McDougal, Littell
GATEWAYS
to Algebra and Geometry
AN INTEGRATED APPROACH

Teacher's Edition

Teacher's Edition

GATEWAYS
to Algebra and Geometry
AN INTEGRATED APPROACH

John Benson
Evanston Township High School
Evanston, Illinois

Sara Dodge
Evanston Township High School
Evanston, Illinois

Walter Dodge
New Trier High School
Winnetka, Illinois

Charles Hamberg
Illinois Mathematics and Science Academy
Aurora, Illinois

George Milauskas
Illinois Mathematics and Science Academy
Aurora, Illinois

Richard Rukin
Evanston Township High School
Evanston, Illinois

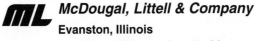

McDougal, Littell & Company
Evanston, Illinois
New York · Dallas · Columbia, SC

Authors

John Benson
Teacher, Evanston Township High School
  Evanston, Illinois
Recipient, Tandy Technology Scholar
  Award, 1991
Recipient, Presidential Award for
  Excellence in Teaching Mathematics,
  Illinois, 1987
Teacher, Center for Talent Development,
  Northwestern University
Head coach, math team,
  Evanston Township High School
Assistant coach, Chicago Area All-Star
  Math Team
Grader, Advanced Placement Mathematics
  Team
Received Distinguished Service Award
  from Luther College, Decorah, Iowa
Previously taught at
  Frederick Douglass High School,
    Atlanta, Georgia

Sara Dodge
Teacher, Evanston Township High School
  Evanston, Illinois
Finalist, Presidential Award for Excellence
  in Teaching Mathematics, Illinois, 1990
  and 1991.
Woodrow Wilson Fellow, 1990
President, Metropolitan Math Club, 1992
Teacher, Center for Talent Development,
  Northwestern University
Delegate, National Council of Teachers of
  Mathematics, 1988–1992
Board Member, Illinois Council of
  Teachers of Mathematics, 1991–1994
Previously taught at
  Wood Oaks Junior High School,
    District 27, Northbrook, Illinois
  Northbrook Junior High School,
    District 28, Northbrook, Illinois

Walter Dodge
Teacher, New Trier High School
  Winnetka, Illinois
Recipient, Presidential Award for
  Excellence in Teaching Mathematics,
  Illinois, 1988
Woodrow Wilson Fellow, 1989
Former grader, table leader, exam leader,
  Writing Committee member, Advanced
  Placement Calculus Test
Past President, Metropolitan Mathematics
  Club
Football and baseball coach
Computer program developer

Charles Hamberg
Teacher, Illinois Mathematics and Science
  Academy, Aurora, Illinois
Recipient, Presidential Award for
  Excellence in Teaching Mathematics,
  Illinois, 1983
Past President, Council of Presidential
  Awardees in Mathematics
Recipient, Golden Apple Award,
  Foundation for Excellence in Teaching
Assistant coach, Chicago Area All-Star Math
  Team
Previously taught at
  Adlai E. Stevenson High School,
    Prairie View, Illinois
  Northern Illinois University
  Laboratory School,
    Dekalb, Illinois (Grades 7 and 8)
  Nathan Hale High School,
    West Allis, Wisconsin
  Marina High School,
    Huntington Beach, California
  Mount Morris High School,
    Mount Morris, Illinois

George Milauskas
Teacher, Illinois Mathematics and Science
  Academy, Aurora, Illinois
Author, *Geometry for Enjoyment and
  Challenge*
Assistant coach, Chicago Area All-Star Math
  Team
Editorial Advisory Board member and
  writer, National Council of Teachers of
  Mathematics Yearbook on Geometry,
  1987
Previously taught at
  Barrington Unit School District 220,
    Barrington, Illinois
  New Trier High School,
    Winnetka, Illinois

Richard Rukin
Teacher, Evanston Township High School
  Evanston, Illinois
Midwest chairperson, American Regional
  Math League
Head Coach, Chicago Area All-Star Math
  Team
Coach, math team, Evanston Township
  High School
Previously taught at
  Senn High School
    Chicago, Illinois

WARNING: No part of this book may be reproduced or transmitted in any form or by any means, electronic or mechanical, including photocopying, recording, or by any information storage and retrieval system without permission in writing from the Publisher.

ISBN 0-8123-8485-7

Copyright © 1994 by McDougal, Littell & Company
Box 1667, Evanston, Illinois 60204

All rights reserved.
Printed in the United States of America

Art and Photography Credits
T4 (bottom) Allan Landau; **T4–T5** Arches National Park. © Gerard Champlong/Image Bank; **T5** (bottom) Allan Landau: **T18–T27** Allan Landau; **T21** (inset) Joe Kay/Hoffman Production; **T28** Computer circuit board. © Bill Gallery/Viesti Associates, Inc.; **T30–T33** Allan Landau; **T34** Maze abstract. © Michel Tcherevkoff/Image Bank; **T36** Allan Landau; **T38–T39** Map. Illinois Bureau of Tourism; **T40** Cartoon. © Sidney Harris; **T44** Cartoon. © Gary Larson

Contents

Letter to Teachers	**T2**
A Walk Through Gateways	**T4**
A Lesson	T6
Key Strands in *Gateways*	T9
Connections	T10
The Spiral Approach	T11
Pacing	T12
Assessment	T13
The *Teacher's Edition*	T14
The *Teacher's Resource File*	T18
Additional Components	T22
Teachers' Workshops	**T24**
Communicating Mathematics	T26
Connections	T28
Cooperative Learning	T30
Calculators and Computers	T32
Problem Solving	T34
Using Manipulatives	T36
Learning Styles	T38
Alternative Assessment	T40
A Word on Math Anxiety	T42
Student's Text with Annotations	
Additional Answers	**A1**

To the Teacher

We welcome you to *Gateways to Algebra and Geometry,* an exciting and innovative text designed to prepare students for success in courses such as algebra and geometry. The main goal of *Gateways* is to allow your students to learn to think mathematically. As they accomplish this goal, students will become problem solvers and will learn to enjoy mathematics more than ever.

A few notes about the structure of *Gateways* might be of assistance to you.

An Integrated Approach
The book is subtitled "An Integrated Approach" for a very good reason: Key strands are woven throughout the text, and problems incorporating each strand appear in nearly every problem set. For example, informal geometry is one of the strands. Instead of being relegated to a separate chapter, however, it is thoroughly integrated, with geometric applications appearing in every section of the book. The ten key strands in *Gateways* are Algebra, Communicating Mathematics, Data Analysis/Statistics, Geometry, Interdisciplinary Applications, Measurement, Number Sense, Patterns and Functions, Probability, and Technology. You will also find that problem solving, critical thinking, and continual review and preview exercises are integrated throughout.

Contents
The Basic assignment guide covers Chapters 1–8, which constitute the core of this text. In these chapters, students will learn to think, use technology, look for patterns, communicate mathematically, and solve problems in such key areas as percent, measurement, ratio, proportion, data analysis, signed numbers, geometry, formulas, and equations. Topics from Chapters 9–12, which appear in the Average and Honors assignment guides, provide you with the flexibility to tailor the course to the needs and abilities of your students.

Chapter 1: Patterns
Studying patterns is a refreshing way to start the year! Besides emphasizing patterns, Chapter 1 introduces students to several key themes that will recur throughout the program: geometry, probability, technology, problem solving, and communicating mathematics.

Lessons
Plan to spend at least two or three days on most lessons. This will give you time to combine various teaching strategies, including discovery, using manipulatives, and cooperative learning, along with lecture and discussion formats.

The Spiral Approach
Students master concepts over time as they experience these concepts in different settings and contexts. Therefore, key topics are previewed before they are formally presented, and after the formal presentation, they are reviewed frequently. This previewing and reviewing takes place throughout the text, being most prominent in the Spiral Learning section of the problem sets.

Problem Sets
The problems constitute the key to success for any mathematics program. In *Gateways,* the emphasis is on variety, applications, and critical thinking, not on drill and practice. Skills are attained—and retained—through continual use, preferably in applied settings.

Each problem set is divided into four parts:
- *Think and Discuss* problems should be done in class orally. They offer you a chance to clarify concepts and assess students' comprehension while at the same time enhancing discourse.
- *Problems and Applications* contains problems that allow students to apply concepts from the lesson in a problem-solving setting while simultaneously reviewing basic arithmetic skills and concepts.
- *Spiral Learning* contains problems that either review previously taught concepts or preview key ideas to come.
- *Investigation* often requires the student to do original work outside the classsroom.

Don't feel you have to assign all of the problems! Refer to the assignment guides, of course, but be sure to tailor assignments to the needs of your particular students.

Multifaceted Problems
Many of the problems throughout the text provide rich learning experiences that help students make connections between various branches of mathematics.

Assessment
We encourage you to employ alternative-assessment strategies regularly. Asking for oral and written explanations, using portfolio assessment, and assessing learning in cooperative settings can contribute significantly to your assessment program; and the *Gateways* supplements package provides a wealth of alternative-assessment ideas, as well as a wide array of more standard tests.

Technology
The concept of variables is essential in algebra, so it is previewed throughout the text. Spreadsheet applications are included for demonstration, for lab work, and for problem solving. We believe that a spreadsheet provides the ultimate demonstration of a variable varying. Students can solve problems via spreadsheets by trying many values for variables quickly and easily. By making use, in whatever way, of the spreadsheet applications throughout the book, you will be empowering your students as problem solvers and preparing them with a marketable skill as you are teaching them the concept of variables.

Use of a calculator—preferably a scientific calculator—is assumed throughout. Remember, though, that while some problems require calculators, others are better done with paper and pencil or mental estimation. Students need to learn when to use and when not to use a calculator. Try not to tell them which problems calculators are needed for; let them tell you.

Gateways students quickly learn that there is more than one way to solve a problem and that there is frequently more than one correct answer. They will be impressed by the utility of mathematics, and they will develop a genuine appreciation for mathematics as they become independent problem solvers. We hope that you and your students enjoy this book. Good luck!

Pupil's and Teacher's Editions

Teacher's Resource File

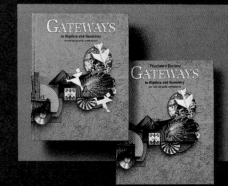

A Walk Through

Also Available

A Lesson

Gateways is all about teaching students to think mathematically. To learn to think mathematically, students need time to investigate, explore, question, and discover important mathematical concepts. Therefore, you will want to spend two or three days on most lessons. The more students are asked to reflect upon their work, the more confident they will become as problem solvers, and the more meaningful mathematics will become in their lives.

Part One: Introduction

The first part of each lesson contains clearly written explanatory material supported by easy-to-follow examples. Graphics and illustrations are often used in the Introduction to address various learning styles.

■ *Gateways* has a unique three-part lesson structure.
1. Introduction
2. Sample Problems
3. Problem Set

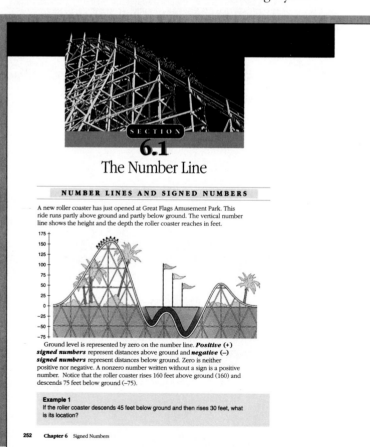

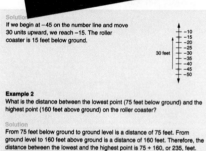

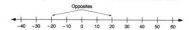

Part Two: Sample Problems

The Sample Problems are applications of lesson concepts and provide excellent models of problem-solving strategies. In addition, these multifaceted problems often preview concepts that will be formally presented later or review previously learned concepts.

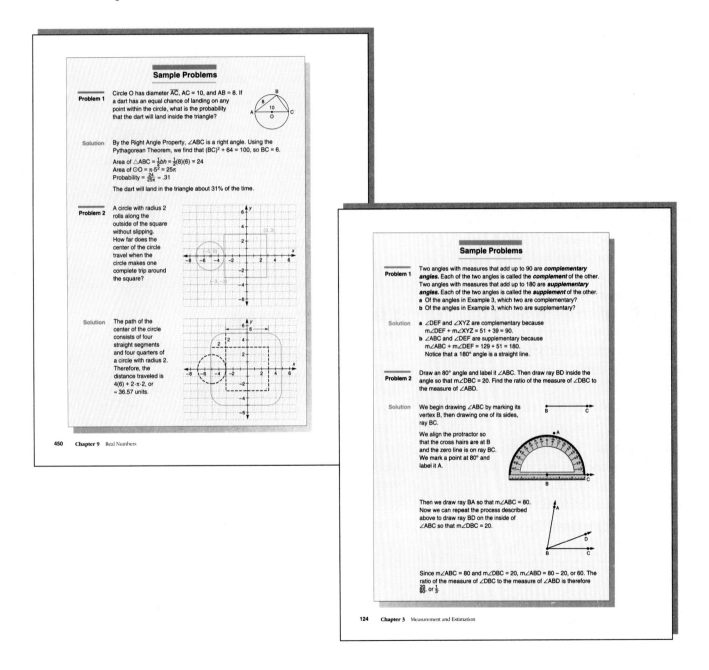

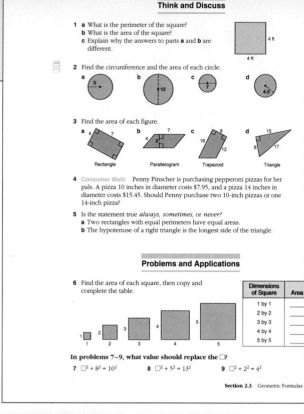

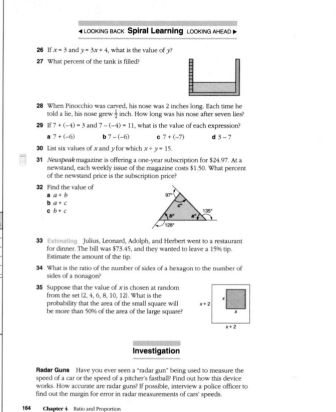

Part Three: Problem Set

The problem sets in *Gateways* offer variety. In this book, you will never see a page with thirty or more similar drill-and-practice problems. Rather, you will find carefully constructed problem sets that implement a spiral approach to promote long-term retention and develop students' critical-thinking and problem-solving skills.

The problem sets in *Gateways* are divided into four parts.

1 Think and Discuss These problems are ideal for enhancing discourse. We encourage teachers to discuss all of these questions with all levels of students.

2 Problems and Applications These problems practice the lesson objectives through multifaceted applications that integrate key strands. While applying concepts from the lesson, students will be continually reviewing previously learned skills.

3 Spiral Learning: Looking Back, Looking Ahead In keeping with the spiral approach, these problems introduce (or preview) concepts that will be formally presented later, and they review previously introduced concepts to promote long-term retention. These problems are identified in the *Teacher's Edition* with an **I** (for Introduce) or an **R** (for Review).

4 Investigation These problems connect mathematics to students' world with a variety of thought-provoking explorations that often require effort outside the classroom.

Key Strands in Gateways

The following examples show how ten key strands are integrated throughout *Gateways*:

Strands	Page	Problem
■ Probability	56	**44** A bag contains four chips, numbered 3, 5, 7, and 8. Three chips are drawn at random. What is the probability that **a** The sum of the numbers on the three chips is even? **b** The product of the numbers on the chips is even?
■ Algebra ■ Interdisciplinary Applications	60	**7** Science If we use v to stand for the speed of sound in feet per second and t to stand for the temperature in degrees Celsius, we can find the speed of sound at sea level with the formula $$v = \frac{1087 \cdot \sqrt{273 + t}}{16.52}$$ What is the speed of sound when the temperature is 25°C?
■ Geometry	61	**20** If $\pi \approx 3.1416$, $r \approx 6.75$, and $d \approx 13.5$, what is the value of **a** πd? **b** $2\pi r$? **c** πr^2?
■ Technology ■ Patterns & Functions	76	**29** Spreadsheets This diagram shows a spreadsheet display, along with formulas that were entered in cells D2, D3, and D4. What formulas should be entered in cells D5 and D6? What numbers will appear in those cells? |	A	B	C	D
---	---	---	---	---
	Value I	Value II	Value III	Average
1				
2	5	7	12	8 ← +(A2+B2+C2)/3
3	9	15	21	15 ← +(A3+B3+C3)/3
4	13	15	26	18 ← +(A4+B4+C4)/3	
■ Measurement ■ Number Sense	97	In problems 15–18, determine which distance is longer. **15** 20 inches or 2 feet **16** 1 mile or 2000 yards **17** 15 millimeters or 1 centimeter **18** 2150 meters or 2 kilometers
■ Communicating Mathematics ■ Data Analysis/ Statistics ■ Interdisciplinary Applications	217	In problems 9–11, refer to the graph. **9** Find the ratio of noodles sold to long goods sold. **10** For every 5 ounces of noodles sold, what quantity of short goods is sold? **11** Why do you think long goods are the most popular type of pasta? **Picking pasta** — Retail sales by shape of pasta: Long goods (spaghetti, linguini) 41%; Short goods (elbows, twists, etc.) 31%; Noodles (egg added) 15%; Specialties (lasagne, jumbo shells, manicotti, flavored shapes) 13%. Source: National Pasta Association/Ron Coddington, USA TODAY

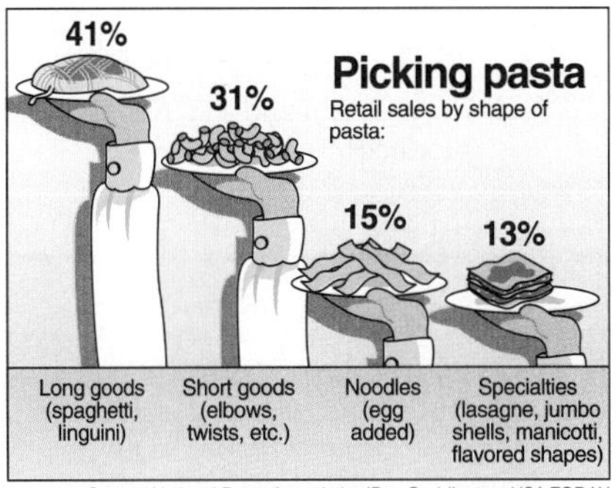

> **Key strands in Gateways:**
> 1. Algebra
> 2. Communicating Mathematics
> 3. Data Analysis/Statistics
> 4. Geometry
> 5. Interdisciplinary Applications
> 6. Measurement
> 7. Number Sense
> 8. Patterns & Functions
> 9. Probability
> 10. Technology

CONNECTIONS

Multifaceted problems help students see connections among various branches of mathematics and between mathematics and the real world.

Page	Problem	Connection
212	**17** In a surveyor's drawing, $\frac{1}{8}$ inch represents 12 feet. What is the actual perimeter of a rectangle if the dimensions in the drawing are 2 feet by $1\frac{1}{2}$ feet?	**Arithmetic** ⇄ **Geometry**
225	**12** Spreadsheets Here are some test scores from Mr. X's math class. 85, 78, 87, 93, 67, 54, 67, 77, 88, 95, 93, 79, 56 **a** Enter the data into cells A1–A13 of a spreadsheet. **b** Calculate the mean in cell A14 by using the function @Avg(A1..A13). **c** Was Mr. X's test a difficult one? Why or why not? **d** Use the spreadsheet to sort the data. What is the median?	**Technology** ⇄ **Data Analysis**
226	**21 a** How many personnel managers were surveyed? **b** How many of them felt that health care will be a major concern in the year 2000? **c** According to the managers surveyed, which issue was a major concern in 1990 but will not be in 2000? Which issues were not major concerns in 1990 but will be in 2000?  Source: Eckerd College 1991 survey of 233 personnel managers/ Elys A. McLean, USA TODAY	**Data Analysis** ⇄ **Business Application**
359	**11** This vertex matrix represents △YES. $\begin{array}{ccc} Y & E & S \\ \begin{bmatrix} -6 & 4 & 10 \\ -10 & -8 & -10 \end{bmatrix} \end{array}$ **a** Multiply the matrix by $\begin{bmatrix} -1 & 0 \\ 0 & -1 \end{bmatrix}$ to produce the vertex matrix of △Y′E′S′. **b** Draw △YES and △Y′E′S′. What transformation did the multiplication produce?	**Algebra** ⇄ **Geometry**

The Spiral Approach

Concepts in *Gateways* are presented using a spiral approach because mastery in learning takes place over an extended period of time. Key concepts are previewed before they are formally introduced. Then after their formal introduction, concepts are practiced and reviewed in a variety of contexts to promote thorough understanding and long-term retention.

Here are some examples that illustrate how the concept of *signed numbers* is spiraled throughout *Gateways:*

	Page	Problem
Introduce (or Preview) **I**	11	**5** Study the pattern shown. What numbers do you think go in the blanks? $3 - 0 = 3$ $3 - 1 = 2$ $3 - 2 = 1$ $3 - 3 = 0$ $3 - 4 =$ ___ $3 - 5 =$ ___
	118	**25** **Spreadsheets** Examine the spreadsheet table. For what value(s) of x does $x^2 - 10 = 3x$? A B C 1 Value of X X^2−10 3X 2 −5 15 −15 3 −4 6 −12 4 −3 −1 −9
	227	**23** Enter 7 [+/−] [×] 8 [=] into your calculator. What did you get? Why?
Formal Introduction	250	**Chapter 6** Signed Numbers
Practice **P**	257	**14** Draw a number line and locate the points with the given coordinates. **a** 2 **b** $\frac{2}{5}$ **c** 7.3 **d** −4.5 **e** $-3\frac{5}{6}$ **f** $-\sqrt{2}$
Review **R**	404	**24** **Science** The high temperatures on the first six days of January were −4°, 6°, 1°, 11°, 7°, and −2°. What must be the high temperature on January 7 in order for the first week of January to have an average high temperature greater than 5°?

PACING

Gateways provides pacing charts with schedules for three ability levels. You may wish to adjust these schedules to meet the individual needs of your class.

Pacing Charts

The **Basic Course** covers chapters 1–8. **Total 149**

Chapter	1	2	3	4	5	6	7	8
Days	16	16	22	21	16	24	17	17

The **Average or Heterogeneous Course** covers chapters 1–9 and 11. **Total 161**

Chapter	1	2	3	4	5	6	7	8	9	11
Days	12	14	19	17	12	18	15	18	17	19

The **Honors Course** covers chapters 1–12. **Total 163**

Chapter	1	2	3	4	5	6	7	8	9	10	11	12
Days	9	13	17	14	11	16	13	14	13	10	19	14

The Pacing Chart

Each chapter opens with a pacing chart to help teachers in class planning.

Days per Section

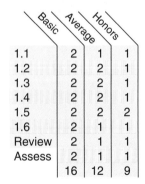

	Basic	Average	Honors
1.1	2	1	1
1.2	2	2	1
1.3	2	2	1
1.4	2	2	1
1.5	2	2	2
1.6	2	1	1
Review	2	1	1
Assess	2	1	1
	16	12	9

The Assignment Guide

Carefully designed assignment guides focus on key strands and include a representative sample of preview, practice, and review problems.

Assignment Guide

Basic
Day 1 6, 9, 14, 15, 24, 28, 32, 36, 39
Day 2 7, 8, 17, 25, 26, 29, 33–35, 38

Average or Heterogeneous
Day 1 6, 7–12, 14, 15, 17, 19
Day 2 20, 21, 24–27, 31, 36, 37

Honors
Day 1 7–14, 17–19, 21–23, 25–27, 30, 31, 35–37, 39, Investigation

ASSESSMENT

Pupil's Edition

A chapter test is provided at the end of each chapter in the student's textbook.

Teacher's Edition

Checkpoint quizzes in every lesson of the *Teacher's Edition* help you determine students' comprehension of lesson concepts.

At the end of each chapter of the *Teacher's Edition* are two important asssessment features.

1 Assessment
Here you will find a reference to all tests in the *Tests and Quizzes Book*.

2 Alternative-Assessment Guide includes:
- *Assessment Tips:* suggestions for assessing projects, group work, and writing
- *Portfolio Projects:* references to problems that make ideal portfolio projects

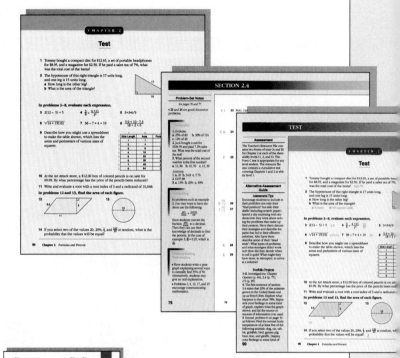

Tests and Quizzes Book

Gateways has a comprehensive test package. *The Tests and Quizzes Book* includes:
- Two forms of three levels of tests per chapter
- One alternative-assessment test per chapter
- Three levels of quizzes per chapter
- Three levels of cumulative tests—quarterlies, mid-year, and final exams

Testing Software

McDougalTest software generates an unlimited variety of tests and worksheets. This software is available for IBM compatible PC, XT, and AT, MAC, and Apple II computers.

T13

The Teacher's Edition

■ **Pacing Chart**
Pacing information for three different levels (basic, average or heterogeneous, and honors) aids teachers in class planning.

■ **Chapter Overview**
Chapter overviews summarize the overall objectives, purpose, and focus of the chapter.

■ **Career Connection**
Career connections capture students' interests by connecting career applications with chapter topics.

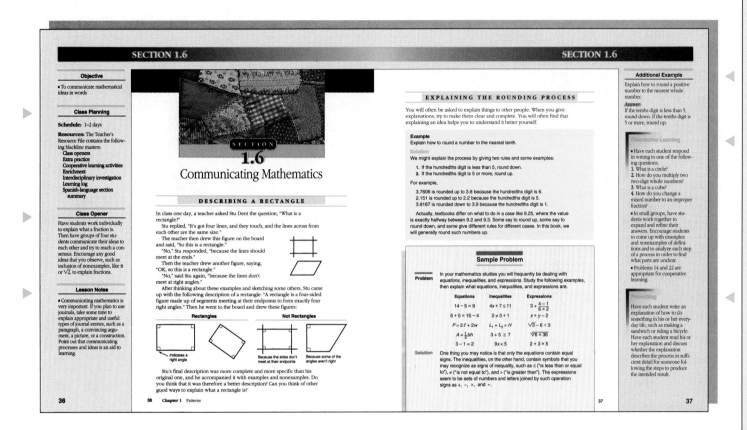

Class Planning
A schedule and list of reference materials from the *Teacher's Resource File* is provided to help you plan and prepare for each class. Feel free to adjust the schedule to meet the specific needs of your class.

Class Opener
Class Openers for every lesson give lessons a "jump start" by getting students on task immediately. Additional Class Openers can be found in the *Teacher's Resource File*.

Lesson Notes
Lesson Notes, written by the authors, provide specific teaching suggestions and strategies for each lesson.

Additional Examples
For each example in the text, an additional parallel example is provided in the *Teacher's Edition* side column.

Cooperative Learning
Cooperative Learning activities in every lesson support a variety of instructional approaches.

Reteaching
This feature contains alternative-instruction suggestions for students who may benefit from having a concept presented from a second perspective.

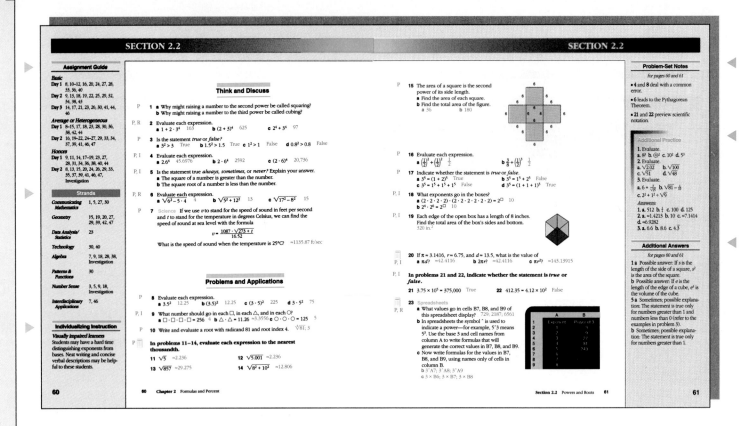

Assignment Guide
Assignment guides are provided for three levels—basic, average or heterogeneous, and honors.

Strands
Ten key strands are integrated throughout the *Gateways* program. Further, numerous multifaceted problems connect these strands to help students develop as real-world problem solvers.

Individualizing Instruction
This feature provides teachers with additional suggestions for tailoring classroom instruction to the individual needs of students.

Problem-Set Notes
Problem-Set Notes provide insight into the problems and help in planning class assignments.

Additional Practice
This feature provides exercises for students who need additional practice with the basic skills encountered in a given lesson.

Additional Answers
Answers that are not annotated directly on the student page appear in the side column where space permits. Overflow answers appear in the back of the *Teacher's Edition*.

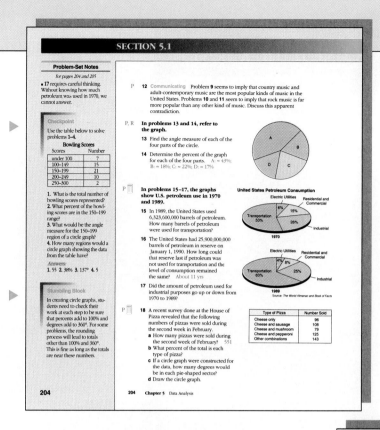

■ Checkpoint
Checkpoints are short quizzes that can be used to determine how well students understand the lesson objectives.

■ Stumbling Block
Stumbling Blocks anticipate areas in which students may have difficulty, and offer suggestions for remediation.

■ Communicating Mathematics
These activities give students opportunities to express their mathematical thoughts.

■ Multicultural Note
These features introduce interesting multicultural perspectives about mathematics.

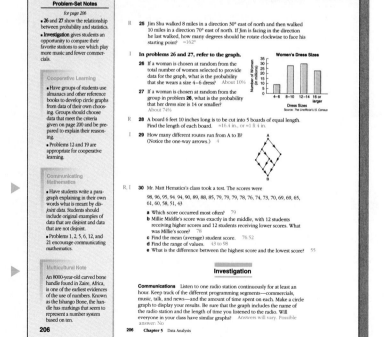

The Teacher's Resource File

Components:
- User's Guide
- Solutions Manual
- Tests and Quizzes Book
- Blackline Masters

User's Guide

The *User's Guide* contains information on how to use the *Teacher's Resource File* and detailed answers for all components of the file.

Solutions Manual

The *Solutions Manual* contains complete step-by-step solutions to all the problems in the student's text.

Tests and Quizzes Book

The *Tests and Quizzes Book* contains two forms of three levels of tests (basic, average or heterogeneous, and honors) per chapter, one alternative assessment test per chapter, three levels of quizzes per chapter, and two quarterlies, one mid-year test, and one final exam for each of the three levels.

Blackline Masters

The following blackline masters accompany *Gateways:*

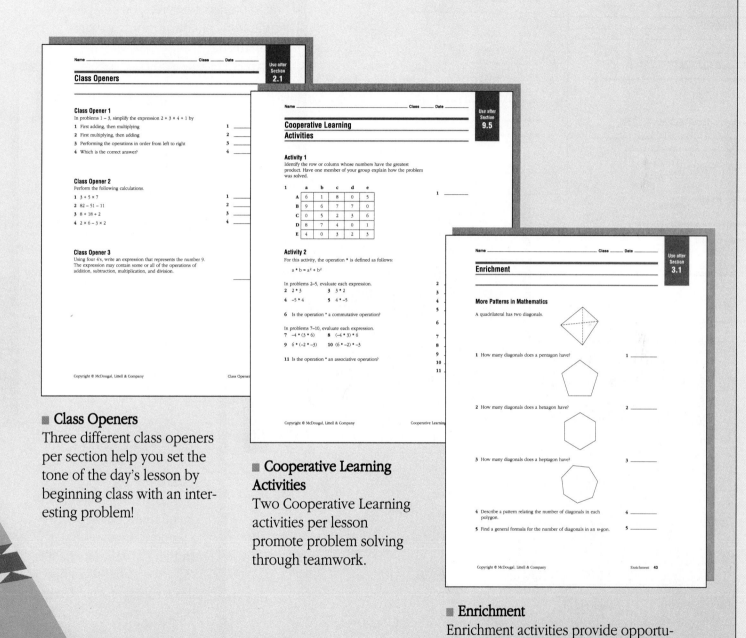

■ Class Openers
Three different class openers per section help you set the tone of the day's lesson by beginning class with an interesting problem!

■ Cooperative Learning Activities
Two Cooperative Learning activities per lesson promote problem solving through teamwork.

■ Enrichment
Enrichment activities provide opportunities for extending ideas and lesson concepts.

Blackline Masters
continued

■ Technology
Technology-related activities provide students with opportunities to discover applications of spreadsheets, calculators, and computers.

■ Spanish-Language Section Summary
These blackline masters summarize the main points of each lesson and provide definitions of key vocabulary words in Spanish.

■ Using Manipulatives
The kinesthetic learner will appreciate how a manipulatives activity can transform an abstract concept into concrete understanding.

■ Learning Log
These blackline masters encourage students to communicate their understanding of lesson concepts and express their feelings about their personal learning experiences.

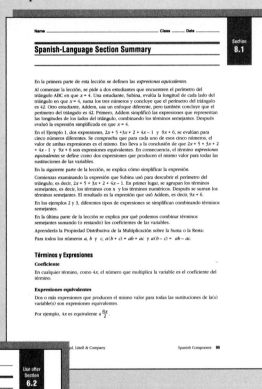

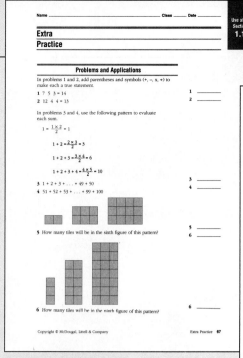
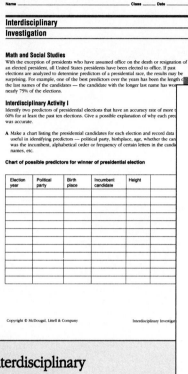
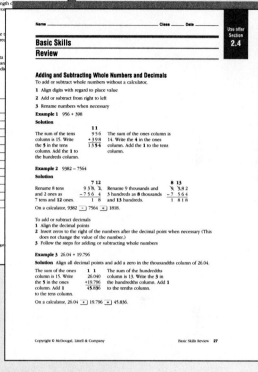

■ Extra Practice
These problems are similar to the problems in the text, and they reinforce the lesson objectives.

■ Interdisciplinary Investigations
These activities allow students to discover mathematics at work in other disciplines.

■ Basic Skills Review
These blackline masters provide numerous worked-out examples and abundant practice in basic arithmetic skills.

Also Available Separately
- Investigations Workbook
- Practice Workbook
- Solutions Manual
- Tests and Quizzes Book

Additional Components

Manipulatives Kit

McDougal, Littell has worked with Educational Teaching Aids (ETA) to design the most complete manipulatives package available for use in secondary mathematics classrooms!

To order or for information, call ETA at 1-800-445-5985.

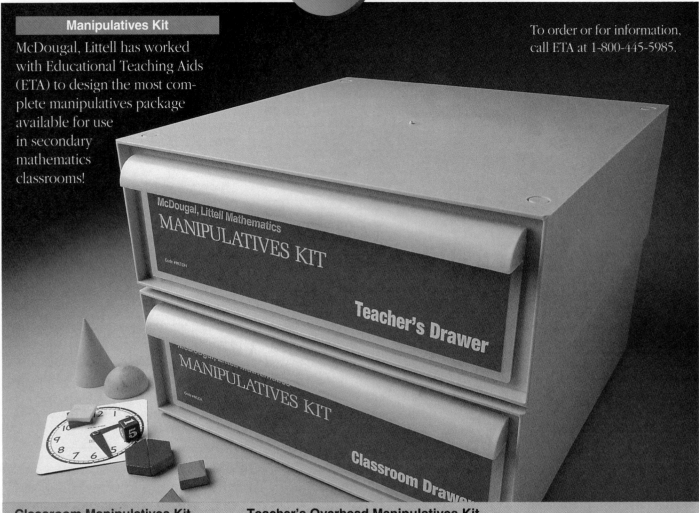

Classroom Manipulatives Kit

Item	Quantity
Angle Rulers (Set/30)	1 Set
Algebra Mods	10 Sets
Number Cubes	5 Sets
Geoboards	10 Each
Clock Dials (Set/10)	1 Set
Color Tiles (Set/400)	1 Set
Protractors	10 Each
Attribute Blocks	4 Sets
Spinners (Set/12)	1 Set
Reflectas	10 Each

+

Teacher's Overhead Manipulatives Kit

Item	Quantity
Angle Ruler	1 Each
Overhead Algebra Mods	1 Set
Metric Grid	1 Each
Overhead Geoboard	1 Each
Grease Pencils	2 Each
Overhead Fraction Squares	1 Set
Overhead Fraction Circles	1 Set
Protractor	1 Each
Overhead Attribute Blocks	1 Set
Overhead Spinner	1 Each
Wood Geometric Solids (Set/12)	1 Set
Overhead Color Tiles	1 Set

= Complete Manipulatives Kit

Software

- **McDougalTest Computer Testing Software**
This software allows teachers to generate an almost limitless number of unique tests and worksheets. Available for IBM compatible PC, XT, and AT, MAC and Apple II computers.

- **Instructional Software**
  - *The Omnifarious Plotter*
  This graphing program developed by author Walter Dodge will graph almost any function or equation.
  - *The Algebra Problem Solver*
  This program contains twelve problem-solving programs for IBM compatible computers.

- **Spreadsheet Software**
ProCalc 3D
This sophisticated yet easy-to-use spread sheet program was developed by Parsons Technology for IBM compatible computers.

Transparencies

Four-color instructional transparencies further complement the *Gateways* program.

Long-Term Investigations

Gateways to Mathematical Investigations consists of six thematic units featuring long-term investigations correlated to the *Gateways to Algebra and Geometry* program. The components include:
- Student Edition
- Teacher's Edition
- Autiotapes that summarize the investigations for students in six languages: Cambodian, Cantonese, English, Hmong, Spanish, and Vietnamese.

Teachers' Workshops

GATEWAYS

WORKSHOP

COMMUNICATING MATHEMATICS

Communication is a process by which information is exchanged. An individual's knowledge of a particular subject is of little interest or value if she or he cannot communicate this knowledge to others. As such, communication skills, whether they are written, oral, or visual, are essential to acquiring, understanding, knowing, and doing mathematics.

According to the *NCTM Teaching and Evaluation Standards,* the goal of teaching mathematics is to develop students' mathematical power, which includes the ability to communicate mathematically. Students should be engaged in mathematical tasks that require reasoning, problem solving, and communicating. Concepts, solutions, and strategies should be discussed and shared; conjectures should be made and validated. During such interactions, ideas are clarified and students find themselves immersed in doing mathematics. Concepts are learned through reading, talking, listening, modeling, and illustrating. Communication enables students to learn mathematics and helps teachers assess whether students are learning what they are trying to teach. Instruction in and assessment of oral and written communication are integral parts of each student's education.

Students should experience a variety of communication models and learn to communicate in a clear, precise manner using appropriate vocabulary and logical reasoning. Students need training in how to read mathematics, how to write mathematics, and how to verbalize

mathematics. These are ongoing activities and should be a strand in weekly lessons. Effective problem solving involves communication. Reading a problem, interpreting the problem, and thinking critically are essential to solving a problem; once the problem is solved, the discussion of solutions and extensions can be the most valuable aspect of the learning experience.

There are many possibilities for communication in a mathematics class, possibilities which we explore in *Gateways*. Every problem set begins with Think-and-Discuss questions to promote discourse. Further, many of the homework problems ask students to explain their answers. These problems can be done by the class as a whole, in cooperative learning groups, or individually.

Teachers should ask questions like "Why?" "What if…?" and "How would you convince someone that…?" At this point in their mathematical growth, students are not familiar with proper mathematical vocabulary, and the emphasis should be on students' using their own words to formulate clear explanations. As the year progresses, growth in the use of mathematical terminology should be encouraged. Students should show the ability to interpret graphs and geometric figures, as well as provide explanations to justify their reasoning.

In *Gateways*, we explored and maximized the use of different communication activities that will not only make a concept clearer to the students but, more importantly, will make it easier for the teacher to assess which concepts the students comprehend and which ones they are having difficulty grasping. *Gateways* makes a student think about different problem situations and real-life applications of mathematical concepts, identify different strategies in solving problems, and communicate how solutions are arrived at and why these solutions are correct. In *Gateways*, students are involved in such activities as describing patterns, writing stories for mathematical formulas, explaining why a solution is correct or incorrect, illustrating how one would go about solving a problem, evaluating how a situation would change if one aspect were altered, communicating why one problem-solving strategy would be better than another, and so on. *Gateways* is rich in activities that promote communication and, as such, it is a valuable and beneficial tool that will develop students' mathematical power.

Resources

Azzolino, Aggie. "Writing as a Tool for Teaching Mathematics: The Silent Revolution." *Teaching and Learning Mathematics in the 1990s.* Ed. Thomas Cooney and Christina Hirsch. Reston: National Council of Teachers of Mathematics, 1990. 92-100.

National Council of Teachers of Mathematics. *Professional Standards for Teaching Mathematics.* Reston: National Council of Teachers of Mathematics, 1991.

Pimm, David. *Speaking Mathematically: Communication in Mathematics Classrooms.* New York: Routledge, 1987.

Silver, Edward, Jeremy Kilpatrick, and Beth Schlesinger. *Thinking Through Mathematics: Fostering Inquiry and Communication in Mathematics Classrooms.* New York: College Entrance Examination Board, 1990.

WORKSHOP
CONNECTIONS

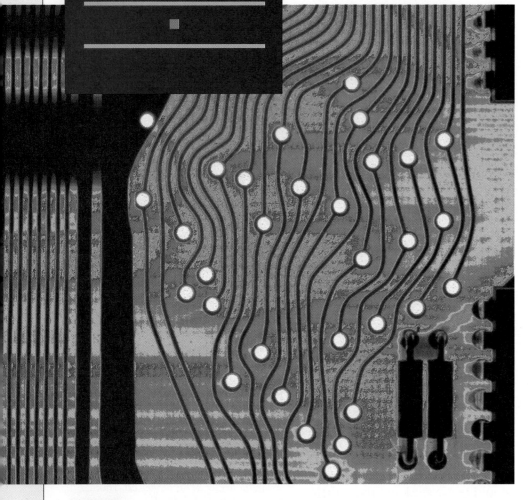

Many students perceive the study of mathematics to be a sequence of disjointed ideas and concepts. The authors of *Gateways* responded to this perception by writing a book with lessons that are designed to lead students toward a discovery and an understanding of the role of connections in the teaching and learning of mathematics.

In *Gateways*, students have numerous opportunities to see and discover many of the natural connections between various branches of mathematics. The Sample Problems and the Problem Sets incorporate mathematical ideas that simultaneously utilize several areas of mathematics. For example, students often must make connections between algebra and geometry, probability and geometry, probability and algebra, arithmetic and geometry, data analysis and arithmetic, and data analysis and algebra.

Students are also given ample opportunities in *Gateways* to view and discover many of the natural connections between mathematics and other disciplines. For example, data analysis problems are integrated throughout the book. They illustrate connections between mathematics and science, business, social science, and agriculture. Problems that require the use of a spreadsheet are also integrated throughout the book and provide valuable insights for bridging mathematics and consumer math, social science, and business.

Gateways provides students with many opportunities to connect mathematics and the real world. Problems throughout the text use statistics from sports, data from newspaper graphs, information from *The World Almanac and Book of Facts,* scientific facts, and health-related information. The Investigation at the end of every lesson also provides an opportunity for students to connect mathematics and the real world. These features ask students to investigate such topics as radar guns, water consumption, and temperature equilibrium of appliances.

The *NCTM Standards* specifically call for students to be able to draw and make connections through a variety of perspectives. The Problem Sets in *Gateways* are a rich source for the development of these connections because the problems enable students to utilize multiple problem-solving approaches. Many of the Examples and Sample Problems offer more than one method of solving the problem at hand. In this way, students are able to make connections while employing equivalent problem-solving strategies, techniques, methods, and processes.

Finally, it is important that students be able to make connections between the concrete and the abstract. In *Gateways*, students explore, investigate, conjecture, and verify their observations and conclusions. Encourage students to communicate, either orally or in writing, the connections they discover while formulating relationships between the concrete and the abstract.

Remember, your students will have a better comprehension of mathematics if they recognize these connections and are able to communicate their understandings.

Resources

Froelich, Gary, Kevin Bartkovich, and Paul Foerster. *Connecting Mathematics*. Addenda Series, Grades 9–12. Reston: National Council of Teachers of Mathematics, 1991.

Gilligan, Carol, Nona Lyons, and Trudy Hanmer. *Making Connections:* Cambridge: Harvard Press, (1990).

Connections are made between:

Various branches of mathematics

19 **a** Write two expressions for the area of rectangle ABCD.
   b Calculate the area of rectangle ABCD.

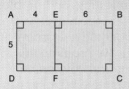

Mathematics and other disciplines

20 Health   To determine your target heart rate for one minute of aerobic exercise, subtract your age from 220 and multiply the result by 75%.
   a Compute your target heart rate.
   b If a represents age, write an expression for target heart rate.
   c Use a spreadsheet to calculate target heart rates for ages 10 through 60.

Mathematics and the real world

25 Geography   Look at a map of the United States. Which states appear to be shaped like quadrilaterals? What kinds of quadrilaterals do they resemble?

Problem-solving approaches

20 **a** What is the area of region A?
   b What is the area of region B?
   c What is the total area of the figure?
   d How does the diagram show that $3(2 + 5) = 3(2) + 3(5)$?

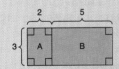

The concrete and the abstract

13 Write an algebraic statement to describe the drawings.

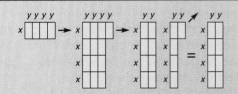

T29

WORKSHOP: COOPERATIVE LEARNING

As professional mathematics educators, we are challenged to make needed changes in both curriculum and instruction to better empower all of our students mathematically. *NCTM Standards for Teaching Mathematics,* published in 1991, calls for mathematics classrooms to become "mathematical communities," where conjecturing, inventing, reasoning, verifying, communicating, and problem solving are the norm. As a teaching tool, cooperative learning is an ideal way to promote this kind of mathematical environment.

Cooperative learning can take many forms. Some teachers use a formal cooperative-learning structure throughout the year, assigning roles like facilitator, timer, recorder, and encourager to students on a rotating basis. Others have a problem of the day on which students work cooperatively. Still others utilize cooperative-group work for projects or reviews.

Regardless of the form of cooperative learning used in the classroom, the basic goals are the same—to foster mathematical communication among students, to encourage problem solving, and to teach social skills in the context of mathematics. Depending on the activity, you may want to set other goals for the day, such as the discovery of a new concept, the mastery of a skill by all group members, the development of connections between concepts, or the preparation of a presentation. As students work through these problems together, the emphasis should be on *process* rather than on the end result. The teacher's job is to facilitate and encourage communication, interacting only to provide encouragement when groups are stuck or discouraged.

The learning that takes place in this group setting can be assessed in several ways. For example, each group may submit one paper, or one member of each group can be randomly chosen to present the group's solution to the class. Teachers may also choose to assess each student individually to insure that all members of the group understand the mathematics. The assessment, whether formal or informal, should also include some discussion about how well the group members worked together.

Students who work in a cooperative setting often gain better understanding of concepts because they must make sincere efforts to communicate what they know, or what they don't know, to their classmates. They learn that by putting their heads together, they can solve problems that may have been difficult to solve on their own. Students not accustomed to being "star" students often contribute a much-needed perspective on a less traditional problem. Open-ended problems can provide a stimulus for more able students to investigate further on their own.

The problem sets in *Gateways to Algebra and Geometry* lead to lively

...the emphasis should be on process rather than on the end result.

mathematical discussions, in which students are encouraged to share and verify not only their solutions to problems but the processes by which they reached these solutions. In addition, the *Teacher's Edition* includes a Cooperative Learning Activity for each lesson, which could serve as an ideal "problem of the day." Generally, such a problem extends a concept presented or practiced in a given lesson. It may be open-ended or more challenging, it may present the concept in another context, or it may be designed to lead students to discover further mathematical connections.

The problem sets and Cooperative Learning Activities in *Gateways* provide materials that can serve as springboards for experimenting with new teaching methods.

Resources

Adams, Dennis, Helen Carson, and Mary Hamm. *Cooperative Learning and Educational Media: Collaborating with Technology and Each Other.* Englewood Cliffs: Educational Technology Publications, 1990.

Artzt, Alice, and Claire Newman. *How to Use Cooperative Learning in the Mathematics Classroom.* Reston: National Council of Teachers of Mathematics, 1990.

Cohen, Elizabeth. *Designing Groupwork: Strategies for the Heterogeneous Classroom.* New York: Teachers College Press, 1986.

Davidson, Neil, and Toni Worsham. *Enhancing Thinking Through Cooperative Learning.* New York: Teachers College Press, 1992.

Slavin, Robert. *Cooperative Learning: Theory, Research, and Practice.* Englewood Cliffs: Prentice Hall, 1990.

WORKSHOP

CALCULATORS AND COMPUTERS

All students will use technology to some extent when they enter the work force. How prepared they are for their future endeavors will depend on their exposure to and comfort level with technology. Although we do not know exactly what kinds of technology students will use in the future, it is our responsibility to give them experience in the technology that is available to us today.

Teachers can take the first step in this direction by incorporating in their lessons whatever technology is available to them. Students will gain confidence in the use of technology over time. They will begin to appreciate the power of technology as they learn how it can be used to help them solve problems. Students will learn to distinguish between problems that can be solved only with the aid of technology and those that are better suited to paper-and-pencil solutions. They will also gain confidence in choosing the best type of technology for solving a particular problem. The most important lesson that students will learn is that although technology is a tool that can help them solve a problem, technology alone cannot solve a problem.

Calculators

Calculators are used by nearly all students, whether sanctioned by the teacher or not. *Gateways* assumes the use of scientific calculators but has many problems that can be solved without a calculator. In the student's text, a calculator symbol is used to identify problems that are best done with a calculator.

However, students vary in their computational abilities. What one student can do correctly without a calculator may be very difficult for another student. For this reason, you may find that some students use calculators for problems that are not identified as calculator problems. Encourage the use of mental math, approximation, and common sense to check answers that are obtained with the aid of a calculator.

Instructions on the use of calculators are essential for students. Throughout *Gateways*, instructions are provided for the use of a scientific calculator. For example, Section 2.1 illustrates the importance of grouping symbols and shows students how to use the parenthesis key on a calculator to replace the vinculum of a fraction. Without such instructions, students may make order-of-operations errors when using calculators. The following section shows students how to use a calculator to evaluate powers and roots. In Section 3.2, students learn how to enter a number in scientific notation and how to interpret a calculator's display of a number in scientific notation. Example 3 in Section 3.4 shows how a calculator can be used to approximate a solution of an equation that students do not know how to solve algebraically. Instruction on calculator use begins early in the text and is followed by plenty of opportunities to use a calculator throughout the rest of the book.

Some other types of calculators, like programmable and graphing calculators, have options that allow students to do more than simple calculating. In Chapter 5, on data analysis, students that have access to a graphing calculator with a

statistics menu can use it to draw bar graphs, line graphs, and scattergrams of data in the problems. Some of these calculators have matrix capabilities and can calculate the mean of a set of data. Sections 6.8 and 9.7 can be greatly enhanced if your students have calculators with drawing and graphing capabilities. Students enjoy being able to draw geometric figures on the screen. As they do, they are becoming familiar with the coordinate plane.

Computers

Spreadsheet software is available for all types of computers. Older computers with less memory have fewer capabilities but may be easier to use. Newer computers support newer versions of spreadsheets, which have many more features.

In *Gateways*, the development of familiarity with spreadsheets is gradual. In the first two chapters, the basics of spreadsheets are presented. Throughout the rest of the book, there are applications and sample problems to help students learn more features of spreadsheets.

A spreadsheet is important as an educational tool because it gives students a real sense of the concept of a variable—it enables students to see variables varying. It can be set up so that students can use it to guess and check solutions, changing values in smaller and smaller increments to find an answer to a specific degree of accuracy.

In Chapter 5, on data analysis, some of the graphing can be done by spreadsheets. We have also incorporated spreadsheets to help students with problem solving as well as the discovery of concepts. If you do not have calculators with iteration capabilities, most of the iteration problems in Chapter 12 can be modeled with spreadsheets. The graphics for the sample problems in Sections 12.1, 12.2, and 12.4 were generated by a spreadsheet. Having students become familiar and comfortable with spreadsheets not only will help them to a better understanding of variables, but will provide them with a marketable skill.

A computer can also be used to demonstrate the dynamic power of geometry software. Programs that allow students to draw, manipulate, and measure figures make Chapter 7 come alive. Their understanding of quadrilaterals and transformations is also greatly enhanced by the use of geometry software. Graphing calculators or graphing software will help students with the coordinate work that is integrated throughout the text.

In addition to helping students bridge the gap between abstract symbolic representation and concrete numbers, technology allows students to explore the concepts of mathematics visually. We hope that in conjunction with *Gateways* you will use throughout the year the technology that you have available.

Resources

Coburn, Terrence. *How to Teach Mathematics Using a Calculator: Activities for Elementary and Middle School.* Reston: National Council of Teachers of Mathematics, 1987.

Fey, James, ed. *Calculators in Mathematics Education.* Reston: National Council of Teachers of Mathematics, 1992.

Hansen, Viggo, ed. *Computers in Mathematics Education.* Reston: National Council of Teachers of Mathematics. 1984.

Langhorne, Mary Jo, et al. *Teaching with Computers: A New Menu for the '90s.* Phoenix: Oryx Press, 1989.

Masalski, William. *How to Use the Spreadsheet as a Tool in the Secondary Mathematics Classroom.* Reston: National Council of Teachers of Mathematics, 1990.

Russell, John. *Spreadsheet Activities in Middle School Mathematics.* Reston: National Council of Teachers of Mathematics, 1992.

WORKSHOP

PROBLEM SOLVING

One of the most misunderstood expressions in mathematics education today is "problem solving." Some people think problem solving means doing word problems. Others think it means doing puzzles. Actually, it means neither. Problem solving simply means that students are being asked to do some thinking for themselves. An equivalent expression, preferred by some people, is "higher-order thinking skills."

An Approach to Teaching

Problem solving can also be thought of as an approach to teaching. For example, ask questions that students do not know how to answer, but are extensions of work they have already done. If a student already knows how to do something, doing it is merely an exercise, not a problem. Or present students with a problem that requires them to use information that is not given in the statement of the problem. For example, they may need to do some measurements or look up some information. The fundamental idea is to get students to make connections and figure out things for themselves. As one educator put it, "It is not the job of the teacher to cover the material; it is the job of the teacher to help students to uncover the material."

Another way to implement a problem-solving approach in your teaching is to keep going even after you've gotten the answer to a problem. Ask students to explain why their method of solving the problem worked. Ask if there are other methods for solving the problem. Ask "What if…" questions, such as "What if the angle had been obtuse?" or "Suppose that the number had been negative. Could the problem still have been worked the same way?" This helps students learn to ask questions themselves.

A Problem-Solving Attitude

It is important that teachers foster a problem-solving attitude in their students. They can start by helping students learn to question —to pursue their own "what ifs"— and to generalize concepts. Eventually, we want students to create and solve problems on their own, individually as well as in groups.

Another way to foster a problem-solving attitude in students is to foster their willingness to try. Students need to develop their own repertoire of approaches and strategies, such as guessing and checking or drawing pictures. Sheila Tobias has written that "the difference between the successful math student and the unsuccessful math student is that the successful student knows what to do when he or she doesn't know what to do."

Many teachers talk of their desire for students to feel good in class— reach a level of comfort. If carried to extremes, this can often be self-defeating. Students know when the work they are given isn't challenging. Furthermore, students need to learn how to deal with difficult problems and problems that take time. Rather, put your efforts into increasing students' threshold of discomfort, and you will begin to develop a problem-solving attitude in them. A certain amount of frustration, when it leads to eventual success, can help students realize the importance of continued effort.

In his *Taxonomy of Educational Objectives,* Benjamin Bloom discusses the various levels of questions we ask students. The highest levels involve analysis and synthesis, and the lowest levels involve knowledge and direct application. To develop a problem-solving attitude in our students, we need to give them an opportunity to work on higher-level questions so that they can put ideas together and take them apart.

> **...the only thing that is always wrong is not to try.**

For us to give students the confidence they need to develop a problem-solving attitude, we have to be sure students are aware that they aren't supposed to know everything in advance. Students need to know that "getting the answer" is not all there is to solving problems. They must know that the only thing that is always wrong is not to try. Even if they don't know how to do something, they must learn to try something intelligent. Teachers need to be encouraging to all students.

Teachers must also demonstrate the problem-solving attitude themselves. We need to be comfortable with the idea that the teacher does not always know everything as we dispense knowledge to the students. As teachers, we must recognize that when students ask questions we don't know the answer to, that's fine. We want students to question and to know that answers will not always be handed to them.

Finally, we need to remember that students still need work on exercises. There are skills they need to know. However, skills are best taught in context. Practice should take place in the context of solving problems. Instead of doing sheets of exercises all the time, students should be doing problems that integrate the skills. This is not only more effective in helping students learn but a lot more exciting for students.

Problem solving is an exciting, interesting, fascinating way to teach. We need to move students away from dull and repetitive exercises and move them toward interesting problems and exciting applications.

Resources

Bloom, Benjamin, ed. *Taxonomy of Educational Objectives.* New York: David McKay, 1956.

Charles, Randall and Edward Silver, eds. *The Teaching and Assessing of Mathematical Problem Solving.* Reston: National Council of Teachers of Mathematics, 1988.

Curcio, Frances, ed. *Teaching and Learning: A Problem-Solving Focus.* Reston: National Council of Teachers of Mathematics, 1987.

Krulik, Stephen. *Problem Solving in School Mathematics.* Reston: National Council of Teachers of Mathematics, 1980.

—, and Jesse Rudnick. *Problem Solving: A Handbook for Teachers.* 2nd ed. Boston: Allyn and Bacon, 1987.

WORKSHOP
USING MANIPULATIVES

Many students, regardless of age, need physical models to help them connect concrete and abstract concepts. This often occurs when students move from arithmetic into algebra and geometry. Manipulatives are now being used at all grade levels to help students move from concrete to abstract levels of thinking. Manipulatives are especially useful for introducing new concepts but can also be used to reinforce previously learned concepts.

We use a very broad definiton of a manipulative: any visual or tactile tool that can model an abstract algebraic or geometric concept, algorithm, or property. Manipulatives can give a dynamic quality to mathematics that cannot be communicated orally or through the printed pages of a textbook. Manipulatives also provide students with an avenue for constructing their own knowledge and understanding of mathematics.

Some manipulatives can be commercially purchased, such as geometric models, algebra mods, reflectas, graphing calculators, and spreadsheets. Others can be simple teacher-constructed items, such as shapes cut from cardboard, string, homemade geoboards, and number lines. Often the best manipulatives are those that are teacher and/or student constructed and designed to introduce the particular lesson for the day.

Ample opportunities are provided in *Gateways* for teachers to use manipulatives with their students. For example, the work with patterns in Chapter 1 is an ideal time to have students use pattern blocks, tiles, and other shapes to construct, create, and extend patterns. Some of the patterns in this chapter are created through rotations. These types of patterns can be demonstrated much better by using manipulative models than by using the static pictures in a textbook.

Spreadsheets are another tool that give students visual images that link arithmetic data, algebraic formulas and graphical representations. By using a "what if" approach to spreadsheets, students see the effects of changing data that are linked via algebraic formulas. In addition, students can dynamically see these effects graphically. Spreadsheets are formally introduced in Chapter 1, further developed in Chapter 2, and then are integrated throughout the rest of the book.

When reviewing the relationships between linear measurements, area measurements, and volume measurements, using blocks can help to clarify these ideas and develop a sense of understanding that is only attainable through the use of a manipulative. In the work with measurement problems in

Gateways, it is assumed that students will actually measure things by using rulers, protractors, angle rulers, and volumetric-measuring devices such as measuring cups. Many problems in Chapter 3 are designed to have students actually measure figures in the text. Students should also be encouraged to do some form of investigation requiring measurement.

In Chapter 4, where the concepts of ratio and proportion are presented, manipulatives can be used to make these ideas concrete to the student. The experience of working with actual maps, blueprints, and other items with a scale will make the abstract ideas of ratio and proportion become more meaningful.

Chapter 5, which deals with data analysis, is full of opportunities for the use of manipulatives in the classroom. Students can construct their own circle graphs, bar graphs, and other charts. They can gather real data, then organize and display the data in some appropriate visual manner. Students will be able to apply what they learn in Chapter 5 to real-world data displays and analyze the quality and fairness of these displays.

In Chapter 6, signed numbers and the operations on these numbers can be effectively modeled with algebra mods or colored tiles.

In Chapter 7, geometry is presented dynamically through stretching, reflecting, translating, and rotating. The use of manipulatives such as geoboards, string, geometric-figure cutouts, reflectas, and individual coordinate systems are ideal for this chapter. Students can manipulate figures on coordinate systems to visualize where the vertices will move under a reflection, a translation, or any other rigid motion. With geoboards, students can conjecture about the invariant properties of a geometric shape. In fact, they can see how one geometric shape can be transformed into another. Software is becoming available that allows students to dynamically work with geometric shapes on a computer. You can use a mouse to pull and adjust the shape of figures and look at the effects on angles, sides, and other attributes of the figure. Manipulating geometry on a computer provides a powerful free-form conjecturing device that aids in the inductive aspects of the study of geometry.

In Chapter 8, algebra mods or colored tiles can be used effectively to introduce operations with variables, such as the distributive property of multiplication over addition, the combining of like terms, the solving of equations, and simple factoring. Using a pan balance with weights is also an effective way to introduce ideas for solving algebraic equations.

In Chapter 10, we recommend that students actually construct Venn diagrams and factor trees and use algebra mods or colored tiles to look for common factors when discussing greatest common factors.

In Chapter 11, constructing solid models will help students see, feel, and discover the various properties of the solid figures that are studied in this chapter.

In Chapter 12, manipulatives can be used to introduce fractals. Have students construct their own fractal pictures. This is an excellent opportunity for students to research an interesting and very modern branch of mathematics.

In addition to helping students bridge the gap between the concrete and the abstract, manipulatives provide mathematical activities that enliven the study of mathematics and bring interest into the mathematics classroom. Students enjoy different approaches to and variety in their learning. Manipulatives provide opportunities to vary the daily routine and make the classroom a more exciting place. In addition, they address the needs of visual-and-tactile learners who have been neglected for too long. We hope you will use manipulatives, where appropriate, throughout the year. They will enhance the students' understanding and help provide needed opportunities to connect students' prior arithmetic work to their later algebraic and geometric studies.

Resources

Denu, Barbara. "The Marvels of the Manipulative: Teaching Mathematics with the Aid of Manipulatives." *Instructor* 101 (1992): 44–45.

Olson, Alton. *Mathematics Through Paper Folding.* Reston: National Council of Teachers of Mathematics, 1975.

Reesink, Carole, ed. *Teacher-Made Aids for Elementary School Mathematics: Readings from the* Arithmetic Teacher. Vol. 2. Reston: National Council of Teachers of Mathematics, 1985.

WORKSHOP

LEARNING STYLES

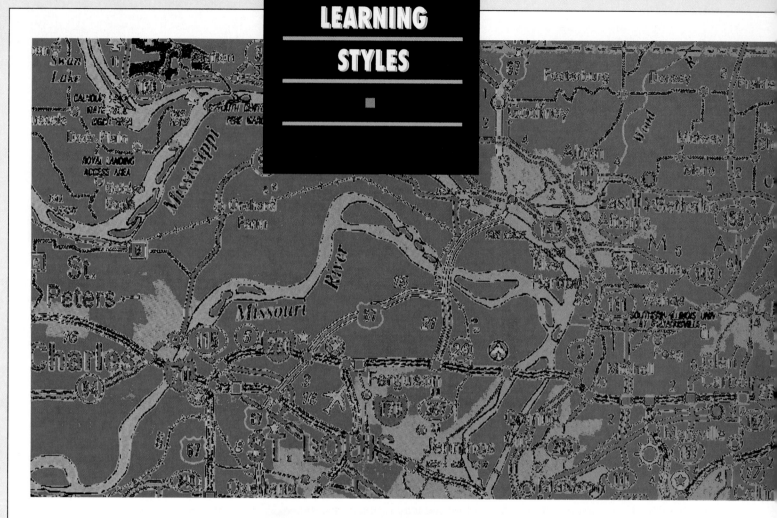

Think about what happens when you ask someone for directions. The person might draw a map, might give specific instructions involving street names and distances, or might give a more general description that includes sentences like "When you pass under the bridge, look for a large church." The same thing happens if you ask people how to do something: one person does it for you, another draws a picture, another writes specific, detailed instructions, and another uses a model to show you how it is done.

What happens is that each person gives directions or explains how to do something in the style that works best for him or her. Each assumes that the method that is the most helpful to him or her will be the method that is most helpful to you. If your learning style matches that particular teaching style, the directions or instructions will be clear and easy to follow. If not, the directions or instructions will not be helpful. A perfectly detailed map will not be of much help in the hands of someone who is not a visual learner.

This phenomenon causes major problems in the classroom. Frequently, the teacher presents material in the style that is the most logical to him or her. When this style matches the learning style of a student, the student understands and thinks the teacher is wonderful. The student across the aisle may have a different learning style, and consequently that student has difficulty understanding. Furthermore, the teacher has probably selected a textbook that matches his or her learning style, so the student who is not receptive to that style experiences the same frustrations from different directions while knowing that students who match the learning style of the teacher are doing quite well. This student begins to think math is impossible to learn and gives up.

What makes the situation even worse is that teachers tend to think that struggling students need more drill and practice, easier problems, and lots of simpler examples. If the student is a visual learner or a global thinker and has trouble generalizing from examples, this one-dimensional approach does not help the situation. The student may be able to finish a work sheet, but there is no progress toward real learning and understanding.

We became acutely aware of this problem as we collaborated on writing *Gateways*. The six of us have different learning styles, and consequently the first draft of a lesson by one author was usually met with many objections from the others. The result is a textbook that represents several learning styles. Most major concepts are addressed by means of concrete examples and models, real-world applications, visual examples and models, pattern recognition, logical argument, and eventually abstract generalization. This is done both in the exposition and in the problem sets. The result is a multidimensional book in which students with various learning styles will find an open door to the ones that suit them best.

There is yet another benefit in providing for a variety of learning styles. A powerful mathematics student needs to develop many learning styles to become a better problem solver, since some problems are intrinsically visual and others are abstract, some are symbolic and others are concrete. Since *Gateways* has such variety, students will have an opportunity to experience a number of different approaches and thus to strengthen their weak areas.

One of the things you should expect from this book is that students will come up with surprising ways to solve problems. A problem you thought was very difficult may be solved in a manner you hadn't thought of by one of your students —frequently not the best student. Be prepared to get excited when you see how students approach problems, and don't underestimate them. After many years in the classroom, we continue to be amazed at how clever students can be if given the opportunity, and *Gateways* gives them plenty of opportunities.

Resources

Claxton, Charles, and Patricia Murrell. *Learning Styles: Implications for Improving Educational Practices*. College Station: Association for the Study of Higher Education, 1988.

Cocking, Rodney, and Jose Mestre, eds. *Linguistic and Cultural Influences on Learning Mathematics*. Hillsdale: Lawrence Erlbaum Associates, 1988.

Schmeck, Ronald, Ed. *Learning Strategies and Learning Styles*. New York: Plenum Press, 1988.

WORKSHOP: ALTERNATIVE ASSESSMENT

"More than half of each semester grade that I assign comes from test scores. The rest comes from quiz scores and homework. Where does the idea of alternative assessment fit into what I've done for years?" If these words could have come from your mouth, read on.

Many new ways of looking at mathematics assessment have been popularized in the past few years because the most common answer to the question, "What is sure evidence that a person knows mathematics?" is "The person effectively uses mathematics in real-life situations." Employers are complaining, however, that graduates of our schools often fall short when called upon to apply mathematics at the workplace.

But let's face it, your classroom is not a place in which real-life mathematics can be perfectly modeled. You have to deal with many students at a time. You have a lot of material to cover. You feel that some of your students probably lack the discipline needed to handle open-ended, real-life problems in a classroom setting. You don't have the extra time needed to grade these real-life mathematical excursions, and, if you ever found the time, you would have no idea how to grade them.

All of these concerns are valid. However, the challenge to bridge the gap between real-life mathematics and what we do in the classroom still exists. In the words of the National Research Council, "We must ensure that tests measure what is of value, not just what is easy to test. If we want students to investigate, explore, and discover, assessment must not measure just mimicry mathematics" (1989). The solution? Compromise. Take a few steps to open your classroom and, more significantly, your grade book to more real-life mathematics, even if the steps are small. *Gateways* offers you a variety of ways to do this. Here are a few:

Portfolio projects

Starting a portfolio system need be no more complicated than having students put some of their work in manila folders and storing the folders in a box. The reasons for initiating a portfolio system are both practical (students know where their work is and spot-checking is easy) and pedagogical (students feel a sense of ownership for their work and they can see their own growth over time). The portfolio can contain projects in progress, projects that were begun then abandoned, and completed projects. Some teachers like to set up separate portfolios for projects in progress and completed projects.

What kind of project goes into a portfolio? The answer will vary according to your teaching style and students. To start with, you may decide to use students' best work from the week as their project: something as simple as "my answer to number 9 in section 4.3" is entirely appropriate for students who have not worked much with open-ended problems. You may also want to refer to the Portfolio Projects section of the *Teacher's Edition* (located on the

"THEN, AS YOU CAN SEE, WE GIVE THEM SOME MULTIPLE CHOICE TESTS."

same page as the Chapter Test). Each of these sections lists about a dozen projects, all of which can be found in, or are related to, the problems found in the chapter.

Acceptable projects will come in a variety of forms: written material, computer programs, graphs, charts, models, spreadsheet applications, class presentations, videos, and artwork are a few examples. Again, if your students are new at this kind of thing, you might start out the year by selecting projects for them, then gradually give them more say in selecting both the projects and the forms they will take.

Responses to these open-ended portfolio projects don't fall into easy-to-score categories. Two students, for example, might take different approaches to a problem and come up with excellent answers that differ in many ways. We recommend, therefore, that you grade students' responses to some of these more open-ended tasks using a holistic method similar to the methods that writing teachers have been using for years to evaluate their students' work. The NCTM publication *Mathematics Assessment: Myths, Models, Good Questions, and Practical Suggestions* (1991) offers some excellent options for scoring student work.

Group projects

Making group work count as part of students' grades reinforces the importance of teamwork—a skill that definitely comes into play in real-world mathematics. Like portfolio projects, group projects can vary in form and scope. Again, a good starting point may be to have students work together on particularly challenging problems in the text. Other suggestions for group work can be found in the Cooperative Learning sections in the *Teacher's Edition* and the Cooperative Learning Activities component of the *Teacher's Resource File*.

Methods for assigning scores to group work include having each group submit one write-up of its work, having each student submit a write-up, or having one or several group members present the results of the group's work to the rest of the class. The key is to get all students involved—a difficult challenge with some students—and to establish an atmosphere in which students are rooting for each other to do well.

Grading and Testing

The problems in *Gateways,* as you probably have noticed, are different from books in which assignments take the form "1–44 evens." Some problems are similar to these "skills" books, but most are multi-faceted. You should not expect, then, that students perform at the level that they would if the book were simply skills-oriented. We recommend, therefore, that if you use a standard 90-80-70-60 scale, you either grade tests and quizzes on a curve or give students partial credit for problems in which their work shows some understanding of the concepts.

The *Gateways Teacher's Resource File* includes six standard tests (Forms A and B at each of three ability levels) and one alternative test (Form C) for each chapter. The Form C tests are open-ended cousins of the Forms A and B tests. They are written to be appropriate for students of any ability level and cover the same material as the Form A and B tests, but with fewer, broader problems. To use an analogy: if this were a literature series, the Form C tests would be the essay tests. To help with the scoring process, we have provided detailed scoring guidelines with the key for each Form C test.

In conclusion, please remember that incorporating new means of assessment into your teaching is a matter of compromise. The steps you take will be dictated by your situation. Many teachers are reporting exciting results as they venture into these new, "untested" waters. It is our wish that, through the use of *Gateways,* you find similar success. Our book offers you many ways to open your classroom and your grade book to new and meaningful forms of assessment.

Resources

Airasian, Peter. *Classroom Assessment.* New York: McGraw-Hill, 1991.

Archbald, Douglas. *Beyond Standardized Testing: Assessing Authentic Academic Achievement in the Secondary School.* Reston: National Association of Secondary School Principals, 1988.

Cain, Ralph, and Patricia Kenney. "A Joint Vision for Classroom Assessment." *Mathematics Teacher* 85 (1992): 612-15.

Clarke, David. "Activating Assessment Alternatives in Mathematics." *Arithmetic Teacher* 39 (1992): 24–29.

Stenmark, Jean. *Assessment Alternatives in Mathematics: An Overview of Assessment Techniques That Promote Learning.* Berkeley: Lawrence Hall of Science, 1989.

—, ed. *Mathematics Assessment: Myths, Models, Good Questions, and Practical Suggestions.* Reston: National Council of Teachers of Mathematics, 1991.

WORKSHOP

A WORD ON MATH ANXIETY

BY SHEILA TOBIAS

Everybody's had it at some time in their lives. Even students who have gone on to do very well in mathematics and math-based fields of study.

Even math teachers.

"Math anxiety" is not a disease, though it can be prevented and cured.

It's a feeling of helplessness accompanied by real panic, when we have to do something we either don't know or have forgotten how to do.

Most students don't have a full-blown case of math anxiety. They just have some anxious moments when it comes to learning math. In ten years of interviewing students who think they could do better at math if they didn't get the shakes, I have learned much about where math anxiety comes from and what students and their teachers can do to prevent it and overcome it.

Time Pressure

For some students math anxiety starts with timed tests and with the elementary school work sheets that have one hundred problems to add, subtract, multiply, or divide in something like ten minutes. No wonder kids get anxious. Who can think while the clock is ticking away? One of the college students interviewed in a math-anxiety clinic said that the only things she ever really learned in elementary school were how to subtract and how to divide. During tests, she said, she kept looking at the clock on her classroom wall, trying to figure out how many minutes she had left for the test and then counting off the number of problems she still had to do. When she realized she had, say, one-half minute for each remaining calculation, she stopped trying altogether and gave up.

It's true that in order to solve problems students need to understand the ideas on which arithmetic, algebra, geometry, and, later, calculus are based. But it is also true that if students are going to be able to concentrate on the difficult stuff, they are going to have to do the "easy" stuff fast. That's why teachers insist that they practice arithmetic, and, later, algebraic skills. You know that the more they practice, the better and faster they can zip through calculations (in arithmetic) and manipulations (in algebra). But students don't know this and become anxious when they have to think under pressure. Teachers need to explain that the better students become at calculations and manipulations, the more they can concentrate on the problem solving that is the focus of this text.

It is mostly at the beginning of every new lesson that students need more time to think. The wise teacher will encourage them to take their time and to develop speed without anxiety.

Time Pressure Exercise

If you think time pressure might be the source of your students' math-anxious moments, try this experiment in class: Move the chairs to face the classroom clock. (If there isn't a clock everyone can see, bring one in.) Then assign the class one-half of their homework under strict time limits. (You might even use an alarm clock that goes off every so many minutes, indicating that time's up.) Then turn the clock to the wall (or cover it with a cloth) and have them do the second half of their homework problems, giving them the feeling of having "all the time in the world" to finish. Watch them work and ask them to compare how they feel under the two artificially constructed situations. Does time pressure make them more nervous? More efficient? Do some of them get more right answers going slowly? Fewer right answers? Perhaps, if time pressure is a major problem, you can make some special accommodations—especially during tests—while students get up to speed.

Solving Problems

Some students do just fine at calculations and manipulations. Their anxious moments come with problems, or real-world applications of what they've learned in school. What makes problems difficult? There is a "Great Divide" between a task like "Go to the chalkboard and multiply 14 times 76" and any of the problems in this text. In the first instance, the student will not know the answer but will surely know what to do. In the multistep problems featured in this text, not only will students not know the answer, they may not even know where to begin. One thing is certain: Until students learn to cope in this kind of environment of uncertainty, anxiety will develop.

The successful student starts to work the problem even before he or she figures out what to do. And the courage to do this is the cure for math anxiety.

Another difficulty students have is in *interpretation.* Problems require the translation of sentences and questions into mathematical statements and symbols. Since there may be more than one interpretation of a sentence (mathematics statements are more precise than English sentences), there is room for doubt and more uncertainty.

One solution to the problem of interpretation is for students to stay with the English sentences a little longer. They may rewrite the problem in their own words, may even write out the two or three interpretations that suggest themselves. If they estimate an answer (roughly calculated) for each interpretation, they may see which one will produce the most reasonable answer. They can be encouraged to do their homework with a friend or to talk on the telephone with a classmate to see whether their interpretations agree. The important thing to communicate to students is that differences in interpretation are possible and that the student who reads a problem differently is not dumb.

One of the great strengths of this text are the Think and Discuss exercises the authors recommend. Some of the exercises are designed particularly to give students the all-important "windows" into their thinking that successful students in mathematics construct for themselves. The exercises ask students to *explain,* to *describe,* and to *discuss* their *start-up decisions* and to *keep track* of their *decision strategies* as they work a problem. This training in self-monitoring and in writing down what they are thinking is very important in giving students a sense of control over their material and over themselves, and can be augmented by the classroom teacher through the following exercise:

Ask students to divide their homework or test page down the middle with a vertical line. On the right-hand side of the line, they should write each problem in mathematical symbols and work the solution. On the left-hand side of the line, they should write down one or more plans of action to solve the problem. Writing down their thoughts has two real benefits: As long as their hands keep moving, their minds won't block. And their notes to themselves will keep them from losing their way.

You will probably not want to collect these jottings, because they are your students' personal notes to themselves. But once in a while some class time can be devoted to talking about the left-hand side of the page. Students need to know that their friends—even the ones who are writing all the time—have difficulties, too.

Getting Unstuck
The book you are about to use is the product of a team of teachers who want your students to succeed. The book is arranged so logically that if, while doing their homework, your students forget how the teacher explained something, they don't need to panic, even if it is the last evening before a test. You can instruct them simply to turn back a few pages and review the previous day's or week's material by themselves.

Still, even the most experienced teachers or the best textbook authors cannot get themselves into every student's mind and figure out exactly what is confusing him or her. Since math anxiety comes from math helplessness and the best cure for math anxiety is getting students back in control, it is important for them to learn how to get themselves unstuck. You can help your students by teaching them, systematically, not simply what to do when they remember what to do, but what to do when they forget. One way to stop the end-of-pencil chewing is to teach them the following exercise to do alone or with a friend:

Getting-Unstuck Exercise
Answer, in writing, the following questions:
- What do I know about the problem?
- What are they asking me to find out?
- What is making this problem difficult for me?
- What can I do to make this problem easier for myself?

Working with friends can be very helpful. Educators are finally discovering that though math is not a spoken language, it helps when students talk to their friends about their homework. Students struggle together better than they struggle alone, and in explaining the examples and sample problems to one another, they learn them well.

Who Can Do Math and Why?
Some teenagers feel uncomfortable doing higher math because of what sociologists call role-stereotyping —that is, because they have the wrong idea that math is for boys

Early stages of math anxiety

and not for girls, for white people and not African Americans and Hispanic Americans. It is true that if you look around, you see fewer girls than boys and fewer minorities than Caucasians doing higher mathematics, engineering, and science. But that's because of past discrimination and lack of opportunity, not because girls and minorities lack some "math gene." Some people disagree, but the evidence is overwhelming: Where opportunities are the same and where there are no prejudices against them, girls, African Americans, and Hispanic Americans do very well. Teachers need to be more than color- and gender-blind. So long as these prejudices exist outside the classroom, teachers will need to counter students' sometimes negative assessment of their own potential and skills.

Another related idea that gets in the way of success in math is that students can't be good in both math and English or math and social studies. Certainly, people have their favorites. But really successful people are good at both. That's because the skills of good writing and the skills of good problem solving overlap. Writing a logical argument is similar to setting up an equation to solve a problem. Combining like terms in an algebraic statement is good practice for editing out unnecessary words and phrases in an English paper.

As a teacher, you can do much to counter these wrong ideas. You can direct students to biographical information about a more varied group of mathematicians and scientists. Interviews with family members or family friends who work with numbers will quickly dispel the notion that such workers do not need to be good with words.

Another wrong idea and source of math anxiety is that memorizing can get students through. While memorizing has its place—it's appropriate for learning times tables, techniques such as "invert and multiply," and even area and perimeter formulas—algebra is the beginning of the end for math memorizers. Memorizing will cause your students more anxious moments than it will cure.

In the atmosphere of a trusting and noncompetitive classroom, led by a teacher who truly believes his or her students can succeed with math, with ample time for thinking about a problem and with tolerance for error, students can learn to cope with their math-anxious moments and even to learn from their mistakes. But where anxiety persists or seems to be ingrained, where parents are discouraging their daughters or putting too much pressure on their sons, the wise teacher will ask a counselor for help or referral. Math-anxiety reduction clinics exist in many locales. My books are helpful, too. Not every student will want to major in mathematics or become professionally involved, but as I wrote in *Suceed with Math,* anyone who wants to lead a rich intellectual and professional life needs to know more math.

GATEWAYS
to Algebra and Geometry
AN INTEGRATED APPROACH

GATEWAYS
to Algebra and Geometry
AN INTEGRATED APPROACH

John Benson
Evanston Township High School
Evanston, Illinois

Sara Dodge
Evanston Township High School
Evanston, Illinois

Walter Dodge
New Trier High School
Winnetka, Illinois

Charles Hamberg
Illinois Mathematics and Science Academy
Aurora, Illinois

George Milauskas
Illinois Mathematics and Science Academy
Aurora, Illinois

Richard Rukin
Evanston Township High School
Evanston, Illinois

McDougal, Littell & Company
Evanston, Illinois
New York   Dallas   Columbia, SC

WARNING: No part of this book may be reproduced or transmitted in any form or by any means, electronic or mechanical, including photocopying, recording, or by any information storage and retrieval system without permission in writing from the Publisher.

ISBN 0-8123-7645-5

Copyright © 1994, 1993 by McDougal, Littell & Company
Box 1667, Evanston, Illinois 60204

All rights reserved.
Printed in the United States of America.

A Special Thank You

A special thank you to the teachers, students, and administrators who helped review and field test **Gateways to Algebra and Geometry**. Their suggestions, comments, and criticisms were invaluable.

Reviewers and Field Test Teachers

Patricia K. Amoroso
Madison High School
Vienna, VA

Dr. Joel L. Arougheti
High School of Art and Design
New York, NY

Larry E. Deis
El Molino High School
Forestville, CA

Dr. Shirley Frye
Numerics, Inc.
Scottsdale, AZ

Madelaine S. Gallin
New York City Board of Education
New York, NY

Elizabeth B. Gilmore
North View Junior High
Brooklyn Park, MN

Dr. Zelda Gold
Los Angeles Unified School
  District
Woodland Hills, CA

Fern Hartman
Wilmette Junior High School
Wilmette, IL

Peg Keller
Roehm Middle School
Berea, OH

Sandy Mackay
Southern Regional Middle School
Manahawkin, NJ

Dr. Stephen E. Moresh
Seward Park High School
New York, NY

Sally J. Nelson
Garrett Junior High School
Boulder City, NV

Dr. Gail Lowe Parrino
Conejo Valley Unified School
  District
Thousand Oaks, CA

Sabina Raab
South Lakes High School
Reston, VA

Glenda Sadler
Columbine High School
Littleton, CO

Guy Sanders
Prospect High School
Saratoga, CA

Sandra M. Schoff
Anchorage School District
Anchorage, AK

Barb Simak
McClure Junior High School
Western Springs, IL

Maureen Sowinski
Roehm Middle School
Berea, OH

Richard L. Stroud
Central Kitsap Junior High
Silverdale, WA

Fred Symonds
Sehome High School
Bellingham, WA

Edward E. Wachtel
Orange High School
Pepper Pike, OH

Mary Wagner
Girard College
Philadelphia, PA

Mary Williams
Elk Grove High School
Elk Grove Village, IL

Multicultural Advisory Committee

Carlos Cumpian
Movimento Artistico Chicano
Chicago, IL

Sandra Mehojah (Kaw/Cherokee)
Office of Indian Education
Omaha Public Schools
Omaha, NE

Alice Kawazoe
Oakland Unified School District
Oakland, CA

Alexs D. Pate
University of Minnesota
Minneapolis, MN

CONTENTS

A Letter to the Student   1

CHAPTER 1: Patterns   2

1.1 A Tiling Problem   4
1.2 Number and Letter Patterns   10
1.3 Practical Patterns   16
     A Family Tree
     Other Applied Patterns
1.4 Probability   23
     What Is Probability?
     Applying Probability
1.5 Tables and Spreadsheets   29
     The Cost of Postage Stamps
     Spreadsheets
1.6 Communicating Mathematics   36
     Describing a Rectangle
     Explaining the Rounding Process

Investigations

Escape Velocity 3   **Tessellation** 9   **Patterns on a Telephone** 15
Technology 22   **Leap Year** 28   **Spreadsheets** 35   **Geometry** 41

Chapter Summary   42
Chapter Review   43
Chapter Test   46
Puzzles and Challenges   47

CHAPTER 2: Formulas and Percent   48

2.1 Operations on Numbers   50
     Which Operation Comes First?
     Operation Symbols
2.2 Powers and Roots   57
     Exponents
     Roots
     Order of Operations
2.3 Geometric Formulas   64
     Area and Circumference Formulas
     The Pythagorean Theorem

2.4 Percent 71
What Is a Percent?
Percent Problems

2.5 More About Spreadsheets 78

Investigations

Selling Prices 49 **Mental Math** 56 **Algebra** 63 **Volume** 70
Decimals 77 **Sports** 85

Chapter Summary 86
Chapter Review 87
Chapter Test 90
Puzzles and Challenges 91

CHAPTER 3: Measurement and Estimation 92

3.1 Linear Measurement 94
Linear Units
Measurement and Polygons

3.2 Scientific Notation 101
Scientific Notation and Large Numbers
Scientific Notation and Small Numbers

3.3 Estimation and Rounding 107
Accuracy
Rounding

3.4 Understanding Equations 114
Solving an Equation
Estimating Solutions
Inverse Operations

3.5 Angle Measurement 121
Rotations
Naming Angles
Measuring Angles

3.6 Constructions with Ruler and Protractor 129

3.7 Geometric Properties 136

Investigations

The Golden Rectangle 93 **The English System** 100 **History** 106
In the News 113 **Merchandising** 120 **Early Civilization** 128
Geometry 135 **Geometry** 141

Chapter Summary	142
Chapter Review	143
Chapter Test	146
Puzzles and Challenges	147

CHAPTER 4: Ratio and Proportion 148

4.1 A Closer Look at Measurement 150
Attributes and Units
Metric Units
Operations with Measurements

4.2 Ratios of Measurements 158
Ratios
Rates
Unit Costs

4.3 More Applications of Ratios 165
Scale and Ratios
Density

4.4 Working with Units 172

4.5 Proportions 178
What Is a Proportion?
Solving Proportions

4.6 Similarity 185
Enlargements and Reductions
Similar Geometric Figures

Investigations

Lift-to-Drag Ratios 149 **Mass and Weight** 157 **Radar Guns** 164
Auto Trip 171 **Water Consumption** 177 **Floor Plans** 184
Similar Polygons 191

Chapter Summary	192
Chapter Review	193
Chapter Test	196
Puzzles and Challenges	197

CHAPTER 5: Data Analysis 198

5.1 Circle Graphs 200
Reading a Circle Graph
Drawing a Circle Graph

5.2 Bar Graphs and Line Graphs 207
Reading Bar Graphs and Line Graphs
Drawing Bar Graphs and Line Graphs

5.3 Other Data Displays 213
Scattergrams
Histograms
Pictographs
Artistic Graphs

5.4 Averages 221
The Mean
The Median
The Mode

5.5 Organizing Data 228
Stem-and-Leaf Plots
Box-and-Whisker Plots

5.6 Matrices 234
Organizing Data in a Matrix
Dimensions of a Matrix
Equal Matrices

Investigations

History 199   **Communications** 206   **Sports** 212   **Data Analysis** 220
Baseball 227   **Statistics** 233   **Language Arts** 241

Chapter Summary 242
Chapter Review 243
Chapter Test 247
Puzzles and Challenges 249

CHAPTER 6: Signed Numbers 250

6.1 The Number Line 252
Number Lines and Signed Numbers
Opposites
Origin and Coordinates

6.2 Inequalities 260
Graphing and Writing Inequalities
And Inequalities and *Or* Inequalities

6.3	**Adding Signed Numbers**	267
	Adding Signed Numbers on the Number Line	
	Using a Calculator to Add Signed Numbers	
6.4	**Multiplying and Dividing Signed Numbers**	275
	Multiplication of Signed Numbers	
	Scalar Multiplication	
	Powers of Signed Numbers	
	Division of Signed Numbers	
6.5	**Matrix Multiplication**	282
6.6	**Subtracting Signed Numbers**	290
	Definition of Subtraction	
	Subtracting Matrices	
6.7	**Absolute Value**	296
	Distance from Zero	
	Distance Between Two Numbers	
6.8	**Two-Dimensional Graphs**	302
	The Rectangular Coordinate System	
	Graphing Geometric Figures	

Investigations

Map Coloring 251   **Temperature Scales** 259   **Equilibrium** 266
Banking 274   **Physics** 281   **Using a Matrix** 289   **Time Differences** 295
Gasoline Mileage 301   **Floor Plan** 309

Chapter Summary	310
Chapter Review	311
Chapter Test	314
Puzzles and Challenges	315

CHAPTER 7: Exploring Geometry   316

7.1	**Line and Angle Relationships**	318
	Parallel and Perpendicular Lines	
	Adjacent and Vertical Angles	
7.2	**Quadrilaterals**	325
	Squares, Rectangles, and Parallelograms	
	Rhombuses and Trapezoids	
7.3	**Symmetry**	332
	What Is Symmetry?	
	Reflection Symmetry	
	Rotation Symmetry	

7.4 Rigid Transformations   339
Reflections
Rotations
Translations

7.5 Discovering Properties   347

7.6 Transformations and Matrices   354
The Vertex Matrix
Translations
Reflections and Rotations

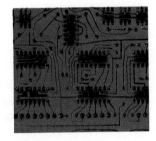

Investigations

Equipment Design 317   **Architecture** 324   **Tessellations** 331
Logos 338   **M. C. Escher** 346   **Playing-Card Symmetries** 353
Origami 361

Chapter Summary	362
Chapter Review	363
Chapter Test	366
Puzzles and Challenges	367

CHAPTER 8: The Language of Algebra   368

8.1 Simplifying Expressions   370
Equivalent Expressions
Adding and Subtracting Like Terms
The Distributive Property

8.2 Using Variables   378
Expressions to Describe Phrases
Expressions to Describe Situations
Equations to Represent Problems

8.3 Solving Equations   385
Inverse Operations
Equations with Several Operations

8.4 Some Special Equations   392
Absolute-Value Equations
Equations with Square Roots
Simplifying to Solve Equations

8.5 Solving Inequalities   399
Adding and Subtracting
Multiplying and Dividing

8.6	**Substitution** Special Substitutions Variables for Expressions	406
8.7	**Functions** What Is a Function? Evaluating Functions	413

Investigations

Fuel Efficiency 369   **The Fibonacci Sequence** 377
Science 384   **Consumer Mathematics** 391   **Science** 398
The Triangle Inequality 405   **Decimals vs. Fractions** 412
Rent Costs 419

Chapter Summary	420
Chapter Review	421
Chapter Test	424
Puzzles and Challenges	425

CHAPTER 9: Real Numbers 426

9.1	**Sets** Describing Sets Subsets	428
9.2	**Intersection and Union** Intersections and Unions of Sets Venn Diagrams	435
9.3	**Important Sets of Numbers** The Rational Numbers The Real Numbers	442
9.4	**Pi and the Circle** The Meaning of π A Property of Circles	448
9.5	**Properties of Real Numbers** Associative Properties Commutative Properties The Zero Product Property	455
9.6	**Two Coordinate Formulas** The Distance Formula The Midpoint Formula	462
9.7	**Graphs of Equations**	469

Investigations

Optical Illusion 427   **Infinite Sets** 434   **Limits** 441
Fractions 447   **Approximating** π 454   **Language** 461
Geography 468   **Data Analysis** 476

Chapter Summary	477
Chapter Review	478
Chapter Test	482
Puzzles and Challenges	483

CHAPTER 10: Topics of Number Theory 484

10.1 Techniques of Counting 486
The Fundamental Counting Principle
Counting Applications

10.2 Primes, Composites, and Multiples 494
Prime Numbers and Composite Numbers
Multiples

10.3 Divisibility Tests 501
Tests for 2, 3, 4, 5, 6, 8, 9, 10, and 12
Divisibility Test for 11

10.4 Prime Factorization 507
Prime-Factorization Form
Prime-Factorization Techniques
General Prime-Factorization Procedure

10.5 Factors and Common Factors 513
Factors of a Natural Number
Abundant, Deficient, and Perfect Numbers
Greatest Common Factors

10.6 LCM and GCF Revisited 521
Using Venn Diagrams
Simplifying Fractions
Equations Containing Fractions

10.7 Number-Theory Explorations 528
Symmetric Primes
Goldbach's Conjectures
Figurate Numbers
Pythagorean Triples

Investigations

Farming 485   **Lotteries** 493   **Tile Arrangements** 500
Numeric Palindrome 506   **Number Game** 512
Relatively Prime Numbers 520   **Lattice Points** 527
Prime Numbers 535

Chapter Summary	536
Chapter Review	537
Chapter Test	540
Puzzles and Challenges	541

CHAPTER 11: Solving Problems with Algebra   542

	Mathematics in Words	544
11.1	**Presenting Solutions to Problems**	546
11.2	**Equations with Variables on Both Sides**	553
11.3	**More Word Problems**	559
11.4	**Equations with Several Variables**	566
11.5	**Polyhedra** Prisms Pyramids	573
11.6	**Solids with Circular Cross Sections** Cylinders Cones Spheres	581
11.7	**Reading Information from Graphs**	589

Investigations

Water Conservation 543   **Language** 552   **Balance** 558
The Teeter-totter Principle 565   **Chemistry** 572   **Polyhedra** 580
Ratio 588   **Graphs** 595

Chapter Summary	596
Chapter Review	597
Chapter Test	600
Puzzles and Challenges	601

CHAPTER 12: Looking Ahead: Iterations and Fractals 602

12.1 Iterations 604
Repeated Calculations
Iteration Diagrams

12.2 More About Iterations 612
Rules Involving the Iteration Counter
Function Notation for Iteration

12.3 Geometric Iterations 620

12.4 Convergent Iterations 625
Convergence
Limiting Values

12.5 Introduction to Fractals 634

Investigations

Music 603   **The Calendar** 611   **Mental Math** 619   **Snowflake Curve** 624
Consumer Math 633   **Technology** 639

Chapter Summary	640
Chapter Review	641
Chapter Test	644
Puzzles and Challenges	645

Handbook of Arithmetic Skills	646
Tables and Charts	660
Glossary	666
Answers to Odd-Numbered Problems	673
Index	695
Credits	703

Letter to the Student

Welcome to *Gateways to Algebra and Geometry: An Integrated Approach*. This book is called *Gateways* because it is an entrance into the world of algebra and geometry. We will build on what you know and will help you learn enough mathematics so you will be well prepared to study algebra and geometry.

Expect to learn something new each day. Some days you may feel confused, but that is OK. Sometimes it is necessary to work on an idea over a period of several days before you understand it. Then it is necessary to keep using the idea to become comfortable with it. All of this takes time. The text and problems are written in a way that requires you to work with important ideas over a period of many weeks.

We have written *Gateways* for you. We expect you to be able to read it and learn from it. You can also use it as a reference and as a source for problems.

Let us explain how the book is organized.

- **Text:** In most sections, new material will be introduced by using a few problems and an explanation that relates to these problems. Often there will be pictures and examples as well. Sample problems will appear before the problem sets. You might find it helpful to model your solutions after the examples and sample problems.

- **Problem Sets:** The Problem Sets are divided into four parts.
    1. The **Think and Discuss** problems will often be done orally in class with your teacher. These questions should help clarify the text you will have just read, and they may raise interesting questions.

2. The **Problems and Applications** will give you a chance to use the new mathematical material in a variety of situations. These problems will provide practice and will help you connect the new ideas to previously learned material. These problems will also help you to connect mathematical ideas with other subjects, such as business, science and geography.
3. The **Spiral Learning** problems will provide additional practice with material from previous sections, and will also prepare you for what you will learn later.
4. The **Investigation** at the end of each section will provide you with open-ended questions to explore and discuss with others. (Your teacher may also choose to work with you on some of the long-term unit investigations available in a softcover book entitled *Gateways to Mathematical Investigations*.)

In addition to learning specific facts and techniques, you will learn to *think mathematically*. You will learn how to approach a problem that you don't know how to do, and you will learn to think critically about problems you have solved.

You will continually be asking yourself questions like these: Should I use my calculator? Is there another way to look at this problem? Should I estimate? Is my answer reasonable? Is there a pattern here? Is there more than one correct answer? Do I need more information? And so on.

You will encounter many interesting and challenging problems in this book.

ENJOY!

John Benson *Charles L. Hamberg*

Sara H. Dodge *George Milauskas*

Walter Dodge *Richard Rukin*

OVERVIEW

LONG-TERM INVESTIGATIONS

See *Investigations* book, Unit A, pages 2-25.

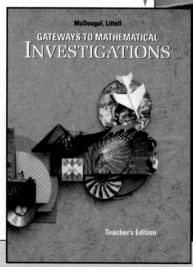

Gateways to Mathematical Investigations is a collection of long-term, real-world investigations that provides students with exciting opportunities to apply the ideas and concepts introduced in *Gateways to Algebra and Geometry: An Integrated Approach*. Unit A: Thought for Food consists of five investigations that develop the theme of opening a restaurant business. This unit is designed so that you can pick and choose the order and number of investigations you would like to do with your students. Each investigation is an exploration that takes several days to complete. Refer to the menu at the right to learn more about the investigations that make up this unit.

As your students work through the Unit A investigations, use Chapters 1 and 2 of *Gateways to Algebra and Geometry: An Integrated Approach* for concept and skill development, and as a source of rich problems, exercises, and applications. When appropriate, refer students to additional supportive sections in *Gateways to Algebra and Geometry* as well.

UNIT A: Thought for Food

MENU OF INVESTIGATIONS

OVERVIEW: Groups apply mathematical concepts and strategies from Chapters 1 and 2 within the context of opening a new, hypothetical restaurant.

Getting Down to Business

(3-4 days)
Students choose and develop strategies for collecting data that will help them decide what kind of restaurant they should open. Students organize, display, and analyze their data, and then decide on a restaurant.
STRANDS
Number, Measurement, Logic and Language, Statistics and Probability
UNIFYING IDEA
Patterns and Generalizations

Designing the Space

(3-5 days)
Students apply the concepts of area and proportional reasoning within the context of designing a restaurant space. They choose a size and shape for their restaurant space. They figure out how to place all the furniture and fixtures within the space. They create a floor plan of the restaurant.
STRANDS
Functions, Algebra, Geometry, Measurement
UNIFYING IDEA
Proportional Relationships

Staffing the Restaurant

(3-5 days)
Students use tables and spreadsheets, and they apply the concepts of percent and basic numeric operations to develop working schedules and payroll tables for restaurant employees.
STRANDS
Functions, Number, Statistics and Probability
UNIFYING IDEA
Patterns and Generalizations

Game Night

(2-3 days)
Students use probability concepts to determine empirical, or experimental, probabilities of winning several games of chance on "Game Night" at the restaurant. They analyze the rules and results of games they play.
STRAND
Statistics and Probability
UNIFYING IDEA
Multiple Representations

A Menu in the Making

(3-5 days)
Students use mathematical tools such as organized lists and tree diagrams to create and organize attractive, competitive restaurant menus. They use tables, spreadsheets, and percents to analyze the effects of various percent discounts on their revenues.
STRANDS
Functions, Number, Statistics and Probability, Discrete Mathematics, Logic and Language
UNIFYING IDEA
Multiple Representations

INVESTIGATIONS FOR UNIT A	SCHEDULE	OBJECTIVES
1. **Getting Down to Business** Pages 5-8*	3-4 Days	• To gather from various sources data about the restaurant business and then organize and analyze the data to make informed decisions about opening a restaurant
2. **Designing the Space** Pages 9-12*	3-5 Days	• To develop and apply area concepts and proportional reasoning within the context of designing a space for a restaurant
3. **Staffing the Restaurant** Pages 13-16*	3-5 Days	• To use tables and spreadsheets to develop working schedules and payroll tables for restaurant employees
4. **Game Night** Pages 17-20*	2-3 Days	• To use probability concepts and to find empirical, or experimental, probabilities of winning a game of chance on "Game Night" at the restaurant
5. **A Menu in the Making** Pages 21-24*	3-5 Days	• To create a menu for the restaurant; and then to combine menu items using various strategies, such as tree diagrams and organized lists, to create special one-price meals offered by the restaurant; to discount menu items using percentages and to compare revenues by varying discounts to establish feasible menu prices

*Pages refer to *Gateways to Mathematical Investigations* Teacher's Edition.

PLANNING
LONG-TERM
INVESTIGATIONS

MATERIALS	RELATED RESOURCES
• Family Letter, 1 per student • Newspapers, including classified ads of businesses for sale and commercial spaces for rent, 1-3 per group • The local telephone company's yellow pages, 1 per group • Chart paper, 1-3 sheets per group • *Gateways to Mathematical Investigations* Student Edition, pp. 1-2, 1 of each per student	**Gateways to Algebra and Geometry** Section 1.3, pp. 16-22: Practical Patterns Section 1.5, pp. 29-35: Tables and Spreadsheets Section 1.6, pp. 36-41: Communicating Mathematics *Additional References* Section 5.2, pp. 207-212: Bar Graphs and Line Graphs Teacher's Resource File: Student Activity Masters
• Calculator; at least 1 per group • Grid paper, several sheets per group • Tape measure, ruler, scissors; 1 of each per group • Compass, protractor; 1 per group • Attribute blocks, square tiles, or cutouts of tiles (optional) • Construction paper, glue (optional) • *Gateways to Mathematical Investigations* Student Edition, pp. 3-4, 1 of each per student	**Gateways to Algebra and Geometry** Section 2.1, pp. 50-56: Operations on Numbers Section 2.2, pp. 57-63: Powers and Roots Section 2.3, pp. 64-70: Geometric Formulas *Additional References* Section 4.2, pp. 158-164: Ratios of Measurements; Section 4.3, pp. 165-171: More Applications of Ratios Teacher's Resource File: Student Activity Masters
• Calculator, at least 1 per group • Grid paper, 1-2 sheets per student • Blank spreadsheet, 1 per student • Compass, protractor, 1 of each per group • Computer, spreadsheet software (optional) • *Gateways to Mathematical Investigations* Student Edition, pp. 5-6, 1 of each per student	**Gateways to Algebra and Geometry** Section 1.5, pp. 29-35: Tables and Spreadsheets Section 2.4, pp. 71-77: Percent Section 2.5, pp. 78-85: More About Spreadsheets *Additional References* Section 5.1, pp. 200-206: Circle Graphs Teacher's Resource File: Student Activity Masters
• Calculator; at least 1 per group • Spinner with 1-6 markings (six 60-degree sectors), 1 per group • Meal cards (20 index cards: write Entree on 4 of the cards; Dessert on 6 of the cards; and Beverage on 10 of the cards.), 1 set per group • *Gateways to Mathematical Investigations* Student Edition, pp. 7-8, 1 of each per student	**Gateways to Algebra and Geometry** Section 1.4, pp. 23-28: Probability *Additional References* Teacher's Resource File: Student Activity Masters
• Calculator, at least 1 per group • Restaurant menus (actual or copies; students may bring in their own), 2 or more per group • Chart paper, 2-4 sheets per group • Blank spreadsheet, 1 per student • Computers, spreadsheet software (optional) • *Gateways to Mathematical Investigations* Student Edition, pp. 9-10, 1 of each per student	**Gateways to Algebra and Geometry** Section 1.4, pp. 23-28: Probability Section 1.5, pp. 29-35: Tables and Spreadsheets Section 2.4, pp. 71-77: Percent Section 2.5, pp. 78-85: More About Spreadsheets *Additional References* Teacher's Resource File: Student Activity Masters

See pages 48A and 48B: Assessment for Unit A.

CHAPTER 1

Contents

1.1 A Tiling Problem
1.2 Number and Letter Patterns
1.3 Practical Patterns
1.4 Probability
1.5 Tables and Spreadsheets
1.6 Communicating Mathematics

Pacing Chart

Days per Section

	Basic	Average	Honors
1.1	2	1	1
1.2	2	2	1
1.3	2	2	1
1.4	2	2	1
1.5	2	2	2
1.6	2	1	1
Review	2	1	1
Assess	2	1	1
	16	12	9

Chapter 1 Overview

- Students are turned off by starting out another year with a dose of whole number and decimal arithmetic. It is important to start the year off with something different, and we felt that patterns were a good choice. Patterns permeate mathematics and are an important strand throughout *Gateways*.

- Chapter 1 gives the students ample experiences with the following notions: 1. Problem solving involves creativity; 2. There is more than one way to solve a problem; and 3. Problems often have more than one correct answer. Students will discover in Chapter 1 that learning to think and solve problems should be their main goal in this course.

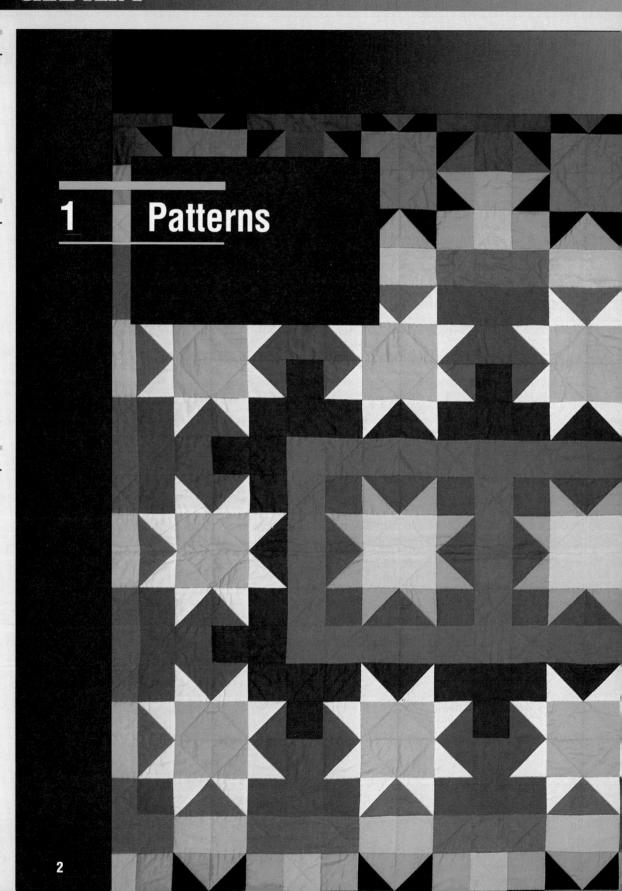

1 Patterns

CHAPTER 1

ASTRONAUT In the past thirty years or so, advancements in science and engineering have opened the way for men and women to begin the exploration of space. United States astronauts are employed by the National Aeronautics and Space Administration (NASA) and are divided into two groups—pilot astronauts, who fly spacecraft, and mission specialists, who maintain the spacecraft and conduct experiments. An applicant for either position must have a college degree in mathematics, science, or engineering. Those who are accepted for astronaut training go to the Johnson Space Center in Houston, Texas, where they undergo further instruction in subjects such as flight mechanics, computer science, mathematics, astronomy, geology, and biology.

INVESTIGATION

Escape Velocity When you throw a ball upward the height it reaches depends on how hard you throw it. If the ball leaves your hand at a speed of 20 feet per second, it will go up only about $6\frac{1}{4}$ feet before it begins to come down. If it is thrown at a speed of 40 feet per second, however, it will reach a height of 25 feet.

At what speed would a ball have to travel to escape the pull of the earth's gravity entirely? Scientists call this speed the earth's *escape velocity*, and this is the speed at which spacecraft must travel to reach outer space. Find out what the earth's escape velocity is, and compare this speed with the speeds of sound and light.

Career Connection

Not all students should aspire to become astronauts; that is not the point of this feature. There are numerous career opportunities that support the space program, and not all of them require engineering degrees or college mathematics. All will, however, involve the ability to think and solve problems.

Video Connection

For more information about the career discussed here, please refer to the Futures video cassette #5, program A: *Putting Man in Space*, featuring Jaime Escalante. For more information or to order a copy of the program, call PBS VIDEO at 800/344-3337.

Investigation Notes

- To escape Earth's gravitational pull, a rocket must reach a speed of about 7 miles per second; if the speed is any less, the rocket will fall back to Earth.

- A ball could never be thrown at a speed anywhere close to the escape velocity. But there is a pattern to the data presented. Notice that when the velocity is doubled (from 20 to 40 ft per sec), the height is quadrupled (from $6\frac{1}{4}$ to 25 feet). In general, the height will be proportional to the *square* of the velocity.

- Seven miles per second is about 0.00004 times the speed of light and about 34 times the speed of sound.

SECTION 1.1

Objective
- To recognize a pattern of geometric figures

Class Planning

Schedule 1–2 days

Resources

In *Chapter 1 Student Activity Masters* Book
Class Openers *1, 2*
Extra Practice *3, 4*
Cooperative Learning *5, 6*
Enrichment *7*
Learning Log *8*
Using Manipulatives *9, 10*
Spanish-Language Summary *11*

Class Opener

1. Draw the next figure in this pattern.

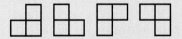

2. Describe the rule that you used to find the figure you drew.
3. Use your rule to draw three more figures in the pattern.

Answers:
1. Answers will vary.
2. Possible answers: Rotate the figure 90° counterclockwise to produce the next figure; shift one tile to produce the next figure. The shifts follow the pattern left, up, right, down.
3. Answers will vary.

SECTION 1.1
A Tiling Problem

If the following pattern is continued, how many tiles will be in the tenth figure?

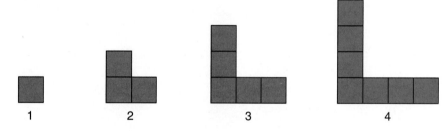

This problem can be solved in several ways.

Method 1: We can draw the next six figures, then count the tiles in the last one.

Method 2: We can list the number of tiles in each figure in a table, then look for a pattern.

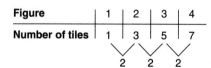

We see that the number of tiles increases by two each time. Let's continue this pattern.

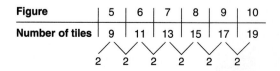

4 Chapter 1 Patterns

Method 3: We can analyze what happens. Two tiles are added to make each new figure—one tile to the top and another to the right side. To make the tenth figure, two tiles must be added six times to the seven tiles of the fourth figure. So there will be 7 + (2 × 6) = 7 + 12, or 19, tiles in the tenth figure.

Method 4: We can rearrange the tiles to make it easier to see the pattern.

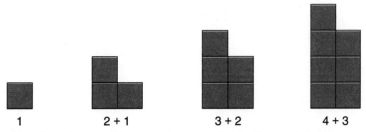

Therefore, there will be 10 + 9, or 19, tiles in the tenth figure.

Method 5: We can add tiles to make complete squares.

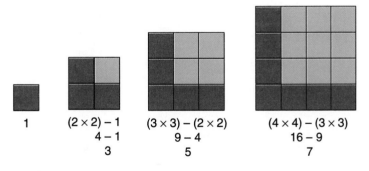

Therefore, the tenth square will contain
(10 × 10) − (9 × 9) = 100 − 81, or 19, red tiles.

Can you think of other ways to solve this problem? You may want to try some of your ideas on the problems that follow.

Think and Discuss

P **1** How many tiles will be in the eighth figure of this pattern? How do you know?

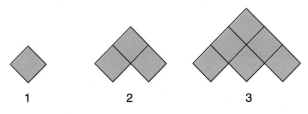

Section 1.1 A Tiling Problem 5

SECTION 1.1

Assignment Guide

Basic
Day 1 7, 8, 10, 16, 18, 20
Day 2 6, 13, 14, 17, 19

Average or Heterogeneous
Day 1 7, 9, 10, 11, 13–15, 18, 19

Honors
Day 1 6, 7, 9–15, 18–20, Investigation

Strands

Communicating Mathematics	7, 8, 13, 16, 20
Geometry	14, 17
Data Analysis/Statistics	10, 13
Technology	10, 15
Probability	18
Algebra	10, 15, 16
Patterns & Functions	1–9, 11–14, 20
Number Sense	4, 16
Interdisciplinary Applications	16, 19

Individualizing Instruction

The kinesthetic learner Hands-on learners will benefit from using paper squares to extend the patterns in problems 8, 9, and 11.

Problem-Set Notes

for pages 5, 6, and 7

■ Think and Discuss problems should elicit discussion from your students. Throughout the book, you will find that lesson concepts are extended in these problems.

■ **1, 3,** and **5** are visual in nature, but **4** is numeric. Students' answers will vary, so communication of their patterns should become the focus of discussion.

R, I **2** Add parentheses and symbols (+, −, ×, ÷) to make a true statement from 7 5 3 = 36. $(7 + 5) \times 3 = 36$

P **3**

  a Draw the next three figures of the pattern shown.
  b Describe a rule that can be used to find the number of red tiles in each figure.
  c Describe a rule that can be used to find the number of blue tiles in each figure.

P, R **4** Use the pattern $1 = \frac{1 \times 2}{2} = 1$

$$1 + 2 = \frac{2 \times 3}{2} = 3$$

$$1 + 2 + 3 = \frac{3 \times 4}{2} = 6$$

$$1 + 2 + 3 + 4 = \frac{4 \times 5}{2} = 10$$

to find the value of 1 + 2 + 3 + . . . + 77 + 78 + 79 + 80. 3240

P, I **5** Find the sum of the numbers on the tiles in the tenth figure of this pattern. 100

Problems and Applications

P, I **6** Draw the next "tree" in this pattern.

 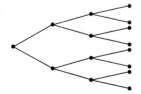

P, R **7** Communicating Add each pair of fractions. Describe any pattern you see.

  a $\frac{1}{2} + \frac{1}{4}$ **b** $\frac{3}{4} + \frac{1}{8}$ **c** $\frac{7}{8} + \frac{1}{16}$ **d** $\frac{15}{16} + \frac{1}{32}$ **e** $\frac{31}{32} + \frac{1}{64}$

Chapter 1 Patterns

SECTION 1.1

P **8** Explain how to make the next figure of this pattern.

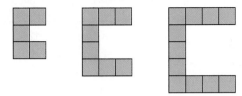

P **9** How many tiles will be in the sixth figure of this pattern?   51

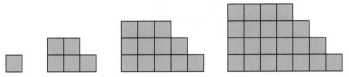

P, I **10** **Spreadsheets** In the table, 2 is in box A1, 3 is in box B2, and 7 is in box B4.
  a Copy and complete the table.
  b Find the value of A1 + B2 + C3 + B4 + A5 + B6 + C7, where A1 means the number in box A1, B2 means the number in box B2, and so forth.   245

	A	B	C
1	2	1	2
2	4	3	12
3	6	5	30
4	8	7	56
5	10	9	90
6	12	11	132
7	14	13	182

P, I **11**

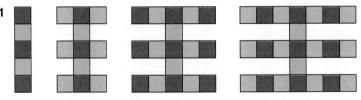

  a Draw the next four figures of the pattern shown.
  b Copy and complete the table.

Number of red tiles	3	3	9	9	15	15	21	21
Number of blue tiles	2	8	8	14	14	20	20	26

P **12** Describe the pattern shown. How many tiles will be in the eighth figure of this pattern?

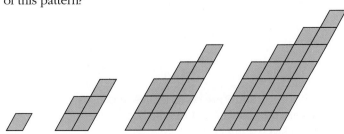

Section 1.1 A Tiling Problem   **7**

■ **2** emphasizes the importance of parentheses. Through discussion, you should be able to evaluate your students' familiarity with order of operations, which is formally presented in Section 2.1. Assess the amount of time necessary to spend on that section.

■ **8, 9, 11,** and **12** all have to do with tile patterns.

■ **6** and **14** are visual patterns but do not involve tiles. In **6,** students may be unfamiliar with the term "tree." Point out the similarity between this tree and a factor tree without numbers.

■ **7** and **13** are numerical patterns involving fractions (discussed in Section 1.2).

■ **10** and **15** preview spreadsheets but does not require the use of one.

Additional Answers

for pages 5, 6, and 7

1 36 tiles; possible answer: The numbers of tiles in the figures seem to follow the pattern 1, 1 + 2, 1 + 2 + 3, 1 + 2 + 3 + 4, . . . , so the eighth figure will be made up of 1 + 2 + 3 + 4 + 5 + 6 + 7 + 8 tiles.

3 a

3 b Possible answer: The red tiles follow the pattern 4, 8, 12, 16, 20, . . . (successive multiples of 4).

c Possible answer: The blue tiles follow the pattern 0, 1, 4, 9, 16, . . . (squares of the whole numbers).

Answers for problems 6, 7, 8, 11 a, and 12 are on page A1.

SECTION 1.1

Problem-Set Notes

for pages 8 and 9

- **13** could be turned into an activity. An interesting logistical question is: Can a strip of paper 12" by 1" be cut out of an 8½" by 11" sheet of paper? (Yes, if cut along a diagonal.)
- **15** previews the idea of function by having students perform operations on an input value.
- **16** does not ask for a numerical answer, just an explanation of how to get one.
- **17** gives a method for determining if an angle is a right angle for students who are not familiar with the term. It also dispels the notion that all right angles must have sides that are parallel to the edges of the page.
- **18** leads into probability.

Additional Practice

1. How many tiles will be in the seventh figure of this pattern?

2. How many tiles will be in the tenth figure of this pattern?

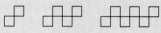

Answers: **1.** 28 **2.** 20

Communicating Mathematics

- Give students the following assignment: Choose the method of solution you like best, from those listed, to solve the opening tiling problem. Write a paragraph describing why you like this method and one or two of its possible drawbacks.

P **13 a** Make a strip of paper a foot long and an inch wide. Experiment with folding it in half repeatedly. Then copy and complete the chart.
b When you unfold the strip, some of the folds are "valleys" and some are "peaks." Describe the pattern of valleys and peaks.

Folds in Half	Numbers of Layers	Approximate Length of Each Layer
0	1	12"
1	2	6"
2	4	3"
3	8	1.5"
4	16	0.75"

P **14** Copy the figure shown, then draw the next triangle in the pattern.

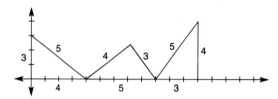

◀ LOOKING BACK **Spiral Learning** LOOKING AHEAD ▶

R, I **15** **Spreadsheets** This table contains seven boxes, A1 through A7. The value in each box depends on the computations performed on INPUT.
a Find the value in each box if INPUT is 12.
b Find the value in each box if INPUT is 0.

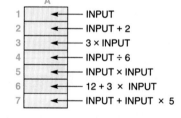

R, I **16** David bought 12 first-class postage stamps and 15 postcard stamps. Explain how he would determine the cost of the stamps.

I **17** Which of these figures do you think are right angles? Check by using the corner of a sheet of paper.   Angles 1 and 4 are right.

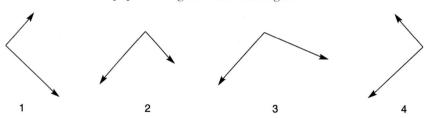

I **18** The numbers 1, 2, 3, 4, 5, 6, 7, 8, and 9 are on nine tiles. If a tile is selected at random, what are the chances that the tile has an even number on it?   Possible answer: 4 out of 9

8 Chapter 1 Patterns

SECTION 1.1

R, I **19** Malik and Marinda sold 27 baskets of fruit to make money for the math club. Each basket was sold for $8.25.
  a How much money did they take in? $222.75
  b How much profit did they earn for the club if each basket cost $5.50 to make? $74.25

P, I **20** **Communicating**   Describe the following pattern. (The three dots mean "and so forth.")

$$Z, X, V, T, R, P, \ldots$$

Possible answer: It is the letters of the alphabet in reverse order, with every other letter left out.

Investigation

Tessellation Look up the meaning of the word *tessellation*. What is a tessellation? Draw an example of a tessellation, and report on some ways in which tessellations are used in everyday life.

Rice servings wrapped in banana leaves cook on a grill in Thailand—a perfect solution to biodegradable packaging.

- Problems 7, 8, 13, 16, and 20 encourage communicating mathematics.

Checkpoint

1. How many tiles will be in the fifth figure of this pattern?

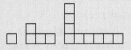

2. How many tiles will be in the eighth figure of this pattern?

Answers: **1.** 25 **2.** 16

Additional Answers

for pages 8 and 9

13 b Possible answer: The difference between the number of peaks and the number of valleys is 1.

14

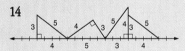

15 a 12, 14, 36, 2, 144, 48, 72
  b 0, 2, 0, 0, 0, 12, 0

16 Possible answer: He could multiply the cost of a first-class stamp by 12, multiply the cost of a postcard stamp by 15, and add the products.

Investigation Tessellation: A covering of a geometric plane by a repeating pattern of shapes, with no gaps or overlaps. Tessellations are evident in wall coverings, floor coverings, and in artwork, most notably some artwork done by M. C. Escher.

SECTION 1.2

Objective
- To extend patterns of numbers and letters

Class Planning

Schedule 1–2 days

Resources

In *Chapter 1 Student Activity Masters* Book
Class Openers 13, 14
Extra Practice 15, 16
Cooperative Learning 17, 18
Enrichment 19
Learning Log 20
Using Manipulatives 21, 22
Technology 23, 24
Spanish-Language Summary 25

Transparency 5

Class Opener

Find the next three items in each list. Explain your thinking.
1. 1, 4, 9, . . .
2. 8, 12, 10, 14, 12, 16, 14, . . .
3. N, NE, E, . . .

Possible answers:
1. 16, 25, 36
2. 18, 16, 20
3. SE, S, SW

Lesson Notes

- In this section, we introduce students to the idea of explaining their thinking. Some students will be more comfortable explaining thoughts orally, other students will prefer writing, and yet others will prefer pictures, diagrams, charts, or cartoons. If you use cooperative groups, it is fun to mix students with these preferences and allow them to "give as many ways as possible to . . . ." Point out that writing is as important in mathematics as it is in other subjects.

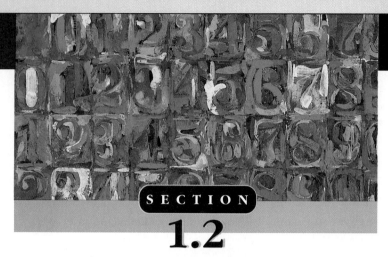

SECTION 1.2
Number and Letter Patterns

Pat Tern claimed that he was an ace at recognizing patterns. One day, his friend Betty Kann challenged Pat to find the next three numbers in the pattern 2, 3, 5, and to explain the pattern.

After thinking for a while, Pat came up with five different possibilities. (You may want to try to find several possibilities before you look at Pat's.) Here is how he explained his thinking:

The differences between consecutive numbers increase by 1.

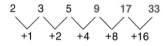

The differences between consecutive numbers double.

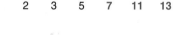

The numbers are consecutive prime numbers.

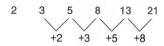

Each number, after the first two, is the sum of the two preceding numbers.

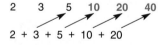

Each number, after the first two, is the sum of all the preceding numbers.

Then Betty challenged Pat to find the next three letters in the pattern T, F, S. Here is what Pat came up with.

T F S **E T T**

The initial letters of *two, four, six, eight, ten, twelve*

T F S **S M T**

The initial letters of *Thursday, Friday, Saturday, Sunday, Monday, Tuesday*

T F S **T F S**

Letters repeating in blocks of three

In these cases, all of Pat's responses were valid, because each was based on a logical explanation. Can you find other valid ways of continuing the patterns?

SECTION 1.2

Patterns play an important role in mathematics. As you continue your studies, keep an eye out for patterns. See if you can come up with logical explanations for the patterns and use the explanations to make predictions about how the patterns might continue.

Think and Discuss

In problems 1 and 2, find the next four items in each pattern.

1 0, 1, 3, 6, 10, 15
21, 28, 36, 45

2 T, F, S, N, E, T
F, S, N, T

3 Study the figures, then copy and complete the table.

Number of rays	1	2	3	4	5
Number of angles	0	1	3	6	10

4 If I have 12 elephants, 15 zebras, 18 spiders, 23 camels, 4 dogs, and a centipede, what fraction of these animals have exactly 4 legs? $\frac{54}{73}$

5 Study the pattern shown. What numbers do you think go in the blanks?

$3 - 0 = 3$
$3 - 1 = 2$
$3 - 2 = 1$
$3 - 3 = 0$
$3 - 4 = \underline{-1}$
$3 - 5 = \underline{-2}$

In problems 6–8, find the next three items in each pattern.

6 $\frac{2}{5}, \frac{4}{10}, \frac{6}{15}, \frac{8}{20}$ $\frac{10}{25}, \frac{12}{30}, \frac{14}{35}$

7 JK, LJ, RN, GF   JC, RR, GB

8 I, III, V, VII
IX, XI, XIII

Problem-Set Notes

for page 11

- **1** and **3** present the triangular numbers. You may want to have students find the triangular numbers on a geoboard or dot paper by building triangles. (See **14**.)
- **6–8** present fractions, presidents' initials, and Roman numerals as patterns.

Stumbling Block

For the pattern "T, F, S, ..." on page 10, students may not realize that these letters could stand for words. Emphasize that a pattern of letters may involve words that are related to each other in some manner.

Reteaching

If students have difficulty analyzing number patterns, encourage them to use the method shown on page 10. Students may find it easier to recognize patterns by drawing lines that show the differences between numbers instead of by just trying to work the patterns out mentally. This strategy can even work for visually represented number patterns like those in problems 14 and 22.

It is important to emphasize that one pattern is as good as another. No pattern is wrong if it works, and no working pattern is best.

Section 1.2 Number and Letter Patterns

SECTION 1.2

Assignment Guide

Basic
Day 1 10, 12, 17, 18, 22, 28, 29, 33
Day 2 14, 19, 20, 24, 26, 30–32

Average or Heterogeneous
Day 1 6, 7, 9–14, 16, 19, 26
Day 2 18, 20, 22, 24, 25, 27, 30, 33

Honors
Day 1 9, 10, 12–16, 18, 20–22, 24–27, 29, 30, 32, 33, Investigation

Strands

Communicating Mathematics	9–11, 23, 31
Geometry	3, 15, 16, 21, 23, 24, 31
Data Analysis/ Statistics	27, 32, 33
Technology	20, 25, Investigation
Algebra	20, 25, 30
Patterns & Functions	1–3, 5–16, 21–24, 27
Number Sense	26, 30, 33
Interdisciplinary Applications	29, 30, 33

Individualizing Instruction

The kinesthetic learner Students may benefit from constructing manipulatives for problems 9 and 10.

Problem-Set Notes

for pages 12 and 13

■ 16 can be linked to 1, 3, 14, and 24. Being able to see and name line segments is an important skill to have in place before doing more advanced geometry. Many of the visual patterns help students interpret geometric figures.

12

Problems and Applications

P **Communicating** In problems 9–11, draw the next figure in each pattern. Then describe the pattern.

9

10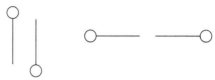

11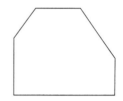

Any seven-sided figure

P **In problems 12 and 13, find the next four items in each pattern.**

12 16, 14, 12, 10, 8, 6
4, 2, 0, –2

13 2, 3, 5, 9, 17, 33
65, 129, 257, 513

P, R **14** How many dots will be in the fifth figure of this pattern? The sixth figure? The seventh figure? The eighth figure?   15; 21; 28; 36

P, I **15** Draw the next two figures in the pattern. (Hint: Cut a right triangle from a sheet of paper and rotate it.)

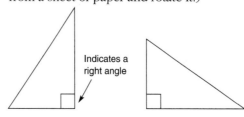 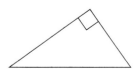

Indicates a right angle

12   Chapter 1   Patterns

SECTION 1.2

P, I **16** Line segments are named by their endpoints. This diagram, for example, shows three segments—$\overline{XY}$, $\overline{YZ}$, and $\overline{XZ}$.

How many segments are in each of the following diagrams?

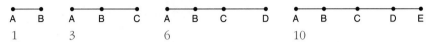

1   3   6   10

R, I **In problems 17–19, subtract the sum of each pair of numbers from their product.**

17 (5, 8)   27   **18** (9, 10)   71   **19** (1.5, 8)   2.5

P, I **20 Spreadsheets** Copy the chart. Use the formulas shown to fill in the rest of column B.

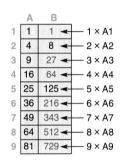

P **21** Draw the next three figures in the pattern.

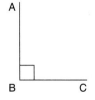 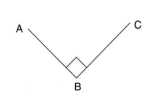 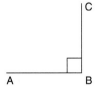

P **22** Find the next number in the pattern.   30

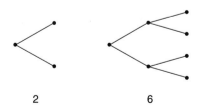

2   6   14

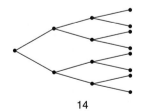

P **23 a** Describe a pattern that can be used to find the number of right angles in the diagram.
  b Use this pattern to find the number of right angles.   64

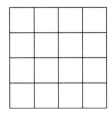

■ **23** is very difficult without developing a pattern. This problem suggests that students process a simpler case, an important problem-solving strategy. Notice that students discover the strategy, rather than being given it.

Checkpoint

Find the next three items in each pattern.
1. $\frac{1}{7}, \frac{3}{9}, \frac{5}{11}, \frac{7}{13}$
2. 52, 50, 47, 43
3. J, F, M, A
4. H, F, G, E

Answers:
1. $\frac{9}{15}, \frac{11}{17}, \frac{13}{19}$ 2. 38, 32, 25
3. M, J, J   4. F, D, E

Additional Answers

for pages 12 and 13

15

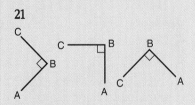

21

23 a Possible answer: Count the smallest squares and multiply by 4.

Section 1.2 Number and Letter Patterns

SECTION 1.2

Problem-Set Notes

for pages 14 and 15

- **24** can be extended by asking for the total number of small triangles and then of all triangles.

Cooperative Learning

- Ask each student to develop the first three numbers or letters of a pattern, list possibilities for the next three items on a separate sheet of paper, and record the rule for each pattern. Collect the patterns and distribute them to small groups of students. Encourage the members of each group to come up with as many possibilities for the next three numbers or letters as they can and to write a rule for each. In a large group, have the small groups share their possibilities and compare them with the patterns that the "author" intended. Then have students discuss the differences between working alone and working in small groups.
- Problem 31 is appropriate for cooperative learning.

Communicating Mathematics

- Have students write a paragraph explaining the differences between the patterns shown in problems 14 and 22. Encourage them to examine the construction of the patterns, and have them explain whether one representation shows the numerical pattern more clearly than the other. Oral explanations may be given instead of written ones.
- Problems 9–11, 23, and 31 encourage communicating mathematics.

P, I **24** Study the pattern, then copy and complete the table.

Number of green tiles	1	3	6	10	15	21	28
Number of blue tiles	0	1	3	6	10	15	21

P, R, I **25 Spreadsheets**
a Find the values that go in boxes B1, B2, B3, B4, B5, B6, and B7.
b Calculate the value
$(B1 + B3 + B5 + B7) \times (B2 + B4 + B6)$.
1,058,400

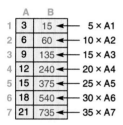

P, R **26** Six numbered bottle caps are arranged in a triangle. The sum of the three corner numbers, $1 + 6 + 5$, is three more than the sum of the remaining numbers. How can you rearrange the bottle caps so that the sum of the corner numbers is

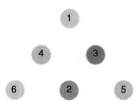

a One more than the sum of the remaining numbers?
b Five more than the sum of the remaining numbers?
c Six more than the sum of the remaining numbers?   Cannot be done.

P **27** Figure out the pattern used to generate the numbers in the chart. Then copy the chart and fill in the empty boxes.

1	2	3	4	5	6
3	5	8	12	17	23
5	10	18	30	47	70
7	17	35	65	112	182
9	26	61	126	238	420
11	37	98	224	462	882
13	50	148	372	834	1716
15	65	213	585	1419	3135

14   Chapter 1   Patterns

SECTION 1.2

◀ LOOKING BACK **Spiral Learning** LOOKING AHEAD ▶

R **28 a** Add 13.5 to 17.4. Multiply the sum by 7.6. What is the result? 234.84
   b Add 13.5 to the product of 17.4 and 7.6. What is the result? 145.74
   c Subtract your answer to part **b** from your answer to part **a**, and divide the difference by 13.5. What is the result? 6.6

R **29** The Samuel Davis Junior High School soccer team won three, lost two, and tied two games during the 1991 season.
   a What fraction of the games did the team win? $\frac{3}{7}$
   b What fraction of the games did the team win or tie? $\frac{5}{7}$

R, I **30 Consumer Math** A plumber's charge of $90 for a service call includes the cost of the first hour of work. For each additional hour or part of an hour, the charge is $45. What is the total charge for two service calls, one lasting 3 hours 15 minutes and the other lasting $4\frac{1}{2}$ hours? $495

P **31** Make five copies of this diagram and try to arrange the five pieces into a five-by-five square. If you succeed, draw a picture of the arrangement. If the pieces cannot form a square, explain why not.
   The pieces cannot form a square. Explanations may vary.

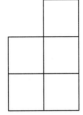

R, I **32** When planning a field trip to the Museum of Lapidary Art, Mr. Boynton decided that he needed to take at least 1 adult for every 10 children. If 74 children signed up for the trip, how many adults did Mr. Boynton need? At least 8 adults

R, I **33 Sports** Samantha ended the baseball season with 6 home runs, 4 triples, 11 doubles, and 35 singles.
   a How many hits did she have? 56
   b How many total bases did she have? 93

Investigation

Patterns on a Telephone Examine the dial or buttons on a telephone to see how letters of the alphabet are matched with numbers. Which numbers have no corresponding letters? Which letters of the alphabet do not appear on the telephone? Find out why some letters are missing and why some numbers are not assigned letters.

Section 1.2 Number and Letter Patterns 15

Additional Practice

Find the next three items in each pattern.
1. 1, 5, 9, 13   2. 64, 56, 48, 40
3. $\frac{3}{9}, \frac{4}{12}, \frac{5}{15}, \frac{6}{18}$ 4. 15, 30, 45, 60
5. J, J, M, A   6. B, D, F, H
7. AC, BD, CE, DF
8. Z, Y, X, W

Answers:
1. 17, 21, 25   2. 32, 24, 16
3. $\frac{7}{21}, \frac{8}{24}, \frac{9}{27}$ 4. 75, 90, 105
5. M, F, J   6. J, L, N
7. EG, FH, GI   8. V, U, T

Multicultural Note

Letter patterns can be interpreted differently in different languages. The pattern U, D, T, C, for example, represents the first letters of the first four cardinal numbers in Spanish (*uno, dos, tres, cuatro*).

Additional Answers

for pages 14 and 15

26 Possible answers:
a ①   b ②
   ② ③   ① ④
   ④ ⑤ ⑥   ⑤ ③ ⑥

31 The pieces cannot be arranged to form a square; possible explanation: When four of the pieces are arranged in a figure that has five squares along each edge, the "hole" in the figure is of a shape different from the fifth piece.

Investigation 1 and 0 have no corresponding letters; Q and Z do not appear on the telephone. Q was probably skipped because it would be easily confused with the letter O. The number 1 has no letters associated with it because it is the access number for long-distance calls.

15

SECTION 1.3

Objective
- To apply numerical patterns to real-world situations

Class Planning

Schedule 1–2 days

Resources

In *Chapter 1 Student Activity Masters* Book
Class Openers 27, 28
Extra Practice 29, 30
Cooperative Learning 31, 32
Enrichment 33
Learning Log 34
Interdisciplinary Investigation 35, 36
Spanish-Language Summary 37

In *Tests and Quizzes* Book
Quiz on 1.1–1.3 (Basic) 1
Quiz on 1.1–1.3 (Average) 2
Quiz on 1.1–1.3 (Honors) 3

Transparencies 5, 6 a–d

Class Opener

1. Make up a story to go with this graph.

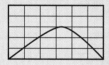

2. Copy the graph and label it to go with your story.

Answers:
1. Possible answers: A boy rides his bicycle on a straight path at a constant rate; he slows, turns around, and rides back at a faster, yet constant, rate to where he started.
2. Answers will vary.

Investigation Connection
Section 1.3 supports Unit A, *Investigation* 1. (See pp. 2A-2D in this book.)

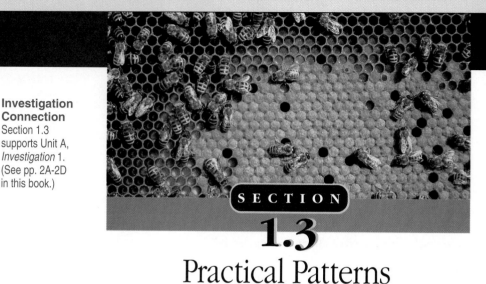

SECTION 1.3
Practical Patterns

A FAMILY TREE

In the preceding section, one of Pat Tern's patterns was 2, 3, 5, 8, 13, 21. Let's take a look at a connection this pattern has with the real world. We will use the symbol ♂ to represent a male bee and the symbol ♀ to represent a female bee. A male bee develops from an unfertilized egg, but a female bee develops from a fertilized egg. Therefore, a male bee has only a mother, whereas a female bee has both a mother and a father.

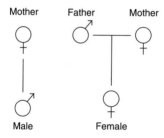

Example 1
Here are parts of the family trees of Mr. and Mrs. Bee.

How many great-grandparents do the Bees have?

16 Chapter 1 Patterns

SECTION 1.3

Solution
Let's continue the family trees.

[family tree diagram showing Great-grandparents, Grandparents, Parents, and Mr. and Mrs. Bee]

Mr. Bee has three great-grandparents, and Mrs. Bee has five. They have a total of eight great-grandparents.

OTHER APPLIED PATTERNS

Sometimes it is difficult to make predictions from a pattern unless you know what real-world situation it represents.

Example 2
What are the next four numbers in the pattern 10, 20, 30, 40, 50?

Solution
It is likely that you would predict the next four numbers to be 60, 70, 80, and 90, since the pattern seems to involve repeated additions of 10. But what if the numbers represented the numbers marking the yardage lines on a football field? In that case, the next four numbers would be 40, 30, 20, and 10. The numbers might also represent the speeds of a car as it accelerates from a standstill and travels along a highway, as shown in the following graph.

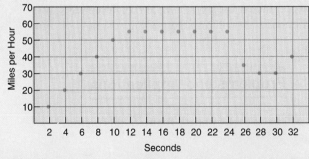

In this case, the next four numbers are 55, 55, 55, and 55. Can you think of any other real-world situations that the pattern 10, 20, 30, 40, 50, might represent? How might the pattern continue in these situations?

Section 1.3 Practical Patterns

Lesson Notes

- Help students realize the importance of patterns in all facets of society. Bringing in speakers from the community or simply having the students share what they and their parents do that involves patterns or prediction from patterns will help to convey this idea.

Additional Examples

Example 1
Every Martian is a gleep, a bop, or a nop. A gleep has one parent, a bop. A bop has two parents, a bop and a nop. A nop has three parents, a nop, a bop, and a gleep. Draw a family tree through the great-grandparents for each kind of Martian.

Answer:
A gleep has 5 great-grandparents; a bop, 11; and a nop, 13.

Example 2
What are the next four numbers in the pattern 5, 10, 15, 20?

Answer:
This series of numbers could represent the outside temperature measured every 3 or 4 hours in a day. The next four numbers might then be 15, 10, 5, 0, not 25, 30, 35, 40, as one might expect.

Reteaching

Give students the following problem: In a human family tree, how many great-grandparents does a person have? How many great-great-grandparents? Students may find it helpful to actually list the names of relatives.

SECTION 1.3

Assignment Guide

Basic
Day 1 6, 9, 14, 15, 24, 28, 32, 36, 39
Day 2 7, 8, 17, 25, 26, 29, 33–35, 38

Average or Heterogeneous
Day 1 6, 7–12, 14, 15, 17, 19
Day 2 20, 21, 24–27, 31, 36, 37

Honors
Day 1 7–14, 17–19, 21–23, 25–27, 30, 31, 35–37, 39, Investigation

Strands

Communicating Mathematics	6, 19, 22, 27, 30
Geometry	21, 31
Data Analysis/Statistics	4, 5, 7, 11, 27
Technology	4, 8, 12, 39, Investigation
Probability	35
Algebra	8, 12, 27–29
Patterns & Functions	1, 2, 5, 6, 8, 9–11, 13–18, 21, 22, 24–27
Number Sense	3, 4, 23, 30, 36
Interdisciplinary Applications	4, 20, 37, 39, Investigation

Individualizing Instruction

The gifted student Students may research topics involving predictions from patterns, such as the stock market, weather forecasting, demographics, energy needs, or business profits. They may present their results to the class.

Think and Discuss

P **1** Refer to the family trees of Mr. and Mrs. Bee in Example 1. What is the total number of the Bees' great-great-grandparents? 13

R **2** The answers to the 20 questions on a multiple-choice test were BACCEDBABCDBADEBADDC. Pat Tern guessed at the answers, using the pattern ABCDEEDCBAABCDEEDCBA. If each answer was worth 5 points, what was Pat's score? 30 points

I **3** Is the statement *true* or *false*?
a 3 + 5 × 2 = 5 + 3 × 2   F   b 3 + 5 × 2 = 2 × 5 + 3   T   c 3 + 5 × 2 = 5 × 2 + 3   T

I **4** Science The Celsius and Fahrenheit scales are two scales used to measure temperature. The chart shows some correspondences between the scales.

Celsius	Fahrenheit
0	32
20	68
40	104
60	140
80	176
100	212

a If the temperature is 50 degrees Celsius, about how many degrees Fahrenheit is it? 122°F
b What Fahrenheit temperature corresponds to 30 degrees Celsius? To 75 degrees Celsius? 86°F; 167°F

R, I **5** A cable TV company was charging customers $48 to view an Old Goats on the Prowl concert. Rocky Rolle decided that he could watch the concert if he invited some friends over to share the cost. He sat down and made the following chart:

Number of viewers	1	2	3	4	5	6	7	8
Cost per viewer	$48.00	$24.00	$16.00	$12.00	$9.60	$8.00	$6.86	$6.00

Copy the chart and continue it for six, seven, and eight viewers. If Rocky invites six friends, what will his share of the cost be? $6.86

Problems and Applications

P **6** How many circles will be in the twentieth figure of this pattern? Explain the pattern.

18 Chapter 1 Patterns

SECTION 1.3

R **7** The chart shows amounts charged by Pam the Painter for painting rooms. What will Pam charge for painting a kitchen, a bathroom, a living room, and
  a One bedroom? $500
  b Two bedrooms? $625
  c Three bedrooms? $750

Room	Cost
Kitchen	$100
Bathroom	75
Living room	200
Bedroom	125

P, I **8 Spreadsheets**
  a Find the values that will be in boxes A3–A6 if A1 is 2 and A2 is 3. 5; 8; 13; 21
  b Predict the values of A7 and A8.
  c What formulas could be written in the blanks to generate these two values?

  b 34; 55   **c** A5 + A6; A6 + A7

P **In problems 9 and 10, draw the next figure in each pattern.**

9

P **10**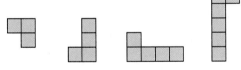

P **11** Nine segments of a pattern are shown. Each of the two shortest segments is one unit long. How many units long will the twentieth segment be? 10 units

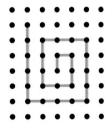

P, R, I **12 Spreadsheets**
  a Determine the value that will be generated in each box if INPUT is 3.
  b Find A1 × A2 × A3 × A4 × A5. 248,832

Problem-Set Notes

for pages 18 and 19

- **3** might require explanation of commutativity and associativity if your students are not familiar with these concepts.

- **4** does not ask for use of a conversion formula, but for estimation (interpolation) using the data given. If students know formulas from science class, more precise answers are acceptable.

Additional Practice

1. One large pizza costs $8.50; each additional pizza is $7.00. How much would five pizzas cost?
2. If a plant grows an average of 5 in. per year, how tall would it be after 24 years?
3. A typist types 60 words per minute. How many words can he type in 14 minutes?

Answers:
1. $36.50 **2.** 120 in. or 10 ft
3. 840 words

Stumbling Block

In problem 27, students may have difficulty recognizing # as an operator. Encourage them to see it as an unknown operation rather than an unknown number.

Additional Answers

for pages 18 and 19

6 210 circles; possible answer: The number of circles in the *n*th figure is the sum of the natural numbers from 1 to *n*.

In problems 13 and 14, draw the next figure in each pattern.

13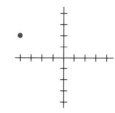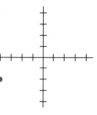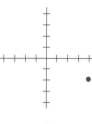

14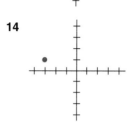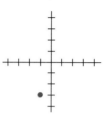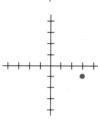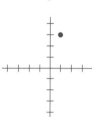

In problems 15–18, find the next four letters or numbers in each pattern. Then explain the pattern.

15 2, 4, 8, 16

16 a, b, b, a

17 128, 111, 96, 83

18 T, T, F, S, E

19 Communicating Describe a situation that this graph might represent.

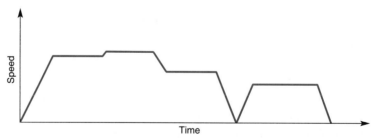

20 Consumer Math A state has a sales tax of 5% on purchased goods. (That is, a purchaser must pay a tax of $0.05 on each dollar of the purchase.)
 a On what purchase amounts would you pay $0.01 in tax? $0.02 in tax? $0.03 in tax? $0.04 in tax? $0.05 in tax?
 b Explain how you can use the answers to part **a** to determine the tax on any purchase.
 c Use your method to compute the tax on purchases of $2.35, $38.59, and $426.87.   Possible answers: $0.12; $1.93; $21.34

21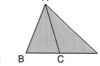

 a How many triangles are in each of the four figures?   1, 3, 6, 10
 b How many triangles will be in the fifth figure? The sixth figure?   15; 21

20   Chapter 1   Patterns

SECTION 1.3

P **22** Write a paragraph explaining the pattern of the green and yellow tiles.

   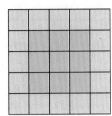

P, R, I **23** In a deep tunnel beneath the city of Rome, the following symbols were discovered on a wall:

(II, III) → VI   (X, III) → XXX   (VII, X) → LXX

The fourth set of symbols was incomplete: (IV, IV) →.
What might have been to the right of the arrow?   XVI

P **In problems 24–26, find the next three items in each pattern.**

24 (1, 5), (2, 9), (3, 13), (4, 17), (5, 21)   (6, 25), (7, 29), (8, 33)

25 (1, 1), (2, 4), (3, 9), (4, 16), (5, 25)   (6, 36), (7, 49), (8, 64)

26 1, 2, 6, 24, 120, 720   5040; 40,320; 362,880

P, R, I **27** The chart indicates that

6 # 4 = 12
4 # 4 = 4
3 # 8 = 24
8 # 6 = 24
7 # 3 = 21

#	3	4	6	7	8
3	3	12	6	21	24
4	12	4	12	28	8
6	6	12	6	42	24
7	21	28	42	7	56
8	24	8	24	56	8

a Copy the chart, filling in the missing numbers.
b Explain how you determined your answers.

---

5. Develop a list of factors other than cost that Rocky should consider as he decides how many people to invite.

Answers:

1.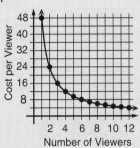

2. Students' descriptions may vary. The graph is hyperbolic.
3. (9, $5.33), (10, $4.80), (11, $4.36), (12, $4.00).
4. The cost approaches $0 but will never be $0.
5. Possible answers: The size of the room; The amount of food available

Additional Answers

for pages 20 and 21

15 Possible answer: 32, 64, 128, 256; each number is double the preceding number.

16 Possible answer: a, b, b, a; the group of four letters repeats.

17 Possible answer: 72, 63, 56, 51; the differences between successive pairs of numbers are successively decreasing odd numbers.

18 Possible answer: T, S, N, T; the letters are the initial letters of the successive prime numbers.

19 Possible answer: The graph might represent the speed of a vehicle as it accelerates, decelerates, and travels at constant speeds.

Answers for problems 20 a and b, 22, and 27 b are on page A1.

Section 1.3 Practical Patterns

SECTION 1.3

Problem-Set Notes

for page 22

- **36** is a number line review, with B = –1.5. A common mistake is B = –2.5.

- **38** requires unit conversion to compare seconds to seconds or minutes to minutes.

Checkpoint

1. If bolts are sold by weight, and each $\frac{1}{2}$ lb costs $0.75, how much would $4\frac{1}{2}$ lb of bolts cost?
2. A chain costs $0.13 per inch. How much would 15 feet cost?
3. Bike rental costs $10.00 per hour. How much would it cost to rent a bike for $3\frac{1}{4}$ hours?

Answers:
1. $6.75 2. $23.40
3. $32.50

Communicating Mathematics

- Have students write a paragraph describing a pattern they know of in the real world. Have them explain whether that pattern is useful for making predictions about the future. Alternatively, students may draw the pattern and explain it orally.
- Problems 19, 22, 27, and 30 all encourage communicating mathematics.

Answers for problem 30 and Investigation are on page A1.

◀ LOOKING BACK **Spiral Learning** LOOKING AHEAD ▶

R, I **In problems 28 and 29, copy the problem and fill in the blank.**

28 _32_ × 7 = 224   29 217 + _343_ = 560

I **30** Arnold and Amy were arguing about Adrian's age. Arnold said that she was 18; Amy said that she was 19. Adrian had lived for 18 years and 11 months. Who was right? Defend your answer.

P, R **31** The diagram shows a square with its two diagonals. How many right angles appear to be in the diagram?   8 right angles

R, I **In problems 32–34, rewrite each fraction as an equivalent fraction with a denominator of 100.**

32 $\frac{2}{5}$ $\frac{40}{100}$ 33 $\frac{8}{5}$ $\frac{160}{100}$ 34 $\frac{5}{5}$ $\frac{100}{100}$

I **35** In how many different ways can a red chip, a blue chip, and a white chip be stacked?   6 ways

I **36** On the number line shown, what numbers are represented by points A, B, C, and D?   –5; ≈–1.5; 0; ≈3.5

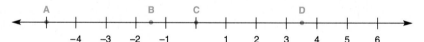

R, I 🖩 **37** **Consumer Math**   The Dry Gulp Water Company charges a customer $5.00 for service each month, plus $1.60 for the first 1000 gallons of water used and $0.002 for each additional gallon. One month Thurston Drinker used 3460 gallons of water. How much was he charged? $11.52

R, I **38** A typical half-hour TV show has 8 commercials. If each commercial is 30 seconds long, what fraction of the show is commercials? $\frac{2}{15}$

R, I **39** **Car Maintenance**   Before driving to Alaska, Biff went to a garage to have antifreeze put in his car's radiator. The mechanic said that Biff would need a mixture of $\frac{1}{3}$ water and $\frac{2}{3}$ antifreeze. The radiator held 2 gallons. How much antifreeze did Biff need? $1\frac{1}{3}$ gal

Investigation

Technology Study the arrangement of the numbers on the buttons of a Touch-Tone telephone. Then study the arrangement of the numbers on the keys of a calculator. Find out why the arrangements are different.

Chapter 1 Patterns

SECTION 1.4

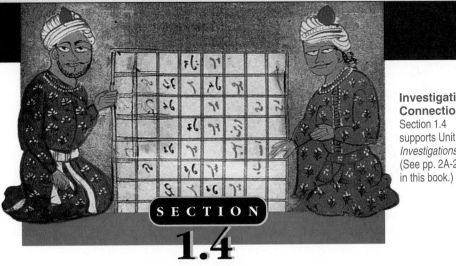

Investigation Connection
Section 1.4 supports Unit A, *Investigations* 4, 5. (See pp. 2A-2D in this book.)

SECTION 1.4
Probability

Objectives
- To understand the meaning of probability
- To calculate probabilities

Vocabulary: probability

Class Planning

Schedule 1–2 days

Resources

In *Chapter 1 Student Activity Masters* Book
Class Openers 39, 40
Extra Practice 41, 42
Cooperative Learning 43, 44
Enrichment 45
Learning Log 46
Using Manipulatives 47, 48
Spanish-Language Summary 49

Class Opener

1. Write your last name.
2. What fraction of the letters are vowels?
3. What fraction of the letters are consonants?
4. What fraction of the letters are capital letters?

Answers:
Answers will vary. Find the students with the least and greatest answers to **2** and **4**.

Lesson Notes

- In example 2, the directed graph representing the results of badminton games is an idea that is used in Chapter 5.

WHAT IS PROBABILITY?

Lois and Lana were playing cards when Lois suddenly said, "Oh, I just remembered! The movie *Superguy* is on TV now. Let's quit playing and watch it."

Lana was not really interested in watching the movie but replied, "I'll tell you what. I'll pick a card at random from the deck. If it's a face card, I'll watch the movie with you."

What was the probability that Lana would watch the movie?

In a regular deck of 52 cards, 12 of the cards (4 jacks, 4 queens, and 4 kings) are face cards. Thus, Lana's chances of watching the movie were 12 out of 52. The probability was $\frac{12}{52}$ or $\frac{3}{13}$.

The **probability** of an event is the fraction of possible outcomes that are favorable to that event ("winners"). In other words, it is the *ratio* of the number of winners to the number of possibilities.

$$\text{Probability} = \frac{\text{number of winners}}{\text{number of possibilities}}$$

APPLYING PROBABILITY

Example 1
If a housefly lands on one of the tiles of this floor, what is the probability that it will land on a blue one?

Solution
There are 25 tiles, and 16 of them are blue. The probability that the fly will land on a blue tile is therefore $\frac{16}{25}$.

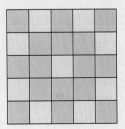

Section 1.4 Probability

SECTION 1.4

Assignment Guide

Basic
Day 1 8, 10, 13, 16, 20, 21, 25, 28
Day 2 9, 11, 12, 17, 18, 23, 26, 27

Average or Heterogeneous
Day 1 8–14, 16, 20
Day 2 17, 19, 22, 23, 25–28

Honors
Day 1 8, 10–15, 17–20, 22, 24–28, Investigation

Strands

Communicating Mathematics	3, 5, 16, 18, 24
Geometry	22, 24
Measurement	21
Data Analysis/ Statistics	10, 13, 20, 27
Technology	17, 26
Probability	1–4, 6, 8, 10–14, 17, 19, 20
Algebra	26, 27
Patterns & Functions	7, 9, 15, 18, 20, 22, 24, 27, 28
Number Sense	19, 23, 25, 27
Interdisciplinary Applications	16, 17, Investigation

Individualizing Instruction

Limited English proficiency
Students may benefit from working through problems such as 4, 10, and 12–14 with a partner.

Problem-Set Notes

for pages 24 and 25

- **3** is the first *always, sometimes,* or *never* problem. It is worth spending time on the categories so that students understand the differences between them. In this case, the probability of an event is *always* less than or equal to

24

Example 2
The diagram shows the games of badminton that were played at a picnic. Each arrow represents a game and points toward the loser—for example, Sue defeated Bill. "Snap" Schott wandered by during the day and took a photo of one of the games. What is the probability that the photo shows a game that Juan won?

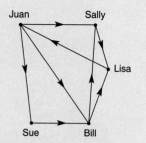

Solution
Eight games were played, and Juan won three. So the probability that the photo shows Juan winning is $\frac{3}{8}$.

Sample Problem

Problem A bag containing five slips of paper—numbered 2, 5, 6, 8, and 9—is used for a game. A player draws two slips and wins if the sum of the two numbers is even. Is the player more likely to win or to lose?

Solution When there aren't too many possible outcomes, we can simply list them and count the possibilities and the winners. To be sure that none are missed, however, we must list them in an orderly way.

This table shows the possible results of drawing two slips in the game described. There are 10 possibilities, and 4 of them (the ones checked) have even sums. The probability that the sum will be even is therefore $\frac{4}{10}$, or $\frac{2}{5}$. Since this probability is less than $\frac{1}{2}$, there is less chance of getting an even sum than of not getting one. The player is more likely to lose.

Numbers	Sum
2, 5	7
2, 6	8 ✓
2, 8	10 ✓
2, 9	11
5, 6	11
5, 8	13
5, 9	14 ✓
6, 8	14 ✓
6, 9	15
8, 9	17

Think and Discuss

P **1** When a coin is tossed, what is the probability that it will land heads up? $\frac{1}{2}$

P **2** What is the probability of rolling an even number with a standard six-sided die? $\frac{1}{2}$

P **3** Is the following statement true *always, sometimes,* or *never?* Explain your answer.

The probability of an event is less than or equal to 1.

24 Chapter 1 Patterns

SECTION 1.4

P   4. Robin Baskin went to an ice-cream parlor and ordered a cone with one scoop of vanilla ice cream, one scoop of chocolate, and one scoop of papaya ripple. In how many ways can the three scoops be arranged on the cone?   6 ways

P   5. What is a ratio?

P   6. A mosquito lands on one of the squares in the diagram shown. What is the probability that it lands on a square containing
     a An even number? $\frac{7}{16}$
     b A multiple of 3? $\frac{5}{16}$
     c A multiple of 5? $\frac{3}{16}$
     d A multiple of both 3 and 5? $\frac{1}{16}$
     e A multiple of 3 or 5? $\frac{7}{16}$

1	2	3	4
5	6	7	8
9	10	11	12
13	14	15	

R   7. What might the following letters represent?

     A, B, B, E, D, C, A, A, C, D, E, A, B, B, E

     Possible answers: The answers to a multiple-choice test; the notes in a song

Problems and Applications

P   8. **a** If a card is drawn at random from a standard 52-card deck, what is the probability that it will be a heart? $\frac{13}{52}$ or $\frac{1}{4}$
     b If the chosen card is a heart, what is the probability that it will be a jack? $\frac{4}{52}$ or $\frac{1}{13}$

R   9. Copy the following statement, filling in the missing numbers.

$$\frac{4}{7} = \frac{16}{28} = \frac{20}{35} = \frac{12}{21} = \frac{140}{245} = \frac{80}{140}$$

P   10. The table represents the possible results of rolling two dice. The numbers at the top represent the possible results for one of the dice, and the numbers at the left represent the possible results for the other.
     a Copy the table, filling in the possible sums of the numbers on the dice.
     b When two dice are rolled, what is the probability that the sum will be a multiple of 2? Of 3? Of 4? Of 5? Of 6? $\frac{1}{2}; \frac{1}{3}; \frac{1}{4}; \frac{7}{36}; \frac{1}{6}$
     c What is the probability that the sum will be a multiple of 2 or 3 or 4 or 5 or 6? $\frac{7}{9}$

+	1	2	3	4	5	6
1	2	3	4	5	6	7
2	3	4	5	6	7	8
3	4	5	6	7	8	9
4	5	6	7	8	9	10
5	6	7	8	9	10	11
6	7	8	9	10	11	12

P   11. If a card is drawn from a standard deck, what is the probability that it will be a face card or an ace? $\frac{4}{13}$

Section 1.4   Probability

---

one. If a student brings up probability as a percent (e.g., 20%), explain that 20 percent means 20 out of 100, or .2.

- **10** is a multitask problem, but it is within reach of most students. They may need to be reminded of the meaning of multiples.

Additional Examples

Example 1
If a housefly lands on a checkerboard, what is the probability that it will land on a black square? **Answer:** $\frac{32}{64}$, or $\frac{1}{2}$

Example 2
This diagram shows phone calls made among some friends. Each arrow points from the caller to the person called. One of the calls was interrupted by a switching error. What is the probability that it was a call made by Fred? **Answer:** $\frac{2}{9}$

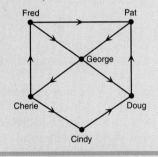

Additional Answers

for pages 24 and 25

3 Always; possible explanation: Since the winners are a subset of the possibilities, the number of winners can never exceed the number of possibilities; thus, the ratio of winners to possibilities can never exceed 1.

5 Possible answers: A fraction; the quotient of two values; a comparison of two values

SECTION 1.4

Problem-Set Notes

for pages 26 and 27

- **16** involves the idea of sampling. Discussion may turn to what makes a viable sample (e.g., if a community is very small, a complete count may be possible from a census).

- **17** may be facilitated by examination of a calculator to determine which bars are lit for which numbers.

- **19** can be modeled by using 27 small cubes with colored tape on each face. Encourage students to visualize the answer.

Cooperative Learning

- Have groups of students develop criteria where, if the names of all the students in the class were put in a bag and one name was drawn at random, the probability that the name would fit the criteria would be:
  a. 1 out of 2 b. 1 c. 0
  d. 1 out of the total number of students
  e. 5 out of the total number of students
  f. 2 less than the total number of students out of the total number of students

Checkpoint

1. Which of the numbers in the set $\{\frac{8}{15}, 0, 27.35, \frac{17}{9}\}$ could represent the probability of choosing a white marble from a bag of white and red marbles?
2. What is the probability of choosing a left shoe from a box of 12 pairs of shoes?

Answers:
1. $\frac{8}{15}$ 2. $\frac{1}{2}$

P   **12** Four boys (Felix, Paul, Alfredo, and Rolf) and one girl (Melissa) want to be on the school debate team. If the debate coach chooses two of them at random, what is the probability that
   a Felix will be on the team? $\frac{2}{5}$
   b Melissa and Alfredo will be on the team? $\frac{1}{10}$
   c Two boys will be chosen for the team? $\frac{3}{5}$

P, I   **13** This diagram shows how some people feel about one another. For example, Sally likes John, but John does not like Sally. Richard doesn't like anyone but is liked by Sally and Chuck.
   a If one of these people is chosen at random, what is the probability that he or she likes none of the others? Likes all of the others? $\frac{1}{4}; \frac{1}{4}$
   b If two of them are chosen at random, what is the probability that they like each other? That only one likes the other? $\frac{1}{6}; \frac{2}{3}$

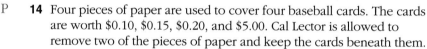

P   **14** Four pieces of paper are used to cover four baseball cards. The cards are worth $0.10, $0.15, $0.20, and $5.00. Cal Lector is allowed to remove two of the pieces of paper and keep the cards beneath them.
   a Find the probability that Cal will get the $5.00 card. $\frac{1}{2}$
   b Why might a baseball card be worth $5.00?

P, R   **15** Copy the diagram, filling in the empty circles with appropriate values.

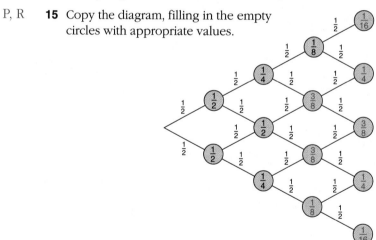

P   **16 Communicating** Describe what you might do to find out what fraction of the people in your community are under the age of 21.

P   **17 Technology** Many calculators use arrays like the one shown to display numbers. By lighting up different parts of the array (lettered A–G in the diagram), a calculator can display any of the ten digits. Suppose that you enter a random digit. What is the probability that
   a Part E will be lit? $\frac{9}{10}$
   b Part D will be lit? $\frac{7}{10}$
   c Parts C and D will both be lit? $\frac{1}{2}$

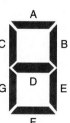

26   Chapter 1   Patterns

SECTION 1.4

R **18** Gus the Guesser wrote the following answers for a true-false quiz:

F, T, T, F, T, F, T, F, F, F, T, F, T, F, F, F, T, F, T, F

Describe a process Gus might have used to make his guesses.

P, I **19** This large cube is cut into 27 smaller cubes. Suppose that you paint the outside of the large cube blue, then put all the small cubes into a bag. If you pick one of the small cubes at random, what is the probability that it will have
  a No blue sides? $\frac{1}{27}$ **b** Exactly one blue side? $\frac{2}{9}$
  c Exactly two blue sides? $\frac{4}{9}$ **d** Exactly three blue sides? $\frac{8}{27}$
  e Exactly four blue sides? 0

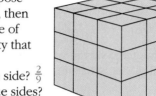

P **20 a** Figure out the pattern used to generate the values in the table. Then copy and complete the table.
  b Find the probability that a number selected randomly from the boxes in the table will be a multiple of 3. $\frac{1}{3}$

#	3	7	11
2	8	16	24
5	11	19	27
8	14	22	30

◀ LOOKING BACK **Spiral Learning** LOOKING AHEAD ▶

P, I **21** A hibernating bear sleeps for one fourth of a year. For how many months of the year is the bear not hibernating? 9 months

R **22**

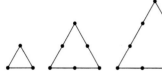

  a Find the number of line segments in each of the three figures. 3; 9; 18
  b How many segments will be in the fourth and fifth figures of the pattern? 30; 45

R, I **23** Mr. Boynton is planning a school field trip. Each of the available buses holds at most 36 people, and there will be 74 students and 8 adults on the trip. How many buses should Mr. Boynton order? 3 buses

R **24** Communicating   Describe a pattern that can be used to find the number of right angles in the diagram. Then determine the number of right angles.

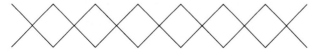

R **25** Arrange $\frac{1}{3}$, 0.3, and 33% in order, from least to greatest value. 0.3, 33%, $\frac{1}{3}$

Section 1.4   Probability   27

SECTION 1.4

Stumbling Block

In example 2 and problem 13, students may have difficulty understanding what the directed graphs indicate. Explain that each directed graph shows "one-way" relationships. For example, only one person can win a badminton game. To show in problem 13 that John and Chuck like each other, two one-way arrows are used.

Additional Practice

1. What is the probability of choosing a white marble from a bag of 16 white and 48 red marbles?
2. What is the probability of choosing a 2 or a 3 from a standard deck of cards?
3. What is the probability of two coins both landing heads up when tossed?

Answers: 1. $\frac{1}{4}$ 2. $\frac{2}{13}$ 3. $\frac{1}{4}$

Additional Answers

for page 28

26 a

A	B
1	3
2	6
4	12
8	24
16	48
32	96
64	65

b

A	B
1	0
2	0
3	0
4	0
5	0
6	0
7	0

Answer for Investigation is on page A1

I **26 Spreadsheets** In the table, the rule for generating the number in each box is shown.
  a Copy the table, putting the number 1 in box A1 and replacing the rules with the numbers that they will generate.
  b Repeat the process, this time putting the number 0 in box A1.

	A	B
1		A1 + A2
2	2 × A1	A2 + A3
3	2 × A2	A3 + A4
4	2 × A3	A4 + A5
5	2 × A4	A5 + A6
6	2 × A5	A6 + A7
7	2 × A6	A7 + A1

R, I **27** Study the way in which the following pairs of numbers generate new numbers. What are the values of INPUT and OUTPUT? 6; 35

(3, 2) → 6   (5, 8) → 40   (7, 2) → 14
(5, 7) → OUTPUT   (5, INPUT) → 30

R **28**

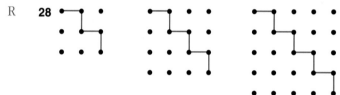

  a How many dots will be connected by segments in the seventh diagram of this pattern? 17
  b In the seventh diagram, how many more dots will be below the segments than above them? 8

Investigation

Leap Year What is a leap year? How can you determine whether a given year is a leap year? If you pick one of the years 1800 to 1899 at random, what is the probability that the year was a leap year?

"You've accounted for leap year, of course?"

28 **Chapter 1** Patterns

SECTION 1.5

1.5 Tables and Spreadsheets

THE COST OF POSTAGE STAMPS

In January 1992, postage for first-class mail was $0.29 for the first ounce and $0.23 for each additional ounce or fraction of an ounce. Calculating the correct postage for every letter, however, is cumbersome. It is much easier to use a **table,** or chart. Here are the first eight columns of such a table.

Single-Piece Letter Rates								
Weight, up to but not exceeding (ounces)	1	2	3	4	5	6	7	8
Cost (dollars)	0.29	0.52	0.75	0.98	1.21	1.44	1.67	1.90

Example 1
Refer to the table of postal rates.
a Calculate the next three columns of the table.
b Neelam paid $1.67 to mail a $6\frac{1}{4}$-ounce letter. Why did she pay the 7-ounce rate rather than the 6-ounce rate?
c What would it cost to mail a $5\frac{1}{2}$-ounce letter?

Solution
a

9	10	11
2.13	2.36	2.59

b Any letter that weighs more than 6 ounces but no more than 7 ounces is charged at the 7-ounce rate.
c $1.44

SECTION 1.5

Additional Examples

Example 1
In Vigesimal City, the sales tax is $0.01 on each purchase of $0.20 or fraction thereof. See the table below.

Price	$0.20	$0.40	$0.60	$0.80
Tax	$0.01	$0.02	$0.03	$0.04

a. Calculate the next 3 columns.
b. Why did Jane pay $0.04 tax on a $0.72 item?
c. What is the tax on a $0.38 item?

Answers:
a. $1.00 $1.20 $1.40
   $0.05 $0.06 $0.07
b. An item costing $0.61–$0.80 has a tax of $0.04. c. $0.02

Example 2
This spreadsheet shows the numbers of red and blue cars produced at a factory.

	A	B	C
1		Red	Blue
2	Mon.	63	13
3	Tues.	21	11
4	Wed.	37	41
5	Total	121	65

a. What formula is probably hidden in cell B5?
b. What cell contains the sum C2 + C3 + C4?

Answers:
a. B2 + B3 + B4   b. C5

Lesson Notes

- Previewing the ideas of incrementing, block copying, using the summing function, and understanding the way that spreadsheets work is important, even if you do not have access to computers or a spreadsheet.

- Spreadsheets focus on data, which is where many real-world problems originate. Many concepts that link data patterns to algebraic formulas and to graphs

---

SPREADSHEETS

So many tables are used in business, science, and mathematics that people in these fields usually find it easiest to use computer programs to create and store the tables and to manipulate the data in them. One sort of table-making program is called a **spreadsheet**. A spreadsheet was used to make the following table, which shows the meal expenses of four people on a one-day trip.

	A	B	C	D	E
1		Andy	Jenny	Diane	George
2					
3	Lunch	$4.29	$3.85	$5.59	$4.33
4	Dinner	$6.75	$6.24	$7.49	$6.85
5	Breakfast	$3.48	$3.95	$2.45	$4.76
6					
7	Total	$14.52	$14.04	$15.53	$15.94

The spreadsheet uses letters to name the columns of a table and numbers to name the rows. Each of the "boxes" in the table is known as a **cell** and is identified by its column letter and row number. In this case, column A and row 1 are used for headings, and rows 2 and 6 are used for spacing. The other cells contain the expense data—for example, cell D5 contains the amount paid by Diane for breakfast.

A spreadsheet can do calculations automatically. In making this table, for example, we "hid" the formula B3 + B4 + B5 in cell B7, so the spreadsheet calculated the sum of the numbers in cells B3, B4, and B5 and displayed the sum in cell B7. If we changed the entry in cell B3 from $4.29 to $4.39, the spreadsheet would automatically change the sum from $14.52 to $14.62. This automatic recalculation makes a spreadsheet a powerful tool for anyone who needs to make and modify tables.

Example 2
Refer to the table of meal expenses.
a What formula is probably hidden in cell C7?
b What cell contains the sum E3 + E4 + E5?

Solution
a Cell C7 contains the sum of the values in cells C3, C4, and C5, so the hidden formula is probably C3 + C4 + C5.
b Cell E7

Spreadsheets can also be used to create patterns of numbers, as the following sample problem shows.

Chapter 1   Patterns

SECTION 1.5

are enhanced by using a spreadsheet. The concept of variable is clearly demonstrated with a spreadsheet, especially in the use of the copy procedure. A spreadsheet can also help students understand the concepts of substitution and recursive thinking.

Sample Problem

Problem This table was created with a spreadsheet. Column A contains the whole numbers 1–9, and column B contains a number pattern. How might the spreadsheet have been used to generate the pattern?

	A	B
1	1	3
2	2	6
3	3	9
4	4	12
5	5	15
6	6	18
7	7	21
8	8	24
9	9	27

Solution There are several possibilities. One way of generating the pattern is to multiply each term in column A by 3. Another way is to enter a 3 in cell B1 and generate each of the other numbers in column B by adding 3 to the number above.

Method 1

	A	B
B1 = 3 × A1	1 →	3
B2 = 3 × A2	2 →	6
B3 = 3 × A3	3 →	9
B4 = 3 × A4	4 →	12

Method 2

	B
B1 = 3	3 ↓
B2 = B1 + 3	6 ↓
B3 = B2 + 3	9 ↓
B4 = B3 + 3	12

In either case, the spreadsheet can copy the pattern after we define the first step. To use Method 1, for instance, we could enter the formula 3 × A1 in cell B1, then copy the formula to the other cells in column B. The program automatically adjusts the formula to 3 × A2, 3 × A3, and so forth, in the other cells.

Assignment Guide

Basic
Day 1 6, 8, 11, 12, 15, 19, 21, 22, 24, 27
Day 2 7, 9, 10, 13, 16, 18, 23, 25, 26, 28

Average or Heterogeneous
Day 1 6, 8, 9, 11, 12, 14–17, 22
Day 2 7, 13, 18–21, 23, 26–28

Honors
Day 1 7–14, 16, 18, 23–25
Day 2 15, 17, 19–22, 26–29, Investigation

Strands

Communicating Mathematics	29
Geometry	3, 14, 16–18, 23
Data Analysis/ Statistics	1, 14, 15, 19, 20, 29
Technology	1, 5, 12, 19, Investigation
Probability	21, 28
Algebra	12
Patterns & Functions	2, 4, 11, 14, 16, 20, 27
Interdisciplinary Applications	1, 15

Individualizing Instruction

The visual learner Students may benefit from actually drawing the rest of the squares in problem 14 before completing the table.

Think and Discuss

P **1** This is part of a spreadsheet display.
  a Name the cells shown in yellow. B2, B3, B4, C4, D4
  b If column A were continued, what number might be in cell A7? In cell A8? In cell A24? 20; 23; 71
  c The formula A1 + 3 is hidden in cell A2. What formula might be hidden in cell A3? In cell A4? In cell A5? A2 + 3; A3 + 3; A4 + 3

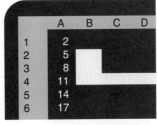

Section 1.5 Tables and Spreadsheets

SECTION 1.5

Problem-Set Notes

for pages 31, 32, and 33

- **1–5** could be done with a spreadsheet, perhaps in groups.
- **16** deals with an area formula which will recur throughout the book.

Stumbling Block

Students may have difficulty with spreadsheet terminology and concepts. Point out that a cell can contain both a hidden formula and the result of that formula, even though only the result is usually visible. Students may not understand that the copy feature of spreadsheets adjusts the cell numbers automatically, so that B2 can become B3, for example. Also, explain that changes in the values entered in certain cells of a spreadsheet can have effects throughout the table because of automatic recalculation.

Additional Answers

for pages 32 and 33

2

Hours	Cost
1	$124.50
2	$174.00
3	$223.50
4	$273.00
5	$322.50
6	$372.00

4b

Month	Amt. (dollars)	Month	Amt. (dollars)
1	100	13	1400
2	200	14	1500
3	325	15	1625
4	425	16	1725
5	525	17	1825
6	650	18	1950
7	750	19	2050
8	850	20	2150
9	975	21	2275
10	1075	22	2375
11	1175	23	2475
12	1300	24	2600

P **2 Consumer Math** Mr. Goodpliers, a mechanic, charges $75.00 for towing and $49.50 for each hour he works on a car. Make a table that shows the total cost of towing plus 1 hour to 6 hours of work, in 1-hour increments.

P, R, I **3 a** Find the area of a square with a side length of 13.   169
 b Find the side length of a square with an area of 49.   7

P **4 a** Bill decides to hide $25.00 in his mattress each week. How much will he have put in the mattress after one month? Two months? Three months? $100; $200; $300
 b Make a table that shows how much money will be in the mattress at the end of each month for two years. Assume that every third month has five weeks.

R **5** In this spreadsheet display, rows 2–6 show the formulas hidden in the cells. Copy the table, replacing the formulas with the numbers they will generate.

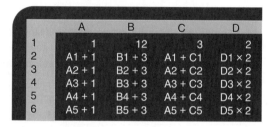

Problems and Applications

P In problems 6 and 7, use the table of postal rates on the first page of this section.

6 What would it cost to mail a letter that weighs $3\frac{1}{2}$ ounces? $0.98

7 What would it cost to mail a letter that weighs $7\frac{3}{4}$ ounces? $1.90

R, I In problems 8–10, find the value generated by adding the second number to three times the first number.

8 (5, 7) 22   **9** (11, 2) 35   **10** $\left(\frac{1}{2}, 5\right)$ $6\frac{1}{2}$

P **11** At Amalgamated Acme, Inc., workers are paid $0.28 per mile when they use their own car for business travel. Make a table showing the amounts paid to workers for trips of 10 miles to 200 miles, in increments of 10 miles.

P **12 Spreadsheets** A spreadsheet is used to create a table. In cell A1 is the number 4. Hidden in cell A2 is the formula A1 + 6. When this formula is copied to cell A3, it automatically changes into A2 + 6.
 a What numbers will appear in cells A2 and A3?   10; 16
 b If you copy the formula in cell A3 to cell A4, what formula will be hidden in cell A4? What number will appear in cell A4?   A3 + 6; 22
 c What formula do you think would be hidden in cell B4 if the formula in cell A4 were copied to it? What number would appear in cell B4?   B3 + 6; 6

SECTION 1.5

R, I **13** Perform the following calculations.
 a 3 × 3   9
 b 3 × 3 × 3   27
 c 3 × 3 × 3 × 3   81
 d 3 × 3 × 3 × 3 × 3   243
 e 3 × 3 × 3 × 3 × 3 × 3   729

P, R, I **14** The area of square 1 is 128 square centimeters. Each of the other squares is formed by connecting the midpoints of the sides of the next larger square.
 a Suppose the pattern of squares continues. Copy and complete the following table.

Square	1	2	3	4	5	6	7	8	9	10	11
Area (cm²)	128	64	32	16	8	4	2	1	0.5	0.25	0.125

 b What will the area of square 25 be? How do you know?
 c If you knew the area of square 49, how could you find the area of square 50?   Possible answer: Divide the area of square 49 by 2.

P, I **15** **History** This graph shows the history of U.S. postal rates.

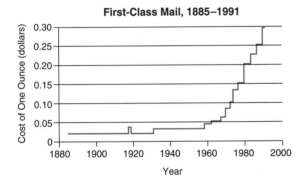

First-Class Mail, 1885–1991

 a When was the increase in postal rates the most rapid?   1980
 b When did the cost of sending a letter decrease?   1919
 c What do you think it will cost to mail a letter in the year 2000? How did you determine this cost?

P **16** Make a table listing the side lengths, perimeters, and areas of squares with whole-number side lengths from 1 to 9.

R, I **17** Find the total length of the fencing used to enclose the four pens shown.   124 ft

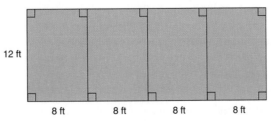

5

	A	B	C	D
1	1	12	3	2
2	2	15	4	4
3	3	18	6	8
4	4	21	9	16
5	5	24	13	32
6	6	27	18	64

11

Mi.	Amt. (dollars)	Mi.	Amt. (dollars)
10	2.80	110	30.80
20	5.60	120	33.60
30	8.40	130	36.40
40	11.20	140	39.20
50	14.00	150	42.00
60	16.80	160	44.80
70	19.60	170	47.60
80	22.40	180	50.40
90	25.20	190	53.20
100	28.00	200	56.00

14 b ≈0.000007629 cm², or ≈7.63 × 10⁻⁶ cm²; possible answer: The area of each square is half the area of the preceding one, so the area of square 25 can be found by multiplying 128 by $\frac{1}{2}$ twenty-four times.

15 c Answers will vary. Accept all answers that students can justify.

16

Side Length	Perimeter	Area
1	4	1
2	8	4
3	12	9
4	16	16
5	20	25
6	24	36
7	28	49
8	32	64
9	36	81

SECTION 1.5

Problem-Set Notes

for pages 34 and 35

- **18** reviews geometric patterns introduced earlier.
- **29** is suitable for journal writing.

Additional Practice

1. Find the value generated by subtracting the second number from 4 times the first number.
a. (7, 4) b. (12, 5) c. ($\frac{1}{4}, \frac{1}{3}$)

2. Column B of a spreadsheet has the formulas B1 = A1 + 4, B2 = A2 + 4, and so on. What numbers appear in column B, if column A contains the numbers 6, 7, 8, 9, 10, 11, and 12?

Answers:
1. a. 24 b. 43 c. $\frac{2}{3}$
2. 10, 11, 12, 13, 14, 15, 16

Checkpoint

1. Refer to the postal-rate table on page 29. What would it cost to mail each of the following letters? a. $3\frac{1}{2}$ oz b. 9 oz c. $7\frac{3}{4}$ oz

2. In the spreadsheet in problem 5, if the pattern of formulas is extended, what number will appear in each of the following cells? a. C8 b. B10 c. A15 d. D8

Answers:
1. a. $0.98 b. $2.13 c. $1.90
2. a. 31 b. 39 c. 15 d. 256

Cooperative Learning

- Have groups of students explore using spreadsheets on a computer. Ask each group to recreate the spreadsheets in the lesson, the sample problem, and problems 5 and 12 to see spreadsheet operation

---

R **18** Find the number of squares in each diagram. (Hint: There are more than four squares in the second diagram.)

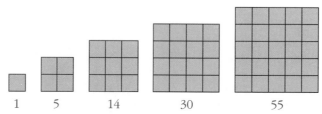

1   5   14   30   55

P **19 Spreadsheets** A spreadsheet was used to create this table, which contains some of the number patterns discussed in Section 1.2. If the numbers in rows 1 and 2 were entered without hidden formulas, what hidden formulas might have been used to generate the numbers in rows 3–6?

	A	B	C	D
1	1	2	2	2
2	2	3	3	3
3	3	5	5	5
4	4	8	8	10
5	5	12	13	20
6	6	17	21	40

R **20**

6 − 1 = 5   10 − 3 = 7   15 − 6 = 9

In the 10 × 10 figure of this pattern, how many more dots will there be below the line than above the line? How many dots will the line connect?   19 dots; 9 dots

◀ LOOKING BACK **Spiral Learning** LOOKING AHEAD ▶

R **21** A beanbag thrown at Mr. Happy Face is equally likely to land in any of the squares. You win a yo-yo if it lands in an eye, a stuffed toy if it lands in the nose, an ice-cream cone if it lands in the mouth, and a radio if it lands in the blue square. Suppose that you are going to throw one beanbag.

a What is the probability that you will win a yo-yo? A stuffed toy? An ice-cream cone? A radio? $\frac{1}{9}; \frac{1}{36}; \frac{1}{12}; \frac{1}{72}$
b What is the probability that you will win no prize? $\frac{55}{72}$

R, I **22** Perform the following calculations.
a 3 + 5 × 2   13   **b** 5 × 2 + 3   13   **c** 3 + 2 × 5   13   **d** 2 × 5 + 3   13

34   Chapter 1   Patterns

R, I 23 **a** Find the area of the square. 49 sq units
      b Find the volume of the cube.
         343 cu units

 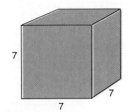

R, I **In problems 24–26, write the first number as many times as the second number indicates. Then multiply together the numbers you have written. For example, (3, 4) → 3 × 3 × 3 × 3 = 81.**

   24 (2, 5) 32   **25** (5, 2) 25   **26** $\left(\frac{1}{2}, 5\right)$ $\frac{1}{32}$

R **27** Draw the next diagram in the following pattern.

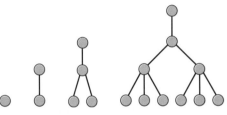

R **28** Wanda bought a box of Plenty Good candy. She found 7 black, 9 white, and 12 pink pieces in the box.
   a If she chooses a piece at random, what is the probability that it will be a pink piece? $\frac{3}{7}$
   b How many more pink pieces would have to be in the box for the probability of selecting a pink piece to be $\frac{1}{2}$? 4 pink pieces

R, I **29** Communicating These two-digit numbers are "cozy":

   13, 20, 24, 35, 43, 46, 53, 57, 64, 66, 76, 77, 88, 89, 97

   These two-digit numbers are not cozy:

   14, 19, 27, 28, 30, 36, 39, 41, 49, 50, 59, 62, 72, 81, 90

   a Which of the following numbers do you think are cozy? 11, 12, 33, 42, 45

       11, 12, 29, 33, 39, 42, 45, 85, 92

   b Explain what makes a two-digit number cozy.
       Two digits are the same or differ by less than 3.

Investigation

Spreadsheets Find an adult—a family member, a neighbor, a friend, a teacher—who uses a spreadsheet at work. Interview that person to find out what he or she uses the spreadsheet for. Be prepared to report your findings to your classmates.

SECTION 1.6

Objective
- To communicate mathematical ideas in words

Class Planning

Schedule 1–2 days

Resources

In *Chapter 1 Student Activity Masters* Book
Class Openers 63, 64
Extra Practice 65, 66
Cooperative Learning 67, 68
Enrichment 69
Learning Log 70
Interdisciplinary
  Investigation 71, 72
Spanish-Language Summary 73

Class Opener

Have students work individually to explain what a fraction is. Then have groups of four students communicate their ideas to each other and try to reach a consensus. Encourage any good ideas that you observe, such as inclusion of nonexamples, like π or $\sqrt{2}$, to explain fractions.

Lesson Notes

- Communicating mathematics is very important. If you plan to use journals, take some time to explain appropriate and useful types of journal entries, such as a paragraph, a convincing argument, a picture, or a construction. Point out that communicating processes and ideas is an aid to learning.

Investigation Connection Section 1.6 supports Unit A, *Investigation* 1. (See pp. 2A-2D in this book.)

SECTION 1.6
Communicating Mathematics

DESCRIBING A RECTANGLE

In class one day, a teacher asked Stu Dent the question, "What is a rectangle?"

Stu replied, "It's got four lines, and they touch, and the lines across from each other are the same size."

The teacher then drew this figure on the board and said, "So this is a rectangle."

"No," Stu responded, "because the lines should meet at the ends."

Then the teacher drew another figure, saying, "OK, so this is a rectangle."

"No," said Stu again, "because the lines don't meet at right angles."

After thinking about these examples and sketching some others, Stu came up with the following description of a rectangle: "A rectangle is a four-sided figure made up of segments meeting at their endpoints to form exactly four right angles." Then he went to the board and drew these figures:

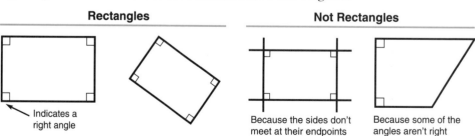

| Rectangles | Not Rectangles |

Indicates a right angle

Because the sides don't meet at their endpoints

Because some of the angles aren't right

Stu's final description was more complete and more specific than his original one, and he accompanied it with examples and nonexamples. Do you think that it was therefore a better description? Can you think of other good ways to explain what a rectangle is?

Chapter 1 Patterns

SECTION 1.6

EXPLAINING THE ROUNDING PROCESS

You will often be asked to explain things to other people. When you give explanations, try to make them clear and complete. You will often find that explaining an idea helps you to understand it better yourself.

Example
Explain how to round a number to the nearest tenth.

Solution
We might explain the process by giving two rules and some examples:

1. If the hundredths digit is less than 5, round down.
2. If the hundredths digit is 5 or more, round up.

For example,

3.7606 is rounded up to 3.8 because the hundredths digit is 6.
2.151 is rounded up to 2.2 because the hundredths digit is 5.
3.9187 is rounded down to 3.9 because the hundredths digit is 1.

Actually, textbooks differ on what to do in a case like 9.25, where the value is exactly halfway between 9.2 and 9.3. Some say to round up, some say to round down, and some give different rules for different cases. In this book, we will generally round such numbers up.

Sample Problem

Problem
In your mathematics studies you will frequently be dealing with *equations, inequalities,* and *expressions*. Study the following examples, then explain what equations, inequalities, and expressions are.

Equations	Inequalities	Expressions
$14 - 5 = 9$	$4x + 7 \leq 11$	$3 + \dfrac{4-1}{6 \times 2}$
$6 + 5 = 15 - 4$	$2 \neq 3 + 1$	$x + y - 2$
$P = 2\ell + 2w$	$L_1 + L_2 > H$	$\sqrt{3} - 6 \div 3$
$A = \tfrac{1}{2}bh$	$3 + 5 \geq 7$	$\sqrt[3]{6 \times 36}$
$3 - 1 = 2$	$9x < 5$	$2 + 3 \times 5$

Solution
One thing you may notice is that only the equations contain equal signs. The inequalities, on the other hand, contain symbols that you may recognize as signs of inequality, such as ≤ ("is less than or equal to"), ≠ ("is not equal to"), and > ("is greater than"). The expressions seem to be sets of numbers and letters joined by such operation signs as +, −, ×, and ÷.

Additional Example
Explain how to round a positive number to the nearest whole number.

Answer:
If the tenths digit is less than 5, round down. If the tenths digit is 5 or more, round up.

Cooperative Learning

- Have each student respond in writing to one of the following questions.
  1. What is a circle?
  2. How do you multiply two two-digit whole numbers?
  3. What is a cube?
  4. How do you change a mixed number to an improper fraction?

- In small groups, have students work together to expand and refine their answers. Encourage students to come up with examples and nonexamples of definitions and to analyze each step of a process in order to find what parts are unclear.

- Problems 14 and 22 are appropriate for cooperative learning.

Reteaching

Have each student write an explanation of how to do something in his or her everyday life, such as making a sandwich or riding a bicycle. Have each student read his or her explanation and discuss whether the explanation describes the process in sufficient detail for someone following the steps to produce the intended result.

SECTION 1.6

Assignment Guide

Basic
Day 1 5, 7, 8, 10, 13, 15, 18, 25, 27, 30
Day 2 6, 9, 11, 12, 16, 19, 26, 28, 29

Average or Heterogeneous
Day 1 5, 7, 8, 10, 11, 14–16, 18–21, 24, 26, 28, 29, 31

Honors
Day 1 7–10, 14–17, 19–24, 26, 29–31, Investigation

Strands

Communicating Mathematics	1, 2, 4, 7, 8, 10, 12, 14, 16–21
Geometry	1, 8, 11, 16, 17, 29, Investigation
Measurement	23
Data Analysis/ Statistics	22, 24
Technology	3, 9, 13, 18–21, 24
Probability	22, 31
Algebra	3, 6, 15
Patterns & Functions	14, 15, 18–22, 26, 29
Number Sense	2, 4, 10, 12, 13, 25
Interdisciplinary Applications	9, 14, 23

Individualizing Instruction

The auditory learner Students may benefit from discussing the text, examples, and problems with other students. They might also record their explanations on tape and listen to the tape in order to study. Encourage students to explain some of their answers to one another.

You will learn more about equations, inequalities, and expressions in later sections of this book, but these examples should make it clear that

1. An equation states that the values before and after the equal sign are equal.

2. An inequality states that the values before and after the inequality sign are either not equal or only sometimes equal.

3. An expression is made up of quantities and the operations that are to be performed on them.

Think and Discuss

P **1** Explain how to find the center of a rectangle. Possible answer: Draw the rectangle's diagonals; their point of intersection is the center.

P, R **2** Is the statement true *always, sometimes,* or *never?* How do you know? Explanations will vary
   a The product of two positive numbers is greater than either number. Sometimes
   b The product of two even numbers is an even number. Always
   c The sum of two odd numbers is an odd number. Never
   d The sum of two positive numbers is greater than either number. Always
   e The sum of two numbers is smaller than their product. Sometimes

R **3** Spreadsheets In a certain spreadsheet display, the number 4 is in cell A1 and the formula 2 × A1 + 5 is hidden in cell A2.
   a What number will appear in cell A2? 13
   b If this formula is copied to cells A3, A4, and A5, what will the formulas in those cells be? What numbers will appear in those cells? $2 \times A2 + 5$; $2 \times A3 + 5$; $2 \times A4 + 5$; 31; 67; 139

P **4** The fractions $\frac{15}{20}$, $\frac{3}{4}$, and $\frac{75}{100}$ are three ways of expressing the same number. Which of these forms do you think is simplest? Why?

Problems and Applications

P **5** Round the number 56.2281 to the nearest hundredth. 56.23

P **6** Write a mathematical expression representing the value of three quarters and five dimes. Possible answer: $3 \times 0.25 + 5 \times 0.10$

P **7** Sung Hi's little brother added $\frac{2}{3}$ and $\frac{3}{2}$ in this way:

$$\frac{2}{3} + \frac{3}{2} = \frac{2+3}{3+2} = \frac{5}{5} = 1$$

How could Sung Hi convince his brother that 1 is not the correct answer?

38 Chapter 1 Patterns

SECTION 1.6

P **8** Explain how to find the center of a circle.
Possible answer: Fold the circle in half in two directions; the intersection of the folds is the center.

R **9 Science**
  a At what temperature does water freeze?   32°F, or 0°C
  b At what temperature does water boil?   212°F, or 100°C

P **10** Explain why you can't divide 12 by 0.

R, I **11 a** Find the area of the square.   64 sq units
  b Find the volume of the cube.
    512 cu units

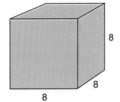

P **12** Explain why $\frac{3}{6}$ is equal to $\frac{1}{2}$. Write as many different explanations as you can.

I **13** Enter 8 $\sqrt{}$ on a scientific calculator. Write down the number that appears in the display. Then clear the display and multiply this number by itself. What do you notice?
Possible answers: 2.828427125; the result is very near 8.

P, R **14** Refer to the table of postal rates in Section 1.5.
  a Why, do you think, is the charge for the first ounce greater than the charge for other ounces?
  b Construct a table for the rates of $0.35 for the first ounce and $0.25 for each additional ounce.
  c What is the cheapest way to send a 4-ounce letter and a 3-ounce letter to the same address? Why?

P, I **15** In the following examples, what rule is applied to the pairs of numbers to generate the results?   Add the second number to twice the first.

  (2, 5) → 9   (3, 5) → 11
  (4, 5) → 13   (4, 6) → 14

P, R, I **16** Explain how to find the area of a square.

P, R, I **17** Explain how to find the volume of a cube.

P, R **In problems 18–21, describe how you could use a spreadsheet to generate each number pattern.**

  18 1, 2, 3, 4, 5, 6, 7
  19 5, 7, 9, 11, 13, 15, 17
  20 0.02, 0.04, 0.08, 0.16, 0.32, 0.64, 1.28
  21 1, 2, 6, 24, 120, 720, 5040

Problem-Set Notes

for pages 38 and 39

- **1–5** stimulate discussion, an important form of communication.
- **6** previews the ideas of coin problems, introduced later.
- **8** can be an activity if you ask students to determine as many ways as possible to find the center of a circle. This could also be a continuing journal entry.
- **13** encourages student explanation of the process.
- **14** encourages students to look for a pattern instead of just calculating repeatedly.
- **16** and **17** can be used to ensure that students explain how to determine the parameters of the formula and how to use them to solve the problem.

Additional Answers

for pages 38 and 39

4 Answers will vary. Of the fractions, $\frac{2}{4}$ is in what is usually called simplest form (lowest terms), but any of the three may be the easiest to use ("simplest") in a particular context.

7 Possible answer: Since $\frac{3}{2}$ is greater than 1, adding another positive quantity to $\frac{3}{2}$ must yield a sum that is also greater than 1.

10 Possible answer: If $12 \div 0$ were equal to some number, then 0 times that number would be equal to 12; but 0 times any number is equal to 0, so there is no value corresponding to $12 \div 0$.

Answers for problems 12, 14, and 16–21 are on page A1.

SECTION 1.6

Additional Practice

1. Explain what a right triangle is.
2. Describe how to reduce $\frac{9}{18}$ to simplest form.

Answers:
1. Possible answer: A polygon with three sides, two of which form an angle of 90°
2. Possible answer: Divide both the numerator and the denominator by their greatest common factor, 9, yielding the fraction $\frac{1}{2}$.

Checkpoint

1. Explain what a triangle is.
2. Explain what a square is.
3. Explain what the center of a circle is.

Answers:
1. Possible answer: A polygon with three sides 2. Possible answer: A polygon with four sides of equal length that meet to form four right angles
3. Possible answer: The point inside the circle that is equidistant from every point on the circle

Communicating Mathematics

- Have students write a paragraph describing the difference between the signs < and ≤.

- Problems 1, 2, 4, 7, 8, 10, 12, 14, and 16–21 encourage communicating mathematics.

P, R, I **22** A two-digit number is cozy if the two digits are the same or if they differ by less than 3. For example, 43, 68, and 88 are cozy numbers, but 52 and 90 are noncozy numbers.
  a Classify 46, 47, 86, and 34 as cozy or noncozy.   Cozy: 46, 86, 34; noncozy: 47
  b List the cozy numbers that contain at least one 5.   35, 45, 53, 54, 55, 56, 57 65, 75
  c List the cozy numbers that contain a 0.   10, 20
  d List the noncozy numbers that contain a 0.   30, 40, 50, 60, 70, 80, 90
  e What is the probability that a two-digit number is cozy? $\frac{41}{90}$
  f How might you define three-digit cozy numbers?

P, R, I **23** **Science** Refer to the thermometer diagram.
  a The *range* of a set of data is the difference between the largest number and the smallest number. The range of the Celsius scale on the thermometer is 100 – 0, or 100. What is the range of the Fahrenheit scale?   180
  b What fraction of the Celsius range is 1 degree? $\frac{1}{100}$
  c Since the range of the Fahrenheit scale is greater, a degree Fahrenheit represents a smaller change in temperature than a degree Celsius. What fraction of a Celsius degree is a Fahrenheit degree? $\frac{5}{9}$
  d How could you use your answer to part **c** to instruct a spreadsheet to list whole-number Celsius temperatures from 0 to 100 degrees along with the corresponding Fahrenheit temperatures?

R, I **24** **Spreadsheets** Suppose you put 4 in cell A1 and 3 in cell B1 of a spreadsheet display. Then you put the formula $A$1 + B1 in cell B2.
  a The number 7 will appear in cell B2. Why? Did the dollar signs have any effect?
  b When the formula in cell B2 is copied to cell B3, the formula in B3 is $A$1 + B2. What effect do the dollar signs seem to have?
  c If the formula in cell B3 is copied to cell B4, what formula do you think will be hidden in cell B4? Why? What number will appear in cell B4?

◀ LOOKING BACK **Spiral Learning** LOOKING AHEAD ▶

R, I **25** Write 5 + 5 + 5 + 5 + 5 + 5 + 5 + 5 as a multiplication problem.   8 × 5 or 5 × 8

R, I **26** Copy this diagram, filling in the missing numbers.

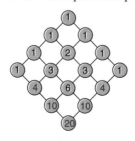

40   Chapter 1   Patterns

SECTION 1.6

R, I **27 a** Add 18 to 37. Multiply the sum by 24. What is the result?   1320
   b Add 18 to the product of 37 and 24. What is the result?   906
   c Subtract your answer to part **b** from your answer to part **a,** then divide the difference by 18. What is the result?   23

R, I **28** Find the value of each expression.
   a $8 + 10 \div 2$ 13   **b** $10 \div 2 + 8$ 13   **c** $8 + \frac{10}{2}$ 13   **d** $\frac{10}{2} + 8$ 13

R, I **29** In the figure, the length of the side of square 1 is 1 cm.
   a Copy and complete the following table.

Square	1	2	3	4	5	6
Side length	1 cm	1 cm	2 cm	3 cm	5 cm	8 cm
Perimeter	4 cm	4 cm	8 cm	12 cm	20 cm	32 cm
Area	1 cm²	1 cm²	4 cm²	9 cm²	25 cm²	64 cm²

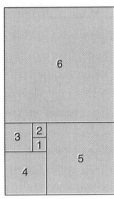

   b If the pattern continues, what will the side length of the tenth square be? The perimeter? The area?
   55 cm; 220 cm; 3025 cm²

R, I **30** In a contest, Lucky Linda won a $100 gift certificate from a local department store. She decided to buy as many pairs of socks as she could. Each pair of socks cost $7. How many pairs could she buy?   14 pairs

R, I **31** If one of the segments connecting two dots is removed at random, what is the probability that the remaining figure will have more than one piece (be a *disconnected graph*)? $\frac{1}{4}$

Investigation

Geometry Draw some figures that are made up of line segments. Then look up the meaning of the word *polygon*. Which of the figures you drew are polygons? If any of them are not polygons, why aren't they?

  Now find out what a convex polygon is. Draw an example of a polygon that is not convex. Is it possible to draw a nonconvex quadrilateral? A nonconvex triangle?

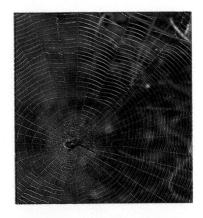

Section 1.6 Communicating Mathematics

SUMMARY

CHAPTER 1

Summary

CONCEPTS AND PROCEDURES

After studying this chapter, you should be able to
- Recognize a pattern of geometric figures (1.1)
- Extend patterns of numbers and letters (1.2)
- Apply numerical patterns to real-world situations (1.3)
- Understand the meaning of probability (1.4)
- Calculate probabilities (1.4)
- Use mathematical tables (1.5)
- Understand how a spreadsheet can be used to create tables and patterns of numbers (1.5)
- Communicate mathematical ideas in words (1.6)

VOCABULARY

cell (1.5)
probability (1.4)
spreadsheet (1.5)
table (1.5)

REVIEW

CHAPTER 1

Review

1. Find the value of each expression.
   a $42.5 + 3.71$ 46.21
   b 42.5×3.71 157.675
   c $42.5 - 3.71$ 38.79
   d $42.5 \div 3.71$ ≈ 11.46

2. How many tiles will be in the eighth figure of this pattern? 64

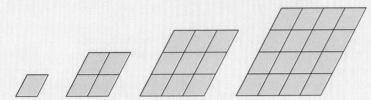

3. Jamie and Adenaky are in charge of their school's fall dance. They plan on spending $50 for rental of the school gym, $200 for custodial service, and $1500 for a disc jockey. What should they charge for each ticket if they estimate that Possible answers: **a** $17.50 **b** $8.75 **c** $5.85 **d** $4.40
   a 100 people will attend the dance?
   b 200 people will attend?
   c 300 people will attend?
   d 400 people will attend?

4. Sales tax in the town of Ripov is 11%. Part of a tax table for the town is shown. Copy the table, filling in the missing amounts. Possible answers shown.

Amount	Tax
$0.78–$0.86	$0.09
0.87– 0.95	0.10
0.96– 1.04	0.11
1.05– 1.13	0.12
1.14– 1.22	0.13
1.23– 1.31	0.14
1.32– 1.40	0.15
1.41– 1.49	0.16
1.50– 1.58	0.17
1.59– 1.67	0.18
1.68– 1.76	0.19
1.77– 1.85	0.20

5. What is a pattern?

6. What is a spreadsheet?

7. List some reasons why it is important for you to be able to explain mathematical ideas in words.

In problems 8–11, find the value of each expression.

8. $\frac{8}{7} \times \frac{1}{2}$ $\frac{4}{7}$

9. $\frac{9}{8} \times \frac{2}{3}$ $\frac{3}{4}$

10. $\frac{10}{9} \times \frac{3}{4}$ $\frac{5}{6}$

11. $\frac{11}{10} \times \frac{4}{5}$ $\frac{22}{25}$

Assignment Guide

Basic
Day 1 1, 2, 4, 7, 9, 13, 14, 18, 23, 24
Day 2 Chapter Test 1–11

Average or Heterogeneous
Day 1 2, 3 (a, c), 4, 7, 8, 10, 12, 14–20, 22–24

Honors
Day 1 2–7, 9, 10, 12, 14–26

Strands

Communicating Mathematics	5–7, 12, 18, 26
Geometry	18, 23
Data Analysis/ Statistics	15, 19
Technology	15, 19
Probability	14, 17
Algebra	3
Patterns & Functions	2, 4, 12, 16, 20–26
Number Sense	13, 26
Interdisciplinary Applications	3, 19

Additional Answers

for page 43

5 Possible answer: A pattern is a group of items in which each item is generated from the preceding one in accordance with a fixed rule.

6 Possible answer: A spreadsheet is a computer program for creating and storing tables and for manipulating the data in them.

7 Possible answer: Because trying to express ideas in words helps you to understand the ideas better; because you need to be able to express ideas to communicate them to other people

REVIEW

Additional Answers
for pages 44 and 45

12

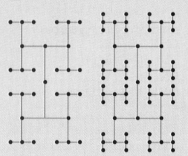

Possible explanation: In each diagram, the points in the preceding diagram that are endpoints of fewer than two segments become the midpoints of new segments drawn at right angles to the segments whose endpoints they were.

15

	A	B	C	D
1	1	1	2	3
2	2	3	5	5
3	3	6	11	16
4	5	11	22	38
5	8	19	41	79

18 Possible answer: A quadrilateral is a figure made up of four line segments meeting in pairs at their endpoints.

24 Possible answers: $\frac{1}{7}, \frac{1}{11}, \frac{1}{13}, \ldots$ (consecutive primes as denominators); $\frac{1}{8}, \frac{1}{13}, \frac{1}{21}, \ldots$ (consecutive Fibonacci numbers as denominators); $\frac{1}{10}, \frac{1}{20}, \frac{1}{40}, \ldots$ (each denominator is the sum of all the preceding ones)

25 Possible answers: H, I, K, ... (capital letters that consist only of straight segments, in alphabetical order); J, K, O, ... (first and last letters of 5-letter blocks of alphabet); A, E, F, ... (three letters repeated)

---

12 Draw the next two figures of the pattern shown. Explain the pattern.

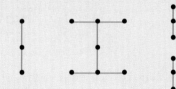

13 Is the statement *true* or *false*?

a $1\frac{3}{8} = 1 + \frac{3}{8}$ True   b $1\frac{3}{8} = 1.375$ True   c $1\frac{3}{8} = \frac{4}{8}$ False

d $1\frac{3}{8} = \frac{11}{8}$ True   e $1\frac{3}{8} = 0.375$ False   f $1\frac{3}{8} = 1 \times \frac{3}{8}$ False

14 Miss Brooks is going to randomly pick two students from a group of five—Bill, Mary, Bob, Gary, and Ellen—to be hall monitors. What is the probability that she will choose

a Two boys? $\frac{3}{10}$ b Two girls? $\frac{1}{10}$ c One boy and one girl? $\frac{3}{5}$

15 Using a spreadsheet, Miriam entered numbers in cells A1 and B1 and hidden formulas in other cells as shown. Copy the table, replacing the formulas with the numbers that they will generate.

	A	B	C	D
1	1	1	A1 + B1	B1 + C1
2	A1 + B1	A2 + B1	A2 + B2	C1 + D1
3	A1 + A2	A3 + B2	B3 + C2	C3 + D2
4	A2 + A3	A4 + B3	B4 + C3	C4 + D3
5	A3 + A4	A5 + B4	B5 + C4	C5 + D4

16 Study the way in which these number pairs generate new numbers. Then find the values of INPUT and OUTPUT. 7; 21

$(1, 2) \to 5$ $(3, 2) \to 11$ $(3, 4) \to 13$

$(5, 1) \to 16$ $(5, 2) \to 17$ $(6, 2) \to 20$

$\left(\frac{1}{3}, 1\right) \to 2$ $(6, 3) \to$ OUTPUT   (INPUT, 5) $\to 26$

17 A card is randomly drawn from a standard deck. What is the probability that the card is

a A heart? $\frac{1}{4}$ b A king? $\frac{1}{13}$ c The king of hearts? $\frac{1}{52}$ d A king or a heart? $\frac{4}{13}$

18 Look at the following examples. Then explain what a quadrilateral is.

Quadrilaterals   **Not Quadrilaterals**

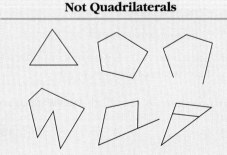

44 Chapter 1 Patterns

19 The Williams family used a spreadsheet to make this table. It shows what they spent to attend a baseball game between the Long Island Metrics and the Quebec Exponents.

	A	B	C	D
1		Lamont	Latoya	Latisha
2	Tickets	$10.00	$10.00	$8.00
3	Soda	$4.50	$3.00	$6.00
4	Hot dogs	$2.75	$2.75	$5.50
5	Peanuts	$2.00	$0.00	$1.00

a How much money did Lamont spend? $19.25
b How much did tickets for the three people cost? $28.00
c How much did Latoya and Latisha spend on food? $18.25

In problems 20–22, find the next three items in the pattern.

20 $\frac{1}{2}, \frac{2}{3}, \frac{3}{4}, \frac{4}{5}, \frac{5}{6}, \frac{6}{7}, \frac{7}{8}, \frac{8}{9}$

21 A, B, a, b, C, D, c   d, E, F

22  , , , ,

23 Copy and complete the chart for the rectangles shown.

Length	Width	Perimeter	Area
6	6	24	36
9	4	26	36
3	12	30	36
18	2	40	36
36	1	74	36

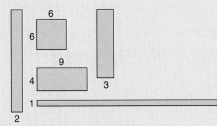

In problems 24 and 25, find as many ways as possible to continue the pattern.

24 $\frac{1}{2}, \frac{1}{3}, \frac{1}{5}, \ldots$

25 A, E, F, . . .

26 Explain the process by which each of these diagrams was generated from the preceding one.

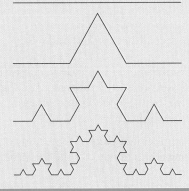

26 Possible answer: Each segment in the preceding diagram is divided into three equal parts, and an equilateral triangle is erected on the central part of each segment, which is then erased.

TEST

Assessment

In *Tests and Quizzes* Book
Basic Level Test 1A 5–7
             Test 1B 8–10
Average Level Test 2A 11–13
              Test 2B 14–16
Honors Level Test 3A 17–19
              Test 3B 20–22
Open-Ended   Test C 23, 24

Alternative-Assessment Guide

Assessment Tips
Portfolios and portfolio projects may be new to some or all of your students. Clearly explain to students what will be expected of them. Discuss the features that all projects will share as well as the various forms that they might take. If possible, have examples on hand to show students what you consider excellent work. Remember to post the standards by which you will evaluate the projects. Explain the purpose of the portfolio, including what should go in it, who will have access to it, and when you will evaluate its contents. Be sure that students are comfortable with all of the ground rules and try to establish an atmosphere in which creativity, risk-taking, and hard word are rewarded.

Portfolio Projects
The following problems are appropriate for use as portfolio projects:
1–6. Investigations: 1.1 (p. 9), 1.2 (p. 15), 1.3 (p. 22), 1.4 (p. 28), 1.5 (p. 35), and 1.6 (p. 41)
7. Five different methods are given for solving the tiling problem on page 4. Come up with another geometric pattern and show at least five different ways to extend the pattern.
8. Answer the same five questions in problem 19 on page 27 for cubes with 2, 4, and 5 blocks to an edge. What patterns do you see? Develop rules for finding the number of cubes with 0, 1, 2, 3,

---

CHAPTER 1

Test

In problems 1 and 2, find the next three items in each pattern.

1. 1, 2, $3\frac{1}{2}$, $5\frac{1}{2}$, 8, 11, . . . $14\frac{1}{2}$, $18\frac{1}{2}$, 23   2. J, F, M, A, M, J, . . . J, A, S

3. A box contains 3 red marbles, 4 white marbles, and 11 black marbles.
   a. What fraction of the marbles are white? $\frac{4}{18}$, or $\frac{2}{9}$
   b. If you pick a marble from the box without looking, what is the probability that it will be white? $\frac{4}{18}$, or $\frac{2}{9}$

4. How many squares will be in the eighth figure of this pattern? Explain the pattern.   22; add three squares to each consecutive figure

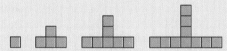

5. Zeke has $15 in his piggy bank. If he drops a quarter into the bank each day, how much money will be in the bank in two weeks? $18.50

6. How many three-digit numbers contain all three of the digits 4, 5, and 6? 6

7. Copy this table, replacing each formula with the appropriate value.

	A	B	C
1	4	A1 × 2	B1 × 3
2	A1 + 3	A2 × 2	B2 × 3
3	A2 + 3	A3 × 2	B3 × 3
4	A3 + 3	A4 × 2	B4 × 3

8. If you toss two coins, what is the probability that one will land heads up and the other will land tails up? $\frac{1}{2}$

9. Explain what a square is.

10. This spreadsheet display shows the monthly salaries of three people.
    a. What formula could you hide in cell B5 so that the spreadsheet would add up the salaries? B1 + B2 + B3
    b. What formula could you hide in cell B6 so that the spreadsheet would calculate the difference between the largest salary and the smallest salary? B3 − B2

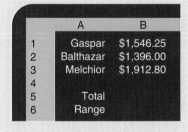

11. The first number in a pattern is 80, and each number in the pattern is half the preceding number. Write the first eight numbers of the pattern.
80, 40, 20, 10, 5, $2\frac{1}{2}$, $1\frac{1}{4}$, $\frac{5}{8}$

Chapter 1 Patterns

PUZZLES AND CHALLENGES

1 A jeweler charges $1 to open a link of a chain and $2 to weld it back together. What is the least that could be charged to join the pieces below into a circular chain?

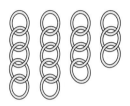

2 How many different triangles are in this figure? What method can you use to make sure that you count them all and do not count any twice?

3 You have 3 bags. One contains blue marbles, one contains green marbles, and one contains both green and blue marbles. Each bag is labeled incorrectly. How could you relabel each bag correctly by looking at just one marble from one bag?

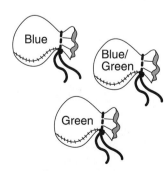

4 Twelve different shapes can be made by joining 5 squares along their sides. Three of the shapes are shown here. How many of the others can you discover? (Shapes that can be made to look the same by flipping or turning them do not count as different.)

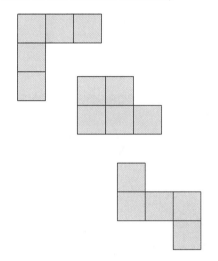

5 The figure shows five 1-by-1 squares constructed with toothpicks.

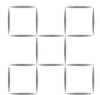

  a. How can you move two toothpicks to make six 1-by-1 squares?
  b. How can you move three toothpicks to make seven 1-by-1 squares?

and 4 blue sides in a cube with edges of n units.

9. Write a computer program that simulates the rolling of a standard 6-sided die. Use the program to "roll the die" 600 times and record the number of 3s that come up. Based strictly on probability, you could expect to roll one hundred 3s. Does this occur when you run your program? If so, how often? Explain what happens as you run the program repeatedly.

10. Extend example 1 on page 29 as follows: Find the current postal rates of at least three other countries. Convert units in order to compare the rates to each other and to rates in the United States. Show your findings in graphic form and explain what the graph shows.

11. Extend problem 31 on page 15 as follows: Come up with all of the different figures made up of 1-by-1 squares that could be used to completely cover a 5-by-5 square. What properties do these figures share? Use your answer to this question to help you come up with all of the different figures made up of 1-by-1 squares that could be used to cover a 6-by-6 square.

Additional Answers

for Chapter 1 Test

7

	A	B	C
1	4	8	24
2	7	14	42
3	10	20	60
4	13	26	78

9 Possible answer: A square is a rectangle in which all four sides are equal in length.

Answers to Puzzles and Challenges are on pages A1 and A2.

UNIT A: Thought for Food

Suggestions for various ways to assess students' performance are integrated throughout the investigations in this unit. These forms of assessment include observing the students while they work, interviewing students, looking at students' self-evaluations and group evaluations, observing peer review/response situations, evaluating students' products, and evaluating students' portfolios.

Refer to the chart below to preview the variety of assessment opportunities that are integrated throughout this unit.

Assessment Opportunities	Before Unit	1	2	3	4	5	End of Unit
Form A: Observation Notes **Bank A/B:** Observation Guide		✖	✖				
Form B: Observation Evaluation Record **Bank A/B:** Observation Guide		✖				✖	
Form C: Interview Record **Bank C:** Interview Guide					✖		
Form D: Product Evaluation Record **Bank D:** Product Evaluation Guide			✖	✖	✖	✖	
Form E: Rubric Chart **Bank E:** Sample Rubric				✖			
Form F: Portfolio Item Description	✖						✖
Form G: Portfolio Notes **Bank G:** Portfolio Evaluation Criteria	✖						✖
Form H: Self-evaluation							✖
Form I: Group Evaluation				✖			
Form J: Peer Review/Response					✖		

Rubrics

Rubrics are used to evaluate students' products. You can develop a scoring rubric for a specific investigation by adapting the blackline masters shown below, which are provided in the *Gateways to Mathematical Investigations* Teacher's Edition. You will probably want to develop a rubric with students at least once per unit. The chart on the opposite page suggests that for Unit A: Thought for Food, you complete a rubric before Investigation 3. Remind students that, at times, they may be asked to revise their work to improve it and to make it complete.

Bank D, page 211

Bank D: Product Evaluation Guide

The statements below can be used as a guide to assess students' products. Adapt these statements as needed to evaluate specific assignments. Record your adaptations on a clean copy of **Form D: Product Evaluation Record** (p. 210). You may also wish to create and record your own criteria on **Form D.** After you have adapted **Form D** for use with a specific assignment, make copies for each student or group as needed.

Mathematical Reasoning Skills

- The student includes examples and/or counter examples to support the mathematical argument.
- The student synthesizes and summarizes own or group's thinking.

Mathematical Understanding

- The student identifies all the important information of the situation or problem.
- The student shows understanding of the situation's mathematical ideas. (List important mathematical

Form E, page 212

Form E: Blank Rubric Chart

NAME (S) ______________ ______________ GROUP ________
DATE ________

STANDARDS FOR STUDENT'S WORK

LEVEL	Standard of performance to be achieved for each specified level
4 Exceptional/ Fully Acceptable	

Bank E, page 213

Bank E: Sample Rubric

STANDARDS FOR STUDENT'S WORK

LEVEL	Standard of performance to be achieved for each specified level
4 Exceptional/ Fully Acceptable	Your presentation fully accomplishes the purpose of the task. Your work is complete, well organized, and clearly explained. You illustrate ideas using clear diagrams and models and communicate them effectively to the audience. You identify all the important information of the situation and show a thorough understanding of the mathematical ideas and procedures. You use strong arguments to support findings, including examples (and/or counterexamples). You demonstrate that you fully grasp the central mathematical ideas developed in this task. You demonstrate an orignial extension or applicatrion of the mathematical concepts.
3 Satisfactory/ Acceptable	Your presentation substantially accomplishes the purpose of the task. Your work is complete, reasonably organized, and explained. You use an appropriate diagram or model and communicate ideas effectively to the audience. You identify most of the important information of the situation and show understanding of the mathmatical ideas and procedures. You present arguments to support findings, including examples (and/or counterexamples). You understand and use the central mathematical ideas developed in this task. You demonstrate and apply the mathematical concepts.
2 Incomplete/ Revise	Your presentation partially accomplishes the purpose of the task. However, your work may be incomplete, or misdirected, or not clearly organized and/or explained. Your diagram or model may be inappropriate or unclear. You may communicate ideas inadequately to the audience. You may identify some of the important information of the situation and you may show only partial and/or limited understanding of the mathematical ideas and procedures. Your arguments to support findings may be weak or incomplete. You may misunderstand or misuse the central mathematical ideas developed in this task. You may have difficulty demonstrating and applying the mathematical tasks. You need to revise this work. (Your teacher will provide suggestions for a satisfactory or competent rating and a due date for your revised work.)
1 Inadequate/ Redo	Your presentation shows little or no understanding of the assignment. Your work is barely, if at all, comprehensible. Your explanations are not understandable. You show little or no understanding of the mathematical ideas and procedures. You made serious computational mistakes and misused mathematical terms. You should meet with your classmates or your teacher to be sure you fully understand this assignment. Then you need to revise this work. (Your teacher will provide suggestions for a satisfactory or competent rating and a due date for your revised work.)

Gateways © McDougal, Littell & Company   Bank E: Sample Rubric   213

Form F, page 214

Form F: Portfolio Item Description

NAME ______________ DATE ______________

Complete this form each time you add an item to your portfolio.

Type of Portfolio Entry: ______________

Why did you choose to place this item in your portfolio?

Form G, page 215

Form G: Portfolio Notes

NAME ______________ CLASS ______________

Date	Observations/Comments	Recommendations

Bank G, page 216

Bank G: Portfolio Evaluation Criteria

At the start of the school year, have a class discussion on the goals and uses of students' portfolios. Discuss with the class some ideas regarding what to include in a portfolio. Then, use the portfolio evaluation criteria which follows to develop a rubric with the class that communicates the criteria you will use to assess their portfolios. Have students record this criteria on copies of **Form E: Rubric Chart.** Have students store this completed form in their portfolios so that they can refer to it throughout the year.

Each time students add items to their portfolios, have them complete copies of **Form F: Portfolio Item Description.**

Use copies of **Form G: Portfolio Notes** to record your observations, comments, and recommendations each time you evaluate students' portfolios.

Ideas for Types of Portfolio Entries

- a portfolio table of contents
- a letter to the reader of the portfolio explaining each item
- teacher-completed checklists/evaluations
- a mathematical autobiography
- notes from interviews, teachers, peers, etc.
- papers that show the student's corrections or errors or misconceptions
- photos or sketches of the student's work with manipulatives or mathematical models
- excerpts from the student's daily journal or reflective writing samples
- tables, charts, and diagrams created by the student
- artwork supporting a problem such as drawings
- designs, coordinate pictures, scale drawings, maps, etc.
- student-created problems with or without solutions
- group projects or individualized reports
- student-developed surveys and results
- self-evaluations

Suggestions for Portfolio Evaluation Criteria

Mathematical Reasoning Skills
The Student:
- analyzes problem situations.
- is flexible and comfortable in using several different problem solving approaches.
- synthesizes and summarizes the individual's or group's thinking.
- provides supporting statements to verify findings.
- shows original and creative thought.

Mathematical Tools and Techniques
The Student:
- uses a variety of diagrams, models, technology, manipulatives, and other resources appropriately.
- uses displays (e.g. graphs, tables, diagrams, sketches, and models) to represent mathematical ideas.

Mathematical Understanding
The Student:
- shows understanding of mathematical ideas and procedures. (List specific skills/concepts on **Form G.**).

Mathematical Communication
The Student:
- organizes work in a logical and clear way.
- expresses mathematical ideas clearly and appropriately.
- uses mathematical language and notation appropriately.

Mathematical Dispositions
The Student:
- shows development as as group participant.
- shows development of a positive attitude.
- shows growth in learning.
- includes revised and complete work.
- shows consistent effort.
- includes a variety of work in his or her portfolio.

216   Assessment   *Gateways* © McDougal, Littell & Company

Portfolios

At the beginning of each unit, you may wish to review with the students the criteria they should use when placing work in their portfolios. Have students complete Form F: Portfolio Item Description each time they wish to place an item in their portfolios.

Periodically, you may wish to evaluate students' portfolios using the blackline masters shown above. For Unit A: Thought for Food, the chart on the opposite page suggests that you evaluate students' portfolios at the end of the unit.

CHAPTER 2

2 Formulas and Percent

Contents

2.1 Operations on Numbers
2.2 Powers and Roots
2.3 Geometric Formulas
2.4 Percent
2.5 More About Spreadsheets

Pacing Chart

Days per Section

	Basic	Average	Honors
2.1	2	2	2
2.2	3	2	2
2.3	2	2	2
2.4	3	2	2
2.5	2	2	2
Review	2	2	1
Assess	2	2	2
	16	14	13

Chapter 2 Overview

■ Chapter 2 is a preview chapter, although it is different from Chapter 1. Much of the content will be review for many students. It is presented with an emphasis on using a calculator to do computation and paves the way for later work. Indeed, presenting the key geometric formulas (area, perimeter, and the Pythagorean Theorem) in Section 2.3 enables students to use them throughout the rest of the text, so geometry can be used in arithmetic, probability, and algebra problems. Likewise, percent, presented in Section 2.4, is such an important topic that it is repeated throughout the book.

■ Spreadsheets, brought up again in Section 2.5, will greatly enhance your students' ability to find patterns and see concepts. They enable variables to actually vary, right there for all to see.

CHAPTER 2

FASHION In the fast-paced world of fashion, a clothing designer needs to be aware of the latest style trends and decide whether to follow them, to introduce a new "look," or to update an older style. There are, however, many other factors that a designer must take into account to create a design that is economical to produce and successful with consumers. Among these are the amount and type of fabric required for the garment, the costs of manufacturing it, its suitability for the regions in which it will be marketed, and its intended selling price. A knowledge of mathematics helps the designer ensure that the price will be low enough to attract buyers while still providing the manufacturer with a decent profit.

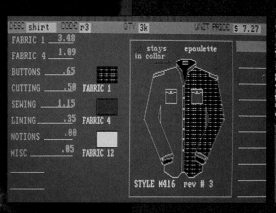

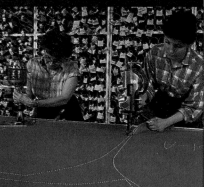

INVESTIGATION

Selling Prices How do the terms *overhead, wholesale price, retail price,* and *markup* apply to the fashion industry? When you buy an article of clothing, how much of the money that you pay reflects the cost of designing and manufacturing the clothing? How much goes for advertising costs and the manufacturer's and merchant's profits?

Career Connection

Many other aspects of creative industries like fashion involve mathematics. Wholesalers and retailers must be able to accurately figure costs, profits, sales, and inventory. Students should recognize that even though many manufacturing industries rely on designers for creative ideas, a great deal of mathematics is required in production.

Video Connection

For more information about the career discussed here, please refer to the Futures video cassette #3, program B: *Fashion,* featuring Jaime Escalante. For more information or to order a copy of the program, call PBS VIDEO at 800/344-3337.

Investigation Notes

■ **Overhead** refers to the general, continuing costs involved in running a business, including costs for labor, machinery, and a building. A store then buys the goods from the manufacturer at a **wholesale price,** which covers the manufacturer's overhead costs plus some profit. The **markup** is the amount the store increases the wholesale price in setting its **retail price,** which is the price consumers pay in the store. The store's profit comes from its markup.

■ Suppose that a manufacturer wants to produce a dress. He figures it costs him $40 in overhead to make the item, plus $20 for the material. He charges a wholesale price of $75 (a $15 profit). A store buys the dress for $75, then marks it at $100 (a $33\frac{1}{3}$% markup) to sell to its patrons.

SECTION 2.1

Objectives

- To recognize the need for a conventional order of operations
- To recognize the symbols used to indicate multiplication and division

Vocabulary: grouping symbol, vinculum

Class Planning

Schedule 2 days

Resources

In *Chapter 2 Student Activity Masters* Book
Class Openers *1, 2*
Extra Practice *3, 4*
Cooperative Learning *5, 6*
Enrichment *7*
Learning Log *8*
Using Manipulatives *9, 10*
Spanish-Language Summary *11*

Class Opener

Perform the following calculations.

1. $3 + 5 \times 7$
2. $82 - 51 - 11$
3. $8 + 18 \div 2$
4. $2 \times 6 - 3 \times 2$

Answers:
1. 38 **2.** 20 **3.** 17 **4.** 6

Lesson Notes

- If your students are inexperienced at using calculators, you might want to spend time familiarizing them with the operation of their calculators. Example 2 is a good problem to have them work through together.
- Sample problem 1 is also a good problem for discussion, as

Investigation Connection
Section 2.1 supports Unit A, *Investigation* 2. (See pp. 2A-2D in this book.)

SECTION 2.1
Operations on Numbers

WHICH OPERATION COMES FIRST?

Esther and her little brother Arthur had just been paid for some odd jobs. "How much money do we have?" asked Arthur.

"Let's see," Esther said, "we have two one-dollar bills and three five-dollar bills. That's seventeen dollars in all." She wrote $2 + 3 \times 5 = 17$ on a sheet of paper.

Arthur was puzzled. "I see that we have seventeen dollars," he said, "but two plus three is five, and when you multiply that by five you get twenty-five. So $2 + 3 \times 5$ should equal 25, not 17."

Who was right—is $2 + 3 \times 5$ equal to 17, or is it equal to 25?

As a rule, multiplication and division come before addition and subtraction:

$$2 + 3 \times 5$$
$$= 2 + 15$$
$$= 17$$

Therefore, Esther was right.

One of the things you will be learning this year is how to use a scientific calculator. To use a calculator to find $2 + 3 \times 5$, enter [AC] 2 [+] 3 [×] 5 [=]. (Before entering a new calculation, you should usually press the "all clear" key as shown here. From now on, this step will not be shown in the calculations in this book.) The display should show 17. If your calculator gives a result different from 17, it is not a scientific calculator.

Example 1
Evaluate (find the value of) each expression.

a $30 - 12 \div 3$ b $4 \times 8 + 6 \div 2$

Chapter 2 Formulas and Percent

SECTION 2.1

Solution

a We divide first, then subtract.

$$30 - 12 \div 3$$
$$= 30 - 4$$
$$= 26$$

b We multiply and divide, then add.

$$4 \times 8 + 6 \div 2$$
$$= 32 + 3$$
$$= 35$$

What about cases in which we need to add or subtract *before* we multiply or divide? In such cases we can use parentheses as **grouping symbols** around the operation that is to be done first:

$$(2 + 3) \times 5$$
$$= 5 \times 5$$
$$= 25$$

On a calculator, you can enter [( ] [2] [+] [3] [ )] [×] [5] [=], and the display will show 25.

Example 2

Evaluate $\frac{14 \div 2 - 1}{9 + 7 \times 3}$.

Solution

A fraction bar is a kind of **vinculum,** which is a grouping symbol, so we evaluate the numerator and the denominator as separate groups.

Method 1

$$\frac{14 \div 2 - 1}{9 + 7 \times 3}$$

In numerator, divide first
In denominator, multiply first
$$= \frac{7 - 1}{9 + 21}$$

Add and subtract
$$= \frac{6}{30}$$

Divide numerator and denominator by 6 to reduce to lowest terms
$$= \frac{1}{5}$$

Method 2 To solve this problem with a calculator, we use parentheses to group the numerator and the denominator. We enter

[( ] [14] [÷] [2] [–] [1] [ )] [÷] [( ] [9] [+] [7] [×] [3] [ )] [=]

          Numerator   **Denominator**

and the display shows 0.2. Since $0.2 = \frac{1}{5}$, our answers are the same.

In general, then, we do mathematical operations in the following order:

1. Operations enclosed by grouping symbols (parentheses, brackets, or vinculums)
2. Multiplication and division from left to right
3. Addition and subtraction from left to right

Section 2.1 Operations on Numbers

---

there are many ways to approach it, and students will probably think about it in different ways. Students with highly visual learning styles will appreciate the opportunity to relate order of operations to a concrete object.

Additional Examples

Example 1
a. Evaluate $15 + 18 \div 6$.

b. Evaluate $8 \times 3 - 14 \div 2$.

Answers: a. 18 b. 17

Example 2
Evaluate $\frac{3 + 5 \div 5}{2 \times 7 + 2}$.

Answer: $\frac{1}{4}$

Example 3
What value would a computer assign to the expression 9∗2–4/4?

Answer: 17

Reteaching

In problems like example 2, students may try to simplify fractions before completing any operations. For example, they may try to reduce example 2 as follows:

$$\frac{\overset{2}{\cancel{14}} \div 2 - 1}{9 + \underset{1}{\cancel{7}} \times 3}$$

To emphasize that the vinculum acts as a grouping symbol, have students place parentheses around the numerator and around the denominator. The parentheses can remain in place until students have simplified the numerator and the denominator completely.

SECTION 2.1

Assignment Guide

Basic
Day 1 9, 13, 15, 17, 19, 22, 25, 32, 36, 38, 42, 45, 46
Day 2 10–12, 14, 18, 24, 27–29, 33, 37, 43, 47

Average or Heterogeneous
Day 1 9, 10, 12–15, 19–21, 25, 28, 29, 33, 34, 36, 37, 44
Day 2 11, 16–18, 22–24, 26, 27, 30–32, 35, 40–42, 45, 46

Honors
Day 1 12, 14, 15, 20, 22, 23, 25, 26, 29, 33, 36, 43, 45
Day 2 10, 11, 13, 17, 19, 24, 27, 28, 30, 32, 34, 35, 37, 44, 46, Investigation

Strands

Communicating Mathematics	1–5, 7, 19, 47
Geometry	25–27, 29, 30, 32, 34, 35
Measurement	20, Investigation
Data Analysis/ Statistics	33
Technology	6, 7, 15, 20, 22, 45
Probability	44
Algebra	8, 33, 35, 36, 42, 46
Patterns & Functions	28, 32
Number Sense	10, 42, 43, Investigation
Interdisciplinary Applications	6, 12, 35, 36

Individualizing Instruction

Visually impaired learners Students may need assistance in operating and/or reading calculators. Make sure that all students

---

OPERATION SYMBOLS

To indicate addition and subtraction, the familiar + and − symbols are almost always used. There are, however, a number of ways of symbolizing multiplication and division.

Ways to Show Multiplication		Ways to Show Division	
Symbol	Example	Symbol	Example
×	3 × 5	÷	12 ÷ 4
*	3*5	/	12/4
·	3 · 5	—	$\frac{12}{4}$
No symbol	3(5) or (3)(5)		

Example 3
Tekkie entered PRINT 14/2*(12−9) on his computer. What value did the computer print?

Solution   14/2*(12−9)
First work inside the parentheses   = 14/2*3
Then divide   = 7*3
Then multiply   = 21

Sample Problems

Problem 1 One way to determine the perimeter of a rectangle is to use the formula

Perimeter = 2(length) + 2(width)

What is the perimeter of the rectangle shown?

Solution Remember, there is usually more than one way of solving a problem correctly.

Method 1 Perimeter = 2(length) + 2(width)
= 2(12) + 2(7)
= 24 + 14
= 38

Method 2 The perimeter is the sum of the lengths of the sides, so it is 12 + 7 + 12 + 7, or 38.

Method 3 Halfway around the rectangle is 12 + 7, or 19. So the entire perimeter is 2(19), or 38.

SECTION 2.1

Problem 2 A formula for changing a temperature from degrees Fahrenheit (°F) to degrees Celsius (°C) is

$$C = \tfrac{5}{9}(F - 32)$$

What is the Celsius equivalent of 212°F?

Solution Since we are given 212°F, we substitute 212 for F in the formula.

$$C = \tfrac{5}{9}(F - 32) = \tfrac{5}{9}(212 - 32)$$
$$= \tfrac{5}{9}(180)$$
$$= \tfrac{900}{9}$$
$$= 100$$

So 212°F is equivalent to 100°C. Do you know what is special about this particular temperature?

Think and Discuss

P, I **1** Why do we need rules to tell us the order in which operations should be performed?

P **In problems 2–5, explain the mistakes that were made in the calculations.**

2 $4 + 12 \div 7 - 6 = 16 \div 1 = 16$

3 $23 - 7 + 6 = 23 - 13 = 10$

4 $\dfrac{4 + 3}{4} = 1 + 3 = 4$

5 $5 + 3(12 - 7) = 8(5) = 40$

R, I **6** **Science** What is the normal temperature of the human body in degrees Celsius? 37°C

P **7** Is the statement *true* or *false?* Explain your answer.

a $(13 + 7) \div 8 = \dfrac{13 + 7}{8}$

b $15*10/4 = (15*10)/4$

P, R **8** Philbert Chesnutt bought 6 pounds of almonds at $2.75 per pound and 4 pounds of macadamia nuts at $9.55 per pound.
a Write a mathematical expression for the total price of the nuts. $6(2.75) + 4(9.55)$
b Find the total price. $54.70

Problems and Applications

9 Evaluate each expression.
a $10 + 3 \cdot 5 + 4$ 29
b $(10 + 3) \cdot 5 + 4$ 69
c $10 + 3 \cdot (5 + 4)$ 37
d $(10 + 3) \cdot (5 + 4)$ 117

Section 2.1 Operations on Numbers

---

have the chance to explore calculator use, perhaps in pairs.

Problem-Set Notes

for page 53

- **8, 12,** and **19** lead naturally into solving more advanced word problems later.
- **9** anticipates common student errors.

Additional Answers

for page 53

1 Possible answer: So that everyone will assign the same value to a mathematical expression

2 The addition and subtraction were performed before the division. (The correct value of $4 + 12 \div 7 - 6$ is $-\tfrac{2}{7}$.)

3 The operations were not performed from left to right. (The correct value of $23 - 7 + 6$ is 22.)

4 The addition was not performed before the division. (The correct value of $\tfrac{4+3}{4}$ is $\tfrac{7}{4}$, or $1\tfrac{3}{4}$.)

5 The numbers 3 and 5 were added. However, 3 and $(12 - 7)$ should be multiplied first. (The correct value of $5 + 3(12 - 7)$ is 20.)

7a True; possible explanation: The parentheses and the vinculum act as grouping symbols in the same way, so the expressions before and after the equal sign are equivalent.

b True; possible explanation: Since multiplication and division are performed from left to right, the parentheses are superfluous—they group the part of the expression that would be evaluated first in any case.

SECTION 2.1

Problem-Set Notes

for pages 54 and 55

■ **29** leads to some interesting questions. What is the perimeter of the rectangle that encloses this figure? What is its area? Would the perimeter get larger if the small rectangular piece were added onto rather than cut out of the 7 × 11 rectangle?

Additional Practice

1. Is each statement *true* or *false*?
a. $14 - 10 \div 4 = \frac{14 - 10}{4}$
b. $(6-1)(2+4) = 6 - 1 \cdot 2 + 4$
c. $(16 \div 2) - 14 = 16 \div 2 - 14$
d. $7 \cdot 4 - 3 \div 6 = [(7 \cdot 4) - 3] \div 6$

2. Evaluate each expression.
a. $(11 + 3)(2)$
b. $132 - 4 + 9 \div 3$
c. $11 + (3 \cdot 2)$
d. $(16 + 4 \div 2) \div (9 - 3 \times 2)$

Answers:
1. a. False b. False c. True d. False
2. a 28 b. 131 c. 17 d. 6

Checkpoint

1. Are the following expressions equal?
a. $6 + (5 \times 9) \div 2$; $(6 + 5) \times 9 \div 2$
b. $(8 \div 2) \times 5$; $8 \div (2 \times 5)$
c. $12 - 4 \times 2$; $12 - (4 \times 2)$
2. If you entered 18/3*(2+5) on a computer, what value would the computer print?
3. Evaluate each expression.
a. $20 - 6 \times 3$
b. $15 \div 3 + 4$

Answers:
1. a. No b. No c. Yes
2. 42 3. a. 2 b. 9

R **10** Indicate whether the statement is *true* or *false*.
a $2\frac{3}{8} = 2 + \frac{3}{8}$ True
b $2 \cdot \frac{3}{8} = 2 + \frac{3}{8}$ False
c $2\frac{3}{8} = 2.375$ True
d $2 \cdot \frac{3}{8} = 2.375$ False

P **11** Evaluate each expression.
a $\frac{3+5}{14+2}$ $\frac{1}{2}$
b $3 + \frac{5}{14} + 2$ $5\frac{5}{14}$
c $\frac{3+5}{14} + 2$ $2\frac{4}{7}$

P, R, I **12 Consumer Math** A plumber charges $52.00 for a service call and $27.50 per hour for labor. How much does a 3-hour service call cost? $134.50

P **In problems 13–18, evaluate each expression.**

13 $3 \cdot 4 \div 2$ 6
14 $3 \div 2 \cdot 4$ 6
15 $16 - 15/5*4 - 4$ 0

16 $\frac{3+5}{2+10}$ $\frac{2}{3}$
17 $\frac{18-6}{3+9}$ 1
18 $3\frac{1}{2} + 2\frac{1}{2}$ 6

P, R, I **19 Communicating** Describe a real-life situation that can be represented by the mathematical expresion $6 + 2 \cdot 5$.

P, I **20** What is the Celsius equivalent of 77°F? 25°C

P **In problems 21–24, evaluate each expression.**

21 $4 + (3 + 9) \div 6$ 6
22 $(15 - 7/10)/2$ 7.15

23 $\frac{24 - 12 \times 2}{13 - 4 \div 2}$ 0
24 $\frac{24 - 12}{13 - 4} \times \frac{12 \times 2}{4 \div 2}$ 16

P **In problems 25–27, find the perimeter of each rectangle.**

25 (2 by 6) 16
26 (3.1 by 12.4) 3.1
27 (3.14 by 6.28) 18.84

P, R **28** Copy the pattern, filling in the missing numbers.

$2 + 3 \cdot 1 = 5$
$2 + 3 \cdot 2 = 8$
$2 + 3 \cdot 3 = \underline{11}$
$2 + 3 \cdot 4 = \underline{14}$
$2 + 3 \cdot 5 = \underline{17}$
$2 + 3 \cdot 6 = \underline{20}$

P **In problems 29 and 30, find the perimeter of each figure.**

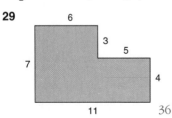

29 36   **30** 24

54 Chapter 2 Formulas and Percent

SECTION 2.1

P, R **31** Evaluate $\frac{3}{5} + \frac{4}{3} \cdot \frac{7}{20}$. Write your answer both as a simplified fraction and as a decimal approximation. $\frac{16}{15}$; ≈1.0667

P, R **32** Draw and label the next three figures in this pattern.

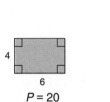
$P = 20$

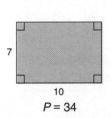
$P = 34$

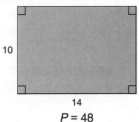
$P = 48$

P, R, I **33 Spreadsheets** This display shows some students' scores on three Latin tests.

	A	B	C	D	E
1		Test 1	Test 2	Test 3	Average
2	Tom	84	75	67	
3	Serena	94	98	84	
4	Jamie	86	74	62	
5	Ivan	93	78	85	
6	Mary	87	64	75	
7					

a What is each student's average score for the tests?
b What formulas could be put in cells E2, E3, E4, E5, and E6 to have the spreadsheet calculate each student's average?
c What is the average of all the scores? ≈80
d What formula could be put in cell E7 so that the spreadsheet would calculate this average? Possible answer: (E2 + E3 + E4 + E5 + E6) ÷ 5

P **34** Contrary Mary wants to put a fence around her flower garden. How many feet of fencing does she need? 42 ft

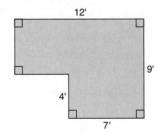

P **35 Consumer Math** A 5-foot-wide gate costs $15.00. A fence post costs $7.95. A 50-foot roll of wire fencing costs $75.00. How much did it cost to enclose this orchard? Possible answer: $209.85

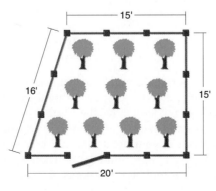

Section 2.1 Operations on Numbers

Cooperative Learning

- Have groups of students put grouping symbols in the following equations so that the equations will be true. Encourage groups to find as many different ways as they can.
  1. $4 + 6 \times 2 \div 7 + 3 = 2$
  2. $16 \div 2 + 2 - 2 \div 2 + 2 = 3.5$
  3. $4 \times 3 + 5 - 1 + 6 \times 3 = 102$
  4. $6 - 3 \times 2 - 4 \div 4 \times 0 = 0$

  Possible answers:
  1. $(4 + 6) \times 2 \div (7 + 3) = 2$
  2. $16 \div (2 + 2) - 2 \div (2 + 2) = 3.5$
  3. $[4 \times (3 + 5 - 1) + 6] \times 3 = 102$
  4. $(6 - 3 \times 2 - 4 \div 4) \times 0 = 0$

- Problem 33 is appropriate for cooperative learning.

Additional Answers

for pages 54 and 55

19 Possible answer: Six 1-dollar bills and two 5-dollar bills

32

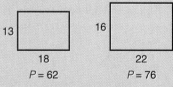

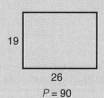

$P = 90$

33 a Tom—75.$\overline{3}$; Serena—92; Jamie—74; Ivan—85.$\overline{3}$; Mary: 75.$\overline{3}$
b (B2 + C2 + D2) / 3; (B3 + C3 + D3) / 3; (B4 + C4 + D4) / 3; (B5 + C5 + D5) / 3; (B6 + C6 + D6) / 3

SECTION 2.1

Problem-Set Notes

for page 56

- 38–41 are appropriate for asking students how many different ways they can think of to solve problems.
- 45 introduces the next section.

Stumbling Block

Students may miss the part of the order of operations that says "left to right" and simply perform multiplications or divisions first in no set order. Point out that all multiplications and divisions, and then all additions and subtractions, are performed from left to right, like the reading and rereading of a line of a book.

Communicating Mathematics

- Have students write a few sentences describing the similarities and differences between the following expressions:
  $(6 + 4) \times 3$ $6 + 4 \times 3$
  $6 + (4 \times 3)$
- Problems 1–5, 7, 19, and 47 encourage communicating mathematics.

Additional Answers

for page 56

Investigation 5°C; 35°C; 91°C; Possible answer: Multiply by 2, then add 30.

P, R, I **36 Consumer Math** Donald and Wendy went into King Castle Burgers and bought three cheeseburgers at $1.50 each, two orders of fries at $1.12 each, a shake at $1.85, and a soft drink at $0.89. There was a 6% sales tax.
  a What was the total cost of the items? $9.48
  b How much tax was charged? $0.57
  c What was the total cost, with tax included? $10.05

◀ LOOKING BACK **Spiral Learning** LOOKING AHEAD ▶

I **37 a** Multiply $7 \cdot 7 \cdot 7 \cdot 7 \cdot 7$. 16,807
  b Multiply six 7's (that is, $7 \cdot 7 \cdot 7 \cdot 7 \cdot 7 \cdot 7$). 117,649
  c Multiply seven 6's. 279,936
  d Multiply eight 3's. 6561

R, I **In problems 38–41, indicate whether the statement is *true* or *false*.**

38 $\frac{2}{3} = \frac{4}{6}$ True   **39** $\frac{9}{11} = \frac{19}{21}$ False   **40** $\frac{5}{8} = \frac{30}{48}$ True   **41** $\frac{3}{4} = \frac{75}{100}$ True

I **42** What number should go in each □, in each △, and in each ○?
  a □ + □ = 81 40.5   **b** △ · △ = 81 9   **c** ○ · ○ · ○ · ○ = 81 3

I **43** Is the following statement true *always, sometimes,* or *never?* Always
  The product of two multiples of 3 is a multiple of 9.

R **44** A bag contains four chips, numbered 3, 5, 7, and 8. Three chips are drawn at random. What is the probability that
  a The sum of the numbers on the three chips is even? $\frac{3}{4}$
  b The product of the numbers on the chips is even? $\frac{3}{4}$

I **45** Most scientific calculators have a key labeled $\boxed{x^y}$, $\boxed{y^x}$, or $\boxed{\wedge}$. Use this key to evaluate each of the following.
  a 7 $\boxed{x^y}$ 2 $\boxed{=}$ 49   **b** 5 $\boxed{x^y}$ 3 $\boxed{=}$ 125   **c** 3 $\boxed{x^y}$ 4 $\boxed{=}$ 81

I **46** If $x = 2$, $y = 3$, and $z = 4$, what is the value of each expression?
  a $x + yz$ 14   **b** $x \cdot y \cdot z$ 24   **c** $(x + y) \cdot z$ 20

I **47 Communicating** Look up the meaning of *pi* (π). Explain what π is. Possible answer: A Greek letter used to symbolize the ratio of a circle's circumference to its diameter

Investigation

Mental Math A rule of thumb for estimating the Celsius equivalent of a temperature expressed in degrees Fahrenheit is to subtract 30 from the Fahrenheit reading, then halve the result. This gives less accurate results than the formula in Sample Problem 2, but it provides a reasonable estimate and can be done in your head. Use this method to estimate the Celsius equivalents of 40°F, 100°F, and 212°F. Then write a rule of thumb for estimating the Fahrenheit equivalents of Celsius readings.

Chapter 2 Formulas and Percent

SECTION 2.2

Investigation Connection
Section 2.2 supports Unit A, *Investigation* 2. (See pp. 2A-2D in this book.)

SECTION 2.2
Powers and Roots

Objectives
- To use a calculator to evaluate powers
- To use a calculator to evaluate roots
- To perform mathematical operations in the conventional order

Vocabulary: base, exponent, power, radicand, root, root index, square root

Class Planning

Schedule 2–3 days

Resources

In *Chapter 2 Student Activity Masters* Book
Class Openers 13, 14
Extra Practice 15, 16
Cooperative Learning 17, 18
Enrichment 19
Learning Log 20
Technology 21, 22
Spanish-Language Summary 23

Transparencies 8 a–d

Class Opener

Use your calculator to evaluate.
1. 5^5 2. 2^5 3. 1^5
4. $(\frac{1}{2})^5$ 5. $(\frac{1}{5})^5$ 6. 0^5

Answers:
1. 3125 2. 32 3. 1
4. 0.03125 5. 0.00032 6. 0

Lesson Notes

- Example 4 will bring out many common calculator errors. Many students overuse the equals key and inadvertently group things that aren't supposed to be grouped.
- Sample problems 1 and 2 are very important for use with the Pythagorean Theorem later. Some students have difficulty knowing

EXPONENTS

M. Ultiply wanted to use his calculator to multiply

$$5 \cdot 5 \cdot 5 \cdot 5 \cdot 5 \cdot 5 \cdot 5 \cdot 5 \cdot 5$$

He had just begun when his friend X. Ponent asked him what he was doing. "I'm multiplying nine fives," he replied. "Don't interrupt me, or I'll lose track."

Mr. Ponent pulled out his calculator. In a moment he said, "I have it! It's 1,953,125."

"How did you do that so fast?" exclaimed Mr. Ultiply. "I'm still multiplying."

"I just calculated five to the ninth power," said Mr. Ponent.

To indicate repeated multiplication of a number, we can use a special notation. Multiplication of nine 5's, for example, can be written 5^9, which is read "five to the ninth ***power***." In this notation, the 5 is called the ***base*** and the 9 is called an ***exponent***.

Notation	Read	Meaning
3^5	Three to the fifth power	$3 \cdot 3 \cdot 3 \cdot 3 \cdot 3$
7^2	Seven to the second power or Seven squared	$7 \cdot 7$
8^3	Eight to the third power or Eight cubed	$8 \cdot 8 \cdot 8$
$(\frac{1}{5})^4$	One fifth to the fourth power	$(\frac{1}{5})(\frac{1}{5})(\frac{1}{5})(\frac{1}{5})$

Section 2.2 Powers and Roots   57

SECTION 2.2

Lesson Notes continued

exactly how to do these on their calculators, so you may want to discuss these problems thoroughly.

■ If your students are interested, you might want to show them how to use a calculator to find roots other than square roots. Some calculators have a root key, but if the one you are using does not, you can still get a root by entering the base, the x^y key, the root you want, the $\frac{1}{x}$ key, and then equals. For example, to find the fifth root of 235, enter 235, x^y, 5, $\frac{1}{x}$, and = and you should get approximately 2.9799815.

■ As students use calculators to explore powers and roots, some results may be given in scientific notation, so the opportunity may present itself to interpret scientific notation answers on the calculators.

Additional Examples

Example 1
Evaluate 6.2^5.
Answer: 9161.3283

Example 2
Evaluate $\sqrt{381.8116}$.
Answer: 19.54

Example 3
Evaluate $\sqrt{68}$.
Answer: ≈8.2462113

Example 4
Evaluate $33 \times 3 - 12 + \frac{8}{\sqrt{16}}$.
Answer: 89

Example 1
Evaluate 3.7^4.

Solution
Scientific calculators have a power key, usually labeled $\boxed{x^y}$, $\boxed{y^x}$, or $\boxed{\wedge}$. We enter 3.7 $\boxed{x^y}$ 4 $\boxed{=}$, and the display shows $\boxed{187.4161}$. Therefore, $3.7^4 = 187.4161$.

ROOTS

Sometimes, instead of finding a power of a base, we need to find the base that corresponds to a given power. For instance, what number raised to the third power is 1000? In this case, the answer is 10, since $10^3 = 10 \cdot 10 \cdot 10 = 1000$. We call 10 the third **root**, or the cube root, of 1000 and write $\sqrt[3]{1000} = 10$.

Similarly, since $5^2 = 25$, we can say that 5 is the second root, or the **square root**, of 25 and write $\sqrt[2]{25} = 5$. And since $2^8 = 256$, we can say that 2 is the eighth root of 256 and write $\sqrt[8]{256} = 2$.

In the notation $\sqrt[8]{256}$, the 8 is called a **root index** and the 256 is called a **radicand**. Because mathematicians work with square roots a lot, they usually omit the root index 2 when they write square roots. Thus, $\sqrt{81}$ means "the square root of 81." (Can you figure out what the value of $\sqrt{81}$ is?)

Example 2
Evaluate $\sqrt{552.7201}$.

Solution
Scientific calculators have a square-root key, labeled $\boxed{\sqrt{\ }}$ or $\boxed{\sqrt{x}}$. We enter 552.7201 $\boxed{\sqrt{\ }}$, and the display shows $\boxed{23.51}$. Therefore, $\sqrt{552.7201} = 23.51$.

It is fairly easy to find exact values of roots such as $\sqrt{25}$ and $\sqrt{81}$. Most of the time, though, we have to be satisfied with approximations.

Example 3
Evaluate $\sqrt{70}$.

Solution
If we enter 70 $\boxed{\sqrt{\ }}$ on a calculator, the display may show $\boxed{8.366600265}$. But 8.366600265 is not the exact value of the square root, since if we calculate 8.366600265^2, we get an answer of 69.99999999, not 70. The calculator could not show all the decimal places of the root, so it showed as many as its display would hold. Usually we round such numbers to a convenient number of decimal places. We might round $\sqrt{70}$ to the nearest hundredth or thousandth, writing $\sqrt{70} \approx 8.37$ or $\sqrt{70} \approx 8.367$. (The symbol ≈ means "is approximately equal to.")

SECTION 2.2

ORDER OF OPERATIONS

In the preceding section, you saw the order in which some mathematical operations should be performed. Now we can add two new operations.

Order of Operations
1. Do operations enclosed by grouping symbols (parentheses, brackets, or vinculums).
2. Evaluate powers and roots.
3. Multiply and divide from left to right.
4. Add and subtract from left to right.

Example 4

Evaluate $2 + 7 \cdot 5^3 - \dfrac{6}{\sqrt{9}}$.

Solution

$$2 + 7 \cdot 5^3 - \dfrac{6}{\sqrt{9}}$$

Evaluate the power and the root $\quad = 2 + 7 \cdot 125 - \dfrac{6}{3}$

Multiply and divide $\quad = 2 + 875 - 2$

Add and subtract $\quad = 875$

Sample Problems

Problem 1 Evaluate $\sqrt{16 + 9}$.

Solution Like a fraction bar, the bar above 16 + 9 is a vinculum. Therefore, we add first.

$$\sqrt{16 + 9} = \sqrt{25} = 5$$

Try solving this problem with a calculator. (Be careful, it's tricky!) Did you get 5 as your answer?

Problem 2 Is the statement $(3 + 5)^2 = 3^2 + 5^2$ *true* or *false*?

Solution Let's evaluate each side of the equation.

Left Side	Right Side
$(3 + 5)^2$	$3^2 + 5^2$
$= 8^2$	$= 9 + 25$
$= 64$	$= 34$

Since 64 is not equal to 34, the statement is false; $(3 + 5)^2 \neq 3^2 + 5^2$.

Stumbling Block

Some students may evaluate an expression such as 3^4 by multiplying 3 by 4. Point out that only numbers that are written on the same level may be multiplied. For example, $3 \times 4 = 12$ and $5^{3 \times 4} = 5^{12}$, but $3^4 \neq 12$. Emphasize that 3^4 means there are 4 factors of 3.

Reteaching

To reteach finding $\sqrt{70}$ in example 3, have students find $\sqrt{64}$ and $\sqrt{81}$. Since $64 < 70 < 81$, $\sqrt{70}$ must be between 8 and 9. Have students find 8.5^2. Ask students if $\sqrt{70}$ will be between 8 and 8.5 or between 8.5 and 9 (Answer: 8 and 8.5). Continue to close in on the value of $\sqrt{70}$ by changing the 8.5 to 8.25, and so on.

Cooperative Learning

- Have groups of students put at least one square root sign and at least one exponent of 2 in the following equations so that they will be true.
  1. $3 + 5 + 4 - 6 + 5 = 25$
  2. $16 + 81 + 2 + (2 + 3) = 42$
  3. $9 \times 4 + 3 - 2 \times 9 = 21$
  4. $25 + 11 + (4 + 3) + 3 = 8$

Possible answers:
  1. $3 + \sqrt{5 + 4} - 6 + 5^2 = 25$
  2. $\sqrt{16} + \sqrt{81} + 2^2 + (2 + 3)^2 = 42$
  3. $9 \times \sqrt{4} + 3^2 - 2 \times \sqrt{9} = 21$
  4. $25 + 11 + (\sqrt{4} + 3)^2 + 3 = 8^2$

- Problem 23 is appropriate for cooperative learning.

SECTION 2.2

Assignment Guide

Basic
Day 1 8, 10–12, 16, 20, 24, 27, 28, 33, 36, 40
Day 2 9, 13, 18, 19, 22, 25, 29, 32, 34, 38, 43
Day 3 14, 17, 21, 23, 26, 30, 41, 44, 46

Average or Heterogeneous
Day 1 8–15, 17, 18, 23, 28, 30, 36, 38, 42, 44
Day 2 16, 19–22, 24–27, 29, 33, 34, 37, 39, 41, 46, 47

Honors
Day 1 9, 11, 14, 17–19, 23, 27, 28, 31, 34, 36, 38, 40, 44
Day 2 8, 13, 15, 20, 24, 26, 29, 33, 35, 37, 39, 41, 46, 47, Investigation

Strands

Communicating Mathematics	1, 5, 27, 30
Geometry	15, 19, 20, 27, 29, 39, 42, 47
Data Analysis/ Statistics	23
Technology	30, 40
Algebra	7, 9, 18, 28, 38, Investigation
Patterns & Functions	30
Number Sense	3, 5, 9, 18, Investigation
Interdisciplinary Applications	7, 46

Individualizing Instruction

Visually impaired learners Students may have a hard time distinguishing exponents from bases. Neat writing and concise verbal descriptions may be helpful to these students.

Think and Discuss

P **1** **a** Why might raising a number to the second power be called squaring?
   b Why might raising a number to the third power be called cubing?

P, R **2** Evaluate each expression.
   a $1 + 2 \cdot 3^4$ 163   **b** $(2 + 3)^4$ 625   **c** $2^4 + 3^4$ 97

P **3** Is the statement *true* or *false*?
   a $3^2 > 3$ True   **b** $1.5^2 > 1.5$ True   **c** $1^2 > 1$ False   **d** $0.8^2 > 0.8$ False

P, I **4** Evaluate each expression.
   a 2.6^4 45.6976   **b** $2 \cdot 6^4$ 2592   **c** $(2 \cdot 6)^4$ 20,736

P, I **5** Is the statement true *always, sometimes,* or *never?* Explain your answer.
   a The square of a number is greater than the number.
   b The square root of a number is less than the number.

P, R **6** Evaluate each expression.
   a $\sqrt{6^2 - 5 \cdot 4}$ 4   **b** $\sqrt{5^2 + 12^2}$ 13   **c** $\sqrt{17^2 - 8^2}$ 15

P **7** **Science** If we use v to stand for the speed of sound in feet per second and t to stand for the temperature in degrees Celsius, we can find the speed of sound at sea level with the formula
$$v = \frac{1087 \cdot \sqrt{273 + t}}{16.52}$$
What is the speed of sound when the temperature is 25°C?   ≈1135.87 ft/sec

Problems and Applications

P **8** Evaluate each expression.
   a 3.5^2 12.25   **b** $(3.5)^2$ 12.25   **c** $(3 \cdot 5)^2$ 225   **d** $3 \cdot 5^2$ 75

P, I **9** What number should go in each □, in each △, and in each ○?
   a $□ \cdot □ \cdot □ \cdot □ = 256$ 4   **b** $△ \cdot △ = 11.26$ ≈3.3556   **c** $○ \cdot ○ \cdot ○ = 125$ 5

P **10** Write and evaluate a root with radicand 81 and root index 4. $\sqrt[4]{81}$; 3

P **In problems 11–14, evaluate each expression to the nearest thousandth.**

11 $\sqrt{5}$ ≈2.236

12 $\sqrt{5.001}$ ≈2.236

13 $\sqrt{857}$ ≈29.275

14 $\sqrt{8^2 + 10^2}$ ≈12.806

Chapter 2 Formulas and Percent

SECTION 2.2

P **15** The area of a square is the second power of its side length.
  a Find the area of each square.
  b Find the total area of the figure.
    a 36   **b** 180

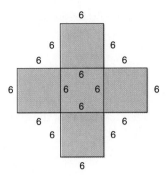

P **16** Evaluate each expression.
  a $\left(\frac{1}{2}\right)^2 + \left(\frac{1}{2}\right)^2$ $\frac{1}{2}$
  b $\frac{3}{8} + \left(\frac{1}{2}\right)^3$ $\frac{1}{2}$

P **17** Indicate whether the statement is *true* or *false*.
  a $3^5 = (1+2)^5$ True
  b $3^5 = 1^5 + 2^5$ False
  c $3^5 = 1^5 + 1^5 + 1^5$ False
  d $3^5 = (1+1+1)^5$ True

P, I **18** What exponents go in the boxes?
  a $(2 \cdot 2 \cdot 2 \cdot 2) \cdot (2 \cdot 2 \cdot 2 \cdot 2 \cdot 2 \cdot 2) = 2^{\square}$ 10
  b $2^4 \cdot 2^6 = 2^{\square}$ 10

P, I **19** Each edge of the open box has a length of 8 inches. Find the total area of the box's sides and bottom.
  320 in.²

P, I **20** If $\pi \approx 3.1416$, $r \approx 6.75$, and $d \approx 13.5$, what is the value of
  a πd? ≈ 42.4116
  b $2\pi r$? ≈ 42.4116
  c πr^2? ≈ 143.13915

P, I In problems 21 and 22, indicate whether the statement is *true* or *false*.

21 $3.75 \times 10^5 = 375{,}000$ True

22 $412.35 = 4.12 \times 10^2$ False

P, R **23** Spreadsheets
  a What values go in cells B7, B8, and B9 of this spreadsheet display?   729; 2187; 6561
  b In spreadsheets the symbol ^ is used to indicate a power—for example, 5^3 means 5^3. Use the base 3 and cell names from column A to write formulas that will generate the correct values in B7, B8, and B9.
  c Now write formulas for the values in B7, B8, and B9, using names only of cells in column B.
    b 3^A7; 3^A8; 3^A9
    c 3 × B6; 3 × B7; 3 × B8

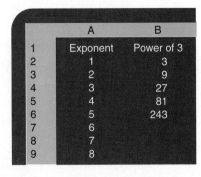

Section 2.2 Powers and Roots

SECTION 2.2

Problem-Set Notes

for pages 62 and 63

- **24–26** and **35** will check to see how well students have mastered the use of their calculators.
- **28** is an important preview as well as a good practice problem.
- **37** should provide some interesting discussion. Be sure to ask students how they thought about this problem.
- **47** might be done as an activity in class.

Multicultural Note

The rules for order of operations are cross-cultural in the sense that mathematicians in all countries adhere to them. Even computers and calculators with algebraic logic follow these rules.

Communicating Mathematics

- Have students describe orally, or write a paragraph describing, the difference between the following expressions: 5^3, 3^5, and 3×5.
- Problems 1, 5, 27, and 30 encourage communicating mathematics.

P **In problems 24–26, evaluate each expression.**

24 $3 + 5 \cdot 8^3 \div \sqrt{16} + \dfrac{10}{\sqrt{25}}$ 645   **25** $5^3 \cdot 5^4 - 5^7$ 0

26 $\sqrt{4 + 5 \cdot 8^3 + 5 \cdot 5^3} + 1$ ≈57.47

I **27** Wanda went from A to B to C. Willie went directly from A to C. Who went the shorter distance? Explain.

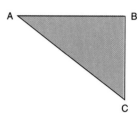

P **28** Indicate whether the statement is *true* or *false*.
  a $5^2 + 5^2 + 5^2 = 3 \cdot 5^2$ True   **b** $\sqrt{9} + \sqrt{9} + \sqrt{9} = 3 \cdot \sqrt{9}$ True
  c $7^5 + 7^5 + 7^5 = 3 \cdot 7^5$ True   **d** $\sqrt{7} + \sqrt{7} + \sqrt{7} = 3 \cdot \sqrt{7}$ True

P **29** The dots in the diagram are 1.8 units apart.
  a Find the area of rectangle ABCD.
  b Find the area of triangle ABC.
    a 25.92 square units
    b 12.96 square units

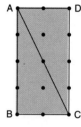

P **30 Communicating**   Enter the given number on a calculator. Then press the $\sqrt{}$ key repeatedly. What do you notice about the numbers displayed?
  a 2   **b** 0.25

P **In problems 31–34, indicate whether the statement is *true* or *false*.**

31 $\sqrt{4+5} = \sqrt{4} + \sqrt{5}$ False   **32** $(3+4)^2 = 3^2 + 4^2$ False

33 $5 \cdot (3+4) = 5 \cdot 3 + 5 \cdot 4$ True   **34** $\sqrt{5^2 - 4^2} = 5 - 4$ False

P **35** Evaluate $\sqrt{\sqrt{2} + \sqrt{3} + \sqrt{5}}$ to the nearest thousandth.   ≈2.320

◀ LOOKING BACK **Spiral Learning** LOOKING AHEAD ▶

R, I **36 a** Change $\tfrac{3}{4}$ to an equivalent fraction with denominator 100. $\tfrac{75}{100}$
  b Change $\tfrac{18}{25}$ to an equivalent fraction with denominator 100. $\tfrac{72}{100}$
  c Which is greater, $\tfrac{3}{4}$ or $\tfrac{18}{25}$? $\tfrac{3}{4}$

R **37** Insert parentheses to make each statement true.
  a $3 + 5 \cdot 6 + 7 = 68$ **b** $3 + 5 \cdot 6 + 7 = 55$ **c** $3 + 5 \cdot 6 + 7 = 104$
    $3 + 5 \cdot (6+7) = 68$ $(3+5) \cdot 6 + 7 = 55$ $(3+5) \cdot (6+7) = 104$

Chapter 2 Formulas and Percent

SECTION 2.2

Checkpoint

1. Evaluate.
   a. 14^5 b. $(\frac{1}{3})^4$
   c. 6.4^3 d. 0.73^2
2. Evaluate.
   a. $\sqrt{7.1}$ b. $\sqrt{0.0381}$
   c. $\sqrt{27}$ d. $\sqrt{80}$
3. Evaluate.
   a. $\frac{(4-1)}{\sqrt{32}}$
   b. $8 \cdot 3 + \sqrt{25}$
   c. $\frac{6}{\sqrt{10}} + 4$

Answers:
1. a. 537,824 b. ≈0.01235
   c. 262.144 d. 0.5329
2. a. ≈2.6646 b. ≈0.1952
   c. ≈5.1962 d. ≈8.9443
3. a. ≈0.53 b. 29 c. ≈5.8974

Additional Answers

for pages 62 and 63

27 Willie; possible explanation: The shortest path between two points is the line segment joining them.

30 a Possible answer: The numbers get smaller (closer to 1), with the calculator eventually displaying 1 each time. **b** Possible answer: The numbers get larger (closer to 1), with the calculator eventually displaying 1 each time.

Investigation 100; The expression $4x^2$ means $4 \cdot x^2$; Square just the variable x, not the 4; it definitely matters! Evaluate the power first, then multiply.

I **38** If $b = 5$ and $h = 6$, what is the value of
  a $b \cdot h$? 30
  b $\frac{1}{2}bh$? 15

R **39** Indicate whether the statement is *true* or *false*.
  a These two figures have the same perimeter. True
  b These two figures have the same area. False

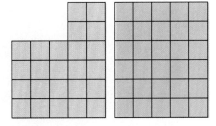

R **40** What is the Celsius equivalent of 90°F? ≈32.2°C

R, I **41** Carya had 5 pounds of pecans and 16 pounds of mixed nuts. By weight, 50 percent of the mixed nuts was pecans. How many pounds of pecans did Carya have in all? 13 lb

R **42** What is the perimeter of this rectangle? $7\frac{1}{4}$

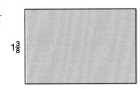

I **In problems 43–45, rewrite the fraction as a decimal.**

43 $\frac{3}{5}$ 0.6   **44** $\frac{7}{4}$ 1.75   **45** $\frac{13}{3}$ $4.\overline{3}$

R, I **46** **Consumer Math** Susie sold seventeen seashells by the seashore for seven dollars and seven cents each. There was also a 6.6% sales tax. How much money did Susie take in? $128.12

I **47** You are given a sheet of paper measuring $8\frac{1}{2}$ inches by 11 inches (a standard sheet of typing paper). Can you cut a 1 inch by 12 inch rectangle from the sheet? Yes

Investigation

Algebra If $x = 5$, what is the value of the expression $4x^2$? Does the expression mean $4 \cdot x^2$, or does it mean $(4x)^2$? That is, should you square just the value of x or square the value $4x$? Or doesn't it matter? See if you can come up with a rule for evaluating expressions like this.

Section 2.2 Powers and Roots

SECTION 2.3

Objectives

- To use formulas to calculate the areas and perimeters of some geometric figures
- To use the Pythagorean Theorem to calculate lengths of sides of right triangles

Vocabulary: hypotenuse, leg, Pythagorean Theorem

Class Planning

Schedule 2 days

Resources

In *Chapter 2 Student Activity Masters* Book
Class Openers 25, 26
Extra Practice 27, 28
Cooperative Learning 29, 30
Enrichment 31
Learning Log 32
Using Manipulatives 33, 34
Interdisciplinary Investigation 35, 36
Spanish-Language Summary 37

In *Tests and Quizzes* Book
Quiz on 2.1–2.3 (Basic) 25
Quiz on 2.1–2.3 (Average) 26
Quiz on 2.1–2.3 (Honors) 27

Transparencies 9–14

Class Opener

Use your calculator to evaluate.
1. $\sqrt{3^2 + 4^2}$ 2. $\frac{1}{2}(4)(7)$
3. $\sqrt{5^2 - 4^2}$ 4. $\sqrt{13^2 - 12^2}$
5. $\frac{1}{2} \cdot (7 + 9) \cdot 3$

Answers:
1. 5   2. 14   3. 3   4. 5   5. 24

Lesson Notes

- If your students have not seen or done some activities to justify the formulas given in this section, take the time to develop them in class. Starting with a rectangle, have students draw a diagonal from one corner to anywhere on an opposite side, then cut along

Investigation Connection
Section 2.3 supports Unit A, *Investigation* 2. (See pp. 2A-2D in this book.)

SECTION 2.3
Geometric Formulas

AREA AND CIRCUMFERENCE FORMULAS

You may already be familiar with formulas that can be used to find the areas and the perimeters of various shapes. Here are some frequently used ones.

Area of a Rectangle or Parallelogram

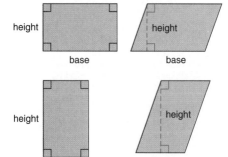

Area = base · height

Area of a Triangle

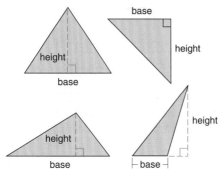

Area = $\frac{1}{2}$ · base · height

Circumference (Perimeter) of a Circle

Circumference = π · diameter

Area of a Circle

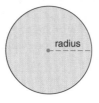

Area = π · (radius)^2

SECTION 2.3

Example 1
Find the area of the parallelogram.

Solution
Area = base · height
    = 7 · 3.4
    = 23.8

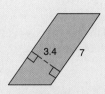

Example 2
Find the area and the circumference of the circle.

Solution
The letter π represents a number that we can only approximate. In this case, we will use the approximation 3.1416 and round the answers to hundredths.

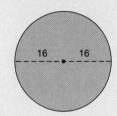

Area = π · (radius)^2 Circumference = π · diameter
    = π · 16^2 = π · 32
    = π · 256   ≈ 3.1416 · 32
    ≈ 3.1416 · 256   ≈ 100.53
    ≈ 804.25

You can find other important formulas in a table in the back of this book.

THE PYTHAGOREAN THEOREM

For many people, one of the most important ideas of geometry is the **Pythagorean Theorem.** It expresses a relationship among the sides of a right triangle—a triangle in which one of the angles is a right (90°) angle.

In a right triangle, the longest side (the side opposite the right angle) is called the **hypotenuse.** The other sides (the sides that meet to form the right angle) are the triangle's **legs.**

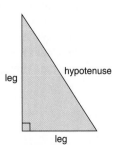

The Pythagorean Theorem states that in any right triangle,

$$(\text{Leg}_1)^2 + (\text{leg}_2)^2 = (\text{hypotenuse})^2$$

Additional Examples

Example 1
Find the area of the parallelogram.

Answer: 32.64

Example 2
Find the area and circumference of the circle.

Answers:
Area ≈ 452.39
Circumference ≈ 75.40

Example 3
What is the length of the other leg of a right triangle with leg 5 and hypotenuse 13?

Answer: 12

---

this line. Move the resulting triangle to the other side of the figure, forming a parallelogram. Point out that the area should be the same as that of the original rectangle. Be sure to discuss that the height of the parallelogram is not the length of the slanted side. The area of a triangle follows readily from the parallelogram. Have students draw one diagonal (connecting opposite corners) and ask them how the area of one of the two identical triangles just formed is related to the area of the parallelogram.

- For the area of a circle, cut a circle into eight equal sections and arrange them to resemble a parallelogram. The dimensions will be about πr and r, which leads to the area formula.

- Point out that we use approximations for the exact value of pi. Ensure that the students do not think that pi equals $\frac{22}{7}$ or 3.14. These are only approximations of pi.

SECTION 2.3

Assignment Guide

Basic
Day 1 6–8, 10–12, 16, 18, 20, 25–27, 31
Day 2 9, 13–15, 23, 28–30, 32–35, 37

Average or Heterogeneous
Day 1 6–9, 11, 12, 14, 15, 18–20, 25–27, 29, 31, 32, 34
Day 2 10, 13, 16, 17, 21–24, 28, 30, 33, 37

Honors
Day 1 6, 8, 11, 12, 15, 18, 19, 21, 25, 26, 30, 33, 36
Day 2 7, 9, 10, 13, 16, 17, 20, 22–24, 31, 32, 34, 35, 37, Investigation

Strands

Communicating Mathematics	1
Geometry	1–6, 10–13, 15–24, 33
Data Analysis/ Statistics	26
Probability	24
Algebra	7–9, 25
Patterns & Functions	6, 26
Number Sense	1, 4, 27, 28, 32, 34–36
Interdisciplinary Applications	4, 20, 33, 37

Individualizing Instruction

The visual learner Some students may have more difficulty than others in working problems in which geometric figures are described only in words. Encourage students to draw diagrams for these problems. This will help them see the relationships among different sides and angles.

66

Example 3

In this right triangle, the length of one leg is 7 and the length of the hypotenuse is 25. What is the length of the other leg?

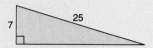

Solution

$$(\text{Leg}_1)^2 + (\text{leg}_2)^2 = (\text{hypotenuse})^2$$
$$(\text{Leg}_1)^2 + 7^2 = 25^2$$
$$(\text{Leg}_1)^2 + 49 = 625$$

Now, since $(\text{leg}_1)^2$ plus 49 is 625, $(\text{leg}_1)^2$ must be 49 less than 625. Thus, $(\text{leg}_1)^2 = 625 - 49 = 576$. But $24^2 = 576$, so the length of the leg is 24.

Sample Problems

Problem 1 The formula for the area of a trapezoid is

Area = $\frac{1}{2} \cdot (\text{base}_1 + \text{base}_2) \cdot \text{height}$

Use this formula to find the area of the trapezoid shown.

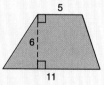

Solution In this trapezoid, the lengths of the bases are 5 and 11, and the height is 6.

Area = $\frac{1}{2} \cdot (\text{base}_1 + \text{base}_2) \cdot \text{height} = \frac{1}{2} \cdot (5 + 11) \cdot 6$
$= \frac{1}{2} \cdot 16 \cdot 6$
$= 48$

Problem 2 A football field is 360 feet long and 160 feet wide. How far is it from one corner of the field to the opposite corner?

Solution The distance we are trying to find is the length of the hypotenuse of a right triangle with legs of 360 feet and 160 feet. We can use the Pythagorean Theorem.

$$(\text{Leg}_1)^2 + (\text{leg}_2)^2 = (\text{hypotenuse})^2$$
$$360^2 + 160^2 = (\text{hypotenuse})^2$$
$$155{,}200 = (\text{hypotenuse})^2$$

Since the length of the hypotenuse squared is 155,200, the length of the hypotenuse must be $\sqrt{155{,}200}$. We can enter 155,200 $\sqrt{}$ on a calculator to find that the distance is approximately 394 feet.

SECTION 2.3

Think and Discuss

P, R **1 a** What is the perimeter of the square? 16 ft
 b What is the area of the square? 16 ft²
 c Explain why the answers to parts **a** and **b** are different.

P **2** Find the circumference and the area of each circle.

≈31.42; ≈78.54 ≈37.70; ≈113.10 ≈21.99; ≈38.48 ≈28.27; ≈63.62

P **3** Find the area of each figure.

 Rectangle Parallelogram Trapezoid Triangle
 28 21 112 60

P, I **4 Consumer Math** Penny Pinscher is purchasing pepperoni pizzas for her pals. A pizza 10 inches in diameter costs $7.95, and a pizza 14 inches in diameter costs $15.45. Should Penny purchase two 10-inch pizzas or one 14-inch pizza?

P **5** Is the statement true *always, sometimes,* or *never?*
 a Two rectangles with equal perimeters have equal areas. Sometimes
 b The hypotenuse of a right triangle is the longest side of the triangle. Always

Problems and Applications

P, R **6** Find the area of each square, then copy and complete the table.

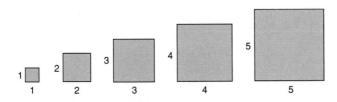

Dimensions of Square	Area
1 by 1	1
2 by 2	4
3 by 3	9
4 by 4	16
5 by 5	25

P, R **In problems 7–9, what value should replace the □?**

 7 $\square^2 + 8^2 = 10^2$ 6 **8** $\square^2 + 5^2 = 13^2$ 12 **9** $\square^2 + 2^2 = 4^2$ $\sqrt{12}$, or ≈3.46

Section 2.3 Geometric Formulas

Problem-Set Notes

for page 67

- **4** should stimulate considerable discussion. Students may be surprised that the two choices are so similar in value.
- **6** is related to **4**.

Cooperative Learning

- Have groups of students consider the following questions, drawing diagrams where needed.
 1. For what size square will the number of perimeter units equal the number of area units? How did you arrive at this answer?
 2. For what size circle will the number of circumference units equal the number of area units? How did you arrive at this answer?

Answers:
1. A square with a side of 4 will have a perimeter of 16 units and an area of 16 square units; answers will vary.
2. A circle with a radius of 2 will have a circumference of 4π units and an area of 4π square units; answers will vary.

- Problem 15 is appropriate for cooperative learning.

Additional Answers

for page 67

1 c Possible answer: Perimeter is a length, whereas area is the measure of a region; therefore, although the answers are numerically equivalent, one is expressed in linear units, and the other is expressed in square units.

4 Possible answer: one 14-inch pizza

SECTION 2.3

Problem-Set Notes

for pages 68 and 69

- **14** should provoke discussion about how it is possible to tell if an answer is exact or not.

- **17** can be extended into finding a segment with length equal to the square root of any positive integer.

- **19** gives a visual justification for the Distributive Property of Multiplication over Addition.

- **21** demonstrates that the area of the ring is more than $1\frac{1}{2}$ times as great as the area of the solid circle in the middle, even though its width is only two-thirds the radius of the circle. Many people are surprised by this counter-intuitive result.

Checkpoint

1. Find the area and perimeter (or circumference) of each figure.

a. b.
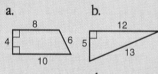

c. d.
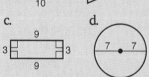

2. Find each missing length to the nearest hundredth.

a. b.
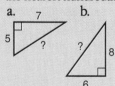

Answers:
1. a. 36; 28 b. 30; 30 c. 27; 24
   d. 49π, or ≈ 153.94; 14π, or ≈ 43.98
2. a. 8.60 b. 10.00

P, I **10** Find the area of the rectangle.   6

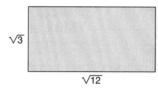

P **In problems 11–13, find the area of the triangle.**

11 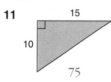 75

12 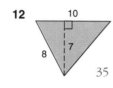 35

13 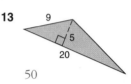 50

P [calc] **14** Indicate whether the statement is *true* or *false*.
  a $\sqrt{33} = 5.744562647$ False
  b $\sqrt{631.5169} = 25.13$ True
  c $\sqrt{2.315} = 1.521512405$ False

P **15** Find the lengths labeled *a, b, c, d,* and *e*. Round to the nearest hundredth if necessary. $a = 10$; $b = 9$; $c \approx 11.18$; $d \approx 8.66$; $e \approx 5.29$

P **16** Find the area of each figure.

a 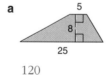 120

b 120

c 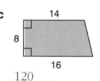 120

d 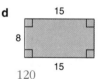 120

P, I **17** Find the lengths labeled *x* and *y*.
  $\sqrt{2}$, or ≈ 1.41; $\sqrt{3}$, or ≈ 1.73

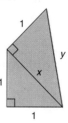

P, R **18** Find the length of the hypotenuse of the right triangle. ≈ 4.19

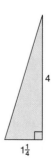

68 **Chapter 2** Formulas and Percent

P, I **19 a** Write two expressions for the area of rectangle ABCD. $5(4+6); 5(4)+5(6)$
b Calculate the area of rectangle ABCD. 50

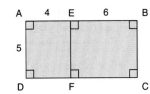

P **20 Sports** How far is it from first base to third base in a straight line? ≈ 127.28 ft

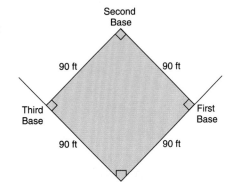

P, I **21 a** Write an expression that stands for the exact area of the shaded region. $5^2\pi - 3^2\pi$
b Find the area of the shaded region to the nearest thousandth. ≈ 50.265 cm^2

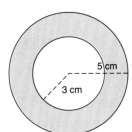

P **22** The dots in the diagram are evenly spaced. Wally walked straight from A to B, then straight from B to C. How far did he walk? 80 ft

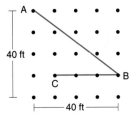

P, I **23 a** How long is segment AD?
b What is the perimeter of rectangle ABCD?
a $\sqrt{10}$ cm, or ≈ 3.16 cm
b $2\sqrt{20} + 2\sqrt{10}$ cm, or ≈ 15.27 cm

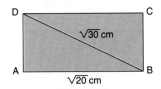

P, I **24** If a dart is thrown so that it lands inside the square, what is the probability that it will land inside the circle? $\approx .79$

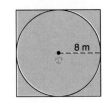

Section 2.3 Geometric Formulas

Additional Practice

1. Find the area of each figure.
   a. A rectangle with sides 4 and 7
   b. A circle with radius 5
   c. A trapezoid with bases 5 and 4 and height 4
   d. A right triangle with height 8 and base 14
2. Find the length of the hypotenuse of a right triangle with the given legs to the nearest hundredth.
   a. 4, 5 **b.** 6.1, 7.8
   c. 9, 14 **d.** 12, 15

Answers:
1. **a.** 28 **b.** 25π, or ≈ 78.54
   c. 18 **d.** 56
2. **a.** 6.4 **b.** 9.90
   c. 16.64 **d.** 19.21

Stumbling Block

Students may try to apply the Pythagorean Theorem to triangles other than right triangles. Emphasize that this relationship is true only for a right triangle. Use a nonright triangle, such as 4-7-9, to show that $4^2 + 7^2 \neq 9^2$.

Reteaching

Use geoboards to help students better visualize how the area formulas are derived. Have students determine the areas of various figures by counting squares directly, then by using the formulas.

SECTION 2.3

Problem-Set Notes

for page 70

- 26 and 34–36 preview percent.

Multicultural Note

The Pythagorean Theorem is attributed to Pythagoras, a Greek mathematician and philosopher who lived in Italy during the fifth century B.C. His followers developed many important mathematical concepts, including concepts of number theory and irrationality.

Communicating Mathematics

- Have students explain how to find the distance a catcher must throw the ball from home plate to second base if the distance between bases along the baselines is 90 feet.
- Problem 1 encourages communicating mathematics.

Additional Answers

for page 70

Investigation The formula is $V = \frac{4}{3}\pi r^3$; $V \approx 4.19$ cm³; 8 times

---

◀ LOOKING BACK **Spiral Learning** LOOKING AHEAD ▶

I **25** Find the value of $x + y - z$.   7

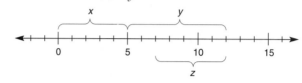

R **26** Copy the pattern, filling in the missing numbers.

15% of 150 is 22.5.
15% of 300 is   45 .
15% of 450 is 67.5 .
15% of 600 is   90 .
15% of 750 is 112.5 .

I In problems 27 and 28, write each number as a common fraction.

27 0.7 $\frac{7}{10}$

28 2.5 $\frac{5}{2}$

R In problems 29–31, evaluate each expression.

29 $5^3 + \sqrt{36}$ 131

30 $\dfrac{3 + 5(4 + \sqrt{9})}{2 \cdot 5^2 - 12}$ 1

31 $3^2 + 4^2 - 2(3)(4)\left(\frac{1}{2}\right)$ 13

I **32 a** Rewrite $\frac{4}{5}$ as a decimal.   0.8

b Rewrite 0.35 as a fraction. $\frac{7}{20}$

R, I **33** **Consumer Math** A rectangular pen measuring 20 feet by 32 feet, with an 8-foot-wide gate, is to be built. The fencing costs $4 per foot, and the material for the gate costs $8 per foot. What will it cost to build the pen? $448

I In problems 34–36, rewrite the fraction as an equivalent fraction having a denominator of 100.

34 $\frac{3}{5}$ $\frac{60}{100}$

35 $\frac{5}{2}$ $\frac{250}{100}$

36 $\frac{35}{500}$ $\frac{7}{100}$

P, I **37** **Consumer Math** Edsel bought a car for $18,500 and paid a sales tax of 15%.

a How much sales tax did Edsel pay? $2775

b What was the total cost of his car? $21,275

Investigation

Volume Look up the formula for the volume of a sphere. Use the formula to find the volume of a marble with a radius of 1 centimeter. If the marble's radius were twice as great, how many times greater would its volume be?

70   **Chapter 2**   Formulas and Percent

SECTION 2.4

Investigation Connection
Section 2.4 supports Unit A, *Investigations* 3, 5. (See pp. 2A-2D in this book.)

SECTION 2.4
Percent

WHAT IS A PERCENT?

According to *The Harper's Index Book,* 22% of the potatoes grown in the United States end up as french fries. What does this statement mean? As you probably know, the symbol % stands for **percent**. The word *percent* means "in each 100." So out of every 100 bushels of potatoes grown in the United States, 22 are prepared as french fries. Other ways of writing 22% are $\frac{22}{100}$ and 0.22.

Example 1
Rewrite each fraction as a percent.

a $\frac{8}{5}$ **b** $\frac{3}{20}$

Solution

a Since 20 × 5 = 100, we can change $\frac{8}{5}$ to hundredths by multiplying both the numerator and the denominator by 20.

$$\frac{8}{5} = \frac{8 \times 20}{5 \times 20} = \frac{160}{100}$$

Therefore, $\frac{8}{5} = 160\%$.

b Let's use a calculator to divide 3 by 20. We enter 3 ÷ 20 =, and the display shows 0.15. Since 0.15 means "15 hundredths," $\frac{3}{20} = 15\%$.

Example 2
Rewrite 24.36% as a fraction and as a decimal.

Solution

$$24.36\% = \frac{24.36}{100} = \frac{2436}{10{,}000} = \frac{2436 \div 4}{10{,}000 \div 4} = \frac{609}{2500}$$

Since $24.36\% = \frac{2436}{10{,}000}$, the decimal form is 0.2436.

Objectives
- To interpret percents
- To solve problems involving percents

Vocabulary: percent

Class Planning

Schedule 2–3 days

Resources

In *Chapter 2 Student Activity Masters* Book
Class Openers 39, 40
Extra Practice 41, 42
Cooperative Learning 43, 44
Enrichment 45
Learning Log 46
Interdisciplinary Investigation 47, 48
Spanish-Language Summary 49

Transparency 15

Class Opener

At the beginning of a game of checkers,
1. What percent of the squares have pieces on them?
2. What percent of the red squares have pieces on them?
3. What percent of the black squares have pieces on them?
4. What percent of the squares are covered by red checkers?
5. What percent of the squares are covered by black checkers?

Answers:
1. 37.5% 2. 0% 3. 75% 4. 18.75%
5. 18.75%

Lesson Notes

- Percent is a difficult topic for students, so sometimes we try to help them through it with overly simple rules, such as *of* means "multiply." While this is an expeditious thing to do, it can eventually lead to difficulties. *Of* doesn't

SECTION 2.4

Lesson Notes continued

always mean "multiply," as in the complement *of* an angle and 20 *of* 30 students. Solving many varied problems over a long period of time will have a much greater impact on a student's understanding than a few rules, which have exceptions that may be overlooked.

- To explain to students why we use percent, ask students if a $50 raise is significant. They should realize that it depends on the base: $50 more than what? Then ask if a 50% raise is significant. This might lead to an interesting discussion about the value of percent.

- Environmental issues provide examples illustrating that raw data are often not as meaningful as percentages. In discussing how much of the available forestland or oil has been used since students were born, the number of acres or barrels is not as informative as a percent.

- Examples from the world of sports can also be motivating. For example, the number of free throws made is not as informative as the percent of free throws made. Other examples include batting averages, winning percentages, completion percentages, and so on.

Additional Examples

Example 1
Rewrite each fraction as a percent.
a. $\frac{4}{5}$ b. $\frac{6}{20}$

Answers: a. 80% b. 30%

Example 2
Rewrite 16.28% as a fraction and as a decimal.

Answer: $\frac{407}{2500}$; 0.1628

Example 3
What percent of 80 is 20?

Solution
When we divide 20 by 80, we get a result of $\frac{1}{4}$, or 0.25. Since $0.25 = \frac{25}{100}$, 20 is 25% of 80.

Some percents are encountered so frequently that it is useful to memorize their fractional and decimal equivalents.

Percent	Fraction	Decimal
1%	$\frac{1}{100}$	0.01
5%	$\frac{1}{20}$	0.05
10%	$\frac{1}{10}$	0.1
20%	$\frac{1}{5}$	0.2
25%	$\frac{1}{4}$	0.25
$33\frac{1}{3}$%	$\frac{1}{3}$	$0.\overline{3}$
50%	$\frac{1}{2}$	0.5

Percent	Fraction	Decimal
60%	$\frac{3}{5}$	0.6
$66\frac{2}{3}$%	$\frac{2}{3}$	$0.\overline{6}$
75%	$\frac{3}{4}$	0.75
80%	$\frac{4}{5}$	0.8
100%	$\frac{1}{1}$	1.0
150%	$\frac{3}{2}$ or $1\frac{1}{2}$	1.5
200%	$\frac{2}{1}$	2.0

PERCENT PROBLEMS

A good deal of the information people encounter — interest rates, election results, forecasts of rain, and so forth—is presented in the form of percentages. Being able to understand these percentages is a necessary skill in everyday life.

Example 4
After having dinner at a restaurant, the members of the Future Accountants of America are given a bill of $350. They vote to leave a 15% tip. How can they calculate how much to leave?

Solution

Method 1 They can write 15% as a fraction.

$$15\% \text{ of } 350 = \frac{15}{100} \cdot 350 = \frac{5250}{100} = \frac{105}{2}$$

Method 2 They can write 15% in decimal form.

$$15\% \text{ of } 350 = 0.15(350) = 52.5$$

Method 3 They can calculate mentally. They know that 10% of 350 is 35. Half of this amount (5% of 350) is $17\frac{1}{2}$. So 15% of 350 is $35 + 17\frac{1}{2}$, or $52\frac{1}{2}$.

The three methods give the same answer in different forms. The tip should be $52.50.

SECTION 2.4

Example 5
While shopping for designer jeans, Calvin found some on sale for 20% off the regular price of $75. How much did the jeans cost?

Solution
The price was reduced by 20% of $75—that is, by 0.2(75), or 15, dollars. The sale price was therefore 75 − 15, or 60, dollars.
  Can you think of any other ways of calculating the price of the pants?

Sample Problems

Problem 1 A pizza was divided into eight slices. Gina got one of the slices. What percent of the pizza did she get?

Solution   There are several ways of solving this problem. Here are two.

  Method 1 Since Gina got one eighth of the pizza and $\frac{1}{8} = 0.125 = \frac{12.5}{100}$, she got 12.5%.

  Method 2 Two slices are one fourth of the pizza, and we know that $\frac{1}{4}$ is 25%. One slice is exactly half as much, or $12\frac{1}{2}\%$.

Problem 2 The label on a bottle of vitamin supplements says that each pill contains 27 milligrams of iron, which is 150% of the recommended daily allowance (RDA). What is the RDA of iron?

Solution   We know that 150% = 1.5. Since 27 milligrams of iron is 1.5 times the RDA, the RDA is 27 ÷ 1.5, or 18, milligrams.

Problem 3 Ira invests $300 at 7% interest, compounded annually.
  a How much will the investment be worth at the end of a year?
  b At the end of the first year, the interest rate is increased to 8%. How much will the investment be worth at the end of the second year?

Solution   **a** Ira starts with $300. During the first year he earns 0.07(300), or 21, dollars in interest. So the investment will be worth 300 + 21, or 321, dollars.
   There is, however, another way of thinking about this problem: Originally, $300 is the whole (100%) of Ira's investment; at year's end 7% more will be added, so the investment will be worth 107% of its original value. Since 107% = 1.07, we can calculate the answer directly by multiplying 1.07(300).

  b At the end of the first year the investment will be worth $321, so at the end of the second year it will be worth 1.08(321) dollars, or $346.68.

Example 3
What percent of 75 is 15?
Answer: 20%

Example 4
A group of friends would like to leave an 18% tip on a restaurant bill of $65. How much should they leave?
Answer: $11.70

Example 5
A coat is on sale for 15% off the regular price of $90. How much does the coat cost?
Answer: $76.50

Cooperative Learning

- Give groups of students newspaper ads and ask them to find examples of items listed at "n% off," with the regular price and the sale price clearly marked. Then ask them to compute the discount and the sale price. Discuss the following questions.
  1. Are your sale prices the same as the ones on the store's list?
  2. If not, why do you think they are not the same? What is the difference in the discount amount?
  3. Why might a store not adhere to proper mathematical rounding procedures?

- Problem 22 is appropriate for cooperative learning.

Stumbling Block

Students may write 3% as 0.3, forgetting to move the decimal two places to the left to get 0.03. Encourage students to think of 0.3 as $\frac{3}{10}$ or $\frac{30}{100}$, which is 30%, not 3%.

Section 2.4 Percent

SECTION 2.4

Assignment Guide

Basic
Day 1 6, 9–11, 13, 16, 21, 23, 26, 32
Day 2 7, 12, 14, 19, 22, 24, 27, 31, 33
Day 3 8, 15, 17, 20, 28–30, 34

Average or Heterogeneous
Day 1 6–12, 16, 17, 19, 20, 23, 27, 30, 31, 33
Day 2 13–15, 18, 21, 22, 24, 25, 29, 32, 34

Honors
Day 1 7, 8, 10, 13, 16, 19, 21, 22, 24, 25, 31
Day 2 9, 11, 12, 14, 17, 18, 20, 23, 27, 29, 30, 33, 34, Investigation

Strands

Communicating Mathematics	2, 4, 12, 17, 25
Geometry	18, 24, 27, 32, 34
Data Analysis/ Statistics	12, 21, 22, 29, 31
Technology	22, 29
Algebra	17, 22, 27, 29
Patterns & Functions	12, 22, 29, 31
Number Sense	1–4, 11, 13–15, 17, 21, Investigation
Interdisciplinary Applications	4, 20, 21, 25

Individualizing Instruction

The visual learner Students may benefit by using grid paper to represent percents. For example, 25% can be shown by shading 25 squares in a 10 × 10 array of squares.

74

Think and Discuss

P **1** Find each value mentally.
   a What is 50% of 90? 45
   b What is 25% of 800? 200
   c What is 46.98% of 100? 46.98
   d What is 100% of 9.4356? 9.4356
   e What percent of 120 is 30? 25%
   f What is $33\frac{1}{3}$% of 6? 2
   g What number is 75 ten percent of? 750

P **2** Coach Roche said, "All my players give 110 percent effort every game." What does her statement mean?

P **3** Is the statement *true* or *false*?
   a 3% of 6% is 18%. False
   b 4% + 8% = 12% True
   c 100% of 3 is equal to 3% of 100. True

P **4** Consumer Math Dr. Drivegood's Car Clinic is having a sale—10% off on brake jobs, 15% off on lube jobs, 20% off on oil changes, 25% off on wheel alignments, and 50% off on car washes. The sign outside the shop reads 120% OFF ON AUTO SERVICES!!! Is the sign accurate? Explain.

P **5** The Chasm marked down its line of sweaters by 40%. A week later there was a storewide sale in which the prices of all merchandise were reduced by 20%.
   a If a sweater's original price was $70, what was its price during the storewide sale? $33.60
   b By what percent were the original prices of the sweaters reduced during the sale? 52%

Problems and Applications

P, R In problems 6–9, indicate what percent of the figure is shaded.

6 25%   **7** 37.5%   **8** 62.5%   **9** $33\frac{1}{3}$%

P **10** Rewrite each fraction as a percent.
   a $\frac{2}{5}$ 40%   **b** $\frac{5}{2}$ 250%   **c** $\frac{9}{5}$ 180%   **d** $\frac{5}{9}$ $55.\overline{5}$%

P **11** List $\frac{2}{3}$, 66%, and 0.6 in order, from least to greatest value.   0.6, 66%, $\frac{2}{3}$

P, R **12** Communicating
   a Copy the pattern, filling in the missing numbers.
   b Explain the pattern.

   10% of 450 is 45.
   20% of 900 is 180.
   40% of 1800 is __720__.
   80% of 3600 is __2880__.
   160% of 7200 is __11,520__.

74 **Chapter 2** Formulas and Percent

SECTION 2.4

In problems 13–15, rewrite each percent as a fraction and as a decimal.

13 15.3% $\frac{153}{1000}$; 0.153   **14** 2.5% $\frac{1}{40}$; 0.025   **15** 60% $\frac{3}{5}$; 0.6

16 At the Café Très Coûteux, a 20% service charge is added to every bill. Find the service charge for each of the following bills.
- **a** $100 $20
- **b** $150 $30
- **c** $200 $40
- **d** $250 $50
- **e** $300 $60
- **f** $350 $70

17 Sue bought a pair of shoes for $34.95 and was charged 6.5% sales tax. Which of the following expressions represents her total bill? Explain your answer.
- **a** 34.95 + 0.065 × 3495
- **b** 3495(0.065)
- **c** (34.95 + 0.065) × 34.95
- **d** 34.95(1 + 0.065)

18 What percent of the area of the square is inside the circle? ≈78.5%

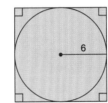

19 At Bali High School, the student government raises money for its activities by selling magazine subscriptions. If the students receive 18% of the total sales, how much money will they raise by selling $4380 worth of subscriptions? $788.40

20 *Finance* A certain investment pays 9% interest, compounded annually. If you invest $860, how much will your investment be worth
- **a** After one year? $937.40
- **b** After two years? $1021.77

21 *Nutrition* Use the information on the label to determine the recommended daily allowance of protein. 52g

```
NUTRITION INFORMATION PER SERVING
SERVING SIZE................10 3/4 OZ. (305 g)
SERVINGS PER CONTAINER.............1
CALORIES..................................200
PROTEIN (GRAMS)........................13
CARBOHYDRATE (GRAMS)..............20
FAT (GRAMS)................................7
SODIUM........................1150 mg/serving
   PERCENTAGE OF U.S. RECOMMENDED
      DAILY ALLOWANCES (U.S. RDA)
PROTEIN..........25   RIBOFLAVIN........10
VITAMIN A.......25   NIACIN................20
VITAMIN C........*   CALCIUM..............2
THIAMINE..........8   IRON.................10
*CONTAINS LESS THAN 2% OF THE U.S. RDA
OF THIS NUTRIENT.
```

22 *Spreadsheets* The spreadsheet printout shows scores of the six members of a bowling team. Column D is for the averages of the bowlers' scores.
- **a** Copy the table, filling in column D.
- **b** What formulas could be used to generate the numbers in column D?

	A	B	C	D
1	120	125	113	
2	146	135	110	
3	151	137	145	
4	129	142	162	
5	144	183	151	
6	137	146	166	

SECTION 2.4

Problem-Set Notes
for pages 76 and 77

- **23** and **24** are good discussion problems.

Checkpoint
1. Evaluate.
   a. 25% of 80   **b.** 30% of 116
   c. 12% of 48
2. Jack bought a suit for $184.59 and paid 7.2% sales tax. What was the total cost of the suit?
3. What percent of the second number is the first number?
   a. 12, 80   **b.** 10, 50   **c.** 12, 30

Answers:
1. **a.** 20 **b.** 34.8 **c.** 5.76
2. $197.88
3. **a.** 15% **b.** 20% **c.** 40%

Reteaching
In problems such as example 3, you may want to have students use the following:

$$\frac{\text{part}}{\text{whole}} = \frac{\text{percent}}{100}$$

Have students convert the fraction $\frac{\text{part}}{\text{whole}}$ to a decimal. Then they can use their knowledge of decimals to find the percent. In the case of example 3, $\frac{20}{80} = 0.25$, which is $\frac{25}{100}$.

Communicating Mathematics

- Have students write a paragraph explaining several ways to mentally find 35% of 50. Alternatively, students may give an oral explanation.
- Problems 2, 4, 12, 17, and 25 encourage communicating mathematics.

P, I   **23** Bob, Carol, and Ted work at Alice's Restaurant. Each was hired at a salary of $350 a week. After having his pay cut by 20%, Bob later got a 20% raise. After receiving a 20% raise, Carol later had her pay cut by 20%. Ted's pay has stayed the same. How much does each person make now?   Bob: $336 a week; Carol: $336 a week; Ted: $350 a week.

P, R   **24** Suppose that the base of this rectangle were increased by 10% and the height were decreased by 10%.
   a By what percent would the perimeter change?   Perim. increases by 6%.
   b By what percent would the area change?   Area decreases by 1%.

P   **25 Environmental Science**   Window insulation can save 12% or more in home heating costs.
   a If heating a home with uninsulated windows costs $1265 a year, how much money might be saved by insulating the windows?
   b Are the windows in your home insulated? If not, how might you determine the amount of money your family could save by insulating them?   Answers will vary.

◀ LOOKING BACK **Spiral Learning** LOOKING AHEAD ▶

R   **26** Evaluate $19 + 9.5^2 \div 15 + 6^3$. $241.01\overline{6}$

I   **27 a** Find lengths x and y (the lengths of segments AB and CD).
   b Find the ratio of x to y—that is, the fraction $\frac{x}{y}$.

   a $\sqrt{125}$, or ≈ 11.18; $\sqrt{5}$, or ≈ 2.24 **b** $\frac{5}{1}$

R   **28** Evaluate each expression.
   a $(3 + 5)^2$ 64   **b** $3^2 + 5^2$ 34

R   **29 Spreadsheets**   This diagram shows a spreadsheet display, along with the formulas that were entered in cells D2, D3, and D4. What formulas should be entered in cells D5 and D6? What numbers will appear in those cells?

	A	B	C	D
1	Value I	Value II	Value III	Average
2	5	7	12	8 +(A2+B2+C2)/3
3	9	15	21	15 +(A3+B3+C3)/3
4	13	15	26	18 +(A4+B4+C4)/3
5	7	11	29	
6	14	38	76	

Chapter 2   Formulas and Percent

SECTION 2.4

R **30** Is the statement 24/3∗5 = 24/(3∗5) *true* or *false*?   False

R, I **31** Study the pattern shown. Then copy the pattern, filling in the missing numbers.

$10^3 = 1000$
$10^2 = 100$
$10^1 = 10$
$10^0 = 1$
$10^{-1} = \underline{0.1}$
$10^{-2} = \underline{0.01}$

R **32** In rectangle ABCD,
  a What is the length of diagonal $\overline{AC}$? 17
  b What is the length of diagonal $\overline{BD}$? 17

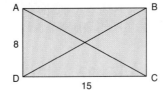

R, I **33** Evaluate each expression.

  a $\dfrac{\sqrt{16}}{\sqrt{9}}$ $\dfrac{4}{3}$ **b** $\sqrt{\dfrac{16}{9}}$ $\dfrac{4}{3}$ **c** $\sqrt{16+9}$ 5   **d** $\sqrt{16}+\sqrt{9}$ 7

R **34** The diagram shows a box without a top. Each of its sides is a square, and the sum of the areas of the five sides is 480.2 cm². What are the dimensions of the box?
9.8 cm × 9.8 cm × 9.8 cm

Investigation

Decimals Perhaps you have noticed that when you rewrite fractions as decimals, some of the decimals terminate and some repeat groups of digits forever. For example, the decimal forms of $\frac{1}{2}$, $\frac{3}{16}$, and $\frac{7}{25}$ terminate, but those of $\frac{1}{3}$, $\frac{5}{6}$, and $\frac{11}{15}$ do not. Using a calculator to explore these and other fractions, try to develop a rule that can be used to determine which sort of decimal form a fraction has.

Section 2.4   Percent

Additional Practice

1. Evaluate.
a. 15% of 75   b. 10% of 50
c. 50% of 250
2. Find the total cost of each item with 5% sales tax added.
a. $125   b. $97.50
c. $474   d. $18.25
3. What percent of the second number is the first number?
a. 35, 70   b. 36, 48   c. 3, 9
4. Evaluate.
a. 75% of 120   b. 35% of 80
c. 90% of 200

Answers:
1. a. 11.25 b. 5 c. 125
2. a. $131.25 b. $102.38
   c. $497.70 d. $19.16
3. a. 50% b. 75% c. $33\frac{1}{3}$%
4. a. 90 b. 28 c. 180

Additional Answers

for pages 76 and 77

25 a $151.80 or more

29 D5: +(A5+B5+C5)/3; D6: +(A6+B6+C6)/3; ≈15.7; ≈42.7

Investigation To decide whether a decimal will terminate or repeat, first reduce the fraction, then look at the denominator. If the prime factors of the denominator consist only of 2s and/or 5s, the decimal will terminate. Otherwise, it will repeat. Thus, $\frac{5}{6}$ repeats because 6 has 3 as a factor, and $\frac{7}{20}$ terminates because $20 = 2^2 \cdot 5$.

SECTION 2.5

Objective
- To use a spreadsheet to construct tables

Class Planning

Schedule 2 days

Resources

In *Chapter 2 Student Activity Masters* Book
Class Openers *51, 52*
Extra Practice *53, 54*
Cooperative Learning *55, 56*
Enrichment *57*
Learning Log *58*
Technology *59, 60*
Spanish-Language Summary *61*

Transparency 7

Class Opener

Use the spreadsheet printout to answer the questions. Suppose that cells A1 and A2 are the only values **not** based on a formula.

1. Write formulas for A3 and A4.
2. Write formulas for cells B2–B4.
3. Write formulas for cells C2–C4.

Possible answers:
1. A3: A2 + 1; A4: A3 + 1
2. B2: A1 * A2; B3: A1 * A3; B4: A1 * A4
3. C2: A2 − B2; C3: A3 − B3; C4: A4 − B4

Lesson Notes

- Spreadsheets can be very important to your students for learning what a variable is and for investigating patterns. Furthermore, many businesses use

SECTION 2.5
More About Spreadsheets

Investigation Connection Section 2.5 supports Unit A, *Investigations* 3, 5. (See pp. 2A–2D in this book.)

Bea Sharp owns a music store at which records, tapes, and compact discs are priced from $3.00 to $16.00, in $1.00 increments. One day Bea was planning a storewide sale and needed to determine how much to reduce her prices. So she sat down at her computer and used a spreadsheet to make a table showing what the price reductions and sale prices would be if she offered a 10% discount. Notice, for example, that for a $6.00 item, the 10% discount would amount to $0.60, so the sale price would be $5.40.

10	Percent	Discount
Reg. Price	Reduction	Sale Price
$3.00	$0.30	$2.70
$4.00	$0.40	$3.60
$5.00	$0.50	$4.50
$6.00	$0.60	$5.40
$7.00	$0.70	$6.30
$8.00	$0.80	$7.20
$9.00	$0.90	$8.10
$10.00	$1.00	$9.00
$11.00	$1.10	$9.90
$12.00	$1.20	$10.80
$13.00	$1.30	$11.70
$14.00	$1.40	$12.60
$15.00	$1.50	$13.50
$16.00	$1.60	$14.40

Then Bea decided to see what the sale prices would be if she offered a 15% discount. All she had to do was change the 10 in the table's title to 15. The computer instantly recalculated and printed out a table for the new discount. As she continued, simply by changing this one number each time, she was able to examine the effect of a variety of discount percentages on the prices of her merchandise. Bea was taking advantage of one of the most useful characteristics of spreadsheets—by immediately showing the results of changes you make, they let you quickly experiment with a variety of possibilities.

15	Percent	Discount
Reg. Price	Reduction	Sale Price
$3.00	$0.45	$2.55
$4.00	$0.60	$3.40
$5.00	$0.75	$4.25
$6.00	$0.90	$5.10
$7.00	$1.05	$5.95
$8.00	$1.20	$6.80
$9.00	$1.35	$7.65
$10.00	$1.50	$8.50
$11.00	$1.65	$9.35
$12.00	$1.80	$10.20
$13.00	$1.95	$11.05
$14.00	$2.10	$11.90
$15.00	$2.25	$12.75
$16.00	$2.40	$13.60

SECTION 2.5

Let's explore the process Bea used to create her spreadsheet. She started by entering the headings and the value of her lowest regular price ($3) as shown. To make each entry, she moved the spreadsheet's "highlight" box to the appropriate cell, typed the entry, and pressed the ENTER key on the computer keyboard. Entering the 10 in a cell by itself allowed her to use it as a value in the spreadsheet's calculations as well as part of the title of the table.

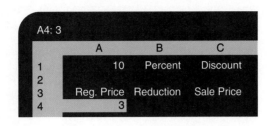

Then Bea entered formulas in cells B4 and C4 to calculate the price reduction and the sale price for a $3 item. She began each formula with a plus sign, which tells the spreadsheet that the entry is a formula rather than a numerical value or a heading. (In some spreadsheets, an equal sign is used for this purpose.) Here are the formulas she used:

In Cell B4

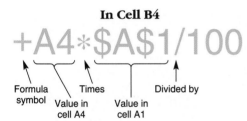

In Cell C4

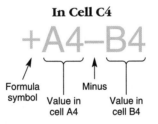

The formula in cell B4 instructs the spreadsheet to take the value in A4 and multiply it by the value in A1 divided by 100. (The division by 100 is needed because the value in cell A1 is a percent.) In this formula, the dollar signs tell the spreadsheet that the reference to cell A1 should not be changed when the formula is copied. The formula in cell C4 instructs the spreadsheet to subtract the value in B4 from the value in A4.

When Bea entered each formula, the value it generated automatically appeared in its cell, as shown. The formula itself is "hidden," but it (or the contents of any other cell) can be seen by simply highlighting the appropriate cell. When a cell is highlighted, whatever was entered in it appears at the top of the display.

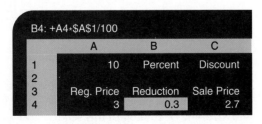

spreadsheets extensively, so students will be using a mathematical tool that they might use someday at work.

■ If two or three students need to share a computer, have them walk through the spreadsheet with Bea, then do problems 3 and 4. The best way to learn about spreadsheets is hands on.

■ If you have only one computer at your disposal, you might still want to walk through Bea's problem and demonstrate problems 3 and 4, but let students anticipate what will happen before you press a key.

■ If you need to, you can simulate a spreadsheet by using templates prepared for use on an overhead projector or have a group of students pretend to be a spreadsheet, as described in the Class Opener on page 29.

Reteaching

Emphasize that the $ means "does not change." For example, $B$1 will always refer to cell B1, wherever it is copied. All other cell references simply point to other cell locations. For example, the formula +A1–2 in cell A2 means "Subtract 2 from the cell above this one" wherever it is copied.

Section 2.5 More About Spreadsheets

SECTION 2.5

Assignment Guide

Basic
Day 1 5–7, 11, 13, 16, 22, 23, 25, 26
Day 2 9, 10, 12, 17–19, 27

Average or Heterogeneous
Day 1 5–7, 9, 12–14, 17–19, 21, 23, 27
Day 2 8, 10, 11, 15, 16, 20, 22, 24–26, 28

Honors
Day 1 5–11, 15, 20, 22, 25–27
Day 2 12–14, 16–18, 23, 24, 28, 29, Investigation

Strands

Communicating Mathematics	3, 16, 17, 25, 26
Geometry	7, 8, 10, 14, 16, 19–21, 23–28
Data Analysis/ Statistics	1, 4, 15
Technology	2–5, 9, 11, 16, 18, 29, Investigation
Patterns & Functions	1, 4, 16, 18, 29
Number Sense	3, 12, 13, 17
Interdisciplinary Applications	11, Investigation

Individualizing Instruction

Visually impaired learners Students may need assistance when using computers. Lighting for computer work should be somewhat dimmer than for normal classroom work to prevent eyestrain. Special display screens that use raised dots are available for students whose vision does not allow them to use a regular computer monitor.

Next, Bea entered the formula +A4+1 in cell A5. By creating a formula that added 1 to the value in the cell above, she would be able to complete column A without typing all the prices. Once the pattern was established, the spreadsheet could do the work, as you will see.

A5: +A4+1

	A	B	C
1	10	Percent	Discount
2			
3	Reg. Price	Reduction	Sale Price
4	3	0.3	2.7
5	4		

Then Bea used the copy feature of the spreadsheet to copy the formulas in cells B4 and C4 to cells B5 and C5. Notice that the formula +A4−B4 changed to +A5−B5 when it was copied. As it copies a formula, the spreadsheet automatically changes the cell names so that they correspond to the formula's new position unless you tell it not to by using dollar signs in the formula.

C5: +A5−B5

	A	B	C
1	10	Percent	Discount
2			
3	Reg. Price	Reduction	Sale Price
4	3	0.3	2.7
5	4	0.4	3.6

Bea was now ready to copy the entries in row 5 to row 6–17. After doing so, she had a list of all her regular prices in column A, along with the corresponding reductions and sale prices in columns B and C. Notice that once again the cell names were updated in the formulas, except for the reference to A1, which the dollar signs indicated should not be changed.

B17: +A17∗$A$1/100

	A	B	C
1	10	Percent	Discount
2			
3	Reg. Price	Reduction	Sale Price
4	3	0.3	2.7
5	4	0.4	3.6
6	5	0.5	4.5
7	6	0.6	5.4
8	7	0.7	6.3
9	8	0.8	7.2
10	9	0.9	8.1
11	10	1	9
12	11	1.1	9.9
13	12	1.2	10.8
14	13	1.3	11.7
15	14	1.4	12.6
16	15	1.5	13.5
17	16	1.6	14.4

Bea's final step was to use the spreadsheet's format function to make it display the table entries in dollars-and-cents form. She then had the table shown at the beginning of this section and was able to experiment with various discounts. Although some commands and symbols differ from spreadsheet to spreadsheet, you can use almost any spreadsheet to do jobs like Bea's. Learn to use one and you will find that you can do amazing things.

Chapter 2 Formulas and Percent

SECTION 2.5

Think and Discuss

P, R **1** Copy and complete the table.

A	B	A ÷ B	Percent
15	20	0.75	75%
3	15	0.2	20%
130	200	0.65	65%
4	25	0.16	16%
18	12	1.5	150%
6	30	0.2	20%

P **2** In this spreadsheet display, Lotus entered 2, 4, 8, and +A1+B1+C1 in row 1.

D1: +A1+B1+C1

	A	B	C	D
1	2	4	8	14
2				
3				
4				

  a If she uses the copy feature to copy row 1 to rows 2–4, what numbers will appear in row 3?
  b If she then changes B3 to 11 and A1 to 6, what number will appear in cell D3? 21
  c If she then changes A3 to +A1, what number will appear in D3? 25
  d If she then changes A1 to 17, which values in the table will change, and what will they become?

P **3** Using a spreadsheet, follow these steps:

  1. Enter any positive value (like 5 or 36 or 7000) in cell A1.
  2. Enter +(A1+3/A1)/2 in cell A2.
  3. Copy the formula in cell A2 to cells A3 through A30.

  What do you notice about the values in column A?

P, R **4** This table deals with right triangles. Using a spreadsheet, copy the entries shown, continue columns A and B to row 50, and fill in cells C2–C50 and D2–D50. (The copy feature will help you fill in the columns.)
Check students' tables.

	A	B	C	D
1	Leg 1	Leg 2	Hypotenuse	Area
2	3	4		
3	4	6		
4	5	8		
5	6	10		
6	7	12		
7	8	14		
8	9	16		
9	10	18		
10	11	20		
11	12	22		
12	13	24		
13	14	26		

Problem-Set Notes

for page 81

■ **3** and **4** are excellent problems for students to work through if they have a spreadsheet.

Multicultural Note

The use of discounts from the regular or list price of an item is common in our society. Students whose parents or grandparents were foreign-born can check with family members to see what the practice is in countries other than the United States.

Stumbling Block

Students may try to type a new formula in each cell of a spreadsheet instead of using the copy function. For example, they may try to enter +A5–B5, +A6–B6, and so on, in column C of a spreadsheet. Emphasize the fact that the copy function automatically changes the formula for each cell.

Additional Answers

for page 81

2a 2; 4; 8; 14

d Besides A1's becoming 17, D1 will change to 29, A3 will change to 17, and D3 will change to 36.

3 Possible answer: The values become smaller, with the spreadsheet eventually listing a value of about 1.73205 (that is, an approximation of $\sqrt{3}$) in each cell.

Section 2.5 More About Spreadsheets

SECTION 2.5

Problems and Applications

P **5** If Wanda uses the copy feature of her spreadsheet to copy row 1 to rows 2–7, what value will appear in cell C6?   8

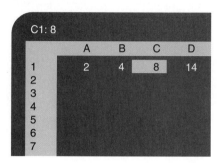

R **6** **a** What is 25% of 76?   19
   b What percent of 46 is 23?   50%
   c What percent of 70 is 8.792?   12.56%

In problems 7 and 8, indicate whether the statement is *true* or *false*.

R **7** The distance from A to C is the same as the distance from B to D.   True

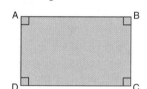

R **8** Length PQ + length PR = length QR   False

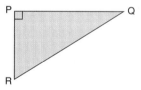

P **9** **Spreadsheets**   Using a spreadsheet, follow these steps:

   1. Enter 1 in cell A1 and 2 in cell B1.
   2. Enter +A1+3*B1 in cell C1.
   3. Enter +2*A1 in cell A2 and +3*B1 in cell B2.
   4. Copy the formula in cell C1 to cell C2.
   5. Use the copy feature to continue the table through row 6.

   What number appears in cell C6?   1490

R, I **10** How many diagonals can be drawn from one vertex of this ten-sided figure? (Two of the diagonals are shown.)   7 diagonals

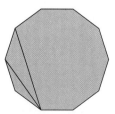

82   Chapter 2   Formulas and Percent

P **11 Consumer Math** Instruct a spreadsheet to display numbers in dollars-and-cents format. Then follow these steps:

1. In column A, list prices from $3 to $21, in $2 increments.
2. In column B, list what the prices would be with a 20% discount.
3. In column C, list what the prices would be with a 25% discount.

I **In problems 12 and 13, indicate whether the statement is true *always*, *sometimes*, or *never*.**

12 The product of two odd numbers is even.   Never

13 The product of an even number and an odd number is even.   Always

R **14** What percent of the perimeter of square ABCD is the perimeter of square EFGH?   ≈79.06%

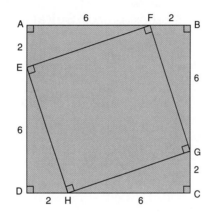

R **15** During the first two weeks of the baseball season, the Missouri Cardinals won 5 out of 8 games, and the Maryland Ordinals won 6 games and lost 4.
   a What fraction of their games did the Cardinals win? $\frac{5}{8}$
   b What fraction of their games did the Ordinals win? $\frac{3}{5}$
   c Which team won a greater fraction of their games?   Cardinals

P, R **16 Spreadsheets** Use a spreadsheet to make a table like this one, showing the circumference and the area of a circle with a radius of 1. (Use formulas to generate the values in cells B3 and C3.)

	A	B	C
1	Radius	Circumference	Area
2			
3	1	6.283185	3.141592
4	15	94.24777	706.8583
5	19	119.3805	1134.114
6	25	157.0796	1963.495

   a Describe the steps you followed to make the table.
   b Use the spreadsheet to calculate the circumferences and the areas of circles with radii of 15 centimeters, 19 inches, and 25 feet.

R **17 Communicating** Mick asked Keith, "How far is it from Los Angeles to San Francisco by car?" Keith replied, "About nine hours." Do you think Keith answered Mick's question? Why or why not?

Section 2.5 More About Spreadsheets

SECTION 2.5

Cooperative Learning

- Refer to problem 14 on page 83. Have small groups of students consider what happens when you reposition the inside square so that its vertices divide each side of the larger square into two segments with lengths that are whole numbers.
1. How many possible figures are there?
2. Draw the possible figures and find the perimeter of the inside square in each case.
3. Make a rule about how the perimeter of the inside square can vary if the vertices fall *between* the whole-number positions.

Answers:
1. There are four possible figures (not including reflected figures or overlapping figures).
2.

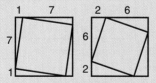

$P \approx 28.28$ units $P \approx 25.30$ units

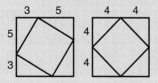

$P \approx 23.32$ units $P \approx 22.63$ units

3. Possible answer: The perimeter will be less than 32 units, but greater than or equal to 22.63 units.

- Problems 4, 16, 18, 29, and the Investigation are appropriate for cooperative learning.

P **18** Use a spreadsheet to make a table like the one shown. (Notice the formula that was used in column B.) Then extend the table to row 20. What number appears in cell B20? −8

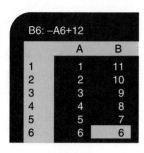

◀ LOOKING BACK **Spiral Learning** LOOKING AHEAD ▶

R **In problems 19–22, find the perimeter or circumference of the figure.**

19

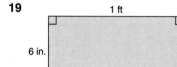

36 in., or 3 ft

20

4 in.

3 in.

12 in.

21

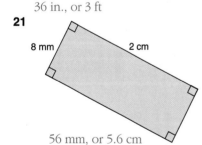

56 mm, or 5.6 cm

22

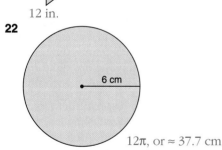

12π, or ≈ 37.7 cm

R, I **23** List $\sqrt{4+9}$, $\sqrt{4} + \sqrt{9}$, and $\sqrt{\frac{13}{2}}$ in order, from least to greatest value.

R **24** Find the area of the trapezoid. 326 in.²

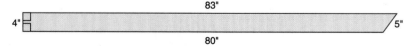

I **In problems 25 and 26, indicate whether the statement is true *always*, *sometimes*, or *never*. Explain your answer.**

25 A square is a rectangle.

26 A rectangle is a square.

R **27** What percent of a rectangle's angles are right angles? 100%

84   Chapter 2   Formulas and Percent

SECTION 2.5

R, I **28 a** What fractional part of the area of the larger circle is the area of the smaller circle? $\frac{1}{4}$
b What fractional part of the circumference of the larger circle is the circumference of the smaller circle? $\frac{1}{2}$

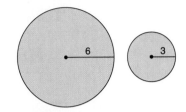

P, I **29 Spreadsheets** Pete wants to make a table of equivalent lengths.

	A	B	C	D
1	Centimeters	Millimeters	Inches	Feet
2				
3	1			
4	2			
5	3			
6	4			
7	5			
8	6			

a There are 10 millimeters in a centimeter. What formula could Pete enter in cell B3 and copy to the rest of column B so that the spreadsheet would calculate appropriate values in that column?   +A3*10
b There are 2.54 centimeters in an inch. What formula could he enter in cell C3 and copy to the rest of column C so that the spreadsheet would calculate appropriate values in that column?   +A3/2.54
c What formula could he enter in cell D3 and copy to the rest of column D so that the spreadsheet would calculate appropriate values in that column?   +C3/12

Investigation

Sports Look in the sports section of a newspaper to find a table showing the standings of the teams in a professional league. If you were using a spreadsheet to generate the table, which numbers would you have to enter as values? Which could you use formulas to calculate?

Checkpoint

These are the first rows of a spreadsheet display.

	A	B	C	D
1	3	6	9	12
2	9	15	21	15
3	24	36	36	24
4	60	72	60	48

1. What formula could be hidden in cell C11?
2. What formula could be hidden in cell D3?
3. Describe the most efficient way to generate the formulas for column C.

Possible answers:
1. +C10+D10 **2.** +D2+A2
3. Enter +C1+D1 in cell C2, then copy the formula into the other cells.

Additional Answers

for pages 84 and 85

23 $\sqrt{\frac{13}{2}}, \sqrt{4+9}, \sqrt{4}+\sqrt{9}$

25 Always; possible explanation: A rectangle is a four-sided figure with four right angles, and a square is a special kind of rectangle, in which all four sides are equal in length.

26 Sometimes; possible explanation: Only those rectangles in which all four sides are equal in length are squares.

Investigation Possible answer: The columns for wins (*W*) and losses (*L*) would be entered as values, but the winning percentage (*P*) can be calculated: $P = \frac{W}{W+L}$.

SUMMARY

CHAPTER 2

Summary

CONCEPTS AND PROCEDURES

After studying this chapter, you should be able to
- Recognize the need for a conventional order of operations (2.1)
- Recognize the symbols used to indicate multiplication and division (2.1)
- Use a calculator to evaluate powers (2.2)
- Use a calculator to evaluate roots (2.2)
- Perform mathematical operations in the conventional order (2.2)
- Use formulas to calculate the areas and perimeters of some geometric figures (2.3)
- Use the Pythagorean Theorem to calculate lengths of sides of right triangles (2.3)
- Interpret percentages (2.4)
- Solve problems involving percentages (2.4)
- Use a spreadsheet to construct tables (2.5)

VOCABULARY

base (2.2)	Pythagorean Theorem (2.3)
exponent (2.2)	radicand (2.2)
grouping symbol (2.1)	root (2.2)
hypotenuse (2.3)	root index (2.2)
leg (2.3)	square root (2.2)
percent (2.4)	vinculum (2.1)
power (2.2)	

CHAPTER 2

Review

Assignment Guide

Basic
Day 1 1–3, 10, 12, 15, 21–23, 26, 30
Day 2 Chapter Test 1–14

Average or Heterogeneous
Day 1 1, 3, 5, 7–10, 15, 18, 21, 29, 30
Day 2 4, 6, 11–14, 17, 20, 22–24, 31–33

Honors
Day 1 1–10, 12, 15, 16, 18–21, 23, 24, 27–29, 31–34

Strands

Communicating Mathematics	21, 31
Geometry	3, 4, 6, 20–24, 29
Measurement	11–14, 19
Technology	21, 33, 34
Probability	32
Algebra	1, 2, 15–17
Patterns & Functions	21, 31
Number Sense	2, 31
Interdisciplinary Applications	5, 15, 18, 30

1 Evaluate each expression.
 a $82 + 18 \div 3 \cdot 6$ 118
 b $(82 + 18) \div 3 \cdot 6$ 200
 c $(82 + 18 \div 3) \cdot 6$ 528
 d $82 + (18 \div 3) \cdot 6$ 118
 e $82 + 18 \div (3 \cdot 6)$ 83
 f $(82 + 18) \div (3 \cdot 6)$ $5.\overline{5}$

2 Write and evaluate a root that has a root index of 5 and a radicand of 759,375. $\sqrt[5]{759,375}$; 15

In problems 3 and 4, find the area and the perimeter of each figure.

3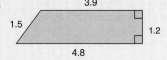

$A = 5.22$; $P = 11.4$

4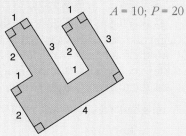

$A = 10$; $P = 20$

5 The CD Music Club is offering compact discs at a price of two for $19.95 instead of the regular price of $12.95 each. What percentage discount is being offered?   ≈23%

6 Rachel wants to make a cube-shaped container that will hold 2700 cubic centimeters of water. What should the length of each edge of the cube be?   ≈13.9 cm

In problems 7–10, find each value.

7 What is 15% of 80?   12

8 What percent of 60 is 90?   150%

9 What is 25% of 30% of 450?   33.75

10 What percent of 75 is 75?   100%

In problems 11–14, what percent of the diagram is shaded?

11
75%

12
32%

13
100%

14
0%

Chapter Review **87**

REVIEW

Additional Answers

for pages 88 and 89

21 Missing values in column C: 12, 18, 24, 30; missing values in column D:, 8, 18, 32, 50; possible answer: First, I entered the headings in row 1 and the values 1 and 2 in cells A2 and B2. Then I entered +2*A2+2*B2 in cell C2, +A2*B2 in cell D2, +A2+1 in cell A3, and +B2+2 in cell B3. Finally, I copied C2 and D2 to C3 and D3, then copied row 3 to rows 4–6.

25 $(4+4) \cdot (4+4) \cdot 4 = 256$

26 $(2+3) \cdot 4 \div 5 = 4$

27 $(8-3) \cdot (4+5) = 45$

28 $(2 \div 4 + 6 - 8) \cdot 10 + 20 = 5$

31 Possible answer: For any (x, y), calculate $x\%$ of y.

33 Possible answer: The values become smaller, with the spreadsheet eventually listing a value of about 1.41421 (that is, an approximation of $\sqrt{2}$) in each cell.

34

	A	B	C	D
1	1	3	6	10
2	2	5	9	16
3	3	7	12	22
4	4	9	15	28

---

15 A law firm charges $185 per hour for legal work. Sue paid for 8 hours of legal work, plus a filing fee of $30.
  a Write a mathematical expression for the amount Sue was charged. $8 \times 185 + 30$
  b Find the total amount Sue paid. $1510

16 Madeline added the cube root of 27 to the fourth power of 3. Adele divided half of 10 by the square root of 20. Darlene computed the product of Madeline's result and Adele's result. What was Darlene's result? $42\sqrt{5}$, or ≈ 93.915

17 Find the value of each expression.
  a $(\sqrt[3]{27} + \sqrt[3]{729})^3$ 1728
  b $27 + 729$ 756
  c $27 + 3(\sqrt[3]{27})^2(\sqrt[3]{729}) + 3(\sqrt[3]{27})(\sqrt[3]{729})^2 + 729$ 1728

18 Ward Roby bought some clothes on sale. Trousers, normally $80, were 25% off. Shirts, normally $46, were 20% off. Shoes, normally $70, were 30% off. Ward bought one pair of trousers, one shirt, and one pair of shoes.
  a What was the total price of the items he bought? $145.80
  b If the sales-tax rate was 7%, how much was the tax? $10.21
  c What was the total cost, including tax? $156.01

19 Mrs. Solemnis had a fever of 101.3°F. Express her temperature in degrees Celsius.   38.5°C

20 a Find the area of the circle.
  b Find the area of triangle ABC.
  c Find the total area of the pieces that would be left if the triangle were removed from the circle.
  a 100π, or ≈ 314.16 **b** 48   **c** ≈ 266.16

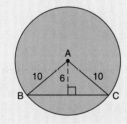

21 Using a spreadsheet, copy this table. Then complete columns C and D to show the perimeters and areas of rectangles with the given dimensions. Describe the procedure you used.

	A	B	C	D
1	Width	Length	Perimeter	Area
2	1	2	6	2
3	2	4		
4	3	6		
5	4	8		
6	5	10		
7				

In problems 22–24, find the area of each triangle.

22

Triangle with sides 8, 15, 17 (right angle)

60

23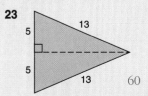

60

24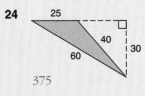

375

---

88   Chapter 2   Formulas and Percent

REVIEW

In problems 25–28, insert parentheses to make each statement true.

25 $4 + 4 \cdot 4 + 4 \cdot 4 = 256$

26 $2 + 3 \cdot 4 \div 5 = 4$

27 $8 - 3 \cdot 4 + 5 = 45$

28 $2 \div 4 + 6 - 8 \cdot 10 + 20 = 5$

29 **a** Find the area of the circle.
b Find the circumference of the circle.
c Find the ratio of the area of the circle to the circumference of the circle (that is, divide the area by the circumference).
a ≈497.18 **b** ≈79.04 **c** 6.29

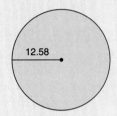

30 A building contractor is penalized 1% of construction charges for each day he or she is late in completing a project. If a contractor charges $3,260,000 to construct a building, how much will the contractor receive if the building is finished 6 days late? $3,064,400

31 Explain the rule by which the pairs of numbers generate new numbers.
$(10, 379) \to 37.9$ $(50, 48) \to 24$
$(100, 97.3) \to 97.3$ $(25, 40) \to 10$

32 If a dart lands on this target, what is the probability that it will land in the shaded area? $\frac{1}{3}$

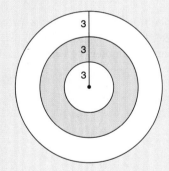

33 Using a spreadsheet, follow these steps:

1. Enter 100 in cell A1.
2. Enter + (A1+2/A1)/2 in cell A2.
3. Copy the formula in cell A2 to cells A4–A20.

What do you notice about the values in column A?

34 Pippin entered the numbers 1, 3, and 6 in cells A1, B1, and C1 of a spreadsheet table. He entered +A1+B1+C1 in cell D1, then entered +A1+1 in cell A2, +B1+2 in cell B2, and +C1+3 in cell C2. After copying D1 to D2, he copied row 2 to rows 3 and 4. What values appeared in the final table?

Chapter Review   89

CHAPTER 2

Test

1. Tommy bought a compact disc for $12.65, a set of portable headphones for $8.95, and a magazine for $2.50. If he paid a sales tax of 7%, what was the total cost of the items? $25.79

2. The hypotenuse of this right triangle is 17 units long, and one leg is 15 units long.
   a. How long is the other leg?
   b. What is the area of the triangle?
      a 8 units   b 60 square units

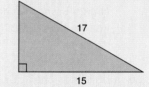

In problems 3–8, evaluate each expression.

3. $2(12 - 5) \div 5$ 2.8

4. $\frac{3}{8} + \frac{4 + 11}{2^4}$ $\frac{21}{16}$

5. $3 + 3*6/9$ 5

6. $\sqrt{14 + 15(16)}$ ≈15.94

7. $38 - 7 \times 4 + 19$ 29

8. $\frac{5.9 + 10 \cdot 7.3}{18 \div 2 + 6}$ 5.26

9. Describe how you might use a spreadsheet to make the table shown, which lists the areas and perimeters of various sizes of squares.

	A	B	C
1	Side Length	Area	Perimeter
2	1	1	4
3	2	4	8
4	3	9	12
5	4	16	16
6	5	25	20
7	6	36	24
8	7	49	28
9	8	64	32

10. At the Art Attack store, a $12.00 box of colored pencils is on sale for $9.99. By what percentage has the price of the pencils been reduced? 16.75%

11. Write and evaluate a root with a root index of 3 and a radicand of 10,648. $\sqrt[3]{10{,}648}$; 22

In problems 12 and 13, find the area of each figure.

12.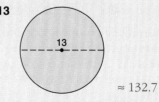
    55

13. ≈ 132.7

14. If you select two of the values 20, 20%, $\frac{1}{5}$, and $\frac{100}{20}$ at random, what is the probability that the values will be equal? $\frac{1}{6}$

Chapter 2 Formulas and Percent

PUZZLES AND CHALLENGES

Each picture below represents a well-known phrase. For example, the picture in box 1A stands for "Fool on the Hill" and box 2A stands for "Stepladder." Figure out the other phrases and try making up some of your own.

graph. Show the temperatures in both Celsius and Fahrenheit.

6. Obtain the nutritional information for either a hamburger, a serving of fries, or a shake from at least two fast-food restaurants. Compare the products' cost, percent of fat and salt (sodium), and number of calories. Show your results in some kind of graph and explain which product you would recommend on the basis of your research.

7. In Section 2.3, the Pythagorean Theorem is introduced. Explore the following questions: Who was Pythagoras? How did he discover his theorem? Why is the side opposite the right angle called the hypotenuse? What is a Pythagorean triple? As part of your explanations, draw a picture that shows how the Pythagorean Theorem works.

8. An algorithm is a step-by-step procedure for solving a problem. When we do long division, we use the algorithm "Divide, multiply, subtract, and bring down." In Section 2.2, we found roots by using a calculator. Find an algorithm for determining the square root of a number. Show several examples of how it works.

9. Extend problem 21 on page 75 as follows: Select two of your favorite cereals and write a report comparing their nutritional values. Be sure to include in your report the percents of fat, sodium, and protein per serving.

Additional Answers

for Chapter 2 Test

9 Enter the headings in row 1. Enter a 1 in cell A2. Enter the formulas +A2^2 in cell B2, +A2*4 in cell C2, and +A2+1 in cell A3. Copy the formulas in B2 and C2 to B3 and C3, then copy row 3 to rows 4–9.

Answers to Puzzles and Challenges are on page A2.

OVERVIEW

LONG-TERM INVESTIGATIONS

See *Investigations* book, Unit B, pages 26-55.

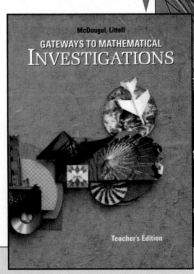

Gateways to Mathematical Investigations is a collection of long-term, real-world investigations that provides students with exciting opportunities to apply the ideas and concepts introduced in *Gateways to Algebra and Geometry: An Integrated Approach*. Unit B: Two-Wheeler Ordeal consists of five investigations that develop the theme of planning a long bicycle trip. This unit is designed so that you can pick and choose the order and number of investigations you would like to do with your students. Each investigation is an exploration that takes several days to complete. Refer to the menu at the right to learn more about the investigations that make up this unit.

As your students work through the Unit B investigations, use Chapters 3 and 4 of *Gateways to Algebra and Geometry: An Integrated Approach* for concept and skill development, and as a source of rich problems, exercises, and applications. When appropriate, refer students to additional supportive sections in *Gateways to Algebra and Geometry* as well.

UNIT B: Two-Wheeler Ordeal

MENU OF INVESTIGATIONS

OVERVIEW: Groups apply mathematical concepts and strategies from Chapters 3 and 4 within the context of planning a long bicycle trip.

A Long and Winding Road

(3-5 days)
Students choose a distant location to reach by bicycle. They plan a route to reach that destination and determine its distance. They describe their route and draw a route map to show it.
STRANDS
Number, Measurement, Logic and Language
UNIFYING IDEAS
Patterns and Generalizations, Proportional Relationships

Getting in Gear

(3-5 days)
Students explore gears and gear ratios and analyze their use in the design and function of a bicycle.
STRANDS
Functions, Algebra, Geometry, Measurement
UNIFYING IDEAS
Proportional Relationships, Multiple Representations

It's About Time

(3-5 days)
Students find the length of time for their trip using rates and ratios while considering factors that might affect their time. They develop an itinerary for their trip.
STRANDS
Functions, Measurement, Algebra, Statistics, Logic and Language
UNIFYING IDEAS
Patterns and Generalizations, Multiple Representations, Proportional Relationships

The Traveling Cyclist Problem

(2-3 days)
Students are introduced to the class of mathematical problems known as "traveling salesperson problems." Groups work together to solve variations of this problem.
STRANDS
Functions, Number, Discrete Mathematics, Logic and Language
UNIFYING IDEAS
Patterns and Generalizations, Multiple Representations

Add Two Vans and Complete Your Plans

(3-5 days)
Students estimate the total cost for the whole class, along with two parents and a teacher driving two supply vans, to take the trip to the destination they have chosen.
STRANDS
Functions, Number, Geometry, Measurement, Logic and Language
UNIFYING IDEAS
Multiple Representations, Proportional Relationships

INVESTIGATIONS FOR UNIT B	SCHEDULE	OBJECTIVES
1. A Long and Winding Road Pages 29-33*	3-5 Days	• To use map-reading skills and concepts of proportions to plan a route and find the distance to a selected destination; to draw a map showing the route
2. Getting in Gear Pages 34-39*	3-5 Days	• To use rates, ratios and proportions, equations, and geometric concepts to investigate gears on bicycles
3. It's About Time Pages 40-44*	3-5 Days	• To apply estimation, rates, and equations to determine how long it will take for the group to cycle to its chosen destination; to develop an itinerary for the trip
4. The Traveling Cyclist Problem Pages 45-50*	2-3 Days	• To determine all possible routes each of which originates at the same point and passes through several specified locations exactly once; to figure out the shortest route
5. Add Two Vans and Complete Your Plans Pages 51-54*	3-5 Days	• To use estimation, rates, ratios, proportions, and equations to figure out the total cost for the trip to the chosen destination

*Pages refer to *Gateways to Mathematical Investigations* Teacher's Edition.

PLANNING LONG-TERM INVESTIGATIONS

MATERIALS	RELATED RESOURCES
• Family Letter, 1 per student • Grid paper, 1 per student • Different-size dot papers, 2-4 per group • Ruler, compass, protractor, 1 of each per pair • String, about 3 ft. per group; proportion sticks, 1 set per group • Calculator, at least one per group • Chart or poster paper, 1 per group; marker, 1-3 per group • Local road map, bicycle map, or road atlas, 1-2 maps per group • *Gateways to Mathematical Investigations* Student Edition, pp. 11-12, 1 of each per student	**Gateways to Algebra and Geometry** Section 3.1, pp. 94-100: Linear Measurement Section 3.3, pp. 107-113: Estimation and Rounding Section 4.3, pp. 165-171: More Applications of Ratios Section 4.5, pp. 178-184: Proportions *Additional References:* Teacher's Resource File: Student Activity Masters
• Compass, protractor, 1 of each per pair • Calculator, 1 per pair • Bicycle brochures, 1-2 per group • Historical photographs of bicycles, at least 3 • 10-speed bicycle, 1 per group • Chalk, 1 piece per pair • *Gateways to Mathematical Investigations* Student Edition, pp. 13-14, 1 of each per student	**Gateways to Algebra and Geometry** Section 3.4, pp. 114-120: Understanding Equations Section 4.2, pp. 158-164: Ratios of Measurements Section 4.3, pp. 165-171: More Applications of Ratio Section 4.4, pp. 172-177: Working with Units Section 4.5, pp. 178-184: Proportions *Additional References* Section 2.3, pp. 64-70: Geometric Formulas Teacher's Resource File: Student Activity Masters
• Calculator, at least 1 per group • Computer, spreadsheet software, 1 per group • Blank spreadsheet, 1 per group • Chart paper, 2-4 per group • Marker, 1-3 per group • Stopwatch or wristwatch, 1 per group • Road maps and student route maps (from Investigation 1) • Campground, motel, and restaurant directories, several • *Gateways to Mathematical Investigations* Student Edition, pp. 15-16, 1 of each per student	**Gateways to Algebra and Geometry** Section 3.3, pp. 107-113: Estimation and Rounding Section 3.4, pp. 114-120: Understanding Equations Section 4.1, pp. 150-157: A Closer Look at Measurement Section 4.2, pp. 158-164: Ratios of Measurements *Additional References* Section 2.5, pp. 78-85: More About Spreadsheets Section 5.1, pp. 200-206: Circle Graphs Teacher's Resource File: Student Activity Masters
• Grid paper, 1-2 per student • Geoboard, at least 1 per group • Calculator, at least 1 per group • Map of school neighborhood, 1 per group • *Gateways to Mathematical Investigations* Student Edition, p. 17, 1 per student	**Gateways to Algebra and Geometry** Section 3.1, pp. 94-100: Linear Measurement Section 3.3, pp. 107-113: Estimation and Rounding *Additional References* Section 9.5, p. 461: #33 Section 10.1, pp. 486-493: Techniques of Counting Teacher's Resource File: Student Activity Masters
• Calculator, at least 1 per group • Computer, spreadsheet software, 1 per group • Blank spreadsheet, 1 per group • Chart paper, 1-2 per group; markers, 1-3 per group • Road maps, at least 1 per group • Student route maps (from Investigation 1) • Campground, motel, and restaurant directories, several • Van owner manual, at least 1 • Compass, protractor, 1 of each per group; ruler, 1 per student • *Gateways to Mathematical Investigations* Student Edition, pp. 18-19, 1 of each per student	**Gateways to Algebra and Geometry** Section 3.3, pp. 107-113: Estimation and Rounding Section 3.5, pp. 121-128: Angle Measurement Section 3.6, pp. 129-135: Const. w/ Ruler & Protractor Section 4.2, pp. 158-164: Ratios of Measurements Section 4.3, pp. 165-171: More Applications of Ratios Section 4.4, pp. 172-177: Working with Units Section 4.5, pp. 172-184: Proportions *Additional References* Section 5.1, pp. 200-206: Circle Graphs Teacher's Resource File: Student Activity Masters

See pages 148A and 148B: Assessment for Unit B.

CHAPTER 3

Contents

3.1 Linear Measurement
3.2 Scientific Notation
3.3 Estimation and Rounding
3.4 Understanding Equations
3.5 Angle Measurement
3.6 Constructions with Ruler and Protractor
3.7 Geometric Properties

Pacing Chart

Days per Section

	Basic	Average	Honors
3.1	2	2	2
3.2	3	3	2
3.3	2	2	2
3.4	3	3	2
3.5	2	2	2
3.6	3	2	2
3.7	2	2	2
Review	2	1	1
Assess	3	2	2
	22	19	17

Chapter 3 Overview

- Science and technology teachers frequently complain that their students come to them with poor measurement skills. Mathematics teachers, well aware of their colleagues' complaints, know that measurement skills are also important in exploring purely mathematical concepts.

- Computing machines have made it essential that students be skilled at estimating (to test for reasonable answers) and rounding (to correctly interpret their answers). Once these two topics have been introduced, we get on with studying geometry, primarily from an experimental point of view.

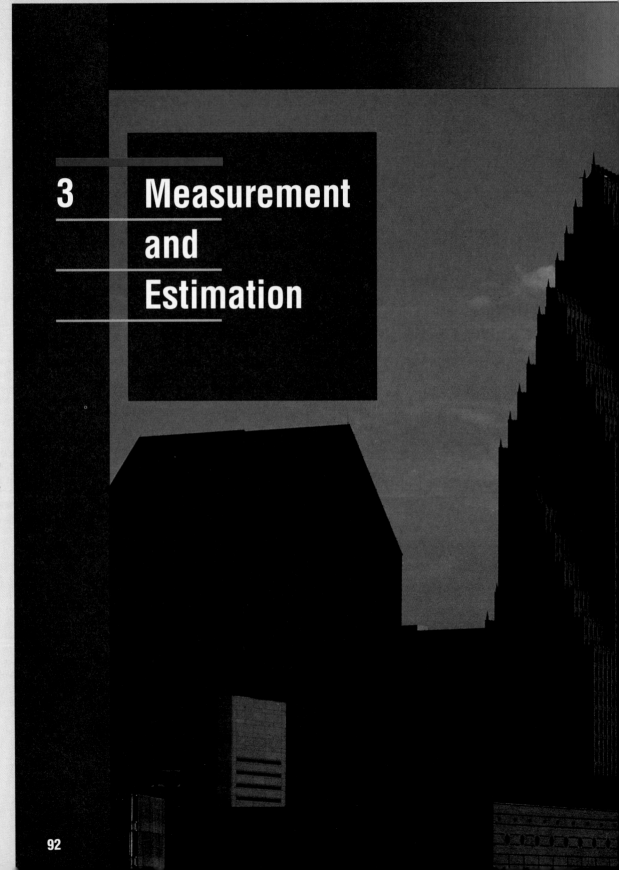

3 Measurement and Estimation

CHAPTER 3

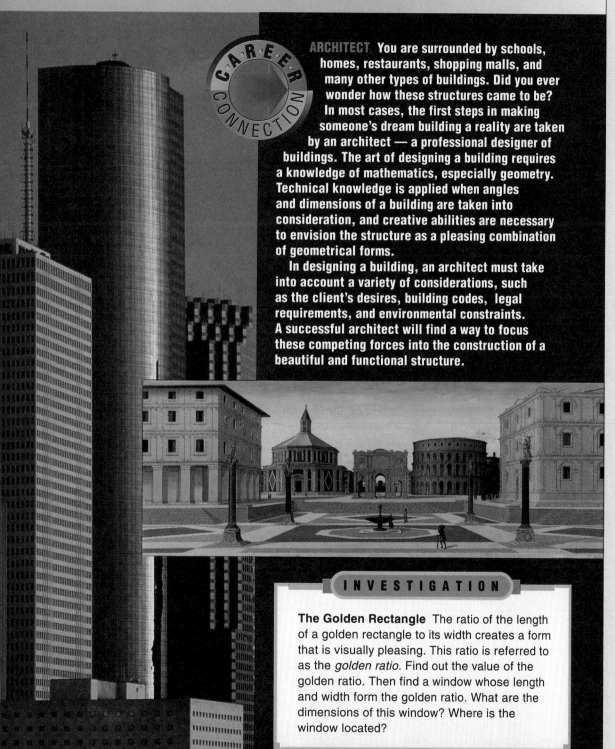

ARCHITECT You are surrounded by schools, homes, restaurants, shopping malls, and many other types of buildings. Did you ever wonder how these structures came to be? In most cases, the first steps in making someone's dream building a reality are taken by an architect — a professional designer of buildings. The art of designing a building requires a knowledge of mathematics, especially geometry. Technical knowledge is applied when angles and dimensions of a building are taken into consideration, and creative abilities are necessary to envision the structure as a pleasing combination of geometrical forms.

In designing a building, an architect must take into account a variety of considerations, such as the client's desires, building codes, legal requirements, and environmental constraints. A successful architect will find a way to focus these competing forces into the construction of a beautiful and functional structure.

INVESTIGATION

The Golden Rectangle The ratio of the length of a golden rectangle to its width creates a form that is visually pleasing. This ratio is referred to as the *golden ratio*. Find out the value of the golden ratio. Then find a window whose length and width form the golden ratio. What are the dimensions of this window? Where is the window located?

Career Connection

Architecture is a good example of a career that has changed due to the development of powerful computers. Computer-aided design (CAD) has made mechanical drawing faster, easier, and more versatile, especially when rendering objects, such as buildings, in three dimensions. Encourage interested students to find out more about CAD and its uses in architecture, in the product design of airplanes and cars, and in the electronics industry. Explain that architecture is a field in which creativity and practicality must be combined to solve aesthetic and engineering problems.

Video Connection

For more information about the career discussed here, please refer to the Futures video cassette #2, program A: *Architecture and Structural Engineering,* featuring Jaime Escalante. For more information or to order a copy of the program, call PBS VIDEO at 800/344-3337.

Investigation Notes

- In a golden rectangle, the length ℓ and width w satisfy the equation $\frac{\ell}{w} = \frac{(\ell + w)}{\ell}$. The ratio $\frac{\ell}{w}$ is called the golden ratio and is equal to about 1.62.

SECTION 3.1

Objectives

- To measure lengths, using both English and metric units
- To recognize various polygons and investigate linear measurements associated with them

Vocabulary: congruent, diagonal, linear, polygon, regular polygon, side, vertex

Class Planning

Schedule 2 days

Resources

In *Chapter 3 Student Activity Masters* Book
Class Openers 1, 2
Extra Practice 3, 4
Cooperative Learning 5, 6
Enrichment 7
Learning Log 8
Interdisciplinary
 Investigation 9, 10
Spanish-Language Summary 11

Transparencies 16, 17 a,b

Class Opener

Estimate the following lengths in both English and metric units:
1. Distance from your location to Chicago
2. Length of the desktop
3. Height of the classroom doorway
4. Length of the chalk ledge of the chalkboard
5. Distance from your classroom to the gym or cafeteria

Answers: Answers will vary.

Lesson Notes

- Many students confuse perimeter, area, and volume, and misuse measurement formulas because they do not adequately understand linear measurement. Allow students plenty of time to

94

Investigation Connection
Section 3.1 supports Unit B, *Investigations* 1, 4. (See pp. 92A-92D in this book.)

SECTION 3.1
Linear Measurement

LINEAR UNITS

In 1954, the British track star Roger Bannister became the first person to run a mile in less than four minutes. Exactly how far is a mile?

A mile is a unit of **linear** measurement—a measure of length or distance. Other examples of linear units are inches, centimeters, meters, yards, kilometers, and microns. There are two basic systems of measurement, the English system and the metric system. The charts show some of the linear units in each system and the units' relationship to each other.

English System	Metric System
12 inches = 1 foot	10 millimeters = 1 centimeter
3 feet = 1 yard	100 centimeters = 1 meter
1760 yards = 1 mile	1000 meters = 1 kilometer
5280 feet = 1 mile	

Did you notice that the metric system is based on powers of ten? Many people think this makes it easier to use than the English system.

Example 1
Measure the butterfly's wingspan
 a Using an English-system ruler
 b Using a metric ruler

Solution
 a Using an English-system ruler, we find that the butterfly's wingspan is about $2\frac{3}{8}$ inches.
 b Using a metric ruler, we find that the butterfly's wingspan is about 6.1 centimeters, or about 61 millimeters.

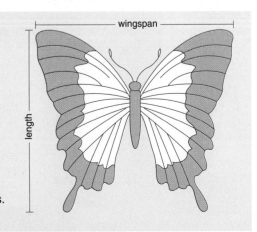

94

SECTION 3.1

MEASUREMENT AND POLYGONS

A **polygon** is a closed figure made up of line segments that intersect at their endpoints. Each point of intersection is a **vertex** of the polygon. (The plural of *vertex* is *vertices*.)

Polygons are usually classified by their number of **sides** (segments).

Number of Sides	Name of Polygon
3	Triangle
4	Quadrilateral
5	Pentagon
6	Hexagon
7	Heptagon
8	Octagon
9	Nonagon
10	Decagon

Example 2
a Identify the type of polygon.
b Explain how to find the perimeter of the polygon.

Solution
a The polygon is a pentagon because it has five sides.
b The perimeter of a polygon is the distance around it. So, to find the perimeter of this pentagon, we would measure the length of each side, then find the sum of all the lengths.

A segment that connects two vertices of a polygon, but is not a side of the polygon, is called a **diagonal**.

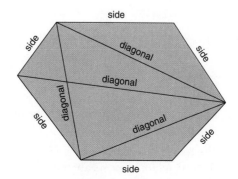

Section 3.1 Linear Measurement   95

---

become numerate in linear measurement in both the English and metric systems.

- Relate the prefixes in the metric system to powers of ten:

kilo-	1000	10^3
hecto-	100	10^2
deka-	10	10^1
	1	10^0
deci-	$\frac{1}{10}$	10^{-1}
centi-	$\frac{1}{100}$	10^{-2}
milli-	$\frac{1}{1000}$	10^{-3}

This pattern may help students understand negative exponents when they encounter them in scientific notation.

Additional Examples

Example 1
Measure the length of the butterfly in example 1:
a. Using an English-system ruler
b. Using a metric ruler

Answers:
a. $2\frac{1}{8}$ in.   **b.** 5.4 cm, or 54 mm

Example 2
a. Identify the type of polygon shown.
b. How many vertices does the polygon have?

Answers:
a. heptagon   **b.** seven

Reteaching

Draw these shapes on the chalkboard.

Have students use the definition of polygon to explain why each is not a polygon.

SECTION 3.1

Assignment Guide

Basic
Day 1 11–13, 19–21, 25, 28, 32, 35–37, 41, 46
Day 2 14–16, 22–24, 31, 33, 34, 38, 42, 45

Average or Heterogeneous
Day 1 11–19, 23, 25, 26, 28, 31, 33, 35, 37, 41, 43
Day 2 20–22, 24, 27, 32, 34, 36, 38, 39, 44, 45, 47

Honors
Day 1 12, 13, 15, 17, 19, 21, 23, 26, 27, 32, 34, 39
Day 2 11, 14, 16, 18, 22, 24, 25, 28–31, 33, 35, 40, 41, 45, 47

Strands

Communicating Mathematics	2, 32, 34
Geometry	1–3, 5, 6, 8, 9, 11–14, 20–22, 27, 34
Measurement	11, 15–18, 20–26, 28–31, 33–35
Data Analysis/ Statistics	14, 39
Technology	10, 39, 47
Probability	25
Algebra	19, 23, 24, 28–30, 33, 36–39, 41, 42, 46
Patterns & Functions	14, 47
Number Sense	2, 4, 7, 15–18, 23, 24, 26, 28–31, 35, 41, 42
Interdisciplinary Applications	23, 33, 40, 45, Investigation

How do the two diagonals of any rectangle compare? Try measuring the diagonals of each of these three rectangles.

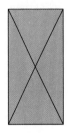 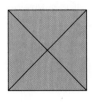 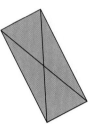

In each case, the two diagonals appear to be the same length. Segments that have the same length are called **congruent.** Do you think the diagonals of all rectangles are congruent?

Sample Problem

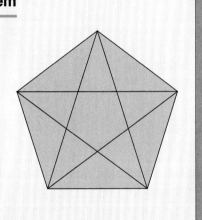

Problem In a *regular polygon,* all sides are congruent and all angles have the same measure. Are all the diagonals of a regular pentagon congruent?

Solution If we measure all five diagonals of the regular pentagon, we find that they have the same length. Therefore, they are congruent. Do you think the diagonals of every regular pentagon are congruent?

Think and Discuss

P **1** How long is the segment? $2\frac{3}{4}$ in.

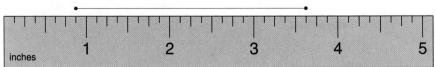

P, R **2** Explain the difference between the perimeter of a polygon and the area of a polygon.

P **3** The fence posts of a fence are 6 feet apart. The fence is 138 feet long. How many fence posts are there? 24 posts

SECTION 3.1

P, R **In problems 4–9, is the statement true *always, sometimes,* or *never*?**

4 A percent of a number is less than the number. Sometimes

5 A rectangle is a parallelogram. Always

6 A parallelogram is a rectangle. Sometimes

7 Every fraction can be written in standard form. Always

8 A rectangle is a quadrilateral. Always

9 A square is a rectangle. Always

I 10 A calculator may express a number as 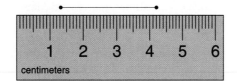, which means 2×10^9.
 a Write 2 E 9 in standard form. 2,000,000,000
 b Use your calculator to multiply 456,789 × 456,789. Express the result in standard form. 208,656,190,500

Problems and Applications

P 11 Measure the length of the butterfly in Example 1
 a Using an English-system ruler About $2\frac{1}{8}$ in.
 b Using a metric ruler About 54 mm, or about 5.4 cm

P 12 How long is the segment?
2.9 cm, or 29 mm

P, I 13 Is the statement *true* or *false*?
 a The diagonals of a rectangle are congruent. True
 b The diagonals of a regular hexagon are congruent. False
 c All radii of a circle are congruent. True
 d The diagonals of a regular octagon are congruent. False

P 14 Copy and complete the table so that it shows the number of diagonals belonging to a polygon with the given number of sides.

Sides	3	4	5	6	7	8	9	10	11	12
Diagonals	0	2	5	9	14	20	27	35	44	54

P, R, I **In problems 15–18, determine which distance is longer.**

15 20 inches or 2 feet 2 ft

16 1 mile or 2000 yards 2000 yd

17 15 millimeters or 1 centimeter 15 mm

18 2150 meters or 2 kilometers 2150 m

Section 3.1 Linear Measurement

SECTION 3.1

Problem-Set Notes

for pages 98 and 99

- **24** emphasizes the need for common units.
- **34** is good for a journal entry.
- **36** and **37** preview scientific notation.

Additional Practice

1. How many sides does each polygon have?
a. octagon   b. quadrilateral
c. pentagon d. hexagon
e. decagon   f. nonagon

2. Which measurement is shorter?
a. 1 km or 175 m
b. 1 m or 1 cm
c. 1760 ft or 1 mi
d. 12 mm or 10 cm
e. 10 yd or 35 ft

Answers: **1. a.** 8 **b.** 4 **c.** 5 **d.** 6
e. 10 **f.** 9 **2. a.** 175 m **b.** 1 cm
c. 1760 ft **d.** 12 mm **e.** 10 yd

Cooperative Learning

- In groups, have students solve the following problem.
- A farmer has 24 posts that will be placed 6 feet apart to support a fence surrounding a rectangular pen.
  a. How much fencing will the farmer need for the pen?
  b. What are the dimensions of the largest pen that the farmer can build?
  c. What are the dimensions of the smallest pen that the farmer can build (and still use all of the posts)?

Answers: **a.** 144 ft
b. 36 ft × 36 ft **c.** 6 ft × 66 ft

- Problems **40** and **47** are appropriate for cooperative learning.

I **19** Find the value of $x^y + y^x$.   23,401

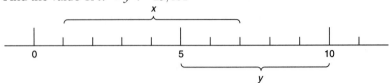

In problems 20–22, draw each polygon.   Check students' drawings.

P **20** A 3-inch-by-5-inch rectangle

P, I **21** A triangle with sides of 4 inches, 5 inches, and 6 inches

P, I **22** A regular hexagon with sides of 3 centimeters

P **23** In football, a team has four downs to gain 10 yards for a first down. The Cats gained $3\frac{1}{2}$ yards on first down, 18 feet on second down, and 2 feet on third down. Did they gain enough yards for a first down? Explain your answer.   Yes. $3\frac{1}{2}$ yd + 18 ft + 2 ft = $3\frac{1}{2}$ yd + 6 yd + $\frac{2}{3}$ yd = $10\frac{1}{6}$ yd

P **24** Cindy Centipede wiggled for 8 centimeters, walked for 35 millimeters, and waddled for 6.3 centimeters. How far did she go?   178 mm, or 17.8 cm

P, R **25** What is the probability that two words chosen from the set {*inch, pound, meter, liter*} both name
a Measures of length? $\frac{1}{6}$
b Metric units of measure? $\frac{1}{6}$
c English units of measure? $\frac{1}{6}$
d Measures of volume? 0

P **26** How many centimeters are in 1 inch?   2.54 cm

P, I **27** The horizontal distance between any two dots is three units, and the vertical distance is four units.
a Find the perimeter of the figure.   86
b Find the area of the enclosed region.   216

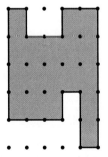

P, R **In problems 28–31, determine which measurement is longer.**

28 50% of 1 yard or 125% of 1 foot   **29** 200% of 1 centimeter or 10% of 1 meter

30 400 yards or 25% of 1 mile   **31** 1 meter or 1 yard   1 m

I **32** **Communicating** An octagon is an eight-sided polygon, an octopus has eight arms, but October is the tenth month. Explain why this is so.

P, R **33** There are two types of socket wrenches—English and metric—for tightening bolts. Juan has a 1.4-centimeter bolt to tighten, but he has only English wrenches in sizes $\frac{7}{16}$, $\frac{1}{2}$, $\frac{9}{16}$, $\frac{5}{8}$, and $\frac{11}{16}$ inch. Which size wrench would work best? $\frac{9}{16}$

Chapter 3 Measurement and Estimation

SECTION 3.1

P, R **34** Thelma and Louise wanted to find the area and the perimeter of a square tile. Thelma measured each side at 4 inches. She concluded that both the area and the perimeter were 16. Louise found the tile to be $\frac{1}{3}$ foot on each side but said that the area was less than the perimeter. Who was correct?

P, I **35 a** Find the distance from A to B. $1\frac{5}{8}$ in.

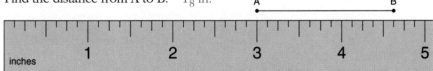

b If each $\frac{1}{4}$ inch on a blueprint represents 12 inches, what length is represented by a $3\frac{3}{4}$-inch segment?   180 in., or 15 ft

◀ LOOKING BACK **Spiral Learning** LOOKING AHEAD ▶

R **36** Evaluate each expression. Use your calculator if you wish.
 a 4^3 **b** 4^6 **c** 4^9 **d** 4^{12} **e** 4^{15} **f** 4^{18}
 g 10^1 **h** 10^3 **i** 10^5 **j** 10^7 **k** 10^9 **l** 10^{11}

I **37** Multiply 3.14159265 by each power of ten.
 a 10   31.4159265   **b** 100   314.159265   **c** 1000   3141.159265
 d $\frac{1}{10}$ 0.314159265   **e** $\frac{1}{100}$ 0.0314159265   **f** $\frac{1}{1000}$ 0.00314159265

R **38** Evaluate each expression.
 a $\sqrt{9} + \sqrt{16}$ 7   **b** $\sqrt{9+16}$ 5
 c $3\sqrt{7} + 9$ 12   **d** $\sqrt{4 \times 9 - 27}$ 3

I **39** Use a scientific calculator to perform the following operations. What do you notice about the results?
 a 4 [x^y] 1 [=]   4   **b** 4 [x^y] 0 [=]   1
 c 4 [x^y] 1 [+/−] [=]   0.25   **d** 4 [x^y] 2 [+/−] [=]   0.0625

R **40 Consumer Math** A mail-order firm charges 5% of the price of an order for shipping and adds 6.5% sales tax (the shipping charge is not taxed). A spreadsheet was used to make this form for the mail-order firm.

	A	B
1	Order	276.84
2	Ship	
3	Tax	
4	Total	

 a What would be the total bill, including shipping and sales tax, for an order of $276.84? $308.68
 b Write an expression to calculate the total bill of an order of any size.
 c Use the cell names to express the formulas that should be placed in cells B2, B3, and B4 to give the correct answer.
 d If the value in B1 is changed to $475.50, what is the total bill? $530.18
 e If the shipping charge is raised to 7% and the order is $245.78, what is the total bill? $278.96

Section 3.1 Linear Measurement 99

Checkpoint

1. On the map of Florida in problem 45 on page 100, measure the distance from Jacksonville to Miami in inches and centimeters.
2. Which polygon listed below has the greatest number of sides? The least number of sides?
triangle, heptagon, quadrilateral, decagon, hexagon

Answers: **1.** $1\frac{3}{16}$ in.; 4.6 cm
2. decagon; triangle

Additional Answers

for pages 98 and 99

28 50% of 1 yd
29 10% of 1 m
30 25% of 1 mi
32 At one time, the Roman calendar consisted of ten months. However, the Romans reorganized the calendar to include two additional months—July, named after Julius Caesar, and August, named after Augustus Caesar. Before the change, October was the eighth month.
34 Neither person was correct. Thelma was incorrect because 16 in. and 16 in.^{2} do not mean the same thing, even though the numbers are identical. Louise made the same mistake by comparing numbers without considering units.
36 a 64 **b** 4096 **c** 262,144
 d 16,777,216 **e** 1,073,741,824
 f 68,719,476,736 **g** 10 **h** 1000
 i 100,000 **j** 10,000,000
 k 1,000,000,000
 l 100,000,000,000
39 The results get smaller as the exponents decrease in value.
40 b +B1+0.05∗B1+0.065∗B1
 c B2: +B1∗0.05; B3: +B1∗0.065;
 B4: +B1+B2+B3

SECTION 3.1

Problem-Set Notes

for page 100

- **42 b** is a concrete example of the Distributive Property of Multiplication over Addition.

Communicating Mathematics

- Have students write a paragraph expressing their preference for the metric system or the English system, discussing the advantages and disadvantages of each.
- Problems 2, 32, and 34 encourage communicating mathematics.

Stumbling Block

In the definition of polygon, students may not understand what is meant by "intersect at their endpoints." Show students these two figures.

In the first figure, all segments intersect at their endpoints. In the second figure, $\overline{EH}$ and $\overline{IG}$ intersect at F, which is not an endpoint of either segment. Have students name the endpoints E, G, H, and I.

Additional Answers

for page 100

46 a 100 **b** 10 **c** 0.01

47 a In each cell, the cell named in the formula changes to the cell above. **b** Place 1 in A1. Enter +A1*2 in A2 and copy down.

Answer for Investigation is on page A2.

100

R **41** Which is greater, 2^6 or 6^2? How much greater? 2^6; greater by 28

I **42** Is the statement *true* or *false?*
   a $729 + 358 \cdot 254$ is less than $(729 + 358) \cdot 254$.   True
   b $277(589 + 816)$ is greater than $277 \cdot 589 + 277 \cdot 816$.   False

R **43 a** What percentage of the length of $\overline{AC}$ is x? ≈224%
   b What percentage of the length of $\overline{BC}$ is x? ≈112%

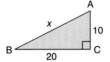

I **44** Harlee David's son can drive 172 miles on 5 gallons of gasoline. How much gasoline must he have to drive 130 miles? ≈3.8 gal

R **45** Geography
   a How far is it from Tampa to Miami? ≈240 mi
   b What is the greatest distance between any two points in Florida?
   Possible answer ≈780 mi

R **46** Copy each equation, filling in the blank to make a true statement.
   a $259.38 = (2.5938)(\_\_\_)$ **b** $81.36 = (8.136)(\_\_\_)$ **c** $0.0529 = (5.29)(\_\_\_)$

R **47** Spreadsheet   Using a spreadsheet, follow these steps:
   1. Enter the number 2 in cell A1.
   2. Enter +A1+3 in cell A2.
   3. Copy this formula to the cells below.

   a What adjustments did the spreadsheet make in your formula?
   b Use a spreadsheet to generate the pattern 1, 2, 4, 8, 16, 32, 64,…

Investigation

The English System   Write a report on the origin of the terms *inch, foot,* and *yard*. In your report, include how and when these terms were derived.

100   Chapter 3   Measurement and Estimation

SECTION 3.2

Scientific Notation

SCIENTIFIC NOTATION AND LARGE NUMBERS

The speed of light is about 300,000,000 meters per second. At that rate, light can travel around the world about seven times in one second!

We can write 300,000,000 as 3×10^8. This form of the number, called ***scientific notation,*** saves space and can simplify certain calculations. A number is expressed in scientific notation when it is expressed as the product of a power of 10 and a number that is greater than or equal to 1 and less than 10.

Example 1
Of 0.42×10^9, 4.2×10^8, and 42×10^7, which is written in scientific notation?

Solution
Since 4.2×10^8 is the only expression in which the multiplier is between 1 and 10, it is the only one correctly expressed in scientific notation.

Example 2
Convert 3.87×10^5 to standard form.

Solution
We move the decimal point five places to the right to multiply by 10^5.
$$3.87 \times 10^5 = 3.87 \times 100{,}000 = 387{,}000$$

Example 3
Convert 2345.6789 to scientific notation.

Solution
We move the decimal point three places to the left so that only one digit is to the left of the decimal point, and multiply by 1000, or 10^3.
$$2345.6789 = 2.3456789 \times 1000 = 2.3456789 \times 10^3$$

Objective
- To use scientific notation to express large and small numbers

Vocabulary: scientific notation

Class Planning

Schedule 2–3 days
Resources
In *Chapter 3 Student Activity Masters* Book
Class Openers *13, 14*
Extra Practice *15, 16*
Cooperative Learning *17, 18*
Enrichment *19*
Learning Log *20*
Technology *21, 22*
Spanish-Language Summary *23*

Class Opener

1. Evaluate each expression.
a. 2^5 b. 2^{15} c. 2^{25} d. 2^{35}
2. Which number is larger?
a. 9.87×10^5 or 1.2×10^6
b. 4.58×10^8 or 45.9×10^7

Answers:
1. a. 32 b. 32,768 c. 33,554,432
d. 3.436×10^{10} 2. a. 1.2×10^6
b. 45.9×10^7

Lesson Notes

- In sample problem 1, students may not know whether a light-year is a measure of time or distance. A light-year converts to centimeters, a unit of distance.

- Where necessary, teach your students how to put their calculators into scientific mode. Also, show them how to enter a number in scientific notation by using the exponent key (EE or EXP). A common oversight is assuming that students will learn on their own how to use a calculator properly.

SECTION 3.2

Assignment Guide

Basic
Day 1 10, 12, 15–18, 26, 29, 32, 34, 39, 43, 52
Day 2 11, 14, 20–22, 27, 33, 35, 40–42
Day 3 13, 19, 23–25, 30, 36, 38, 44, 46, 51

Average or Heterogeneous
Day 1 10–15, 18–19, 23, 31, 34, 38, 50
Day 2 16, 17, 20, 21, 25, 27, 30, 39–44, 53, 56
Day 3 22, 24, 28, 29, 32, 33, 35, 46, 47, 49, 55

Honors
Day 1 11, 13, 14, 16, 18, 19, 24, 26, 27, 30, 31, 34, 38, 49
Day 2 10, 12, 15, 17, 20, 22, 23, 28, 29, 35, 40, 42, 47, 48, 53, 54, 56

Strands

Communicating Mathematics	1, 4, 52
Geometry	15, 31, 34–36, 46, 52
Measurement	30, 34–36, 38
Technology	1, 24, 56
Probability	23, 53
Algebra	9, 26–30, 32, 33, 37, 39–44, 49, 51, 54, 55
Patterns & Functions	47
Number Sense	2–6, 10–13, 16–21, 22, 37, 39–44, 46, 54
Interdisciplinary Applications	14, 23, 24, 30, 45, Investigation

Individualizing Instruction

Visually impaired learners
Students may have difficulty accurately counting the number

Your calculator probably has a key labeled EE, EXP, or EEX. This key is used to enter numbers in scientific notation.

Example 4
Evaluate $(4.267 \times 10^6)(6.34 \times 10^{11})$.

Solution
If we enter 4.267 EE 6 × 6.34 EE 11 =, the display shows 2.705278 18, which means 2.705278×10^{18}.

SCIENTIFIC NOTATION AND SMALL NUMBERS

We can also use scientific notation to express very small numbers.

Example 5
Evaluate $0.00025 \div 5{,}000{,}000$.

Solution
On your calculator, enter 0.00025 ÷ 5,000,000 =. The calculator will display 5. –11. This means 5×10^{-11}. The negative exponent (−11) indicates that we are dividing by a power of 10 instead of multiplying by it. So $5 \times 10^{-11} = 5 \times \frac{1}{10^{11}} = \frac{5}{10^{11}} = 0.00000000005$.

Sample Problems

Problem 1 A light-year is the distance that light travels in one year. About how many centimeters are in a light-year?

Solution The speed of light is about 3×10^{10} centimeters per second. To approximate the number of centimeters in a light-year, we multiply the number of seconds in a year by the speed of light.
　　Since there are 365 days in a year, 24 hours in a day, 60 minutes in an hour, and 60 seconds in a minute, there are (365)(24)(60)(60), or 31,536,000, seconds in one year.

$$(31{,}536{,}000)(3 \times 10^{10}) = 9.4608 \times 10^{17}$$
$$\approx 9.5 \times 10^{17}$$

Light travels about 9.5×10^{17} centimeters in one year.

Problem 2 Evaluate $987{,}234.4 \div 10^8$, and express the answer in scientific notation.

SECTION 3.2

> **Solution** Multiplying by 10^{-8} gives the same result as dividing by 10^8. To multiply by 10^{-8}, we move the decimal point eight places to the left. Notice that it is necessary to insert zeros to the left of the first digit in order to do this.
> $$987{,}234.4 \div 10^8 = 987{,}234.4 \times 10^{-8}$$
> $$= 0.009872344$$
> $$= 9.872344 \times 10^{-3}$$

Think and Discuss

P **1** Why do we use scientific notation?

P **2** What is true about all the expressions in Example 1?   All are equal to 420,000,000.

P **3** Which number is greater?
   a -2 or -3 -2 **b** 10^{-2} or 10^{-3} 10^{-2} **c** 9.87×10^5 or 1.2×10^6 1.2×10^6

P, R **In problems 4–6, is the statement *true* or *false*?**

  4 $3591 \times 10^2 = 359.1$ False   **5** $0.00123 = \frac{1.23}{1000}$ True   **6** $5 \times 10^2 = 50^2$ False

P **7** Express 43,000 in scientific notation. 4.3×10^4

P **8** Express 8.135×10^9 in standard form. 8,135,000,000

P, R **9** Evaluate each expression.
   a $10^5 + 10^7$ 10,100,000   **b** $10^5 \cdot 10^7$ **c** $10^7 \div 10^5$ 100
   d $10^5 \div 10^7$ $\frac{1}{100}$ **e** $10^5 - 10^7$ $-9{,}900{,}000$ **f** $10^7 - 10^5$ 9,900,000

Problems and Applications

P **In problems 10–13, express each number in scientific notation.**

  10 93.4   **11** 385.12   **12** 0.01234   **13** 5647

P **14** **Science** In chemistry, a mole is a unit of measure equal to 6.02×10^{23} molecules of a substance. Write this number in standard form. 602,000,000,000,000,000,000,000

R **15** Which side of the triangle is the longest? $\overline{BC}$

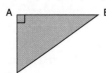

P **In problems 16 and 17, write each number in scientific notation.**

  16 52×10^6 5.2×10^7 **17** 0.52×10^6 5.2×10^5

---

of decimal places in numbers in scientific notation. Assist these students with more verbal guidance, such as reading the numbers aloud.

Additional Examples

Example 1
Which number, 16×10^4, 0.16×10^7, or 1.6×10^5, is correctly written in scientific notation?
Answer: 1.6×10^5

Example 2
Convert 7.12×10^3 to standard form.
Answer: 7120

Example 3
Convert 46215.4231 to scientific notation.
Answer: 4.62154231×10^4

Example 4
Evaluate $(1.243 \times 10^4)(4.692 \times 10^{13})$.
Answer: 5.8322×10^{17}

Example 5
Evaluate $0.0004 \div 16{,}000{,}000$.
Answer: 2.5×10^{-11}

Additional Answers

for page 103

1 Possible answer: Scientific notation is a space-saving method of writing very large and very small numbers. Scientific notation also makes it possible to express these numbers on a calculator.

9 b 1,000,000,000,000
10 9.34×10^1
11 3.8512×10^2
12 1.234×10^{-2}
13 5.647×10^3

SECTION 3.2

Problem-Set Notes

for pages 104 and 105

- **22** illustrates the need for scientific notation to compare more numbers easily.
- **23** emphasizes the importance of converting to scientific notation first.
- **26–29** should be done with a calculator.
- **47** may be discussed or given as a writing assignment. Using strictly volume and dividing 27 into 1000, we seem to be able to get thirty-seven 3" cubes. However, we can only get twenty-seven 3" cubes. Some students may need to use manipulatives.

Checkpoint

1. Express the following numbers in scientific notation.
   a. 45,600,000 b. 13,900
   c. 0.00056 d. 32.9
2. Evaluate $(6.2 \times 10^{12})(4.3 \times 10^9)$.
3. Express these numbers in standard form.
   a. 2.01×10^3 b. 7.432×10^4
   c. 4×10^5 d. 9.842×10^6

Answers:
1. a. 4.56×10^7 b. 1.39×10^4
c. 5.6×10^{-4} d. 3.29×10^1
2. 2.666×10^{22}
3. a. 2010 b. 74,320 c. 400,000
d. 9,842,000

Cooperative Learning

- Have groups of students do the following four problems on a calculator.
   a. $(3 \times 10^2)(2 \times 10^5)$
   b. $(2.5 \times 10^5)(3 \times 10^4)$
   c. $(1.2 \times 10^8)(4 \times 10^3)$
   d. $(1.6 \times 10^3)(3.4 \times 10^4)$

P **In problems 18–21, write each expression in standard form.**

18 3.05×10^4 30,500 **19** 23.4×10^{-2} 0.234

20 3.15648×10^3 3156.48 **21** 8.7643×10^1 87.643

P, R **22** Arrange the numbers 130×10^7, 92.5×10^8, and 8.27×10^9 in order, from least to greatest. $130 \times 10^7, 8.27 \times 10^9, 92.5 \times 10^8$

R **23** In poker, there are 2,598,960 possible 5-card hands. Of these, 4 are royal flushes. What is the probability of getting a royal flush? Express your answer in scientific notation. $\approx 1.539 \times 10^{-6}$

P **24** Science How long does it take sunlight to reach the Earth? About 8 minutes 20 seconds

P, R **25** a What percent of 8×10^{30} is 2×10^{30}? 25%
b What percent of 8×10^{-2} is 8×10^{-3}? 10%
c What percent of 10^{23} is 10^{21}? 1%

P **In problems 26–29, find the value of each expression. Express each answer in scientific notation.**

26 $(3.5 \times 10^{14})(8.13 \times 10^{27})$ **27** $\dfrac{9.52 \times 10^{44}}{4.12 \times 10^{12}}$ $\approx 2.310679612 \times 10^{32}$

28 $5.1 \times 10^3 + 6.2 \times 10^4$ 6.71×10^4 **29** $95{,}476 \div (3 \times 10^8)$ $3.1825\overline{3} \times 10^{-4}$

P **30** Environmental Science In 1990, about 50 million tons of paper was used in the United States. The U.S. population in 1990 was about 250 million. What was the average number of pounds of paper used by each person? 400 lb

P, I **31** a What is the length of $\overline{AB}$?
b Find the area of $\triangle ABC$.
c Find the perimeter of $\triangle ABC$.

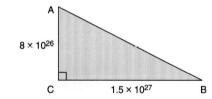

◀ LOOKING BACK **Spiral Learning** LOOKING AHEAD ▶

I **In problems 32 and 33, find a number that makes the equation true.**

32 $\square^3 = 64$ 4   **33** $5 + 4 \cdot \triangle^2 = 41$ 3

I **In problems 34–36, use a ruler to draw each polygon.**

34 A rectangle measuring 4.5 centimeters by 5.6 centimeters

35 A rectangle with one 5-centimeter side and one 7-centimeter diagonal

36 A triangle with sides of 6 inches, 8 inches, and 10 inches

Chapter 3 Measurement and Estimation

SECTION 3.2

Answers:
a. 6×10^7 **b.** 7.5×10^9
c. 4.8×10^{11} **d.** 5.44×10^7

- After considering both the factors and the products in these problems, ask students to develop a rule for multiplying two numbers written in scientific notation.
- After students have time to develop a rule, ask them to consider these three problems.
e. $(3 \times 10^5)(5 \times 10^4)$
f. $(6 \times 10^{10})(7 \times 10^3)$
g. $(2.5 \times 10^7)(8 \times 10^5)$

Answers:
e. 1.5×10^{10} **f.** 4.2×10^{14}
g. 2×10^{13}

- Ask students: Do these problems follow the rule you have written? Why or why not? How are these problems different from the first four given?
- Problem 56 is appropriate for cooperative learning.

I **37** Write a square-root expression that has a value of 0.36. $\sqrt{0.1296}$

I **38** A map legend indicates that 1 inch = 8.5 miles. What is the distance between two cities that are 6 inches apart on the map? 51 mi

I **In problems 39–44, estimate the answer without using a calculator.**

39 $\sqrt{85}$ **40** $(9.9)^3$ **41** $\sqrt{48}$

42 $4\sqrt{65}$ **43** $\dfrac{18}{\sqrt{80}}$ **44** $(9.1 - 4)^2$

R **45** **Health** In the state of Health, high school students are not allowed to wrestle unless 7% of their weight is body fat. Matt weighs 154 pounds, 10 pounds of which is body fat. Will Matt be allowed to wrestle, or will he need to increase his body fat? Matt will need to increase his body fat from its current 6.5%.

I **46** Which of the expressions represent the area of rectangle ABCD? Why?
a $3 \cdot 4 + 3 \cdot 2 + 2 \cdot 4 + 2 \cdot 2$
b $3 + 2(4 + 2)$
c $(3 + 2)(4 + 2)$
d $4(3 + 2) + 2(3 + 2)$

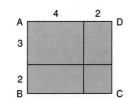

I **47** When this 10" × 10" × 10" cube is cut into 2" cubes as shown, exactly 5 · 5 · 5, or 125, of the small cubes are produced. Copy the following table, indicating the greatest number of cubes of each size that can be cut from a 10" × 10" × 10" cube.

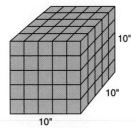

Size of cubes	1"	2"	3"	4"	5"	6"	7"	8"	9"
Number of cubes	1000	125	27	8	8	1	1	1	1

R **48** The pizza that Yogi, Ellie, and Mickey bought was cut into 12 slices. Yogi ate 6 slices, Ellie ate 3 slices, and Mickey ate 2 slices.
a What percent of the pizza did each person eat?
b What is the total percent of the pizza the three men ate? ≈91.7%
c What percent of the pizza was not eaten? ≈8.3%

I **49** What is the value of x if the perimeter of this polygon is 112 inches? 10 in.

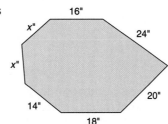

Additional Practice

1. Express each number in scientific notation.
a. 46,842,000 **b.** 0.00346
c. 1.5432 **d.** 794,577
2. Express these numbers in standard form.
a. 1.43×10^{-3}
b. 8.954288×10^6
c. 6.02×10^4
3. Evaluate.
a. $3 \times 10^5 + 6 \times 10^4$
b. $(4.2 \times 10^8)(7.7 \times 10^3)$
c. $5.9 \times 10^2 - 1.5 \times 10^2$

Answers:
1. a. 4.6842×10^7
b. 3.46×10^{-3} **c.** 1.5432×10^0
d. 7.94577×10^5 **2. a.** 0.00143
b. 8,954,288 **c.** 60,200
3. a. 3.6×10^5 **b.** 3.234×10^{12}
c. 4.4×10^2

Answers for problems 26, 31, 34–36, 39–44, 46, and 48 are on page A2.

Section 3.2 Scientific Notation

SECTION 3.2

Stumbling Block

Students may be confused by the negative exponent in example 5. Point out that the statement "dividing by a power of 10" means dividing by a *positive* power of 10. Multiplying by 10 to a negative exponent is the same as dividing by 10 to a positive exponent. Give students several examples, such as $2 \times 10^{-3} = 2 \div 10^3$.

Communicating Mathematics

- Have students write a paragraph explaining, or have them explain orally, how to order the following numbers from least to greatest: 63.5×10^3; $.786 \times 10^4$; 4628000×10^{-2}; 5×10^5.

- Problems 1, 46, and 52 encourage communicating mathematics.

Additional Answers

for page 106

52 A trapezoid has one pair of parallel sides, and a parallelogram has two pairs of parallel sides.

55 Possible answers:
a $4 + 12 + (7 \times 9)$
b $4 \times (12 - 7) + 9$
c $(4 + 12) \div (7 + 9)$
d $4 \div 12 \times 7 \times 9$

56 The population of Movile approaches 75,000; the population of Transitory approaches 100,000.

Investigation About 4.6×10^7 min

106

I **50** Use the digits 2, 3, 4, and 5 to make two 2-digit numbers whose product is as large as possible. Use each digit only once. 52×43

I **51** Evaluate each expression for $a = 4$, $b = 6$, and $c = 2$.
 a b^2 36
 b $4ac$ 32
 c $b^2 - 4ac$ 4
 d $\sqrt{b^2 - 4ac}$ 2

I **52 Communicating** What is the difference between a trapezoid and a parallelogram?

R **53** Margaret tosses a ring at random onto one of the numbered pegs on this pegboard.
 a Find the probability that the number of the peg is a prime number. $\frac{3}{8}$
 b Find the probability that the number of the peg is a multiple of 2 or 3. $\frac{2}{3}$

I **54** Arrange the numbers 4, 5, and 6 in the expression $\triangle \cdot \triangle^\triangle$ to make the greatest possible number. $4 \cdot 5^6$

R **55** Add parentheses and symbols $(+, -, \times, \div)$ between the numbers 4 12 7 9 in the given order so that the value of the expression is
 a 79
 b 29
 c 1
 d 21

R **56 Spreadsheets** Each year, 40% of the population of the town of Movile moves to Transitory, and the other 60% remains in Movile. Each year, 30% of the population of the town of Transitory moves to Movile, and the other 70% stays in Transitory. In 1960, the population of Movile was 95,000, and the population of Transitory was 80,000. Use a spreadsheet to make a table like the one shown.

	A	B	C
1	Year	Movile's Population	Transitory's Population
2	1960	95,000	80,000
3	1961		
4	1962		

Continue the spreadsheet through the present year. What happens to the populations over the years? (Hint: The formula for cell B3 is +0.6*B2+0.3*C2. Do you see why?)

Investigation

History Approximately how many minutes passed between the signing of the Declaration of Independence and the end of the American Civil War?

106 Chapter 3 Measurement and Estimation

SECTION 3.3

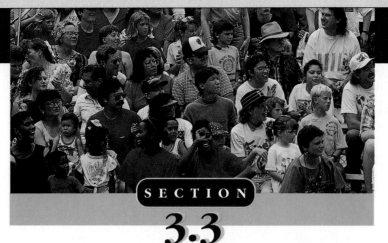

3.3 Estimation and Rounding

Investigation Connection
Section 3.3 supports Unit B, *Investigations* 1, 3, 4, 5. (See pp. 92A–92D in this book.)

ACCURACY

The 1990 census gave the United States population as 248,709,873. Since the number is close to 249,000,000, we can say that in 1990, the population of the United States was 249 million, to the nearest million.

We often estimate a quantity to a reasonable accuracy. What is reasonable depends on the situation. For example, with a metric ruler, you can measure to within a millimeter, or one tenth of a centimeter. If a ruler has English units, you can round off to the nearest sixteenth of an inch. One general rule is that our result can be only as accurate as the least accurate measurement.

Example 1
What would be the most reasonable way to measure the height of a basketball player—to the nearest meter, to the nearest centimeter, or to the nearest millimeter?

Solution
Measuring to the nearest meter would not give a very accurate impression of the player's height, since all basketball players are about 2 meters tall. However, the height could not be measured accurately to the nearest millimeter. So in this case, the nearest centimeter is reasonable.

ROUNDING

We round a number to a specific place value. For example, rounding to the nearest ten-thousandth means leaving four numbers to the right of the decimal point.

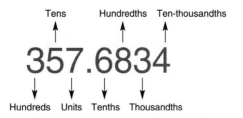

Section 3.3 Estimation and Rounding   107

Objectives
- To recognize that different situations require different degrees of accuracy in measurement
- To round a number to a specified number of decimal places
- To express results to a reasonable degree of accuracy

Vocabulary: acute angle, obtuse angle, right angle

Class Planning

Schedule 2 days
Resources
In *Chapter 3 Student Activity Masters* Book
Class Openers 25, 26
Extra Practice 27, 28
Cooperative Learning 29, 30
Enrichment 31
Learning Log 32
Interdisciplinary
  Investigation 33, 34
Spanish-Language Summary 35

In *Tests and Quizzes* Book
Quiz on 3.1–3.3 (Basic) 53
Quiz on 3.1–3.3 (Average) 54
Quiz on 3.1–3.3 (Honors) 55

***Transparency* 18**

Class Opener

Perform the following calculations.
1. 24 × 32   **2.** 23.5 × 31.5
3. 24.5 × 32.5
4. 24.5 × 32.5 − 24 × 32
5. (24.5 × 32.5 − 24 × 32) ÷ (24 × 32)

Answers: **1.** 768 **2.** 740.25
3. 796.25 **4.** 28.25 **5.** ≈0.037

Lesson Notes

- Students have a tendency to copy all of the place values that they can, especially if they use calculators. Point out that results of calculations should only be as

107

SECTION 3.3

Lesson Notes continued

accurate as the original measurements, so calculated measurements must usually be rounded.

■ You may want to discuss exact versus estimated values. Some students may think that a multitude of place values makes an answer exact. Use $\frac{5}{17}$ and 0.2941176460588 as examples. The fraction is an exact value, whereas the decimal is only an approximation.

■ Example 3 and sample problem 1 show how to be reasonable when rounding. Remind students to use the rounding rules only when it makes sense to do so.

Additional Examples

Example 1
Which of the following would be the most reasonable way to measure the change in a baby's weight over the course of a month: to the nearest kilogram, gram, or milligram?

Answer: gram

Example 2
Round 426.1493 to the nearest
a. Ten   **b.** Hundredth

Answers: **a.** 430 **b.** 426.15

Example 3
There are 350 books to be mailed. If each box can hold up to 18 books, how many boxes are needed?

Answer: 350 ÷ 18 = 19.4. Round up to 20.

Additional Answers

for page 109

9 Possible answer: To make the measurement look as if it were not an estimate.

108

To help us decide whether to round a number up or down, we look at the value in the next lower decimal place and use the following rules.

1. If the value in the next decimal place is 5 or more, round up.
2. If the value in the next decimal place is less than 5, round down.

Example 2
Round 357.6834 to the nearest
a Ten   **b** Hundred   **c** Hundredth

Solution
a Since 7, the units digit, is greater than 5, we round up. Rounded to the nearest ten, 357.6834 is 360.
b Since the tens digit is 5, we round up. Rounded to the nearest hundred, 357.6834 is 400.
c Since 3, the thousandths digit, is less than 5, we round down. Rounded to the nearest hundredth, 357.6834 is 357.68.

However, there are some cases when it is necessary to revise the rules for rounding. Always make sure your answer is reasonable.

Example 3
There are 410 people going on a field trip. If the seating capacity of each bus is 44, how many buses are needed?

Solution
Since 410 ÷ 44 ≈ 9.32, we must round to 9 or 10. According to the rules for rounding, 9.32 should be rounded down to 9. However, if we do that, some people will not have a seat on the bus. In this case, it is necessary to revise the rules and round 9.32 up to 10. Therefore, 10 buses must be ordered.

Sample Problems

Problem 1 Barry has $10.00. How many burgers can he order?

Solution If we divide 10 by 1.7, we get about 5.88. In this case, we cannot round 5.88 up to 6 because Barry doesn't have enough money to buy the sixth burger. So Barry can buy only 5 burgers.

SECTION 3.3

Problem 2 Maddy is painting the gutters that surround her house. How long is the ladder she is using?

Solution When rounding a number, always remember to round to a reasonable degree of accuracy. The top of the ladder is 11.25 feet above the ground, and the base of the ladder is 3.5 feet from the house. We can use the Pythagorean Theorem to solve this problem.

$$11.25^2 + 3.5^2 = (\text{hypotenuse})^2$$
$$126.5625 + 12.25 = (\text{hypotenuse})^2$$
$$138.8125 = (\text{hypotenuse})^2$$

So the ladder's length is $\sqrt{138.8125}$, or approximately 11.78187167, feet. This answer, however, suggests that we can determine the length of a ladder to the nearest hundred-millionth of a foot. Is this reasonable? Not even the most accurate ruler could be that precise. Furthermore, we cannot be sure that the original measurements were exact. It is more reasonable to round and say that the ladder is about 12 feet long.

Think and Discuss

P **1** Have someone record the time it takes you to perform a task. Then estimate how long it took. How accurate were you?   Answers will vary.

P, R **2** Have several people use rulers to measure the width of your teacher's desk. Then compare measurements. How accurate were you?   Answers will vary.

P, R **3** Estimate the height of the ceiling in your classroom. Then try to get an accurate measurement. How close was your estimate?   Answers will vary.

P **In problems 4–7, round each number to the nearest unit.**

4 47.239   47   **5** 2.45   2   **6** 9.5   10   **7** 17.76   18

P, R **8** Jerome bought a pair of Atmosphere basketball shoes that cost $89.95. The sales tax was 7%. Estimate the cost of the shoes to the nearest dollar. $96

P **9** A survey in the 1800's determined the height of Mount Everest to be 29,000 feet (to the nearest foot). However, when the statistics were later published, the height was listed as 29,002 feet. Why do you think this was done?

Section 3.3 Estimation and Rounding

Problems and Applications

P In problems 10–12, round each number to the nearest ten-thousandth.

10 3.29681   3.2968
11 4.26   4.2600
12 16.15936828   16.1594

P, R 13 If you have $5.00, how many of each can you purchase?
  a Packages of gum at 65¢ each
  b Boxes of candy at 50¢ each
  c Bags of chips at $1.29 each
  d Newspapers at 45¢ each

I 14 A *right angle* is exactly 90°, an *acute angle* is less than 90°, and an *obtuse angle* is between 90° and 180°. Classify each angle as *right, acute,* or *obtuse.*

  a Acute
  b Right
  c Acute
  d Obtuse

P 15 **Consumer Math** The price of granola bars is 3 for $1. Gary bought one granola bar and was charged 34¢. How can you justify this method of rounding?

P In problems 16–21, round each number to the nearest hundredth.

16 385.2865   385.29
17 475.391112 475.39
18 253.496   253.50
19 132.62   132.62
20 3.006   3.01
21 263.109   263.11

P, R 22 The shelves in Mr. Ditto's bookcase are 18 inches long. How many books can he put on each shelf if each book is
  a $\frac{3}{4}$ inch thick?   24
  b $1\frac{1}{2}$ inches thick? 12
  c $1\frac{1}{3}$ inches thick? 13
  d $\frac{3}{8}$ inch thick?   48

P, R 23 **Estimation** Approximately how many seconds have you been alive? c
  a 40,000
  b 4,000,000
  c 400,000,000
  d 400,000,000,000

P, I 24 **Communicating** Felix computed the length of the hypotenuse of the triangle to be 64.03124237 feet. Do you agree with his calculation? No. The answer cannot be more accurate than the input.

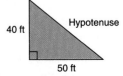

P, R 25 Discount Music Sales sells compact discs for $9.75 each. The sales tax is 6%. How many discs can Juanita buy for $50.00?   4

110   Chapter 3   Measurement and Estimation

SECTION 3.3

P 26 How many of the small cubes will fit inside the large cube? 27 cubes

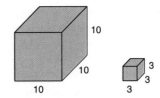

P 27 A newspaper reported that 53,000 people attended a ballgame. Assume that this number was rounded to the nearest thousand. What might have been the exact attendance? Give a range, from lowest to highest. 52,500 to 53,499

P, R 28 Kurt needs $2500.00 to buy a used car. He earns $9.75 per hour and works 40 hours per week. Assuming no pay deductions, how long would Kurt have to work in order to pay for the car? ≈6.4, or about ≈7, weeks

P 29 Consumer Math Victoria's Video has two plans for renting movies.

Plan 1 Rent one movie for one night for $3.
Plan 2 Pay an annual membership fee of $96; rent one movie for one night for $1.

How many movies would you need to rent annually to make the membership beneficial? 49 movies

P, R, I 30 How many 16-ounce glasses of water will it take to fill a 10-gallon aquarium? 80 glasses of water

P, I 31 Science A patient of Dr. J's has a heart rate of 36 beats every half minute.
  a What is the average amount of time between two beats? ≈0.8 sec
  b How many beats occur per minute? 72 beats per min

P 32 Estimation The bases on a baseball diamond are 90 feet apart.
  a Approximately how far did Babe Ruth run around the bases during his 714 home runs? About 280,000 ft
  b Explain how you arrived at your estimated answer. Multiply 700 × 400.

P 33 Using cruise control, Eddie can drive at a constant speed of 63 miles per hour. At that speed, his car gets 32.4 miles per gallon. If gasoline costs $1.32 per gallon and he travels 442 miles along an interstate highway,
  a How long will the trip take? ≈7 hr
  b How much gasoline will he use? ≈13.6 gal
  c How much will the gasoline cost? $17.95

P 34 How many gallons of water are needed to fill a 24-inch-by-9-inch-by-18-inch aquarium if one cubic foot holds approximately 7.5 gallons of water? About 16.9 gal

P, R 35 Spreadsheets Use a spreadsheet to create a table in which the numbers 1, 0.5, 0.25, 0.125, 0.0625, … are in the cells of column A. In column B, use formulas to list 30% of each value in column A.
  a When does the spreadsheet begin to display the numbers in scientific notation?
  b Set your spreadsheet to round to four decimal places. What happens to the entries in the table?

Communicating Mathematics

- Have students write a paragraph describing how a number line can be used to visually show how to round numbers to the nearest unit.

- Problems 9, 15, and 37 encourage communicating mathematics.

Multicultural Note

The height of Mt. Everest, referred to in problem 9, has been measured several different ways, including by using barometric altimeters, geometry, and trigonometry. Climatic conditions can affect instrument readings, however, and different methods produce different results. A 1954 Indian government survey measured Everest at 29,028 ft, but a widely accepted unofficial figure is 29,141 ft.

Additional Answers

for pages 110 and 111

13 a 7 packages **b** 10 boxes **c** 3 bags **d** 11 newspapers
15 Possible answer: The purpose of the sale is to sell three granola bars at a time. If the buyer does not take advantage of the sale, the seller keeps the extra penny.
35 a Possible answer: A: line 34; B: line 32 **b** The numbers are no longer expressed in scientific notation. When numbers become too small to be expressed in four digits, they are rounded to zero.

SECTION 3.3

36 A rectangular container measuring 28 feet by 28 feet by 10 feet is filled with gasoline.
  a How much did it cost to fill the container if each cubic foot holds approximately 7.5 gallons of gasoline and each gallon costs $1.30?   ≈$76,440
  b How many miles could be driven using this gasoline if the drivers average 25 miles per gallon?   ≈1,470,000 mi

37 According to the U.S. Geological Survey, the Caspian Sea is 3363 feet deep. What does this mean?

◀ LOOKING BACK **Spiral Learning** LOOKING AHEAD ▶

38 Find the area of the trapezoid.   578

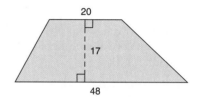

39 Stand and face north. Turn 90 degrees to the right. What direction are you facing?   East

40 Is each statement *true* or *false*?
  a $5 + 4^2 = 13$ False
  b $5(4^2) = 400$ False
  c $5 \cdot 4^2 = 40$ False
  d $5 \cdot 4^2 = 80$ True

41 The Furd Family owned two cars, five bicycles, two tricycles, and a motorcycle. What fraction of their tires belongs to vehicles with motors? (Don't forget that each car has a spare tire.) $\frac{3}{7}$

42 Draw a triangle with sides of the following lengths.
  a 5 centimeters, 6 centimeters, 7 centimeters
  b 5 centimeters, 5 centimeters, 3 centimeters
  c 5 centimeters, 5 centimeters, 5 centimeters

43 Spreadsheets

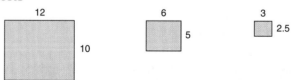

  a What is the perimeter of each figure?   44, 22, 11
  b Explain the pattern.
  c A spreadsheet was used to calculate the perimeters. What formula was probably entered in cell B4 to calculate the value in B4 in terms of the value in B3.   +B3/2

	A	B	C
1	Length	Width	Perimeter
2	12	10	44
3	6	5	22
4	3	2.5	11

Chapter 3   Measurement and Estimation

SECTION 3.3

I **44** Which angle is larger? ∠B

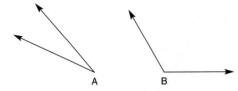

R **45** Find the value of each expression.

a $(3.7 \times 10^4) + (5.2 \times 10^3)$ 42,200
b $\dfrac{3.7 \times 10^4}{5.2 \times 10^3}$ ≈ 7.1154
c $(3.7 \times 10^4) - (5.2 \times 10^3)$ 31,800
d $(3.7 \times 10^4)(5.2 \times 10^3)$ 1.924×10^8

R **46** Find the distance from A to B.   ≈7.1

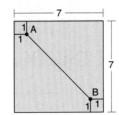

R **In problems 47–49, write each number in standard form.**

47 5.23×10^1 52.3   **48** 5.23×10^{-4} 0.000523   **49** 5.23×10^6 5,230,000

I **In problems 50 and 51, is each statement *true* or *false*?**

50 If $1 + a = 5$, then $a^2 + a = 30$. False   **51** If $1 + x = 4$, then $x^x = 256$. False

R **52** Communicating

a If $c = \sqrt{9^2 + 12^2}$, what's the value of c?   15
b Explain why this formula works.

I **In problems 53–56, is each statement *true* or *false*?**

53 If $x = 5$, then $x^x = 3125$. True   **54** If $y = 2$, then $4 + 3y = 14$. False
55 If $z = 9$, then $\sqrt{z} = 3$. True   **56** $3.5^{3.5} \approx 80.21$ True

Investigation

In the News Look through a newspaper and find several examples in which a rounded number is used to describe a situation. Then find several examples in which an exact number is used. Also find an example in which you cannot determine whether the number has been rounded or is exact.

SECTION 3.4

Objectives
- To identify solutions of equations
- To estimate the values of solutions of equations
- To use inverse operations to solve equations

Vocabulary: solution, variable

Class Planning

Schedule 2–3 days

Resources

In *Chapter 3 Student Activity Masters* Book
Class Openers 37, 38
Extra Practice 39, 40
Cooperative Learning 41, 42
Enrichment 43
Learning Log 44
Using Manipulatives 45, 46
Spanish-Language Summary 47

Class Opener

1. Find the number that goes in each □.
   a. $246 + 492 = \square$; $\square - 246 = 492$
   b. $1269 \div 27 = \square$; $\square \cdot 27 = 1269$
   c. $43^2 = \square$; $\sqrt{\square} = 43$
2. What do you notice about each pair of equations?

Answers:
1. a. 738; 738 b. 47; 47
   c. 1849; 1849
2. Possible answer: Each pair seems to illustrate opposite operations.

Lesson Notes

- You may want to emphasize that the word *variable* has as its stem the word *vary*. Since the concept of a variable is central to algebra, students need to become comfortable with using variables

---

Investigation Connection
Section 3.4 supports Unit B, *Investigations* 2, 3. (See pp. 92A-92D in this book.)

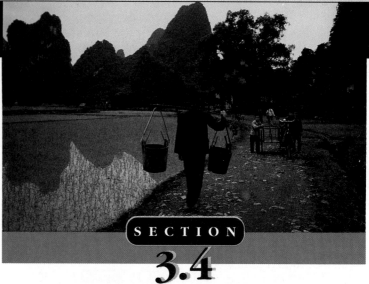

SECTION 3.4
Understanding Equations

SOLVING AN EQUATION

We have solved many problems similar to $\square + 7 = 15$. In this equation, $\square$ is a **variable.** A variable is a symbol that represents a number. If we replace the $\square$ with 8, the equation will be true. By making this replacement, we have solved the equation. This type of equation is more commonly written as $x + 7 = 15$, with solution $x = 8$. In this case, x is the variable.

Example 1
Solve the equation $3x + 7 = 25$, choosing numbers from the set {4, 5, 6}.

Solution
We will substitute each number for the x to determine whether that number makes the equation true.

Try $x = 4$		Try $x = 5$		Try $x = 6$	
Left Side	Right Side	Left Side	Right Side	Left Side	Right Side
$= 3(4) + 7$	$= 25$	$= 3(5) + 7$	$= 25$	$= 3(6) + 7$	$= 25$
$= 19$		$= 22$		$= 25$	
False		False		True	

We say that 6 is a **solution** of the equation.

Example 2
Solve $c = 0.75 + 6(0.40)$ for c.

Solution
$c = 0.75 + 6(0.40)$
$ = 0.75 + 2.40$
$ = 3.15$

```
SINKIN DONUT SHOP
Doughnuts . . . . . . . 40¢ each
Coffee  . . . . . . . . . 75¢
```

Chapter 3 Measurement and Estimation

SECTION 3.4

ESTIMATING SOLUTIONS

One good way to solve an equation is to guess and check—make a guess, check to see if the number is too large or too small, and revise the guess.

Example 3
Solve the equation $x^x = 8$.

Solution
Our goal is to find a number that when raised to itself equals 8. We know that $2^2 = 4$ and $3^3 = 27$, so the number we are looking for must be between 2 and 3.

Let's try $x = 2.4$. Enter 2.4 $\boxed{x^y}$ 2.4 $\boxed{=}$ on your calculator. Since $2.4^{2.4} \approx 8.175$, 2.4 is too large. Let's try 2.3. Since $2.3^{2.3} \approx 6.79$, 2.3 is too small. So the number we are looking for must be between 2.3 and 2.4, probably closer to 2.4. Let's try some more values.

$$2.38^{2.38} \approx 7.875$$
$$2.39^{2.39} \approx 8.0236$$

We now know that x is between 2.38 and 2.39. If we continue this process, we will get an even closer approximation.

INVERSE OPERATIONS

Inverse operations—operations that reverse the effect of each other—are commonly used to solve equations.

- Addition and subtraction are inverse operations.
- Multiplication and division are inverse operations.
- Taking the square root and squaring are inverse operations.

Example 4
Solve the equation $12x = 54$.

Solution
The first thing we may recognize is that the variable is being multiplied by 12. Therefore, to solve the equation, we will use the inverse operation of dividing by 12.

$$12 \cdot x = 54$$
$$x = 54 \div 12$$
$$= 4.5$$

It is always important to check the solution to see if it works. Since $12(4.5) = 54$, $x = 4.5$ is the answer.

---

in different contexts. The following types of applications are particularly important:

Arithmetic word problems: A hamburger costs $1.79. How much will 5 hamburgers cost? 7 hamburgers? 12? 20?

Geometric applications: The area of a circle varies as the radius changes.

Spreadsheet applications: Find the maximum area of a rectangle that has a perimeter of 75 feet.

Expressions: Evaluate $2a + 5b - 3$ for $a = 3$ and $b = 7$.

Equations: Solve for x: $2x - 6 = 9$.

- There are two important ideas in example 3: (1) The guess-and-check method is an important one for helping students understand what a variable is; and (2) some problems can only be solved by numerical analysis methods like guess and check.

- One way to introduce the concept of inverse operations is to discuss some nonmathematical examples, such as screwing in a light bulb and removing it, putting on a shoe and taking it off, and changing a tire. Contrast these with some nonexamples, such as pounding in a nail and removing it (the nail most likely cannot be used again) and peeling an orange.

- Sample problem 1 is an example of an equation with more than one solution.

Additional Examples

Example 1
Solve the equation $4 + 6x = 28$, choosing numbers from the set $\{2, 4, 6\}$.

Answer: $x = 4$

SECTION 3.4

Example 2
Solve $a = 0.50 + 3(1.15)$ for a.
Answer: $a = 3.95$

Example 3
Solve the equation $y^y = 7$.
Answer: y is between 2.31 and 2.32.

Example 4
Solve the equation $8b = 84$.
Answer: $b = 10.5$.

Example 5
The area of the square is 130. Find the length of each side.
Answer: ≈ 11.4

Problem-Set Notes

for page 117

- **2** and **4** (one solution), **3** (many solutions), and **8** (two solutions) are good for discussion of numbers of solutions.

Checkpoint

1. Solve the following equations, choosing from the set $\{1, 2, 5, 6\}$.
   a. $9x + 18 = 36$ b. $5x = 25$
   c. $\frac{1}{3}x + 1 = 3$
2. Which is the best estimate of the solution of
   $\frac{x}{49.08} = 16.72 - 4.82$?
   50, 15, 500, 400
3. Solve each equation for y.
   a. $11.56y = 108.664$
   b. $25.4y = 81.28$
   c. $47y = 94$

Answers:
1. a. 2 b. 5 c. 6
2. 500 3. a. 9.4 b. 3.2 c. 2

Example 5
The area of the square is 72. Find the length, ℓ, of each side.

Solution
Recall that taking the square root of a number is the inverse operation of squaring the number. Therefore, if $\ell^2 = 72$, $\ell = \sqrt{72} \approx 8.5$.

Sample Problems

Problem 1   Solve the equation $x^2 + 15 = 8x$, choosing numbers from the set $\{1, 3, 5, 7\}$.

Solution   We will substitute each number for x to determine whether that number makes the equation a true statement.

Try $x = 1$		Try $x = 3$	
Left Side	**Right Side**	**Left Side**	**Right Side**
$1^2 + 15$	$= 8(1)$	$3^2 + 15$	$= 8(3)$
$= 1 + 15$	$= 8$	$= 9 + 15$	$= 24$
$= 16$		$= 24$	
False		True	

Try $x = 5$		Try $x = 7$	
Left Side	**Right Side**	**Left Side**	**Right Side**
$5^2 + 15$	$= 8(5)$	$7^2 + 15$	$= 8(7)$
$= 25 + 15$	$= 40$	$= 49 + 15$	$= 56$
$= 40$		$= 64$	
True		False	

We can say that both 3 and 5 are solutions of the equation.

Problem 2   Solve $y = 2 + 3x$ for y if $x = 7$.

Solution   If we substitute 7 for x in the equation, we get $y = 2 + 3(7) = 2 + 21 = 23$.

Problem 3   The base of a triangle with area 84 is 20. Find the height, h, of the triangle.

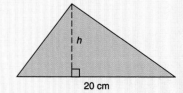

SECTION 3.4

Solution Recall the formula for the area of a triangle.

$$\text{Area} = \tfrac{1}{2} \cdot \text{base} \cdot \text{height}$$
$$84 = \tfrac{1}{2} \cdot 20 \cdot h$$
$$84 = 10 \cdot h$$
$$8.4 = h$$

Since the variable is being multiplied by 10, we used the inverse operation of dividing by 10 to solve the equation. Since $84 = \tfrac{1}{2}(20)(8.4)$, the height of the triangle is 8.4.

Problem 4 Four blocks were put together to form a box. Find length ℓ.

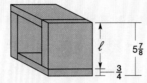

Solution The length of the box can be represented by the equation

$$\ell + \tfrac{3}{4} = 5\tfrac{7}{8}$$

Since we are adding $\tfrac{3}{4}$ to the variable, we use the inverse operation, subtracting $\tfrac{3}{4}$, to solve for ℓ.

$$5\tfrac{7}{8} - \tfrac{3}{4} = 5\tfrac{7}{8} - \tfrac{6}{8} = 5\tfrac{1}{8}$$

Since $5\tfrac{1}{8} + \tfrac{3}{4} = 5\tfrac{7}{8}$, $\ell = 5\tfrac{1}{8}$ is the solution of the equation.

Think and Discuss

P, I **1** Solve the equation $4x + 10 = 38$, choosing numbers from the set {7, 9, 11}. 7

P, I **2** Solve for y if $y = 4 + 6x$ and $x = 3$. 22

P, I **3** List several pairs of values of x and y that make the equation $x + y = 10$ true. Answers will vary. Possible answers: (0,10), (1, 9), (5, 5)

P **In problems 4–7, solve each equation.**

4 $x - 12 = 57$ 69

5 $x + 12 = 57$ 45

6 $\tfrac{x}{12} = 57$ 684

7 $x(12) = 57$ 4.75

P, I **8** Solve the equation $x^2 + 12 = 7x$, choosing numbers from the set {1, 2, 3, 4}. 3, 4

P, R **9** Find the value of w that will make the perimeter of the rectangle 11.2. 1.8

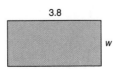

Section 3.4 Understanding Equations

Assignment Guide

Basic
Day 1 10, 13, 15, 18, 27, 28, 31, 36, 41
Day 2 11, 14, 17, 19, 21, 25, 30, 38, 42
Day 3 12, 16, 20, 24, 32–34, 40, 44

Average or Heterogeneous
Day 1 11–17, 22, 23, 26, 27, 34, 38, 39
Day 2 10, 18–21, 29, 32, 33, 36, 37, 40
Day 3 24, 25, 28, 30, 31, 35, 41–44

Honors
Day 1 10, 12, 14, 16, 17, 19, 22, 25, 29, 36, 37, 41
Day 2 11, 13, 15, 18, 20, 24, 27, 30, 32, 34, 38, 40, 44

Strands

Communicating Mathematics	28
Geometry	9, 19, 22, 23, 26, 29, 35, 36, 38–44
Measurement	29, 39
Data Analysis/ Statistics	10, 25
Technology	25
Probability	29
Algebra	1–21, 24, 25, 27, 30–35, 37
Patterns & Functions	25
Number Sense	28
Interdisciplinary Applications	Investigation

Individualizing Instruction

Learning disabled students Students may have difficulty distinguishing between variables and words. They may confuse the variable *a* with the word *a*. Also, students may get caught up in the

SECTION 3.4

Individualizing Instruction continued

fact that some variables stand for words (e.g., *h* for height) while others do not. Emphasize that the letter selected need not relate to any word. We could, in fact, use ℓ to stand for *height*.

Problem-Set Notes

for pages 118 and 119

- **21** can be used to challenge students to find more than the six answers involving whole numbers. Encourage answers involving fractions, decimals, square roots, and negatives.

- **26** can be done by guess and check or by using the inverse operation (i.e., taking the cube root).

- **34** can be done by first substituting 11 for $a + b$ and solving for c.

- **37** can be solved quickly by noting that $(x - x) = 0$, so the entire product is 0. The substitution and computation need not be done.

Reteaching

To reteach how to solve equations by using inverse operations, have students show that the inverse operation is performed on both sides of the equation. In example 4, students should show:
$$12x = 54$$
$$\frac{12x}{12} = \frac{54}{12}$$
$$x = 4.5$$
In example 5, they should show:
$$\ell^2 = 72$$
$$\sqrt{\ell^2} = \sqrt{72}$$
$$\ell = \sqrt{72} \approx 8.5$$

Problems and Applications

P **10** Solve for *a*, *b*, and *c*. $a = 66, b = 9.4, c = 2.6$

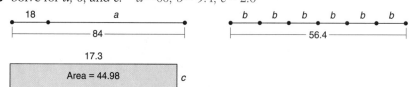

P **In problems 11–16, solve each equation.**

11 $x = 3 + 5(4)$ 23   **12** $x + 387 = 4129$ 3742   **13** $\frac{x}{54} = 96$ 5184

14 $18.4x = 437.92$ 23.8   **15** $x - 23.5 = 90.1$ 113.6   **16** $12.3 = x - 11.2$ 23.5

P, R **17** Find the value of *x* that will make the perimeter 53.4.   9.4

P, I **18** If $x = 3$ and $y = 3x + 4$, solve for *y*.   13

P **19** Find the length of each side of a square that has an area of 40.   ≈6.3

P, I **20** If $x = 7$, $y = 5$, and $z = 16x - 8y$, solve for *z*.   72

P, I **21** Find as many values of *x* and *y* as you can for which $xy = 6$.

I **22** The height of a triangle with area 125 is 5. Find the length of the base of the triangle.   50

P, I **23** The area of a rectangle is 22,215.97 square inches. One side has length 684.20 inches. Find the length of the other side.   32.47 in.

P, I **24** Solve the equations $19 + 8.4 = x$, $x + 13 = y$, and $\frac{y}{4} + \frac{x}{17} = z$ for *x*, *y*, and *z*.
$x = 27.4, y = 40.4, z \approx 11.7$

P, R **25** Spreadsheets Examine the spreadsheet table. For what value(s) of *x* does $x^2 - 10 = 3x$?
−2 and 5

	A	B	C
1	Value of X	X^2−10	3X
2	−5	15	−15
3	−4	6	−12
4	−3	−1	−9
5	−2	−6	−6
6	−1	−9	−3
7	0	−10	0
8	1	−9	3
9	2	−6	6
10	3	−1	9
11	4	6	12
12	5	15	15
13	6	26	18

118   Chapter 3   Measurement and Estimation

SECTION 3.4

26 The volume of a cube is 200. How long is each edge? ≈5.85

27 Solve $5w = w + 40$ for w. 10

28 **Communicating** The grape juice Jamie's little brother spilled on Jamie's homework covered part of this problem. Do you think $x = 1$ is a good estimate for the solution of this problem? Explain your answer. Yes: $5(1) + 19 = 24$

$5x + 19.\ \ \ = 24.$

29 The lengths of two sides of a triangle with a perimeter of 18 centimeters are 4.5 centimeters and 6 centimeters.
   a Find the length of the third side. 7.5 cm
   b Draw the triangle accurately. Check students' drawings.
   c If one angle of the triangle is selected at random, find the probability that it is acute. $\frac{2}{3}$

30 Solve each equation for x.
   a $\frac{1}{x} = \frac{3}{5}$ $\frac{5}{3}$
   b $\frac{1}{x} = 6$ $\frac{1}{6}$
   c $\frac{1}{x} = \frac{1}{3}$ 3

In problems 31 and 32, approximate the solution of each equation to the nearest hundredth.

31 $x^x = 25$ 2.96

32 $8^x = 897$ 3.27

33 Solve for N, M, and W.
   a $78N + 4 = 160$ 2
   b $52M - 8 = 5$ $\frac{1}{4}$
   c $512 - W = 437$ 75

34 Solve the equations $a + b + c = 24$, $a + b = 11$, $a + c = 16$, and $b + c = 21$ for a, b, and c. $a = 3$, $b = 8$, and $c = 13$

35 Find a value of x so that the area of the triangle is
   a 20% of the area of the rectangle 2
   b 30% of the area of the rectangle 6
   c 40% of the area of the rectangle No such value

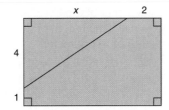

◀ LOOKING BACK **Spiral Learning** LOOKING AHEAD ▶

36 Arrange angles A, B, and C in order, from smallest to largest. B, A, C

37 Evaluate the expression $(v - x)(w - x)(x - x)(y - x)(z - x)$ for $v = 21$, $w = 19$, $x = 17$, $y = 15$, and $z = 13$. 0

Section 3.4 Understanding Equations 119

SECTION 3.4

Communicating Mathematics

- Have students list several ways in which a spreadsheet could be helpful and several ways in which it could be limiting in the process of finding the solution set of an equation.
- Problem 28 encourages communicating mathematics.

Stumbling Block

When solving an equation by testing numbers from a replacement set, students might not test all members of the replacement set. That is, they might stop after finding one solution. Emphasize that all members of the replacement set should be tested, since many types of equations have more than one solution. Sample problem 1 is one example.

Additional Answers

for page 120

39 a 4.2 cm; ≈6.1 cm; ≈6.1 cm
b Check students' drawings.
c ≈20.6 cm

Investigation Answers will vary. Many stores have markups in the 25%–50% range, depending on the type of store and the types of goods sold. Surprisingly, businesses usually express the markup as a percent of the selling price, not as a percent of the cost; so if a store pays $100 for a suit and sells it for $200, the markup is considered to be 50%, not 100%.

120

I **38** A right angle is exactly 90°, an acute angle is less than 90°, and an obtuse angle is between 90° and 180°. Classify each angle as *acute, right,* or *obtuse*.

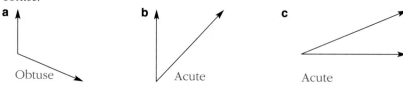

R, I **39** The area of the rectangle is 25.6 square centimeters.
  a Find the lengths of the other three sides.
  b Draw the rectangle accurately.
  c Find the perimeter of the rectangle.

I **40** Every day Jo Ger runs from his house to a circular track. The diameter of this track is 40 yards. He runs around the track 15 times, then back home. All together he runs about 3000 yards. How far is it from his home to the track? 558 yd

R **In problems 41 and 42, solve each triangle for the length of the third side.**

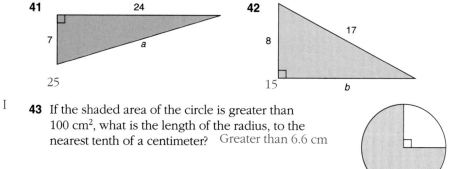

I **43** If the shaded area of the circle is greater than 100 cm², what is the length of the radius, to the nearest tenth of a centimeter? Greater than 6.6 cm

R **44** What percent of 8×10^7 is 3.2×10^6? 4%

Investigation

Merchandising Interview someone who works in a store. Find out what the average percent markup is on merchandise in that store. Give examples of two or three items, stating the cost to the store and the list price. Is the percent markup a percent of the cost or of the list price?

Chapter 3 Measurement and Estimation

SECTION 3.5

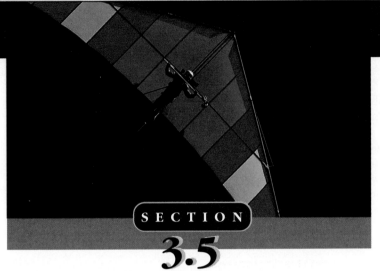

Angle Measurement

Investigation Connection
Section 3.5 supports Unit B, *Investigation* 5. (See pp. 92A–92D in this book.)

Objectives
- To interpret an angle as a rotation
- To name angles
- To use a protractor to measure angles

Vocabulary: angle, complement, complementary angles, degree, final ray, initial ray, protractor, side, supplement, supplementary angles, vertex

Class Planning

Schedule 2 days

Resources

In *Chapter 3 Student Activity Masters* Book
Class Openers 49, 50
Extra Practice 51, 52
Cooperative Learning 53, 54
Enrichment 55
Learning Log 56
Using Manipulatives 57, 58
Spanish-Language Summary 59

In *Tests and Quizzes* Book
Quiz on 3.4–3.5 (Basic) 56
Quiz on 3.4–3.5 (Average) 57

***Transparencies* 18, 19**

ROTATIONS

If you face any direction and turn all the way around in a complete circle so that you are facing the same direction as when you started, you have rotated 360°. If you turn halfway around (a half circle), you have rotated 180°. If you make a quarter turn, you have rotated 90°.

To represent a rotation, we begin with a **vertex,** the center of rotation. The **initial ray** shows the initial direction, and the rotation is represented by the curved arrow. The final position is shown by the **final ray.** Together, the two rays form an **angle.**

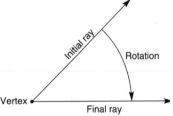

Class Opener

Draw diagrams that meet the given requirements.

1. Two acute angles that, together, make a right angle

2. Two acute angles that together, make an obtuse angle

3. Two obtuse angles that, together, make a right angle

4. Three acute angles that, together, make a straight angle

Answers:
Check students' drawings.

Example 1
What angle is formed by the hands of a clock at 5:00?

Solution
Recall that there are 360° in a circle. The numbers on the clock divide the 360° of the circular clock into twelve 30° intervals (360 ÷ 12 = 30). The rotation could be 150° (that is, 30° × 5) or 210° (that is, 30° × 7). We will agree to the following: When the problem does not say what the direction of the rotation is, we will give the smaller value. Therefore, we say that the hands of a clock form a 150° angle at 5:00.

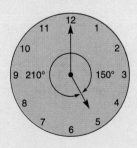

Section 3.5 Angle Measurement 121

SECTION 3.5

Lesson Notes

- It is important to assess your students' abilities to use a protractor correctly. Be careful not to assume that all students have this skill. If some, but not all, of your students already have this skill, group or peer tutoring may be a way to help those that do not.

- If you have access to the computer program Logo, having students move the turtle around the screen reinforces the concept of an angle defined as a rotation.

Additional Examples

Example 1
What angle is formed by the hands of a clock at 8:00?

Answer: Angle of rotation could be 240° or 120°.

Example 2
Name three angles in the diagram.

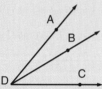

Answer: ∠ADC, ∠ADB, ∠BDC

Example 3
Find the measure of ∠PSR in example 2 on page 122.

Answer: 35°

Multicultural Note

Many cultures, such as the Mayans, Incas, and ancient Egyptians, relied on measuring the angles of the sun and stars to determine dates, seasons, agricultural plans, and festivals. Mayan observatories, which still survive, were often constructed on pyramids.

122

NAMING ANGLES

The symbol ∠ is used to represent the word *angle*. This angle can be named in three different ways. The vertex can be used to name the angle (∠A). Or one point from each ray, or **side,** of the angle can be used along with the vertex (∠BAC or ∠CAB). When three points are used, the vertex is always named in the middle.

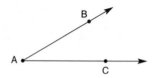

Example 2
Name three angles in the diagram.

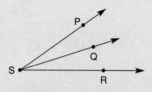

Solution
There are two smaller angles: ∠PSQ and ∠QSR. The largest angle is ∠PSR.

MEASURING ANGLES

A **protractor** is an instrument used to measure angles. Here are some guidelines on how to use a protractor to measure an angle.

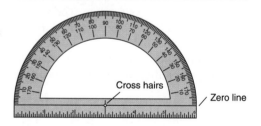

1. Estimate the measure of the angle. Is it acute, right, or obtuse? This is an acute angle.

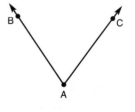

2. Locate the vertex and place the cross hairs on the vertex.

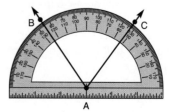

122   Chapter 3   Measurement and Estimation

SECTION 3.5

3. Rotate the protractor, keeping the cross hairs on the vertex, until one of the zero lines aligns with one side of the angle.

4. Read the angle measurement by determining the place where the other side of the angle crosses the row of numbers that corresponds to the zero line. We used the zero line on the right side of the protractor, so we will read the bottom row of numbers. (If we had used the zero line on the left, we would have read the top row of numbers.) Since a protractor measures angles in units called ***degrees,*** we can say that ∠A is a 72° angle. We can also write m∠A = 72, which is read "the measure of ∠A is equal to 72."

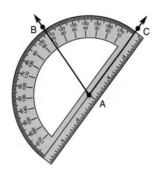

Example 3
Find the measure of each angle.

Solution

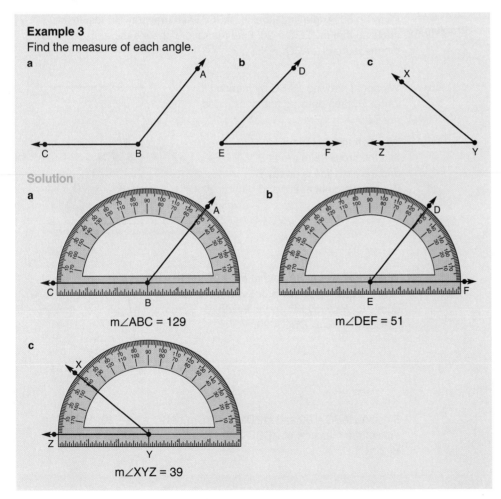

m∠ABC = 129

m∠DEF = 51

m∠XYZ = 39

Checkpoint

1. What angle is formed by the hands of a clock at 3:00?
2. Give two names for an angle with vertex P and rays $\overrightarrow{AP}$ and $\overrightarrow{BP}$.
3. Measure ∠ADB in problem 21 on page 126 with a protractor.

Answers:
1. 90° 2. ∠APB, ∠BPA 3. 40°

Additional Practice

1. Measure ∠BDC and ∠ADC in problem 21 on page 126 with a protractor.
2. What is the complement of each angle?
a. 15° b. 6° c. 71° d. 55°
3. What is the supplement of each angle?
a. 34° b. 161° c. 85° d. 102°

Answers:
1. 64°; 104°
2. a. 75° b. 84° c. 19° d. 35°
3. a. 146° b. 19° c. 95° d. 78°

Section 3.5 Angle Measurement 123

SECTION 3.5

Assignment Guide

Basic
Day 1 12–15, 17–19, 21, 24, 30, 39, 41, 45, 47
Day 2 16, 20, 23, 25–28, 37, 43, 46, 49

Average or Heterogeneous
Day 1 12–17, 21, 24, 27, 28, 30, 33, 37, 40, 47–50
Day 2 18, 20, 22, 23, 25, 26, 32, 34, 36, 38, 39, 41–45, 46b

Honors
Day 1 12, 15, 18, 22, 23, 25, 28, 32, 35, 37, 39, 46
Day 2 13, 16, 21, 24, 26, 27, 30, 31, 33, 34, 36, 40, 45, 50

Strands

Communicating Mathematics	3, 6, 28
Geometry	1–15, 21–25, 29–32, 34–36, 45, 50
Measurement	8–11, 13–20, 24–27, 29, 35
Data Analysis/ Statistics	39
Technology	20, 28
Probability	18
Algebra	24, 28, 30–32, 34, 36, 37, 45, 47–49
Number Sense	3, 6, 33, 38, 50
Interdisciplinary Applications	40

Individualizing Instruction

Physically impaired learners
Students may need assistance reading or manipulating protractors. Once you recognize this difficulty, you might have students work in pairs to facilitate learning protractor skills.

Sample Problems

Problem 1 Two angles with measures that add up to 90 are **complementary angles.** Each of the two angles is called the **complement** of the other. Two angles with measures that add up to 180 are **supplementary angles.** Each of the two angles is called the **supplement** of the other.
  a Of the angles in Example 3, which two are complementary?
  b Of the angles in Example 3, which two are supplementary?

Solution
  a ∠DEF and ∠XYZ are complementary because
     m∠DEF + m∠XYZ = 51 + 39 = 90.
  b ∠ABC and ∠DEF are supplementary because
     m∠ABC + m∠DEF = 129 + 51 = 180.
     Notice that a 180° angle is a straight line.

Problem 2 Draw an 80° angle and label it ∠ABC. Then draw ray BD inside the angle so that m∠DBC = 20. Find the ratio of the measure of ∠DBC to the measure of ∠ABD.

Solution We begin drawing ∠ABC by marking its vertex B, then drawing one of its sides, ray BC.

We align the protractor so that the cross hairs are at B and the zero line is on ray BC. We mark a point at 80° and label it A.

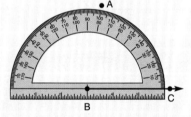

Then we draw ray BA so that m∠ABC = 80. Now we can repeat the process described above to draw ray BD on the inside of ∠ABC so that m∠DBC = 20.

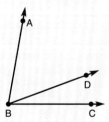

Since m∠ABC = 80 and m∠DBC = 20, m∠ABD = 80 − 20, or 60. The ratio of the measure of ∠DBC to the measure of ∠ABD is therefore $\frac{20}{60}$, or $\frac{1}{3}$.

SECTION 3.5

Think and Discuss

P **1** If your initial direction is south and you rotate 90° to the right, what is your final direction?   West

P, R **2** Is the statement true *always, sometimes,* or *never*?
a An acute angle has a supplement.   Always
b An acute angle has a complement.   Always
c An obtuse angle has a complement.   Never
d An obtuse angle has a supplement.   Always

P **3** Why is it necessary to use three letters to name each angle in Example 2?

P **4** What is another name for each of the three angles in Example 2?

P, I **5** Identify the initial ray and the final ray of this angle.

Either ray can be considered the initial ray. The other is the final ray.

P **6** Name some situations in which it is important to know the measure of an angle.

P **7** If you rotated 120° to your left, then 70° to your right, at what angle, with respect to your original position, would you be facing?   50° to the left.

P, R **In problems 8–11, find the measure of each angle. Then classify the angle as *acute, right,* or *obtuse*. (When you are asked to measure an angle, it is helpful to trace the angle onto another piece of paper and extend the angle's sides.)**

8
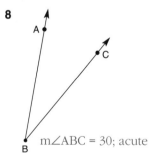
m∠ABC = 30; acute

9
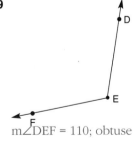
m∠DEF = 110; obtuse

10
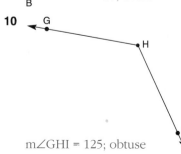
m∠GHI = 125; obtuse

11
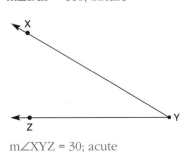
m∠XYZ = 30; acute

Problem-Set Notes

for page 125

■ **1** helps students relate angle measurement to North-South-East-West orientation.

Reteaching

Refer to example 1. Use the following reasoning: Once around the clock is 12 hours. The difference in the positions of the hands is 5 hours. This is $\frac{5}{12}$ of the distance around the clock; $\frac{5}{12} \times 360° = 150°$

Additional Answers

for page 125

3 Possible answer: S is the vertex of each of the angles in the diagram.

4 ∠QSP, ∠RSQ, ∠RSP

5 Initial ray: $\overrightarrow{EF}$; final ray: $\overrightarrow{ED}$; or initial ray: $\overrightarrow{ED}$; final ray: $\overrightarrow{EF}$

6 Possible answers: navigation, architecture, sports

Section 3.5   Angle Measurement

Problems and Applications

P   **12** Find the angle formed by the hands of the clock at 4:00.   120°

P, R   In problems 13–16, find the measure of each angle. Then classify the angle as *acute*, *right*, or *obtuse*.

P   **13** 90; right

14 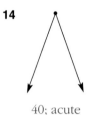 40; acute

15 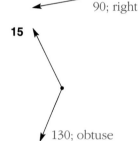 130; obtuse

16  107; obtuse

P   In problems 17–20, draw an angle of the given number of degrees.

17 15°   **18** 105°   **19** 90°   **20** 180°

P   **21** Name three angles in the diagram.
∠ADB, ∠BDC, and ∠ADC
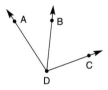

P   **22** If you are facing west and rotate 90° to the left, then 180° to the right, in what direction are you facing?   North

P, R   **23** If m∠A = 64,
     a What is the measure of the supplement of ∠A?   116
     b What is the measure of the complement of ∠A?   26

P, I   **24** Find the value of *n*.
34

P, I   **25** Draw an obtuse angle. Then draw a ray that divides it into two angles with equal measures.   Check students' drawings.

126   Chapter 3   Measurement and Estimation

SECTION 3.5

P, R   **26 a** Find the measure of ∠ABC.   150
            b Draw an angle that is 80% of the measure of ∠ABC.

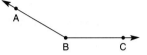

P, R   **27** Draw a triangle with three 60° angles. Check students' drawings.

P, I   **28 Communicating**
            a If $a = 14°$, what is b?   166°
            b If $a = 88°$, what is b?   92°
            c Explain how to find b if you are given a.

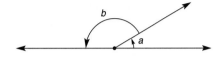

P, R, I   **29** Draw a rectangle with dimensions 7 centimeters by 15 centimeters.

P, I   **30** Find the value of $a + b + c$.
            360°

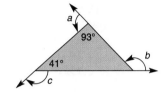

P, I   **31** In triangle ABC, m∠A = 78 and the measure of ∠B is half the measure of ∠A. What is the measure of ∠C?   63

P, I   **32** If m∠ABC = 126, what is the value of x?   74

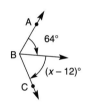

R   **33 Estimating Costs** The blue segments represent pipelines to the cities of Liberty and Vernon. The cost of every two miles of pipeline is $10,000,000.
            a Which pumping station, A, B, or C, will be the least expensive to pump from if costs for both cities are combined?   B
            b Estimate the cost of the pipelines from that station.   About $45,000,000

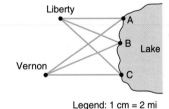

Legend: 1 cm = 2 mi

P, I   **34 a** Find the value of a.   168°
            b Find the value of b.   78°

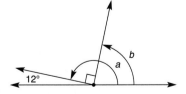

P, I   **35** Draw a square with vertices A, B, C, and D, in that order. Put point E midway between C and D. Connect A to E. Find the measure of ∠EAD.   30

P, R, I   **36** Find the measure of each of the supplementary angles.
            40; 140

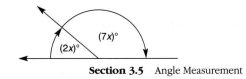

Section 3.5 Angle Measurement

Cooperative Learning

- Have each student in a group draw two triangles of different types. For each triangle, the student should extend the sides to show the exterior angles as in problem 30. Students should then exchange triangles and should measure each of the interior angles and the exterior angles of the triangles they are given.
- After comparing and checking the measurements, the groups should list as many generalizations as possible about the interior and exterior angles of a triangle.
- Answers will vary. Some possible generalizations students may make are: the three interior angles of a triangle add up to 180; the three exterior angles of a triangle add up to 360; each interior angle and its exterior angle add up to 180; each exterior angle equals the sum of the two remote interior angles.
- Problem 28 is appropriate for cooperative learning.

Additional Answers

for pages 126 and 127

17–20 Check students' drawings.

26 b Check students' drawings.

28 c Subtract the measure of the given angle from 180.

29 Check students' drawings.

SECTION 3.5

◀ LOOKING BACK **Spiral Learning** LOOKING AHEAD ▶

R **37** Solve each equation for x.
  a $x + 6 = 91.8$ 85.8
  b $6x = 91.8$ 15.3
  c $x - 6 = 91.8$ 97.8
  d $\frac{x}{6} = 91.8$ 550.8

I **38** Given that $\sqrt[3]{64} = 4$ and $\sqrt[3]{125} = 5$, estimate $\sqrt[3]{80}$ to the nearest thousandth.   4.309

P, I **39** What is the mean (average) measure of the angles of a triangle?   60

R, I **40** The Sunbirds have won 15 of their first 27 games. Find the least number of additional games they need to play and win so that they will have won 75% of their games.   21 games

R **In problems 41–44, write each number in scientific notation.**
  41 0.0392   **42** 41.2   **43** 3156   **44** 0.00002345

R **45** The perimeter of the pentagon is 71. Find the value of x.   13

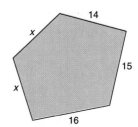

R, I **46** Evaluate each expression.
  a $\frac{1}{2} \cdot \frac{2}{3} \cdot \frac{3}{5} \cdot \frac{5}{7}$ $\frac{1}{7}$
  b $\frac{3}{65} \cdot \frac{92}{7} \cdot \frac{65}{11} \cdot \frac{7}{31} \cdot \frac{31}{92}$ $\frac{3}{11}$

R **In problems 47–49, solve each equation for y.**
  47 $y - 312 = 587$ 899   **48** $\frac{y}{2} = 36$ 72   **49** $100y = 839.7$ 8.397

R, I **50 Estimating** The perimeter of the right triangle is 100 inches. Estimate, to the nearest tenth, the length of the shortest side of the triangle.   27.3 in.

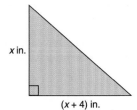

Investigation

Early Civilization Who decided that there were 360° in a circle? Why was that number chosen?

SECTION 3.6

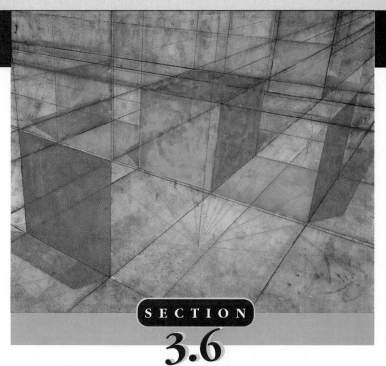

Investigation Connection
Section 3.6 supports Unit B, *Investigation* 5. (See pp. 92A-92D in this book.)

SECTION 3.6
Constructions with Ruler and Protractor

In this section, you will learn how to draw figures accurately with a ruler and a protractor.

Example 1
Draw a triangle with a 65° angle between sides with lengths of 3.7 centimeters and 5.5 centimeters.

Solution
We can begin by drawing one of the sides of the triangle. In this case, we will draw a segment that is 3.7 centimeters long.

Then, using one endpoint of the segment as the vertex, we can use a protractor to construct a 65° angle.

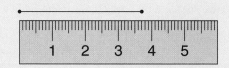

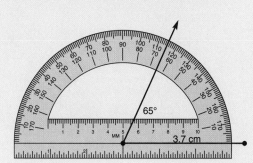

Objective
- To use a ruler and a protractor to make accurate drawings of geometric figures

Class Planning

Schedule 2–3 days
Resources
In *Chapter 3 Student Activity Masters* Book
Class Openers *61, 62*
Extra Practice *63, 64*
Cooperative Learning *65, 66*
Enrichment *67*
Learning Log *68*
Using Manipulatives *69, 70*
Spanish-Language Summary *71*

In *Tests and Quizzes* Book
Quiz on 3.4–3.6 (Honors) *58*

Transparency *20*

Class Opener
Construct a triangle that meets the given requirements.

	sides	angles
1.	2 cm	
	4 cm	
	5 cm	
2.	3 cm	30°
	1 cm	
3.	3 cm	55°
		50°
4.		35°
		65°
		80°

Answers: Students' drawings for **1** should be congruent and for **4**, similar. Drawings for **2** and **3** will vary.

Lesson Notes
- In the opening section, students use simple construction tools to discover some properties of geometric figures, rather than memorizing or being told. This tactile

SECTION 3.6

Lesson Notes continued

learning promotes investigation skills and retention.

- In example 2, the line drawn through points A and B will intersect the two parallel lines and is called a *transversal*. You may introduce this term here.

Additional Examples

Example 1
Draw a triangle with a 40° angle between sides with lengths of 2.4 cm and 4.6 cm.

Answer: See students' diagrams.

Example 2
Label two points A and B. Draw $\overleftrightarrow{AB}$. Construct a line $\overleftrightarrow{CD}$ parallel to $\overleftrightarrow{AB}$.

Answer: See students' diagrams.

Multicultural Note

The Danish mathematician Georg Mohr (1640–1697) showed that any construction that can be performed with a compass and a straightedge can be performed with a compass alone. If one regards a line as known whenever two points on it are known, then a straightedge is not necessary.

Reteaching

In example 2, reteach by having students place the cross hairs at B along $\overrightarrow{AB}$, then mark the 58° angle from B. This method makes use of congruent alternate interior angles, rather than congruent corresponding angles.

Now we can draw a segment 5.5 centimeters long for the other side of the angle. Then if we connect the ends of the segments, the triangle will be complete.

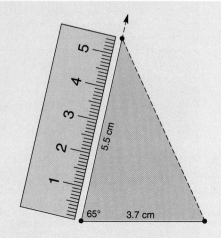

Example 2
Draw two parallel lines—one through point A and one through point B.

Solution
We begin by drawing any line through point A. Then we measure the angle.

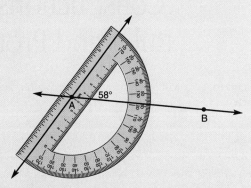

Now we construct the same size angle at the other point and draw the second line. The two lines are parallel. Can you figure out why?

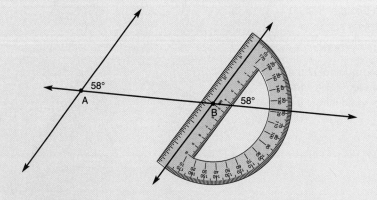

130 **Chapter 3** Measurement and Estimation

SECTION 3.6

Sample Problem

Problem Make an accurate drawing of this parallelogram.

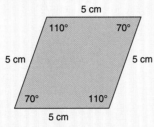

Solution We will begin by drawing one of the sides. In this case, we will draw a horizontal segment 5 centimeters long. Then we will construct a 110° angle at an endpoint of this segment.

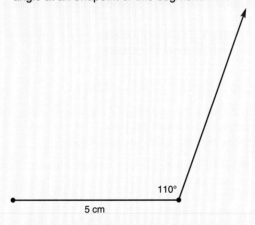

On the side of the angle, we can mark off a segment that is 5 centimeters long. Then we can construct a 70° angle at the other endpoint of the horizontal segment and mark off a 5-centimeter segment on the side of this angle.

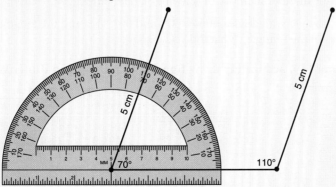

Assignment Guide

Basic
Day 1 10, 13, 17, 19, 24, 27, 31
Day 2 11, 14, 16, 22, 30, 34
Day 3 12, 18, 21, 25, 32, 33

Average or Heterogeneous
Day 1 10, 11, 14, 17, 18, 22, 23, 27, 29, 31
Day 2 12, 13, 15, 16, 19–21, 24, 26, 30, 33, 34

Honors
Day 1 8, 12, 14, 16, 20, 21, 23, 24, 28, 30
Day 2 9, 11, 13, 15, 17, 18, 22, 26, 27, 29, 31, 33, 34, Investigation

Strands

Communicating Mathematics	1, 4–6, 12, 16, Investigation
Geometry	1–18, 21, 24–27, 31, Investigation
Measurement	2, 6–15, 18, 32
Technology	23, 30
Probability	21
Algebra	22, 24–28, 34
Patterns & Functions	29
Number Sense	16, 20, 30

Individualizing Instruction

The kinesthetic learner Students will benefit from using geoboards or sticks of varying lengths to construct geometric figures with specific measures.

SECTION 3.6

Problem-Set Notes

for pages 132 and 133

- **1** has as another answer in that the sum of the interior angles of both is 360°. But while both pairs of opposite sides of the parallelogram are congruent, a trapezoid can have at most one pair of opposite sides congruent.

- **10 c** is a counterexample of the rule that *of* always means "multiply."

- **15** reviews linear units, both English and metric.

Checkpoint

1. Draw triangles with the following angles.
a. 45°, 78°, 57°
b. 34°, 38°, 108°
c. 55°, 50°, 75°

2. Draw a parallelogram with one pair of sides 4.6 cm long and one pair of angles measuring 115°.

Answers: **1. a–c.** Check students' drawings. **2.** Check students' drawings.

Additional Practice

1. Draw triangles with the following angles or sides.
a. 45°, 45°, 90°
b. 3 cm, 4 cm, 5 cm
c. 30°, 60°, 90°
d. 5 cm, 7 cm, 9 cm
e. 59°, 61°, 60°

2. Draw parallelograms with the following side and angle measures.
a. 4 cm, 2 cm; 90°, 90°, 90°, 90° b. 3 cm, 1.5 cm; 55°, 55°, 125°, 125°

Answers: **1. a–e.** Check students' drawings.
2. a. and **b.** Check students' drawings.

132

Finally, we connect the endpoints of the two segments to complete the parallelogram.

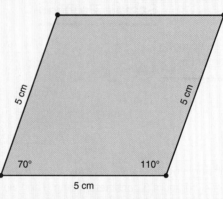

Think and Discuss

P, I **1** How is a trapezoid like a parallelogram? How is it different?

P, I **2** Draw a parallelogram with a 4-centimeter side, a 7-centimeter side, and a 60° angle between these sides. Draw a second parallelogram with the same dimensions. How do the two parallelograms compare?

P **3** Draw a trapezoid ABCD. Be sure $\overline{AB}$ and $\overline{CD}$ are parallel. Measure the four angles. What did you discover about
a ∠A and ∠D? *They are supplementary.*
b ∠C and ∠B? *They are supplementary.*

P **4** Explain how to draw a circle with a radius that is 6 centimeters long.

P **5** A method for drawing parallel lines is given in Example 2.
a Explain why this method works.
b Describe another method for drawing parallel lines.

P, I **6** Draw a triangle with angles of 50°, 50°, and 80°. Measure the lengths of the sides of the triangle. What did you discover?

P **In problems 7–9, accurately draw figures with the given dimensions.**

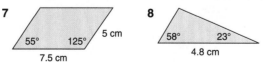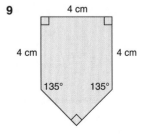

In problems 7–9, check students' drawings.

132 Chapter 3 Measurement and Estimation

SECTION 3.6

Problems and Applications

P, R 10 Draw an angle congruent to the one shown.
Use a protractor to construct
  a Its complement   55°
  b Its supplement   145°
  c The supplement of its complement   125°
  In parts **a–c**, check students' drawings.

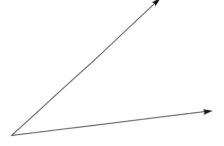

P 11 Construct a parallelogram with four congruent angles and with sides of 2.3 centimeters and 4.9 centimeters.   Check students' drawings.

P, I 12 At least two sides of an isosceles triangle are equal in length. Draw an isosceles triangle and measure its three angles. What did you discover?

P, I 13 Accurately draw this figure using the given dimensions. Is each statement *true* or *false?* (≅ means "is congruent to.")
  a $\overline{AC} \cong \overline{BD}$ True
  b $\angle A \cong \angle B$ True

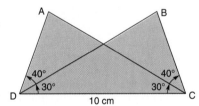

P 14 Make an accurate drawing of the figure, using the given dimensions. How many times longer is $\overline{AC}$ than $\overline{AB}$?   2.5 times longer

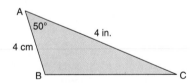

P, R 15 Draw a rectangle that has a length of 5 inches and a width of 12.7 centimeters.
  a Find the ratio of the rectangle's length to its width.   1:1
  b What is a special name for this kind of rectangle?   Square

P, R 16 Communicating The area of the rectangle is 5050 square inches. The width is 68.3 inches. When the teacher asked for the length, Hal calculated that it is about 74 inches. Damon said it is 73.938506558 inches. With whom do you agree? Explain why.

P, R 17 Find the ratio of the complement of a 70° angle to the supplement of a 70° angle. $\frac{2}{11}$

P, R 18 A stick 24 centimeters long is to be cut into four pieces to form a rectangle. The lengths of the sides are to be whole numbers. Draw and label the lengths of the sides of all such rectangles. Find the area of each rectangle.

Additional Answers

for pages 132 and 133

1 Possible answer: Both have four sides. Only one pair of opposite sides of a trapezoid are parallel.

2 Possible answer: Parallelograms have the same area.

4 Possible answer: Use a compass opened to a radius of 6 cm.

5 a Answers will vary. The method is based on the geometric theorem that states that if two lines are cut by a transversal so that the corresponding angles are congruent, the lines are parallel.
b Answers will vary. Students may use methods based on congruent alternate interior angles, congruent alternate exterior angles, or perpendicular lines.

6 Possible answer: The sides opposite the 50° angles are the same length.

12 At least two of the angles are equal in measure.

16 Hal; Damon's answer is too precise for the information given in the problem.

18 1 cm by 11 cm, 11 cm²; 2 cm by 10 cm, 20 cm²; 3 cm by 9 cm, 27 cm²; 4 cm by 8 cm, 32 cm²; 5 cm by 7 cm, 35 cm²; 6 cm by 6 cm, 36 cm²

SECTION 3.6

◀ LOOKING BACK **Spiral Learning** LOOKING AHEAD ▶

R **19** Evaluate each expression.
  a $\sqrt{16+9}$ 5
  b $\sqrt{16} + \sqrt{9}$ 7

R **20** **Estimating** Sue buys four cans of soup at 89¢ per can, one loaf of bread at 97¢ per loaf, and three candy bars at 37¢ each.
  a Estimate whether $5 is enough to pay for her purchase. No
  b If there is a 5% sales tax, will $6 be enough? Yes

P, R, I **21** A circular region is divided into eight equal parts.
  a Find the ratio of the shaded area to the total area. $\frac{3}{8}$
  b What percent of the circular region is shaded? 37.5%
  c If a point is chosen at random inside the circle, what is the probability that it is in the shaded area? $\frac{3}{8}$

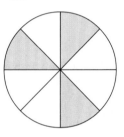

I **22** Solve each equation for n.
  a $\frac{2}{5} = \frac{n}{40}$ 16
  b $\frac{3}{11} = \frac{15}{n}$ 55
  c $\frac{n}{68} = \frac{8}{17}$ 32
  d $\frac{n}{17} = \frac{25}{85}$ 5

R **23** **Spreadsheets** A radiator contains four gallons of a mixture of water and antifreeze.
  a If the mixture is 50% antifreeze, how many gallons are water and how many are antifreeze? 2; 2
  b If the mixture is 60% antifreeze, how many gallons are water and how many are antifreeze? 1.6; 2.4
  c Use a spreadsheet to make a table showing how many gallons of antifreeze and water are in the radiator for mixtures containing from 40% to 100% antifreeze. Use intervals of 10% and a total amount of four gallons.

In problems 24–27, find the length of the third side of each triangle.

R **24**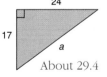
About 29.4

25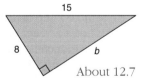
About 12.7

26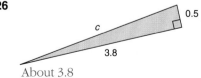
About 3.8

27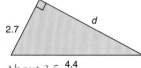
About 3.5

134 **Chapter 3** Measurement and Estimation

SECTION 3.6

R, I **28** Tom set his cruise control at 57 miles per hour when he got on the interstate at 1:00 P.M. By 3:15 P.M. he had reached only the halfway point of his trip. How fast would he have to go to reach his destination by 5:00 P.M.?   73 mph

R, I **29** Continue the pattern.
$$\left(\tfrac{1}{2}, \tfrac{1}{8}\right) \rightarrow \tfrac{3}{8}$$
$$\left(\tfrac{5}{12}, \tfrac{1}{4}\right) \rightarrow \tfrac{1}{6}$$
$$\left(\tfrac{2}{3}, \tfrac{1}{6}\right) \rightarrow \tfrac{1}{2}$$
$$\left(\tfrac{4}{5}, \tfrac{3}{10}\right) \rightarrow \tfrac{1}{2}$$

R **30** **Spreadsheets** Using a spreadsheet, enter the formula @SQRT(A1^2+B1^2) in cell C1.
  a Try entering the values (3, 4), (5, 12), and (7, 24) in cells (A1, B1). What appears in cell C1 for each pair?
  b What does the formula do? What might A1, B1, and C1 represent?
  c What formula could you use to have the spreadsheet find A1 in terms of B1 and C1?

R **31** Is the statement true *always, sometimes,* or *never?*
  a The supplement of an obtuse angle is acute.   Always
  b The complement of an obtuse angle is acute.   Never
  c The supplement of an acute angle is acute.   Never
  d The supplement of an acute angle is obtuse.   Always
  e The complement of an acute angle is acute.   Always

R, I **32** Is each statement *true* or *false?*
  a 4 inches = 4 feet   False
  b 3 feet + 1 foot = 4 feet   True

I **33** Write each number in lowest terms.
  a $\tfrac{15}{20}$ $\tfrac{3}{4}$
  b $\tfrac{25}{20}$ $\tfrac{5}{4}$
  c $\tfrac{9}{10}$ $\tfrac{9}{10}$

R, I **34** List several pairs of values of x and y for which the equation $3x + 4y = 48$ is true.   Answers will vary. Possible answers: $x = 16, y = 0$; $x = 0, y = 12$; $x = 12, y = 3$

Investigation

Geometry In classical geometry constructions, what are the only two tools you are permitted to use? How is a straightedge different from a ruler? Learn how to bisect an angle by using a classical geometry construction. Demonstrate this construction to the class.

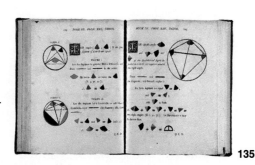

SECTION 3.7

Objective
- To discover some of the geometric properties of a figure

Vocabulary: convex, equiangular, equilateral

Class Planning

Schedule 2 days
Resources
In *Chapter 3 Student Activity Masters* Book
Class Openers 73, 74
Extra Practice 75, 76
Cooperative Learning 77, 78
Enrichment 79
Learning Log 80
Technology 81, 82
Spanish-Language Summary 83

Transparencies 17 a,b, 21

Class Opener

1. Draw a quadrilateral.
2. Measure each interior angle.
3. Find the sum of the four angles.

Answers:
1. Check students' drawings.
2. Answers will vary.
3. All sums should be about 360.

Lesson Notes

- In example 1, more triangles can be tested if your class uses geometry software, like Geometric Supposer.

- In example 2, students might measure the angles in a hexagon and obtain a sum of 720° before noting that the hexagon can be divided into four triangles.

- In the sample problem, other observations that may be noted include: the sum of the angles is 360°; the area is equal to the base

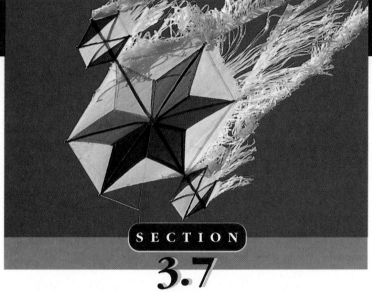

SECTION 3.7
Geometric Properties

Draw a triangle, then tear off the corners of the triangle and rearrange them as shown below.

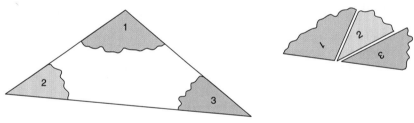

What did you discover?

Example 1
What is true about the sum of the measures of the angles of any triangle?

Solution
Let's draw some triangles and look for a pattern.

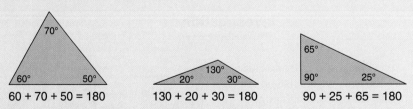

60 + 70 + 50 = 180   130 + 20 + 30 = 180   90 + 25 + 65 = 180

The sum of the measures of the angles in each triangle is 180. Do you think that this means that the sum of the measures of the angles of *any* triangle is 180°?

136   Chapter 3   Measurement and Estimation

SECTION 3.7

Example 2
What is the sum of the measures of the angles of a hexagon?

Solution
The diagonals drawn from one vertex of a hexagon divide the hexagon into four triangles. Therefore, the sum of the measures of the angles of the hexagon is 4 × 180, or 720.

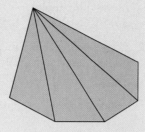

Example 3
An *equilateral* triangle has all sides congruent. What is true about the angles of an equilateral triangle?

Solution
If we measure the angles of several equilateral triangles, we find in each triangle that each angle has a measure of 60.

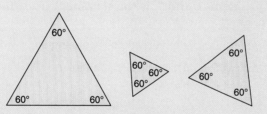

Figures like this, in which all angles are congruent, are called *equiangular*.

Sample Problem

Problem Refer to the diagrams. What conclusions can you draw about parallelograms?

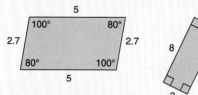 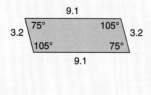

Solution Here are some possible conclusions:
- Opposite sides of a parallelogram are congruent.
- Opposite angles of a parallelogram have equal measures.
- Consecutive angles of a parallelogram are supplementary.

Can you draw any more conclusions?

---

times the height; and the diagonals bisect each other.

Additional Examples

Example 1
What is true about the sum of the measures of the angles of any quadrilateral?

Answer: The sum of the measures of the angles is 360°.

Example 2
What is the sum of the measures of the angles of an octagon?

Answer: 1080°

Example 3
A square has all sides congruent. What is true about the angles of a square?

Answer: Each angle measures 90°, and all angles are congruent.

Reteaching

To reteach equilateral versus equiangular, give students the following examples:

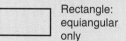 Square: equilateral and equiangular

Rectangle: equiangular only

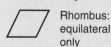 Rhombus: equilateral only

Students should see that having congruent sides does not guarantee having congruent angles, and vice versa.

SECTION 3.7

Assignment Guide

Basic
Day 1 8–10, 17, 20, 22, 25, 31
Day 2 11, 13–15, 21, 23, 28, 34

Average or Heterogeneous
Day 1 8–12, 14–16, 18, 22, 23, 25, 27
Day 2 13, 17, 19–21, 24, 28, 29, 31, 34

Honors
Day 1 8–13, 15, 17, 20, 22, 24–26, 31, 34
Day 2 14, 16, 18, 19, 21, 23, 27–29, 32, 33, Investigation

Strands

Communicating Mathematics	14, 30
Geometry	1–21, 25, 29–31, 34, Investigation
Measurement	13, 14, 22, 24, 34, Investigation
Data Analysis/ Statistics	2, 16
Technology	18
Probability	33
Algebra	19, 26, 31–33
Patterns & Functions	2
Number Sense	24, 25, Investigation

Individualizing Instruction

Limited English proficiency
Students may need help with terms like *equilateral* and *equiangular*. Use several examples when explaining these terms.

138

Think and Discuss

P **1** The measures of two angles of a triangle are given. What is the measure of the third angle?
  a 90, 45   45   **b** 23, 72   85   **c** 106, 41   33   **d** 90, 53   37

P, R **2** Copy and complete the chart by filling in the sum of the measures of the interior angles of a polygon with the given number of sides.

Sides	3	4	5	6	7	8	9	10	11
Sum of angles	180	360	540	720	900	1080	1260	1440	1620

P **In problems 3–5, find the sum of the measures of the angles of each polygon.**

  3 Quadrilateral   360   **4** Pentagon   540   **5** Octagon   1080

P **In problems 6 and 7, is the statement true *always, sometimes,* or *never*?**

  6 An equilateral quadrilateral is equiangular.   Sometimes

  7 An equilateral triangle is equiangular.   Always

Problems and Applications

P, R **In problems 8–11, is the statement true *always, sometimes,* or *never*?**

  8 Two angles of a right triangle are acute.   Always

  9 If one angle of a triangle is obtuse, then the other two angles must be acute.   Always

  10 Two angles of a triangle are complementary.   Sometimes

  11 Two angles of a triangle are supplementary.   Never

P **12** The sides of the drawbridge rise to a 60° angle from horizontal. If these two sides are extended, what angle would be formed where they meet?   60°

P, I **13** Draw △ABC so that m∠A = 55, m∠B = 65, and m∠C = 60.
  a Which side is the longest? $\overline{AC}$ **b** Which side is the shortest? $\overline{BC}$

138   Chapter 3   Measurement and Estimation

SECTION 3.7

P **14 Communicating** Draw several different isosceles triangles and measure their angles. What did you discover? There are always at least two congruent angles.

P, R **15** Is the statement *true* or *false?*
  a The diagonals of a rectangle are congruent. True
  b The diagonals of any parallelogram are congruent. False
  c A square is the only quadrilateral with four congruent sides. False
  d Opposite angles of a parallelogram are congruent. True
  e Each angle of a regular pentagon is 110°. False

P, R **16** What is the mean (average) measure of the angles of a pentagon? 108

P, R, I **17** Find the number of degrees in ∠1, ∠2, and ∠3.
∠1: 77; ∠2: 68; ∠3: 33

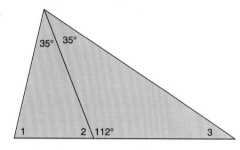

P, R **18 Spreadsheets**
  a Use a spreadsheet to make a two-column table—one column with the numbers of sides of various polygons, the other column with the sum of the measures of the interior angles of each polygon.
  b What is the sum of the measures of the interior angles of a 15-sided polygon? 2340

R, I **19** Lines *a* and *b* are parallel. What are the measures of the eight acute and obtuse angles in this figure?

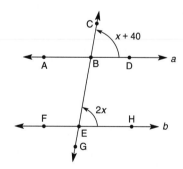

I **20** A polygon is **convex** if any two points in the interior of the polygon can be connected by a segment that lies entirely in the polygon's interior. Which of these polygons are convex? **a** and **c**

  a   **b**   **c**  **d**

P, I **21** Draw two line segments that have different lengths and divide each other in half. Connect the endpoints of the segments to make a quadrilateral. What kind of quadrilateral is it? Parallelogram

Cooperative Learning

- Have students work together to complete the chart in problem 2. A modified version of the chart may be helpful.

Sides	3	4	5	...	n
No. of Triangles	1	2	3		
Sum of Angles	180	360	540		

- Have students work together to develop a formula for finding the sum of the measures of the interior angles of a polygon with *n* sides.

Answer: $S_n = 180(n-2)$

- Problem 18 is appropriate for cooperative learning.

Multicultural Note

In the United States, the shape of a stop sign is an octagon that is equiangular and equilateral. Do stop signs in other countries also have this shape? Students from other countries may be able to answer this question.

Additional Answers

for pages 138 and 139

18 a

No. of Sides	Sum of Int. Angles	No. of Sides	Sum of Int. Angles
3	180	10	1440
4	360	11	1620
5	540	12	1800
6	720	13	1980
7	900	14	2160
8	1080	15	2340
9	1260		

19 m∠CBA = 100; m∠CBD = 80; m∠ABE = 80; m∠DBE = 100; m∠FEB = 100; m∠BEH = 80; m∠FEG = 80; m∠HEG = 100

SECTION 3.7

Problem-Set Notes

for pages 140 and 141

- **24** illustrates the difference between density and weight.

- **28** can be solved by the Pythagorean Theorem.

- **30** can be solved in a number of ways. You might want to keep track of these on an ongoing basis either in journals or on a bulletin board.

- **32** has many pairs of numbers that are correct. Encourage students to find fractions, negative numbers, and decimal answers, as well as whole number answer pairs.

- **33** should be solved by trial and error, not by an algebraic solution of the quadratic equation.

Checkpoint

Is each statement *true* or *false*?
1. Isosceles triangles are always equilateral.
2. Parallelograms have opposite angles congruent.
3. The diagonals of any rectangle are congruent.
4. The diagonals of any quadrilateral are congruent.
5. The average of the measures of the angles in a hexagon is 120°.
6. Two angles of a right triangle are always acute.

Answers: 1. false 2. true
3. true 4. false 5. true 6. true

Communicating Mathematics

- Have students write a few sentences explaining why an obtuse triangle may have only one obtuse angle in it.

◀ LOOKING BACK **Spiral Learning** LOOKING AHEAD ▶

I   **22** Match each item in the first column with appropriate items in the other two columns.

A	2.54 centimeters	I	1 ton	1	Volume	A, IV, 2
B	144 square inches	II	1 cup	2	Length	B, III, 4
C	2000 pounds	III	1 square foot	3	Weight	C, I, 3
D	250 milliliters	IV	1 inch	4	Area	D, II, 1

R   **23** Evaluate each expression.

     a $\frac{3}{5} \cdot \frac{5}{7} \cdot \frac{7}{11} \cdot \frac{11}{23} \cdot \frac{23}{41}$ $\frac{3}{41}$ **b** $4 \cdot \frac{3}{4} \cdot \frac{2}{3} \cdot \frac{7}{2}$ 7

R, I   **24** Jack loosely packed 8 ounces of feathers inside a box with a volume of 1 cubic foot (each edge was 1 foot long). Portia took the feathers and compressed them to fit into a cubical box with a volume of 27 cubic inches.
     a How much do the feathers weigh in the smaller box?   8 oz
     b How many of the smaller boxes can fit into the larger box?   64

I   **25** Draw each triangle and find its area. Which triangle has the greater area?
The areas are the same.

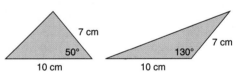

R 🖩   **26** Solve each equation.
     a $(0.78)x = 94$ About 120.51   **b** $\frac{2}{3} + x = \frac{8}{9}$ $\frac{2}{9}$
     c $x^2 = 75$ About 8.7   **d** $x^x = 900$ About 4.51

I   **27** Use the diagram to find each ratio.
     a $\frac{BN}{NC}$ $\frac{2}{3}$ **b** $\frac{BT}{TM}$ $\frac{2}{3}$

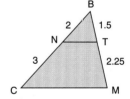

R   **28** A large tree was struck by lightning and was broken as shown. About how tall was the tree before it was broken? ≈43.9 ft

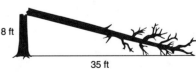

I   **29**   **a** Find the ratio of m∠ABD to m∠DBC. $\frac{3}{2}$
     b Find the ratio of m∠ABD to m∠ABC. $\frac{3}{5}$
     c What percent of m∠DBC is m∠ABD?   150%
     d What percent of m∠ABC is m∠ABD?   60%

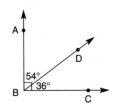

R   **30** **Communicating** Explain how to find the center of a circle by using a ruler and protractor.

140   **Chapter 3**   Measurement and Estimation

SECTION 3.7

R, I **31** Determine the value of x.
15

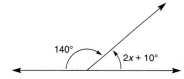

I **32** **a** Find some values of x and y for which $5x = 3y$.
  b For each pair of values, compute $\frac{x}{y}$. $\frac{3}{5}$

R, I **33** If x is chosen from the set $\{0, 1, 2, 3, 4, 5\}$, what is the probability that x is a solution of $x^2 + 15 = 8x$? $\frac{1}{3}$

R **34** Draw a triangle with sides 5 centimeters long, 7 centimeters long, and 8 centimeters long. Find the area of this triangle. $\approx 17.3 \text{ cm}^2$

Investigation

Geometry Determine whether the three given lengths can be the lengths of the sides of a triangle. If they can, draw the triangle as accurately as possible. If they cannot, explain why not.

a 5 in., 8 in., 7 in.
b 10 cm, 20 cm, 8 cm
c 2 in., 5 in., 3 in.
d 220 mm, 70 mm, 190 mm

State a general rule for determining whether three lengths can be the lengths of the sides of a triangle.

SUMMARY

CHAPTER 3

Summary

CONCEPTS AND PROCEDURES

After studying this chapter, you should be able to
- Measure lengths, using both English and metric units (3.1)
- Recognize various polygons and investigate linear measurements associated with them (3.1)
- Use scientific notation to express large and small numbers (3.2)
- Recognize that different situations require different degrees of accuracy in measurement (3.3)
- Round a number to a specified number of decimal places (3.3)
- Express results to a reasonable degree of accuracy (3.3)
- Identify solutions of equations (3.4)
- Estimate the values of solutions of equations (3.4)
- Use inverse operations to solve equations (3.4)
- Interpret an angle as a rotation (3.5)
- Name angles (3.5)
- Use a protractor to measure angles (3.5)
- Use a ruler and a protractor to make accurate drawings of geometric figures (3.6)
- Discover some of the geometric properties of a figure (3.7)

VOCABULARY

acute angle (3.3)	obtuse angle (3.3)
angle (3.5)	polygon (3.1)
complement (3.5)	protractor (3.5)
complementary angles (3.5)	regular polygon (3.1)
congruent (3.1)	right angle (3.3)
convex (3.7)	scientific notation (3.2)
degree (3.5)	side (3.1, 3.5)
diagonal (3.1)	solution (3.4)
equiangular (3.7)	supplement (3.5)
equilateral (3.7)	supplementary angles (3.5)
final ray (3.5)	variable (3.4)
initial ray (3.5)	vertex (3.1, 3.5)
linear (3.1)	

CHAPTER 3

Review

1 Solve each equation for x.
  a $18 + x = 58.5$ 40.5
  b $18x = 58.5$ 3.25
  c $x - 18 = 58.5$ 76.5
  d $\frac{x}{18} = 58.5$ 1053

In problems 2 and 3, express each number in scientific notation.

2 743,617 7.43617×10^5

3 0.0041758 4.1758×10^{-3}

4 Construct a parallelogram with each side 4.7 centimeters long and with at least one 52° angle.   Check students' drawings.

5 Find the length of $\overline{BC}$.   15

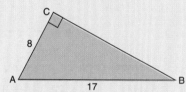

In problems 6 and 7, express each number in standard form.

6 3.86×10^{-2} 0.0386

7 6.44×10^5 644,000

8 Is $x = 4$ a solution of the equation?
  a $x^2 = 3x + 4$ Yes
  b $2x + x^2 = 6x$ Yes
  c $x^x = 16x$ No
  d $x^x = 256$ Yes

9 List some properties of a parallelogram.

10 What percent of the circular region is in the shaded sector?   5.6%

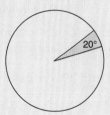

11 The measures of three angles of a pentagon are 60, 80, and 100. What is the sum of the measures of the other angles?   300

12 What percent of 6.5×10^7 is 3.25×10^6?   5%

In problems 13–16, estimate the value of each expression.

13 $(4.1)(2.9)$ ≈ 12

14 $3.01^{1.99}$ ≈ 9

15 $\frac{512}{10}$ ≈ 50

16 $788 \div 19$ ≈ 40

Assignment Guide

Basic
Day 1   1–6, 9, 13, 19, 22, 26, 28, 34, 38
Day 2   Chapter Test 1–12

Average or Heterogeneous
Day 1   1–3, 5, 6, 9, 10, 12, 17, 19–22, 26, 29, 31, 37, 40, 42

Honors
Day 1   1–12, 14, 15, 17, 19–29, 31, 33, 35, 37–41

Strands

Communicating Mathematics	33, 42
Geometry	4, 5, 9–11, 17, 18, 20, 26, 31–33, 35–37, 39, 41, 42
Measurement	4, 18, 27, 30, 32, 33, 36, 39, 42
Algebra	1, 8, 19, 28, 29, 31, 34
Number Sense	2, 3, 6, 7, 13–16, 38, 42

Additional Answers

for page 143

9 Possible answers: Opposite sides of a parallelogram are congruent. Opposite angles of a parallelogram are congruent. Consecutive angles of a parallelogram are supplementary. Opposite sides of a parallelogram are parallel.

REVIEW

17 Express the volume of the box in scientific notation. 9.3×10^8 cm^3

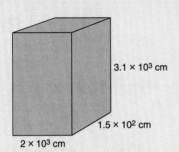

18 Draw three different triangles, each with a perimeter of 18 centimeters.

19 Solve for y if $x = 8$ and $y = 3x - 2$. 22

20 What are the measures of ∠A and ∠BCD? 30; 125

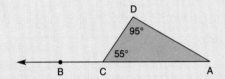

In problems 21–23, evaluate each expression.

21 $(10^6)(10^3)(10^{-4})$ 10^5

22 $10^4 + 10^3 + 10^2 + 10^1$ 11,110

23 $10^{-3} + 10^{-2} + 10^{-1}$ 0.111

In problems 24 and 25, evaluate each expression.

24 60% of 50% of 40% of $250 $30

25 12% of $250 $30

26 Draw a line ℓ and locate a point P that is not on the line. Use a ruler and protractor to construct a line through point P parallel to line ℓ. Check students' drawings.

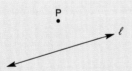

27 Give some examples of situations in which an approximation is preferable to an exact answer.

In problems 28 and 29, estimate the solution of each equation.

28 $x^2 = 18$ ≈4.2

29 $x^x = 200$ ≈3.9

30 Measure the top and the side of a piece of paper in centimeters. Find the ratio of the length of the top to the length of the side. Repeat the same procedure, using inches. What did you discover? The ratios are the same.

31 Find the value of x if m∠ABC = 60. 15

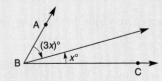

144 Chapter 3 Measurement and Estimation

REVIEW

32 The length of one side of a triangle is 4.8 centimeters, the length of another side is 5.9 centimeters, and the perimeter of the triangle is 17 centimeters. Draw the triangle. Check students' drawings.

33 Draw a parallelogram in which the measures of one pair of opposite angles are 60. Measure the other angles. What did you discover?

34 Solve each equation for x.
 a $x + 7 = 10$ 3
 b $x + 10 = 7$ –3

35 Use the figure to compute each ratio.
 a $\dfrac{a}{d}$ $\dfrac{1}{2}$
 b $\dfrac{c}{f}$ $\dfrac{1}{2}$

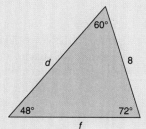

36 Draw a quadrilateral, other than a rectangle, with each side 2 inches long. Find the area of that quadrilateral.

37 Find the measure of each angle.
 a $\angle AOB$ 45
 b $\angle COD$ 22.5
 c $\angle DOB$ 135

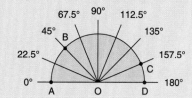

38 Which is greater, 0.036 or 3.6×10^{-3}?   0.036

39 Draw a line and pick a point on that line. Draw a 42° angle, using the point as its vertex. What is the measure of the obtuse angle in your diagram?   138

40 What is the ratio of the length of a shoelace that is 24 inches long to the length of a belt that is 3 feet long? $\dfrac{2}{3}$

41 If r represents the radius of the circle and A represents its area, then $A = \pi r^2$, where $\pi \approx 3.1416$.
 a If $r = 7$, what is A?   ≈153.9384
 b If $r = 3.27$, what is A?   ≈33.59
 c If $A = 153.86$, what is r?   ≈6.998217242

42 Draw two line segments, each 4 inches long, that intersect at their midpoints. Connect the endpoints to form a quadrilateral.
 a Find the perimeter of the quadrilateral.
 b How can you do this problem so that the perimeter would be less or greater?   Change the measure of the angles between the segments.

Chapter Review

Additional Answers

for pages 144 and 145

18 Check students' drawings.

26 Students should follow the procedure described in Section 3.6.

27 Possible answer: The 1990 census (an exact count would be too expensive)

33 The measure of the other pair of opposite angles is 120°.

36 Answers will vary depending on the measure of the angles between the adjacent sides.

42 a Answers will vary depending on the measure of the angle drawn between the two segments.

CHAPTER 3

Test

1. Use a protractor to draw a triangle with angles of 27°, 68°, and 85°.

In problems 2–4, rewrite each number in scientific notation.

2. 23,450 2.3450×10^4
3. 103.45 1.0345×10^2
4. 0.00791 7.91×10^{-3}

5. Carefully trace a copy of the parallelogram shown. On the copy, label each angle with its measure and each side with its length. What do you notice about the angles and sides?   Opposite angles are congruent; opposite sides are congruent.

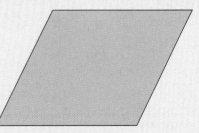

6. Solve the equation $\frac{4}{5} = \frac{x}{40}$ for x. Explain each step of your solution.

7. If you are facing east and you turn 90° to the left, 120° to the right, and 75° to the left, in what direction will you be facing?   Northeast

8. If the perimeter of the triangle is $26\frac{1}{4}$ feet, what is the value of p?   13

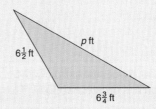

9. Which is shorter,
   a. 30 yards or 100 feet?   30 yd
   b. $\frac{1}{2}$ mile or 100,000 inches? $\frac{1}{2}$ mi
   c. 3.5×10^4 centimeters or 5.1×10^{-1} kilometer? 3.5×10^4 cm

10. Find the measures of ∠BCA and ∠BCD.
    35; 145

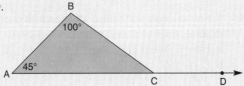

11. Estimate the solution of each equation.
    a. $y^2 = 240$ ≈15.5
    b. $2^y = 240$ ≈7.9

12. Rachel wants to draw a rectangle that has an area of 0.5 square meters. She starts by using a meterstick to draw a segment 80 centimeters long. How long should the other three sides of the rectangle be?   80 cm, 62.5 cm, 62.5 cm

Chapter 3   Measurement and Estimation

PUZZLES AND CHALLENGES

Draw each figure without lifting your pencil from the paper and without tracing any line twice.

1

2

3

How many triangles are in each figure?

4

5

6 At the firm Dewey, Cheatem, and Howe, Inc., someone has been embezzling money. Each of the three executives made a statement to the police. Exactly one of them lied. Given the statements below, who was the guilty party?

  Dewey: Cheatem did it.
  Cheatem: Howe did it.
  Howe: Cheatem lied when he said
     I did it.

for other types of signs? Look up the traffic signs from one other country. Compare how this country uses polygons for its signs with how the United States uses them.

8. Extend problem 45 on page 105 as follows: Use a spreadsheet to make a table that shows the least number of pounds of body fat that a wrestler could have and still be allowed to wrestle. Have the weights range from 100–180 lb. Find out why many states have rules about the percent of body fat in wrestlers.

9. Use a ruler and a protractor to construct the following regular polygons: triangle, quadrilateral, pentagon, octagon, and nonagon. Besides having sides with the same length and angles with the same measure, what other special properties do the regular polygons have?

10. Extend problem 38 on page 99 as follows: Compare your answers to parts **a** and **b**. Now compare $\sqrt{4+9}$ and $\sqrt{4}+\sqrt{9}$, and $\sqrt{36+64}$ and $\sqrt{36}+\sqrt{64}$. What conclusions can you draw about how the square root of a sum compares to the sum of the square roots? Make up other examples if you need to. Now compare $\sqrt{9 \times 16}$ and $\sqrt{9} \times \sqrt{16}$ and $\sqrt{4 \times 9}$ and $\sqrt{4} \times \sqrt{9}$. What conclusions can you draw about how the square root of a product compares to the product of the square roots?

Additional Answers

for Chapter 3 Test

1 Check students' drawings.

6 $x = 32$; Multiply both sides of the equation by 40.

Answers for Puzzles and Challenges are on page A2.

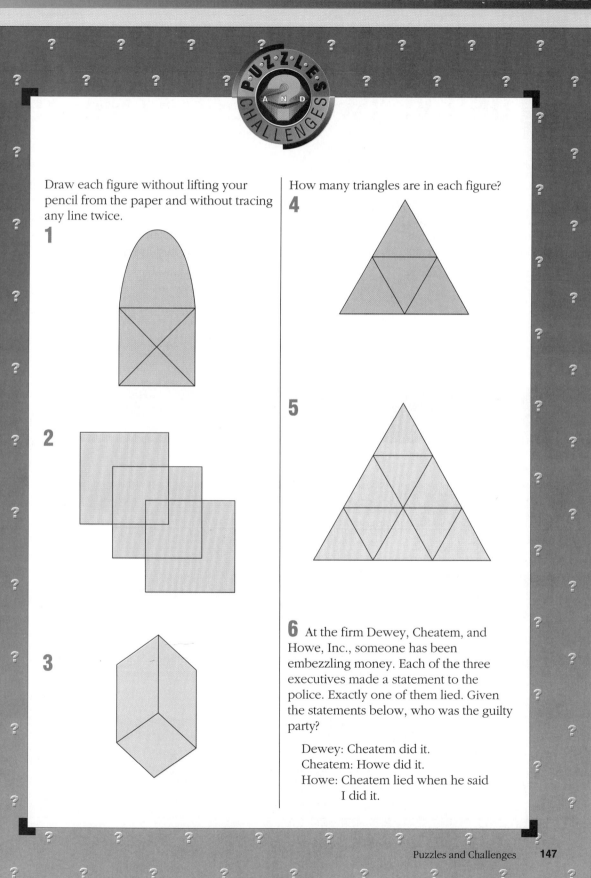

ASSESSMENT LONG-TERM INVESTIGATIONS

UNIT B: Two-Wheeler Ordeal

Suggestions for various ways to assess students' performance are integrated throughout the investigations in this unit. These forms of assessment include observing the students while they work, interviewing students, looking at students' self-evaluations and group evaluations, observing peer review/response situations, evaluating students' products, and evaluating students' portfolios.

Refer to the chart below to preview the variety of assessment opportunities that are integrated throughout this unit.

Assessment Opportunities	Investigations						
	Before Unit	1	2	3	4	5	End of Unit
Form A: Observation Notes **Bank A/B:** Observation Guide		✖			✖		
Form B: Observation Evaluation Record **Bank A/B:** Observation Guide			✖				
Form C: Interview Record **Bank C:** Interview Guide				✖			
Form D: Product Evaluation Record **Bank D:** Product Evaluation Guide		✖	✖	✖	✖	✖	
Form E: Rubric Chart **Bank E:** Sample Rubric						✖	
Form F: Portfolio Item Description	✖						✖
Form G: Portfolio Notes **Bank G:** Portfolio Evaluation Criteria	✖						✖
Form H: Self-evaluation			✖		✖		✖
Form I: Group Evaluation				✖		✖	
Form J: Peer Review/Response			✖				

148A

Rubrics

Rubrics are used to evaluate students' products. You can develop a scoring rubric for a specific investigation by adapting the blackline masters shown below, which are provided in the *Gateways to Mathematical Investigations* Teacher's Edition. You will probably want to develop a rubric with students at least once per unit. The chart on the opposite page suggests that for Unit B: Two-Wheeler Ordeal, you complete a rubric before Investigation 5. Remind students that, at times, they may be asked to revise their work to improve it and to make it complete.

Bank D, page 211

Bank D: Product Evaluation Guide

The statements below can be used as a guide to assess students' products. Adapt these statements as needed to evaluate specific assignments. Record your adaptations on a clean copy of **Form D: Product Evaluation Record** (p. 210). You may also wish to create and record your own criteria on **Form D.** After you have adapted **Form D** for use with a specific assignment, make copies for each student or group as needed.

Mathematical Reasoning Skills

- The student includes examples and/or counter examples to support the mathematical argument.
- The student synthesizes and summarizes own or group's thinking.

Mathematical Understanding

- The student identifies all the important information of the situation or problem.
- The student shows understanding of the situation's mathematical ideas. (List important mathematical concepts and skills for the assignment on Form D.)

Form E, page 212

Form E: Blank Rubric Chart

NAME (S) ______________ ______________ GROUP ______
______________ ______________ DATE ______

STANDARDS FOR STUDENT'S WORK

LEVEL	Standard of performance to be achieved for each specified level
4 Exceptional/ Fully Acceptable	

Bank E, page 213

Bank E: Sample Rubric

STANDARDS FOR STUDENT'S WORK

LEVEL	Standard of performance to be achieved for each specified level
4 Exceptional/ Fully Acceptable	Your presentation fully accomplishes the purpose of the task. Your work is complete, well organized, and clearly explained. You illustrate ideas using clear diagrams and models and communicate them effectively to the audience. You identify all the important information of the situation and show a thorough understanding of the mathematical ideas and procedures. You use strong arguments to support findings, including examples (and/or counterexamples). You demonstrate that you fully grasp the central mathematical ideas developed in this task. You demonstrate an orignial extension or applicatrion of the mathematical concepts.
3 Satisfactory/ Acceptable	Your presentation substantially accomplishes the purpose of the task. Your work is complete, reasonably organized, and explained. You use an appropriate diagram or model and communicate ideas effectively to the audience. You identify most of the important information of the situation and show understanding of the mathmatical ideas and procedures. You present arguments to support findings, including examples (and/or counterexamples). You understand and use the central mathematical ideas developed in this task. You demonstrate and apply the mathematical concepts.
2 Incomplete/ Revise	Your presentation partially accomplishes the purpose of the task. However, your work may be incomplete, or misdirected, or not clearly organized and/or explained. Your diagram or model may be inappropriate or unclear. You may communicate ideas inadequately to the audience. You may identify some of the important information of the situation and you may show only partial and/or limited understanding of the mathematical ideas and procedures. Your arguments to support findings may be weak or incomplete. You may misunderstand or misuse the central mathematical ideas developed in this task. You may have difficulty demonstrating and applying the mathematical tasks. You need to revise this work. (Your teacher will provide suggestions for a satisfactory or competent rating and a due date for your revised work.)
1 Inadequate/ Redo	Your presentation shows little or no understanding of the assignment. Your work is barely, if at all, comprehensible. Your explanations are not understandable. You show little or no understanding of the mathematical ideas and procedures. You made serious computational mistakes and misused mathematical terms. You should meet with your classmates or your teacher to be sure you fully understand this assignment. Then you need to revise this work. (Your teacher will provide suggestions for a satisfactory or competent rating and a due date for your revised work.)

Gateways © McDougal, Littell & Company   Bank E: Sample Rubric   213

Form F, page 214

Form F: Portfolio Item Description

NAME ______________ DATE ______________

Complete this form each time you add an item to your portfolio.

Type of Portfolio Entry: ______________

Why did you choose to place this item in your portfolio?

Form G, page 215

Form G: Portfolio Notes

NAME ______________ CLASS ______________

Date	Observations/Comments	Recommendations

Bank G, page 216

Bank G: Portfolio Evaluation Criteria

At the start of the school year, have a class discussion on the goals and uses of students' portfolios. Discuss with the class some ideas regarding what to include in a portfolio. Then, use the portfolio evaluation criteria which follows to develop a rubric with the class that communicates the criteria you will use to assess their portfolios. Have students record this criteria on copies of **Form E: Rubric Chart.** Have students store this completed form in their portfolios so that they can refer to it throughout the year.

Each time students add items to their portfolios, have them complete copies of **Form F: Portfolio Item Description.**

Use copies of **Form G: Portfolio Notes** to record your observations, comments, and recommendations each time you evaluate students' portfolios.

Ideas for Types of Portfolio Entries

- a portfolio table of contents
- a letter to the reader of the portfolio explaining each item
- teacher-completed checklists/evaluations
- a mathematical autobiography
- notes from interviews, teachers, peers, etc.
- papers that show the student's corrections or errors or misconceptions
- photos or sketches of the student's work with manipulatives or mathematical models
- excerpts from the student's daily journal or reflective writing samples
- tables, charts, and diagrams created by the student
- artwork supporting a problem such as drawings
- designs, coordinate pictures, scale drawings, maps, etc.
- student-created problems with or without solutions
- group projects or individualized reports
- student-developed surveys and results
- self-evaluations

Suggestions for Portfolio Evaluation Criteria

Mathematical Reasoning Skills

The Student:

- analyzes problem situations.
- is flexible and comfortable in using several different problem solving approaches.
- synthesizes and summarizes the individual's or group's thinking.
- provides supporting statements to verify findings.
- shows original and creative thought.

Mathematical Tools and Techniques

The Student:

- uses a variety of diagrams, models, technology, manipulatives, and other resources appropriately.
- uses displays (e.g. graphs, tables, diagrams, sketches, and models) to represent mathematical ideas.

Mathematical Understanding

The Student:

- shows understanding of mathematical ideas and procedures. (List specific skills/concepts on **Form G.**).

Mathematical Communication

The Student:

- organizes work in a logical and clear way.
- expresses mathematical ideas clearly and appropriately.
- uses mathematical language and notation appropriately.

Mathematical Dispositions

The Student:

- shows development as as group participant.
- shows development of a positive attitude.
- shows growth in learning.
- includes revised and complete work.
- shows consistent effort.
- includes a variety of work in his or her portfolio.

216   Assessment   *Gateways* © McDougal, Littell & Company

Portfolios

At the beginning of each unit, you may wish to review with the students the criteria they should use when placing work in their portfolios. Have students complete Form F: Portfolio Item Description each time they wish to place an item in their portfolios.

Periodically, you may wish to evaluate students' portfolios using the blackline masters shown above. For Unit B: Two-Wheeler Ordeal, the chart on the opposite page suggests that you evaluate students' portfolios at the end of the unit.

CHAPTER 4

Contents

4.1 A Closer Look at Measurement
4.2 Ratios of Measurement
4.3 More Applications of Ratios
4.4 Working with Units
4.5 Proportions
4.6 Similarity

Pacing Chart

Days per Section

	Basic	Average	Honors
4.1	2	2	2
4.2	2	2	2
4.3	3	3	2
4.4	3	2	2
4.5	3	3	2
4.6	3	2	2
Review	2	1	1
Assess	3	2	1
	21	17	14

Chapter 4 Overview

Ratio and proportion are at the center of many relationships among real-world objects. Therefore, there are many ways to approach the subject. We will emphasize ratios between different units of measurement. This extends the material from Chapter 3 while introducing new concepts. Many of the applications of ratio and proportion are treated in the problems. Density, however, is formally discussed in Section 4.3 because it is such a rich application and an important link to science. We also continue to emphasize the relationship between geometry and arithmetic, culminating in a section on similarity.

4 Ratio and Proportion

CHAPTER 4

AIRCRAFT DESIGNER Improvements are continually being made in the design of civilian and military aircraft. Aircraft are flying higher and faster and are becoming more automated. People who design these aircraft must be familiar with every aspect of aircraft construction and operation, including mechanics, guidance and control systems, propulsion systems, and communications. An aircraft designer must have a background in engineering, with studies in mathematics, physics, and materials science. Most aircraft designers are employed by private aerospace firms.

INVESTIGATION

Lift-to-Drag Ratios One important aspect of aircraft design is the creation of an efficient wing shape. The flow of air over an airplane's wing produces a force, called lift, that enables the plane to fly: the faster the flow, the greater the lift. On the other hand, any object that moves through the air is affected by a force known as drag, which slows down the object. Aircraft designers try to design wing shapes that will give planes a high *lift-to-drag ratio*. Do some research to find out about lift-to-drag ratios and how they are affected by the shapes of wings. What are the lift-to-drag ratios of various types of airplanes?

Career Connection

Although it is not expected that many students will become aircraft designers, many techniques of aircraft design, like computer-aided design, are used in other industries. Help students to see that understanding the uses of mathematics in familiar products and activities can increase understanding about and appreciation of those products and reduce anxiety about learning new technology. Airline pilots, for example, must now be more knowledgeable than ever before about the computers on their aircraft, such as satellite navigation systems.

Video Connection

For more information about the career discussed here, please refer to the Futures videocassette #1, program B: *Aircraft Design*, featuring Jaime Escalante. For more information or to order a copy of the program, call PBS VIDEO at 800/344-3337.

Investigation Notes

Encourage students to explore the influence an airplane's intended use has on its wing and engine design. All aircraft need a lift-to-drag ratio that is greater than one in order to fly. A small fighter plane, which must fly and climb at great speed and be very maneuverable, needs a higher lift-to-drag ratio than a large cargo plane, which generally flies more slowly.

SECTION 4.1

Objectives

- To identify measurable attributes of objects and some of the units in which measurements of the attributes may be expressed
- To recognize metric units of length, mass, and capacity
- To add, subtract, multiply, and divide measurements of like attributes

Vocabulary: attribute, *centi-*, gram, *kilo-*, meter, *milli-*

Class Planning

Schedule 2 days

Resources

In *Chapter 4 Student Activity Masters* Book
Class Openers 1, 2
Extra Practice 3, 4
Cooperative Learning 5, 6
Enrichment 7
Learning Log 8
Interdisciplinary Investigation 9, 10
Spanish-Language Summary 11

Class Opener

List various ways in which you could measure the following objects:
1. A strand of hair
2. The moon
3. Hot chocolate in a mug
4. Your math book

Possible answers:
1. Length, weight, volume
2. Volume, mass, surface area, circumference
3. Volume, temperature
4. Length, width, height, volume, weight

Lesson Notes

- Students usually know many units, but not always when and

150

Investigation Connection
Section 4.1 supports Unit B, *Investigation* 3. (See pp. 92A-92D in this book.)

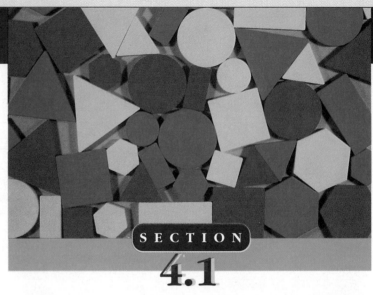

SECTION 4.1
A Closer Look at Measurement

ATTRIBUTES AND UNITS

Young children are sometimes given blocks called attribute tiles to help them learn about shapes, colors, and sizes. An **attribute** is a characteristic, a quality that distinguishes an object from other objects. Some attributes, such as size, can be measured. In the preceding chapter, you learned about measuring one sort of size—length or distance. Size may also mean *area,* usually expressed in square units. Another sort of size is *volume,* the amount of space that an object takes up. It is usually expressed in cubic units.

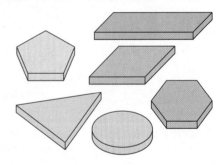

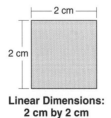
Linear Dimensions:
2 cm by 2 cm

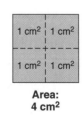
Area:
4 cm²

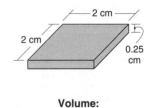
Volume:
1 cm³

Notice how the exponents 2 and 3 are used to indicate square units and cubic units.

Units of capacity, such as quarts and liters, are used to measure the volume of liquids that a container can hold.

150   **Chapter 4**   Ratio and Proportion

Many other attributes can be measured—mass, weight, time, temperature, speed, and cost are but a few—and a variety of units can be used to measure each. To avoid confusion, it is useful to ask yourself the following questions whenever you need to decide what unit you should use to express a measurement.

1. **Does the unit represent the attribute being measured?** It is impossible, for instance, to measure a distance in pints, since pints are units of capacity, not length. Similarly, we cannot measure temperatures in meters, lengths of time in ounces, or angles in square feet.
2. **Does the size of the unit fit the quantity being measured?** Although you could express your height in miles (it's about 0.001 mile), it is more reasonable to use a smaller unit, such as feet or centimeters. It would likewise be awkward (but not wrong) to express the weight of a dog in tons or the age of a friend in seconds.

Sometimes tradition makes a particular unit appropriate. For example, we measure speeds of ships in knots, lengths of horse races in furlongs, amounts of petroleum in barrels, and the area of land in acres.

METRIC UNITS

In Section 3.1, you learned about some metric units of length. The metric system is now the standard of measurement in most countries. It is also the basis for the units used in science. In the metric system, the basic unit of length is the meter, the basic unit of mass is the gram, and the basic unit of capacity is the liter. Larger and smaller units are obtained by multiplying and dividing these units by powers of 10. The following table shows how various prefixes—kilo-, centi-, etc. are used. You will use the units shown in bold face most frequently.

Attribute	Basic Unit	Other Units	Meaning
Length	**Meter**	**Kilometer**	1000 meters
		Hectometer	100 meters
		Dekameter	10 meters
		Decimeter	0.1 meter
		Centimeter	0.01 meter
		Millimeter	0.001 meter

The same prefixes are also used with grams and liters.
To see how metric units compare with English units, refer to the table of equivalent measurements in the back of this book.

---

how to use them appropriately. This section will heighten their awareness of units and expand their knowledge base.

▪ This unit might be one to coordinate with science, shop, music, art, or social studies teachers to show that mathematics relates to many areas. For example, a whole note in music is held for four times the length of a quarter note, but not necessarily for one second. Another example might be the use of scale drawings in social studies to determine distances.

▪ Be sure that students understand why there are no units in the answer to example 2.

Additional Examples

Example 1
On top of a 0.75 m high box is a box that is 0.25 m high. On top of that box is a box that is 50 cm high. How high is the stack of boxes?

Answer: 1.5 m or 150 cm

Example 2
Evaluate each ratio of measurements.
a. $\dfrac{6 \text{ yards}}{2 \text{ yards}}$
b. $\dfrac{3 \text{ yards}}{2 \text{ feet}}$

Answers: a. 3 b. $\dfrac{9}{2}$

SECTION 4.1

Assignment Guide

Basic
Day 1 14–16, 18, 22–24, 32, 34, 37, 41, 47, 48
Day 2 17, 19–21, 25–28, 33, 40, 43, 44, 46

Average or Heterogeneous
Day 1 14–19, 21, 22, 25, 26, 30, 32, 38, 39, 41, 42
Day 2 20, 23, 24, 27–29, 31, 34, 36, 37, 43, 45, 48

Honors
Day 1 14–17, 19, 20, 23–26, 28, 29, 34, 37, 42, 45
Day 2 18, 21, 22, 27, 31, 32, 35, 36, 39, 40, 48, Investigation

Strands

Communicating Mathematics	10, 13, 28, 33, 37, 40, 44, Investigation
Geometry	18, 21, 31, 36, 42
Measurement	1–11, 14–19, 21–23, 26–31, 33–36, 38, Investigation
Data Analysis/ Statistics	30, 45
Technology	12, 39, 41, 45
Algebra	14, 31, 33, 41, 43, 47
Patterns & Functions	39, 45
Number Sense	1–11, 13, 21, 22, 28, 34, 37, 48
Interdisciplinary Applications	13, 20, 30, 32, 37, Investigation

Individualizing Instruction

The learning disabled student Students can benefit from working with tangible examples of different units of measure. Using visual and manual representations of various units, such as

152

OPERATIONS WITH MEASUREMENTS

When we want to compare or combine measurements of the same attribute, we usually need to express the measurements in the same units.

Example 1
On top of a 1-meter-high box is a box that is half a meter high. On top of that is a box that is 75 centimeters high. How high is the stack of boxes?

Solution
In order to add the heights, we need to express them in the same units. The easiest way to do this is to rewrite centimeters as meters or rewrite meters as centimeters.

Method 1 1 m + 0.5 m + 75 cm
= 1 m + 0.5 m + 0.75 m
= 2.25 m

Method 2 1 m + 0.5 m + 75 cm
= 100 cm + 50 cm + 75 cm
= 225 cm

Since 2.25 m = 225 cm, the answers are the same.

Example 2
Evaluate each ratio of measurements.

a $\dfrac{9 \text{ feet}}{6 \text{ feet}}$ b $\dfrac{6 \text{ feet}}{2 \text{ yards}}$

Solution
Notice that in part **a** the units are the same, so they cancel. In part **b** we need to change the measurements to the same units.

a $\dfrac{9 \text{ feet}}{6 \text{ feet}} = \dfrac{3}{2}$ b $\dfrac{6 \text{ feet}}{2 \text{ yards}} = \dfrac{6 \text{ feet}}{6 \text{ feet}} = 1$

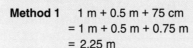

Sample Problems

Problem 1 Express 2 hours 12 minutes as a number of hours.

Solution We need to change 12 minutes to hours. There are 60 minutes in an hour; so 1 minute is $\frac{1}{60}$ hour, and 12 minutes is $\frac{12}{60}$, or $\frac{1}{5}$, hour. Therefore, 2 hours 12 minutes is $2\frac{1}{5}$ hours.

SECTION 4.1

Problem 2 If a speedboat travels at 65 kilometers per hour, how far can it travel in 2 hours 12 minutes?

Solution The boat travels 65 kilometers every hour. In the preceding problem, we found that 2 hours 12 minutes is equivalent to $2\frac{1}{5}$ hours.

$$1 \text{ hr} + 1 \text{ hr} + \tfrac{1}{5} \text{ hr} = 2\tfrac{1}{5} \text{ hr}$$
$$\downarrow \quad\quad \downarrow \quad\quad \downarrow$$
$$65 \text{ km} + 65 \text{ km} + \tfrac{1}{5}(65) \text{ km}$$
$$65 \text{ km} + 65 \text{ km} + 13 \text{ km} = 143 \text{ km}$$

Think and Discuss

P In problems 1–8, choose from the box the units that would be most useful in measuring each attribute.

Box: Quarts, Cents, Millimeters, Tons, Feet, Grams, Square centimeters, Fluidounces, Knots, Acres

1 The area of a postage stamp sq cm

2 The height of a building ft

3 The length of a housefly mm

4 The size of a lake Acres

5 The amount of oil in a car's engine qt

6 The amount of scrap iron in a truck Tons

7 The mass of a kernel of corn g

8 The amount of root beer in a cup fl oz

P, R **9** Refer to the table of metric units in this section. What power of 10 seems to be indicated by each of these prefixes?
  a Hecto- 10^2
  b Deci- 10^{-1}
  c Milli- 10^{-3}
  d Kilo- 10^3
  e Centi- 10^{-2}
  f Deka- 10^1

P **10** List as many attributes of this book as you can. Which of these attributes can be measured? What units might be used?

P **11** Would you use a 12-inch ruler, a meterstick, a 100-foot tape measure, or a car's odometer to measure
  a A vegetable garden?
  b The length of a phone cord?
  c A baseball diamond?
  d The distance between two cities?
  e A pencil?
  f A poster hung on the wall?

---

metersticks, measuring cups, and clocks can be very helpful.

Problem-Set Notes

for page 153

- 1–8 and 11 reinforce the use of appropriate units.

Communicating Mathematics

- Have students write a paragraph describing the process used to add
5 minutes 43 seconds,
16 minutes 37 seconds, and
3 minutes 27 seconds.

- Problems 10, 13, 28, 33, 37, 40, 44, and Investigation encourage communicating mathematics.

Additional Answers

for page 153

10 Answers will vary. Students may mention such attributes as weight (measurable in ounces, pounds, grams, etc.), color (measurable only with electronic instruments), and linear dimensions (measurable in inches, centimeters, etc.).

11 Most likely answers: **a** Tape measure **b** Meterstick, ruler, or tape measure **c** Tape measure **d** Odometer **e** Ruler **f** Meterstick or ruler

Section 4.1 A Closer Look at Measurement

SECTION 4.1

Problem-Set Notes

for pages 154 and 155

- 15–17, 23–25, and 28 reinforce how important a correct unit is to an answer.
- 21 illustrates an important concept: relating linear, square, and cubic units.

Multicultural Note

The metric system originated with Gabriel Mouton, a French clergyman, who proposed a decimal measurement system in 1670. In the original metric system, the meter was 0.0000001 of the distance from the North Pole to the equator along the line of longitude near Dunkerque, France, and Barcelona, Spain. A surveying error led to a meter that was slightly shorter than the intended length. The metric system is officially used by every country except the United States, Brunei, and Myanmar.

Additional Practice

1. Which is a unit of length: acre, quart, gram, or inch?
2. Which is a unit of volume: kilogram, cubic foot, ounce, or centimeter?
3. Which is a unit of capacity: liter, gram, acre, or meter?
4. Find each difference.
   a. 1.4 m – 30 cm
   b. 11 ft – 72 in.
   c. 23 quarts – 6 quarts
   d. 2.4 kg – 128 g

Answers:
1. Inch 2. Cubic foot 3. Liter
4. a. 110 cm or 1.10 m b. 5 ft
c. 17 quarts d. 2272 g or 2.272 kg

P, R   **12** Mr. Snackmeister can eat peanuts at a rate of 3 ounces per minute. If peanuts cost $2.88 per pound, how long would it take him to eat $4.50 worth of nuts? $8.\overline{3}$ min, or 8 min 20 sec

P   **13** A famous book by Jules Verne is titled *Twenty Thousand Leagues Under the Sea*. What do you think this title means?

Problems and Applications

P, R   **14** Determine the value of n that makes the statement true.
   a n cm = 9 m   900   **b** 5 yd = n in.   180   **c** 5 yd = n ft   15

P   **In problems 15–17, evaluate each ratio of measurements.**

   15 $\dfrac{2 \text{ feet}}{6 \text{ inches}}$ $\dfrac{4}{1}$ **16** $\dfrac{3 \text{ inches}}{1 \text{ foot}}$ $\dfrac{1}{4}$ **17** $\dfrac{4 \text{ feet}}{2 \text{ yards}}$ $\dfrac{2}{3}$

P, R   **18** Find the area of a rectangle with the given dimensions. Be sure to include the proper units in your answer.
   a 3 cm by 5 cm   15 cm²   **b** 4 in. by 6 in.   24 in.²   **c** 5 m by 7 m   35 m²

P, R   **19 a** What is the total length, in feet and inches, of segment AC?   15 ft 3 in.   (A to B: 5 ft 8 in., B to C: 9 ft 7 in.)
   b Evaluate the sum $5\frac{8}{12} + 9\frac{7}{12}$. $15\frac{3}{12}$, or $15\frac{1}{4}$

P, R   **20 Consumer Math**   If the cost of running an air conditioner is 10.4 cents per hour, how much does it cost to run the air conditioner for 250 hours? $26

P, R   **21** Use the diagram to help you answer the following questions.
   a What fraction of a yard is a foot? $\frac{1}{3}$
   b What fraction of a square yard is a square foot? $\frac{1}{9}$
   c What fraction of a cubic yard is a cubic foot? $\frac{1}{27}$

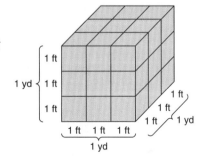

P   **22** List two things you might measure with each unit.
   a Miles per hour   **b** Meters per day   **c** Inches per year

P, R   **In problems 23–25, evaluate each ratio of amounts of money.**

   23 $\dfrac{3 \text{ quarters}}{5 \text{ dimes}}$ $\dfrac{3}{2}$ **24** $\dfrac{6 \text{ nickels}}{3 \text{ dimes}}$ $\dfrac{1}{1}$ **25** $\dfrac{4 \text{ quarters}}{15 \text{ dimes}}$ $\dfrac{2}{3}$

Chapter 4   Ratio and Proportion

SECTION 4.1

P, R **26** A gallon is equivalent to 4 quarts.
  a What percent of a gallon is a quart?   25%
  b What percent of a quart is a gallon?   400%

P, R **27** Find the difference between the heights of the buildings shown.   46 ft, or $15\frac{1}{3}$ yd

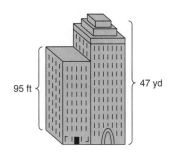

95 ft   47 yd

P, R **28 Communicating**
  a Which is greater, 3 or 8?   8
  b Which is longer, 3 meters or 8 centimeters?   3 m
  c Explain why your answers to parts **a** and **b** were the same or were different.

R, I **29** Find the volume of a rectangular solid that measures 12 inches by 5 inches by 5 inches.   300 in.^{3}

P, R **30** The criminal lawyer Dee Fender charges $125 per hour for her services. Use the picture of her billing notes to determine how much she earned on October 25. $1156.25

P, R, I **31** Find the value of w for which the perimeter of the rectangle is 10 yards.   9 ft, or 3 yd

6 ft

P **32 Consumer Math**   A 12-ounce can of Meadowlark lemonade costs 60¢. How much does Meadowlark cost per ounce? $0.05

I **33 Communicating**
  a Find the value of x. $\frac{2}{3}$
  b Describe the situation that could be represented by the diagram.

42 mi	42 mi	28 mi
1 hr	1 hr	x hr

Stumbling Block

Students may not understand what is meant by the term *attribute*. One way to teach the concept is to name an object in the room and give examples of attributes of the object (e.g., length, color, weight). Then have students try to define the term from your example. Have a student then name an object and ask other students to list attributes of that object.

Reteaching

Reteach example 2b by converting feet to yards, rather than yards to feet.

$$\frac{6 \text{ feet}}{2 \text{ yards}} = \frac{2 \cancel{\text{yards}}}{2 \cancel{\text{yards}}} = 1$$

Point out that the ratio is the same, no matter which measurement is converted.

Additional Answers

for pages 154 and 155

13 Possible answer: The title refers to a 60,000-mile undersea journey in a submarine (since a league is equivalent to 3 miles).

22 Possible answers: **a** The speed of a car; the speed of a pitched baseball **b** The speed of a lava flow; the rise of flood waters **c** The growth of a tree; the movement of a continent

28 c Possible explanation: Even though the number 3 is less than the number 8, 3 m is longer that 8 cm because a centimeter is a much smaller unit of measure than a meter (100 cm = 1 m).

33 b Possible answer: The distance covered in 2 hr 40 min by a car traveling at 42 mph

Section 4.1   A Closer Look at Measurement

SECTION 4.1

Problem-Set Notes

for pages 156 and 157

- **40** introduces some units which are less commonly known. You might ask students to add to the list and have them share their findings with the class.

Cooperative Learning

Have groups of students work on the following problems:
1. Draw two diagrams similar to the one in problem 21, but instead of showing feet and yards, show
   a. inches and feet
   b. millimeters and centimeters
2. Use your diagrams to help you write a fraction for each relationship.
   a. $\frac{ft}{in.}$ b. $\frac{square\ ft}{square\ in.}$
   c. $\frac{cubic\ ft}{cubic\ in.}$ d. $\frac{cm}{mm}$
   e. $\frac{square\ cm}{square\ mm}$ f. $\frac{cubic\ cm}{cubic\ mm}$
3. What pattern do you notice in the fractions as you go from linear to square to cubic units?

Answers: 1. See students' drawings. 2. a. $\frac{1}{12}$ b. $\frac{1}{144}$ c. $\frac{1}{1728}$ d. $\frac{1}{10}$ e. $\frac{1}{100}$ f. $\frac{1}{1000}$
3. If the ratio of linear units is $\frac{1}{n}$, the ratio of square units is $\frac{1}{n^2}$, and the ratio of cubic units is $\frac{1}{n^3}$.

- Problems 10, 13, and 40 are appropriate for cooperative learning.

P   **34** Mickey is trying to estimate how much a trip to Florida will cost. One of the things he needs to calculate is the amount of gasoline he will need to buy. What kinds of measurements and units will he probably use in this calculation?

P, R   **35** Di has 4 quarters, 3 dimes, and 2 nickels. Nick has 3 quarters, 5 dimes, and 2 nickels.
    a Determine the ratio of the number of coins that Di has to the number of coins that Nick has. $\frac{9}{10}$
    b Determine the ratio of the value of Di's coins to the value of Nick's coins. $\frac{28}{27}$

R, I   **36** Six 1-by-1 squares can be joined edge to edge in various ways to form a figure with an area of 6. (Three possibilities are shown, but there are others.)
    a Find the perimeter of each figure shown.   10; 14; 12
    b What is the greatest possible perimeter of such a figure?   14
    c What is the least possible perimeter of such a figure?   10

 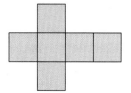 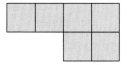

P   **37 Consumer Math**
    a A 19-inch television set is not 19 inches long or wide. To what might "19-inch" refer?
    b A spool of 6-pound-test fishing line does not weigh 6 pounds. To what might "6-pound" refer?

P, R   **38** Rachel rode her bike for 2 hours at a speed of 12 mi/hr, then slowed down to 10 mi/hr for the next $1\frac{1}{2}$ hours. What was the total distance she traveled?   39 mi

R   **39 Spreadsheets**   The spreadsheet display shows that 1 mile per hour is equal to about 1.609 kilometers per hour. How could you use the copy feature of the spreadsheet to list speeds from 1 mi/hr to 100 mi/hr along with their equivalents in kilometers per hour?

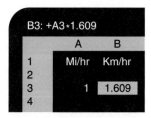

P   **40** Choose three of the units listed below, and write a description of each. Be sure to explain what the unit is used to measure, and give some examples of situations in which the unit might be used. (Look up the units in a dictionary or an encyclopedia if you need to.)
    a Angstrom   **b** Bushel   **c** Calorie   **d** Carat
    e Cord   **f** Decibel   **g** Fathom   **h** Furlong
    i Lumen   **j** Peck   **k** Roentgen   **l** Watt

156   **Chapter 4**   Ratio and Proportion

SECTION 4.1

◀ LOOKING BACK **Spiral Learning** LOOKING AHEAD ▶

R **41** Solve each equation for *x*.
 a $x + 534 = 961.2$ **b** $x - 534 = 961.2$ **c** $534x = 961.2$
 d $\frac{x}{534} = 961.2$ **e** $x + 961.2 = 534$ **f** $x + 534 = -961.2$

R **42** Which angle is greater, ∠ACE or ∠BCD?
 They are the same angle and therefore
 equal in measure.

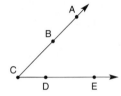

I **43** Evaluate each expression.
 a $\frac{2}{3} \cdot \frac{3}{5} \cdot \frac{5}{7}$ $\frac{2}{7}$
 b $\frac{a}{b} \cdot \frac{b}{c} \cdot \frac{c}{d}$ $\frac{a}{d}$

I **44 Communicating** Greta added the fractions $\frac{1}{4}$ and $\frac{1}{6}$ in the following way.
 $$\frac{1}{4} + \frac{1}{6} = \left(\frac{1}{4} \cdot \frac{3}{3}\right) + \left(\frac{1}{6} \cdot \frac{2}{2}\right) = \frac{3}{12} + \frac{2}{12} = \frac{5}{12}$$
 Explain why she multiplied $\frac{1}{4}$ by $\frac{3}{3}$ and multiplied $\frac{1}{6}$ by $\frac{2}{2}$.

R **45 Spreadsheets** Use a
 spreadsheet to make a
 table like the one shown.
 Continue the pattern to find
 the year when the ratio of
 Wally's age to Nancy's age
 was equal to 3.   1979

Year	Wally's Age	Nancy's Age	Ratio
1991	48	24	2
1990	47	23	2.043
1989	46	22	2.09

I **46** Given that $\frac{3}{5} = \frac{6}{10}$, indicate whether each statement is *true* or *false*.
 a $\frac{3}{10} = \frac{6}{5}$ False   **b** $3 \cdot 10 = 5 \cdot 6$ True   **c** $3 \cdot 6 = 5 \cdot 10$ False

I **47** Evaluate each expression.
 a $\frac{3}{8} \cdot 8$ 3   **b** $\frac{3 \cdot 8}{8}$ 3   **c** $\frac{x}{y} \cdot y$ x **d** $\frac{x \cdot y}{y}$ x

R **48** Arrange the four numbers 3.24×10^{-6}, 2×10^5, 2.13×10^{-7}, and 1.2×10^3 in
 order, from least to greatest. 2.13×10^{-7}, 3.24×10^{-6}, 1.2×10^3, 2×10^5

Investigation

Mass and Weight Find out the scientific definitions of *mass* and *weight*.
How do the attributes of mass and weight differ? In what kinds of situations
would each be measured?

Section 4.1   A Closer Look at Measurement   157

Checkpoint

1. Which of the following units would be best to use for each measurement?

grams, tons, feet,
dollars, acres, quarts

a. The size of a cornfield
b. The mass of a pencil
c. The amount of water in a bucket
d. The length of a ladder
e. The price of a car
f. The mass of an adult elephant

2. Find each sum.
a. 1 ft 4 in. + 3 ft 7 in.
b. 14 g + 16 g
c. 15 cm + 0.15 m
d. 12 yd + 4 ft

Answers:
1. a. Acres **b.** Grams **c.** Quarts
d. Feet **e.** Dollars **f.** Tons
2. a. 4 ft 11 in. **b.** 30 g
c. 30 cm or 0.30 m
d. 13 yd 1 ft

Additional Answers

for pages 156 and 157

34 Possible answer: Distance to be traveled (in miles) and gas mileage of his car (in miles per gallon)

37 Possible answers: **a** The distance from one corner of the TV screen to the opposite corner (the diagonal measure of the screen) **b** The amount of pull that the line has consistently borne (in controlled tests) without breaking

39 Possible answer: Enter +A3+1 in cell A4, and copy B3 to B4. Then copy A4–B4 to A102–B102.

Answers for problems 40, 44, and Investigation are on page A2.

157

SECTION 4.2

Objectives
- To identify and evaluate ratios
- To identify and evaluate rates
- To calculate unit costs

Vocabulary: base unit, rate, ratio, unit cost

Class Planning

Schedule 2 days

Resources

In *Chapter 4 Student Activity Masters* Book
Class Openers 13, 14
Extra Practice 15, 16
Cooperative Learning 17, 18
Enrichment 19
Learning Log 20
Using Manipulatives 21, 22
Spanish-Language Summary 23

Transparency 22

Class Opener

Reduce each fraction to lowest terms.
1. $\frac{5}{25}$ 2. $\frac{10}{70}$ 3. $\frac{68}{12}$ 4. $\frac{35}{75}$ 5. $\frac{114}{54}$

Answers:
1. $\frac{1}{5}$ 2. $\frac{1}{7}$ 3. $\frac{17}{3}$ 4. $\frac{7}{15}$ 5. $\frac{19}{9}$

Lesson Notes

- A *rate* is defined in the text in the most general of terms: a ratio that compares different attributes. When discussing rates, use varied examples, not just rates that involve time. Discuss whether a unit cost is a rate (it is) and how other factors besides cost per unit of weight are important in determining better buys (see problem 5).

Investigation Connection
Section 4.2 supports Unit B, *Investigations* 2, 3, 5. (See pp. 92A-92D in this book.)

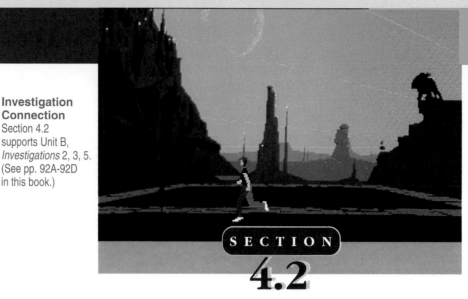

SECTION 4.2
Ratios of Measurements

RATIOS

A **ratio** is a way of comparing two values. The ratio of any value a to any value b can be written either in fraction form (that is, as $\frac{a}{b}$) or in the form $a:b$.

Example 1
While playing the video game Astroblaster, Luke fired 450 shots and destroyed 300 asteroids. What was the ratio of the number of destroyed asteroids to the number of shots fired?

Solution
We divide the number of destroyed asteroids by the number of shots. The ratio is $\frac{300}{450}$, or $\frac{2}{3}$, which can also be expressed in the form 2:3. This means that Luke destroyed, on the average, 2 asteroids with every 3 shots.

In Section 1.4 you saw that probabilities are ratios, but you may not realize that percents, speeds, averages, and many other quantities you encounter in your daily life are also based on ratios.

RATES

A ratio that compares measurements of two different attributes is called a **rate**. The word *per* is often a signal that a rate is being discussed.

SECTION 4.2

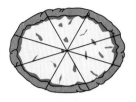  

There are 4 pieces of pizza per person.

A bicycle wheel may make 100 revolutions per minute.

The gold sells for $396 per ounce.

A rate is usually expressed in terms of a **base unit** by rewriting the rate as a fraction with a denominator of 1. A rate of $\frac{135 \text{ miles}}{3 \text{ hours}}$, for example, can be reduced to $\frac{45 \text{ miles}}{1 \text{ hour}}$ (usually written "45 miles per hour," "45 mi/hr," or "45 mph").

Example 2
A boat traveled 126 kilometers in 4 hours. What was its average speed for the trip?

Solution
Speed is a kind of rate. It is calculated by evaluating the ratio $\frac{\text{distance}}{\text{time}}$.

$$\text{Speed} = \frac{\text{distance}}{\text{time}} = \frac{126 \text{ km}}{4 \text{ hr}} = \frac{31.5 \text{ km}}{1 \text{ hr}} = 31.5 \text{ km/hr}$$

The boat traveled at an average speed of 31.5 kilometers per hour. Notice that the units do not cancel when we evaluate a rate, since the measurements cannot be rewritten in terms of a common unit.

UNIT COSTS

People frequently want to compare similar products or services to determine which is cheapest. To do so, they may compare the items' **unit costs**—the costs of equal amounts of the items.

Example 3
If the box of Shreds costs $3.36 and the box of Flakes costs $2.64, which cereal is the better buy?

Additional Examples

Example 1
In a basketball practice, Jenny took 200 shots and made 120 baskets. What was the ratio of baskets to shots?
Answer: $\frac{3}{5}$

Example 2
A car traveled 410 km in 5 hr. What was its average speed for the trip?
Answer: 82 km/hr

Example 3
A 32-oz bag of Graino rice costs $3.74. A 26-oz bag of Uncle Sven's rice costs $3.30. Which rice is the better buy?
Answer: Graino (11.7¢ per oz)

Multicultural Note

An example of a ratio is a currency exchange rate (the value of one country's money in comparison to another's). On August 10, 1992, for example, $1.00 U.S. could be exchanged for 1.4704 German marks, 128 Japanese yen, 3102 Mexican pesos, or 0.51942 British pound.

Stumbling Block

Students may think that a rate such as $\frac{23 \text{ mi}}{2 \text{ hr}}$ is in lowest terms, since the fraction $\frac{23}{2}$ is in lowest terms. Emphasize that $\frac{23 \text{ mi}}{2 \text{ hr}}$ can still be written in terms of a base unit. Both numerator and denominator can be divided by 2. In terms of the base unit "miles per hour," the rate is $\frac{11.5 \text{ mi}}{1 \text{ hr}}$, or 11.5 mi/hr. Point out that the numerator can contain a decimal or a fraction.

SECTION 4.2

Assignment Guide

Basic
Day 1 7–9, 13, 14, 17, 24, 29, 30, 32, 33
Day 2 10–12, 18–20, 26–28, 34, 35

Average or Heterogeneous
Day 1 7, 9, 11, 12, 16, 18–20, 23, 24, 28, 31
Day 2 8, 10, 13, 15, 17, 22, 25–27, 32, 33, 35

Honors
Day 1 7, 8, 11, 12, 15, 16, 18–20, 22, 24, 25, 29, 34, 35
Day 2 9, 10, 13, 14, 17, 21, 23, 27, 28, 31–33, Investigation

Strands

Communicating Mathematics	4–6, 23
Geometry	10, 19, 32, 34, 35
Measurement	1, 2, 5, 8–10, 13, 15, 17, 18, 22, 28
Data Analysis/ Statistics	4, 14, 20, 25, 27
Technology	12, 14, 20, 24, 31, Investigation
Probability	35
Algebra	26, 30, 32, 35
Number Sense	3, 5, 7, 33
Interdisciplinary Applications	4, 5, 12, 14, 16, 22, 23, 31, 33, Investigation

Individualizing Instruction

The kinesthetic learner For students having difficulty understanding unit cost, have them create paper models of items that are commonly divided up into equal portions, such as pizzas. Students can then determine the cost of each portion. For example, if an 8-slice pizza costs $12.00, each slice costs $1.50.

160

Solution
Since the boxes contain different amounts of cereal, it is difficult to make a direct comparison of the prices. We can, however, calculate the cost per ounce of each cereal.

$$\text{Shreds: } \frac{\$3.36}{16 \text{ oz}} = \frac{\$0.21}{1 \text{ oz}}, \text{ or } \$0.21 \text{ per ounce}$$

$$\text{Flakes: } \frac{\$2.64}{12 \text{ oz}} = \frac{\$0.22}{1 \text{ oz}}, \text{ or } \$0.22 \text{ per ounce}$$

An ounce of Shreds costs 1 cent less than an ounce of Flakes, so in terms of unit cost, Shreds is the better buy.

Sample Problems

Problem 1 Tito Tenspeed left home on his bicycle, traveling at a rate of 8 mi/hr. Half an hour later, his sister Tina rode off after him. If she averages 10 mi/hr, how long will it take her to catch Tito?

Solution It is important to begin by thinking about the situation rather than just starting to do calculations. Since Tito travels 8 miles in one hour, he goes 4 miles in the half hour before Tina starts.

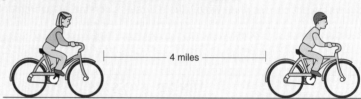

Tina's speed of 10 mi/hr is 2 mi/hr faster than Tito's speed. She will therefore catch up to him by 2 miles every hour. Since Tito's head start is 4 miles, it will take Tina $\frac{4}{2}$, or 2, hours to overtake him.

Problem 2 The Dash Telephone Company's charge for a call between New York City and Chicago is $0.48 for the first minute and $0.19 for each additional minute. For the same call, the TT&T Company charges $0.75 for the first minute and $0.16 for each additional minute.
a Which company charges less for a 5-minute call between New York and Chicago?
b Which company charges less for a 20-minute call?

160

SECTION 4.2

Solution

a A 5-minute call includes charges for the first minute and for 4 additional minutes.

Dash: $0.48 + 4($0.19) = $0.48 + $0.76 = $1.24
TT&T: $0.75 + 4($0.16) = $0.75 + $0.64 = $1.39

Dash charges $0.15 less for the call.

b A 20-minute call includes charges for the first minute and for 19 additional minutes.

Dash: $0.48 + 19($0.19) = $0.48 + $3.61 = $4.09
TT&T: $0.75 + 19($0.16) = $0.75 + $3.04 = $3.79

In this case, TT&T's charge is $0.30 less. So one company is less expensive for a 5-minute call, but the other company is less expensive for a 20-minute call.

Think and Discuss

P **1** Reduce each rate to a base unit.
  a $\frac{200 \text{ miles}}{5 \text{ hours}}$ 40 mi/hr **b** $\frac{75 \text{ dollars}}{5 \text{ books}}$ $15/book **c** $\frac{720 \text{ miles}}{30 \text{ gallons}}$ 24 mi/gal

P **2** The U.S. Postal Service handled 106,311,062,000 pieces of mail in 1980. There were about 227 million people in the United States at that time. Find the ratio of the number of pieces of mail to the number of people.   ≈468:1

P **3** Is the statement *true* or *false*?
  a Distance = speed + time   False
  b Speed · distance = time   False
  c $\frac{\text{Distance}}{\text{Speed}}$ = time   True
  d $\frac{\text{Distance}}{\text{Time}}$ = speed   True

P **4 Consumer Math** The makers of car B advertise that their car goes farther on a tank of gas than cars A, C, and D.
  a Does the advertisement mean that car B is the most economical to drive? Why or why not?
  b Which of the cars is the most economical to drive?   Car A
  c Which is the least economical to drive?   Car D

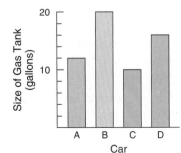
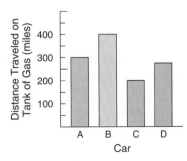

Section 4.2 Ratios of Measurements

SECTION 4.2

Problem-Set Notes

for pages 161, 162, and 163

- **4** represents the type of critical thinking that students should be guided to do. "Going farther on a tank of gas" may simply mean that the gas tank is larger, not that the car gets better gas mileage.
- **6** points out that flat amounts may be misleading; ratios give more information. An $80 per month raise is certainly more significant if Federico earns $800 per month than if he earns $2000 per month.
- **7** may be approached in a variety of ways. Have students discuss the different approaches they employed.
- **17** makes sure that students do not equate 6.15 hr with 6 hr 15 min.
- **20** highlights the fact that average speed is a rate, defined as

$$\frac{\text{total distance}}{\text{total time}}$$

Additional Practice

1. Reduce each rate to a base unit.
   a. $\frac{80 \text{ mi}}{2 \text{ hr}}$ b. $\frac{64 \text{ gal}}{8 \text{ min}}$
   c. $\frac{65 \text{ ft}}{13 \text{ sec}}$ d. $\frac{24 \text{ oz}}{4 \text{ days}}$
2. Find the average speed of each airplane trip.
   a. 1170 mi in 3 hr
   b. 800 mi in 4 hr
   c. 4950 mi in 9 hr
3. Calculate each unit cost.
   a. 14 kg for $35
   b. 16 oz for $14.24
   c. 25 pieces for $43.75
   d. 10 boxes for $120

Answers:
1. a. 40 mi/hr b. 8 gal/min
   c. 5 ft/sec d. 6 oz/day
2. a. 390 mi/hr b. 200 mi/hr
   c. 550 mi/hr 3. a. $2.50/kg
   b. $0.89/oz c. $1.75/piece
   d. $12/box

P 5 Consumer Math In Example 3, on page 159, which cereal had the lower unit cost? What are some factors other than unit cost that help you determine the "better buy"?

P 6 Federico was bragging that he had received a raise of $80 a month. Katia said that she would not be impressed until she knew the ratio of his new salary to his old salary. Why?

Problems and Applications

P, R 7 Arrange the ratios $\frac{2}{3}, \frac{3}{5}, \frac{4}{7}$, and $\frac{5}{9}$ in order, from least to greatest. $\frac{5}{9}, \frac{4}{7}, \frac{3}{5}, \frac{2}{3}$

P, R 8 Letitia can walk at a speed of 6.9 kilometers per hour. At that rate, how far could she walk in
   a 3 hours? 20.7 km
   b 2 hours 20 minutes? 16.1 km

P, R 9 At Ed's Foods, the price of 10 pounds of potatoes is $4.50.
   a How much do the potatoes cost per pound? $0.45
   b How much do 25 pounds of potatoes cost? $11.25

P, I 10 a Determine the ratio $\frac{AC}{AB}$. $\frac{4}{3}$
   b Determine the ratio $\frac{AC}{BC}$. $\frac{6}{7}$
   c What is the perimeter of $\triangle ABC$?
      35 ft, or 11 yd 2 ft

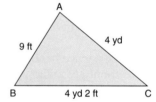

P, R 11 Bernice the bricklayer can set about 96 bricks per hour. If her workday is $7\frac{1}{2}$ hours long, about how many bricks can Bernice set in a day? ≈720 bricks

P 12 Consumer Math If a 12-ounce bottle of shampoo costs $2.80 and an 18-ounce bottle of the same shampoo costs $4.00, which is the better buy? The 18-oz bottle

P 13 A plane flew for 2 hours 45 minutes at a speed of 550 miles per hour. How far did the plane fly? 1512.5 mi

P 14 History The first United States census was taken in 1790. According to this census, there were 3,172,006 white people and 757,208 black people. What was the ratio of black people to white people in the United States in 1790? ≈1:4.2

P 15 Farmer Fox raises corn on 120 acres of land. This year he expects a yield of about 185 bushels of corn per acre. About how many bushels of corn does he expect to harvest? About 22,200 bushels

P, R 16 Consumer Math In the Ticonderoga Boutique is a sign reading Erasers: 6 for a Dollar. How much would you expect to pay for one eraser? Possible answer: $0.17

162 Chapter 4 Ratio and Proportion

SECTION 4.2

P **17** Los Angeles and San Francisco are about 400 miles apart. If the speed limit on the highway connecting these cities is 65 miles per hour, what is the shortest time in which a person could drive from San Francisco to Los Angeles without breaking the law?   ≈6.15 hr, or about 6 hr 9 min

P **18** Billy Backstroke kept his head shaved while he was a member of his school's swimming team. After swimming season was over, he decided to let his hair grow out again. How long did it take his hair to reach a length of 4 inches? (Human hair grows about 0.01 inch per day.)

P, I **19** Find the ratio of m∠C to m∠B.   4:5

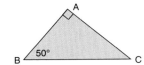

P **20** Annette drives 200 miles from her home to Bingo Beach in 4 hours. When she drove home, the trip took her 5 hours.
 a What was her average speed when she drove to the beach?   50 mi/hr
 b What was her average speed when she drove home?   40 mi/hr
 c What was her average speed for the round trip? $44.\overline{4}$ mi/hr

P, R **21** Two cars are directly opposite each other on a 1-mile oval track. One is traveling at 80 mph, the other at 95 mph. How long will it take the faster car to overtake the slower one?   2 min

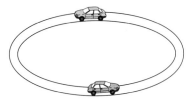

P, R **22** **Consumer Math** A 32-ounce bottle of Teucer cleaner costs $4.68. A 10-ounce bottle of Teucer concentrate costs $5.39. If it takes only a third as much concentrate as cleaner to do a cleaning job, which is the better buy?   The cleaner

P, R **23** When a person buys CD's from the Mayflower Compact Disc Club, the first disc costs $15.00, the second costs $7.50, and each additional disc costs $6.00. The Laser Trax store sells CD's for $9.75 each. If Aaron wants to buy three discs, where will he get the best price? Justify your answer.

P, R **24** **Spreadsheets** Use a spreadsheet to make a table showing the times needed to make a trip of 450 miles at speeds of 25, 30, 35, 40, 45, 50, 55, 60, and 65 miles per hour.

P, R **25** The graph shows numbers of magazine subscriptions sold during the annual fund-raising drive of the Yoknapatawpha School District. What is the ratio of the number sold by students in grades 9–12 to the number sold by students in grades 7–8? $\frac{309}{173}$

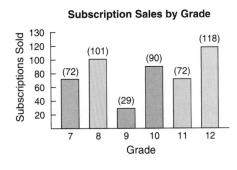

Cooperative Learning

- Have students develop advertising slogans for each of the cars in problem 4 relative to the other cars. Groups can present and justify their slogans to the class. Discuss how the slogans might be misleading.

Possible answers: Car A achieves the most miles per gallon. Car B goes farthest on a tank of gas. Car C costs the least to fill up. Car D uses less than one tenth of a gallon of gas per mile traveled.

- Problem 24 and the Investigation are appropriate for cooperative learning.

Additional Answers

for pages 162 and 163

5 Shreds; possible answers: Ingredients of a product; quality of a product; loyalty to a particular brand

6 Possible answer: Because the raise can be properly assessed only in comparison with the amount of his former salary (if Federico's salary had been $4000 per month, for example, his raise would be one of only 2%)

18 ≈400 days, or about a year and 5 weeks

23 Mayflower; since 15 + 7.5 + 6 < 3(9.75)

24
Speed (mi/hr)	Time (hr)
25	18
30	15
35	≈12.86
40	11.25
45	10
50	9
55	≈8.18
60	7.5
65	≈6.92

SECTION 4.2

Problem-Set Notes

for page 164

- **28** introduces the concept that a linear function has constant rate of change; here, $\frac{1}{2}$ inch per lie.

Checkpoint

1. Reduce each rate to a base unit.
a. $\frac{150 \text{ km}}{3 \text{ hr}}$ b. $\frac{175 \text{ mi}}{5 \text{ gal}}$ c. $\frac{86 \text{ in.}}{4 \text{ days}}$

2. Find the average speed for each car trip.
a. 330 mi in 6 hr
b. 224 km in 2.8 hr
c. 56 mi in 1.4 hr

3. Calculate each unit cost.
a. 3 lb for $5.84
b. 14 g for $9.66
c. 12 gal for $54

Answers: **1. a.** 50 km/hr
b. 35 mi/gal **c.** 21.5 in./day
2. a. 55 mi/hr **b.** 80 km/hr
c. 40 mi/hr
3. a. $1.95/lb **b.** $0.69/g
c. $4.50/gal

Additional Answers

for page 164

30 Possible answers: 15, 1; 30, 2; 7.5, 0.5; 5, $\frac{1}{3}$; 150, 10; 1, $\frac{1}{15}$

Investigation In a radar gun, Doppler radar is used to make speed measurements. It works on the basis of the Doppler effect—a change in wave frequency caused by motion. An outgoing electromagnetic wave strikes an object that is approaching the radar gun, and the wave is reflected at a higher frequency than it was sent out. When an object is moving away from the gun, it is bounced back at a lower frequency. By measuring the difference in frequency, Doppler radar determines the speed of the object. Radar guns are often accurate to within one mile per hour.

◀ LOOKING BACK **Spiral Learning** LOOKING AHEAD ▶

I **26** If $x = 3$ and $y = 3x + 4$, what is the value of y?   13

R **27** What percent of the tank is filled?   30%

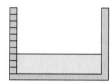

P, I **28** When Pinocchio was carved, his nose was 2 inches long. Each time he told a lie, his nose grew $\frac{1}{2}$ inch. How long was his nose after seven lies? $5\frac{1}{2}$ in.

I **29** If $7 + (-4) = 3$ and $7 - (-4) = 11$, what is the value of each expression?
a $7 + (-6)$ 1   **b** $7 - (-6)$ 13   **c** $7 + (-7)$ 0   **d** $3 - 7$ -4

I **30** List six values of x and y for which $x \div y = 15$.

R **31** *Newspeak* magazine is offering a one-year subscription for $24.97. At a newsstand, each weekly issue of the magazine costs $1.50. What percent of the newsstand price is the subscription price?   ≈32%

I **32** Find the value of
a $a + b$ 97
b $a + c$ 128
c $b + c$ 135

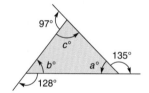

R **33** **Estimating** Julius, Leonard, Adolph, and Herbert went to a restaurant for dinner. The bill was $73.45, and they wanted to leave a 15% tip. Estimate the amount of the tip.   Possible answer: $11

P, R, I **34** What is the ratio of the number of sides of a hexagon to the number of sides of a nonagon?   2:3

R, I **35** Suppose that the value of x is chosen at random from the set {2, 4, 6, 8, 10, 12}. What is the probability that the area of the small square will be more than 50% of the area of the large square? $\frac{2}{3}$

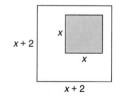

Investigation

Radar Guns Have you ever seen a "radar gun" being used to measure the speed of a car or the speed of a pitcher's fastball? Find out how this device works. How accurate are radar guns? If possible, interview a police officer to find out the margin for error in radar measurements of cars' speeds.

Chapter 4 Ratio and Proportion

SECTION 4.3

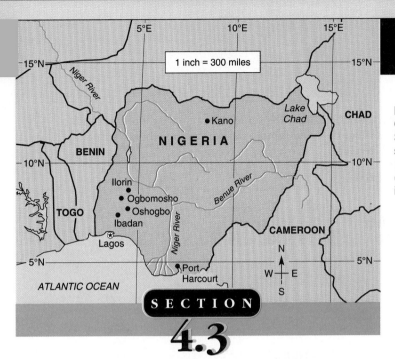

Investigation Connection
Section 4.3 supports Unit B, *Investigations* 1, 2, 5. (See pp. 92A–92D in this book.)

4.3
More Applications of Ratios

SCALE AND RATIOS

Most maps contain a legend telling how lengths on the map correspond to lengths in the real world. Each inch on the map above, for example, represents a length of 300 miles. The **scale** of a map is the ratio of distances on the map to real-world distances, so the scale of this map is 1 in. : 300 mi.

When units are the same, such as 1 ft:72 ft, you can give the scale without units, as in the following example.

Example 1
Orville built a 1:72 scale model of a jet plane. If the wingspan of the model is $9\frac{1}{2}$ inches, what is the wingspan of the real plane?

Solution
The scale of 1:72 means that each dimension of the real jet is 72 times the corresponding dimension of the model. Since the wingspan of the model is 9.5 inches, the wingspan of the jet is 9.5(72), or 684, inches. This is equivalent to 684 ÷ 12, or 57, feet.

Objectives
- To solve problems involving scale ratios
- To solve problems involving densities

Vocabulary: density, scale

Class Planning

Schedule 2–3 days

Resources

In *Chapter 4 Student Activity Masters* Book
Class Openers 25, 26
Extra Practice 27, 28
Cooperative Learning 29, 30
Enrichment 31
Learning Log 32
Interdisciplinary
  Investigation 33, 34
Spanish-Language Summary 35

In *Tests and Quizzes* Book
Quiz on 4.1–4.3 (Basic) 85
Quiz on 4.1–4.3 (Average) 86
Quiz on 4.1–4.3 (Honors) 87

Class Opener

Refer to the map on page 165.

1. Measure the following distances in inches:
   a. Kano to Port Harcourt
   b. Ilorin to Ibadan
   c. Port Harcourt to Oshogbo

2. Convert each distance to miles, using the scale provided.

Answers:
1. a. $1\frac{3}{4}$ in. b. $\frac{5}{16}$ in. c. $\frac{7}{8}$ in.
2. ≈525 mi; ≈94 mi; ≈263 mi

Lesson Notes

- Some students may be familiar with the idea of scaling from models that they have made. Let these students share what knowledge they already possess.

SECTION 4.3

Lesson Notes continued

- The concept of density may be unfamiliar to many students. You may want to demonstrate that the same volume of two materials with different densities will have different masses.

Additional Examples

Example 1
Ernest built a 1:50 scale model of a boat. If the length of the model is 11 in., what is the length, in feet, of the real boat?
Answer: ≈45.8 ft

Example 2
If a 15 cm³ piece of oak has a mass of 10.8 g, what is the density of the oak?
Answer: 0.72 g/cm³

Problem-Set Notes

for pages 166 and 167

- **3** requires that students measure a side of the house drawing in centimeters and compare that to either 12 m or 8.4 m to get the 1:240 ratio. Students may give different but equivalent ratios, such as $\frac{5 \text{ cm}}{12 \text{ m}}$, $\frac{2 \text{ in.}}{12 \text{ m}}$, or $\frac{1 \text{ in.}}{6 \text{ m}}$. These equivalent ratios can be used to preview of the discussion of proportions in Section 4.5.

Stumbling Block

In example 1, students may assume that 1:72 means 1 inch : 72 feet. Emphasize that if units are not included in a ratio, this means that like units are being compared.

DENSITY

Two objects can be exactly the same size but have different weights, or mass. That is because one is more *dense* than the other. A substance's **density** is the ratio of its mass to its volume.

$$\text{Density} = \frac{\text{mass}}{\text{volume}}$$

Example 2
If a 9-cubic-inch piece of granite has a mass of $14\frac{1}{2}$ ounces, what is the density of the granite?

Solution
We use the formula given above.

$$\text{Density} = \frac{\text{mass}}{\text{volume}} = \frac{14.5 \text{ oz}}{9 \text{ in.}^3} \approx \frac{1.61 \text{ oz}}{1 \text{ in.}^3}, \text{ or } 1.61 \text{ oz/in.}^3$$

The density of the granite is about 1.61 ounces per cubic inch.

This example shows that density is a kind of rate, expressed in units of mass per unit of volume. The table shows the densities of a few common substances in grams per cubic centimeter and in pounds per cubic foot. More densities are listed in the table of densities in the back of the book.

Substance	Density g/cm³	Density lb/ft³
Liquid water	1.0	62.4
Ice	0.922	57.5
Steel	7.8	486.7
Aluminum	2.7	168.5
Wood (pine)	0.56	34.9

Sample Problems

Problem 1 Use the map of Nigeria on page 165 to find the distance, in miles, between Kano and Lagos.

Solution If you use a ruler to measure the distance between Kano and Lagos on the map, you will find that they are about $1\frac{3}{4}$ inches apart. Since 1 inch on the map represents 300 miles, the cities are about $1\frac{3}{4} \cdot 300$, or 525, miles apart.

Problem 2 The cube on the left is made of ice.
The cube on the right is made of steel.
Which is heavier?

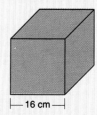

SECTION 4.3

Solution The volume of the ice cube is 16^3, or 4096, cm^3. The volume of the steel cube is 8^3, or 512, cm^3. According to the table on page 166, the density of ice is 0.922 g/cm^3 and the density of steel is 7.8 g/cm^3.

Ice:

$$\text{Density} = \frac{\text{mass}}{\text{volume}}$$

$$0.922 \text{ g/cm}^3 = \frac{\text{mass}}{4096 \text{ cm}^3}$$

$$\frac{0.922 \text{ g}}{1 \text{ cm}^3}(4096 \text{ cm}^3) = \text{mass}$$

$$3777 \text{ g} \approx \text{mass}$$

Steel:

$$\text{Density} = \frac{\text{mass}}{\text{volume}}$$

$$7.8 \text{ g/cm}^3 = \frac{\text{mass}}{512 \text{ cm}^3}$$

$$\frac{7.8 \text{ g}}{1 \text{ cm}^3}(512 \text{ cm}^3) = \text{mass}$$

$$3994 \text{ g} \approx \text{mass}$$

The two cubes actually weigh about the same, with the steel cube being slightly heavier.

Think and Discuss

P **1 Estimating** Estimate the distance, by plane, between Thomasville and St. Henry.
Possible answer: About 200 mi.

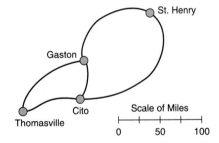

P **2** Why does ice float in water? (Hint: What are the densities of ice and water?) Because frozen water is less dense than liquid water

P, R **3 a** What is the scale of this house plan? 1:240
b What are the dimensions of the kitchen? Of the master bedroom?
$\approx$2.9m × $\approx$4.8m; $\approx$3.6m × $\approx$4.6m

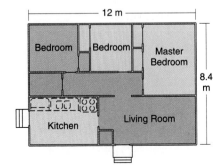

Assignment Guide

Basic
Day 1 6, 7, 10, 12, 14, 18, 28, 29, 34, 35
Day 2 8, 11, 17, 19, 23, 27, 31, 36, 38
Day 3 9, 13, 15, 20, 26, 30, 33, 37

Average or Heterogeneous
Day 1 6, 7, 10–13, 15, 16, 24, 28, 31
Day 2 8, 9, 14, 17–19, 22, 25, 29, 34
Day 3 20, 21, 23, 26, 27, 30, 32, 33, 35–38

Honors
Day 1 9, 11, 13, 15–17, 21, 22, 25, 26, 29, 30, 33, 35
Day 2 7, 8, 10, 12, 14, 18, 19, 23, 24, 27, 28, 31, 32, 34, 36, 38, Investigation

Strands

Communicating Mathematics	2, 19, 38
Geometry	25, 30, 32, 38
Measurement	1–28, 31, 33, Investigation
Data Analysis/ Statistics	5, 27, 34
Technology	5, 13, 23, 27
Algebra	30, 35–37
Patterns & Functions	21
Number Sense	1, 15, 17, 22, 29, 32, 34, 38, Investigation
Interdisciplinary Applications	2, 5, 13, 15, 18, 19, 24, 26–28, 33, Investigation

Individualizing Instruction

The kinesthetic learner Students may benefit from demonstrations of density, such as feeling the weights of same-size cubes of different materials or discovering which of these cubes float in water.

SECTION 4.3

Problem-Set Notes

for pages 168 and 169

- 5 might include a discussion about the relevance of the figure 264.5 people/mi². What is its meaning? Could you find a square mile with 265 people in it? With 264.5 people?
- 17 may not be solved correctly by some students because the smaller ratio yields a larger map. Some discussion might be helpful.
- 19 is a good application of density. Some students will know from experience that oil floats on water.

Additional Practice

1. If a 100-meter bridge appears as the following lengths on a set of maps, what is the scale of each map?
   a. 10 cm   b. 5 cm
   c. 2 cm   d. 50 cm
2. Calculate the following densities.
   a. 500 lb per 50 ft³
   b. 64 kg per 200 cm³
   c. 78 g per 13 cm³
3. Use the table on page 661 to calculate the weight of the following.
   a. 60 cm³ of gold
   b. 100 ft³ of helium
   c. 35 cm³ of pine
   d. 16 ft³ of gasoline
   e. 400 cm³ of glass

Answers:
1. a. 1 cm : 10 m b. 1 cm : 20 m
   c. 1 cm : 50 m d. 1 cm : 2 m
2. a. 10 lb/ft³ b. 0.32 kg/cm³
   c. 6 g/cm³
3. a. 1158 g, or 1.158 kg
   b. 1.123 lb c. 19.6 g
   d. 659.2 lb
   e. 1040 g, or 1.040 kg

P, R   **4** Balsa wood has a density of about 130 kg/m³.
      a How much does a 4-cubic-meter block of balsa wood weigh? ≈520 kg
      b How many grams does a cubic centimeter of balsa wood weigh? ≈0.13 g

P, R   **5** The land area of the state of Ohio is 41,004 square miles. In 1990, the population of Ohio was about 10,847,000. What was Ohio's population density (in people per square mile) in 1990? ≈264.5 people/mi²

Problems and Applications

P   In problems 6–8, find the density of a substance with the given mass and volume.

  6 Mass: 150 g   **7** Mass: 4 lb   **8** Volume: 2.3 m³
     Volume: 40 cm³   Volume: 29 in.³   Mass: 0.01 kg

P   **9** The scale of a model airplane is 1:32. The tail fin of the model is 4 inches high. How high, in feet and inches, is the tail fin of the actual plane? 10 ft 8 in.

P   In problems 10–12, rewrite each map scale as a ratio of numbers.

  10 1 in. : 50 mi   **11** 1 cm : 100 km   **12** $1\frac{1}{2}$ in. : 100 ft

P, R   **13** Science   Gasoline has a density of 660 kilograms per cubic meter. A cubic meter of gasoline contains about 264 gallons. How much does a gallon of gasoline weigh? ≈2.5 kg

P   **14** The distance between two cities is 150 miles. On a map, the cities are represented by dots 3 inches apart. What is the scale of the map?

P, R   **15** Estimating   On the map shown, 1 inch represents 250 miles. Use the map to estimate the distance between
    a Milwaukee and Louisville ≈360 mi
    b Indianapolis and Chicago ≈165 mi

Chapter 4 Ratio and Proportion

SECTION 4.3

P, R **16** This 2-inch segment represents a distance of 5280 feet.

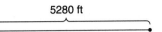

   a Draw, to the same scale, a segment that represents 1760 yards.
   b What is the ratio of the length of the segment shown to the length of the segment you drew in part **a**?   **a** Segment should be 2 in. long.   **b** 1:1

P **17** In Miss McPherson's classroom are two maps of the United States. The scale of one is 1:5,000,000, and the scale of the other is 1:3,500,000. Which map is larger?   The one with a scale of 1:3,500,000

P **18** The ratio between the height of the Empire State Building and its height in this picture is about 11,200:1. How tall is the Empire State Building?   ≈1400 ft

P **19** **Science**   Water has a density of 1 g/cm^3, and lubricating oil has a density of 0.9 g/cm^3.
   a Does oil float on water, or does water float on oil? Justify your answer.
   b A container weighs 300 grams and has a capacity of 2.1×10^3 cm^3. How much does the container weigh when filled with water? When filled with oil?   2400 g, or 2.4 kg; 2190 g, or 2.19 kg

P **20** A map has a scale of 1 inch : 60 miles. What distance on the map represents 165 miles? $2\frac{3}{4}$ in.

P **21** Phil wants to enlarge a photo that is 3 inches wide and 5 inches high to make a 24-inch-wide poster. How high will the poster be?   40 in.

P, R **22** **Estimating**   Estimate the distance between Veronica Lodge and Pinepole Lodge.
Possible answer: 13 mi.

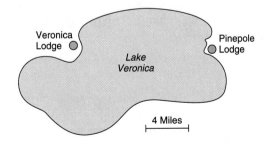

P, R, I **23** The Abstract Concrete Company makes concrete blocks measuring 30 cm by 14 cm by 20 cm. If the concrete has a density of 2000 kg/m^3, how much does a pallet of 128 blocks weigh?   2150.4 kg

P, R **24** **Architecture**   Lynn is making her niece Katy a dollhouse that is a scale model of Katy's real house. The living room of the dollhouse measures 18 inches by 12 inches. Katy's living room measures 24 feet by 16 feet. If the fireplace in Katy's living room is 8 feet wide and 5 feet high, what size should the fireplace in the dollhouse be?   6 in. by $3\frac{3}{4}$ in.

Reteaching

Refer to the map on page 165. You may wish to use the following method to reteach changing map scales to like units.

$$\frac{1 \text{ in.}}{300 \text{ mi}} \times \frac{1 \text{ mi}}{5280 \text{ ft}} \times \frac{1 \text{ ft}}{12 \text{ in.}}$$

$$= \frac{1}{19,008,000}$$

Using this method, students can more easily see the cancellation of like units of measurement and whether 5280 and 12 belong in the numerator or the denominator.

Multicultural Note

Population density, mentioned in problems 5 and 27, varies greatly in different cities, as shown by the following data: Hong Kong: 247,004 persons per square mile (ppm^2); Bombay: 120,299 ppm^2; Rome: 43,556 ppm^2; Mexico City: 38,528 ppm^2; New York City: 11,473 ppm^2. Asia has the highest continental population density at 168 ppm^2, and Australia has the lowest at 5 ppm^2.

Additional Answers

for pages 168 and 169

6 3.75 g/cm^3

7 ≈0.14 lb/in.^{3}

8 ≈0.004 kg/m^3

10 1:3,168,000

11 1:10,000,000

12 1:800

14 1 in. : 50 mi, or 1:3,168,000

19 a Oil floats on water, since the density of oil is less than that of water.

SECTION 4.3

Problem-Set Notes

for pages 170 and 171

- **25** may surprise some students. Any ball with a 1 ft diameter might seem liftable regardless of its material.
- **34** is an important example, mathematically and physically. Twice as fast does not mean twice as far. Discuss this.

Checkpoint

1. How long would a 10-mile lake appear on maps with the following scales?
   a. 1 in. : 30 mi b. 1 cm : 5 mi
   c. 1 in. : 100 mi d. 1 cm : 1 mi
2. Calculate the following densities.
   a. 24 lb per 6 ft^3
   b. 55 g per 25 cm^3
   c. 18 oz per 36 in.^{3}
3. Use the table on page 661 to determine the denser substance.
   a. Glass or oak
   b. Gasoline or milk
   c. Air or helium

Answers:
1. a. $\frac{1}{3}$ in. b. 2 cm c. 0.1 in.
   d. 10 cm 2. a. 4 lb/ft^3
   b. 2.2 g/cm^3 c. 0.5 oz/in.^{3}
3. a. Glass b. Milk c. Air

Communicating Mathematics

- Have students imagine they are cartographers. Have them write a letter in response to a student who wonders why map scales are sometimes given in unitless form (for example, 1:19,008,000 for the map of Niger on page 165).
- Problems 2, 19, and 38 encourage communicating mathematics.

P, I **25** Arnold is given a set of six balls, each having a radius of $\frac{1}{2}$ foot. The balls are made of different materials—steel, aluminum, copper, gold, oak, and glass. If the greatest weight Arnold can lift is 200 pounds, which of the balls could he lift? (Hint: Use the table of densities in the back of the book and the fact that the volume of a sphere with radius r is equal to $\frac{4}{3}\pi r^3$.) The aluminum, oak, and glass balls

P, R **26 Consumer Math** Tony's car gets 24 miles to the gallon. If gas costs $1.20 per gallon, how much will Tony save by taking the direct route from Franklin to Accarda instead of going by way of Bolen? $1.45

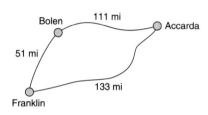

P **27**

State	Population	Land Area (square miles)
Nevada	1,202,000	109,872
Connecticut	3,287,000	4,872

a Use the table to determine the population densities of Nevada and Connecticut. ≈11 people/mi^2; ≈675 people/mi^2
b What would the population of Nevada have to be for its population density to be the same as Connecticut's? ≈74,200,000
c The Nevada cities of Las Vegas and Reno have populations of 258,000 and 134,000. If all the people in these two cities moved to Connecticut, what would the population densities of Nevada and Connecticut be?
≈7 people/mi^2; ≈755 people/mi^2

◄ LOOKING BACK **Spiral Learning** LOOKING AHEAD ►

R **28 Consumer Math** A gallon of apple juice costs $4.25. A quart of apple juice costs $1.65. Which is a better buy? The gallon

R **29** On a 75-question test, Mandy answered 62 questions correctly. On a 70-question test, Sandy answered 59 questions correctly. Which student has the greater percent of correct answers? Sandy

I **30** Use a base and an exponent to write an expression for each measurement.
a The length of one side of the cube 8^1
b The area of one face of the cube 8^2
c The volume of the cube 8^3

R **31** What is the ratio of the value of ten 29¢ stamps to the value of fifteen 25¢ stamps? 290:375, or 58:75

170 **Chapter 4** Ratio and Proportion

SECTION 4.3

I **32** Which of the figures below represent figures the same shape as the 3-by-5 rectangle shown? **b** and **d**

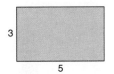

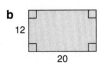
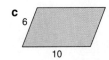
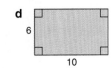

R, I **33** A freight train leaves New Orleans at noon, headed for Phoenix at 45 mph. At 2 P.M. an express train leaves New Orleans for Phoenix, traveling at 60 mph. At what time will the express train pass the freight train? 8 P.M.

I **34** Stopping distances of a car with good brakes on dry pavement are given in the graph. Estimate the stopping distance at a speed of
 a 20 mph ≈50 ft
 b 40 mph ≈125 ft
 c 60 mph ≈225 ft

I In problems 35–37, solve each equation for x.

35 $\frac{1}{x} = \frac{3}{5}$ $x = \frac{5}{3}$ **36** $\frac{1}{x} = 6$ $x = \frac{1}{6}$ **37** $\frac{1}{x} = \frac{1}{3}$ $x = 3$

I **38** **Communicating** Four students used four different methods to find the area of the unshaded part of this figure. Their calculations are shown below. Describe the method each student used, and explain why the method works.
Anna: 12(15) – 5(5) – 5(5) = 180 – 25 – 25 = 130
Bert: 12(15) – 2(5)(5) = 180 – 50 = 130
Carl: 10(5) + 2(5) + 10(7) = 50 + 10 + 70 = 130
Dana: 12(5) + 10(7) = 60 + 70 = 130

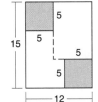

Investigation

Auto Trip Holly Day is planning an automobile trip from Atlanta, Georgia, to Chicago, Illinois, to Minneapolis, Minnesota, to Portland, Oregon, to Los Angeles, California, to New Orleans, Louisiana, to Miami, Florida, then back to Atlanta. What information would you need in order to calculate her fuel expenses for the trip? Make some assumptions and estimate how much money she will spend on gasoline.

SECTION 4.4

Objective
- To use dimensional analysis to change the units in which measurements are expressed

Vocabulary: dimensional analysis

Class Planning

Schedule 2–3 days

Resources

In *Chapter 4 Student Activity Masters* Book
Class Openers *37, 38*
Extra Practice *39, 40*
Cooperative Learning *41, 42*
Enrichment *43*
Learning Log *44*
Using Manipulatives *45, 46*
Spanish-Language Summary *47*

Class Opener

Suppose that a major-league pitcher can throw a baseball at a speed of 90 miles per hour.

1. Convert 90 miles per hour into miles per minute.

2. Convert your answer to **1** to miles per second.

3. Convert your answer to **2** to feet per second (1 mi = 5280 ft).

4. The distance from the pitcher's mound to home plate is 60 ft 6 in. How long does it take the pitch to reach home plate?

Answers:
1. 1.5 mi/min 2. 0.025 mi/sec
3. 132 ft/sec 4. ≈0.46 sec

Investigation Connection
Section 4.4 supports Unit B, *Investigations* 2, 5. (See pp. 92A-92D in this book.)

SECTION 4.4
Working with Units

By now, you can probably figure out how many feet are in 5 yards or how many minutes are in 2 hours without too much trouble. But do you know how to convert from liters to gallons? Do you know the number of gallons in a cubic foot?

Example 1
The owner's manual for Mercedes's new sports car says that the car's gas tank holds 50 liters. The local gas station, however, sells gasoline by the gallon, not by the liter. How many gallons of gas does the tank hold?

Solution
We need to express 50 liters as a number of gallons. Looking at the table of equivalent measurements in the back of the book, we find that 1 gallon is about 3.785 liters, so $\frac{1 \text{ gallon}}{3.785 \text{ liters}} \approx 1$. We can therefore multiply 50 liters by this ratio to obtain an equivalent measurement.

$$50 \text{ liters} \approx 50 \text{ liters} \cdot \frac{1 \text{ gallon}}{3.785 \text{ liters}} \approx 13.2 \text{ gallons}$$

Notice that liters "canceled" and we were left with the desired units, gallons, after multiplying by a fraction equal to 1. This is an example of **dimensional analysis.**

Example 2
How many times does a person's heart beat in one day if it beats an average of 63 times per minute?

Solution
We are given that the person's heart beats 63 times per minute, and we know that there are 60 minutes in an hour and 24 hours in a day. Since we want our

SECTION 4.4

answer to be in beats per day, we can multiply by appropriate conversion factors so that all units will cancel except beats/day.

$$63 \text{ beats/minute} = \left(\frac{63 \text{ beats}}{1 \text{ minute}}\right)\left(\frac{60 \text{ minutes}}{1 \text{ hour}}\right)\left(\frac{24 \text{ hours}}{1 \text{ day}}\right)$$

$$= \frac{63 \cdot 60 \cdot 24 \text{ beats}}{1 \text{ day}}$$

$$= 90{,}720 \text{ beats/day}$$

Be sure to include units in all your calculations involving measurements. If the units don't come out right in the answer, go back and look for an error. A common error is for one of the fractions to be upside down. Since the numerator and the denominator of a conversion factor are equal, the fraction can be "flipped over" so that the correct units will be on the top.

Sample Problems

Problem 1 How many gallons of lime gelatin does it take to completely fill the wading pool shown?

(Diagram: cylindrical wading pool, 6 ft diameter, $1\frac{1}{2}$ ft height)

Solution The pool is cylindrical, and we need to know the volume of space it encloses to determine the amount of liquid it will hold. According to the table of geometric formulas in the back of the book, the volume of a cylinder is equal to $\pi r^2 h$, where r represents the cylinder's radius and h represents its height. This pool has a radius of 3 feet and a height of 1.5 feet.

$$\text{Volume} = \pi \cdot r^2 \cdot h$$

$$= \pi \cdot 3^2 \cdot 1.5$$

Your calculator shows 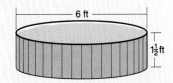 42.41150082, so the pool will hold about 42 cubic feet of lime gelatin. From the table of equivalent measurements in the back of the book, we find that 1 gallon is equal to 231 cubic inches. Furthermore, since there are 12 inches in a foot, there are 12^3, or 1728, cubic inches in a cubic foot. We can now convert the volume from cubic feet to cubic inches to gallons.

$$42 \text{ ft.}^3 = (42 \text{ ft}^3)\left(\frac{1728 \text{ in.}^3}{1 \text{ ft}^3}\right)\left(\frac{1 \text{ gal}}{231 \text{ in.}^3}\right)$$

$$= \frac{42 \cdot 1728}{231} \text{ gal}$$

$$\approx 314 \text{ gal}$$

It will take about 314 gallons of lime gelatin to fill the pool.

Lesson Notes

- Some of your students may have gotten into the habit of simply multiplying or dividing by a conversion factor when converting units (e.g., to convert 24 inches to feet). When problems involve more than one conversion, however, the techniques of dimensional analysis (i.e., multiplying by one conversion factor at a time and crossing out units) will prove most helpful.

Additional Examples

Example 1
A pitcher is advertised to hold 1.5 liters. How many quarts can the pitcher hold?
Answer: ≈1.59 qt

Example 2
How many times do a hummingbird's wings beat in one minute if they beat 70 times per second?
Answer: 4200

Cooperative Learning

- Have students work together to invent a system of linear measurement with a basic unit, two related units, and conversion factors for changing one unit to another and to English units. Have students express the following measurements in their systems: the width of a desk, the width of a math book, the height of a doorway, the length of the classroom, and the height of each student in the group.
- Problems 6, 25, and Investigation are appropriate for cooperative learning.

SECTION 4.4

Assignment Guide

Basic
Day 1 7, 8, 12, 16, 20, 25, 28, 31, 33
Day 2 9, 11, 13, 18, 22, 24, 29, 32, 37
Day 3 10, 14, 15, 19, 23, 26, 30, 34, 35

Average or Heterogeneous
Day 1 7–10, 12, 14, 15, 19, 21, 23, 26, 28, 31, 34, 37
Day 2 11, 13, 16–18, 20, 22, 24, 25, 29, 32, 33, 35, 36

Honors
Day 1 7, 8, 12, 14, 16, 17, 20, 21, 26, 28, 31, 37
Day 2 9, 10, 13, 18, 19, 22, 25, 27, 32–34, Investigation

Strands

Communicating Mathematics	2, 5, 26, 37
Geometry	15
Measurement	1–6, 8–25, 27–30, 33, 34, 36, Investigation
Technology	9, 15, 16, 37
Algebra	5, 7, 8, 12, 13, 31, 32, 35
Number Sense	2, 4, 6, 27, 37, Investigation
Interdisciplinary Applications	2, 6, 17, 20, 22, 34, 36, 37, Investigation

Individualizing Instruction

Learning-disabled students Some students may have difficulty canceling the correct units in dimensional analysis problems. It might be helpful for them to use different-colored pencils or markers to cancel similar units.

Problem 2 How many miles did a car travel in 3 hours if it was traveling at a speed of 50 km/hr?

Solution The speed of the car is given as 50 kilometers per hour. In 3 hours the car traveled $(3 \text{ hr})\left(50 \frac{\text{km}}{\text{hr}}\right)$, or 150 km. But we need the answer in miles, so we multiply by 1 in the form $\frac{1 \text{ mi}}{1.61 \text{ km}}$, since 1 mi ≈ 1.61 km.

$$150 \text{ km} \approx 150 \text{ km} \cdot \frac{1 \text{ mi}}{1.61 \text{ km}}$$

$$\approx \frac{150}{1.61} \text{ mi}$$

$$\approx 93.2 \text{ mi}$$

The car traveled about 93.2 miles.

Think and Discuss

P, R **1** Which of the following ratios are equal to 1? **a, c**, and **d**
  a $\frac{1 \text{ yd}}{3 \text{ ft}}$ **b** $\frac{5 \text{ gal}}{10 \text{ qt}}$ **c** $\frac{288 \text{ in.}^2}{2 \text{ ft}^2}$ **d** $\frac{100 \text{ cm}}{1 \text{ m}}$

P, R **2** *Estimating* The speed of sound is approximately 1100 feet per second.
  a About how many miles will the sound of thunder travel in 5 seconds? 1 mi
  b Explain how to estimate how far away a storm is by counting the seconds between seeing the lightning and hearing the thunder.

P, R **3 a** How many seconds are there in an hour? 3600 sec
  b If it takes a snail 10 seconds to travel an inch, how far could the snail travel in an hour? 360 in.

P, R **4** Match each metric unit with the closest customary unit.
  a quart II   I meter
  b inch III   II liter
  c mile IV   III centimeter
  d yard I   IV kilometer

P **5** Copy each expression, filling in the missing number. Then simplify the expression, being sure to include the proper units, and describe a situation in which you might use the expression.

  a $\frac{4 \text{ mi}}{1 \text{ min}} \cdot \frac{3 \text{ min}}{\underline{} \text{ sec}}$ **b** $\frac{34 \text{ mi}}{1 \text{ gal}} \cdot \frac{2 \text{ gal}}{1 \text{ hr}} \cdot \frac{\frac{1}{2} \text{ hr}}{\underline{} \text{ min}}$

P, R **6** In 1985, the A. C. Nielsen Company found that in the United States the average child spends 3 hours 54 minutes watching television each day. Estimate the number of hours of TV programs that the average child watches in a year. Possible answer: ≈1460 hr

Chapter 4 Ratio and Proportion

SECTION 4.4

Problems and Applications

7 Evaluate each product.
  a $\frac{7}{3} \cdot \frac{4}{4}$ $\frac{7}{3}$
  b $\frac{7}{3} \cdot \frac{1.2}{1.2}$ $\frac{7}{3}$
  c $\frac{7}{3} \cdot \frac{x}{x}$ $\frac{7}{3}$
  d $\frac{7}{3} \cdot \frac{x^5}{x^5}$ $\frac{7}{3}$

8 Simplify each expression, being sure to include the correct units.
  a $\frac{6 \text{ dollars}}{1 \text{ hour}} \cdot \frac{5 \text{ hours}}{1 \text{ workday}}$ $30/workday
  b $\frac{3 \text{ pounds}}{1 \text{ minute}} \cdot \frac{7 \text{ minutes}}{1 \text{ person}}$ 21 lb/person

9 If a canoe is 17 feet long, how many canoe lengths are in a mile? ≈310.6

10 Lynda is 5 feet 3 inches tall, and Don is 6 feet 3 inches tall. What is the ratio of Lynda's height to Don's height? 21:25

11 Use dimensional analysis to express
  a 5 gallons as a number of quarts 20 qt
  b 6 gallons as a number of pints 48 pt
  c $2\frac{1}{2}$ pounds as a number of ounces 40 oz

In problems 12 and 13, simplify each expression.

12 $\frac{4 \text{ ft}}{1 \text{ sec}} \cdot \frac{120 \text{ sec}}{2 \text{ min}}$ 240 ft/min

13 $\frac{5 \text{ yd}}{1 \text{ min}} \cdot \frac{1 \text{ min}}{60 \text{ sec}} \cdot \frac{36 \text{ in.}}{1 \text{ yd}}$ 3 in./sec

14 Express 32 pounds as a number of kilograms. ≈14.5 kg

15 What is the volume of the cube in cubic feet? ≈551.8 ft³

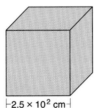
2.5×10^2 cm

16 The printer that Carlo uses with his computer prints 160 characters per second. How many characters can the printer print in 10 minutes? 96,000

17 Maple syrup is delivered to Dr. Terror's House of Pancakes in 1-gallon cans. The syrup is served to customers in 8-ounce pitchers. How many pitchers can be filled from one can? 16 pitchers

18 Danielle earns $5.50 per hour. She works $5\frac{1}{2}$ hours per day, 24 days per month.
  a How much does she earn per month? $726
  b How much does she earn for each 8 hours she works? $44

19 Olivia can maintain an average speed of 14 miles per hour when riding her bicycle.
  a How far can she cycle in 8 hours? 112 mi
  b If she cycles 8 hours a day, how long will it take her to travel 2000 miles? ≈17 da 7 hr

Section 4.4 Working with Units

Stumbling Block

In example 1, some students may not know whether to multiply by $\frac{1 \text{ gallon}}{3.785 \text{ liters}}$ or by $\frac{3.785 \text{ liters}}{1 \text{ gallon}}$. Point out that 50 liters is equivalent to $\frac{50 \text{ liters}}{1}$, so we must multiply by $\frac{1 \text{ gallon}}{3.785 \text{ liters}}$ in order for the term *liters* to cancel.

Reteaching

Make a set of 20 cards, each with a measurement on it. Use either metric or English units. Have students pick pairs of cards. Then have them make a ratio of the measures they chose and try to come up with a real-life situation in which they might encounter this ratio.

Additional Answers

for pages 174 and 175

2 b You can estimate how far away a storm is (in miles) by dividing the number of seconds between seeing the lightning and hearing the thunder by 5.

5 a Missing expression: 180 sec; ≈0.07 mi/sec; possible situation: Wanting to compute the speed in miles per second of a race car that travels at 4 mi/min for 3 min

b Missing expression: 30 min; ≈1.13 mi/min; possible situation: Wanting to compute the speed in miles per minute of a car that gets 34 miles per gallon of gasoline and is consuming gas at a rate of 2 gal/hr for a period of $\frac{1}{2}$ hr

SECTION 4.4

Problem-Set Notes

for pages 176 and 177

- **20, 24,** and **26** are problems that represent real applications.
- **37** refers to how computers use the binary, or base-two, system. The number 1024 is 2^{10}.

Checkpoint

1. Convert each length to meters.
   a. 6 ft   b. 3 ft 5 in.
   c. 4 yd   d. 1 mi
2. Convert each measure to English units.
   a. 3 g   b. 23 cm
   c. 4 kg   d. 3 m
3. Which measure is greater?
   a. 2 kg or 2 lb
   b. 100 m or 100 yd
   c. 10 in. or 10 cm
   d. 5 g or 5 oz

Answers:
1. a. ≈1.8288 m b. ≈1.0414 m
   c. ≈3.6576 m d. ≈1610 m
2. a. ≈11 oz b. ≈9.1 in.
   c. ≈8.83 lb d. ≈9.8 ft
3. a. 2 kg b. 100 m
   c. 10 in. d. 5 oz

Communicating Mathematics

- Have students use dimensional analysis to change 346,289 in. to mi and 346,289 cm to km. Then have students write a paragraph explaining which process seems easier. In the students' opinions, is dimensional analysis more useful in the metric system or the English system? Students should support their answers.

- Problems 2, 5, 26, and 37 encourage communicating mathematics.

P, R   **20** Mr. McCall knows that his truck, including its contents, weighs 6 tons. If he comes to a bridge in Canada that has a posted load limit of 5000 kilograms, should he drive across the bridge?   No

P, R   **21** Pauline drove from Bodega Bay to Santa Mira by way of Hadleyville. She returned to Bodega Bay by way of Bedford Falls. How many kilometers long was the round trip?   ≈306 km

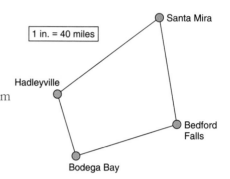

P   **22** The directions on a can of powdered lemonade mix say to use two scoops of mix to make an 8-ounce glass of lemonade. How many scoops of mix are needed to make
   a A quart?   8 scoops   **b** A 12-ounce glass?   3 scoops
   c A 10-ounce glass? $2\frac{1}{2}$ scoops   **d** A liter?   ≈$8\frac{1}{2}$ scoops

P, R   **23** The scale of a map is 1 inch : 60 miles. Boris measured a journey on the map and found it to be $4\frac{1}{2}$ inches long. If Boris drives at an average speed of 50 miles per hour, how long will it take him to make the journey?   5.4 hr, or 5 hr 24 min

P   **24** The rules of a 500-mile automobile race restrict each car's fuel supply to a maximum of 110 gallons.
   a How many miles per gallon must a car average for its driver to complete the race?   ≈4.55 mi/gal
   b If a car averages 6 miles per gallon, how much of its fuel supply will be left at the end of the race? $26\frac{2}{3}$ gal

P, R   **25** Hilda planned to go hiking in a national forest. The government map of the region was too big to carry, however, so she decided to make her own map. On the government map, 6 inches represented 25 miles, and the distance from Jailhouse Rock to Blueberry Hill was 8 inches. If Hilda drew her map to a scale of 2 inches : 25 miles, how far apart were Jailhouse Rock and Blueberry Hill on her map? $2\frac{2}{3}$ in.

P, R   **26 Communicating**   A car going 60 miles per hour is traveling 88 feet per second. How can you use this information to compute the number of feet in a mile?

P   **27** In the land of Decimalia, each day is divided into 10 gours, each gour into 10 linutes, and each linute into 10 reconds.
   a If a Decimalian tells you that it is 4 linutes past 6, what time is it according to our system?   3:21:36 P.M.
   b How would a Decimalian express 10:15 A.M.?   ≈4:2:7

176   **Chapter 4**   Ratio and Proportion

SECTION 4.4

◀ LOOKING BACK **Spiral Learning** LOOKING AHEAD ▶

P, R **In problems 28–30, find each ratio of lengths.**

28 $\frac{8 \text{ feet}}{4 \text{ yards}}$ $\frac{2}{3}$ 29 $\frac{9 \text{ feet}}{3 \text{ yards}}$ 1   30 $\frac{36 \text{ inches}}{1 \text{ yard}}$ 1

R, I **In problems 31 and 32, solve each equation for x.**

31 $\frac{x}{4} = \frac{9}{12}$ $x = 3$ 32 $\frac{x}{18} = \frac{3}{2}$ $x = 27$

R 33 Jasmine is 5'8" tall. Althea's shoulders are 4'9" off the ground. When Jasmine stands on Althea's shoulders, how far off the ground is the top of Jasmine's head? 10 ft 5 in.

R 34 **Science** Diamonds burn at temperatures between 1400°F and 1670°F. Express this range of temperature in degrees Celsius. Between 760°C and 910°C

I 35 Use these three equations to find the values of x, y, and z.

$19 + 8(4) = x$ $x = 51$ $x + 13 = y$ $y = 64$ $z = \frac{y}{4} + \frac{x}{17}$ $z = 19$

I 36 The Sears Tower in Chicago is 1454 feet high. What is the scale of this picture of the Sears Tower? ≈1:8724

I 37 **Technology** Computer memory can be measured in kilobytes, but in this case *kilo-* does not mean exactly "one thousand." A kilobyte is actually 1024 bytes.
  a How many bytes are in 640 kilobytes? 655,360 bytes
  b Why might computer memory be measured in multiples of 1024 rather than multiples of 1000?

Investigation

Water Consumption How much water do you think you use in a day? Try to estimate the amount of water used by an average person on an average day. Then use your estimate to calculate, both in gallons and in cubic feet, the amount of water used in the United States during a year.

Additional Practice

1. Convert each length to meters.
  a. 8 in.   b. 2 ft 6 in.
  c. 3 yd
2. Convert each length to feet.
  a. 40 cm   b. 0.64 m
  c. 0.89 km
3. Convert each measurement to the indicated units.
  a. 6.25 m to ft
  b. 23 in. to cm
  c. 0.16 km to ft
  d. $23\frac{1}{5}$ hr to min

Answers:
1. a. ≈0.2032 m b. ≈0.762 m
  c. 2.742 m
2. a. ≈1.31 ft b. ≈2.1 ft
  c. ≈2920 ft
3. a. ≈20.5 ft b. 58.42 cm
  c. ≈524.9 ft d. 1392 min

Additional Answers

for pages 176 and 177

26 Possible answer: By using the two rates, along with the fact that there are 60 seconds in a minute and 60 minutes in an hour, to write the expression

$\frac{1 \text{ hr}}{60 \text{ mi}} \times \frac{88 \text{ ft}}{1 \text{ sec}} \times \frac{60 \text{ sec}}{1 \text{ min}} \times \frac{60 \text{ min}}{1 \text{ hr}}$

then simplifying the expression to obtain $\frac{5280 \text{ ft}}{1 \text{ mi}}$.

37 b Possible answer: Because computers handle information in binary (base-2) form, and 1024 is a power of 2 (it is 2^{10}) that is close to 1000

Investigation The average person may use about 100 gallons of water per day, including drinking, cooking, washing, flushing, and other uses. Estimating the U.S. population at 250 million, that is about 9×10^{12} gallons (or about 1.2×10^{12} cubic feet) of water used in the United States in a year.

Section 4.4 *Working with Units*

SECTION 4.5

Objectives
- To recognize proportions
- To use proportions to solve problems

Vocabulary: cross multiply, proportion

Class Planning

Schedule 2–3 days

Resources

In *Chapter 4 Student Activity Masters* Book
Class Openers 49, 50
Extra Practice 51, 52
Cooperative Learning 53, 54
Enrichment 55
Learning Log 56
Technology 57, 58
Spanish-Language Summary 59

Class Opener

Tell whether the following ratios are *equal* or *not equal*.
1. $\frac{1}{2}, \frac{2}{4}$ 2. $\frac{3}{9}, \frac{6}{15}$ 3. $\frac{4}{7}, \frac{8}{16}$ 4. $\frac{10}{50}, \frac{1}{5}$ 5. $\frac{5}{9}, \frac{9}{5}$

Answers:
1. Equal 2. Not equal
3. Not equal 4. Equal
5. Not equal

Lesson Notes

- Students have been using proportions throughout this chapter, so only terminology is being introduced, not the concept.
- In both examples, students do not need to use cross multiplication to solve the proportions. Multiplying both sides of the proportion by 3 in example 1 leads to $n = \frac{15}{7}$. In example 2, multiplying $\frac{4}{5}$ by $\frac{7}{7}$ leads to $\frac{28}{35}$, so g must be 35. Point out that there are alternative methods of solving proportions.

178

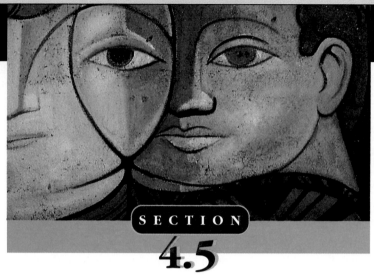

Investigation Connection
Section 4.5 supports Unit B, *Investigations* 1, 2, 5. (See pp. 92A-92D in this book.)

SECTION 4.5
Proportions

WHAT IS A PROPORTION?

A ***proportion*** is an equation stating that two ratios are equal. The two forms of the proportion shown can both be read "*a* is to *b* as *c* is to *d*."

$$\frac{a}{b} = \frac{c}{d} \qquad a:b = c:d$$

If a line parallel to one side of a triangle intersects the other two sides of the triangle, it divides those two sides proportionally. For example, in triangle ABC the arrows indicate that $\overline{DE}$ is parallel to $\overline{AC}$. The ratios of the labeled parts of sides $\overline{BA}$ and $\overline{BC}$ are equal $\left(\frac{8}{6} = \frac{12}{9}\right)$, since each ratio is equal to $\frac{4}{3}$. Notice also that $8 \cdot 9 = 72$ and $6 \cdot 12 = 72$, so $8 \cdot 9 = 6 \cdot 12$.

SOLVING PROPORTIONS

Proportions are useful in solving many types of problems. When you want to solve a proportion, it is frequently useful to rewrite the equation by ***cross multiplying***—that is, multiplying the numerator of each ratio by the denominator of the other.

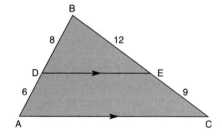

$$\frac{a}{b} = \frac{c}{d} \rightarrow a \cdot d = b \cdot c$$

178   Chapter 4   Ratio and Proportion

SECTION 4.5

Example 1
Find the value of n for which $\frac{n}{3} = \frac{5}{7}$.

Solution
We can use cross multiplication.

$$\frac{n}{3} = \frac{5}{7} \rightarrow 7n = 3(5) \rightarrow 7n = 15$$

Since n is multiplied by 7 in the rewritten equation, we can use the inverse operation, division, to solve the equation.

$$7n = 15$$
$$n = \frac{15}{7}, \text{ or about } 2.14$$

Example 2
The ratio of boys to girls at a party was 4:5. If there were 28 boys at the party, how many girls were at the party?

Solution
Let's use g to represent the number of girls. Since $\frac{\text{number of boys}}{\text{number of girls}} = \frac{4}{5}$ and there were 28 boys, we can write the proportion $\frac{28}{g} = \frac{4}{5}$. To solve it, you can cross multiply.

$$\frac{28}{g} = \frac{4}{5} \rightarrow 28(5) = 4g \rightarrow 140 = 4g$$
$$\frac{140}{4} = g$$
$$35 = g$$

There were 35 girls at the party. Notice that the proportion is written so that the numerators and denominators of the ratios correspond. If $\frac{\text{boys}}{\text{girls}}$ is on one side of the equation, the ratio on the other side must also be in the form $\frac{\text{boys}}{\text{girls}}$.

Sample Problems

Problem 1 The town council of Springfield decided to decorate a large wall in the center of town with a mural, so a contest was held to obtain a design for the artwork. Gloria won the contest with this 20-inch-by-28-inch drawing. The wall on which the mural is to be painted is 25 feet high and 42 feet wide. What is the largest mural that can be based on Gloria's drawing without changing its proportions?

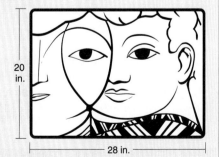

Additional Examples

Example 1
Find the value of n for which $\frac{n}{6} = \frac{8}{5}$.
Answer: $n = 9.6$

Example 2
The ratio of peanuts to raisins in a party mix is 3:2. If there are 75 peanuts in a bowl of the mix, how many raisins are in the bowl?
Answer: 50

Stumbling Block

Students sometimes try to apply cross multiplication to problems that are not proportions. Emphasize that there can be only one term, a ratio, on each side of the equal sign in order to use cross multiplication. Problem 4 brings up another potential misapplication of cross multiplication: expressions in the form $\frac{a}{b} + \frac{c}{d}$.

Reteaching

In sample problem 1, have students solve the proportion $\frac{20 \text{ in.}}{28 \text{ in.}} = \frac{x \text{ ft}}{42 \text{ ft}}$, as suggested at the end of the solution. Students will find that $x = 30$ ft. Have students explain why 30 ft is not an acceptable value for the mural's height.

SECTION 4.5

Assignment Guide

Basic
Day 1 7–9, 14, 18, 21, 29, 33, 34
Day 2 10–12, 19, 22, 26, 30–32
Day 3 13, 15–17, 23–25, 35

Average or Heterogeneous
Day 1 7–9, 13, 16, 17, 20, 24, 29, 31, 32
Day 2 10–12, 18, 19, 21, 26, 27, 33, 34, 37
Day 3 14, 15, 22, 23, 25, 28, 30, 35

Honors
Day 1 8, 9, 14, 15, 18, 20, 21, 26, 27, 31, 36, 38
Day 2 10, 12, 13, 17, 19, 22–24, 28, 30, 32, 33, 37

Strands

Communicating Mathematics	4, 6, 35
Geometry	3, 22, 23, 25, 32, 34, 35, 37, Investigation
Measurement	13, 14, 20, 26–28, 30, 31, 33, 34, 38, Investigation
Data Analysis/Statistics	13, 15, 22
Technology	16–19, 36
Probability	21, 36
Algebra	2, 7–13, 16–19, 22, 24, 29
Patterns & Functions	2, 5, 35
Number Sense	1, 4, 6, 15, 35, 38
Interdisciplinary Applications	14, 15, 26–28, 34, 38

Individualizing Instruction

The visual learner To demonstrate proportional amounts, you might produce two different-sized mixtures of water and food coloring. For example, five drops

180

Solution We set up and solve a proportion.

$$\frac{\text{Height of drawing}}{\text{Width of drawing}} = \frac{\text{height of mural}}{\text{width of mural}}$$

$$\frac{20 \text{ in.}}{28 \text{ in.}} = \frac{25 \text{ ft}}{x \text{ ft}}$$

$$\frac{5}{7} = \frac{25}{x}$$

$$5x = 7(25)$$

$$5x = 175$$

$$x = \frac{175}{5} = 35$$

The largest possible mural is one 25 feet high by 35 feet wide. Since the wall is 42 feet wide, 7 feet of its width will remain unpainted. Can you figure out why we used 25 feet for the mural's height and x for its width instead of using x for its height and 42 feet for its width?

Problem 2 The recipe for a fruit punch says to mix 2 pints of fruit concentrate with enough water to make 3 gallons of the punch. How much concentrate is needed to make 10 gallons of punch?

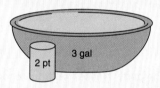

Solution Once again, we can solve this problem by using a proportion to compare like quantities. Notice that both ratios are in the form "concentrate:punch" and in the form "pints:gallons."

$$\frac{2 \text{ pints of concentrate}}{3 \text{ gallons of punch}} = \frac{x \text{ pints of concentrate}}{10 \text{ gallons of punch}}$$

$$\frac{2}{3} = \frac{x}{10}$$

$$20 = 3x$$

$$\frac{20}{3} = x$$

Since x represents the number of pints of concentrate that are needed, it would take $\frac{20}{3}$, or $6\frac{2}{3}$, pints of concentrate to make 10 gallons of punch. This result can also be obtained by solving the proportion $\frac{2 \text{ pints of concentrate}}{x \text{ pints of concentrate}} = \frac{3 \text{ gallons of punch}}{10 \text{ gallons of punch}}$. Do you see why?

Chapter 4 Ratio and Proportion

SECTION 4.5

Think and Discuss

P **1** Maude remembered that three of the four numbers in a proportion were 1, 3, and 4. She could not, however, remember the other number or the position of each number in the proportion. Help Maude to find all possible values of the missing number. $\frac{3}{4}, \frac{4}{3}$, or 12

P, R **2** Solve each proportion for *m*. (Hint: Look for a pattern.)

a $\frac{8}{m} = \frac{2}{5}$ $m = 20$ **b** $\frac{8}{m} = \frac{4}{5}$ $m = 10$ **c** $\frac{8}{m} = \frac{8}{5}$ $m = 5$

d $\frac{8}{m} = \frac{16}{5}$ $m = 2\frac{1}{2}$ **e** $\frac{8}{m} = \frac{32}{5}$ $m = 1\frac{1}{4}$ **f** $\frac{8}{m} = \frac{64}{5}$ $m = \frac{5}{8}$

P **3** $\overline{PQ}$ is parallel to $\overline{BC}$, AP = 4 in., PB = 8 in., and AQ = 5 in. Find length QC.   10 in.

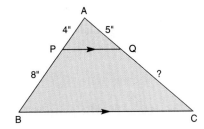

P, R **4** Explain why you cannot use the cross-multiplication technique to calculate $\frac{1}{5} + \frac{2}{3}$.

P **5** A photo measuring 6 inches by 8 inches is enlarged into a poster. The longer side of the poster is 3 feet long. How long is the shorter side?   27 in., or $2\frac{1}{4}$ ft

P **6** On this bicycle, the pedal gear has 48 teeth, and the gear on the rear wheel has 18 teeth. Suppose we use *P* to stand for the number of times the pedals turn and *R* to stand for the number of times the rear wheel turns. Write a valid proportion involving 48, 18, *P*, and *R*. Explain why your proportion is valid.

Wheel gear   Pedal gear

Problems and Applications

P, R **In problems 7–12, solve each proportion for *x*.**

7 $\frac{12}{x} = \frac{3}{4}$ $x = 16$ **8** $\frac{12}{x} = \frac{4}{3}$ $x = 9$ **9** $\frac{12}{x} = \frac{6}{5}$ $x = 10$

10 $\frac{12}{x} = \frac{2}{5}$ $x = 30$ **11** $\frac{12}{x} = \frac{1}{5}$ $x = 60$ **12** $\frac{12}{x} = \frac{12}{5}$ $x = 5$

Section 4.5 Proportions   181

SECTION 4.5

Checkpoint

1. Solve each proportion for x.
a. $\frac{6}{x} = \frac{2}{7}$ b. $\frac{x}{3} = \frac{8}{4}$
c. $\frac{9}{4} = \frac{x}{12}$ d. $\frac{4}{16} = \frac{5}{x}$

2. A model house is 6 inches wide and 3 inches high. The real house is 60 feet wide. How high is the real house?

3. An ant weighing 0.25 g can lift about 20 times its own weight. If a 60-kg human had proportional strength, how much could the person lift?

Answers:
1. a. 21 **b.** 6 **c.** 27 **d.** 20
2. 30 ft **3.** 1200 kg

Additional Practice

1. Solve each proportion for y.
a. $\frac{3}{y} = \frac{11}{22}$ b. $\frac{y}{4} = \frac{24}{6}$
c. $\frac{13}{3} = \frac{y}{6}$ d. $\frac{5}{15} = \frac{5}{y}$

2. The ratio of engineers to technicians in a factory is 3:11. If there are 42 engineers, how many technicians are there?

3. Solve each proportion for b.
a. $\frac{b}{16} = \frac{4}{8}$ b. $\frac{b}{8} = \frac{11}{4}$
c. $\frac{b}{7} = \frac{9}{3}$

Answers:
1. a. 6 **b.** 16 **c.** 26 **d.** 15
2. 154 **3. a.** 8 **b.** 22 **c.** 21

P, R 13 The ratio of the height of bar A to the height of bar B is 2:5.
a What is the value of x?   25
b What is the ratio of the height of bar C to the height of bar B?   8:25

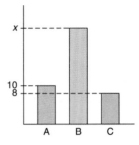

P 14 The directions on a can of lemonade mix say to use 8 tablespoons of mix to make 2 quarts of lemonade. Heathcliff wants to make 7 quarts of lemonade. How many tablespoons of mix should he use?   28 tbsp

P, R 15 Estimating Use the gas gauge to estimate the number of gallons of gasoline left in the motorist's 20-gallon tank.   Possible answer: ≈9 gal

In problems 16–19, solve the proportions for w, x, y, and z.

P, R 16 $\frac{3.5}{8.4} = \frac{5.7}{w}$ **17** $\frac{2\frac{1}{2}}{3\frac{3}{4}} = \frac{4\frac{4}{5}}{x}$ **18** $\frac{2.75}{5.5} = \frac{y}{4.8}$ **19** $\frac{40\%}{60\%} = \frac{8}{z}$

P 20 For safety, a swimming pool should contain 2 parts of liquid chlorinator for each 1,000,000 parts of water. How much chlorinator should be added to a pool that contains 50,000 gallons of water? $\frac{1}{10}$ gal, or ≈13 fl oz

P, R 21 In a recent year, the ratio of men to women in the United States armed forces was about 9 to 1.
a If there were 90 men in a random group of military personnel, how many women would you expect to find in the group?   Possible answer: 10
b In a group of 90 military personnel, how many of them would you expect to be women?   Possible answer: 9
c What percent of armed-forces personnel were women in that year?   ≈10%

P, R, I 22 If the ratio of x to y in the diagram is 7:8, what are the values of x and y?   28; 32

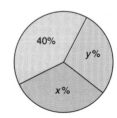

P, R 23 The ratio of the width to the length of a rectangle is 3:7. The shorter side is 9 inches long. What is the area of the rectangle?   189 in.²

P, I 24 Solve $\frac{x+4}{6} = \frac{9}{12}$ for x. (Hint: First solve $\frac{n}{6} = \frac{9}{12}$ for n.) $x = \frac{1}{2}$

Chapter 4 Ratio and Proportion

SECTION 4.5

P **25** In the diagram, DI = 7, DA = 22, and NE = 8. Find AN. ≈17.14

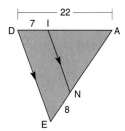

P, I **26** At Wassamatta U. the admissions office rejects nine applicants for every two that are accepted. If 1408 people applied for admission last year, how many were accepted? 256 people

P, R **27** The North Shore Blood Center has a goal of 3500 units of blood in its yearly blood drive. By September 1, the center has received 2000 units. If donations continue at the same rate, will the center reach its goal or fall short of it? By how much? Will fall short by 500 units

P **28** The ratio of officers to enlisted personnel in the United States Air Force is about 9:41. If there are about 606,000 people in the air force, how many of them are officers? ≈109,000

P, I **29** Solve each proportion for y.
  a $\frac{y}{6} = \frac{6}{y}$ 6   **b** $\frac{y}{2} = \frac{18}{y}$ 6   **c** $\frac{y}{4} = \frac{16}{y}$ 8   **d** $\frac{y}{3} = \frac{15}{y}$ 6.71

P **30** At 3 P.M. on a sunny day, a 2-foot-high stick casts a shadow 3 feet long. At the same time, a telephone pole casts a shadow 50 feet long. How tall is the telephone pole? (Hint: Use a proportion to calculate the answer.) 33 ft 4 in.

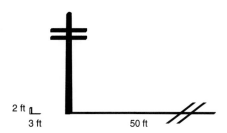

◀ LOOKING BACK **Spiral Learning** LOOKING AHEAD ▶

R **31** Find the ratio of 2 yards 2 feet 4 inches to 1 foot 8 inches. 5:1

I **32** In ⊙B, m∠ABC = 120. What percentage of the circular region is shaded? 33.$\overline{3}$%

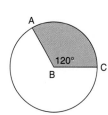

R **33** Jacob left his house, walking south at 4 miles per hour. Three hours later, his sister Danielle followed him on her bicycle at 12 miles per hour. How long did she ride before she caught up to him? $1\frac{1}{2}$ hr

Section 4.5 Proportions 183

Communicating Mathematics

- Have students consider the following problem: A cyclist can travel 25 miles in 3 hours. How far can she travel in 5 hours? Students should first solve this problem by setting up a proportion, then by finding and using the cyclist's rate of speed. Have students then choose the method they think is easier or more efficient and write a letter to you to persuade you that this method is the best method to use.

- Problems 4, 6, and 35 encourage communicating mathematics.

Cooperative Learning

- Bring in a recipe for punch or some other treat and have groups of students figure out how much of each ingredient they would need in order to make enough for the entire class.

- Problem 34 is appropriate for cooperative learning.

Additional Answers

for pages 182 and 183

16 $w = 13.68$
17 $x = 7\frac{1}{5}$
18 $y = 2.4$
19 $z = 12$

SECTION 4.5

Additional Answers

for page 184

34 a

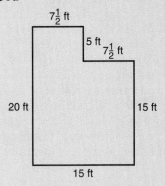

35 Possible answer: The angle with its vertex at the chosen point is a 90° angle, and the sum of the measures of the other two angles is 90.

Investigation Answers will vary.

R **34 Architecture** On this floor plan, 1 in. = 10 ft.
  a Make a sketch showing the lengths of all six sides of the actual room.
  b Find the area of the floor plan in square inches.   2.625 in.²
  c Find the area of the actual room in square feet.   262.5 ft²
  d Find the ratio of the area you found in part **b** to the area you found in part **c**.   1:14,400

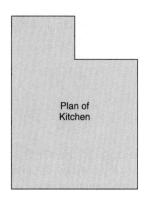

Plan of Kitchen

R **35 Communicating** Draw a circle and one of its diameters. Pick a point on the circle that is not an endpoint of the diameter. Draw segments connecting this point to each endpoint of the diameter. Then measure the angles formed. What do you notice?

R **36** The chances that a certain state lottery ticket costing $2 will be a winning ticket are 4 out of 11,543,672.
  a What is the probability that a ticket will be a winning one? (Express the probability in decimal form.)   ≈.00000035
  b If you spend $100 on these lottery tickets, what is the probability that you will win?   ≈.000017

P, I **37** ∠R is a right angle, and $\overline{QS} \parallel \overline{PT}$ (that is, $\overline{QS}$ is parallel to $\overline{PT}$).
  a Find PQ. $1.\overline{6}$
  b Find QS. 13
  c Find PT. $17.\overline{3}$

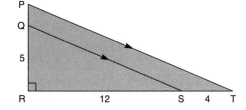

38 Estimate the distance from the tee to the flagstick on the golf hole shown.   Possible answer: 450 yd

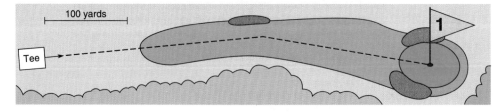

Investigation

Floor Plans Draw, to scale, a floor plan of your home. Be sure that the plan includes a legend explaining the scale that you used. What is the total floor space, in square feet, represented by the plan?

184   **Chapter 4**   Ratio and Proportion

SECTION 4.6

SECTION 4.6
Similarity

Objectives
- To understand the concept of similarity
- To solve problems involving similar geometric figures

Vocabulary: scale factor, similar

Class Planning

Schedule 2–3 days
Resources
In *Chapter 4 Student Activity Masters* Book
Class Openers 61, 62
Extra Practice 63, 64
Cooperative Learning 65, 66
Enrichment 67
Learning Log 68
Using Manipulatives 69, 70
Technology 71, 72
Spanish-Language Summary 73

ENLARGEMENTS AND REDUCTIONS

When a photographic slide is projected, the image on the screen is the same as the image on the slide, only larger. Maps, on the other hand, are smaller than what they represent.

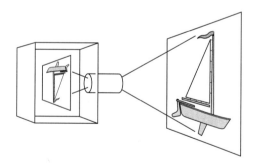
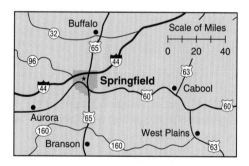

The slide and its image are *similar;* the map and the area it represents are also *similar.*

Let's reconsider the drawing and the mural discussed in Sample Problem 1 of the preceding section. These similar pictures show two characteristics that are shared by all pairs of similar objects.

1. The ratios of all dimensions of the drawing to the corresponding dimensions of the mural are the same.

$$\frac{\text{Width of drawing}}{\text{Width of mural}} = \frac{\text{height of drawing}}{\text{height of mural}}$$

$$\frac{20 \text{ in.}}{25 \text{ ft}} = \frac{28 \text{ in.}}{35 \text{ ft}}$$

$$\frac{4 \text{ in.}}{5 \text{ ft}} = \frac{4 \text{ in.}}{5 \text{ ft}}$$

Class Opener

Which of the following pairs represent enlargements or reductions?
1. Cat, kitten
2. Blueprint, house
3. Document, microfilm
4. Airplane, model airplane
5. Lake, ocean
6. Bush, shrub
7. Negative, 5" × 8" photo
8. House, skyscraper
9. Map, Ohio

Answers:
2, 3, 4, 7, 9

Lesson Notes

- The applications of proportions emphasized in this section are (1) scaling for enlargement and reduction and (2) similar geometric figures. The ratios of

Section 4.6 Similarity 185

SECTION 4.6

Lesson Notes continued
corresponding sides of similar figures are equal, so unknown lengths of sides can be calculated by using proportions. Recognizing which sides correspond is not always easy for students. Some practice and help may be needed.

Additional Example

Example
In the diagram, △MNO ~ △XYZ.

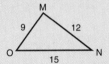

a. Find the following ratios:
$\frac{MN}{XY}$, $\frac{NO}{YZ}$, and $\frac{OM}{ZX}$

b. What can we conclude about the angle measures of the triangles?

Answers:
a. 3; 3; 3 b. The measures of the corresponding angles are equal.

Cooperative Learning

- Have students work together to determine the least amount of information necessary to determine if two triangles are similar.

Answers: (1) Two pairs of corresponding congruent angles, (2) equal ratios of three corresponding sides, (3) equal ratios of two corresponding sides and congruence of the included angles

- Problems 16, 23, and Investigation are appropriate for cooperative learning.

In other words, any 4-inch length on the drawing will become a 5-foot length on the mural. If we express the ratio in terms of a common unit, we find that $\frac{4 \text{ in.}}{5 \text{ ft}} = \frac{4 \text{ in.}}{60 \text{ in.}} = \frac{1}{15}$. The drawing is therefore $\frac{1}{15}$ the size of the mural, and the mural is $\frac{15}{1}$, or 15, times the size of the drawing. The scale factor of the enlargement is 15:1.

2. The ratio of any two dimensions of the drawing is equal to the ratio of the corresponding dimensions of the mural.

$$\frac{\text{Height of drawing}}{\text{Width of drawing}} = \frac{\text{height of mural}}{\text{width of mural}}$$

$$\frac{20 \text{ in.}}{28 \text{ in.}} = \frac{25 \text{ ft}}{35 \text{ ft}}$$

$$\frac{5}{7} = \frac{5}{7}$$

Whenever a model or copy differs from its original in size but not in shape, the two objects are said to be **similar.** The ratio of linear measurements of the objects corresponds to the amount of enlargement or reduction and is sometimes called the **scale factor** of the change in size.

SIMILAR GEOMETRIC FIGURES

Similarity of polygons is like similarity of other objects. When polygons have exactly the same shape, they are similar even if their sizes are different.

   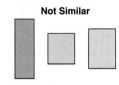

In any pair of similar polygons, the ratios of the lengths of corresponding sides are equal, and the measures of corresponding angles are equal. The symbol ~ is used to indicate similarity.

Example
In the diagram, △ABC ~ △PQR.
a Find the ratios $\frac{AB}{PQ}$, $\frac{BC}{QR}$, and $\frac{AC}{PR}$.
b What can we conclude about the angle measures of the triangles?

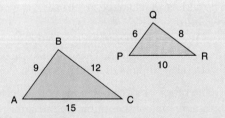

Solution
a $\frac{AB}{PQ} = \frac{9}{6} = \frac{3}{2}$ $\frac{BC}{QR} = \frac{12}{8} = \frac{3}{2}$ $\frac{AC}{PR} = \frac{15}{10} = \frac{3}{2}$

Notice that all three ratios are the same. In this pair of similar triangles, the ratio of corresponding lengths is 3:2.

SECTION 4.6

b Since the triangles are similar, the measures of corresponding angles are equal: m∠A = m∠P, m∠B = m∠Q, and m∠C = m∠R.

Whenever each angle of one triangle is equal to the corresponding angle of another triangle, the two triangles are similar. (Although this is true of triangles, it usually is *not* true of other kinds of polygons.)

Sample Problem

Problem Given that $\overline{AC} \perp \overline{BD}$ (segment AC is perpendicular to segment BD), m∠BAC = m∠DEC, and m∠B = m∠D, find length CD.

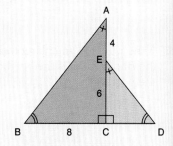

Solution $\overline{AC}$ is perpendicular to $\overline{BD}$, so ∠ACB and ∠ACD both have measures of 90. Since m∠BAC = m∠DEC and m∠B = m∠D, each angle of △ABC is equal to the corresponding angle of △EDC. Therefore, △ABC is similar to △EDC. Since $\overline{AC}$ corresponds to $\overline{EC}$ and $\overline{CB}$ corresponds to $\overline{CD}$,

$$\frac{AC}{EC} = \frac{CB}{CD}$$

$$\frac{10}{6} = \frac{8}{x}$$

$$10x = 48$$

$$x = \frac{48}{10} = 4.8$$

$\overline{CD}$ is 4.8 units long.

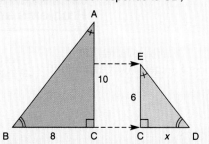

Think and Discuss

P **1** Felicia drew two triangles and said, "These two triangles are similar." Connie said that she didn't think they were similar. How could they decide?

P **2** Which of these rectangles are similar?   I and II, III and IV

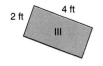

Section 4.6 Similarity

Assignment Guide

Basic
Day 1 6, 12–14, 20, 24–26
Day 2 8–10, 15, 21–23, 27, 31
Day 3 7, 11, 16–18, 28–30

Average or Heterogeneous
Day 1 6, 7, 9, 13, 14, 18–20, 22, 23
Day 2 8, 10–12, 15–17, 21, 25–27, 30, 31

Honors
Day 1 6, 8, 9, 14, 15, 18, 20, 21, 26, 27, 31, 36, 38
Day 2 10, 12, 13, 17, 19, 22–24, 28, 30, 32, 33, 37

Strands

Communicating Mathematics	1, 4, 16, 25
Geometry	1–9, 11–15, 17–19, 26, Investigation
Measurement	7, 10, 12, 13, 15, 16, 18, 22, 24, 30, Investigation
Data Analysis/ Statistics	21, 31
Technology	16, 18, 22, 23, 27–29, 31
Algebra	3, 5, 20, 21, 26–30
Patterns & Functions	16, 23
Number Sense	24
Interdisciplinary Applications	10, 16, 22, 25, 31

Individualizing Instruction

The visual learner Students may benefit from using a pantograph to enlarge and reduce figures. This may help them see how to use proportions to construct similar figures.

SECTION 4.6

Problem-Set Notes

for pages 187, 188, and 189

- **2** points out that similar relationships must be expressed in terms of the same unit.
- **4** shows that equal ratios are not the only criterion for similar figures—the angles must have the same measures.
- **13** uses the Pythagorean Theorem to determine YZ, then a proportion to solve for TK.

Reteaching

To reteach similarity of polygons other than triangles, show students the following pairs of figures and ask them to explain why the figures are or are not similar.

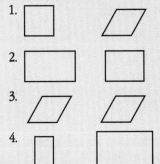

Stumbling Block

In the sample problem, students may have difficulty determining which sides are corresponding sides. Have students trace the triangles, mark the vertices on the insides of the figures, then cut the triangles out. Students can then flip one triangle over, copy the letters of the vertices from the other side, and overlap the triangles. Have students write the pairs of corresponding sides from the overlapped triangles.

188

P, R   **3** If △ABC ~ △DEF, what are the values of *x* and *y*?   12; 15

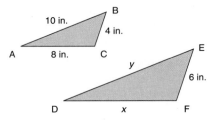

P, R   **4 a** Are the ratios of corresponding sides of these two figures equal?
      b Are the two figures similar? Explain your answer.   **a** yes

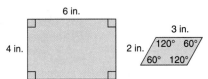

P, R   **5** If the poster is an enlargement of the photo, what is the height of the poster?   2 ft 1 in.

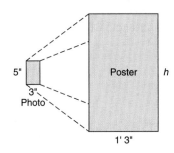

Problems and Applications

P, R   **6** △ABC is similar to △PQR. Find lengths AC, QR, and PR.   5 cm; 16 cm; 20 cm

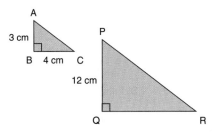

P, R   **7** Draw a rectangle measuring 2 inches by 3 inches. Then draw another rectangle, measuring 2 inches by 6 inches.
      a Do the two rectangles have equal angles?   Yes
      b Do the two rectangles have the same shape? (That is, are they similar?)   No
      c Are any two geometric figures with equal angles similar?   No

P, R   **8** If $\overline{BC}$ is parallel to $\overline{DE}$, what is the ratio of AC to CE?   5:6

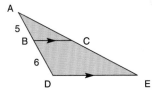

188   Chapter 4   Ratio and Proportion

SECTION 4.6

P, R **9** △ABC ~ △AFG, and AB = BD = DF.
  a Find the ratio AB:AF.   1:3
  b Find the ratio AC:AG.   1:3
  c Find the ratio BC:FG.   1:3
  d If the area of △ABC is 1, what is the area of △AFG?   9

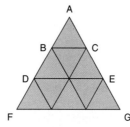

P, R **10 Consumer Math** A 4-acre parcel of property has an assessed value of $90,000. The taxes on the property are $3000 per year. A second parcel has an assessed value of $12,000. How much are the taxes on the second property if both are taxed at the same rate? $400 per year

P, R **11** Compare the larger rectangle with the smaller rectangle. What is
  a The ratio of corresponding side lengths?   4:3
  b The ratio of the rectangles' areas?   16:9
  c The ratio of the rectangles' perimeters?   4:3

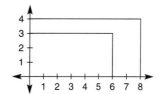

P, R **12** △RMW ~ △FAZ, RW = 18 ft, RM = 24 ft, MW = 12 ft, and FZ = 48 ft.
  Find FA and AZ.   64 ft; 32 ft

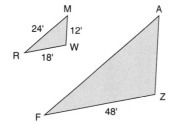

P, R **13** Rectangle SUZY is similar to rectangle KATE.
  a Find YZ.   6 in.
  b Find TK to the nearest hundredth.   ≈16.67 in.

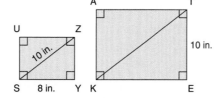

P, R **14** $\overline{DF}$ is an enlargement of $\overline{AC}$. If $\overline{AC}$ is 9 units long and $\overline{DF}$ is 30 units long, how long is $\overline{DE}$?   10 units

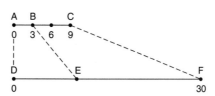

P, R **15** △BIG is similar to △TOP.
  a How long is $\overline{OP}$?   12 in.
  b How long is $\overline{TP}$?   16 in.
  c If $\overline{TO}$ were 10 feet long instead of 10 inches long, how long would $\overline{OP}$ and $\overline{TP}$ be?   12 ft; 16 ft

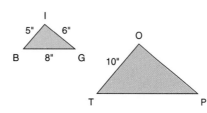

Section 4.6 Similarity

Checkpoint

1. Are the following pairs of rectangles similar?
a. Sides 4 in. and 6 in., sides 12 in. and 18 in.
b. Sides 11 in. and 5 in., sides 15 in. and 5 in.
c. Sides 9 in. and 2 ft, sides 13.5 in. and 3 ft

2. A photo with height 3" and width 5" is enlarged into a 6-ft-high poster. How wide is the poster?

3. △FAR has sides 12, 16, and 21. △GBS is similar to △FAR, with shortest side 26. What are the other two sides to the nearest hundredth?

Answers:
1. a. Yes **b.** No **c.** Yes **2.** 10 ft
3. 34.67 and 45.50

Additional Answers

for pages 187, 188, and 189

1 Possible answer: They could measure the angles of the triangles to determine whether corresponding angles are equal, or they could measure the sides to determine whether the ratios of the lengths of corresponding sides are equal.

4 b No; possible explanation: Although the ratios of corresponding sides are equal, the corresponding angles of the figures are not equal (the figures' shapes are different).

SECTION 4.6

Problem-Set Notes

for pages 190 and 191

- **19** requires that students recognize that the smaller blue triangles are similar to the larger ones. Students will likely use the Pythagorean Theorem to solve the problem, but a simpler solution is represented by the expression 220 − 147 − 3.

- **21** offers a preview of Chapter 5 and emphasizes that 100% is the whole.

Additional Practice

1. Are the following pairs of triangles similar?
   a. Sides 9, 12, 15; sides 7, 11, 13
   b. Sides 6, 8, 12; sides 6, 8, 13
   c. Sides 13, 15, 21; sides 18, 20, 26
   d. Sides 5, 12, 13; sides 7, 13, 15

2. A computer screen with height 9" and width 10" is projected onto a wall so that the image is 6 ft high. How wide is the image?

3. △BYE has sides 10, 13, and 16. △ERN has sides 20, 26, and 32. Are the triangles similar?

Answers:
1. a. No b. No c. No d. No
2. $6\frac{2}{3}$ ft 3. yes

Communicating Mathematics

- Have students respond to the following question in a paragraph: If you want to know if two triangles are similar, must you measure one, two, or all three pairs of corresponding angles? Explain.

- Problems 1, 4, 16, and 25 encourage communicating mathematics.

P, R **16** **Spreadsheets** A dollhouse is constructed so that each inch of the dollhouse corresponds to $2\frac{1}{2}$ feet of a real house.
   a Use a spreadsheet to make a table listing "dollhouse lengths" of 1 in. to 20 in., in increments of $\frac{1}{4}$ in., along with the equivalent lengths in the real house.
   b Now use the spreadsheet to make a table listing lengths of 1 ft to 20 ft, in increments of 1 ft, along with the equivalent dollhouse lengths.
   c How might these conversion tables be useful?

P, R **17** △GOC ~ △ART, GO = 7, and AR = 3.
   a If CO = 8, what is RT?   ≈3.43
   b If AT = 5, what is GC?   ≈11.67

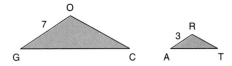

P, R **18** Recall that the volume of a cylinder with radius r and height h is equal to $\pi r^2 h$.
   a What is the volume of the cylindrical can?   ≈35.34 in.³
   b If a similar can has a diameter of 4 inches, what is its height? What is its volume? $6.\overline{6}$ in.; ≈83.78 in.³
   c What is the ratio of the volumes of the cans in parts **a** and **b**?   27:64

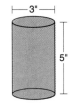

P, R **19** ABCD and EFGH are rectangles, AE = 3, AH = 1, and DG = 7. What is the area of rectangle EFGH?   70

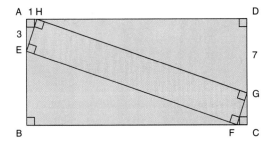

◀ LOOKING BACK   **Spiral Learning**   LOOKING AHEAD ▶

R **20** Solve each proportion for y.
   a $\frac{y}{12} = \frac{3}{4}$ $y = 9$ **b** $\frac{12}{y} = \frac{3}{4}$ $y = 16$ **c** $\frac{12}{4} = \frac{3}{y}$ $y = 1$ **d** $\frac{12}{3} = \frac{4}{y}$ $y = 1$

I **21** In the circle graph shown, what is the value of x?   15

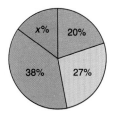

190   Chapter 4   Ratio and Proportion

SECTION 4.6

R 22. Aluminum has a density of 2700 kg/m³. If Billy cannot lift a weight greater than 23 kilograms, what is the size of the largest cube of aluminum that he can lift?

R 23. **Spreadsheets** Consider the proportion $\frac{a}{b} = \frac{3}{7}$. Use a spreadsheet to make a table showing the values of b that correspond to whole-number values of a from 1 to 10.

R 24. In this diagram, 1 in. = 66 ft. How far apart are the house and the oak tree?   ≈145 ft

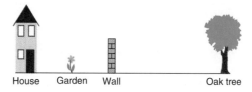

R 25. **Consumer Math** A pack of ProStar baseball cards contains 16 cards and costs $0.59. A pack of Slugger cards contains 14 cards and costs $0.49. Which is the better buy? Why?

P, I 26. What is the value of y in the diagram? $2\frac{2}{3}$

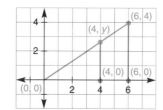

R **In problems 27–29, solve each proportion for w.**

27. $\frac{w}{3.7} = \frac{8.9}{10.4}$ $w \approx 3.17$ 28. $\frac{7.6}{w} = \frac{3.14}{180}$ $w \approx 435.67$ 29. $\frac{11.7}{12.4} = \frac{w}{8.2}$ $w \approx 7.74$

R, I 30. Suppose that $\frac{n^2}{9} = \frac{7}{3}$ and that $n > 0$. Find the value of n to the nearest hundredth.

I 31. **Environmental Science** If 6323.6 million barrels of petroleum were used in the United States in 1989, how much petroleum is represented by each part of the circle graph?

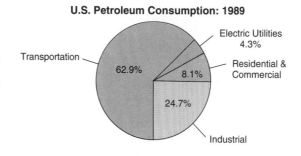

Investigation

Similar Polygons Draw two triangles that are similar, then draw two triangles that are not similar. Label the sides of each figure with their lengths, and label the angles of each figure with their measures. Repeat this process for quadrilaterals and for pentagons. What conclusions can you draw from your figures? Are all squares similar to one another? Why or why not?

Section 4.6 Similarity

SUMMARY

CHAPTER 4

Summary

CONCEPTS AND PROCEDURES

After studying this chapter, you should be able to
- Identify measurable attributes of objects and some of the units in which measurements of the attributes may be expressed (4.1)
- Recognize metric units of length, mass, and capacity (4.1)
- Add, subtract, multiply, and divide measurements of like attributes (4.1)
- Identify and evaluate ratios (4.2)
- Identify and evaluate rates (4.2)
- Calculate unit costs (4.2)
- Solve problems involving scale ratios (4.3)
- Solve problems involving densities (4.3)
- Use dimensional analysis to change the units in which measurements are expressed (4.4)
- Recognize proportions (4.5)
- Use proportions to solve problems (4.5)
- Understand the concept of similarity (4.6)
- Solve problems involving similar geometric figures (4.6)

VOCABULARY

attribute (4.1)
base unit (4.2)
centi– (4.1)
cross multiply (4.5)
density (4.3)
dimensional analysis (4.4)
gram (4.1)
kilo– (4.1)
meter (4.1)

milli– (4.1)
proportion (4.5)
rate (4.2)
ratio (4.2)
scale (4.3)
scale factor (4.6)
similar (4.6)
unit cost (4.2)

CHAPTER 4

Review

1. What units might you use to express   Possible answers given.
   a The amount of cloth needed to make a dress?   Yards
   b The amount of baking soda needed to make a cake?   Teaspoons
   c The amount of oil in an automobile's crankcase?   Quarts
   d The amount of soda pop in a bottle?   Fluidounces

2. Evaluate the ratio $\frac{84 \text{ in.}}{2\frac{1}{2} \text{ yd}}$. $\frac{14}{15}$

3. Which is the better buy, a dozen daisies for $3 or 20 daisies for $5?

4. A storage tank holds 185 liters of liquid. How many gallons does it hold?   ≈48.87 gal

In problems 5–8, solve each proportion for x.

5. $\frac{x}{8} = \frac{6}{5}$ $x = 9.6$ 6. $\frac{x}{8} = \frac{5}{6}$ $x = 6.\overline{6}$ 7. $\frac{x}{6} = \frac{8}{5}$ $x = 9.6$ 8. $\frac{x}{6} = \frac{5}{8}$ $x = 3.75$

9. A model limousine is 10 inches long and 3 inches wide. The scale of the model is 1:21. How long is the real car?   210 in., or 17 ft 6 in.

10. How many centimeters are in 5 kilometers?   500,000 cm

11. Find the value of x. $6.\overline{6}$

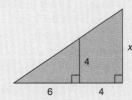

12. A photograph measuring 3 inches by 5 inches is to be enlarged so that the longer side of the enlargement will measure 11 inches. How long will the shorter side of the enlargement be?   6.6 in.

13. Evaluate the ratio $\frac{1 \text{ hour}}{1 \text{ year}}$. Express your answer as a fraction, then rewrite the fraction in scientific notation.   Possible answer: $\frac{1}{8760}$; ≈1.14×10^{-4}

14. A car traveled at a speed of 60 miles per hour for 100 miles, then at 40 miles per hour for 50 miles. How long did the trip take?   ≈2.92 hr, or 2 hr 55 min

15. Express the area of the shaded region
    a In square meters   38 m²
    b In square feet (Hint: 1 ft² ≈ 0.093 m².)
    ≈408.6 ft²

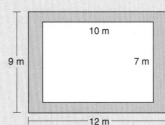

Chapter Review   193

REVIEW

Additional Answers

for pages 194 and 195

17 Possible answers: Height, texture, weight, color, volume, girth; height, weight, volume, girth

26 Possible answer: The 14-in. pizza, since it costs less per square inch

29 Possible answer: Because the two photographs are not similar—the image in the smaller photograph must have been cropped to fit in the 8" × 10" format

16 A pack of eight 16-ounce bottles of Coola Cola costs $5.60. A pack of six 12-ounce cans of Coola Cola costs $3.60. Which pack gives you more cola for your money?   The pack of bottles

17 List six attributes of a tree. Which of these attributes can be measured?

18 Given that $\overline{RA}$ is parallel to $\overline{DM}$, ER = 12, RD = 16, and EA = 15, find length EM.   35

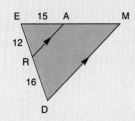

19 Bernie's car has a 12-gallon gas tank. He estimates that he can drive approximately 320 miles on a tank of gas. If his gas gauge indicates that his tank is about three-eighths full, about how far has he driven since he last filled his tank?   ≈200 mi

20 At a temperature of 25°C, a maximum of 36.2 grams of salt will dissolve in 100 milliliters of water. How much water is needed to dissolve 300 grams of salt at 25°C.   ≈829 ml

21 The Belmont Stakes is a $1\frac{1}{2}$-mile horse race. In 1941, Whirlaway won the race by running the course in 2 minutes 31 seconds. What was Whirlaway's average speed in miles per hour?   ≈35.76 mi/hr

22 Rectangle BIRD is similar to rectangle BATH. If BI = 16, BD = 24, and BH = 18, what fraction of rectangle BIRD is shaded? $\frac{7}{16}$

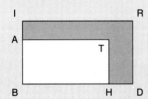

23 An ice sculptor is using a block of ice measuring 3 feet by 4 feet by 5 feet.
  a What is the volume of the block?   60 ft³
  b How many pounds does the block weigh?   3450 lb

24 Find the height of a building that casts a 75-foot shadow when a 4-foot fence post casts a 5-foot shadow.   60 ft

25 The diagram represents the floor of a room. The scale of the diagram is 1:216. If carpeting sells for $15 per square yard, what is the price of the carpeting needed to cover the floor? $810

Chapter 4 Ratio and Proportion

REVIEW

26 At Pies 'R' Round, a pizza 10 inches in diameter costs $6.50, and a pizza 14 inches in diameter costs $12.50. Which is the better buy?

27 △MTV ~ △VHI, TM = 5, MV = 8, and TV = 10. If HI = 15, what is the perimeter of △VHI?   34.5

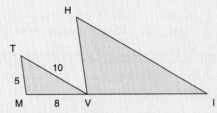

28 The maximum speed at which Tess Turtle can crawl is 20 feet per hour. Sidney Slug's top speed is 1 foot per hour. They decide to have a race, with Sidney getting a 100-foot head start because of his sluggishness. If both race at top speed, how long will it take Tess to catch Sidney?   ≈ 5.26, or 5 hr 16 min

29 Why is an 8-inch-by-10-inch "enlargement" of a $3\frac{1}{2}$-inch-by-5-inch photograph not a true enlargement?

30 Pam is planning to tile the floor of this L-shaped room. The tiles she wants to use measure 20 cm by 20 cm and are sold in boxes of 40 tiles. If each box costs $25, how much should Pam expect to spend on tiles? $300

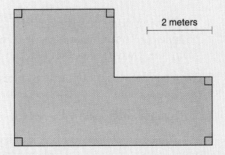

31 An acre is equal to 4840 square yards.
  a How many acres are in a rectangular lot that is 227 feet wide and 450 feet long?   ≈2.35 acres
  b The property tax on a 12.3-acre plot of land is $654. What is the tax rate per square yard?   ≈$0.011/yd^2

32 Les Cargo the snail can travel at a speed of $\frac{1}{5}$ inch per minute. How long would it take Les to finish a 2500-mile cross-country snail race if he never stopped to rest? 7.92×10^8 min, or ≈1506 yr

33 The ratio of AB to BC is 4:9. The total length of $\overline{AC}$ is 195. Use a spreadsheet to make a table like the one shown, and extend the pattern to find the lengths of $\overline{AB}$ and $\overline{BC}$.   60; 135

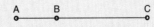

AB	BC	Total
4	9	13
8	18	26
12	27	39
.	.	.
.	.	.
.	.	.

Chapter Review

Assessment

In *Tests and Quizzes* Book
Basic Level   Test 1A   89–91
                Test 1B   92–94
Average Level   Test 2A   95–97
                Test 2B   98–100
Honors Level   Test 3A   101–103
                Test 3B   104–106
Open-Ended   Test C
  (for all ability levels)   107, 108
Midyear Test 1 (Basic)   109–114

Alternative-Assessment Guide

Assessment Tips
Have students select their best project and share it with the rest of the class in a short oral presentation. Some students may need help selecting their best work. Encourage students to be creative in their presentations. They may do demonstrations or use handouts, props, or an overhead projector.

Portfolio Projects
The following problems are appropriate for use as portfolio projects:
1–5. Investigations: Chapter Opener (p. 149), 4.1 (p. 157), 4.2 (p. 164), 4.3 (p. 171), 4.5 (p. 184)
6. Extend problem 36 on page 156 as follows: Find as many shapes as you can that can be formed by joining six 1-by-1 squares edge to edge. Be careful not to count as a new figure one that could be made by flipping or rotating another figure. Find the perimeter of each of these figures and display the data in a graph. Explain what the graph shows.
7. Extend sample problem 2 on page 160 as follows: Find the rates for making a long-distance phone call for at least two different companies. How do the rates vary by time of day? Do they also vary according to where the call is made? Compare the costs of making a 5-minute phone call and a 20-minute phone call at various times of day. Which

196

CHAPTER 4

Test

1 List as many attributes of a compact disc as you can think of. Which of these attributes can be measured?   Answers will vary.

2 Which length is longer, 30 cm or 2 m?   2 m

3 For what value of x is the perimeter of this rectangle 12 yards?   1

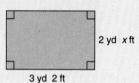

4 For the triangle shown, what is the ratio $\frac{AB}{BC}$? $\frac{22}{25}$

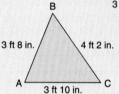

5 The driving distance between two towns is 250 miles. If the speed limit is 55 miles per hour, what is the least time in which a person could drive from one town to the other without breaking the law? $4\frac{6}{11}$ hr, or ≈4 hr 33 min

6 Which is the better buy, four 8-ounce cans of tomato paste for $4.12 or six 6-ounce cans of tomato paste for $4.50?   Six 6-oz cans

7 Rewrite the map scale 1 in. : 25 mi as a ratio of numbers.   1:1,584,000

8 If a 200-cubic-centimeter block of a substance has a mass of 450 grams, what is the density of the substance?   2.25 g/cm^3

9 If a typist can type an average of 65 words per minute, how many words can he type in 8 hours?   31,200 words

10 Solve the proportion $\frac{4}{m} = \frac{20}{25}$ for m. $m = 5$

11 If △XYZ ~ △MNO, what are the values of p and q?   20; 25

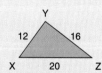 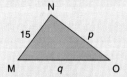

12 △ABC ~ △DEC, AB = 8, BC = 7, AC = 6, and DC = 4. Find the ratio of the perimeter of △ABC to the perimeter of △DEC.   3:2

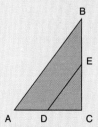

196   **Chapter 4**   Ratio and Proportion

PUZZLES AND CHALLENGES

1 Using four straight lines, connect the nine dots without lifting your pencil from the paper.

2 Use exactly six toothpicks, each the same length, to create four equilateral triangles, each the same size.

3 Rearrange the letters in the word *elation* to spell a part of the body.

4 Figure out what rule was used to determine which letters in the diagram are above the line and which letters are below the line.

A E F H I K L M N T V W Y X Z
―――――――――――――――――――――――――
B C D G J O P Q R S U

5 Write a ten-digit number in which the first digit tells how many zeros are in the number, the next digit tells how many ones are in the number, the next digit tells how many twos, and so on.

phone company would you recommend?

8. Extend example 3 on page 159 as follows: Go to a grocery store with a notebook and a pen and record the costs and the weights of at least 30 boxes of dry cereal (for example, bran cereals, "sugar" cereals), and whether or not the cereal is on sale. Write a report summarizing your findings. Include a graph or two to support your conclusions.

9. Extend sample problem 1 on page 173 as follows: Obtain a straw from at least three fast-food restaurants or convenience stores. (Remember to label each one so that you don't confuse them!) Record lengths and diameters and calculate the volume of each straw. Make a graph showing the results. Try to explain any significant differences between the straws.

Additional Answers

for Chapter 4 Test

1 Possible answers: weight, radius, color; weight and radius

for Puzzles and Challenges

1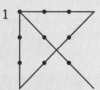

2 A pyramid with a triangular base

3 Toenail

4 The letters above the line are made up of straight segments only.

5 6,210,001,000

OVERVIEW

LONG-TERM INVESTIGATIONS

See *Investigations* book, Unit C, pages 56-79.

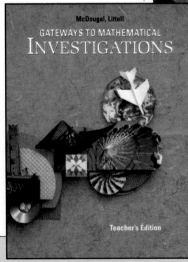

Gateways to Mathematical Investigations is a collection of long-term, real-world investigations that provides students with exciting opportunities to apply the ideas and concepts introduced in *Gateways to Algebra and Geometry: An Integrated Approach*. Unit C: The Sporting Life consists of five investigations that apply mathematical concepts and strategies within the context of sports and games. This unit is designed so that you can pick and choose the order and number of investigations you would like to do with your students. Each investigation is an exploration that takes several days to complete. Refer to the menu at the right to learn more about the investigations that make up this unit.

As your students work through the Unit C investigations, use Chapters 5 and 6 of *Gateways to Algebra and Geometry: An Integrated Approach* for concept and skill development, and as a source of rich problems, exercises, and applications. When appropriate, refer students to additional supportive sections in *Gateways to Algebra and Geometry* as well.

UNIT C: The Sporting Life

MENU OF INVESTIGATIONS
OVERVIEW: Groups apply mathematical concepts and strategies from Chapters 5 and 6 within the context of sports and games.

Where We Stand
(3-4 days)
Students conduct a weighted survey of their classmates to find out their favorite sports. They interpret the data they collect and use a variety of graphs to display those data. Students then analyze their graphs.
STRANDS
Number, Statistics, Geometry
UNIFYING IDEAS
Multiple Representations, Proportional Relationships

Faster, Farther, and Higher
(3-4 days)
Students examine data about Olympic track and field results, analyze changes over the years in those data, and then draw conclusions about those changes.
STRANDS
Statistics, Functions, Number, Measurement
UNIFYING IDEAS
Proportional Relationships, Multiple Representations, Patterns and Generalizations

Hoops du Jour
(2-3 days)
Students use stem-and-leaf plots, box-and-whisker plots, matrices, and integers to analyze basketball data.
STRANDS
Statistics, Functions, Number, Algebra
UNIFYING IDEAS
Multiple Representations, Patterns and Generalizations

Hole in One
(4-5 days)
Students design miniature golf courses on rectangular coordinate systems and play "integer golf" games using their courses.
STRANDS
Statistics, Number, Measurement, Logic and Language
UNIFYING IDEAS
Multiple Representations, Proportional Relationships

Hit the Bull's Eye
(3-4 days)
Students simulate archery tournaments by using operations with integers to make coordinate pairs that "land" on the target. They analyze the results of their tournaments.
STRANDS
Number, Statistics, Measurement, Logic and Language
UNIFYING IDEAS
Multiple Representations, Proportional Relationships

INVESTIGATIONS FOR UNIT C	SCHEDULE	OBJECTIVES
1. **Where We Stand** Pages 59-62*	3-4 Days	• To collect and interpret data about the class; to use a variety of graphs to display those data; to analyze the graphs
2. **Faster, Farther, and Higher** Pages 63-66*	3-4 Days	• To use graphs, statistics, and integers to interpret and analyze Olympic track and field results
3. **Hoops du Jour** Pages 67-70*	2-3 Days	• To use plots, matrices, statistics, and integers to interpret and analyze data about basketball
4. **Hole in One** Pages 71-74*	4-5 Days	• To use a rectangular coordinate system to design a miniature golf course; to use operations with integers to estimate the distance of the course; to play a game on the course
5. **Hit the Bull's Eye** Pages 75-78*	3-4 Days	• To use operations with integers and graphing on a rectangular coordinate system to simulate an archery tournament; to apply several of the mathematical ideas from the unit to analyze the results of a game

*Pages refer to *Gateways to Mathematical Investigations* Teacher's Edition.

PLANNING
LONG-TERM
INVESTIGATIONS

MATERIALS	RELATED RESOURCES
<ul><li>Family Letter, 1 per student</li><li>Calculator, at least 1 per group</li><li>Grid paper, 1-3 sheets per student</li><li>Compass, protractor, ruler or straightedge, 1 of each per student</li><li>*Gateways to Mathematical Investigations* Student Edition, p. 20, 1 per student</li></ul>	**Gateways to Algebra and Geometry** Section 5.1, pp. 200-206: Circle Graphs Section 5.2, pp. 207-212: Bar Graphs and Line Graphs Section 5.3, pp. 213-220: Other Data Displays Section 5.4, pp. 221- 227: Averages Section 5.5, pp. 228-233: Organizing Data *Additional References:* Teacher's Resource File: Student Activity Masters
<ul><li>Calculator, at least 1 per group</li><li>Grid paper, 1-3 sheets per student</li><li>Ruler or straightedge, 1 per student</li><li>A recent almanac for data about Summer Olympics track and field winners, 2-3 sets of data per group</li><li>Data about other events in the Summer Olympics, for example, swimming (optional)</li><li>*Gateways to Mathematical Investigations* Student Edition, pp. 21-22, 1 of each per student</li></ul>	**Gateways to Algebra and Geometry** Section 5.2, pp. 207-212: Bar Graphs and Line Graphs Section 5.3, pp. 213-220: Other Data Displays Section 5.4, pp. 221-227: Averages Section 6.1, pp. 252-259: The Number Line Section 6.3, pp. 267-274: Adding Signed Numbers Section 6.6, pp. 290-295: Subtracting Signed Numbers *Additional References* Section 2.4, pp. 71-77: Percent Section 4.1, pp. 150-157: A Closer Look at Measurement Teacher's Resource File: Student Activity Masters
<ul><li>Calculator, at least 1 per group</li><li>Sports sections from a Sunday newspaper (for college basketball statistics), 1 copy per student</li><li>*Gateways to Mathematical Investigations* Student Edition, pp. 23-24, 1 of each per student</li></ul>	**Gateways to Algebra and Geometry** Section 5.5, pp. 228-233: Organizing Data Section 5.6, pp. 234-241: Matrices Section 6.3, pp. 267-274: Adding Signed Numbers Section 6.4, pp. 275-281: Mult. and Div. Signed Numbers Section 6.6, pp. 290-295: Subtracting Signed Numbers *Additional References* Teacher's Resource File: Student Activity Masters
<ul><li>Coordinate grid paper, 3 sheets per student</li><li>Spinner marked 1 to 4, 1 per group</li><li>Spinner marked -1 to -4 , 1 per group</li><li>Ruler or straightedge, 1 per student</li><li>Two colors of counters (optional)</li><li>Miniature golf scorecard/course layout (optional)</li><li>*Gateways to Mathematical Investigations* Student Edition, pp. 25-26, 1 of each per student</li></ul>	**Gateways to Algebra and Geometry** Section 6.3, pp. 267-274: Adding Signed Numbers Section 6.4, pp. 275-281: Multiplying and Dividing Signed Numbers Section 6.6, pp. 290-295: Subtracting Signed Numbers Section 6.7, pp. 296-301: Absolute Value Section 6.8, pp. 302-309: Two-Dimensional Graphs *Additional References* Section 8.2, pp. 378-384: Using Variables Teacher's Resource File: Student Activity Masters
<ul><li>Coordinate grid paper, 1-2 sheets per student</li><li>Set of 9 index cards numbered -4 to 4, 1 set per group</li><li>Archery target (optional)</li><li>*Gateways to Mathematical Investigations* Student Edition, pp. 27-28, 1 of each per student</li></ul>	**Gateways to Algebra and Geometry** Section 5.6, pp. 234-241: Matrices Section 6.2, pp. 260-266: Inequalities Section 6.3, pp. 267-274: Adding Signed Numbers Section 6.4, pp. 275-281: Multiplying and Dividing Signed Numbers Section 6.6, pp. 290-295: Subtracting Signed Numbers Section 6.7, pp. 296-301: Absolute Value Section 6.8, pp. 302-309: Two-Dimensional Graphs *Additional References* Teacher's Resource File: Student Activity Masters

See pages 250A and 250B: Assessment for Unit C.

CHAPTER 5

Contents

5.1 Circle Graphs
5.2 Bar Graphs and Line Graphs
5.3 Other Data Displays
5.4 Averages
5.5 Organizing Data
5.6 Matrices

Pacing Chart

Days per Section

	Basic	Average	Honors
5.1	2	1	1
5.2	2	2	1
5.3	2	2	1
5.4	2	1	2
5.5	2	2	2
5.6	2	2	2
Review	2	1	1
Assess	2	1	1
	16	12	11

Chapter 5 Overview

Representing and analyzing data is an increasingly important task for people, both at home and at work. Computers, with their increasingly user-friendly spreadsheets, data bases, and graphing capabilities, make the organization and display of data accessible to virtually anyone. This chapter emphasizes learning to intelligently represent and communicate data. Along the way, students will have the opportunity to apply and integrate many previously learned concepts. They will also be able to bring their own perspective to the subject—both in the graphs they create and in their interpretation of those graphs.

5 Data Analysis

CHAPTER 5

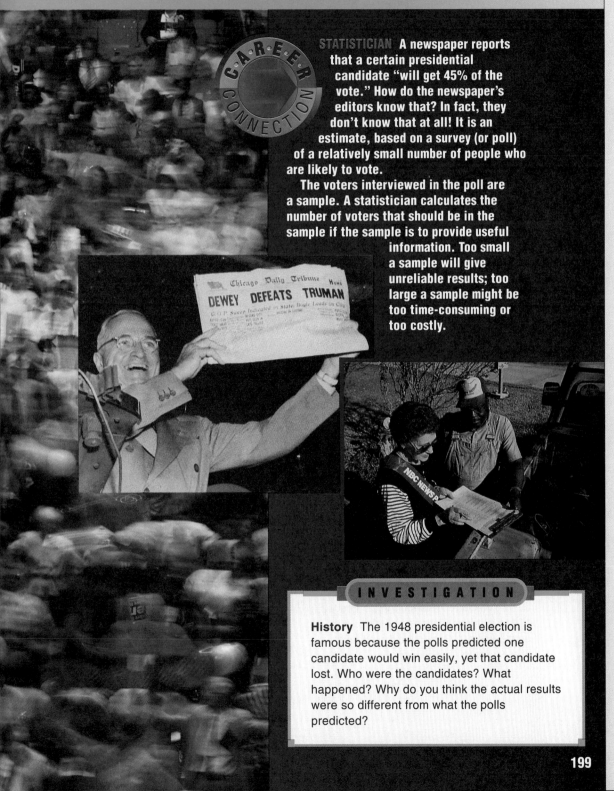

STATISTICIAN A newspaper reports that a certain presidential candidate "will get 45% of the vote." How do the newspaper's editors know that? In fact, they don't know that at all! It is an estimate, based on a survey (or poll) of a relatively small number of people who are likely to vote.

The voters interviewed in the poll are a sample. A statistician calculates the number of voters that should be in the sample if the sample is to provide useful information. Too small a sample will give unreliable results; too large a sample might be too time-consuming or too costly.

INVESTIGATION

History The 1948 presidential election is famous because the polls predicted one candidate would win easily, yet that candidate lost. Who were the candidates? What happened? Why do you think the actual results were so different from what the polls predicted?

Career Connection

Students should be aware of the great influence statistics has in government, business, and education. Encourage them to find examples of polls, statistical samples, and statistical predictions in newspapers, magazines, and books. Point out that all statistical calculations are subject to errors and that statistical conclusions are not facts but interpretations and extrapolations. Discuss how statistics can be misleading and what students can do to learn to recognize statistics that are misleading.

Video Connection

For more information about the career discussed here, please refer to the Futures videocassette #6, program A: *Statistics* featuring Jaime Escalante. For more information or to order a copy of the program, call PBS VIDEO at 800/344-3337.

Investigation Notes

In 1948, Thomas E. Dewey, the Republican governor of New York, ran for President against the incumbent Democrat, Harry S Truman. Opinion polls and most observers predicted that Dewey would win easily, and Dewey campaigned quietly. Truman, however, campaigned vigorously, charming the public with his warmth and humor and traveling the country tirelessly. As a result, he won the election with about 24,000,000 popular votes and 303 electoral votes to Dewey's approximately 22,000,000 popular votes and 189 electoral votes.

SECTION 5.1

Objective

- To read and draw circle graphs

  Vocabulary: circle graph, data analysis, disjoint, pie chart

Class Planning

Schedule 1–2 days

Resources

In *Chapter 5 Student Activity Masters* Book
Class Openers *1, 2*
Extra Practice *3, 4*
Cooperative Learning *5, 6*
Enrichment *7*
Learning Log *8*
Technology *9, 10*
Spanish-Language Summary *11*

Class Opener

Make a circle graph for each set of data. Estimate the size of each pie-shaped sector. Do not use a protractor.
1. Hair color of 100 students: red, 5; blond, 20; black, 25; brown, 50.
2. Grades of 50 students: A, 5; B, 10; C, 25; D, 5; F, 5.

Answers:
Approximate central angle measures of each sector are given.
1. red, 18°; blond, 72°; black 90°; brown 180° **2.** A, 36°; B, 72°; C, 180°; D, 36°; F, 36°

Lesson Notes

- Circle graphs help students not only to organize data but also to conceptualize area, especially portions of a circle. An important theme of this chapter is addressed on page 200: What is the best way to display information? For circle graphs, we are taking the total amount of something and examining fractional parts of it.

200

Investigation Connection
Section 5.1 supports Unit C, *Investigation* 1. (See pp. 198A-198D in this book.)

SECTION
5.1
Circle Graphs

READING A CIRCLE GRAPH

Through a process called **data analysis,** we can collect, organize, display, and interpret data. To be meaningful, data must be collected in a fair and impartial manner, and the presentation of the data should be clear, accurate, and not misleading.

A **circle graph,** often referred to as a **pie chart,** relates portions of data to the total number of data. A circle represents the total number of data, and pie-shaped sectors represent portions of the data.

A circle graph is an appropriate graph to use when the data have the following attributes:

1. The sum of the data is known.
2. The groups of data are **disjoint**—that is, divisible into categories that do not overlap.

Example
The circle graph shows the results of a survey conducted at Hoday Junior High School. One of the survey questions asked students to list their favorite food. Of the total number of students surveyed, 154 students responded to the question and 23 did not. Determine the number of students in each category.

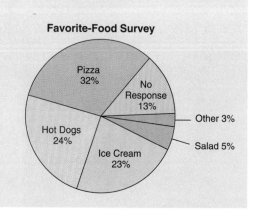

200   Chapter 5   Data Analysis

SECTION 5.1

Solution

Since 154 students responded to the favorite-food question and 23 did not, we know that a total of 177 students took the survey. We use this total and the percent from the graph to determine the number of students in each category.

Favorite Food	Percentage	Number of Students
Pizza	32% of 177 = 0.32 × 177 = 56.64	57
Hot dogs	24% of 177 = 0.24 × 177 = 42.48	42
Ice cream	23% of 177 = 0.23 × 177 = 40.71	41
Salad	5% of 177 = 0.05 × 177 = 8.85	9
Other	3% of 177 = 0.03 × 177 = 5.31	5
No response	13% of 177 = 0.13 × 177 = 23.01	23

DRAWING A CIRCLE GRAPH

To draw a circle graph, we need to

1. Identify the whole
2. Determine the number of parts
3. Calculate the percent of the circle represented by each part
4. Determine the measures of the angles that correspond to the percents
5. Draw and title the graph, labeling each pie-shaped sector with a category and a percent

Sample Problem

Problem Look at a map of the United States. Determine the number of states that border each state. (States that share only corners are not considered bordering states.) Make a circle graph, with each pie-shaped sector showing the percent of states with a given number of bordering states.

Solution
1. The entire set of data is the 50 states.
2. The number of states bordering each state is shown in the map at the top of the next page. We can see that some states are bordered by as many as eight states, and some states are not bordered by any other state. Therefore, the circle will be divided into nine parts.

■ Having students create circle graphs from data that you provide—or better yet, that they collect—is a wonderful review of percent and area of circles. Creating circle graphs is a good group activity.

Additional Example

Example
Use the circle graph on page 200. Assume 200 students were surveyed. How many students are in each category?

Answer: Pizza, 64; hot dogs, 48; ice cream, 46; salad, 10; other, 6; no response, 26

Additional Practice

Use the table below to solve problems 1–4.

Weekly Allowance Survey

Weekly Allowance	Number of Students
under $5.00	63
$5.00–9.99	47
$10.00–14.99	40
$15.00 or over	21
no allowance	79

1. Find the total number of students represented.
2. Find the percent each entry represents of the total.
3. Find the angle measure of each region of a circle graph.
4. Make a circle graph for the data.

Answers: **1.** 250 **2.** 25%, 19%, 16%, 8%, 32% **3.** 90°, 68°, 58°, 29°, 115°
4.

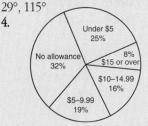

SECTION 5.1

Assignment Guide

Basic
Day 1 9–14, 20–22, 25
Day 2 15–18, 23, 26, 28, 30

Average or Heterogeneous
Day 1 9–11, 13–19, 22–28, 30

Honors
Day 1 9–14, 18, 19, 23, 24, 27–30, Investigation

Strands

Communicating Mathematics	1, 2, 5, 6, 12, 17, 21
Geometry	3, 13, 22
Technology	15, 16, 18, 19
Probability	26, 27
Algebra	22, 24
Data Analysis/ Statistics	5, 8–10, 15–19, 26, 27, 30
Measurement	13, 28
Patterns & Functions	20, 21
Number Sense	21
Interdisciplinary Applications	9–12, 15–19, 23, Investigation

Individualizing Instruction

Emotionally handicapped learners Students may not have the patience to work completely through problems such as 18 or 19. You may need to provide some extra support and encouragement to these students, and you might choose to allow them to work in pairs, so that the amount of work they need to do per problem is reduced.

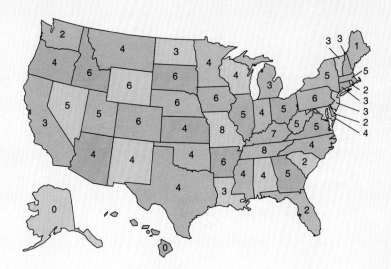

The table summarizes the data displayed on the map.

Number of bordering states	0	1	2	3	4	5	6	7	8
Number of states	2	1	5	8	13	9	9	1	2

3 and 4. We can now calculate the percent and angle measures.

Number of Bordering States	Percent of States with the Given Number(s) of Bordering States	Angle Measure
0 or 8	$\frac{2}{50}(100) = 4\%$	4% of 360° = 0.04 × 360 ≈ 14°
1 or 7	$\frac{1}{50}(100) = 2\%$	2% of 360° = 0.02 × 360 ≈ 7°
2	$\frac{5}{50}(100) = 10\%$	10% of 360° = 0.10 × 360 = 36°
3	$\frac{8}{50}(100) = 16\%$	16% of 360° = 0.16 × 360 ≈ 58°
4	$\frac{13}{50}(100) = 26\%$	26% of 360° = 0.26 × 360 ≈ 94°
5 or 6	$\frac{9}{50}(100) = 18\%$	18% of 360° = 0.18 × 360 ≈ 65°

5. Here is the completed circle graph with all labels.

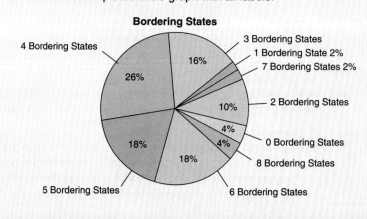

SECTION 5.1

Think and Discuss

P   **1** What does a circle graph represent?   An entire quantity divided into groups.

P, I   **2** Explain how to make a circle graph.

P, I   **3** What is the sum of the measures of all the angles that make up the pie-shaped sectors of a circle graph?   360

P, R   **4** What is the sum of the percent represented by the pie-shaped sectors into which a circle graph is divided?   100%

P, R, I   **In problems 5–7, refer to the example on pages 200–201.**

  5 Why is the number of "no responses" included on the graph?

  6 List possible reasons that students did not respond to the question.

  7 Can we conclude that less than one fourth of the students identified hot dogs as their favorite food?   Answers may vary. Discuss $\frac{42}{177} \approx 24\%$ versus $\frac{42}{154} \approx 27\%$.

P   **8** Poll your class to determine students' favorite lunch foods. Make a pie chart representing the data.   Answers will vary.

Problems and Applications

P   **9** In a given year, U.S. radio stations were polled to determine the most popular kinds of programming. The data are displayed on the graph. Can you conclude that country music and adult-contemporary music were the most popular kinds of programming that year?   No

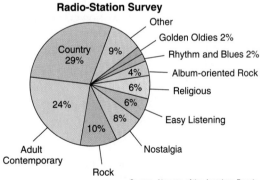

Source: *Almanac of the American People*

P, R   **In problems 10 and 11, refer to the survey that shows how U.S. consumers spent money to purchase albums, tapes, and compact discs.**

  10 Make a circle graph to display this information.

  11 Compare the circle graph you drew for problem **10** to the one in problem **9**.

Type of Music	Percent of Money Spent
Rock	46.6%
Pop	14.2%
Disco	10.1%
Country	9.7%
Classical	5.9%
Jazz	3.9%
Gospel	3.0%
Other	6.6%

Source: *Almanac of the American People*

Section 5.1 Circle Graphs   203

Problem-Set Notes

for page 203

■ **5** and **6** are important questions when analyzing data. You might ask how a graph would look if 60% of the students did not respond or if you used only the 40% that did respond. How would the example look if the 23 no-responses were not included? Would this graph show the same information?

■ **9** should lead to excellent discussions. Many students will conclude that the answer is yes because a majority of the stations play country and adult-contemporary music. The real issue is how many people listen to the music. It is possible that the rock stations have a much bigger audience per station and so are more popular. Thinking about what the graph really says takes a lot of practice and requires discussion. It is essential that students learn to think carefully about statistics.

Additional Answers

for page 203

2 First, identify the whole quantity. Then determine the number of parts into which the whole will be divided. Calculate the percent of the circle represented by each part. Determine the angle measure that corresponds to the percent. Draw the circle and label each pie-shaped sector with a category and a percent. Give the graph a title.

5 The number of *no responses* must be included so that the percents add up to 100.

6 Possible answers: Students may have been absent; students did not want to participate.

Answers for problems 10 and 11 are on page A3.

SECTION 5.1

Problem-Set Notes

for pages 204 and 205

- 17 requires careful thinking. Without knowing how much petroleum was used in 1970, we cannot answer.

Checkpoint

Use the table below to solve problems 1–4.

Bowling Scores

Scores	Number
under 100	7
100–149	15
150–199	21
200–249	10
250–300	2

1. What is the total number of bowling scores represented?
2. What percent of the bowling scores are in the 150–199 range?
3. What would be the angle measure for the 150–199 region of a circle graph?
4. How many regions would a circle graph showing the data from the table have?

Answers:
1. 55 2. 38% 3. 137° 4. 5

Stumbling Block

In creating circle graphs, students need to check their work at each step to be sure that percents add to 100% and degrees add to 360°. For some problems, the rounding process will lead to totals other than 100% and 360°. This is fine as long as the totals are near these numbers.

P **12 Communicating** Problem **9** seems to imply that country music and adult-contemporary music are the most popular kinds of music in the United States. Problems **10** and **11** seem to imply that rock music is far more popular than any other kind of music. Discuss this apparent contradiction.

P, R **In problems 13 and 14, refer to the graph.**

13 Find the angle measure of each of the four parts of the circle.

14 Determine the percent of the graph for each of the four parts.   A: ≈ 43%; B: ≈ 18%; C: ≈ 22%; D: ≈ 17%

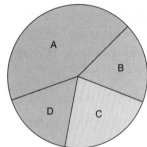

P **In problems 15–17, the graphs show U.S. petroleum use in 1970 and 1989.**

15 In 1989, the United States used 6,323,600,000 barrels of petroleum. How many barrels of petroleum were used for transportation?

16 The United States had 25,900,000,000 barrels of petroleum in reserve on January 1, 1990. How long could that reserve last if petroleum was not used for transportation and the level of consumption remained the same?   About 11 yrs

17 Did the amount of petroleum used for industrial purposes go up or down from 1970 to 1989?

United States Petroleum Consumption

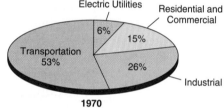

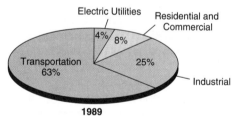

Source: *The World Almanac and Book of Facts*

P **18** A recent survey done at the House of Pizza revealed that the following numbers of pizzas were sold during the second week in February.
  a How many pizzas were sold during the second week of February?   551
  b What percent of the total is each type of pizza?
  c If a circle graph were constructed for the data, how many degrees would be in each pie-shaped sector?
  d Draw the circle graph.

Type of Pizza	Number Sold
Cheese only	96
Cheese and sausage	108
Cheese and mushroom	79
Cheese and pepperoni	125
Other combinations	143

204 Chapter 5 Data Analysis

SECTION 5.1

P, R **19 Spreadsheets** The table displays the ages of cars being used in the United States.
  a Enter the two columns of data on a spreadsheet.
  b Find the total number of cars represented. 139,500,000
  c Make a third column on the spreadsheet for the percent each entry represents. (Hint: Use a formula and the copy function.)
  d Make a fourth column to show the angle measure of each portion of a pie chart. (Hint: Use a formula and the copy function.)
  e Make a pie chart of the data by hand or with the spreadsheet.

Age of Car	Number of Cars on the Road
Less than 1 year	7,812,000
1–5 years	47,569,500
5–10 years	42,687,000
10–15 years	27,760,500
15 years or older	13,671,000

Source: *The Unofficial U.S. Census*

◀ LOOKING BACK **Spiral Learning** LOOKING AHEAD ▶

R **20** Continue the pattern.
$5 + 3 \cdot 6 = 23$
$8 + 4 \cdot 6 = \underline{32}$
$11 + 5 \cdot 6 = \underline{41}$
$14 + 6 \cdot 6 = \underline{50}$
$17 + 7 \cdot 6 = \underline{59}$

R **21** Explain the pattern in problem **20**. Why does it work?

R **22** $\triangle ACB \sim \triangle DEF$. Determine x and y.
$x = 39; y = 15$

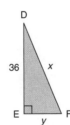

R, I **23 Sports** The Mildcats' football team had the ball at their 28 yard line. On the first play, they gained 7 yards. On the next play, they gained 4 more yards. On the third play, they lost 8 yards. On the fourth play, they gained 2 yards. What was the team's net gain after the four plays? 5 yds

I **24** Find lengths a, b, and c.
$a = 66, b = 9.4, c = 2.6$

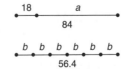

Section 5.1 Circle Graphs 205

Additional Answers

for pages 204 and 205

12 Possible answer: Different listening audiences.

13 A: ≈155°; B: ≈65°; C: ≈80°; D: ≈60°

15 About 3,983,868,000 barrels

17 The answer cannot be determined. Even though the percentage went down, the total amount consumed in 1970 is not known.

18 b Cheese only ≈17.4%
Cheese and sausage ≈19.6%
Cheese and mushroom ≈14.3%
Cheese and pepperoni ≈22.7%
Other combinations ≈26.0%
  c Cheese only ≈63°
Cheese and sausage ≈71°
Cheese and mushroom ≈51°
Cheese and pepperoni ≈82°
Other combinations ≈94°
  d

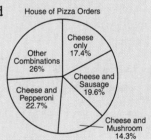

19 a See students' work.
  c Less than 1 yr: 5.6%;
1–5 yr: 34.1%; 5–10 yr: 30.6%;
10–15 yr: 19.9%; 15 yr or older: 9.8%
  d Less than 1 yr: 20°; 1–5 yr: 123°; 5–10 yr: 110°; 10–15 yr: 72°; 15 yr or older: 35°
  e

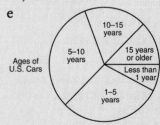

21 Possible answer: Each number is 9 more than the preceding number.

SECTION 5.1

Problem-Set Notes

for page 206

- **26** and **27** show the relationship between probability and statistics.
- **Investigation** gives students an opportunity to compare their favorite stations to see which play more music and fewer commercials.

Cooperative Learning

- Have groups of students use almanacs and other reference books to develop circle graphs from data of their own choosing. Groups should choose data that meet the criteria given on page 200 and be prepared to explain their reasoning.
- Problems 12 and 19 are appropriate for cooperative learning.

Communicating Mathematics

- Have students write a paragraph explaining in their own words what is meant by *disjoint data*. Students should include original examples of data that are disjoint and data that are not disjoint.
- Problems 1, 2, 5, 6, 12, and 21 encourage communicating mathematics.

Multicultural Note

An 8000-year-old carved bone handle found in Zaire, Africa, is one of the earliest evidences of the use of numbers. Known as the Ishango Bone, the handle has markings that seem to represent a number system based on ten.

206

R **25** Jim Shu walked 8 miles in a direction 30° east of north and then walked 10 miles in a direction 70° east of north. If Jim is facing in the direction he last walked, how many degrees should he rotate clockwise to face his starting point? ≈162°

I **In problems 26 and 27, refer to the graph.**

26 If a woman is chosen at random from the total number of women selected to provide data for the graph, what is the probability that she wears a size 4–6 dress? About 10%

27 If a woman is chosen at random from the group in problem **26**, what is the probability that her dress size is 14 or smaller? About 74%

R **28** A board 6 feet 10 inches long is to be cut into 5 boards of equal length. Find the length of each board. ≈16.4 in., or ≈1 ft 4 in.

I **29** How many different routes run from A to B? (Notice the one-way arrows.) 4

R, I **30** Mr. Matt Hematics's class took a test. The scores were

98, 96, 95, 94, 94, 90, 89, 88, 85, 79, 79, 79, 78, 76, 74, 73, 70, 69, 69, 65, 61, 60, 58, 51, 43

  a Which score occurred most often? 79
  b Millie Middle's score was exactly in the middle, with 12 students receiving higher scores and 12 students receiving lower scores. What was Millie's score? 78
  c Find the mean (average) student score. 76.52
  d Find the range of values. 43 to 98
  e What is the difference between the highest score and the lowest score? 55

Investigation

Communications Listen to one radio station continuously for at least an hour. Keep track of the different programming segments—commercials, music, talk, and news—and the amount of time spent on each. Make a circle graph to display your results. Be sure that the graph includes the name of the radio station and the length of time you listened to the radio. Will everyone in your class have similar graphs? Answers will vary. Possible answer: No

206 Chapter 5 Data Analysis

SECTION 5.2

Investigation Connection
Section 5.2 supports Unit C, *Investigations* 1, 2. (See pp. 198A-198D in this book.)

5.2
Bar Graphs and Line Graphs

READING BAR GRAPHS AND LINE GRAPHS

A ***bar graph*** is a graph in which data can be displayed as vertical or horizontal parallel bars of appropriate length. Each of the axes of a bar graph is labeled with appropriate units. Bar graphs allow us to make quick comparisons.

Example 1

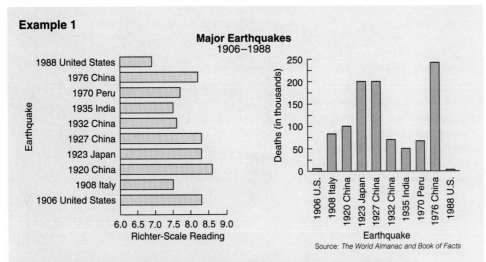

Source: *The World Almanac and Book of Facts*

a Where and when did the quake that caused the fewest number of deaths occur? What was the Richter-scale reading for this quake?
b Where and when did the quake with the highest Richter-scale reading occur? How many deaths resulted from that quake?
c What country was hit by four major quakes? What is the difference between the largest and the smallest number of resulting deaths?
d Is there a relationship between the Richter-scale magnitude of an earthquake and the number of resulting deaths?

Objectives

- To read and draw bar graphs and line graphs

  Vocabulary: bar graph, line graph

Class Planning

Schedule 1–2 days

Resources

In *Chapter 5 Student Activity Masters* Book
Class Openers *13, 14*
Extra Practice *15, 16*
Cooperative Learning *17, 18*
Enrichment *19*
Learning Log *20*
Using Manipulatives *21, 22*
Interdisciplinary
  Investigation *23, 24*
Spanish-Language Summary *25*

Transparencies *23 a,b*

Class Opener

1. Make a bar graph with vertical bars for the following data. Favorite pet: dog, 20; cat, 10; bird, 15; other, 5

2. Make a bar graph with horizontal bars for the following data. Miles jogged each week: week 1, 10; week 2, 8; week 3, 12; week 4, 8

Answers:

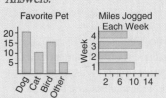

Lesson Notes

- Example 1 is particulary interesting because it contains a surprising result. You might war ask students before they look . the graphs if they expect to find that earthquake casualties go up

Section 5.2 Bar Graphs and Line Graphs

SECTION 5.2

Lesson Notes continued

as Richter-scale readings increase. Many of them will not consider that a large earthquake in a sparsely populated area will injure fewer people than a smaller earthquake in a densely populated area. If anyone raises that point in class, commend them for their careful reasoning.

■ Bar graphs are useful when comparing things over time or when comparing quantities when the sum of *all* the quantities is unknown. Line graphs and bar graphs are generally interchangeable. A line graph, though, is preferable when recording single events over time, such as world records or test results. A bar graph might be used to show total sales for a given item, company, or time period and to compare them with total sales for other items, companies, or time periods.

■ Example 2 introduces the concept of extrapolation.

Additional Examples

Example 1
Refer to the bar graphs on page 207.
a. What was the Richter-scale reading for the earthquake that occurred in Japan in 1923?
b. How many deaths occurred in that earthquake?
c. For which of the earthquakes that occurred in China is the ratio of the number of deaths to the Richter-scale reading the largest?

Answers: **a.** 8.3 **b.** 200,000
c. the 1976 earthquake

Example 2
Refer to the line graph on page 208. What do you think the women's Olympic high-jump record might be in 1996?

Answer: ≈83 inches

208

Solution

a The quake occurred in the United States in 1988. The Richter-scale reading was about 6.9.
b The quake occurred in China in 1920. There were about 100,000 deaths.
c China was hit by four major quakes. The difference between the largest and the smallest number of resulting deaths is about 170,000.
d There seems to be very little relationship between the Richter-scale magnitude of a quake and the number of resulting deaths. However, a smaller quake would probably cause fewer casualties. We need more data to draw a valid conclusion.

A **line graph** is a graph in which data are displayed as points that are connected by segments. Both axes on a line graph are also labeled. A line graph allows us to show trends.

Example 2
Do you think the women's Olympic high-jump record will reach 7 feet by the year 2000? Why?

Women's Olympic High-Jump Winners

Source: The World Almanac and Book of Facts

Solution
If we lay a ruler along the graph and estimate an average rate of increase, it appears that the women's Olympic high-jump record will have reached 7 feet by the year 2000.

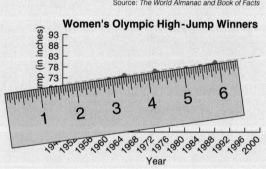

DRAWING BAR GRAPHS AND LINE GRAPHS

To make a bar graph or a line graph,

1. Decide what data to graph along each axis
2. Find the range of values
3. Label each axis with appropriate units
4. Draw the bars for a bar graph, or plot and connect the points for a line graph
5. Title the graph

208   Chapter 5   Data Analysis

SECTION 5.2

Sample Problem

Problem Mrs. Kur E. Uss surveyed her class to find out the types and numbers of pets her students owned. The table summarizes the results of the survey.

Pet	Number of Students Owning Pets	Total Number of Pets
Cat	8	13
Dog	10	12
Fish	6	68
Gerbil	5	8
Bird	3	4
Guinea pig	2	2
Other	2	4
No pet	3	0

Make a bar graph showing the total number of each type of pet owned by students.

Solution
1. We will label the horizontal axis with the number of pets and the vertical axis with the types of pets.
2. The number of pets ranges from 0 ("no pet") to 68 ("fish").
3. We will label the horizontal axis in increments of 10—each division of this axis will represent 10 pets. Since there are 8 types of pets, 1 type will be named in each of the 8 divisions of the vertical axis.
4 and 5.

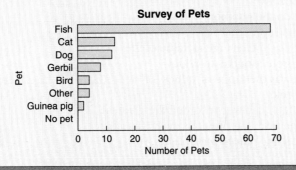

Think and Discuss

P **1** What would you need to do to change the horizontal bar graph in Example 1 into a vertical bar graph?

P **2** What would you need to do to change the vertical bar graph in Example 1 into a horizontal bar graph?

Assignment Guide

Basic
Day 1 6, 7, 10, 15–17, 21
Day 2 8, 11–13, 18–20, 22

Average or Heterogeneous
Day 1 6, 7, 9, 12, 14, 16, 19, 21
Day 2 8, 10, 11, 13, 15, 17, 18, 20, 22

Honors
Day 1 6–8, 10–13, 15, 21, 22, Investigation

Strands

Communicating Mathematics	1–5, 7, 12–14, 20
Geometry	16, 19, 20
Algebra	18, 20
Data Analysis/ Statistics	3, 4, 6–15, 22
Technology	21
Measurement	17, 19, 21
Interdisciplinary Applications	6, 8, 11, 12, 21, 22, Investigation

Individualizing Instruction

Limited English proficiency Before attempting to write a story for problem 13, students might benefit from talking about the graph with you or with a peer. Describing statistical data requires precise word choice, so LEP students will need extra help.

Additional Answers

for page 209

1 The Richter-scale readings would appear along the vertical axis, and the countries would appear along the horizontal axis.

2 The number of deaths would appear along the horizontal axis, and the countries would appear along the vertical axis.

SECTION 5.2

Problem-Set Notes

for pages 210 and 211

- **4** and **5** are important in helping students determine how best to express their data. Be sure to allow students to express their opinions about these problems.

- **13** could have as many different responses as there are students. Allow some time for students to share their stories with the class.

Additional Practice

The chart shows the daily high temperatures during a week in a particular city.

Day	Temperature
Sunday	83
Monday	89
Tuesday	80
Wednesday	72
Thursday	81
Friday	85
Saturday	92

1. Make a line graph to represent the data.
2. How many decreases in temperature occurred?
3. Between what days did the greatest change occur?
4. Describe the weather pattern at the end of the week.

Answers:

1.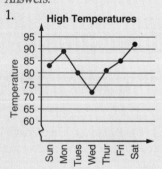

2. 2   3. Monday–Tuesday and Wednesday–Thursday
4. A warming trend

P   **3** What conclusion was suggested by the data in Example 2? How certain are you about your conclusion?

P, R   **4** Why is a circle graph not appropriate for the data in the sample problem on page 209?   The data overlap. A student may have both a fish and a dog.

P, R   **5 a** What kinds of data are best displayed in a circle graph?
    b What kinds of data are best displayed in a bar graph?

Problems and Applications

P, R   **6 Science**   Make a line graph using the following data.

Days after planting	0	10	20	30	40	50	60	70	80	90
Height of plant (in inches)	0	0	4	9	20	40	64	80	84	84

P   **7** Refer to the sample problem on page 209.
   a Make a bar graph that shows the number of students owning each type of pet.
   b Is a fish or a cat the more popular pet among the students in Mrs. Kur E. Uss's class?   Cat

P   **8 Business**   The value of Solar Plexus stock was $39.00 on Monday. The value of the stock rose by $1.50 on Tuesday, fell by $4.00 on Wednesday, rose by $1.75 on Thursday, and fell by $5.50 on Friday. Display this information in a line graph.

P, R   **9** The chart displays weekly statistics for BCA stock.

Day	Value at Close
Monday	$35\frac{7}{8}$
Tuesday	$33\frac{1}{2}$
Wednesday	$35\frac{1}{4}$
Thursday	$37\frac{1}{8}$
Friday	$36\frac{3}{4}$

   a What do the numbers in the value column represent?
   b What was the change in value of the stock from Monday to Friday?
   c What was the range of values for the week? $37\frac{1}{8}$ to $33\frac{1}{2}$
   d On which day was recorded the greatest change from the previous day?   Tuesday
   e Make a line graph of this data.

P   **10** The following is a list of test scores for a geology test:

97, 83, 65, 82, 99, 81, 89, 77, 63, 100, 100, 91, 82, 88, 88, 97, 92, 96, 94, 84, 88, 89, 73, 78, 87, 82, 71, 73, 97, 52, 68, 84, 86, 81, 92

The given grading scale is used. Make a bar graph that shows the number of students that earned each grade.

Grade	Range
A	100–93
B	92–85
C	84–77
D	76–68
F	67–0

210   Chapter 5   Data Analysis

SECTION 5.2

P, R **11 Social Science**
  a What percent of people got married in 1960? 0.85%
  b What percent of people got married in 1970? 1.06%
  c What percent of people got married in 1975? 1%
  d Did more people get married in 1970 or in 1975?
  e What is the ratio of marriages to divorces, rounded to the nearest hundredth, in 1960? 1965? 1975? 1980? 1988? 3.86; 3.72; 2.08; 2.04; 2.02

P **12** A study of deaths over a 25-year period revealed data on the average number of annual deaths from various weather-related conditions.
  a Display the data in a bar graph.
  b Explain why a circle graph is not an appropriate way to display the data.

Weather Condition	Average Annual Deaths
Heat	200
Floods	151
Lightning	94
Tornadoes	87
Hurricanes	31

P, I **13 Communicating** Write a story about what the graph might represent.

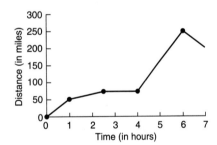

P **14 Communicating**
  a About how many times larger were farms in 1989 than in 1940? ≈3 times
  b What may have happened to the number of farms from 1940 to 1989?
  c Explain your response to part **b**.

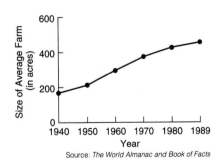

P, R **15** For a basketball player to slam dunk, the sum of the vertical reach and the vertical jump of the player must be 10 feet 6 inches. Copy and complete the bar graph to show the vertical jump needed to slam dunk.

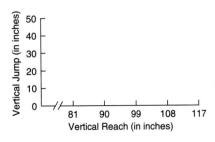

Section 5.2 Bar Graphs and Line Graphs 211

Checkpoint

Mr. Marks surveyed his class on the first day of school. The results are given below.

Length of Family Vacation	Number of Students
no vacation	3
weekend	10
1 week	12
2 weeks	7
longer than 2 weeks	2

1. What was the total number of responses?
2. What was the most frequent length for a family vacation?
3. How many students were on vacation for less than a week?
4. Make a bar graph for the data.

Answers: **1.** 34 **2.** 1 week **3.** 13
4.

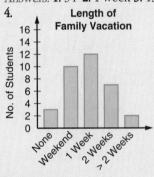

Cooperative Learning

- Have groups of students choose a line or bar graph from a newspaper, textbook, or magazine. Have groups write three questions involving at least two data points from the graph and one question focusing on a trend or prediction indicated by the graph. Groups should then exchange graphs and answer each others' questions.

- Problem 15 is appropriate for cooperative learning.

Answers for problems 3, 5, 6, 7 a, 8, 9 a, b, and e, 10, 11 d, 12, 13, 14 b and c, and 15 are on page A3.

211

SECTION 5.2

Problem-Set Notes

for page 212

- **18** can be done easily without a calculator. Ask the class if anyone realized this.
- **22** relates bar graphs and circle graphs. Ask students which is better for these data and why.

Additional Answers

for page 212

20 b One leg of the triangle is much longer than the other leg.

22 b

c You cannot determine the total number of phones because the number of phones in the "5 or more" category is unknown.

Communicating Mathematics

- Have students write a paragraph or give an oral presentation describing a recent experiment performed in science class. They should describe the data they collected and discuss what kind of graph (circle graph, bar graph, or line graph) is best suited to displaying those data. They should give reasons for their choice.

- Problems 1–4, 12–14, and 20 encourage communicating mathematics.

◀ LOOKING BACK **Spiral Learning** LOOKING AHEAD ▶

R   **16**   Classify each angle as *acute, right,* or *obtuse*.

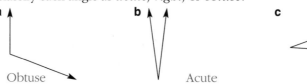

Obtuse   Acute   Acute

R   **17**   In a surveyor's drawing, $\frac{1}{8}$ inch represents 12 feet. What is the actual perimeter of a rectangle if the dimensions in the drawing are 2 feet by $1\frac{1}{2}$ feet?   672 ft

I   **18**   If $v = 21$, $w = 19$, $x = 17$, $y = 15$, and $z = 13$, what is the value of $(v - x)(w - x)(x - x)(y - x)(z - x)$?   0

R, I   **19**   The small rectangles are congruent. How many gallons does the red region represent?   7.5 gal

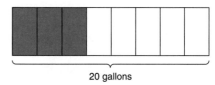

20 gallons

R, I   **20**   **a** Find the value of x. $\approx 4.7001 \times 10^8$

       b Why is the value of x close to the length of one of the legs of the triangle?

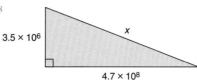

R   **21**   **Science**   Ice (at 0°C) has a density of 922 kg/m³.

       a If the edge of a cube of ice is 4 cm, what is the mass of the cube? ≈ 0.059 kg

       b Check your answer to part **a** by actually weighing such an ice cube.

P, R   **22**   **a** How many U.S. homes have four or fewer phones? ≈ 216 million

        b Make a circle graph of the data.

        c Based on the given data, what is the total number of phones in the U.S.?

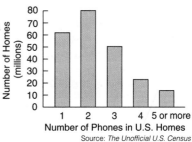

Source: *The Unofficial U.S. Census*

Investigation

Sports   The winning women's high jump in the 1968 Olympics in Mexico City was surprisingly low. Research the history of the 1968 Olympic Games to find possible explanations.   Answers will vary. Possible answer: High altitude

Chapter 5   Data Analysis

SECTION 5.3

Investigation Connection
Section 5.3 supports Unit C, *Investigations* 1, 2. (See pp. 198A–198D in this book.)

SECTION 5.3
Other Data Displays

SCATTERGRAMS

In a **scattergram,** data are graphed along vertical and horizontal axes but cannot be connected by a line. The data in a scattergram often follow a pattern from which we can derive a trend or draw some conclusions.

Example 1
Ms. Anna Liss gathered data on the length of time that students studied for a science test and the resulting test scores. What trend do you see?

Study Time (minutes)	Scores
0–15	0, 20, 24, 30
16–30	20, 30, 40
31–45	36, 38, 40, 42, 46
46–60	34, 38, 40, 42, 44, 50
More than 60	38, 40, 44, 46, 48, 49, 50

Solution
We can draw a scattergram with the test scores graphed along the vertical axis and the lengths of study time graphed along the horizontal axis.

The scattergram shows a trend of better scores for students who studied longer.

Objectives

- To interpret and draw scattergrams, histograms, pictographs, and artistic graphs

Vocabulary: artistic graph, histogram, line plot, pictograph, scattergram

Class Planning

Schedule 1–2 days

Resources

In *Chapter 5 Student Activity Masters* Book
Class Openers *27, 28*
Extra Practice *29, 30*
Cooperative Learning *31, 32*
Enrichment *33*
Learning Log *34*
Using Manipulatives *35, 36*
Spanish-Language Summary *37*

In *Tests and Quizzes* Book
Quiz on 5.1–5.3 (Basic) *115*
Quiz on 5.1–5.3 (Average) *116*

Transparencies *23 a,b, 24*

Class Opener

Make one graph that shows the data from the two graphs on page 207. Show the number of casualties on one axis and the Richter-scale rating on the other. Show each earthquake as a dot on the graph. Do not label or connect the dots.

Answer:

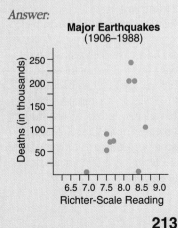

Section 5.3 Other Data Displays

SECTION 5.3

Lesson Notes

- A scattergram is appropriate when one variable will yield several different outcomes or when several variables will yield the same outcome. A scattergram is also useful when looking for a cause and effect, or correlation, between two different things. Example 1 shows that more time spent studying will generally produce higher scores, although certainly not in every case, as there are other variables to consider.
- Direct students to the sample problem for an example of a histogram. Most students have used a variation of a histogram to keep score, only they probably made vertical lines instead of dots.
- Pictographs and artistic graphs are generally just fancier versions of bar graphs, line graphs, circle graphs, or histograms. They seem to be growing in popularity despite their potential for being misleading (note the last paragraph at the bottom of page 215). You might have students bring in some examples of pictographs or artistic graphs and discuss whether the art is necessary and how else the information might have been presented.

Additional Examples

Example 1
Refer to the scattergram on page 213.
a. Did any student whose study time was 0–15 minutes score as high as a student whose study time was 31–45 minutes? Explain.
b. Did any student whose study time was more than 60 minutes score lower than a student whose study time was 60 minutes or less?

HISTOGRAMS

A **histogram** (or **line plot** if data are graphed on a number line) indicates the number of times that a value or range of values occurs in a set of data. Histograms are similar to bar graphs with respect to the type of information they can display. A histogram can be drawn vertically or horizontally, but in line plots the number line is usually horizontal.

Example 2
Display the scores from Ms. Anna Liss's class in a line plot.

Solution

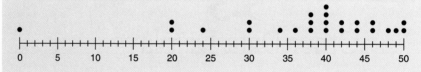

PICTOGRAPHS

A **pictograph** is used to represent data that would be difficult to display in a histogram. Pictographs use symbols related to the nature of the data.

Example 3
How much petroleum did the United States use in 1989?

Petroleum Consumption in 1989

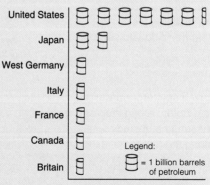

Source: Energy Information Administration / The World Almanac and Book of Facts

Solution
The legend at the bottom of the graph indicates that each barrel symbol on the pictograph represents 1 billion barrels of petroleum. Therefore, one half of a barrel symbol represents 500 million barrels. The $6\frac{1}{3}$ barrel symbols indicate that the United States used about 6,300,000,000 barrels of oil in 1989.

Chapter 5 Data Analysis

SECTION 5.3

ARTISTIC GRAPHS

Sometimes information can be represented by an ***artistic graph***. Artistic graphs vary greatly and usually manage to depict a great deal of information within a small space. Sometimes artistic graphs give impressions that are not exactly correct.

Example 4
Much information about the depletion of the world's rain forests is presented in this artistic graph. What might you conclude by looking at this graph?

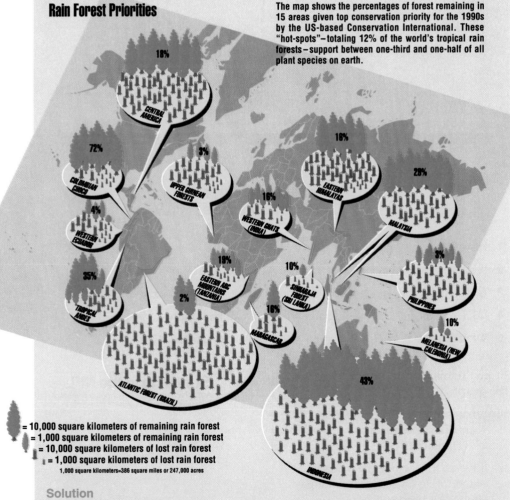

Rain Forest Priorities

The map shows the percentages of forest remaining in 15 areas given top conservation priority for the 1990s by the US-based Conservation International. These "hot-spots"—totaling 12% of the world's tropical rain forests—support between one-third and one-half of all plant species on earth.

🌲 = 10,000 square kilometers of remaining rain forest
🌲 = 1,000 square kilometers of remaining rain forest
🌲 = 10,000 square kilometers of lost rain forest
🌲 = 1,000 square kilometers of lost rain forest
1,000 square kilometers=386 square miles or 247,000 acres

Solution
We might conclude that most of the world's rain forests have been heavily depleted. However, if we read carefully, we learn that only 12% of the world's rain forests are depicted in this artistic graph and that these are the 15 rain forests that need the most help. If the artist had included all rain forests, our impressions might be different.

Answers: **a.** No. The highest score in the 0–15 interval is 30; the lowest score in the 31–45 interval is 36. **b.** Yes

Example 2
Refer to the line plot on page 214. What was the most frequent score?
Answer: 40

Example 3
Refer to the pictograph on page 214. Compare the oil consumption of the United States with the total consumption of the other six countries on the graph.
Answer: U.S. consumption was ≈1.7 billion barrels more than the others combined (6.3 − 4.6).

Example 4
Refer to the artistic graph on page 215.
a. What percent of the world's tropical rain forests are *not* represented in this graph?
b. What fraction of the earth's plant species are not found in the rain forests represented?

Answers: **a.** 88% **b.** $\frac{1}{2}$ to $\frac{2}{3}$

Additional Practice

1. Refer to the histogram for problem 8. How would you represent four field goals and two free throws?

2. In a pictograph, one symbol represents 10 bushels. How many symbols would you need to represent 55 bushels?

3. Refer to the graph for problems 15–18 on page 218. What years are represented on the graph?

Answers:
1. ●●●● ○○
2. $5\frac{1}{2}$ symbols 3. 1970–1990

SECTION 5.3

Assignment Guide

Basic
Day 1 9–11, 19–22, 24, 26, 33
Day 2 12, 13, 15–18, 23, 25, 30, 32

Average or Heterogeneous
Day 1 9–11, 14–18, 20–22
Day 2 12, 13, 19, 23–27, 30–33

Honors
Day 1 9–12, 14–19, 23–25, 27, 30, 31, 33, Investigation

Strands

Communicating Mathematics	1–5, 11, 21, 30
Geometry	26, 27, 29, 31
Probability	25, 33
Algebra	22, 29, 32
Data Analysis/ Statistics	6–10, 12–20, 30, 33, Investigation
Technology	14
Measurement	31
Number Sense	30
Interdisciplinary Applications	6–11, 13–19, 23 33

Individualizing Instruction

The gifted student Students with special artistic talents have an opportunity to use these gifts when constructing pictographs. Make sure they keep in mind that a good pictograph is not necessarily a fancy one, but one that communicates the correct information in an interesting and effective way.

Problem-Set Notes

for pages 216 and 217

■ **1–5** are important discussion questions about determining which representation best suits various types of data.

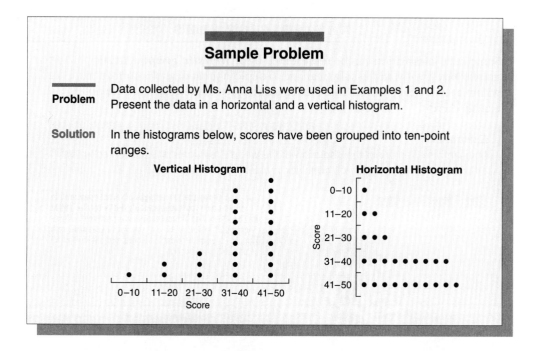

Sample Problem

Problem Data collected by Ms. Anna Liss were used in Examples 1 and 2. Present the data in a horizontal and a vertical histogram.

Solution In the histograms below, scores have been grouped into ten-point ranges.

Think and Discuss

P, R **1** How does a histogram differ from a bar graph?

P **2** What is the difference between a histogram and a line plot?

P, R **3** How would you decide whether to use a scattergram or a bar graph?

P, R **4** What advantages does a pictograph have over a bar graph?

P, R **5** What advantage does a bar graph have over a pictograph?

P **In problems 6 and 7, refer to the graph.**

6 What is the average number of years of experience for all 28 NFL coaches? 6 yr

7 If we do not include the 6 head coaches listed, what is the average tenure of a head coach? About 3.2 yr

SECTION 5.3

8 This is a histogram of the scoring of the Chicago Bulls in their first home game of the 1991 season against the Philadelphia 76ers.
   a How many points did Scottie Pippen score? 23
   b How many points did Michael Jordan score? 26
   c What was the team score? 110
   d The 76ers scored 90 points. Which team won? The Bulls

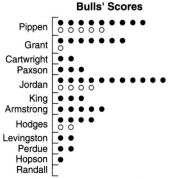

Problems and Applications

In problems 9–11, refer to the graph.

9 Find the ratio of noodles sold to long goods sold. 15:41, or $\frac{15}{41}$

10 For every 5 ounces of noodles sold, what quantity of short goods is sold?

11 Why do you think long goods are the most popular type of pasta?
   Answers may vary.

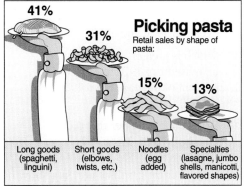

12 Adam rolled a pair of dice 116 times. Each time he rolled the dice, he wrote down the sum of the dots on the faces. The table summarizes the results. Represent the data in a histogram.

Sum of dots	2	3	4	5	6	7	8	9	10	11	12
Number of times	3	5	9	14	16	22	18	14	8	5	2

13 **Science** Draw a pictograph that displays the data in this table.

City	Snowfall (inches)
Juneau, Alaska	103
Buffalo, New York	92
Burlington, Vermont	78
Duluth, Minnesota	77
Portland, Maine	72
Denver, Colorado	60

Section 5.3 Other Data Displays

SECTION 5.3

Problem-Set Notes

for pages 218 and 219

- **14** might be replicated with your class if you have some way to have students timed in a 100-meter dash. The results might prove interesting.

- **27** provides a good visual connection between area and graphs.

Additional Answers

for pages 218 and 219

14 a

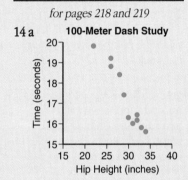

d The greater the hip height, the shorter the time to run the 100-meter dash.

15 About 1.36×10^{10} gal; About 1.72×10^{10} gal

19

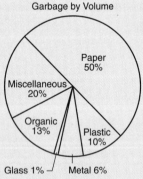

20 Answers will vary. See students' graphs.

21 Answer depends on graph drawn in problem 20.

26 a and **b** Answers will depend on individual student's diagram. Students should find that the ratios are equal.

218

---

14 **Science** Mr. Lewis had each student in his class run the 100-meter dash. The students' times were measured in seconds, and their hip heights were measured in inches. The data are displayed in a chart.

Hip height	30	31	28	34	26	29	32	22	26	33	32
Time	16.2	16.0	18.3	15.6	19.1	17.4	16.1	19.9	18.7	15.7	16.3

a Present the data in a scattergram.
b Find the average hip height of the students in Mr. Lewis's class. About 29.4 in.
c Find the average time it takes the students in Mr. Lewis's class to run the 100-meter dash. 17.2 sec
d Describe a relationship between hip height and the time it takes to run the 100-meter dash.

Agriculture In problems 15–18, refer to the graph, which shows the annual milk production of U.S. dairy cows.

15 How many gallons of milk were produced in 1970? In 1990?

16 What is the percent of increase in annual milk production per cow between 1970 and 1990? 50%

17 Which is greater, the percent of increase from 1970 to 1980 or the percent of increase from 1980 to 1990? 1970 to 1980

18 How much milk per day did cows give in 1970? In 1990? 3.1 gal; 4.67 gal

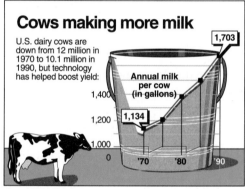

19 Make a circle graph of the data showing the distribution of landfill garbage by volume.

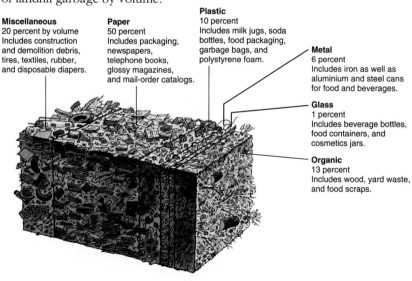

218 Chapter 5 Data Analysis

SECTION 5.3

◀ LOOKING BACK **Spiral Learning** LOOKING AHEAD ▶

P, R In problems 20 and 21, refer to the table, which displays the high temperatures in Frostbite Falls during the first week in January.

20 Display these data in a pictograph or an artistic graph.

21 *Communicating* Explain your choice of graph in problem **20**.

Day	Temperature
Monday	−27° F
Tuesday	−18°
Wednesday	−7°
Thursday	−12°
Friday	−19°
Saturday	−21°
Sunday	−25°

R, I 22 Solve each equation for x.
  a $13x = 1352$ 104
  b $13 + x = 1352$ 1339

R, I 23 Halfback Crazy Legs carried the ball seven times in last week's game. The following changes in yardage took place: +4, −7, −3, +5, −2, 0, +3. What was his average gain?   0

R, I 24

How many units apart are
  a A and B? 2   **b** B and C? 2   **c** A and C? 4   **d** A and D? 7

R, I 25 If you study the signs of the zodiac, you will see that $\frac{4}{12}$ of the signs represent people and $\frac{7}{12}$ of the signs represent animals. What fraction of the signs represent neither people nor animals? $\frac{1}{12}$

R 26 Draw $\triangle ABC$. Draw $\overline{PQ}$ parallel to $\overline{BC}$, where P is on $\overline{AB}$ and Q is on $\overline{AC}$. Measure $\overline{AP}$, $\overline{PB}$, $\overline{AQ}$, and $\overline{QC}$.
  a Find the ratio of the length of $\overline{AP}$ to the length of $\overline{PB}$.
  b Find the ratio of the length of $\overline{AQ}$ to the length of $\overline{QC}$.

R 27 **a** Calculate the area of each polygon to the left of the numbers 1 through 7.
  b Make a bar graph of the data.

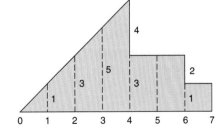

R 28 Evaluate each of the following.
  a 42.63×10^3 42,630
  b $523,609 \div 10^4$ 52.3609

---

27 a The integer is listed, followed by the area: $1, \frac{1}{2}$; 2, 3; 3, $\frac{15}{2}$; 4, 14; 5, 17; 6, 20; 7, 21

b

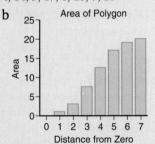

Cooperative Learning

■ Have groups of students choose two easily measured physical characteristics and collect data about their classmates. Examples include foot length and height, arm length and height, and thumb and ankle circumferences. Have students plot their data in a scattergram and write a short paragraph explaining what trends, if any, they discovered and what predictions they might make.

■ Problem 19 and the Investigation are appropriate for cooperative learning.

Communicating Mathematics

■ After students have drawn a circle graph of the data displayed in problem 19, have them write a paragraph explaining which graph they feel best displays the data and why. Students should suggest situations where the circle graph might be best, where the given artistic graph might be best, and where another type of graph might be best.

■ Problems 11, 21, and 30 encourage communicating mathematics.

Section 5.3 *Other Data Displays*

SECTION 5.3

R, I 29. a. Compare the perimeters of the two rectangles.
b. Compare the areas of the two rectangles.

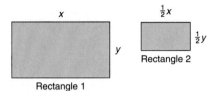
Rectangle 1 / Rectangle 2

I 30. a. List these seven people in order of height. Who is in the middle? Fionnuala
b. Explain your answer to part **a**. There are three people taller than Fionnuala and three people shorter than Fionnuala.

Alan — 5'4"
Cindy — 5'6"
Ellen — 5'3"
Greg — 5'7"
Bob — 5'8"
David — 5'2"
Fionnuala — 5'5"

R 31. a. Measure each section of the pie chart with a protractor and use your results to calculate the approximate number of dogs in each category.
b. Each year of a dog's life is equivalent to about seven years of a human's life. Calculate the equivalent human ages for each category.

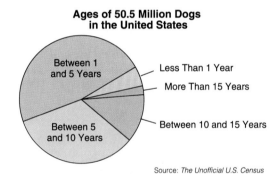

Ages of 50.5 Million Dogs in the United States

Source: *The Unofficial U.S. Census*

I 32. Solve for n, m, and w.
a. $78n + 4 = 160$ 2
b. $52m - 8 = 5$ $\frac{1}{4}$
c. $512 - w = 437$ 75

R 33. The data for this problem were gathered from a 1991 survey.
a. In 1991, if a science teacher were selected at random, what is the probability the teacher would have had a science degree? $\frac{61}{100}$
b. In 1991, if a mathematics teacher were selected at random, what is the probability the teacher would have had a mathematics degree? $\frac{21}{50}$

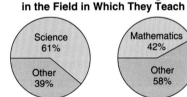

Teachers Who Have a Degree in the Field in Which They Teach

All high school science teachers / All high school mathematics teachers

Source: *National Center for Education Statistics*

Investigation

Data Analysis Find several examples of graphs from magazines or newspapers. Describe the main point(s) being made by each graph. Is some of this information misleading? Explain how graphs are sometimes used to mislead the reader. Produce examples if possible.

220   Chapter 5   Data Analysis

SECTION 5.4

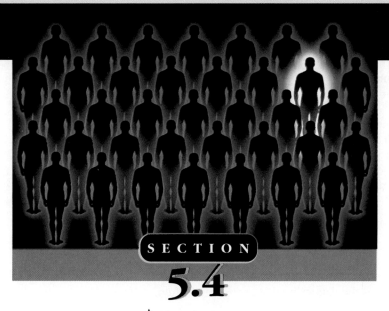

Investigation Connection
Section 5.4 supports Unit C, *Investigations* 1, 2. (See pp. 198A-198D in this book.)

SECTION 5.4
Averages

THE MEAN

An average tells us something about an entire set of data we are trying to analyze. As you will see, there are several kinds of averages. One kind, with which you are already familiar, is called the **mean.** In everyday life, when people talk about an average, they are usually referring to a mean. The mean is actually a ratio—the sum of the data values divided by the number of values.

Example 1
Here is a set of test scores from Miss Riley's third-period class. What is the mean test score?

65	76	84	93	69
87	95	92	79	72
82	86	88	94	66
82	77	74	82	95
97	82	91	89	73

Solution

$$\text{Mean} = \frac{\text{sum of test scores}}{\text{number of test scores}} = \frac{2070}{25} = 82.8$$

The mean test score is 82.8.

THE MEDIAN

The **median,** another kind of average, is the value in the middle of the data. To determine the median, you may put the data in order, either from least to greatest or from greatest to least. The value in the middle is the median. If the number of data items is even, the median is the mean of the two middle values.

Objective
- To find the mean, the median, and the mode of a set of data

Vocabulary: mean, measure of central tendency, median, mode

Class Planning

Schedule 1–2 days
Resources
In *Chapter 5 Student Activity Masters* Book
Class Openers *39, 40*
Extra Practice *41, 42*
Cooperative Learning *43, 44*
Enrichment *45*
Learning Log *46*
Using Manipulatives *47, 48*
Spanish-Language Summary *49*

In *Tests and Quizzes* Book
Quiz on 5.1–5.4 (Honors) *117*

Class Opener
Use the following groups of data to answer the questions.
A: 7, 6, 5, 6, 9, 100, 8, 15, 96
B: 5, 9, 8, 15, 6, 6, 7
1. Arrange the numbers in each group from greatest to least.
2. Find the middle value in each ordered group.
3. Find the average of the values in each group.
4. Find the value that occurs most often in each group.

Answers:
1. 100, 96, 15, 9, 8, 7, 6, 6, 5; 15, 9, 8, 7, 6, 6, 5 **2.** 8; 7 **3.** 28; 8 **4.** 6; 6

Lesson Notes
- As the opener makes evident, determining which statistical average best represents a set of data is not trivial. Some discussion

SECTION 5.4

Lesson Notes continued

should focus on when each average is preferable. As the opener shows, the median is less affected by extreme high or low scores than the mean. This is also illustrated in sample problem 1.

- Some calculators allow the user to compute the mean just by entering the data, instead of by having to press the addition button each time and then divide. You might demonstrate this if such a calculator is available.

- Is there a difference between the average height of your students and the height of the average student? You might ask your class how they would determine who the average student in the class is. We suspect they will conclude that there is no such person.

Additional Examples

Example 1
Here is a set of test scores from a mathematics placement test: 42, 84, 88, 70, 75, 68, 79, 62, 91, 57, 73, 94, 82, 85, 74, 68, 88, 67, 88, 71. What is the mean test score?

Answer: 75.3

Example 2
What is the median of the data?

Answer: 74.5

Example 3
What is the mode of the data?

Answer: 88

Reteaching
Have students give situations in which each of the three measures of central tendency would be preferred. Have students create data that illustrate each situation, and find the appropriate measures of central tendency.

Example 2
Find the median of the test scores in Example 1.

Solution
We will organize the data from least to greatest, then find the middle test score. Since there are 25 scores, the middle value of the data is the thirteenth score, 82. Therefore, the median is 82.

65	66	69	72	73
74	76	77	79	82
82	82	82	84	86
87	88	89	91	92
93	94	95	95	97

THE MODE

The **mode** is the data value that occurs most often.

Example 3
What is the mode of the data in Example 2?

Solution
In the set of scores, 82 occurs four times, more than any other score. The mode is 82.

The three averages—mean, median, and mode—are also called **measures of central tendency** because they are different ways of locating the center of a set of data.

Sample Problems

Problem 1 The Micro-Bitts Computer Company (MBCC) has 20 employees with the following annual salaries. Find the average salary at MBCC.

```
 1 President .......................$100,000
 2 Vice-presidents..............$ 80,000 each
 1 Plant Manager ................$ 40,000
16 Assembly workers ..........$ 15,000 each
```

Solution To answer this question, we will find three kinds of averages—the mean, the median, and the mode.

$$\text{Mean} = \frac{\text{total of salaries}}{\text{number of employees}}$$

$$= \frac{\$100{,}000 + 2(\$80{,}000) + \$40{,}000 + 16(\$15{,}000)}{20}$$

$$= \frac{\$540{,}000}{20}$$

$$= \$27{,}000$$

There are 20 employees, so the median will be the mean of the tenth and eleventh salaries. The amounts of the tenth and eleventh salaries are each $15,000 because 16 of the 20 salaries are $15,000.

$$\text{Median} = \frac{\text{sum of the tenth and eleventh values}}{2}$$

$$= \frac{\$15,000 + \$15,000}{2}$$

$$= \$15,000$$

The mode is $15,000, the salary that occurs most often.

Problem 2 The list contains the sales volume for Top Ten compact discs sold at Seady's Record Store last week. Find the mode of the sales.

Pistols and Posies.....15	Snow Leopards 32
Twigs........................19	Cellophane Rappers...45
Yukon II....................25	Madora 50
Nebraska..................25	Jack Michaelson.........52
Keen Teens...............28	A and J 52

Solution There are two modes in this set of data, 25 and 52.

Problem 3 The mean of Jeremy's first three test scores is 88 points. How many points must he earn on the fourth test to have a mean score of 90?

Solutions **Method 1**
Because the mean of the first three scores is 88, the total number of points Jeremy scored must be the same as if he had received an 88 on each test. Here is one way to determine the number of points he needs on the fourth test.

Calculate points earned:	$88 \times 3 = 264$
Calculate necessary points:	$90 \times 4 = 360$
Find the difference:	$360 - 264 = 96$

Jeremy needs 96 points.

Method 2
We can also determine the number of points Jeremy needs on the fourth test by using a graph. On the first three tests, Jeremy's scores were a total of 6 points below a mean of 90. Therefore, he would need to score 90 + 6, or 96, on the last test to have a score of 90 on all four tests.

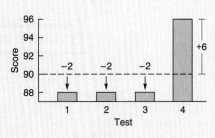

SECTION 5.4

Problem-Set Notes

for pages 224 and 225

- **7** demonstrates the usefulness of spreadsheets for computing means and medians. Have students compare doing the problem by hand with using the built-in spreadsheet functions.
- **10** and **15** illustrate why there are different kinds of averages. Students could create an advertisement for entering the Big Raffle and an advertisement against entering the Big Raffle.
- **16** should be interesting to students and create discussion about how it is possible to get more than 100%. Again, it is easy to use these data to misinform.

Cooperative Learning

- Have small groups of students create sets of data that meet the following conditions:
  1. The mean, median, and mode are all equal.
  2. The median and mode are equal and less than the mean.
  3. The mode and mean are equal and greater than the median.
- Students might begin by creating relatively simple sets with few data. Then they should try to create some more complex examples.
- Problem 27 is appropriate for cooperative learning.

Additional Answers

for pages 224 and 225

2 d The data are symmetric at about 500, so the averages are all the same.

Think and Discuss

P **1** Find the mode, the median, and the mean of 13, 21, 35, 35, 48, 56, 65, 74, and 88. $35; 48; 48\frac{1}{3}$

P **2 a** Find the mean of the data.   500
  b Find the median of the data.   500
  c Find the mode of the data.   500
  d Explain the results of parts **a–c**.

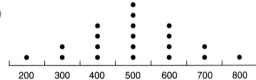

P **3** Sample Problem 1 on page 222 refers to an average. Why might it be important to specify which of the three averages is to be calculated?

P, I **4** The measures of the angles of a pentagon are 108, 108, 108, 78, and 138.
  a Find the mean measure of the angles.   108
  b What is the measure of any angle of a regular pentagon?   108

P **5** Tenna Pin bowled 120, 125, 130, 136, and 170 in the first five games of the Latin Club bowling tournament. She bowled 212 in the sixth game.
  a Find the mean and the median of the first five games.   136.2; 130
  b Find the mean and the median of all six games.   ≈148.8; 133
  c What is the change in the mean and the median between the first five games and all six games?   Change in mean: 12.6; change in median: 3

P **6** During the first nine weeks of the summer, Chip earned an average of $35 per week mowing lawns. How much would he need to earn during the tenth week to have a mean weekly income of $40? $85

P **7 Spreadsheets**   Here are some test scores from Mr. X's math class.

38, 43, 98, 96, 55, 95, 95, 56, 62, 63, 89, 87, 65

  a Find the mean. (Use a spreadsheet if one is available.)   72.46
  b Find the median. (If you use a spreadsheet, try the sorting function.)   65
  c What is the mode of the data? Is the mode a good measure of central tendency for the data? Explain your response.
  d Can we use the data to draw any meaningful conclusions?

Problems and Applications

P, R **8** Find the mean and the median for each set of data.
  a 3, 8, 24, 35, 90   32; 24
  b 0, 20, 20, 40, 80   32; 20
  c 30, 30, 30, 30, 40   32; 30
  d 0, 0, 50, 50, 60   32; 50

P, I **9 a** Find the mean of 56, 94, 62, 75, and 83.   74
  b Find the sum of all the differences of the data from the mean in part **a**.   0

224   Chapter 5   Data Analysis

SECTION 5.4

P   **10** The table shows the income of each member of the Mint family. Which of the following statements is correct?

Person	Income
S. Pierre Mint	$110,000
"Pepper" Mint	50,000
Merry Mint	35,000
Bela Mint	10,000
Horace Mint	10,000

    a The average income of the Mint family is $43,000.
    b The average income of the Mint family is $10,000.
    c The average income of the Mint family is $35,000.

P, I   **11** Here are Pete's scores on video games.

21, 18, 19, 39, 26, 36, 28, 17, 19, 34, 36, 18, 21, 19, 29, 28, 24, 21, 13, 18, 29, 36, 29, 22, 29, 26, 27, 28, 21, 38, 23, 39, 32, 37, 39, 18, 12, 28, 31, 27, 32, 35, 37, 18

    a How many scores are between 10 and 19?   8
    b How many scores are between 20 and 29?   15
    c How many scores are between 30 and 39?   11

P, R   **12 Spreadsheets**   Here are some test scores from Mr. X's math class.

85, 78, 87, 93, 67, 54, 67, 77, 88, 95, 93, 79, 56

    a Enter the data into cells A1–A13 of a spreadsheet.
    b Calculate the mean in cell A14 by using the function @Avg(A1..A13).   ≈78.38
    c Was Mr. X's test a difficult one? Why or why not?
    d Use the spreadsheet to sort the data. What is the median?   79

P   **13** The table displays the results of a survey taken to determine the average number of pieces of jewelry worn by women. What is the average?

Number of pieces	0	1	2	3	4	5 or more
Percent of women	30%	5%	14%	19%	14%	18%

P, R, I   **14 Communicating**   Copy the "ruler" onto your paper. Mark 44 and 47. How did you determine where the two marks should go?

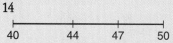

P   **15** In the Big Raffle, 125 people won nothing, 4 people won $50 each, and 1 person won $1000.
    a Find the mean of the people's winnings.   ≈$9.23
    b Find the median of the people's winnings.   0
    c Find the mode of the people's winnings.   0

P, R   **16** The graph displays ten popular adult activities and the percent of adults that participate in them.
    a What is the sum of the percents?
    b Why can the sum of the percent exceed 100%?
    c Is it possible to determine the most popular activity for the average person in the United States?

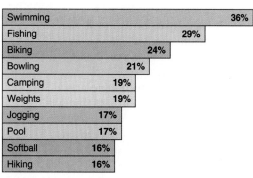

Section 5.4   Averages

---

3 Since each of the three averages may yield a different number, it is important to identify the average so that the data are not incorrectly interpreted.

7 c Possible answer: 95; No, 9 of the 13 test scores are well below 95.

d Possible answer: No. The data items vary greatly in value. The averages are very different from each other.

10 All of the statements are correct. The mean salary is $43,000, the median salary is $35,000, and the mode of the salaries is $10,000.

12 c Possible answer: The test was challenging but was not impossible to pass.

13 Since you don't know the number of pieces in the "5 or more" category, the mean can't be found. The median is three. The mode is zero.

14

```
├─────┼─────┼─────┤
40    44    47   50
```

Possible explanation: Mark the midpoint of the segment at 45. Divide the length of the segment by 10 to find out how far apart each number is. Subtract this distance from 45 to mark 44, then add two times this distance to 45 to mark 47.

16 a 214%

b Many people in the United States do more than one activity.

c Swimming is possibly the most popular activity, but we do not know if some popular activities were not included in the survey.

SECTION 5.4

Problem-Set Notes

for pages 226 and 227

- **22** previews stem-and-leaf plots, which are discussed in the next section.

Additional Answers

for pages 226 and 227

17 b and **c**

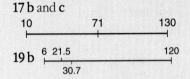

19 b 6 21.5   120
         ├──┼──────────────────────────────┤
              30.7

21 c Health-care costs; Skill levels and aging workforce

22 a 99, 97, 93, 92, 87, 87, 87, 84, 81, 79, 70, 68, 65, 64

23 −56; (−7)(8) = −56

25 a The larger the wrist size, the larger the neck size.

b ≈25 cm

c A scale beginning at zero would not be a feasible set of numbers.

27 a No

b Answers may vary. Possible answer: Other American League teams in Oakland's division may be weak.

c Cincinnati won the World Series in 1975, 1976, and 1990. Los Angeles won the World Series in 1981 and 1988.

Investigation Batting average is the ratio of hits to at-bats, usually rounded to 3 decimal places, and often expressed without the decimal point. (One hit in 3 at-bats is 0.333, or sometimes, 333.) A pitcher's ERA is the average number of earned runs allowed per 9 innings pitched, usually rounded to two decimal places:

ERA = runs ÷ ($\frac{1}{9}$ · innings)

So a pitcher who gives up 4 earned runs in 18 innings has

P, I   **17**   10   22   35   53 57   72 78   85   95   102   113   130

       a Find the mean of the numbers on the number line.   71
       b Copy the number line onto your paper, marking only the extremes, 10 and 130.
       c Mark the mean in the appropriate place.

P, R   **18** During lunch hour at a compact-disc store, the employees kept track of the number of discs bought by each customer who entered the store. The results are listed in the table.

Number of customers	2	3	10	8	16	27
Number of discs purchased	5	4	3	2	1	0

       a How many discs were sold?   84
       b How many discs did the average person buy?   1.27
       c If the average price of a disc is $14.00, how much money did the store make? $1176.00
       d How much did the average customer spend? $17.82

P, I   **19**   6 10 14 18 22 27   60   120

               16 21 25 29

       a Find the mean and the median of the numbers plotted.   ≈ 30.7; 21.5
       b Copy the number line onto your paper. Plot 6, 120, the mean, and the median.

◀ LOOKING BACK   **Spiral Learning**   LOOKING AHEAD ▶

R, I   **20** Find the measure of ∠D in pentagon ABCDE.   125

R   **21 a** How many personnel managers were surveyed?   233
       b How many of them felt that health care will be a major concern in the year 2000?   ≈21
       c According to the managers surveyed, which issue was a major concern in 1990 but will not be in 2000? Which issues were not major concerns in 1990 but will be in 2000?

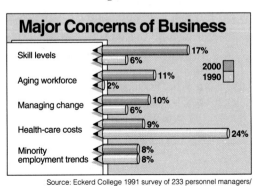

226   **Chapter 5**   Data Analysis

SECTION 5.4

I **22** Mr. E. Ficcient organized his test scores into this table. Four of the scores, 92, 93, 97, and 99, are in the 90's.
   a List all the scores.
   b Find the mean of the scores. ≈82.4
   c Find the median score. 85.5
   d Find the mode of the scores. 87

Tens	Units
9	2 3 7 9
8	1 4 7 7 7
7	0 9
6	4 5 8

I **23** Enter 7 [+/−] [×] 8 [=] into your calculator. What did you get? Why?

R **24** Find the ratio of the values of six quarters to five dimes. 3:1

R **25** **Communicating**
   a What trend does the scattergram show?
   b If someone had a neck size of about 53 centimeters, what would you predict the person's wrist size to be?
   c Explain why the scales do not start with zero.

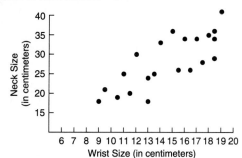

R **26 a** How many paths run from A to C, traveling downward? 8
   b What percent of the paths go through B? 75%

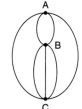

R **27 a** Does the graph mean that Oakland has been the best team since 1969?
   b What does the fact that Oakland has won so many division titles suggest?
   c Cincinnati and Los Angeles have each been 5-2 in play-off series. Do some research and find out how they did in the World Series.

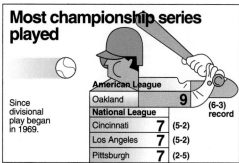

Investigation

Baseball How is a batting average calculated in baseball? How is a pitcher's earned run average calculated? Are these averages examples of a mean, a median, or a mode? Explain your answer. Then write formulas that can be used to calculate a batting average and an earned run average.

Section 5.4 Averages

---

an ERA of $4 \div (\frac{1}{9} \cdot 18)$, or 2.00. Both BA and ERA are examples of means.

Communicating Mathematics

- Have students write a paragraph describing a public relations situation where it would be more beneficial to cite the mean as the average salary at MBCC in sample problem 1. Students might write a second paragraph describing a situation where it would be more beneficial to cite the median or mode.

- Problems 3, 14, and 25 encourage communicating mathematics.

Checkpoint

1. Find the mode, median, and mean for these data.
84, 93, 72, 86, 95, 86, 75, 81, 89, 97

2. The mean of Jenna's first four test scores is 89. What must she earn on the fifth test to have a mean score of 91?

Answers:
1. 86; 86; 85.8 **2.** 99

Additional Practice

Find the mean, the median, and the mode for each set of data.
1. 8.6, 9.1, 7.3, 8.4, 5.9
2. 143, 179, 106, 129, 134, 106, 125, 163, 119, 124
3. 84, 76, 92, 73, 86, 49, 97, 89, 72, 69, 76, 76, 84, 92, 73

Answers:
1. 7.86; 8.4; no mode **2.** 132.8; 127; 106 **3.** 79.2; 76; 76

227

SECTION 5.5

Objectives
- To interpret and draw stem-and-leaf plots and box-and-whisker plots

Vocabulary: box-and-whisker plot, extremes, lower quartile, midquartile range, outlier, range, stem-and-leaf plot

Class Planning

Schedule 1–2 days

Resources

In *Chapter 5 Student Activity Masters* Book
Class Openers *51, 52*
Extra Practice *53, 54*
Cooperative Learning *55, 56*
Enrichment *57*
Learning Log *58*
Technology *59, 60*
Spanish-Language Summary *61*

Transparencies 25 a–c

Class Opener

Refer to the data in example 1 of Section 5.4. Make a table for the data like the one in problem 22 of that section.

Answer:

Tens	Units
6	5 6 9
7	2 3 4 6 7 9
8	2 2 2 2 4 6 7 8 9
9	1 2 3 4 5 5 7

Lesson Notes

- A stem-and-leaf plot is a useful way to begin when there are data in a random order. It is not particularly powerful, but it is relatively easy to do and can be a first step towards understanding the data. You might discuss how to decide what to use for the stems and what to use for the leaves.

Investigation Connection
Section 5.5 supports Unit C, *Investigations* 1, 3. (See pp. 198A-198D in this book.)

SECTION 5.5
Organizing Data

STEM-AND-LEAF PLOTS

When data are collected, they are often unsorted or are sorted in a way that does not meet our needs. In this section, we will use a **stem-and-leaf plot** to organize data.

Example 1
Below is a list of the score margins for the top 25 college football teams on October 18, 1991. Two teams did not play, so we ignore them. Four teams lost, so their margins are negative.

Florida State 32	Florida..............16	Alabama............46
Washington.......54	California............3	Illinois................21
Miami................37	Ohio State.........15	Texas A & M 23
Tennessee ....—	Pittsburgh............4	Georgia............15
Michigan 19	Syracuse.........–32	Mississippi 21
Oklahoma 21	N.C. State 7	Arizona State 6
Notre Dame 16	Iowa ....–19	Auburn ....–1
Baylor................17	Nebraska............—	
Penn State 17	Clemson..........–15	

Organize these data in a stem-and-leaf plot.

Solution
The greatest and least values in a set of data are called the **extremes** of the data, and the difference between these values is the **range** of the data. The extremes of these data are 54 and –32, so the range of the data is 86. We will use the tens digits of the numbers –32 through 54 as the stems in our plot.

228   **Chapter 5**   Data Analysis

SECTION 5.5

We add leaves to the stems by writing each unit's digit next to the corresponding stem. For example, Alabama had a score margin of 46, so we write a 6 to the right of 4. Similarly, we write a 1 next to –0 for Auburn's margin of –1.

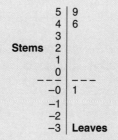

Here is the completed stem-and-leaf plot. There are 23 data values, so the median is the twelfth value in the list (the value with 11 values greater and 11 values less). The median of these data is 16.

```
        5 | 4
        4 | 6
        3 | 2 7
Stems   2 | 1 1 1 3
        1 | 5 5 6 6 7 7 9
        0 | 3 4 6 7
       ---|---
       -0 | 1
       -1 | 5 9
       -2 |
       -3 | 2   Leaves
```

BOX–AND–WHISKER PLOTS

A **box-and-whisker plot** is a visual representation of data that shows the middle 50% of the data relative to the extremes.

Example 2
Draw a box-and-whisker plot of the data in Example 1.

Solution
First we draw a segment whose length represents the range of the data. The range of the data in Example 1 is 86, so we can draw a segment 86 millimeters long. Each 1 millimeter of the segment will represent 1 point of a score margin. Next, we plot the median of the data. The median of these data is 16, and the difference between 54 and 16 is 38, so we plot the median 38 millimeters from the 54.

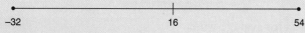

The median divides the data into two equal parts. The median of the upper part is called the **upper quartile** of the data. In this case, the upper quartile is 21, the value from the top of the stem-and-leaf plot (or the sixth value above the median). It should be plotted 5 millimeters to the right of the median, since 21 – 16 = 5. Similarly, the **lower quartile,** 4, is the median of the lower part of the data. It should be plotted 12 millimeters to the left of the median.

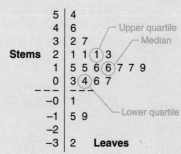

- A box-and-whisker plot gives an immediate visual picture of how the data are distributed. It shows the median and the quartiles, which can be helpful. If most of the numbers are bunched in the middle, the plot appears different than if they are evenly spread out. Don't allow students to get lost in the process of how to create the box and whiskers and miss the significance of what they are able to see after the picture has been drawn.

Additional Examples

Example 1
Mr. McTrack measured the heights of his 4th period physical education class. Arrange the results in a stem-and-leaf plot.
43, 52, 70, 56, 64, 61, 59, 49, 55, 51, 67

Answer:
```
4 | 3 9
5 | 1 2 5 6 9
6 | 1 4 7
7 | 0
```

Example 2
Draw a box-and-whisker plot for the data in example 1.

Answer:

Communicating Mathematics

- Have students write a paragraph describing what the position and shape of the box in a box-and-whisker plot tells about the data shown in the graph. Alternately, they might draw a box-and-whisker plot and explain it orally.

- Problems 1, 4–6, 9, and 11 encourage communicating mathematics.

Section 5.5 Organizing Data

SECTION 5.5

Assignment Guide

Basic
Day 1 8, 9, 12–14
Day 2 10, 11, 15, 16

Average or Heterogeneous
Day 1 8, 9, 12, 13
Day 2 10, 11, 14–16, Investigation

Honors
Day 1 8, 9, 12–14
Day 2 10, 11, 15, 16, Investigation

Strands

Communicating Mathematics	1, 4–6, 9, 11
Data Analysis/ Statistics	1–12, 16, Investigation
Technology	11
Number Sense	12, 14, 15
Interdisciplinary Applications	10, 11, 16, Investigation

Individualizing Instruction

Learning-disabled students
Students may benefit from copying data, such as those listed in problem 8, onto separate pieces of paper so they can be more easily arranged from least to greatest.

Problem-Set Notes

for pages 230 and 231

■ **1** may lead to a discussion of outliers and why we throw them out. If a data set is large, one or two outliers will give a misleading picture of the overall average, just as one student who does a lot better (or worse) than all of the others can give an unrealistic picture of how well a class did as a whole.

■ **4** and **6** focus on students being able to interpret a box-and-whisker plot. Students may find this difficult. Some discussion here might be helpful.

230

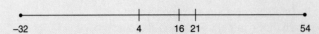

To complete the box-and-whisker plot, we draw a box around the **midquartile range**—the part of the segment between the upper and lower quartiles.

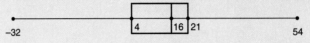

A box-and-whisker plot gives a pictorial view of a central tendency of a set of data. The midquartile range, corresponding to the middle 50% of the data, can be used to locate **outliers.** Outliers are data values that are more than 1.5 times the midquartile range above the upper quartile or below the lower quartile.

Sample Problem

Problem Use the box-and-whisker plot in Example 2 to determine whether the value 54 is an outlier of the data presented in the plot.

Solution The midquartile range of the data is the difference between the upper quartile and the lower quartile, so it is 21 − 4, or 17. If we multiply 17 by 1.5, we get a result of 25.5. Since 54 is more than 25.5 units above the upper quartile (it is 54 − 21, or 33, units above the quartile), it is an outlier of the set of data.

Think and Discuss

P **1 a** Organize the data for average wind speeds into a stem-and-leaf plot.
   b Which of the data values is an obvious outlier? Explain why this number is an outlier.

Location	Average Speed (mph)
Mt. Washington, N.H.	35.3
New Orleans, La.	8.2
New York, N.Y.	9.4
Omaha, Neb.	10.6
Philadelphia, Pa.	9.5
Phoenix, Ariz.	6.3
Pittsburgh, Pa.	9.1
Portland, Ore.	7.9
St. Louis, Mo.	9.7
Salt Lake City, Ut.	8.8
San Diego, Cal.	6.9
Seattle, Wash.	9.0
Washington, D.C.	9.3

Source: *World Almanac and Book of Facts, 1991*

230 **Chapter 5** Data Analysis

SECTION 5.5

P **2** Is the value −32 an outlier of the data in Example 1? Yes

P **3** What percent of the data values are in each of the four sections of a box-and-whisker plot? 25%

P **4** Coach Bradley recorded the number of seconds it took her students to do 50 sit-ups. Then she drew a box-and-whisker plot to represent the data she collected.

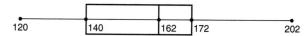

The next day she did the same thing and drew this box-and-whisker plot.

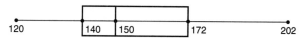

Explain how the students' performances changed.

P **5** In a box-and-whisker plot, data are divided into quarters, which are separated by quartiles. Standardized-test data are often divided into percentiles. Explain the difference between percentiles and quartiles.

P **6**

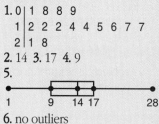

Explain how the test data in the first box-and-whisker plot differ from the data in the second.

P, R **7**
a Find the range of the data in the stem-and-leaf plot. 83
b Find the median of the data. 69
c Find the mode of the data. 68
d Find the lower quartile of the data. 60
e Find the upper quartile of the data. 84
f Make a box-and-whisker plot for the data.

```
0 |
1 | 5
2 |
3 | 2
4 | 3 5      Stems
5 | 1 7 8
6 | 2 3 5 8 8 8 9
7 | 0 0 4 7 9
8 | 2 4 4 8
9 | 3 4 5 8 8
      Leaves
```

Problems and Applications

P **8** 56, 68, 15, 98, 84, 38, 100, 95, 61, 46, 99, 81, 63, 49, 55, 100, 75, 91, 84, 56, 99, 71, 84, 64, 63, 84, 56

a Find the median, M, of the given test scores. 71
b Find the median of the data less than M. What is this number called?
c Find the median of the data greater than M. What is this number called?

Checkpoint

Use the following data for problems 1–6.

8, 12, 16, 21, 14, 17, 14, 15, 8, 12, 9, 28, 1, 17, 12

1. Make a stem-and-leaf plot.
2. Find the median.
3. Find the upper quartile.
4. Find the lower quartile.
5. Make a box-and-whisker plot.
6. Find any outliers.

Answers:

1. 0 | 1 8 8 9
   1 | 2 2 2 4 4 5 6 7 7
   2 | 1 8

2. 14 3. 17 4. 9
5.
   ```
   •——[ | ]——•
   1   9 14 17   28
   ```
6. no outliers

Additional Answers

for pages 230 and 231

1 a Answer on page A4.
b 35.3; It lies far beyond where the rest of the data is bunched.

2 Yes, because −32 is more than 22.5 units below the lower quartile.

4 The students in the middle group improved their time the second day. The best students and the poorest students did not improve their time.

5 Percentiles divide data into 100 parts, and quartiles divide data into 4 parts.

6 The highest and the lowest scores did not change, but there was a drop in the average score.

7 f
```
•——[ | ]——•
15   60 69 84   98
```

8 b 56; lower quartile
c 91; upper quartile

SECTION 5.5

Problem-Set Notes

for pages 232 and 233

- 16 shows an example of how a matrix can be used to organize some information. This is the topic of the next section.

Cooperative Learning

- Obtain some almanacs and have groups of students create stem-and-leaf plots, then box-and-whisker plots, using as data the ages of U.S. Presidents when they took office. Discuss the implications of what the plots show.
- Problem 11 is appropriate for cooperative learning.

Stumbling Block

When finding the median of data arranged in a stem-and-leaf plot, students might make the following errors:
- counting stems as though they were additional data
- when counting data from the top down, counting leaves from left to right instead of from right to left

Additional Practice

Use the following data for problems 1–6.

72, 59, 83, 91, 13,
85, 79, 56, 83, 85

1. Make a stem-and-leaf plot.
2. Find the median.
3. Find the upper quartile.
4. Find the lower quartile.
5. Make a box-and-whisker plot.
6. Find any outliers.

P **9** **a** Make a stem-and-leaf plot for the data.
   b Find the median, the lower quartile, the upper quartile, the extremes, and the range for the data.
   c Make a box-and-whisker plot for the data.
   d Are there any outliers? Justify your answer.

Forest and Range Fires: Acres Burned, 1981–90 (in millions)

Year	Acreage	Year	Acreage
1981	4.8	1986	3.3
1982	2.3	1987	4.2
1983	5.1	1988	7.4
1984	2.3	1989	3.3
1985	4.4	1990	5.5

Source: Boise Interagency Fire Center, 1992

P **10** **Science**

Major California Earthquakes, 1900–1989

Year	Location	Richter-scale reading	Year	Location	Richter-scale reading	Year	Location	Richter-scale reading
1906	San Francisco	8.3	1980	Eureka	7.0	1971	San Francisco	6.4
1952	Tehachapi-Bakersfield	7.8	1940	Imperial Valley	6.7	1933	Long Beach	6.3
1927	Offshore San Luis Obispo	7.7	1911	Coyote	6.6	1925	Santa Barbara	6.3
1923	North Coast	7.2	1980	Mammoth Lakes	6.6	1984	Morgan Hill	6.2
1989	San Francisco	7.0	1983	Coalinga	6.5	1987	Los Angeles	6.1
			1979	Imperial Valley	6.4	1986	Palm Springs	6.0
			1968	Anza-Borrego Mountains	6.4			

a Make a stem-and-leaf plot for the data.
b Find the median of the data.   6.55
c Find the lower and upper quartiles.   6.3; 7.0
d Make a box-and-whisker plot for the data.

P **11** **Communicating**
   a Arrange the data values from least to greatest.
   b Find the median, the lower quartile, the upper quartile, and the extremes of the data.
   c Compare the mean and the median of the data. Why is the median a better choice as an average of these data?
   d Draw a box-and-whisker plot for the data.
   e What are the outliers? Explain why these numbers are outliers.
   f Why, do you think, is much more fish caught in Alaska than in any other state?

Fisheries—Quantity of Catch, by State, (millions of pounds, live weight)

State	1988
U.S. total	7,155
Alabama	24
Alaska	2,639
California	496
Connecticut	9
Delaware	6
Florida (east coast)	56
Florida (west coast)	126
Georgia	17
Hawaii	21
Louisiana	1,356
Maine	157
Maryland	80
Massachusetts	287
Mississippi	336
New Hampshire	11
New Jersey	113
New York	38
North Carolina	191
Oregon	149
Rhode Island	106
South Carolina	16
Texas	96
Virginia	651
Washington	174

Source: *Universal Almanac*

Chapter 5 Data Analysis

SECTION 5.5

◀ LOOKING BACK **Spiral Learning** LOOKING AHEAD ▶

R **12 a** Find the mean and the median of 3, 6, 8, 9, 10, 12, and 15. 9; 9
   b What number would you add to the list of numbers in part **a** to make 30 the mean of the data? 177
   c Find the median of the new list of numbers. 9.5

R **13** The legs of a right triangle are 2.5 centimeters long and 4 centimeters long. How long is the hypotenuse? ≈4.7 cm

R **14 Communicating** C.D. can press the < symbol on her tape player to rewind the tape and the > symbol to fast-forward the tape. C.D. was surprised when she saw the expressions 15 > 9 and 9 < 15 in her math book. What do the symbols > and < represent in her math book?

R **15** Determine whether each statement is *true* or *false*.
   a $8 + 5 > 9 + 4$ False   **b** $8 + 5 \leq 9 + 4$ True   **c** $8 + 5 < 9 + 4$ False
   d $8 + 5 \geq 9 + 4$ True   **e** $8 + 5 \neq 9 + 4$ False

I **16 Business** Here is an inventory sheet from Suzhanna's Dress Shop. A zero (0) entry means there are no dresses of the given size from that manufacturer in the store. A dash (–) entry means that dresses in the given size are not available.

Manufacturer Account Number

	#1023	#1203	#1312	#1332
Size 4	2	1	0	–
Size 6	2	2	1	1
Size 8	3	5	4	2
Size 10	3	4	6	5
Size 12	2	5	4	6
Size 14	1	3	6	5
Size 16	0	2	4	6
Size 18	–	0	2	3

a How many dresses from #1312 are there in size 12? 4
b What manufacturer does not produce size-4 dresses? #1332
c What two manufacturers produce all of the sizes? #1203 and #1312
d How many size-8 dresses are in stock? 14
e What sizes are most heavily stocked?

Investigation

Statistics Gather data on how your favorite athlete performed during a three-year period. Organize the data, and present the data to the class. Decide whether or not the athlete's performance is improving, and explain your decision.

Section 5.5 *Organizing Data* 233

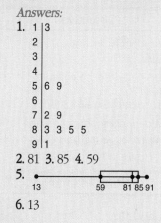

Answers:
1.
```
1 | 3
2 |
3 |
4 |
5 | 6 9
6 |
7 | 2 9
8 | 3 3 5 5
9 | 1
```
2. 81 3. 85 4. 59
5. [box plot with values 13, 59, 81, 85, 91]
6. 13

Reteaching

To reteach finding the median and the upper and lower quartiles, have students write the data, in order, on one line. Have students place their fingers on the least data point and the greatest data point and move each finger inward one data point at a time until they find the median. Repeat the process using the least data point and the median to find the lower quartile. Use the median and the greatest data point to find the upper quartile.

Multicultural Note

Suppose a grid consisting of 1-mile squares were laid over a country and the number of people living in each square were counted. Have students find countries for which the mode and mean of the resulting data would differ greatly and countries for which these averages would be roughly equal.

Answers for 9, 10 a and d, 11, 14, 16 e, and Investigation are on page A4.

233

SECTION 5.6
Matrices

ORGANIZING DATA IN A MATRIX

One of the most common ways of organizing data is in a **matrix.** This is a two-dimensional table in which the rows (the horizontal groups of entries) and the columns (the vertical groups of entries) are labeled to indicate the meaning of the data. The plural form of *matrix* is *matrices*.

The following matrix contains data about the U.S. recording industry in the years 1986–1989. The entries represent approximate numbers of records, CD's, and tape cassettes shipped to stores, in millions.

	Records	CD's	Cassettes
1986	125	53	345
1987	107	102	410
1988	72	150	450
1989	35	207	446

The entries in the first row, for example, indicate how many of each item were shipped in 1986. The entries in the first column represent the number of records shipped each year. What do you think the entry 150 (in the third row and the second column) represents?

DIMENSIONS OF A MATRIX

The size of a matrix is expressed by giving its **dimensions**—the number of rows by the number of columns. The matrix above, for instance, is a 4-by-3, or 4 × 3, matrix. (The number of rows is always given first.) The individual entries can be referred to by their row and column numbers—in the matrix above, the entry in row 2, column 3, is 410. Sometimes subscripts are used to identify entries: The entry in row 2, column 3, can be called $e_{2,3}$, and we can write $e_{2,3} = 410$.

234   Chapter 5   Data Analysis

SECTION 5.6

Example
This matrix gives information about some rectangles.

	Length	Width	Perimeter	Area
Rectangle 1	17	1	36	17
Rectangle 2	16	2	36	32
Rectangle 3	15	3	36	45
Rectangle 4	14	4	36	56
Rectangle 5	13	5	36	65

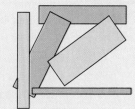

a What is the size of the matrix?
b What is entry $e_{2,4}$?
c What is the location of the entry that represents an area of 45?
d What are the dimensions of the rectangle with the greatest area?

Solution
a 5 by 4, or 5 × 4
b 32
c $e_{3,4}$
d 13 by 5, or 13 × 5

EQUAL MATRICES

A capital letter is often used to name a matrix. Let's take a look at two matrices—matrix A and matrix B. Do you notice anything special about the entries in these two matrices?

$$A = \begin{bmatrix} 1 & \frac{1}{2} & \frac{1}{3} & \frac{1}{4} & \frac{1}{5} \\ \frac{1}{6} & \frac{1}{8} & \frac{1}{9} & \frac{1}{10} & \frac{1}{20} \\ \frac{2}{3} & \frac{3}{4} & \frac{2}{5} & \frac{3}{5} & \frac{4}{5} \\ \frac{3}{8} & \frac{5}{8} & \frac{7}{8} & \frac{3}{10} & \frac{3}{20} \end{bmatrix}$$

$$B = \begin{bmatrix} 1 & 0.5 & 0.\overline{3} & 0.25 & 0.2 \\ 0.1\overline{6} & 0.125 & 0.\overline{1} & 0.1 & 0.05 \\ 0.\overline{6} & 0.75 & 0.4 & 0.6 & 0.8 \\ 0.375 & 0.625 & 0.875 & 0.3 & 0.15 \end{bmatrix}$$

We observe that corresponding entries in matrix A and matrix B have equal values. For example, $a_{1,4}$ of matrix A and $b_{1,4}$ of matrix B are equal—$\frac{1}{4} = 0.25$. Matrices whose corresponding entries are equal are called **equal matrices**.

Lesson Notes
- With the advent of spreadsheets, working with matrices is becoming an increasingly important skill. We begin by looking at a matrix as a table. Because students are so familiar with tables, the arithmetic and analysis that follows emerges in a natural way. Our objective is to help students get comfortable with matrices and see why they are useful. We suggest that you don't dwell on mastery of new terminology; rather, use the terminology frequently so students gradually become comfortable with it. The connection between a matrix and a spreadsheet should become self-evident.

Additional Example

Example
This matrix gives information about some isosceles right triangles.

Hypotenuse	Leg	Area
$\sqrt{2}$	1	$\frac{1}{2}$
$\sqrt{8}$	2	2
$\sqrt{18}$	3	$\frac{9}{2}$
$\sqrt{50}$	5	$\frac{25}{2}$

a. What is the size of the matrix?
b. What is the entry $e_{3,2}$?
c. What is the location of the entry that represents an area of $\frac{1}{2}$?
d. What is the measure of the hypotenuse of the triangle with the greatest area?

Answers: **a.** 4 × 3 **b.** 3
c. $e_{1,3}$ **d.** $\sqrt{50}$

SECTION 5.6

Assignment Guide

Basic
Day 1 8, 10–12, 15, 19–21, 25–27, 32, 36, 37
Day 2 9, 13, 14, 16, 22, 28–30, 34, 38–40

Average or Heterogeneous
Day 1 8, 9, 12, 13, 15, 18, 20, 22, 24, 35
Day 2 11, 14, 16, 17, 19, 21, 25, 28, 29, 31, 36

Honors
Day 1 8, 9, 12–14, 18, 20–22, 25–27
Day 2 11, 15–17, 19, 23, 24, 28–31, 33–36

Strands

Communicating Mathematics	1, 2, 4, 6, Investigation
Geometry	11, 22, 36
Algebra	3–5, 7–9, 11–13, 20, 21, 31–34
Data Analysis/ Statistics	6, 16, 23, 24
Technology	36
Probability	23
Patterns & Functions	21
Number Sense	35
Interdisciplinary Applications	6, 10, 16, Investigation

Individualizing Instruction

The kinesthetic learner Make cards representing the entries of various sizes of matrices (e.g., 3 × 3 or 2 × 4). Give one card to each student in a group and have the students form a human matrix. By assigning an extra student to each group and having students pass the cards to each other, this activity can be played as a game similar to musical chairs.

236

Sample Problems

Problem 1 In this matrix of tennis matches of players A, B, C, D, and E, each entry represents the number of games the player listed in the row won against the player listed in the column.

Player C won 6 matches against player B.

$$\begin{array}{c|ccccc} & A & B & C & D & E \\ \hline A & - & 4 & 6 & 7 & 3 \\ B & 6 & - & 4 & 2 & 1 \\ C & 4 & 6 & - & 5 & 9 \\ D & 3 & 8 & 5 & - & 6 \\ E & 7 & 9 & 1 & 4 & - \end{array}$$

a Why are there five missing entries in the matrix?
b How many times did each pair of players meet?
c Who is the best player?

Solution
a A player cannot play against himself or herself.
b For any pair of players, the total number of wins is ten. Therefore, each pair of players met ten times.
c Who is best is a matter of interpretation. Player C won the most games—24. However, player A beat player C 6 out of 10 times, and player D won half of the matches with C.

Problem 2 In this diagram, called a *digraph*, a route is defined as a road that travels through a city at most one time. Construct a matrix that shows the number of routes from the city listed in the row to the city listed in the column.

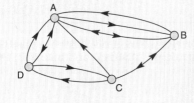

Solution

$$\begin{array}{c|cccc} & A & B & C & D \\ \hline A & 0 & 3 & 3 & 3 \\ B & 5 & 0 & 3 & 4 \\ C & 5 & 7 & 0 & 4 \\ D & 5 & 7 & 5 & 0 \end{array}$$

Think and Discuss

P **1** What is a matrix?

P **2** Why is a matrix used? A matrix is used to organize data.

236 Chapter 5 Data Analysis

SECTION 5.6

P **3** What are the dimensions of the matrix $\begin{bmatrix} 3 & 6 & 8 \\ 4 & 5 & 1 \end{bmatrix}$? 2 by 3

P **4** Is matrix A equal to matrix B? Explain your answer.

$$A = \begin{bmatrix} 2 & 3 \\ 4 & 5 \end{bmatrix} \qquad B = \begin{bmatrix} 3 & 2 \\ 5 & 4 \end{bmatrix}$$

P **5** Construct a matrix with entries $e_{1,1} = 12$, $e_{2,1} = 15$, $e_{1,2} = 18$, and $e_{2,2} = 20$. $\begin{bmatrix} 12 & 18 \\ 15 & 20 \end{bmatrix}$

P **6** **Sports** Refer to the table showing 1990 National League records.
  a Which team had the best record? Pittsburgh
  b What was the total number of games won by the teams in the Western Division? 477
  c Which team scored the least number of runs? Houston
  d Which team scored the greatest number of runs?
  e How is the table similar to a matrix?

National League Records in 1990 Final Standings

Eastern Division	W	L	Runs Avg	Runs vs.
Pittsburgh	95	67	4.5	3.8
New York	91	71	4.8	3.8
Montreal	85	77	4.1	3.7
Philadelphia	77	85	4.0	4.5
Chicago	77	85	4.3	4.8
St. Louis	70	92	3.7	4.3

Western Division	W	L	Runs Avg	Runs vs.
Cincinnati	91	71	4.3	3.7
Los Angeles	86	76	4.5	4.2
San Francisco	85	77	4.4	4.4
Houston	75	87	3.5	4.0
San Diego	75	87	4.2	4.2
Atlanta	65	97	4.2	5.1

Source: *World Almanac and Book of Facts*

P, I **7** Construct a 4-by-4 matrix in which $e_{a,b} = a + b$.

Problems and Applications

P **8** Consider the matrix $\begin{bmatrix} 5 & 2 & 1 \\ 2 & 15 & 10 \\ 4 & 21 & 22 \end{bmatrix}$
  a What are the dimensions of the matrix? 3 by 3
  b What is entry $e_{3,2}$? 21
  c What is the sum of the elements in column 2? 38
  d What is the sum of the elements in row 3? 47

P, I **9** Consider the matrix $\begin{bmatrix} 3 & -4 & 8 \\ -9 & 5 & -4 \end{bmatrix}$.
  a In what locations are the entries negative?
  b In what locations are the entries even?

P **10** **Consumer Math** Create a matrix that displays your weekly expenses. Label the rows with the days of the week. For the columns, choose labels like *Food*, *Clothes*, *Entertainment*, and *Savings*.

Additional Practice

1. Complete the matrix by finding the perimeter and area of a square with a side of each given length.

Length	Perimeter	Area
2	__	__
5	__	__
7	__	__

2. What are the dimensions of the matrix in problem 1?

3. What values of w, x, y, and z will make these matrices equal?

$$\begin{bmatrix} x+7 & 5 \\ w-7 & 4z \end{bmatrix} = \begin{bmatrix} 9 & 15y \\ -4 & 16 \end{bmatrix}$$

Answers:
1. 8, 4; 20, 25; 28, 49
2. 3 × 3   3. $w = 3$, $x = 2$, $y = \frac{1}{3}$, $z = 4$

Additional Answers

for pages 236 and 237

1 A matrix is a two-dimensional table of rows and columns.

4 No, corresponding entries are not equal.

6 d New York
 e The table has rows and columns. The data are organized so they are easy to read.

7 Possible answer:
$$\begin{bmatrix} 2 & 3 & 4 & 5 \\ 3 & 4 & 5 & 6 \\ 4 & 5 & 6 & 7 \\ 5 & 6 & 7 & 8 \end{bmatrix}$$

9 a $e_{1,2}$, $e_{2,1}$, $e_{2,3}$
 b $e_{1,2}$, $e_{1,3}$, $e_{2,3}$

10 Check students' matrices.

Section 5.6 Matrices

SECTION 5.6

P, R   **11**   Complete the matrix by finding the perimeter and the area of a rectangle with each given length and width.

$$\begin{bmatrix} \text{Length} & \text{Width} & \text{Perimeter} & \text{Area} \\ 1 & 2 & 6 & 2 \\ 2 & 4 & 12 & 8 \\ 3 & 6 & 18 & 18 \\ 4 & 8 & 24 & 32 \\ 5 & 10 & 30 & 50 \\ 6 & 12 & 36 & 72 \end{bmatrix}$$

P, I   **12**   Create one 4-by-4 matrix in which
1. The only entries are 1, 2, 3, and 4
2. Every column contains 1, 2, 3, and 4
3. Every row contains 1, 2, 3, and 4

P, R, I   **13**   What values of $w, x, y, z,$ and t will make these matrices equal?   6, 9, 6, −1, 2

$$\begin{bmatrix} 2 & x-2 & w \\ 3y & z+5 & 4-t \end{bmatrix} = \begin{bmatrix} 2 & 7 & 6 \\ 18 & 4 & 2 \end{bmatrix}$$

P, I   **14**   The labels of the rows and columns of the matrix below correspond to the vertices of the digraph. The matrix has an entry of 1 if there is an arrow leaving the row vertex and going to the column vertex. A zero is placed in all other entry locations. Complete the matrix.

$$\begin{array}{c|ccccc} & A & B & C & D & E \\ \hline A & 0 & 1 & 0 & 0 & 1 \\ B & 0 & 0 & 1 & 0 & 1 \\ C & 0 & 0 & 0 & 0 & 0 \\ D & 0 & 0 & 1 & 0 & 0 \\ E & 0 & 0 & 0 & 1 & 0 \end{array}$$

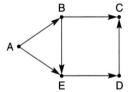

P, I   **15**   Create a matrix in which $e_{a,b} = a \cdot b$.

P   **16**   **Science**   In the windchill table, the rows represent wind speed and the columns represent temperature.

Determining the Windchill Factor																	
Wind speed	Actual temperature (°F)																
	35°	30°	25°	20°	15°	10°	5°	0°	−5°	−10°	−15°	−20°	−25°	−30°	−35°	−40°	−45°
5 mph	33°	27°	21°	16°	12°	7°	0°	−5°	−10°	−15°	−21°	−26°	−31°	−36°	−42°	−47°	−52°
10 mph	22	16	10	3	−3	−9	−15	−22	−27	−34	−40	−46	−52	−58	−64	−71	−77
15 mph	16	9	2	−5	−11	−18	−25	−31	−38	−45	−51	−58	−65	−72	−78	−85	−92
20 mph	12	4	−3	−10	−17	−24	−31	−39	−46	−53	−60	−67	−74	−81	−88	−95	−102
25 mph	8	1	−7	−15	−22	−29	−36	−44	−51	−59	−66	−74	−81	−88	−96	−103	−110
30 mph	6	−2	−10	−18	−25	−33	−41	−49	−56	−64	−71	−79	−86	−93	−101	−109	−116
35 mph	4	−4	−12	−20	−27	−35	−43	−52	−58	−67	−74	−82	−89	−97	−105	−113	−120
40 mph	3	−5	−13	−21	−29	−37	−45	−53	−60	−69	−76	−84	−92	−100	−107	−115	−123
45 mph	2	−6	−14	−22	−30	−38	−46	−54	−62	−70	−78	−85	−93	−102	−109	−117	−125

a   If the temperature is 0°F and the wind is blowing 35 miles per hour, what is the windchill?   −52°

b   A windchill below −40°F is very dangerous. What combinations of wind speed and temperature are dangerous?

238   Chapter 5   Data Analysis

SECTION 5.6

P, I **17** Use the following matrix to determine who should be assigned jobs A, B, and C so that the jobs will be done in the minimum time. (The people cannot share jobs, and each must do one of the jobs.)

	Job A	Job B	Job C
Arti	6 hr	5 hr	4 hr
Rita	5 hr	4 hr	2 hr
Tria	6 hr	3 hr	5 hr

P **18** The matrix represents a digraph. What is a possible graph for this matrix?

	A	B	C	D	E
A	0	1	1	1	1
B	0	0	1	0	0
C	0	0	0	0	0
D	0	0	0	0	1
E	0	0	1	1	0

P **19** Create a matrix showing the number of routes from point to point.

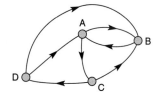

◀ LOOKING BACK **Spiral Learning** LOOKING AHEAD ▶

R, I **20** If $x = 7$, $y = 5$, and $z = 16x - 8y$, what is the value of z? 72

P, R **21** The salary schedule for employees at Dominique's Food Store is summarized in the matrix. (The entries represent dollars per hour.)

Years Experience	Bagger	Stockboy	Cashier
1	4.50	4.75	7.35
2	4.90	5.25	8.00
3	5.35	5.85	8.70
4	5.70	6.15	9.50
5	6.25	6.75	10.45

In the new contract signed by the grocer's union, the salary in each category is increased by 6.7%. What is the new salary matrix?

R, I **22** Find the measures $\angle 1$, $\angle 2$, and $\angle 3$.
70, 110, 40

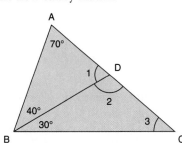

Section 5.6 Matrices 239

Additional Answers

for pages 238 and 239

12 Possible answer:
$$\begin{bmatrix} 1 & 2 & 3 & 4 \\ 4 & 1 & 2 & 3 \\ 3 & 4 & 1 & 2 \\ 2 & 3 & 4 & 1 \end{bmatrix}$$

15 Possible answer:
$$\begin{bmatrix} 1 & 2 & 3 \\ 2 & 4 & 6 \\ 3 & 6 & 9 \end{bmatrix}$$

16 b 5°, more than 30 mph; 0°, more than 25 mph; –5°, more than 20 mph; –10°, more than 15 mph; –15°, more than 10 mph; –20°, more than 10 mph; –30°, more than 10 mph; –35°, more than 5 mph.

17 Job A: Arti; Job B: Tria; Job C: Rita

18 Possible answer:

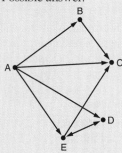

19

	A	B	C	D
A	0	3	1	1
B	1	0	1	1
C	3	3	0	1
D	2	3	2	0

21

Years Exp.	Bagger	Clerk	Cashier
1	4.80	5.07	7.84
2	5.23	5.60	8.54
3	5.71	6.24	9.28
4	6.08	6.56	10.14
5	6.67	7.20	11.15

SECTION 5.6

Problem-Set Notes

for pages 240 and 241

- **24** reviews making a bar graph, previews the addition of matrices, provides some arithmetic practice, and is an interesting application of matrices.

Stumbling Block

Students may not realize that the definition of *equal matrices* requires that equal matrices have the same dimensions. For each element in one of the matrices, there must be a corresponding element in the other matrix. For example,

$A = \begin{bmatrix} 0 & 1 \\ 2 & 3 \end{bmatrix}$ and

$B = \begin{bmatrix} 0 & 1 & 6 \\ 2 & 3 & 7 \end{bmatrix}$

are not equal. While for each element of A there is an equal corresponding element in B, the reverse is not true.

Communicating Mathematics

- Have students imagine what their lives will be like in 20 years. In a letter to you, have them describe their career and the type of data displays and analyses they encounter most frequently in their daily work. Encourage imagination and creativity. Students might present their letters orally.

- Problems 1, 2, 4, 6, and Investigation encourage communicating mathematics.

R   **23** Ernest had an urn that contained some red marbles and some green marbles. He picked 20 marbles out of the urn, counted the number of red marbles, then returned all marbles to the urn. He did this ten times. Here are the numbers of red marbles he chose: 8, 4, 10, 9, 6, 7, 9, 6, 11, 8.
     a Make a box-and-whisker plot of these data.
     b Find the mean of the data, rounded to the nearest whole number.   8
     c What fraction of the marbles probably are red? $\frac{2}{5}$, or 0.4
     d If Ernest's urn contains 300 marbles, about how many are red?   about 120

24 Business

P, R   **a** The matrix represents data on the first week's sales of compact discs at Abby Rhodes's Music Store. Represent this data in a bar graph.

	Mon.	Tues.	Wed.	Thurs.	Fri.	Sat.	Sun.
Rock	18	11	13	19	16	38	8
Classical	6	5	8	11	9	22	3
Country	14	18	9	10	7	10	4
Jazz	4	2	0	3	4	6	1
Blues	9	1	3	0	2	7	4

     b This matrix represents data on the second week's sales of compact discs at Abby Rhodes's Music Store. Construct a matrix representing the total sales for both weeks.

	Mon.	Tues.	Wed.	Thurs.	Fri.	Sat.	Sun.
Rock	23	8	15	17	12	33	15
Classical	8	11	7	8	12	28	6
Country	12	15	12	8	6	15	7
Jazz	8	3	2	5	2	3	0
Blues	7	4	2	8	5	0	0

R, I   In problems 25–30, use the number line to find the distance between each pair of numbers.

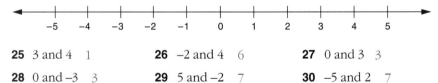

25 3 and 4   1   **26** −2 and 4   6   **27** 0 and 3   3

28 0 and −3   3   **29** 5 and −2   7   **30** −5 and 2   7

P, R, I   **31**

           Inventory Matrix

	Kix	Trox	Soggies
Store A	10	5	30
Store B	9	14	6
Store C	12	12	7
Store D	25	3	18

           Price Matrix

	Cost
Kix	3.25
Trox	2.75
Soggies	4.25

     a What is the total dollar amount of inventory at Stores A–D of Kix? Of Trox? Of Soggies? $182.00; $93.50; $259.25
     b What is the value of the cereal inventory at each store?
     c Which store has the greatest dollar amount of cereal inventory?   Store A

Chapter 5   Data Analysis

SECTION 5.6

R, I **In problems 32–34, solve each equation. Round answers to the nearest hundredth.**

32 $17x = -543$ ≈ -31.94 **33** $853y = 97$ ≈ 0.11 **34** $\dfrac{z}{-15.2} = 0.913$ ≈ -13.88

I **35** Complete the matrix with the least common multiples (LCM) of the numbers in columns 1 and 2.

$$\begin{array}{ccc} \text{Value} & \text{Value} & \text{LCM} \\ \begin{bmatrix} 5 & 8 & 40 \\ 6 & 10 & 30 \\ 1 & 6 & 6 \\ 14 & 21 & 42 \\ 5 & 2 & 10 \\ 6 & 8 & 24 \end{bmatrix} \end{array}$$

R **36** What percent of the area of the rectangle is the area of the triangle?   20%

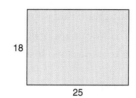

In problems 37–40, evaluate each expression.

R **37** $6 + 3 \div 9 \cdot 6$ 8   **38** $\dfrac{6+3}{9 \cdot 6}$ $\dfrac{1}{6}$ **39** $6 + \dfrac{3}{9} \cdot 6$ 8   **40** $\dfrac{6+3}{9} \cdot 6$ 6

Investigation

Language Arts The word *matrix* has different meanings in different fields. Find out what a matrix is in each of the following disciplines: model making, geology, printing, and anatomy. Do these uses of the word have anything in common? Do they relate to the mathematical meaning of matrix? Explain your answer.

CHAPTER 5

Summary

CONCEPTS AND PROCEDURES

After studying this chapter, you should be able to
- Read a circle graph (5.1)
- Draw a circle graph (5.1)
- Read a bar graph and a line graph (5.2)
- Draw a bar graph and a line graph (5.2)
- Interpret scattergrams, histograms, pictographs, and artistic graphs (5.3)
- Draw a scattergram, a histogram, a pictograph, and an artistic graph (5.3)
- Find the mean, the median, and the mode of a set of data (5.4)
- Interpret a stem-and-leaf plot and a box-and-whisker plot (5.5)
- Draw a stem-and-leaf plot and a box-and-whisker plot (5.5)
- Identify the entries in a matrix (5.6)
- Determine the dimensions of a matrix (5.6)
- Identify equal matrices (5.6)

VOCABULARY

artistic graph (5.3)
bar graph (5.2)
box-and-whisker plot (5.5)
circle graph (5.1)
data analysis (5.1)
digraph (5.6)
dimensions (5.6)
disjoint (5.1)
equal matrices (5.6)
extremes (5.5)
histogram (5.3)
line graph (5.2)
line plot (5.3)
lower quartile (5.5)

matrix (5.6)
mean (5.4)
measure of central tendency (5.4)
median (5.4)
midquartile range (5.5)
mode (5.4)
outlier (5.5)
pictograph (5.3)
pie chart (5.1)
range (5.5)
scattergram (5.3)
stem and leaf plot (5.5)
upper qaurtile (5.5)

CHAPTER 5

Review

1 a How many days (on average) did each of the top 32 fifth-grade students miss? ≈4.2
b How many days (on average) did each of the bottom 32 fifth-grade students miss? ≈7.5
c Find the ratio of days missed by the bottom 32 seventh-grade students to days missed by the top 32 seventh-grade students. 2:1
d What conclusion would you draw from the data?

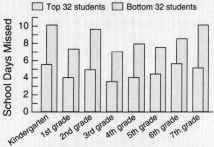

Average number of days missed over the last eight years by students who are now eighth graders
Source: *Chicago Sun-Times*

2 Mr. Ed's students scored 75, 80, 85, 74, 49, 63, 41, 91, 98, 71, 85, 81, 91, 66, 95, 91, 88, 91, and 83 on a test about horses.
a Make a stem-and-leaf plot for the data.
b Find the mean of the data. ≈78.8
c Find the median of the data. 83
d Find the mode of the data. 91
e Find the lower and upper quartiles of the data. 71; 91
f Make a box-and-whisker plot for the data.

3 Construct a matrix with entries $e_{1,1} = 5$, $e_{1,2} = 7$, $e_{2,1} = 9$, and $e_{2,2} = 11$.

4 a What percent does each pie-shaped sector of the circle graph represent?
b Find the number of degrees for each pie-shaped sector of the circle graph.
c Use a protractor to measure each pie-shaped sector of the circle graph. Do the measurements agree with the calculations you made in part **b**?

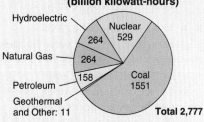

Source: Energy Information Administration

5 Watch a half-hour television news broadcast. Keep track of how much time is devoted to international news, national news, local news, sports, weather, special features, commercials, and other items (such as introduction, logo, and entertainment). Draw a circle graph to represent the data you have collected. Check students' graphs.

Chapter Review   243

Assignment Guide

Basic
Day 1 1–4, 6, 7, 10, 11, 14
Day 2 Chapter Test 1–11

Average or Heterogeneous
Day 1 1–4, 6–8, 10–12, 14, 15

Honors
Day 1 1–4, 6–12, 14, 15

Strands

Communicating Mathematics	1, 8, 10, 11, 13
Geometry	4
Data Analysis/ Statistics	1, 2, 4–12, 14, 15
Technology	12
Probability	15
Algebra	3, 7
Patterns & Functions	1, 8, 12
Interdisciplinary Applications	1, 4–6, 8, 11

Additional Answers

for page 243

1 d Answers will vary. Possible answer: Students do better in school if their attendance is better.

2 a
```
9 | 1 1 1 1 5 8
8 | 0 1 3 5 5 8
7 | 1 4 5
6 | 3 6
5 |
4 | 1 9
```

f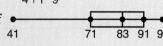
  41   71 83 91 98

3 $\begin{bmatrix} 5 & 7 \\ 9 & 11 \end{bmatrix}$

Answer for problem 4 is on the following page.

243

REVIEW

Additional Answers
for pages 243, 244, and 245

4 a Nuclear: 19%; Hydroelectric: 9.5%; Natural Gas: 9.5%; Petroleum: 6%; Geothermal and Other: 0.4%; Coal: 56%
b Nuclear: 68°; Hydroelectric: 34°; Natural Gas: 34°; Petroleum: 22°; Geothermal and Other: 1.4°; Coal: 202°
c Answers may vary depending on the degree of students' accuracy.

6 a
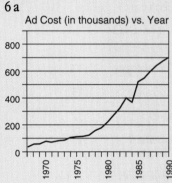

b

c
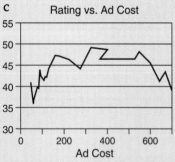

7 a Bill: $188.75; Sue: $219.00; Harry: $246.25
b 6 hr as a bagger, 7 hr as a clerk, or 4 hr as a cashier

---

6 The table displays Super Bowl television ratings and the costs of advertising during the Super Bowl game. Use the data in the chart to draw a line graph in which
 a The year is graphed against ad costs
 b The year is graphed against ratings
 c The ratings are graphed against ad costs

Year	Rating	Ad Cost (30 seconds)
1967	41.1	$40,000
1968	36.8	$54,500
1969	36.0	$55,000
1970	39.7	$78,200
1971	39.9	$72,500
1972	44.2	$86,700
1973	42.7	$88,100
1974	41.6	$103,500
1975	42.4	$107,000
1976	42.3	$110,000
1977	44.4	$125,000
1978	47.2	$162,300
1979	47.1	$185,000
1980	46.3	$222,000
1981	44.4	$275,000
1982	49.1	$324,300
1983	48.6	$400,000
1984	46.4	$368,200
1985	46.4	$525,000
1986	48.3	$550,000
1987	45.8	$600,000
1988	41.9	$645,000
1989	43.5	$675,000
1990	39.0	$700,000

Source: *Daily Herald*

7 Each week, Bill, Sue, and Harry work the given numbers of hours in each of three different jobs.

	Bagger	Clerk	Cashier
Hourly wage	[5.25	4.75	8.25]

	Bill	Sue	Harry
Bagger	10	5	4
Clerk	20	18	4
Cashier	5	13	25

 a How much money does each person make per week?
 b How many additional hours per week would Bill need to work in order to earn more than Sue?

8 a What percent of degrees earned in 1959–60 were bachelor's degrees? In 1969–70? In 1979–80? In 1984–85? In 1985–86?
 b Why, do you think, does the number of advanced degrees seem to level off after 1979–80?

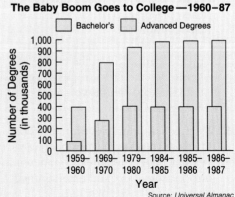

Source: *Universal Almanac*

244 Chapter 5 Data Analysis

REVIEW

9 Telly surveyed his classmates to find out how many hours of television they watched on a given day. Here are their responses. 3, 2.5, 4, 3, 2.5, 1.5, 0, 1, 0, 4, 2.5, 1, 2, 1, 0, 2.5, 3, 1, 0, 2.5, 3, 4, 2 Make a line plot of the data.

10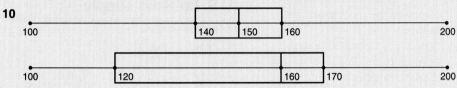

Explain the differences between the sets of data represented by the two box-and-whisker plots.

11 a Which team had the best home record in each division?
    b Which team had the best away record in each division?
    c Which team scored fewer runs than its opposition but still won more than half of its games?
    d Which team won the American League title?

American League Records in 1990 Final Standings

Eastern Division	W	L	Home	Away	Runs Avg	Runs vs.
Boston	88	74	51-30	37-44	4.3	4.1
Toronto	86	76	44-37	42-39	4.7	4.1
Detroit	79	83	39-42	40-41	4.6	4.6
Cleveland	77	85	41-40	36-45	4.5	4.5
Baltimore	76	85	40-40	36-45	4.2	4.3
Milwaukee	74	88	39-42	35-46	4.5	4.7
New York	67	95	37-44	30-51	3.7	4.6

Western Division	W	L	Home	Away	Runs Avg	Runs vs.
Oakland	103	59	51-30	52-29	4.5	3.6
Chicago	94	68	49-31	45-37	4.2	3.9
Texas	83	79	47-35	36-44	4.2	4.3
California	80	82	42-39	38-43	4.3	4.4
Seattle	77	85	38-43	39-42	3.9	4.2
Kansas City	75	86	45-36	30-50	4.4	4.4
Minnesota	74	88	41-40	33-48	4.1	4.4

American League Championship Series

Oakland 9, Boston 1   Oakland 4, Boston 1
Oakland 4, Boston 1   Oakland 3, Boston 1

Source: *The World Almanac and Book of Facts*

12 A town's population of 40,000 is decreasing by 8% each year. Of the total population, 1.5% is junior-high age, and 60% of these students attend Antelope Valley Junior High School.
    a Use a spreadsheet to copy and complete the table showing the change in the population over the next 15 years.
    b What will the population of Antelope Junior High School be in 2007? ≈103

Year	Town Population	Junior- High Age	Attend Antelope Valley
1992	40,000		
1993			
1994			
.			
.			
.			
2007			

---

8 a 1959–60: ≈83%; 1969–70: 74.5%; 1979–80: 69.9%; 1984–85: 71.3%; 1985–86: 71.4%
  b Possible answer: There was less demand in the job market for people with an advanced degree.

9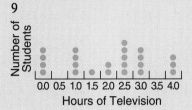

10 Possible answer: The upper half has several higher scores the second time. The lower half has several lower scores the second time. The median increased by ten. The midquartile range is much greater the second time, which indicates more of a spread of the middle 50% of the data.

11 a East: Boston; West: Oakland
   b East: Toronto; West: Oakland
   c Texas
   d Oakland

12 a

Year	Town Population	Junior High Age	Attend Antelope Valley
1992	40,000	600	360
1993	36,800	552	331
1994	33,856	508	305
1995	31,148	467	280
1996	28,656	430	258
1997	26,363	395	237
1998	24,254	364	218
1999	22,314	335	201
2000	20,529	308	185
2001	18,886	283	170
2002	17,376	261	156
2003	15,985	240	144
2004	14,707	221	132
2005	13,530	203	122
2006	12,448	187	112
2007	11,452	172	103

Chapter Review

REVIEW

Additional Answers

for page 246

13 Answers will vary.

14 a Mobile homes and trailers
b 53,692,000   **c** 60,357,000
d Increase   **e** Decrease

13 Look at newspapers and magazines. What kinds of data displays are the most prevalent? Why do you suppose this is true?

14 a Which housing category increased the most from 1980 to 1990?
b How many detached houses were there in 1980?
c How many detached houses were there in 1990?
d Did the number of detached houses increase or decrease?
e Did the percent of detached houses increase or decrease?

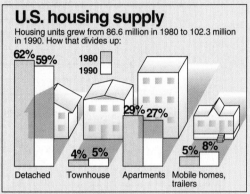

15 The map shows the 12 states with the greatest potential for wind-generated electricity. The numbers represent the percent of potential U.S. wind-generated electricity for each state.

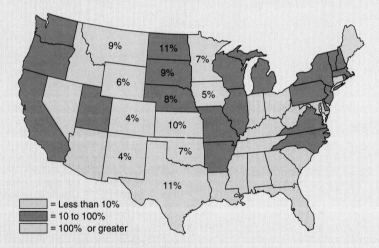

a Which states have the greatest potential for wind-generated electricity?   North Dakota and Texas
b What fraction of potential wind-generated electricity is in Iowa? $\frac{1}{20}$
c Which states each have $\frac{1}{25}$ of the U.S. potential for wind-generated electricity?   Colorado and New Mexico

Chapter 5   Data Analysis

Test

1. In a survey, 450 people identified their favorite color. Draw a circle graph to represent the data in the table.

Color	Number of People
Red	82
Blue	92
Yellow	66
Green	58
Purple	70
Pink	82

2. The stem-and-leaf plot shows the numbers of books read by the students in Mr. Reed's class during the first semester.
   a. What are the extremes of the data? 15 and 35
   b. Draw a box-and-whisker plot for the data.
   c. Between what two numbers are 50% of the data values located? Possible answer: 18 and 27

```
1 | 5 5 6 8 9
2 | 0 0 3 4 4 5 7
3 | 0 2 5
```

3. a. What is the mean number of hours worked by the students each week? 8.6 hr
   b. What is the median number of hours worked by the students each week? 8 hr
   c. What is the mode of the data? 5 hr

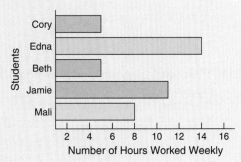

4. Consider the matrix $A = \begin{bmatrix} 2 & -1 & 3 \\ 4 & 2 & 6 \end{bmatrix}$

   a. What are the dimensions of matrix A 2 by 3
   b. What is $e_{1,3}$? 3

5. a. What percent of the households surveyed have more than three telephones? About 25%
   b. How many households were surveyed? 355

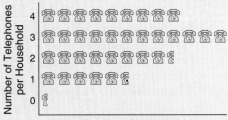

TEST

Alternative-Assessment Guide continued

determine the number of countries that border it. Make a circle graph using the method detailed in the sample problem. Write a brief report comparing your graph to the graph at the bottom of page 202.

10. Extend problem 9 on page 210 as follows: Select a stock from the newspaper and follow its trading value over the course of one week. Summarize the results in a table and answer the five parts of problem 9 using your table. Do you have enough information to conclude whether the stock you picked would have been a good investment?

11. Look up the final results of the 1992 Summer Olympics. Make a line plot in which each dot represents the total number of medals (gold, silver, or bronze) won by a country. Find the mean, median, and mode of the data and compare the three statistics. Which statistic best represents the results of a country selected at random? If your data did not include countries that participated in the Olympics but did not win a medal, how would this affect your results?

12. Search for some data that you think are interesting and create an artistic graph to display them. Write a paragraph describing ways that someone viewing your graph might view the data differently than someone who studied them in a nonartistic form.

6 a Draw a vertical histogram using the data, which consist of the scores from Mr. Miske's history test.

Score	50	55	60	65	70	75	80	85	90	95	100
Number	2	1	3	2	5	6	8	4	3	1	2

b What percent of the students received a score of 90% or better?   About 16%

7 Solve $\begin{bmatrix} 2 & 13-w \\ x+7 & 3y \end{bmatrix} = \begin{bmatrix} 2 & 5 \\ 14 & 9 \end{bmatrix}$ for w, x, and y. $w = 8$; $x = 7$; $y = 3$

8 Draw a pictograph to represent the number of strokes Debbie took per hole during a round of golf.

Hole	1	2	3	4	5	6	7	8	9
Strokes	2	4	4	6	4	8	2	6	4

9 a Draw a line graph for the data.
b What trend do you see in the graph?
The greatest number of students is absent on Mondays and Fridays, and the least number of students is absent on Wednesday.

Day	Number of Absences
Monday	56
Tuesday	28
Wednesday	18
Thursday	21
Friday	49

10 What conclusion can you draw from this scattergram?   As the number of free throws attempted increases, the number of free throws made increases.

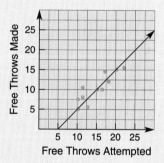

11 Mr. Litt E. Rature gave an English test. His students' test scores were 42, 87, 62, 94, 88, 82, 91, 85, 86, 92, 91, 95, 81, 79, 74, 65, 33, 98, 91
a Organize the data into a stem-and-leaf plot.
b Find the range of scores.   65
c Find the median.   86
d Find the lower quartile.   74
e Make a box-and-whisker plot.
f Which values are outliers?   33 and 42

Chapter 5 Data Analysis

PUZZLES AND CHALLENGES

1 Using operation signs, we can arrange four 4's to obtain various numbers. For example:

$$\frac{44}{44} = 1$$

$$\frac{4}{4} + \frac{4}{4} = 2$$

$$\frac{4+4+4}{4} = 3$$

$$\frac{4}{\sqrt{4}} + \frac{4}{\sqrt{4}} = 4$$

$$4 + \sqrt{4} - \frac{4}{4} = 5$$

Come up with ways to use four 4's to obtain 6, 7, 8, and 9.

2 Here is one way to use six 9's to obtain 100.

$$9 + 9 + 9 \times 9 + \frac{9}{9}$$

Find another way to use six 9's to obtain 100.

3 Arrange the numbers 1 through 9 in the squares so that the sums of the numbers in all the rows, columns, and diagonals are the same. This is called a *magic square*.

4 Find a two-digit number in which
  a. The number is twice the product of its digits
  b. The number is three times the sum of its digits
  c. The number is the square of its units digit
  d. The number exceeds its reversal by 20%
  e. The number and its reversal add to a perfect square

Additional Answers
for Chapter 5 Test

1
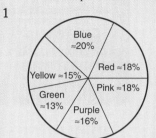

2 b
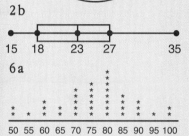

6 a
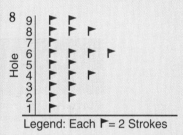

8
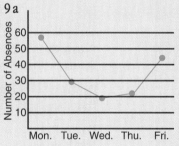
Legend: Each ⚑ = 2 Strokes

9 a
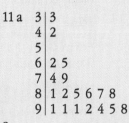

11 a
```
3 | 3
4 | 2
5 |
6 | 2 5
7 | 4 9
8 | 1 2 5 6 7 8
9 | 1 1 1 2 4 5 8
```

e
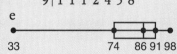

Answers for Puzzles and Challenges are on page A4.

ASSESSMENT
LONG-TERM INVESTIGATIONS

UNIT C: The Sporting Life

Suggestions for various ways to assess students' performance are integrated throughout the investigations in this unit. These forms of assessment include observing the students while they work, interviewing students, looking at students' self-evaluations and group evaluations, observing peer review/response situations, evaluating students' products, and evaluating students' portfolios.

Refer to the chart below to preview the variety of assessment opportunities that are integrated throughout this unit.

Assessment Opportunities	Before Unit	1	2	3	4	5	End of Unit
Form A: Observation Notes **Bank A/B:** Observation Guide		✖				✖	
Form B: Observation Evaluation Record **Bank A/B:** Observation Guide				✖	✖		
Form C: Interview Record **Bank C:** Interview Guide				✖			
Form D: Product Evaluation Record **Bank D:** Product Evaluation Guide		✖	✖		✖	✖	
Form E: Rubric Chart **Bank E:** Sample Rubric			✖				
Form F: Portfolio Item Description	✖						✖
Form G: Portfolio Notes **Bank G:** Portfolio Evaluation Criteria	✖						✖
Form H: Self-evaluation		✖		✖			✖
Form I: Group Evaluation			✖			✖	
Form J: Peer Review/Response					✖		

Rubrics

Rubrics are used to evaluate students' products. You can develop a scoring rubric for a specific investigation by adapting the blackline masters shown below, which are provided in the *Gateways to Mathematical Investigations* Teacher's Edition. You will probably want to develop a rubric with students at least once per unit. The chart on the opposite page suggests that for Unit C: The Sporting Life, you complete a rubric before Investigation 2. Remind students that, at times, they may be asked to revise their work to improve it and to make it complete.

Bank D, page 211

Bank D: Product Evaluation Guide

The statements below can be used as a guide to assess students' products. Adapt these statements as needed to evaluate specific assignments. Record your adaptations on a clean copy of **Form D: Product Evaluation Record** (p. 210). You may also wish to create and record your own criteria on **Form D**. After you have adapted **Form D** for use with a specific assignment, make copies for each student or group as needed.

Mathematical Reasoning Skills

- The student includes examples and/or counter examples to support the mathematical argument.
- The student synthesizes and summarizes own or group's thinking.

Mathematical Understanding

- The student identifies all the important information of the situation or problem.
- The student shows understanding of the situation's mathematical ideas. (List important mathematical concepts and skills for the assignment on Form D.)

Form E, page 212

Form E: Blank Rubric Chart

NAME (S) ____________ ____________ GROUP ______
____________ ____________ DATE ______

STANDARDS FOR STUDENT'S WORK

LEVEL	Standard of performance to be achieved for each specified level
4 Exceptional/ Fully Acceptable	

Bank E, page 213

Bank E: Sample Rubric

STANDARDS FOR STUDENT'S WORK

LEVEL	Standard of performance to be achieved for each specified level
4 Exceptional/ Fully Acceptable	Your presentation fully accomplishes the purpose of the task. Your work is complete, well organized, and clearly explained. You illustrate ideas using clear diagrams and models and communicate them effectively to the audience. You identify all the important information of the situation and show a thorough understanding of the mathematical ideas and procedures. You use strong arguments to support findings, including examples (and/or counterexamples). You demonstrate that you fully grasp the central mathematical ideas developed in this task. You demonstrate an orignial extension or applicatrion of the mathematical concepts.
3 Satisfactory/ Acceptable	Your presentation substantially accomplishes the purpose of the task. Your work is complete, reasonably organized, and explained. You use an appropriate diagram or model and communicate ideas effectively to the audience. You identify most of the important information of the situation and show understanding of the mathmatical ideas and procedures. You present arguments to support findings, including examples (and/or counterexamples). You understand and use the central mathematical ideas developed in this task. You demonstrate and apply the mathematical concepts.
2 Incomplete/ Revise	Your presentation partially accomplishes the purpose of the task. However, your work may be incomplete, or misdirected, or not clearly organized and/or explained. Your diagram or model may be inappropriate or unclear. You may communicate ideas inadequately to the audience. You may identify some of the important information of the situation and you may show only partial and/or limited understanding of the mathematical ideas and procedures. Your arguments to support findings may be weak or incomplete. You may misunderstand or misuse the central mathematical ideas developed in this task. You may have difficulty demonstrating and applying the mathematical tasks. You need to revise this work. (Your teacher will provide suggestions for a satisfactory or competent rating and a due date for your revised work.)
1 Inadequate/ Redo	Your presentation shows little or no understanding of the assignment. Your work is barely, if at all, comprehensible. Your explanations are not understandable. You show little or no understanding of the mathematical ideas and procedures. You made serious computational mistakes and misused mathematical terms. You should meet with your classmates or your teacher to be sure you fully understand this assignment. Then you need to revise this work. (Your teacher will provide suggestions for a satisfactory or competent rating and a due date for your revised work.)

Gateways © McDougal, Littell & Company   Bank E: Sample Rubric   213

Form F, page 214

Form F: Portfolio Item Description

NAME ____________ DATE ____________

Complete this form each time you add an item to your portfolio.

Type of Portfolio Entry: ____________

Why did you choose to place this item in your portfolio?

____

Form G, page 215

Form G: Portfolio Notes

NAME ____________ CLASS ____________

Date	Observations/Comments	Recommendations

Bank G, page 216

Bank G: Portfolio Evaluation Criteria

At the start of the school year, have a class discussion on the goals and uses of students' portfolios. Discuss with the class some ideas regarding what to include in a portfolio. Then, use the portfolio evaluation criteria which follows to develop a rubric with the class that communicates the criteria you will use to assess their portfolios. Have students record this criteria on copies of **Form E: Rubric Chart.** Have students store this completed form in their portfolios so that they can refer to it throughout the year.

Each time students add items to their portfolios, have them complete copies of **Form F: Portfolio Item Description.**

Use copies of **Form G: Portfolio Notes** to record your observations, comments, and recommendations each time you evaluate students' portfolios.

Ideas for Types of Portfolio Entries

- a portfolio table of contents
- a letter to the reader of the portfolio explaining each item
- teacher-completed checklists/evaluations
- a mathematical autobiography
- notes from interviews, teachers, peers, etc.
- papers that show the student's corrections or errors or misconceptions
- photos or sketches of the student's work with manipulatives or mathematical models
- excerpts from the student's daily journal or reflective writing samples
- tables, charts, and diagrams created by the student
- artwork supporting a problem such as drawings
- designs, coordinate pictures, scale drawings, maps, etc.
- student-created problems with or without solutions
- group projects or individualized reports
- student-developed surveys and results
- self-evaluations

Suggestions for Portfolio Evaluation Criteria

Mathematical Reasoning Skills

The Student:
- analyzes problem situations.
- is flexible and comfortable in using several different problem solving approaches.
- synthesizes and summarizes the individual's or group's thinking.
- provides supporting statements to verify findings.
- shows original and creative thought.

Mathematical Tools and Techniques

The Student:
- uses a variety of diagrams, models, technology, manipulatives, and other resources appropriately.
- uses displays (e.g. graphs, tables, diagrams, sketches, and models) to represent mathematical ideas.

Mathematical Understanding

The Student:
- shows understanding of mathematical ideas and procedures. (List specific skills/concepts on **Form G.**).

Mathematical Communication

The Student:
- organizes work in a logical and clear way.
- expresses mathematical ideas clearly and appropriately.
- uses mathematical language and notation appropriately.

Mathematical Dispositions

The Student:
- shows development as as group participant.
- shows development of a positive attitude.
- shows growth in learning.
- includes revised and complete work.
- shows consistent effort.
- includes a variety of work in his or her portfolio.

216   Assessment   *Gateways* © McDougal, Littell & Company

Portfolios

At the beginning of each unit, you may wish to review with the students the criteria they should use when placing work in their portfolios. Have students complete Form F: Portfolio Item Description each time they wish to place an item in their portfolios.

Periodically, you may wish to evaluate students' portfolios using the blackline masters shown above. For Unit C: The Sporting Life, the chart on the opposite page suggests that you evaluate students' portfolios at the end of the unit.

CHAPTER 6

6 Signed Numbers

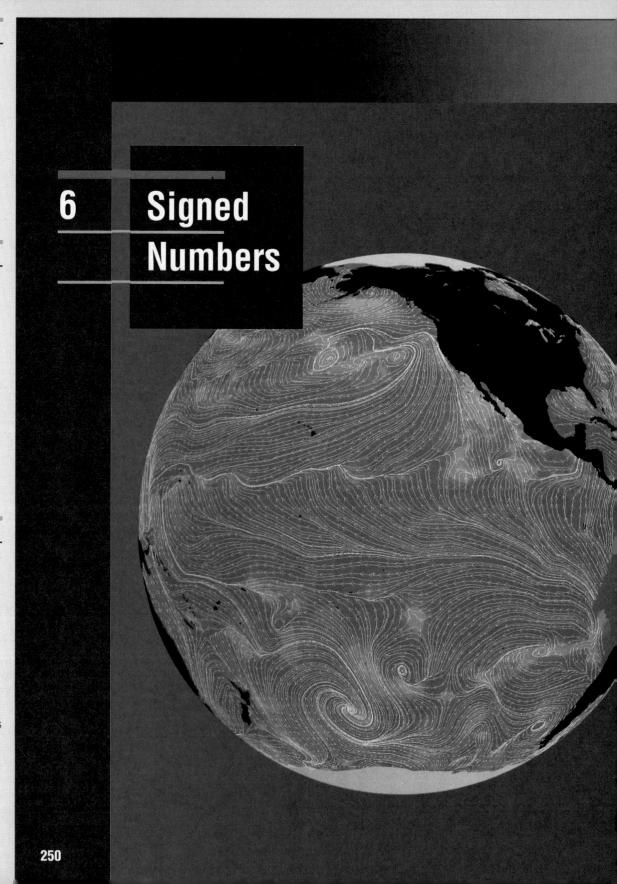

Contents

6.1 The Number Line
6.2 Inequalities
6.3 Adding Signed Numbers
6.4 Multiplying and Dividing Signed Numbers
6.5 Matrix Multiplication
6.6 Subtracting Signed Numbers
6.7 Absolute Value
6.8 Two-Dimensional Graphs

Pacing Chart

Days per Section

	Basic	Average	Honors
6.1	2	1	1
6.2	2	2	2
6.3	2	2	2
6.4	3	2	2
6.5	2	2	2
6.6	2	2	2
6.7	2	2	1
6.8	3	2	1
Review	3	1	1
Assess	3	2	2
	24	18	16

Chapter 6 Overview

A firm grasp of signed numbers is critical for success in algebra. We use real-world applications of signed numbers as the basis of each lesson and make liberal use of number lines. Students are given opportunities to practice computation with signed numbers in a variety of contexts. Multiplication of matrices, in particular, provides a considerable amount of practice. The emphasis on graphing inequalities visually reinforces the meaning of signed numbers, while spreadsheet applications reinforce the results by helping students discover numerical patterns. The final

CHAPTER 6

CARTOGRAPHY Cartographers are people who make maps. There are many kinds of maps. Some show the physical features of the earth, such as hills, forests, and ravines. Other maps show political boundaries, or population densities, or even information about climate. Most cartographers work extensively with computer graphics.

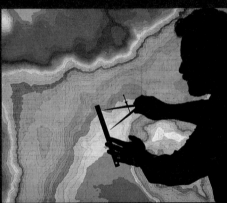

INVESTIGATION

Map Coloring Four families own an island. They plan to divide the entire island into four separate lots by building fences at the boundaries. They have agreed to the following conditions:
- Each lot must share a boundary with each of the other three lots.
- Each lot must have its own beach.
- The lots can be irregular in shape and need not be equal in size.

The families need your help. Draw a map of an island, and construct boundaries according to the conditions stated above. Make each lot a different color. Explain your results to the class.

section on two-dimensional graphs reinforces the geometrical aspects of signed numbers and sets the stage for the next chapter.

Career Connection

Point out that many maps are made from aerial and satellite photographs. The increasing sophistication of computerized map production has allowed cartographers to respond to rapidly changing political boundaries. For example, in 1990 the National Geographic Society was able to alter a map to show a reunified Germany only a short time after reunification. This would have taken months longer without computer-aided mapmaking.

Video Connection

For more information about the career discussed here, please refer to the Futures videocassette #3, program A: *Cartography*, featuring Jaime Escalante. For more information or to order a copy of the program, call PBS VIDEO at 800/344-3337.

Investigation Notes

- The four families are trying to do something that is impossible. While it is possible for all four plots to border each other, one of the four must be landlocked.

- The four-color map theorem, proved in the 1970s after more than a century of failed attempts, states that four different colors are sufficient to color any map on a plane surface. If the situation described in the problem were possible, then five regions—the four plots plus the water—would all border one another, contradicting the four-color theorem.

SECTION 6.1

Objectives

- To associate signed numbers with points on a number line
- To determine the opposite of a signed number
- To graph signed numbers on a number line

Vocabulary: coordinate, integer, negative signed number, opposite, origin, positive signed number

Class Planning

Schedule 1–2 days

Resources
In *Chapter 6 Student Activity Masters* Book
Class Openers 1, 2
Extra Practice 3, 4
Cooperative Learning 5, 6
Enrichment 7
Learning Log 8
Technology 9, 10
Spanish-Language Summary 11

Transparency 24

Class Opener

Give the opposite of each expression.
1. Withdrawing $60 from your bank account
2. –60
3. Taking an elevator up 20 floors
4. 20
5. 0

Answers:
1. Depositing $60 in your bank account 2. 60 3. Taking an elevator down 20 floors 4. –20 5. 0

Lesson Notes

- Number lines are usually thought of only as horizontal lines. Three types—horizontal, vertical, and circular—are

252

Investigation Connection
Section 6.1 supports Unit C, *Investigation* 2. (See pp. 198A-198D in this book.)

SECTION 6.1
The Number Line

NUMBER LINES AND SIGNED NUMBERS

A new roller coaster has just opened at Great Flags Amusement Park. This ride runs partly above ground and partly below ground. The vertical number line shows the height and the depth the roller coaster reaches in feet.

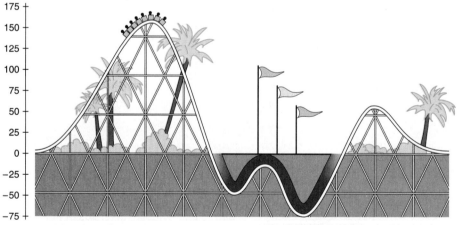

Ground level is represented by zero on the number line. **Positive (+) signed numbers** represent distances above ground and **negative (–) signed numbers** represent distances below ground. Zero is neither positive nor negative. A nonzero number written without a sign is a positive number. Notice that the roller coaster rises 160 feet above ground (160) and descends 75 feet below ground (–75).

Example 1
If the roller coaster descends 45 feet below ground and then rises 30 feet, what is its location?

252   Chapter 6   Signed Numbers

SECTION 6.1

Solution
If we begin at −45 on the number line and move 30 units upward, we reach −15. The roller coaster is 15 feet below ground.

Example 2
What is the distance between the lowest point (75 feet below ground) and the highest point (160 feet above ground) on the roller coaster?

Solution
From 75 feet below ground to ground level is a distance of 75 feet. From ground level to 160 feet above ground is a distance of 160 feet. Therefore, the distance between the lowest and the highest point is 75 + 160, or 235, feet.

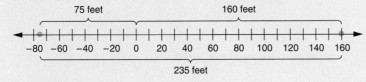

You may have noticed that on a horizontal number line the negative numbers are to the left of zero and the positive numbers are to the right of zero. To save space, we usually use a horizontal number line rather than a vertical one.

The circular number line used on some outdoor thermometers is another type of number line. Notice that the negative numbers are arranged in a counter-clockwise sequence and the positive numbers in a clockwise sequence.

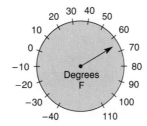

OPPOSITES

Two nonzero numbers are **opposites** if the numbers are the same distance from zero on the number line but one of the numbers is positive and the other is negative. For example, −30° and 30° are opposites on the thermometer scale, as are −20 and 20 on this horizontal number line.

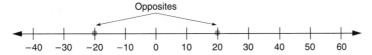

---

discussed in this section. Spend enough time with the examples for the students to be familiar with all three types.

- Make sure that students know the appropriate calculator keystrokes for entering the opposite of a number. Some calculators require that the opposite sign be entered after the number; others use the more natural before-number entry. In either case, the subtraction key is usually not a good choice.

- In example 4, stress the placement of the numbers on the number line, especially the three negative numbers.

Reteaching
To reteach locating mixed numbers on a number line, have pairs of students use their rulers to draw a number line 8 inches long. Have them label a zero point at the 4-inch mark, then use two rulers to show positive and negative scales (one ruler will be upside-down). Have students locate numbers such as $2\frac{1}{2}$, $-1\frac{1}{4}$, $5\frac{7}{8}$, and $-4\frac{7}{16}$. Use rulers marked in tenths of an inch for numbers such as 2.7 and −1.9.

Stumbling Block
Students may get hung up on long-held notions, such as "The sum of two numbers is a bigger number," that have been reinforced through experience with whole numbers. Expand students' concepts of addition by giving them plenty of practice with problems like the roller-coaster problem in example 1 and with manipulatives representing positives and negatives.

SECTION 6.1

Assignment Guide

Basic
Day 1 10–12, 15, 16, 18, 21, 26, 29, 32, 37
Day 2 13, 14, 17, 19, 22–24, 27, 31, 35

Average or Heterogeneous
Day 1 10–18, 20, 22–25, 27, 28, 30, 31, 33, 34, 36

Honors
Day 1 11–14, 16, 17, 20, 22–26, 30, 31, 33, 36

Strands

Communicating Mathematics	2, 3, 8, 35, Investigation
Geometry	26, 27, 34
Measurement	5, 8, 11, 23, 28, 34, Investigation
Data Analysis/Statistics	20, 22, 23, 30, 31
Technology	6, 18, 24, 29, 37
Probability	34
Algebra	1, 4–7, 9–30, 32–36
Patterns & Functions	6, 18, 19, 29
Number Sense	3, 16–18, 20–22, Investigation
Interdisciplinary Applications	5, 8, 11–13, 20, 28, 33

Individualizing Instruction

The auditory learner Students may benefit from reading problems such as 28 aloud to each other before attempting to solve them. You might have students work together to assist each other.

Example 3
Use a calculator to
a Add 33 and –33
b Subtract –33 from 33

Solution
Many calculators have a $\boxed{+/-}$ key for entering signed numbers.
a To add, enter 33 $\boxed{+}$ 33 $\boxed{+/-}$ $\boxed{=}$. The display shows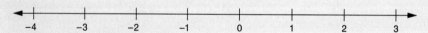
b To subtract, enter 33 $\boxed{-}$ 33 $\boxed{+/-}$ $\boxed{=}$. The display shows

ORIGIN AND COORDINATES

The number associated with a point on a number line is the **coordinate** of the point. The point on the number line assigned to 0 is called the **origin.** When we draw a number line, we usually label the divisions of the line with **integers**—... , –3, –2, –1, 0, 1, 2, 3, ...

Example 4
Draw a number line and locate the points with the coordinates –3.5, $-1\frac{7}{8}$, 2.6, and $-\pi$.

Solution
First we draw a number line and label the origin and several integer coordinates.

Then we label the four points.

1. The point with the coordinate –3.5 is located halfway between –3 and –4.
2. To locate the point with the coordinate $-1\frac{7}{8}$, we subdivide the interval between –1 and –2 into eight equal parts and locate $-1\frac{7}{8}$ at the mark nearest to –2.
3. To locate the point with the coordinate 2.6, we subdivide the interval between 2 and 3 into ten equal parts and locate 2.6 at the sixth mark to the right of 2.
4. Since $-\pi \approx -3.14$, we can find an approximate location for it by subdividing the interval between –3 and –4 into ten equal parts and locating $-\pi$ near the middle of the second subdivision to the left of –3.

Chapter 6 Signed Numbers

SECTION 6.1

Sample Problem

Problem Tellie Fone calls friends in the other five time zones of the United States from her home in San Francisco. To reach her friends at the right times, she needs a fast way to calculate the time of day in each location.

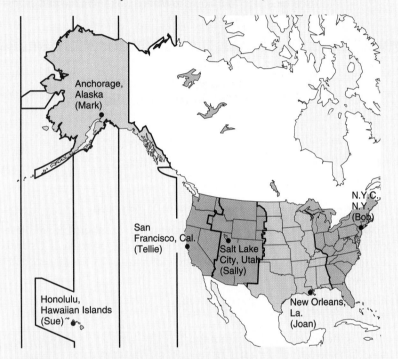

a Use a number line to show the time differences of the six zones.
b If Tellie's clock reads 7:00 P.M., how can she find the time in Bob's time zone? In Sue's time zone?
c Which of her friends are in time zones with opposite values?
d Mark needs to get up at 6:00 A.M. At what time must Bob set his own alarm if he has to give Mark a wake-up call?

Solution **a** We will place Tellie at the origin. To the east of her, the time becomes one hour later in each successive zone. To the west, the time becomes one hour earlier in each successive zone. We therefore place zones to the east of Tellie at positive integer coordinates and zones to the west at negative integer coordinates. Each coordinate shows how much a friend's time differs from Tellie's.

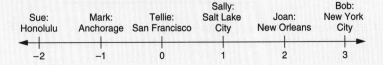

Additional Examples

Example 1
If the roller coaster shown on page 252 is 75 feet below the ground and then goes up 130 feet, what is its location?

Answer: 55 feet above the ground

Example 2
The roller coaster descends how many feet during the first drop?

Answer: About 200 feet

Example 3
Use your calculator to
a. Add 54 and –54
b. Subtract –54 from 54

Answers: **a.** 0 **b.** 108

Example 4
Draw a number line and label four points with the coordinates 2.7, $-\frac{\pi}{2}$, $-\frac{4}{5}$, and 1.1.

Answer:

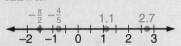

Additional Practice

What is the opposite of each number?
1. 7 **2.** –4 **3.** 5 **4.** –91

Draw a number line and locate each point at the given coordinate.
5. Point A: –5 **6.** Point B: 3.7
7. Point C: $-2\frac{1}{3}$ **8.** Point D: $4\frac{1}{3}$

Which number is greater?
9. –3 or –4 **10.** –9 or 4
11. 0 or –7 **12.** 5 or –5

Answers:
1. –7 **2.** 4 **3.** –5 **4.** 91
5–8.

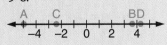

9. –3 **10.** 4 **11.** 0 **12.** 5

SECTION 6.1

Problem-Set Notes

for pages 256 and 257

- **9** is a good visual problem that will help the discussion of absolute value in Section 6.7.
- **11** and **12** check to see if students understood the reading.
- **16** and **17** offer students a way to visualize a comparison of two numbers.
- **19** illustrates for students who get hung up on the often-taught rule "You can't subtract a larger number from a smaller one" that it is indeed possible. Arranging the problems vertically might help students to see a pattern.

Cooperative Learning

- Have students create a number line that includes all 24 time zones, using the time zone they are in as the zero point. Have them list under each coordinate two or three major cities that are in that time zone. Have them find the difference in hours between their time zone and the time zones in which the following cities are located.
  a. Paris, France
  b. Mexico City, Mexico
  c. Sydney, Australia
  d. Cairo, Egypt
  e. Tokyo, Japan

Answers: Answers are based on eastern time. Add 1 hour for each successive time zone west of the eastern time zone.
  a. +6 hours   b. –1 hour
  c. +15 hours   d. +7 hours
  e. +14 hours

- Problem 28 is appropriate for cooperative learning.

---

b Bob lives three time zones to the east, so Tellie must add three hours to her time to determine the time in Bob's time zone (10:00 P.M.). Sue lives two time zones to the west, so Tellie must subtract two hours to determine the time in Sue's time zone (5:00 P.M.).

c Mark (–1) and Sally (1) are in time zones with opposite values on the number line. So are Sue (–2) and Joan (2).

d Bob is four time zones to the east of, or four hours later than, Mark. So Bob must add four hours to Mark's wake-up time and set his own alarm for 10:00 A.M.

Think and Discuss

P **1** What is the sum of two opposite numbers? 0

P **2** Why do we need signed numbers?

P, I **3** What are some ways to define *integers?*

P **4** On a number line, what is the coordinate of the origin?

P **5** **Science** In a countdown for a space flight, the numbers before blast-off are negative and the numbers after blast-off are positive. If T stands for the time of blast-off, how many seconds pass from T – 15 seconds to T + 30 seconds? 45 sec

P, R **6** Use a calculator to evaluate each sum.
  a –55 + 55   0   **b** –55 + 56   1   **c** –55 + 57   2

P **7** What is the opposite of each number?
  a –3.6   3.6   **b** 102   –102   **c** 0   0

P **8** If –8°F is really cold, is 8°F really hot? Explain your response.

P, R, I **9** Suppose point A on the number line is moved five units and point B is moved four units.

  a Find all possible locations for points A and B after they are moved.
  b How far apart are the points in each possible pair of locations?

Problems and Applications

P **10** Draw a number line and locate each point at the given coordinate.
  a Point A: 3.5   **b** Point B: $-1\frac{2}{3}$
  c Point C: $-1\frac{1}{2}$ **d** Point D: –4

256   **Chapter 6** Signed Numbers

SECTION 6.1

P **11** The temperature rose from −6°F to 10°F. What was the change in temperature?   16°

P **12** Refer to the sample problem on page 255. Sue wants to call Bob to wish him a happy birthday. If she wants to reach him at 6:00 P.M., what time should she place her call?   1 P.M.

P **13** There are 15 levels below ground in a mine shaft. The levels are numbered from top to bottom as follows: −1, −2, −3, . . ., −15.
  a A miner is at level −3 and goes down 8 levels. On what level is she?   −11
  b If a mine inspector starts at level −13 and goes up 3 levels, then up 3 more levels, at what level is she?   −7

P, R **14** Draw a number line and locate the points with the given coordinates.
  a 2   **b** $\frac{2}{5}$ **c** 7.3   **d** −4.5   **e** $-3\frac{5}{6}$ **f** $-\sqrt{2}$

P **15** Write the opposite of each number.
  a 14 −14   **b** π −π   **c** $-\sqrt{2}$ $\sqrt{2}$ **d** 0 0   **e** −1 1   **f** $-\frac{3}{4}$ $\frac{3}{4}$

P, R, I **16** For each pair of numbers, determine which number is greater (further to the right on a number line).
  a 8 or 0   8   **b** 3.6 or 3.7   3.7   **c** −3.6 or −3.7   −3.6
  d −7 or 0   0   **e** −1 or −3   −1   **f** $-2\frac{1}{2}$ or $-2\frac{5}{8}$ $-2\frac{1}{2}$

P, R, I **17** For each pair of numbers, determine which number is less (further to the left on a number line).
  a 4 or 2   2   **b** 0 or −9   −9   **c** −8 or 1   −8
  d −4 or −2   −4   **e** −3.4 or −3.5   −3.5   **f** $-4\frac{5}{9}$ or $-4\frac{2}{3}$ $-4\frac{2}{3}$

P 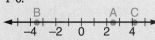 **18** Use a calculator to find each sum.
  a 12 + (−8)   4   **b** 12 + (−9)   3   **c** 12 + (−10)   2   **d** 12 + (−11)   1
  e 12 + (−12)   0   **f** 12 + (−13)   −1   **g** 12 + (−14)   −2   **h** 12 + (−15)   −3

P **19** Find each difference.
  a 5 − 3   2   **b** 5 − 4   1   **c** 5 − 5   0   **d** 5 − 6   −1
  e 5 − 7   −2   **f** 5 − 8   −3   **g** 5 − 9   −4

P **20 Sports** The data shown represent goals scored by our hockey team and the opposition during the time each of our players was on the ice.

Our Player	Our Goals	Opponents' Goals	Our Player Index
#1	35	41	−6
#2	16	8	+8
#3	19	17	+2
#4	47	43	+4
#5	40	52	−12
#6	29	31	−2
#7	38	32	+6
#8	22	17	+5

  a Determine the player index for the remaining players.
  b In a close game, which player(s) should be used?

Section 6.1   The Number Line   257

Checkpoint

What is the opposite of each number?
1. 3   **2.** $-4\frac{1}{2}$ **3.** 0

Draw a number line and locate each point at the given coordinate.
4. Point A: 2.5   **5.** Point B: −3.6
6. Point C: $4\frac{1}{4}$

Which number is less?
7. −3 or 3   **8.** 2 or −9
9. −5 or −4   **10.** 0 or −6

Answers: **1.** −3 **2.** $4\frac{1}{2}$ **3.** 0
4–6.

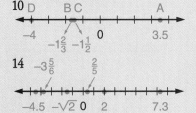

7. −3 **8.** −9 **9.** −5 **10.** −6

Additional Answers

for pages 256 and 257

2 Possible answer: Signed numbers can be used to distinguish between gain and loss, hot and cold temperatures, east and west, or north and south.

3 Possible answer: The integers consist of the natural numbers, their opposites, and zero.

4 0

8 No. There is a difference of only 16°.

9 a A: −2 or 8; B: 11 or 19
b 13 units (−2 to 11); 21 units (−2 to 19); 3 units (8 to 11); 11 units (8 to 19)

10

```
   D           B C                    A
◄──┼──┼──┼──┼──┼──┼──┼──┼──┼──┼──►
  −4        −1⅔ −1½  0            3.5
```

14

```
       −3⅚                ⅖
◄──┼──┼──●──┼──┼──●──┼──┼──┼──┼──┼──►
    −4.5  −√2  0   2              7.3
```

20 b In order: #2, #7, #8, #4, #3, #6, #1, and #5

Problem-Set Notes

for pages 258 and 259

- **24** may need to be modified for the calculators your students are using.
- **36** provides a good review of percent as well as more practice with number lines.
- **37** suggests a feature of spreadsheets with which students may not yet be familiar. Note that some spreadsheets display negative numbers by means other than parentheses.
- **Investigation** Students might be interested to know that the zero point of the Fahrenheit scale was set as the temperature at which a saturated solution of salt in water freezes, in order to avoid negative temperatures.

Communicating Mathematics

- Have students write a paragraph explaining how positive and negative numbers could be used to represent the amount of money earned in a year, the amount spent in a year, and the amount saved in a year.
- Problems 2, 3, 8, 35, and Investigation encourage communicating mathematics.

P **21** Locate $\frac{-3}{4}$, −1, 6, −2.43, 2.3, −π, 4.7, $\frac{-\sqrt{2}}{2}$, 0, and −7 on a number line.

P, R **22** **a** Locate points 6, 8, 10, 17, 22, and 26 on a number line.
b Locate the mean and median of the numbers in part **a** on the same number line. Mean: 14.8; median: $13\frac{1}{2}$

P **23** At 9:00 A.M. the temperature was 11°C. The table describes the temperature changes over the next 13 hours.

Time	Temperature Behavior
9:00 — 10:00 A.M.	Dropped 4°
10:00 — 2:00 P.M.	Remained steady
2:00 — 4:00 P.M.	Dropped 3°
4:00 — 6:00 P.M.	Dropped 5°
6:00 — 10:00 P.M.	Dropped 3°

a Graph the temperatures from 9:00 A.M. to 10:00 P.M.
b What was the temperature at 10:00 P.M.? −4°C

P **24** Perform the following operations on a calculator. Record your results.
a 9 [+/−] [+] 15 [+/−] [=] −24
b 9 [+/−] [×] 15 [+/−] [=] 135
c 9 [+/−] [−] 15 [+/−] [=] 6
d 9 [+/−] [÷] 15 [+/−] [=] 0.6

P, I **25** Suppose that Bill stands on square 0 and tosses a die seven times. On the first, third, fifth, and seventh tosses, he moves to the right the number of squares indicated by the die. On the second, fourth, and sixth tosses, he moves to the left the number of squares indicated by the die.

−6	−5	−4	−3	−2	−1	0	1	2	3	4	5	6
						START						

a If Bill's first three tosses are 3, 6, and 1, on which square will he be standing? −2
b If Bill's next four tosses are 1, 2, 5, and 4, on which square will he be standing? −2

P, I **26**

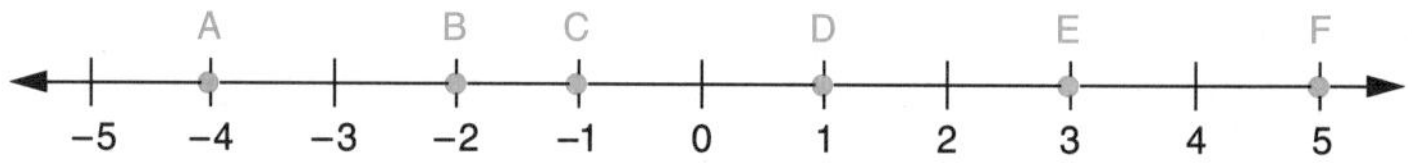

Find the length of the segment from
a D to F 4
b C to D 2
c A to B 2
d A to F 9
e B to E 5
f B to C 1

P, I **27** Point A is the midpoint of segment BC. What is the coordinate of A if the coordinates of B and C are
a 16 and 36? 26
b 0 and 20? 10
c −12 and 8? −2
d $-\frac{2}{3}$ and $4\frac{2}{3}$? 2

P **28** Every day, Roger tries to run a mile in 4 minutes 30 seconds. He records the difference between his target time and his actual time in seconds, using a + if it takes more time and a − if it takes less time. His goal is for the sum of the numbers for each week to be 0. One week his records showed −3, +5, +7, −2, −1, and 0 for the first six days. What must his running time be on the seventh day if he is to meet his goal? 4 min 24 sec

SECTION 6.1

◀ LOOKING BACK **Spiral Learning** LOOKING AHEAD ▶

I **29** Use your calculator to find each product.
   a (–3)(–2) 6   **b** (–3)(–1) 3   **c** (–3)(0) 0
   d (–3)(1) –3   **e** (–3)(2) –6   **f** (–3)(3) –9

R **30**
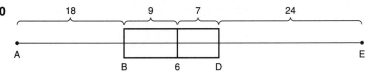

The coordinate of the median in the box-and-whisker plot is 6.
Determine the coordinates of A, B, D, and E. –21; –3; 13; 37

P, R **31** Find the mean of –8, –6, –3, 0, 3, 6, and 8. 0

R, I **32** Evaluate each expression for $x = 5$ and $y = 3$.
   a $11x$ 55   **b** $4x + 7x$ 55   **c** $17y$ 51
   d $9y + 8y$ 51   **e** $2x + 3y + 4x + 5y$ 54   **f** $6x + 8y$ 54

R **33** **Consumer Math** Stocky Dealer bought 40 shares of Widget stock at $57.50 a share. He later sold 25 of these shares at $49.25 a share. How much money did Stocky lose on these 25 shares of stock? $206.25

R **34** The perimeter of a triangle is 18 centimeters. The lengths of two sides of the triangle are 4.5 centimeters and 6 centimeters.
   a Find the length of the third side of the triangle. 7.5 cm
   b Draw an accurate picture of this triangle.
   c If one angle of the triangle is selected at random, what is the probability that the angle is acute? $\frac{2}{3}$

R, I **35 a** If $x = 8$ and $y = 24$, what is the value of $\frac{x+y}{2}$? 16
   b If x, y, and your answer to part **a** are placed on a number line, what conclusion can you draw about the answer to part **a**?

R, I **36** Point B is 20% of the way from A to C. Find the coordinate of B. 20

R **37** **Technology** When a spreadsheet is instructed to display numbers in dollars and cents, how does it show negative amounts of money? In parentheses

Investigation

Temperature Scales Who were Fahrenheit and Celsius? Where and when did they live? Which of the two temperature scales do you find more convenient, and why? If you could invent your own temperature scale, how might it be different?

Section 6.1 The Number Line 259

Additional Answers
for pages 258 and 259

21
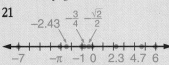

22 a See students' drawings.

23 a
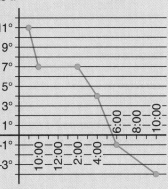

34 b See students' drawings.

35 b 16 is the midpoint of the segment connecting x to y.

Investigation Gabriel Daniel Fahrenheit (1686–1736) was a German physicist who made thermometers more accurate by using mercury instead of alcohol. Anders Celsius (1701–1744) was a Swedish astronomer who invented the Celsius scale (also sometimes called the Centigrade scale). The freezing point of pure water is 32°F or 0°C, and the boiling point of pure water is 212°F or 100°C. Other answers may vary. Students may find it more convenient to use the Fahrenheit scale because they are more familiar with it. Students may invent a scale that uses multiples of 10, with 0° as the freezing point and 200° as the boiling point.

SECTION 6.2

Objectives

- To write inequalities and graph them on a number line
- To write and graph inequalities joined by *and* or *or*

Vocabulary: inequality

Class Planning

Schedule 2 days

Resources

In *Chapter 6 Student Activity Masters* Book
Class Openers 13, 14
Extra Practice 15, 16
Cooperative Learning 17, 18
Enrichment 19
Learning Log 20
Interdisciplinary Investigation 21, 22
Spanish-Language Summary 23

Transparency 24

Class Opener

List five examples of numbers that satisfy the given condition or conditions.
1. Numbers less than 3
2. Numbers greater than −7
3. Numbers greater than or equal to −2 and less than 1
4. Numbers less than −2 or greater than 1
5. Numbers less than 10 and greater than 20

Answers:
1–4. Answers will vary.
5. No numbers meet both conditions.

Lesson Notes

- Students will benefit from drawing a number line for any problem involving compound inequalities expressed in words or symbols. It's important for students to see what's going on with

260

Investigation Connection
Section 6.2 supports Unit C, *Investigation* 5. (See pp. 198A-198D in this book.)

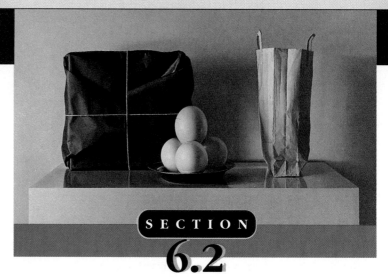

SECTION 6.2
Inequalities

GRAPHING AND WRITING INEQUALITIES

United Shipping People (USP) will ship a package only if it meets the following three conditions:

Condition 1: The package is no more than 70 pounds in weight.
Condition 2: The package is no more than 130 inches in combined length and girth.
Condition 3: The package is no more than 108 inches in length.

These three conditions can be described graphically.

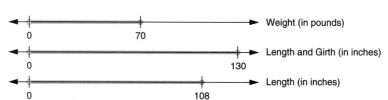

A solid endpoint indicates that the coordinate of that point is included in the graph. An open endpoint indicates that the coordinate is not included in the graph. The shaded region indicates that all the numbers between the endpoints are included in the graph. The maximum values allowed by USP are represented by the solid endpoints at 70, 130, and 108. Why are all the endpoints at 0 open?

Example 1
A package has a length of 84 inches. Graph the possible values of the girth.

260   **Chapter 6**   Signed Numbers

SECTION 6.2

Solution
A package must have a girth greater than zero, but the sum of the girth and the length cannot exceed 130 inches. Since the length of the package is 84 inches, the girth can be at most 130 − 84, or 46 inches. The graph of the solution is

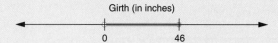

For example, a 36-inch girth is acceptable, but a 60-inch girth is not.

We can also express the USP conditions in mathematical symbols, as an ***inequality***—a statement containing one of the following symbols.

Symbol	Meaning
<	is less than
>	is greater than
≤	is less than or equal to
≥	is greater than or equal to
≠	is not equal to

Example 2
Express each USP condition as an inequality.

Solution
Condition 1:
$0 < \text{weight} \leq 70$

Condition 2:
$0 < \text{length} + \text{girth} \leq 130$

Condition 3:
$0 < \text{length} \leq 108$

AND INEQUALITIES AND *OR* INEQUALITIES

Graphs that are shaded between two numbers can be represented by two inequalities joined by the word *and*. For example, the inequality that describes condition 1 of the USP guidelines for mailing can be written as

$$0 < \text{weight} \text{ and } \text{weight} \leq 70.$$

This statement is read "0 is less than the weight *and* the weight is less than or equal to 70."

Example 3
A computer needs at least 90 volts of electricity to work. It is damaged, however, if the voltage is 140 volts or more.
a Express the safe voltage range as an inequality.
b Graph the inequality in part **a**.

---

these problems, even when a number line is not called for.

Additional Examples

Example 1
A package has a length of 42 inches. Graph the possible values for the girth if $0 < \text{length} + \text{girth} \leq 130$.

Answer:

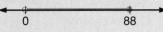

Example 2
Express the following package conditions as inequalities.
a. Weight less than 50 pounds
b. Length plus girth less than 88 inches
c. Length no more than 40 inches

Answers: **a.** $0 < \text{weight} < 50$
b. $0 < \text{length} + \text{girth} < 88$
c. $0 < \text{length} \leq 40$

Example 3
A plant needs at least 50 days of sunlight during the year to bloom. It will not bloom, however, if it gets 200 days or more of sunlight.
a. Express the range of days of sunlight needed as an inequality.
b. Graph the inequality in **a**.

Answers: **a.** $50 \leq \text{sunlight} < 200$
b.

Example 4
Refer to the set "all numbers that are less than 4 units from 2 on the number line."
a. Draw a graph that represents this set.
b. Describe the set, using an inequality.

Answers:

a.

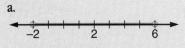

b. $-2 < \text{number} < 6$

SECTION 6.2

Example 5
a. Write an inequality that describes the graph.
b. What could this graph represent?

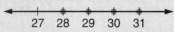

Answers:
a. $28 \leq$ whole numbers ≤ 31
b. Possible answer: the set of possible last dates of the month

Communicating Mathematics

- Have students write a paragraph explaining how the graphs of the following inequalities are the same and how they are different:
$0 \leq$ whole numbers ≤ 23 and
$0 \leq$ numbers ≤ 23

- Problems 2–8, 11, 17, 20, 23, and 25 encourage communicating mathematics.

Cooperative Learning

- Have groups of students contact airlines or parcel moving companies to find out the price structure they use when determining the cost of sending packages. Have each group create a poster that displays this information, using inequalities.

- Problems 24, 28, and 33 are appropriate for cooperative learning.

Solution

a Here are two ways to write the inequality:

1. Voltage ≥ 90 and voltage < 140
2. $90 \leq$ voltage < 140

b This is the graph of the inequality.

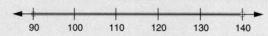

Some graphs have two or more separate sections. The inequalities that represent these graphs are joined by the word *or*.

Example 4
Refer to the set of "All numbers that are more than three units from zero on the number line."
a Draw a graph that represents this set.
b Describe the set using an inequality.

Solution

a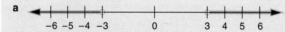

The numbers 3 and –3 are not included in the graph because they are not more than three units from zero on the number line—they are *exactly* three units from zero. We indicate this with the open endpoints. The arrowheads indicate that the numbers go on forever—all the numbers greater than 3 and all the numbers less than –3 are included in the graph.

b Number > 3 or number < -3

This inequality is read "The number is greater than 3 or the number is less than –3."

Example 5

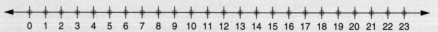

a Write an inequality that corresponds to this graph.
b What could this graph represent?

Solution

a $0 \leq$ whole number ≤ 23
b This graph could represent the set of possible numbers in the hour portion of a 24-hour digital clock display.

Chapter 6 Signed Numbers

SECTION 6.2

Sample Problem

Problem For what values of N is it possible to compute the value of $\sqrt{N+7}$?

Solution Let's try some values for N. A spreadsheet is useful.

Although we are trying only a few numbers, we can see a pattern. It appears that no number less than −7 will work. A number less than −7 would result in the square root of a negative number, which we cannot compute. Therefore, N must be greater than or equal to −7.

	A	B
1	N	SQRT(N + 7)
2		
3	3	3.1622776602
4	2	3
5	1	2.8284271247
6	0	2.6457513111
7	−1	2.4494897428
8	−2	2.2360679775
9	−3	2
10	−4	1.7320508076
11	−5	1.4142135624
12	−6	1
13	−7	0
14	−8	ERR
15	−9	ERR

Using mathematical symbols, we can express all possible values of N as $N \geq -7$. We can also express the answers in a graph.

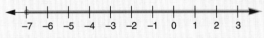

Think and Discuss

P, R **1** Match each set of numbers with the appropriate graph.
  a The nonnegative numbers 3
  b The negative integers 5
  c Integers between −4 and 4 1
  d The positive numbers 4
  e All numbers between −4 and 4 2

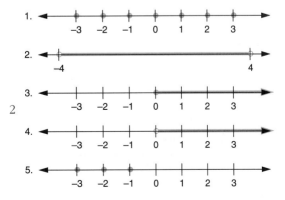

Assignment Guide

Basic
Day 1 8–12, 16, 21, 26, 29–31
Day 2 13–15, 18–20, 22, 23, 27, 32, 34

Average or Heterogeneous
Day 1 8–11, 12 (a, c), 14–16, 22, 26–28, 32, 34
Day 2 12 (b, d), 13, 17–21, 23, 24, 29, 30, 31, 33

Honors
Day 1 8, 9, 11–13, 14 (a, b), 18, 19, 23, 25, 29, 30, 34
Day 2 10, 14 (c, d), 15–17, 20–22, 24, 26–28, 31–33

Strands

Communicating Mathematics	2–8, 11, 17, 20, 23, 25
Geometry	19, 27, 30, 33
Measurement	11, 14, 30, 33, Investigation
Data Analysis/Statistics	14, 29
Technology	24, 31
Algebra	1–7, 9, 10, 12, 13, 15–19, 21–27, 33, 34
Patterns & Functions	28, 31
Number Sense	1, 9, 10, 13, 14, 18, 26, 32
Interdisciplinary Applications	8, 11, 14, 20, Investigation

Individualizing Instruction

The auditory learner Have a group of students form a human number line, with each student representing an integer value. Read to these students various compound inequalities. For each inequality, have the students whose values make the inequality true raise their hands.

Section 6.2 Inequalities

SECTION 6.2

Problem-Set Notes

for pages 264 and 265

- **7** requires students to make a distinction that is difficult to communicate. One explanation might be to say that *or* inequalities require that only one of two conditions be met, while *and* inequalities require that both conditions be satisfied.

- **19** requires students to recognize that $L + W$ is half the perimeter. It is necessary to include zero as the lower bound because perimeter cannot be negative.

- **27** makes a nice connection between prior knowledge and the concepts introduced in this section.

Checkpoint

Graph each inequality.
1. $x < 3$ 2. $-2 < x \leq 4$
3. $x \geq -1$ 4. $x < -1$ or $x \geq 3$

Write an inequality for each graph.

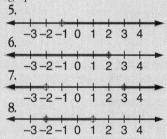

Answers:

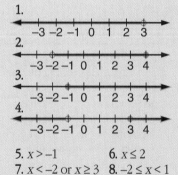

5. $x > -1$ 6. $x \leq 2$
7. $x < -2$ or $x \geq 3$ 8. $-2 \leq x < 1$

P **2** What is the difference between an equation and an inequality?

P **3** What does a shaded endpoint on the graph of an inequality represent?

P **4** What does an open endpoint on the graph of an inequality represent?

P **5** What does an arrowhead on the graph of an inequality represent?

P **6** What is the difference between the symbols < and ≤?

P **7** What is the difference between an *and* inequality and an *or* inequality?

Problems and Applications

P **8 Communicating**
  a Why do you think USP uses girth and length instead of length, width, and height?
  b Why do you think USP sets a maximum weight?
  c Why do you think USP sets a maximum length?

P **9 a** What is the first integer to the left of −3.8 on the number line?   −4
    b Is your answer to part **a** less than or greater than −3.8?   Less than

P **10** Graph all integers between −3.4 and 3.9.

P **11 Communicating** Will USP deliver a pair of cross-country skis that are 215 centimeters long? Explain your answer.

P, I **12** Graph each inequality.
  a $x > -4$ **b** $x < 2$ **c** $x \geq -6$ **d** $x \leq 1$

P, R **13 a** Between which two integers is $\sqrt{39}$ located on the number line?   6 and 7
    b Between which two integers is $-\sqrt{39}$ located on the number line?   −7 and −6

P **14** Assuming each box weighs less than 70 pounds, which ones will USP deliver?   **a** and **d**

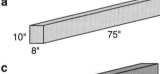 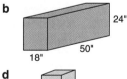

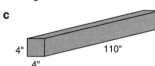

P **15** Graph each inequality.
  a $x < -2$ or $x \geq 3$ **b** $x \geq -2$ and $x < 3$ **c** $-1 < x$ or $-4 > x$

P **16** Write an inequality for each graph.

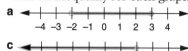 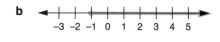

264   **Chapter 6** Signed Numbers

SECTION 6.2

P **17 Communicating** The inequality 3 < number < −3 represents this graph.

a Why is this the wrong inequality? **b** What is the correct inequality?

P, I **18** If A < 7, 12 < B, and C > A, which inequalities must be true? **a and d**
  a A < B **b** C < 7 **c** C < B **d** B > 7 **e** C > B

P, R **19** The perimeter of the rectangle is less than 50. Write an inequality for L + W. $0 < L + W < 25$

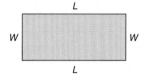

P **20 Communicating** Will USP accept a package with a girth of 10 inches and a combined girth and length of 130 inches? Why or why not?

P, I **21** Write an inequality for each graph.

a  Possible answer: $-2 < x \leq 3$

b  Possible answer: $y \geq -1$

c  Possible answer: $z < 1$

P, R **22** Graph each inequality on a number line.
  a $-3 < x < 4$ **b** $-3 < x < 4$, and x is an integer
  c $-3 < x < 4$, and x is a whole number

P **23** What is the difference between the graph of $x < -2$ or $x > -1$ and the graph of $x < -2$ and $x > -1$?

R **24 Spreadsheets**
  a Use a spreadsheet to evaluate the expression $\sqrt{9 - x^2}$ for integer values of x ranging from −6 to 6.
  b What numbers seem to work? $-3 \leq x \leq 3$
  c Draw a number line that shows all the possible values of x.

◀ LOOKING BACK **Spiral Learning** LOOKING AHEAD ▶

P, R **25 Communicating** Explain whether each number line is properly constructed.
  a 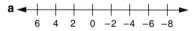 **b**

R, I **26** For $x = 7$, determine whether each statement is *true* or *false*.
  a $3x + x = 4x$ True **b** $5x + 6x = 11x^2$ False **c** $4x^2 + 8x^2 = 12x^2$ True
  d $5x \cdot 3x = 15x$ False **e** $5x \cdot 3x = 15x^2$ True

P, R, I **27** Draw a number line that shows all the possible measures of an obtuse angle.

Section 6.2 Inequalities 265

Reteaching

To help students make the connection between phrases and mathematical symbols, consider sentences such as "All students scored at least 70 on the test," and "No one scored higher than 98." Ask students what possible test scores could be. If they don't mention 70 or 98, be sure to ask whether these values should be included.

Additional Practice

Graph each inequality.
1. $x \geq 2$ 2. $x < -3$
3. $-4 \leq x < 1$ 4. $x < -2$ or $x \geq 1$

Write an inequality for each graph.

5.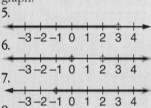
6.
7.
8.

Answers:
1.
2.
3.
4.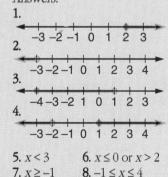

5. $x < 3$ 6. $x \leq 0$ or $x > 2$
7. $x \geq -1$ 8. $-1 \leq x \leq 4$

Answers for problems 2–8, 10–12, 15–17, 20, 22, 23, 24 (a, c), 25, and 27 are on page A4.

265

SECTION 6.2

Stumbling Block

When graphing a compound inequality, students are sometimes unsure about which parts of the number line should be shaded. Suggest that they replace the variable in the inequality with numbers. If a number satisfies the inequality, that part of the number line should be shaded.

Additional Answers

for page 266

30 a See students' drawings.
32 1, 2, 3, 6, 9, 18
33 a

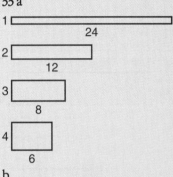

b

The *Investigation* Answers will vary from one appliance to another, and will also depend on how the appliance's thermostat is calibrated. A home-heating furnace, for example, is designed to kick on when the temperature falls below the thermostat setting, and kick off when the temperature goes above the setting. But how far below or above the setting? The variation could be half a degree or even less, or in an older or less efficient unit, the variation could be several degrees above or below the setting.

R **28** How many paths are there from A to G? Follow the directions indicated by the arrows. 6

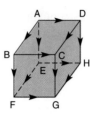

R **29** All 27 students in a class took a test, and no two students got the same score. The box-and-whisker plot represents the scores.

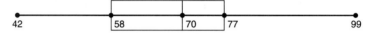

  a How many students scored over 70? 13
  b How many students scored at least 58 but less than 70? 7
  c How many students scored more than 77? 6

R **30 a** Draw a parallelogram with one side 3 centimeters long, another side 5 centimeters long, and one 70° angle.
  b Measure the other three angles of the parallelogram and record your results. 110°, 70°, and 110°

R, I **31** Complete the following pattern, using a spreadsheet if possible.

  $6 \cdot 4 = 24$ $6 \cdot 0 = \underline{0}$
  $6 \cdot 3 = 18$ $6 \cdot (-1) = \underline{-6}$
  $6 \cdot 2 = \underline{12}$ $6 \cdot (-2) = \underline{-12}$
  $6 \cdot 1 = \underline{6}$ $6 \cdot (-3) = \underline{-18}$

R, I **32** Find all whole numbers, N, for which $\frac{18}{N}$ is a whole number.

R **33** Ty Isle is making a rectangular kitchen counter using all 24 of his tiles. He wants a design where *no* tiles have to be cut.
  a Sketch the rectangular designs he could possibly use.
  b Graph all the possible lengths and widths.

R **34** Refer to a number line.
  a Start at 3 and move 7 units to the left. On which number do you land? −4
  b Start at −2 and move 5 units to the right. On which number do you land? 3
  c Start at −18 and move 39 units to the right. On which number do you land? 21

Investigation

Equilibrium Investigate the operation of a refrigerator, furnace, air conditioner, or other appliance that is designed to maintain a particular temperature. How is the temperature equilibrium maintained? How many degrees above or below the preset temperature does the appliance tolerate before the unit kicks on or off? Report your findings.

SECTION 6.3

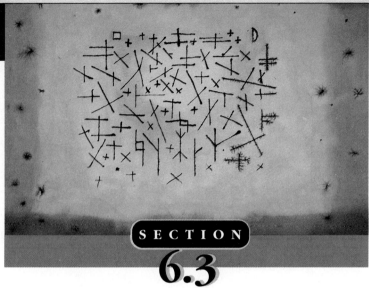

Investigation Connection
Section 6.3 supports Unit C, *Investigations* 2, 3, 4, 5. (See pp. 198A-198D in this book.)

SECTION 6.3
Adding Signed Numbers

Objectives
- To use a number line to model the addition of signed numbers
- To use a calculator to add signed numbers
- To add matrices

Class Planning

Schedule 2 days
Resources
In *Chapter 6 Student Activity Masters* Book
Class Openers 25, 26
Extra Practice 27, 28
Cooperative Learning 29, 30
Enrichment 31
Learning Log 32
Using Manipulatives 33, 34
Spanish-Language Summary 35

In *Tests and Quizzes* Book
Quiz on 6.1–6.3 (Basic) 145
Quiz on 6.1–6.3 (Average) 146
Quiz on 6.1–6.3 (Honors) 147

Transparencies 24, 27

ADDING SIGNED NUMBERS ON THE NUMBER LINE

We can use a number line to model the addition of signed numbers. When signed numbers are added, the signs indicate the directions to move on the number line. On horizontal number lines, a positive sign (+) indicates a move to the right and a negative sign (−) indicates a move to the left.

Example 1
Add two positive numbers, (+3) + (+6).

Solution
We start at zero and move three units to the right. Then we go six more units to the right. The sum is +9.

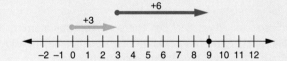

Example 2
Add a positive and a negative number, (+4) + (−6).

Solution
If we study the pattern shown, we see that as consecutively smaller integers are added to +4, the sums decrease by one. We can continue the pattern until we see that (+4) + (−6) = −2.

+4 + (+3) = +7	+4 + (−2) = +2
+4 + (+2) = +6	+4 + (−3) = +1
+4 + (+1) = +5	+4 + (−4) = 0
+4 + (0) = +4	+4 + (−5) = −1
+4 + (−1) = +3	+4 + (−6) = −2

Class Opener

Use a number line to show these movements. Start at 0 and describe your final position.
1. 3 units right; 4 units right
2. 5 units right; 4 units left
3. 8 units left; 2 units right
4. 7 units left; 1 unit left
5. 3 units right; 7 units left
6. 5 units left; 5 units right

Answers:
1. 7 2. 1 3. −6 4. −8 5. −4 6. 0

Lesson Notes

- Rather than simply giving students rules for adding integers, let students form their own rules by studying the patterns in example 2.

Section 6.3 Adding Signed Numbers 267

SECTION 6.3

Lesson Notes continued

- The process of adding two matrices can give students a lot of practice in adding signed numbers. Whenever possible, matrices should have labels, like those in the sample problem, to help students relate matrix addition to real-world contexts.

Additional Examples

Example 1
Add (+8) + (+11).

Answer: +19

Example 2
Add (+3) + (−7).

Answer: −4

Example 3
On Tuesday the temperature was −15°C. By Wednesday the temperature had risen 10 degrees. What was the temperature on Wednesday?

Answer: −5°C

Example 4
At 8 A.M. the level in a water tank was $11\frac{1}{4}$ ft. By 3 P.M. the level had dropped by $2\frac{1}{2}$ ft due to use. By 11 P.M. the tank had been filled, and the level had risen by $7\frac{1}{8}$ ft. What was the final water level?

Answer: $15\frac{7}{8}$, or 15.875, ft

Cooperative Learning

- Have students write five survey questions that can be answered with this scale:
  - −2 Strongly disagree
  - −1 Disagree somewhat
  - 0 Neither agree nor disagree
  - 1 Agree somewhat
  - 2 Strongly agree

Have students record each respondent's answers in a 5 × 1 matrix. Have them use matrix addition to summarize

---

To use a number line to show this addition, we begin at the point with coordinate 0 and move four units to the right to the point with coordinate +4. Then, from +4 we move six units to the left. The sum is −2.

USING A CALCULATOR TO ADD SIGNED NUMBERS

A calculator can also be used to add signed numbers.

Example 3
At 8:00 P.M. the temperature was −2° Celsius. By 7:00 the next morning, the temperature had dropped 15 degrees. What was the temperature at 7 A.M.?

Solution
The expression (−2) + (−15) models the problem. We can use a number line to help us add.

We can also use a calculator:

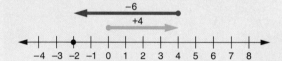

Using either method, we find that the temperature at 7 A.M. was −17°C.

Example 4
BCA computer stock sold for $18\frac{7}{8}$ dollars per share on Tuesday morning, but by noon its value had dropped $1\frac{1}{2}$ dollars per share. By closing time, its value had risen $\frac{5}{8}$ dollar per share. What was the closing value of the stock?

Solution
If you choose to use a calculator to solve this problem, it may be helpful to first write an expression that models the problem.

$$+18\frac{7}{8} + \left(-1\frac{1}{2}\right) + \left(+\frac{5}{8}\right)$$

If your calculator has a fraction key, you can enter the problem directly. If not, you could use the decimal equivalents for the fractions. If you enter

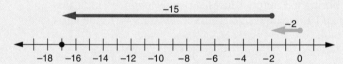

, the display shows 18. Notice that the +/− key is used to obtain the negative number.

268   Chapter 6   Signed Numbers

SECTION 6.3

Sample Problem

Problem Tiana and Ping each had money in a piggy bank, in a checking account, and in a savings account. The money they had at the end of November is represented by this 2 × 3 matrix.

	Piggy Bank	Checking Account	Savings Account
Tiana	37.50	125.00	368.27
Ping	18.74	145.00	547.19

In December, Ping deposited money (+) in one account, and both Tiana and Ping withdrew money (−) from other accounts. Again, the data are represented in a 2 × 3 matrix.

	Piggy Bank	Checking Account	Savings Account
Tiana	−24.25	−76.87	0
Ping	−10.47	−165.43	+125

How much money did each person have in each of the three accounts at the end of December?

Solution To answer this question, we need to add the two matrices. Two or more matrices can be added only if all the matrices have the same dimensions. To add matrices, we add corresponding entries.

$$\begin{bmatrix} 37.50 & 125.00 & 368.27 \\ 18.74 & 145.00 & 547.19 \end{bmatrix} + \begin{bmatrix} -24.25 & -76.87 & 0 \\ -10.47 & -165.43 & 125 \end{bmatrix}$$

$$= \begin{bmatrix} 13.25 & 48.13 & 368.27 \\ 8.27 & -20.43 & 672.19 \end{bmatrix}$$

The answer matrix gives the amounts of money in each person's accounts at the end of December. Why should Ping transfer some money from his savings account into his checking account?

Think and Discuss

P **1** In the sample problem above, two 2-by-3 matrices were added. Can you add a 4-by-3 matrix and a 3-by-4 matrix? Explain.

P **2** When a number line is used to add signed numbers, why is the endpoint of the first arrow at zero?

P, I **3** If your calculator has the fraction key [a b/c], use this key to help you find the sum $2\frac{2}{3} + \left(-3\frac{1}{2}\right) + 1\frac{5}{8}$. $\frac{19}{24}$

---

the results, then interpret the results.

- Problems 3 and 41 are appropriate for cooperative learning.

Assignment Guide

Basic
Day 1 9–11, 14–16, 22, 25, 33, 35, 38
Day 2 12, 17–20, 26, 29–31, 37, 39

Average or Heterogeneous
Day 1 9–11, 13–18, 21, 23, 24, 30 (a, c), 33
Day 2 12, 19, 20, 22, 26–29, 30 (b, d), 31, 36, 37b, 38, 40

Honors
Day 1 11–13, 15, 19, 20, 27, 29, 31, 32, 37
Day 2 9, 14, 16–18, 22–24, 26, 28, 30, 33–36, 38, 40, 41

Strands

Communicating Mathematics	1, 2, 4, 5, 8, 21, 26, 27, 35, 38, 41
Geometry	32, 37
Data Analysis/ Statistics	8, 17, 29, 34, 41
Technology	3, 24, 29, 38, 39
Probability	19, 23, 40
Algebra	1–23, 25–28, 30–33, 35, 36, 38–40
Patterns & Functions	5, 18, 24
Number Sense	5, 11, 13, Investigation
Interdisciplinary Applications	8, 27, 29, 34, 41, Investigation

Individualizing Instruction

The gifted learner Students may extend problem 29 by making a spreadsheet that models an actual checkbook record.

SECTION 6.3

Problem-Set Notes

for pages 269, 270, and 271

- **1** emphasizes that in order to be added, matrices must have the same dimensions.
- **10** consists of problems that students can check by using their calculators.
- **13** may need some discussion for clarification.

Additional Practice

Find each sum. Use a number line if necessary.
1. $2 + 5$
2. $-3 + 7$
3. $4 + (-3)$
4. $-2 + (-3)$

Find the sum of the matrices.

5. $\begin{bmatrix} -1 & 2 & 4 \\ 5 & 7 & -8 \end{bmatrix} + \begin{bmatrix} 4 & -6 & 8 \\ -2 & -5 & 1 \end{bmatrix}$

6. $\begin{bmatrix} 3 \\ -4 \\ -2 \end{bmatrix} + \begin{bmatrix} -1 \\ -6 \\ 7 \end{bmatrix}$

Answers:
1. 7   2. 4   3. 1   4. −5
5. $\begin{bmatrix} 3 & -4 & 12 \\ 3 & 2 & -7 \end{bmatrix}$ 6. $\begin{bmatrix} 2 \\ -10 \\ 5 \end{bmatrix}$

Communicating Mathematics

- Have students write a list of rules for adding signed numbers. Students might compare their rules to explore different ways of thinking about adding signed numbers.
- Problems 1, 2, 4, 5, 8, 21, 26, 27, 35, 38, and 41 encourage communicating mathematics.

P, I **4** In the expression $(+2) + (-5) - (-7)$, how is the use of the plus and minus symbols outside the parentheses different from the use of the plus and minus symbols within the parentheses?

P **5** Add four pairs of positive signed numbers. Add four pairs of negative signed numbers. Can you draw any conclusions about the addition of numbers with the same sign?

P **6** What addition problem is suggested by each of the following diagrams?

a 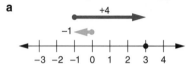   **b**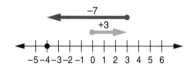

c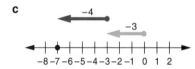

P **7** Use a number line to find each sum.
   a $3 + 5$ 8   **b** $-3 + 6$ 3   **c** $-3 + (-4)$ −7   **d** $5 + (-7)$ −2

P **8** **Budgeting** Sue and Juan's budgeted money (+) and actual expenses (−) for a month are arranged by categories in two matrices.

	Budget	**Expenses**
Food	250.00	−275.00
Housing	745.00	−745.00
Utilities	140.00	−130.00
Auto	210.00	−300.00
Entertainment	75.00	−10.00
Miscellaneous	100.00	−50.00

a Add the two matrices. Then explain the meaning of the answer matrix.
b Did Sue and Juan stay within their budget? Explain your answer.

Problems and Applications

P, I **9** What addition problem is represented by the diagram? Find the answer to the addition problem. $-3 + 8 + (-2) = 3$

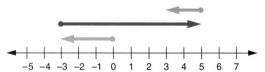

P **10** Find each sum.
   a $7 + 91$ 98
   b $-4 + (-16)$ −20
   c $-86 + 35$ −51
   d $-42 + (-73) + 2$ −113
   e $19 + (-30) + 79$ 68
   f $19 + (-30) + (-19)$ −30
   g $15 + (-5) + 12 + (-10) + (-12)$ 0

270 Chapter 6 Signed Numbers

SECTION 6.3

P, I   **11** Is the sum of a nonzero number and itself farther from zero on the number line than the number itself?   Yes

P   **12** Find each sum.
    a $(-8) + (-3)$ -11
    b $-7 + 12 + 7 + (-12)$ 0
    c $19 + (-5) + (-19)$ -5
    d $38.6 + (-15.2) + (-29) + (-38.6) + 29$ -15.2
    e $-93.92 + 15 + 38.6 + (-15) + (-93.92)$ -149.24

P, I   **13** Is the statement true *always*, *sometimes*, or *never*?
    a The sum of two negative numbers is greater than either number.   Never
    b The sum of two positive numbers is greater than either number.   Always
    c The sum of two numbers is equal to one of the numbers.   Sometimes

P, R   **14** Find each sum.
    a $3\frac{1}{5} + \left(-4\frac{3}{5}\right)$ $-1\frac{2}{5}$ **b** $-6.8 + (-9.3)$ -16.1 **c** $-4\frac{1}{2} + 6\frac{1}{4}$ $1\frac{3}{4}$

R   **15** Find the sum of the matrices.
$$\begin{bmatrix} -4 & -16 & 10 \\ -22 & 9 & 16 \end{bmatrix} + \begin{bmatrix} -7 & 6 & -5 \\ -17 & 12 & -3 \end{bmatrix} \quad \begin{bmatrix} -11 & -10 & 5 \\ -39 & 21 & 13 \end{bmatrix}$$

P, I   **16** Fill in the blank with a number that makes the equation true.
    a $19 + \underline{\quad} = -2$ -21 **b** $-15 + \underline{\quad} = -13$ 2 **c** $-11 + \underline{\quad} = -35$ -24

P, R   **17** Find the mean of $2\frac{2}{5}, -4\frac{5}{8}, 7\frac{1}{2}, 2.78, -3.69,$ and $4\frac{7}{8}$. 1.54

R, I   **18** Evaluate the expression $3 + x$ for each number in the set $\{-1, -2, -3, -4, -5\}$. $\{2, 1, 0, -1, -2\}$

P, R   **19** If one of the expressions $3 + (-9), -10 + 2, -6 + 12, 15 + (-14.3),$ $19 + (-21), 75.5 + (-75.6), 18 + (-18),$ and $-4 + 4$ is chosen at random, what is the probability that the value of the expression is
    a Positive? $\frac{1}{4}$ **b** Negative? $\frac{1}{2}$ **c** Neither positive nor negative?

P, R   **20** Find the sum of the matrices.
    a $\begin{bmatrix} 19 & 3.5 \\ -6 & -8 \end{bmatrix} + \begin{bmatrix} -7 & 4.3 \\ 6 & -12 \end{bmatrix}$ **b** $\begin{bmatrix} 11 & -4 \\ -8 & -3.1 \\ 6.3 & -10 \end{bmatrix} + \begin{bmatrix} -15 & 6 \\ -2.5 & 4.2 \\ -9 & 10 \end{bmatrix}$

P   **21 Communicating** Perform the indicated operations. What did you discover?
    a $18 + (-3)$ 15 **b** $18 - 3$ 15
    c $24.5 + (-18.2)$ 6.3 **d** $24.5 - 18.2$ 6.3

P   **22** Find the sum of the matrices.
$$\begin{bmatrix} -2 & 3 \\ 5 & 0 \\ -9 & -1 \end{bmatrix} + \begin{bmatrix} -6 & -7 \\ -2 & -8 \\ 3 & 4 \end{bmatrix} + \begin{bmatrix} -2 & -1 \\ -7 & 3 \\ 2 & -1 \end{bmatrix} \quad \begin{bmatrix} -10 & -5 \\ -4 & -5 \\ -4 & 2 \end{bmatrix}$$

P, R   **23** If one of the expressions $12.47 + (-8.53), -\frac{2}{3} + \left(-\frac{5}{8}\right), -1\frac{2}{3} + 4\frac{5}{6},$ and $-2.5 + 3.95$ is chosen at random, what is the probability that the value of the expression is greater than 2? $\frac{1}{2}$

Section 6.3   Adding Signed Numbers

Additional Answers
for pages 269, 270, and 271

1 No, the dimensions of the matrices must be the same.

2 Every addition starts with a zero sum before the numbers are added to each other.

4 The + and – symbols outside the parentheses represent the operations of addition and subtraction. The + and – symbols inside the parentheses represent the signs of the numbers, positive or negative.

5 Possible answer: The sum of two positive numbers is positive; the sum of two negative numbers is negative.

6 a $-1 + 4 = 3$
   b $3 + (-7) = -4$
   c $-3 + (-4) = -7$

8 a
Food	-25
Housing	0
Utilities	10
Auto	-90
Entertainment	65
Miscellaneous	50

Shows overspent or underspent by type.

b Since the sum of the items in the matrix is +10, they met their budget goal.

19 c $\frac{1}{4}$

20 a $\begin{bmatrix} 12 & 7.8 \\ 0 & -20 \end{bmatrix}$

b $\begin{bmatrix} -4 & 2 \\ -10.5 & 1.1 \\ -2.7 & 0 \end{bmatrix}$

21 Adding a negative number gives the same result as subtracting the opposite of the number.

SECTION 6.3

Problem-Set Notes

for pages 272 and 273

- **26** should be followed up with some discussion after students have written their explanations.

Checkpoint

Find each sum.
1. $-18 + 24$
2. $2\frac{5}{8} + (-3\frac{1}{2})$
3. $18 + (-3) + (-6)$
4. $21.9 + (-8.6) + 5.7$
5. $-4.3 + (-1.2)$
6. $-4\frac{1}{3} + 2\frac{1}{9}$
7. $\begin{bmatrix} 2 & -1 \\ 4 & 7 \end{bmatrix} + \begin{bmatrix} 3 & 5 \\ -6 & -11 \end{bmatrix}$
8. $\begin{bmatrix} -1 & 5 & 7 \\ -2 & 4 & -8 \end{bmatrix} + \begin{bmatrix} -5 & -1 & 2 \\ -4 & 7 & 9 \end{bmatrix}$

Answers:
1. 6 2. $-\frac{7}{8}$ 3. 9
4. 19 5. -5.5 6. $-2\frac{2}{9}$
7. $\begin{bmatrix} 5 & 4 \\ -2 & -4 \end{bmatrix}$ 8. $\begin{bmatrix} -6 & 4 & 9 \\ -6 & 11 & 1 \end{bmatrix}$

Additional Answers

for pages 272 and 273

25 a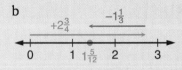

b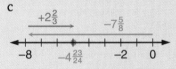

c

$+2\frac{2}{3}$ $-7\frac{5}{8}$

-8 $-4\frac{23}{24}$ -2 0

26 b To add -4215 to 2363 on a number line, you move to the left.

27 No change. $-45 - 25 + 40 + 30 = 0$

P, R **24 Spreadsheets** The spreadsheet display shows some of the formulas hidden in the cells. The pattern of these formulas is continued in the rows below.

	A	B	C
1	–4	18	–20
2	A1 + 1	B1 + (–5)	C1 + (–2.5)
3	A2 + 1	B2 + (–5)	C2 + (–2.5)

a What formula is in cell A5? In cell B5? In cell C5? A4 + 1; B4 + (–5); C4 + (–2.5)
b What number will appear in cell A5? In cell B5? In cell C5? 0; –2; –30

P **25** Perform the indicated operation. Then graph each sum or difference on a number line.
a $\frac{7}{3} + \frac{8}{21}$ $\frac{57}{21}$, or ≈ 2.7 **b** $2\frac{3}{4} - 1\frac{1}{3}$ $\frac{17}{12}$, or ≈ 1.4 **c** $-7\frac{5}{8} + 2\frac{2}{3}$ $\frac{-119}{24}$, or ≈ -4.96

P **26 Communicating**
a Find the sum.
$$\begin{array}{r} -4215 \\ +\;\;2363 \\ \hline \end{array} \quad -1852$$
b Explain why you actually *subtract* to find the sum of the numbers in part **a**.
c Find the sum.
$$\begin{array}{r} 2363 \\ +\;-4215 \\ \hline \end{array} \quad -1852$$

P **27 Communicating** Joey owed Mandy 45 cents. He borrowed another quarter from her and later paid her back four dimes. The next day he forgot what he owed Mandy and gave her six nickels. How much change should Mandy give him back? Explain your answer.

P, R **28** Find the sum of the matrices.

$$\begin{bmatrix} -\frac{3}{8} & \frac{7}{3} \\ \frac{5}{12} & \frac{-14}{15} \end{bmatrix} + \begin{bmatrix} \frac{2}{5} & \frac{-4}{3} \\ \frac{2}{3} & \frac{-5}{21} \end{bmatrix} \quad \begin{bmatrix} \frac{1}{40} & 1 \\ 1\frac{1}{12} & -1\frac{6}{35} \end{bmatrix}$$

P, R **29 Banking** The balance of Tom's checking account was $140.00. Then he wrote some checks.
a Create a third column that shows the balance of the checking account after each check was written. You may want to use a spreadsheet.
b Make a bar graph to display the data.

Check	Amount
105	$74.47
106	32.53
107	24.18
108	16.42
109	5.49
110	19.25

◀ LOOKING BACK **Spiral Learning** LOOKING AHEAD ▶

P, R, I **30** Solve each equation for x.
a $x + 98.3 = 0$ -98.3
b $-15.27 + x = 0$ 15.27
c $93.16 + x = 0$ -93.16
d $x + (-113.96) = 0$ 113.96

272   **Chapter 6** Signed Numbers

SECTION 6.3

P, I **31** Find all numbers on the number line that are 8.4 units away from 3.1.   11.5, −5.3

R **32 a** Solve for h. ≈5 ft 11 in.
   b Solve for h when the base of the ladder is 2 feet from the wall. ≈5 ft 8 in.

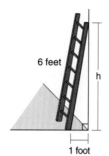

R, I **33** Matrix addition is needed to change the Start triangle into the Finish triangle. What matrix should be added to the ordered pairs (x, y) of the Start matrix to give the ordered pairs of the Finish matrix?

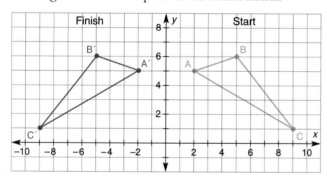

$$\text{Start} \quad + \quad ? \quad = \quad \text{Finish}$$

$$\begin{array}{c} \\ \mathbf{A} \\ \mathbf{B} \\ \mathbf{C} \end{array} \begin{bmatrix} x & y \\ 2 & 5 \\ 5 & 6 \\ 9 & 1 \end{bmatrix} + \begin{bmatrix} -4 & 0 \\ -10 & 0 \\ -18 & 0 \end{bmatrix} = \begin{array}{c} \\ \mathbf{A'} \\ \mathbf{B'} \\ \mathbf{C'} \end{array} \begin{bmatrix} x & y \\ -2 & 5 \\ -5 & 6 \\ -9 & 1 \end{bmatrix}$$

R **34** The table contains time and temperature data for Denver, Colorado.
   a Display the data in a bar graph.
   b What was the greatest temperature change in a two-hour period?
   A drop of 16°

Time	Temperature (°F)
6 A.M.	−12
8	−4
10	−2
12 P.M.	7
2	12
4	18
6	12
8	−4
10	−6
12 A.M.	−9

R **35 Communicating** Is the number line drawn correctly? Explain your answer.

```
←┼┼┼┼┼┼┼┼┼┼┼┼→
 −1−2−3−4−5−6 0 1 2 3 4 5 6
```

R **36** Graph each inequality.
   a $w \geq -2$ **b** $w < -2$ **c** $-2 < w$ **d** $-2 \geq w$

---

29 a Balance
$ 65.53
33.00
8.82
−7.60
−13.09
−32.34

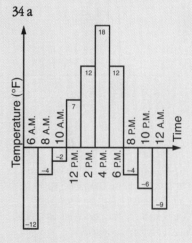

b

34 a

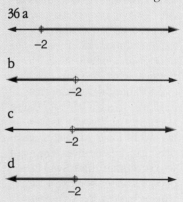

35 No; the negative numbers should increase from left to right.

36 a
```
←————•————→
    −2
```
b
```
←————◦————→
    −2
```
c
```
←————◦————→
    −2
```
d
```
←————•————→
    −2
```

Section 6.3 Adding Signed Numbers

SECTION 6.3

Additional Answers

for page 274

38 e The quotient of two numbers with the same sign is positive. The quotient of two numbers with different signs is negative.

41 a

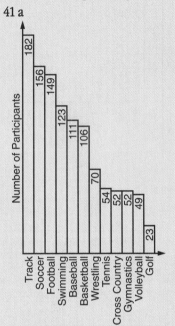

b Some athletes may have participated in more than one sport. A circle graph may be used only when each item belongs to exactly one category.

Investigation In financial statements, minus signs are often not used. Instead, negative numbers are shown in parentheses. From the standpoint of a bank depositor, the terms *credit, receipt, deposit,* and *interest* are usually positive, while *debit, payment,* and *withdrawal* can be seen as negative. To a banker, profits and assets are usually seen as positive, while losses, liabilities, and overhead are negatives. For many words, though, the answer depends on the context. To a banker, interest received from borrowers is positive, while interest paid to depositors is negative.

R **37** Use the Pythagorean Theorem to find the length of the third side of each triangle.

a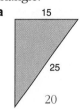
 15, 25, 20

b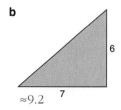
 6, 7, ≈9.2

c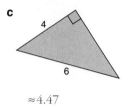
 4, 6, ≈4.47

I **38** Use a calculator to find each quotient.

a $\frac{-42}{-7}$ 6   **b** $\frac{-42}{-21}$ 2   **c** $\frac{42}{-7}$ -6   **d** $\frac{42}{21}$ 2

e What do you think is the rule for determining the sign of the quotient of signed numbers?

I **39** Use your calculator to perform the indicated operations in parts **a–f**.

a $-7 + (-7) + (-7) + (-7)$ -28
b $(-4) + (-4) + (-4) + (-4) + (-4) + (-4)$ -24
c $2 + 2 + 2 + 2 + 2 + 2 + 2 + 2 + 2 + 2$ 20
d $(-7)(4)$ -28
e $(-4)6$ -24
f $(2)(10)$ 20

R, I **40** If a number, x, is chosen at random so that $-5 \leq x \leq 6$, what is the probability that $-2 < x < 1$? $\frac{3}{11}$

R **41 Communicating** The table indicates the sports in which Antelope High School students participate and the number of participants in each sport.
a Make a bar graph to display the data.
b Why is a circle graph not an appropriate way to display the data?

Sport	Number of Participants
Track	182
Soccer	156
Football	149
Swimming	123
Baseball	111
Basketball	106
Wrestling	70
Tennis	54
Cross country	52
Gymnastics	52
Volleyball	49
Golf	23

Investigation

Banking How are signed numbers used in fields such as accounting, banking, and investing? Find out the meaning and use of the terms *credit, debit, payment, receipt, deposit, withdrawal, profit, loss, asset, liability, loan, interest,* and *overhead*. Which of these terms could be described as "positive," and which could be described as "negative"?

274   Chapter 6   Signed Numbers

SECTION 6.4

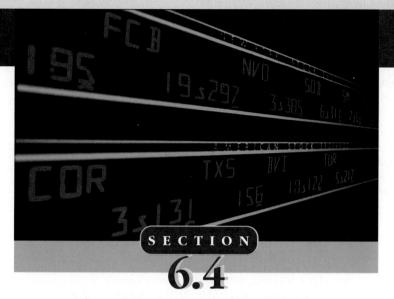

Investigation Connection
Section 6.4 supports Unit C, *Investigations* 3, 4, 5. (See pp. 198A-198D in this book.)

SECTION 6.4
Multiplying and Dividing Signed Numbers

MULTIPLICATION OF SIGNED NUMBERS

The patterns shown below will help us discover the rules for multiplying signed numbers.

$3 \cdot 4 = 12$	$3 \cdot (-4) = -12$
$3 \cdot 3 = 9$	$2 \cdot (-4) = -8$
$3 \cdot 2 = 6$	$1 \cdot (-4) = -4$
$3 \cdot 1 = 3$	$0 \cdot (-4) = 0$
$3 \cdot 0 = 0$	$-1 \cdot (-4) = 4$
$3 \cdot (-1) = -3$	$-2 \cdot (-4) = 8$
$3 \cdot (-2) = -6$	$-3 \cdot (-4) = 12$
$3 \cdot (-3) = -9$	$-4 \cdot (-4) = 16$
$3 \cdot (-4) = -12$	$-5 \cdot (-4) = 20$

The patterns suggest three rules for multiplying signed numbers.

1. The product of two positive numbers is positive.
2. The product of a positive number and a negative number is negative.
3. The product of two negative numbers is positive.

Example 1
Find each product.
a $4 \cdot (-7)$ **b** $-5 \cdot 3$ **c** $-7 \cdot (-3)$

Solution
a $4 \cdot (-7) = -28$ **b** $-5 \cdot 3 = -15$ **c** $-7 \cdot (-3) = 21$

Section 6.4 Multiplying and Dividing Signed Numbers

Objectives
- To multiply signed numbers
- To multiply matrices by scalars
- To calculate powers of signed numbers
- To divide signed numbers

Vocabulary: scalar, scalar multiplication

Class Planning

Schedule 2–3 days
Resources
In *Chapter 6 Student Activity Masters* Book
Class Openers 37, 38
Extra Practice 39, 40
Cooperative Learning 41, 42
Enrichment 43
Learning Log 44
Interdisciplinary Investigation 45, 46
Spanish-Language Summary 47

Transparency 24

Class Opener

Evaluate each expression.
1. $5 + 5 + 5 + 5$
2. $(-3) + (-3) + (-3) + (-3) + (-3) + (-3)$
3. $(-2) + (-2) + (-2) + (-2) + (-2)$
4. $8 + 8 + 8 + 8 + 8 + 8 + 8$
5. $(-7) + (-7) + (-7)$

Answers:
1. 20 2. –18 3. –10 4. 56 5. –21

Lesson Notes

- The purpose of the patterns of products shown on page 275 is to give students a chance to develop on their own the rules for multiplying signed numbers.

- Scalar multiplication should be introduced via an application problem, such as sample problem 2. This problem can be done

275

SECTION 6.4

Lesson Notes continued

in one step by using 1.23 as the scalar rather than using 0.23 as the scalar and adding the result to the original matrix. Be sure to discuss this alternative method.

- Be sure that students know how to evaluate the two expressions in example 4 by hand as well as on a calculator.

Additional Examples

Example 1
Find each product.
a. $-6 \cdot 2$ b. $-4 \cdot (-3)$ c. $8 \cdot (-1)$

Answers: a. -12 b. 12 c. -8

Example 2
Solve the equation $5x = -35$ for x.

Answer: $x = -7$

Example 3
Multiply $-2 \cdot \begin{bmatrix} 6 & -1 & -3 \\ 2 & -2 & 4 \end{bmatrix}$.

Answer: $\begin{bmatrix} -12 & 2 & 6 \\ -4 & 4 & -8 \end{bmatrix}$

Example 4
Evaluate each expression.
a. $(-4)^6$ b. -4^6

Answers: a. 4096 b. -4096

Example 5
Find each quotient.
a. $-6.9 \div 3$ b. $\frac{-15}{-5}$

Answers: a. -2.3 b. 3

Communicating Mathematics

- Have students write a paragraph explaining the difference between the terms *reciprocal* and *opposite*.
- Problems 3, 5–9, and 14 encourage communicating mathematics.

Example 2
Solve the equation $-8x = 40$ for x.

Solution
We can experiment with our calculator. First we try 5 and find that $-8(5) = -40$. It then seems logical to try -5. Since $-8(-5) = 40$, the solution of the equation is -5. On a calculator, 8 [+/−] [×] 5 [+/−] [=] results in a display of 40 . Therefore, $x = -5$.

SCALAR MULTIPLICATION

To multiply a matrix by a number, we multiply each element of the matrix by the number. This is called **scalar multiplication,** and the number we multiply by is called a **scalar.**

Example 3
Multiply $-3 \begin{bmatrix} 5 & -1 & 2 \\ -2 & 3 & 6 \end{bmatrix}$.

Solution
$$-3 \begin{bmatrix} 5 & -1 & 2 \\ -2 & 3 & 6 \end{bmatrix} = \begin{bmatrix} (-3)(5) & (-3)(-1) & (-3)(2) \\ (-3)(-2) & (-3)(3) & (-3)(6) \end{bmatrix} = \begin{bmatrix} -15 & 3 & -6 \\ 6 & -9 & -18 \end{bmatrix}$$

POWERS OF SIGNED NUMBERS

We apply the rules for multiplying signed numbers when we evaluate powers of signed numbers.

Example 4
Evaluate each expression.
a $(-3)^4$ b -3^4

Solution
a In the expression $(-3)^4$, the base is -3 and the exponent is 4. Therefore, -3 is used as a factor four times.

$$(-3)^4 = (-3)(-3)(-3)(-3) = 9 \cdot 9 = 81$$

b In the expression -3^4, the base is 3 and the exponent is 4. The value of this expression is the opposite of the number we get when we use 3 as a factor four times.

$$-3^4 = -(3)(3)(3)(3) = -81$$

Chapter 6   Signed Numbers

SECTION 6.4

DIVISION OF SIGNED NUMBERS

Dividing by a number gives the same result as multiplying by the reciprocal of the number. Keep this in mind as you study the next example.

Example 5
Find each quotient.
a $-4.5 \div 2$
b $\frac{-12}{-8}$

Solution
a $-4.5 \div 2 = -4.5 \cdot \frac{1}{2} = -4.5 \cdot 0.5 = -2.25$

b $\frac{-12}{-8} = -12 \cdot \left(\frac{1}{-8}\right) = -12 \cdot (-0.125) = 1.5$

The equation in Example 2 on the previous page, $-8x = 40$, can also be solved by using division, the inverse operation of multiplication. For example, if $-8x = 40$, then $x = 40 \div (-8)$.
Using a calculator, we get the same result.

$$40 \;\boxed{\div}\; 8 \;\boxed{+/-}\; \boxed{=} \qquad -5$$

Since dividing has the same result as multiplying by a reciprocal, the rules for dividing are the same as those for multiplying.

1. The quotient of two positive numbers is positive.
2. The quotient of a positive and a negative number is negative.
3. The quotient of two negative numbers is positive.

Sample Problems

Problem 1 Wally Street has 157 shares of Zenox stock. The stock's value was $32.75 per share on Tuesday morning. At the close of Tuesday's trading, the stock was down $1.50 per share.
a How much money did Wally lose on his Zenox stock on Tuesday?
b By what percent did the stock decrease in value?

Solution **a** We indicate "down $1.50" by -1.5 and multiply to find Wally's total loss on the stock. Since $(-1.5) \cdot 157 = -235.50$, Wally's loss was $235.50.

b Method 1
From part **a** we know that the total loss was $235.50 out of a total value of 157($32.75), or $5141.75. Therefore, the percent of change was

$$\frac{-235.50}{5141.75} \approx -0.046, \text{ or } \approx -4.6\%$$

Therefore, the value of the stock decreased 4.6%.

Checkpoint

Evaluate.
1. $-4(7)$ 2. $5(-9)$ 3. $-5(-6)$
4. $\frac{-56}{-8}$ 5. $\frac{28}{-4}$ 6. $\frac{-36}{6}$
7. $(-4)^3$ 8. -6^2

Find the product by using scalar multiplication.

9. $3 \begin{bmatrix} 1 & -1 & 9 \\ 2 & 4 & -6 \end{bmatrix}$

10. $-2 \begin{bmatrix} 7 & -1 & 5 & 9 \\ 1 & -3 & 2 & -4 \end{bmatrix}$

Answers:
1. -28 2. -45 3. 30
4. 7 5. -7 6. -6
7. -64 8. -36
9. $\begin{bmatrix} 3 & -3 & 27 \\ 6 & 12 & -18 \end{bmatrix}$

10. $\begin{bmatrix} -14 & 2 & -10 & -18 \\ -2 & 6 & -4 & 8 \end{bmatrix}$

Reteaching

You might use the idea that multiplication is repeated addition to help students see that a positive number times a negative number has to be negative. An example such as $4 \cdot (-3)$ can be thought of as $(-3) + (-3) + (-3) + (-3)$. Any repeated movements to the left of 0 on a number line must always result in a negative number.

Stumbling Block

Students may confuse -3^4 and $(-3)^4$. Remind students that unless parentheses have been included, an exponent applies only to the number base, not the negative sign.

SECTION 6.4

Assignment Guide

Basic
- Day 1 Section 6.3: 13, 21, 28, 36
         Section 6.4: 9, 11, 12, 18, 21
- Day 2 10, 14, 17, 19, 20, 23, 24, 27
- Day 3 13, 16, 22, 25, 26, 28

Average or Heterogeneous
- Day 1 9–12, 14, 15, 18, 19a, 20, 21, 23 (a, c), 27
- Day 2 13, 16, 17, 19b, 22, 23 (b, d), 24–26, 28

Honors
- Day 1 11, 12, 15, 16, 21, 23 (a–d), 24, 25
- Day 2 10, 13, 14, 17–20, 22, 23 (e–h), 26–28, Investigation

Strands

Communicating Mathematics	3, 5–9, 14
Geometry	17, 24, 27
Measurement	17, 27
Data Analysis/ Statistics	14, 15, 22
Technology	22
Probability	18, 25
Algebra	1, 2, 4–28, Investigation
Patterns & Functions	9, 14
Number Sense	2, 5, 8, 10, 12, 14
Interdisciplinary Applications	15, 16, 21, 28, Investigation

Individualizing Instruction

Limited English proficiency In problems 6 and 7 students may need help with the meaning of the term *reciprocal*. They may benefit from answering problem 7 aloud before attempting to write an answer.

Method 2
The percent of change for all 157 shares of stock is the same as the percent of change for a single share of the stock. Therefore, we can divide the loss per share by the value of one share of the stock.

$$\frac{-1.5}{32.75} \approx -0.046, \text{ or } \approx -4.6\%$$

Problem 2 The matrix shows the average number of miles per gallon of gasoline for three Major Motors cars. Included are figures for both city and highway driving.

Miles per Gallon

	City	Highway
Guzzle Mobile	14	17
Family Auto	24	30
Mini Mite	35	47

If an engineering breakthrough raises mileages on all three cars for both types of driving by 23%, what will each car's new mileage figures be?

Solution To determine the result of the 23% increase, we first use scalar multiplication to find 23% of each number in the matrix, then add these results to the original numbers in the matrix.

$$0.23 \cdot \begin{bmatrix} 14 & 17 \\ 24 & 30 \\ 35 & 47 \end{bmatrix} + \begin{bmatrix} 14 & 17 \\ 24 & 30 \\ 35 & 47 \end{bmatrix} = \begin{bmatrix} 0.23(14) & 0.23(17) \\ 0.23(24) & 0.23(30) \\ 0.23(35) & 0.23(47) \end{bmatrix} + \begin{bmatrix} 14 & 17 \\ 24 & 30 \\ 35 & 47 \end{bmatrix}$$

$$\approx \begin{bmatrix} 3.22 & 3.91 \\ 5.52 & 6.90 \\ 8.05 & 10.8 \end{bmatrix} + \begin{bmatrix} 14 & 17 \\ 24 & 30 \\ 35 & 47 \end{bmatrix}$$

$$= \begin{bmatrix} 17.22 & 20.91 \\ 29.52 & 36.9 \\ 43.05 & 57.8 \end{bmatrix} \approx \begin{bmatrix} 17 & 21 \\ 30 & 37 \\ 43 & 58 \end{bmatrix}$$

Think and Discuss

P **1** Find each product.
   a –3(6) –18   **b** 3(–6) –18   **c** 6(–3) –18   **d** –6(3) –18

P **2** Condense the three rules for multiplying signed numbers into two rules—one for multiplying numbers with the same sign and one for multiplying numbers with different signs.

P **3 a** Write a definition of the term *scalar*.
      b Why do you think the term *scalar* is used?

SECTION 6.4

P **4** Find the products by using scalar multiplication.

a $-2 \cdot \begin{bmatrix} 2 & -4 \\ -6 & 8 \end{bmatrix}$ $\begin{bmatrix} -4 & 8 \\ 12 & -16 \end{bmatrix}$ **b** $\frac{1}{2} \cdot \begin{bmatrix} -2 & -4 \\ 6 & 8 \end{bmatrix}$ $\begin{bmatrix} -1 & -2 \\ 3 & 4 \end{bmatrix}$

P, I **5 a** Evaluate $(-2)^2$. 4 **b** Evaluate -2^2. -4
c Explain why the answers in parts **a** and **b** have different signs.
d Evaluate $(-2)^3$. -8 **e** Evaluate -2^3. -8
f Explain why the answers in parts **d** and **e** have the same sign.

P, R, I **6** What is a reciprocal? Find the reciprocals of these numbers.

a 2 $\frac{1}{2}$ **b** $-\frac{1}{3}$ -3 **c** $\frac{3}{4}$ $\frac{4}{3}$ **d** -2.5 $-\frac{2}{5}$

R **7** What is the relationship between reciprocals and division?

P, I **8** Find each quotient, then explain why $-\frac{a}{b} = \frac{-a}{b} = \frac{a}{-b}$.

a $-\frac{20}{5}$ -4 **b** $\frac{-20}{5}$ -4 **c** $\frac{20}{-5}$ -4

Problems and Applications

P **9** **Communicating** Find each product. Then graph the products on a number line. What pattern do you see in the graph?

a $(-5)4$ **b** $(-5)3$ **c** $(-5)2$ **d** $(-5)1$ **e** $(-5)0$
f $(-5)(-1)$ **g** $(-5)(-2)$ **h** $(-5)(-3)$ **i** $(-5)(-4)$

P, R **10** Perform each operation. Arrange the answers in order, from least to greatest.

a $(-5)(-7)$ 35 **b** $\frac{-50}{-5}$ 10 **c** $\frac{-50}{5}$ -10 **d** $(-5)(7)$ -35
e $\frac{20}{-4}$ -5 **f** $(9)(-5)$ -45 **g** $3(4)$ 12

P, R, I **11** Solve each equation for x.

a $-4x = 32$ -8 **b** $-4x = -32$ 8 **c** $4x = 32$ 8 **d** $4x = -32$ -8

P **12** Is the statement true *always, sometimes,* or *never?*
a A negative number added to a negative number is a positive number. Never
b A negative number added to a positive number is a positive number. Sometimes
c A negative number multiplied by a negative number is a positive number. Always
d A negative number divided by a negative number is a positive number. Always
e A negative number added to a positive number is a negative number. Sometimes

P, I **13** Evaluate each expression.   **a** $\frac{(-5)^2(-3) + 4(-3)}{(-1)(-3)}$ -29 **b** $\frac{-12^2}{-24 + (-2)(-4)}$ 9

P, I **14** **Communicating** Study the pattern in column A. Then determine which powers in column B will be negative. Explain your answer.

Column A	Column B
$(-2)^2 = 4$	$(-2)^{77}$
$(-2)^3 = -8$	$(-2)^{88}$
$(-2)^4 = 16$	$(-2)^{36}$
$(-2)^5 = -32$	$(-2)^{100}$

Section 6.4 Multiplying and Dividing Signed Numbers

Problem-Set Notes

for pages 278 and 279

- **12** reinforces the rules for signed numbers with all operations except subtraction.

Additional Answers

for pages 278 and 279

2 The product of two numbers with the same sign is positive. The product of two numbers with different signs is negative.

3 a A scalar is a number by which a matrix is multiplied.
b Because a scalar changes the "scale" of a matrix by increasing or decreasing each entry by the same factor

5 c In part **a**, -2 is squared. In part **b**, only the 2 is.
f Because the cube of -2 and the opposite of the cube of 2 are both negative

6 The reciprocal of a number is a number such that when the number and the reciprocal are multiplied, the result is 1.

7 Division by a number is equivalent to multiplication by the number's reciprocal.

8 Possible answer: All three result in an equivalent quotient.

9 a -20 **b** -15 **c** -10 **d** -5
e 0 **f** 5 **g** 10 **h** 15 **i** 20

$-20 \quad -10 \quad 0 \quad 10 \quad 20$

Possible answer: The products increase in increments of 5.

10 Answers arranged in order, from least to greatest: **f** -45, **d** -35, **c** -10, **e** -5, **b** 10, **g** 12, **a** 35

14 Only $(-2)^{77}$ is negative; possible explanation: When a negative number is raised to an even power, the result is positive; when it is raised to an odd power, the result is negative.

SECTION 6.4

Cooperative Learning

■ Using the coordinate matrix given in problem 24, have students determine the vertices of the new triangles for scalars of 3, 5, and ½. Have students draw the resulting triangles on the same coordinate grid, determine the area of each triangle, and write a rule relating the change in area and to the scalar.

Answers:

Scalar	Area	Coordinate Matrix
3	144	$\begin{bmatrix} -6 & -6 & 18 \\ 15 & 3 & 3 \end{bmatrix}$
5	400	$\begin{bmatrix} -10 & -10 & 30 \\ 25 & 5 & 5 \end{bmatrix}$
½	4	$\begin{bmatrix} -1 & -1 & 3 \\ 2\tfrac{1}{2} & \tfrac{1}{2} & \tfrac{1}{2} \end{bmatrix}$

The area changes by a factor equal to the square of the scalar.

■ Problems 12 and 18 are appropriate for cooperative learning.

Additional Practice

Evaluate.
1. $-9(-7)$ 2. $4(-3)$ 3. $\tfrac{-18}{6}$
4. $\tfrac{-24}{-4}$ 5. $-7(-3)$ 6. $\tfrac{27}{-9}$
7. -2^4 8. $(-1)^{78}$

Perform each scalar multiplication.

9. $-3 \begin{bmatrix} -1 & 2 \\ -5 & 7 \\ 1 & 4 \end{bmatrix}$

10. $2.5 \begin{bmatrix} 2 & -4 & 1 \\ 3 & -7 & -6 \end{bmatrix}$

Answers: 1. 63 2. –12 3. –3
4. 6 5. 21 6. –3 7. –16 8. 1
9. $\begin{bmatrix} 3 & -6 \\ 15 & -21 \\ -3 & -12 \end{bmatrix}$ 10. $\begin{bmatrix} 5 & -10 & 2.5 \\ 7.5 & -17.5 & -15 \end{bmatrix}$

P, R 15 Four employees at Joe's Burger Barn will receive a 7% raise. The matrix shows their current wages.
  a Write a matrix problem that can be used to determine the new hourly wage for each employee.
  b Compute the new salary for each of the four employees.

$$\begin{array}{r} \text{Current} \\ \text{Hourly Wage} \end{array}$$
$$\begin{array}{r} \text{Raul} \\ \text{Pablo} \\ \text{Sue} \\ \text{Melissa} \end{array} \begin{bmatrix} 4.95 \\ 5.65 \\ 4.85 \\ 5.75 \end{bmatrix}$$

R 16 **Business** Jan has 47 shares of Libom Oil stock. One share was worth $37.75 on Tuesday and $36.85 on Wednesday.
  a How much did she lose on a single share of the stock? $0.90
  b What was the percent of loss per share? ≈ 2.4%
  c How much did she lose on all 47 shares? $42.30
  d What was the percent of loss on all 47 shares? ≈ 2.4%

P, R, I 17 The area of the rectangle is 320. Solve for x. -4

P, R 18 Five wooden squares are numbered as indicated.

If two of the squares are chosen at random, what is the probability that
  a Their product is negative? .6   b Their product is positive? .4
  c Their sum is negative? .3

P 19 Perform each scalar multiplication.

  a $-2 \begin{bmatrix} -3 & 4 \\ -9 & 18 \\ 6 & -2 \end{bmatrix}$ $\begin{bmatrix} 6 & -8 \\ 18 & -36 \\ -12 & 4 \end{bmatrix}$ b $-3.5 \begin{bmatrix} -1.2 & -6 \\ 3.4 & -9 \\ 11.1 & 0 \end{bmatrix}$ $\begin{bmatrix} 4.2 & 21 \\ -11.9 & 31.5 \\ -38.85 & 0 \end{bmatrix}$

P, R, I 20 Solve each equation.
  a $-6x = 30$ -5 b $-7x = -49$ 7   c $-4x + 3 = 31$ -7 d $12 - 18 = -2x$ 3

P, R 21 In 1990 the dollar values of five homes were $175,000, $98,000, $112,000, $145,000, and $205,000. In 1991 the value of each home decreased 14%.
  a Write a scalar matrix multiplication problem that represents the values of the homes at the end of 1991.
  b Solve the problem you wrote in part **a**.

P, R 22 **Spreadsheets** The students' scores in Mrs. Quadratic's algebra class are 97, 84, 69, 42, 79, 84, 86, 77, 62, 71, and 55. If possible, use a spreadsheet to answer these questions.
  a What are the new scores for all the students if Mrs. Quadratic increases each student's score by 5%? By 10%? By 12%? By 15%?
  b What percent of increase is necessary to have no scores below 60? What would be the highest score after that increase? ≈43%; 139

SECTION 6.4

◀ LOOKING BACK **Spiral Learning** LOOKING AHEAD ▶

R, I **23** Graph each inequality.
  a $n > 3$ **b** $n \geq 3$ **c** $n < 3$ **d** $n \leq 3$
  e $3 > n$ **f** $3 \leq n$ **g** $3 \geq n$ **h** $3 < n$

P, I **24** The x-coordinates, and the y-coordinates, of vertices A, B, and C of the triangle are arranged in a matrix.

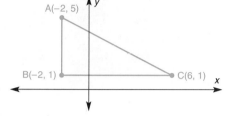

  a If this matrix is multiplied by 2, what is the new matrix?
  b Determine the vertices of the triangle represented by the matrix in part **a**, and draw the triangle.
  c Compare the area of the original triangle with the area of the new triangle.

P, R **25** If one of the following expressions is selected at random, what is the probability that the expression has a negative answer? $\frac{4}{6} = \frac{2}{3}$
  a $-1732 + 1492$ **b** $(-81)(759)$ **c** $\frac{-1001}{-13}$
  d $-8118 + 1881$ **e** $-4224 + (-8448)$ **f** $-37 \cdot [847 + (-847)]$

R **26** Find the sum of the matrices.
$$\begin{bmatrix} 7 & 12 & 15 \\ 8 & 10 & 19 \\ 6 & 14 & 8 \end{bmatrix} + \begin{bmatrix} -8 & -13 & -16 \\ -7 & -9 & -18 \\ -7 & -13 & -7 \end{bmatrix} \begin{bmatrix} -1 & -1 & -1 \\ 1 & 1 & 1 \\ -1 & 1 & 1 \end{bmatrix}$$

R **27** A rectangular field with dimensions of 30 meters by 50 meters is to be surrounded by a fence. Posts are placed every 2 meters.
  a How many posts are needed? 80 posts
  b How many posts are needed if one 50-meter side is not fenced? 56 posts
  c How many posts are needed if one 30-meter side is not fenced? 66 posts

R **28** **Consumer Math** If a hamburger costs $2.00, an order of fries costs $0.75, and a soft drink costs $0.60, how much is each person's bill?

	Hamburgers	Fries	Drinks
Alf	4	3	5
Betty	2	4	2
Grandma	1	3	5
Del	3	2	4

Investigation

Physics Find out the meaning of the term *half-life* as it applies to nuclear science. Draw a graph to help you explain the term to the class.

Section 6.4 Multiplying and Dividing Signed Numbers

Problem-Set Notes

for pages 280 and 281

- **21** can be done by multiplying the matrix by 0.14 and subtracting from the original matrix, or by using 0.86 as the scalar.
- **22** demonstrates to students the power of a spreadsheet.
- **28** previews matrix multiplication, the topic of the next section.

Additional Answers

for pages 280 and 281

15 a Possible answer:

$$1.07 \begin{bmatrix} 4.95 \\ 5.65 \\ 4.85 \\ 5.75 \end{bmatrix}$$

b Raul: $5.30; Pablo: $6.05; Sue: $5.19; Melissa: $6.15

21 a
$$0.86 \begin{bmatrix} 175,000 \\ 98,000 \\ 112,000 \\ 145,000 \\ 205,000 \end{bmatrix}$$

b
$$\begin{bmatrix} 150,500 \\ 84,280 \\ 96,320 \\ 124,700 \\ 176,300 \end{bmatrix}$$

23

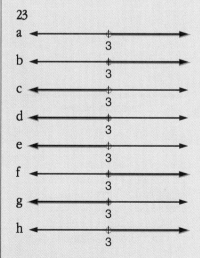

Answers for problems 22 a, 24, 28, and Investigation are on page A5.

SECTION 6.5

Objective
- To multiply two matrices

Class Planning

Schedule 2 days

Resources

In *Chapter 6 Student Activity Masters* Book
Class Openers *49, 50*
Extra Practice *51, 52*
Cooperative Learning *53, 54*
Enrichment *55*
Learning Log *56*
Technology *57, 58*
Spanish-Language Summary *59*

Transparencies 28 a–e

Class Opener

Rosa bought 8 plates and 4 bowls. Ralph bought 6 plates and 6 bowls. Plates cost $3.50 each and bowls cost $2.00 each.
1. How much did Rosa spend?
2. How much did Ralph spend?
3. How much was spent for plates?
4. How much was spent for bowls?

Answers:
1. $36.00 2. $33.00 3. $49.00
4. $20.00

Lesson Notes

- Have students discover the algorithm for matrix multiplication by working through the bills for several of the customers in example 1. Matrix multiplication should be approached via a real-world application, or else it quickly becomes a tedious exercise.

- In example 1, it is important for students to notice that the labels of the columns in the quantity matrix match the labels of the

SECTION 6.5
Matrix Multiplication

In the last section we looked at scalar multiplication—multiplying a matrix by a number. In this section you will learn how to multiply two matrices.

Example 1
Alpha, Beta, Gamma, and Delta want to buy burgers, soft drinks, and fries. The quantity matrix shows the number of items each person wants to buy.

Quantity Matrix

	Burgers	Soft Drinks	Fries
Alpha	4	4	4
Beta	6	1	1
Gamma	2	10	2
Delta	3	2	4

The price of each item at three fast-food restaurants—McDuck's, Burger Barn, and Wanda's—is shown in the price matrix.

Price Matrix

	McDuck's	Burger Barn	Wanda's
Burger	$1.69	$1.85	$1.75
Drink	0.79	0.59	0.65
Fries	0.69	0.75	0.59

a How much will each person pay at McDuck's? At Burger Barn? At Wanda's?
b Which fast-food restaurant has the best prices?

Solution
a There are 12 values to calculate because each of the 4 people could buy from any of the 3 restaurants. To calculate any of these values, we first choose a row from the quantity matrix and a column from the price matrix. Let's try the Alpha row and the Wanda's column to find out how much Alpha would pay at Wanda's.

4 burgers × $1.75 each + 4 drinks × $0.65 each + 4 fries × $0.59 each
= $7.00 + $2.60 + $2.36
= $11.96

The total cost for Alpha to purchase four burgers, four soft drinks, and four fries at Wanda's is $11.96.

282 Chapter 6 Signed Numbers

SECTION 6.5

The $11.96 it costs Alpha to buy at Wanda's is placed in the product matrix at the intersection of row 1 and column 3.

Product Matrix

	McDuck's	Burger Barn	Wanda's
Alpha			$11.96
Beta			
Gamma			
Delta			

Since each of the four people can make her purchase at any of the three fast-food restaurants, we will repeat this procedure $4 \cdot 3$, or 12, times.

Product Matrix

	McDuck's	Burger Barn	Wanda's
Alpha	$12.68	$12.76	$11.96
Beta	11.62	12.44	11.74
Gamma	12.66	11.10	11.18
Delta	9.41	9.73	8.91

b Determining the fast-food restaurant with the best prices depends on what items are ordered. Some people might say that Wanda's has the best price because it has the lowest-priced fries and the lowest combined price for the three menu items. The entries in the product matrix indicate that Alpha should go to Wanda's, Beta should go to McDuck's, Gamma should go to Burger Barn, and Delta should go to Wanda's.

Notice that in the preceding example it was important that the number of elements in a row of the first matrix matched the number of elements in a column of the second matrix. Also notice that the values in the product matrix were not found by simply multiplying corresponding elements of the two matrices.

Example 2

Multiply $\begin{bmatrix} -12 & 2.5 & -3 \\ 7 & -5 & 4.2 \end{bmatrix} \cdot \begin{bmatrix} 2 & -6 \\ 4 & 3 \\ -5 & -1 \end{bmatrix}$.

Solution

Row 1 $\begin{bmatrix} -12 & 2.5 & -3 \end{bmatrix} \cdot \begin{bmatrix} 2 \\ 4 \\ -5 \end{bmatrix}$ Column 1

$(-12)(2) + 2.5(4) + (-3)(-5) = \boxed{1}$

Row 1 $\begin{bmatrix} -12 & 2.5 & -3 \end{bmatrix} \cdot \begin{bmatrix} -6 \\ 3 \\ -1 \end{bmatrix}$ Column 2

$(-12)(-6) + 2.5(3) + (-3)(-1) = \boxed{82.5}$

Row 2 $\begin{bmatrix} 7 & -5 & 4.2 \end{bmatrix} \cdot \begin{bmatrix} 2 \\ 4 \\ -5 \end{bmatrix}$ Column 1

$7(2) + (-5)(4) + 4.2(-5) = \boxed{-27}$

Row 2 $\begin{bmatrix} 7 & -5 & 4.2 \end{bmatrix} \cdot \begin{bmatrix} -6 \\ 3 \\ -1 \end{bmatrix}$ Column 2

$7(-6) + (-5)(3) + 4.2(-1) = \boxed{-61.2}$

The product matrix is $\begin{bmatrix} 1 & 82.5 \\ -27 & -61.2 \end{bmatrix}$ Row 1 / Row 2

---

rows in the price matrix (Burgers, Soft Drinks, Fries) and that these labels do not appear in the product matrix. Encourage students to see how these units "cancel" each other.

Additional Examples

Example 1
Mary, José, and Suki want to buy some fruit. The amounts they want to buy are given in the quantity matrix, and the prices for the items at three stores are given in the price matrix.

Quantity Matrix

	Apples	Oranges	Bananas
Mary	3	5	11
José	6	1	4
Suki	2	3	8

Price Matrix

	Sav-Rite	Fruiteria	Healthmart
Apple	0.15	0.21	0.09
Orange	0.29	0.23	0.13
Banana	0.10	0.12	0.18

Show the cost of each person's produce at each store in a product matrix.

Answer:

	Sav-Rite	Fruiteria	Healthmart
Mary	3.00	3.10	2.90
José	1.59	2.01	1.39
Suki	1.97	2.07	2.01

Example 2
Multiply.

$\begin{bmatrix} 3 & -5 & 2 \\ 8 & 6 & 1 \end{bmatrix} \cdot \begin{bmatrix} 4 & 2 \\ -1 & 7 \\ -4 & 5 \end{bmatrix}$

Answer: $\begin{bmatrix} 9 & -19 \\ 22 & 63 \end{bmatrix}$

SECTION 6.5

Assignment Guide

Basic
Day 1 7–9, 15, 19–21, 26, 27, 29
Day 2 11–14, 22–25, 30, 31

Average or Heterogeneous
Day 1 7, 9, 12 (a, c), 14–16, 19–22, 24, 29
Day 2 8, 11, 12 b, 13, 17, 18, 23, 25–27, 30, 31

Honors
Day 1 8, 9, 12, 13, 16, 19–21, 25–27
Day 2 7, 10, 11, 14, 15, 17, 18, 22–24, 28–31

Strands

Communicating Mathematics	1, 2, 10, 14, 15, 19, Investigation
Geometry	25
Data Analysis/ Statistics	8, 16–19, 28
Technology	18, 19
Algebra	1–27, 29–31
Number Sense	13, 20, 24
Interdisciplinary Applications	16–19, 28, Investigation

Individualizing Instruction

The kinesthetic learner Students might more easily remember the procedure for multiplying two matrices if they practice touching an index finger to the entries of two matrices in the order in which the operations will occur. Have students explain as they point, and have them point to the proper location of each result in the product matrix.

Sample Problems

Problem 1 Suppose that $A = \begin{bmatrix} 2 & 4 \\ -5 & 3 \end{bmatrix}$ and $B = \begin{bmatrix} -6 & 12 \\ 3 & -7 \end{bmatrix}$. Find $A \cdot B$ and $B \cdot A$.

Solution

$$A \cdot B = \begin{bmatrix} 2 & 4 \\ -5 & 3 \end{bmatrix} \cdot \begin{bmatrix} -6 & 12 \\ 3 & -7 \end{bmatrix} = \begin{bmatrix} 0 & -4 \\ 39 & -81 \end{bmatrix}$$

$$B \cdot A = \begin{bmatrix} -6 & 12 \\ 3 & -7 \end{bmatrix} \cdot \begin{bmatrix} 2 & 4 \\ -5 & 3 \end{bmatrix} = \begin{bmatrix} -72 & 12 \\ 41 & -9 \end{bmatrix}$$

By looking at these results, we can see that the order in which the matrices are multiplied is important. Sometimes matrices that can be multiplied in one order cannot even be multiplied when their order is reversed. You will see an example of such matrices in the Think and Discuss problems of this section.

Problem 2 The shareholders' matrix gives the number of shares of three stocks owned by four stockholders.

Shareholders' Matrix

		Juan	Ali	Samir	Rachel
	ADE	15	5	9	7
Stock	CAR	12	13	4	34
	ING	9	24	7	5

On Tuesday, stock ADE lost $5.00 per share, stock CAR gained $2.00 per share, and stock ING lost $2.50 per share.

a Write a matrix multiplication problem to find the total losses or gains for each person's portfolio.

b Evaluate the matrix product you wrote in part **a**. How much did each person gain or lose on his or her stocks?

Solution **a** The dollar gain or loss for each type of stock is displayed in the 1 × 3 gain/loss matrix. The product of the gain/loss matrix and the shareholders' matrix will be a matrix that displays the total gain or loss in each person's portfolio.

Gain/Loss Matrix

ADE	CAR	ING
$\begin{bmatrix} -5$	2	$-2.5 \end{bmatrix}$

Shareholders' Matrix

Juan	Ali	Samir	Rachel
15	5	9	7
12	13	4	34
9	24	7	5

SECTION 6.5

b To evaluate the product matrix we wrote in part **a**, we multiply each column of the second matrix by the single row of the first matrix.

Juan: $-5(15) + 2(12) + (-2.5)(9) = -73.5$
Ali: $-5(5) + 2(13) + (-2.5)(24) = -59$
Samir: $-5(9) + 2(4) + (-2.5)(7) = -54.5$
Rachel: $-5(7) + 2(34) + (-2.5)(5) = 20.5$

Shareholders' Gain/Loss Matrix

Juan	Ali	Samir	Rachel
$[-73.5$	-59	-54.5	$20.5]$

Note that everyone but Rachel lost money on Tuesday.

Think and Discuss

P **1** Find the product in parts **a** and **b**.

  a $\begin{bmatrix} 3 & -1 \\ 2 & -4 \end{bmatrix} \cdot \begin{bmatrix} -2 & 3 \\ 1 & -2 \end{bmatrix}$ $\begin{bmatrix} -7 & 11 \\ -8 & 14 \end{bmatrix}$

  b $\begin{bmatrix} -2 & 3 \\ 1 & -2 \end{bmatrix} \cdot \begin{bmatrix} 3 & -1 \\ 2 & -4 \end{bmatrix}$ $\begin{bmatrix} 0 & -10 \\ -1 & 7 \end{bmatrix}$

  c What conclusion can you draw about matrix multiplication from the results in parts **a** and **b**?

P **2** Find the product in parts **a** and **b**.

  a $\begin{bmatrix} 2 & 1 \\ 1 & -1 \end{bmatrix} \cdot \begin{bmatrix} 4 & 3 & 2 \\ -2 & 4 & 5 \end{bmatrix}$ $\begin{bmatrix} 6 & 10 & 9 \\ 6 & -1 & -3 \end{bmatrix}$

  b $\begin{bmatrix} 4 & 3 & 2 \\ -2 & 4 & 5 \end{bmatrix} \cdot \begin{bmatrix} 2 & 1 \\ 1 & -1 \end{bmatrix}$ No product

  c What conclusion can you draw about matrix multiplication from the results in parts **a** and **b**?

P **3** Give an example of dimensions of two matrices that can be multiplied.

P **4** Give an example of dimensions of two matrices that cannot be multiplied.

P **5** If you multiply a 4 × 6 matrix by a 6 × 3 matrix, what will the dimensions of the product matrix be? 4 × 3

P **6** In matrix multiplication, if you multiply row 3 of the first matrix by column 5 of the second matrix, where is the result placed in the product matrix? Row 3, column 5

Problem-Set Notes

for page 285

- **1** and **2** are problems that you'll probably want to discuss only after students are comfortable with the process of multiplying matrices.

Reteaching

Some students may find it helpful to compare solutions obtained through matrix multiplication with those obtained through more familiar or intuitive methods. For example, students needed their intuition to solve the class opener on page 282. Show them how matrix multiplication can be used to model the same problem, allowing students to make the connections between their logic and the process of matrix multiplication.

Additional Answers

for page 285

1c Possible answer: The order in which matrices are multiplied affects the answer.

2c Possible answer: For two matrices to be multiplied, the number of columns in the first matrix must be the same as the number of rows in the second matrix.

3 Possible answer: 2 × 3 and 3 × 4

4 Possible answer: 3 × 2 and 4 × 3

SECTION 6.5

Problem-Set Notes

for pages 286 and 287

- **13** is true sometimes. For example, if C is a 2 × 2 matrix with all entries 1 and D is a 2 × 2 matrix with all entries 2, then C · D and D · C will both be 2 × 2 matrices with all entries 4.

Communicating Mathematics

- Have students write a paragraph explaining whether matrix multiplication is commutative. Students should include an original example in their explanation.
- Problems 1, 2, 10, 14, 15, 19, and Investigation encourage communicating mathematics.

Checkpoint

Use the matrices below to find each product.

$A = \begin{bmatrix} 3 & -4 \end{bmatrix}$ $B = \begin{bmatrix} 1 \\ 5 \end{bmatrix}$

$C = \begin{bmatrix} 0 & 2 \\ 3 & -6 \end{bmatrix}$

1. A · B
2. B · A
3. A · C
4. C · A
5. B · C
6. C · B

Answers:
1. $\begin{bmatrix} -17 \end{bmatrix}$
2. $\begin{bmatrix} 3 & -4 \\ 15 & -20 \end{bmatrix}$
3. $\begin{bmatrix} -12 & 30 \end{bmatrix}$
4. No product
5. No product
6. $\begin{bmatrix} 10 \\ -27 \end{bmatrix}$

Problems and Applications

P, R **7** Find each product.

  a $\begin{bmatrix} -3 & -4 \end{bmatrix} \cdot \begin{bmatrix} 2 \\ -7 \end{bmatrix}$ $\begin{bmatrix} 22 \end{bmatrix}$

  b $\begin{bmatrix} 2 \\ -7 \end{bmatrix} \cdot \begin{bmatrix} -3 & -4 \end{bmatrix}$ $\begin{bmatrix} -6 & -8 \\ 21 & 28 \end{bmatrix}$

P **8** Possible dimensions of matrix A and matrix B are given. Copy and complete the table, determining the dimensions of each product matrix A · B.

	Matrix B		
×	2 × 4	2 × 2	2 × 5
3 × 2	3 × 4	3 × 2	3 × 5
2 × 2	2 × 4	2 × 2	2 × 5
5 × 2	5 × 4	5 × 2	5 × 5

(Matrix A labels the row headers)

P, R **9** Find each product.

  a $\begin{bmatrix} 3 & 1 & 5 \\ 2 & 0 & 6 \end{bmatrix} \cdot \begin{bmatrix} 4 \\ 8 \\ 9 \end{bmatrix}$ $\begin{bmatrix} 65 \\ 62 \end{bmatrix}$

  b $\begin{bmatrix} 3 & -1 & -5 \\ -2 & 0 & 6 \end{bmatrix} \cdot \begin{bmatrix} -4 \\ 8 \\ -9 \end{bmatrix}$ $\begin{bmatrix} 25 \\ -46 \end{bmatrix}$

P **10** **Communicating** Write a procedure for multiplying two matrices. Read it to someone and have that person use your procedure to multiply two matrices.

P, R **11** The product of two matrices is given, but three of the elements are missing. Fill in the missing elements so that the product matrix will be correct.

$\begin{bmatrix} 2 & 2 \\ \underline{1} & 4 \end{bmatrix} \cdot \begin{bmatrix} 5 & -3 \\ -2 & \underline{-3} \end{bmatrix} = \begin{bmatrix} 6 & -12 \\ -3 & -15 \end{bmatrix}$

P, I **12** Find each product.

  a $\begin{bmatrix} 1 & 0 \\ 0 & 1 \end{bmatrix} \cdot \begin{bmatrix} 3 & 2 \\ 3 & 9 \end{bmatrix}$

  b $\begin{bmatrix} 3 & 2 \\ 13 & 9 \end{bmatrix} \cdot \begin{bmatrix} 1 & 0 \\ 0 & 1 \end{bmatrix}$

  c $\begin{bmatrix} 3 & 2 \\ 13 & 9 \end{bmatrix} \cdot \begin{bmatrix} 9 & -2 \\ -13 & 3 \end{bmatrix}$

P, R, I **13** Is the following statement true *always, sometimes,* or *never?*

For matrix C and matrix D, C · D = D · C.   Sometimes

P **14** Matrix A has 5 rows and 9 columns. Matrix B has 12 columns. We want to find the product A · B.
  a How many rows must matrix B have? Explain your answer.
  b If matrix B has the number of rows indicated in your answer to part **a**, how many rows does the product matrix have?   5
  c How many columns does the product matrix have?   12

P **15** Find each product. What conclusion about matrix multiplication do your results suggest?

  a $\begin{bmatrix} 3 & 1 \\ 4 & 2 \end{bmatrix} \cdot \begin{bmatrix} 5 & 6 \\ 9 & 8 \end{bmatrix}$ $\begin{bmatrix} 24 & 26 \\ 38 & 40 \end{bmatrix}$

  b $\begin{bmatrix} 3 & -1 \\ -4 & 2 \end{bmatrix} \cdot \begin{bmatrix} 5 & -6 \\ -9 & 8 \end{bmatrix}$ $\begin{bmatrix} 24 & -26 \\ -38 & 40 \end{bmatrix}$

Chapter 6 Signed Numbers

SECTION 6.5

P **16. Consumer Math** The shares matrix gives the number of shares of MBI stock and PI stock held by Peter, Kathy, and Bob. MBI stock lost $2.25 per share and PI gained $1.50 per share.

Shares Matrix

	Peter	Kathy	Bob
MBI	72	58	39
PI	14	47	81

  a. Write a matrix multiplication problem that can be used to determine how much money each person's portfolio gained or lost.
  b. Find the product of the matrices in part **a** to determine the actual gains and losses.
  c. Who lost the most?

P **17. Business** The wage matrix represents the hourly wage Clyde and Chloe earn working at Mickey Dees and Burger Barn. The time matrix shows that each person worked 10 hours at Mickey Dees and 8 hours at Burger Barn. Use matrix multiplication to find the total amount each person earned.

Wage Matrix

	Mickey Dees	Burger Barn
Clyde	$3.80	$4.10
Chloe	4.00	3.90

Time Matrix

	Hours
Mickey Dees	10
Burger Barn	8

Clyde: $70.80
Chloe: $71.20

P 📱 **18. Environmental Science** Four conservationists want to purchase three types of evergreen trees. The trees are available at three nurseries.

Quantity Matrix

	Spruce	Red Pine	White Pine
George	25	35	14
Wally	45	50	32
Chuck	18	35	75
Bill	20	40	18

Price Matrix

	Nursery A	B	C
Spruce	1.45	1.53	1.62
Red pine	1.50	1.42	1.38
White pine	1.25	1.60	1.35

  a. How much would it cost each person to purchase the given numbers of trees at each of the nurseries?
  b. Which nursery offers the best overall deal for each person?

P 📱 **19.** The number matrix for the Alsip Expressway's three toll booths gives the throughput numbers for three types of vehicles in one week. The toll matrix lists the toll fee per vehicle.

Number Matrix

	Cars	Trailers	Trucks
Booth 1	72,000	8,000	12,500
Booth 2	84,000	9,756	8,750
Booth 3	68,000	6,540	9,984

Toll Matrix

	Fee per Vehicle
Car	0.40
Trailer	0.60
Truck	0.95

  a. Find the total amount collected at each toll booth for the week.
  b. What was the total amount collected during the week at all the toll booths combined?
  c. Why do you think tollways charge more for trucks and for trailers than for cars?

Additional Answers

for pages 286 and 287

10. Possible answer: Multiply each element in each column of the second matrix by the corresponding element in each row of the first matrix. Then add these numbers to determine the corresponding entry of the product matrix.

12. a. $\begin{bmatrix} 3 & 2 \\ 3 & 9 \end{bmatrix}$ **b.** $\begin{bmatrix} 3 & 2 \\ 13 & 9 \end{bmatrix}$

  c. $\begin{bmatrix} 1 & 0 \\ 0 & 1 \end{bmatrix}$

14. a. 9 rows, so that every row entry in matrix A has a matching column entry in matrix B

15. Possible answer: Altering the signs of the entries in the factor matrices affects the product matrix.

16. a. $\begin{bmatrix} -2.25 & 1.50 \end{bmatrix} \cdot \begin{bmatrix} 72 & 58 & 39 \\ 14 & 47 & 81 \end{bmatrix}$

  b. $\begin{bmatrix} -141 & -60 & 33.75 \end{bmatrix}$
  c. Peter, by quite a lot.

18. a.

	A	B	C
George	106.25	110.35	107.70
Wally	180.25	191.05	185.10
Chuck	172.35	197.24	178.71
Bill	111.50	116.20	111.90

  b. Nursery A

19. a. Booth 1 $45,475.00
    Booth 2 $47,766.10
    Booth 3 $40,608.80

  b. $133,849.90

  c. Possible answer: Because they are responsible for more wear on the roadway

SECTION 6.5

Cooperative Learning

- Have groups of students write problems that could be solved by using matrix multiplication. Be sure to require that each group provide an answer to their problem. Then have groups trade problems and solve them.

- Problems 10, 16, 18, and 19 are appropriate for cooperative learning.

Stumbling Block

Because product matrices, such as the one for problem 18, often require many calculations, students may have difficulty organizing their work. You might encourage students to write the answer to each intermediate step on paper or, if students are using calculators, provide guidance in using calculators to perform the operations. Scientific calculators make matrix multiplication relatively painless because they are programmed to use the standard order of operations.

Additional Practice

Find each product.
1. $[2 \ -1] \cdot \begin{bmatrix} 1 \\ 3 \end{bmatrix}$
2. $\begin{bmatrix} 1 \\ 3 \end{bmatrix} \cdot [2 \ -1]$
3. $[2 \ -1] \cdot \begin{bmatrix} 1 & -3 \\ 2 & 4 \end{bmatrix}$
4. $\begin{bmatrix} 1 \\ -6 \end{bmatrix} \cdot [1 \ -1 \ 4]$

288

◀ LOOKING BACK **Spiral Learning** LOOKING AHEAD ▶

R **20** Find the missing numbers so that each side of the triangle will have the same sum.

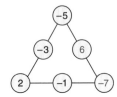

R **21** Evaluate each expression.
   a $-16^2 + (-10)^2$ -156 **b** $(-16)^2 - 10^2$ 156

R, I **22** Solve each equation for x.
   a $-8x = -120$ 15 **b** $5x = -60$ -12 **c** $-1.5x = 90$ -60

R, I **23** Compute the product. Express your answer in scientific notation.
$$(-3.67 \times 10^{18})(4.00 \times 10^{10}) \quad -1.468 \times 10^{29}$$

R **24** Which is greater, 20% of -6 or 40% of -8? 20% of -6

R **25** Find the area of the circle.
   36π, or ≈ 113.1

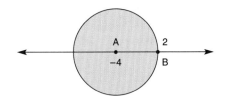

R **26** Write a mathematical expression that represents each phrase. Then evaluate the expression.
   a The sum of -18, 7, and -4 $-18 + 7 + (-4) = -15$
   b The product of -5, 3, and -2 $(-5)(3)(-2) = 30$
   c The sum of 3 times -4 and -2 times -5 $3(-4) + (-2)(-5) = -2$
   d The product of the sum of 3 and -8 and the sum of -7 and -8
   e The product of -8 and the sum of 3 and -7 $-8 \cdot [3 + (-7)] = 32$

R, I **27** Evaluate each expression for $x = -3$ and $y = -5$.
   a $3x + 2y$ -19 **b** $x \cdot y^2$ -75 **c** $-5x + x \cdot y$ 30

R **28** **Sports** The table shows the field-goal statistics in the Fumbler's Football League for last year. The data for attempted and completed field goals is broken down by the distance of the kicks.

Distance (yards)	Number Attempted	Number Made
Under 20	42	39
20–29	477	456
30–39	626	454
40–49	625	319
50–59	167	59
60 or more	5	0

   a Make a circle graph using the distance and number attempted.
   b Make a circle graph using the distance and the number made.

288   Chapter 6   Signed Numbers

SECTION 6.5

R, I **29** Solve for a. 10

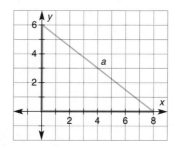

R **30** Solve each equation for n.

a $\dfrac{n}{-2} = \dfrac{12}{-3}$ 8

b $\dfrac{7}{n} = \dfrac{14}{-6}$ -3

R **31** Evaluate each expression.

a $\dfrac{(-48)(-3)}{-6}$ -24

b $-9 + (-5) \cdot 3$ -24

c $[-9 + (-5)] \cdot 3$ -42

Investigation

Using a Matrix Interview one or more of the following: a scientist, an engineer, a computer programmer, or a business owner. Discuss at least one example of how matrices are used in a job situation. Explain your findings.

5.
$\begin{bmatrix} -1 & 2 & 4 \\ 2 & 0 & 3 \end{bmatrix} \cdot \begin{bmatrix} -3 \\ 1 \\ 2 \end{bmatrix}$

Answers: 1. $\begin{bmatrix} -1 \end{bmatrix}$ 2. $\begin{bmatrix} 2 & -1 \\ 6 & -3 \end{bmatrix}$

3. $\begin{bmatrix} 0 & -10 \end{bmatrix}$ 4. $\begin{bmatrix} 1 & -1 & 4 \\ -6 & 6 & -24 \end{bmatrix}$

5. $\begin{bmatrix} 13 \\ 0 \end{bmatrix}$

Additional Answers

for pages 288 and 289

26 d $[3 + (-8)] \cdot [-7 + (-8)]$
$= (-5)(-15) = 75$

28 a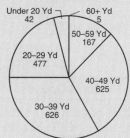

b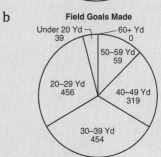

Investigation Answers will vary. A businessperson running a clothing store might record inventory in the form of a matrix: Column 1 might be the item identification number; column 2, the different sizes; column 3, the quantity on hand for each size; column 4, the unit cost; and column 5, the current list price.

SECTION 6.6

Objectives
- To subtract signed numbers
- To subtract matrices

Class Planning

Schedule 2 days

Resources

In *Chapter 6 Student Activity Masters* Book
Class Openers *61, 62*
Extra Practice *63, 64*
Cooperative Learning *65, 66*
Enrichment *67*
Learning Log *68*
Spanish-Language Summary *69*

In *Tests and Quizzes* Book
Quiz on 6.4–6.6 (Basic) *148*
Quiz on 6.4–6.6 (Average) *149*
Quiz on 6.4–6.6 (Honors) *150*

Class Opener

Find the next three numbers in each pattern. What number is being *subtracted* to get the next number in the pattern?
1. 15, 11, 7, 3
2. 10, 8, 6, 4
3. –3, –5, –7, –9
4. –8, –5, –2, 1
5. –2, 0, 2, 4

Answers:
1. –1, –5, –9; 4 2. 2, 0, –2; 2
3. –11, –13, –15; 2 4. 4, 7, 10; –3
5. 6, 8, 10; –2

Lesson Notes

- Sample problem 2 illustrates two ways that students can think about subtraction and reinforces the idea that there's often more than one way to solve a problem. Combine practice in subtraction with practice in multiplication, division, and addition so that students do not confuse the rules.

Investigation Connection
Section 6.6 supports Unit C, *Investigations* 2, 3, 4, 5. (See pp. 198A-198D in this book.)

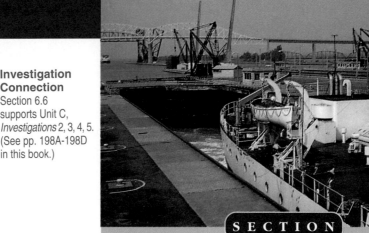

SECTION 6.6
Subtracting Signed Numbers

DEFINITION OF SUBTRACTION

The number -4 can be read "the opposite of the number 4." Similarly, $-(-4)$ can be interpreted as the opposite of -4.

Let's take a look at two patterns—one in which numbers are subtracted from 10 and one in which the opposites of these numbers are added to ten.

Subtraction	Addition
$10 - 7 = 3$	$10 + (-7) = 3$
$10 - 8 = 2$	$10 + (-8) = 2$
$10 - 9 = 1$	$10 + (-9) = 1$
$10 - 10 = 0$	$10 + (-10) = 0$
$10 - 11 = -1$	$10 + (-11) = -1$
$10 - 12 = -2$	$10 + (-12) = -2$
$10 - 13 = -3$	$10 + (-13) = -3$
$10 - 14 = -4$	$10 + (-14) = -4$
$10 - 15 = -5$	$10 + (-15) = -5$

The patterns help us understand why subtracting a signed number is the same as adding its opposite.

▶ **For any two numbers a and b, $a - b = a + (-b)$.**

Example 1
Find each difference.
a $5 - 9$ b $-5 - 9$ c $5 - (-9)$ d $-5 - (-9)$

Chapter 6 Signed Numbers

Solution
a $5 - 9 = 5 + (-9) = -4$
b $-5 - 9 = -5 + (-9) = -14$
c $5 - (-9) = 5 + [-(-9)]$
$\quad = 5 + 9$
$\quad = 14$
d $-5 - (-9) = -5 + [-(-9)]$
$\quad = -5 + 9$
$\quad = 4$

Example 2
Friday at 5:00 P.M. the temperature was −4°F. At 5:00 A.M. Saturday the temperature was −22°F. What was the change in temperature?

Solution
To find the change in temperature, we subtract the 5:00 P.M. temperature from the 5:00 A.M. temperature.

$$-22 - (-4) = -22 + 4$$
$$= -18$$

The temperature decreased by 18 degrees Fahrenheit.

If we use a number line to represent the problem in Example 2, we see that the definition of subtraction makes sense.

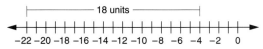

The number line shows that −22 is 18 units to the left of −4.

SUBTRACTING MATRICES

To subtract matrices, we subtract corresponding entries.

Example 3
Subtract $\begin{bmatrix} 2 & -4 \\ -5 & 9 \end{bmatrix} - \begin{bmatrix} -14 & 12 \\ -4 & -6 \end{bmatrix}$.

Solution
We will subtract each entry of the second matrix from the corresponding entry of the first matrix.

$$\begin{bmatrix} 2 & -4 \\ -5 & 9 \end{bmatrix} - \begin{bmatrix} -14 & 12 \\ -4 & -6 \end{bmatrix} = \begin{bmatrix} 2 - (-14) & -4 - 12 \\ -5 - (-4) & 9 - (-6) \end{bmatrix}$$

$$= \begin{bmatrix} 2 + 14 & -4 + (-12) \\ -5 + 4 & 9 + 6 \end{bmatrix}$$

$$= \begin{bmatrix} 16 & -16 \\ -1 & 15 \end{bmatrix}$$

Additional Examples

Example 1
Find each difference.
a. $6 - 8$ b. $-6 - 8$
c. $6 - (-8)$ d. $-6 - (-8)$

Answers: a. −2 b. −14 c. 14 d. 2

Example 2
Monday at 12 P.M. the temperature was −5°C. Tuesday at 8 A.M. the temperature was 14°C. What was the change in temperature?

Answer: 19°C

Example 3
Subtract.
$$\begin{bmatrix} -2 & -3 \\ 4 & 7 \end{bmatrix} - \begin{bmatrix} -5 & 4 \\ 3 & -11 \end{bmatrix}$$

Answer: $\begin{bmatrix} 3 & -7 \\ 1 & 18 \end{bmatrix}$

Cooperative Learning

- Have groups of students create four 3 × 3 magic squares with the following series of numbers: −3 to 5, −2 to 6, −1 to 7, and 0 to 8. Remind students that in a magic square, the sum of the numbers in each row, column, and diagonal is the same. For each magic square, have them find the sum, and have them find a rule for determining the sum for any magic square of 9 consecutive integers.

Answer: If a and b are the least and greatest integers, the sum equals $1.5(a + b)$.

- Problem 30 is appropriate for cooperative learning.

SECTION 6.6

Assignment Guide

Basic
Day 1 9–11, 15, 18, 20, 21, 25, 28
Day 2 12–14, 16, 22, 23, 29–31

Average or Heterogeneous
Day 1 9–14, 17, 18, 20, 21 (a, c), 22, 25, 27, 31
Day 2 15, 16, 19, 21 (b, d), 23, 24, 26, 28–30

Honors
Day 1 11, 12, 14–16, 18, 20, 23, 24, 27, 30
Day 2 9, 10, 13, 17, 19, 21, 22, 25, 26, 28, 29, 31

Strands

Communicating Mathematics	1, 3, 4, 6, 28
Geometry	28
Data Analysis/ Statistics	11, 27
Algebra	1–31, Investigation
Patterns & Functions	10, 11, 19
Number Sense	4, 6, 10, 18, 19, 30
Interdisciplinary Applications	27, Investigation

Individualizing Instruction

The kinesthetic learner Students may find it useful to tape two rulers together, with one upside-down so that the zero points are in the middle, to form a number line. Encourage these students to use their number-line rulers to help with computations involving negative numbers.

Problem-Set Notes

for pages 292 and 293

- 4 can be discussed by small groups before you discuss it with the whole class.

292

Sample Problems

Problem 1 How many integers are in the set $\{-15, -14, -13, \ldots, 5, 6\}$?

Solution **Method 1**
From -15 to -1 there are   15 numbers.
Counting 0 adds   1 number.
From 1 to 6 there are   6 numbers.
Total: 22 numbers

Method 2
We can also subtract the number with the smallest value from the number with the largest value, then add 1.

$$6 - (-15) = 6 + [-(-15)] = 6 + 15 = 21$$

Adding 1, we have $21 + 1$, or 22 numbers.

Problem 2 Sally and Tom were asked to solve the equation $x + 12 = 19$. Sally said she would subtract 12 from both sides of the equation to solve it. Tom said he would add -12 to both sides of the equation to solve it. Who was right? Explain your answer.

Solution Let's try Sally's approach:
If $x + 12 = 19$, then $x + 12 - 12 = 19 - 12$ and $x = 7$.
If we replace x with 7, we see that Sally is correct because $7 + 12 = 19$.
Now let's try Tom's approach:
If $x + 12 = 19$, then $x + 12 + (-12) = 19 + (-12)$ and $x = 7$.
Tom's answer is the same as Sally's.
Both Sally's approach and Tom's approach are correct. Subtracting 12 gives the same result as adding -12.

Think and Discuss

P, R   **1** How do we find the opposite of a number?

P, R   **2** What is the opposite of 5? Of 0? Of -7? $-5, 0, 7$

P   **3** **a** How would you read the expression $-(-100)$?
b Evaluate the expression in part **a**.   100

P   **4** Which difference has the smallest value? Which has the largest value? How could you arrange the differences from smallest to largest without doing any computation?
a $-6 - 8$ **b** $6 - 8$ **c** $6 - (-8)$ **d** $-6 - (-8)$

292 Chapter 6 Signed Numbers

SECTION 6.6

P, R   **5** Match each expression in the column on the left with an equivalent expression in the column on the right.

a $4 - (-3)$ 1. $-4 + (-3)$ a 3
b $-4 - 3$ 2. $-4 - (-3)$ b 1
c $4 + (-3)$ 3. $4 + 3$ c 4
d $-4 + 3$ 4. $4 - 3$ d 2

P   **6** Refer to Sample Problem 1. Does method 2 work if the set has only positive numbers? Only negative numbers? Explain your answers.

P   **7** Subtract $\begin{bmatrix} -4 & 6 & 0 \\ 3 & -2 & 5 \end{bmatrix} - \begin{bmatrix} 2 & 5 & -7 \\ -3 & 0 & 7 \end{bmatrix}$. $\begin{bmatrix} -6 & 1 & 7 \\ 6 & -2 & -2 \end{bmatrix}$

P   **8** How many units apart on the number line are the numbers in each pair?
a -12 and 7 19   b -18 and -11 7   c 8 and -17 25

Problems and Applications

P   **9** Find each difference.
a $5 - 8$ -3 b $-5 - 8$ -13 c $5 - (-8)$ 13   d $-5 - (-8)$ 3

P, R   **10** Find the numbers that will complete the pattern.

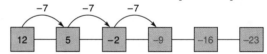

P, I   **11** Copy and complete the table if $y = 4 - x$.

x	-8	-6	-4	-2	0	2	4	6	8
y	12	10	8	6	4	2	0	-2	-4

P, R   **12** For each diagram, write a subtraction expression that represents length of segment AB.

a    b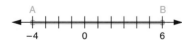

c

P, I   **13** Evaluate the expression $x - y$ for the given values.
a $x = 6, y = 2$ 4   b $x = -2, y = 4$ -6 c $x = 35, y = -2$ 37
d $x = -11, y = -6.5$ -4.5 e $x = -8.36, y = 4.2$ -12.56 f $x = 19.31, y = -4.58$ 23.89

P   **14** Subtract $\begin{bmatrix} -4 & 7 & -1 \\ 3 & -8 & 0 \end{bmatrix} - \begin{bmatrix} 8 & -2 & -5 \\ 9 & -3 & -11 \end{bmatrix}$. $\begin{bmatrix} -12 & 9 & 4 \\ -6 & -5 & 11 \end{bmatrix}$

P, I   **15** Solve each equation for n.
a $n - 23 = -12$ 11   b $n - (-5) = 3$ -2 c $n + 18 = 6$ -12

Section 6.6 Subtracting Signed Numbers

■ 8 should be contrasted with sample problem 1.

Stumbling Block

Some students may prefer to subtract signed numbers by changing problems to addition problems. However, watch that, in doing so, they don't change the sign of the wrong number. Use a number line to explain why the sign of the number being subtracted must be changed to the opposite sign.

Communicating Mathematics

■ Have students write a paragraph or give an oral presentation describing how to use one of the visual models for adding signed numbers to solve problems involving the subtraction of signed numbers.

■ Problems 1, 3, 4, 6, and 28 encourage communicating mathematics.

Additional Answers

for pages 292 and 293

1 Possible answer: By multiplying the number by -1

3 a The opposite of negative 100

4 Smallest: $-6 - 8 = -14$; largest: $6 - (-8) = 14$; possible answer: Use a number line.

6 Yes; yes; the method works for any set of integers.

12 a $12 - 7 = 5$
b $6 - (-4) = 10$
c $-1 - (-14) = 13$

SECTION 6.6

Problem-Set Notes

for pages 294 and 295

- **14, 16,** and **22** provide quite a bit of practice with subtraction by means of matrix subtraction problems.

- **27** reinforces the idea that setting up a matrix multiplication problem requires matching the labels of the columns of the first matrix with the labels of the rows of the second.

Additional Practice

Find each difference.
1. $8 - (-3)$ 2. $4 - 7$
3. $12 - 26$ 4. $-8 - 3$
5. $-9 - (-4)$ 6. $-4.6 - (-9.3)$
7. $4\frac{1}{2} - 9\frac{1}{6}$ 8. $-3.4 - 5.2$
9. $\begin{bmatrix} 8 & -1 & 9 \\ -2 & 4 & 7 \end{bmatrix} - \begin{bmatrix} -1 & 3 & -5 \\ 7 & 5 & 0 \end{bmatrix}$
10. $\begin{bmatrix} 1.3 & -8.9 \\ 2.4 & 7.6 \end{bmatrix} - \begin{bmatrix} 8.1 & -1.3 \\ 4.9 & 2.6 \end{bmatrix}$

Answers: **1.** 11 **2.** –3
3. –14 **4.** –11 **5.** –5
6. 4.7 **7.** $-4\frac{2}{3}$ **8.** –8.6
9. $\begin{bmatrix} 9 & -4 & 14 \\ -9 & -1 & 7 \end{bmatrix}$
10. $\begin{bmatrix} -6.8 & -7.6 \\ -2.5 & 5.0 \end{bmatrix}$

Checkpoint

Find each difference.
1. $8 - 9$ 2. $4 - (-6)$
3. $-2 - 3$ 4. $-4 - (-9)$
5. $\begin{bmatrix} 1 & -3 \\ 2 & 6 \end{bmatrix} - \begin{bmatrix} 3 & -5 \\ 7 & 4 \end{bmatrix}$
6. $\begin{bmatrix} 1 & -2 & 3 \end{bmatrix} - \begin{bmatrix} 4 & -6 & 9 \end{bmatrix}$

Answers: **1.** –1 **2.** 10 **3.** –5 **4.** 5
5. $\begin{bmatrix} -2 & 2 \\ -5 & 2 \end{bmatrix}$ **6.** $\begin{bmatrix} -3 & 4 & -6 \end{bmatrix}$

P **16** Evaluate $\begin{bmatrix} -2 & 0 \\ -4 & 5 \\ 3 & -6 \end{bmatrix} - \begin{bmatrix} 10 & -8 \\ 7 & 5 \\ -2 & 3 \end{bmatrix} \cdot \begin{bmatrix} -12 & 8 \\ -11 & 0 \\ 5 & 9 \end{bmatrix}$

P **17** How many numbers are in each set?
  a $\{-7, -6, -5, \ldots, 4\}$ 12
  b $\{-7, -6, -5, \ldots, 12\}$ 20
  c $\{-100, -99, -98, \ldots, 100\}$ 201
  d $\{0, 1, 2, 3, \ldots, 100\}$ 101

P, R **18** Arrange these differences in order from least value to greatest value.
  a $-8579 - 3654$
  b $-8579 - (-3654)$
  c $8579 - 3654$
  d $8579 - (-3654)$ **a, b, c, d**

P **19** Each number in the second row is obtained by subtracting the number to the right above it from the number to the left above it. Use this rule to form additional rows until there is a row at the bottom with only one number.

  1   –7   21   –35   35   –21   7   –1
    8   –28   56   –70   56   –28   8

P **20** Find each difference.
  a $3.7 - 4.8$ –1.1 **b** $4.95 - (-15.28)$ 20.23 **c** $3\frac{1}{2} - 4\frac{7}{8}$ $-1\frac{3}{8}$ **d** $-4\frac{3}{5} - \left(-2\frac{5}{8}\right)$ $-1\frac{39}{40}$

P, R **21** Evaluate each expression.
  a $-4 + (-3)(-2) - (-5)(-6)$ –28
  b $[-4 + (-3)](-2) - (-5)(-6)$ –16
  c $-4 + (-3)[-2 - (-5)] - 6$ –19
  d $-4 + (-3)[-2 - (-5)](-6)$ 50

P **22** Subtract $\begin{bmatrix} 2.9 & -4.7 \\ -3.6 & 4.58 \end{bmatrix} - \begin{bmatrix} 6.9 & 2.4 \\ -4.6 & -2.76 \end{bmatrix} \cdot \begin{bmatrix} -4.0 & -7.1 \\ 1.0 & 7.34 \end{bmatrix}$

P **23** Tommy Toad hopped from point A to point B. How far did he hop? 40 units

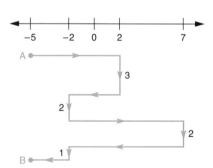

P, R **24** Refer to the set $\{10, 11, 12, \ldots, 20, 21, 22\}$.
  a How many numbers are in this set? 13
  b Subtract 14 from all numbers in the set, and list the resulting set. $\{-4, -3, -2, \ldots, 8\}$
  c Find the sum of the numbers in the original set. 208
  d Using your answers in parts **a** and **c**, find the sum of all the numbers in the set in part **b**. 26

P, I **25** Find the distance between these numbers on the number line.
  a 0 and 35 35 **b** 0 and –35 35 **c** 12 and 0 12 **d** –12 and 0 12

P, R, I **26** If the set $\{-11, -10, -9, -8, \ldots, x\}$ contains 34 integers, what is x? 22

Chapter 6 Signed Numbers

SECTION 6.6

◀ LOOKING BACK **Spiral Learning** LOOKING AHEAD ▶

R **27** The first matrix gives the number of grades earned last quarter by four students in five grade categories. The second matrix gives the points awarded for each grade.

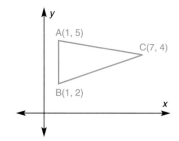

a Use matrix multiplication to determine the number of points each person earned for the quarter.
b Use your results from part **a** to determine each student's grade-point average.

R, I **28 a** Subtract 5 from each x-coordinate and -3 from each y-coordinate of the triangle. Then graph and label the resulting triangle.
b How has the position of the triangle shifted from its original placement?

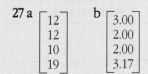

R, I **29** List all pairs of integers that have a product of -18.

R **30** The Game of -12 involves using one or more operation symbols $(+, -, \times, \div)$ along with three numbers to write an expression having a value of -12. For example, for the numbers 2, 42, and 3 the expression would be $2 - 42 \div 3 = -12$. Play this game with
a 1, -3, 4   **b** 3, 8, -2 **c** 2, 1, 7   **d** -15, 2, 9

R **31** If $a = -5$ and $b = -3$, are the statements below *true* or *false*? All are true.
a $2a + 7a = 9a$ **b** $9a + 6b + 4b + 3a = 12a + 10b$
c $14a - 3a - 2a = 9a$ **d** $9a + 4b + 6a - b = 15a + 3b$

Investigation

Time Differences   List a dozen or so major cities from around the world. Assign to each city a positive or negative integer to indicate the time difference from your home.

Additional Answers
for pages 294 and 295

19
36 -84 126 -126 84 -36
120 -210 252 -210 120
330 -462 462 -330
792 -924 792
1716 -1716
3432

27 a $\begin{bmatrix} 12 \\ 12 \\ 10 \\ 19 \end{bmatrix}$ **b** $\begin{bmatrix} 3.00 \\ 2.00 \\ 2.00 \\ 3.17 \end{bmatrix}$

28 a

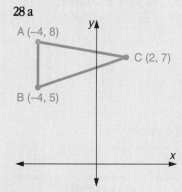

b The triangle has shifted to the left 5 units and up 3 units.

29 $(-1)(18), (-2)(9), (-3)(6),$
$(-6)(3), (-9)(2), (-18)(1)$

30 a $(-3) \times 4 \times 1$
b $3 \times [8 \div (-2)]$
c $2 \times (1 - 7)$
d $\frac{(-15) - 9}{2}$

Investigation Answers will vary. In general, cities that are to the west of your location and to the east of the international date line will be represented by negative numbers; other cities, by positive numbers. Cities in your time zone should be represented by 0. Students may note that the times in some cities differ from their own time by nonintegral numbers of hours.

SECTION 6.7

Objectives

- To determine the absolute values of numbers
- To recognize that the distance between two numbers on the number line is the absolute value of the difference of the numbers

Vocabulary: absolute value

Class Planning

Schedule 1–2 days

Resources

In *Chapter 6 Student Activity Masters* Book
Class Openers *71, 72*
Extra Practice *73, 74*
Cooperative Learning *75, 76*
Enrichment *77*
Learning Log *78*
Using Manipulatives *79, 80*
Spanish-Language Summary *81*

Transparency 24

Class Opener

On a number line, how many units apart are
1. 6 and 19?
2. –3 and –10?
3. –2 and 9?
4. –5 and –6?
5. –3 and 4?

Answers:
1. 13 2. 7 3. 11 4. 1 5. 7

Lesson Notes

- The content of this section is an extension of what students have done previously with number lines. Sample problems 1 and 2 should be solved as applications of the definition on page 297, not algebraically.

Investigation Connection
Section 6.7 supports Unit C, *Investigations* 4, 5. (See pp. 198A-198D in this book.)

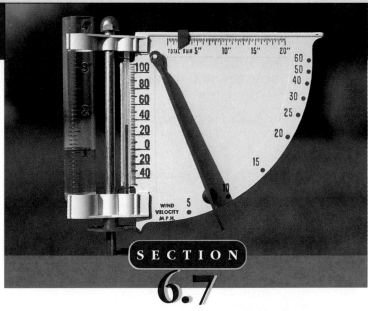

6.7 Absolute Value

DISTANCE FROM ZERO

Two temperatures, 6°C and –6°C, are both 6 degrees from 0° Celsius. If we use a number line to display this information, we see that –6 and 6 are each six units from 0 but located in opposite directions.

We call the distance of a number from zero, regardless of its direction from zero, the **absolute value** of the number. The mathematical symbol | | (vertical bars) represents the absolute value of a number. This symbol is read "the absolute value of."

Example 1
Evaluate each expression.
a |6| b |–6|

Solution
a Since the number 6 is six units from 0 on the number line, |6| = 6.
b Since the number –6 is six units from 0 on the number line, |–6| = 6.

Example 2
Solve each absolute-value equation for all values of *x*.
a |x| = 12   b |x| = –2

296   **Chapter 6**   Signed Numbers

SECTION 6.7

Solution

a To solve the equation $|x| = 12$, we need to determine what values of x have an absolute value of 12—that is, what numbers are 12 units away from zero on the number line. The numbers -12 and 12 fit both descriptions. Notice that the solutions to the equation are opposites.

b The absolute value of a number is its distance from 0 on a number line. Since this distance is either a positive number or 0, the equation $|x| = -2$ has no solutions.

DISTANCE BETWEEN TWO NUMBERS

Now let's see what happens when we put the difference of two numbers within the absolute-value symbol. The absolute-value symbol is a grouping symbol.

$$|10 - 4| = |6| = 6$$
$$|4 - 10| = |-6| = 6$$

The result, 6, is the same in both cases. Notice that 6 is also the number of units between the numbers 10 and 4 on the number line.

▶ **The distance between two numbers, a and b, on the number line is equal to $|a - b|$ or $|b - a|$.**

Example 3
Find the distance between -20 and 6 on the number line.

Solution
We can evaluate the expression $|-20 - 6|$.

$$|-20 - 6| = |-20 + (-6)| = |-26| = 26$$

Or we can evaluate the expression $|6 - (-20)|$.

$$|6 - (-20)| = |6 + 20| = |26| = 26$$

Both expressions result in the same answer. The numbers are 26 units apart on the number line.

Section 6.7   Absolute Value   297

Additional Examples

Example 1
Evaluate each expression.
a. $|12|$ b. $|-12|$

Answers: a. 12 b. 12

Example 2
Solve each equation for all values of x.
a. $|x| = 3$ b. $|x| = -5$

Answers: a. 3, -3 b. No solutions

Example 3
Find the distance between -11 and 12 on the number line.

Answer: 23 units

Additional Practice

Evaluate each expression.
1. $|-14|$ 2. $|12|$
3. $|8 - (-4)|$ 4. $|8| - |-10|$
5. $|-3(4)|$ 6. $|-3| \times |4|$

Write an absolute-value inequality to represent each graph.

7.
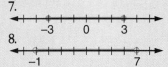

8.

9. Graph the solution set of $|x - 1| > 2$ on a number line.

10. Use absolute-value symbols to write the distance between -1 and 9 in two ways.

Answers: **1.** 14 **2.** 12 **3.** 12
4. -2 **5.** 12 **6.** 12 **7.** $|x| \leq 3$
8. $|x - 3| < 4$
9.
10. $|9 - (-1)|$, $|-1 - 9|$

SECTION 6.7

Assignment Guide

Basic
Day 1 12–14, 19, 21–23, 26, 32, 36, 41
Day 2 15–18, 25, 33, 34, 37, 39, 40

Average or Heterogeneous
Day 1 12–14, 16 (a, c, e), 17, 19, 21, 22, 24 (a, b), 26, 31, 32, 34
Day 2 15, 16 (b, d, f), 18, 20, 24 c, 27, 29, 33, 36 (a, c, e), 37, 38, 40

Honors
Day 1 15, 17, 19, 20, 22, 24, 26, 28, 30, 31, 38, 40

Strands

Communicating Mathematics	7–9, 11
Geometry	22, Investigation
Measurement	27, Investigation
Data Analysis/ Statistics	33, 41
Probability	26, 30
Algebra	1–37, 39–41
Patterns & Functions	19, 35, 38
Number Sense	4, 15, 23, 27
Interdisciplinary Applications	38, Investigation

Individualizing Instruction

Visually impaired learners
Students may have difficulty distinguishing absolute-value bars from 1's or from parentheses. If these students are making errors, it may not be due to a lack of understanding.

Problem-Set Notes

for pages 298 and 299

■ **8** has more than one answer. Be sure that students understand why.

298

Sample Problems

Problem 1 Solve the equation $|x - 6| = 7$ for all possible values of x.

Solution The solutions of the equation are the numbers that are 7 units from 6 on the number line.

Check If $x = 13$, $|x - 6| = |13 - 6| = |7| = 7$.
If $x = -1$, $|x - 6| = |-1 - 6| = |-1 + (-6)| = |-7| = 7$.
Therefore, the solutions to the equation are 13 and –1.

Problem 2 Write an inequality to represent the diagram.

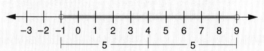

Solution The midpoint of the shaded region is the point with a coordinate of 4. The shaded region represents all numbers that are less than five units away from 4. The distance from any number x to 4 can be written as $|x - 4|$. Since this distance must be less than 5 units, the inequality that describes the graph is $|x - 4| < 5$.

Think and Discuss

P **1** Evaluate each expression.
  a $|-5|$ 5   **b** $|0|$ 0   **c** $|6.23|$ 6.23   **d** $\left|-3\tfrac{2}{3}\right|$ $3\tfrac{2}{3}$

P **2** What points are 22 units from zero on the number line?   22, –22

P **3** Solve each absolute-value equation.
  a $|x| = 10$ 10, –10   **b** $|x| = 1.3$ 1.3, –1.3   **c** $|x| = 0$ 0

P **4** Is the statement true *always, sometimes,* or *never*?
  a The absolute value of a number is negative.   Never
  b Opposite numbers have the same absolute value.   Always

P **5** Evaluate each expression.   **a** $|-12 + 8|$ 4   **b** $|-12| + |8|$ 20

P **6** Use absolute-value symbols to express the distance between 3 and 6 in two ways.
  $|3 - 6|$; $|6 - 3|$

298 **Chapter 6** Signed Numbers

SECTION 6.7

P, I **7** Explain why $|x - y| = |y - x|$.

P **8** Point A is five units from zero on the number line and point B is three units from zero on the number line.
  a How far apart are point A and point B? Explain your answer.   2 or 8 units
  b What is the coordinate of the midpoint of segment AB?   −4, −1, 1, or 4

P **9** Find a number for each box so that the inequality represents the graph. Explain how you solved each problem.
  a $|x - 7| \leq \boxed{4}$

  [number line from 3 to 11]

  b $|x - \boxed{-8}| < 6$

  [number line from −14 to −2]

P **10** Solve for x.   **a** $|x - 2| = 8$ 10, −6   **b** $|x + 6| = 3$ −3, −9

P **11** Graph $|x - 3| = 2$, $|x - 3| < 2$, and $|x - 3| > 2$. Discuss the differences among these three expressions.

Problems and Applications

P **12** Evaluate each expression.
  a $|-7|$ 7   **b** $|24|$ 24   **c** $\left|\frac{3}{4}\right|$ $\frac{3}{4}$ **d** $|8.3|$ 8.3

P **13** Solve each equation.
  a $|x| = 8$ −8, 8   **b** $|x| = 0$ 0   **c** $|x| = -9$ No solution

P **14** Evaluate each expression.
  a $|8 - 15|$ 7   **b** $|8 + 15|$ 23   **c** $|-8 - 15|$ 23   **d** $|15 - 8|$ 7

P **15** Arrange the expressions $|18 + (-7)|$, $|-18 + 7|$, $|7 - 18|$, $|-7 - 18|$, $|-7 - (-18)|$, and $|-7| - |18|$ in order from smallest value to largest value.

P **16** Find the value of each expression.
  a $|6| - |2|$ 4   **b** $|-8| - |-10|$ −2   **c** $|-8 - (-10)|$ 2
  d $|-8| - (-|10|)$ 18   **e** $-|8| - |-10|$ −18   **f** $-|8 - (-10)|$ −18

P **17** **a** If $|\text{number}| = 5$, what are the possible values of the number? −5 or 5
  b If $|x - 9| = 5$, what are the possible values of $x - 9$? −5 or 5
  c If $|x - 9| = 5$, what are the possible values of x? 4 or 14

P, R **18** Find all numbers that are $5\frac{1}{4}$ units away from $-7\frac{1}{2}$ on the number line. $-12\frac{3}{4}$ and $-2\frac{1}{4}$

P, R **19** Evaluate each expression for $a = -3$, $b = 4$, and $c = -6$.
  a $|ab|$ 12   **b** $|a||b|$ 12   **c** $|ac|$ 18   **d** $|a||c|$ 18

P, R **20** Write an absolute-value inequality to represent the graph.
  a [number line −4 to 4] $|x| < 4$ **b** [number line 1 to 9] $|x - 5| < 4$
  c [number line −6 to 2] $|x - (-2)| < 4$

Section 6.7 Absolute Value

- **9** could yield some interesting explanations from students.
- **15** has some equal expressions, so as students order them, answers may vary but still be correct.
- **19** suggests that the absolute value of a product is the same as the product of the absolute values of the factors. Ask students whether this will always be true.

Additional Answers
for pages 298 and 299

7 Both expressions stand for the distance between the points whose coordinates are x and y. This distance is unique, so the expressions must be equal.

8 a If the coordinate of A is 5, the coordinate of B could be 3 or −3. If the coordinate of A is −5, the coordinate of B could be 3 or −3.

9 a 4; 7 is 4 units from 11 and 4 units from 3
b −8; −8 is 6 units from −14 and 6 units from −2

11
$|x - 3| = 2$ [points at 1 and 5]
$|x - 3| < 2$ [between 1 and 5]
$|x - 3| > 2$ [outside 1 and 5]

The solutions of $|x - 3| = 2$ are 1 and 5. The solutions of $|x - 3| < 2$ are all the numbers between 1 and 5. The solutions of $|x - 3| > 2$ are all the numbers less than 1 or greater than 5.

15 $|-7| - |18|$, $|18 + (-7)|$
$= |-18 + 7| = |7 - 18|$
$= |-7 - (-18)|$, $|-7 - 18|$

SECTION 6.7

Problem-Set Notes

for pages 300 and 301

- **23** is similar to **19**, but it involves addition rather than multiplication. You might discuss this problem with the class.
- **38** can be solved by copying the diagram and writing on each vertex the number of paths there are to that vertex. The number of paths to any vertex is the sum of the numbers of paths to the vertices immediately above and to the left of it.

Communicating Mathematics

- Have students write a paragraph explaining the similarities and differences between the graphs of $|x| > 7$ and $|x| < 7$.
- Problems 7–9 and 11 encourage communicating mathematics.

Cooperative Learning

- Have students work in groups to write general rules for the values of a and b that will make the following true.
  1. $|a + b| = |a| + |b|$
  2. $|a + b| < |a| + |b|$
  3. $|a + b| > |a| + |b|$
  4. $|a - b| = |a| - |b|$
  5. $|a - b| < |a| - |b|$
  6. $|a - b| > |a| - |b|$

Answers: 1. True if a and b are both positive or both negative or if either value is zero 2. True if one value is positive and one is negative 3. Never true 4. True if both a and b are positive and $a > b$, if $b = 0$, or if both values are negative and $b > a$ 5. Never true 6. True if $a = 0$, if both values are negative and $a > b$, or if

---

P   **21**   Which of the numbers $-7, -4, -2, 0, 3$, and 5 are solutions of $|x + 2| < 3$? $-4, -2, 0$

P   **22**   The midpoint of segment AB is point C. If you slide segment AB along the number line until the coordinate of C is 0, what will be the new coordinates of points A and B? -3 and 3

(Number line: A at -17, C, B at -11)

P, I   **23**   Substitute numbers for a and b so that the statement is true. Repeat this process three times for different values of a and b.
    a $|a + b| = |a| + |b|$ **b** $|a + b| < |a| + |b|$ **c** $|a + b| > |a| + |b|$

P, R, I   **24**   Graph each solution on a number line.
    a $|x| > 5$ **b** $|x - 3| > 5$
    c $|x + 3| > 5$ [Hint: Remember that $x + 3 = x - (-3)$.]

P   **25**   If $|4 - y| = 11$, solve for all possible values of y. -7 and 15

P   **26**   If one of the numbers $|-15|, |5 - 8|, -|-21|, |0|$, and $|0 - 1|$ is selected at random, what is the probability that the number is negative? $\frac{1}{5}$

P, R   **27**   Find, to the nearest tenth, the distance between $\sqrt{31}$ and 5.5. 0.1

P, R   **28**   Consider the line segment on the number line.

    a What percent of the segment contains solutions of the inequality $|x| \leq 8$? 80%
    b What percent of the segment contains solutions of the inequality $|x - 5| \leq 3$? 30%

P   **29**   Point M is midway between -12 and 30.

    a What is the coordinate of M? 9
    b What is the distance between M and -12? Between M and 30? $21; 21$
    c Use the results of parts **a** and **b** to write an absolute-value equation that has the solutions -12 and 30. $|x - 9| = 21$

P, R   **30**   If x is a number between 6 and -8 inclusive, what is the probability that $|x| < 3$? $\frac{3}{7}$

P   **31**   Write an absolute-value equation to describe each statement.
    a The distance between 3 and a number x is 12. $|x - 3| = 12$
    b The distance between -5 and a number x is 8. $|x + 5| = 8$

◀ LOOKING BACK   **Spiral Learning**   LOOKING AHEAD ▶

R   **32**   Subtract $\begin{bmatrix} 3 & -6 \\ -2 & -4 \\ 10 & 0 \end{bmatrix} - \begin{bmatrix} 5 & -2 \\ 4 & 3 \\ 0 & -7 \end{bmatrix} \cdot \begin{bmatrix} -2 & -4 \\ -6 & -7 \\ 10 & 7 \end{bmatrix}$

R   **33**   Find the mean, the median, and the mode of $-12, -34, -12, 6, -26, -1, -10, 18, 2, 5, -3$, and 0.   Mean: ≈ -5.58, median: -2; mode: -12

300   Chapter 6   Signed Numbers

SECTION 6.7

R **34** Evaluate the expression $(-3) \cdot \begin{bmatrix} -5 & 4 \\ -2 & -6 \end{bmatrix} + (-2) \cdot \begin{bmatrix} 6 & -8 \\ 7 & 3 \end{bmatrix} \cdot \begin{bmatrix} 3 & 4 \\ -8 & 12 \end{bmatrix}$

R **35** In the following pattern, each number in the second row is found by subtracting the number to its right in the row above from the number to its left in the row above. Continue this pattern until there is a row at the bottom with only one number.

$$-25 \qquad 15 \qquad -10 \qquad 5 \qquad -20$$
$$-40 \qquad 25 \qquad -15 \qquad 25$$

R, I **36** Write a mathematical expression that represents each phrase. Then evaluate each expression.
  a The sum of −4 and −6 and 2 $-4 + (-6) + 2; -8$
  b The product of −4 and −3 and −2 $(-4)(-3)(-2); -24$
  c Subtract −7 from 8 $8 - (-7); 15$
  d 3 times −5 is subtracted from −6 $-6 - 3(-5); 9$
  e Subtract the sum of −3 and −1 from −10 $-10 - [(-3) + (-1)]; -6$

R **37** Multiply $\begin{bmatrix} 3 & -4 \\ -2 & 6 \end{bmatrix} \cdot \begin{bmatrix} -4 \\ 5 \end{bmatrix} \cdot \begin{bmatrix} -32 \\ 38 \end{bmatrix}$

R **38** Using only movements to the right and down, how many routes are there from
  a A to B? 3
  b A to C? 10
  c A to D? 35

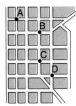

R, I **39** Solve for P.
  a $-7 \cdot P = 28$ -4 **b** $\frac{P}{-4} = \frac{5}{2}$ -10 **c** $-7 - P = 5$ -12

R **40** Solve $\frac{n}{-8} = \frac{21}{4}$ for n. -42

R **41** Copy and complete the multiplication table.

*	−3	−7	+4	−2
+2	−6	−14	8	−4
−5	15	35	−20	10
−6	18	42	−24	12

Investigation

Gasoline Mileage Choose a make and model car. Find out how many gallons of gas the tank holds and the average miles per gallon for this car. Obtain a map of your region and indicate the approximate location of your home. Now mark the areas you could reach on one tankful of gas or less. What shape is your travel region? What places in this region would you like to visit?

- one value is positive and one is negative
- Problem 38 is appropriate for cooperative learning.

Stumbling Block

In an expression such as $|6-4|$, students may try to change the minus sign to a plus sign, thinking that absolute value makes everything positive. Point out that the absolute-value symbol is a type of grouping symbol, like parentheses. What appears within absolute-value symbols should be simplified to a single number, if possible, before the definition of absolute value is applied.

Checkpoint

Evaluate each expression.
1. $|0|$ 2. $|3.6|$ 3. $|-9|$
4. Use absolute-value symbols to represent the distance between 9 and 15 in two ways.

Solve each absolute-value equation.
5. $|x| = 9$ 6. $|x| = -2$

Write an absolute-value inequality to represent each graph.
7.

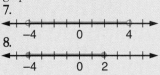

8.

Answers: 1. 0 2. 3.6 3. 9
4. $|15-9|, |9-15|$ 5. 9, −9
6. No solution 7. $|x| < 4$
8. $|x + 1| \leq 3$

Answers for problems 23, 24, 35, and Investigation are on page A5.

SECTION 6.8

Objectives

- To create a rectangular coordinate system
- To graph geometric figures on a rectangular coordinate system

Vocabulary: midpoint, ordered pair, quadrant, rectangular coordinate system, x-axis, x-coordinate, y-axis, y-coordinate

Class Planning

Schedule 1–3 days

Resources

In *Chapter 6 Student Activity Masters* Book
Class Openers 83, 84
Extra Practice 85, 86
Cooperative Learning 87, 88
Enrichment 89
Learning Log 90
Using Manipulatives 91, 92
Spanish-Language Summary 93

Transparencies 23 a,b

Class Opener

Describe the location of each point in relation to point O.

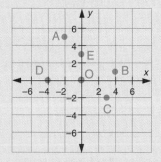

Answers: A is two units left and five units up. B is four units right and one unit up. C is three units right and two units down. D is four units left. E is three units up.

Investigation Connection
Section 6.8 supports Unit C, *Investigations* 4, 5. (See pp. 198A-198D in this book.)

SECTION 6.8
Two-Dimensional Graphs

THE RECTANGULAR COORDINATE SYSTEM

A **rectangular coordinate system** is formed by drawing two number lines that are perpendicular to each other and that intersect at their zero points. This point of intersection is the origin of the coordinate system.

The horizontal number line is the **x-axis,** with positive numbers to the right. The vertical number line is the **y-axis,** with positive numbers upward. The two number lines divide the coordinate system into four regions, called **quadrants,** numbered I, II, III, and IV as in the diagram.

Each point on the coordinate plane corrresponds to a unique **ordered pair** of numbers (x, y). For example, the point $(-4, 5)$ is located 4 units to the left of the y-axis and 5 units above the x-axis. The point $(5, -4)$, on the other hand, is located 5 units to the right of the y-axis and 4 units below the x-axis. The first number in the ordered pair (the **x-coordinate**) indicates how far to move to the left or right of the y-axis. The second number (the **y-coordinate**) indicates how far to move above or below the x-axis.

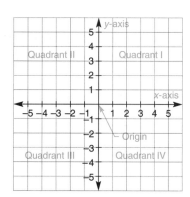

302   **Chapter 6**   Signed Numbers

SECTION 6.8

Example 1
What are the coordinates of points A and B in the coordinate system shown?

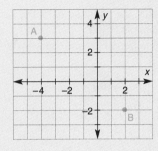

Solution
To reach point A from the origin, we can move 4 units to the left along the x-axis, then 3 units upward. The point's coordinates are therefore (−4, 3). (Notice that point A is directly above the −4 point of the x-axis and directly to the left of the 3 point of the y-axis.) Point B is 2 units to the right of the y-axis and 2 units below the x-axis, so its coordinates are (2, −2).

Example 2
a Plot points C(1, 3) and D(−4, −2) on a coordinate system.
b What are the coordinates of the point that is 3 units to the left of and 2 units below point C?

Solution
a To locate point C, we start at the origin and move 1 unit to the right, then 3 units up. We locate point D in a similar manner—starting at the origin and moving 4 units to the left, then 2 units down.

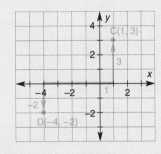

b Method 1: If we start at point C and move 3 units to the left, then 2 units down, we end up at the point with coordinates (−2, 1).
Method 2: We can add −3 (3 units in the negative direction) to the x-coordinate of point C and add −2 (2 units in the negative direction) to the y-coordinate of point C.

$$[1 + (-3), 3 + (-2)] = (-2, 1)$$

GRAPHING GEOMETRIC FIGURES

Drawing geometric figures on a coordinate system can often help us to identify some of the characteristics of the figures.

Lesson Notes

- A basic understanding of the coordinate system is key to students' success in Chapter 7. Be sure that all students are at least able to plot points and recognize coordinates.

- Counting squares is an acceptable method for finding the area in example 3.

- Most computer languages allow you to plot points on-screen through use of a modified coordinate system. If any students are familiar with such a language, you may have them describe to the class how the plotting function works.

Reteaching

To reteach the quadrants of a coordinate plane, remind students that ordered pairs in Quadrant I take the form (+, +) and that the other quadrants follow in a counterclockwise direction. Give students practice with these ideas by naming several points and having students tell which quadrant each point lies in.

Multicultural Note

Rectangular coordinates were used as long ago as 2750 B.C. in Africa. The coordinate system was used for locating a point by giving the horizontal distance and the vertical distance from fixed perpendicular axes.

SECTION 6.8

Assignment Guide

Basic
Day 1 7, 9, 11, 12, 15, 20, 24, 29, 35
Day 2 10, 13, 16, 19, 22, 25, 30, 34, 36
Day 3 8, 14, 18, 23, 32, 33, 37, 38

Average or Heterogeneous
Day 1 7–12, 14, 18–20, 22, 25, 26, 30, 38
Day 2 13, 15–17, 21, 23, 24, 27, 28, 31, 34, 35–37, 39

Honors
Day 1 8, 9, 11–14, 16, 17, 19, 20, 22, 23, 25, 34

Strands

Communicating Mathematics	2, 3, 6, 11, 17, Investigation
Geometry	5, 10, 15, 16, 20, 23, 24
Measurement	15, 20, 23, Investigation
Technology	25
Algebra	1–5, 7–40, Investigation
Patterns & Functions	3, 24, 34
Number Sense	8, 34
Interdisciplinary Applications	Investigation

Individualizing Instruction

Learning-disabled students
Students may need some extra patience and monitoring in this section. Working with a coordinate system requires constant attention to accuracy, which might be too much for some students to handle without extra guidance.

Example 3
a What are the coordinates of the vertices of rectangle RECT?
b Find the perimeter and the area of RECT.

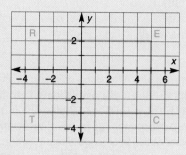

Solution
a The vertices are R(–3, 2), E(5, 2), C(5, –3) and T(–3, –3).
b Since $\overline{RE}$ is horizontal, its length is the difference between the x-coordinates of R and E. This length is | –3 – 5 |, or 8, units. Since $\overline{RT}$ is vertical, its length is the difference between the y-coordinates of R and T. This is | 2 – (–3) |, or 5, units. The perimeter is 2(RE) + 2(RT) = 2(8) + 2(5) = 26 units, and the area is (RE)(RT) = 8(5) = 40 square units. (Try counting the unit squares inside RECT. Are there 40 of them?)

Sample Problem

Problem   In the figure shown, point J is the **midpoint** of $\overline{GH}$—that is, it is halfway between points G and H.
a Draw the figure on a coordinate system, labeling F, G, H, and K with their coordinates.
b What are the coordinates of point J?
c What is length FJ?

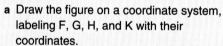

Solution a There are many ways of placing the figure on a coordinate system. We will put F at the origin, so that $\overline{FG}$ lies on the y-axis and $\overline{FH}$ lies on the x-axis. From the given lengths, we can see that the coordinates of F, G, H, and K are (0, 0), (0, 5), (6, 0), and (3, 0).

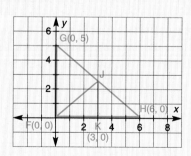

SECTION 6.8

b When a point is the midpoint of a segment on the coordinate plane, the point's x-coordinate is the mean of the x-coordinates of the segment's endpoints, and the point's y-coordinate is the mean of the y-coordinates of the segment's endpoints.

Coordinates of $J = \left(\dfrac{0+6}{2}, \dfrac{5+0}{2}\right) = (3, 2.5)$

c Since $\overline{JK}$ is vertical, its length is $|2.5 - 0|$, or 2.5, units. $\overline{FJ}$ is the hypotenuse of right triangle FJK, so we can use the Pythagorean Theorem to find the length of $\overline{FJ}$.

$$(FK)^2 + (JK)^2 = (FJ)^2$$
$$3^2 + 2.5^2 = (FJ)^2$$
$$9 + 6.25 = (FJ)^2$$
$$15.25 = (FJ)^2$$

Since $(FJ)^2 = 15.25$, length FJ must be $\sqrt{15.25}$, or ≈ 3.9, units.

Think and Discuss

P **1** What are the coordinates of points A, B, C, D, E, and F? (2, 4); (0, 1); (–2, 2); (–3, 0); (–3, –4); (3, –1)

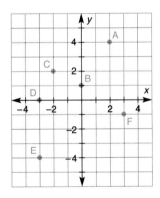

P, I **2** Why, do you think, are pairs of coordinates, such as (5, 3) and (3, 5), called *ordered* pairs?

P **3** How can you tell what quadrant a point is in just by looking at the signs of its coordinates? Write a rule that can be used to identify the quadrant of any given ordered pair.

P, R **4** Find the lengths labeled *a*, *b*, and *c* in the diagram. 3; 4; 5

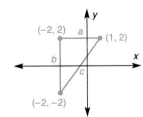

Additional Examples

Example 1
What are the coordinates of a point 3 units to the left of the origin? 3 units above the origin?
Answer: (–3, 0), (0, 3)

Example 2
Plot points Q(–5, 2), U(1, 2), A(1, –2), and D(–5, –2) on a coordinate system. Connect the points.
Answer:

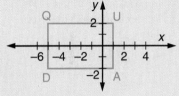

Example 3
Find the perimeter and the area of QUAD in example 2.
Answer: Perimeter = 20 units, area = 24 square units

Additional Answers

for page 305

2 The order (position) of the numbers in the ordered pairs matters.

3 Possible answer: Quadrant I ordered pairs have the form (+, +); Quadrant II ordered pairs, (–, +); Quadrant III ordered pairs, (–, –); and Quadrant IV ordered pairs, (+, –).

Section 6.8 Two-Dimensional Graphs

SECTION 6.8

Problem-Set Notes

for pages 306 and 307

- **8** might be reinforced by having students draw an example and a counterexample.
- **13–16** preview some ideas in Chapter 7.

Additional Practice

1. What are the coordinates of the center of the circle shown in problem 20? Where does the circle intersect the x-axis? The y-axis?

2. Graph the points E(0, –3), F(2, –1), G(–1, –3), and H(–2, 4).

Tell which quadrant each point is in.
3. (–1, 4) **4.** (2, 7)
5. (0, 6) **6.** (–5, –9)

The endpoints of a line segment are given. Find the coordinates of each midpoint.
7. (2, 6), (8, 16)
8. (–1, 4), (7, 10)

Answers: **1.** (3, 3); (3, 0); (0, 3)
2.

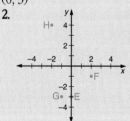

3. II **4.** I **5.** None **6.** III
7. (5, 11) **8.** (3, 7)

Additional Answers

for pages 306 and 307

6 Possible answer: Place F at the origin, G at (0, –5), H at (6, 0), and K at (3, 0). The method given in the sample problem is best because all coordinates are positive.

306

P, R   **5** If point Y is the midpoint of $\overline{XZ}$, what are the coordinates of Y? (2.5, 1)

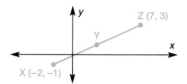

P   **6** In this section's sample problem, a figure was placed on a coordinate system in a specific manner. Discuss some other ways of positioning the figure on a coordinate system. Which way is best? Why?

Problems and Applications

P   **7** Create a rectangular coordinate system on graph paper. Then locate and label each of the following points and indicate which quadrant the point is in.
  a A(6, 2)   I
  b B(8, 0)   No quadrant
  c C(4, –6)   IV
  d D(–1, 3)   II
  e E(–3, –4)   III
  f F(0, –2)   No quadrant

P   **8** Is the following statement true *always, sometimes,* or *never*?   Sometimes
  A segment with one endpoint in Quadrant I and the other endpoint in Quadrant III passes through Quadrant II.

P   **9** What are the coordinates of points G, H, J, K, and L?   (–5, 1); (–3, 2); (1, 1); (0, –3); (–4, –2)

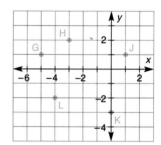

P   **10** Point N is the midpoint of horizontal segment MP, and N is 15 units to the right of point M.
  a What are the coordinates of N and P?
  b What are lengths MN, NP, and MP?

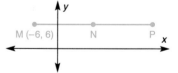

P   **11** Communicating   Create a coordinate system and locate the points with coordinates (3, 2), (3, –2), (–3, 2), and (–3, –2). Describe how the points are related to one another.

P, R   **12** Each segment in the diagram is either vertical or horizontal. What are the coordinates of points T, U, F, E, and N?
  (–3, –1); (–1, –1); (–1, 2); (4, 2); (4, 4)

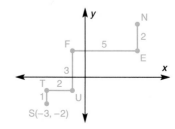

306   Chapter 6 Signed Numbers

SECTION 6.8

P **13** Locate points S(–4, 8), T(–4, –4), and U(1, –4) on a rectangular coordinate system. Connect the points to form △STU. What are the lengths of sides $\overline{ST}$, $\overline{TU}$, and $\overline{SU}$? 12; 5; 13

P, R **14** If $\overline{QR}$ is moved 3 units horizontally to the right, what will be the new coordinates of points Q and R? (1, 3); (7, –1)

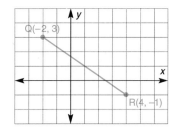

P, R **15** RECT is a rectangle.
  a What are the coordinates of the vertices of RECT? R(–6, 5); E(2, 5); C(2, –1); T(–6, –1)
  b What is the area of RECT? 48
  c What is the length of diagonal $\overline{ET}$? 10

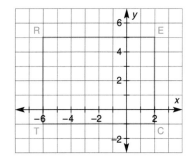

P **16 a** What are lengths TR and RY? 2; 12
  b What are the coordinates of the midpoints of $\overline{TR}$ and $\overline{TY}$? (–2, 5); (4, 4)

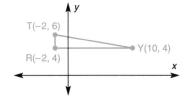

P **17 Communicating**
  a Describe three ways in which this figure can be placed on a coordinate system.
  b Which of the three ways do you think is most convenient? Why?
  c Position the figure on a coordinate system in one of the ways you described, labeling each vertex with its coordinates.

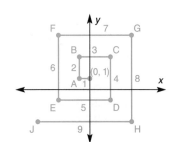

P **18** Each segment in the diagram is either horizontal or vertical. Find the coordinates of each lettered point. A(–1, 1), B(–1, 3), C(2, 3), D(2, –1), E(–3, –1), F(–3, 5), G(4, 5), H(4, –3), J(–5, –3)

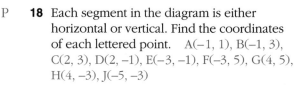

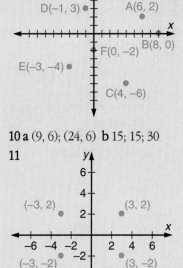

7

10 a (9, 6); (24, 6) **b** 15; 15; 30

11

Possible answer: The points are the vertices of a rectangle.

17 a Possible answers: With the bottom side and the left side lying on the x- and the y-axis; with the bottom side lying on the x-axis and the origin at its midpoint; with the origin at the figure's center

b Possible answer: With the bottom side and the left side lying on the x- and the y-axis

c

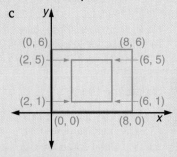

Section 6.8 Two-Dimensional Graphs **307**

SECTION 6.8

Checkpoint

1. What are the coordinates of the vertices of the triangle shown in problem 20?

Problems 2–4 refer to the following points: E(–1, 4), F(–5, 0), G(2, –1), and H(–4, –1).
2. Graph each point on a coordinate system.
3. Name the quadrant in which each point lies.
4. Find the midpoint of $\overline{EF}$.

Answers: 1. (–4, –1), (–1, –1), and (–2, 7)
2.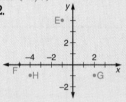
3. E: II; F: None; G: IV; H: III
4. (–3, 2)

Cooperative Learning

- Have pairs of students draw pictures on the coordinate plane and then write directions for re-creating the pictures, including sets of coordinates. Students should then try to re-create another group's drawing by following the directions.
- Investigation is appropriate for cooperative learning.

Stumbling Block

Students may graph (3, 7) as (7, 3). Remind students that the horizontal component is first, just as it is in the naming of matrices and in the multiplication of matrices (i.e., row times column).

P **19** If this figure is placed on a coordinate system so that point P is at (–29, –4) and point A is at (–13, –4), what will be the coordinates of point R? (–19, 4)

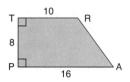

P, R **20** What are the areas of the circle and the triangle?
9π, or ≈ 28.27; 12

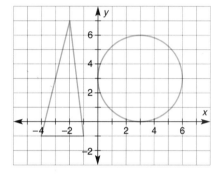

P **21** Find the coordinates of points A, B, and C.
(5, –2); (8, –2); (8, –6)

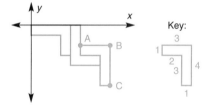

P **22** QUAD is a square. If point U is moved 3 units up and D is moved 3 units down,
  a What will be the new coordinates of U and D? (2, 5); (–2, –5)
  b What will lengths QD, UA, QU, and DA become? 7; 7; 5; 5

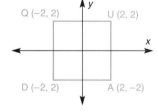

P, R, I **23** The shaded region in the diagram represents all ordered pairs (x, y) in which $-3 \le x \le 8$ and $-4 \le y \le 4$.
  a Find the area of rectangle RECT. 88
  b What percentage of the shaded region is in Quadrant II? ≈13.6%

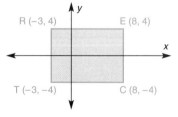

P, I **24** List five ordered pairs (x, y) in which $x - y = 2$. On a rectangular coordinate system, locate the points corresponding to these ordered pairs. What do you notice?

308   Chapter 6   Signed Numbers

SECTION 6.8

◀ LOOKING BACK **Spiral Learning** LOOKING AHEAD ▶

25. Express 60 mi/hr as a number of feet per second.   88 ft/sec

26. How many numbers are in each set?
    a $\{-4, -3, -2, \ldots, 10\}$ 15   b $\{0, 1, 2, 3, \ldots, 14\}$ 15   c $\{-8, -6, -4, \ldots, 20\}$ 15

In problems 27 and 28, graph each inequality.

27. $|x| < 8$
28. $|x - 3| < 8$

29. Each point of $\triangle A'B'C'$ is the same distance from the y-axis as the corresponding point of $\triangle ABC$. What are the coordinates of A', B', and C'? $(-1, 3)$; $(-5, -1)$; $(-2, -2)$

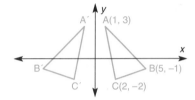

In problems 30–33, solve each equation for z.

30. $z + 1432 = 976$ $z = -456$
31. $\dfrac{z}{-93} = 58$ $z = -5394$
32. $z - 349.7 = 41.26$ $z = 390.96$
33. $25.1z = 921.17$ $z = 36.7$

34. Copy and complete the diagram so that the sums of the numbers lying along the five segments are the same.

In problems 35–38, solve each equation for x.

35. $|x| = 2$
36. $|x| = -2$
37. $|x - 1| = 3$
38. $|2x| = 3$

In problems 39 and 40, write an absolute-value inequality that is represented by the graph.

39.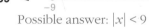
Possible answer: $|x| < 9$

40.
Possible answer: $|x + 2| < 9$

Investigation

Floor Plan Using a coordinate system, draw to scale a floor plan of your classroom. Label several important positions in the floor plan (doors, windows, and so on) with their coordinates. Use the floor plan to determine some distances that you have not measured, then check your findings. What scale did you choose for your floor plan, and why?

Section 6.8   Two-Dimensional Graphs

SUMMARY

CHAPTER 6

Summary

CONCEPTS AND PROCEDURES

After studying this chapter, you should be able to
- Associate signed numbers with points on a number line (6.1)
- Determine the opposite of a signed number (6.1)
- Graph signed numbers on a number line (6.1)
- Write inequalities and graph them on a number line (6.2)
- Write and graph inequalities joined by *and* or *or* (6.2)
- Use a number line to model the addition of signed numbers (6.3)
- Use a calculator to add signed numbers (6.3)
- Add matrices (6.3)
- Multiply signed numbers (6.4)
- Multiply matrices by scalars (6.4)
- Calculate powers of signed numbers (6.4)
- Divide signed numbers (6.4)
- Multiply two matrices (6.5)
- Subtract signed numbers (6.6)
- Subtract matrices (6.6)
- Determine the absolute values of numbers (6.7)
- Recognize that the distance between two numbers on the number line is the absolute value of the difference of the numbers (6.7)
- Create a rectangular coordinate system (6.8)
- Graph geometric figures on a rectangular coordinate system (6.8)

VOCABULARY

absolute value (6.7)
coordinate (6.1)
inequality (6.2)
integer (6.1)
midpoint (6.8)
negative signed number (6.1)
opposite (6.1)
ordered pair (6.8)
origin (6.1)

positive signed number (6.1)
quadrant (6.8)
rectangular coordinate system (6.8)
scalar (6.4)
scalar multiplication (6.4)
x-axis (6.8)
x-coordinate (6.8)
y-axis (6.8)
y-coordinate (6.8)

REVIEW

CHAPTER 6

Review

1 Perform each operation.
  a $-246 - (-93)$ -153
  b $\frac{9432}{-12}$ -786
  c $|-16| - (-42)$ 58
  d $|-3 - (-9)|$ 6
  e $(413)(-6)(-2)$ 4956
  f $-15 - (-4 - 3)$ -8

2 The area of square AUQS is 400.
  a Find the coordinates of points S and Q. $S(0, 20)$ $Q(20, 20)$
  b Move point Q along diagonal $\overline{AQ}$ to coordinates (12, 12). (Remember to move points S and U.) By what percent has the area of the square been reduced? 64%

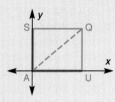

3 How far from zero on the number line is each number?
  a -8 8 units
  b $-3\frac{1}{2}$ $3\frac{1}{2}$ units
  c 7 7 units
  d -5.34 5.34 units

4 List all pairs of integers that have a product of 24.

5 Write an inequality that represents the number line.
  a 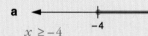 $x \geq -4$
  b 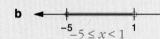 $-5 \leq x < 1$

6 If -7 is subtracted from each of the numbers in the set $\{-10, -5, -1, 1, 5, 10\}$, what is the probability that a number selected at random from the new set will be greater than 0? $\frac{5}{6}$

7 A scuba diver swims to a depth of -42 feet. How many feet must she rise to reach a depth of -24 feet? 18 ft

8 Find the mean of each set of numbers.
  a $2, 4, 6, 8$ 5
  b $-8, -6, -4, -2$ -5
  c $8, -6, -4, 2$ 0

9 Evaluate each expression for $x = -8$, $y = -2$, and $z = 4$.
  a $x + y$ -10
  b $x - y$ -6
  c xy 16
  d $\frac{x}{y}$ 4
  e $x + z$ -4
  f $x - z$ -12
  g x^z 4096
  h $\frac{x}{z}$ -2

10 Is the statement *true* or *false*? If it is false, give a counterexample.
  a The sum of a negative number and a positive number is a negative number.
  b The product of two integers is an integer.
  a False; Possible answer: $-4 + 5 = 1$ **b** True

11 Refer to rectangle ECTR. Solve for w, x, y, and z. $w = -80$; $x = 100$; $y = -9$; $z = 9$

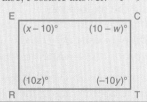

Assignment Guide

Basic
Day 1 1–5, 10, 11, 14, 18–20, 25–27
Day 2 7–9, 13, 15–17, 21–23, 28, 30
Day 3 Chapter Test 1–8, 10–18

Average or Heterogeneous
Day 1 1, 2, 4, 5, 7–13, 15, 16, 17 (a, c), 18, 20, 23, 24, 27, 29 (a, c), 30

Honors
Day 1 2, 4, 6–8, 9 (a–d, g), 10–13, 15–21, 23, 24, 27, 29, 30

Strands

Geometry	2, 11, 19, 24
Measurement	2, 7
Data Analysis/ Statistics	8, 12
Probability	6
Algebra	1–27, 29, 30
Patterns & Functions	28
Number Sense	4, 6, 10, 12, 16, 21, 23, 26, 27
Interdisciplinary Applications	12

Additional Answers

for page 311

4 $(-1)(-24)$, $1 \cdot 24$, $(-2)(-12)$, $2 \cdot 12$, $(-3)(-8)$, $3 \cdot 8$, $(-4)(-6)$, $4 \cdot 6$

REVIEW

Additional Answers

for pages 312 and 313

12 a

$$\text{Degree-Day Matrix}$$

	Cold	Warm
Monday	33	4
Tuesday	50	0
Wednesday	31	−4
Thursday	26	14
Friday	35	−8
Saturday	40	−21
Sunday	58	6

b Negative numbers represent days with weather warmer than 68°. Positive numbers represent days with weather colder than 68°.

c Cold Town: 273; Warm Ville: −9

d Approximately 4 weeks

13 a −27 **b** −64 **c** −125 **d** 81 **e** 256 **f** 625

16 Possible answers:

a True: 1 + 2; False: 1 + (−2)

b True: 1 + 2; False: −1 + (−2)

c True: 2 − 1; False: 1 − 2

d True: 2 × 3; False: 2 × (−1)

e True: 6 ÷ 3; False: 6 ÷ 2

f True: −1 × 1; False: 1 × 2

18 a

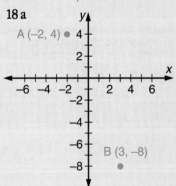

---

12 Companies that deliver heating oil calculate "degree days" to estimate when their customers will require an oil delivery. The degree-day number is calculated each day by subtracting the average temperature for the day from 68°. For example, if the average temperature on a given day was 35°, the calculation yields 68 − 35, or 33 degree days. The matrix gives the average temperatures over a one-week period in two towns.

a Use the average-temperature matrix to write a degree-day matrix.

b What do negative numbers represent in the degree-day matrix? What do positive numbers represent?

c How many degree days did each town have for the week?

d If the oil company fills the tanks of all customers after 1000 degree days have accumulated, estimate the length of time between fills for the customers in Cold Town. Assume all weekly data weeks are similar to the data that you have for one week.

Average-Temperature Matrix		
	Cold Town	Warm Ville
Monday	35	64
Tuesday	18	68
Wednesday	37	72
Thursday	42	54
Friday	33	76
Saturday	28	89
Sunday	10	62

13 If $(-2)^3 = -8$ and $(-2)^4 = 16$, evaluate each expression.

a $(-3)^3$ **b** $(-4)^3$ **c** $(-5)^3$ **d** $(-3)^4$ **e** $(-4)^4$ **f** $(-5)^4$

14 Evaluate each expression.

a $3 \cdot (-4) - 5 \cdot (-3)$ 3 **b** $8 \cdot (-6) - 9 \cdot (-7)$ 15 **c** $-4^2 - (-4)^2$ −32

15 Find the values of u, v, x, y, and z.

$$\begin{bmatrix} 2 & -5 & -4 \\ 1 & -6 & z \end{bmatrix} + \begin{bmatrix} -5 & -8 & 12 \\ -2 & y & 10 \end{bmatrix} = \begin{bmatrix} x & u & v \\ -1 & 0 & -10 \end{bmatrix}$$
$u = -13$, $v = 8$, $x = -3$, $y = 6$, $z = -20$

16 Find a pair of numbers for which each of the following statements is true and a pair of numbers for which each is false.

a The sum of the two numbers is a positive number.

b The sum of the two numbers is greater in value than either number.

c The difference of the two numbers is a positive number.

d The product of the two numbers is greater in value than either number.

e The quotient of the two numbers is smaller in value than either number.

f The product of the two numbers is equal to −1.

17 Write a mathematical expression that represents each phrase. Then evaluate the expression.

a Multiply the sum of 5 and 2 by (−6). $-6(5 + 2) = -42$

b Add the product of −9 and 5 to the product of −9 and 2. $-9 \cdot 5 + -9 \cdot 2 = -63$

c The product of −3 and the sum of 4 and −9. $-3[4 + (-9)] = 15$

18 a Plot points A(−2, 4) and B(3, −8)

b Find the coordinates of the midpoint of segment AB $(\frac{1}{2}, -2)$

312 Chapter 6 Signed Numbers

REVIEW

19 a Determine the area of △ABC. 45
b If △A′B′C′ is the result of reflecting △ABC over the x-axis, what are the coordinates of A′, B′, and C′?
A′ (−3, 1), B′ (−3, 11), C′ (6, 6)

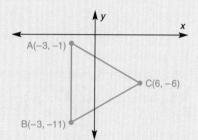

20 Multiply $\begin{bmatrix} -2 & 6 \\ 4 & 3 \end{bmatrix} \cdot \begin{bmatrix} 9 & -3 \\ -2 & -1 \end{bmatrix} \cdot \begin{bmatrix} -30 & 0 \\ 30 & -15 \end{bmatrix}$

21 The sum of two numbers is −29, and one of the numbers is 13. What is the other number? −42

22 Find values for x and y that make each equation true. Then find values for x and y that make each equation false.
a $|x + y| = |x| + |y|$ **b** $|x| = -(x)$ **c** $|x - y| = |x + y|$

23 Is the statement *true* or *false*? If the statement is false, give a counterexample.
a The product of two negative numbers is a negative number.
b The sum of two negative numbers is a positive number.

24 a In rectangle ABCD each coordinate of each vertex is increased by 20%. Find the new coordinates of A, B, C, and D.
b If the new coordinates are called A′, B′, C′, and D′, respectively, compare the areas of rectangle ABCD and rectangle A′B′C′D′.

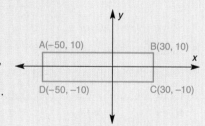

25 Evaluate each expression.
a $(-1)(2^6)$ −64 **b** -2^6 −64 **c** $(-2)^6$ 64 **d** $(-1)^6 (2)^6$ 64

26 a Plot $-6, \pi, -4.3, \frac{22}{7}, \sqrt{11}, -\sqrt{35},$ and $-4\frac{1}{3}$ on a number line.
b Arrange the numbers in part **a** in order, from least to greatest in value.

27 If $y = -4$, solve for x.
a $-2x = 7y$ 14 **b** $x - 3y = 4$ −8 **c** $12 - x = y$ 16

28 Find the next three terms in each pattern.
a $\frac{8}{9}, \frac{4}{3}, 2,$ ___, ___, ___ $3, \frac{9}{2}, \frac{27}{4}$ **b** 81, 27, 9, ___, ___, ___ $3, 1, \frac{1}{3}$

29 Solve for w.
a $|w - 5| = 8$ 13, −3 **b** $|w + 5| = 8$ −13, 3 **c** $|w - 5| \leq 8$ $-3 \leq w \leq 13$

30 Evaluate $b^2 - 4ac$, for $a = 2, b = -7,$ and $c = -3$. 73

---

22 Possible answers:
a True: $x = 2, y = 4$; False: $x = 2, y = -4$
b True: $x = 0$; False: $x = 7$
c True: $x = -9, y = 0$; False: $x = 3; y = 5$
23 a False; $-2 \cdot (-3) = 6$
b False; $-2 + -3 = -5$
24 a A′: (−60, 12), B′: (36, 12); C′: (36, −12); D′: (−60, −12)
b Area of A′B′C′D′ is $(1.2)^2$ times area of ABCD.
26 a

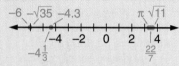

b $-6, -\sqrt{35}, -4\frac{1}{3}, -4.3, \pi, \frac{22}{7}, \sqrt{11}$

Chapter Review **313**

CHAPTER 6

Test

1. Which number is greater in value, −8 or 6? 6

2. What is the opposite of the given number?
   a −5   5
   b −7   7
   c 0   0

3. Point A is the midpoint of $\overline{BC}$. What is the coordinate of point A on the number line if the coordinate of point B is −8 and the coordinate of point C is 16? 4

4. Graph each inequality.
   a $x \leq -3$ or $x \geq 4$
   b $x \geq -4$ and $x < 5$
   c $-2 < x < 5$, x is an integer

5. Perform the indicated operation.
   a $28 + 56 + (-43)$ 41
   b $-18.7 + 9(-29.6)$ -285.1
   c $3(-15)(2)$ -90
   d $18(-3)(0)$ 0
   e $53 - 96$ -43
   f $-42 - (-18)$ -24
   g $\frac{-56}{-8}$ 7
   h $\frac{-3(-4)^2}{2}$ -24

6. Perform the indicated matrix operation.
   a $\begin{bmatrix} 2 & -3 & 0 \\ 4 & 5 & -9 \end{bmatrix} + \begin{bmatrix} -6 & 12 & 15 \\ 4 & 3 & -1 \end{bmatrix}$
   b $\begin{bmatrix} 2 & 3 \\ -1 & 4 \end{bmatrix} \cdot \begin{bmatrix} 1 \\ 6 \end{bmatrix}$
   c $-4 \begin{bmatrix} -3 & 2 & -6 \\ 0 & 4 & -7 \end{bmatrix}$

7. Solve $-9x = 180$ for x. -20

8. Evaluate $|-14 - (-9)|$. 5

9. Graph the solution of $|x - 2| > 3$ on a number line.

10. Draw a number line and label the points with the given coordinates.
    a -4
    b 0
    c 5
    d $-1\frac{3}{4}$
    e $2\frac{1}{3}$

11. Write the inequality that describes the graph. $x \leq 2$

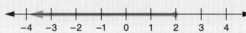

12. NCB stock sold for $19\frac{1}{2}$ dollars per share on Monday morning, but by noon its value had dropped $2\frac{1}{2}$ dollars per share. By closing time the same day, the stock has risen $\frac{3}{4}$ of a dollar per share. What was the closing value of the stock? $17.75

In problems 13–16, evaluate each expression.

13. -4^2 -16
14. $(-4)^2$ 16
15. $|-19|$ 19
16. $|0|$ 0

17. What is the distance between -214 and 16 on the number line?   230

18. Locate points A(−3, 2), B(−3, −5), and C(1, −5) on a coordinate plane. Connect the points to form a triangle. Find the length of $\overline{AB}$, $\overline{BC}$, and $\overline{AC}$.   7, 4, ≈ 8

PUZZLES AND CHALLENGES

1 If 78 players enter a singles tennis tournament, how many matches must be played for a winner to be determined?

2 A bear left its den and went due south for a mile. Then it turned 90° to the left and walked straight ahead for some distance. Then the bear turned 90° to the left and walked for one mile. It had arrived back at its den. What color is the bear?

3 In chess, a knight moves in an L-shaped manner, two spaces in one direction and one space in a direction perpendicular to the first direction.

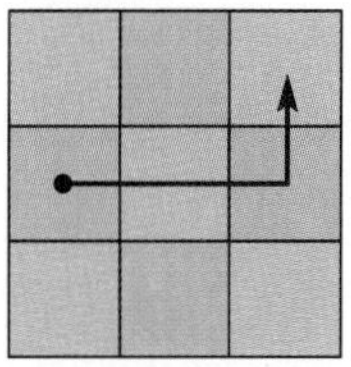

In the 3-by-3 square shown, we have numbered the squares to show the order in which the knight moved from square to square, ultimately landing in 8 of the 9 squares but never visiting the same square twice.

1	4	7
6		2
3	8	5

In the 4-by-4 square, the knight starts in the upper left-hand corner and visits 15 of the 16 squares, again never visiting the same square twice.

1	4	11	8
10	7	14	3
5	2	9	12
	13	6	15

In the 5-by-5 square, how can the knight start in the upper left-hand corner and visit at least 22 of the 25 squares, never visiting the same square twice?

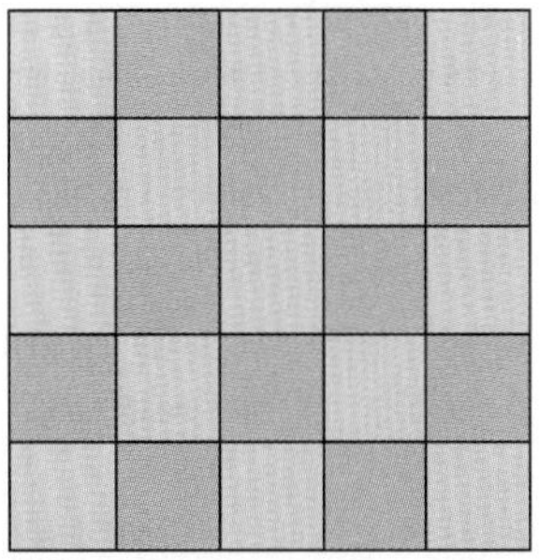

your findings in line graphs and write a paragraph comparing the graphs to each other.

12. Extend sample problem 2 on page 278 as follows: Design a spreadsheet application to solve the problem. Check your answers with those in the book and make any necessary adjustments. Then explore what would happen if Major Motors discovered that the engineering breakthrough raised the mileage by only 9% instead of 23%. Is there an advantage to using a spreadsheet for this problem? Explain.

13. Extend problem 19 on page 287 as follows: Design a spreadsheet application for multiplying the two matrices given in the problem. Suppose that officials for the Alsip Expressway decided to increase all tolls by 10% (rounded to the nearest $0.05). Use your spreadsheet to calculate the amount that each toll booth would have collected with a 10% increase in tolls. Compare these totals with the totals before the increase.

Additional Answers

for Chapter 6 Test

4 a

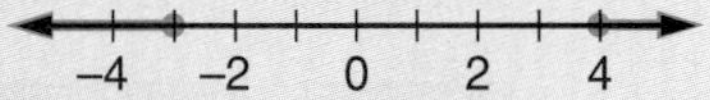

b

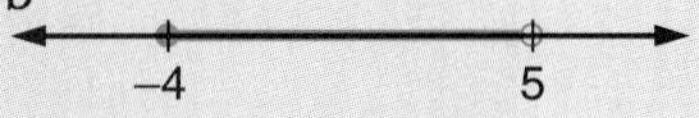

c

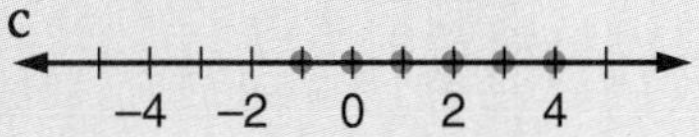

6 a $\begin{bmatrix} -4 & 9 & 15 \\ 8 & 8 & -10 \end{bmatrix}$ b $\begin{bmatrix} 20 \\ 23 \end{bmatrix}$

Answers for test problems 6 c, 9, 10, and 18 are on page A5.

Answers for Puzzles and Challenges are on page A5.

OVERVIEW

LONG-TERM INVESTIGATIONS

See *Investigations* book, Unit D, pages 80-

Gateways to Mathematical Investigations is a collection of long-term, real-world investigations that provides students with exciting opportunities to apply the ideas and concepts introduced in *Gateways to Algebra and Geometry: An Integrated Approach*. Unit D: Patterns with Shapes consists of five investigations that apply mathematical concepts and strategies within the context of tessellations and patterns. This unit is designed so that you can pick and choose the order and number of investigations you would like to do with your students. Each investigation is an exploration that takes several days to complete. Refer to the menu at the right to learn more about the investigations that make up this unit.

As your students work through the Unit D investigations, use Chapters 7 and 8 of *Gateways to Algebra and Geometry: An Integrated Approach* for concept and skill development, and as a source of rich problems, exercises, and applications. When appropriate, refer students to additional supportive sections in *Gateways to Algebra and Geometry* as well.

UNIT D: Patterns with Shapes

MENU OF INVESTIGATIONS

OVERVIEW: Groups apply mathematical concepts and strategies from Chapters 7 and 8 within the context of creating and analyzing tessellations and other patterns.

Covering Up

(4-5 days)
Students share prior knowledge about tessellations, predict which figures will tessellate, and test their predictions using manipulatives. Using what they know about the measures of angles of regular polygons, they determine which figures or combinations of figures will tessellate. They create tessellations.
STRANDS
Geometry, Measurement, Functions, Algebra
UNIFYING IDEAS
Multiple Representations, Patterns and Generalizations

Pentomino Puzzles

(2-3 days)
Students create and identify all twelve pentominoes. They determine which are symmetrical and which tessellate. They solve and create pentomino and tangram puzzles.
STRANDS
Geometry, Measurement, Logic and Language
UNIFYING IDEAS
Patterns and Generalizations

Another Way to Look At It

(4-5 days)
Students create tessellations and model them using coordinate graphing, rigid transformations, and matrices.
STRANDS
Geometry, Functions, Measurement
UNIFYING IDEAS
Multiple Representations, Patterns and Generalizations

Stretching the Boundaries

(4-5 days)
Students find function rules for transformations of a rectangle. They represent transformations, including dilations, using coordinate graphing, matrices, and function equations. Students analyze the transformations by creating tables and/or spreadsheets that use algebraic expressions and formulas for length, width, area, and perimeter. They create tiling patterns using transformations.
STRANDS
Functions, Algebra, Geometry, Measurement
UNIFYING IDEAS
Proportional Relationships, Multiple Representations, Patterns and Generalizations

Designers at Work

(3-4 days)
Students use the "take-and-put" technique with polygons that can tessellate to create interesting and whimsical figures that when translated and/or rotated make artistic tessellations.
STRANDS
Geometry, Measurement, Logic and Language
UNIFYING IDEAS
Multiple Representations, Patterns and Generalizations

INVESTIGATIONS FOR UNIT D	SCHEDULE	OBJECTIVES
1. **Covering Up** Pages 83-86*	4-5 Days	• To investigate the concept of tessellation and to determine which polygons tessellate
2. **Pentomino Puzzles** Pages 87-91*	2-3 Days	• To explore pentominoes; to determine which pentominoes are symmetrical and which tessellate; to solve and create pentomino puzzles
3. **Another Way to Look At It** Pages 92-96*	4-5 Days	• To use coordinate geometry, rigid transformations, and matrices to create and describe a tessellation
4. **Stretching the Boundaries** Pages 97-102*	4-5 Days	• To describe transformations using function equations, graphs, and matrices; to explore dilations and compare them to rigid transformations; to create tiling patterns
5. **Designers at Work** Pages 103-106*	3-4 Days	• To create artistic patterns using figures developed from polygons that tessellate and using rigid transformations

*Pages refer to *Gateways to Mathematical Investigations* Teacher's Edition.

PLANNING LONG-TERM INVESTIGATIONS

MATERIALS	RELATED RESOURCES
• Family Letter, 1 per student • Examples of tessellations, 1 per student • Pattern blocks or tiles, 1 set per group • Polygon patterns, 6-8 copies per student • Overhead projector; transparencies, markers, 1-2 per group • Grid paper, dot paper, 1-3 sheets per student • Construction paper, 1-2 sheets per student • Scissors, at least 1 per group; envelopes, 1 per student • Examples of artistic tessellations (M.C. Escher prints), (optional) • *Gateways to Mathematical Investigations* Student Edition, pp. 29-30, 1 of each per student	**Gateways to Algebra and Geometry** Section 7.1, pp. 318-324: Line and Angle Relationships Section 7.2, pp. 325-331: Quadrilaterals *Additional References* Section 3.7, pp. 136-141: Geometric Properties. Teacher's Resource File: Student Activity Masters
• Grid paper, 2-4 sheets per student • Color tiles; tangram squares (all 7 pieces), 1 set per student • Scissors, 1 per pair; sets of tangram pieces (optional) • *Gateways to Mathematical Investigations* Student Edition, pp. 31-32, 1 of each per student	**Gateways to Algebra and Geometry** Section 7.3, pp. 332-338: Symmetry Section 7.4, pp. 339-346: Rigid Transformations *Additional References* Teacher's Resource File: Student Activity Masters
• Coordinate grid paper, 3-5 sheets per student • Ruler or straightedge, 1 per student • Tracing paper, 3-5 sheets per student • Overhead projector; transparencies, markers, 1-3 per group • Colored construction paper, at least 1 sheet per student • Scissors, 1 per pair; envelopes, 1 per student • Compasses, protractors (optional) • Computer, LOGO software (optional) • *Gateways to Mathematical Investigations* Student Edition, pp. 33-34, 1 of each per student	**Gateways to Algebra and Geometry** Section 7.4, pp. 339-346: Rigid Transformations Section 7.6, pp. 354-361: Transformations and Matrices *Additional References* Section 6.3, pp. 267-274: Adding Signed Numbers Section 6.4, pp. 275-281: Mult. and Div. Signed Numbers Section 6.5, pp. 282-289: Matrix Multiplication Section 6.6, pp. 290-295: Subtracting Signed Numbers Section 6.8, pp. 302-309: Two-Dimensional Graphs Teacher's Resource File: Student Activity Masters
• Coordinate grid paper, 2-4 sheets per student • Protractor, ruler, envelope, 1 per student • Pattern blocks, color tiles, or students' cutout shapes; several per student • Scissors, 1 per group; markers, 2-5 per student • Blank spreadsheet, 1 per student • Computer, spreadsheet software, 1 per group • 1 Cardboard box; index cards, 24 (optional) • *Gateways to Mathematical Investigations* Student Edition, pp. 35-36, 1 of each per student	**Gateways to Algebra and Geometry** Section 7.1, pp. 318-324: Line and Angle Relationships Section 7.4, pp. 339-346: Rigid Transformations Section 8.2, pp. 378-384: Using Variables Section 8.3, pp. 385-391: Solving Equations Section 8.6, pp. 406-412: Substitution Section 8.7, pp. 413-419: Functions *Additional References* Section 1.5, pp. 29-35: Tables and Spreadsheets Section 2.5, pp. 78-85: More About Spreadsheets Section 6.4, p. 276: Mult. and Div. Signed Numbers Teacher's Resource File: Student Activity Masters
• Examples of tessellations, 1 per student • Examples of artistic tessellations (M.C. Escher prints), (optional) • Grid paper, 2-4 sheets per student; oaktag paper, 1 sheet per student • Blank paper, at least 3 sheets per student • Markers, 3-5 per student; scissors, ruler, 1 of each per student • Clear tape, 1 roll per group • Compass, protractor, at least 1 of each per group • Pattern blocks or color tiles (optional) • Transparency of coordinate grid axis and tessellations (optional) • *Gateways to Mathematical Investigations* Student Edition, pp. 37-38, 1 of each per student	**Gateways to Algebra and Geometry** Section 7.1, pp. 318-324: Line and Angle Relationships Section 7.2, pp. 325-331: Quadrilaterals Section 7.4, pp. 339-346: Rigid Transformations Section 7.5, pp. 347-353: Discovering Properties *Additional References* Teacher's Resource File: Student Activity Masters

See pages 368A and 368B: Assessment for Unit D.

CHAPTER 7

Contents

7.1 Line and Angle Relationships
7.2 Quadrilaterals
7.3 Symmetry
7.4 Rigid Transformations
7.5 Discovering Properties
7.6 Transformations and Matrices

Pacing Chart

Days per Section

	Basic	Average	Honors
7.1	2	2	1
7.2	3	2	1
7.3	2	2	2
7.4	2	2	2
7.5	2	2	2
7.6	2	2	2
Review	2	1	1
Assess	2	2	2
	17	15	13

Chapter 7 Overview

■ This chapter is designed to be a hands-on, activity-based discovery unit. You might want to use geoboards at the beginning of the chapter, as some discoveries are made by stretching figures into new ones and observing the results. Reflective tools (such as the Mira) and cutouts of figures can be used for hands-on practice with symmetry and transformations. To complement the hands-on work, matrices are introduced as a purely mathematical means of modeling transformations.

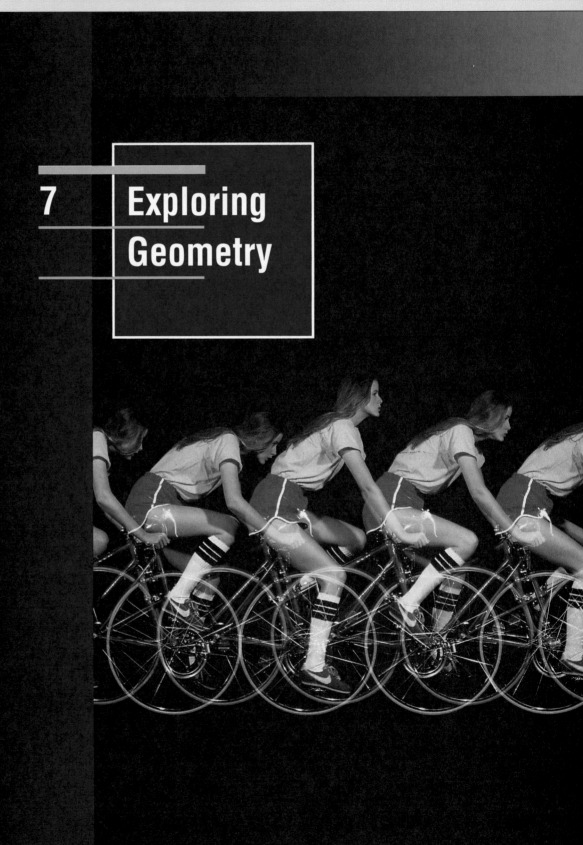

7 Exploring Geometry

CHAPTER 7

SPORTS PERFORMANCE A spherical basketball flies through the air in a symmetrical arc above a rectangular court and into a circular hoop with 1.6 seconds left in the game. Then, even as the 17,308 fans are still cheering, the players' statistics are updated. Mathematics pervades the world of sports!

In recent years, mathematical studies of the geometry of motion and the performance of the human body have led to improved training techniques, helping athletes move faster, jump higher, and throw farther. A lot of mathematics also goes into the design of sports equipment. A racing bicycle, for example, must be sturdy but light. By analyzing the properties of various materials and making refinements in shape and construction, designers have been able to develop bicycles that go faster with less effort than ever before.

INVESTIGATION

Equipment Design Choose a sport and explain in detail how changes in the design of equipment have had an impact on that sport.

Career Connection

Many students may be familiar with the use of statistics in sports. Examples include the earned run average and batting average in baseball, the field-goal and pass-completion percentages in football, and the free-throw percentage in basketball. Point out that computer simulations can provide athletes, coaches, and product designers with information about the effectiveness of new techniques and equipment without risking injury or building expensive prototypes.

Video Connection

For more information about the career discussed here, please refer to the Futures videocassette #6, program B: *Sports Performance,* featuring Jaime Escalante. For more information or to order a copy of the program, call PBS VIDEO at 800/344-3337.

Investigation Notes

Examples could include:

Baseball—bigger gloves, aluminum bats

Basketball—better shoes, hoops that can withstand slam dunks

Football—more streamlined football that is easier to pass

Golf—graphite or steel club shafts instead of wood, dimpled golf balls

Pole Vault—fiberglass poles instead of metal

Tennis—oversized graphite or aluminum rackets instead of smaller wooden ones

SECTION 7.1

Objectives

- To recognize parallel lines and perpendicular lines
- To recognize adjacent angles and vertical angles

Vocabulary: adjacent angles, parallel, perpendicular, vertical angles

Class Planning

Schedule 1–2 days

Resources

In *Chapter 7 Student Activity Masters* Book
Class Openers *1, 2*
Extra Practice *3, 4*
Cooperative Learning *5, 6*
Enrichment *7*
Learning Log *8*
Interdisciplinary Investigation *9, 10*
Spanish-Language Summary *11*

Transparencies *18, 21*

Class Opener

Use a ruler and a protractor to sketch an example of each of the following:

1. Two lines that do not meet (parallel lines)
2. Two lines that meet at right angles
3. Four angles with a common vertex
4. Three lines, each of which intersects the other two
5. Two lines that are not parallel yet do not meet

Answers:
1–4. Check students' drawings. Drawings for 2 and 3 may be identical.
5. Not possible in two dimensions.

Investigation Connection
Section 7.1 supports Unit D, *Investigations* 1, 4, 5. (See pp. 316A-316D in this book.)

SECTION 7.1
Line and Angle Relationships

Geometry is concerned with the size, shape, and location of objects. These attributes are often determined by the intended use of the object. Consider, for example, a loudspeaker. If the loudspeaker is intended for use at a rock concert, the speaker will have to be quite large. Small speakers cannot produce loud, distortion-free sound. If speakers are to be placed on a bookshelf, loudness is not critical, so the speakers can be smaller, but appearance and shape become important. Most people do not want unattractive objects on their bookshelves. If speakers are designed to fit in a person's ear, appearance becomes less important, but shape and size become critical.

PARALLEL AND PERPENDICULAR LINES

Picture a line and another line beside it. If the two lines are the same distance apart, so that they never meet, the lines are **parallel.** The symbol ∥ means "is parallel to." The rails of a straight railroad track are parallel.

If one of a pair of parallel lines is rotated 90°, the lines become **perpendicular.** The symbol ⊥ means "is perpendicular to." The top and side of a rectangular doorway are perpendicular.

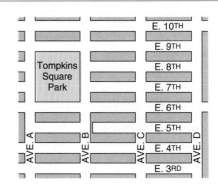

In the portion of New York City shown, the numbered streets are parallel to one another, and the lettered avenues are parallel to one another. Each of the streets is perpendicular to each of the avenues.

Chapter 7 Exploring Geometry

SECTION 7.1

Example 1
Refer to the map of a portion of New York City on the previous page.
a Suppose Juan is standing at the corner of B Avenue and 4th Street and René is standing at the corner of D Avenue and 10th Street. If they have agreed to meet at the corner of C Avenue and 7th Street, who will walk farther?
b From the corner of C Avenue and 7th Street, Juan will walk to the corner of A Avenue and 6th Street and René will walk to the corner of D Avenue and 10th Street. Who will walk farther?

Solution
a Both Juan and René will walk one long block and three short blocks. Since the streets are parallel and the avenues are parallel, all of the long blocks are the same distance and all of the short blocks are the same distance. Therefore, Juan and René will walk the same distance.
b Juan will walk two long blocks and one short block. René will walk one long block and four short blocks. Since we don't have a scale for the map, we are unable to answer the question. If we estimate that the length of three short blocks is equal to the length of one long block, Juan walks seven short blocks and René walks seven short blocks, so assume they will travel the same distance.

ADJACENT AND VERTICAL ANGLES

Consider three rays—$\vec{PA}$, $\vec{PB}$, and $\vec{PC}$—that have a common endpoint, P. Three angles are formed: ∠APB, ∠BPC, and ∠APC. ∠APB and ∠BPC are **adjacent angles**, since (1) they have the same vertex, (2) they have a side ($\vec{PB}$) in common, and (3) neither has a side that lies within the other.

Now consider these two intersecting lines and the four numbered angles. Two pairs of these angles—angles 1 and 3 and angles 2 and 4—are not adjacent. These pairs of angles are called **vertical angles.** Vertical angles have the same vertex, and their sides form two straight lines.

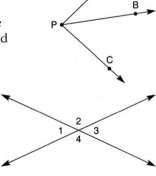

Example 2
Are ∠AFT and ∠AFM adjacent angles?

Solution
No, ∠AFT and ∠AFM are not adjacent angles because $\vec{FT}$ is inside ∠AFM. However, ∠AFT and ∠TFM are adjacent angles.

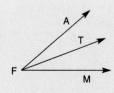

Lesson Notes
In the sample problem, students should draw and measure pairs of vertical angles and conclude that vertical angles are congruent. In this text we consider only angles greater than 0° and less than or equal to 180°.

Additional Examples

Example 1
If $\overline{CD}$ is parallel to the x-axis and $\overline{DE}$ is perpendicular to the x-axis, what is the value of p? The value of q?

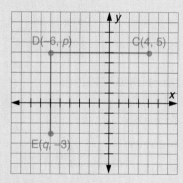

Answers: $p = 5$; $q = -6$

Example 2
Look at the rectangle in problem 11 on page 350. Are the following pairs of angles *adjacent, vertical,* or *neither*?
a. ∠VPE and ∠EPL
b. ∠VPO and ∠EPL
c. ∠OPL and ∠EVO
d. ∠EVP and ∠OVP
e. ∠OPL and ∠EPV

Answers: **a.** Adjacent **b.** Vertical **c.** Neither **d.** Adjacent **e.** Vertical

SECTION 7.1

Assignment Guide

Basic
Day 1 5–8, 13, 15, 17, 20, 21, 24, 27
Day 2 9, 10, 14, 16, 18, 22, 25, 28, 29

Average or Heterogeneous
Day 1 5–9, 11, 15, 16, 19, 21, 24, 25, 28
Day 2 10, 12–14, 17, 18, 20, 22, 23, 26

Honors
Day 1 5–9, 11–14, 16–20, 23–26, 28

Strands

Communicating Mathematics	1, 4–7, Investigation
Geometry	1–15, 17–19, 23, 25, 26, Investigation
Measurement	2, 3, 8, 9, 17–19, 23, 26
Data Analysis/ Statistics	19, 25
Technology	26
Probability	20
Algebra	2, 3, 8–11, 13 15–17, 20–24, 28, 29
Patterns & Functions	25, 26
Number Sense	27–29
Interdisciplinary Applications	1, 4, 19, Investigation

Individualizing Instruction

Limited English proficiency Students may need help understanding the vocabulary in this section, as the words are long and some have nonmathematical meanings. You may need to devote some extra time to helping these students master the vocabulary.

Sample Problem

Problem In the diagram, m∠SOT = 57. What is the measure of
 a ∠SOP?   **b** ∠TOX?   **c** ∠POX?

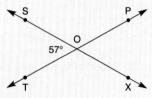

Solution **a** $\overleftrightarrow{TP}$ is a straight line, so ∠TOP is a 180° angle. Therefore, m∠SOP = 180 − 57 = 123.
 b $\overleftrightarrow{SX}$ is another straight line, so m∠TOX = 180 − 57 = 123.
 c Since adjacent angles SOP and POX form a 180° angle, m∠POX = 180 − 123 = 57.

Like two segments with the same length, two angles with the same measure are said to be congruent. The symbol ≅ means "is congruent to," so we can write ∠SOT ≅ ∠POX and ∠SOP ≅ ∠TOX. Draw a few other pairs of vertical angles and measure them. What can you conclude about vertical angles?

Think and Discuss

P **1** Find several examples of parallelism and perpendicularity in your classroom. Are there reasons why the objects involved are parallel or perpendicular?

P **2** $\overline{DE}$ is parallel to the y-axis, and $\overline{EF}$ is parallel to the x-axis.
 a What are the coordinates of point E? (−3, −4)
 b How far is it from D to E? 7 units
 c How far is it from E to F? 7 units
 d How long is $\overline{DF}$? $\sqrt{98}$, or ≈9.9, units

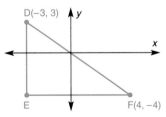

P **3** If m∠4 = 33, what is the measure of
 a ∠2? 33   **b** ∠1? 147   **c** ∠3? 147

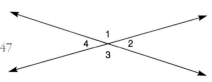

320   **Chapter 7** *Exploring Geometry*

P **4** The shape and size of most objects is determined by their intended use. Think about the following questions and explain your answers.
  a Why are books rectangular?
  b Why are soft-drink cans round and cereal boxes rectangular?
  c Why are most houses and office buildings rectangular and most large stadiums round?
  d Why are computer disks, compact discs, and records round, yet stored in square cases?
  e Why do road signs have different shapes (circular, rectangular, triangular, and so on)?

Problems and Applications

P **Communicating** In problems 5–7, identify the pair of angles indicated by blue arrows as *adjacent* or *not adjacent*. If the angles are not adjacent, explain why they are not.

5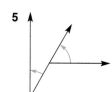

6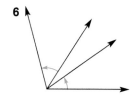

7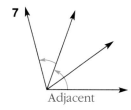
Adjacent

P **8** In the diagram, m∠VZW = 23, m∠WZX = 45, and m∠XZY = 31.
What is
  a m∠VZX?   68
  b m∠WZY?   76
  c m∠VZY?   99

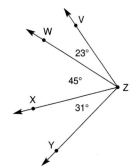

P **9** If m∠LJM = 129, what is
  a m∠KJL?   51
  b m∠KJN?   129
  c m∠NJM?   51

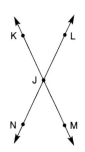

Section 7.1 Line and Angle Relationships

SECTION 7.1

Problem-Set Notes

for pages 322 and 323

- 14 might be shown as nonadjacent sections of a circle graph or as nonadjacent angles that share a ray (such as ∠s AFT and AFM in example 2).

Checkpoint

1. Graph the points N(5, 4) and M(–3, –2) on a coordinate plane. Place point O in Quadrant IV so that NO is parallel to the y-axis, and MO is parallel to the x-axis.
2. What are the coordinates of point O?
3. How far is it from N to O?
4. How far is it from M to O?
5. How long is MN?

Problems 6–8 refer to the angles shown in problem 8 on page 321. Classify the angles as *adjacent, vertical,* or *neither*.
6. ∠WZX and ∠WZV
7. ∠XZY and ∠WZV
8. ∠WZX and ∠WZY

Answers:
1.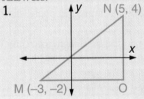

2. (5, –2) 3. 6 units 4. 8 units
5. 10 units 6. Adjacent
7. Neither 8. Neither

Stumbling Block

Students may assume that all angles that appear to be right angles are right angles. Remind students that a right angle in a diagram will either be marked as such or identified as such in directions.

322

P 10 ABCD is a rectangle with its sides parallel to the axes. What are the coordinates of points A and C? (–3, 4); (9, –2)

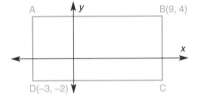

P, R, I 11 On a rectangular coordinate system, draw the line that passes through (–2, 5) and (6, 5). Label this line ℓ. Then draw a line that is parallel to line ℓ, a line that is perpendicular to line ℓ, and a line that is neither parallel nor perpendicular to line ℓ. Label each line by identifying the coordinates of two points on it.

P, R 12 **a** List all the pairs of adjacent angles in the figure.
b List all the pairs of complementary angles in the figure. ∠MSW and WST
c List all the pairs of supplementary angles in the figure.

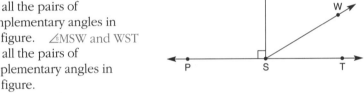

P, I 13 Carefully copy this graph of points A, B, C, and D. Draw AB and CD. Are they parallel? No

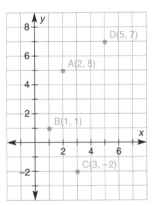

P 14 Can two supplementary angles have the same vertex without being adjacent? If so, draw an example of two such angles.

P 15 List all the pairs of parallel segments and all the pairs of perpendicular segments in polygon SQRE.

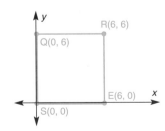

322   Chapter 7   Exploring Geometry

SECTION 7.1

P, R **16** What is the length of $\overline{AB}$? 10

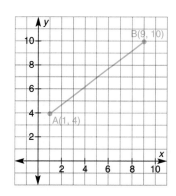

P, R **17** On a rectangular coordinate system, locate the points A(2, 5), B(–1, 3), C(6, 1), and D(6, –1). Then use a protractor to find
 a m∠BAD 90 **b** m∠BAC ≈101 **c** m∠BDA ≈27 **d** m∠ABD ≈63

P **18** What is the measure of the supplement of the complement of a 35° angle? 125

P **19 a** Find the measures of the eight numbered angles.
 b Which of the angles are congruent?
 c Which of the angles are complementary?
 d Which of the angles are supplementary?

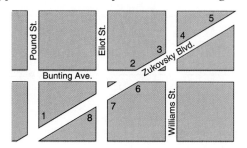

◀ LOOKING BACK **Spiral Learning** LOOKING AHEAD ▶

R, I **20** If one of the points in the diagram is picked at random, what is the probability that it is
 a In Quadrant I? $\frac{2}{9}$
 b In Quadrant III? $\frac{1}{9}$
 c On the x-axis? $\frac{1}{3}$

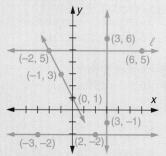

R **21**

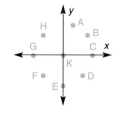

 a How many units apart are the numbers –10 and –4? 6 units
 b Evaluate –4 – (–10). 6
 c How many units apart are the numbers –8 and 12? 20 units
 d Evaluate 12 – (–8). 20

R **22** Evaluate $\begin{bmatrix} -7 & 2 & 4 \\ 15 & 9 & -31 \end{bmatrix} + \begin{bmatrix} -6 & -8 & -1 \\ -18 & -12 & 31 \end{bmatrix}$. $\begin{bmatrix} -13 & -6 & 3 \\ -3 & -3 & 0 \end{bmatrix}$

Section 7.1 Line and Angle Relationships 323

Cooperative Learning

- Refer to problem 25. Have groups of students develop a general rule for the number of angles formed by *n* rays with a common endpoint. Encourage them to expand the table above and below what is given if necessary.
 Answer: $T_n = \frac{n(n-1)}{2}$
- Problems 4 and 26 are appropriate for cooperative learning.

Additional Answers

for pages 322 and 323

11 Possible graph:

12 a ∠PSM and ∠MSW, ∠PSM and ∠MST, ∠PSW and ∠WST, ∠MSW and ∠WST **b** ∠MSW and ∠WST **c** ∠PSM and ∠MST, ∠PSW and ∠WST

14 Yes; possible example:

15 $\overline{QR} \parallel \overline{SE}$; $\overline{QS} \parallel \overline{RE}$; $\overline{QR} \perp \overline{RE}$; $\overline{RE} \perp \overline{ES}$; $\overline{ES} \perp \overline{SQ}$; $\overline{SQ} \perp \overline{QR}$

19 a m∠1 = 60; m∠2 = 150; m∠3 = 120; m∠4 = 60; m∠5 = 30; m∠6 = 150; m∠7 = 120; m∠8 = 60 **b** ∠1 ≅ ∠4 ≅ ∠8, ∠2 ≅ ∠6, ∠3 ≅ ∠7 **c** ∠1 and 5, ∠4 and 5, ∠8 and 5 **d** ∠1 and 3, ∠1 and 7, ∠2 and 5, ∠3 and 4, ∠3 and 8, ∠4 and 7, ∠5 and 6, ∠7 and 8

323

SECTION 7.1

Problem-Set Notes

for page 324

- 25 encourages students to consider solving a simpler problem, then looking for a pattern in order to solve a more difficult problem.
- 26 can be adapted into a classroom activity using geoboards.

Communicating Mathematics

Have students describe a situation in which groups of people form parallel lines and a situation in which groups of people form perpendicular lines.

- Problems 1, 4–7, and Investigation encourage communicating mathematics.

Additional Practice

Refer to ∠1–4 shown on page 319. If m∠1 = 46, what are the measures of the angles named in problems 1–3?

1. ∠2 2. ∠3 3. ∠4
4. Name an acute angle adjacent to ∠2.
5. Name two obtuse angles that are vertical angles.

Answers:
1. 134 2. 46 3. 134
4. ∠1 or ∠3 5. ∠2 and ∠4

Answers for problem 26 and Investigation are on page A6.

R, I **23** What are the lengths of $\overline{AB}$, $\overline{BC}$, $\overline{CD}$, and $\overline{DA}$? 5; 5; 5; 5

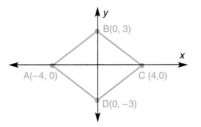

R **24** Evaluate $\begin{bmatrix} -6 & 4 \\ 3 & -6 \end{bmatrix} \cdot \begin{bmatrix} 12 & -3 \\ -5 & -2 \end{bmatrix} \cdot \begin{bmatrix} -92 & 10 \\ 66 & 3 \end{bmatrix}$

R **25** Study the figures shown. Then copy and complete the table.

Figure	Number of Rays	Number of Angles
A	3	3
B	4	6
C	5	10
D	6	15
E	12	66

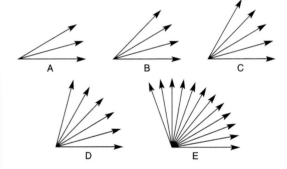

R **26** **Spreadsheets** Use a spreadsheet to make a table listing (1) integer values of a from 1 to 20, (2) the lengths of $\overline{AC}$, $\overline{CB}$, and $\overline{AB}$ for each value of a, (3) the perimeter of $\triangle ABC$ for each value of a, and (4) the area of $\triangle ABC$ for each value of a.

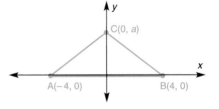

R **In problems 27–29, name the two consecutive integers that the given number is between.**

27 $\frac{16}{3}$ 5 and 6   **28** -8.2 -9 and -8 **29** $-\frac{12}{5}$ -3 and -2

Investigation

Architecture When a skyscraper is built with vertical sides, the opposite sides are almost parallel, but not quite. Why is this? Draw a picture or build a model to show why the sides must be designed to be a little farther apart at the top than at the bottom.

Chapter 7 Exploring Geometry

SECTION 7.2

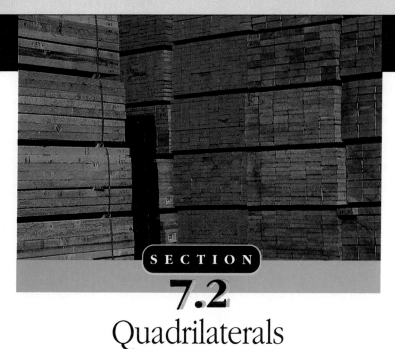

Investigation Connection
Section 7.2 supports Unit D, *Investigations* 1, 5. (See pp. 316A-316D in this book.)

SECTION 7.2
Quadrilaterals

SQUARES, RECTANGLES, AND PARALLELOGRAMS

As you saw in Chapter 3, a **quadrilateral** is a polygon with four sides. By using the concepts of parallelism, perpendicularity, and congruence, we can identify some special kinds of quadrilaterals. You are already familiar with squares, like the one shown here. All four sides of a square are congruent. Adjacent sides are perpendicular and opposite sides are parallel.

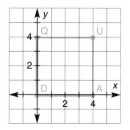

Example 1
Suppose that side $\overline{UA}$ of square QUAD above is moved to the right in such a way that its length is not changed, it is kept parallel to side $\overline{QD}$, and sides $\overline{QU}$ and $\overline{DA}$ are allowed to stretch. What kind of figure will QUAD become?

Solution
Let's try moving $\overline{UA}$ three units to the right, so that vertex U moves from (4, 4) to (7, 4) and vertex A moves from (4, 0) to (7, 0). All four angles are still right angles, and the opposite sides are still congruent and parallel. The only change is that adjacent sides are no longer congruent. The result is a rectangle.

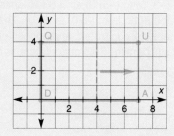

Objective
- To identify squares, rectangles, parallelograms, rhombuses, and trapezoids

Vocabulary: consecutive vertices, diagonal, opposite vertices, parallelogram, quadrilateral, rhombus, trapezoid

Class Planning

Schedule 1–3 days

Resources

In *Chapter 7 Student Activity Masters* Book
Class Openers 13, 14
Extra Practice 15, 16
Cooperative Learning 17, 18
Enrichment 19
Learning Log 20
Technology 21, 22
Spanish-Language Summary 23

Transparencies 29 a,b, 30 a–f

Class Opener

For problems 1–4, draw two four-sided figures that both have
1. Just one set of parallel sides
2. Only right angles
3. Two sets of parallel sides but no right angles
4. At least one right angle

Answers will vary.

Lesson Notes

- The definitions of several quadrilaterals are presented by starting with a square and stretching it to form new figures. The use of geoboards might be helpful. If a computer is available, software such as the Geometer's Sketchpad can be used to effectively demonstrate the concepts in this section.

Section 7.2 Quadrilaterals   325

SECTION 7.2

Additional Examples

Examples 1–4 refer to square QUAD on page 325.

Example 1
Suppose that side $\overline{QU}$ is moved up so that its length remains the same, it is kept parallel to side $\overline{DA}$, and sides $\overline{QD}$ and $\overline{UA}$ are allowed to stretch. What kind of figure will QUAD become?

Answer: Rectangle

Example 2
Suppose that side $\overline{DA}$ is moved to the left, remaining parallel to side $\overline{QU}$, and sides $\overline{QD}$ and $\overline{UA}$ are allowed to stretch. What kind of figure will QUAD become?

Answer: Parallelogram

Example 3
If vertex U is moved *n* units to the right and *n* units up, vertex D is moved *n* units down and *n* units to the left, and all sides are allowed to stretch, what kind of figure will QUAD become?

Answer: Rhombus

Example 4
If vertex D is moved to the left along the x-axis while the other vertices remain fixed and sides $\overline{QD}$ and $\overline{DA}$ are allowed to stretch, what kind of figure will QUAD become?

Answer: Trapezoid

Communicating Mathematics

Have students describe in writing or orally how a square can be stretched to form a rectangle, a parallelogram, a trapezoid, and a rhombus. Have students start by drawing a square and labeling its vertices so that they can refer to specific points.

Example 2
Consider the rectangle produced in Example 1. Suppose that side $\overline{DA}$ is moved to the right along the x-axis. If its length is not changed and sides $\overline{QD}$ and $\overline{UA}$ are allowed to stretch, what kind of figure will QUAD become?

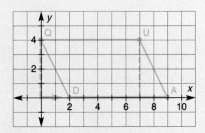

Solution
Let's try moving $\overline{DA}$ two units to the right, so that D moves from (0, 0) to (2, 0) and A moves from (7, 0) to (9, 0). The opposite sides remain parallel and congruent, but the figure's angles are no longer right angles. The quadrilateral has become a **parallelogram**.

RHOMBUSES AND TRAPEZOIDS

There are two other kinds of special quadrilaterals you should know about.

Example 3
Figure RMBS is a square. Suppose that vertices R and B are moved the same distance in opposite directions along the x-axis, with all four sides being allowed to stretch. What kind of figure will RMBS become?

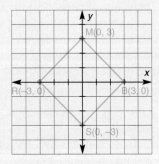

Solution
Let's move R two units to the left, from (−3, 0) to (−5, 0), and B two units to the right, from (3, 0) to (5, 0). The four sides are still congruent, and opposite sides are still parallel. This quadrilateral is a **rhombus**.

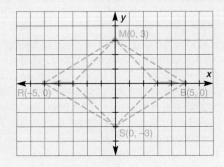

Chapter 7 Exploring Geometry

SECTION 7.2

Example 4
Suppose that vertex A of parallelogram DRAG is moved to the right along the x-axis. If the other vertices remain fixed and sides $\overline{GA}$ and $\overline{RA}$ are allowed to stretch, what kind of figure will DRAG become?

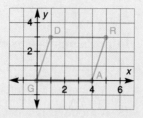

Solution
Let's move A three units to the right, to (7, 0). $\overline{DR}$ and $\overline{GA}$ are still parallel, but $\overline{DG}$ and $\overline{RA}$ are not. A quadrilateral such as this, with one pair of parallel sides, is called a **trapezoid**.

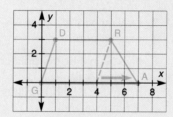

Quadrilaterals have four vertices. The quadrilateral on the right has vertices A, B, C, and D. If two vertices are the endpoints of an edge (A and B), they are said to be **consecutive vertices**. If they are not consecutive, like (A and C), they are **opposite vertices**. If two opposite vertices are connected, the segment formed is a **diagonal**. So far our discussion has focused on the sides and angles of quadrilaterals.

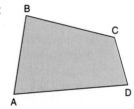

Sample Problem

Problem
a What kind of quadrilateral is figure AREH?
b What is the area of AREH?
c What is the length of $\overline{AR}$?
d What is the perimeter of AREH?

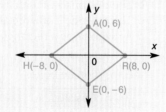

Solution
a AREH is a rhombus.
b Since OA = 6 and OR = 8, the area of $\triangle$ AOR is $\frac{1}{2} \cdot 6 \cdot 8$, or 24. Similarly, each of the triangles ROE, EOH, and HOA has an area of 24, so the area of AREH is 4(24), or 96.
c We use the Pythagorean Theorem.

$$(AR)^2 = 6^2 + 8^2 = 36 + 64 = 100$$

Since $(AR)^2 = 100$, AR must be $\sqrt{100}$, or 10.
d Since a rhombus has four congruent sides, the perimeter of AREH is 4(10), or 40.

■ Problems 7, 8, and 19 encourage communicating mathematics.

Assignment Guide

Basic
Day 1 12, 13, 15, 21, 24, 29, 30, 35
Day 2 14, 17, 19, 22, 27, 32, 36
Day 3 16, 18, 20, 25, 28, 33, 34

Average or Heterogeneous
Day 1 15, 16, 19, 21, 24, 25, 27, 29
Day 2 12–14, 18, 20, 22, 23, 26, 28, 35

Honors
Day 1 12–14, 19–21, 23–25, 27–29, 35

Strands

Communicating Mathematics	7, 8, 19
Geometry	1–28, 35, 36, Investigation
Measurement	14, 20, 23, 24, 26, 35
Technology	26
Algebra	8, 9, 14, 19–21, 24, 26, 28–36
Patterns & Functions	18, Investigation
Interdisciplinary Applications	22, 25, Investigation

Individualizing Instruction

Learning-disabled students Students may have difficulty with classification problems like 1–6 and 10. These students will need physical representations and extra guidance to draw proper conclusions.

SECTION 7.2

Problem-Set Notes

for pages 328 and 329

- 1–6 could be enhanced by having students give examples and counterexamples to justify their answers.
- 7 might be somewhat tricky, in that a square is not a trapezoid.
- 12 and 13 reinforce some of the new vocabulary.

Stumbling Block

Students may have a tough time identifying nonisosceles trapezoids as trapezoids. Emphasize that it is the condition of having only one pair of parallel sides that makes a quadrilateral a trapezoid. Use the following examples to expand students' concepts of a trapezoid.

Cooperative Learning

- Have groups of students develop a Venn diagram for the family of figures that includes quadrilaterals, squares, rhombuses, rectangles, trapezoids, and parallelograms.

Answer:

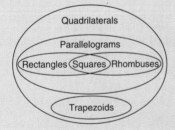

- Problems 19 and 25 are appropriate for cooperative learning.

Think and Discuss

P **In problems 1–6, is the statement true *always*, *sometimes*, or *never*?**

1 A square is a rectangle.   Always
2 A rectangle is a square.   Sometimes
3 A rhombus is a trapezoid.   Never
4 A rhombus is a parallelogram.   Always
5 A rhombus is a rectangle.   Sometimes
6 A rhombus is a square.   Sometimes

P **7** Explain why a square is a rectangle, a rhombus, and a parallelogram, but not a trapezoid.

P **8 a** Explain why quadrilateral DQUA is a rhombus.
  b Explain why quadrilateral DQUA is not a square.

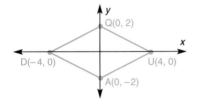

P **9 a** What kind of quadrilateral is KLMN?
  b How could points K and N be moved so that KLMN would become a rectangle that is not a square?
  c How could points K and L be moved so that KLMN would become a rectangle that is not a square?

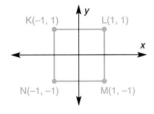

P **10** Is the statement true *always, sometimes,* or *never*?
  a The diagonals of a rectangle are perpendicular.   Sometimes
  b The diagonals of a rectangle are congruent.   Always
  c The diagonals of a rectangle are parallel.   Never
  d A diagonal of a rectangle is perpendicular to a side of the rectangle.   Never

P **11** Copy the following table, then classify the polygons below by checking the appropriate boxes.

	Quadrilateral	Trapezoid	Parallelogram	Rhombus	Rectangle
a	✓				
b	✓		✓		✓
c	✓		✓	✓	
d	✓	✓			

a   b 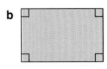   c   d

328   Chapter 7   Exploring Geometry

SECTION 7.2

Problems and Applications

In problems 12 and 13, refer to quadrilateral WHAT.

P **12** Classify each pair of vertices as *consecutive* or *opposite*.
   a H and T   Opp.   **b** A and T   Consec.
   c A and H   Consec.   **d** A and W   Opp.

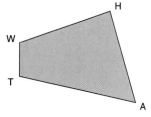

P **13** Classify each pair of sides as *adjacent* or *opposite*.
   a $\overline{TW}$ and $\overline{AT}$ Adj.   **b** $\overline{TW}$ and $\overline{HW}$ Adj.
   c $\overline{TA}$ and $\overline{HW}$ Opp.   **d** $\overline{HA}$ and $\overline{HW}$ Adj.

P **14** ABCD is a square. If you move $\overline{BC}$ to the right, keeping it parallel to $\overline{AD}$ and allowing $\overline{AB}$, $\overline{DB}$, and $\overline{DC}$ to stretch, what will happen to the value of
   a w?   **b** z?   **c** w + z?

   a Will decrease   **b** Will increase
   c Will remain the same (90)

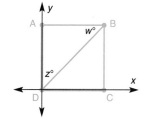

P **In problems 15–18, sketch an example of each figure described. If no such figure exists, write *impossible*.**

   15 A rectangle that is not a trapezoid

   16 A rectangle that is not a parallelogram   Impossible

   17 A rectangle that is not a rhombus

   18 A rectangle that is not a square

P **19** **Communicating**
   a What kind of quadrilateral is QUAD?
   b How could Q and U be moved to make QUAD a parallelogram that is not a rectangle?
   c How could U and A be moved to make QUAD a parallelogram that is not a rectangle?
   d How could D and U be moved to make QUAD a parallelogram that is not a rectangle?

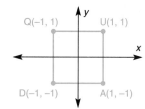

P, R **20** Determine the area of trapezoid TRAP.   35

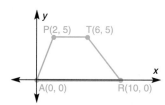

Reteaching

Have students attempt to define various quadrilaterals in terms of other quadrilaterals. For example, a square is a rhombus with four right angles. Have students explore the following question: Which figure can be defined in terms of the most other figures?

Additional Answers

for pages 328 and 329

7 A square is a rectangle because adjacent sides are perpendicular. A square is a rhombus because all four sides are congruent. A square is a parallelogram because opposite sides are parallel. A square is not a trapezoid because it has two pairs of parallel sides.

8 a Because its opposite sides are parallel and all four of its sides are congruent **b** Because its angles are not right angles

9 a Square **b** Possible answer: They could be moved the same distance to the left. **c** Possible answer: They could be moved the same distance upward.

15 Any rectangle

16 Impossible

17 Any rectangle that does not have 4 equal sides

18 Any rectangle that does not have 4 equal sides

19 a Square **b** Possible answer: They could be moved the same distance to the right or to the left. **c** Possible answer: They could be moved the same distance up or down. **d** Possible answer: D could be moved to the left, and U could be moved the same distance to the right.

SECTION 7.2

Problem-Set Notes

for pages 330 and 331

- **27** and **36** are activity-based problems.

Additional Practice

In problems **1–4**, is the statement true *always, sometimes,* or *never?*
1. A rectangle is a rhombus.
2. A parallelogram has only one pair of parallel sides.
3. A quadrilateral with a right angle is a rectangle.
4. A rhombus with a right angle is a square.

Problems **5** and **6** refer to rhombus QUAD in problem 8 on page 328.
5. Classify each pair of vertices as *consecutive* or *opposite*.
   a. Q and U   b. D and U
   c. U and A   d. A and Q
6. Classify each pair of sides as *adjacent* or *opposite*.
   a. $\overline{QD}$ and $\overline{UA}$ b. $\overline{DQ}$ and $\overline{UQ}$
   c. $\overline{DA}$ and $\overline{AU}$ d. $\overline{QU}$ and $\overline{DA}$

Answers:
1. Sometimes   2. Never
3. Sometimes   4. Always
5. a. Consecutive   b. Opposite
   c. Consecutive   d. Opposite
6. a. Opposite   b. Adjacent
   c. Adjacent   d. Opposite

Checkpoint

In problems **1–4**, is the statement true *always, sometimes,* or *never?*
1. A square is a parallelogram.
2. A trapezoid is a rectangle.
3. A rectangle is a parallelogram.
4. A parallelogram is a trapezoid.

P, I **21** If BLUE is a parallelogram, what are the coordinates of point L?   (−3, 4)

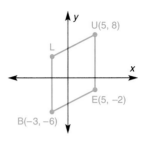

P **22** **Language Arts** Look up the words *quadrilateral, rectangle,* and *parallelogram* in a dictionary. What languages are the sources of these words? To what characteristics of the three polygons do the words refer?

P **23** Use a protractor to measure ∠F, ∠R, ∠E, and ∠D of this parallelogram. (You may need to trace the figure and extend its sides.) What do you notice?

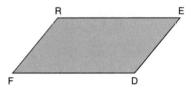

P, R **24**
a What is the area of SPQR?   36
b What is the perimeter of SPQR?   24
c Draw a new quadrilateral by moving S to (10, 6) and R to (4, 6).
d What is the area of the new quadrilateral?   36
e What is the perimeter of the new quadrilateral?   ≈26.42

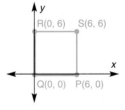

P **25** **Geography** Look at a map of the United States. Which states appear to be shaped like quadrilaterals? What kinds of quadrilaterals do they resemble?

P, R **26** **Spreadsheets** Use a spreadsheet to make a table listing (1) integer values of *a* from 1 to 20, (2) the lengths of $\overline{AB}$, $\overline{BC}$, $\overline{AD}$, and $\overline{DC}$ for each value of *a*, (3) the perimeter of quadrilateral ABCD for each value of *a*, and (4) the area of ABCD for each value of *a*. For what values of *a* is AB = DC?   8 and 12

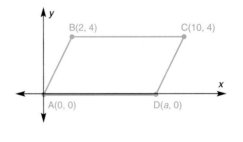

P **27** On a sheet of paper, draw a triangle with sides of different lengths. Then put another sheet of paper under your drawing and cut along the segments you drew so that you get two identical triangles. How many different parallelograms can you make by putting the two triangles together?   3 parallelograms

330 Chapter 7 *Exploring Geometry*

SECTION 7.2

P, I **28** If you slide points Y and R so that quadrilateral GRAY becomes a square, what will the new coordinates of Y and R be?   (–3, 0); (5, 0)

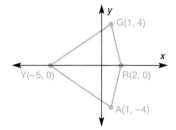

◀ LOOKING BACK   **Spiral Learning**   LOOKING AHEAD ▶

R **29 a** What is 20% of –40?   –8
  b What percent of –45 is –90?   200%
  c If a value decreases from 40 to 30, by what percent has the value decreased?   25%

R **In problems 30 and 31, write an absolute-value equation or inequality that is represented by the graph.**

30    31

R **In problems 32–34, solve the proportions for a, b, and c.**

32 $\dfrac{a}{-12} = \dfrac{-6}{-9}$ $a = -8$ **33** $\dfrac{-14}{-3} = \dfrac{b}{-6}$ $b = -28$ **34** $\dfrac{-2 + 5}{-6 - 1} = \dfrac{-15}{c}$ $c = 35$

R, I **35** Suppose that ∠FOG and ∠GOH are complementary, and m∠GOH = 34. Find the measures of ∠FOH and ∠FOG.   90; 56

I **36** Imagine that the shaded rectangular region is balanced on the origin, so that it can be spun around like a phonograph record. If it is turned 90° clockwise, what will the coordinates of its vertices be?
(–2, 4), (2, 4), (2, –4) (–2, –4)

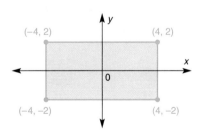

Investigation

Tessellations   From a sheet of paper, cut a square, a rectangle, a rhombus, a parallelogram, a trapezoid, and a quadrilateral having no parallel sides. If you had a large number of copies of each figure, which could you use to tile a floor, without leaving gaps or overlapping tiles? What other kinds of polygons could be used to tile floors?

Section 7.2 Quadrilaterals   331

Problems 5 and 6 refer to figure FRED on page 330.
5. Identify each pair of vertices as *consecutive* or *opposite*.
  a. F and R   **b.** R and D
  c. E and D   **d.** D and F
6. Identify each pair of sides as *adjacent* or *opposite*.
  a. $\overline{FR}$ and $\overline{RE}$ **b.** $\overline{RE}$ and $\overline{FD}$
  c. $\overline{ED}$ and $\overline{DF}$ **d.** $\overline{ED}$ and $\overline{FR}$

Answers:
1. Always   2. Never
3. Always   4. Never
5. a. Consecutive b. Opposite
   c. Consecutive d. Consecutive
6. a. Adjacent b. Opposite
   c. Adjacent d. Opposite

Additional Answers

for pages 330 and 331

22 Answer on page A6.

23 m∠F = 50, m∠R = 130, m∠E = 50, m∠D = 130; possible answer: Opposite angles of a parallelogram are congruent; consecutive angles of a parallelogram are supplementary.

24 c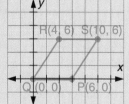

25 Possible answer: Wyoming and Colorado appear to be shaped like rectangles.

26 Answer on page A6.

30 $|x| = 5$

31 $|x - 2| < 4$

Answer to Investigation is on page A6.

SECTION 7.3

Objectives

- To identify figures that possess reflection symmetry
- To identify figures that possess rotation symmetry

Vocabulary: line of symmetry, reflection symmetry, rotation symmetry, symmetry

Class Planning

Schedule 2 days

Resources

In *Chapter 7 Student Activity Masters* Book
Class Openers 25, 26
Extra Practice 27, 28
Cooperative Learning 29, 30
Enrichment 31
Learning Log 32
Using Manipulatives 33, 34
Spanish-Language Summary 35

In *Tests and Quizzes* Book
Quiz on 7.1–7.3 (Basic) 181
Quiz on 7.1–7.3 (Average) 182

Transparencies 31 a,b

Class Opener

Name as many symmetrical objects as you can that appear in the classroom. Be prepared to show the line or center of symmetry for each object you name.

Lesson Notes

- Students may need some hands-on practice with symmetrical figures to find lines of symmetry. Pieces of cardboard and pushpins will allow students to manipulate figures to discover rotation symmetry.

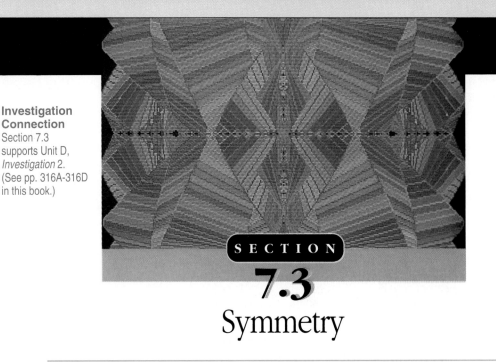

Investigation Connection
Section 7.3 supports Unit D, *Investigation* 2. (See pp. 316A-316D in this book.)

SECTION 7.3
Symmetry

WHAT IS SYMMETRY?

In the pictures below, the objects on the left have an attribute that the objects on the right do not. That attribute is **symmetry.**

Symmetric   **Not Symmetric**

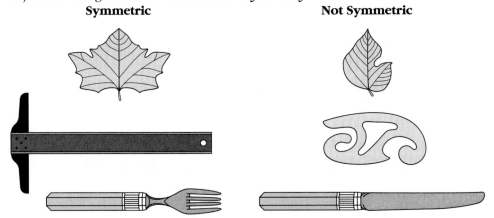

Basically, symmetry involves balance. The objects on the left appear balanced because parts of their shape correspond exactly to other parts. This is not true of the objects on the right.

Symmetry is widespread in nature, and it also plays an important role in manufacturing and design. Here are a few examples:

- Artists, architects, and photographers make use of symmetry (and of carefully planned departures from symmetry) to make visually pleasing designs and to draw attention to certain parts of their works.
- Containers are often made to be symmetric so that they are easier to handle and to stack.

Chapter 7 Exploring Geometry

SECTION 7.3

- Symmetric tools can be used as easily by left-handed people as by right-handed people, but nonsymmetric tools (most kinds of scissors, for example) are usually designed for right-handed people and are difficult for lefties to use.
- Mathematicians find symmetry useful because it helps them understand properties of complex figures and it can reduce the effort required to solve difficult problems.

REFLECTION SYMMETRY

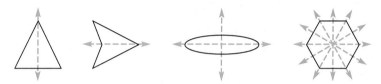

Each of these figures possesses **reflection symmetry.** A figure has this kind of symmetry if you can draw a line dividing the figure into two halves that are mirror images of each other. The figure is said to be symmetric about this **line of symmetry.** Figures may have only one line of symmetry, or they may have two, or they may have many.

A rectangle is drawn on a sheet of stiff paper, as shown on the right. If we fold the paper along one of the rectangle's lines of symmetry, the two halves of the rectangle will fit together perfectly. And if we cut the rectangle out, the line of symmetry will serve as a "balance line"—the rectangle will balance on a ruler placed along the line.

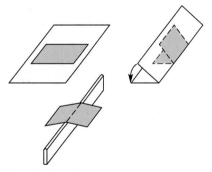

Example 1
A rectangle has vertices R(–3, 4), E(3, 4), C(3, 0), and T(–3, 0). Locate the lines of symmetry of the rectangle.

Solution
First, we draw the figure described. We notice that for each point of the rectangle on one side of the y-axis, there is a corresponding point the same distance on the other side of the y-axis. Therefore, the y-axis is a line of symmetry of the rectangle. For the same reason, a line parallel to the x-axis and two units above it is a line of symmetry of the rectangle.

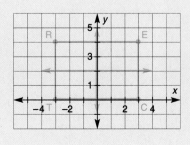

Additional Examples

Example 1
A square has vertices S(0, 3), P(3, 3), Q(3, 0) and R(0, 0). Locate the lines of symmetry of the square.

Answer:
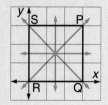

Example 2
Identify the rotation symmetries of the figure shown.

Answer: 90°, 180°

Cooperative Learning

- Have groups of students classify the lowercase letters of the English alphabet or the uppercase letters of the Greek alphabet according to the kinds of symmetry they have.
- Answers will vary, depending on how students form letters. Possible answers: **c, i, l, o, v, t, w,** and **x** have reflection symmetry; **l, o, s, x,** and **z** have rotation symmetry. A, B, Δ, E, H, Θ, I, K, Λ, M, Ξ, O, Π, Σ, T, Y, Φ, X, Ψ, and Ω have reflection symmetry. Z, H, Θ, I, N, Ξ, O, Φ, and X have rotation symmetry.
- Problems 29 and 33 are appropriate for cooperative learning.

SECTION 7.3

Assignment Guide

Basic
Day 1 12, 13, 16–18, 24–27, 30, 34–36, 45
Day 2 14, 15, 19–23, 28, 32, 37–39, 43, 44

Average or Heterogeneous
Day 1 12–15, 17, 18, 21–24, 28, 33, 35–38
Day 2 16, 19, 20, 25–27, 29, 31, 32, 39–43, 45

Honors
Day 1 13, 16–18, 21–24, 30, 39, 41
Day 2 12, 14, 15, 19, 20, 25–29, 32, 33, 40, 42

Strands

Communicating Mathematics	11, 16, 29, 33
Geometry	1–33, 39–42, Investigation
Measurement	28, 31, 43–46
Algebra	28, 31, 34–38, 43–46
Number Sense	34–38
Interdisciplinary Applications	6, 29, 33, Investigation

Individualizing Instruction

The gifted student Students could do some research to find examples of symmetry in sculpture. Have them include works from the earliest known times, as well as works created by contemporary artists, and present their findings to the class.

Stumbling Block

Students may look for only one line of symmetry or one angle of rotational symmetry. Remind them that many figures are symmetrical in more than one way.

334

ROTATION SYMMETRY

  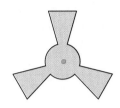 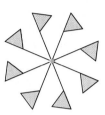

These figures all possess **rotation symmetry.** When each figure is rotated a certain number of degrees about a center point, the figure fits perfectly on top of its original position. (For each figure, try turning this book until the figure looks exactly as it does now. How much do you have to turn the book? Is it the same amount in each case?) If the amount of rotation is greater than zero and less than or equal to 180°, the figure has rotation symmetry.

Example 2
Identify the rotation symmetry of the figure shown.

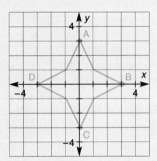

Solution
The figure's center of rotation is the origin. When the figure is rotated 90° about this point, it lies on its original position. The figure therefore has 90° rotation symmetry about the origin. Notice that the figure can also be rotated 180° and 270° to obtain the original figure.

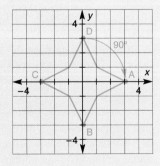

334   Chapter 7   Exploring Geometry

SECTION 7.3

Sample Problem

Problem What kinds of symmetry does each figure have?

a   b 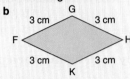 c

Solution a The square has reflection symmetry about both diagonals and about lines through the midpoints of opposite sides. It also has rotation symmetry about the point where its diagonals intersect.

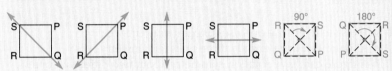

b The rhombus has reflection symmetry about both diagonals. It also has 180° rotation symmetry about the point where its diagonals intersect.

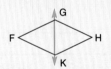 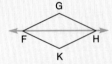 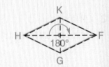

c Do you see that the circle has infinitely many line symmetries, since it is symmetrical about *any* diameter? It also has rotation symmetry, since no matter how much it is rotated about its center, it will always lie on its original position.

Think and Discuss

P **In problems 1–4, indicate whether the figure has reflection symmetry, rotation symmetry, both kinds of symmetry, or neither kind of symmetry.**

1  2  3   4

Additional Practice

In problems **1–4**, indicate whether the figure has reflection symmetry, rotation symmetry, both kinds of symmetry, or neither kind of symmetry.

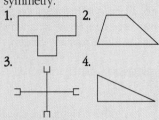

In problems **5** and **6**, draw all the lines of symmetry.

Answers:
1. Reflection   2. Neither
3. Both   4. Neither
5.   6.

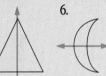

Communicating Mathematics

Have students write a paragraph on the following questions: Which type of symmetry—reflection or rotation—is easier for you to spot in a figure? Why do you think this is the case?

■ Problems 11, 16, 29, and 33 encourage communicating mathematics.

Section 7.3 Symmetry

SECTION 7.3

Problem-Set Notes

for pages 336 and 337

- 7–10 and 17–20 could be made into hands-on activities by having students copy and cut out each figure.

- 6, 29, 33, and **Investigation** point out that examples of symmetry can be found in many contexts.

Checkpoint

In problems **1** and **2**, indicate whether the figure has reflection symmetry, rotation symmetry, both kinds of symmetry, or neither kind of symmetry.

1.   2.

In problems **3** and **4**, draw all the figure's lines of symmetry.

3.   4.

5. Draw a figure that has
   a. Exactly three lines of symmetry
   b. Only one line of symmetry
   c. Rotation symmetry

Answers:
1. Both   2. Both
3.   4.

5. Check students' drawings.

Multicultural Note

Sofya Kovalevskaya, the daughter of a Russian nobleman, is regarded by historians as one of the greatest mathematicians of the nineteenth

336

P   **5** Is the statement true *always, sometimes,* or *never?*
   a A balance line is a line of symmetry.   Sometimes
   b A line of symmetry is a balance line.   Always
   c A center of rotation symmetry is on a balance line.   Always

P   **6** Sports   Discuss the kinds of symmetry possessed by
   a A football field   **b** A baseball field
   c A tennis court   **d** A golf course

P   **In problems 7–10, copy each figure. Then draw all the figure's lines of symmetry.**

7   8   9   10

P   **11** Is the statement *true* or *false?* Explain your answer.
   a A figure can have exactly three lines of symmetry.
   b A figure can have rotation symmetry about two different centers.

Problems and Applications

P   **12** Of lines a, b, c, d, and e, which are lines of symmetry of the rectangle?   b and d

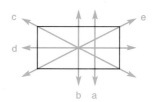

P   **In problems 13 and 14, each diagram shows part of a figure that has reflection symmetry about the dashed line. Copy the diagram and complete the figure.**

13   14

P   **15** What kind of symmetry, if any, does figure JKLMNOPQ have?

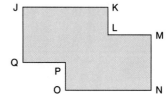

336   **Chapter 7**   Exploring Geometry

SECTION 7.3

P **16 a** How would you go about finding the center of symmetry of parallelogram PARL?
b How many degrees must PARL be rotated before it lies atop its original position? 180°
c When PARL is rotated that number of degrees, which parts of the rotated figure will lie on the positions now occupied by $\overline{PL}$, $\overline{RL}$, and ∠A? $\overline{RA}$, $\overline{PA}$, and ∠L

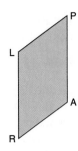

P, I **In problems 17–20, sketch each figure and draw all of the polygon's lines of symmetry.**

17 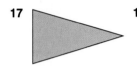   18    19 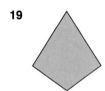   20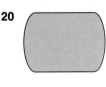

P **In problems 21–27, sketch an example of the figure described. If no such figure exists, write *impossible*.**

21 A trapezoid without reflection symmetry
22 A rectangle without reflection symmetry Impossible
23 A rectangle without rotation symmetry Impossible
24 A parallelogram without reflection symmetry
25 A parallelogram without rotation symmetry Impossible
26 A rhombus without reflection symmetry Impossible
27 A rhombus without rotation symmetry Impossible

P **28** Quadrilateral ABCD has 180° rotation symmetry about point P. If m∠A = 58, AB = 8 cm, and AD = 6 cm, what is
a DC? 8 cm   **b** BC? 6 cm   **c** m∠C? 58

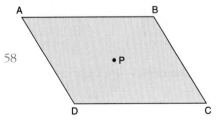

P **29** *Industrial Arts* Many bolt heads and nuts are hexagonal in shape.
a What advantage does a hexagonal nut have over a square nut?
b Why are octagonal nuts not used?
c Why are pentagonal nuts not used?
d See if you can find an example of a pentagonal bolt head.

Section 7.3 Symmetry 337

---

century. One important contribution was her mathematical approach to the conclusion that the rings of Saturn are egg-shaped ovals with one line of symmetry rather than ellipses with two lines of symmetry.

Additional Answers

for pages 335, 336, and 337

1 Both   **2** Rotation
3 Rotation   **4** Neither
6 *Answer on page A6.*
7   **8** No lines of symmetry

9   **10**

11 *Answer on page A6.*
13   **14**

15 180° rotation symmetry about the midpoint of $\overline{PL}$

16 a Find the point of intersection of the diagonals.

17   **18**

19   **20**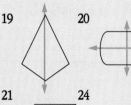

21   **24**

Answer for problem 29 is on page A6.

337

SECTION 7.3

Reteaching

Reteach line symmetry by having students use a device, such as a Mira, that is both transparent and reflective. Students can use a Mira both to verify that a given figure has a line of symmetry and to draw the mirror image of a figure.

Additional Answers

for page 338

30 Answers may vary, depending on how students form letters. Possible answer: A, B, C, D, E, K, M, T, U, V, W, and Y have reflection symmetry; N, S, and Z have rotation symmetry; H, I, O, and X have both kinds of symmetry; and F, G, J, L, P, Q, and R are not symmetric.

31 a (–1, 1); (–4, 0); (–1, –1); (0, –4); (1, –1)

33 Possible answer: Many kinds of flowers have radial (rotation) symmetry; some other kinds, as well as many kinds of leaves, have bilateral (reflection) symmetry. The arrangement of leaves on plant stems is often bilaterally symmetric. Some lower animals (starfish, coral polyps) have radial symmetry, whereas most higher animals (including human beings) are bilaterally symmetric.

39 Possible answer: They are congruent right triangles.

40 Possible answer: They are noncongruent.

41 Possible answer: They are congruent isosceles triangles.

42 Possible answer: They are congruent.

Investigation Answers will vary.

338

P **30** Classify the 26 capital letters of the alphabet according to the kinds of symmetry they have.

P **31** Both axes are lines of symmetry of the figure.
  a What are the coordinates of points A, B, C, D, and E?
  b Does the figure have rotation symmetry? If so, what point is the center of rotation, and what is the least rotation that will make the figure lie atop its original position? Yes; (0, 0); 90°

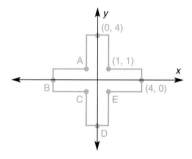

P **32** What kinds of quadrilateral always have reflection symmetry over each of their diagonals? Squares and rhombuses

P **33** **Communicating** Discuss some examples of symmetry that occur in nature.

◀ LOOKING BACK **Spiral Learning** LOOKING AHEAD ▶

R, I **34** Of the three expressions $9 - 8(5 + 3)$, $9 - 8(5) + 8(3)$, and $9 - 8(5) - 8(3)$, which are equal? $9 - 8(5 + 3)$ and $9 - 8(5) - 8(3)$

R **In problems 35–38, evaluate each expression.**

35 $-8(-7)$ 56   **36** $-8 + (-7)$ -15 **37** $\dfrac{-42}{7}$ -6 **38** $3 - (-1)$ 4

R, I **A diagonal of a quadrilateral divides the quadrilateral into two triangles. In problems 39–42, what can be concluded about the triangles formed by the diagonals of each quadrilateral?**

39 Rectangle   **40** Trapezoid   **41** Rhombus   **42** Parallelogram

R **In problems 43–46, express each length as a number of meters.**

43 245 cm 2.45 m   **44** 1.33 km 1330 m   **45** 10 ft ≈3.05 m   **46** 2 mi ≈3220 m

---

Investigation

Logos Many corporate logos (the symbols the companies use on their products and in their advertising) are symmetric. One that you may be familiar with is the CBS "eye." Find some other examples of symmetric logos—both ones with reflection symmetry and ones with rotation symmetry. Then design a logo that you can use as your personal symbol.

338 **Chapter 7** Exploring Geometry

SECTION 7.4

Investigation Connection
Section 7.4 supports Unit D, *Investigations* 2, 3, 4, 5. (See pp. 316A-316D in this book.)

SECTION 7.4

Rigid Transformations

REFLECTIONS

Triangle ABC has been "flipped" over line ℓ to produce triangle A´B´C´. Each point of the original figure has a corresponding point in triangle A´B´C´.

This "flip" is an example of a **reflection**. The new figure has the same size and shape as the original figure, but the new figure has a different position.

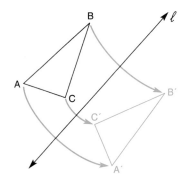

Example 1
Reflect figure TERASF over the x-axis.

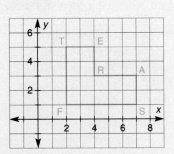

Objectives
- To transform a figure by means of reflection, rotation, and translation

Vocabulary: image, preimage, reflection, rotation, slide, translation, translation numbers

Class Planning

Schedule 2 days

Resources

In *Chapter 7 Student Activity Masters* Book
Class Openers 37, 38
Extra Practice 39, 40
Cooperative Learning 41, 42
Enrichment 43
Learning Log 44
Using Manipulatives 45, 46
Spanish-Language Summary 47

In *Tests and Quizzes* Book
Quiz on 7.1–7.4 (Honors) 183

Transparencies 32–36

Class Opener

On graph paper, draw △ABC, with A(2, 1), B(2, 5), and C(7, 1).

1. Draw △DEF with D(2, –1), E(2, –5), and F(7, –1). Describe how you would get △ABC to fit exactly on △DEF.

2. Draw △GHI with G(–1, 5), H(–1, 9), and I(4, 5). Describe how you would get △ABC to fit exactly on △GHI.

3. Draw △JKL with J(1, –2), K(5, –2), and L(1, –7). Describe how you would get △ABC to fit exactly on △JKL.

Answers:
1. Flip △ABC over the x-axis.
2. Slide △ABC 3 units left and 4 units up.
3. Turn △ABC 90° clockwise about the origin.

Section 7.4 Rigid Transformations

SECTION 7.4

Lesson Notes

- Students should have an opportunity to physically model each transformation. This will help their understanding of the effect each transformation has on a figure. You might have students verify the solutions to the examples and the sample problem by cutting out each figure from graph paper, and then performing the transformation.

Additional Examples

Example 1
Refer to figure TERASF on page 339. Reflect the figure over the y-axis.

Answer: Figure with vertices T'(–2, 5), E'(–4, 5), R'(–4, 3), A'(–7, 3), S'(–7, 1), F'(–2, 1)

Example 2
Refer to figure ABCD at the top of page 341. Rotate the figure 90° counterclockwise about the point (1, 1).

Answer: Rectangle with vertices A'(1, 2), B'(1, 1), C'(–1, 1), D'(–1, 2)

Example 3
Refer to △STU on page 342. Translate the triangle 3 units up.

Answer: Triangle with vertices S'(–2, 2), T'(–3, 7), U'(1, 6)

Cooperative Learning

- Refer to problem 15 on page 345. Assume that PMRV is the result of the following three transformations, performed in the given order: (1) a 90° clockwise rotation about the origin, (2) a translation of <–3, 2>, and (3) a reflection over the y-axis.

Solution

Notice that when TERASF is reflected over the x-axis, another figure, T´E´R´A´S´F´, is formed. Each point of the original figure is transformed into a point of the new figure. Each new point is called the ***image*** of an original point. Each original point is called the ***preimage*** of a new point. In order to distinguish between the image and the preimage, we will label an image point with the same letter as the preimage point and the additional symbol ´. In the example, the image of T is T´, read "T prime."

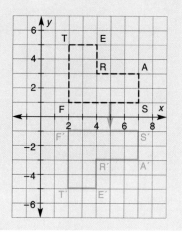

ROTATIONS

Let's revisit figure TERASF from Example 1.

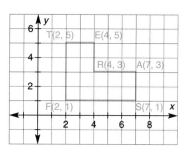

Select a point, in this case point S, and rotate the figure 90° in a clockwise direction about S. This movement transforms the figure into T´E´R´A´S´F´. The transformation is a ***rotation*** about S. To rotate a figure, we need a point about which the figure is rotated, a direction of rotation (clockwise or counterclockwise), and the number of degrees the figure is rotated. We then rotate all the points of the figure by that number of degrees around the point of rotation.

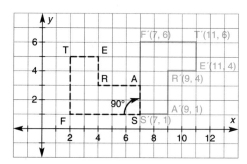

340   Chapter 7 Exploring Geometry

SECTION 7.4

Example 2
Draw the image of the figure after a rotation of 90 degrees in a clockwise direction about the point (2, 3), and label its vertices using prime notation. Give the coordinates of the image's vertices.

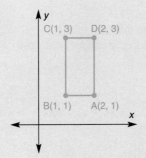

Solution

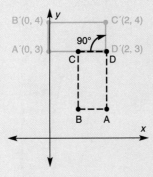

TRANSLATIONS

A third kind of transformation is a **translation,** or **slide.**

Without turning or flipping, each point of the figure has been moved the same distance in the same direction, or translated. If we do the translation on a coordinate plane, we can find the image of a point by adding the **translation numbers** to the coordinates of the point being translated.

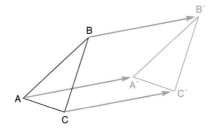

Translation Numbers

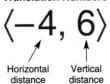

Horizontal distance   Vertical distance

Section 7.4 Rigid Transformations

What were the coordinates of the original figure?

Answers: (0, 9), (0, 4), (4, 2), (4, 11)

■ Problems 2 and 4 are appropriate for cooperative learning.

Assignment Guide

Basic
Day 1 6, 8–10, 15, 19, 22–24
Day 2 7, 11, 12, 17, 20, 21, 25–27

Average or Heterogeneous
Day 1 6–10, 12, 15, 16, 18, 21–23, 25
Day 2 11, 13, 14, 17, 19, 20, 24, 26, 27

Honors
Day 1 6–9, 12, 13, 16, 17, 20–22
Day 2 10, 11, 14, 15, 18, 19, 27

Strands

Communicating Mathematics	2, 4, 5, 21, 22
Geometry	1–20, Investigation
Measurement	1, 3, 4, 8, 9, 11, 13, 15, 16, 18
Algebra	2–4, 6–12, 14, 15, 17–27
Patterns & Functions	Investigation
Number Sense	21–26
Interdisciplinary Applications	18, Investigation

Individualizing Instruction

Visually impaired learners
Students may find this section difficult, as there are many movements on a graph that need to be exact and they may have trouble locating points on graph paper or a grid with small squares. Have them use graph paper with squares that are 1 cm × 1 cm or 0.5 in. × 0.5 in.

SECTION 7.4

Problem-Set Notes

for page 343

- **2** might be extended by asking if there is any triangle whose reflection over the y-axis could also be accomplished with a translation. The answer is no, though students might argue that it is possible with an isosceles or equilateral triangle. Point out that while the images overlap, the vertices do not match.

- **3** could be done as a group activity.

Communicating Mathematics

Have students draw on a piece of graph paper a triangle and a reflection of that triangle over the x-axis. Then have them connect each vertex of the triangle to its image. Have students explain, in writing, how the image points are related to the preimage points, using the words *segment, perpendicular,* and *midpoint.*

- Problems 2, 4, 5, 21, and 22 encourage communicating mathematics.

Additional Answers

for page 343

2 No; A translation affects the coordinates of every point. The points on $\overline{AC}$ are not affected by the reflection.

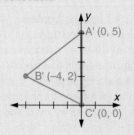

Example 3
Translate △STU
a 3 units to the right
b 1 unit to the left and 4 units down

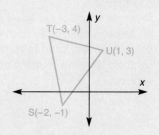

Solution
We write translation numbers and add them to the corresponding coordinates of the vertices.

Slide	Translation Numbers
a +3 units to the right 0 units up	<3, 0>

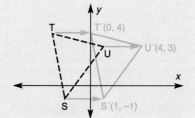

b 1 units to the left 4 units down	<–1, –4>

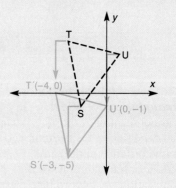

Sample Problem

Problem The coordinates of the vertices of a triangle are A(2, 5), B(6, 1) and C(4, –3). Find the coordinates of the image after a translation of <–6, 4>.

SECTION 7.4

Solution You can use a matrix to help you organize the information.

$$\begin{matrix} & \text{A} & \text{B} & \text{C} \\ x & \begin{bmatrix} 2 & 6 & 4 \\ 5 & 1 & -3 \end{bmatrix} \end{matrix} + \begin{matrix} \text{Translation} \\ \text{Numbers} \\ \begin{bmatrix} -6 & -6 & -6 \\ 4 & 4 & 4 \end{bmatrix} \end{matrix} = \begin{matrix} \text{A}' & \text{B}' & \text{C}' \\ \begin{bmatrix} -4 & 0 & -2 \\ 9 & 5 & 1 \end{bmatrix} \end{matrix}$$

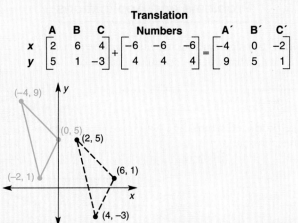

Think and Discuss

P **1** If points A, B, and C are rotated 120° clockwise about the center of the clock, where will their images be? At 6, 8, and 9, respectively

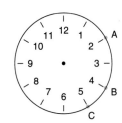

P **2** △ABC has vertices A(0, 5), B(4, 2), and C(0, 0). Reflect this triangle over the y-axis to produce △A´B´C´. Is there a translation that would slide △ABC onto △A´B´C´? Why or why not?

P **3** Copy figure TERASF in Example 1. Then draw its image after each of the following transformations. Label the vertices of each image with their coordinates.
   a A reflection over the y-axis
   b A 180° clockwise rotation about the origin
   c A 90° counterclockwise rotation about point T
   d A translation of <−4, −8>

P **4** Suppose that a triangle with vertices D(0, 8), R(6, 3), and Y(0, 0) is rotated 180° clockwise about the origin to produce the image △D´R´Y´.
   a If △DRY had been rotated 180° counterclockwise about the origin, would the image have been the same? Why or why not?
   b Is there a translation that will slide △DRY onto △D´R´Y´? Why or why not?

P, R **5** How is symmetry related to transformations?

3 a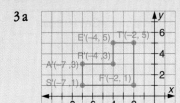

b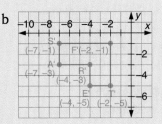

c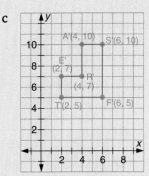

d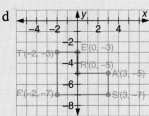

4 a Yes; possible explanation: A rotation of 180° has the same effect whether it is performed clockwise or counterclockwise.

b No; possible answer: R is to the right of segment DY. After the rotation, R' is to the left of segment D'Y'.

5 Possible answer: A figure is symmetric if it can undergo a reflection or a rotation of 180° or less without changing position.

Section 7.4 Rigid Transformations

SECTION 7.4

Problem-Set Notes

for pages 344 and 345

- **6** is an example of a one-dimensional translation.

- **7** and **14** will be done by some students in their heads, while others will need models to manipulate.

- **13** makes a connection with supplementary angles.

Additional Practice

Refer to figure BONE at the top of page 345. Draw an image of BONE after each of the following transformations and label the vertices of each image with their coordinates.
1. A reflection over the y-axis
2. A reflection over the x-axis
3. A 180° counterclockwise rotation about the origin
4. A translation of <3, –1>

Answers:
Coordinates of the vertices of each image:
1. B′(2, 5), O′(0, 0), N′(6, 0), E′(8, 5)
2. B′(–2, –5), O′(0, 0), N′(–6, 0), E′(–8, –5)
3. B′(2, –5), O′(0, 0), N′(6, 0), E′(8, –5)
4. B′(1, 4), O′(3, –1), N′(–3, –1), E′(–5, 4)

Additional Answers

for pages 344 and 345

7a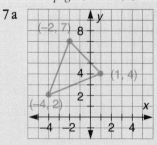

Problems and Applications

P, R   **6** Suppose that $\overline{AB}$ is translated along the number line to form $\overline{A'B'}$. If the coordinate of A′ is 4, what will the coordinate of B′ be?   8

P   **7** Draw the image produced when the triangle shown is translated
     a <–3, 5>   **b** <2, 6>

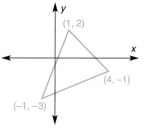

P   **8** What are the coordinates of the image of P(–2, 7) after
     a A reflection over the x-axis?   (–2, –7)
     b A reflection over the y-axis?   (2, 7)
     c A 90° clockwise rotation about the origin?   (7, 2)

P   **9** Sketch the image produced when this "propeller" is rotated 60° clockwise about the origin.

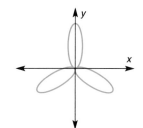

P   **10** If △ARM is reflected over the x-axis to produce △A′R′M′, what will lengths RR′ and A′M′ be?   12; 8

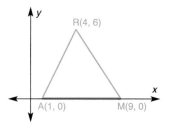

P   **11** If rectangle RECT is rotated 90° clockwise about the origin, what will be the coordinates of the vertices of the image?
R′(4, 2), E′(4, –3), C′(–3, –3), T′(–3, 2)

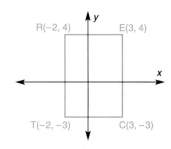

344   **Chapter 7**   Exploring Geometry

SECTION 7.4

P **12** Draw the image produced when quadrilateral BONE is reflected over the y-axis. Label the vertices of the image with their coordinates.

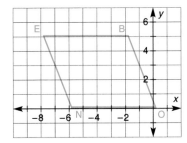

P, R **13** If $\overrightarrow{AB}$ is rotated 50° clockwise about point A to form $\overrightarrow{A'B'}$, what will the measure of ∠DA'B' be? 130

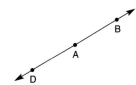

P **14** If figure MANDIBLE is translated <−4, −6>, what will the coordinates of the image's vertices be?
M′(−5, −5), A′(−5, −3), N′(−4, −3), D′(−4, −4), I′(−2, −4), B′(−2, −6), L′(−3, −6), E′(−3, −5)

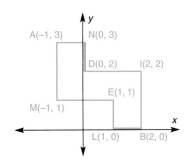

P, R **15** Rotate PMRV 180° clockwise about the origin to produce P′M′R′V′. What kind of quadrilateral is PV′P′V? Parallelogram

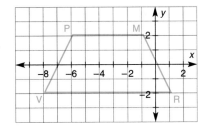

P, R **16** If line ℓ is a 180° rotation of line m about point A, what are m∠1, m∠2, m∠3, and m∠4?
70; 110; 70; 110

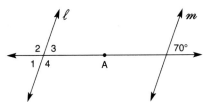

P, R **17** The matrix shows the coordinates of the vertices of △LEG.
  a Draw △LEG on a coordinate system.
  b Translate △LEG <4, −7> and label the vertices of the image, △L′E′G′, with their coordinates.
  c What matrix could you add to the given matrix to produce a matrix containing the coordinates of L′, E′, and G′?

$$\begin{array}{c} \\ \text{x-coordinate} \\ \text{y-coordinate} \end{array} \begin{array}{c} \text{L} \quad \text{E} \quad \text{G} \\ \begin{bmatrix} 5 & 3 & -1 \\ 9 & -2 & 7 \end{bmatrix} \end{array}$$

b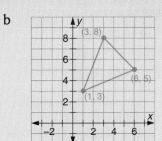

9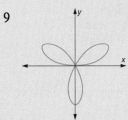

12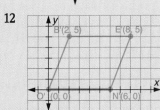

17 a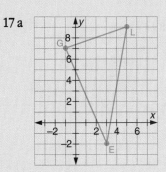

b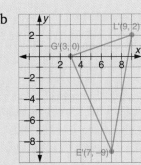

c $\begin{bmatrix} 4 & 4 & 4 \\ -7 & -7 & -7 \end{bmatrix}$

Section 7.4 Rigid Transformations

SECTION 7.4

Checkpoint

Refer to △ARM in problem 10 on page 344. Draw an image of △ARM after each of the following transformations and label the vertices of each image with their coordinates.
1. A reflection over the y-axis
2. A reflection over the x-axis
3. A 90° counterclockwise rotation about the origin
4. A translation of <2, –1>

Answers:
Coordinates of the vertices of each image:
1. A'(–1, 0), R'(–4, 6), M'(–9, 0)
2. A'(1, 0), R'(4, –6), M'(9, 0)
3. A'(0, 1), R'(–6, 4), M'(0, 9)
4. A'(3, –1), R'(6, 5), M'(11, –1)

Additional Answers

for page 346

18 a

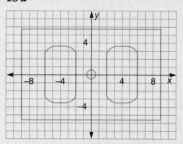

20 d Possible answer: It shows that either the expression 3(2 + 5) or the expression 3(2) + 3(5) can be used to calculate the area of the figure.

21 True; possible answer: 5 less than 20 is 15. 10 is less than 15.

22 False; possible answer: 5 less than 20 is equal to 15, not greater than 15

Investigation Escher used geometric concepts of translation, rotation, and reflection to produce the tessellation designs for which his work is known.

346

P, R **18 Design** When drawing plans for a symmetrical object, a designer sometimes draws only part of the object, since reflections can be used to complete the plans accurately. This diagram shows one fourth of an electrical-outlet cover.

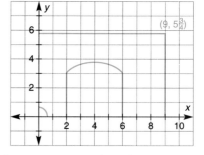

  a Copy the diagram, reflect the figure over the x-axis, then reflect the figure and its image over the y-axis to complete the cover.
  b If each unit of the coordinate system represents $\frac{1}{4}$ inch, what are the dimensions of the cover? $4\frac{1}{2}$ in. by $2\frac{7}{8}$ in.

P, R **19 a** If point B were moved 2 units to the right and point C were moved 2 units down, by what percent would length BC increase? 0%
  b If point B were moved 3 units to the right and point C were moved 3 units down, by what percent would length BC increase? ≈ 2.96%

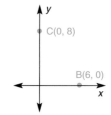

◀ LOOKING BACK **Spiral Learning** LOOKING AHEAD ▶

R, I **20 a** What is the area of region A? 6
  b What is the area of region B? 15
  c What is the total area of the figure? 21
  d How does the diagram show that 3(2 + 5) = 3(2) + 3(5)?

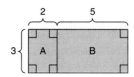

I **Communicating** In problems 21 and 22, indicate whether the statement is *true* or *false*. Explain your answer.

  21 10 is less than 5 less than 20.   **22** 5 less than 20 is greater than 15.

R, I In problems 23–26, indicate whether the statement is *true* or *false*.

  23 7(3 – 5) = 7(3) – 7(5) True   **24** $\sqrt{25-9} = \sqrt{25} - \sqrt{9}$ False
  25 –2(7 + 11) = –2(7) + (–2)(11) True   **26** $(3+5)^2 = 3^2 + 5^2$ False

R **27 a** Evaluate the ratio $\frac{2x}{5x}$ for $x = 4$. $\frac{2}{5}$ **b** Evaluate the ratio $\frac{3k}{4k}$ for $k = -7$. $\frac{3}{4}$

Investigation

M. C. Escher Find a book containing some examples of the works of the Dutch artist Maurits Cornelis Escher. Examine the pictures carefully. What are some of the ways in which Escher made use of transformations?

346 **Chapter 7** Exploring Geometry

SECTION 7.5

Investigation Connection
Section 7.5 supports Unit D, *Investigation* 5. (See pp. 316A-316D in this book.)

Discovering Properties

Recognizing symmetries and performing transformations can help you find useful properties of geometric figures. The three examples in this section show some ways in which properties can be discovered. In the problems that follow, you will have an opportunity to apply these properties and discover some other properties.

Example 1
A triangle with two congruent sides is called an ***isosceles triangle***. △ABC is isosceles, with congruent legs $\overline{AB}$ and $\overline{BC}$ and base $\overline{AC}$. From the symmetry of this triangle, what can you conclude about base angles A and C?

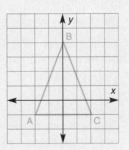

Solution
△ABC possesses reflection symmetry about the y-axis. When we reflect the triangle about its line of symmetry, the image (△A′B′C′) matches the preimage exactly. We observe that ∠C ≅ ∠A′ and that ∠A ≅ ∠C′. So, ∠A ≅ ∠C. Since every isosceles triangle has reflection symmetry, we can generalize this conclusion in the statement below.

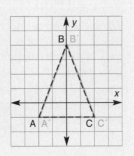

▶ **In an isosceles triangle, the base angles are congruent.**

Section 7.5 Discovering Properties

Objective
- To discover and apply properties of polygons

Vocabulary: bisect, isosceles triangle

Class Planning

Schedule 2 days
Resources
In *Chapter 7 Student Activity Masters* Book
Class Openers *49, 50*
Extra Practice *51, 52*
Cooperative Learning *53, 54*
Enrichment *55*
Learning Log *56*
Interdisciplinary
 Investigation *57, 58*
Spanish-Language Summary *59*

Transparencies *37 a,b*

Class Opener

Draw a rectangle and label it ABCD. Describe a way to transform ABCD so that ∠A overlaps
1. ∠B
2. ∠C
3. ∠D

Answers:
Answers will vary; but answers for **1** and **3** should involve reflections over lines of symmetry, and the answer for **2** should involve a rotation of 180°.

Lesson Notes

Suggest to students that they try all of the examples with cutout figures. You may even do this as a class activity before assigning the reading. We would encourage you to structure the lesson so that students discover the properties instead of being told of them.

347

SECTION 7.5

Additional Examples

Example 1
△ABC is isosceles, with ∠A ≅ ∠C. From the symmetry of this triangle, what can you conclude about sides $\overline{AB}$ and $\overline{BC}$?

Answer: They are congruent.

Example 2
Think about the symmetries of a rectangle. What parts of the rectangle are congruent?

Answer: Both pairs of opposite sides, all angles

Example 3
Draw a circle and any two of its radii. Think of the rotation symmetry of a circle about its center. What can you conclude about any two radii of a circle?

Answer: They are congruent.

Additional Answers

for page 349

1 Possible answer: All four sides are congruent, so ABCD is a rhombus.

2 a Possible answer: ∠T ≅ ∠R and ∠P ≅ ∠A **b** Possible answer: The diagonals are congruent.

3 a Reflect △NST over $\overline{TN}$, forming △T'S'N'; △TSS' will be equilateral.

4 a Possible answers: They are perpendicular; they bisect the four angles of the rhombus; they bisect each other
b Possible answers: Opposite angles are congruent; consecutive angles are supplementary.
c All sides are congruent.

5 Possible answers: One diagonal bisects the other; the diagonals are perpendicular; one diagonal bisects a pair of opposite angles;

348

Example 2
Think about the symmetry of a parallelogram. What parts of the parallelogram are congruent?

Solution
First, we draw a general parallelogram, GRAM. We know that the opposite sides of GRAM are congruent and that GRAM has a rotation symmetry of 180° about the intersection of its diagonals. When we rotate the parallelogram GRAM 180° about its center of symmetry, we see that each side matches its opposite side exactly. We also notice that ∠A′ matches ∠G, ∠M′ matches ∠R, ∠G′ matches ∠A, and ∠R′ matches ∠M. Since there is nothing special about this particular parallelogram, we generalize our observation in the statement below.

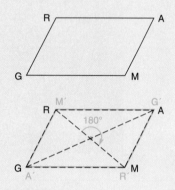

▶ **Opposite angles of a parallelogram are congruent.**

If we include the diagonals in the rotation, we see that $\overline{P'A'}$ matches $\overline{PG}$ and $\overline{P'M'}$ matches $\overline{PR}$. This means that $\overline{PA} \cong \overline{PG}$ and $\overline{PM} \cong \overline{PR}$. When a segment is divided into two congruent parts, it is said to be **bisected**, so we can draw the following general conclusion:

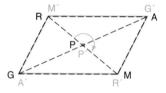

▶ **The diagonals of a parallelogram bisect each other.**

Since squares, rectangles, and rhombuses are kinds of parallelograms, these properties of parallelograms apply to them, too. Do the rotations we performed suggest any other congruences?

Example 3
Draw a rectangle and its diagonals. From the reflection symmetry of the rectangle, what can you conclude about its diagonals?

Solution
When we draw rectangle RECT and reflect it over one of its lines of symmetry, the image R'E'C'T' matches the original rectangle exactly. Notice also that the image of diagonal $\overline{RC}$ ($\overline{R'C'}$) matches $\overline{ET}$ and that the image of diagonal $\overline{TE}$ ($\overline{T'E'}$) matches $\overline{CR}$. The diagonals are therefore congruent. Again, we generalize the conclusion in the statement below.

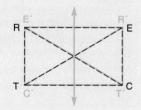

▶ **The diagonals of a rectangle are congruent.**

348 Chapter 7 Exploring Geometry

SECTION 7.5

Think and Discuss

P **1** Suppose that the diagonals of a quadrilateral bisect each other and are perpendicular to each other, like $\overline{AC}$ and $\overline{BD}$ in the diagram. By using reflections, what can you conclude about quadrilateral ABCD?

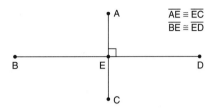

P, I **2** An *isosceles trapezoid* is a trapezoid in which the nonparallel sides are congruent. In isosceles trapezoid TRAP, $\overline{TR} \parallel \overline{PA}$ and $\overline{TP} \cong \overline{RA}$.
  a Reflect TRAP over line of symmetry $\overleftrightarrow{EZ}$. What properties do you notice?
  b If diagonals $\overline{PR}$ and $\overline{TA}$ are drawn, what additional properties do you notice from the reflection in part **a**?

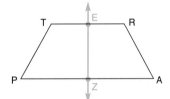

P, I **3** In your mathematics studies, you will frequently encounter triangles with angles of 30°, 60°, and 90°.
  a Describe how, with one reflection of △TSN, you can create an equilateral triangle.
  b What can you conclude about the lengths of $\overline{TS}$ and $\overline{SN}$? $SN = \frac{TS}{2}$

P **4** From the symmetries of a rhombus, what conclusions can you draw about
  a Its diagonals?
  b Its angles?
  c Its sides?

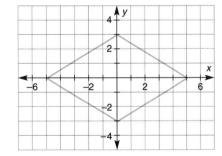

P **5** A quadrilateral that has two pairs of congruent adjacent sides is called a *kite*. In the kite shown, $\overline{EK} \cong \overline{KI}$ and $\overline{ET} \cong \overline{TI}$. Reflect kite KITE over its line of symmetry, $\overleftrightarrow{KT}$. What properties of the kite do you see?

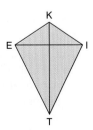

Section 7.5 Discovering Properties

SECTION 7.5

Problem-Set Notes

for pages 350 and 351

- 13–15 require a protractor. You may want to do these in class.

Multicultural Note

Little is known about the life and background of the great mathematician Euclid. It is known that he lived and taught in Alexandria, in Egypt, around 300 B.C., when that city had become the center of learning in the Western world. He is famous primarily because he compiled all of the geometry known in his time into thirteen books that are called the *Elements*.

Checkpoint

From the symmetries of a square, what conclusion can you draw about
1. Its diagonals?
2. Its angles?
3. Its sides?
4. If JKLM is a rectangle and JL = 15, what is MK?

Answers:

1. Bisect each other, equal in length 2. Opposite angles are equal, consecutive angles are equal, so all angles are equal. 3. Opposite sides are equal, adjacent sides are equal, so all sides are equal. 4. 15

Stumbling Block

The symmetries of a figure can be used to conclude that certain parts of the figure are congruent. However, students must use care in observing where the image of each part of the figure is. For example,

Problems and Applications

P **6** If AMRV is a rectangle and AR = 24, what is length VM? 24

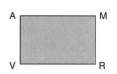

P **7** FAST is a parallelogram, and m∠A = 74. What is m∠T? 74

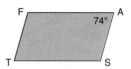

P, R **8** In this triangle, $\overline{AB} \cong \overline{AC}$ and m∠B = 52. What are m∠A and m∠C? 76; 52

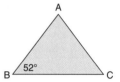

P **9** $\overline{AL}$ and $\overline{PF}$ are parallel to the x-axis, and $\overline{PA}$ and $\overline{FL}$ are parallel to the y-axis. What are the coordinates of points P and L? (5, 7), (11, 2)

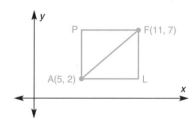

P, R **10** If this trapezoid is symmetric about the y-axis, what are the coordinates of points C and W? (–6, 3); (–8, –4)

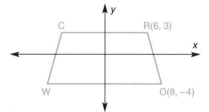

P **11** VOLE is a rectangle, and VL = 18. What is length PO? 9

P **12** Parallelogram WREN can be translated as shown so that point W′ lies on N and point R′ lies on E. From this translation, what can be concluded about the consecutive angles of a parallelogram? They are supplementary.

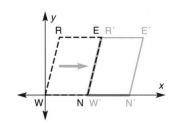

350   **Chapter 7** Exploring Geometry

SECTION 7.5

In problems 13–16, carefully draw each of the polygons described.

P, R

13 A quadrilateral that has angles of 35°, 35°, 145°, and 145° but is not a parallelogram

14 A parallelogram that has angles of 35°, 35°, 145°, and 145° but is not a rhombus

15 A quadrilateral that has angles of 35°, 35°, 145°, and 145° and is both a parallelogram and a rhombus

16 A rhombus that has angles of 35°, 35°, 145°, and 145° but is not a parallelogram   No such figure

P **17** SMOU is a parallelogram. If MU = 31 and EO = 17, what are lengths EU and SE?   15.5; 17

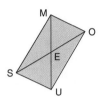

P **18 a** If points C, O, N, and Y are connected by segments to form a quadrilateral, what kind of quadrilateral will CONY be?   Square
  b If the quadrilateral is translated <3, –2>, what will be the coordinates of the vertices of the image?   C′(3, 4), O′(3, –2), N′(9, –2), Y′(9, 4)
  c What kind of quadrilateral will the image be?   Square

P **19** If $\overline{TC} \cong \overline{AC}$ and m∠C = 44, what are m∠A and m∠T?   68; 68

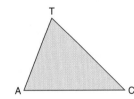

P, I **20** Find the area of △BUG for each of the values of *k*.
  a –6   25   **b** 0   25   **c** 4   25

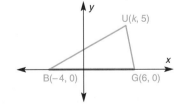

P, R **21** PICA is a parallelogram, with PA = 18 and CA = 7. What is the perimeter of PICA?   50

---

when a parallelogram is rotated 180°, each side or angle is matched with the opposite side or angle. Therefore, these parts are congruent. However, each diagonal is matched with itself, not with the other diagonal. Therefore, congruence of diagonals is not an appropriate inference.

Additional Practice

1. Refer to the figure in problem 6 on page 350. If AV = 16.5, what is MR?
2. Refer to the figure in problem 7 on page 350. If m∠F = 106°, what is m∠S?
3. Refer to the figure in problem 10 on page 350. What are the coordinates of the points at which the trapezoid intersects the y-axis?
4. Refer to the figure in problem 12 on page 350. If m∠NWR = 75°, what is m∠WRE?

Answers:
1. 16.5   2. 106°
3. (0, 3) and (0, –4)   4. 105°

Additional Answers

for pages 350 and 351

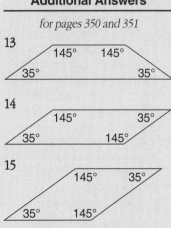

Section 7.5 Discovering Properties

SECTION 7.5

Problem-Set Notes

for pages 352 and 353

- **24** results in the creation of a regular hexagon and should allow students to come up with properties of the figure more easily.

- **26** is a review problem, but an essential one for the next section.

Reteaching

All the examples in this section can be demonstrated more concretely by using overhead transparencies of the given figures and actually modeling the transformations. Students can be asked to describe which parts of each figure match with other parts of the figure.

Cooperative Learning

- Small groups of students can use paper folding to investigate line symmetry and rotation symmetry. For example, they can fold a piece of paper in half and then cut a pattern from the paper, keeping the folded edge intact. The unfolded pattern will have line symmetry. Next, they can fold a piece of paper in half twice in succession and cut a pattern, keeping the folded edges intact. The unfolded pattern will have rotation symmetry. Finally, students can use the concepts of the lesson to explain why these paper-folding activities yielded the results they observed.

- Problems 13–16 are appropriate for cooperative learning.

P, I **22** What kind of quadrilateral does square SQUA become if
  a S is moved to (4, 4) and U is moved to (–4, –4)?   Parallelogram
  b S is moved to (9, 0) and U is moved to (–9, 0)?   Rhombus
  c S is moved to (2, 0) and U is moved to (–2, 0)?   Rhombus
  d S is moved to (2, 3) and Q is moved to (–2, 7)?   Parallelogram

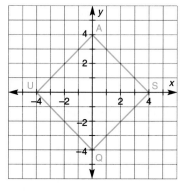

P **23** Parallelogram JETS has 180° rotation symmetry about point P. If JE = 20, JS = 21.2, JP = 17, and SP = 12, what are JT, SE, ET, and ST?
  34; 24; 21.2; 20

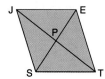

P, I **24 a** Draw an equilateral triangle with side length 1 inch.
  b Keeping one vertex stationary, reflect the triangle over a side five times in a counterclockwise pattern. What kind of polygon is formed by the original triangle and the reflections?   Hexagon
  c What is the perimeter of the new figure?   6 in.
  d List some properties of the new figure.

◀ LOOKING BACK   **Spiral Learning**   LOOKING AHEAD ▶

R **25** In this diagram, what are the coordinates of points D and E?
  (10, 0); (0, 8)

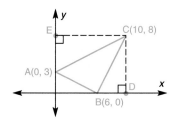

R **26 a** Evaluate $\begin{bmatrix} 1 & 0 \\ 0 & -1 \end{bmatrix} \cdot \begin{bmatrix} 3 & 2 & 7 \\ 5 & 4 & 1 \end{bmatrix}$. $\begin{bmatrix} 3 & 2 & 7 \\ -5 & -4 & -1 \end{bmatrix}$

  b Evaluate $\begin{bmatrix} -1 & 0 \\ 0 & 1 \end{bmatrix} \cdot \begin{bmatrix} 3 & 2 & 7 \\ 5 & 4 & 1 \end{bmatrix}$. $\begin{bmatrix} -3 & -2 & -7 \\ 5 & 4 & 1 \end{bmatrix}$

  c Describe the effects of multiplication by $\begin{bmatrix} 1 & 0 \\ 0 & -1 \end{bmatrix}$ and of multiplication by $\begin{bmatrix} -1 & 0 \\ 0 & 1 \end{bmatrix}$ on a matrix having two rows.

Chapter 7   Exploring Geometry

SECTION 7.5

P, R **27** Of a square, a rectangle, a rhombus, a parallelogram, and a trapezoid, which figures always are symmetric about a line through the midpoints of opposite sides?   Square and rectangle

R **28**

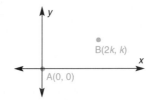

Suppose that on the number line, points A and B are reflected over D to form A′ and B′, then points A′, B′, and D are reflected over C to form A″, B″, and D′. What will the coordinates of A″, B″, and D′ be?   −12; −9, −11

R, I **29** Spreadsheets   Use a spreadsheet to make a table showing the distances between A and B for k = 1, 2, 3, … , 24.

R Science   **The formula $F = \frac{9}{5}C + 32$ can be used to calculate the Fahrenheit equivalent of a temperature expressed in degrees Celsius. In problems 30–33, convert each temperature to degrees Fahrenheit.**

30 20°C   68°F   **31** 0°C   32°F   **32** −8.4°C   16.88°F   **33** −32°C   −25.6°F

I **34** Without using a calculator, evaluate the expressions on both sides of the equal sign. Which expression is easier to evaluate?

17(34) + 17(66) = 17(34 + 66)   1700; 1700; the expression on the right

R **35** Communicating
a What is the length of $\overline{PQ}$?
b What is the area of rectangle RSTU?
c Would it make sense to calculate the sum of your answers to parts **a** and **b**? Why or why not?
a 20 mm   **b** 100 mm²

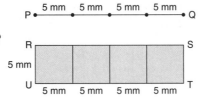

Investigation

Playing-Card Symmetries   Look closely at the playing cards in a standard 52-card deck. What kind of symmetry do most of the cards have? Why are they designed to have this kind of symmetry? Which of the cards are not symmetric? Why aren't these cards symmetric?

SECTION 7.6

Objectives
- To represent a figure by means of a vertex matrix
- To use matrix addition to model translations
- To use matrix multiplication to model reflections and rotations

Vocabulary: encasement, vertex matrix

Class Planning

Schedule 2 days

Resources

In *Chapter 7 Student Activity Masters* Book
Class Openers 61, 62
Extra Practice 63, 64
Cooperative Learning 65, 66
Enrichment 67
Learning Log 68
Technology 69, 70
Spanish-Language Summary 71

Transparency 38

Class Opener

Evaluate.

1. $\begin{bmatrix} -1 & 0 \\ 0 & 1 \end{bmatrix} \cdot \begin{bmatrix} 2 & 5 & 3 \\ 1 & 3 & -2 \end{bmatrix}$

2. $\begin{bmatrix} 1 & 0 \\ 0 & -1 \end{bmatrix} \cdot \begin{bmatrix} 2 & 5 & 3 \\ 1 & 3 & -2 \end{bmatrix}$

3. $\begin{bmatrix} 2 & 5 & 3 \\ 1 & 3 & -2 \end{bmatrix} + \begin{bmatrix} 2 & 2 & 2 \\ -1 & -1 & -1 \end{bmatrix}$

Answers:

1. $\begin{bmatrix} -2 & -5 & -3 \\ 1 & 3 & -2 \end{bmatrix}$ 2. $\begin{bmatrix} 2 & 5 & 3 \\ -1 & -3 & 2 \end{bmatrix}$

3. $\begin{bmatrix} 4 & 7 & 5 \\ 0 & 2 & -3 \end{bmatrix}$

354

Investigation Connection
Section 7.6 supports Unit D, *Investigation* 3. (See pp. 316A–316D in this book.)

SECTION 7.6
Transformations and Matrices

THE VERTEX MATRIX

In this section, you will see how your knowledge of matrix addition and multiplication can help you transform geometric figures. As you study the examples and the sample problem, look for patterns and try to figure out why the methods presented here work.

Example 1
Create a matrix to represent this quadrilateral.

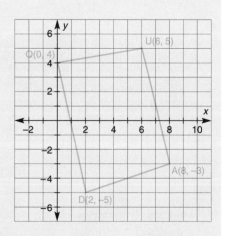

Solution
The vertices of the quadrilateral are Q(0, 4), U(6, 5), A(8, −3), and D(2, −5). Here are three possible ways of organizing the coordinates of the vertices in a matrix:

$\begin{array}{c} \\ x \\ y \end{array} \begin{array}{c} \text{Q U A D} \\ \begin{bmatrix} 0 & 6 & 8 & 2 \\ 4 & 5 & -3 & -5 \end{bmatrix} \end{array} \quad \begin{array}{c} \\ x \\ y \end{array} \begin{array}{c} \text{A D Q U} \\ \begin{bmatrix} 8 & 2 & 0 & 6 \\ -3 & -5 & 4 & 5 \end{bmatrix} \end{array} \quad \begin{array}{c} \\ x \\ y \end{array} \begin{array}{c} \text{A U Q D} \\ \begin{bmatrix} 8 & 6 & 0 & 2 \\ -3 & 5 & 4 & -5 \end{bmatrix} \end{array}$

Chapter 7 Exploring Geometry

Each of these matrices is called a **vertex matrix** of the quadrilateral. Notice that the x-coordinates are always in the first row and the y-coordinates are always in the second row. The coordinates of the points are entered in order, going clockwise or counterclockwise around the figure from any of its vertices.

TRANSLATIONS

After writing a vertex matrix for a figure, you can use matrix addition to model a translation of the figure.

Example 2
The diagram shows a triangle being translated <3, −4>—that is, 3 units to the right and 4 units down. Use matrix addition to find the coordinates of A′, B′, and C′.

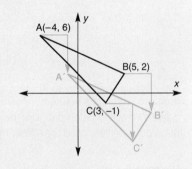

Solution
To translate the vertices, we need to add 3 to each x-coordinate and −4 to each y-coordinate.

$$\begin{array}{c} \text{Vertices of Preimage} \\ \begin{array}{cc} & \begin{array}{ccc} A & B & C \end{array} \\ \begin{array}{c} x \\ y \end{array} & \begin{bmatrix} -4 & 5 & 3 \\ 6 & 2 & -1 \end{bmatrix} \end{array} \end{array} + \begin{array}{c} \text{Translation Matrix} \\ \begin{bmatrix} 3 & 3 & 3 \\ -4 & -4 & -4 \end{bmatrix} \end{array} = \begin{array}{c} \text{Vertices of Image} \\ \begin{array}{cc} & \begin{array}{ccc} A' & B' & C' \end{array} \\ \begin{array}{c} x \\ y \end{array} & \begin{bmatrix} -1 & 8 & 6 \\ 2 & -2 & -5 \end{bmatrix} \end{array} \end{array}$$

Notice that in the translation matrix we entered the translation numbers vertically so that we could add them to the corresponding x- and y-coordinates. The vertices of the image are A′ (−1, 2), B′ (8, −2), and C′ (6, −5).

REFLECTIONS AND ROTATIONS

To model reflections, we use matrix multiplication. Two key reflection matrices are

$$\begin{bmatrix} -1 & 0 \\ 0 & 1 \end{bmatrix} \qquad \begin{bmatrix} 1 & 0 \\ 0 & -1 \end{bmatrix}$$

For a reflection over the y-axis   **For a reflection over the x-axis**

Lesson Notes

- This section combines the work with matrices taught earlier and the geometry of this chapter. Students who may not necessarily find the hands-on approach beneficial may prefer this section, which is more theoretical.

- In the sample problem encasement is used to find the area of a triangle on the coordinate plane. This strategy might be helpful to students entered in MathCounts or other contests.

Additional Examples

Example 1
Create a matrix to represent quadrilateral JHIK.

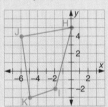

Answer:
$$\begin{array}{c} \begin{array}{cccc} J & H & I & K \end{array} \\ \begin{array}{c} x \\ y \end{array} \begin{bmatrix} -6 & 0 & -2 & -5 \\ 4 & 5 & -2 & -3 \end{bmatrix} \end{array}$$

Example 2
JHIK is being translated 2 units to the left and 3 units down. Use matrix addition to find the coordinates of J′, H′, I′, and K′.

Answer:
$$\begin{array}{c} \begin{array}{cccc} J' & H' & I' & K' \end{array} \\ \begin{array}{c} x \\ y \end{array} \begin{bmatrix} -8 & -2 & -4 & -7 \\ 1 & 2 & -5 & -6 \end{bmatrix} \end{array}$$

Example 3
Use matrix multiplication to locate the vertices of the image produced by reflecting JHIK over the x-axis.

Answer:
$$\begin{array}{c} \begin{array}{cccc} J' & H' & I' & K' \end{array} \\ \begin{array}{c} x \\ y \end{array} \begin{bmatrix} -6 & 0 & -2 & -5 \\ -4 & -5 & 2 & 3 \end{bmatrix} \end{array}$$

SECTION 7.6

Additional Examples continued

Example 4
Locate the vertices of the image produced by rotating JHIK 90° counterclockwise about the origin.

Answer:
$$\begin{array}{c} \;\;J'\;\;H'\;\;I'\;\;K' \\ x \\ y \end{array} \begin{bmatrix} -4 & -5 & 2 & 3 \\ -6 & 0 & -2 & -5 \end{bmatrix}$$

Cooperative Learning
Extend problem 29 by having groups of students complete the following:
1. Choose another point, Q, and translate it repeatedly, using the translation <4, 2>.
2. How does the result of this translation compare with the translation of P?
3. Now choose a point A on the line through the translations of P. Find translation numbers for A so that the line through the translations is perpendicular to the line containing P.

Answers:
1. See students' drawings.
2. The points lie on two parallel lines.
3. Possible answers:
<1, −2>, <−1, 2>,
<2, −4>, <−2, 4>

■ Problems 4 and 28 are appropriate for cooperative learning.

Communicating Mathematics

■ Have students write a paragraph explaining how to determine the area of a figure by encasement. Have them give an example with their explanation.

Example 3
Use matrix multiplication to locate the vertices of the image produced by reflecting △DEF over the y-axis.

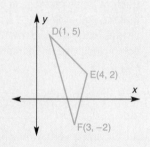

Solution
The reflection is over the y-axis, so we multiply the vertex matrix of △DEF by
$\begin{bmatrix} -1 & 0 \\ 0 & 1 \end{bmatrix}$

$$\begin{bmatrix} -1 & 0 \\ 0 & 1 \end{bmatrix} \cdot \begin{matrix} D & E & F \\ \begin{bmatrix} 1 & 4 & 3 \\ 5 & 2 & -2 \end{bmatrix} \end{matrix} = \begin{matrix} D' & E' & F' \\ \begin{bmatrix} -1 & -4 & -3 \\ 5 & 2 & -2 \end{bmatrix} \end{matrix}$$

The vertices of the image are D′ (−1, 5), E′ (−4, 2), and F′ (−3, −2). Draw the image to check this solution.

Matrix multiplication can also be used to rotate figures, but the mathematics involved is complicated for all but the simplest rotations.

Example 4
Locate the vertices of the image produced by rotating △GHK 90° counterclockwise about the origin.

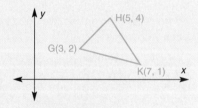

Solution
To rotate 90° counterclockwise, multiply the vertex matrix by $\begin{bmatrix} 0 & -1 \\ 1 & 0 \end{bmatrix}$.

The vertices of the image are G′ (−2, 3), H′ (−4, 5), and K′ (−1, 7).

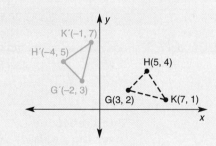

356 Chapter 7 Exploring Geometry

SECTION 7.6

Sample Problem

Problem Suppose △DIL has vertex matrix $\begin{array}{c} x \\ y \end{array} \begin{bmatrix} \overset{D}{0} & \overset{I}{5} & \overset{L}{7} \\ 3 & 8 & 2 \end{bmatrix}$.

Solution Find the area of △DIL.

We will begin by graphing △DIL. Then we will encase the triangle in a rectangle. The area of △DIL is the area of the rectangle minus the area of the three shaded triangles: △A_1, △A_2, and △A_3. This procedure is called **encasement**.

Area of rectangle: 7 · 6, or 42

Area of △A_1: $\frac{1}{2}$ · 5 · 5, or 12.5

Area of △A_2: $\frac{1}{2}$ · 2 · 6, or 6

Area of △A_3: $\frac{1}{2}$ · 7 · 1, or 3.5

Area of △DIL: 42 − (12.5 + 6 + 3.5), or 20

The area of △DIL is 20 square units.

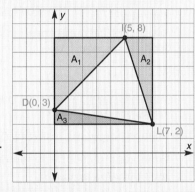

Think and Discuss

P **1** Which of these matrices is not a vertex matrix for quadrilateral MIRO? **c**

  a $\begin{bmatrix} -5 & -6 & 0 & 1 \\ 5 & -1 & 0 & 6 \end{bmatrix}$ **b** $\begin{bmatrix} -6 & -5 & 1 & 0 \\ -1 & 5 & 6 & 0 \end{bmatrix}$

  c $\begin{bmatrix} 0 & -5 & 1 & -6 \\ 0 & 5 & 6 & -1 \end{bmatrix}$ **d** $\begin{bmatrix} 0 & 1 & -5 & -6 \\ 0 & 6 & 5 & -1 \end{bmatrix}$

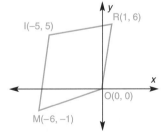

P **2** What is the area of △ABC? 28

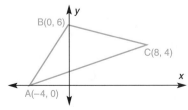

Section 7.6 Transformations and Matrices

- Problems 9 and 28 encourage communicating mathematics.

Assignment Guide

Basic
Day 1 5, 7–9, 14–17, 21, 23
Day 2 6, 10, 11, 13, 20, 22, 24–27

Average or Heterogeneous
Day 1 5–7, 11, 14, 15, 17, 19, 21, 23, 25, 27
Day 2 8–10, 16, 18, 20, 24, 26, 29

Honors
Day 1 5, 6, 8, 10, 13, 15, 18, 21–23, 26, 27
Day 2 7, 9, 11, 12, 14, 17, 19, 24, 25, 29

Strands

Communicating Mathematics	9, 28
Geometry	2–4, 6–12, 13–19, 28, 29, Investigation
Measurement	2, 3, 8, 10, 12, 14, 17, 23
Data Analysis/ Statistics	22
Technology	24–27
Algebra	1–27, 29
Interdisciplinary Applications	Investigation

Individualizing Instruction

The auditory learner Students may benefit from working in pairs to verbalize each step of the process for multiplying two matrices. You might suggest that students pretend to be teaching the process to a younger student so that their descriptions have enough detail.

SECTION 7.6

Problem-Set Notes

for pages 358 and 359

- **4** is a good problem to do in class and then discuss the results.
- **10** can be completed by finding the areas of triangles.
- **12** can be extended by asking students to compare the areas.

Additional Answers

for pages 358 and 359

3 a $\begin{bmatrix} 0 & 10 & 14 \\ 6 & 16 & 4 \end{bmatrix}$ **b** 80 **c** 1:4

4 a Possible answer: $\begin{bmatrix} 0 & 0 & 4 \\ 8 & 0 & 0 \end{bmatrix}$

b Sum matrix: $\begin{bmatrix} 5 & 5 & 9 \\ 14 & 6 & 6 \end{bmatrix}$

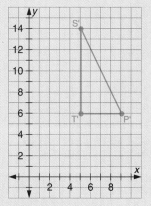

Translation of <5, 6>

c Product matrix: $\begin{bmatrix} 0 & 0 & -4 \\ -8 & 0 & 0 \end{bmatrix}$

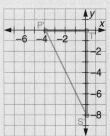

Rotation of 180° about origin

P, R   **3** Refer to △DIL in the sample problem.
 a Multiply the vertex matrix of △DIL by $\begin{bmatrix} 2 & 0 \\ 0 & 2 \end{bmatrix}$ to find the vertex matrix of another triangle: △D′I′L′.
 b Find the area of △D′I′L′.
 c Find the ratio of the area of △DIL to the area of △D′I′L′.

P   **4** Refer to the diagram of △STP.
 a Write a vertex matrix for the triangle.
 b Add $\begin{bmatrix} 5 & 5 & 5 \\ 6 & 6 & 6 \end{bmatrix}$ to the vertex matrix and draw the figure represented by the sum. What transformation did the addition produce?
 c Multiply the vertex matrix of △STP by $\begin{bmatrix} -1 & 0 \\ 0 & -1 \end{bmatrix}$ and draw the figure represented by the product. What transformation did the multiplication produce?
 d Multiply the vertex matrix of △STP by $\begin{bmatrix} -1 & 0 \\ 0 & 1 \end{bmatrix}$ and draw the figure represented by the product. What transformation did the multiplication produce?
 e Multiply the vertex matrix of △STP by $\begin{bmatrix} 1 & 0 \\ 0 & -1 \end{bmatrix}$ and draw the figure represented by the product. What transformation did the multiplication produce?

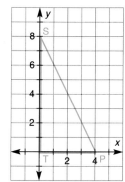

Problems and Applications

P   **5** Write a vertex matrix for the triangle.

Possible answer: $\begin{matrix} & P & Q & T \\ & -3 & 4 & -4 \\ & 3 & -1 & -2 \end{matrix}$

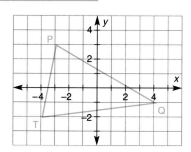

P   **6 a** Write a vertex matrix for △MUD.
 b Multiply the matrix by $\begin{bmatrix} 1 & 0 \\ 0 & -1 \end{bmatrix}$ and draw the image represented by the product.

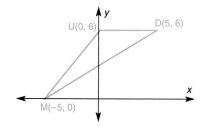

Chapter 7 Exploring Geometry

SECTION 7.6

P   **7** A triangle has vertices A(0, –6), B(4, –4), and C(6, 6). Use matrix addition to find the coordinates of the vertices of △A′B′C′, the image of △ABC after a translation of <0, –1>.   A′(0, –7), B′(4, –5), C′(6, 5)

P, R   **8** What is the area of quadrilateral SQUA?   200

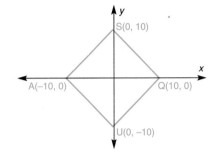

P   **9 Communicating**
    a Describe what happens to the x- and y-coordinates of F, U, and R when △FUR is reflected over the y-axis.
    b Evaluate $\begin{bmatrix} -1 & 0 \\ 0 & 1 \end{bmatrix} \cdot \begin{bmatrix} -6 & 5 & -8 \\ 4 & 3 & 2 \end{bmatrix}$.

P, R   **10** If points B and C are both moved 4 units to the right to form a new quadrilateral, by how much will the area of the new quadrilateral differ from the area of the one shown? It will not differ.

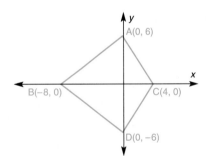

P   **11** This vertex matrix represents △YES.
$\begin{matrix} Y & E & S \\ \begin{bmatrix} -6 & 4 & 10 \\ -10 & -8 & -10 \end{bmatrix} \end{matrix}$
    a Multiply the matrix by $\begin{bmatrix} -1 & 0 \\ 0 & -1 \end{bmatrix}$ to produce the vertex matrix of △Y′E′S′.
    b Draw △YES and △Y′E′S′. What transformation did the multiplication produce?

P, R   **12** Suppose that points Y and X are moved to the right so that the lengths of $\overline{BY}$ and $\overline{OX}$ increase by 20%. By what percent will the lengths of the figure's diagonals change?   Will increase by ≈10.45%

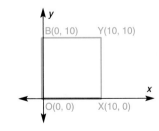

d Product matrix: $\begin{bmatrix} 0 & 0 & -4 \\ 8 & 0 & 0 \end{bmatrix}$

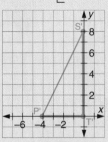

Reflection over y-axis

e Product matrix: $\begin{bmatrix} 0 & 0 & 4 \\ -8 & 0 & 0 \end{bmatrix}$

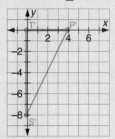

Reflection over x-axis

6 a $\begin{bmatrix} -5 & 0 & 5 \\ 0 & 6 & 6 \end{bmatrix}$

b Product matrix: $\begin{bmatrix} -5 & 0 & 5 \\ 0 & -6 & -6 \end{bmatrix}$

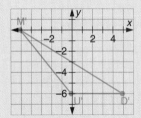

9 a The x-coordinates change to their opposites; the y-coordinates remain the same.
b Product matrix: $\begin{bmatrix} 6 & -5 & 8 \\ 4 & 3 & 2 \end{bmatrix}$

11 a $\begin{matrix} Y' & E' & S' \\ \begin{bmatrix} 6 & -4 & -10 \\ 10 & 8 & 10 \end{bmatrix} \end{matrix}$

Answer for 11b is on page A7.

Section 7.6 Transformations and Matrices

SECTION 7.6

Problem-Set Notes

for pages 360 and 361

- 15–17 could be done as in-class projects.
- 28 provides some review for the chapter test.

Reteaching

To reteach the use of matrices

$$\begin{bmatrix} -1 & 0 \\ 0 & 1 \end{bmatrix} \text{ and } \begin{bmatrix} -1 & 0 \\ 0 & -1 \end{bmatrix}$$

to perform reflections over the axes, have students consider the effect that multiplying by each matrix has on a matrix representing a single point. For example,

$$\begin{bmatrix} -1 & 0 \\ 0 & 1 \end{bmatrix} \cdot \begin{bmatrix} 2 \\ 3 \end{bmatrix} = \begin{bmatrix} -2 \\ 3 \end{bmatrix}$$

Graph the point and its image and have students explain how the multiplication affected the coordinates of the point. Then move on to working with a vertex matrix.

Additional Practice

1. Write a vertex matrix for a triangle with vertices C(–3, –2), A(4, 5), N(7, –1).
2. Multiply the matrix from problem 1 by $\begin{bmatrix} 1 & 0 \\ 0 & -1 \end{bmatrix}$ and draw the figure represented by the product.
3. Perform the translation <–4, –1> on the triangle from problem 1 and sketch the image.

Answers:

1. C   A   N
$$\begin{bmatrix} -3 & 4 & 7 \\ -2 & 5 & -1 \end{bmatrix}$$

360

P **13 a** Write a vertex matrix for △SAM.
b If △S′A′M′ is the image of △SAM after a translation of <p, q>, what are the values of p and q and the coordinates of S′ and A′?

a $\begin{bmatrix} 4 & 2 & 7 \\ 9 & 6 & 5 \end{bmatrix}$

b –3; –11; (1, –2); (–1, –5)

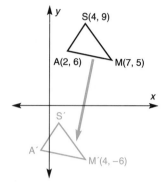

P, I **14** Using the vertex matrix shown, determine the area of △XYZ for
a k = 0   32   **b** k = 4   48

$$\begin{array}{c c c} X & Y & Z \\ \begin{bmatrix} k+2 & 0 & k+10 \\ 0 & k+8 & 0 \end{bmatrix} \end{array}$$

P **In problems 15–17, refer to the diagram of △HAT.**

15 Use matrix addition to locate the vertices of the image of △HAT after a translation of <–3, 5>.

16 Use matrix multiplication to locate the vertices of the image of △HAT after a reflection over the y-axis.

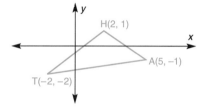

17 Use matrix multiplication to locate the vertices of the image of △HAT after a 90° counterclockwise rotation about the origin.

P, R **18** For what values of w and z will the quadrilateral represented by the following vertex matrix be a square?   w = 3 and z = 3 or w = 13 and z = –7

$$\begin{bmatrix} -8+w & 0 & 2+z & 0 \\ 0 & -5 & 0 & 5 \end{bmatrix}$$

P, R, I **19 a** Write a vertex matrix for △REV.
b Multiply the vertex matrix by $\begin{bmatrix} -1 & 0 \\ 0 & 1 \end{bmatrix}$ and draw the image represented by the product.
c Multiply the product matrix from part **b** by $\begin{bmatrix} 1 & 0 \\ 0 & -1 \end{bmatrix}$ and draw the new image.
d Evaluate $\begin{bmatrix} -1 & 0 \\ 0 & 1 \end{bmatrix} \cdot \begin{bmatrix} 1 & 0 \\ 0 & -1 \end{bmatrix}$.
e Multiply the original vertex matrix of △REV by your answer to part **d**. What do you notice?

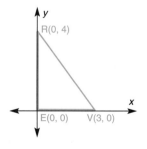

360   Chapter 7 *Exploring Geometry*

SECTION 7.6

◀ LOOKING BACK **Spiral Learning** LOOKING AHEAD ▶

R, I **20** Indicate whether the statement is an *equation*, an *expression*, or an *inequality*.
  a $9x - 3y + 7\sqrt{3}\,z^5$ Expression
  b $9x = 3y + 7\sqrt{3}\,z^5$ Equation
  c $9x - 3y + 7 = \sqrt{3}\,z^5$ Equation
  d $9x - 3y < 7\sqrt{3}\,x^5$ Inequality
  e $9x - 3y = 7\sqrt{3}\,z^5$ Equation

R **21** Evaluate $\begin{bmatrix} 1 & 0 \\ 0 & 1 \end{bmatrix} \cdot \begin{bmatrix} 2 & -1 & 1 \\ 3 & -4 & 6 \end{bmatrix}$. What do you notice about the product?

R **22** Elsie, Lacie, and Tillie are sisters. Their mean age is 11, and their median age is 10. If Elsie is 15 years old, how old are the other two girls?

R **23**

  A at −12, B at 6 on a number line.

  Suppose that $\overline{AB}$ is translated along the number line to form $\overline{A'B'}$. If the coordinate of A' is 0,
  a What will the coordinate of B' be? 18
  b What will the length of $\overline{A'B'}$ be? 18

R **In problems 24–27, solve each equation for x.**

  24 $x + 31.8 = 47.136$ $x = 15.336$
  25 $9.6x = -34.416$ $x = -3.585$
  26 $x - 39 = -42$ $x = -3$
  27 $\dfrac{x}{-4.36} = -91.3$ $x = 398.068$

R **28 Communicating** PENTA is a regular pentagon.
  a Describe the reflection symmetries of the pentagon.
  b Describe the rotation symmetries of the pentagon.
  c List as many properties of the pentagon and its diagonals as you can, justifying each by a rotation or reflection.

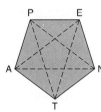

R, I **29** Suppose that point $P(-6, -5)$ is translated $\langle 4, 2 \rangle$ to P', then the translation is repeated four more times to produce P'', P''', P'''', and P'''''. What are the coordinates of each of the images?

Investigation

Origami Do you know what origami is? Try to find a person (perhaps an art teacher) or a book that can teach you about origami, then construct an origami figure of your own. How are symmetries and transformations important in origami?

---

2. C' A' N'
$\begin{bmatrix} -3 & 4 & 7 \\ 2 & -5 & 1 \end{bmatrix}$
Drawings should show a triangle with vertices C'(−3, 2), A'(4, −5), N'(7, 1).

3. C' A' N'
$\begin{bmatrix} -7 & 0 & 3 \\ -3 & 4 & -2 \end{bmatrix}$
Drawings should show a triangle with vertices C'(−7, −3), A'(0, 4), N'(3, −2).

Checkpoint

1. Write a vertex matrix for the triangle with vertices S(−2, 1), T(4, 8), U(7, 3).

2. Multiply the matrix from problem 1 by $\begin{bmatrix} -1 & 0 \\ 0 & 1 \end{bmatrix}$ and describe the transformation.

3. Use a translation matrix to perform the transformation $\langle 5, -2 \rangle$ on the triangle in problem 1.

Answers:

1. S T U
$\begin{bmatrix} -2 & 4 & 7 \\ 1 & 8 & 3 \end{bmatrix}$

2. S' T' U'
$\begin{bmatrix} 2 & -4 & -7 \\ 1 & 8 & 3 \end{bmatrix}$
Reflection over y-axis

3. S T U
$\begin{bmatrix} -2 & 4 & 7 \\ 1 & 8 & 3 \end{bmatrix} + \begin{bmatrix} 5 & 5 & 5 \\ -2 & -2 & -2 \end{bmatrix}$
  S' T' U'
$= \begin{bmatrix} 3 & 9 & 12 \\ -1 & 6 & 1 \end{bmatrix}$

Answers for problems 15–17, 19, 21, 22, 28, 29, and Investigation are on page A7.

SUMMARY

CHAPTER 7

Summary

CONCEPTS AND PROCEDURES

After studying this chapter, you should be able to
- Recognize parallel lines and perpendicular lines (7.1)
- Recognize adjacent angles and vertical angles (7.1)
- Identify squares, rectangles, parallelograms, rhombuses, and trapezoids (7.2)
- Identify figures that possess reflection symmetry (7.3)
- Identify figures that possess rotation symmetry (7.3)
- Transform a figure by means of reflection (7.4)
- Transform a figure by means of rotation (7.4)
- Transform a figure by means of translation (7.4)
- Discover and apply properties of polygons (7.5)
- Represent a figure by means of a vertex matrix (7.6)
- Use matrix addition to model translations (7.6)
- Use matrix multiplication to model reflections and rotations (7.6)

VOCABULARY

adjacent angles (7.1)
bisect (7.5)
consecutive vertices (7.2)
diagonal (7.2)
encasement (7.6)
image (7.4)
isosceles triangle (7.5)
line of symmetry (7.3)
opposite vertices (7.2)
parallel (7.1)
parallelogram (7.2)
perpendicular (7.1)
preimage (7.4)

quadrilateral (7.2)
reflection (7.4)
reflection symmetry (7.3)
rhombus (7.2)
rotation (7.4)
rotation symmetry (7.3)
slide (7.4)
symmetry (7.3)
translation (7.4)
translation numbers (7.4)
trapezoid (7.2)
vertex matrix (7.6)
vertical angles (7.1)

CHAPTER 7

Review

1. One angle of a triangle is a 70° angle. The other two angles have equal measures. What are the measures of those two angles? Each is 55

2. Which of the following pairs of angles are adjacent? a and c
   a ∠PBQ and ∠SBQ
   b ∠RBS and ∠QBP
   c ∠RBQ and ∠SBR
   d ∠PBR and ∠QBR

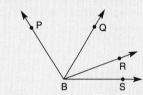

3. If points D, E, F, and G are reflected over the y-axis, what will the coordinates of their images be? (−2, 5); (3, 1); (1, 0); (0, −5)

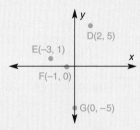

In problems 4–7, copy each polygon, indicating whether it has rotation symmetry, reflection symmetry, or both. If it has reflection symmetry, draw all of its lines of symmetry. If it has rotation symmetry, indicate the center of symmetry and the angle(s) of rotation for the symmetry.

4   5 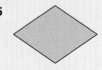   6   7

8. The coordinates of the vertices of a triangle are E(−6, −8), F(−2, 10), and G(4, 0). Suppose that △EFG is translated <−5, 3> to form △E´F´G´. Write a vertex matrix for △E´F´G´.

9. a What kind of quadrilateral is QUAD? Rectangle
   b How could $\overline{UA}$ be moved to make QUAD a square? Move 4 units left
   c How could $\overline{QU}$ be moved to make QUAD a square? Move 4 units up
   d How could $\overline{DA}$ be moved to make QUAD a square? Move 4 units down

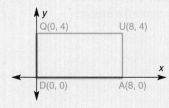

Assignment Guide

Basic
Day 1 1–4, 6, 9, 12, 15–18, 21, 23
Day 2 Chapter Test 1–10

Average or Heterogeneous
Day 1 1–4, 6, 8, 9, 11, 12, 15–18, 21–23

Honors
Day 1 1–3, 5–9, 11–14, 16–18, 20–22, 24

Strands

Geometry	1–24
Measurement	1, 4–7, 10–12, 14, 17, 18, 20, 21, 24
Algebra	1, 3, 8–13, 15–18, 20–24

Additional Answers

for page 363

4 Both

5 Both

6 Rotation

7 Both

8 $\begin{matrix} E' & F' & G' \end{matrix}$
$\begin{bmatrix} -11 & -7 & -1 \\ -5 & 13 & 3 \end{bmatrix}$

REVIEW

Additional Answers

for pages 364 and 365

11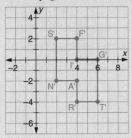

15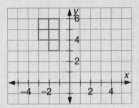

16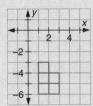

17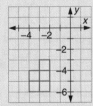

19 a Similarities: four congruent sides; diagonals are perpendicular bisectors of each other. Differences: The angles of a rhombus need not be right angles; the diagonals of a rhombus need not be congruent.

b Similarities: Opposite sides are congruent and parallel; opposite angles are congruent; consecutive angles are supplementary; diagonals bisect each other. Differences: The angles of a parallelogram need not be right angles; the diagonals of a parallelogram need not be congruent.

10 a Copy the diagram and reflect $\overline{RN}$ over the x-axis to form $\overline{R'N'}$. What are the coordinates of R' and N'? (−2, −1); (22, −8)
  b What are the lengths of $\overline{RN}$ and $\overline{R'N'}$? 25; 25

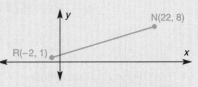

11 Draw the image of figure TRANSFIG after a 180° rotation about point F. Use prime notation to label the vertices of the image.

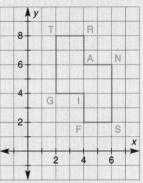

12 If $\overline{MT} \cong \overline{MV}$ and m∠T = 72, what are m∠M and m∠V? 36; 72

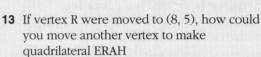

13 If vertex R were moved to (8, 5), how could you move another vertex to make quadrilateral ERAH
  a A rectangle? Move A to (8, 0)
  b A parallelogram? Move A to (8, 0), move H to (−3, 0), or move E to (3, 5)

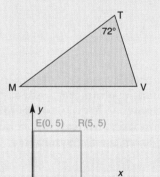

14 GRAM is a parallelogram. If $\overline{GR}$ is 12 cm long and $\overline{RA}$ is 8 cm long, how long are $\overline{GM}$ and $\overline{MA}$? 8 cm; 12 cm

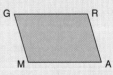

In problems 15–17, draw the image of this figure after the given transformation.

15 Reflection over the y-axis

16 Reflection over the x-axis

17 Rotation of 180° counterclockwise about the origin

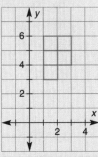

Chapter 7 Exploring Geometry

REVIEW

18 If △TRY is rotated 180° clockwise about the origin, what will be the coordinates of the vertices of the image?
T′ (3, 0), R′ (0, −5), Y′ (−7, 0)

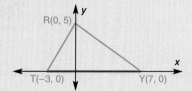

19 Describe the similarities and differences between
  a A square and a rhombus
  b A rectangle and a parallelogram

20 If point F is translated 3 units to the right to form F′ and point L is translated 5 units to the left to form L′,
  a What will the coordinates of F′ and L′ be? (−3, 8); (1, 8)
  b What will the area of trapezoid F′L′EX be? 128

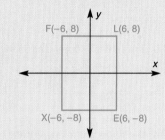

21 What is the area of △JST? 100

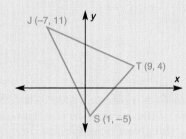

22 a Write a vertex matrix for rectangle RECT.
  b Multiply the vertex matrix by $\begin{bmatrix} -1 & 0 \\ 0 & -1 \end{bmatrix}$.
  c Draw the quadrilateral represented by your answer to part **b**. What transformation did the multiplication produce?

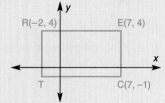

23 A quadrilateral has vertices K(−1, 5), I(1, 7), T(3, 5), and E(1, 0). Use matrix multiplication to locate the vertices of the image of KITE after a reflection over
  a The x-axis
  b The y-axis

24 FIRM is a rectangle, $\overline{FI}$ is 6 inches long, and $\overline{FM}$ is 8 inches long. How long are $\overline{IR}$, $\overline{MR}$, $\overline{FR}$, and $\overline{IM}$?   8 in.; 6 in.; 10 in.; 10 in.

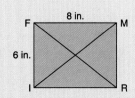

22 a $\begin{bmatrix} -2 & 7 & 7 & -2 \\ 4 & 4 & -1 & -1 \end{bmatrix}$

  b $\begin{bmatrix} 2 & -7 & -7 & 2 \\ -4 & -4 & 1 & 1 \end{bmatrix}$

  c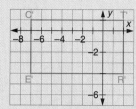

A rotation of 180° about the origin

23 a K′(−1, −5), I′(1, −7), T′(3, −5), E′(1, 0) **b** K′(1, 5), I′(−1, 7), T′(−3, 5), E′(−1, 0)

Chapter Review 365

CHAPTER 7

Test

In problems 1–3, refer to quadrilateral ABCD.

1. Identify all pairs of parallel segments and all pairs of perpendicular segments in the quadrilateral.

2. Describe all the symmetries of the quadrilateral.

3. What kind of quadrilateral is ABCD?   Kite

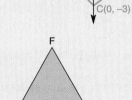

4. △EFG is equilateral. Copy the triangle and draw all its lines of symmetry.

5. What are the coordinates of vertex H of parallelogram HJKL?   (−2, −5)

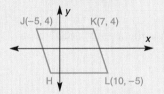

In problems 6–8, refer to △MNP. Find the coordinates of the vertices of the triangle's image after the given transformation.

6. A reflection over the y-axis

7. A translation of <−5, 2>

8. A 90° counterclockwise rotation about the origin

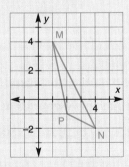

9. What is the area of △QRS?   30

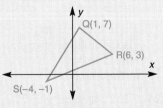

10. TUVW is a rhombus, with WV = 4.5 and m∠V = 72. What are UV, TW, m∠T, and m∠U?   4.5; 4.5; 72; 108

PUZZLES AND CHALLENGES

1 A **word chain** is a list of words, in order, where each word is made from the previous word by changing one letter. All words in a chain must be legitimate English words.

Example: A word chain can change JACK to KING with the following list: JACK, BACK, BANK, RANK, RINK, RING, KING

Find word chains for the following:

a Change MAST to SPAR.
b Change CENT to DIME.
c Change FAST to SLOW.
d Change WORK to PLAY.
e Change SHAVE to CLEAN.
f Change MEAT to FISH.
g Change WOK to FRY.
h Change WISH to HOPE.

2 Put 3 pennies, 3 dimes, 1 quarter, and 1 nickel into a 3-by-3 grid as shown. Slide the coins around, one at a time and one to a square. A coin can be moved only into an adjacent open space. Each move may be only left, right, up, or down. The idea is to end up with the same configuration, except that the nickel and quarter are switched. What is the fewest number of moves it takes for you to do this?

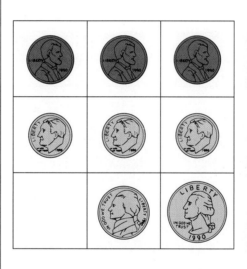

rectangle, a rhombus, a parallelogram, a kite, and a trapezoid. Organize the information in a chart that other students could use to help them distinguish between the figures.

10. Section 7.3 presents symmetry. Find several examples of reflection and rotation symmetry in the objects you deal with everyday. For each object, describe its symmetry and try to explain why it possesses the type of symmetry that it does.

11. Create a spreadsheet application that produces the coordinates of a figure after a translation. Use problems 3 d, 7, 14, and 17 b on pages 343–345 to test your spreadsheet application. Write an explanation of how you set up the spreadsheet to perform translations.

12. Extend problem 18 on page 346 as follows: Find another object besides an electrical-outlet cover that could be diagrammed in the way described in the problem. Make a scale drawing of part of this object in the first quadrant of a coordinate system. Explain how your diagram could be used to construct a complete diagram of the object.

Additional Answers

for Chapter 7 Test

1 $\overline{DA}$ and $\overline{DC}$ are perpendicular; no parallel segments

2 It has reflection symmetry over the x-axis.

4

6 M'(−1, 4), N'(−4, −2), P'(−2, −1)
7 M'(−4, 6), N'(−1, 0), P'(−3, 1)
8 M'(−4, 1), N'(2, 4), P'(1, 2)

Answers for Puzzles and Challenges are on page A7.

ASSESSMENT LONG-TERM INVESTIGATIONS

UNIT D: Patterns with Shapes

Suggestions for various ways to assess students' performance are integrated throughout the investigations in this unit. These forms of assessment include observing the students while they work, interviewing students, looking at students' self-evaluations and group evaluations, observing peer review/response situations, evaluating students' products, and evaluating students' portfolios.

Refer to the chart below to preview the variety of assessment opportunities that are integrated throughout this unit.

Assessment Opportunities	Before Unit	1	2	3	4	5	End of Unit
Form A: Observation Notes **Bank A/B:** Observation Guide		✘	✘		✘		
Form B: Observation Evaluation Record **Bank A/B:** Observation Guide						✘	
Form C: Interview Record **Bank C:** Interview Guide			✘			✘	
Form D: Product Evaluation Record **Bank D:** Product Evaluation Guide		✘		✘	✘		
Form E: Rubric Chart **Bank E:** Sample Rubric				✘			
Form F: Portfolio Item Description	✘						✘
Form G: Portfolio Notes **Bank G:** Portfolio Evaluation Criteria	✘						✘
Form H: Self-evaluation				✘		✘	✘
Form I: Group Evaluation		✘				✘	
Form J: Peer Review/Response				✘			

Rubrics

Rubrics are used to evaluate students' products. You can develop a scoring rubric for a specific investigation by adapting the blackline masters shown below, which are provided in the ***Gateways to Mathematical Investigations*** Teacher's Edition. You will probably want to develop a rubric with students at least once per unit. The chart on the opposite page suggests that for Unit D: Patterns with Shapes, you complete a rubric before Investigation 3. Remind students that, at times, they may be asked to revise their work to improve it and to make it complete.

Bank D, page 211

Bank D: Product Evaluation Guide

The statements below can be used as a guide to assess students' products. Adapt these statements as needed to evaluate specific assignments. Record your adaptations on a clean copy of **Form D: Product Evaluation Record** (p. 210). You may also wish to create and record your own criteria on **Form D.** After you have adapted **Form D** for use with a specific assignment, make copies for each student or group as needed.

Mathematical Reasoning Skills

- The student includes examples and/or counter examples to support the mathematical argument.
- The student synthesizes and summarizes own or group's thinking.

Mathematical Understanding

- The student identifies all the important information of the situation or problem.
- The student shows understanding of the situation's mathematical ideas. (List important mathematical concepts and skills for the assignment on Form D.)

Form E, page 212

Form E: Blank Rubric Chart

NAME (S) ______________ ______________ GROUP ________
DATE ________

STANDARDS FOR STUDENT'S WORK

LEVEL	Standard of performance to be achieved for each specified level
4 Exceptional/ Fully Acceptable	

Bank E, page 213

Bank E: Sample Rubric

STANDARDS FOR STUDENT'S WORK

LEVEL	Standard of performance to be achieved for each specified level
4 Exceptional/ Fully Acceptable	Your presentation fully accomplishes the purpose of the task. Your work is complete, well organized, and clearly explained. You illustrate ideas using clear diagrams and models and communicate them effectively to the audience. You identify all the important information of the situation and show a thorough understanding of the mathematical ideas and procedures. You use strong arguments to support findings, including examples (and/or counterexamples). You demonstrate that you fully grasp the central mathematical ideas developed in this task. You demonstrate an orignial extension or applicatrion of the mathematical concepts.
3 Satisfactory/ Acceptable	Your presentation substantially accomplishes the purpose of the task. Your work is complete, reasonably organized, and explained. You use an appropriate diagram or model and communicate ideas effectively to the audience. You identify most of the important information of the situation and show understanding of the mathmatical ideas and procedures. You present arguments to support findings, including examples (and/or counterexamples). You understand and use the central mathematical ideas developed in this task. You demonstrate and apply the mathematical concepts.
2 Incomplete/ Revise	Your presentation partially accomplishes the purpose of the task. However, your work may be incomplete, or misdirected, or not clearly organized and/or explained. Your diagram or model may be inappropriate or unclear. You may communicate ideas inadequately to the audience. You may identify some of the important information of the situation and you may show only partial and/or limited understanding of the mathematical ideas and procedures. Your arguments to support findings may be weak or incomplete. You may misunderstand or misuse the central mathematical ideas developed in this task. You may have difficulty demonstrating and applying the mathematical tasks. You need to revise this work. (Your teacher will provide suggestions for a satisfactory or competent rating and a due date for your revised work.)
1 Inadequate/ Redo	Your presentation shows little or no understanding of the assignment. Your work is barely, if at all, comprehensible. Your explanations are not understandable. You show little or no understanding of the mathematical ideas and procedures. You made serious computational mistakes and misused mathematical terms. You should meet with your classmates or your teacher to be sure you fully understand this assignment. Then you need to revise this work. (Your teacher will provide suggestions for a satisfactory or competent rating and a due date for your revised work.)

Gateways © McDougal, Littell & Company   Bank E: Sample Rubric   213

Form F, page 214

Form F: Portfolio Item Description

NAME ______________ DATE ________

Complete this form each time you add an item to your portfolio.

Type of Portfolio Entry: ______________

Why did you choose to place this item in your portfolio?

Form G, page 215

Form G: Portfolio Notes

NAME ______________ CLASS ________

Date	Observations/Comments	Recommendations

Bank G, page 216

Bank G: Portfolio Evaluation Criteria

At the start of the school year, have a class discussion on the goals and uses of students' portfolios. Discuss with the class some ideas regarding what to include in a portfolio. Then, use the portfolio evaluation criteria which follows to develop a rubric with the class that communicates the criteria you will use to assess their portfolios. Have students record this criteria on copies of **Form E: Rubric Chart.** Have students store this completed form in their portfolios so that they can refer to it throughout the year.

Each time students add items to their portfolios, have them complete copies of **Form F: Portfolio Item Description.**

Use copies of **Form G: Portfolio Notes** to record your observations, comments, and recommendations each time you evaluate students' portfolios.

Ideas for Types of Portfolio Entries

- a portfolio table of contents
- a letter to the reader of the portfolio explaining each item
- teacher-completed checklists/evaluations
- a mathematical autobiography
- notes from interviews, teachers, peers, etc.
- papers that show the student's corrections or errors or misconceptions
- photos or sketches of the student's work with manipulatives or mathematical models
- excerpts from the student's daily journal or reflective writing samples
- tables, charts, and diagrams created by the student
- artwork supporting a problem such as drawings
- designs, coordinate pictures, scale drawings, maps, etc.
- student-created problems with or without solutions
- group projects or individualized reports
- student-developed surveys and results
- self-evaluations

Suggestions for Portfolio Evaluation Criteria

Mathematical Reasoning Skills
The Student:
- analyzes problem situations.
- is flexible and comfortable in using several different problem solving approaches.
- synthesizes and summarizes the individual's or group's thinking.
- provides supporting statements to verify findings.
- shows original and creative thought.

Mathematical Tools and Techniques
The Student:
- uses a variety of diagrams, models, technology, manipulatives, and other resources appropriately.
- uses displays (e.g. graphs, tables, diagrams, sketches, and models) to represent mathematical ideas.

Mathematical Understanding
The Student:
- shows understanding of mathematical ideas and procedures. (List specific skills/concepts on **Form G.**).

Mathematical Communication
The Student:
- organizes work in a logical and clear way.
- expresses mathematical ideas clearly and appropriately.
- uses mathematical language and notation appropriately.

Mathematical Dispositions
The Student:
- shows development as as group participant.
- shows development of a positive attitude.
- shows growth in learning.
- includes revised and complete work.
- shows consistent effort.
- includes a variety of work in his or her portfolio.

216   Assessment   *Gateways* © McDougal, Littell & Company

Portfolios

At the beginning of each unit, you may wish to review with the students the criteria they should use when placing work in their portfolios. Have students complete Form F: Portfolio Item Description each time they wish to place an item in their portfolios.

Periodically, you may wish to evaluate students' portfolios using the blackline masters shown above. For Unit D: Patterns with Shapes, the chart on the opposite page suggests that you evaluate students' portfolios at the end of the unit.

CHAPTER 8

Contents

8.1 Simplifying Expressions
8.2 Using Variables
8.3 Solving Equations
8.4 Some Special Equations
8.5 Solving Inequalities
8.6 Substitution
8.7 Functions

Pacing Chart

Days per Section

	Basic	Average	Honors
8.1	3	2	1
8.2	3	2	1
8.3	2	2	1
8.4	*	2	2
8.5	*	3	2
8.6	2	2	2
8.7	2	2	2
Review	3	1	1
Assess	2	2	2
	17	18	14

* Optional at this level

Chapter 8 Overview

Most students will probably take a formal algebra course following this course. This chapter introduces and reinforces many of the central ideas of the first part of that course. Topics are presented concretely, geometrically, and abstractly. The major objectives of this chapter are to get students comfortable with representing quantities with variables and to help them skillfully manipulate the variables. Whenever possible, students are called on to use both skills. In order to emphasize understanding rather than rote manipulation, students are asked to solve inequalities and equations with absolute value, roots,

8 The Language of Algebra

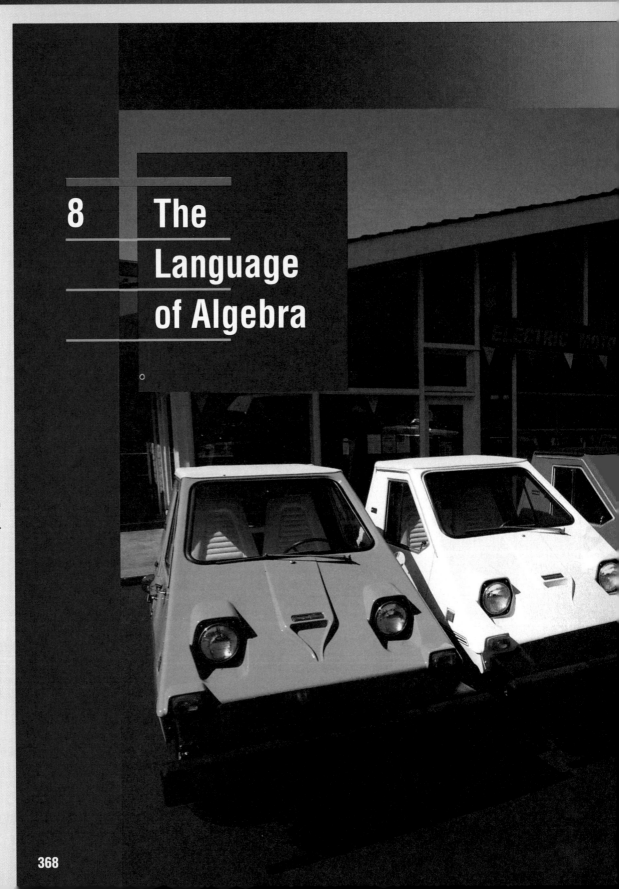

CHAPTER 8

AUTOMOTIVE DESIGN Why are today's cars so different from earlier models? Were changes made over the years just to make cars *look* different?

Some design changes are strictly for appearance, but others are for safety. At one time, for example, few cars had either seat belts or safety glass, and dashboards were made of steel. Air bags are a relatively recent safety feature.

Automotive designers use computers that can display three-dimensional designs. These designs can be altered by a few keystrokes. In addition to affecting safety and appearance, decisions about design can help to determine how long the car will last, how often it will need repairs, how much it will cost, and how much gas mileage it will get.

INVESTIGATION

Fuel Efficiency Ask a friend or relative how many miles per gallon his or her car gets. Does the type of driving affect the mileage? Explain the steps you would follow in order to determine a car's gas mileage. Will you get the same answer every time you follow these steps? Explain.

---

and exponents. The concept of function is presented as a way of applying equations and relating them to formulas.

Career Connection

Computer models are used to test many features of cars, and they have reduced the time from conception to production of new models from over five years to under two years. Explain to students that mathematical modeling is used in engineering to test the effects of different conditions on final results and that such modeling has helped designers build safer, more efficient, cars. An automobile designer needs to be able to combine creativity and practical use of mathematics.

Video Connection

For more information about the career discussed here, please refer to the Futures videocassette #2, program B: *Automotive Design,* featuring Jaime Escalante. For more information or to order a copy of the progam, call PBS VIDEO at 800/344-3337.

Investigation Notes

Stop-and-go city driving yields lower gas mileage than highway driving. Figure gas mileage this way: (1) fill the tank and record the mileage, (2) drive, (3) fill the tank again and note the mileage and the number of gallons in the second fill-up, and (4) divide the number of miles driven between fill-ups by the number of gallons in the second fill-up. Repeated calculations on the same car may produce different results, because different driving conditions, such as hills or air conditioner use, affect gas mileage.

SECTION 8.1

Objectives

- To recognize equivalent expressions
- To add and subtract like terms
- To use the Distributive Property of Multiplication over Addition and Subtraction

Vocabulary: equivalent expressions, like terms, numerical term, simplify

Class Planning

Schedule 1–3 days

Resources

In *Chapter 8 Student Activity Masters* Book
Class Openers 1, 2
Extra Practice 3, 4
Cooperative Learning 5, 6
Enrichment 7
Learning Log 8
Using Manipulatives 9, 10
Spanish-Language Summary 11

Class Opener

A quadrilateral has sides of length $3x + 1$, $6x - 2$, $4x + 5$, and $7x - 4$. Find its perimeter for each value of x.
1. $x = 3$ 2. $x = 4$
3. $x = 10$ 4. $x = 100$

Answers:
1. 60 2. 80 3. 200 4. 2000

Lesson Notes

- The class opener and the first two pages of the section are meant to motivate students to see why simplification is important. You can amaze your students by asking them for a value for x in the class opener and then calculating the perimeter simply by multiplying x by 20.

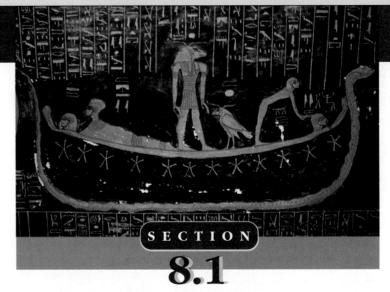

SECTION 8.1
Simplifying Expressions

EQUIVALENT EXPRESSIONS

Ms. Perry asked her class to find the perimeter of triangle ABC for $x = 4$.

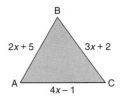

Subina wrote the expression for the length of each side and substituted 4 for x.

AB	= $2x + 5$	BC	= $3x + 2$	AC	= $4x - 1$
	= $2(4) + 5$		= $3(4) + 2$		= $4(4) - 1$
	= 13		= 14		= 15

She added the lengths of the sides.   Perimeter = 13 + 14 + 15
                                                  = 42

Addem wrote an expression for   Perimeter = $9x + 6$
the perimeter and substituted   Perimeter = $9(4) + 6$
4 for x.   = 42

Both students used different methods and found the same perimeter.

Example 1
Evaluate the two expressions $2x + 5 + 3x + 2 + 4x - 1$ and $9x + 6$ for five different values of x. What do you observe?

370   **Chapter 8**   The Language of Algebra

SECTION 8.1

Solution

We can set up a table and choose five values.

Value of x	$2x + 5 + 3x + 2 + 4x - 1$	$9x + 6$
2.5	28.5	28.5
10	96	96
2	24	24
3.5	37.5	37.5
7	69	69

It isn't practical to substitute every possible value for x. The five values we used show that the two expressions $2x + 5 + 3x + 2 + 4x - 1$ and $9x + 6$ produce the same value for a replacement of x.

If two expressions produce the same value for all replacements of the variable(s), the two expressions are called **equivalent expressions**. We indicate this by connecting the two expressions with =, an equal sign.

The expression $9x + 6$ is shorter and simpler than $2x + 5 + 3x + 2 + 4x - 1$. When we work with an expression, we usually try to simplify the expression by finding its simplest equivalent form.

ADDING AND SUBTRACTING LIKE TERMS

How did Addem arrive at the simpler form $9x + 6$?

Rewrite the expression Subina used. $2x + 5 + 3x + 2 + 4x - 1$

Group the x-terms and the terms
without x, the **numerical terms**. $2x + 3x + 4x + 5 + 2 - 1$

We call terms that have the same variable raised to the same power **like terms**. In this problem, $2x$, $3x$, and $4x$ are like terms. Like terms can be added or subtracted.

$$(2 + 3 + 4)x + (5 + 2 - 1)$$
Add the like terms. $= 9x + 6$

The result is the expression Addem used. Addem **simplified** the expression.

Example 2
Simplify the expression $3xy + 8xy - 5xy$.

Solution

Method 1 In this expression, all the terms are like terms.
They have the same variable, xy.

Add and subtract the coefficients of xy
$$3xy + 8xy - 5xy$$
$$= (3 + 8 - 5)xy$$
$$= 6xy$$

- Students are prone to several kinds of errors. One is interpreting $3x$ as "three equals x" rather than "three times x." A second error is failing to distribute a factor to both terms, as in problems similar to example 5 and sample problem 2. A third error is failing to equate the expression x with 1 times x. This error is addressed in sample problem 1.
- The model on page 372 will be helpful to the visual learner. Some discussion will help make the model more accessible.

Additional Examples

Example 1
Evaluate the two expressions
$3x - 1 + 5x + 4 - 2x - 7$ and $6x - 4$
for $x = 1, 2, 3, 4,$ and 5. What do you observe?

Answer: 2, 2; 8, 8; 14, 14; 20, 20; 26, 26; answers are the same and successive answers are 6 more than the previous answer.

Example 2
Simplify the expression
$4ab + 6ab - 3ab$.

Answer: $7ab$

Example 3
Simplify the following expressions.
a. $3y^2 + 4x - 2y^2 + 3y + 5y^2 - 12x$
b. $6x + 3y - 4x^2$

Answers: a. $6y^2 - 8x + 3y$
b. $6x + 3y - 4x^2$

Example 4
Multiply the quantity
$27(3x + 2y)$.

Answer: $81x + 54y$

Example 5
Multiply the quantity
$-7(4x - 3y)$.

Answer: $-28x + 21y$

Section 8.1 Simplifying Expressions

SECTION 8.1

Stumbling Block

Students may simplify an expression such as $3x + 4x + x$ by adding only the numerical coefficients that they can see. Remind students that the term x has an understood coefficient of 1.

Reteaching

Reteach combining like terms by using physical models. Cut out several of the following shapes.

ones   x x^2 y

Have students model problems such as $x^2 + 3x - 4 + y + 2x^2 - 5x + 9$. For this problem, the result would be:

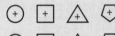
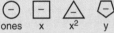

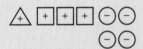

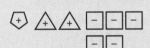

Figures with the same shape should then be placed together. A pair of identical shapes with one plus sign and one minus sign add to 0, so these pairs are removed. Have students write the simplified expression, $3x^2 - 2x + y + 5$, from the remaining figures.

Method 2 We can model the problem using rectangles with sides x and y and area xy.

Start with $3xy$.

Add $8xy$.

Subtract $5xy$.

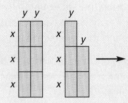

There are $6xy$ left.

Example 3
Simplify the following expressions.
a $14x + 9x^2 - 3x^2 + 13y - 18x - 2x^2 + 8y$
b $3x + 8y + 5x^2$

Solution
a This expression contains x-terms, x^2-terms, and y-terms.

$$14x + 9x^2 - 3x^2 + 13y - 18x - 2x^2 + 8y$$

First, group like terms $= 14x - 18x + 9x^2 - 3x^2 - 2x^2 + 13y + 8y$

Combine like terms $= (14 - 18)x + (9 - 3 - 2)x^2 + (13 + 8)y$

$= -4x + 4x^2 + 21y$

b The expression $3x + 8y + 5x^2$ cannot be simplified because it has no like terms. The simplest form is $3x + 8y + 5x^2$.

SECTION 8.1

THE DISTRIBUTIVE PROPERTY

You have simplified expressions such as $-3x + 5x$ and $14y - 6y$.

$$-3x + 5x = (-3 + 5)x \text{ and } 14y - 6y = (14 - 6)y$$

The general form of these expressions is $ax + bx = (a + b)x$ or $ax - bx = (a - b)x$. These statements are true for any values of a, b, and x and are examples of one of the most important properties in algebra.

▶ **The Distributive Property of Multiplication over Addition or Subtraction:** For all numbers a, b, and c, $a(b + c) = ab + ac$ and $a(b - c) = ab - ac$.

Example 4
Multiply the quantity $34(2x + 6y)$.

Solution
We distribute the multiplication by 34.

$$34(2x + 6y)$$
$$= 34(2x) + 34(6y)$$

Multiply $2x$ by 34 and $6y$ by 34
$$= (34)(2)x + (34)(6)y$$
$$= 68x + 204y$$

Example 5
Multiply the quantity $-3(5x - 7y)$.

Solution
We distribute the multiplication by (-3). Be careful to keep track of the negative and subtraction signs.

$$-3(5x - 7y)$$
$$= (-3)(5x) - (-3)(7y)$$
$$= (-3)(5)x - (-3)(7)y$$
$$= -15x - (-21)y$$
$$= -15x + 21y$$

Sample Problems

Problem 1 Simplify the expression $6x + x$.

Solution Remember that $x = 1x$. Therefore, $6x + x = 7x$.

Assignment Guide

Basic
Day 1 11, 13, 16, 18, 23, 28–30
Day 2 12, 15, 17, 21, 24, 31, 35
Day 3 14, 19, 20, 22, 26, 32–34

Average or Heterogeneous
Day 1 11, 13, 14, 16, 20, 21, 24–26, 31, 32
Day 2 12, 15, 17, 19, 22, 23, 27–30, 34, 35

Honors
Day 1 12, 13, 15–17, 19–21, 23, 25–28, 30–32, 34, Investigation

Strands

Communicating Mathematics	1, 3–10, 21, 28
Geometry	6, 16, 19, 23, 25
Measurement	6, 16, 19, 23, 25
Technology	25, Investigation
Algebra	1–7, 9–33, 35
Patterns & Functions	20, 25, 33, 34, Investigation
Number Sense	8, 10, 15, 21, 26, 28, 35

Individualizing Instruction

The visual learner Students may be better able to explain their answers to problems 3 and 4 if they draw diagrams to represent the equations. Students' representations will vary. Have them share and discuss their drawings with the class.

SECTION 8.1

Problem-Set Notes

for pages 374 and 375

- **6** provides a concrete, visual representation of the ideas presented in this section.
- **7** leads to interesting questions. The simpler method of evaluation depends on the numbers involved.
- **9** is a rather deep question. The first equation is always true, while the second requires us to find all values for which it is true. The discussion can become much more involved, but for now this is probably a satisfactory answer.

Additional Practice

Simplify each expression.
1. $3x - 2y + 7x + 6y$
2. $3x^2 - 2x + 5x^2 + 9x - 4$
3. $5(3x - 7y)$
4. $4(2x - 3y) - 2(3x + 7y)$
5. $2x - 5x + 6x^2$
6. $8(2x + 3y) - 5(3x + 4y)$
7. $3xy + 2x - 4y - xy$
8. $3x^2 + 8xy - x + 2xy + 3x - x^2$

Answers:
1. $10x + 4y$ 2. $8x^2 + 7x - 4$
3. $15x - 35y$ 4. $2x - 26y$
5. $-3x + 6x^2$ 6. $x + 4y$
7. $2xy + 2x - 4y$
8. $2x^2 + 10xy + 2x$

Problem 2 Simplify the expression $12(3x - 5y) - 14(2 - 8y)$.

Solution

$12(3x - 5y) - 14(2 - 8y)$

Use the distributive property $= 12(3x) - 12(5y) - 14(2) - 14(-8y)$
Multiply $= 36x - 60y - 28 + 112y$
$= 36x - 60y + 112y - 28$
Combine like terms $= 36x + 52y - 28$

Problem 3 Find the area of the figure.

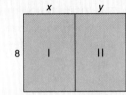

Solution

Method 1 The area of rectangle I is $8x$.
The area of rectangle II is $8y$.
The total area will be $8x + 8y$.

Method 2 The total length of the top is $x + y$.
The area is $8(x + y)$.

Because $8x + 8y$ and $8(x + y)$ are equivalent, both forms of the answer are equally valid.

Think and Discuss

P **1** Are $4x + 3 + 2x$ and $6x - 3$ equivalent expressions? Explain your answer.

P **2** Simplify $5x + 2 - x^2 + 3x^2 - 4x + 3$. $2x^2 + x + 5$

P **3** Explain why $8x + x = 9x$.

P **4** Explain why $7(5x) = 35x$.

P **5** Johnny said, "Terms with x and terms with x^2 should be like terms. They both include x's." Susie said, "They aren't like terms. If x represented length, x^2 would represent area." Who do you think was right and why?

P, R **6 a** Express the total area of the figure in two different ways. $5(x + 3y + 8); 5x + 15y + 40$
b Are the expressions equivalent?   Yes
c Defend your answer to the question in part **b**.   The distributive property has been used.

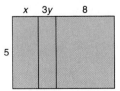

374   Chapter 8   The Language of Algebra

SECTION 8.1

P, R **7** Use the equation $a(b + c) = ab + ac$.
  a Evaluate each side for $(a, b, c) = (5, 20, 7)$. $135 = 135$
  b Which side was easier to evaluate?   Right
  c Evaluate each side for $(a, b, c) = (27, 13, 7)$. $540 = 540$
  d Which side was easier to evaluate?   Left
  e When will the left side be easier to evaluate? When will the right side be easier to evaluate?

P, R **8** Is it always appropriate to change $\frac{9}{12}$ to $\frac{3}{4}$? Explain your answer.

P, R **9** The = sign in $3x + 4x = 7x$ means something different from the = sign in $x + 5 = 8$. Explain the difference between these two equations.

P, R **10** Nikita said, "Subtracting a positive number is just like adding a negative number." Explain why you think Nikita is or isn't correct.

Problems and Applications

P **11** Simplify each expression.
  a $8x + 5x + x$ $14x$
  b $5x + 3x - 8x$ 0
  c $3x + 6y + 9x + y - x$ $11x + 7y$
  d $4x - 9 - 10x + 6$ $-6x - 3$
  e $4x^2 + x - 8x^2 - 6x$ $-4x^2 - 5x$
  f $3(2x + 5y) - 8(3y - 5x)$ $46x - 9y$

P **12** Write each expression in simplest form by combining like terms.
  a $x + x + x$ $3x$
  b $2xy + 9xy - xy$ $10xy$
  c $xy + x + y$ $xy + x + y$
  d $ab - ab$ 0
  e $2xy + 5xy - x$ $7xy - x$

P, R **13** Write an algebraic statement to describe the drawings.
  $4xy + 9xy - 5xy = 8xy$

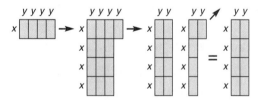

P **14** Match each item in the column on the left with an item in the column on the right.
  a x vi
  b $2x + 3x$ iii
  c $9x + 11x$ vii
  d $9(11x)$ i
  e $3(x + 6)$ iv
  f $3x + 6$ ii
  g $10(3x - x)$ vii

  i $99x$
  ii $3(x + 2)$
  iii $5x$
  iv $3x + 18$
  v $9(x + 11)$
  vi $1x$
  vii $30x - 10x$

P **15** Is the statement *true* or *false* for $x = 4$ and $y = 9$?
  a $2x + 5x = 7x^2$ False
  b $3y + 5y = 8y$ True
  c $36x + 14x = 50x$ True
  d $2x + 7y + 9x + 11y = 29xy$ False

Cooperative Learning

- Have students work in groups to draw pictures like those in example 2 (method 2) and problem 6 to illustrate how to simplify the following expressions.
  1. $3x + 5 + 4x + 3 + 2x + 4$
  2. $(9x + 7) - (3x + 5)$
  3. $2xy + 5xy + 3x^2 + 4y^2 + 3y^2 + 2y^2$

Answers: Check students' drawings.

- Problem 28 and Investigation are appropriate for cooperative learning.

Additional Answers

for pages 374 and 375

1 No; $4x + 3 + 2x = 6x + 3$

3 $8x + x = (8 + 1)x = 9x$

4 $7(5x) = (7 \cdot 5) \cdot x = 35 \cdot x = 35x$

5 Susie; like terms have the same variable and the same power.

7 e When it is easier to add first; when it is easier to multiply first.

8 It is always appropriate to reduce the fraction, but the unreduced fraction may be a more representative form of the number in some cases.

9 $3x + 4x = 7x$ is true for every value of x, but $x + 5 = 8$ is true only when $x = 3$.

10 Nikita is correct. Both methods have the effect of moving left on the number line.

Section 8.1 Simplifying Expressions

SECTION 8.1

Problem-Set Notes

for pages 376 and 377

- **19** shows another way to visualize the concepts in this section.
- **20** motivates students to simplify expressions first.
- **26** might be fun to do in groups. Have students explain their reasoning.
- **27** should generate some class discussion. How many pairs of numbers work? Do all pairs of numbers work? Are there other sets of equations that have many solutions?

Communicating Mathematics

- Have students write a few sentences explaining why $2x^4 + 3x^4$ does not equal $5x^8$.
- Problems 1, 3–10, 21, and 28 encourage communicating mathematics.

Checkpoint

Simplify each expression.
1. $2x^2 - 3x + 4x^2 + 7x$
2. $3x - 4y - x + 6y$
3. $3(2x - 5y)$
4. $4(3x + 2y) - 3(2x + 5y)$
5. $3x + 7x - 2x^2$
6. $-(x + 2y) - 2(-x - 3y)$

Answers:
1. $6x^2 + 4x$ 2. $2x + 2y$
3. $6x - 15y$ 4. $6x - 7y$
5. $10x - 2x^2$ 6. $x + 4y$

P, R   **16** Write an expression in simplest form to represent

     a The area of the rectangle

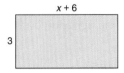

$3x + 18$

     b The perimeter of the triangle

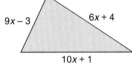

$25x + 2$

     c The coordinates of points A, B, and C

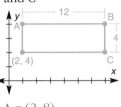

$A = (2, 8)$,
$B = (14, 8)$, $C = (14, 4)$

P, R   **17** Simplify each expression.

     a $\frac{-3}{4} + \frac{1}{2}$ $-\frac{1}{4}$ **b** $\frac{-3}{4}x + \frac{1}{2}x$ $-\frac{1}{4}x$ **c** $\frac{-3}{4}xy + \frac{1}{2}x^2$

P   **18** Simplify each expression.

     a $3x + 6y$ **b** $9x^2 - 6x$ **c** $6x^2 + 3x - 4y + 2$

P, R   **19** The lengths of line segments AB and CD are equal.

     a Write an expression for the length of segment AB. $4x + 4$

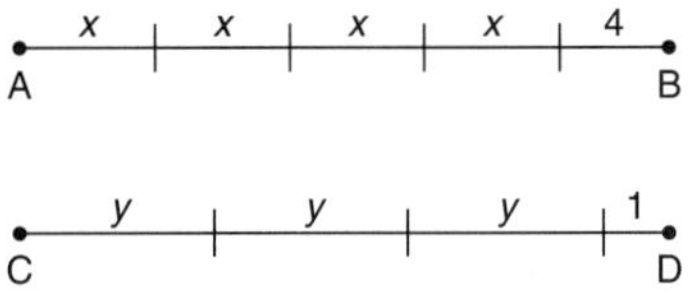

     b Write an expression for the length of segment CD. $3y + 1$

     c If the lengths of segments AB and CD are each 16, what are the values of x and y? $x = 3$, $y = 5$

     d What would be the lengths of segments AB and CD if $(x, y) = (6, 9)$?

P, I   **20** Complete the table.

x	-8	-4	-2	0	1	3	5
$9x + 6 - 3x + 2 - 4x - 7$	-15	-7	-3	1	3	7	11

P   **21 Communicating**   Explain why we write $3x - 3x = 0$, rather than $3x - 3x = 0x$.

P, R   **22** Simplify each expression.

     a $8(6.5) - 16\left(\frac{3}{4}\right) + 100(3.15)$ 355 **b** $14(2x - 6y) - 9(4 - 3x)$ $55x - 84y - 36$

     c $19x + (-6x)$ $13x$

P, R   **23 a** Write an expression for the length of the segment. $13x + 3$

     [segment labeled $x + 5$, $2x - 3$, $10x + 1$]

     b Find the length of the segment for $x = 0.1$. 4.3

P   **24** Simplify each expression.

     a $-2(3x - 8y) - (4y + 2x)$ $-8x + 12y$ **b** $5(2x - 6y^2) + 3(5x - 8x^2)$ $25x - 24x^2 - 30y^2$

P, R   **25 a** Write an expression to represent the perimeter of triangle TRI. $11x - 2$

     b Use the values $x \geq 1$. Create a spreadsheet to calculate the length of each side of the triangle, the sum of the sides, and the simplified form of the expression for the perimeter.

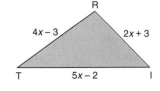

376   Chapter 8   The Language of Algebra

SECTION 8.1

P, R **26** **Mental Math** Without using a calculator or a pencil and paper, evaluate each expression.
  a $37 \cdot 2 + 37 \cdot 5 + 37 \cdot 3$ 370   **b** $8 \cdot 23 + 23 \cdot 3 - 23 \cdot 11$ 0
  c $137 \cdot 93 - 37 \cdot 93$ 9300

P, I **27** Find the value of x and y if $5x + 5y = 30$ and $x + y = 6$.

P **28** **Communicating** The two expressions $18 + (-5)$ and $18 + -5$ mean the same thing. Why might someone want to use one form or the other?

◀ LOOKING BACK **Spiral Learning** LOOKING AHEAD ▶

R **29** Indicate whether each statement is *true* or *false*.
  a If $x = 8$, then $x^2 - x = 56$.   True   **b** If $x = 1$, then $x^0 + x^1 = 1$.   False

R **30** Write an expression to represent each phrase.
  a 5 more than 27   **b** The product of 7 and 11   **c** 3 less than 56 $56 - 3$

P, I **31** Evaluate $2x - y$ for the following values.
  a $(x, y) = (-5, 7)$ -17 **b** $(x, y) = (-3, -8)$ 2

P, I **32** Solve each equation for x.
  a $x - 12 = 57$ $x = 69$ **b** $\frac{x}{12} = 57$ $x = 684$ **c** $12x = 57$ $x = 4.75$

R **33** Evaluate each expression and divide the result by -9.
  a $-859 + 958$ 99; -11 **b** $-813 + 318$ -495; 55   **c** $-512 + 215$ -297; 33

R **34** Find the next three terms in the pattern. $\frac{1}{3}, \frac{3}{5}, \frac{5}{7}, \frac{7}{9}$ $\frac{9}{11}, \frac{11}{13}, \frac{13}{15}$

R **35** Fill in each blank with one of the symbols <, >, or =.
  a $3(5)$ _____ $3(4)$ >   **b** $0(5)$ _____ $0(4)$ =   **c** $-2(5)$ _____ $-2(4)$ <

Investigation

The Fibonacci Sequence The first two numbers in the following pattern are both 1. Each number after that is the sum of the two numbers that precede it. Continuing in this manner, you form the Fibonacci sequence.

1, 1, 2, 3, 5, 8, 13, 21,…

You can generate some Fibonacci-like sequences by starting with two numbers that aren't both 1. For example, write several terms of the sequences whose first two terms are

1. 3, 3,…   **2.** 2, 1,…   **3.** x, x,…

Try some of your own. Then, for both the Fibonacci sequence and for your Fibonacci-like sequences, use your calculator to investigate the pattern of the ratios of one term to the next.

Additional Answers

for pages 376 and 377

17 c Cannot be simplified.
18 a Cannot be simplified.
  b Cannot be simplified.
  c Cannot be simplified.
19 d AB = 28, CD = 28
21 $0x = 0$; so 0 is the simpler form.
25 b

	A	B	C	D	E
1	x	TR	RI	TI	Sum
2	1	1	5	3	9
3	2	5	7	8	20
4	3	9	9	13	31
5	4	13	11	18	42
6	5	17	13	23	53
7	6	21	15	28	64
8	7	25	17	33	75
9	8	29	19	38	86
10	9	33	21	43	97
11	10	37	23	48	108

27 $x = 6 - y$; $y = 6 - x$

28 Possible answer: The first form makes it less likely that either the + or − will be overlooked; the second takes less time to write.

30 a $27 + 5$ **b** $(7)(11)$

Investigation
1. 3, 3, 6, 9, 15, 24, …
2. 2, 1, 3, 4, 7, 11, …
3. x, x, $2x$, $3x$, $5x$, $8x$, …
Remarkably, for all Fibonacci-like sequences generated in this manner, the ratio of one term to the next gets closer and closer to approximately 0.618.

SECTION 8.2

Objectives
- To use expressions to describe phrases and situations
- To use equations to represent problems

Class Planning

Schedule 1–3 days

Resources

In *Chapter 8 Student Activity Masters* Book
Class Openers *13, 14*
Extra Practice *15, 16*
Cooperative Learning *17, 18*
Enrichment *19*
Learning Log *20*
Interdisciplinary Investigation *21, 22*
Spanish-Language Summary *23*

In *Tests and Quizzes* Book
Quiz on 8.1–8.2 (Basic) *205*

Transparency *39*

Class Opener

Write an expression for each phrase and give the value of the expression.
1. 4 more than 12
2. 6 less than 10
3. The product of 4 and 7
4. The quotient of 10 and 1 more than 4
5. 2 less than the product of 6 and 8

Answers:
1. $12 + 4$; 16   2. $10 - 6$; 4
3. 4×7; 28   4. $\frac{10}{1+4}$; 2
5. $6 \times 8 - 2$; 46

Lesson Notes

- One of the major goals of studying variables is to learn to use them to represent relationships in a situation. This may be difficult for many students, but it is central to success in algebra. Don't

Investigation Connection Section 8.2 supports Unit D, *Investigation* 4. (See pp. 316A–316D in this book.)

SECTION 8.2
Using Variables

EXPRESSIONS TO DESCRIBE PHRASES

Many math problems start as a situation described in words. The first step in solving these problems is to translate the words into symbols.

The phrase "2 more than 47" can be written as $47 + 2$.
The phrase "2 less than a number" can be written as $n - 2$.
The phrase "half of a number" can be written as $\frac{n}{2}$.

Example 1
Translate each phrase into an expression.
a The product of 4 and a number
b The quotient of 16 divided by a number

Solution
We can use any variable to represent the number.
a $4 \cdot x$ or $4x$
b $16 \div y$ or $\frac{16}{y}$

EXPRESSIONS TO DESCRIBE SITUATIONS

Many problems are based on physical situations, such as a drawing. In these problems, using symbols to represent the situation is the first step in solving the problem.

378 Chapter 8 The Language of Algebra

SECTION 8.2

Example 2
a Write an expression for the area of each rectangle.
b Write an expression for the perimeter of each rectangle.

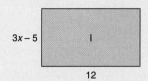 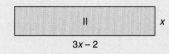

Solution
a The formula for the area of a rectangle is $A = \ell w$.

For rectangle I, $A = 12(3x - 5)$
For rectangle II, $A = x(3x - 2)$

b A formula for the perimeter is $P = 2\ell + 2w$.

For rectangle I, $P = 2(3x - 5) + 2(12) = 6x + 14$
For rectangle II, $P = 2(x) + 2(3x - 2) = 8x - 4$

EQUATIONS TO REPRESENT PROBLEMS

If we know that two or more expressions represent equal quantities, we use an equation to find out more about the problem.

Example 3
In the previous example, suppose the area of rectangle I is equal to the area of rectangle II. Write an equation to represent this equality.

Solution
From the previous example, we know the following:

$$\text{Area of rectangle I} = 12(3x - 5)$$
$$\text{Area of rectangle II} = x(3x - 2)$$

Since the areas are equal, $12(3x - 5) = x(3x - 2)$.

Example 4
Write an equation suggested by the figure.

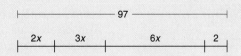

Solution
The total length is described in two ways.

$$\text{Total length} = 97$$
$$\text{Total length} = 2x + 3x + 6x + 2$$

An equation is $2x + 3x + 6x + 2 = 97$.

---

expect students to immediately become proficient at this skill; it will develop over time as they work on problems.

- Some students will not recognize the difference between an equation and an expression. In addition to bringing up the obvious "equations have equal signs," it is probably worthwhile to discuss how equations and expressions are related—in particular, how we often create an equation by setting two expressions equal to each other.

- Sample problem 2, with its three methods of solution, is worth discussing. The equation yielded by the first method is easy to generate but cannot be solved, while the equation yielded by the last method is rather tough to generate but can be solved.

Additional Examples

Example 1
Translate each phrase into an expression.
a. The product of 7 and a number
b. The quotient of a number and 5

Answers: a. $7x$ b. $\frac{x}{5}$

Example 2
Write an expression for the area and perimeter of each rectangle.
a. length 9, width $2x - 3$
b. length $5x - 2$, width x

Answers: a. $9(2x - 3)$; $4x + 12$
b. $x(5x - 2)$; $12x - 4$

Example 3
In the previous example, suppose the two rectangles have the same area. Write an equation to represent this.

Answer: $9(2x - 3) = x(5x - 2)$

SECTION 8.2

Additional Examples continued

Example 4
Write an equation suggested by the figure.

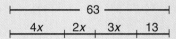

Answer: $4x + 2x + 3x + 13 = 63$

Problem-Set Notes

for page 381

- **2** and **16** involve easily misunderstood phrases such as "6 less than a number" and "6 is less than a number." Students have to learn to read for meaning, not rely on key words, since these words often have more than one meaning.

- **5** and **13** are problems for which, if students have trouble finding a correct equation, they should try substituting numbers and looking for a pattern.

Additional Answers

for page 381

2 Possible answers:
a Four more than a number
b A number subtracted from 9
c Nine less than a number
d Nine is less than a number
e Any number less than 9
f Five more than 3 times a number

3 The sentence is true when you substitute any value for x.

5 b $3h + (0.12)d$

7 $(2x + 1) + (8x - 3) + (15x + 2)$ or $25x$; $25x$ is simpler because it is shorter

---

Sample Problems

Problem 1 Lydia has drawn and labeled a line segment.

$$\begin{array}{|c|c|c|} \hline 3x & x-5 & 6x-8 \\ \hline \end{array}$$

Write an expression to represent the segment.

Solution The length of the segment is $3x + x - 5 + 6x - 8$. This expression can be simplified to $10x - 13$.

Problem 2 In Friday's basketball game, Al made six more baskets than Barry. Barry made twice as many baskets as Carl. Together, the three players made 56 baskets. Write an equation to represent this situation.

Solution **Method 1** Let A = the number of baskets Al made, B = the number of baskets Barry made, and C = the number of baskets Carl made. An equation is $A + B + C = 56$.

This equation has three variables. For this problem, it will be easier to use only one variable.

Method 2 Start with the equation $A + B + C = 56$. We know that Al made six more baskets than Barry ($A = B + 6$). Barry made twice as many baskets as Carl ($B = 2C$). Substitute equal quantities:

$$A = B + 6$$
$$B = 2C$$
Combine like terms

$$A + B + C = 56$$
$$B + 6 + B + C = 56$$
$$2C + 6 + 2C + C = 56$$
$$(2 + 2 + 1)C + 6 = 56$$
$$5C + 6 = 56$$

Method 3 We know how many baskets Carl made, and we can work backward to find the number of baskets Bob and Al made.

Let x = the number of baskets Carl made.
Barry made twice as many baskets as Carl.
Let $2x$ = the number of baskets Barry made.
Al made six more baskets than Barry.
Let $2x + 6$ = the number of baskets Al made.
The three made a total of 56 baskets.

$$x + 2x + (2x + 6) = 56$$
Combine like terms $\quad 5x + 6 = 56$

Chapter 8 The Language of Algebra

SECTION 8.2

Think and Discuss

P, R **1** Write an expression to represent each phrase. Simplify the expression.
   a The sum of 8 and its opposite $8 + (-8); 0$
   b The product of $\frac{2}{3}$ and its reciprocal $\frac{2}{3} \cdot \frac{3}{2}; 1$
   c The sum of x and its opposite $x + (-x); 0$
   d The product of y and its reciprocal (assume $y \neq 0$) $(y)(\frac{1}{y}); 1$

P **2** Write a phrase that represents each expression.
   a $x + 4$ **b** $9 - x$ **c** $x - 9$
   d $9 < x$ **e** $x < 9$ **f** $3x + 5$

P **3** Demonstrate that $5x - 12 = 5x - 12$ is true for all values of x.

P, R **4** The area of the rectangle is equal to the area of the triangle.

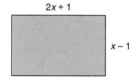
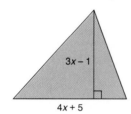

   a Write an expression to represent the area of the rectangle. $(2x + 1)(x - 1)$
   b Write an expression to represent the area of the triangle. $\frac{1}{2}(4x + 5)(3x - 1)$
   c Write an equation to express the equality. $(2x + 1)(x - 1) = \frac{1}{2}(4x + 5)(3x - 1)$

P, R **5** **Business** Sally sells seashells at Seashore Susie's. Sally's salary is $3 per hour plus a 12% commission on her sales.
   a If Sally works 8 hours and sells shells worth $60, what is her pay? $31.20
   b If Sally works h hours and sells seashells worth d, what is her pay?

P, R **6** A box has length ℓ, width w, and height h.
   a Write an expression to represent the volume of the box. $V = \ell w h$
   b Write an expression to represent the surface area of the box. $SA = 2(\ell w + \ell h + w h)$

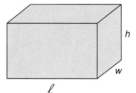

P, R **7** Write several expressions to represent the total length of the segment. Explain which form is simplest.

   | 2x + 1 | 8x – 3 | 15x + 2 |

P **8** Match each phrase with its algebraic representation.
   a $2(N - 3)$ iii
   b $3 - N$ vi
   c $3 > 2N$ ii
   d $2N + 3$ i
   e $N - 3$ iv
   f $3N$ v

   i 3 more than twice a number
   ii 3 is greater than twice a number
   iii Twice the difference of 3 and a number
   iv 3 less than a number
   v The product of 3 and a number
   vi 3 decreased by a number

Section 8.2 Using Variables

Assignment Guide

Basic
Day 1 9, 10, 14, 16, 19, 21, 25
Day 2 11, 15, 17, 24, 26–28, 31
Day 3 12, 13, 18, 20, 29, 30, 32

Average or Heterogeneous
Day 1 9–12, 15–17, 20, 21, 26, 30, 32
Day 2 13, 14, 18, 19, 22, 23, 25, 27–29, 31

Honors
Day 1 10–12, 14–18, 21–23, 26, 27, 30, 32

Strands

Communicating Mathematics	7, 16, 31
Geometry	4, 6, 7, 10, 12, 14, 17, 19, 20, 22, 23, 27, 30, Investigation
Measurement	4–7, 10, 12, 14, 18–23, 27, 30
Algebra	1–32
Patterns & Functions	21
Number Sense	24, 29, 32
Interdisciplinary Applications	5, 11, 13, 15, 18, 21, Investigation

Individualizing Instruction

Learning-disabled students Students may get hung up on the fact that some variables stand for a word or set of words (e.g., h for height) while others do not. Have students compare the use of variables in problems 4 and 6. Could the variable h replace x in problem 4? (Yes, but it wouldn't stand for height.) Could the variables x, y, and z replace the variables h, ℓ, and w in problem 6? (Yes, but it's more convenient to use h, ℓ, and w because they represent *height*, *length*, and *width*.)

SECTION 8.2

Problem-Set Notes

for pages 382 and 383

- **15** is comparable to sample problem 2.
- **18** and **21** introduce rates of change.

Checkpoint

In problems **1–6**, match each phrase with the algebraic expression it represents.

Phrases:
i. Two more than 3 times a number
ii. The product of 3 and a number
iii. Three times the sum of a number and 2
iv. Three is greater than twice a number.
v. Two decreased by a number
vi. Two less than a number

Algebraic expressions:
1. $3(N+2)$ 2. $2-N$
3. $3 > 2N$ 4. $3N+2$
5. $N-2$ 6. $3N$

Answers: **1.** iii **2.** v **3.** iv **4.** i **5.** vi **6.** ii

Stumbling Block

Students may write an expression such as "4 less than a number" as $4-n$. To overcome this tendency, have students pick a number for the variable, such as 10. They can then reason: "I know that 4 less than 10 is 6. I can write that as $10-4=6$. So 4 less than any number must be the number minus 4 or, using a variable, $n-4$."

Problems and Applications

P   **9** Translate each phrase into an algebraic expression and each algebraic expression into a phrase.
  a 6 less than a number $n-6$
  b $\frac{x}{2}$ Half of a number
  c Twice the sum of a number and 3
  d $5-n$ A number subtracted from 5

P, R   **10** Segment AB is equal in length to segment CD. Write an equation to represent this equality. $x + 2x + 1 + 3x - 2 = 3x - 5 + 18 + x + 3$; or $6x - 1 = 4x + 16$

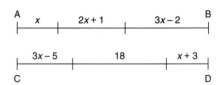

P, R   **11 Consumer Math**   Rich bought an item that cost P and had a 6% sales tax added to the price. Write a formula for the total cost C of the item. $C = P + 0.06P$ or $C = 1.06P$

P, R, I   **12** In the figure, $\ell \parallel m$. Write an equation suggested by the figure. $2x + 8y = 3(x + 2y)$

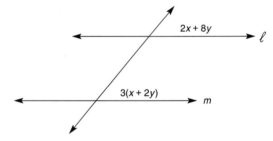

P, R   **13** Andy is paid $4 per hour plus 9% commission on his sales.
  a If he works 6 hours and sells $850 worth of goods, write an expression for his earnings. $\$4(6) + 0.09(\$850)$, or $\$100.50$
  b If he works H hours and sells D dollars of goods, write an expression for his income. $4(H) + 0.09(D)$

P, R, I   **14** Write at least two equations that can represent the parallelogram.

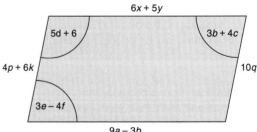

P, I   **15** Yogi, Mickey, and Casey have a total of 63 hits. In the baseball season to date, Mickey has 14 more hits than Yogi. Yogi has 3 times as many hits as Casey. Use one or more variables and write one or more equations.

P, R   **16 Communicating**   Explain the difference between the phrases "13 more than 10" and "13 is more than 10."

382   Chapter 8   The Language of Algebra

SECTION 8.2

P, R **17** Write an equation for each right triangle.

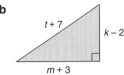

P, R **18** There are 2 inches of snow on the ground. Then it starts snowing at a rate of $\frac{1}{2}$ inch per hour. Write an expression for the depth of the snow after it has been snowing for
a 1 hour   **b** 2 hours   **c** 3 hours   **d** h hours

P, R **19** Write an equation to represent that AB = CD. $4x + 17 = 2x + 21$

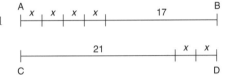

P, R, I **20** In △ABC, the length of $\overline{AC}$ is twice the length of $\overline{AB}$ and the length of $\overline{BC}$ is 9 more than the length of $\overline{AB}$.

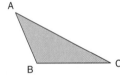

a Write an expression for the perimeter of △ABC using only one variable. $2x + (x + 9) + x$, or $4x + 9$
b If the perimeter of the triangle is 91, write an equation that represents the perimeter of the triangle in terms of the expression you wrote for part **a**. $91 = 4x + 9$

P, R **21** A plane is at 33,000 feet. It descends 400 feet per minute. Find an expression for the plane's altitude after the plane has descended for
a 1 minute $33,000 - 400(1)$ **b** 2 minutes $33,000 - 400(2)$
c 3 minutes $33,000 - 400(3)$ **d** 4 minutes $33,000 - 400(4)$
e m minutes $33,000 - 400(m)$

P, R **22** Write an expression for
a The area of the base of the box $4(x + 7)$
b The volume of the box $6(4)(x + 7)$
c The perimeter of the base of the box $2(4 + (x + 7))$

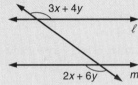

P, R, I **23** The length of one diagonal of a quadrilateral is $2x + 4y$ and the length of the other diagonal is $6x + 3$.
a If the figure is a rectangle, can an equation describe the diagonals? If so, what is the equation?
b If the figure is a trapezoid, can an equation describe the diagonals? If so, what is the equation?   No, not unless the trapezoid is isosceles.

P, R **24** If A = apples, B = basket and P = pear, the B had a total of $10A$ and at least one P. Carlos took $2A$ and $1P$. There are 12 pieces of fruit left. How many pears were in the basket originally?   5

Section 8.2   Using Variables   **383**

Additional Practice

In problems **1–6**, write a phrase for each expression.
1. $x + 3$ **2.** $2x - 3$ **3.** $x \leq 3$
4. $x - 3$ **5.** $3 - x$ **6.** $4x + 1$

7. Write and simplify an expression that represents the length of the segment.

8. In the figure, $\ell \parallel m$. Write an equation suggested by the figure.

Answers:
1. 3 more than a number
2. 3 less than twice a number
3. x is less than or equal to 3
4. 3 less than a number
5. 3 less a number   **6.** 1 more than 4 times a number
7. $(2x - 1) + (3x + 1) + (x + 6)$; $6x + 6$ **8.** $3x + 4y = 2x + 6y$

Additional Answers

for pages 382 and 383

9 c $2(n + 3)$

14 $(6x + 5y) = (9a - 3b)$, $10q = (4p + 6k)$, $(3e - 4f) = (3b + 4c)$, $(5d + 6) = 180 - (3b + 4c)$, $(5d + 6) = 180 - (3e - 4f)$

15 Possible answer: $M = 14 + Y$, $Y = 3C$, $63 = C + (3C) + (14 + 3C)$

16 Answers will vary. "13 more than 10" represents a single value, 23; "13 is more than 10" represents an inequality, $13 > 10$.

17 a $a^2 = b^2 + c^2$
b $(t + 7)^2 = (k - 2)^2 + (m + 3)^2$

18 a $2 + 0.5(1)$ or 2.5
b $2 + 0.5(2)$ or 3
c $2 + 0.5(3)$ or 3.5
d $2 + 0.5(h)$

23 a Yes; $2x + 4y = 6x + 3$

383

SECTION 8.2

Problem-Set Notes

for page 384

- 32 will help students understand the full power of the substitution principle.

Cooperative Learning

- Give groups of students the following expressions and have then come up with verbal equivalents.
  a. $2n + 5$ b. $2(n + 5)$
  c. $\frac{n+5}{2}$ d. $\frac{n}{2} + 5$
  e. $2n + 5n$

Answers will vary somewhat. Have groups share their answers and discuss whether their phrases have the same meaning as the expressions.

- Problems 15 and 24 are appropriate for cooperative learning.

Communicating Mathematics

- Have students write a paragraph explaining the following analogy: "A phrase is to an expression as a sentence is to an equation or inequality."
- Problems 7, 16, and 31 encourage communicating mathematics.

Additional Answers

for page 384

31 a Substitute 4 for x, multiply 3 times 4, add 7 to 12.

b Subtract 5 from 35 to get 30. Divide 30 by 2 to get 15.

Investigation Both are spherical. A sphere contains the greatest volume for a given amount of surface area. Other examples include the planets and stars, which are nearly spherical.

384

◀ LOOKING BACK **Spiral Learning** LOOKING AHEAD ▶

R   **25** Simplify each expression.
   a $x - (3x + 5)$ $-2x - 5$
   b $3(x + y) + 12(x + y) - 4(x + y)$ $11x + 11y$

R, I   **26 a** If $7(a + b) = 84$, what is $a + b$?   12   **b** If $p + q = 5$, what is $7(p + q) - 8$?   27

R, I   **27** Find the perimeter of the quadrilateral if $(x, y) = (3, -2)$.   55

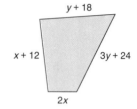

R   **28** Evaluate each expression.
   a $-3(5 - 12)$ 21   **b** $-3(5) - 12$ -27 **c** $-35 - 12$ -47

R   **29** Fill in each blank with one of the symbols <, >, or =.
   a -3 ___ 15 <
   b $-3 + 2$ ___ $15 + 2$ <
   c $(-3)(2)$ ___ $(15)(2)$ <
   d $-3 - 2$ ___ $15 - 2$ <
   e $(-3)(-2)$ ___ $(15)(-2)$ >
   f $\frac{-3}{-2}$ ___ $\frac{15}{-2}$ >

R, I   **30** Draw a pentagon with the segments of the sides extended as shown. Measure the five marked angles. Find the sum of the measures of the five angles.   360

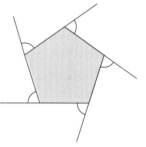

R, I   **31 Communicating** Explain in detail the steps that must be taken to
   a Evaluate $3x + 7$ for $x = 4$
   b Solve the equation $2x + 5 = 35$ without using a calculator or pencil and paper   15

R, I   **32** If $2x + 2y = 12$, what is the value of $6x + 6y$?   36

Investigation

Science What shape is a single soap bubble as it floats in the air? What shape is a raindrop as its falls through the air? Find out why these shapes are formed naturally. Give some other examples.

SECTION 8.3

Investigation Connection
Section 8.3 supports Unit D, *Investigation*, 4. (See pp. 316A-316D in this book.)

SECTION 8.3
Solving Equations

INVERSE OPERATIONS

In Section 3.4, we learned that we could use inverse operations to solve equations.

Example 1
Solve the equation $x + 57 = 126$ for x.

Solution
Since 57 added to x equals 126, we can find x by subtracting 57 from 126.

$$126 - 57 = 69$$

To see if 69 is the correct answer, check:

$$\text{Does } 69 + 57 = 126? \text{ Yes.}$$

So the answer is $x = 69$.

Example 2
Solve $y - 34.9 = 66.6$ for y.

Solution
Since 34.9 is subtracted from y, add 34.9 to each side of the equation. On the next page, you will see two different forms that can be used to solve the equation.

Objectives
- To use inverse operations to solve equations
- To solve equations with several operations

Class Planning

Schedule 1–2 days

Resources

In *Chapter 8 Student Activity Masters* Book
Class Openers *25, 26*
Extra Practice *27, 28*
Cooperative Learning *29, 30*
Enrichment *31*
Learning Log *32*
Using Manipulatives *33, 34*
Spanish-Language Summary *35*

In *Tests and Quizzes* Book
Quiz on 8.1–8.3 (Average) *206*
Quiz on 8.1–8.3 (Honors) *207*

Class Opener

For each sentence, write an equation. Then give a solution to the equation.
1. When 4 is added to a number, the result is 11.
2. When 12 is subtracted from a number, the result is 25.
3. When a number is multiplied by 9, the result is –36.
4. When a number is divided by –3, the result is –6.
5. When a number is multiplied by 4, with the product decreased by 1, the result is 27.

Answers:
1. $x + 4 = 11$, $x = 7$
2. $x - 12 = 25$, $x = 37$
3. $9x = -36$, $x = -4$
4. $\frac{x}{-3} = -6$, $x = 18$
5. $4x - 1 = 27$, $x = 7$

Section 8.3 Solving Equations **385**

SECTION 8.3

Lesson Notes

- Some students are more comfortable with the vertical form of solving equations than with the horizontal form, so it might be best to use both as you work the problems. Some students will not want to write down all the steps. If these students can still solve the equations, you might be flexible about how much they write. As the problems get tougher, though, the value of writing each step will become evident.

- The check is an important part of every problem, and it will only be seen as critical if students see you check every equation. It not only helps students locate errors but also reminds them what it means to solve an equation. In algebra, students will encounter solutions that are extraneous, so the check becomes essential. For now, it is just a sound habit.

- In some situations, the second method in example 4 can save a considerable amount of work.

Additional Examples

Example 1
Solve the equation $x + 38 = 95$ for x.
Answer: $x = 57$

Example 2
Solve $y - 28 = 65$ for y.
Answer: $y = 93$

Example 3
Write an equation to represent the diagram. Find the value of x.

|— 14 —|——— 9x ———|
|————— 86 —————|

Answer: $14 + 9x = 86$; $x = 8$

Example 4
Solve the equation $5y - 11 = 49$.
Answer: $y = 12$

Horizontal Form

$y - 34.9 = 66.6$
$y - 34.9 + 34.9 = 66.6 + 34.9$
$y = 101.5$

Check the answer. $y - 34.9 = 101.5 - 34.9$
                              $= 66.6$

The correct answer is $y = 101.5$.

Vertical Form

$$\begin{aligned} y - 34.9 &= 66.6 \\ + 34.9 & \quad +34.9 \\ \hline y + 0 &= 101.5 \\ y &= 101.5 \end{aligned}$$

In general, using inverse operations can help you solve many equations. In Example 1, 57 is added to x, so you subtracted 57 from both sides of the equation. In Example 2, 34.9 is subtracted from y, so you added 34.9 to both sides of the equation.

EQUATIONS WITH SEVERAL OPERATIONS

In many equations, more than one operation is used with the variable. When this happens, we need to remember the order of operations we learned in Section 2.1.

Example 3
Write an equation to represent the diagram, and find the value of x.

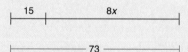

Solution
We can write the equation $15 + 8x = 73$ and solve for x.

Remember the order of operations. First, the variable x is multiplied by 8. Then, 15 is added to $8x$. To solve the equation, we will do the inverse operations in reverse order. First subtract 15 from each side of the equation, then divide each side of the equation by 8.

First, subtract 15

$$\begin{aligned} 15 + 8x &= 73 \\ -15 & \quad -15 \\ \hline 8x &= 58 \end{aligned}$$

Then divide by 8 $\dfrac{8x}{8} = \dfrac{58}{8}$

$x = 7.25$

Check the answer $15 + 8x = 15 + 8(7.25)$
                           $= 15 + 58$
                           $= 73$

The answer is $x = 7.25$.

386   Chapter 8   The Language of Algebra

SECTION 8.3

Example 4
Solve the equation $6y - 12 = 42$.

Solution
Method 1

$$6y - 12 = 42$$

Add 12 to each side
$$ +12 \quad +12$$
$$6y = 54$$

Divide each side by 6
$$\frac{6y}{6} = \frac{54}{6}$$
$$y = 9$$

Check the answer
$$6 \cdot 9 - 12 = 54 - 12 = 42$$

Method 2
$$6y - 12 = 42$$

First let's divide each side by 6
$$\frac{6y - 12}{6} = \frac{42}{6}$$
$$\frac{6y}{6} - \frac{12}{6} = \frac{42}{6}$$
$$y - 2 = 7$$

Add 2 to each side
$$ +2 \quad +2$$
$$y = 9$$

We arrive at the same solution using either method.

Sample Problem

Problem Assume the measure of each of the 15 angles is k. Find k.

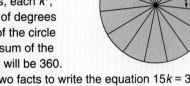

Solution There are 15 congruent angles, each $k°$, in the circle. The total number of degrees in all the angles in the center of the circle is $15k$. We also know that the sum of the measures of the fifteen angles will be 360. Therefore, we can use these two facts to write the equation $15k = 360$. Since k is multiplied by 15, we use the inverse operation, dividing by 15, to solve the equation.

Divide by 15
$$\frac{15k}{15} = \frac{360}{15}$$

$$\frac{15}{15} = 1$$
$$1k = k$$

$$1k = 24$$
$$k = 24$$

Check the answer
$$15k = 15(24)$$
$$= 360$$

The measure of each central angle k is 24.

Assignment Guide

Basic
Day 1 7, 10–12, 17, 22, 25
Day 2 8, 9, 18, 21, 23, 28, 30

Average or Heterogeneous
Day 1 7, 9, 11 (a, c, e, g, i, k), 12, 14, 15, 18, 20, 24, 27
Day 2 8, 10, 11 (b, d, f, h, j, l), 13, 16, 17, 19, 21, 23, 26, 29

Honors
Day 1 8–10, 12–14, 16–18, 20, 21, 26, 27, 29

Strands

Communicating Mathematics	3, 4, 6, 10, 12, 14, 18, 24, Investigation
Geometry	4, 10, 13, 15, 17, 19, 21, 24, 26
Measurement	4, 13–15, 17, 19, 21, 26, 29
Technology	11, 18, 20
Probability	21
Algebra	1–23, 25–30, Investigation
Number Sense	22, 28
Interdisciplinary Applications	1, 12, 14, 16, 18, 20, Investigation

Individualizing Instruction

The gifted learner Students could brainstorm several different ways to demonstrate, without verbal explanation, the meaning of an equal sign and present their ideas to the class.

SECTION 8.3

Problem-Set Notes

for pages 388 and 389

- **2 and 11** help to illustrate the difference between an equation and an expression. Here is another situation where the check reinforces understanding.

- **4** is an example of a problem in which the algebraic solution makes no sense in the context of the problem. You might ask students before they attempt an algebraic solution whether they can tell if there is a value of x that makes sense in the problem. You might also ask if students can come up with a situation in which the equation $2x + 25 = 19$ does make sense. (Possible answer: a decrease in temperature of 3 degrees over each of two hours.)

- **10** differs from problem 4 in that even though the value of x is negative, this value is still a solution because the lengths are positive.

- **14** can preview slope if you show this problem visually.

Checkpoint

Solve each equation for x.
1. $x + 8.6 = 19.7$
2. $x - 3 = 27$
3. $5 = 3x + 4 + 2x + 6$
4. $2x - 1 = 19$
5. $3x + 4 = 22$
6. $3x - 5 + x = 11$

Answers:
1. $x = 11.1$ 2. $x = 30$
3. $x = -1$ 4. $x = 10$
5. $x = 6$ 6. $x = 4$

388

Think and Discuss

P   **1** To get dressed, we put on socks, then shoes. Later, we remove the shoes first, then the socks. This is an example of using inverse operations in reverse order. Give two more examples.

P, R   **2** Identify each statement as an equation or an expression. If it is an equation, solve it. If it is an expression, simplify it.
  a $3x - 82 + 53$ $3x - 29$ **b** $3x - 82 = 53$ 45
  c $12 - 5x = -18$ 6 **d** $12 - 5x - 18$ $-6 - 5x$
  e $12 = 5x - 18$ 6 **f** $6 - (x - 2)$ $8 - x$

P   **3** Each equation in the first column can be solved by using one of the steps in the second column. Match each equation with its step. Explain why you made each match and how you would apply the step.
  a $x + 17 = 34$ ii   **i** add 17
  b $x - 17 = 34$ i   **ii** subtract 17
  c $17x = 34$ iv   **iii** multiply by 17
  d $\frac{x}{17} = 34$ iii   **iv** divide by 17   Each is the inverse operation.

P, R   **4 a** Write an equation that represents that the length of $\overline{AB}$ is equal to the length of $\overline{CD}$. $2x + 25 = 19$
  b Solve the equation in part **a**. $x = -3$
  c Does your answer make sense? Explain why.

```
A    x    x        25          B
|----|----|--------------------|

C           19              D
|---------------------------|
```

R   **5** Write each statement as an equation or as an expression.
  a Seven is three more than a number $7 = n + 3$
  b Seven more than a number $n + 7$ or $7 + n$
  c Seven more than three times a number $3n + 7$
  d Seven more than three times a number is twelve $3n + 7 = 12$
  e Seven more than three is a number $3 + 7 = n$

P, R   **6** When Moe and Curly solved the equation $3x + 15 = 57$, they each found that $x = 14$. Explain the different forms they used.

Moe's Solution	Curly's Solution
$3x + 15 = 57$	$3x + 15 = 57$
$3x + 15 = 57$	$\dfrac{3x + 15}{3} = \dfrac{57}{3}$
$\underline{-15\ \ -15}$	
$3x = 42$	$x + 5 = 19$
	$\underline{-5\ \ -5}$
$\dfrac{3x}{3} = \dfrac{42}{3}$	$x = 14$
$x = 14$	

388   **Chapter 8**   The Language of Algebra

SECTION 8.3

Problems and Applications

P, R 7. Solve each equation for x.
   a. $12x = 66$ $x = 5.5$
   b. $3x + 5x = 7(6)$ $x = 5.25$
   c. $3x + 2y = -6$, if $y = 9$ $x = -8$

P, R 8. Solve each equation for y.
   a. $-19 - 6y = -43$ $y = 4$
   b. $7y - 3 = 53$ $y = 8$
   c. $19 + 8y = 47$ $y = 3.5$
   d. $3y + 42 = 21$ $y = -7$

P, R 9. Solve each equation or simplify each expression.
   a. $2 + 5x - 3 - 2x + 8$ $3x + 7$
   b. $2 = 5x - 3 - 2x + 8$ $x = -1$
   c. $2 + 5x - 3 - 2x = 8$ $x = 3$
   d. $2 - 5x - 3 - 2(x + 8)$ $-17 - 7x$

P, R 10. **Communicating** Segment AB is equal in length to segment CD.
   a. Write an equation that describes the equality.
   b. Solve the equation for x.
   c. Does the answer make sense? Explain why you think it does or does not.

   A ⊢ $x + 18$ ⊢ $13 - x$ ⊢ $7 - x$ ⊢ B

   C ⊢ 42 ⊢ D

P, R 11. Solve each equation or simplify each expression.
   a. $19x = 85.5$ $x = 4.5$
   b. $x + 349 = 156$ $x = -193$
   c. $a + b + 3a - 4b$ $4a - 3b$
   d. $y - 43.25 = 103.26$ $y = 146.51$
   e. $92 = 4t - 16$ $t = 27$
   f. $24k = 463.2$ $k = 19.3$
   g. $45 = 19 - 2k$ $k = -13$
   h. $m - 392 + m$ $2m - 392$
   i. $5x - 10x^2$ $5x - 10x^2$
   j. $d + d + d = 963$ $d = 321$
   k. $6 - 5t = -129$ $t = 27$
   l. $10 - (3x - 5)$ $15 - 3x$

P 12. Karen taped a key to an index card and placed it in an envelope. Then she put the envelope in a box. Next she wrapped the box in paper and mailed it to Kitty. Describe the correct order of the process Kitty must use to get the key.

R 13. This figure is a regular octagon. All of its angles have the same measure. The sum of the measure of the angles is 1080. Find the measure of each angle. 135

P, R 14. **Aviation** An airplane is at 14,000 feet when the pilot is told by air traffic control to ascend to 24,000 feet. The pilot puts the plane into a climb rate of 425 feet/minute.
   a. Write an expression for the height of the plane after it has ascended for t minutes. $14{,}000 + 425t$
   b. Write an equation for the time it takes the plane to reach its assigned height. $24{,}000 = 14{,}000 + 425t$
   c. Solve the equation in part **b**. What does your answer represent?

Additional Practice

Solve each equation and simplify each expression.
1. $18x = 23.4$
2. $m + m = 842$
3. $7k - 3 + 10k - 4$
4. $m + 92 = 106$
5. $2d + 3 = 19$
6. $7x + 4y - 8x + 9y$
7. $r + 3 + 2r = 12$
8. $12 - (2v + 1)$

Answers:
1. $x = 1.3$ 2. $m = 421$
3. $17k - 7$ 4. $m = 14$
5. $d = 8$ 6. $-x + 13y$
7. $r = 3$ 8. $11 - 2v$

Additional Answers

for pages 388 and 389

1 Possible answers: Raising the shade and opening the window, then closing the window and lowering the shade; putting a gift in a box and wrapping the box, then unwrapping the box and taking out the gift.

4 c No. A segment cannot have negative length.

6 Curly first divided both sides by 3, but Moe subtracted first. Both methods are correct.

10 a $(x + 18) + (13 - x) + (7 - x) = 42$ or $-x = 4$

b $x = -4$

c Answers will vary. The answer doesn't seem to make sense, but it is mathematically correct—the length of the segment is 42.

12 Kitty unwrapped the box, opened the box, took out the envelope, opened the envelope, removed the index card from the envelope, and took the key off the card.

14 c $t = 23.5$; 23.5 is the approximate number of minutes it took the plane to climb 10,000 ft.

Section 8.3 Solving Equations

SECTION 8.3

Problem-Set Notes

for pages 390 and 391

- **15** can also be solved with the equation $9(4 + x) - 2x = 57$ or the equation $7(4 + x) + 8 = 57$.

- **Investigation** might generate some interesting discussion. It can be shown algebraically that $(1.05x)(0.80) = (0.80x)(1.05)$, so the price to the consumer is the same in either case. However, which method might the store prefer? The tax collector? Why?

Cooperative Learning

- Have pairs of students come up with short sets of instructions (3 or 4 steps) that can be acted out. They might perform an action, for example, or hide an object. Have each group perform its action or instructions, then have a second group attempt to do it in reverse order. Discuss what changes need to be made to the directions in order to undo what has been done, and relate these changes to using inverses to solve equations.

- Problem 20 and Investigation are appropriate for cooperative learning.

Communicating Mathematics

- Have students write a poem, rap, or song parody that encourages other students to check their solutions to an equation by substituting the value back into the original equation. Tell students to be sure to give reasons why checking is important. Students may perform their work for the class.

P, R   **15** The area of the figure is 57 square units.
   a Write an equation that describes the area of the figure. $57 = 36 + 7x$
   b Solve the equation for x. $x = 3$

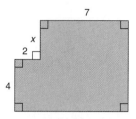

P, R   **16** **Business** Juan works at the Higgly-Piggly. He is paid $5.75 per hour Monday through Saturday. On Sunday he is paid time and a half (one and one half times his hourly pay). Let w = the number of Juan's Monday-through-Saturday hours and s = the number of Juan's Sunday hours.
   a Write an expression for his total pay for the week. $5.75w + 8.625s$
   b If Juan worked 35 hours Monday through Saturday and had a total weekly pay of $280, how many hours did he work on Sunday? ≈ 9 hr

P, R   **17** In $\triangle ABC$, $\angle A = (12x + 5)°$, $\angle B = (9x + 15)°$, and $\angle C = (2x - 1)°$.
   a Write an equation to describe the sum of the measures of the angles. (Hint: The sum of the angles in a triangle is 180°.) $23x + 19 = 180$
   b Solve the equation for x. $x = 7$
   c Find the measures of angles A, B, and C. $m\angle A = 89$, $m\angle B = 78$, $m\angle C = 13$

P, R   **18** **Communicating** Bertha buys a box of beeswax and pays 6% sales tax. The total cost is $25.97.
   a Explain what each term of the equation $x + 0.06x = 25.97$ represents.
   b Solve the equation for x. What does x represent?

P, R   **19** ABCD is a parallelogram.
   a Write an equation that can be used to determine a. $a + 4 = 15$
   b Solve the equation for a. $a = 11$
   c The perimeter of ABCD is 34. How long is $\overline{AD}$? How long is $\overline{BC}$? $AD = 5$, $BC = 5$
   d What is the value of b? $b = 6$

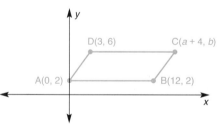

P, R   **20** **Health** To determine your target heart rate for one minute of aerobic exercise, subtract your age from 220 and multiply the result by 75%.
   a Compute your target heart rate.   Answers will vary.
   b If a represents age, write an expression for target heart rate.
   c Use a spreadsheet to calculate target heart rates for ages 10 through 60.

P, R   **21** The probability a dart that lands on the figure will be in the shaded square is .4.
   a Write an equation, in terms of x, that describes this situation. $\frac{x^2}{36} = .4$
   b Find the value of x. $x \approx 3.8$

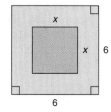

390   Chapter 8   The Language of Algebra

SECTION 8.3

◀ LOOKING BACK **Spiral Learning** LOOKING AHEAD ▶

R **22** Evaluate each expression.
   a $\sqrt{36}$ 6   **b** $\sqrt{25 + 144}$ 13   **c** $\sqrt{5}$ ≈ 2.236   **d** $\sqrt{25} + \sqrt{144}$ 17

R **23** Evaluate each expression.
   a $|8| - |5|$ 3   **b** $|5 - 8|$ 3   **c** $|8 - 5|$ 3   **d** $|5| - |8|$ -3

R, I **24 Communicating** Sticks can be used to make a "raft." Discuss the symmetries of the "raft" shown.

R **25** Evaluate each expression.
   a $-2(6 - 3 - 5)$ 4   **b** $\frac{-3}{8}(-8 + 16 - 32 + 64 - 128)$ 33

R **26** An angle has measure x. Write an expression for the measure of
   a The complement of the angle $90 - x$ **b** The supplement of the angle $180 - x$

R, I **27** Graph each inequality.
   a $-2 < x$ **b** $x \geq 7$ **c** $-5 < x \leq 2$

P, I **28** If $8x - 8y = 12$, find the value of $6x - 6y$. $6x - 6y = 9$

P, R, I **29** Copy the diagram and shade each region described.
   a $4 \leq x \leq 7$ using ▨
   b $2 \leq y \leq 4$ using ▧
   c What is the area of the region that is shaded both ways ▩? 6

R **30** Perform the indicated operation.
   a $8.7 - 9.16$ -0.46 **b** $\frac{3}{5} + \frac{1}{2}$ $\frac{11}{10}$ or $1\frac{1}{10}$ **c** $\frac{3}{5} - \frac{1}{2}$ $\frac{1}{10}$ **d** $\frac{1}{3} - \frac{1}{2}$ $-\frac{1}{6}$

Investigation

Consumer Mathematics A store advertises a sale with a sign that reads, "20% off!" However, the store must add 5% tax to the price of your purchase. Will you pay more if the clerk calculates the tax first or the discount first? Use expressions from algebra to support your argument.

Section 8.3 Solving Equations

SECTION 8.4

Objectives

- To solve absolute-value equations
- To solve equations with square roots
- To simplify expressions to solve equations

Class Planning

Schedule 2 days

Resources

In *Chapter 8 Student Activity Masters* Book
Class Openers 37, 38
Extra Practice 39, 40
Cooperative Learning 41, 42
Enrichment 43
Learning Log 44
Technology 45, 46
Spanish-Language Summary 47

Class Opener

For each sentence, find the number or numbers that make it true.
1. It is 6 units away from 2 on a number line.
2. Its square is 25.
3. Its square root, minus 2, is 3.
4. Its square added to its absolute value is 56.

Answers:
1. $-4, 8$ 2. $5, -5$
3. 25 4. $7, -7$

Lesson Notes

- This section gives us an opportunity to work with equations that have two answers or no answers. It also provides problems for which guessing is a little tougher and checking is essential.

- Students have been simplifying expressions for a few days now,

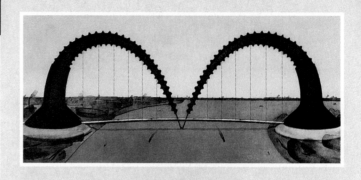

SECTION 8.4
Some Special Equations

ABSOLUTE-VALUE EQUATIONS

To solve an equation that contains an absolute value, we apply the definition of absolute value.

Example 1
Solve the equation $|x| + 8 = 36$ for x.

Solution
Subtract 8 from each side or add -8 to each side.
$$\begin{array}{r} |x| + 8 = 36 \\ -8 -8 \\ \hline |x| = 28 \end{array}$$

In solving an equation, we try to find all values that work. There are two numbers, -28 and 28, that have an absolute value of 28. The equation has two solutions, $x = 28$ or $x = -28$.

Example 2
Solve the equation $|q| = -35$ for q.

Solution
Recall that the absolute value of a number cannot be negative. This equation has no solution.

EQUATIONS WITH SQUARE ROOTS

Squaring and finding the square root are inverse operations. You can use this idea in Examples 3 and 4.

392   **Chapter 8** The Language of Algebra

SECTION 8.4

Example 3
Solve the equation $\sqrt{r} = 4$ for r.

Solution
The inverse operation of taking the square root is squaring (raising to the second power).

Square both sides of the equation
$$\sqrt{r} = 4$$
$$(\sqrt{r})^2 = (4)^2$$
$$r = 16$$

Check to see that the answer is correct.

Example 4
Solve the equation $y^2 = 25$ for y.

Solution
In the equation, y is squared, and the inverse of squaring is taking the square root. So you can find one answer by taking the square root of 25.

If $y^2 = 25$, then $y = 5$.

But is 5 the only answer? What if y is negative? Since $(-5)^2$ also equals 25, the equation has two answers.

If $y^2 = 25$, then $y = 5$ or $y = -5$.

Always look for two answers in an equation with an x^2 term..

SIMPLIFYING TO SOLVE EQUATIONS

Sometimes one or both sides of an equation have expressions that must be simplified before we can solve the equation.

Example 5
Solve the equation $3(-5n + 8) = 237$ for n.

Solution
To solve the equation in Example 5, start by distributing multiplication over addition on the left side of the equation.

---

so they should find it natural to apply the skill to solving equations.

Additional Examples

Example 1
Solve the equation $|x| + 7 = 24$ for x.

Answer: $x = 17$ or $x = -17$

Example 2
Solve the equation $|z| = -22$ for z.

Answer: no solution

Example 3
Solve the equation $\sqrt{p} = 9$ for p.

Answer: $p = 81$

Example 4
Solve the equation $n^2 = 36$ for n.

Answer: $n = 6$ or $n = -6$

Example 5
Solve the equation $4(-3m + 5) = 80$ for m.

Answer: $m = -5$

Reteaching

For problems such as example 5, reteach using the following method.

$$\frac{3(-5n + 8)}{3} = \frac{237}{3}$$
$$-5n + 8 = 79$$
$$\underline{-8 \quad -8}$$
$$\frac{-5n}{-5} = \frac{71}{-5}$$
$$n = -\tfrac{71}{5}, \text{ or } -14.2$$

Section 8.4 Some Special Equations

SECTION 8.4

Assignment Guide

Basic
Optional; skip to Section 8.6

Average or Heterogeneous
Day 1 7–9, 11, 12 (a, c, e, g), 13, 16 (a, c), 18, 20, 27, 29
Day 2 10, 12 (b, d, f, h), 14, 15, 16 (b, d), 17, 21, 25, 28, 30, 34

Honors
Day 1 7, 9, 10, 12, 13, 17, 18, 22, 29
Day 2 8, 11, 14–16, 19–21, 23, 25, 26, 28, 30, 33, Investigation

Strands

Communicating Mathematics	1, 3, 4, 28, Investigation
Geometry	11, 13, 15, 17, 18, 21, 29, 33
Measurement	11, 13, 15, 17, 18, 21, 29, Investigation
Data Analysis/Statistics	19, 25
Technology	15, 22, 25
Algebra	1–34
Patterns & Functions	25
Number Sense	5, 6, 9, 12, 19, 20, 24, 25, 27, 34
Interdisciplinary Applications	11, 15, Investigation

Individualizing Instruction

The gifted learner Have students use logical reasoning to solve the following equations.
1. $2|x| = 2x$
2. $2|x| = |2x|$
3. $\sqrt{x} = \sqrt{|x|}$
4. $|x| = |-x|$
5. $-|x| = x$

Answers: **1.** all nonnegative numbers **2.** all numbers **3.** all nonnegative numbers **4.** all numbers **5.** all nonpositive numbers

---

Subtract 24 on each side

Divide both sides by −15

$$3(-5n + 8) = 237$$
$$-15n + 24 = 237$$
$$\underline{-24 \quad -24}$$
$$-15n = 213$$

$$\frac{-15n}{-15} = \frac{213}{-15}$$

$$n = -14.2$$

If we check the answer, we will see that $n = -14.2$ is the correct answer.

Sample Problems

Problem 1 Solve $|x - 3| = 5$ for x. Graph the solution on the number line. Explain where the points are in relation to the number 3.

Solution We know that the absolute value of the number $x - 3$ is 5, so $x - 3$ can be either 5 or −5. Let's try both numbers.

$$x - 3 = 5 \quad \text{or} \quad x - 3 = -5$$
$$x = 8 x = -2$$

Check the answers
$$|8 - 3| = |5| = 5$$
$$|-2 - 3| = |-5| = 5$$

The solutions of the equation are 8 and −2.

$$\leftarrow|\!+\!|\!+\!|\!+\!|\!+\!|\!+\!|\!+\!|\!+\!|\!+\!|\!+\!|\!+\!|\!+\!|\rightarrow$$
$$-3\;-2\;-1\;0\;1\;2\;3\;4\;5\;6\;7\;8$$

Both points are exactly five units from 3 on the number line.

Problem 2 Solve the equation $\sqrt{x} + 8 = 19$ for x.

Solution

Subtract 8 from both sides
$$\sqrt{x} + 8 = 19$$
$$\underline{-8 \quad -8}$$

Square both sides
$$\sqrt{x} = 11$$
$$(\sqrt{x})^2 = 11^2$$
$$x = 121$$

Check the answer
$$\sqrt{x} + 8$$
$$= \sqrt{121} + 8$$
$$= 11 + 8$$
$$= 19$$

The correct answer is $x = 121$.

Chapter 8 The Language of Algebra

SECTION 8.4

Think and Discuss

P, R   **1** Solve each equation for x.
     a $|x| + 4 = 21$ $x = 17$ or -17 **b** $|x + 4| = 21$ $x = 17$ or -25
     c $|x + 4| = 0$ $x = -4$ **d** $|x| + 4 = 0$ No solution
     e Explain the difference between parts **a** and **b**.
     f Explain why there is only one answer to part **c**.

P   **2** Is the equation $x(x + 3) = x^2 + 3x$ true for
     a $x = 1$?   Yes   **b** $x = 4$?   Yes   **c** $x = -6$?   Yes
     d $x = \frac{3}{5}$?   Yes   **e** $x = 0$?   Yes
     f For how many values of x can the equation $x(x + 3) = x^2 + 3x$ be true?

P, R   **3** Ms. E. Kwayshun gave her class the problem
$$3x + 5x + 9 - 2 = 5 + 8$$
Kwami said, "First, we combine the like terms on the left and get $8x + 7$. Then, we do the arithmetic on the right to get 13. Now it's $8x + 7 = 13$." Gyo argued, "That can't be right. We didn't do the same thing on each side." Who was right, Kwami or Gyo? Why?

P   **4** Discuss what types of equations have more than one solution.

P, R   **5** Is the statement true *always, sometimes,* or *never*?
     a $|x| = x$ Sometimes   **b** $|x| = -x$ Sometimes
     c $|x| \geq 0$ Always   **d** $|x - 10| < 0$ Never

P   **6** Solve each equation for y.
     a $-4(\sqrt{y} - 6) = 12$ 9   **b** $y^2 = 49$ $y = 7$ or -7
     c $\sqrt{y} = -6$ No solution

Problems and Applications

P, R   **7** For each equation, tell how many values of x are solutions. Find the solutions.
     a $\sqrt{x} = 16$ 1; 256   **b** $x^2 = 16$ 2; 4 and -4
     c $x^2 = x$ 2; 1 or 0   **d** $2(x + 4) = 2x + 8$ All values

P   **8** Solve each equation for k.
     a $|k| = -3$ No solution   **b** $|k| - 9 = -2$ $k = 7$ or -7
     c $|k - 1| = 5$ $k = 6$ or $k = -4$

P   **9** Solve each equation for x.
     a $\sqrt{x} = 9$ $x = 81$ **b** $\sqrt{x} = -9$ No solution   **c** $x^2 = 9$ $x = 3$ or $x = -3$

P   **10** Solve each equation for n.
     a $5(3n - 7) = 55$ $n = 6$ **b** $5(3n - 7) = 4$ $n = 2.6$

Problem-Set Notes

for page 395

- **1** is very important. Students often miss the significance of what is contained in the absolute-value bars.

- **3** should help clarify what it means to operate on both sides of an equation.

- **5** may be difficult for some students, especially **a** and **b**. The tendency to think of absolute value as "drop the minus sign" and of x as "positive x" make this a hard problem to grasp. This problem will provoke some class discussion.

Stumbling Block

In sample problem 1, students may think that the problem to be solved is $x + 3 = 5$ or $x + 3 = -5$. Point out that absolute value symbols do not change a subtraction expression to an addition expression.

Additional Answers

for page 395

1e In part **a** we found the absolute value of x before adding 4. In **b** we found the absolute value of the sum $x + 4$.
f 4 has only one additive inverse, -4.

2f The equation is true for all real numbers.

3 Kwami; the first step is to simplify each side of the equation.

4 Some equations with absolute values, equations with a squared term, and equations that are identities.

Section 8.4   Some Special Equations

SECTION 8.4

Problem-Set Notes

for pages 396 and 397

- **14** gives students a lot of practice with solving equations.
- **25** may surprise some students. In some branches of mathematics, $|x|$ is defined to be $\sqrt{x^2}$.
- **26** represents one of the applications of absolute value.

Checkpoint

Solve each equation for x.
1. $|x| = 8$
2. $|x| + 4 = 9$
3. $\sqrt{x} = 5$
4. $x^2 = 36$
5. $4(x - 5) = 36$
6. $3(x + 2) = 19$
7. $\sqrt{x} = 25$
8. $x^2 = 49$

Answers:
1. $x = 8, -8$ 2. $x = 5, -5$
3. $x = 25$ 4. $x = 6, -6$
5. $x = 14$ 6. $x = 4\frac{1}{3}$
7. $x = 625$ 8. $x = 7, -7$

Multicultural Note

Chu Shih-chieh was the last and greatest mathematician of the golden age of Chinese mathematics during the Song Dynasty. In the title of one of his most important books, *Precious Mirror of the Four Elements*, written in 1303, the four elements—heaven, earth, man, and matter—represent four unknowns in a single equation.

P, R **11** Jeff wants to enlarge a room so that it will have an area of 162 square feet, as in the diagram. He needs to find the value of x. He wrote the formula $A = 12(8 + x)$ for the area.
  a Use Jeff's formula and solve for x. $x = 5.5$ ft
  b Find a way to find the size of the extension without using Jeff's area formula. Answers will vary. Possible answer: Divide 162 by 12 and subtract 8

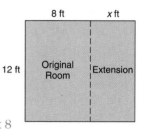

P, R **12** Solve each equation or simplify each expression.
  a $3(7n - 38) = 138$ $n = 12$
  b $|a - 5| = 13$ $a = 18$ or -8
  c $\sqrt{x} + 3\sqrt{x} - 5$ $4\sqrt{x} - 5$
  d $\sqrt{x} - 3 = 5$ $x = 64$
  e $2\sqrt{y} - 9 = 13$ $y = 121$
  f $3x^2 - 6(4 - x^2)$ $9x^2 - 24$
  g $3x^2 + 6x^2 = 36$ $x = 2$ or -2
  h $4x - (3 - x)$ $5x - 3$

P, R **13** Find the value of x that makes 50 the perimeter of the triangle. $x = 2$

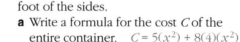

P, R **14** Find the value of w, x, y, and z in the equation $\begin{bmatrix} 2x - 10 & 3y \\ \frac{1}{2}w & 4 - z \end{bmatrix} = \begin{bmatrix} 22 & -24 \\ 40 & 30 \end{bmatrix}$ $w = 80$, $x = 16$, $y = -8$, $z = -26$

P, R **15 Business** The cost of producing an open-top cubical container is \$5 per square foot of base and \$8 per square foot of the sides.
  a Write a formula for the cost C of the entire container. $C = 5(x^2) + 8(4)(x^2)$
  b Find the difference in the cost of a container with edges that are 3 feet long and the cost of a container with edges that are 4 feet long. \$259

P, R **16** Solve each equation or simplify each expression.
  a $3[4 + (3)(-2)] = 3x - (x - 10)$ $x = -8$
  b $3[4 + 5(-6)] + 3x - (x - 10)$ $2x - 68$
  c $15 - (9 - p) = 14\left(\frac{1}{2}\right) - 3\left(\frac{2}{3}\right)$ $p = -1$
  d $\frac{1}{4}(8x) - 2(15 - \frac{1}{2}x)$ $3x - 30$
  e $\frac{2}{3}(6 - 12y) + 5\left(\frac{1}{5} + 2y\right)$ $2y + 5$
  f $5(-3m + 6) = 660$ $m = -42$
  g $3a + 6 = 90$ $a = 28$

P, R **17** The area of rectangle ABCD is 42. Find the value of a. $a = 7$ or $a = -7$

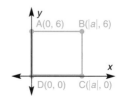

396   **Chapter 8**   The Language of Algebra

SECTION 8.4

P, R **18** Find the value of x so that the perimeter of the square is equal to the perimeter of the triangle. $x = 14$

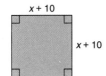
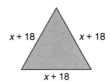

R **19** Refer to the expressions $x + 75$, $x + 5$, $x - 10$, x, and $x + 100$.
a Arrange the data from smallest to largest. $x - 10, x, x + 5, x + 75, x + 100$
b What is x if the median is 200? 195

P, R **20** Use a number line to graph the values of x for which $|x| = x$.

P, R **21** The diagonals of a rhombus are 12 and 16. Determine the value of x if the length of each side is $4x - 2$. $x = 3$

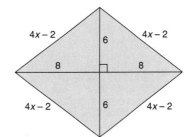

P **22** Solve the equation $(0.2)(0.3)(0.4)x = (0.5)(0.6)$. $x = 12.5$

R **23** Write an inequality to represent each graph.

a    b

R **24 a** List the operations, in order, that should be used to evaluate the expression $\sqrt{x^2 - 25}$ for $x = 13$.
b List the operations, in order, that should be used to solve the equation $\sqrt{x^2 - 9} = 5$.

P, R **25 Spreadsheets** Set up a spreadsheet like the one shown, using values for x from -10 to 10. (Note: Different spreadsheets use different notations for square root and absolute value.) What do you notice about columns $|x|$ and $\sqrt{x^2}$?

The values of $|x|$ and $\sqrt{x^2}$ are the same.

	A	B	C	D
1	x	x^2	$|x|$	$\sqrt{x^2}$
2				
3	-10			
4	-9			
	.			
	.			
	.			
	.			
	.			
23	10			

R **26** Find the value(s) of t that describe the indicated length of the number line. $t = -26$ or $t = 62$

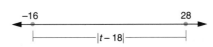

Additional Practice

Solve each equation and simplify each expression.
1. $|x| = 4$
2. $\sqrt{r} = 9$
3. $|x| - 1 = 3$
4. $5(2y - 9) = 25$
5. $4(3x - 7) + 2(3x + 1)$
6. $\sqrt{m} = 6$
7. $y^2 - 1 = 35$
8. $5a + 3 = 22$
9. $3t^2 - 7t + 4t^2 + 9t$
10. $|v| - 4 = 6$

Answers:
1. $x = 4, -4$ 2. $r = 81$
3. $x = 4, -4$ 4. $y = 7$
5. $18x - 26$ 6. $m = 36$
7. $y = 6, -6$ 8. $a = 3\frac{4}{5}$
9. $7t^2 + 2t$ 10. $v = 10, -10$

Additional Answers

for pages 396 and 397

20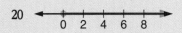

23 a $|x - 2| \leq 5$ or $-3 \leq x \leq 7$
b $-5 < x \leq 5$

24 a Substitute 13 for x, square 13, subtract 25, and take the square root of the result.

b Square both sides of the equation, add 9 to both sides of the equation, then take the square root of the result.

Section 8.4 Some Special Equations

SECTION 8.4

◀ LOOKING BACK **Spiral Learning** LOOKING AHEAD ▶

R **27** Indicate whether the statement is *true* or *false*.
  a $15 > 10$ True   **b** $15 + 7 > 10 + 7$ True   **c** $15 - 7 > 10 - 7$ True
  d $15(7) > 10(7)$ True   **e** $15(-7) > 10(-7)$ False   **f** $\frac{15}{-5} > \frac{10}{-5}$ False

R **28 Communicating**   Explain why $5 > 2$, but $-5 < -2$.

P, R, I **29 a** Find the area of rectangle ABCD.   40
  b Move C and D three units to the right and three units down. What are the coordinates of the new points? $C' = (11, 2)$, $D' = (3, 2)$
  c Find the area of the new quadrilateral.   16

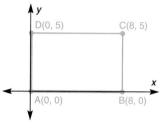

R, I **30** Find the value of x, y, and z if $2x = 12$, $y = x + 8$, and $3z - 1 = y$.

R, I **31** Solve each equation for x.
  a $x(b + c) = 2(b + c)$ $x = 2$ **b** $3x + ax = 5(a + 3)$ $x = 5$

R **32** Simplify each expression.
  a $-42 + 16$ -26 **b** $-11 - 16$ -27 **c** $(-11)(-16)$ 176
  d $(-11) - (-16)$ 5   **e** $\frac{-11}{16} + \frac{1}{2}$ $-\frac{3}{16}$ **f** $\frac{-8}{-4} - \frac{-4}{-2}$ 0

P, R, I **33** Find the coordinates of the images of points A, B, and C when $\triangle ABC$ is reflected over line ℓ.
  $A' = (-1, 5)$; $B'(2, -2)$; $C' = (10, 2)$

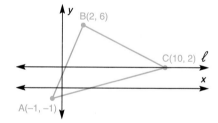

R **34** Find the number that will make the equation true if the number replaces the box.
  a $\frac{1}{8} + \square = \frac{3}{4}$ $\frac{5}{8}$ **b** $\frac{3}{4} - \frac{\square}{3} = \frac{1}{12}$ 2   **c** $\frac{5}{6} + \square = \frac{5}{4}$ $\frac{5}{12}$

Investigation

Science   What is the absolute, or Kelvin, temperature scale? What is absolute zero? Explain any connection between this use of the word *absolute* and the concept of absolute value. Write a formula showing how to convert a temperature from Celsius (C) to Kelvin (K).

SECTION 8.5

SECTION 8.5
Solving Inequalities

ADDING AND SUBTRACTING

What happens when we add the same value to each side of the inequality $12 > -9$? What happens when we subtract the same value from each side?

Add 3 to each side.
$12 + 3 > -9 + 3$

Add −3 to each side.
$12 + (-3) > -9 + (-3)$

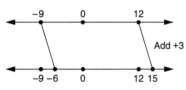

$15 > -6$ True

$9 > -12$ True

Subtract 3 from each side.
$12 - 3 > -9 - 3$

Subtract −3 from each side.
$12 - (-3) > -9 - (-3)$

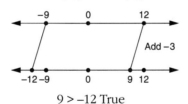

$9 > -12$ True

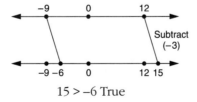

$15 > -6$ True

When the same value was added to or subtracted from both sides of the inequality, the order of the numbers stayed the same. This result leads to a rule for inequalities.

▶ **Addition and Subtraction Rule for Solving Inequalities**
If the same number is added to or subtracted from both sides of an inequality, an equivalent inequality results.

Section 8.5 Solving Inequalities **399**

Objectives
- To use addition and subtraction to solve an inequality
- To use multiplication and division to solve an inequality

Class Planning

Schedule 2 days
Resources
In *Chapter 8 Student Activity Masters* Book
Class Openers *49, 50*
Extra Practice *51, 52*
Cooperative Learning *53, 54*
Enrichment *55*
Learning Log *56*
Interdisciplinary
 Investigation *57, 58*
Spanish-Language Summary *59*

In *Tests and Quizzes* Book
Quiz on 8.4–8.5 (Average) *209*
Quiz on 8.4–8.5 (Honors) *210*

Transparency *24*

Class Opener

Write < or > in the blank.
1. -1 ___ 2
2. $-1 + 1$ ___ $2 + 1$
3. $-1 - 3$ ___ $2 - 3$
4. -1×2 ___ 2×2
5. $-1 \div 4$ ___ $2 \div 4$
6. -1×-3 ___ 2×-3
7. $-1 + (-2)$ ___ $2 + (-2)$
8. $-1 \div (-\frac{1}{2})$ ___ $2 \div (-\frac{1}{2})$

Answers: **1.** < **2.** < **3.** < **4.** < **5.** < **6.** > **7.** < **8.** >

Lesson Notes

- The double number line pictured in the text may make more sense to your students after they watch you demonstrate a few problems. This visual method of teaching the multiplication and division rules should help your students remember why the rules work.

399

SECTION 8.5

Lesson Notes continued

- Emphasize the similarities between solving equations and inequalities. The only real difference is what happens when you multiply or divide by a negative number.
- Encourage students to rewrite the inequality when they divide by a negative number, as in the next-to-the-last line in sample problem 2.
- Placing answers to inequalities on number lines will help students make sense of the solutions.

Additional Examples

Example 1
Solve the inequality $x + 36 > 19$ for x.

Answer: $x > -17$

Example 2
Solve the inequality $-8x > 20$ for x.

Answer: $x < -2.5$

Additional Practice

For problems **1–4**, solve each inequality, then graph the solution on a number line.
1. $x + 5 \geq 9$
2. $2x - 11 < 15$
3. $15 > 3x - 5$
4. $4 - 3x \geq 16$
5. If $t \not> 4$, what do we know about t?

Answers:
1. $x \geq 4$
2. $x < 13$
3. $x < 6\frac{2}{3}$
4. $x \leq -4$
5. $t \leq 4$

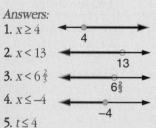

Inequalities involving addition and subtraction are solved in a way similar to the way we solve equations. We use the inverse operation.

Example 1
Solve the inequality $x + 315 > 296$ for x.

Solution
Subtract 315 from each side

$$\begin{array}{r} x + 315 > 296 \\ -315 -315 \\ \hline x > -19 \end{array}$$

Any number greater than -19 (-18.4, 234, 0, -15, or many others) makes the original inequality true. You may want to check some values.

MULTIPLYING AND DIVIDING

What happens to the inequality $12 > -9$ when we multiply or divide each side of the inequality by the same value?

Multiply each side by 3.
$3(12) > 3(-9)$

Multiply each side by -3.
$-3(12) > -3(-9)$

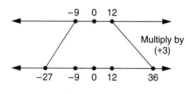

$36 > -27$ True

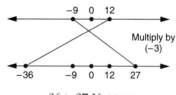

$-36 > 27$ Not true
$-36 < 27$ True

Divide each side by 3.
$\dfrac{12}{3} > \dfrac{-9}{3}$

Divide each side by -3.
$\dfrac{12}{-3} > \dfrac{-9}{-3}$

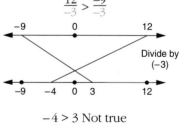

$4 > -3$ True

$-4 > 3$ Not true
$-4 < 3$ True

When we multiply or divide by a negative number, both numbers are reflected over the origin on the number line, and the order (direction of inequality) is reversed. This illustrates an important rule for solving inequalities.

Chapter 8   The Language of Algebra

SECTION 8.5

▶ **Multiplication and Division Rules for Solving Inequalities**

If both sides of an inequality are multiplied or divided by a positive number, the result is an equivalent inequality with the same order.

If both sides of an inequality are multiplied or divided by a negative number, the result is an inequality with the order reversed.

Example 2
Solve the inequality $-4x > 34$ for x.

Solution
When we divide by -4, we reverse the order of the inequality.

$$-4x > 34$$
$$\frac{-4x}{-4} < \frac{34}{-4}$$
$$x < -8.5$$

Sample Problems

Problem 1 Solve the inequality $-19 + 8p \geq 37$ for p.

Solution The symbol $\geq$ means "is greater than or equal to." It is a way to combine two separate statements: the equation $-19 + 8p = 37$ and the inequality $-19 + 8p > 37$.

Add 19 to each side
$$\begin{array}{r} -19 + 8p \geq 37 \\ +19 +19 \\ \hline 8p \geq 56 \end{array}$$

Divide each side by 8
$$\frac{8p}{8} \geq \frac{56}{8}$$
$$p \geq 7$$

Problem 2 Solve the inequality $5 - 8y \leq -35$ for y and graph the solution.

Solution First, subtract 5 from each side of the inequality. This has the same effect as adding -5 to each side.

$$\begin{array}{r} 5 - 8y \leq -35 \\ -5 -5 \\ \hline -8y \leq -40 \end{array}$$

Divide by -8
$$\frac{-8y}{-8} \geq \frac{-40}{-8}$$

Change $\leq$ to $\geq$
$$y \geq 5$$

Graph this solution on a number line.

Assignment Guide

Basic
Optional; skip to Section 8.6

Average or Heterogeneous
Day 1 9, 10 (a, b), 11 (a, c), 13 (a, b), 14 (a, c, e), 15, 19, 30
Day 2 10 (c, d), 12, 13 (c, d), 14 (b, d, f), 16, 20
Day 3 14 (g, h), 17, 18, 21, 22, 25, 28, 31, 33, 34

Honors
Day 1 10, 11, 13, 15, 18, 19, 21, 25, 30
Day 2 8, 9, 12, 14, 16, 17, 20, 22, 24, 28, 29, 34

Strands

Communicating Mathematics	1–3, 5, 29, Investigation
Geometry	17, 18, 21, 33–35, Investigation
Measurement	17, 18, 21, 24, 29, 33, Investigation
Data Analysis/ Statistics	12, 24, 28
Algebra	1–33, Investigation
Number Sense	4, 6–9, 11, 16, 20, 27, 30
Interdisciplinary Applications	22, 24, 25, 29

Individualizing Instruction

Learning-disabled students Work through a number of examples with double number lines to develop the multiplication and division rules for solving inequalities.

401

SECTION 8.5

Problem-Set Notes

for pages 402 and 403

- **5** might lead into a discussion of *and* inequalities if someone notices that $4(7 - 2x)$ must be greater than zero for the problem to make sense. **15** has similar possibilities.

- **10** and **11** differ in that **10** includes all real numbers, not only integers. Some students think of solution sets as containing only integers.

- **18** is an application of inequalities to geometry.

Cooperative Learning

- Have groups of students consider the following:

1. In △ABC, AB = 8 and BC = 14. Write an inequality that describes the possible lengths of AC if
a. AC < AB
b. AB < AC < BC
c. AC > BC

2. In △DEF, DE = 5 and EF = 12. Write an inequality that describes the possible lengths of DF if
a. DF < DE
b. DE < DF < EF
c. DF > EF.

3. Why are the answers to these two problems different?

Answers: **1. a.** 6 < AC < 8 **b.** 8 < AC < 14 **c.** 14 < AC < 22
2. a. impossible **b.** 7 < DF < 12 **c.** 12 < DF < 17 **3.** 5 is less than half of 12, so there cannot be a side shorter than DE in problem 2.

- Problems 35 and Investigation are appropriate for cooperative learning.

Problem 3 Write an inequality to represent the diagram.

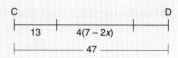

Solution The length of $\overline{CD}$ is $13 + 4(7 - 2x)$ plus an unknown amount. The length is also 47. An inequality that represents this is $13 + 4(7 - 2x) < 47$.

Think and Discuss

P **1** Which of the following operations require reversing the order of an inequality? Explain your answer.
a Add a positive number to each side
b Add a negative number to each side
c Multiply by a positive number on each side
d Multiply by a negative number on each side
e Subtract a positive number from each side
f Subtract a negative number from each side
g Divide by a positive number on each side
h Divide by a negative number on each side

P, R **2** Explain the difference between $x > 7$ and $x \geq 7$.

P, R **3** Why does subtracting 5 have the same effect as adding -5?

R **4** If $y \not< 8$ (y is not less than 8) and $y \neq 8$, what do we know about y? $y > 8$

P **5** Use the inequality $13 + 4(7 - 2x) < 47$ from Sample Problem 3.
a What are the possible values for x? $x > -0.75$
b Explain why $13 + 4(7 - 2x) < 47$ represents the diagram in Sample Problem 3.

P, R **6** Is the statement true *always, sometimes,* or *never?*
a If $x < 0$, then $x < -2$ b If $x > 3$, then $x > 0$ c $|x| < 0$

R **7** Is the inequality $x > -34.8$ *true* or *false* for the following values of x?
a -40 False   b -34.08 True   c -34.80 False   d -34 True

Problems and Applications

R **8** Is the inequality $x \geq 31.7$ *true* or *false* for the following values of x?
a -1 False   b 31.70 True   c 31.71 True   d 3.17 False
e 31 False   f 31.07 False   g 37.1 True   h -30.7 False

R **9** If $z \not< 5$, what do we know about z? $z \geq 5$

402 **Chapter 8** The Language of Algebra

SECTION 8.5

P, R, I   **10** Solve each inequality. Graph the solution on a number line.
  a $2x - 15 < 37$
  b $-12 - 3x \geq -48$
  c $x + 43.6 > -91.7$
  d $14 \geq 9x - 13$

P, R   **11** Find the smallest integer that solves each inequality.
  a $3x - 6 > 23$ 10
  b $2x + 5 \geq -22$ -13
  c $10 - 3x \leq 56$ -15

P, R   **12** The mean of $3x$, $2x + 5$, $4x - 11$, and $18 - x$ is less than 21.
  a Write an inequality to represent this. $\frac{8x+12}{4} < 21$
  b Solve the inequality for x. $x < 9$

P   **13** Solve each inequality.
  a $3x + 52 > 13$ $x > -13$
  b $-14 - 6x \leq -98$ $x \geq 14$
  c $-2x - 53 > 47$ $x < -50$
  d $9 > 3 + 2x$ $x < 3$

P, R, I   **14** Solve each equation or inequality. Simplify each expression.
  a $3 - 8x > 27$ $x < -3$
  b $3 - 8x + 27$ $30 - 8x$
  c $-5(x + 3) > -10$ $x < -1$
  d $-5(x + 3) - 10$ $-5x - 25$
  e $-4(3 - 2x) - (6 - 8x)$ $16x - 18$
  f $-3(6 - 2x) \leq 15$ $x \leq 5.5$
  g $4(3 - 2x) = 62$ $x = -6.25$
  h $\frac{-2}{3}x < \frac{-4}{5}$ $x > 1.2$

P, R, I   **15** Refer to the drawing.
  a Write an inequality to describe the figure. $3(x - 2) + 8 < 39$ or $3x + 2 < 39$
  b Solve for x. $x < \frac{37}{3}$ or $x < 12.\overline{3}$
  c Graph the solutions for x on a number line.

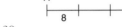

P, R   **16** Copy the figure. Multiply $-\frac{1}{2}$ by the coordinate of each point shown on the top number line. Plot the results on the bottom number line, and draw segments connecting each point on the bottom number line with its corresponding point on the top number line. An example is shown.

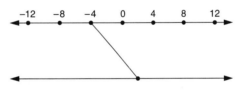

P, R   **17** The perimeter of the square is at least 70% of the perimeter of the triangle. Find the value of x. $x \geq 8.75$

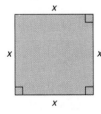
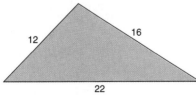

P, R   **18** If $10° \leq y \leq 40°$, what is the range of x? $90° \leq x \leq 120°$

Stumbling Block

Watch for students who start reversing the inequality sign when adding or subtracting a negative number to an inequality. Use double number lines to show these students why the sign should not be reversed when adding or subtracting a negative.

Additional Answers

for pages 402 and 403

1 Parts **d** and **h** are the only operations that reverse order. Multiplying or dividing by a negative number requires reversing the order of the inequality.

2 In $x \geq 7$, x can equal 7. In $x > 7$, x cannot equal 7.

3 Possible answer: Each operation decreases the value by 5.

5 b Possible answer: The length 13 added to the length $4(7 - 2x)$ is shorter than 47. Therefore, $13 + 4(7 - 2x)$ is less than 47.

6 a Sometimes
b Always
c Never

10 a $x < 26$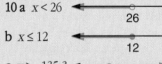

b $x \leq 12$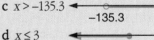

c $x > -135.3$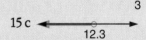

d $x \leq 3$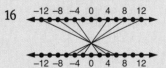

15 c

16

Section 8.5 Solving Inequalities   403

SECTION 8.5

Problem-Set Notes

for pages 404 and 405

- 21, 22, 24, and 25 explore some situations in which inequalities are useful.
- 26 will help to clarify this section for students.
- 29 provides a physical model illustrating the addition property for inequalities.
- **Investigation** reinforces an earlier Investigation (Section 3.7) and extends it by having students write a compound inequality.

Communicating Mathematics

- Have each student write a paragraph describing how the graphs of the solutions sets of $3x + 4 > 10$ and $3x + 4 \leq 10$ are the same and how they are different.
- Problems 1–3, 5, 29, and Investigation encourage communicating mathematics.

Reteaching

Have students get in the habit of checking solutions to inequalities. A quick and easy way to check that the inequality symbol is pointing in the right direction is to substitute zero for the variable in the original equation. If zero is not a solution, then it should make the original inequality false. If zero is a solution, it should make the original inequality true.

P   **19** Solve each inequality for k.
    a $2(4k + 3k) > 56$ $k > 4$ **b** $-2(4k + 3k) > 84$ $k < -6$
    c $2k - 5k < -24$ $k > 8$

R   **20** For each statement, decide which symbol, <, >, or =, should replace the box.
    a $-6 \;\square\; 4$ <   **b** $\frac{-5}{8} \;\square\; \frac{(-5)(1)}{(8)(-1)}$ <
    c $(-3)(-6) \;\square\; (-3)(4)$ >   **d** $-\frac{-3}{-7} \;\square\; \frac{-3}{7}$ =

P, R, I   **21** Find the value(s) of x. $x \geq 6$

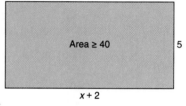

P, R, I   **22 Business** Jon's sales commission ranges from 2% to 7% depending on the items he sells. What is the range of possible commissions on the sale of a $650 item? $\$13.00 \leq c \leq \$45.50$

R   **23** The coordinates of the set of points shown are each divided by -4. Show the new graph on a number line.

P, R, I   **24 Science** The high temperatures on the first six days of January were $-4°$, $6°$, $1°$, $11°$, $7°$, and $-2°$. What must be the high temperature on January 7 in order for the first week of January to have an average high temperature greater than 5°? $t > 16°$

R   **25 Business** Yang receives an $8 bonus for every CD player she sells. Her CD sales bonus for December exceeded $150. What can we conclude about the number of CD players she sold in December?   She sold more than 18 CD players.

P, R, I   **26** Write an inequality having the form $nx \geq 20$ and having a solution given by the graph.

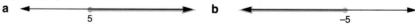

P, R   **27** Is the inequality $|x| > 5$ *true* or *false* for the following values of x?
    a 5.1   True   **b** -5.1 True
    c 5   False   **d** -5 False
    e 4.8   False   **f** -4.8 False
    g 7   True   **h** -7 True

P, R   **28** The mean of -10, 6, x, 15, and 31 is greater than 5. Find the value of x. $x > -17$

P, R   **29 Communicating** Penny had a balance scale, but the left side was heavier than the right. She put the same type of coin on each side, and the left side remained heavier than the right side. She put more coins, the same number and the same type on each side. The left side was still heavier than the right side. Explain why Penny didn't get the scale to change position.

404   Chapter 8   The Language of Algebra

SECTION 8.5

◀ LOOKING BACK **Spiral Learning** LOOKING AHEAD ▶

R **30** Solve each equation for x.
  a $|x| - 3 = 15$ $x = 18$ or -18
  b $|x - 3| = 15$ $x = 18$ or -12
  c $\sqrt{x} = 36$ $x = 1296$
  d $x^2 = 36$ $x = 6$ or -6

R **31** Multiply $\begin{bmatrix} -4 & 13 \\ 16 & -2 \end{bmatrix} \cdot \begin{bmatrix} -2 & 3 \\ -1 & 1 \end{bmatrix}$. $\begin{bmatrix} -5 & 1 \\ -30 & 46 \end{bmatrix}$

R, I **32 a** Solve $3y + 18 = 12$ for y. $y = -2$
  b Solve $3(a + b) + 18 = 12$ for $a + b$. $a + b = -2$
  c Solve $3(2x + 8) + 18 = 12$ for $2x + 8$. $2x + 8 = -2$
  d Solve $3(2x + 8) + 18 = 12$ for x. $x = -5$

R **33** Segment AB is 8 units long. It is to be placed on a number line. Either A or B could be on the left.

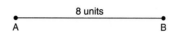

  a If A is placed at -2, where is B? $B = 6$ or $B = -10$
  b If $\overline{AB}$ is centered at the origin, what are the coordinates of A and B?
  $A = -4$ or $A = 4$, $B = -4$ or $B = 4$

R **34** Which of the following are possible?
  a A trapezoid with rotation symmetry
  b A parallelogram with reflection symmetry
  c A rectangle with rotation symmetry
  d A rhombus with rotation symmetry
  e A trapezoid with reflection symmetry   **b, c, d, e**

R **35** ABCD is a square with its diagonals drawn as shown.
  a Name all pairs of segments that are parallel. $\overline{AB} \parallel \overline{CD}$, $\overline{AD} \parallel \overline{BC}$
  b Name all pairs of segments that are perpendicular.
  $\overline{AB} \perp \overline{BC}$ and $\overline{AD}$, $\overline{DC} \perp \overline{DA}$ and $\overline{CB}$, $\overline{AC} \perp \overline{DB}$

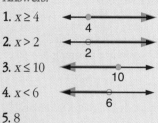

Investigation

The Triangle Inequality

1 Can you draw a triangle that has sides with the following dimensions? Explain your answer.
  a 4 in., 2 in., 7 in.
  b 8 cm, 5 cm, 15 cm
  c 9 in., 1 in., 9 in.

2 The sides of one triangle are 3, 5, and x. Write an inequality that describes the possible values of x.

3 Find out what is meant by the triangle inequality in geometry.

Section 8.5 Solving Inequalities 405

Checkpoint

For problems 1–4, solve each inequality, then graph the solution set on a number line.
1. $4x - 1 \geq 15$
2. $3 - 2x < -1$
3. $x - 3.9 \leq 6.1$
4. $15 > 3x - 3$

5. Find the smallest integer that satisfies the inequality $4x - 7 > 21$.

Answers:

1. $x \geq 4$
2. $x > 2$
3. $x \leq 10$
4. $x < 6$
5. 8

Multicultural Note

Sonya Kovalevsky, a nineteenth-century Russian mathematician, is associated with the Cauchy-Kovalevsky Theorem. The theorem can be used to determine whether certain types of equations have one or no solutions.

Additional Answers

for pages 404 and 405

23

26 a $4x \geq 20$ b $-4x \geq 20$

29 She always added the same amount to each side.

Investigation 1 a No **b** No **c** Yes
In parts **a** and **b**, one side is too short for the triangle to be a closed figure.
2 $2 < x < 8$
3 The sum of the lengths of any two sides of the triangle is greater than the length of the third side.

SECTION 8.6

Objectives

- To use substitution to solve equations
- To use a quantity as a single term

Class Planning

Schedule 2 days

Resources

In *Chapter 8 Student Activity Masters* Book
Class Openers *61, 62*
Extra Practice *63, 64*
Cooperative Learning *65, 66*
Enrichment *67*
Learning Log *68*
Technology *69, 70*
Spanish-Language Summary *71*

In *Tests and Quizzes* Book
Quiz on 8.3 & 8.6 (Basic) *208*

Class Opener

For each equation, find y if $x = 4y - 1$. The first step has been completed for you.
1. $ 3x + y = 10$
   $3(4y - 1) + y = 10$
2. $ 2x + 4(y + 1) = 8$
   $2(4y - 1) + 4(y + 1) = 8$
3. $ 3y - x = 7$
   $3y - (4y - 1) = 7$

Answers: **1.** 1 **2.** $\frac{1}{2}$ **3.** −6

Lesson Notes

- This section allows us to continue solving equations in one variable, while extending the topic to two variables. While we have not called this solving systems of equations, that is what students are doing in examples 2 and 3.

Investigation Connection
Section 8.6 supports Unit D, *Investigation* 4. (See pp. 316A-316D in this book.)

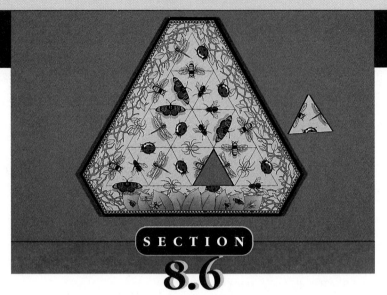

SECTION 8.6
Substitution

SPECIAL SUBSTITUTIONS

We have substituted values for variables in earlier sections. Now let's look at some special kinds of substitution.

Example 1
If $(x, y) = (12, -7)$, find the area of the parallelogram.

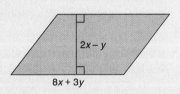

Solution
The base of the parallelogram is
$ 8x + 3y$
$ = 8(12) + 3(-7)$
$ = 75$

The height is
$ 2x - y$
$ = 2(12) - (-7)$
$ = 31$

The area is (base) × (height)
$ = (75)(31)$
$ = 2325$

Notice that when we substituted values for variables in Example 1, we used parentheses ( ). You will find parentheses to be a valuable tool when doing substitutions. Sometimes we substitute expressions for variables in order to solve equations.

Chapter 8 The Language of Algebra

SECTION 8.6

Example 2
If $3x + 5y = 47$ and $x = 2y + 1$, find x and y.

Solution

$$3x + 5y = 47$$

Substitute $(2y + 1)$ for x $3(2y + 1) + 5y = 47$
Use the distributive property $6y + 3 + 5y = 47$
Combine like terms $11y + 3 = 47$
Subtract 3 $11y = 44$
Divide by 11 $y = 4$

Now that we know $y = 4$, we can find x.

$$x = 2y + 1$$
$$= 2(4) + 1$$
$$= 9$$

The answers are $x = 9$ and $y = 4$.

In Example 2, we again used parentheses around $2y + 1$ when we substituted for x. If we had not used parentheses, we could have distributed the multiplication incorrectly.

VARIABLES FOR EXPRESSIONS

You have simplified expressions like $3x + 5x = 8x$. The variable, x, can be any value.

Example 3
Simplify the expression $4(3x - 2) - 7(3x - 2) + 6(3x - 2)$.

Solution

Method 1
Using familiar techniques,

$$4(3x - 2) - 7(3x - 2) + 6(3x - 2)$$
$$= 12x - 8 - 21x + 14 + 18x - 12$$
$$= 12x - 21x + 18x - 8 + 14 - 12$$
$$= 9x - 6$$

Method 2
Think of $3x - 2$ as a single value.

$$4(3x - 2) - 7(3x - 2) + 6(3x - 2)$$
$$= (4 - 7 + 6)(3x - 2)$$
$$= 3(3x - 2)$$
$$= 9x - 6$$

Treating an expression as a single value can make some problems easier to solve.

- You might point out that substitution is the basic idea behind using a formula.

- Thinking of an expression as a single value (as in example 3) is an efficient technique for solving equations as well as a generalization of the concept of variable. Students will need a detailed explanation of this method.

- The second method in sample problem 1 is particularly useful.

Additional Examples

Example 1
Find the area of a parallelogram whose base is $7x + 2y$ and height is $3x - y$ if (x, y) is $(6, -4)$.

Answer: 748

Example 2
If $2x + 3y = 25$ and $y = 3x + 1$, find x and y.

Answer: $x = 2$ and $y = 7$

Example 3
Simplify the expression
$5(5x - 3) + 4(5x - 3) - 7(5x - 3)$.

Answer: $10x - 6$

Example 4
Simplify the expression
$8(VAR) - 11(VAR) + 7(VAR)$.

Answer: $4(VAR)$

Communicating Mathematics

- Give students the following two equations to consider: $A = \pi d$ and $A = 3.14d$. Have them write a paragraph to explain whether this is a proper use of substitution. Students should be sure to defend their answers.

- Problems 2–5, 16, 19, and Investigation encourage communicating mathematics.

SECTION 8.6

Assignment Guide

Basic
Day 1 11, 12, 15, 18, 21 (a, b), 28, 31 (a, b), 37
Day 2 14, 17, 19, 21d, 23, 30, 31 (e, f), 35

Average or Heterogeneous
Day 1 11, 12, 14–16, 19, 20, 21 (a, b, c), 22, 24, 29 (a, b), 30, 37
Day 2 17, 18, 21 (d, e), 23, 25, 28, 29 (c, d), 31 (a, b, c), 33–35

Honors
Day 1 11, 12, 14, 18, 22, 23, 25, 29, 32, 33
Day 2 15–17, 19–21, 24, 26, 27, 30, 31, 34, 37

Strands

Communicating Mathematics	2–5, 16, 19, Investigation
Geometry	6, 10, 17, 18, 20, 23, 28, 32, 34, 36, 37
Measurement	4, 6, 10, 17, 18, 20, 23, 24, 28, 32, 34–36
Data Analysis/ Statistics	30, 33
Technology	5, 27, 30
Algebra	1, 2, 4, 6–33, 35, 36, Investigation
Patterns & Functions	30, 37
Number Sense	15, 16, Investigation
Interdisciplinary Applications	24, 27, 33, 35

Individualizing Instruction

Learning-disabled students Students may have difficulty with the attention to detail needed in this section. Care is needed to make proper substitutions as well as to avoid losing negative signs.

Example 4
Simplify the expression $4(NUM) - 7(NUM) + 6(NUM)$.

Solution

$$4(NUM) - 7(NUM) + 6(NUM)$$
$$= (4 - 7 + 6)NUM$$
$$= 3(NUM)$$

Sample Problems

Problem 1 Find two numbers that have a ratio of 1:3 and a sum of 92.

Solution **Method 1** Call the numbers a and b. We know two facts: $\frac{a}{b} = \frac{1}{3}$ and $a + b = 92$. Using cross multiplication, $b = 3a$.
In the equation $a + b = 92$, we substitute $3a$ for b.

$$a + b = 92$$
Substitute $\quad a + 3a = 92$
$$4a = 92$$
$$a = 23$$

If $a = 23$ and $b = 3a$, $b = 3(23)$, or 69.
The numbers are 23 and 69.

Method 2 Since the numbers are in the ratio 1:3, we can call the numbers $1x$ and $3x$. The sum of the numbers is 92, so

$$1x + 3x = 92$$
$$4x = 92$$
$$x = 23$$
$$3x = 69$$

Using either method, the numbers are 23 and 69.

Problem 2 The angles of a triangle are in the ratio 2:3:4. Find their measures.

Solution If the numbers are in the ratio 2:3:4, we can call the angle measures $2x$, $3x$, and $4x$. We also know that the sum of the measures of the angles of a triangle are 180. We can write an equation.

$$2x + 3x + 4x = 180$$
$$9x = 180$$
$$x = 20$$

The measures of the angles are 2(20), or 40; 3(20), or 60; and 4(20), or 80.

408   **Chapter 8**   The Language of Algebra

SECTION 8.6

Think and Discuss

P **1** Solve each equation.
   a Solve $3x + 8 = 26$ for x. $x = 6$
   b Solve $3 \cdot NUM + 8 = 26$ for NUM. $NUM = 6$
   c Solve $3(a + b) + 8 = 26$ for $a + b$. $a + b = 6$
   d Solve $3(5x - 9) + 8 = 26$ for $5x - 9$. $5x - 9 = 6$
   e Solve $3(5x - 9) + 8 = 26$ for x. Use at least two methods. $x = 3$

P **2 a** Describe how to solve for x and y if $2x - 6y = -8$ and $x = 9y - 6$.
   b Find x and y. $x = -3$, $y = \frac{1}{3}$, or 0.3

P **3** The word *substitute* is commonly used in conversation, as in a substitute teacher, a substitute player in sports, a substitute ingredient in a recipe, or a substitute part in a machine.
   a Give several other examples of substitutes.
   b Discuss similarities and differences between these uses of the word *substitute* and the mathematical use of the word.

P, R **4** The following procedure converts 8 feet to inches.

$$8 \text{ ft} = 8 \text{ ft} (1)$$
$$= 8 \text{ ft} \left(\frac{12 \text{ in.}}{1 \text{ ft}} \right)$$
$$= 96 \text{ inches}$$

Explain the substitution that took place.

P, R **5** Ben and Frank were evaluating the expression $\frac{1}{3} \cdot 10$. They both reached for their calculators, entered 1 ÷ 3 =, and got the result 0.3333333. Ben left the number on his screen, then entered × 10 = and got 3.3333333. Frank cleared his calculator, entered 0.3333333 × 10 = and got 3.333333. Explain why they got different results.

P, R **6** Triangle EQU is equilateral. Find the length of a side and find the perimeter. 20; 60

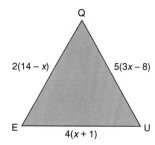

P **7** If $-2x + 3y = 12$ and $y = 6$, find the value of x. $x = 3$
P **8** If $5(x + 2y) + 3(x + 2y) - 2(x + 2y) = 36$, find the value of $x + 2y$. $x + 2y = 6$
P, R **9** Two numbers, a and b, have a ratio $a{:}b = 5{:}2$. Find a if $6a + 10b = 200$. $a = 20$

Section 8.6 Substitution

Problem-Set Notes

for page 409

■ **4** and **19** are common to students' experiences, but students probably haven't thought of them as substitution problems.

Checkpoint

1. If $3x - 4y = 24$ and $y = 3$, find the value of x.
2. If $-2x + 3y = 12$ and $x = 9$, find the value of y.
3. If $3(2x + y) + 4(2x + y) - 2(2x + y) = 35$, find the value of $2x + y$.
4. Solve $5(2x + 3y) - 9 = 51$ for $2x + 3y$.
5. Two numbers, m and n, have a ratio $m{:}n = 3{:}7$. Find m if $4m + 6n = 210$.

Answers:
1. $x = 12$ 2. $y = 10$
3. $2x + y = 7$
4. $2x + 3y = 12$ 5. $m = \frac{35}{3}$

Additional Answers

for page 409

2 a Substitute $9y - 6$ for x, solve as usual.

3 a Possible answers: A store substitutes one brand of product for another; a store substitutes a generic prescription drug for a brand-name prescription drug.

b A nonmathematical substitute can take the place of something and not be equivalent. In mathematics, the substituted value must be equivalent to the value it replaces.

4 $\frac{12 \text{ in.}}{1 \text{ ft}} = 1$, an equal quantity was substituted for an equal amount.

5 Possible answer: The calculator stored at least one more decimal place than it showed.

SECTION 8.6

Problem-Set Notes

for pages 410 and 411

- **26** previews the formula for slope.

Additional Practice

1. If $5x - 2y = 20$ and $y = 5$, find the value of x.
2. If $7(3x + y) - 4(3x + y) + 5(3x + y) = 48$, find the value of $3x + y$.
3. Two numbers, a and b, have a ratio $a{:}b = 4{:}3$. Find b if $9a + 10b = 660$.
4. Find the value of x and y if $4x + 5y = 20$ and $y = x - 8$.
5. Find two numbers that have a ratio of 5:7 and a sum of 36.
6. Two complementary angles have a ratio of 4:5. Find the measure of the smaller angle.

Answers:
1. $x = 6$ 2. $3x + y = 6$ 3. $b = 30$
4. $x = \frac{20}{3}$, $y = -\frac{4}{3}$ 5. 15 and 21
6. 40°

Reteaching

In method 2 of example 3, have students write a substitution for $3x - 2$. Let $y = 3x - 2$. Then $4(3x - 2) - 7(3x - 2) + 6(3x - 2) = 4y - 7y + 6y = 3y$. Then substitute $3x - 2$ for y: $3y = 3(3x - 2) = 9x - 6$.

Stumbling Block

In sample problem 2, some students may stop when they have solved the equation for x. Remind students to read the problem one more time to determine whether they have found what the problem is asking for. In this case, the problem asks for angle measures, not the value of x.

P, R **10** If $(b_1, b_2, h) = (10, 24, 6)$, find the area of the trapezoid. $A = 102$

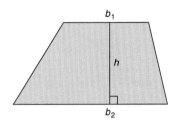

Problems and Applications

P **11** Find the value of x if $4x + y = 8$ and $y = 10$. $x = -0.5$

P **12** Find the value of x and y if $2x + 3y = 8$ and $y = x - 9$. $x = 7, y = -2$

P, R **13** Simplify the expression $9(2x - 7) - 15(2x - 7) + 11(2x - 7)$ in two different ways. $5(2x - 7)$, or $10x - 35$

P, R **14** Find two numbers that have a ratio of 3:5 and a sum of 52. 19.5, 32.5

P, R **15** Which of the following values can be substituted for 25%?
  a 0.025   **b** $\frac{5}{20}$ **c** 0.25%   **d** $\frac{1}{4}$
  e $\frac{2}{5}$ **f** $\frac{75}{300}$ **g** 0.25   **h** $\frac{25}{100}$ b, d, f, g, h

P, R **16** Communicating Explain why $5 \cdot \left(\frac{3}{8}\right) = 5 \cdot (0.375)$. $0.375 = \frac{3}{8}$

P, R **17** The angles of a quadrilateral are in the ratio 2:3:3:4. Find their measures.

P **18** Two complementary angles have a ratio 2:3. Find the measure of the larger angle. 54

P, R **19** Communicating Explain how Edna used substitution to simplify the fraction $\frac{18}{24}$. Edna substituted 1 for $\frac{6}{6}$, then multiplied.
$\frac{18}{24} = \frac{6 \cdot 3}{6 \cdot 4} = \frac{6}{6} \cdot \frac{3}{4} = 1 \cdot \frac{3}{4} = \frac{3}{4}$

P, I **20** The lengths of the sides of the triangle are in the ratio $a{:}b{:}c = 5{:}6{:}7$. The perimeter is less than 135. What are the possible lengths of sides a, b, and c? $0 < a < 37.5, 0 < b < 45, 0 < c < 52.5$

P, R **21** Solve each equation or inequality and simplify each expression.
  a $3(x + 5) + 8(x + 5) = 11$ $x = -4$ **b** $3(x + 5) + 8(x + 5) + 11$ $11x + 66$
  c $3(x + 5) + 8(x + 5) \geq 11$ $x \geq -4$ **d** $3(x + 5) + 8(x + 5) - 11(x + 5)$ 0
  e $3(x + 5) + 8(x + 5) = 11(x + 5)$ All values of x

P, R, I **22** On the planet Sram, a dyme is worth 12 centts and a nickall is worth 6 centts. On Sram, how many packages of guum, which cost 28 centts each, can be purchased with a total of 11 dymes and 9 nickalls? 6 packages

P, R **23** The lengths of segments AB and BC have the ratio AB:BC = 3:2, and $AC = 60$. Find the length of segment AB. $AB = 36$

410 Chapter 8 The Language of Algebra

SECTION 8.6

P, R **24** Franklin remembered that a rod is a unit of measurement equal to about $16\frac{1}{2}$ feet. On a canoe trip, the map showed a portage that was 13 rods long. How many feet long was the portage? 214.5 ft

P, R **25** Find the coordinate of point P on the number line if $a:b = 1:4$. P = 1

P, R, I **26** If $(x_1, y_1) = (3, -9)$ and $(x_2, y_2) = (1, 7)$, evaluate the expression $\frac{y_2 - y_1}{x_2 - x_1}$.

P, R **27 Business** A company had a profit of $78,000 in a given year. The three owners of the business divided the profits in a ratio of 4:4:5. How much did each owner receive? $24,000, $24,000, and $30,000

P, R **28** In triangle ABC, AB = AC. The ratio of $\angle B$ to $\angle A$ is 2:1. Find the measure of $\angle C$. $\angle C = 72°$

P, R **29** Solve each inequality.
  a $30 - 8x \leq 126$ $x \geq -12$
  b $15 + 2x > 9$ $x > -3$
  c $x + 382.16 \leq -15.05$ $x \leq -397.21$
  d $15 - 6x \geq -18$ $x \leq 5.5$

P, R **30 Spreadsheets** Make a spreadsheet to complete the table.

x	y	2x + 3y
4	2	
3	1	
2	0	
.	.	.
.	.	.
.	.	.
-8	-10	
-9	-11	

P, R **31** Simplify each expression and solve each equation or inequality.
  a $2(x + 1) + 5$ $2x + 7$
  b $2(y + 1) = 5$ $y = 1.5$
  c $2(z - 2) - (3 - z)$ $3z - 7$
  d $2 < 8 - 4a$ $a < 1.5$
  e $10(q + 10q)$ $110q$
  f $10 = q + 10q$ $q \approx 0.91$

P, R **32** In the figure, $\angle ABC \cong \angle DEF$. Find the measure of $\angle DEG$. $\angle DEG = 124$

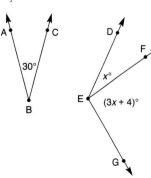

Cooperative Learning

- Refer to problem 4. The ratio $\frac{12 \text{ in.}}{1 \text{ ft}}$ is called a *scale factor* and is used to convert a measurement expressed in inches to one expressed in feet. Have students refer to tables of weights and measures to develop other scale factors. For example, $\frac{1 \text{ ft}}{12 \text{ in.}}$ is used to convert feet to inches and $\frac{16 \text{ oz}}{1 \text{ lb}}$ to convert ounces to pounds. Small groups might make a chart or poster showing their results.

- Problems 20 and 27 are appropriate for cooperative learning.

Additional Answers

for pages 410 and 411

13 $(9 - 15 + 11)(2x - 7) =$
$5(2x - 7) = 10x - 35; 18x - 63$
$- 30x + 105 + 22x - 77 = 10x - 35$

16 The decimal 0.375 is equal to the fraction $\frac{3}{8}$ and, therefore, can be substituted for it.

17 60°, 90°, 90°, 120°

19 Edna substituted 1 for $\frac{6}{6}$, then multiplied.

26 −8

30

x	y	2x + 3y
4	2	14
3	1	9
2	0	4
1	-1	-1
0	-2	-6
-1	-3	-11
-2	-4	-16
-3	-5	-21
-4	-6	-26
-5	-7	-31
-6	-8	-36
-7	-9	-41
-8	-10	-46
-9	-11	-51

SECTION 8.6

◀ LOOKING BACK **Spiral Learning** LOOKING AHEAD ▶

R **33** Ace earned the same score on every one of his five tests. Bee scored 81, 68, 96, 91, and 89 on her five tests. Bee and Ace had the same mean score. What did Ace score on each of his tests? 85

R **34** Use a ruler to find the area of the parallelogram. What dimensions do you need to measure?
The base and the height

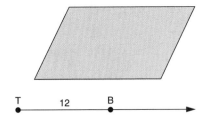

R, I **35** Tom starts riding his bicycle at point T. Bill, who is 12 miles east of Tom, starts at point B. Tom rides 18 miles per hour and Bill rides 13 miles per hour. They both start riding east at the same time. Let x = the number of hours they have been riding.
  a Write an expression for Tom's distance from point T.
  b Write an expression for Bill's distance from point T.
  c Write an equation for when the distances in parts **a** and **b** are equal. What does this mean?
  d Solve for x. What does this solution mean?

P, R, I **36** The length of the sides of a square is x. The perimeter of the square is between 12 and 28. What are the possible values of x? $3 < x < 7$

R **37 a** How many angles are in each figure? 0; 1; 3; 6
  b If a similar figure has five rays, how many angles would there be? 10

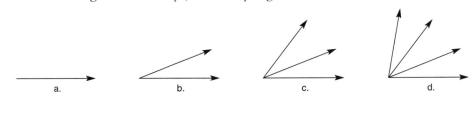

a.   b.   c.   d.

Investigation

Decimals vs. Fractions Why doesn't $\frac{1}{3} + \frac{2}{3}$ equal the same quantity as $0.333\ldots + 0.666\ldots$, or does it? Explain.

$$\begin{aligned} \frac{1}{3} &= 0.333\ldots \\ +\frac{2}{3} &= +0.666\ldots \\ \hline 1 &= 0.999\ldots \end{aligned}$$

Chapter 8 The Language of Algebra

Problem-Set Notes

for page 412

- **35** previews a common word problem in algebra.

Multicultural Note

Seki Kowa was the greatest Japanese mathematician of the seventeenth century. He improved upon the Chinese methods of solving higher-degree equations and developed the use of determinants in solving simultaneous equations. He is credited with awakening the scientific spirit that continues to thrive in modern Japan. In 1907 the Japanese Emperor honored Seki Kowa posthumously for his efforts to create a better understanding and knowledge of mathematics in Japan.

Additional Answers

for page 412

35 a $18x$ **b** $13x + 12$
c $18x = 13x + 12$; Tom caught up with Bill
d $x = 2.4$; Tom will catch up with Bill in 2 hr 24 min.

Investigation They are the same; $1 = 0.999\ldots$ because the 9's repeat infinitely. To prove this, let $n = .999\ldots$, so $10n = 9.999\ldots$. Now subtract the first equation from the second, and you have $9n = 9$, or $n = 1$. Similarly, $2 = 1.\overline{9}$, $0.5 = 0.4\overline{9}$, etc.

SECTION 8.7

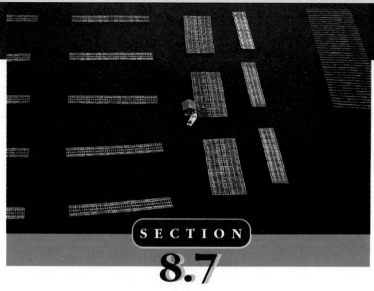

Investigation Connection
Section 8.7 supports Unit D, *Investigation* 4. (See pp. 316A–316D in this book.)

SECTION 8.7
Functions

WHAT IS A FUNCTION?

In order to find the area of a rectangle, we need to know the rectangle's length and width. To find out how long a trip will take, we need to know the distance (how far we will travel), and the speed (how fast we will travel).

A *function* describes a relationship in which one value depends on the other values. In a rectangle, the area depends on the length and the width. The area is a function of the length and the width. For a trip, the time depends on the distance and the speed. The time is a function of the distance and the speed.

We write a function by naming the function, placing parentheses around what we need to know, and giving a formula, equation, or a set of instructions that shows how to find what we need to know. We call the information in parentheses the **input**, and the result of the formula or equation the **output**.

Example 1
Express the perimeter of the rectangle as a function.

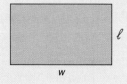

Solution
The perimeter of the rectangle depends on the length and the width. We can write a function for the perimeter.

$$\text{Perimeter (Length, Width)} = 2 \cdot \text{Length} + 2 \cdot \text{Width}$$

The function would be much shorter if we used single variables like P for perimeter, ℓ for length, and w for width.

$$P(\underbrace{\ell, w}_{\text{Inputs}}) = \underbrace{2\ell + 2w}_{\substack{\text{Formula to}\\ \text{compute outputs}}}$$

(Name of function)

Objectives
- To recognize a function
- To calculate the value of a function

Vocabulary: function, input, output

Class Planning

Schedule 2 days
Resources
In *Chapter 8 Student Activity Masters* Book
Class Openers 73, 74
Extra Practice 75, 76
Cooperative Learning 77, 78
Enrichment 79
Learning Log 80
Using Manipulatives 81, 82
Spanish-Language Summary 83

Class Opener

1. If a regular hexagon has sides of length x, write a formula for its perimeter.
2. Find the perimeter of the hexagon if
a. $x = 4$ **b.** $x = 7$ **c.** $x = 10$
d. $x = 25$ **e.** $x = 100$

Answers: **1.** $P = 6x$ **2. a.** 24 **b.** 42 **c.** 60 **d.** 150 **e.** 600

Lesson Notes

- Functions are commonplace in mathematics, although we don't always identify them as such. The concept of function is central to all of mathematics beyond algebra. It is also useful to have the vocabulary of functions because it enables us to clarify difficult concepts with simple language.

- Our introduction will be informal and intuitive, focusing on the idea of input and output.

SECTION 8.7

Lesson Notes continued

Formulas are usually functions, as in examples 1 and 2 and sample problem 4.

■ Some students will have trouble with the notation because they are used to problems like (5)(7) = 35, in which the parentheses mean multiplication. Now they mean something else. One way to explain the notation is to make a list on the board as follows:

Price of 1	=	$0.50
Price of 5	=	$0.50 · 5
Price(8)	=	$0.50 · 8
Pr(12)	=	$0.50 · 12
P(15)	=	$0.50 · 15
P(x)	=	$0.50x

The notation is quite natural if approached in this way.

Additional Examples

Example 1
Express the perimeter of a square with a side length of x as a function.

Answer: $P(x) = 4x$

Example 2
Express the area of the window as a function.

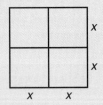

Answer: $A(x) = 4x^2$

Example 3
Using the function from example 2, evaluate $A(6)$.

Answer: $A(6) = 144$

Example 4
If $f(x) = 3x^2 - 4x$, evaluate $f(3)$.

Answer: $f(3) = 15$

Example 2
Write a function for the surface area of a cube.

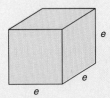

Solution

Let S represent the surface area of a cube and e represent the length of each edge of a cube. The length of an edge is the only input we need. Each face of the cube has area e^2, and there are six faces, so the formula will be $6e^2$.

The function for the surface area of a cube is $S(e) = 6e^2$.

Example 3
Using the function from Example 2, evaluate $S(5)$. What does $S(5)$ mean?

Solution

$S(5)$ means that $e = 5$ in the formula $6e^2$. We use the function and the input 5 to find the output.

Use 5 as an input for e
$$S(e) = 6e^2$$
$$S(5) = 6 \cdot 5^2$$
$$= 6 \cdot 25$$
$$= 150$$

For the input 5, the output is 150. This means that 150 is the surface area of a cube with an edge length of 5.

EVALUATING FUNCTIONS

Letters like f and g are often used to name functions. In the case of perimeter or area, we will use P or A to name a function. Usually, x or y is used for inputs. It makes sense to use ℓ for length, w for width, or e for edge.

Example 4
If $f(x) = 4x^2 + 7x$, evaluate $f(5)$.

Solution

For the function $f(x) = 4x^2 + 7x$, $f(5)$ means that the value of x is 5. To evaluate the function at 5, we substitute 5 in place of x in the formula or equation.

$$f(5) = 4(5)^2 + 7(5)$$
$$= 4(25) + 7(5)$$
$$= 100 + 35$$
$$= 135$$
$$f(5) = 135$$

We read the result as "f of 5 equals 135."

SECTION 8.7

Sample Problems

Problem 1 Evaluate $g(x, y) = -3x - 7y$ for $g(-4, 11)$.

Solution Substitute -4 for x and 11 for y in the formula.
$$g(x, y) = -3x - 7y$$
$$g(-4, 11) = -3(-4) - 7(11)$$
$$= 12 - 77$$
$$= -65$$

Problem 2 If $f(x) = 8x - 43$, find x so that $f(x) = 97$.

Solution
$$f(x) = 8x - 43$$
Find x so that $\quad f(x) = 97$
Substitute $8x - 43$ for $f(x)$ $\quad 8x - 43 = 97$
and solve $\quad\quad\quad\quad\quad +43 \;\; +43$
$$8x = 140$$
$$\frac{8x}{8} = \frac{140}{8}$$
When $x = 17.5$, $f(x) = 97$. $\quad x = 17.5$

Problem 3 How much is the total simple interest received if $5000 is invested for 3 years at an annual interest rate of 8%?

Solution The total interest, I, depends on the amount of principal invested, P; the interest rate, r; and the number of years, t. The formula is $I = Prt$. As a function we write $I(P, r, t) = Prt$.

In this problem, $\quad I(5000, 0.08, 3) = (5000)(0.08)(3)$
$$= 1200$$

The total interest is $1200.

Problem 4 Tandy is a salesperson at TV Shack. She is paid $4 per hour and an 8% commission on whatever she sells. Write a function to express Tandy's earnings.

Solution Since we are computing earnings, let's call this function E. We need two inputs to compute the salary.

Let h = the number of hours Tandy works and
let d = the total dollar amount of Tandy's sales.

To compute Tandy's earnings, we use the formula $4h + 0.08d$.
Now we can write the function $E(h, d) = 4h + 0.08d$.

Assignment Guide

Basic
Day 1 7, 8, 13, 20, 24, 29, 32, 34, 37
Day 2 9–11, 17, 19, 26–28, 30, 35

Average or Heterogeneous
Day 1 7–9, 12–14, 18, 20, 21, 27, 29, 30, 32, 36, 37
Day 2 10, 11, 17, 19, 22, 24, 26, 28, 31, 33, 34, 38

Honors
Day 1 7, 8, 11, 14, 16, 18, 21, 22, 26, 27
Day 2 9 (a–c), 10, 12, 13, 17, 19, 23, 24, 29, 30, 36–38

Strands

Communicating Mathematics	1, 3, 4, 10, 14, 15, Investigation
Geometry	13, 17, 26, 31, 33, 36, 38
Measurement	4, 13, 15, 17, 33, 36
Data Analysis/ Statistics	14
Technology	6, 25, 29, 33
Algebra	1–29, 32–36, 38
Patterns & Functions	1–4, 6–10, 13–15, 17–28
Number Sense	30, 37
Interdisciplinary Applications	8, 10–12, 16, 18, 20, Investigation

Individualizing Instruction

Learning-disabled students
Students may mistakenly think of the name of a function, such as f, as a variable. Emphasize that the names of functions are not variables in that they don't stand for number values. Have students consider the parallels between using f as a function name and x as a variable and P as a function name (where P stands for *perimeter*) and h as a variable (where h stands for *height*).

SECTION 8.7

Problem-Set Notes

for pages 416 and 417

- **3** helps to clarify the uses of parentheses that students encounter.
- **14** can lead to the graphing of functions and the concept of slope.
- **19** deals with a concept that often confuses students.

Checkpoint

1. Given the function $f(x) = 4x + 1$, evaluate $f(3)$.
2. Use the function $f(x, y) = 2x^2 - 3x + 4y$ to find $f(1, -3)$.
3. Given the function $f(x) = x^2 - 3x + 1$, evaluate $f(-2)$.
4. Evaluate the function $f(x, y) = 3x + 4y$ for $f(2, 7)$.
5. Write a function for the perimeter of a rhombus.

Answers:
1. 13   2. –13   3. 11   4. 34
5. Possible answer: $P(a) = 4a$

Additional Practice

1. Given the function $f(x) = 3x^2 - 4x + 5$, evaluate $f(4)$.
2. Use the function $f(x, y) = x^2 + 4y - 2$ to find $f(-1, 4)$.
3. Given the function $f(x) = x^2 + 4x - 9$, evaluate $f(2)$.
4. Evaluate the function $f(x, y) = 2x^3 - 3y + 8$ for $f(2, -4)$.
5. Write a function for the area of a rectangle.

Answers:
1. 37   2. 15   3. 3   4. 36
5. Possible answer: $A(\ell, w) = \ell w$

Think and Discuss

P   **1** What is the difference between a formula and a function?

P   **2** Given the function $f(x) = 2x + 7$, evaluate
     a $f(3)$ 13   **b** $f(6)$ 19   **c** $f(9)$ 25   **d** $f(3 + 6)$ 25

P, R   **3** Explain the different meanings of parentheses in each of the following.
     a $3(5)$ **b** $f(5)$ **c** $4(5 + 2)$

P, R   **4** For the figure shown, $V(r, h) = \pi r^2 h$.
     a What do V, r, and h represent?
     b Compute $V(3, 4)$. 36π, or ≈ 113.1
     c Compute $V(4, 3)$. 48π, or ≈ 150.8
     d Explain the meaning of the result in parts **b** and **c**.
     The results are the volume.

P, R   **5** Define the operation $\otimes$ as $a \otimes b = a + 2b - 3$. Evaluate $5 \otimes 7$.   16

P   **6** Use the function $f(x) = x + 2x + 3x + 4x + 5x + 6$ and find
     a $f(0)$ 6   **b** $f(1)$ 21   **c** $f(2)$?₅   **d** $f(7.36)$ 116.4

Problems and Applications

P   **7** Use the function $f(x, y) = x^2 + y$ and find
     a $f(1, 0)$ 1   **b** $f(0, 1)$ 1   **c** $f(2, 3)$ 7   **d** $f(3, 2)$ 11

P   **8 Sports** Play Skicker is a field-goal and extra-point kicker for his football team. He tries F field goals and E extra points during the season. He is 65% successful on field goals and 98% successful on extra points. Write a function P for his point total during the season. [Note: A field goal is worth three points and an extra point is worth one point.] $P(F, E) = 0.65(3F) + 0.98E$

P, R   **9** Evaluate the functions $f(x) = 3x + 4$ and $g(x) = x^2 + x$ for each of the following values.
     a $f(5)$ 19   **b** $g(5)$ 30   **c** $f(1) + g(3)$ 19   **d** $g(1) + f(3)$ 15
     e $g(1.5)$ 3.75   **f** $f\left(\frac{1}{3}\right)$ 5   **g** $g(0)$ 0   **h** $f(0)$ 4

P, R   **10** If *Batting Average (At Bats, Hits)* = $\frac{Hits}{At\ Bats}$, find the value of *Batting Average* (20, 7). Explain what your answer means.

P, R   **11** Find the interest received if $6000 is invested for one year at 6% simple interest. $360

P   **12 Finance** How much simple interest is received from an investment of $8000 for 4 years at an annual interest rate of 8%? $2560

416   Chapter 8   The Language of Algebra

SECTION 8.7

P, R **13** Write a function for the perimeter of each figure. **b** $P(\ell, w) = 2\ell + 2w$
    a Square $P(s) = 4s$ **b** Rectangle   **c** Parallelogram $P(b, s) = 2b + 2s$

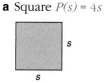

    d Kite $P(s_1, s_2) = 2s_1 + 2s_2$ **e** Isosceles trapezoid $P(b_1, b_2, s) = b_1 + b_2 + 2s$

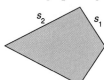
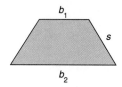

P, R, I **14** Complete the following table for $f(x) = 3x + 5$.

x	−5	−4	−3	−2	−1	0	1	2	3	4	5	6
f(x)	−10	−7	−4	−1	2	5	8	11	14	17	20	23

Explain the pattern in the results.

P, R **15** Write a function for the volume of a box. Be sure to explain what is represented by any letters you use.

P, R **16** **Business** How much is the commission on $1700 of sales if the commission rate is 4%? $68

P, R **17** Write a function for the area of each figure.
    a Square $A(s) = s^2$ **b** Triangle $A(b, h) = \frac{1}{2}bh$ **c** Parallelogram $A(b, h) = bh$

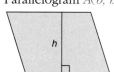
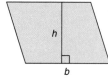

    d Kite $A(d_1, d_2) = \frac{1}{2}d_1 \cdot d_2$ **e** Trapezoid $A(b_1, b_2, h) = \frac{1}{2}h(b_1 + b_2)$

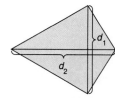
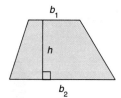

P **18** Biff is a waiter at the Let Us Eat Lettuce restaurant. He is paid $5.50 per hour and receives tips that average 18% of each bill. If he works h hours and earns t dollars in tips, write a function for how much he earns.

P **19** For the functions $f(x) = 3x^2$ and $g(x) = (3x)^2$,
    a Evaluate $f(4)$ and $g(4)$ $f(4) = 48$, $g(4) = 144$
    b Evaluate $f\left(\frac{1}{3}\right)$ and $g\left(\frac{1}{3}\right)$ $f\left(\frac{1}{3}\right) = \frac{1}{3}$, $g\left(\frac{1}{3}\right) = 1$
    c Evaluate $f(-2)$ and $g(-2)$ $f(-2) = 12$, $g(-2) = 36$
    d Find the value(s) of x so that $f(x) = g(x)$ 0

Stumbling Block

Students may have trouble making substitutions properly. They may replace the coefficient or lose it entirely, or they may not use parentheses around a substitution of two or more terms, preventing them from seeing that the distributive property needs to be used. Modeling some of the more difficult problems for students will be helpful.

Additional Answers

for pages 416 and 417

1 A function uses a formula to assign an output value to an input value.

3 a The parentheses indicate multiplication.
b The parentheses enclose the input of a function.
c The parentheses indicate the order of operations—addition before multiplication.

4 a V = volume, r = radius, h = height

10 0.35; It is a ratio—for every 20 times at bat, there will be 7 hits.

13 b $P(\ell, w) = 2\ell + 2w$

14 Possible answer: Each answer is three greater.

15 $V(\ell, w, h) = \ell w h$, where ℓ represents length, w represents width, h represents height

18 $F(h, t) = 5.50h + t$

SECTION 8.7

Problem-Set Notes

for pages 418 and 419

- 21 relates pattern recognition with function ideas.
- 28 will allow you to see if students understand the notation.

Reteaching

Students may not readily accept that $f(x)$ does not mean f times x. You might find it helpful to define $f(x)$ as "the function f evaluated for x." So, for $f(x) = 3x + 2$, $f(4)$ means "$3x + 2$ evaluated for $x = 4$," which is a problem students should recognize.

Communicating Mathematics

- Have students write a few sentences to answer the following questions: What is a function? What does the notation $g(x)$ mean?
- Problems 1, 3, 4, 10, 14, 15, and Investigation encourage communicating mathematics.

Multicultural Note

Galileo is given credit for realizing that past scientific thinking had not advanced science very far by the sixteenth century. He proposed to replace such questions as "Why does a ball fall?" with quantitative descriptions, such as "the distance a ball falls from its starting point increases as time elapses from the moment it was dropped." This way of representing a relationship between variables is an example of a function.

P   **20** Write a formula to represent the function Value(nickels, dimes, quarters). $V(n, d, q) = 0.05n + 0.10d + 0.25q$

P, R, I   **21** For each of the following inputs, the function f gives the following outputs. Write function f for input x and output y.

Inputs	Output	Inputs	Output
(3, 4)	10	(1, 5)	7
(0, 4)	4	(0, 11)	11
(6, 1)	13	(5, 0)	10

$f(x, y) = 2x + y$

P, R   **22** Evaluate $f(2)$ if $f(x) = \begin{bmatrix} 2x & 3x-6 \\ x^2 & 5-x \end{bmatrix} \begin{bmatrix} 4 & 0 \\ 4 & 3 \end{bmatrix}$

P   **23** Solve each of the following for y if $g(y) = 3y - 11$.
    a $g(y) = 49$ $y = 20$ **b** $g(y) > 4$ $y > 5$

P   **24** If $f(x) = 3x - 2$ and $g(x) = 4^x$, find
    a $f(5)$ 13   **b** $g(3)$ 64   **c** $f(5) + g(3)$ 77

P   **25** Evaluate the function $f(x) = 8(3^x)$ for each input.
    a $f(2)$ 72   **b** $f(0)$ 8
    c $f(1.7)$ ≈ 51.8 **d** $f(2.6)$ ≈ 139.2

P, R   **26** Write a function $L(a, b)$ for the length of segment PQ.
$L(a, b) = \sqrt{a^2 + b^2}$

P, R   **27** If $A(b, h) = \frac{1}{2} bh$, and $A(b, 6) = 84$, find b. $b = 28$

P   **28** The function $M(x + 2) = 3x$. Evaluate $M(5)$.   9

◀ LOOKING BACK   **Spiral Learning**   LOOKING AHEAD ▶

R   **29** **Spreadsheets**   Make a spreadsheet that resembles the diagram.

Use the spreadsheet to solve $x^2 + 88 = 21x - 20$ by copying down until the values in columns B and C are equal.
(Hint: There are two answers.)

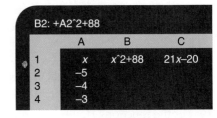

R, I   **30** List the set of all even integers between -7 and 3. $I = \{-6, -4, -2, 0, 2\}$

418   Chapter 8   The Language of Algebra

SECTION 8.7

R **31** What set of points lies in both triangle ABE and quadrilateral DEBC?

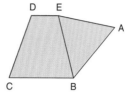

R **32** Write an expression for
  a The sum of p and q $p+q$
  b The product of p and q pq
  c The quotient of p and q $\frac{p}{q}$
  d The difference of p and q $p-q$ or $q-p$

P, R, I **33** The sides of a quadrilateral have a ratio of 3:6:8:9. The quadrilateral's perimeter is 91. Find the lengths of the sides.   10.5, 21, 28, 31.5

P, R, I **34** If $8x-3y=1$ and $y=3x-2$, find x and y. $x=5, y=13$

R **35** Evaluate each expression.
  a $\frac{-3}{8} \cdot \frac{5}{7} \cdot \frac{-7}{11}$ $\frac{15}{88}$
  b $\frac{-3}{5} + \frac{1}{2}$ $-\frac{1}{10}$
  c $-8 \cdot \frac{-4}{5} \cdot \frac{-1}{16}$ $-\frac{2}{5}$
  d $\frac{-2}{3} - \frac{3}{4}$ $-\frac{17}{12}$ or $-1\frac{5}{12}$

R, I **36** The perimeter of the figure is less than 25. Write an inequality for x and solve the inequality. $25 > 18 + x; 0 < x \leq 7$

R, I **37** How many integers are between $-\sqrt{51}$ and $\sqrt{55}$ on the number line?   15

P, R, I **38** $\triangle ABC$ is reflected across the y-axis. What are the coordinates of the new vertices?
  $A' = (-2, 7), B' = (-2, 3), C' = (-5, 7)$

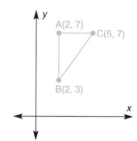

Investigation

Rent Costs What is the monthly rent for an apartment in your area? The answer certainly depends on the size of the apartment. (We can say that the monthly rent is a function of its size.) But the cost also depends on many other factors.

Report on the range in rents for typical apartments or houses in your area. Include evidence such as newspaper ads. List all the factors you can think of that influence rent prices.

Section 8.7 Functions

SUMMARY

CHAPTER 8

Summary

CONCEPTS AND PROCEDURES

After studying this chapter, you should be able to
- Recognize equivalent expressions (8.1)
- Add and subtract like terms (8.1)
- Use the Distributive Property of Multiplication over Addition and Subtraction (8.1)
- Use expressions to describe phrases and situations (8.2)
- Use equations to represent problems (8.2)
- Use inverse operations to solve equations (8.3)
- Solve equations with several operations (8.3)
- Solve absolute-value equations (8.4)
- Solve equations with square roots (8.4)
- Simplify expressions to solve equations (8.4)
- Use addition and subtraction to solve an inequality (8.5)
- Use multiplication and division to solve an inequality (8.5)
- Use substitution to solve equations (8.6)
- Use a quantity as a single term (8.6)
- Recognize a function (8.7)
- Calculate the value of a function (8.7)

VOCABULARY

equivalent expressions (8.1)
function (8.7)
input (8.7)
like terms (8.1)
numerical term (8.1)
output (8.7)
simplify (8.1)

REVIEW

CHAPTER 8

Review

1. Evaluate each expression for $x = -7$ and $y = 12$.
   a. $3(x + 2y) - 2(x - 3y)$ 137
   b. $x + 12y$ 137
   c. $2x^2 + y$ 110
   d. $\dfrac{(2x)^2}{y}$ $\dfrac{49}{3}$

2. Solve each equation and simplify each expression.
   a. $3y - (2 - 3y) = 40$ $y = 7$
   b. $3y - (2 - 3y) - 40$ $6y - 42$
   c. $\dfrac{1}{3}(6a - 5) + \dfrac{2}{3}$ $2a - 1$
   d. $\dfrac{1}{3}(6a - 5) = 7$ $a = \dfrac{13}{3}$ or ≈ 4.3

3. Add $\begin{bmatrix} 4x & 2y & 3z \\ 5x & y & -z \\ 7x & 3y & 2z \end{bmatrix} + \begin{bmatrix} x & 3y & 2z \\ 0 & 4y & 6z \\ -2x & 2y & 2z \end{bmatrix}$ $\begin{bmatrix} 5x & 5y & 5z \\ 5x & 5y & 5z \\ 5x & 5y & 4z \end{bmatrix}$

4. Complete the table.

x	−18	−15	−13	−5	0	3	8
8x − 4x + 2x − 6x	0	0	0	0	0	0	0

5. If $x = -3$ and $y = \dfrac{2}{3}$, which of the following are true?   b, c, d
   a. $3x + 4y = 12xy$
   b. $x + 2x + 3x + 4x = 10x$
   c. $2x + 2y = 2(x + y)$
   d. $xy = 1 + x$

6. Find x. $x = 3$

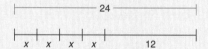

7. Simplify each expression.
   a. $3 + x + 4x + 5$ $5x + 8$
   b. $2x^2 + x + x^2 + 3$ $3x^2 + x + 3$

8. Write an expression, an equation, or an inequality for each of the following.
   a. 17 less than four times a number $4x - 17$
   b. The square of a number is less than 25 $x^2 < 25$
   c. The average of x and 12 $\dfrac{x + 12}{2}$
   d. The area of a given rectangle is numerically the same as its perimeter.

9. Write a sentence or phrase to represent each of the following.
   a. $3x + 9$ 9 more than 3 times a number
   b. $3(x + 9)$ 3 times the sum of a number and 9
   c. $\dfrac{2}{3} = \dfrac{5}{x}$ $\dfrac{2}{3}$ is 5 divided by a number

10. Carlo sells cameras at Photo World. He earns $15 per hour plus a 10% commission on camera purchases. Find his earnings if he
    a. Works 8 hours and sells $350 worth of cameras $155
    b. Works b hours and sells $$d$ worth of cameras $15b + 0.10d$

Assignment Guide

Basic
Day 1 1–3, 7–9, 11, 16, 22, 25, 31
Day 2 4–6, 10, 12, 18, 28, 29, 32
Day 3 Chapter Test 1–10, 16–22

Average or Heterogeneous
Day 1 1 (a, c), 2 (a, c), 3, 6–12, 14, 16, 18, 19, 21 (b, c), 22, 24 (a, b), 27, 29, 32

Honors
Day 1 1–4, 6, 8–19, 21–24, 26, 27, 30, 32–34

Strands

Communicating Mathematics	29, 33
Geometry	8, 11, 16, 17, 19, 22, 23, 33
Measurement	11, 13, 14, 16, 17, 19, 22, 23, 33
Data Analysis/ Statistics	8, 12, 15
Algebra	1–34
Patterns & Functions	4, 12, 19, 22, 31, 32
Number Sense	24, 25, 34
Interdisciplinary Applications	10, 13, 14

Additional Answers

for page 421

8 d $\ell w = 2(\ell + w)$

Chapter Review

REVIEW

Additional Answers

for pages 422 and 423

16 a Possible answer:
$6x + 20 + 40 = 180$

29 Treat $(2x + 4)$ as a unit and combine like terms, or multiply to remove parentheses, then combine like terms.

30 It must be greater than 9.

11 Find the perimeter of the triangle in terms of x and y.
$P = 6x + 18y + 20$

12 Write a function $Ave(a, b, c)$ to find the mean average of the numbers a, b, and c. $Ave(a, b, c) = \frac{a+b+c}{3}$

13 The appliance repairer in Pleasantville charges $30 for a service call plus $20 per hour. Find the total cost for a service call if the repairer works
   a $2\frac{1}{2}$ hours $80
   b h hours $30 + 20h$

14 A train travels 45 miles per hour for t hours and 50 miles per hour for 4 hours. Find the value of t if the entire trip covered 520 miles. ≈ 7.1

15 The mean of 80, 64, 94, 96, and x is 84. Find the value of x.   86

16 a Write an equation to represent the measures of angles ABD and DBC.
   b Solve for x. $x = 20$
   c Determine the measure of $\angle ABD$.   140

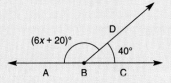

17 The measures of the four angles of a quadrilateral have a ratio of 3:4:5:6. Find the measures of the angles.   60, 80, 100, 120

18 Solve each equation.
   a $40 - 4x = -32$ $x = 18$
   b $\frac{8}{5}x + \frac{4}{5} = \frac{2}{5}$ $x = -0.25$
   c $-6(y + 4) = 39$ $y = -10.5$

19 In the triangle, AB = BC. Write a function, in simplest form, for the perimeter of the triangle.
$P(AB, AC, BC) = 13x + 11$

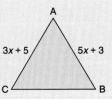

20 Solve $9(3y - 2) + 4(3 - 6y) = \frac{14+22}{4}$ for y. $y = 5$

21 Solve each equation for x.
   a $|x| + 3 = 8$ $x = 5$ or $x = -5$
   b $|x + 3| = 8$ $x = 5$ or $x = -11$
   c $|x| + 8 = 3$ No solution
   d $|x + 8| = 3$ $x = -5$ or $x = -11$

22 Write a function V for the volume of a cone with base radius r and height h. $V(r, h) = \frac{1}{3}\pi r^2 h$

422   Chapter 8   The Language of Algebra

REVIEW

23 The perimeter of the triangle is 40. Find the value of x. $x = 6$ or $x = -6$

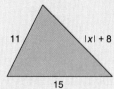

24 Solve each equation for x.
  a $\sqrt{x} = 81$ $x = 6561$ **b** $x^2 = 81$ $x = 9$ or $x = -9$
  c $\sqrt{x} = -81$ No real solution   **d** $x^2 = -81$ No real solution

25 Write the symbol, >, <, or =, that should be in each ☐.
  a $\frac{7}{8}$ ☐ $\frac{8}{7}$ <   **b** $\frac{-3}{5}$ ☐ $\frac{-5}{3}$ >
  c 0.125 ☐ $\frac{1}{8}$ =   **d** 40% ☐ $\frac{3}{5}$ <

26 Solve $3x + y > -10$ for x if $y = 5$. $x > -5$

27 Refer to the diagram.
  a Write an inequality to represent the diagram. $2x + 33 > 99$
  b Solve the inequality. $x > 33$

28 Solve $3x + 8y = 39$ for y if $x = 5$. $y = 3$

29 What two different methods can be used to simplify the expression $5(2x + 4) - 18(2x + 4) + 7(2x + 4) + 9(2x + 4)$?

30 Two numbers have a ratio of 4:3. The sum of the two numbers is greater than 21. What must be true about the smaller of the two numbers?

31 Evaluate the functions $f(x, y) = 4x - 7y$, $g(x) = x^2$, and $h(x) = 3^x$ for each of the following inputs.
  a $f(3, 1)$ 5   **b** $f(1, 3)$ -17 **c** $g(4)$ 16   **d** $g(-4)$ 16
  e $h(2)$ 9   **f** $h(-2)$ $\frac{1}{9}$ **g** $f(7, 4)$ 0   **h** $f(4, 7)$ -33

32 The function $f(x, y) = \begin{bmatrix} 6 \\ 2 \end{bmatrix} + \begin{bmatrix} x \\ y \end{bmatrix}$. Evaluate $f(3, -4)$. $\begin{bmatrix} 9 \\ -2 \end{bmatrix}$

33 Describe what happens to the value of a if b is increased by 3 and the perimeter stays the same. a is decreased by 6

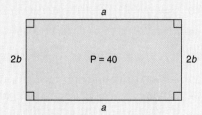

34 a If $x < \frac{2}{3}$, is it always true that $\frac{1}{x} > \frac{2}{3}$?   No

  b If $x > \frac{2}{3}$, is it always true that $\frac{1}{x} < \frac{2}{3}$?   No

Chapter Review

CHAPTER 8

Test

In problems 1–3, simplify each expression.

1. $4x + 6 - 3x - 9$ $x - 3$
2. $3(2x + 5) + 4(7x - 1)$
3. $-4(3x - 2y)$ $-12x + 8y$

4. Add $\begin{bmatrix} 3xy & -2x & -4y \\ -5x & 5xy & -6y \end{bmatrix} + \begin{bmatrix} -4xy & -6x & 3y \\ 7y & -3xy & 6x \end{bmatrix}$. $\begin{bmatrix} -xy & -8x & -y \\ -5x+7y & 2xy & -6y+6x \end{bmatrix}$

5. Write an algebraic expression that represents the phrase "three less than a number." $n - 3$

6. Write an equation that shows that the area of the rectangle is equal to the area of the triangle. $(x+6)(x+1) = \frac{1}{2}(x+2)(4x)$

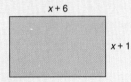
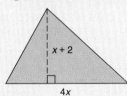

In problems 7–12, solve each equation.

7. $17x = 20.4$ $x = 1.2$
8. $2b + 1 = 5b - 9$ $b = \frac{10}{3}$
9. $2x + x = 15$ $x = 5$
10. $5(3x + 4) = 125$ $x = 7$
11. $|a - 2| = 8$
12. $\sqrt{m} + 2 = 9$ $m = 49$

In problems 13–15, solve each inequality.

13. $3x + 1 < 10$ $x < 3$
14. $8 - 3x \geq 17$ $x \leq -3$
15. $5(2x - 7) > 120$ $x > 15.5$

16. If Megan sells from 10 to 20 items at $50 each, she earns a 6% commission. What is the possible range of Megan's commission in this situation? $\$30 \leq c \leq \$60$

17. If $3x - 4y = 18$ and $y = -3$, solve for x. $x = 2$

18. If $m:n = 5:3$, solve the equation $2m + 3n = 76$ for m. $m = 20$

19. If $3x + 4y = 17$ and $y = x - 1$, solve for x and y. $x = 3; y = 2$

20. If $f(x, y) = x^2 - 2xy + y^2$, evaluate $f(2, -1)$. 9

21. A $4000 investment is made for two years at a 6% annual interest rate.
    a. Write a function using one variable for this situation.   Possible answer: $I(P) = 0.12P$
    b. What will be the actual dollar amount of interest earned? $480

22. Find the value of x that makes the perimeter of the rectangle 40. $x = 3$

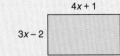

PUZZLES AND CHALLENGES

1 Find an English word in which all five vowels (a, e, i, o, u) appear in alphabetical order, and each vowel appears only once.

2 Find a common English word in which three pairs of double letters appear one after another.

3 Continue this pattern for two more rows.

```
          1
        1   1
       1  2  1
      1  3  3  1
     1  4  6  4  1
    1  5 10 10  5  1
```

4 Carl is standing by a river with a 7-quart bottle and a 4-quart bottle. Explain how he can bring home exactly 5 quarts of water.

5 Interpret the phrase each puzzle suggests.

a

b

c S r P a A i l n N

d SAFE
   FIRST

8. Extend problem 15 on page 396 as follows: Design a spreadsheet application to find the cost of producing open-top cubical containers with lengths ranging from 1 to 6 feet in increments of 1 foot. Find the volume of each container. Which container has the lowest cost per cubic foot of storage? Explain your answer.

9. Savings can be expressed as the following function:
$S(i, e) = i - e$, where i is income and e is expenses. Use this function to determine your savings at the end of one month. To do this, you'll have to keep exact records of your income (from allowance, jobs, etc.) and your expenses (for movies, school supplies, etc). Show the results of your record keeping in some sort of graph. Does your research indicate any need for you to change your spending habits?

10. Write a function that can be used to calculate the score of a football game. Describe what each of the variables stands for and the possible values the variable can have. Use your function to explore the following questions: Of the numbers from 1 to 20, which could not be football scores? Which numbers are not likely to be football scores? Explain your answers.

Additional Answers

for Chapter 8 Test

2 $34x + 11$

11 $a = 10$ or $a = -6$

Answers for Puzzles and Challenges are on page A8.

OVERVIEW

LONG-TERM INVESTIGATIONS

See *Investigations* book, Unit E, pages 108-133.

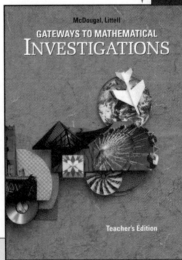

Gateways to Mathematical Investigations is a collection of long-term, real-world investigations that provides students with exciting opportunities to apply the ideas and concepts introduced in *Gateways to Algebra and Geometry: An Integrated Approach*. Unit E: Playing with Numbers consists of five investigations that explore properties of real numbers. This unit is designed so that you can pick and choose the order and number of investigations you would like to do with your students. Each investigation is an exploration that takes several days to complete. Refer to the menu at the right to learn more about the investigations that make up this unit.

As your students work through the Unit E investigations, use Chapters 9 and 10 of *Gateways to Algebra and Geometry: An Integrated Approach* for concept and skill development, and as a source of rich problems, exercises, and applications. When appropriate, refer students to additional supportive sections in *Gateways to Algebra and Geometry* as well.

UNIT E: Playing with Numbers

MENU OF INVESTIGATIONS

OVERVIEW: Groups apply mathematical concepts and strategies from Chapters 9 and 10 within the context of exploring the properties of real numbers.

Prime Time Mathematics

(2-3 days)
Working in groups, students investigate prime numbers to study how they occur, and to test conjectures about primes.
STRAND
Number
UNIFYING IDEA
Patterns and Generalizations

Billiard Table Mathematics

(3-5 days)
Working on grid paper, students investigate the behavior of a billiard ball bouncing around a billiard table and find rules explaining the patterns that a ball traces on the table.
STRANDS
Number, Geometry
UNIFYING IDEAS
Patterns and Generalizations, Multiple Representations, Proportional Relationships

Terminators and Repeaters

(4-6 days)
Groups investigate the nature of fractions and their decimal representations and identify how and why some decimals terminate while others repeat.
STRAND
Number
UNIFYING IDEA
Patterns and Generalizations

Building Irrational Numbers

(2-3 days)
Working in groups, students construct spirals of right triangles as a way of physically creating irrational numbers. As they build, students apply the Pythagorean Theorem to calculate measures and use rulers and calculators to understand their results.
STRANDS
Number, Measurement, Algebra, Geometry, Functions
UNIFYING IDEAS
Patterns and Generalizations, Multiple Representations

Taxicab Geometry

(3-5 days)
Students calculate distances based on the real world geometry of a city street as well as on the abstract geometry of the Euclidean plane. Connections are made between the two geometries as a way to strengthen understandings.
STRANDS
Algebra, Geometry, Measurement
UNIFYING IDEAS
Patterns and Generalizations, Multiple Representations

INVESTIGATIONS FOR UNIT E	SCHEDULE	OBJECTIVES
1. **Prime Time Mathematics** Pages 111-114*	**2-3 Days**	• To generate prime numbers using the sieve of Eratosthenes; to develop conjectures regarding prime numbers
2. **Billiard Table Mathematics** Pages 115-118*	**3-5 Days**	• To use grid paper to investigate how billiard balls move around a billiard table to form conjectures about rectangles and numbers; to discover mathematical ideas regarding factors and multiples.
3. **Terminators and Repeaters** Pages 119-123*	**4-6 Days**	• To construct tables displaying the decimal representations for the first 50 unit fractions, 1/1, 1/2, . . . 1/50, and their decimal representations; to study patterns in the decimal representations; to formulate and test conjectures on other unit fractions
4. **Building Irrational Numbers** Pages 124-127*	**2-3 Days**	• To use the Pythagorean Theorem, a protractor, a compass, and a ruler as the means for constructing and approximating irrational numbers
5. **Taxicab Geometry** Pages 128-131*	**3-5 Days**	• To use a compass and point plotting to study the concept of distance in the standard Euclidean geometry and in taxicab geometry

*Pages refer to *Gateways to Mathematical Investigations* Teacher's Edition.

PLANNING LONG-TERM INVESTIGATIONS

MATERIALS	RELATED RESOURCES
• Family Letter, 1 per student • Calculator, 1 per group • *Gateways to Mathematical Investigations* Student Edition, pp. 39-40, 1 of each per student	**Gateways to Algebra and Geometry** Section 10.2, pp. 494-500: Primes, Composites, and Multiples Section 10.4, pp. 507-512: Prime Factorization Section 10.7, pp. 528-535: Number-Theory Explorations *Additional References* Teacher's Resource File: Student Activity Masters
• Grid paper, 3 or more sheets per student • Ruler, 1 per student • Poster board, protractor, 1 per student (optional) • Overhead projector (optional) • Transparency, 1 per group (optional) • *Gateways to Mathematical Investigations* Student Edition, pp. 41-43, 1 of each per student	**Gateways to Algebra and Geometry** Section 10.2, pp. 494-500: Primes, Composites, and Multiples Section 10.3, pp. 501-506: Divisibility Tests Section 10.4, pp. 507-512: Prime Factorization Section 10.5, pp. 513-520: Factors and Common Factors Section 10.6, pp. 521-527: LCM and GCF Revisited *Additional References* Teacher's Resource File: Student Activity Masters
• Calculator, at least 1 per group • Computer, spreadsheet software (optional) • Blank spreadsheet, 1 per student • *Gateways to Mathematical Investigations* Student Edition, pp. 44-45, 1 of each per student	**Gateways to Algebra and Geometry** Section 9.3, pp. 442-447: Important Sets of Numbers Section 10.2, pp. 494-500: Primes, Composites and Multiples Section 10.4, pp. 507-512: Prime Factorization Section 10.5, pp. 513-520: Factors and Common Factors *Additional References* Teacher's Resource File: Student Activity Masters
• Calculator, protractor, 1 per group • Ruler, compass, 1 per student • Poster board, 1 per group (optional) • Markers, 1-3 per group (optional) • Overhead projector (optional) • Transparencies, 2-3 per group (optional) • *Gateways to Mathematical Investigations* Student Edition, pp. 46-47, 1 of each per student	**Gateways to Algebra and Geometry** Section 9.1, pp. 428-434: Sets Section 9.2, pp. 435-441: Intersection and Union Section 9.3, pp. 442-447: Important Sets of Numbers Section 10.7, pp. 528-535: Number-Theory Explorations *Additional References* Section 2.3, pp. 64-70: Geometric Formulas Section 3.6, pp. 129-135: Constructions with Ruler and Protractor Teacher's Resource File: Student Activity Masters
• Grid paper, several per student • Ruler, 1 per student • Compass, 1 per student • *Gateways to Mathematical Investigations* Student Edition, pp. 48-49, 1 of each per student	**Gateways to Algebra and Geometry** Section 9.1, pp. 428-434: Sets Section 9.4, pp. 448-454: Pi and the Circle Section 9.6, pp. 462-468: Two Coordinate Formulas Section 10.1, pp. 486-493: Techniques of Counting *Additional References* Teacher's Resource File: Student Activity Masters

See pages 484A and 484B: Assessment for Unit E.

CHAPTER 9

Contents

9.1 Sets
9.2 Intersection and Union
9.3 Important Sets of Numbers
9.4 Pi and the Circle
9.5 Properties of Real Numbers
9.6 Two Coordinate Formulas
9.7 Graphs of Equations

Pacing Chart

Days per Section

	Basic	Average	Honors
9.1		*	1	1
9.2		*	2	1
9.3		*	2	1
9.4		*	2	2
9.5		*	2	2
9.6		*	2	1
9.7		*	3	2
Review	–	1	1
Assess	–	2	2
	–	17	13

* Optional at this level

Chapter 9 Overview

- Set theory allows us to clarify many important ideas by referring to more general, underlying ideas. The terms *intersection*, *union*, and *subset* make clearer the concepts of solution sets of inequalities, greatest common factor, and least common multiple. Set notation enables us to avoid verbosity by saying precisely what we mean.

- The application of set theory to the real numbers is significant because many problems involve only a subset of the real numbers. Students need to be familiar with the terminology in order to understand how the statement of a problem places restrictions on the solution.

9 Real Numbers

CHAPTER 9

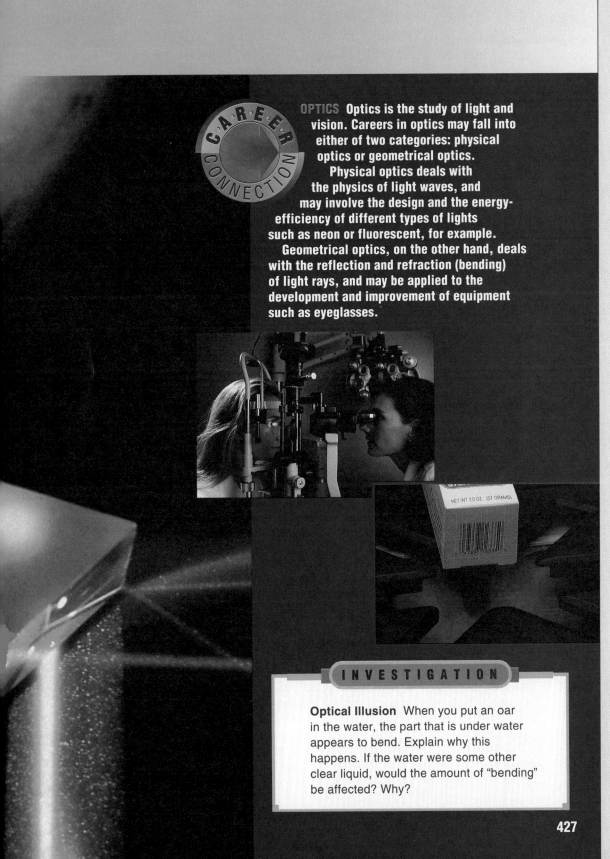

CAREER CONNECTION

OPTICS Optics is the study of light and vision. Careers in optics may fall into either of two categories: physical optics or geometrical optics. Physical optics deals with the physics of light waves, and may involve the design and the energy-efficiency of different types of lights such as neon or fluorescent, for example. Geometrical optics, on the other hand, deals with the reflection and refraction (bending) of light rays, and may be applied to the development and improvement of equipment such as eyeglasses.

INVESTIGATION

Optical Illusion When you put an oar in the water, the part that is under water appears to bend. Explain why this happens. If the water were some other clear liquid, would the amount of "bending" be affected? Why?

■ While discussing irrational numbers, we deal with perhaps the most famous one, π. We review some number properties to enable us to discuss them within a problem-solving context, and we conclude by introducing graphs of equations.

Career Connection

There are many careers involving optics, including optometry and lighting design. Have students discuss how optics affects their daily lives, mentioning items such as eyeglasses, contact lenses, full-spectrum fluorescent lighting, and high-intensity mercury- and sodium-vapor streetlights. Even though students are not involved in manufacturing these products, they will probably encounter them at some point and may need to know how to choose among various optical products.

Video Connection

For more information about the career discussed here, please refer to the Futures videocassette #4, program B: *Optics*, featuring Jaime Escalante. For more information or to order a copy of the program, call PBS VIDEO at 800/344-3337.

Investigation Notes

The image appears to bend due to refraction. When a light beam enters a more dense substance, one side of the beam is slowed down before the other side, causing the light to bend. Liquids with different densities bend the light different amounts.

SECTION 9.1

Objectives

- To represent a set of numbers by listing, by using set-builder notation, and by graphing
- To identify sets as subsets of other sets

Vocabulary: element, empty set, set, set-builder notation, solution set, subset

Class Planning

Schedule 1 day

Resources

In *Chapter 9 Student Activity Masters* Book
Class Openers 1, 2
Extra Practice 3, 4
Cooperative Learning 5, 6
Enrichment 7
Learning Log 8
Interdisciplinary Investigation 9, 10
Spanish-Language Summary 11

Class Opener

List the items described by each set. Use braces around your list.
1. {whole numbers between $3\frac{1}{2}$ and $8\frac{1}{2}$}
2. {the four seasons}
3. {the first four presidents of the United States}
4. {multiples of 10 between 17 and 77}
5. {integers between –9 and –3, inclusive}

Answers:
1. {4, 5, 6, 7, 8}
2. {spring, summer, autumn, winter}
3. {Washington, Adams, Jefferson, Madison}
4. {20, 30, 40, 50, 60, 70}
5. {–9, –8, –7, –6, –5, –4, –3}

428

Investigation Connection
Section 9.1 supports Unit E, *Investigations* 4, 5. (See pp. 426A–426D in this book.)

SECTION 9.1
Sets

DESCRIBING SETS

A **set** is any collection of objects. Usually we will be considering sets of numbers or sets of geometric figures. Capital letters are often used to name sets. A single member of a set is called an **element** of the set. The set, I, of integers can be written as follows:

$$I = \{\ldots, -3, -2, -1, 0, 1, 2, 3, \ldots\}$$

Example 1
A cassette tape deck has a three-digit tape counter. The tape counter was set at 200 at the start of a tape, and it reads 486 at the end of the tape. In four different ways, describe the set of numbers that appeared on the tape counter during the playing of the tape.

Solution
1. In words: The counter showed every integer from 200 to 486 inclusive.
2. By listing the elements: {200, 201, 202, ... , 486}
3. By using **set-builder notation**: {x:200 ≤ x ≤ 486 and x is an integer}. This notation is read "the set of all values of x such that x is greater than or equal to 200 and less than or equal to 486 and is an integer."
4. By graphing:

198 199 200 201 202 203   484 485 486 487 488 489

Example 1 shows several different ways of representing a set. We can describe a set in words. We can make a list or a partial list that shows how the set continues. We can make a rule. We can draw a graph.

428   Chapter 9   Real Numbers

SECTION 9.1

Example 2
A circle has its center at the origin of the coordinate system, and the circle has a radius of 2 units. In three different ways, describe the set of points where the circle intersects the axes of the coordinate system.

Solution
Here are three possible ways:

1. By graphing: The points are shown on the coordinate system.
2. By listing: {(2, 0), (0, 2), (–2, 0), (0, –2)}
3. In words: The set of points on the coordinate axes are 2 units from the origin.

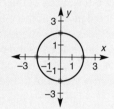

SUBSETS

Set B is a **subset** of a set A if every element in set B is also an element of set A. We write this as $B \subset A$, which is read, "B is a subset of A." Or we can write $\{1, 2\} \subset \{1, 2, 3\}$, which is read, "The set containing the elements 1 and 2 is a subset of the set containing the elements 1, 2, and 3." Every set is a subset of itself.

Example 3
Amy Planter purchases 20 feet of fence to border a rectangular garden. She wants the length of each side of the garden to be a whole number of feet.
a Find the set of possible areas of the garden.
b Find the set of possible areas for which the garden will be at least 3 feet wide.
c Find the set of possible areas greater than 30 square feet.

Solution
a The figure shows the rectangles that have dimensions that are whole numbers of feet and perimeters of 20 feet. The set of areas is {9, 16, 21, 24, 25}.
b The set of areas for which the garden is at least 3 feet wide is {21, 24, 25}.
c There are no dimensions for which the area is greater than 30 square feet. The set with no elements in it is called the **empty set**. The empty set can be written { }. The empty set is a subset of every set.

Lesson Notes

- The symbol ø, commonly used as a symbol for the empty (null) set, can easily be confused with zero, especially among computer science students who have learned to cross zeros to distinguish them from letter O's. Therefore, we will use empty braces as the symbol for the empty set. Be sure to reiterate that the empty set and the set itself are both subsets of every set.

- Parentheses around ordered pairs are different from brackets around sets. (2, 0) and (0, 2) are not equal as ordered pairs, but {2, 0} and {0, 2} are equal sets.

Additional Examples

Example 1
A digital thermometer that displays temperatures to the nearest whole number is placed in a container of chemicals. It reads 50°C at the start of heating and 175°C at the end of the heating. Use set-builder notation to describe the set of numbers that appears on the thermometer display during heating.

Answer: $\{x: 50 \leq x \leq 175,$ and x is an integer$\}$

Example 2
A square with sides of 4 units has its center at the origin. List the set of points where the square intersects the axes of the coordinate system.

Answer: {(0, 2), (0, –2), (2, 0), (–2, 0)}

Example 3
Matt Ricks has to arrange 30 chairs in at least 2 rows, with equal numbers of chairs in all rows. Find the set, B, of the possible

SECTION 9.1

Additional Examples continued

numbers of rows containing at least 3 chairs.

Answer: B = {2, 3, 5, 6, 10}

Example 4
Set A = {factors of 4}. If B is a subset of A, what is the probability that B contains

a. 2?
b. 1 and 4?

Answers: a. $\frac{1}{2}$ b. $\frac{1}{4}$

Assignment Guide

Basic
Optional

Average or Heterogeneous
Day 1 7, 9–23, 27–29, 38, 40, 42, 44, 46

Honors
Day 1 8, 11, 13–17, 19–22, 24–27, 30, 32, 33, 35, 37, 38, 41, 45

Strands

Communicating Mathematics	1–5, 7, 31, 36, 38, 47, Investigation
Geometry	9, 12, 17, 25, 30, 36
Measurement	25, 33, 35
Algebra	1–12, 24, 25, 27, 28, 32–38, 40–47, Investigation
Patterns & Functions	6, 42
Number Sense	8, 21, 26
Interdisciplinary Application	19, 20, 35, 41, 43, 45

Individualizing Instruction

Limited English proficiency Have students read their answers to problems 7, 31, 36, and 38 to you. Coach them on proper word choice and sentence structure.

430

Example 4

a By listing coordinates, identify the set of vertices of this figure.
b Reflect the figure over the x-axis. Which vertices of the image form a subset of the vertices of the preimage?
c Reflect the figure over the y-axis. Which vertices of the image form a subset of the vertices of the preimage?
d Translate the figure 1 unit to the left and 2 units up. Which vertices of the image form a subset of the vertices of the preimage?

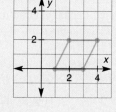

Solution

a {(1, 0), (3, 0), (4, 2), (2, 2)}   b {(1, 0), (3, 0)}

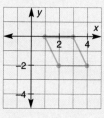

c { }   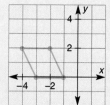   d {(2, 2)}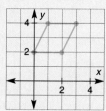

Sample Problems

Problem 1 List all the subsets of the set {1, 2, 3}. How many are there?

Solution { }, {1}, {2}, {3}, {1, 2}, {1, 3}, {2, 3}, {1, 2, 3}
There are a total of eight subsets.

Problem 2 The perimeter of the figure is less than 20 units. Describe all possible values of *x* as a set. (This is known as a **solution set**.)

430

SECTION 9.1

Solution
$2(x + 2) + 2x < 20$
$2x + 4 + 2x < 20$
$4x + 4 < 20$
$4x < 16$
$x < 4$

We know that the lengths of the rectangle's sides must be positive. This means that $x > 0$. Here are two ways to show the set:

1. Graph: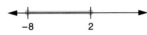

2. Set-builder notation: $\{x : 0 < x < 4\}$

Think and Discuss

P **1** A set of numbers is represented on the number line. Describe the set of numbers in words and in set-builder notation. Why can't you list the elements in the set?

P **2** Suppose that $P = \{-2, 17, \frac{3}{5}, -\frac{4}{9}, \pi, \sqrt{5}, 0\}$. Explain why listing is the only method that can be used to describe set P.   There is no rule or pattern.

P **3** If $R = \{x : 200 \leq x \leq 486\}$, is the statement *true* or *false*? Explain your answer.
   a 300 is an element of R.   True   **b** −300 is an element of R.   False
   c 203.5 is an element of R.   True   **d** 486 is an element of R.   True

P, R **4** Describe the shaded set in several ways. Which method seems easiest? Why?   Possible answers: $\{(x, y) : 1 \leq x \leq 2 \text{ and } 1 \leq y \leq 3\}$; all points on or inside the rectangle with vertices (1, 1), (2, 1), (1, 3), and (2, 3); set-builder notation is more concise.

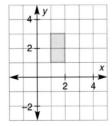

P, R **5** If $A = \{x : x > 4\}$, is 5 the smallest element of A? Why or why not?

P **6 a** List all subsets of each of the sets {a}, {a, b}, and {a, b, c}.
   b Do you see a pattern that can be used to determine the number of subsets a set has if you know the number of elements in the set? What is the pattern?   They increase in powers of 2.
   c How many subsets do you think {a, b, c, d} has?   16

Problem-Set Notes

for page 431

■ **6** can be explained by pointing out that when a new element is added to a set, you still have all the subsets you had before it was added, and you form new subsets by attaching the new element to each of the old subsets. This is twice the number that you had before; so each time a new element is added, the number of subsets doubles.

Additional Practice

1. If $R = \{x : 100 < x \leq 250\}$, is each statement *true* or *false*?
a. 100 is an element of R.
b. 200.5 is an element of R.

2. Indicate whether the statement is true or false.
a. $\{\ \} \subset \{1, 2, 3\}$
b. $\{1, 2, 3, 4\} \subset \{1, 3, 5, 7, 9\}$

Answers:
1. a. False **b.** True
2. a. True **b.** False

Additional Answers

for page 431

1 The set of all points on the number line between −8 and 2 inclusive; $\{x : -8 \leq x \leq 2\}$; because there are infinitely many elements

3 a True; 300 is less than 486 and greater than 200.

b False; −300 is less than 486 but is not greater than or equal to 200.

c True; 203.5 is greater than 200 and less than 486.

d True; 486 is greater than 200 and less than or equal to 486.

5 No, since the set contains all real numbers greater than 4

6 a { }, {a}; { }, {a}, {b}, {a, b}; { }, {a}, {b}, {c}, {a, b}, {b, c}, {a, c}, {a, b, c}

SECTION 9.1

Problem-Set Notes

for pages 432 and 433

- **11** looks at **6** in a different way: Do you need to know all of the subsets in order to know the set? Which ones could be left out? If you knew that there were exactly eight subsets, how many would you have to see in order to be certain that you had the entire set? **39** also addresses these ideas.

- **31** is fairly easy to answer, but rather tough to explain.

Cooperative Learning

- Refer to example 3. Give each group a number from the set {14, 18, 22, 26}. The number represents the amount (in feet) of fence available to border the garden. Remind groups that the gardens must be rectangular and have sides with whole-number lengths.
  1. Sketch the gardens you could make.
  2. Find the set of possible areas of the garden.
  3. Suppose that the lengths of the sides were not limited to whole numbers. Use set-builder notation to describe the possible lengths, widths, and areas.

Answers:
Answers for 26 ft. of fence are given:
1. $1 \times 12, 2 \times 11, 3 \times 10, 4 \times 9, 5 \times 8, 7 \times 6$
2. {12, 22, 30, 36, 40, 42}
3. {length: $0 < \text{length} \leq 6.5$}, {width: $0 < \text{width} \leq 6.5$}, {area: $0 < \text{area} \leq 42.25$}

- Problems 6 and 39 are appropriate for cooperative learning.

Problems and Applications

P   **7 Communicating** Suppose that A = {–4, –2, 0, 2, 4, 6, 8, 10, 12, 14, 16}.
  a Describe A in words.
  b Represent A by graphing.
  c Describe A in set-builder notation.
  d Which is the best method?

P   **8** If B = {$b: -5 < b < 5$ and b is an integer}, which of the following values of b are in B?   Only **c**
  a $b = \frac{1}{2}$ **b** $b = -5$ **c** $b = 0$

P, I   **9** A circle with center at (2, 0) has a radius of 2 units. Describe in two ways the set of points where the circle intersects the coordinate axes.

P, R   **10** Undistributed Middle School is expecting between 80 and 100 students for a ski trip. Students will be transported by minivans that hold 7 students each. Use set-builder notation to show the set of possible numbers of minivans that will be needed for the trip.

P   **11** If all of the subsets of set X are { }, {5}, {10}, {a}, {5, 10}, {5, a}, {10, a}, and {5, 10, a}, what are the elements of X?   {5, 10, a}

P, R   **12 a** By listing coordinates, identify the set of vertices of the figure shown.
  b Reflect the figure over the y-axis. Which vertices of the image form a subset of the vertices of the preimage?
  c Reflect the figure over the x-axis. Which vertices of the image form a subset of the vertices of the preimage?
  d Translate the figure ⟨6,2⟩. Which vertices of the image form a subset of the vertices of the preimage?   {(3, 1)}

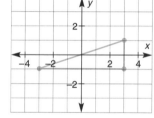

P   In problems 13–16, indicate whether the statement is *true* or *false*.

13 {1, 2, 3} ⊂ {1, 3, 5, 7, 9, 11}   False   **14** {4, 7} ⊂ {4, 7}   True

15 {1, 3} ⊂ {0, 1, 3, 5, 7}   True   **16** {1, 3, 5, 7, 9} ⊂ {2, 4, 6, 8, 10}   False

P, R   **17** List the set of all angles in the figure.
{∠NAT, ∠TAP, ∠PAC, ∠NAP, ∠TAC, ∠NAC}

P   In problems 18–23, determine the number of elements in each set.

18 {1, 2, 3, 4, ... , 41}   41   **19** {United States senators}   100

20 {Phases of matter}   3   **21** {Even prime numbers}   1

22 {0}   1   **23** { }   0

432   Chapter 9   Real Numbers

SECTION 9.1

P, R, I **24** For what values of x chosen from $\{-4, -3, \ldots, 4, 5, 6\}$ is $2x - 1 < 3$?

P, R **25** If the perimeter of the parallelogram is less than 43 units, what is the set of all possible integer values of x? $\{-1, 0, 1, 2, 3\}$

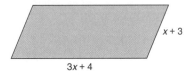

P, R **26 a** List the set of perfect squares less than or equal to 100.
b List the subset of the set in part **a** that consists of even numbers.
c What else do you notice about all the numbers in part **b**?

P **In problems 27–29, determine the number of elements in each set.**

27 $\{y : y^2 = 9\}$ 2 **28** $\{y : y = 1, 2, 3, 4, 5\}$ 5

29 $\{(x, y) : x = 1, 2, \text{ or } 3; y = 1, 2, \text{ or } 3\}$ 9

P, R **30** Suppose that A = {squares}, B = {rectangles}, C = {rhombuses}, and D = {parallelograms}. Indicate whether each statement is *true* or *false*.
a $A \subset B$ True **b** $A \subset C$ True **c** $A \subset D$ True **d** $B \subset C$ False
e $B \subset D$ True **f** $C \subset D$ True **g** $B \subset A$ False **h** $C \subset A$ False
i $D \subset A$ False **j** $C \subset B$ False **k** $D \subset B$ False **l** $D \subset C$ False

P **31 Communicating** If $A \subset B$ and $B \subset D$, is it true that $A \subset D$? Explain your answer.

P, R **32** Write an equation for which the solution set is $\{-5, 5\}$.

P, R, I **33** In the figure, $AB = x + 5$. Find the set of integers that are possible values of x.

P **34** Suppose that $A = \{x : -8 < x \leq 6\}$, $B = \{x : -6 \leq x \leq 3\}$, and $C = \{x : -7 < x < 11\}$. Indicate which of the sets, if any, contains the given number.
a -10 None **b** -7 A **c** -6 A, B, C **d** 10 C
e 7 C **f** 2 A, B, C **g** -6.5 A, C **h** 5.4 A, C
i 8.2 C **j** 0 A, B, C **k** -7.6 A **l** -8 None

P, R **35 Architecture** An office building is to be built with a height between 100 and 120 feet. Depending on the type of construction and the heights of the ceilings, each story will be between $9\frac{1}{2}$ and 12 feet high. List the possible numbers of stories that the building can have. $\{9, 10, 11, 12\}$

P, R, I **36 Communicating** Describe and draw on a coordinate system the set of points that are
a 4 units from the point (1, 2) **b** Less than 4 units from (1, 2)
c More than 4 units from (1, 2)

P **37** List the elements of each set.
a {Numbers that are factors of 64} **b** $\{x : x = 2^n, \text{ where } n = 1, 2, 3, 4, 5, 6\}$

P **38 Communicating** Describe in words $\{x : -4 < x < 4\}$.

P **39** Some of the subsets of S are $\{15\}$, $\{3, 6, 9\}$, $\{9, 12\}$, $\{6, 9, 12\}$, and $\{9, 18\}$.
a What is the smallest possible number of elements of S? 6
b List as many subsets of S as you can for which $\{3, 6, 9, 12\}$ is a subset.

Section 9.1 Sets 433

Additional Answers

for pages 432 and 433

7 a The set of even integers between -4 and 16, inclusive

b [number line from -4 to 16 with even integers marked]

c $\{x : -4 \leq x \leq 16, \text{ and } x \text{ is an even integer}\}$

d Possible answer: graphing

9 Possible answer: $\{(0, 0), (4, 0)\}$; the set of points consisting of the origin and (4, 0)

10 $\{x : 12 \leq x \leq 15, \text{ and } x \text{ is an integer}\}$

12 a $\{(3, 1), (3, -1), (-3, -1)\}$
b $(-3, -1)$ and $(3, -1)$
c $(3, -1)$ and $(3, 1)$

24 $\{-4, -3, -2, -1, 0, 1\}$

26 a $\{1, 4, 9, 16, 25, 36, 49, 64, 81, 100\}$
b $\{4, 16, 36, 64, 100\}$
c Possible answer: They are divisible by 4.

31 Yes. If every member of A is in B and every member of B is in D, then every member of A is in D.

32 Possible answers: $x^2 = 25$; $|x| = 5$

33 $\{x : -5 < x < 15, \text{ and } x \text{ is an integer}\}$

36 In parts **a–c**, see students' drawings.

a A circle with center (1, 2) and radius 4 units

b All points on the interior of a circle with center (1, 2) and radius 4 units

c All points on the exterior of a circle with center (1, 2) and radius 4 units

37 a $\{1, 2, 4, 8, 16, 32, 64\}$
b $\{2, 4, 8, 16, 32, 64\}$

38 All numbers greater than -4 and less than 4

39 b $\{3, 6, 9, 12, 15\}, \{3, 6, 9, 12, 15, 18\}, \{3, 6, 9, 12, 18\}, \{3, 6, 9, 12\}$

SECTION 9.1

Problem-Set Notes

for page 434

- 41, 43, and 45 are examples of questions that often appear in math contests. Making a diagram can be useful.

Communicating Mathematics

- Have students write a paragraph explaining which kinds of sets are best described by listing the members and which are best described by using set-builder notation.
- Problems 1–5, 7, 31, 36, 38, 47, and Investigation encourage communicating mathematics.

Checkpoint

1 If R = {x: 100 < x ≤ 250, and x is an integer}, is each statement true or false?
a. 250 is an element of R.
b. 100.5 is an element of R.

2. Describe the shaded set using words.

3. List all of the subsets of {*, ©, w}.

4. Tell whether the statement is true or false.
a. {3} ⊂ {4, 7, 9}
b. {1, 2, 3} ⊂ {1, 2, 3}

Answers:
1. a. True b. False 2. The set of all points at most 2 units from the origin 3. { }, {*}, {©}, {w}, {*, ©}, {*, w}, {©, w}, {*, ©, w}
4. a. False b. True

Answer for Investigation is on page A8.

◀ LOOKING BACK **Spiral Learning** LOOKING AHEAD ▶

R **40** Evaluate $\begin{bmatrix} 4 & -1 & 6 \\ -6 & 19 & -7 \end{bmatrix} - \begin{bmatrix} 5 & -7 & -3 \\ -3 & -14 & -7 \end{bmatrix} \cdot \begin{bmatrix} -1 & 6 & 9 \\ -3 & 33 & 0 \end{bmatrix}$

R **41** There are 40 people at a costume party. Twenty-eight are wearing masks, 31 are wearing belts, and 23 are wearing both masks and belts. How many are wearing neither masks nor belts? 4

R **42** If $f(t) = 3t + 8$ and $g(t) = 4t^2 - t$, what is
  a $f(4)$? 20   **b** $g(4)$? 60   **c** $f(0)$? 8
  d $g(0)$? 0   **e** $f(-3)$? -1   **f** $g(-3)$? 39

P, R **43** At Elisha Cook Junior High School, the math club had 30 members and the science club had 21 members. (No one was a member of both clubs.) Some members left each of these clubs and formed a computer club. All three clubs now have the same number of members. How many members from the math club and how many from the science club transferred to the computer club? 13 from math; 4 from science

P **44** If A = {1, 2, 3, 4} and B = {3, 5, 7, 9}, is the statement *true* or *false*?
  a 2 is an element of A. True   **b** 2 is an element of B. False
  c 3 is an element of A. True   **d** 3 is an element of B. True
  e 2 is an element of A or 2 is an element of B. True
  f 3 is an element of A or 3 is an element of B. True
  g 2 is an element of A and 2 is an element of B. False
  h 3 is an element of A and 3 is an element of B. True

P, R **45** Students at Rocky Mountain High School can win tickets to a ball game by having both an A average and perfect attendance. There are 366 students in the school. One hundred students have perfect attendance, 120 have an A average, and 186 have neither. How many students earned tickets to a ball game? 40

R, I **46** If $3t - 6y = 18$ and $t = -5$, what is the value of y? $y = -5.5$

P, R **47** Communicating Suppose that X = {multiples of 4}, Y = {multiples of 6}, and Z = {multiples of 8}.
  a Which set is a subset of another set? Z ⊂ X
  b Describe the set containing all the numbers that are in both X and Y. {Multiples of 12}

Investigation

Infinite Sets Write down some examples of sets that have a finite number of elements. Then give some examples of sets that have an infinite number of elements. Write a convincing argument to show that there is an infinite set of points on any line segment.

434 Chapter 9 Real Numbers

SECTION 9.2

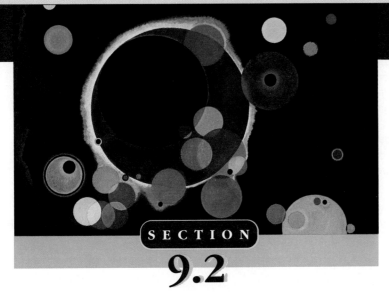

9.2 Intersection and Union

INTERSECTIONS AND UNIONS OF SETS

Sometimes the solution of a problem requires numbers that are in one set and also in another set. For instance, to be hired by the police department in Center City, a male candidate must be at least 66 inches tall but not more than 79 inches tall. His height must fall into both sets $A = \{x: x \geq 66\}$ and $B = \{x: x \leq 79\}$. The set $A \cap B$ is called the ***intersection*** of sets A and B. $A \cap B$ consists of all the numbers or objects the two sets have in common.

At other times a solution calls for things that are either in one set or in another set. For example, at the time of a flu epidemic, the local health department recommends flu shots for people over the age of 60 or people who have traveled outside the country in the past six months.

If we let C = {those older than 60} and D = {those who have traveled outside the country in the past six months}, then any person who is in set C or in set D or in both sets should have the flu shot. The set $C \cup D$ is called the ***union*** of sets C and D. $C \cup D$ consists of all elements that are in set C or in set D. (An element that is in both of two sets is an element of both the union and the intersection of the sets.)

Example 1
Let A be the set of points on the circle and B be the set of points on the square.
What are the elements of
a $A \cap B$?   b $A \cup B$?

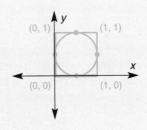

Objectives
- To determine the intersection and the union of two sets
- To use Venn diagrams to represent intersections and unions

Vocabulary: complement, intersection, union, Venn diagram

Class Planning

Schedule 1–2 days

Resources
In *Chapter 9 Student Activity Masters* Book
Class Openers *13, 14*
Extra Practice *15, 16*
Cooperative Learning *17, 18*
Enrichment *19*
Learning Log *20*
Using Manipulatives *21, 22*
Spanish-Language Summary *23*

Transparencies *40 a–d*

Class Opener

Use the diagram to answer problems **1–5**.

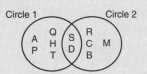

1. List the elements in Circle 1.
2. List the elements in Circle 2.
3. List the elements that are in Circle 1 but not in Circle 2.
4. List the elements that are in Circle 2 but not in Circle 1.
5. List the elements that are in both Circle 1 and Circle 2.

Answers:
1. {A, Q, P, H, T, S, D}
2. {R, M, C, B, S, D}
3. {A, Q, P, H, T}
4. {R, M, C, B}
5. {S, D}

Section 9.2 Intersection and Union   435

SECTION 9.2

Lesson Notes

- Intersection and union are concepts that help us to be precise in our language. Both concepts have arithmetic, algebraic, and geometric applications.
- Venn diagrams are quite useful in clarifying sets, as well as in solving problems like 41, 43, and 45 in the preceding section.
- The calendar problem in example 3 is an example of modular arithmetic. A student might investigate modular arithmetic as a special project.

Additional Examples

Example 1
Refer to the diagram on page 435. Describe the set $\overline{A} \cap B$.

Answer: All points on the square except $(0, \frac{1}{2})$, $(\frac{1}{2}, 0)$, $(1, \frac{1}{2})$, and $(\frac{1}{2}, 1)$.

Example 2
Let set M be defined as $\{x: x = 4n - 3$, where n is a natural number less than 7$\}$ and set P be defined as $\{$odd numbers less than 12$\}$.
a. List the elements in each set.
b. Find $M \cap P$.
c. Find $M \cup P$.

Answers: **a.** $M = \{1, 5, 9, 13, 17, 21\}$; $P = \{1, 3, 5, 7, 9, 11\}$
b. $M \cap P = \{1, 5, 9\}$
c. $M \cup P = \{1, 3, 5, 7, 9, 11, 13, 17, 21\}$

Example 3
Refer to example 3. Use set-builder notation to describe the set of dates in April that Gerry will work.

Answer: $\{x: x = 4n + 2$, where n is a whole number less than 8$\}$

Solution
a The intersection consists of the points belonging to both figures. According to the graph, $A \cap B = \{(0.5, 0), (1, 0.5), (0.5, 1), (0, 0.5)\}$.
b $A \cup B$ consists of an infinite number of points. We can describe $A \cup B$ in words with the phrase "all the points on the square or on the circle." We can also describe the set by referring to the graph. The whole of the figure shown in blue represents the union of sets A and B.

Example 2
Suppose we define set A as $\{x:x = 3n + 1$, where n is a natural number less than 7$\}$ and we define set B as $\{$prime numbers less than 20$\}$.
a List the elements in each set.
b Find $A \cap B$.
c Find $A \cup B$.

Solution
a $A = \{4, 7, 10, 13, 16, 19\}$, $B = \{2, 3, 7, 11, 13, 17, 19\}$
b $A \cap B = \{7, 13, 19\}$
c $A \cup B = \{2, 3, 4, 7, 10, 11, 13, 16, 17, 19\}$

VENN DIAGRAMS

A **Venn diagram** is a way to show the intersection or union of two or more sets.

The Venn diagram at the right shows the intersection of sets A and B from Example 2.

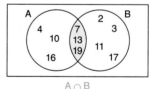

$A \cap B$

The Venn diagram at the right shows the union of sets A and B from Example 2.

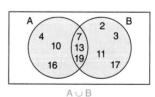

$A \cup B$

Example 3
Gerry Booker will work in the library every fourth day, starting on Monday, April 2. Draw a calendar for the month of April and circle the days he works. Underline the weekend days. If S is the set of days he will work and T is the set of weekend days, list the elements of S, T, and $S \cap T$.

436   Chapter 9   Real Numbers

SECTION 9.2

Solution
In the calendar, the circled dates represent the elements of set S. The underlined dates represent the elements of set T.
S = {2, 6, 10, 14, 18, 22, 26, 30}
T = {1, 7, 8, 14, 15, 21, 22, 28, 29}
S∩T = {14, 22}

S	M	T	W	T	F	S
1	②	3	4	5	⑥	7
8	9	⑩	11	12	13	⑭
15	16	17	⑱	19	20	21
㉒	23	24	25	㉖	27	28
29	㉚					

In Example 3, we talked about the set T of weekend days. We might also refer to the set of weekdays. This is the set of all days that are not weekend days. It is designated by $\overline{T}$ and is called the complement of T.

▶ The **complement** of a set A is the set of all numbers or elements under discussion that are not in A. The complement of A is written $\overline{A}$.

Sample Problems

Problem 1 Given the Venn diagram for sets A and B shown, draw Venn diagrams for
  a A∪B   **b** A∩B

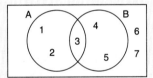

Solution **a** 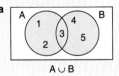   A∪B

b 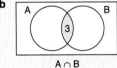   A∩B

Problem 2 Copy this Venn diagram and shade the region(s) representing
  a A∩B∩C   **b** (A∪B)∩C   **c** $\overline{A}$

Solution **a** 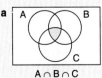 A∩B∩C   **b** (A∪B)∩C   **c** 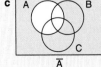 $\overline{A}$

Assignment Guide

Basic
Optional

Average or Heterogeneous
Day 1 8, 10, 12, 15, 20, 21, 24 (a, c, e), 27, 28, 30
Day 2 11, 14, 16, 22, 23, 24 (b, d, f), 26, 29, 31, 33

Honors
Day 1 8–11, 13, 14, 17, 21 (a, b), 24 (a–d), 25 (a–d, h), 30–32

Strands

Communicating Mathematics	1, 3–5, 32, Investigation
Geometry	15, 17, 19, 20, 23, 28, 30
Measurement	15, 23, 28, 30
Technology	16, Investigation
Algebra	2–4, 6–25, 27–31, 33
Number Sense	11, 16, 26
Interdisciplinary Applications	5

Individualizing Instruction

The visual learner Students might be more likely to discover the different interpretations of the statement in problem 1 if they draw Venn diagrams. They may also find it helpful to draw Venn diagrams for problems 8 and 25 before attempting to answer.

Stumbling Block

Students may have difficulty finding the intersection or union of sets when all of the members are not listed. Example 2 shows a good strategy: list all of the sets' elements before trying to find their intersection or union.

437

SECTION 9.2

Problem-Set Notes

for pages 438 and 439

- 1–5 should help to clarify the differences between the mathematical usages of *or* and *and* and the everyday usages of these words.

- 16 can be done without a calculator or spreadsheet if students notice that the units digit of the square of a number will always be the units digit of the square of the units digit of the number, since there is no regrouping involved in finding the units digit. For example, any number ending in 7 will have a square ending in 9, since $7^2 = 49$.

Communicating Mathematics

- Ask students to consider various subsets of students in the school (for example, members of a homeroom, band, or sports team). Have them come up with subsets that meet each of the following requirements:
  a. $A \cap B = \{\}$
  b. $C \cap D = D$
  c. $E \cap F$ has one member.
  d. $G \cup H$ is the entire student body.

- Have students compare their answers and explain their reasoning.

- Problems 1, 3–5, 32, and Investigation encourage communicating mathematics.

Additional Answers

for pages 438 and 439

1 It might refer to the members of the band and the members of the orchestra, or it might refer only to those who are members of both the band and the orchestra.

438

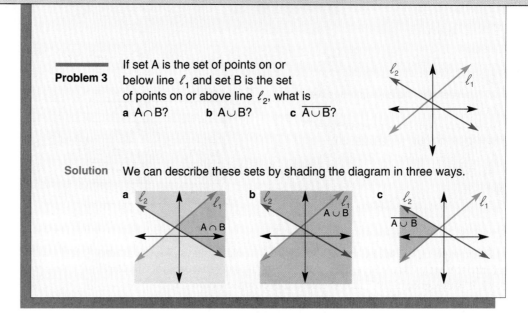

Problem 3 If set A is the set of points on or below line ℓ_1 and set B is the set of points on or above line ℓ_2, what is
a $A \cap B$?   b $A \cup B$?   c $\overline{A \cup B}$?

Solution We can describe these sets by shading the diagram in three ways.

Think and Discuss

P **1** Explain the possible confusion in the statement, "Members of the band and the orchestra are to meet in the auditorium."

P **2** If B is the set of members of the band and O is the set of members of the orchestra, what is
a $B \cup O$? Those in band or orchestra b $B \cap O$? Those in both band and orchestra

P **3** An announcement says that all eighth graders with perfect attendance may watch a movie after lunch. Does this refer to a union or an intersection of sets? Explain your answer.

P **4** How do the uses of the word *or* differ in the following statements:

I left my book in my locker or in the cafeteria.
The number 8 belongs to $A \cup B$—that is, it is in set A or in set B.

P **5** **Language Arts** In everyday language, we talk about the intersection of two roads. How does this use of the word *intersection* relate to the intersection of two sets? It is a place that is part of both roads.

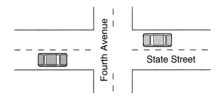

P **6** Under what circumstances will each statement be true?
a $A \cap B = A$ b $A \cup B = A$ c $A \cap B = B \cap A$

P **7** Refer to the diagram and list the numbers in each set.
a A $\{1, 2, 3\}$ b $\overline{A}$ $\{4, 5, 6, 7\}$
c $A \cup B$ $\{1, 2, 3, 4, 5\}$ d $A \cap B$ $\{3\}$
e $\overline{A \cup B}$ $\{6, 7\}$

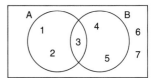

438   Chapter 9   Real Numbers

SECTION 9.2

Problems and Applications

P **8** Given that A = {1, 3, 5, 7, 9}, B = {2, 3, 5, 7, 11, 13}, C = {5, 7, 9}, and D = {6, 8, 10}, list the elements of each intersection or union.
  a A∩B **b** A∩C **c** B∩C **d** A∪B
  e A∪C **f** B∪C **g** C∩D **h** C∪D

P **9** Use symbols to describe the shaded set. B∩C
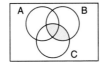

P **10** Draw four Venn diagrams like the one shown. Shade regions for
  a A∩B **b** A∪B
  c $\overline{A∩B}$ **d** $\overline{A∪B}$

P **11**
  a List the elements of {x:x is a multiple of 3 and 1 < x < 41}.
  b List the elements of {x:x is a multiple of 4 and 1 < x < 41}.
  c List the elements of the intersection of the sets in parts **a** and **b**.
  d Use set-builder notation to represent your answer to part **c**.

P, I **12** Sets A, B, and C are graphed on the number lines shown. Draw a graph of
  a A∩B **b** A∩C
  c B∩C **d** A∪B
  e A∪C **f** B∪C

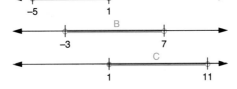

P, R **13** If A = {..., −4, −3, −2, −1, 0, 1, 2, 3,...} and B = {0, 1, 2, 3, 4, ...}, what is
  a A∩B? B   **b** A∪B? A

P **14** There are 71 students on the track team and 53 students on the swimming team. Sixteen of the students are on both teams. Find the total number of students on the two teams. 108

P, R **15** The two rectangular regions are congruent.
  a Find the area of their intersection. 27
  b Find the area of the complement of their intersection. 141

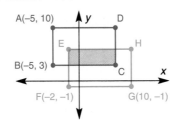

P, R **16** Use a spreadsheet or a calculator to find the set of squares of the integers from 1 to 50. Then find the subset of the squares that have
  a A units digit of 1 **b** A units digit of 5 **c** A units digit of 3

---

3 Intersection; it means those who meet both conditions.

4 The book can be in only one place; the 8 can be in both sets.

6 a A is a subset of B.
b B is a subset of A.
c The statement is always true.

8 a {3, 5, 7} **b** {5, 7, 9} **c** {5, 7}
d {1, 2, 3, 5, 7, 9, 11, 13} **e** {1, 3, 5, 7, 9} **f** {2, 3, 5, 7, 9, 11, 13}
g { } **h** {5, 6, 7, 8, 9, 10}

10 a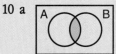

b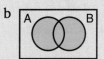

c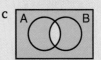

d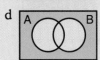

11 a {3, 6, 9, 12, 15, 18, 21, 24, 27, 30, 33, 36, 39}
b {4, 8, 12, 16, 20, 24, 28, 32, 36, 40}
c {12, 24, 36}
d Possible answer: {x: x = 12n, where n = 1, 2, 3}

12 a ⟵┼┼┼●┼┼┼┼┼●┼┼┼┼┼⟶
         −3 1

b ⟵┼┼┼┼┼┼●┼┼┼┼┼┼⟶
              1

c ⟵┼┼┼┼┼┼●┼┼┼┼●┼┼┼┼⟶
              1   7

d ⟵┼●┼┼┼┼┼┼┼┼●┼┼┼⟶
       −5   7

e ⟵┼●┼┼┼┼┼┼┼┼┼┼●┼⟶
       −5   11

f ⟵┼┼┼●┼┼┼┼┼┼┼┼●┼⟶
          −3   11

16 a {1, 81, 121, 361, 441, 841, 961, 1521, 1681, 2401}
b {25, 225, 625, 1225, 2025}
c { }

SECTION 9.2

Problem-Set Notes

for pages 440 and 441

- 18 is a nice thought exercise. It should be discussed in class to see how students approached it.

Additional Practice

Problems **1–8** refer to the diagram. List the numbers in each set.

1. A
2. B
3. $\overline{A}$
4. $\overline{B}$
5. A ∪ B
6. A ∩ B
7. $\overline{A} \cap \overline{B}$
8. $\overline{A \cup B}$

9. List the elements of {x: x is a multiple of 6 and 3 < x < 41}.

10. Sets A, B, and C are graphed on the number lines shown. Draw a graph of (A ∩ B) ∪ C.

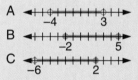

Answers:
1. {1, 2, 7, 9, 12} 2. {5, 9, 12, 13, 14} 3. {5, 8, 13, 14, 15}
4. {1, 2, 7, 8, 15} 5. {1, 2, 5, 7, 9, 12, 13, 14} 6. {9, 12} 7. {8, 15}
8. {1, 2, 5, 7, 8, 13, 14, 15}
9. {6, 12, 18, 24, 30, 36}
10. ◄─┼┼┼┼┼┼┼┼┼┼┼►
      −6   3

Cooperative Learning

- Borrow some almanacs from the library and have groups of students make Venn diagrams that show the following sets of U.S. presidents: A = {those who served as vice-president}, B = {those who served more than 4 years}, C = {those who died during their presidencies}.

P, R   **17 a** Find the intersection of the interiors of rectangle ACFH and rectangle BDEG.   Interior of rect. BCFG
      b Find the union of the interiors of rectangle ACFH and rectangle BGED.   Interior of rect. ADEH

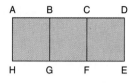

P   **18** Give an example of two sets, A and B, such that A ⊂ B and B ⊂ A.

P, R   **19** Reflect $\overline{AB}$ over the y-axis to form $\overline{A'B'}$. Use set-builder notation to represent
    a $\overline{A'B'} \cap \overline{AB}$ {x: −4 ≤ x ≤ 4}
    b $\overline{A'B'} \cup \overline{AB}$ {x: −6 ≤ x ≤ 6}

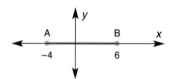

P, R   **20** Q is the set of all quadrilaterals. Name at least four subsets of this set.

P, R, I   **21** Set A is the shaded region.   Set B is the shaded region.

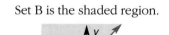

On a coordinate system, draw
  a A ∩ B   **b** A ∪ B   **c** $\overline{A \cup B}$

P, I   **22** ℓ is the line through (−3, 2) and (3, 2).
m is the line through (−4, −4) and (4, 4).
A is the region above ℓ. B is the region below m.
Of (8, 5), (0, 5), (−5, 0), and (5, 0), which are in
  a A ∩ B?   (8, 5)   **b** A ∪ B?   (8, 5), (0, 5), (5, 0)

P, R   **23** Suppose that A = {rectangles whose side lengths are a whole number of centimeters and whose perimeter is 10 centimeters} and D = {rectangles whose side lengths are a whole number of centimeters and whose area is 6 square centimeters}. What is
  a A ∩ D?   {Rectangles 2 cm by 3 cm}   **b** A ∪ D?   {Rectangles 1 × 6, 2 × 3, 1 × 4}

P   **24** Draw Venn diagrams like the one shown and shade regions to show
  a A ∩ B   **b** A ∪ B
  c B ∩ C   **d** A ∩ B ∩ C
  e (A ∩ B) ∩ $\overline{C}$ **f** (A ∪ B) ∩ $\overline{C}$

P   **25** Suppose that A = {1, 4, 7, 10}, B = {2, 4, 6}, and C = {1, 3, 5}. List the elements of each of the following sets.
  a A ∩ B   **b** (A ∩ B) ∪ C   **c** A ∪ C   **d** (A ∪ C) ∩ B
  e A ∩ B ∩ C   **f** A ∪ B ∪ C   **g** (A ∪ C) ∩ (B ∪ C)   **h** (A ∩ B) ∪ (C ∩ B)

P, R   **26** List all the subsets of the set of prime numbers less than 10.

440   Chapter 9   Real Numbers

SECTION 9.2

◀ LOOKING BACK **Spiral Learning** LOOKING AHEAD ▶

R   **27** Rewrite each number as a fraction:
     a 0.3 $\frac{3}{10}$ **b** 0.3333... $\frac{1}{3}$ **c** 15% $\frac{3}{20}$ **d** 9.2 $\frac{46}{5}$

R   **28**   **a** Find the length of $\overline{AB}$.   6
     b Find the length of $\overline{AC}$.   8
     c Find the length of $\overline{BC}$.   10
     d Find the area of $\triangle ABC$.   24
     e Find the perimeter of $\triangle ABC$.   24

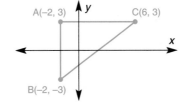

R   **29** Find the solution set of each inequality.
     a $3x - 8 < 13$ $\{x : x < 7\}$ **b** $10 - 4x \leq 14$ $\{x : x \geq -1\}$

R   **30**   **a** Find the area and the circumference of the circle inscribed in the square.
     b Find the area of the region that is outside the circle but inside the square.
     $36 - 9\pi \approx 7.73$

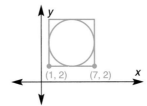

R, I   **31** Find the solution set of each equation.
     a $x^2 = 36$ $\{6, -6\}$ **b** $\sqrt{x} = 36$ $\{1296\}$ **c** $x^2 = -36$ $\{\}$ **d** $\sqrt{x^2} = x$ $\{x : x \geq 0\}$

P   **32 Communicating**   Describe the relationship between sets A and B. $B \subset A$

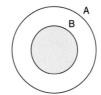

R, I   **33**   **a** Find several possible values of x, not necessarily integers.
     b Using set-builder notation, represent all values of x such that AB < BC.

Investigation

Limits   Using a calculator, enter any positive number greater than 1, then press the $\boxed{\sqrt{}}$ key repeatedly. Describe what happens. Now enter any positive number less than 1 and press the $\boxed{\sqrt{}}$ key repeatedly. Describe what happens. Try the procedure with some other numbers. Can you explain your results?

Section 9.2   Intersection and Union   441

Answers:
A = {J. Adams, Jefferson, Van Buren, Tyler, Fillmore, A. Johnson, Arthur, T. Roosevelt, Coolidge, Truman, L. Johnson, Nixon, Ford, Bush}, B = {Washington, Jefferson, Madison, Monroe, Jackson, Grant, Cleveland, McKinley, T. Roosevelt, Wilson, Coolidge, F. Roosevelt, Truman, Eisenhower, L. Johnson, Nixon, Reagan}, C = {W. Harrison, Taylor, Lincoln, Garfield, McKinley, Harding, F. Roosevelt, Kennedy}.

■ Problems 16 and 23 are appropriate for cooperative learning.

Checkpoint

Problems 1–6 refer to the diagram. List the numbers in each set.

1. A   2. B   3. $\overline{A}$
4. $A \cup B$ 5. $\overline{A} \cap \overline{B}$ 6. $A \cap B$

7. Use symbols to describe the shaded set.

8. List the elements of $\{x : x \text{ is a multiple of 5, and } 3 < x < 41\}$.

Answers:
1. {1, 3, 5, 7, 9}   **2.** {2, 4, 5, 6, 7, 8}   **3.** {2, 4, 6, 8, 12, 15}   **4.** {1, 2, 3, 4, 5, 6, 7, 8, 9}   **5.** {12, 15}   **6.** {5, 7}   **7.** $A \cap B$ **8.** {5, 10, 15, 20, 25, 30, 35, 40}

Answers for problems 18, 20, 21, 24–26, 30, 33, and Investigation are on page A8.

SECTION 9.3

Objective
- To categorize numbers according to the number sets to which they belong

  Vocabulary: disjoint sets, irrational number, natural number, rational number, real number

Class Planning

Schedule 1–2 days

Resources

In *Chapter 9 Student Activity Masters* Book
Class Openers 25, 26
Extra Practice 27, 28
Cooperative Learning 29, 30
Enrichment 31
Learning Log 32
Technology 33, 34
Spanish-Language Summary 35

In *Tests and Quizzes* Book
Quiz on 9.1–9.3 (Average) 241
Quiz on 9.1–9.3 (Honors) 242

Transparencies 41 a–f

Class Opener

Evaluate each fraction to a maximum of five decimal places.
1. $\frac{10}{2}$ **2.** $-\frac{21}{7}$ **3.** $\frac{15}{4}$ **4.** $\frac{2}{9}$ **5.** $\frac{3}{11}$ **6.** $\frac{1}{8}$

Answers: **1.** 5 **2.** −3 **3.** 3.75
4. 0.22222 **5.** 0.27272 **6.** 0.125

Lesson Notes

- References to the sets of numbers discussed in this section will be made throughout your students' studies in mathematics. To help students become more comfortable with the language, use the terms frequently in class and have students put numbers in categories.

- Some students may assume that every radical number is irrational.

Investigation Connection
Section 9.3 supports Unit E, *Investigations* 3, 4. (See pp. 426A–426D in this book.)

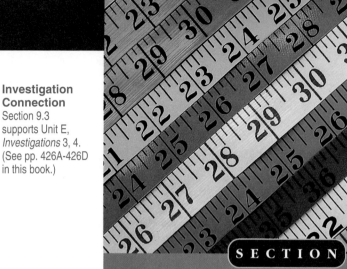

SECTION 9.3
Important Sets of Numbers

THE RATIONAL NUMBERS

In mathematics we try to be as clear and precise as possible with vocabulary and notation. When you refer to a triangle, a circle, or a square, other people know just what you mean. In a similar way, we define important sets of numbers. We build these sets by starting with the ***natural numbers,*** which are also called the counting numbers.

Name	Set	Graph
Natural numbers	{1, 2, 3, …}	number line with −3 to 3, points at 1, 2, 3
Whole numbers	{0, 1, 2, 3, …}	number line with −3 to 3, points at 0, 1, 2, 3
Integers	{…, −3, −2, −1, 0, 1, 2, 3, …}	number line with −3 to 3, points at all integers

On the graph of the integers, there are many points that can be paired with numbers that are not integers. Points for numbers like $\frac{1}{2}$, −3.25, and 8.1 lie between the points representing the integers. These numbers, along with the integers, make up a fourth set of numbers.

▶ A ***rational number*** is any number that can be written in the form $\frac{a}{b}$, where a and b are integers and $b \neq 0$. In set-builder notation,
$R = \left\{ \frac{a}{b} : a \text{ and } b \text{ are integers}, b \neq 0 \right\}$.

442 **Chapter 9** Real Numbers

SECTION 9.3

Example 1
Which of the following are rational numbers?

$2, \frac{3}{4}, -\frac{5}{8}, 3\frac{6}{7}, 2.54, 0, \sqrt{49}, 0.\overline{3}$

Solution
All of the numbers are rational numbers. We can show that each number can be written as a ratio of two integers. The numbers $\frac{3}{4}$ and $-\frac{5}{8}$ are already in fraction form. The other fraction forms are

$2 = \frac{2}{1}$ $3\frac{6}{7} = \frac{27}{7}$ $2.54 = \frac{254}{100}$ $0 = \frac{0}{1}$ $\sqrt{49} = \frac{7}{1}$ $0.\overline{3} = \frac{1}{3}$

Any fraction can be converted to a decimal. Thus, any rational number can be written as a decimal. The decimal form of a rational number either terminates or repeats. In numbers such as $0.4545...$ and $3.2\overline{716}$, the ellipsis dots and the bar mean that the groups of digits repeat, in the same pattern, without end.

THE REAL NUMBERS

Not all of the points on the number line represent rational numbers. There is another set of numbers on the number line in addition to the rationals. This set contains numbers that cannot be written as a ratio of two integers and so cannot be written as terminating or repeating decimals.

▶ An ***irrational number*** is a number that is the coordinate of a point on the number line but is not a rational number. Some irrational numbers are $\sqrt{2}, \sqrt[3]{6}, -\sqrt{24}, 2\sqrt{3}, \pi,$ and $0.010010001...$

Since no irrational number can be written as a ratio of two integers, the set of rational numbers and the set of irrational numbers have no numbers in common. In other words, {rational numbers} ∩ {irrational numbers} = { }. When two sets have an intersection that is empty, the two sets are called ***disjoint sets.***

Rational	Irrational
2 $\frac{13}{4}$	$\sqrt{2}$ $\sqrt[3]{6}$
−5 2.54	
$\frac{-5}{8}$	$-\sqrt{24}$
$\sqrt{16}$	π
$3.\overline{6}$	$0.01001...$

▶ The union of the set of rational numbers and the set of irrational numbers is the set of ***real numbers.***

Every point on the number line has a coordinate that is a real number.

Stress to students that they must try to simplify every radical. Since $\sqrt{25} = 5$, and $5 = \frac{5}{1}$, $\sqrt{25}$ is a counting number, a whole number, an integer, and a rational number.

■ If students want to know why any rational number can be converted to a repeating or terminating decimal, have them consider what can happen when they divide the numerator of a rational number by the denominator. If one of the remainders repeats, the division process repeats, and since there are only a finite number of possible remainders (1 less than the denominator), the division must either repeat or terminate.

■ If students want proof that the number the calculator gives as the square root of 2 isn't quite exact, ask them to square the number. The last digit will not be 0, since the square of a digit other than zero never ends in zero, so the square of whatever number they have cannot be exactly 2.

■ Sample problem 1 shows how to convert a repeating decimal to a fraction. Students should become proficient with this skill.

Additional Examples

Example 1
Which of the following are rational numbers?
$\frac{1}{4}, 13, \sqrt{11}, 0.125, 3.145$

Answer: $\frac{1}{4}, 13, 0.125, 3.145$

Example 2
On the number line, locate $\sqrt{6}, \frac{\pi}{3}$, and $-4\sqrt{2}$.

Answer:

SECTION 9.3

Assignment Guide

Basic
Optional

Average or Heterogeneous
Day 1 10–16, 18, 19, 23, 25, 26, 29, 30
Day 2 17, 20, 22, 24, 27, 28, 31

Honors
Day 1 10, 11 (a, b), 12–14, 16–20, 28, 30, 32

Strands

Communicating Mathematics	8, 9, 29, Investigation
Geometry	16, 24, 25
Measurement	16, 24, 25
Technology	4, 6, 7, 10, 14, 15, 24, 28, Investigation
Probability	22
Algebra	1, 8–10, 12, 16, 17, 22, 24–28, 30–32
Patterns & Functions	Investigation
Number Sense	1–9, 11–21, 23–26, 29, 31, 32, Investigation
Interdisciplinary Applications	28

Individualizing Instruction

The auditory learner Before students are asked to work on their own, have them pair off and discuss problems 1 and 18 with each other.

Problem-Set Notes

for pages 444 and 445

■ 1 is really about closure of the sets, but it also checks to see if students have learned the sets of numbers.

444

Example 2
On a number line, locate $\sqrt{5}$, π, and $-2\sqrt{3}$.

Solution
Using a calculator, we find that $\sqrt{5} \approx 2.2$, $\pi \approx 3.1$, and $-2\sqrt{3} \approx -3.5$.

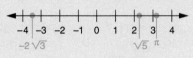

Sample Problems

Problem 1 Write $-2.\overline{57}$ as a ratio of two integers.

Solution Let $n = -2.5757\ldots$

Multiplying by 100 provides a way to eliminate the repeating decimal

$$100n = -257.5757\ldots$$
$$n = -2.5757\ldots$$

Subtract
$$99n = -255$$

Divide and simplify
$$n = \frac{-255}{99} = \frac{-85}{33}$$

Therefore, $-2.\overline{57} = \frac{-85}{33}$.

Problem 2 Indicate the sets of numbers containing each of the following:

$$8, -2, 0, \sqrt{25}, \tfrac{3}{4}, 2\pi, -5.2, 3.1\overline{7}, -\sqrt{5}$$

Solution
Natural numbers: 8, $\sqrt{25}$
Whole numbers: 8, $\sqrt{25}$, 0
Integers: 8, $\sqrt{25}$, 0, –2
Rational numbers: 8, $\sqrt{25}$, 0, –2, $\tfrac{3}{4}$, –5.2, 3.1$\overline{7}$
Irrational numbers: 2π, $-\sqrt{5}$
Real numbers: 8, $\sqrt{25}$, 0, –2, $\tfrac{3}{4}$, –5.2, 3.1$\overline{7}$, 2π, $-\sqrt{5}$

Think and Discuss

P 1 Indicate whether the statement is *true* or *false*.
   a The sum of two natural numbers is always a natural number. True
   b The difference of two natural numbers is always a natural number. False
   c The difference of two integers is always an integer. True
   d The quotient of two integers is always an integer. False

Chapter 9 Real Numbers

SECTION 9.3

P, R **2** What is the intersection of the set of natural numbers and the set of whole numbers? The natural numbers

P **3** Which of the following numbers are integers? **a, b, d,** and **e**
  a −764,002   **b** $\frac{48}{6}$ **c** 4.2   **d** $\frac{1.2}{0.4}$ **e** $\sqrt{64}$

P, R **4** For which of the following will a calculator provide the exact value? **a** and **e**
  a $2 \div 5$ **b** $\sqrt{3}$ **c** $\frac{4}{7}$ **d** $\frac{353}{120}$ **e** $\frac{17}{32}$

P **5** Which of the following are rational numbers? **a, b, c,** and **d**
  a 2.54789   **b** −2.1717…   **c** $0.\overline{6}$ **d** 0.66   **e** 0.20020002…

P **6** Does a calculator display the exact values of all rational numbers? No

P **7** Does a calculator display the exact values of all irrational numbers? No

P, R **8** Explain the difference between $\{x:0 < x < 10\}$ and $\{x:0 < x < 10 \text{ and } x \text{ is an integer}\}$.

P **9** You know that $\frac{1}{3} = 0.\overline{3}$ and $\frac{2}{3} = 0.\overline{6}$. Explain why $1 = 0.\overline{9}$.

Problems and Applications

10 Write each number as a decimal.
  a $\frac{7}{8}$ 0.875   **b** $1\frac{4}{11}$ $1.\overline{36}$ **c** $-\frac{5}{6}$ $-0.8\overline{3}$ **d** $\frac{3}{64}$ 0.046875

P, R **11** Write each number in fraction form.
  a 0.35 $\frac{7}{20}$ **b** $0.\overline{3}$ $\frac{1}{3}$ **c** 0.5555… $\frac{5}{9}$ **d** $0.\overline{6}$ $\frac{2}{3}$

P **12** Write each number as a ratio of two integers.
  a $0.\overline{15}$ $\frac{5}{33}$ **b** −2.456 $-\frac{307}{125}$ **c** 4.111… $\frac{37}{9}$ **d** 6 $\frac{6}{1}$

P **13** Which of the following are rational numbers? **a, b,** and **c**
  a $\sqrt{3} \cdot \sqrt{3}$ **b** $\sqrt{7} - \sqrt{7}$ **c** $\sqrt{14} \div 2\sqrt{14}$ **d** $2\sqrt{14}$

P **14** Locate the following values on a number line.
  a $2\sqrt{7}$ **b** $\sqrt{11} - 1$ **c** −1.2   **d** −0.55…

P, R **15** Arrange the following numbers in order, from smallest to largest.
  $\frac{11}{10}, \frac{100}{99}, \sqrt{5} \div 2, 1.111\ldots \frac{100}{99}, \frac{11}{10}, 1.\overline{1}, \sqrt{5} \div 2$

P, R **16** Find the values of **a, b,** and **c**. Indicate whether each is *rational* or *irrational*.

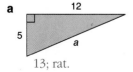
13; rat.

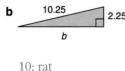
10; rat

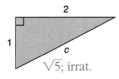
$\sqrt{5}$; irrat.

P, I **17** What points (x, y) satisfy the equation $x + y = 10$ if x and y must be natural numbers? (1, 9), (2, 8), (3, 7), (4, 6), (5, 5), (9, 1), (8, 2), (7, 3), (6, 4)

■ **9** is hard for some students to accept. It seems to them that $0.\overline{9}$ is close to 1, but not equal to 1. Interesting discussions should follow.

Cooperative Learning

■ A set is closed under an operation if performing that operation with any two numbers from that set results in another number from that set. For example, the set of whole numbers is closed under addition because adding two whole numbers always gives you another whole number. For each set listed below, have students indicate whether it is closed or not closed under addition, subtraction, multiplication, and division.

a. natural numbers
b. whole numbers
c. integers
d. rational numbers
e. irrational numbers
f. real numbers

Answers:

	a	b	c	d	e	f
+	c	c	c	c	nc	c
−	nc	nc	c	c	nc	c
×	c	c	c	c	nc	c
÷	nc	nc	nc	c	nc	c

■ The Investigation is appropriate for cooperative learning.

Additional Answers

for pages 444 and 445

8 The first includes all real numbers in the interval; the second includes only integers.

9 Possible answer: Because $1 = \frac{1}{3} + \frac{2}{3} = 0.\overline{3} + 0.\overline{6} = 0.\overline{9}$

14 −1.2 −0.5̄ √11−1   2√7
  ◄─┼─●─●─┼──┼─●─┼──┼─●─┼─►
   −2   0   2   4   6

SECTION 9.3

Problem-Set Notes

for pages 446 and 447

- **18–20** motivate students to be sure they understand the relationships between the sets of numbers.

- **24** is an interesting geometry question. What about numbers larger than 12? Does the number of triangles grow according to a pattern as we allow the dimensions to increase?

- **28** can be extended to include letters that can be formed by lighting up certain parts. How many lighted parts are needed to create various letters of the alphabet?

- **32** might be extended by asking students how many different answers they can come up with.

Additional Practice

In problems **1–6,** write the set or sets to which the number belongs: naturals, wholes, integers, rationals, irrationals, reals.
1. –3   2. 14   3. $\sqrt{9}$
4. $\sqrt{7}$ 5. 1.3   6. 2.6

In problems **7–9,** write each number as a decimal.
7. $1\frac{6}{7}$ 8. $2\frac{5}{16}$ 9. $-3\frac{1}{9}$

In problems **10–12,** write each number as a ratio of two integers.
10. 9   11. $-2.\overline{3}$ 12. 0.45

Answers:
1. integer, rational, real 2. natural, whole, integer, rational, real 3. natural, whole, integer, rational, real 4. irrational, real 5. rational, real 6. rational, real 7. $1.\overline{857142}$ 8. 2.3125 9. $-3.\overline{1}$ 10. $\frac{9}{1}$ 11. $-\frac{7}{3}$ 12. $\frac{9}{20}$

P 18 Is the statement true *always, sometimes,* or *never?*
   a An integer is a rational number. Always
   b A whole number is a natural number. Sometimes
   c A rational number is an irrational number. Never
   d An irrational number is a real number. Always
   e A rational number is an integer. Sometimes

P 19 Suppose that N = {natural numbers}, I = {integers}, Q = {rational numbers}, and R = {real numbers}. Is the statement *true* or *false?*
   a $N \subset I$ True   **b** $I \cap Q = Q$ False   **c** $I \cup Q = R$ False   **d** $I \cap R = I$ True

P 20 Copy the table, writing *yes* or *no* in each space to indicate the sets to which the numbers belong.

	Natural	Whole	Integer	Rational	Irrational	Real
13	Yes	Yes	Yes	Yes	No	Yes
$\sqrt{5}$	No	No	No	No	Yes	Yes
–7	No	No	Yes	Yes	No	Yes
$-\frac{7}{2}$	No	No	No	Yes	No	Yes
0	No	Yes	Yes	Yes	No	Yes
$\sqrt{-4}$	No	No	No	No	No	No
$-\frac{6}{2}$	No	No	Yes	Yes	No	Yes

P, R 21 Draw a Venn diagram showing the relationship between the six sets of numbers discussed in this section.

P, R, I 22 a Use set-builder notation to describe the set of points shown.
   b If a point on this graph is selected at random, what is the probability that it will be in the third quadrant? .125

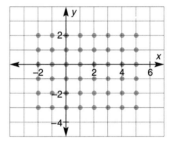

P, R 23 Which of the following pairs of sets are disjoint? **a, b,** and **d**
   a {Even integers} and {odd integers}
   b {Positive integers} and {negative integers}
   c {Prime numbers} and {even integers}
   d {Positive multiples of 5 less than 5} and {triangles with 5 sides}

P, R 24 Spreadsheets Suppose that *x* and *y* are both integers less than or equal to 12. Use a spreadsheet to find the set of all pairs (x, y) for which the hypotenuse of the triangle is an integer.

446   Chapter 9   Real Numbers

◀ LOOKING BACK **Spiral Learning** LOOKING AHEAD ▶

R, I **25** **a** Find the radius of the smallest circle that has its center at the origin and contains all of the points shown, either in its interior or on its circumference. $\sqrt{13}$

b What is the smallest radius that is an integer? 4

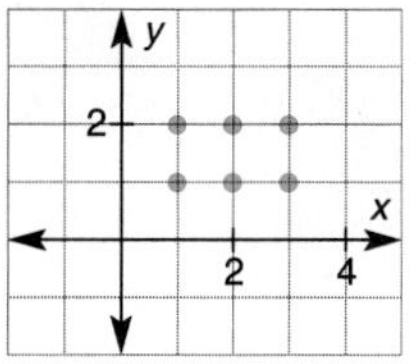

R **26** Evaluate each expression.

a $(-3)(5 \times 7)$ −105 **b** $(-3)(5)(7)$ −105 **c** $19 - 4$ 15 **d** $4 - 19$ −15

R **27** Evaluate $\begin{bmatrix} 3 & -4 \\ -1 & 6 \\ 9 & -5 \end{bmatrix} \cdot \begin{bmatrix} -2 & 4 \\ 3 & -5 \end{bmatrix}$. $\begin{bmatrix} -18 & 32 \\ 20 & -34 \\ -33 & 61 \end{bmatrix}$

R **28** **Technology** Many calculators use arrays like the one shown to display numbers. By lighting up different parts of the array (lettered A–G in the diagram), a calculator can display any of the ten digits. For example, the set of lit parts for a 3 is {A, B, D, E, F}. What set of parts corresponds to each of the following?

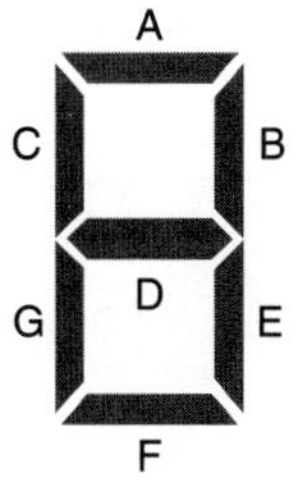

a 2 {A, B, D, F, G} **b** $3 \cap 2$ {A, B, D, F}

c 4 {B, C, D, E} **d** $1 \cap 5$ {E}

R **29** **Communicating** Suppose that A = {positive multiples of 4} and B = {positive multiples of 6}.

a Describe in words $A \cap B$. The set of positive multiples of 12

b Find the least element in $A \cap B$. {12}

R **30** Is the statement true *always, sometimes,* or *never?*

a If $A \subset B$, then $B \subset A$. Sometimes

b If $A \cap B = \{\ \}$, then A and B are disjoint. Always

R **31** If W = {w: w is a multiple of 3} and Z = {z: z is a factor of 60}, what is $W \cap Z$? {3, 6, 12, 15, 30, 60}

R **32** Given the irrational number 0.010010001..., find another irrational number that you can add to this number so that the sum will be a rational number. Possible answer: 0.101101110...

Investigation

Fractions Use a calculator to investigate the decimal forms of fractions that have a denominator of 9. Then extend your investigation to fractions with denominators of 99, of 999, of 9999, and so forth. Write a description of the patterns you notice.

Checkpoint

In problems **1–6,** write the set or sets to which the number belongs: naturals, wholes, integers, rationals, irrationals, reals.

1. 8 **2.** 0 **3.** $4\frac{1}{2}$

4. $8.\overline{6}$ **5.** $\sqrt{3}$ **6.** π

In problems **7–9,** write each number as a decimal.

7. $\frac{5}{8}$ **8.** $1\frac{6}{11}$ **9.** $\frac{11}{3}$

In problems **10–12,** write each number as a ratio of two integers.

10. $0.\overline{23}$ **11.** 5.39 **12.** $2.1\overline{6}$

Answers:

1. natural, whole, integer, rational, real **2.** whole, integer, rational, real **3.** rational, real **4.** rational, real **5.** irrational, real **6.** irrational, real **7.** 0.625 **8.** $1.\overline{54}$ **9.** $3.\overline{6}$ **10.** $\frac{23}{99}$ **11.** $\frac{539}{100}$ **12.** $\frac{13}{6}$

Communicating Mathematics

- Have students write a paragraph explaining why $\sqrt{25}$ and $\sqrt{0.25}$ are rational numbers but $\sqrt{2.5}$ is not.
- Problems 8, 29, and Investigation encourage communicating mathematics.

Additional Answers

for pages 446 and 447

21

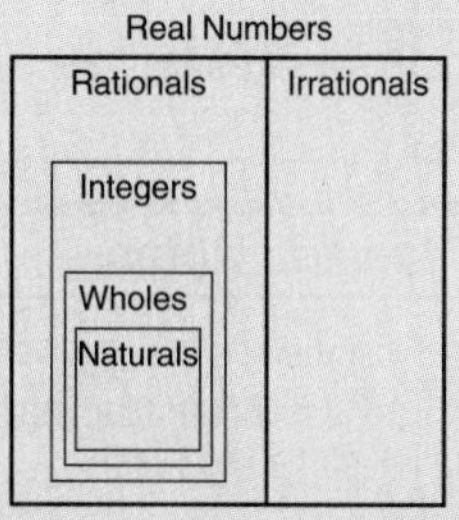

Answers for problems 22 a, 24, and Investigation are on page A8.

SECTION 9.4

Objectives

- To recognize that π is the ratio of a circle's circumference to its diameter
- To recognize that if the vertex of an angle is on a circle and the angle's rays pass through the endpoints of a diameter of the circle, the angle is a right angle

Vocabulary: chord

Class Planning

Schedule 2 days

Resources

In *Chapter 9 Student Activity Masters* Book
Class Openers 37, 38
Extra Practice 39, 40
Cooperative Learning 41, 42
Enrichment 43
Learning Log 44
Using Manipulatives 45, 46
Spanish-Language Summary 47

Class Opener

Use a compass to draw two circles of different sizes. Draw a diameter in each. Choose three points on each circle and, with your ruler, draw the two segments connecting each point with the endpoints of the diameter. Measure each angle formed by the pairs of segments you drew. What do you notice?

Answer: All are right angles.

Lesson Notes

- The irrationality of π is provable only with advanced mathematics, so at this point it must be taken on faith. There have been many attempts to compute as many digits of π as possible. The purpose

Investigation Connection
Section 9.4 supports Unit E, *Investigation* 5. (See pp. 426A–426D in this book.)

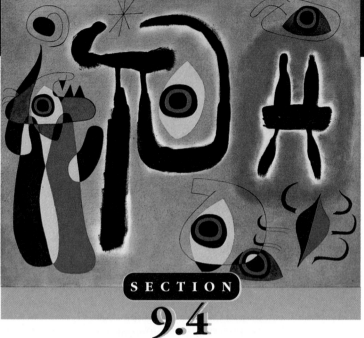

SECTION 9.4
Pi and the Circle

THE MEANING OF π

A well known irrational number is π (pi). You have used this number to find circumferences and areas of circles. The symbol ⊙O (read "circle O") identifies a circle with center at point O.

$\overline{OA}$ is a radius of the circle. $\overline{CD}$ is a **chord.** The longest chord passes through the center and is a diameter of the circle. A diameter of a circle is twice as long as a radius of the circle. $\overline{RS}$ is a diameter.

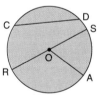

The people of many ancient civilizations discovered that dividing the circumference of a circle by the circle's diameter always results in the same number, regardless of the size of the circle. This ratio of the circumference to the diameter of a circle is the number π.

You can estimate the value of π by collecting several cans of different sizes. Use a string to measure their circumferences and diameters as accurately as possible. Divide each circumference by the corresponding diameter. Take the average of the ratios that you calculate. It should be close to 3.14.

448 Chapter 9 Real Numbers

SECTION 9.4

The number π cannot be written as a ratio of two integers or as a terminating or repeating decimal. A close approximation of π can be found by pressing the π key on the calculator.

Example 1
If the distance around a circle is 34.7 centimeters, what is the circle's radius?

Solution
Based on the definition of π, $\frac{34.7}{\text{diameter}} = \pi$. Therefore,

$$\pi \times \text{diameter} = 34.7$$
$$\text{diameter} = \frac{34.7}{\pi} \approx 11.0 \text{ cm}$$

The radius is half the diameter, so the radius is ≈5.5 centimeters

Example 2
a Find the circumference of ⊙P.
b Find the area of ⊙P.

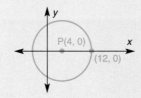

Solution
Since (4, 0) and (12, 0) are 8 units apart, the radius of the circle is 8.
a Circumference = $2\pi r = 2 \cdot \pi \cdot 8 = 16\pi \approx 50.27$
b $A = \pi r^2 = \pi \cdot 8^2 = 64\pi \approx 201.06$
Sometimes an exact answer, such as 16π, is more useful than an approximation, such as 50.27, and sometimes not.

A PROPERTY OF CIRCLES

Circles have a number of interesting and useful properties. Draw several circles of different sizes, along with their diameters. Then draw a triangle in each circle such that the diameter is one side and the vertex opposite the diameter is on the circumference of the circle. In each figure, measure the angle opposite the diameter. You will find that the angle is a right angle. In the figure, $\overline{AC}$ is a diameter and ∠B is a right angle.

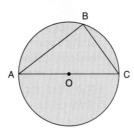

▶ **Right Angle Property:** If $\overline{AC}$ is a diameter of a circle and B is any point of the circle other than A or C, then ∠ABC is a right angle.

---

has not been to see if it repeats but rather to look for nonrepeating patterns within the digits. Computing digits of π depends on results from calculus.

- Emphasize to students that since π is irrational, numbers such as $\frac{22}{7}$, 3.14, and even 3.141592654 are only approximations.
- An easy way to create a right angle is to pick any point on a semicircle and connect it to the endpoints of the diameter.

Additional Examples

Example 1
The circumference of a circle is 56.26 cm. What is the circle's radius?
Answer: ≈8.95 cm

Example 2
a. Find the circumference of ⊙H.
b. Find the area of ⊙H.

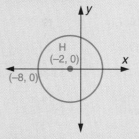

Answers:
a. 12π, or ≈37.7 **b.** 36π, or ≈113

Communicating Mathematics

- Have students write a letter to the principal of your school explaining what π is and how it can be used.
- Problems 1, 2, 4–6, 10, and 14 encourage communicating mathematics.

SECTION 9.4

Assignment Guide

Basic
Optional

Average or Heterogeneous
Day 1 11–14, 16, 19, 21, 22, 24, 27, 30, 32
Day 2 15, 17, 18, 20, 23, 25, 26, 28, 31, 33

Honors
Day 1 11, 14, 16, 17, 19, 21, 24, 27, 30
Day 2 12, 13, 15, 18, 20, 22, 23, 25, 28

Strands

Communicating Mathematics	1, 2, 4–6, 10, 14
Geometry	5, 6, 8–27
Measurement	9, 11–27
Data Analysis/Statistics	29
Technology	3, 4, 24, Investigation
Probability	9, 22, 28
Algebra	1, 2, 8–10, 12–24, 26, 27, 31–34
Patterns & Functions	8, 14, Investigation
Number Sense	1–4, 6, 7, 14, 25, 27, 30, Investigation
Interdisciplinary Applications	24–26, 29, Investigation

Individualizing Instruction

Limited English proficiency
Students may be more likely to understand and remember the meaning of *concentric*, introduced in problem 9, if they draw concentric circles. Also, you may have them consider the Latin roots of the word: *com-* (together) + *centrum* (center).

450

Sample Problems

Problem 1 Circle O has diameter $\overline{AC}$, AC = 10, and AB = 8. If a dart has an equal chance of landing on any point within the circle, what is the probability that the dart will land inside the triangle?

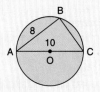

Solution By the Right Angle Property, ∠ABC is a right angle. Using the Pythagorean Theorem, we find that $(BC)^2 + 64 = 100$, so BC = 6.

Area of △ABC = $\frac{1}{2}bh = \frac{1}{2}(8)(6) = 24$
Area of ⊙O = $\pi \cdot 5^2 = 25\pi$
Probability = $\frac{24}{25\pi} \approx .31$

The dart will land in the triangle about 31% of the time.

Problem 2 A circle with radius 2 rolls along the outside of the square without slipping. How far does the center of the circle travel when the circle makes one complete trip around the square?

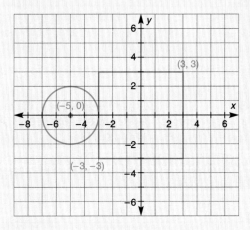

Solution The path of the center of the circle consists of four straight segments and four quarters of a circle with radius 2. Therefore, the distance traveled is $4(6) + 2 \cdot \pi \cdot 2$, or ≈ 36.57 units.

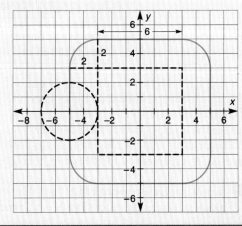

450 Chapter 9 Real Numbers

SECTION 9.4

Think and Discuss

P, R **1** Can the ratio of two irrational numbers be a rational number? Explain your answer. Yes; for example, $\frac{\pi}{\pi} = 1$.

P, R **2** Can the ratio of two rational numbers be an irrational number? Explain your answer. No. Such a fraction can always be simplified: $\frac{a}{b} \div \frac{c}{d} = \frac{ad}{bc}$

P **3** Press the π key on a scientific calculator. Does the display show a rational number or an irrational number? Rational

4 Three values sometimes used for π are $\frac{22}{7}$, 3.14, and $\frac{355}{113}$.
P, R **a** Are any of these the exact value of π? Why or why not? No. They are rational.
  b Which of the three is the closest to π? $\frac{355}{113}$

P, R **5** In the diagram shown, AB = CD, BC = AD, and $\overline{AC}$ is a diameter.
  a What type of figure is ABCD? Why?
  b What is a way to find the center of a circle?

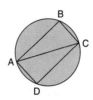

P **6** Explain what the letter π stands for.

P **7** Is the statement *true* or *false*?
  a $7\pi = 22$ False **b** $100\pi = 314$ False **c** $\pi > 3.14$ True

P, I **8** Let r = radius of a circle, d = diameter of the circle, A = area of the circle, and C = circumference of the circle. Write a function for
  a d in terms of r **b** r in terms of d **c** A in terms of r
  d A in terms of d **e** C in terms of r **f** C in terms of d
  g d in terms of C **h** r in terms of C **i** r in terms of A

P, R **9** These two circles are *concentric*— that is, they have the same center. If a dart has an equal chance of landing on any point within the outer circle, what is the probability that the dart will land within the inner circle? $\frac{1}{16}$

P, R **10** In solving problem 9, Laura could not remember the value of π, so she used 5. Why did she still get the correct answer? π divides out of the ratio.

Problems and Applications

P **11** Determine the
  a Area of a circle with radius 7 49π, or ≈ 153.9
  b Circumference of a circle with radius 8 16π, or ≈ 50.3
  c Circumference of a circle with diameter 9 9π, or ≈ 28.3
  d Area of a circle with diameter 10 25π, or ≈ 78.5

Section 9.4 Pi and the Circle   451

SECTION 9.4

Problem-Set Notes

for pages 452 and 453

- 21 and 24 point out that the circumference of the circle tells us how far it would roll in one revolution.

Reteaching

To reteach sample problem 2, have students cut out a circle and square from cardboard. Have them tape the square to a piece of graph paper, poke a hole in the center of the circle, and use a pencil to run the circle around the square. Students could cut up the figure generated and verify that it forms figures congruent to the original square and circle. This activity could also be done using Spirograph pieces.

Additional Practice

In problems 1–4, determine each value.
1. The area of a circle with radius 5.
2. The circumference of a circle with radius 9.
3. The circumference of a circle with diameter 7.
4. The area of a circle with diameter 12.

In problems 5–9, use ⊙F.

5. What is the diameter of ⊙F?
6. What is the radius of ⊙F?
7. What is the area of ⊙F?
8. What is the circumference of ⊙F?
9. Find the area of the triangle.

452

P **12** What is this circle's
  a Diameter? 10
  b Radius? 5
  c Area? 25π, or ≈ 78.5
  d Circumference? 10π, or ≈ 31.4

P **13** Circle O rolls along the line from A to B. How far does point O travel? 20 ft

P, R **14** **Communicating** If $C(r) = 2\pi r$ and $A(r) = \pi r^2$, what are $C(5)$ and $A(4)$? Why does $C(-2)$ not make sense?

15 The circle rolls along the outside of the rectangle without slipping. How far has the center of the circle traveled after the circle has made one complete trip around the rectangle? $56 + 8\pi$, or ≈ 81

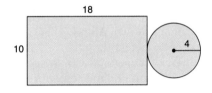

P **16** The circumference of a juice can is about 8.5 inches. Find the diameter of the can. ≈ 2.7 in.

P, R **17** $\overline{AB}$ is a diameter of the circle. Find the coordinates of the endpoints of the diameter that is perpendicular to $\overline{AB}$. $(12, 6); (12, -6)$

P **18** $\overline{AC}$ is a diameter of ⊙O. $AC = 26$ and $AB = BC$.
  a Find the area of $\triangle ABC$. 169
  b Find AB. ≈ 18.4

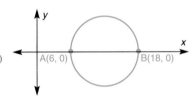

P **19** The circle is inscribed in the triangle.
  a Find the radius of the circle. 5
  b Which is greater, AB or the circumference of the circle? AB

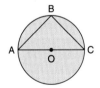

P **20** Circle O has diameter $\overline{BD}$. $AB = 15$, $AD = 20$, and $DC = 7$. Find BD and BC.
  25; 24

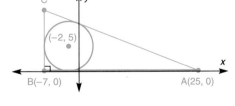

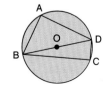

452 Chapter 9 Real Numbers

SECTION 9.4

P **21** A wheel with a diameter of 3 feet rolls along a surface, making one complete revolution. How far does the center of the wheel travel? 3π, or ≈9.4 ft

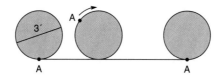

P, R **22** For the circle shown,
  a Find the length of the diameter ≈5.7 in.
  b Find the area of the circle ≈25.1 in.^{2}
  c If a dart has an equal chance of landing on any point within the circle, what is the probability that the dart will land in the square? ≈.64

P **23** The circle rolls along the outside of the triangle without slipping. Find the distance that the center of the circle has traveled after one complete trip around the triangle. $24 + 6\pi$, or ≈42.8

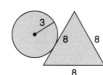

P **24** A bicycle wheel has a diameter of 27 inches.
  a When the wheel makes one revolution, how far does the bike travel? ≈84.82 in.
  b How far is this in feet? ≈7.07 ft
  c How many revolutions of the wheel are needed for the bike to travel one mile? (Hint: There are 5280 feet in a mile.) 747

P **25** Science The pedal gear of a bike has 48 teeth. The gear on the rear wheel has 14 teeth.
  a Which completes a revolution faster, the pedal gear or the gear on the rear wheel? Rear-wheel gear
P
  b How many revolutions does the rear wheel make for each revolution of the pedals? $\frac{48}{14}$, or ≈3.4, rev.

P **26** In problem 25, if the diameter of the rear wheel is 27 inches, how many revolutions of the pedals are needed to travel one mile? 218

Answers:
1. 25π, or ≈78.5 2. 18π, or ≈56.5 3. 7π, or ≈22.0 4. 36π, or ≈113.1 5. 10 6. 5 7. 25π, or ≈78.5 8. 10π, or ≈31.4 9. 24

Checkpoint

In problems **1–4**, determine each value.
1. The area of a circle with radius 17.
2. The circumference of a circle with radius 12.
3. The circumference of a circle with diameter 11.
4. The area of a circle with diameter 12.

In problems **5–9**, use ⊙D.

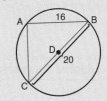

5. What is the diameter of ⊙D?
6. What is the radius of ⊙D?
7. What is the area of ⊙D?
8. What is the circumference of ⊙D?
9. Find the length of $\overline{AC}$.

Answers:
1. 289π, or ≈907.9 2. 24π, or ≈75.4 3. 11π, or ≈34.6 4. 36π, or ≈113.1 5. 20 6. 10 7. 100π, or ≈314 8. 20π, or ≈62.8 9. 12

Additional Answers

for pages 452 and 453

14 10π, or ≈31.4; 16π, or ≈50.3; length of the radius cannot be negative

◀ LOOKING BACK **Spiral Learning** LOOKING AHEAD ▶

R, I **27** Find the largest integer value of x for which the figure will have a perimeter of less than 27 inches. 4

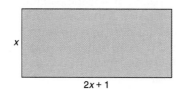

R **28** In how many ways can eight students be arranged in a row of eight seats? 40,320

Section 9.4 Pi and the Circle

SECTION 9.4

Multicultural Note

The Egyptians identified π, the constant ratio of the circumference of any circle to its diameter, around 4000 B.C. At that time, they determined the value of π to be 3.16, which is very close to its actual value.

Stumbling Block

Students may confuse radius and diameter. Simple memory joggers such as "the radius radiates from the center" will help. Also, varying the given information in practice problems (i.e., using radius one time, diameter the next) keeps students on their toes.

Additional Answers

for page 454

29 Favorite Fitness Activities

31 a

b

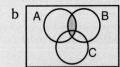

c

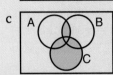

Investigation After evaluating the expression for the first three fractions, the resulting value is ≈2.86; check students' spreadsheets for accuracy; more than one hundred

454

R **29** The following favorite fitness activities were reported in a survey of 397 people: running, 150; biking, 92; aerobics, 73; and walking, 82. Draw a circle graph to represent this information. Include an appropriate percent in each section.

R **30** Copy the table and complete it by writing *yes* or *no* in each space to indicate the sets to which each number belongs.

	Natural	Whole	Integer	Rational	Irrational	Real
−2.3	No	No	No	Yes	No	Yes
π	No	No	No	No	Yes	Yes
$-\frac{6}{2}$	No	No	Yes	Yes	No	Yes
3	Yes	Yes	Yes	Yes	No	Yes
$2.\overline{3}$	No	No	No	Yes	No	Yes
0	No	Yes	Yes	Yes	No	Yes
$\sqrt{2}$	No	No	No	No	Yes	Yes

R **31** Draw three Venn diagrams like the one shown. Shade the regions that represent
  a $A \cup B$ **b** $A \cap B$ **c** $(A \cap B) \cup C$

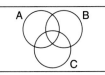

R, I **32** Solve each equation and simplify each expression.
  a $3(x-2) - 4(2-x)$ $7x - 14$
  b $3(x+4) - 2x = 12$ $x = 0$
  c $9 - 3(4 - x) = 18$ $x = 7$
  d $-10 - 3x = -76$ $x = 22$
  e $9 - 4(1 + 3x) - 76$ $-12x - 71$
  f $4x - 2(x-3) - 6$ $2x$

R **33** Write $12.\overline{39}$ as a fraction. $\frac{409}{33}$

R **34** Write $4\frac{7}{11}$ as a decimal. $4.\overline{63}$

Investigation

Approximating π   Mathematicians have discovered a number of expressions that can be used to approximate the value of π. One of them is

$$\sqrt{6\left(\frac{1}{1^2} + \frac{1}{2^2} + \frac{1}{3^2} + \frac{1}{4^2} + \ldots\right)}$$

The farther the pattern inside the parentheses is extended, the closer the value of the expression is to π. Use a calculator to evaluate the expression for the first three fractions in the pattern. How close is the resulting value to π? Then see if you can figure out a way to use a spreadsheet to continue the pattern. How many fractions do you have to use to approximate π to the nearest hundredth?

454   Chapter 9   Real Numbers

SECTION 9.5

Properties of Real Numbers

Objectives
- To apply the associative properties of real numbers
- To apply the commutative properties of real numbers
- To apply the Zero Product Property

Vocabulary: associative, commutative

Class Planning

Schedule 2 days

Resources

In *Chapter 9 Student Activity Masters* Book
Class Openers 49, 50
Extra Practice 51, 52
Cooperative Learning 53, 54
Enrichment 55
Learning Log 56
Using Manipulatives 57, 58
Spanish-Language Summary 59

Class Opener

Evaluate each side of the equation. Is the statement *true* or *false*?
1. $(3 + 10) + 8 = 3 + (10 + 8)$
2. $(3 - 10) - 8 = 3 - (10 - 8)$
3. $(8 \cdot 4) \cdot 2 = 8 \cdot (4 \cdot 2)$
4. $(8 \div 4) \div 2 = 8 \div (4 \div 2)$
5. $10 \cdot 5 = 5 \cdot 10$
6. $10 \div 5 = 5 \div 10$
7. $4 + 7 = 7 + 4$
8. $4 - 7 = 7 - 4$

Answers: 1. True 2. False 3. True
4. False 5. True 6. False 7. True
8. False

Lesson Notes

- Most students take the associative and commutative properties for granted—they have used them many times without realizing it. Giving these properties

ASSOCIATIVE PROPERTIES

Example 1
Find the area of the triangle.

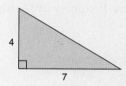

Solution
The formula for the area is $A = \frac{1}{2}bh$.
Method 1 $\frac{1}{2}(4 \cdot 7) = \frac{1}{2} \cdot 28 = 14$
Method 2 $(\frac{1}{2} \cdot 4)7 = 2 \cdot 7 = 14$

The answer is the same for both methods. It doesn't matter which two numbers you multiply first. We say that multiplication is **associative**.

▶ **Associative Property of Multiplication:** $a(bc) = (ab)c$

Addition has a similar property: $4 + (5 + 7) = (4 + 5) + 7 = 16$. It doesn't matter which two numbers you add first.

▶ **Associative Property of Addition:** $a + (b + c) = (a + b) + c$

Example 2
Evaluate each pair of expressions.
a $36 \div (6 \div 2)$ and $(36 \div 6) \div 2$
b $12 - (7 - 2)$ and $(12 - 7) - 2$

Solution
a $36 \div (6 \div 2) = 36 \div 3 = 12$
   $(36 \div 6) \div 2 = 6 \div 2 = 3$

b $12 - (7 - 2) = 12 - 5 = 7$
   $(12 - 7) - 2 = 5 - 2 = 3$

Section 9.5 Properties of Real Numbers 455

SECTION 9.5

Lesson Notes continued

names makes it easier for us to talk about them and to identify situations where they are used.

- Examples 2 and 4 emphasize that subtraction and division are neither commutative nor associative.

- The Zero Product Property plays an important role in several algebraic situations. Sample problem 2 shows an application of this property. Students may have difficulty with the property in all but its most simple applications.

Additional Examples

Example 1
Write two different expressions for the area of the triangle. Evaluate both expressions.

Possible answers:
$\frac{1}{2}(6 \cdot 9) = \frac{1}{2} \cdot 54 = 27$
$(\frac{1}{2} \cdot 6) \cdot 9 = 3 \cdot 9 = 27$

Example 2
Evaluate each pair of expressions.
a. $(15 - 4) - 1$ and $15 - (4 - 1)$
b. $60 \div (10 \div 2)$ and $(60 \div 10) \div 2$

Answers:
a. 10 and 12   b. 12 and 3

Example 3
Write two different expressions for the volume of the box. Evaluate each expression.

Possible answers:
$3 \cdot 5 \cdot 7 = 105$; $7 \cdot 3 \cdot 5 = 105$

456

Example 2 shows that division and subtraction are not associative.

COMMUTATIVE PROPERTIES

Example 3
Find the length of the hypotenuse in the figure shown.

Solution
Method 1 $8^2 + 15^2 = 64 + 225 = 289$; $\sqrt{289} = 17$
Method 2 $15^2 + 8^2 = 225 + 64 = 289$; $\sqrt{289} = 17$

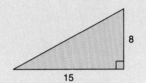

The order in which the numbers are added does not matter. We say that addition is *commutative.*

▶ **Commutative Property of Addition:** $a + b = b + a$

The order of multiplication also does not affect a result. The area of an 8-by-3 rectangle can be found by evaluating either $8 \cdot 3$ or $3 \cdot 8$.

▶ **Commutative Property of Multiplication:** $ab = ba$

Example 4
Evaluate each pair of expressions.
a $42.3 \div 6.2$ and $6.2 \div 42.3$ b $31.7 - 58.5$ and $58.5 - 31.7$

Solution
a $42.3 \div 6.2 = 6.8$ b $31.7 - 58.5 = -26.8$
   $6.2 \div 42.3 = 0.15$ $58.5 - 31.7 = 26.8$

Example 4 shows that division and subtraction are not commutative.

THE ZERO PRODUCT PROPERTY

Example 5
Find all values of x that solve each equation.
a $6x = 0$ b $x(5 - 9) = 0$ c $x(x - 2) = 0$ d $(x - 3)(x + 4) = 0$

Solution
The only number that can make a product equal to 0 is 0.
a $6x = 0$ will be true only if $x = 0$.

456   Chapter 9   Real Numbers

SECTION 9.5

b $x(5 - 9) = x(-4) = 0$ will be true only if $x = 0$.
c $x(x - 2) = 0$ will be true if $x = 0$ or if $x - 2 = 0$.
Therefore, there are two solutions: $x = 0$ and $x = 2$.

Check: $0(0 - 2) = 0$; $2(2 - 2) = 2(0) = 0$

d $(x - 3)(x + 4) = 0$ will be true if $x - 3 = 0$ or if $x + 4 = 0$.
Therefore, $x = 3$ and $x = -4$ are solutions.

Check: $(3 - 3)(3 + 4) = 0(7) = 0$
$(-4 - 3)(-4 + 4) = (-7)0 = 0$

▶ **Zero Product Property:** If $ab = 0$, then either $a = 0$ or $b = 0$.

Sample Problems

Problem 1 In finding the circumference of the circle, Walker and Maria used different formulas but got the same answer.

Walker: $C = 2\pi r = 2\pi(8) \approx 50.27$
Maria: $C = \pi d = \pi(16) \approx 50.27$

Use number properties to show that the formulas are equivalent.

Solution
Commutative Property of Multiplication $2\pi r = \pi(2)(r)$
Associative Property of Multiplication $= \pi(2r)$
                                                        $= \pi d$

Problem 2
a For what values of x will the area of the rectangle equal 0?
b On a number line, graph the possible values of x in this figure.

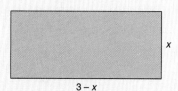

Solution
a Area of a rectangle $= bh$ $x(3 - x) = 0$
Zero Product Property $x = 0$ or $3 - x = 0$
                                         $x = 0$ or $x = 3$

b The area of a rectangle must be positive. Testing values less than 0, between 0 and 3, and greater than 3 shows that a positive product is found only when x is between 0 and 3.

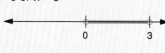

Example 4
Evaluate each pair of expressions.
a. $39.8 \div 5.1$ and $5.1 \div 39.8$
b. $27.6 - 8.4$ and $8.4 - 27.6$
Answers: **a.** ≈ 7.8, ≈ 0.13
b. 19.2, -19.2

Example 5
Find all values of x that are solutions of each equation.
a. $3x = 0$
b. $x(3 - 7) = 0$
c. $x(x + 4) = 0$
d. $(x + 6)(x - 1) = 0$
Answers: **a.** $x = 0$ **b.** $x = 0$
c. $x = 0, x = -4$ **d.** $x = -6, x = 1$

Cooperative Learning

■ Use a group activity to facilitate students' discovery of the Zero Product Property. Give each group the following table and ask them to find the row or column with the greatest product. Have students share how they solved the problem.

	A	B	C	D	E
1	6	1	8	0	5
2	9	6	7	7	0
3	0	5	2	3	6
4	8	7	4	0	1
5	4	0	3	2	3

Answer: Column C

■ Problem 34 is appropriate for cooperative learning.

Communicating Mathematics

■ Have students write a paragraph explaining how we use, or might use, the associative properties when doing mental math.

■ Problems 7, 8, 17, 27, and Investigation encourage communicating mathematics.

SECTION 9.5

Assignment Guide

Basic
Optional

Average or Heterogeneous
Day 1 9–11, 15, 16, 20, 22 (b, d), 27, 31, 32
Day 2 13, 14, 17–19, 23, 24, 28, 29, 34

Honors
Day 1 9, 10, 14, 16–18, 20, 22, 23, 27
Day 2 12, 13, 15, 19, 21, 24, 26, 28, 29, 32, 33

Strands

Communicating Mathematics	7, 8, 17, 27, Investigation
Geometry	1, 17, 23, 26–31, 34
Measurement	17, 23, 26–31, 34
Data Analysis/ Statistics	18
Technology	33, 34
Probability	32, 34
Algebra	1–17, 19–33, 35, Investigation
Patterns & Functions	23, 33
Number Sense	3, 4, 11, 13, 18, 24
Interdisciplinary Applications	14, 32, Investigation

Individualizing Instruction

The kinesthetic learner Students may benefit from using manipulatives to verify the associative properties of addition and multiplication. Problem 1 provides a good model for multiplication.

Think and Discuss

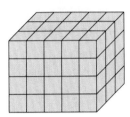

P 1 Rachel wants to know how many small cubes are in the large rectangular solid. Should she evaluate $(5 \times 4)3$ or $5(4 \times 3)$?

P 2 Match each equation with the property it illustrates.
  a $4(3 \cdot 5) = (4 \cdot 3) \cdot 5$ 1
  b $3 + 5 = 5 + 3$ 4
  c $(9 + 4) + 2 = (4 + 9) + 2$ 4
  d $(4 + 7) \cdot 3 = 4 \cdot 3 + 7 \cdot 3$ 5
  e $(4 + 7) \cdot 3 = 3 \cdot (4 + 7)$ 3
  f $5 + (2 + 6) = (5 + 2) + 6$ 2

  1. Associative Property of Multiplication
  2. Associative Property of Addition
  3. Commutative Property of Multiplication
  4. Commutative Property of Addition
  5. Distributive Property of Multiplication over Addition

P, R 3 Without using a calculator, indicate whether $<$, $>$, or $=$ goes in the blank.
  a $(-81)(-59)$ ____ $(-59)(-81)$ $=$
  b $-81 \div 3$ ____ $3 \div -81$ $<$
  c $-18 - 57$ ____ $-57 - (-18)$ $<$
  d $(-49) + 87$ ____ $87 + (-49)$ $=$

P, I 4 For what set of values of x is $(x - 100)(x - 99)(x - 98)(x - 97) = 0$ a true equation? $\{97, 98, 99, 100\}$

P, R 5 Use several sets to determine if, for the intersection and union of sets,
  a $\cap$ is commutative   Yes
  b $\cup$ is commutative   Yes
  c $\cap$ is associative   Yes
  d $\cup$ is associative   Yes

P, I 6 If the empty set, $\{\ \}$, is like 0 and intersection is an operation, is there a zero intersection property—if $A \cap B = \{\ \}$, then is $A = \{\ \}$ or $B = \{\ \}$? If not, give a counterexample.   No. If $A = \{1, 2\}$ and $B = \{3, 4\}$, $A \cap B = \{\ \}$.

P, I 7 Is the converse of the zero intersection property discussed in problem 6 true? (If $A = \{\ \}$ or $B = \{\ \}$, is $A \cap B = \{\ \}$?) Explain your answer.

P, I 8 If $a \wedge b$ is defined to mean a^b, is the operation symbolized by $\wedge$ commutative? Why or why not?   No. For example, $2^3 \neq 3^2$.

Problems and Applications

P 9 Identify the property illustrated.
  a $(3 \cdot 4) \cdot 7 = (4 \cdot 3) \cdot 7$ Com. Mult.
  b $(3 + 4) + 10 = 3 + (4 + 10)$ Assoc. Add.
  c $7x = 0$, so $x = 0$ Zero Prod.
  d $4(x + 2) = 4x + 8$ Distr.
  e $4(x + 3) = 4(3 + x)$ Com. Add.

P, R 10 Solve each equation for x.
  a $(x)6 = 0$ $x = 0$
  b $(x - 4)6 = 0$ $x = 4$
  c $(x - 4)(6x) = 0$ $x = 4, x = 0$
  d $(x - 2)(x + 9) = 0$ $x = 2, x = -9$

458   Chapter 9   Real Numbers

SECTION 9.5

P   **11** Without using a calculator, indicate whether <, >, or = goes in the blank.
      a $91 + (73 + 87)$ _____ $(91 + 73) + 87$ =
      b $77 + (41 - 82)$ _____ $(77 + 41) - 82$ =
      c $59 - (72 - 77)$ _____ $(59 - 72) - 77$ >
      d $37 - (18 + 15)$ _____ $(37 - 18) + 15$ <

P   **12** State a property or operation for each step.

$\frac{1}{2} \cdot 4(7 + 6) = (\frac{1}{2} \cdot 4)(7 + 6)$ Associative Property of Multiplication
$= 2(7 + 6)$ Multiplication
$= 14 + 12$ Distributive Property
$= 26$ Addition

P   **13** The product of three consecutive integers is 0. What are the possible values of these three integers? $\{-2, -1, 0\}, \{-1, 0, 1\}, \{0, 1, 2\}$

P, R   **14** **Consumer Math**   For a $340 television set, Goodbuy offers a 10% clearance discount followed by a 15% holiday discount. On the same item, Cheapo offers a 15% seasonal discount followed by a 10% overstocked discount. How do the TV prices of the two stores compare?   They are the same.

P, R   **15** Suppose that $A = \begin{bmatrix} 2 & 3 \\ 4 & 5 \end{bmatrix}$ and $B = \begin{bmatrix} -1 & 5 \\ 3 & 7 \end{bmatrix}$.
      a Is $A + B = B + A$?   Yes
      b Is $AB = BA$?   No
      c Do you think matrix addition is commutative?   Yes
      d Do you think matrix multiplication is commutative?   No

P   **16** Use the number properties to evaluate each expression. Do not use a calculator.
      a $(0.06 \cdot 8)100$ 48   **b** $(15 \cdot 7) \cdot \frac{2}{5}$ 42
      c $(432 + 450) + 18$ 900   **d** $98 + (576 + 2)$ 676

P, R   **17** **Communicating**   Two students computed the area of the trapezoid in different ways. Explain why each method works.

Mandy: $A = \frac{1}{2} \cdot 6(4 + 7) = 3(4 + 7) = 3 \cdot 11 = 33$
Carlos: $A = \frac{1}{2} \cdot 6(4 + 7) = \frac{1}{2} \cdot 6 \cdot 11 = \frac{1}{2}(66) = 33$

P   **18** A "combination" is determined by multiplying all of the numbers in a row or column. Without multiplying, determine which row or column will have the greatest combination.   Row D

	1	2	3	4	5
A	0	9	17	4	3
B	6	21	0	2	19
C	18	0	5	3	6
D	4	11	12	8	2
E	1	18	6	4	0

P   **19** If $r \# s = r\sqrt{s}$.
      a Is the operation symbolized by # commutative?   No
      b Is there a "zero property" for the operation?   Yes

Section 9.5   Properties of Real Numbers

SECTION 9.5

Problem-Set Notes

for pages 460 and 461

- **24** looks formidable at first glance, but is actually rather straightforward. Use some deductive reasoning.

Multicultural Note

An illustration of a small ear of corn was used to symbolize zero in the Aztec culture. Since corn, or maize, was the Aztecs' major crop, we can surmise that the concept of zero must have been important to them.

Additional Practice

Identify the property illustrated.
1. $8 + (9 + 10) = 8 + (10 + 9)$
2. $4(7 + 101) = (7 + 101)4$
3. $8 + (9 + 11) = (8 + 9) + 11$
4. $9x = 0$, so $x = 0$
5. $8(19 \cdot 21) = (8 \cdot 19)21$
6. $15 + 31 = 31 + 15$

Solve each equation for x.

7. $8x = 0$
8. $x(x + 9) = 0$
9. $(x + 1)(x - 3) = 0$
10. $(x + 5)(x + 4) = 0$

Use number properties to evaluate each expression. Do not use a calculator.

11. $(0.09 \cdot 4)100$
12. $(25 \cdot 3) \cdot \frac{3}{5}$

Answers: 1. Comm. Prop. of Add. 2. Comm. Prop. of Mult. 3. Assoc. Prop. of Add. 4. Zero Product Prop. 5. Assoc. Prop. of Mult. 6. Comm. Prop. of Add. 7. $x = 0$ 8. $x = 0$ or $x = -9$ 9. $x = -1$ or $x = 3$ 10. $x = -5$ or $x = -4$ 11. 36 12. 45

460

P   **20** If $a @ b = a + b - 2$, is the operation symbolized by @ commutative?   Yes

P, R   **21** For 2×2 matrices, the zero matrix is $\begin{bmatrix} 0 & 0 \\ 0 & 0 \end{bmatrix}$.

    a Evaluate $\begin{bmatrix} 2 & 2 \\ 0 & 0 \end{bmatrix} \cdot \begin{bmatrix} 0 & -2 \\ 0 & 2 \end{bmatrix}$. $\begin{bmatrix} 0 & 0 \\ 0 & 0 \end{bmatrix}$

    b Is there a zero product property for matrices?   No

P, I   **22** Solve each equation for x.
    a $2x(3x - 1) = 0$ $x = 0, x = \frac{1}{3}$ **b** $(3x - 4)(2x + 7) = 0$ $x = \frac{4}{3}, x = -\frac{7}{2}$
    c $5x(x - 7)(x + 11) = 0$ **d** $3(x + 2) + 0 = 15$ $x = 3$

P   **23 a** Write a function $A(x)$ for the area of the figure. $A(x) = (6 - x)(x - 2)$
    b What values of x give an area of zero? $x = 6, x = 2$
    c What is the set of possible values of x? $\{x : 2 < x < 6\}$

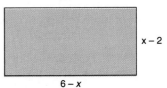

P, R   **24** The digits of a three-digit number are all different. The product of the digits is 0 and the sum is 7. If the number is divided by 10, the result is a prime number. What might the three-digit number be?   430 or 610

P   **25** Use the number properties to simplify each expression.
    a $\frac{2}{3}\left(5 \cdot \frac{3}{2}\right)$ 5   **b** $\left(\frac{1}{7} \cdot 14\right)35$ 70
    c $\frac{7}{5}\left(\frac{5}{7} \cdot 9\right)$ 9   **d** $0.85 - \left(\frac{1}{2} + \frac{17}{20}\right)$ $-\frac{1}{2}$

P   **26** In the squares shown, lengths x and y are related by the equations $(6y - 3x) \cdot 10 = 0$ and $y = x - 2$. Find the perimeter of the smaller square.   8

◀ LOOKING BACK   **Spiral Learning**   LOOKING AHEAD ▶

R   **27** For the points $(4, 9)$ and $(-2, 7)$,
    a Find the mean of the x-coordinates and the mean of the y-coordinates.   1 : 8
    b Graph the two given points and the point found in part **a**.
    c Discuss the relationships among the three points.   (1, 8) is the midpoint.

R   **28** Find lengths a, b, and c.
    $a = 8, b = 17, c = 15$

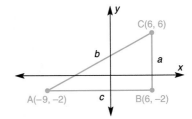

460   **Chapter 9** Real Numbers

SECTION 9.5

R **29** The measures of the angles of a triangle are in the ratio 2:5:8. Find the measures of the angles.   24, 60, 96

R **30** Copy and complete the table for the formula $A = \pi r^2$. Round to the nearest tenth.

r	2	4	6	8	10
A	12.6	50.3	113.1	201.1	314.2

R **31** $\overline{AC}$ is a diameter and AB = 7 inches. The area of the triangle is 14 square inches. Find
  a BC   4
  b AC   ≈8.06
  c The circumference of the circle   ≈25.32

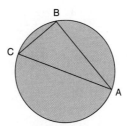

R **32** The Acme computer can be purchased with any one of 6 different disk drives, 9 different memory sizes, and 5 different monitors. How many different types of computer configurations are possible? $6 \times 9 \times 5 = 270$

I
  33 **Technology** Most scientific calculators have a [x!] key, called a *factorial* key. What are the results of the following calculator operations?
  a 2 [x!]   2
  b 1 [×] 2 [=]   2
  c 3 [x!]   6
  d 1 [×] 2 [×] 3 [=]   6
  e 4 [x!]   24
  f 1 [×] 2 [×] 3 [×] 4 [=]   24
  g 5 [x!]   120
  h 1 [×] 2 [×] 3 [×] 4 [×] 5 [=]   120

R, I **34 a** The square has side 4. A dart lands somewhere inside the square. What is the probability the dart has landed inside the circle?   ≈.785
  b Repeat part **a** for a square of side 8 and a circle that is proportionately larger than the circle in part **a**.   ≈.785

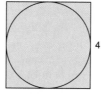

R, I **35** Find the values of *x* and *y* that solve the equations $x = 19 + 4y$ and $2x + 5y = 12$. $x = 11, y = -2$

Investigation

Language What meanings of the words *commute* and *associate* are most familiar to you? How are the everyday meanings of these words related to the mathematical meanings of *commutative* and *associative?* Consult a dictionary to find out the meanings of the Latin words that are the ancestors of these terms. Does knowing the meanings of the Latin words help you see how the English meanings are related?

Checkpoint

Match each equation with the property it illustrates.

Equations
1. $x(x + 2) = 0$, so $x = 0$ or $x = -2$
2. $5 + (6 + 9) = (5 + 6) + 9$
3. $5(8 + 3) = (8 + 3)5$
4. $4(8 \cdot 3) = (4 \cdot 8)3$
5. $8 + 9 = 9 + 8$
6. $8 + (6 + 7) = (6 + 7) + 8$

Properties
a. Associative Property of Multiplication
b. Associative Property of Addition
c. Commutative Property of Multiplication
d. Commutative Property of Addition
e. Zero Product Property

Answers: 1. e 2. b 3. c 4. a 5. d 6. d

Additional Answers

for pages 460 and 461

22 **c** $x = 0, x = 7, x = -11$

27 **b**

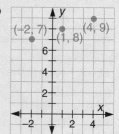

Answer for Investigation is on page A8.

Section 9.5 Properties of Real Numbers   461

SECTION 9.6

Objective
- To apply the distance and midpoint formulas

Class Planning

Schedule 1–2 days

Resources

In *Chapter 9 Student Activity Masters* Book
Class Openers 61, 62
Extra Practice 63, 64
Cooperative Learning 65, 66
Enrichment 67
Learning Log 68
Interdisciplinary
  Investigation 69, 70
Spanish-Language Summary 71

In *Tests and Quizzes* Book
Quiz on 9.4–9.6 (Average) 243
Quiz on 9.4–9.6 (Honors) 244

Transparencies 42, 43

Class Opener

Graph the points A(–5, 1), B(7, 1), and C(7, 6).
1. Use the Pythagorean Theorem to find AC.
2. Find the coordinates of the point halfway between A and B.
3. Find the coordinates of the point halfway between B and C.
4. Use a ruler to find the coordinates of the point halfway between A and C.

Answers:
1. 13   2. (1, 1)   3. $(7, 3\frac{1}{2})$ 4. $(1, 3\frac{1}{2})$

Lesson Notes

- Point out that the distance formula is just the Pythagorean Theorem and that the midpoint formula is just the mean of the coordinates.

- Subscripted variables sometimes confuse students. You might do

---

Investigation Connection
Section 9.6 supports Unit E *Investigation* 5. (See pp. 426A–426D in this book.)

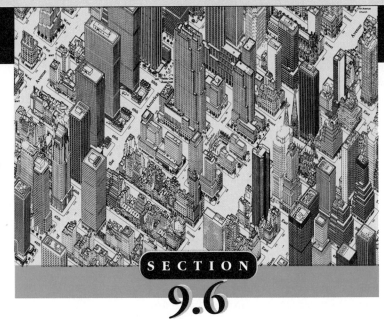

SECTION 9.6
Two Coordinate Formulas

THE DISTANCE FORMULA

In the city of Fairview, the avenues run east-west and the streets run north-south—except for Main Street, which cuts diagonally across many streets and avenues. Fairview Middle School is located at the intersection of Second Avenue, Fifth Street, and Main Street. Fairview High School is located at the intersection of Eighth Avenue, Ninth Street, and Main Street. The distance between two consecutive streets or avenues is 0.1 mile.

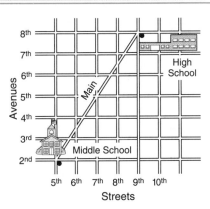

You can find the distance from the middle school to the high school by drawing a right triangle. The distance is 0.4 mile east-west and 0.6 mile north-south.

$$d = \sqrt{0.6^2 + 0.4^2} \approx 0.7 \text{ mi}$$

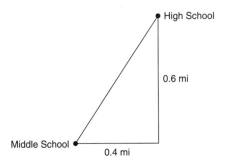

462   **Chapter 9**   Real Numbers

SECTION 9.6

Example 1
Find the distance between points A and B.

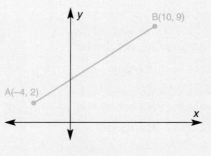

Solution
Form a right triangle. Find the lengths of the legs.

Horizontal leg: $|10 - (-4)| = 14$
Vertical leg: $|9 - 2| = 7$

By the Pythagorean Theorem,
$AB = \sqrt{14^2 + 7^2} \approx 15.7$

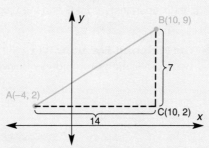

It is possible to find the distance by working directly with the coordinates. In Example 1, $AB = \sqrt{[10 - (-4)]^2 + (9 - 2)^2}$

▶ **The Distance Formula:** If (x_1, y_1) and (x_2, y_2) are the coordinates of any two points, then the distance between these two points is equal to

$$\sqrt{(x_2 - x_1)^2 + (y_2 - y_1)^2}$$

THE MIDPOINT FORMULA

Look again at the map of Fairview. Can you determine what intersection is midway between the middle school and the high school?

Example 2
Find the coordinates of the point midway between A(−3, 8) and B(15, 2).

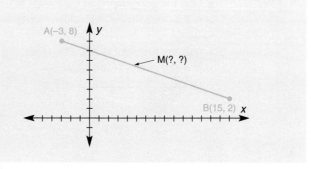

Section 9.6 Two Coordinate Formulas

---

some examples with and without the subscripts to help students along, but, ultimately, it is important for students to become familiar with these variables.

- Sample problem 2 is an application of the distance and midpoint formulas to geometry.

Additional Examples

Example 1
Refer to example 2 on page 463. Find the distance between points A and B.

Answer: $AB \approx 18.97$

Example 2
Refer to example 1 on page 463. Find the coordinates of the point midway between points A and B.

Answer: (3, 5.5)

Reteaching

We can represent a general ordered pair as (x, y). The distance and midpoint formulas, however, require that we find some way to represent a second general ordered pair. We could use other letters, but then we would have to keep track of which were x-coordinates and which were y-coordinates. It would be nice to use x and y again, but how? Subscripts are the answer. They allow us to use x and y for both ordered pairs. The general points (x_1, y_1) and (x_2, y_2) can stand for any two ordered pairs. We read (x_1, y_1) as "x sub 1, y sub 1."

SECTION 9.6

Assignment Guide

Basic
Optional

Average or Heterogeneous
Day 1 8 (a, c), 9 (a, c), 10, 12, 13 (a, c), 16, 22, 25, 26, 30
Day 2 8 (b, d), 9 (b, d), 11, 14, 15, 19, 21, 27, 29, 31

Honors
Day 1 8 (a, c), 9 (a, c), 10, 11, 13, 15, 18, 19, 22, 23, 30–32

Strands

Communicating Mathematics	1, 3, 5, 7, Investigation
Geometry	1, 2, 5–23, 32, Investigation
Measurement	4–8, 10–19, 21–23
Data Analysis/ Statistics	24
Technology	13, 15
Algebra	1–21, 23–27, 29–32
Patterns & Functions	7, 11, 24, 32
Number Sense	1, 2, 27, 28
Interdisciplinary Applications	13, 25, Investigation

Individualizing Instruction

The visual learner Although students may use formulas to find the answers to problems 8, 9, and 12, they may benefit from drawing graphs first.

Problem-Set Notes

for page 465

■ **1** and **2** confuse many students. In the midpoint formula, the order of the coordinates is not important, since addition is commutative. In the distance formula,

464

Solution
First, draw a right triangle.
Find the coordinates of P by averaging the y-coordinates.

$$\left(-3, \frac{8+2}{2}\right) = (-3, 5)$$

Find the coordinates of Q by averaging the x-coordinates.

$$\left(\frac{-3+15}{2}, 2\right) = (6, 2)$$

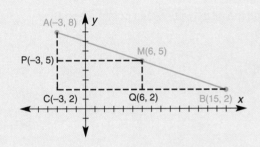

Using the x-coordinate from Q and the y-coordinate from P gives the midpoint M(6, 5).

▶ **The Midpoint Formula:** If (x_1, y_1) and (x_2, y_2) are the coordinates of two points, the point midway between the two points has the coordinates

$$\left(\frac{x_1 + x_2}{2}, \frac{y_1 + y_2}{2}\right)$$

Sample Problems

Problem 1
a Find the circumference of Circle O.
b Find the area of Circle O.

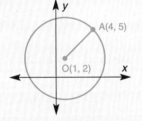

Solution Find the radius of the circle by using the distance formula.

$$r = \sqrt{(4-1)^2 + (5-2)^2} = \sqrt{9+9} \approx 4.24$$

a $C = 2\pi r = 2\pi(4.24) \approx 26.6$ b $A = \pi r^2 = \pi(4.24)^2 \approx 56.5$

Problem 2 The endpoints of a diameter of a circle are A(−4, −3) and B(6, 15).
a Find the coordinates of the center of the circle.
b Is the point P(5, −3.4) inside or outside the circle?

Solution
a M is the midpoint of $\overline{AB}$. Therefore, M is the center of the circle.

$$M\left(\frac{-4+6}{2}, \frac{-3+15}{2}\right) = M(1, 6)$$

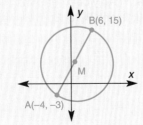

464

SECTION 9.6

b The radius of the circle can be found by using the distance formula with B(6, 15) and M(1, 6).

$$r = \sqrt{(6-1)^2 + (15-6)^2} = \sqrt{5^2 + 9^2} \approx 10.296$$

We next want to find the distance from M(1, 6) to P(5, –3.4).

$$MP = \sqrt{(1-5)^2 + [6-(-3.4)]^2} = \sqrt{(-4)^2 + 9.4^2} \approx 10.216$$

Since the distance from M to P is less than the radius of the circle, point P lies inside the circle.

Think and Discuss

P **1** When you use the distance formula, does it matter which point you use as (x_1, y_1) and which you use as (x_2, y_2)? Why or why not?

P **2** Does the order of points matter in the midpoint formula? No

P, R **3** Explain the connection between averages and the midpoint formula.

P **4** Find the coordinates of some points that are 5 units from the origin.

P **5** Knowing an endpoint, A, and the midpoint, M, of a segment, Jason used the figure shown to find the other endpoint.
  a What are the coordinates of the other endpoint, B? (8, –3)
  b Why does the figure work?

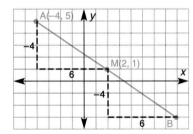

P **6** Given points A(–10, 2) and B(6, –10), find points that divide segment $\overline{AB}$ into four congruent segments. (–6, –1), (–2, –4), (2, –7)

P, I **7 a** Find the distance AB for $a = 2$ and $b = 3$. ≈ 9.90
  b Find AB for $a = 7$ and $b = 1$. ≈ 9.90
  c Find AB for $a = -1$ and $b = 6$. ≈ 9.90
  d Explain the results of parts **a, b,** and **c.**

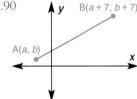

Problems and Applications

P **8** Find the distance between each pair of points.
  a (–5, 6) and (8, 6) 13
  b (4, –1) and (4, 12) 13
  c (5, 0) and (0, –12) 13
  d (6, –3) and (3, 1) 5

---

even though subtraction is not commutative, the squares of both differences are equal.

■ **5** emphasizes the conceptual underpinnings of the midpoint formula.

■ **6** extends the midpoint formula. As a challenge problem, you might ask students to divide the segment into three equal parts.

Communicating Mathematics

■ Have students write a paragraph explaining how the Pythagorean Theorem is related to the distance formula.

■ Problems 1, 3, 5, 7, and Investigation encourage communicating mathematics.

Additional Answers

for page 465

1 No, because $(x_2 - x_1)^2$ and $(y_2 - y_1)^2$ have the same value no matter which order the coordinates are substituted in.

3 The coordinates of the midpoint are the mean of the x-coordinates and the mean of the y-coordinates.

4 There are infinitely many. Possible answers: (5, 0), (3, 4), (–3, 4)

5 b Possible answer: Since the right triangles are congruent, AM = MB.

7 d No matter what the values of a and b, the difference between a and $a + 7$ and between b and $b + 7$ is always 7.

SECTION 9.6

Problem-Set Notes

for pages 466 and 467

- **19** reviews the idea of a vertex matrix and provides practice with the midpoint formula.

Cooperative Learning

Have students work in groups on the following problems.

1. Make a table of all the lattice points (points with integer coordinates) in problem 4 that are five units from the origin. Use the Pythagorean Theorem to find a pattern or rule associated with these numbers.

2. Make another table of all the lattice points that are 13 units from the origin. What rule do these points follow?

3. Try to write a general rule for circles.

Answers:
1. (0, 5), (3, 4), (4, 3), (5, 0), (4, –3), (3, –4), (0, –5), (–3, –4), (–4, –3), (–5, 0), (–4, 3), (–3, 4); $(x, y): x^2 + y^2 = 25$
2. $(x, y): x^2 + y^2 = 169$
3. $(x, y): x^2 + y^2 = r^2$, where r is the radius of the circle.

- Problem 13 and Investigation are appropriate for cooperative learning.

Multicultural Note

The Babylonians knew the relationship among the sides of a right triangle long before it was named the Pythagorean Theorem. The theorem is not totally misnamed, however, because it was probably first proved by a Greek mathematician about the time that Pythagoras lived.

P **9** Find the midpoint of the segment joining each pair of points.
  a (9, –4) and (15, –10)   (12, –7)   **b** (–8, 10) and (16, 5)   (4, 7.5)
  c (1.3, 5.8) and (2.5, –6)   (1.9, –0.1)   **d** (3.9, 2) and (–4.7, 2)   (–0.4, 2)

P **10** M is the midpoint of $\overline{AB}$. Find the coordinates of B.   (8, 5)

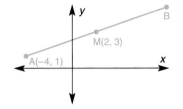

P, R **11** For the given circle, find
  a AO   5   **b** BO   5
  c CO   5   **d** FO   5

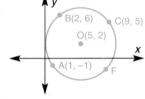

P **12** Given points A(–8, –3) and B(4, 13), find
  a The length of $\overline{AB}$ 20
  b The midpoint of $\overline{AB}$ (–2, 5)
  c The distance from A to the midpoint   10

P **13** **Surveying** A land surveyor for a housing development drew the grid shown. The side length of each small square represents 1000 feet. Find, to the nearest foot, the distance between points
  a B2 and B5   3000 ft   **b** C3 and F6   4243 ft
  c G1 and F5   4123 ft   **d** A2 and D6   5000 ft

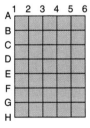

P **14** Find the distance from C to the midpoint of $\overline{AB}$.   ≈ 8.1

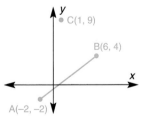

P **15** Find AB, BC, and AC. What kind of triangle is △ABC? $\sqrt{32}, \sqrt{80}, \sqrt{80}$; isosceles

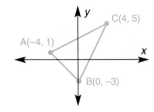

466   Chapter 9   Real Numbers

SECTION 9.6

P **16** Pick three points on the coordinate plane that do not all lie on a straight line, and connect them to form a triangle. Locate the midpoints of two sides of the triangle. *Answers will vary in parts* **a** *and* **b.**
  a Find the length of the segment joining the two midpoints.
  b Find the length of the side opposite the segment used in part **a.**
  c What is the ratio of the lengths found in parts **a** and **b**? 1:2

P **17** Find the midpoint of the segment joining each pair of points.
  a $A(\sqrt{2}, 3)$ and $B(-\sqrt{2}, 5)$ (0, 4)   **b** $C(0.5, -0.5)$ and $D(-0.5, 1.5)$ (0, 0.5)

P, R **18** For circle O, find
  a The radius 5
  b The diameter 10
  c The area 25π, or ≈ 78.5
  d The circumference 10π, or ≈ 31.4

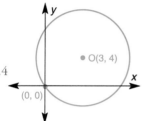

P, R **19** Draw a triangle with the given vertex matrix. Connect the midpoints of the sides of the triangle and find the vertex matrix for the smaller triangle. Could you have found the matrix without drawing the smaller triangle?

$$\begin{array}{c} x \\ y \end{array} \begin{bmatrix} 3 & -4 & 8 \\ -9 & 5 & -4 \end{bmatrix}$$

P, R, I **20** If A$B means finding the midpoint of $\overline{AB}$, is A$B = B$A? Yes

P **21** Locate the points A(3, 9), B(−5, 6), C(−3, −5), and D(5, −2) on a coordinate system.
  a Draw and name the quadrilateral having these four points as vertices.
  b Find the midpoint of $\overline{AC}$. (0, 2)
  c Find the midpoint of $\overline{BD}$. (0, 2)
  d Find the lengths of $\overline{AB}$ and $\overline{CD}$. $\sqrt{73}$, or ≈ 8.5
  e Find the lengths of $\overline{BC}$ and $\overline{AD}$. $\sqrt{125}$, or ≈ 11.2

P, R **22** Find the area of the shaded region, given that ABCD is a square.
$36\pi - 72$, or ≈ 41.1, in.^{2}

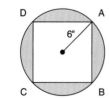

P **23 a** Find the coordinates of the midpoint of $\overline{AC}$. (5, 2.5)
  b Find the coordinates of the midpoint of $\overline{BD}$. (5, 2.5)
  c What do you notice?
    They are the same.

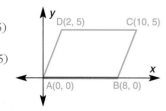

Additional Practice

Find the distance between each pair of points.
1. (8, 6) and (12, −1)
2. (4, 6) and (−2, 3)
3. (7, 12) and (4, 9)

Find the midpoint of the segment joining each pair of points.
4. (7, 9) and (3, 15)
5. (2, −1) and (7, 14)
6. (2.6, 3.9) and (3.8, 4.7)

Answers:
1. $\sqrt{65}$ 2. $\sqrt{45}$ 3. $\sqrt{18}$
4. (5, 12) 5. (4.5, 6.5)
6. (3.2, 4.3)

Additional Answers

for pages 466 and 467

19
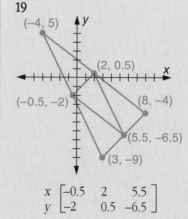

$$\begin{array}{c} x \\ y \end{array} \begin{bmatrix} -0.5 & 2 & 5.5 \\ -2 & 0.5 & -6.5 \end{bmatrix}$$

21 a
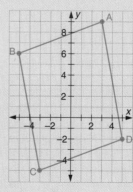

Section 9.6 Two Coordinate Formulas 467

SECTION 9.6

Problem-Set Notes

for page 468

- 32 is a review of functions, but is also a preview of the material in Chapter 12.

Checkpoint

Find the distance between each pair of points.
1. (4, 3) and (7, 9)
2. (–1, 2) and (8, 5)
3. (4, –6) and (–9, –2)

Find the midpoint of the segment joining each pair of points.
4. (8, 3) and (16, 7)
5. (–4, 3) and (–2, –1)
6. (8.7, 1.6) and (3.9, 4.6)

Answers:
1. $\sqrt{45}$ 2. $\sqrt{90}$ 3. $\sqrt{185}$
4. (12, 5) 5. (–3, 1)
6. (6.3, 3.1)

Additional Answers

for page 468

24 a
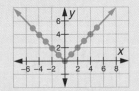

28 b $2 \times 2 \times 3 \times 3 \times 5$

31 f Commutative Property of Multiplication and Associative Property of Addition

Investigation On the globe, distances are measured along a "great circle" route, that is, a circle whose center is the center of the earth. Therefore, a trip from New York to Japan will take you close to the North Pole, rather than on an east-west route. Longitude lines are great circles; so is the equator, although other latitude lines are not.

468

◀ LOOKING BACK **Spiral Learning** LOOKING AHEAD ▶

I   **24**   **a** Plot the points represented in the table below and connect them.
           b What function F could have produced the values in the table—that is, if $y = F(x)$, what is $F(x)$? $F(x) = |x|$

x	–5	–4	–3	–2	–1	0	1	2	3	4	5
y	5	4	3	2	1	0	1	2	3	4	5

R   **25**   **Consumer Math**   Mark bought six sandwiches at $1.64 each, four pieces of pie at $1.05 each, and four juices at $0.98 each. There is a 7% sales tax. Find the total bill. $19.22

R   **26**   Solve each equation, and simplify each expression.
           a $9x(2x – 3) = 0$ $x = 0, x = \frac{3}{2}$ **b** $9x + 6x – 3 = 2$ $x = \frac{1}{3}$
           c $5x + 8 – 3(9 – 2x)$ $11x – 19$ **d** $–3x – 2(4 – x)$ $–x – 8$

R   **27**   Evaluate each expression.
           a $9 – (3 – 7)$ 13 **b** $2 – 3^2$ $–7$
           c $\frac{18 – 6}{–3} – (5)(–2)$ 6 **d** $\sqrt{3^2} + \sqrt{4^2} – \sqrt{3^2 + 4^2}$ 2

R   **28**   Express each number as a product of prime numbers.
           a 46 2×23 **b** 180   **c** 363 $3 \times 11 \times 11$ **d** 462 $2 \times 3 \times 7 \times 11$

R, I   **29**   If $(x, y) = (3, 8)$, $a = \frac{4}{3}$, and $y = ax + b$, what is the value of b? $b = 4$

R   **30**   If A = {9, 12, 15} and B = {–6, 0, 6, 12, 18}, what is
           a $A \cup B$?   {–6, 0, 6, 9, 12, 15, 18}   **b** $A \cap B$?   {12}

R   **31**   Name the property illustrated by each equation.
           a $a + 7 = 7 + a$ Com. Add.   **b** $(ab)c = a(bc)$ Assoc. Mult.
           c $(a + b) + c = (b + a) + c$ Com. Add.   **d** $p(q + r) = pq + pr$ Distr.
           e $3ab = 3ba$ Com. Mult.   **f** $a + (2 \cdot 3 + x) = (a + 3 \cdot 2) + x$

R, I   **32**   Suppose that 4 segments make 1 square, 7 segments make 2 squares, 10 segments make 3 squares, and 13 segments make 4 squares. Write a function for the number of segments needed to make n squares. $f(n) = 3n + 1$

Investigation

Geography   On a flat surface, like the coordinate plane, distances are measured along straight line segments. How are distances measured on curved surfaces, like the surface of the earth? Select two cities that are at roughly the same latitude and are on different continents. Use a globe and a piece of string to find the length of the shortest route between the cities.

468   Chapter 9   Real Numbers

SECTION 9.7

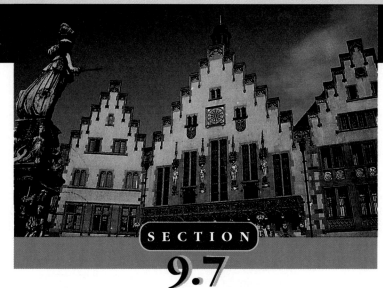

9.7 Graphs of Equations

There are a number of ways to show mathematical relationships. Pairs of values can be listed in a table. A graph can be used to provide a visual presentation of a direction or trend. An equation can be used to show the relationship between the values of two or more variables.

Li and Josh went for a 5-hour hike. They walked at a rate of 3 miles an hour. If we let h stand for hours and m for miles, then we can make a table showing the relationship between time and distance.

h	1	2	3	4	5
m	3	6	9	12	15

We can use the pairs of numbers as coordinates, and we can use the coordinates to plot points. Connecting the points will give the graph shown.

To find the number of miles traveled, you multiply the number of hours by 3. This can be shown by the equation $m = 3h$.

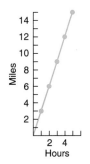

Example 1
Make a table and a graph for the equation $y = 2x - 5$.

Solution
We can select any values to substitute for x to find y values. We will substitute $\{-1, 0, 1, 2, 3, 4\}$.

Objective
- To recognize that the solution sets of many equations can be represented by straight lines on a coordinate system

Class Planning

Schedule 2–3 days

Resources

In *Chapter 9 Student Activity Masters* Book
Class Openers 73, 74
Extra Practice 75, 76
Cooperative Learning 77, 78
Enrichment 79
Learning Log 80
Technology 81, 82
Spanish-Language Summary 83

Class Opener

1. Generate five pairs of numbers in the form (x, y) that satisfy each equation. (Hint: Select any number for x, then solve for y.)

a. $x + y = 3$
b. $y = 2x - 3$

2. Plot each set of ordered pairs on a coordinate plane. What do you observe?

Answers:
1. Points will vary. 2. Points for each equation lie in a straight line.

Lesson Notes

- Graphing equations is the link between algebra and geometry. When we can visualize a formula or an equation, we have insight into its behavior. The ability to graph equations and functions is an important skill, but the advent of graphing calculators and computer graphics tools has made

SECTION 9.7

Lesson Notes continued

understanding the relationship between a graph and what it represents the critical skill.

- Encourage students to observe the numeric patterns in their tables of values and the geometric patterns in their graphs and to realize that each is a different way of expressing the same thing.

- A student who thoroughly understands examples 1, 2, and 3 and sample problems 1 and 2 will have a good intuitive feel for the concept of *slope*. Slope is an important idea in both algebra and calculus. The foundation is being carefully laid here!

Additional Practice

1. Do these points lie on a line? (8, 6), (5, 2), (14, 10)

2. Complete the table for the equation $y = 2x - 3$.

x	-2	-1	0	1	2
y					

3. Draw a graph for the table. Then write an equation for the relationship.

x	-2	-1	0	1	2
y	5	7	9	11	13

4. Make a table for the equation $y = \frac{1}{2}x + 1$.

Answers:
1. No 2. -7, -5, -3, -1, 1
3. $y = 2x + 9$

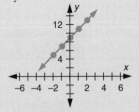

4. Possible answer:

x	-2	0	2
y	0	1	2

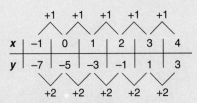

Note that as the *x* values increase by 1, the *y* values increase by 2. The coefficient of *x* in the equation is 2 and causes this increase in *y*.

Also note that when $x = 0$, $y = -5$. The constant term in the equation is also -5.

We can use the pattern of *x* and *y* values to help us draw the graph. We start at any point in the table, say (0, -5), and move to the right 1 unit (thus increasing *x* by 1). Then we move upward 2 units (thus increasing *y* by 2) to arrive at another point on the graph, (1, -3). We repeat this process several times to locate several points on the graph, then draw a straight line through the points.

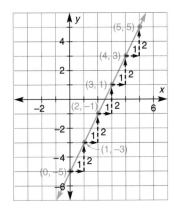

Example 2
Use the table to draw a graph. Then write an equation for the relationship.

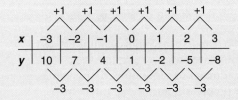

Solution
To draw the graph, we use the pattern from the table.

In determining the equation, we notice that each change of +1 in *x* corresponds to a change of -3 in *y*. So *x* must be multiplied by -3 in the equation. When $x = 0$, $y = 1$, so the constant term is 1. The equation is $y = -3x + 1$.

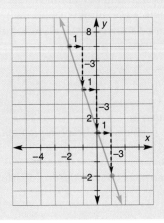

Chapter 9 Real Numbers

SECTION 9.7

Example 3
Use the graph to create a table. Then write an equation for the relationship.

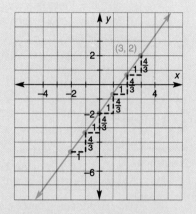

Solution
We notice that the point (3, 2) is on the graph. Also, for every increase of 1 in the value of x, there is an increase of $\frac{4}{3}$ in the value of y. With this information we can create the table shown.

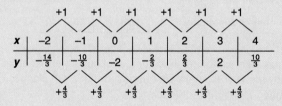

Using the pattern for the change in y values and the value of y when $x = 0$, we have the equation $y = \frac{4}{3}x + (-2)$.

Sample Problems

Problem 1 Write an equation that represents the graph.

(The solution to this problem appears on the next page.)

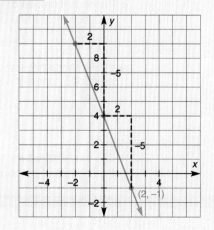

Additional Examples

Example 1
Make a table and graph for the equation $y = 3x - 1$.

Possible answers:

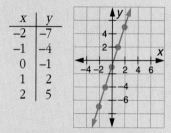

Example 2
Use the table to draw a graph. Then write an equation for the relationship.

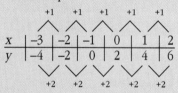

Answers:

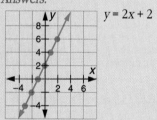

$y = 2x + 2$

Example 3
Use the graph to create a table. Then write an equation for the relationship.

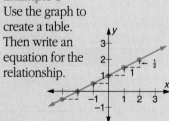

Answers:

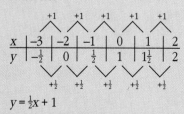

$y = \frac{1}{2}x + 1$

SECTION 9.7

Assignment Guide

Basic
Optional

Average or Heterogeneous
Day 1 8, 9, 11, 13, 15, 17, 21 a, 24, 27, 28
Day 2 10 (a, b), 12 a, 14, 16, 20, 21 b, 22, 25, 31
Day 3 10 a, 12 b, 18, 19, 23, 26, 29, 30, 32, 33

Honors
Day 1 8, 11, 12, 14, 18, 20, 21, 24, 27
Day 2 9, 10, 13, 15–17, 22, 28, 32

Strands

Communicating Mathematics	5, Investigation
Geometry	1, 3, 4, 6, 22, 24–27, 33
Measurement	17, 20, 22, 24, 25, 27, Investigation
Data Analysis/ Statistics	6, 11, 15–17, Investigation
Algebra	1–30, 33, Investigation
Patterns & Functions	4, 6–12, 14–16, 20, 22, 23, 29, 32
Number Sense	28, 30–32
Interdisciplinary Applications	7, 8, 17, 20, Investigation

Individualizing Instruction

The gifted student After completing probem 17, encourage students to research other sports statistics in order to graph the data and predict possible achievements for the year 2000. Have them share their results with the class.

472

Solution For each change of 2 in x, there is a change of -5 in y. So for a change of 1 in x, there is a change of -2.5 in y. The equation must have the form $y = -2.5x +$ constant. To find the constant, we substitute the coordinates, $(2, -1)$.

$$-1 = -2.5(2) + \text{constant}$$
$$-1 = -5 + \text{constant}$$
$$4 = \text{constant}$$

The equation is $y = -2.5x + 4$.

Problem 2 The math class at Algebra High is planning a car wash to raise money for a $450 computer. The students' plans are based partly on the following:

  Expenses (polish, cloths, sponges): $100
  Income: washes, $3 each; polishes, $4 each

The students estimate that half of the customers will want a polish.
a What is the average income expected per customer?
b Complete the table.

Number of customers	0	10	20	30	40
Profit					

c Write an equation relating number of customers to profit.
d How many customers are needed for the math class to earn enough money to purchase the computer?

Solution **a** Income for every two customers:

  2 washes + 1 polish = $3 + $3 + $4 = $10

Therefore, the average income per customer is $5.

b

Number of customers	0	10	20	30	40
Profit	-100	-50	0	50	100

c For each increase of 10 customers, there is a $50 increase in profit. For each increase of 1 customer, there is a $5 increase in profit.

  Profit = 5(customers) + (-100)

d For the students to purchase the computer, the ordered pair $(c, 450)$ is needed, where c is the number of customers.

$$450 = 5c - 100$$
$$550 = 5c$$
$$110 = c$$

The students will earn enough money to purchase the computer if they have 110 or more customers.

472 Chapter 9 Real Numbers

SECTION 9.7

Think and Discuss

P **1** Indicate which of the three groups of points lie on a line. **a and c**
   a (1, 4), (2, 5), (3, 6), (4, 7) **b** (−1, 1), (0, 0), (1, 1), (2, 4)
   c (0.5, 3.5), (4, 21), (7.5, 38.5)

P **2** In the equation $y = 2x + 3$,
   a Which number shows how much y changes for each increase of 1 in x? 2
   b Which number is called the constant? 3

P **3** If you know that the graph of an equation is a straight line, how many points should you plot before you draw the line? Possible answer: 3 to be safe

P **4** Copy the table twice and fill in values for y such that
   a The points lie on a line
   b The points do not lie on a line

x	−4	−3	−2	−1	0	1	2	3
y								

P, I **5** For the equation $y = 5x + 1$, explain why when x increases by 1, y increases by 5. Each value of x is multiplied by 5.

P **6** The points represented by the values in the table all lie on the same line except one. Which coordinates represent a point not on the line? How can you change the coordinates so that the point is on the line?

x	−3	−2	−1	0	1	2	3	4	5
y	−8	−5	−2	1	4	8	10	13	16

P **7** **Business** The students of the Math Club are selling mugs to raise money. They buy 500 mugs at $1.25 each. They are selling the mugs for $4.00 each.
   a What is the total expense? $625
   b What is the income per sale of each mug? $4
   c Copy and complete the table.

Mugs sold	0	1	2	3	4	5	6
Profit (dollars)	−625	−621	−617	−613	−609	−605	−601

   d Write a profit equation. $p = 4m − 625$
   e How many mugs must be sold to make a profit of $400? 257

Problems and Applications

P **8** **Business** Suppose that potatoes cost $0.59 a pound.
   a Copy and complete the table.

Pounds	0	1	2	3	4	5
Cost	0	0.59	1.18	1.77	2.36	2.95

   b Write an equation for the cost, c, of p pounds of potatoes. $c = 0.59p$

Section 9.7 Graphs of Equations 473

Problem-Set Notes

for pages 473, 474, and 475

- **2** previews the idea of slope.
- **7** relates a change in values to an amount of profit.
- **8–23** illustrate many different ways of looking at rates of change.

Checkpoint

1. Do these points lie on a line? (4, 7), (5, 9), (6, 11)

2. Complete the table for the equation $y = -3x + 1$.

x	−2	−1	0	1	2
y					

3. Draw a graph for the table. Then write an equation for the relationship.

x	−4	−2	0	2	4
y	−1	0	1	2	3

4. Make a table for the equation $y = -2x + 5$.

Answers:
1. Yes **2.** 7, 4, 1, −2, −5
3. $y = \frac{1}{2}x + 1$

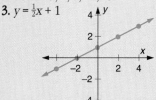

4. Possible answer:

x	−2	−1	0	1	2
y	9	7	5	3	1

Additional Answers

for page 473

4 a Possible answer:

x	−4	−3	−2	−1	0	1	2	3
y	−2	−1	0	1	2	3	4	5

b Possible answer:

x	−4	−3	−2	−1	0	1	2	3
y	4	3	2	1	0	1	2	3

6 (2, 8); possible answer: change the 8 to 7

SECTION 9.7

Communicating Mathematics

- Have students write a paragraph explaining how the coefficient of the *x* term in an equation such as $y = 4x + 2$ affects the graph of the equation. Alternatively, students might describe this orally.
- Problem 5 and Investigation encourage communicating mathematics.

Additional Answers

for pages 474 and 475

10 a Possible answer:

x	-3	-2	-1	0	1	2	3
y	-9	-7	-5	-3	-1	1	3

b Possible answer:

x	-3	-2	-1	0	1	2	3
y	-8	-5	-2	1	4	7	10

c Possible answer:

x	-4	-2	0	2	4	6
y	-1	0	1	2	3	4

11 a

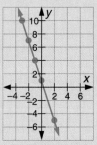

b

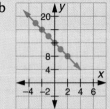

12 a

x	-1	0	1	2	3	4
y	-3	-2	-1	0	1	2

P 9 Copy and complete the table for the equation $y = 4x + 7$.

x	-5	-4	-3	-2	-1	0	1	2	3	4
y	-13	-9	-5	-1	3	7	11	15	19	23

P 10 Make a table for each equation.
  a $y = 2x - 3$ **b** $y = 3x + 1$ **c** $y = \frac{1}{2}x + 1$

P 11 Draw a graph for each table.

  a
x	-3	-2	-1	0	1	2
y	10	7	4	1	-2	-5

  b
x	-3	-2	-1	0	1	2
y	18	16	14	12	10	8

P 12 Make a table and draw a graph for each equation.
  a $y = x - 2$ **b** $y = -\frac{2}{3}x + 1$

P 13 For the graph shown,
  a Find the increase in the value of *y* for each increase of 1 in *x*. 0.5
  b Find the value of *y* when $x = 0$. -0.5
  c Write an equation for the graph.
  $y = 0.5x - 0.5$

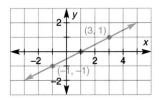

P 14 a Continue the pattern in the table and then graph the data.

x	-9	-7	-5	-3	-1	1	3	5
y	-5	-1	3	7	11	15	19	23

  b How much does *y* change for each increase of 1 in *x*? 2

P 15 For each table, write an equation of the form $y = ax + b$.

  a
x	-3	-2	-1	0	1	2
y	-3	0	3	6	9	12

  b
x	-3	-2	-1	0	1	2
y	7	4	1	-2	-5	-8

P 16 Which of the following tables have graphs that are straight lines? Only **a**

  a
x	-3	-2	-1	0	1	2
y	9	5	1	-3	-7	-11

  b
x	-3	-2	-1	0	1	2
y	9	4	1	0	1	4

  c
x	-3	-2	-1	0	1	2
y	$-\frac{1}{3}$	$-\frac{1}{2}$	-1	—	1	$\frac{1}{2}$

P, R 17 Sports The winning heights in the men's Olympic high jump from 1948 through 1988 were as follows. Heights are in inches.

Year	'48	'52	'56	'60	'64	'68	'72	'76	'80	'84	'88
Height	78	80.3	83.25	85	85.75	88.24	87.75	88.5	92.75	92.5	93.5

  a Plot the points, using the x-axis for years and the y-axis for inches.
  b Draw a line to approximate the data.
  c Predict the winning height in 2000. Possible answer: About 96 or 97 in.

P, I 18 Given the equation and point, determine the value of the constant. Then graph each equation.
  a $y = \frac{2}{3}x + $ constant; (6, -2) -6   **b** $y = -\frac{1}{4}x + $ constant; (8, 5) 7

474   **Chapter 9** Real Numbers

SECTION 9.7

P **19** Using x values from -3 to 3, make tables for
    a $y = 5x$ **b** $y = 9$

P **20** **Science** A plane is descending at a rate of 640 feet per minute. Right now (at time zero) its altitude is 30,000 feet. Negative time means minutes just past. Positive time means minutes to come.
    a Copy and complete the table.

Time	-3	-2	-1	0	1	2	3
Altitude	31,920	31,280	30,640	30,000	29,360	28,720	28,080

    b Write an equation for this table. $a = -640t + 30{,}000$
    c When will the plane be at 5,000 feet?   In about 39.1 min

P **21** Write an equation for each graph.

    a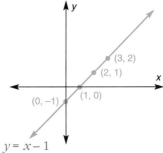
    $y = x - 1$

    b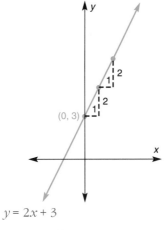
    $y = 2x + 3$

P **22** **a** Suppose that y is the area of the shaded portion of the rectangle. Use the given values of x to complete the table.

x	4	5	6	7	8	9	10
y	20	26	32	38	44	50	56

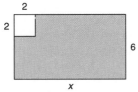

    b Using the pattern from the table, complete the function
        Shaded area(x) = _____ $6x - 4$
    c Draw a graph of the data.
    d What is the set of possible values of x? $\{x : x > 2\}$

P **23** Copy and complete the table, and draw a graph, for the equation $y = 1 + \frac{3}{4}x$.

        A   B   C

x	-4	-3	-2	-1	0	1	2	3	4	5	6
y	-2	-1.25	-0.5	0.25	1	1.75	2.5	3.25	4	4.75	5.5

Compute the distance between points A and B and between C and B.

---

12 a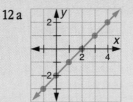

b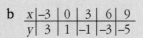

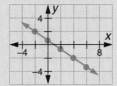

14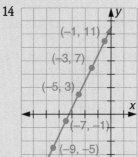

15 a $y = 3x + 6$ b $y = -3x - 2$

17 a, b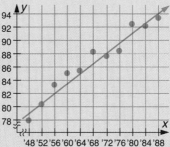

18 a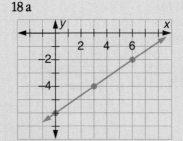

Answers for problems 18 b, 19, 22 c and 23 are on the following page.

Section 9.7 Graphs of Equations   **475**

SECTION 9.7

◀ LOOKING BACK **Spiral Learning** LOOKING AHEAD ▶

R **24** If PQ = 18, find the area of the two semicircles with centers at P and Q. 81π, or ≈ 254.5

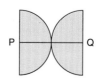

R, I **25 a** Find the distance between the midpoints of $\overline{AB}$ and $\overline{BC}$. ≈ 5.31
b Compare your answer in part **a** with the length of $\overline{AC}$. It is half of AC.

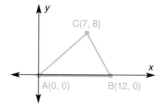

R **26** Given A $(-4, -2)$ and B $(10, -6)$, find the coordinates of
a The point one fourth of the way from A to B $\left(-\frac{1}{2}, -3\right)$
b The point three fourths of the way from A to B $\left(\frac{13}{2}, -5\right)$

R **27** Find the area of the triangle that has the vertex matrix shown. 32

$$\begin{array}{c} x \\ y \end{array} \begin{bmatrix} 3 & -1 & 5 \\ 7 & 1 & -6 \end{bmatrix}$$

R, I **28** Find the values of $(x-4)(x-3)(x-2)(x-1)(x)(x+1)(x+2)$ when x is taken from the set $\{-2, -1, 0, 1, 2, 3, 4\}$. All are 0.

R **29** Evaluate $f(x) = 7 - 2(-4 + 3x)$ for $x = 5$. -15

R **30** Write $3.\overline{571428}$ as a ratio of integers. Reduce your answer to lowest terms. $\frac{25}{7}$

R **31** Suppose that A = {whole-number divisors of 20}, B = {whole-number divisors of 36}, and C = {whole-number divisors of 30}. Make a Venn diagram of sets A, B, and C.

R **32** Find the next three numbers in the pattern $\sqrt{5}, \sqrt{20}, \sqrt{45}, \sqrt{80}, \ldots$

R **33** Point A has coordinates $(-3, 4)$. It is translated $\langle 5, -8 \rangle$ to produce A′, then A′ is translated $\langle 5, -8 \rangle$ to produce A″.
a What are the coordinates of A′ and A″? A′$(2, -4)$, A″$(7, -12)$
b What are the coordinates of the midpoint of $\overline{AA''}$? $(2, -4)$

Investigation

Data Analysis Find out the heights and the shoe sizes of at least eight people, all girls or all boys. Then draw a graph in which one axis represents height and the other represents shoe size. Plot the data you gathered. Does the graph suggest a linear relationship between height and shoe size? Try combining your data with your classmates' data in a class graph. Does the relationship (or lack of relationship) become clearer?

476   Chapter 9   Real Numbers

SUMMARY

CHAPTER 9

Summary

CONCEPTS AND PROCEDURES

After studying this chapter, you should be able to
- Represent a set of numbers by listing, by using set-builder notation, and by graphing (9.1)
- Identify sets as subsets of other sets (9.1)
- Determine the intersection and the union of two sets (9.2)
- Use Venn diagrams to represent intersections and unions (9.2)
- Categorize numbers according to the number sets to which they belong (9.3)
- Recognize that π is the ratio of a circle's circumference to its diameter (9.4)
- Recognize that if the vertex of an angle is on a circle and the angle's rays pass through the endpoints of a diameter of the circle, the angle is a right angle (9.4)
- Apply the associative properties of real numbers (9.5)
- Apply the commutative properties of real numbers (9.5)
- Apply the Zero Product Property (9.5)
- Apply the distance and midpoint formulas (9.6)
- Recognize that the solution sets of many equations can be represented by straight lines on a coordinate system (9.7)

VOCABULARY

associative (9.5)
commutative (9.5)
chord (9.4)
complement (9.2)
disjoint sets (9.3)
element (9.1)
empty set (9.1)
intersection (9.2)
irrational number (9.3)

natural number (9.3)
rational number (9.3)
real number (9.3)
set (9.1)
set-builder notation (9.1)
solution set (9.1)
subset (9.1)
union (9.2)
Venn diagram (9.2)

22 c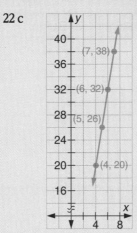

23
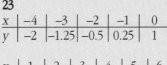

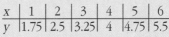

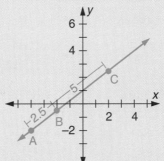

31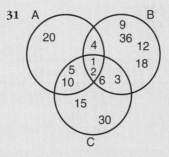

32 $\sqrt{125}, \sqrt{180}, \sqrt{245}$

Investigation Normally there will be evidence of a positive correlation between height and shoe size; that is, in general, taller people have bigger feet. It may not be clear with a sample of just eight, but it will be clearer as you combine data.

Review

Assignment Guide

Basic
Optional

Average or Heterogeneous
Day 1   1 a, 2, 6, 8, 9, 13, 14, 16, 19, 24 (a, c), 26, 28, 33

Honors
Day 1   1–10, 12, 14, 15, 17, 19–23, 24 (a, c), 26, 29, 30, 32, 33

Strands

Geometry	6, 11, 13, 18, 20, 25, 27, 28, 31, 32
Measurement	3, 11, 13, 18, 20, 25, 27, 28, 31, 32
Data Analysis/ Statistics	21
Probability	25
Algebra	1–7, 9, 11–14, 16, 18–21, 23–29, 31–33
Patterns & Functions	3, 7, 14, 21, 32, 33
Number Sense	3, 4, 9, 10, 16, 17, 19, 22–24, 29
Interdisciplinary Applications	8, 14

Additional Answers

for pages 478 and 479

2 a {(0, 0), (0, 1), (0, 2), (0, 3), (1, 0), (1, 1), (1, 2), (1, 3)}
b {(1, 1), (2, 1), (3, 1), (4, 1), (5, 1), (1, 2), (2, 2), (3, 2), (4, 2), (5, 2), (1, 3), (2, 3), (3, 3), (4, 3), (5, 3)}
c {(1, 1), (1, 2), (1, 3)}
4 f True

CHAPTER 9

Review

1. Using set-builder notation, express the solution set of each inequality.
   a $3x - 4 < 11$ $\{x : x < 5\}$ **b** $-5x + 8 > 7$ $\{x : x < 0.2\}$ **c** $12x + 4 \geq 16x$ $\{x : x \leq 1\}$

2. Consider the points shown on the graph.
   If A = {points such that $x \geq 2$ and $y \leq 5$} and
   B = {points on the axes}, what points are in
   the following sets?
   a $\overline{A}$ **b** $\overline{B}$ **c** $\overline{A \cup B}$

   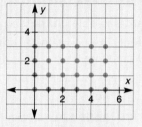

3. Suppose that M = {distances from O to each of the other points}.
   a List all of the rational numbers in M.   {1}
   b If you were to continue the pattern by adding point F, what would be its distance from O? $\sqrt{26}$

   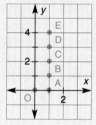

4. Indicate whether the statement is *true* or *false*.
   a $\pi = 3.14$ False
   b $\pi = \frac{22}{7}$ False
   c $\pi > 3.14$ True
   d $\pi < 3.15$ True
   e $\pi > \frac{22}{7}$ False
   f π is an irrational number.

5. Give examples to show that   Answers will vary.
   a There is no associative property of subtraction
   b There is no associative property of division
   c There is no commutative property of subtraction
   d There is no commutative property of division

6. For points A (4, −3) and B (x, y), the midpoint of $\overline{AB}$ is (6, 4). Find (x, y).   (8, 11)

7. Each table gives coordinates of points on a straight line. Copy and complete the tables.

   a
x	−3	−2	−1	0	1	2
y	4	6	8	10	12	14

   b
x	−3	−2	−1	0	1	2
y	5	2	−1	−4	−7	−10

8. There are 270 students at Algebra High. On Friday, 194 students went to the basketball game, 125 went to the dance held after the game, and 80 went to both the game and the dance. Use a Venn diagram to determine how many students did not go to either the game or the dance.   31

Chapter 9   Real Numbers

REVIEW

9 Find the set of all values of x such that AB is a positive even integer. {1, 3, 5, 7}

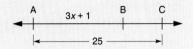

10 Is each statement *true* or *false*?
   a 0 is a rational number. True
   b $-\sqrt{16}$ is a natural number. False
   c The intersection of the rational and the irrational numbers is {0}. False
   d The intersection of the integers and the whole numbers is the natural numbers. False

11 Figure ABCD is a rectangle. AD = 6 and DC = 8. If ABCD is rotated 90° clockwise about the center of the circle, what is the probability that a randomly selected point inside the circle will be inside both the preimage and the image? $\frac{36}{25\pi}$ or $\approx .46$

12 For real numbers a, b, and c, is each statement true *always*, *sometimes*, or *never*?
   a If $abc = 0$, then $a = 0$. S
   b $a(b + c) = ab + ac$ A
   c $a(b + c) = ab + bc$ S
   d $(a + a)^b = a^b + a^b$ S
   e If $ab = 0$, then $a = 0$ or $b = 0$. A
   f $|ab| = |a| \cdot |b|$ A

13 For circle A, find
   a The length of the diameter 13
   b The length of the radius 6.5
   c The area of the circle ≈ 132.7
   d The circumference of the circle ≈ 40.8

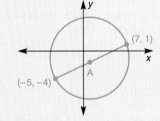

14 A boat can be rented for a fee of $23 plus $17 an hour.
   a Copy and complete the table.

Hours	0	1	2	3	4	5	6
Cost (dollars)	23	40	57	74	91	108	125

   b Write an equation for the cost in terms of hours. $c = 17h + 23$

15 a List all subsets of {□, @}   {□, @}, {□}, {@}, { }
    b List all subsets of {B, E, A, R}

16 A = {$x : 2x < 21$ and x is a natural number}
    B = {$x : x^2 < 25$ and x is an integer}
    C = {$x : 4x = 16$}
    a List the numbers in each of the sets.
    b Is $C \subset (A \cap B)$? Yes
    c Is $C \cup B = B$? Yes
    d Is $(C \cap B) \subset A$? Yes

---

5 Possible answers:
a $12 - (7 - 2) \neq (12 - 7) - 2$
b $36 \div (6 \div 2) \neq (36 \div 6) \div 2$
c $7 - 4 \neq 4 - 7$
d $81 \div 3 \neq 3 \div 81$

15 b { }, {B}, {E}, {A}, {R}, {B, E}, {B, A}, {B, R}, {E, A}, {E, R}, {A, R}, {B, E, A}, {B, E, R}, {B, A, R}, {E, A, R}, {B, E, A, R}

16 a A = {1, 2, 3, 4, 5, 6, 7, 8, 9, 10}
B = {−4, −3, −2, −1, 0, 1, 2, 3, 4}
C = {4}

Chapter Review

REVIEW

17 Copy the Venn diagram and place each of following numbers in the proper section.

$-2, 0, \frac{3}{2}, \sqrt{49}, \sqrt{61}, \frac{-15}{6},$
$\frac{1}{\sqrt{2}}, 5\pi, 2.76, 0.010010001\ldots,$
$3.14, -15, -2\sqrt{3}$

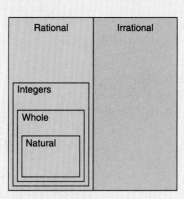

18 $\overline{AC}$ is a diameter of the circle shown. $AB = 6$, $BC = 8$, and $AD = 9$. Find CD. $\sqrt{19}$, or ≈ 4.36

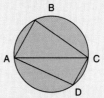

19 Find the set of all values of x for which $5x(x-4)(2x-3)(3x+7) = 0$.

20 P is a point on circle O.
  a Find the radius of the circle. ≈ 9.4
  b Find the area of the circle. ≈ 279.6

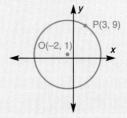

21 Use the relationship shown in the table to create another table in which the x values increase by 1.

x	−4	−1	0	2	5		−4	−3	−2	−1	0
y	$\frac{8}{3}$	$\frac{11}{3}$	4	$\frac{14}{3}$	$\frac{17}{3}$		$\frac{8}{3}$	3	$\frac{10}{3}$	$\frac{11}{3}$	4

22 List the set of all times for which the digits are consecutive integers. (Hint: The order of the integers can be either smallest to largest or largest to smallest.)

23 Suppose that A = {positive-integer divisors of 36} and B = {positive-integer divisors of 30}.
  a Describe in words A ∩ B.   {Positive-integer divisors of 6}
  b Find the greatest number in A ∩ B.   6

24 Convert each decimal to a fraction.
  a 2.76 $\frac{69}{25}$ **b** 3.94 $\frac{197}{50}$ **c** $5.87\overline{12}$ $\frac{775}{132}$

Additional Answers

for pages 480 and 481

17

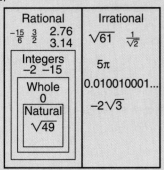

19 $\{0, 4, \frac{3}{2}, -\frac{7}{3}\}$

22 {1:23, 2:34, 3:45, 4:56, 2:10, 3:21, 4:32, 5:43, 6:54, 12:34}

30 a

b

33 a $y = 2x + 10$
  b $y = -1.5x + 0.5$

REVIEW

25 In rectangle PMTK, PK = 12, and the area of rectangle PMTK is 192 square units. If a point inside the circle is selected at random, what is the probability that it will be inside the rectangle? ≈.611

26 Match each property with the appropriate equation.
 a Commutative Property of Addition 4
 b Commutative Property of Multiplication 5
 c Associative Property of Addition 1
 d Distributive Property 2
 e Associative Property of Multiplication 3

 1. $9 + (7 + 4) = (9 + 7) + 4$
 2. $9(7 + 4) = 9 \cdot 7 + 9 \cdot 4$
 3. $9(7 \cdot 4) = (9 \cdot 7)4$
 4. $9 + (7 + 4) = 9 + (4 + 7)$
 5. $9(7 \cdot 4) = 9(4 \cdot 7)$

27 The midpoint of $\overline{AB}$ is on circle B. Find the area of the circle. 13π

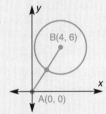

28 If $(-8, 11)$ and $(-2, -3)$ are the endpoints of a diameter of a circle, is $(1, 1)$ inside, on, or outside the circle? Inside the circle

29 How many elements are in each set?
 a $\{(x, y): x + y = 10 \text{ and } x \text{ and } y \text{ are whole numbers}\}$ 11
 b $\{(x, y): x + y = 10 \text{ and } x \text{ and } y \text{ are integers}\}$ Infinitely many

30 Copy the Venn diagram and shade the area representing
 a $A \cup (B \cap C)$ b $\overline{A} \cap \overline{B}$

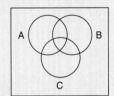

31 The circumference of circle A is 50 cm. The area of circle B is 50 cm². Find the ratio of the radius of circle A to the radius of circle B. About 2:1

32 Point P is translated to P′ using the translation numbers <8, 15>. Find the distance from P to P′ if P has the coordinates
 a $(0, 0)$ 17   b $(5, 8)$ 17   c $(-11, 1)$ 17

33 Copy and complete each table, assuming that the entries represent points on a line. Then write an equation for the table.

a
x	-3	-2	-1	0	1	2	3
y	4	6	8	10	12	14	16

b
x	-3	-2	-1	0	1	2	3
y	5	3.5	2	0.5	-1	-2.5	-4

Chapter Review 481

TEST

Assessment

In Tests and Quizzes Book
Average Level Test 2A 245–247
                Test 2B 248–250
Honors Level   Test 3A 251–253
                Test 3B 254–256
Open-Ended   Test C
  (for both ability levels) 257, 258
Third Qtr. Test 3 (Hnrs.) 259–261

Alternative-Assessment Guide

Assessment Tips
Point out to students concerned about their performance on standardized tests that the skills developed in working on portfolio projects are the skills called for in the new wave of standardized mathematics tests. Specifically, students will need to be able to use various problem-solving strategies and to solve novel problems from a variety of contexts.

Portfolio Projects
The following problems are appropriate for use as portfolio projects:
1–8. Investigations: Chapter Opener (p. 427), 9.1 (p. 434), 9.2 (p. 441), 9.3 (p. 447), 9.4 (p. 454), 9.5 (p. 461), 9.6 (p. 468), 9.7 (p. 476)

9. Create a Venn diagram that shows the following sets: (a) U.S. states that begin with a vowel; (b) U.S. states that border a body of salt water; (c) U.S. states that contain at least one city with a population of over 1 million; (d) the first thirteen U.S. states; (e) U.S. states with an area of over 100,000 square miles; and (f) U.S. states with fewer than 10 counties. Are any of the fifty states not represented in your diagram?

10. Problem 21 on page 453 suggests a way to estimate the value of π: measure the distance that a

CHAPTER 9

Test

1. If V = {$x : -4 < x \le 20$}, is -5 an element of set V?   No

2. List all the subsets of {2, 4, 6}.   { }, {2}, {4}, {6}, {2, 4}, {2, 6}, {4, 6}, {2, 4, 6}

3. If values of x are chosen from {$-3, -2, -1, \ldots, 6, 7$}, for what subset of the values is $3x + 5 < 24$?   {$-3, -2, -1, \ldots, 6$}

4. Suppose that A = {1, 3, 5, 7, 9}, B = {1, 4, 9, 16}, and C = {2, 3, 5, 7}. List the elements of each of the following sets.
   a $A \cap B$ {1, 9}
   b $B \cup C$ {1, 2, 3, 4, 5, 7, 9, 16}

5. Out of 90 freshman students, 47 are involved in sports and 53 are involved in clubs. How many students are involved in both sports and clubs?   10 students

6. If N = {natural numbers}, I = {integers}, Q = {rational numbers}, and R = {real numbers}, is the statement *true* or *false*?
   a $N \subset Q$ True
   b $I \cap R = R$ False

7. Write $1.\overline{23}$ as a ratio of two integers. $\frac{122}{99}$

8. What is the area of a circle with a diameter of 8 meters? 16π, or ≈ 50.27, m²

9. What is the circumference of a circle with a radius of 10 inches?

10. If the circle rolls once around the rectangle without slipping, how far will the center of the circle travel? $42 + 4\pi$, or ≈ 54.57

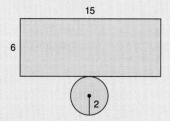

11. Solve $(x + 4)(x - 2) = 0$ for x. $x = -4$ or $x = 2$

12. Use number properties to evaluate $0.75 - (8.15 + 0.75)$ mentally. -8.15

13. What is the distance between the points A(7, 5) and B(–1, 12)? $\sqrt{113}$, or ≈ 10.63

14. Given C(–8, 12) and D(2, –8), find the coordinates of the midpoint of $\overline{CD}$. $(-3, 2)$

15. **a** In the following table, each pair (x, y) represents a point on a line. Copy and complete the table.

x	–3	–2	–1	0	1	2	3
y	–9	–7	–5	–3	–1	1	3

    b Draw a graph showing the relationship between x and y.
    c Write an equation that represents the graph you drew. $y = 2x - 3$

PUZZLES AND CHALLENGES

1 Divide this shape into four parts, each exactly congruent to the other three.

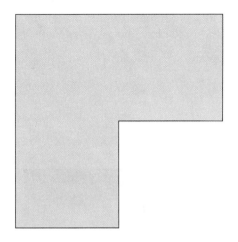

2 Point O is the center of this circle. OA = 3 and OB = 4. Find the radius of the circle.

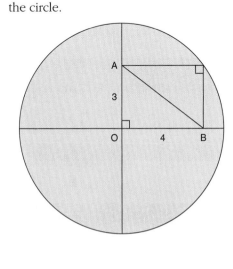

3 Who says that two WRONG's can't make a RIGHT? Solve the alphametic, where the letter O stands for the digit 0 and H stands for 8.

$$\begin{array}{r} \text{WRONG} \\ + \text{WRONG} \\ \hline \text{RIGHT} \end{array}$$

4 Mr. and Mrs. Smith went to a party with three other married couples. Several people shook hands with some other people. No one shook hands with himself or herself and no one shook hands with his or her spouse. No one shook hands with the same person twice. After all this handshaking took place, Mr. Smith asked each person, "How many handshakes did you make?" Each person gave him a different answer. How many handshakes did Mrs. Smith make?

wheel travels to make one revolution, then divide this distance by the diameter of the wheel. Use a tape measure, some chalk (to mark starting points), and several different-sized bicycles to obtain an estimate for π. Find the average of your estimates of π. How close to the actual value is your estimate? Explain your results.

11. In March of 1987, Hideyaki Tomoyori of Japan recited π from memory to 40,000 places. Estimate how long it took him to complete this feat. Explain how you made your estimate. Then use a computer to generate a page of random digits. How many such pages would be needed to show 40,000 digits? Explain how you made your estimate.

12. Extend sample problem 1 on page 444 as follows: Explain why $n = -2.5757\ldots$ is multiplied by 100 as part of converting $-2.\overline{57}$ to a fraction. Explore what happens if you try multiplying the equation by 10 or 1000. Will multiplying by 100 work if you want to convert $3.\overline{123}$ to a fraction? Why or why not? What about $5.\overline{6}$? Use these examples to write a general rule about how to convert any repeating decimal to a fraction.

Additional Answers

for Chapter 9 Test

9 20π, or ≈ 62.83, in.

15 b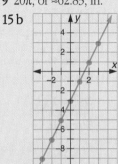

Answers for Puzzles and Challenges are on page A8.

ASSESSMENT

LONG-TERM INVESTIGATIONS

UNIT E: Playing with Numbers

Suggestions for various ways to assess students' performance are integrated throughout the investigations in this unit. These forms of assessment include observing the students while they work, interviewing students, looking at students' self-evaluations and group evaluations, observing peer review/response situations, evaluating students' products, and evaluating students' portfolios.

Refer to the chart below to preview the variety of assessment opportunities that are integrated throughout this unit.

Assessment Opportunities	Investigations						
	Before Unit	1	2	3	4	5	End of Unit
Form A: Observation Notes **Bank A/B:** Observation Guide			✖		✖	✖	
Form B: Observation Evaluation Record **Bank A/B:** Observation Guide		✖				✖	
Form C: Interview Record **Bank C:** Interview Guide		✖					
Form D: Product Evaluation Record **Bank D:** Product Evaluation Guide			✖	✖	✖		
Form E: Rubric Chart **Bank E:** Sample Rubric				✖			
Form F: Portfolio Item Description	✖						✖
Form G: Portfolio Notes **Bank G:** Portfolio Evaluation Criteria	✖						✖
Form H: Self-evaluation						✖	✖
Form I: Group Evaluation				✖			
Form J: Peer Review/Response			✖				

Rubrics

Rubrics are used to evaluate students' products. You can develop a scoring rubric for a specific investigation by adapting the blackline masters shown below, which are provided in the *Gateways to Mathematical Investigations* Teacher's Edition. You will probably want to develop a rubric with students at least once per unit. The chart on the opposite page suggests that for Unit E: Playing with Numbers, you complete a rubric before Investigation 3. Remind students that, at times, they may be asked to revise their work to improve it and to make it complete.

Bank D, page 211

Bank D: Product Evaluation Guide

The statements below can be used as a guide to assess students' products. Adapt these statements as needed to evaluate specific assignments. Record your adaptations on a clean copy of **Form D: Product Evaluation Record** (p. 210). You may also wish to create and record your own criteria on **Form D**. After you have adapted **Form D** for use with a specific assignment, make copies for each student or group as needed.

Mathematical Reasoning Skills

- The student includes examples and/or counter examples to support the mathematical argument.
- The student synthesizes and summarizes own or group's thinking.

Mathematical Understanding

- The student identifies all the important information of the situation or problem.
- The student shows understanding of the situation's mathematical ideas. (List important mathematical

Form E, page 212

Form E: Blank Rubric Chart

NAME (S) ____________ ____________ GROUP ________
____________ ____________ DATE ________

STANDARDS FOR STUDENT'S WORK

LEVEL	Standard of performance to be achieved for each specified level
4 Exceptional/ Fully Acceptable	

Bank E, page 213

Bank E: Sample Rubric

STANDARDS FOR STUDENT'S WORK

LEVEL	Standard of performance to be achieved for each specified level
4 Exceptional/ Fully Acceptable	Your presentation fully accomplishes the purpose of the task. Your work is complete, well organized, and clearly explained. You illustrate ideas using clear diagrams and models and communicate them effectively to the audience. You identify all the important information of the situation and show a thorough understanding of the mathematical ideas and procedures. You use strong arguments to support findings, including examples (and/or counterexamples). You demonstrate that you fully grasp the central mathematical ideas developed in this task. You demonstrate an orignial extension or applicatrion of the mathematical concepts.
3 Satisfactory/ Acceptable	Your presentation substantially accomplishes the purpose of the task. Your work is complete, reasonably organized, and explained. You use an appropriate diagram or model and communicate ideas effectively to the audience. You identify most of the important information of the situation and show understanding of the mathmatical ideas and procedures. You present arguments to support findings, including examples (and/or counterexamples). You understand and use the central mathematical ideas developed in this task. You demonstrate and apply the mathematical concepts.
2 Incomplete/ Revise	Your presentation partially accomplishes the purpose of the task. However, your work may be incomplete, or misdirected, or not clearly organized and/or explained. Your diagram or model may be inappropriate or unclear. You may communicate ideas inadequately to the audience. You may identify some of the important information of the situation and you may show only partial and/or limited understanding of the mathematical ideas and procedures. Your arguments to support findings may be weak or incomplete. You may misunderstand or misuse the central mathematical ideas developed in this task. You may have difficulty demonstrating and applying the mathematical tasks. You need to revise this work. (Your teacher will provide suggestions for a satisfactory or competent rating and a due date for your revised work.)
1 Inadequate/ Redo	Your presentation shows little or no understanding of the assignment. Your work is barely, if at all, comprehensible. Your explanations are not understandable. You show little or no understanding of the mathematical ideas and procedures. You made serious computational mistakes and misused mathematical terms. You should meet with your classmates or your teacher to be sure you fully understand this assignment. Then you need to revise this work. (Your teacher will provide suggestions for a satisfactory or competent rating and a due date for your revised work.)

Gateways © McDougal, Littell & Company Bank E: Sample Rubric 213

Form F, page 214

Form F: Portfolio Item Description

NAME ____________ DATE ____________

Complete this form each time you add an item to your portfolio.

Type of Portfolio Entry: ____________

Why did you choose to place this item in your portfolio?

Form G, page 215

Form G: Portfolio Notes

NAME ____________ CLASS ____________

Date	Observations/Comments	Recommendations

Bank G, page 216

Bank G: Portfolio Evaluation Criteria

At the start of the school year, have a class discussion on the goals and uses of students' portfolios. Discuss with the class some ideas regarding what to include in a portfolio. Then, use the portfolio evaluation criteria which follows to develop a rubric with the class that communicates the criteria you will use to assess their portfolios. Have students record this criteria on copies of **Form E: Rubric Chart.** Have students store this completed form in their portfolios so that they can refer to it throughout the year.

Each time students add items to their portfolios, have them complete copies of **Form F: Portfolio Item Description.**

Use copies of **Form G: Portfolio Notes** to record your observations, comments, and recommendations each time you evaluate students' portfolios.

Ideas for Types of Portfolio Entries

- a portfolio table of contents
- a letter to the reader of the portfolio explaining each item
- teacher-completed checklists/evaluations
- a mathematical autobiography
- notes from interviews, teachers, peers, etc.
- papers that show the student's corrections or errors or misconceptions
- photos or sketches of the student's work with manipulatives or mathematical models
- excerpts from the student's daily journal or reflective writing samples
- tables, charts, and diagrams created by the student
- artwork supporting a problem such as drawings
- designs, coordinate pictures, scale drawings, maps, etc.
- student-created problems with or without solutions
- group projects or individualized reports
- student-developed surveys and results
- self-evaluations

Suggestions for Portfolio Evaluation Criteria

Mathematical Reasoning Skills
The Student:
- analyzes problem situations.
- is flexible and comfortable in using several different problem solving approaches.
- synthesizes and summarizes the individual's or group's thinking.
- provides supporting statements to verify findings.
- shows original and creative thought.

Mathematical Tools and Techniques
The Student:
- uses a variety of diagrams, models, technology, manipulatives, and other resources appropriately.
- uses displays (e.g. graphs, tables, diagrams, sketches, and models) to represent mathematical ideas.

Mathematical Understanding
The Student:
- shows understanding of mathematical ideas and procedures. (List specific skills/concepts on **Form G.**).

Mathematical Communication
The Student:
- organizes work in a logical and clear way.
- expresses mathematical ideas clearly and appropriately.
- uses mathematical language and notation appropriately.

Mathematical Dispositions
The Student:
- shows development as as group participant.
- shows development of a positive attitude.
- shows growth in learning.
- includes revised and complete work.
- shows consistent effort.
- includes a variety of work in his or her portfolio.

216 Assessment *Gateways* © McDougal, Littell & Company

Portfolios

At the beginning of each unit, you may wish to review with the students the criteria they should use when placing work in their portfolios. Have students complete Form F: Portfolio Item Description each time they wish to place an item in their portfolios.

Periodically, you may wish to evaluate students' portfolios using the blackline masters shown above. For Unit E: Playing with Numbers, the chart on the opposite page suggests that you evaluate students' portfolios at the end of the unit.

CHAPTER 10

10 Topics of Number Theory

Contents

10.1 Techniques of Counting
10.2 Primes, Composites, and Multiples
10.3 Divisibility Tests
10.4 Prime Factorization
10.5 Factors and Common Factors
10.6 LCM and GCF Revisited
10.7 Number-Theory Explorations

Pacing Chart

Days per Section

	Basic	Average	Honors
10.1	*	*	1
10.2	*	*	1
10.3	*	*	1
10.4	*	*	1
10.5	*	*	2
10.6	*	*	1
10.7	*	*	1
Review	–	–	1
Assess	–	–	1
	–	–	10

* Optional at these levels

Chapter 10 Overview

The chapter begins with a discussion of counting methods and a preview of sophisticated methods to be presented in later courses. We progress to a detailed examination of primes, composites, and multiples, including greatest common factors, least common multiples, prime-factorization form, and divisibility tests. All of these will enable students to work with integers quickly and accurately. Because these are review topics, we reintroduce them in what we anticipate are novel ways in order to broaden students' perspec-

CHAPTER 10

AGRICULTURE If you pursue a career in agriculture, you may be growing crops for food or other uses, or you may be raising livestock. Either way, there will be plenty of mathematics involved. For crops, you will need to calculate yield per acre or the profit per bushel. You may want to weigh the costs and benefits of using fertilizers or weed killers.

A person raising livestock will need to be able to estimate the amount of feed to keep on hand, the costs of shipping animals to market, the number of animals that can be supported per acre, and the best time to sell.

INVESTIGATION

Farming Choose a crop that is grown in your area. Find out the approximate yield per acre for that crop and the current or projected selling price for that crop. Also list some of the major costs involved in producing it.

tives. The emphasis in the later part of the chapter is on material that foreshadows algebraic work and on some special topics that take students beyond the customary treatment of integers.

Career Connection

Agriculture is a familiar way to show how mathematics affects daily life. Many agricultural workers in the United States work on farms, but many others work in industry, education, and government. There are many agricultural careers, including those in agronomy (soil science), agricultural economy, nutrition, and government regulation. Mathematical skills in accounting and statistics are usually required.

Video Connection

For more information about the career discussed here, please refer to the Futures videocassette #1, program A: *Agriculture*, featuring Jaime Escalante. For more information or to order a copy of the program, call PBS VIDEO at 800/344-3337.

Investigation Notes

Some widely grown crops that students might choose are corn, wheat, rice, and soybeans. The major costs of production may include farm machinery, labor, seeds, fertilizer, and irrigation equipment. Students may be able to obtain information about yields by contacting a state agriculture department.

SECTION 10.1

Objective

- To apply the Fundamental Counting Principle to solve problems

Vocabulary: Fundamental Counting Principle, tree diagram

Class Planning

Schedule 1 day

Resources

In *Chapter 10 Student Activity Masters* Book
Class Openers *1, 2*
Extra Practice *3, 4*
Cooperative Learning *5, 6*
Enrichment *7*
Learning Log *8*
Interdisciplinary Investigation *9, 10*
Spanish-Language Summary *11*

Transparencies 44, 45

Class Opener

List all possible results for each activity.

1. Two coins are tossed.
2. A coin is tossed and a die is rolled.
3. One letter and one number are selected from {A, B, C} and {1, 2, 3}.

Answers:
1. HH, HT, TH, TT **2.** H1, H2, H3, H4, H5, H6, T1, T2, T3, T4, T5, T6
3. A1, A2, A3, B1, B2, B3, C1, C2, C3

Lesson Notes

- Systematic counting methods are becoming increasingly important in mathematics, due in part to the use of computers and calculators. The Fundamental Counting Principle is aptly named: it is the principle upon

Investigation Connection
Section 10.1 supports Unit E, *Investigation* 5. (See pp. 426A–426D in this book.)

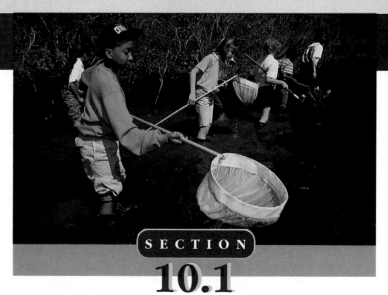

SECTION 10.1
Techniques of Counting

THE FUNDAMENTAL COUNTING PRINCIPLE

We have worked with different sets of numbers. In the study of number theory we will deal primarily with the set of natural or counting numbers—{1, 2, 3, 4, 5, ...}—and related subsets.

Example 1
Every camper attending Timber Lake Summer Camp is required to select two activities each day, one activity from Group I and one activity from Group II.

Group I	Group II
Scuba Diving (S)	Archery (A)
Horseback Riding (H)	Tennis (T)
Fishing (F)	Dance (D)
	Photography (P)

How many combinations of activities are possible?

Solution
There are three ways of selecting an activity from Group I and four ways of selecting an activity from Group II. Each activity in Group I can be paired with each activity in Group II. Thus, there are (3)(4), or 12, possible combinations for each camper.

Example 1 is a direct application of the **Fundamental Counting Principle** stated below:

▶ If Event I can happen in m ways and Event II can happen in n ways, then the number of ways that both Event I and Event II can happen is $m \cdot n$.

486 **Chapter 10** Topics of Number Theory

SECTION 10.1

There are other ways we can view the solution to Example 1. Using letter abbreviations, we can list the possible arrangements in a display. For example, SA will represent the scuba diving and archery combination.

$$\text{Group I} \begin{array}{c} \\ \end{array} \begin{bmatrix} \text{SA} & \text{ST} & \text{SD} & \text{SP} \\ \text{HA} & \text{HT} & \text{HD} & \text{HP} \\ \text{FA} & \text{FT} & \text{FD} & \text{FP} \end{bmatrix}$$

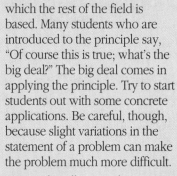

Notice that the list of all possibilities forms a 3 by 4 matrix. We see that there are (3)(4), or 12, entries in the display.

We can also use a horizontal or a vertical ***tree diagram.***

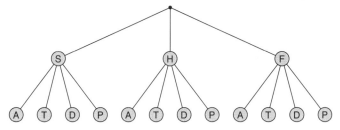

We can count 12 paths along the branches of the trees in each tree diagram. Again we see that a camper can choose from 12 different schedules.

COUNTING APPLICATIONS

Counting techniques can be used to solve a number of problems.

Example 2
How many three-digit even numbers can be formed using the digits 1, 2, 3, 4, and 5?

Solution
There are three blanks to be filled in a three-digit number:

$$\underline{\quad ? \quad} \quad \underline{\quad ? \quad} \quad \underline{\quad ? \quad}$$
$$\text{(hundreds)} \quad \text{(tens)} \quad \text{(units)}$$

Since there are two even digits, 2 and 4, we have two choices for filling in the units blank. There are five different numbers that we can select for both the tens and the hundreds blanks. Thus, we have 5 · 5 · 2, or 50, possible three-digit even numbers.

---

which the rest of the field is based. Many students who are introduced to the principle say, "Of course this is true; what's the big deal?" The big deal comes in applying the principle. Try to start students out with some concrete applications. Be careful, though, because slight variations in the statement of a problem can make the problem much more difficult.

■ Example 1 illustrates how a tree diagram can be used to help solve a counting problem. Tree diagrams are also useful in solving probability problems, as shown in the sample problem. Using them in this manner should be encouraged. Many of the problems in the problem set can best be solved using a tree diagram. Other methods of solution include making an organized list and directly applying the counting principle through multiplication. Allow students to employ the strategy that makes the most sense to them.

Communicating Mathematics

■ Have students answer the following questions: How many different seven-digit phone numbers are mathematically possible in any one area code? Is this the same as the number of seven-digit phone numbers that would be practical to assign in any one area code? Why or why not?

■ Problems 3, 7, 10, 25, 33, and Investigation encourage communicating mathematics.

SECTION 10.1

Assignment Guide

Basic
Optional

Average or Heterogeneous
Optional

Honors
Day 1 12–14, 16–18, 20–22, 24, 26, 28–30, 35

Strands

Communicating Mathematics	3, 7, 10, 25, 33, Investigation
Geometry	27, 35
Data Analysis/ Statistics	18, 21, 25
Technology	21, 30, 33
Probability	1, 2, 4–19, 21–23, 26, 28, 29, 34, Investigation
Algebra	20
Patterns & Functions	33
Number Sense	7, 20, 24, 30, 31, 36
Interdisciplinary Applications	2, 4, 5, 14–16, 18, 19, 21, 23, 26, 30, Investigation

Individualizing Instruction

The kinesthetic learner Students may benefit from experimenting with manipulatives to compare probabilities with actual results. You may have them use coins in problems 9 and 22 and dice in problem 29.

Problem-Set Notes

for page 489

- 4–6 should be done in class to see how well students are understanding the ideas.

488

Example 3
Donisha has 12 songs on his compact disc. How many different ways can he program his compact disc player to play 3 of the 12 songs?

Solution
There are 12 choices for the first song, 11 choices for the second song, and 10 choices for the third song. By the Fundamental Counting Principle there are 12 · 11 · 10, or 1320, different ways 3 out of 12 songs can be programmed on the compact disc player.

Sample Problem

Problem When Jordanne Michaels shoots two free throws, she makes 70% of her first free-throw attempts and 80% of her second free-throw attempts.
  a What percent of the time does she make both of the free-throw attempts?
  b What percent of the time does she make exactly one free-throw attempt?

Solution We can draw a tree diagram to model Jordanne's shooting of two free throws.

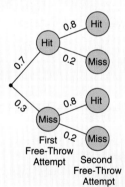

  a To determine the percent of times Jordanne makes both free throws, we follow the path where both shots are made and multiply

  (0.7)(0.8) = 0.56

  She makes both shots 56% of the time.

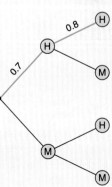

488

SECTION 10.1

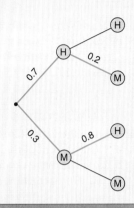

b To determine the percent of times Jordanne makes exactly one free throw, we follow both paths where one shot is made and one shot is missed. We multiply (0.7)(0.2) and (0.3)(0.8) and add the results for the two parts.

$$(0.7)(0.2) + (0.3)(0.8) = 0.14 + 0.24$$
$$= 0.38$$

Jordanne makes one shot 38% of the time.

Think and Discuss

P **1** Suppose a camper in Example 1 was not allowed to choose horseback riding. How many choices would he or she have?   8

P **2** Pat can go either to the movies or to a concert on Friday night and can babysit or mow the lawn on Saturday. How many combinations of activities are possible?   4

P **3** State in your own words the Fundamental Counting Principle.

P, R **4** The weather forecast for the weekend is for a 60% chance of rain on Saturday and a 60% chance of rain on Sunday. Find the probability of
 a Rain on both Saturday and Sunday   36%
 b No rain all weekend   16%
 c Rain on Saturday but no rain on Sunday   24%
 d No rain on Saturday but rain on Sunday   24%
 e Some rain all weekend   84%

P, R **5** On each of three days, the probability of rain is listed as 50%.
 a Find the probability that on two days it rains and on one day it does not rain.   37.5%
 b Find the probability that on two days it does not rain and on one day it does rain.   37.5%

P, R **6** In the sample problem, what is the probability that Jordanne will make the first shot? The second shot? Both shots? Neither shot?   .7; .8; .56; 0.06

P **7 a** Multiply each number in column A by each number in column B. How many different products are possible?   12
 b Do the same problem for columns C and D.   9
 c Explain why your answers are different.

A	B	C	D
2	11	1	3
3	13	2	5
5	17	4	6
7		8	

■ **7** is an application of the Fundamental Counting Principle that demonstrates that the pairs must yield distinct results for the principle to make sense.

Additional Examples

Example 1
A diner gets a choice of one soup and one sandwich from the following menu for a lunch special.

Soups	Sandwiches
Lentil	Tuna salad
Mushroom	Cheese
Chicken	Ham
Barley	Vegetarian

How many combinations of soups and sandwiches are there?

Answer: 16

Example 2
How many three-digit odd numbers can be formed using the digits 5, 6, 7, 8, and 9?

Answer: 75

Example 3
Ten teams enter a holiday basketball tournament. In how many different ways can the first-, second-, and third-place trophies be awarded?

Answer: 720

Additional Answers

for pages 488 and 489

3 Answers will vary.

7 c Possible answer: In **a**, each pair of numbers is prime, so there is a total of 4 × 3, or 12, different products, but in **b**, the answers aren't unique (e.g., 2 × 6 = 4 × 3), so there are fewer distinct products (9).

SECTION 10.1

Problem-Set Notes

for pages 490 and 491

- 12–17 provide many situations in which the counting principle can be applied. Students may need some guidance to see that these are fundamentally the same problem.
- 20 involves matrices, primes, and counting. This problem is one for which the best strategy is to directly apply the counting principle, since listing all of the possibilities would be quite an exhaustive task.

Additional Practice

1. At camp, Dawn must choose one of five activities for the morning, and one of six activities for the afternoon. In how many ways can she arrange her day?
2. A school cafeteria offers three different drinks, two different main courses, and four different desserts. How many lunches are available if one item from each category is chosen?
3. How many four-digit numbers can be formed from the digits 0, 1, 2, 3, 4, and 5?
4. Use a tree diagram to find the probability of getting four tails when a coin is tossed four times.
5. The probability of Arthur's making a basket is .4. What is the probability that Arthur will make exactly two out of three baskets?

Answers:
1. 30   2. 24   3. 1080   4. .0625
5. .288

P   **8 a** What is the probability of getting a head on one toss of a coin?   0.5
   b Tokahama tossed a coin four times in a row and it came up heads each time. What is the probability the coin will come up heads on the next toss?   0.5

P, R   **9** Use a tree diagram to find the probability of getting heads exactly three times when a coin is tossed four times.   .25

P, R   **10** Refer to the sample problem on pages 488–489.
   a What percent of the time does Jordanne miss both free throw attempts?   6%
   b Jordanne makes both shots 56% of the time, makes exactly one out of two 38% of the time, and misses both shots 6% of the time. What is the sum of these three percents? Explain your answer.   100%

Problems and Applications

P   **11** Copy the tree diagram onto your paper. Then, complete the diagram.

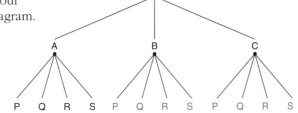

P, R   **12** Set $A_1 = \{a, b, c, \ldots, w, x, y, z\}$ and set $A_2 = \{A, B, C, \ldots, X, Y, Z\}$. How many ways can we pair one element from set A_1 with one element from set A_2?   676

P   **13** How many line segments can be drawn between the points on the left and the points on the right?   12

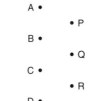

P   **14** Suzanne has five blouses and three skirts that can be mixed and matched. How many different outfits can she make?   15

P   **15** At a family restaurant, diners are given their choice of soup or salad; asparagus, carrots, or beets; and pasta, chicken, beef, or pork. How many different meals are possible if we assume one choice is made from each of the three categories?   24

P   **16** The Savory Sandwich Shop allows customers to select one of five kinds of meat and one of three kinds of cheeses. How many different sandwiches can a person have if he or she orders a meat-and-cheese sandwich?   15

490   **Chapter 10**   Topics of Number Theory

SECTION 10.1

P **17** How many three-digit numbers can be formed using the digits 0, 1, 2, 3, 4, 5, 6, 7, 8, and 9?   900

P **18** **Sports** The tree diagram models Tom's shooting of two free throws. Find the probability that he will make exactly one free throw.   .44

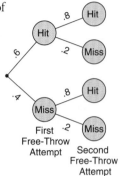

P **19** The coach of the girls' softball team is selecting a catcher and a pitcher for the starting lineup. How many combinations of catcher and pitcher can the coach select if there are three catchers and five pitchers on the team?   15

P, R **20** How many two-by-two matrices can be written if each entry is a single-digit prime number?   256

P **21** **Sports** Scottie Pitten has a 64% free throw average. What is the probability that he will make four free throws in a row?   ≈16.8%

P **22** A coin is tossed three times. What is the probability that it lands heads twice and tails once? $\frac{3}{8}$ or .375

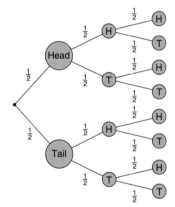

P **23** **Business** A clothing store is selling sweatshirts at 50% off. The store guarantees it will have a wide selection of sizes and colors. The colors are black, white, blue, red, green, brown, yellow, and purple. The sizes are small, medium, large, and extra-large. How many choices are available?   32

I **24** The notation 7!, read "7 factorial," means $7 \cdot 6 \cdot 5 \cdot 4 \cdot 3 \cdot 2 \cdot 1$. Evaluate the expression $4! + 3!$.   30

Reteaching

To reteach the sample problem, have students approach it as an application of the Fundamental Counting Principle. Suppose that Jordanne shoots 20 free throws. The outcomes, based on the percentages given, would be:

Free throw 1	Free throw 2
hit	hit
hit	hit
hit	hit
hit	hit
hit	hit
hit	hit
hit	hit
miss	hit
miss	miss
miss	miss

By the counting principle, there are 10 × 10, or 100, possible outcomes. Of these, 7 × 8, or 56, will be (hit, hit). The probability, then, of Jordanne hitting two free throws is $\frac{56}{100}$, or 56%.

Stumbling Block

Students may have difficulty deciding how the Fundamental Counting Principle can be applied to a given problem. In particular, distinguishing between problems in which items can be repeated (as in example 2) and those in which items cannot be chosen more than once (as in examples 1 and 3) is difficult. Model a variety of applications of the counting principle.

Additional Answers

for pages 490 and 491

10 b Possible explanation: Since the three percentages represent all possible outcomes, their total must be 100%.

SECTION 10.1

Problem-Set Notes

for pages 492 and 493

- 25 is a nice variation from what students are used to: It asks that they interpret a tree diagram rather than make one.
- 34 is related to 7.
- 35 previews one way we develop the concept of primes in the next section.

Cooperative Learning

- Have students hypothesize in writing about the probability of having 2 boys and 2 girls in a family of 4 children. Then have students work in groups to determine the correct answer.

Answer: $\frac{6}{16}$

- Problem 33 is appropriate for cooperative learning.

Checkpoint

1. Judy has four pairs of pants, five blouses, and three vests that she can mix and match. How many outfits does she have available if she wears one of each item?
2. How many three-digit odd numbers can be formed from the digits 1, 2, 3, 4, and 5?
3. How many three-digit odd numbers can be formed from the digits 0, 1, 2, 3, and 4?
4. If the probability of Greg's making a putt is .65, what is the probability he will make his first putt on each of the first two holes?
5. Use a tree diagram to find the probability of getting heads exactly two times when a coin is tossed four times.

492

P **25 Communicating** Write a problem that can be represented by the tree diagram.

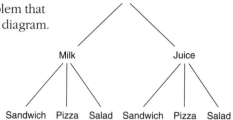

P **26 Science** The weatherman says that the probability for rain today is 70% and for tomorrow is 40%.
  a What is the probability that it rains both days? .28
  b What is the probability that it rains only one of the days? .54
  c What is the probability that it does not rain either day? .18

P, I **27** Suppose a line segment joins A to one of the other points on the circle. Then a third point is selected to be joined to the first two points. How many triangles are possible? 15

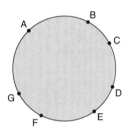

P, R **28** In a video game, a Victory missile has an 80% chance of hitting a Scum missile. Three Scum missiles are observed to be honing in on a target. A Victory missile is fired at each of the Scum missiles. Find the probability that the Victory missiles hit
  a Exactly one Scum missile .096   b At least one Scum missile .992
  c At least two Scum missiles .896   d No Scum missiles .008

P **29 a** What is the probability of rolling a 5 on one toss of one fair die? $\frac{1}{6}$
  b What is the probability of rolling an even number on one toss of one fair die? $\frac{1}{2}$
  c What is the probability of rolling a 5 and then an even number if a fair die is tossed twice? $\frac{1}{12}$
  d What is the probability of rolling an even number and then a 5 if a fair die is tossed twice? $\frac{1}{12}$
  e What is the probability of rolling a 5 and an even number if a fair die is tossed twice? $\frac{1}{6}$
  f What is the probability of rolling an even number or a 5 if a fair die is tossed twice? $\frac{8}{9}$

492   Chapter 10   Topics of Number Theory

SECTION 10.1

P, R, I **30 Technology** A binary digit is either a zero or one. A binary number used by computers consists of a string of binary digits. For example, 100,110 is a six-digit binary number. One K is defined as "the number of different ten-digit binary numbers."
  a How large is 1 K?   1024 digits
  b How large is 640 K?   655,360 digits
  c One meg is 1000 K. How large is 1 meg?   1,024,000 digits
  d A computer has a 40-meg hard drive. How large is 40 meg?   40,960,000 digits

◀ LOOKING BACK **Spiral Learning** LOOKING AHEAD ▶

R **31** Each ticket for the school play costs $6.00. George totaled the sales for one night and wrote the number on a slip of paper. When he looked for this slip of paper, he found five slips of paper, with the numbers $576.00, $584.00, $596.00, $564.50, and $591.00. Which number represents the ticket sales for the school play? $576.00

R **32** Perform the indicated operation.
  a $\frac{1}{4} - \frac{1}{6}$ $\frac{1}{12}$
  b $\frac{1}{2} - \frac{1}{3} - \frac{1}{4}$ $-\frac{1}{12}$

I **33 Spreadsheets** Copy machine A makes 8 copies per minute, and copy machine B makes 12 copies per minute.
  a Suppose each machine runs for an hour. Use a spreadsheet to list, minute by minute, the total number of copies made by each machine.
  b List the numbers that appear in both columns.
  c Why do these numbers appear in both columns?

R, I **34** How many different fractions can be made if the numerator of the fraction is selected from {1, 3, 5, 7} and the denominator of the fraction is selected from {1, 4, 7, 10}?   15

R **35** A solid rectangle is formed by using 20 tiles. Find all possible perimeters of the rectangles.   18, 24, 42

I **36** Find the smallest natural number that is divisible by
  a 16 and 24   48
  b 7 and 11   77

Investigation

Lotteries Does your state have a state lottery? If so, find out how it is played. If not, check out a lottery in some other state. What is the probability of winning? What percent of the money spent on tickets is actually paid out to the winners? On average, is playing the lottery profitable? Explain.

Answers:
1. 60 **2.** 75 **3.** 40 **4.** .4225
5. .375

Multicultural Note
Arthur Cayley (1821–1895) was one of the first to study matrices. His discovery of the rules for their manipulation secured his place in mathematics as one of the earliest investigators of abstract algebra.

Additional Answers
for pages 492 and 493

25 Possible answer: How many choices does Jane have for lunch if she can choose milk or juice and either a sandwich, pizza, or a salad?

33 a Column 1, Minutes: 1, 2, 3, . . . 60
Column 2, Machine A: 8, 16, 24, 32, . . . 480
Column 3, Machine B: 12, 24, 36, 48, . . . 720
b 24, 48, 72, 96, 120, 144, 168, 192, 216, 240, 264, 288, 312, 336, 360, 384, 408, 432, 456, 480
c They are the common multiples of 8 and 12.

Investigation Answers will vary. A lottery game that involves matching all 6 numbers (out of 54) gives you less than a 1-in-25-million chance of hitting the jackpot. Typically, states pay out to winners only about 40% to 60% of ticket sales.

SECTION 10.2

Objectives

- To identify prime numbers and composite numbers
- To identify natural-number multiples of a number
- To identify the least common multiple of natural numbers

Vocabulary: composite number, factor, least common multiple, prime number

Class Planning

Schedule 1 day

Resources

In *Chapter 10 Student Activity Masters* Book
Class Openers *13, 14*
Extra Practice *15, 16*
Cooperative Learning *17, 18*
Enrichment *19*
Learning Log *20*
Using Manipulatives *21, 22*
Spanish-Language Summary *23*

Transparencies *40 a–d*

Class Opener

Cut out twelve identical squares. Find the number of rectangles you can make using all of the following numbers of squares.
a. 3   b. 6   c. 7   d. 8   e. 10   f. 12

Answers:
a. 1   b. 2   c. 1   d. 2   e. 2   f. 3

Lesson Notes

- The concept of prime numbers is often an abstract one, so we develop it through concrete materials. It might be helpful for some students to experiment with actual tiles, as in the class opener.
- We assume that students have had considerable work with primes, composites, multiples,

494

Investigation Connection
Section 10.2 supports Unit E, *Investigations* 1, 2, 3. (See pp. 426A-426D in this book.)

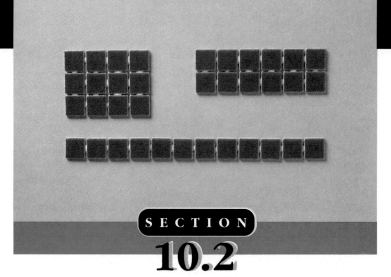

SECTION 10.2
Primes, Composites, and Multiples

PRIME NUMBERS AND COMPOSITE NUMBERS

We can use tiles to learn about prime numbers and composite numbers.

Example 1
a How many different rectangles can be formed using 12 tiles?
b How many different rectangles can be formed using 7 tiles?

Solution
a We can arrange the 12 tiles into three different rectangles.

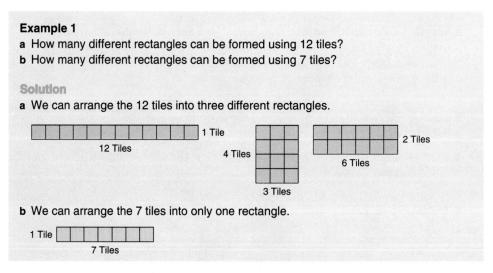

b We can arrange the 7 tiles into only one rectangle.

The example shows that the numbers 1, 12, 2, 6, 3, and 4 are ***factors*** of 12. Each number divides into 12 leaving a remainder of zero. The only factors of 7 are 1 and 7. A number greater than one is a ***prime number*** if it has exactly two factors, 1 and itself. Seven is a prime number. A natural number greater than one is a ***composite number*** if it has more than two factors. The number 12 is a composite number, since it has factors other than 1 and 12. Every prime number and every composite number have at least two different factors. The number 1 has only one factor; therefore, it is neither a prime number nor a composite number.

494   **Chapter 10**   Topics of Number Theory

SECTION 10.2

We can represent the natural numbers using a Venn diagram.

Natural Numbers

Prime Numbers	1	Composite Numbers
2, 3, 5, 7, 11, ...	(Neither prime nor composite)	4, 6, 8, 9, 10, 12, ...

Example 2
List the factors of each number and classify each number as *prime* or *composite*.
a 28   b 31   c 64

Solution
a The factors of 28 are 1, 2, 4, 7, 14, and 28. The number 28 is composite.
b The factors of 31 are 1 and 31. The number 31 is prime.
c The factors of 64 are 1, 2, 4, 8, 16, 32, and 64. The number 64 is composite.

MULTIPLES

Let's take a look at these two patterns.

$1 \cdot 4 = 4$ $1 \cdot 6 = 6$

$2 \cdot 4 = 8$ $2 \cdot 6 = 12$

$3 \cdot 4 = 12$ $3 \cdot 6 = 18$

$4 \cdot 4 = 16$ $4 \cdot 6 = 24$

$5 \cdot 4 = 20$ $5 \cdot 6 = 30$

$6 \cdot 4 = 24$

$7 \cdot 4 = 28$

$8 \cdot 4 = 32$

The set A = {4, 8, 12, 16, 20, 24, 28, 32} contains the first eight natural-number multiples of 4. The set B = {6, 12, 18, 24, 30} contains the first five natural-number multiples of 6. The numbers 12 and 24 are common multiples of 4 and 6. The number 12 is the **least common multiple (LCM)**—the common multiple with the smallest value—of 4 and 6.

We can represent set A and set B using a Venn diagram.

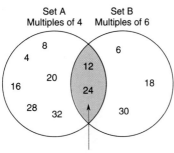

Section 10.2 Primes, Composites, and Multiples

---

and factors prior to this class. It is included here because an understanding of primes and composites will enable students to efficiently solve a surprising number of problems.

■ Many algebraic concepts are simply generalizations of this material. In some cases the language is even the same (as in prime polynomials). The subject also provides many challenging questions for students who are simply interested in numbers.

■ Use the terms *least common multiple* and *greatest common factor* rather than the abbreviated forms LCM and GCF when speaking about the concepts. Students quickly lose track of what the letters stand for when we speak in acronyms. The words are descriptive and will help students remember what it is they are doing.

■ Example 3 gives a visual model of multiples. This method of visualization will be helpful in many of the problems that require multiples for their solution.

Additional Examples

Example 1
a. How many different rectangles can be formed using 5 tiles?
b. How many different rectangles can be formed using 16 tiles?

Answers: **a.** 1 **b.** 3

Example 2
List the factors of each number and classify each number as prime or composite.
a. 37 **b.** 46 **c.** 24

Answers: **a.** 1, 37; prime **b.** 1, 2, 23, 46; composite **c.** 1, 2, 3, 4, 6, 8, 12, 24; composite

SECTION 10.2

Additional Examples continued

Example 3

Draw a number line and graph all the natural numbers less than 80 that are

a. Multiples of 8
b. Multiples of 18
c. Common multiples of 8 and 18

Answers:

```
  8  16  24  32  40  48  56  64  72
←─┼──┼──┼──┼──┼──┼──┼──┼──┼──→
      18       36       54       72
```

Problem-Set Notes

for page 497

■ **6** and **7** will stimulate discussion. They are good problems for small groups.

■ **11** shows another way to find the least common multiple.

■ **12** should elicit many responses. For example, 1 is the only number that is neither prime nor composite, 12 is the only number with more than one prime factor and is the only two-digit number, 9 is the only perfect square, 5 is the only odd prime, and 2 is the only even prime. Your students will probably come up with many more.

■ **13** points out that there are many common multiples, but only one that is least.

Additional Answers

for page 497

5 All prime natural numbers and 1.

7 4, 6, 8, 9, 10, 14

Example 3

Draw a number line and graph all the natural numbers less than 100 that are

a Multiples of 9   **b** Multiples of 12   **c** Common multiples of 9 and 12

Solution

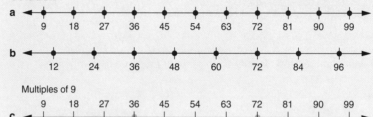

Notice that the least common multiple (LCM) of 9 and 12 is 36.

Sample Problems

Problem 1   We will define LCM(m, n) to be the least common multiple of m and n. Find LCM(4, 6) + LCM(6, 9).

Solution   We first evaluate LCM(4, 6) and then evaluate LCM(6, 9). The least common multiple of 4 and 6 is 12 and the least common multiple of 6 and 9 is 18. Thus, LCM(4, 6) + LCM(6, 9) = 12 + 18 = 30.

Problem 2   Find the value of n if $n = \frac{7}{8} - \left(\frac{1}{2} - \frac{1}{3}\right)$.

Solution   We will use the LCM of the denominators of the fractions as the common denominator of the fractions.

$$n = \frac{7}{8} - \left(\frac{1}{2} - \frac{1}{3}\right)$$

Rewrite $\frac{1}{2}$ as $\frac{3}{6}$ and $\frac{1}{3}$ as $\frac{2}{6}$
$$n = \frac{7}{8} - \left(\frac{3}{6} - \frac{2}{6}\right)$$

$$n = \frac{7}{8} - \frac{1}{6}$$

Rewrite $\frac{7}{8}$ as $\frac{21}{24}$ and $\frac{1}{6}$ as $\frac{4}{24}$
$$n = \frac{21}{24} - \frac{4}{24}$$

$$n = \frac{17}{24}$$

Note that the LCM of 2 and 3 is 6 and the least common denominator of $\frac{1}{2}$ and $\frac{1}{3}$ is 6. Also, the LCM of 6 and 8 is 24, and the least common denominator of $\frac{1}{6}$ and $\frac{7}{8}$ is 24.

Chapter 10   Topics of Number Theory

SECTION 10.2

Think and Discuss

P   **1** What are the first five prime numbers?   2, 3, 5, 7, 11

P   **2** What are the first five natural-number multiples of 3?   3, 6, 9, 12, 15

P   **3** What is the least common multiple of 2 and 5?   10

P   **4** Complete the table for the dimensions of all the rectangles that can be formed with each given number of tiles.

Number of Tiles	Dimensions of Rectangles
1	1 × 1
2	1 × 2
3	1 × 3
4	1 × 4; 2 × 2
5	1 × 5
6	1 × 6; 2 × 3
7	1 × 7
8	1 × 8; 2 × 4
9	1 × 9; 3 × 3
10	1 × 10; 2 × 5
11	1 × 11
12	1 × 12; 2 × 6; 3 × 4

P   **5** For what natural numbers of tiles can exactly one rectangle be formed?

P   **6** Ms. Monroe's class of ten students was using tiles to build a rectangle. Each of the students had a different number of tiles and each was able to build only one rectangle. Ms. Monroe's class did this with the least number of tiles. How many tiles did the students have?   101

P   **7** Mr. Quincy's class was using tiles to build rectangles. He had six different groups, and each group had a different number of tiles. Each group was able to build exactly two rectangles. Mr. Quincy's class did this using the least number of tiles. How many tiles did each group use?

P   **8** A given number of tiles can be arranged to form a 3-by-6 rectangle. What are the dimensions of the other rectangles that can be formed using the same tiles?   18 × 1; 9 × 2

P   **9** List the dimensions of the rectangles that can be formed using
    a 10 tiles   10 × 1; 5 × 2   **b** 11 tiles   11 × 1   **c** 9 tiles   9 × 1; 3 × 3

P   **10** Find four numbers of tiles that will produce exactly
    a One rectangle 1, 2, 3, 5   **b** Two rectangles 4, 6, 8, 9   **c** Three rectangles 12, 18, 20, 28

P   **11** **a** List the first three multiples of 16.   16, 32, 48
     b List the first three multiples of 24.   24, 48, 72
     c What is the LCM of 16 and 24?   48

P   **12** Which of the numbers 1, 2, 5, 9, and 12 is different from the others? Explain your answer.   Possible answer: 1; It is neither prime nor composite.

P   **13** Find the five numbers with the smallest values that are divisible by 10 and 15.   30, 60, 90, 120, 150

Assignment Guide

Basic
Optional

Average or Heterogeneous
Optional

Honors
Day 1 15, 16, 18, 19, 21–24, 26–28, 30–32, 34, 35, 39, Investigation

Strands

Communicating Mathematics	12
Geometry	4–10, 22, 25, 26
Measurement	26, 32, 35, 38, 39, 43
Data Analysis/ Statistics	18, 30
Technology	34
Probability	28, 37, 42
Algebra	24, 29, 31, 33, 35, 36, 38, 40, 41, 43
Patterns & Functions	11, 13, 20, 24, 27, 34
Number Sense	1, 2, 5–11, 13–25, 27–34, 36, 37, 40–42
Interdisciplinary Applications	30, 32, 38, 39

Individualizing Instruction

The gifted learner Students should recognize that rearranging a certain number of tiles into different rectangles keeps the area constant. But how is the perimeter affected? Have students find out which arrangement has the greatest perimeter and which has the least perimeter.

Section 10.2   Primes, Composites, and Multiples

SECTION 10.2

Problem-Set Notes

for pages 498 and 499

- **14 a** and **b** require some thought and some trial and error; **c** and **d** can be answered using the definition of composite numbers.
- **30** is an application of number theory.

Cooperative Learning

- Have students explore the Sieve of Eratosthenes:
  1. Write all the numbers from 1 to 100.
  2. Cross out the number 1. We know it's not prime.
  3. Circle 2, then cross out all of its multiples.
  4. Circle 3, then cross out all of its multiples.
  5. Circle 5, the next number that is not crossed out. Then cross out all of its multiples.
  6. Continue in the same way by circling numbers and crossing out multiples.
- Problems 6 and 7 are appropriate for cooperative learning.

Additional Practice

In problems 1–4, classify each number as *prime* or *composite*.
1. 29 2. 91 3. 15 4. 43

In problems 5 and 6, list the first six natural-number multiples of the given number.
5. 13   6. 17

In problems 7 and 8, find the least common multiple of each pair of numbers.
7. 9 and 24   8. 48 and 60

9. Find all natural numbers so that $\frac{18}{n}$ is a natural number.
10. Find $\frac{4}{9} + \frac{5}{24}$.

498

P **14** Is the statement true *always, sometimes,* or *never?*
  a The sum of two prime numbers is a prime number. Sometimes
  b The sum of two composite numbers is a composite number. Sometimes
  c The product of two prime numbers is a composite number. Always
  d The product of two composite numbers is a composite number. Always

Problems and Applications

P **15** Classify each number as *prime* or *composite*.
  a 31 Prime   **b** 51 Composite   **c** 72 Composite   **d** 43 Prime

P **16** List the first six natural-number multiples of
  a 4   4, 8, 12, 16, 20, 24   **b** 7   7, 14, 21, 28, 35, 42

P **17** Find the least common multiple of each pair of prime numbers.
  a 5 and 7   35   **b** 2 and 3   6   **c** 11 and 13   143

P **18** What three numbers with the smallest values belong in II?   56, 112, 168

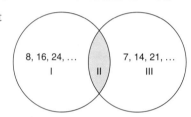

P **19** Find the least common multiple of
  a 5 and 7   35   **b** 12 and 3   12   **c** 8 and 6   24

P **20** List the first five common multiples of 6 and 8.

P **21 a** Find the least common multiple of 12 and 8.   24
       b Find the least common multiple of 10 and 12.   60

P **22** How many solid rectangles can be formed using 40 tiles?   4

P, R **23** Find all natural numbers n so that $\frac{12}{n}$ is a natural number.   1, 2, 3, 4, 6, 12

P, R **24** Draw a number line and graph all positive multiples of
  a 4 that are less than 50   **b** 3 that are less than 50
  c 3 and 4 that are less than 50

P **25** Draw the other rectangles that can be made with these tiles.   20 × 1; 10 × 2

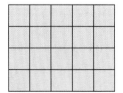

P, R **26** Toby arranged a set of tiles in a single row. The perimeter of the rectangle formed was 100. What is the perimeter of the square that can be constructed with these tiles?   28

498   Chapter 10   Topics of Number Theory

SECTION 10.2

P, R **27** a List the positive multiples of 7. 7, 14, 21, 28, ...
  b List the whole-number multiples of 7. 0, 7, 14, 21, 28, ...
  c List the integer multiples of 7. ...−14, −7, 0, 7, 14, 21, ...

P, R **28** Two dice are tossed. The table is filled in by determining the LCM of the numbers on the faces of the two dice.
  a Copy and complete the table.
  b Find the probability that the LCM of the numbers on the faces of the two dice is a multiple of 6. $\frac{5}{12}$
  c Find the probability that the LCM of the numbers on the faces of the two dice is a multiple of 5. $\frac{11}{36}$

LCM	1	2	3	4	5	6
1	1					
2		2				
3				12		
4						
5						
6						

P **29** Let $a \# b$ = LCM $\{a, b\}$. Find the values of
  a $(4 \# 6) \# 8$ 24
  b $4 \# (6 \# 8)$ 24

P, R **30** Science Gear A rotates in a clockwise direction while Gear B rotates in a counter-clockwise direction. How many complete revolutions of Gear A and Gear B are necessary in order for the gears to align themselves at 0? 5 of A; 4 of B

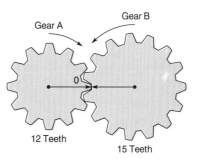
Gear A Gear B
12 Teeth 15 Teeth

P, R **31** a Find LCM (9, 12). 36
  b Use your answer in part a to write equivalent fractions for $\frac{1}{9}$ and $\frac{1}{12}$ that have a common denominator. $\frac{4}{36}; \frac{3}{36}$
  c Use your answer to part b to find $\frac{1}{9} - \frac{1}{12}$. $\frac{1}{36}$

P, R **32** Phyllis can run once around a certain field in 8 minutes. Tyred can run once around the same field in 12 minutes. If they begin at the same time and run in the same direction for an hour, when will Phyllis pass Tyred?

P, R **33** a Find the LCM (18, 48). 144
  b Use your answer in part a to rewrite $\frac{17}{18} - \left(\frac{5}{18} + \frac{11}{48}\right)$ as an expression with a common denominator. $\frac{136}{144} - \left(\frac{40}{144} + \frac{33}{144}\right)$
  c Evaluate the new expression you wrote in part b. $\frac{7}{16}$

P, R **34** Spreadsheets
  a Write a spreadsheet to generate the first 100 multiples of 3 and the first 100 multiples of 5. How would you describe the common multiples in these two lists?
  b Repeat part a for 6 and 8.

P, R **35** Doralee rides 14 miles per hour on her bike. Annabelle rides 19 miles per hour. At noon, they begin to ride toward each other. If they start 80 miles apart, at what time will they meet? ≈2:25 P.M.

Section 10.2 Primes, Composites, and Multiples 499

Answers:
1. Prime 2. Composite
3. Composite 4. Prime
5. 13, 26, 39, 52, 65, 78
6. 17, 34, 51, 68, 85, 102 7. 72
8. 240 9. 1, 2, 3, 6, 9, 18 10. $\frac{47}{72}$

Communicating Mathematics

- Have students write a paragraph explaining why least common multiple is important in working with fractions.
- Problem 12 encourages communicating mathematics.

Multicultural Note

Eratosthenes, one of history's greatest thinkers, was born in what is now Libya around 274 B.C. In the area of mathematics, he is best known for a method of generating prime numbers, called the Sieve of Eratosthenes, which is presented in the Cooperative Learning feature for this section.

Additional Answers

for pages 498 and 499

20 24, 48, 72, 96, 120

24 a–c
4, 8, 12, 16, 20, 24, 28, 32, 36, 40, 44, 48
3, 6, 9, 12, 15, 18, 21, 24, 27, 30, 33, 36, 39, 42, 45, 48

28 a

LCM	1	2	3	4	5	6
1	1	2	3	4	5	6
2	2	2	6	4	10	6
3	3	6	3	12	15	6
4	4	4	12	4	20	12
5	5	10	15	20	5	30
6	6	6	6	12	30	6

32 After 24 min and after 48 min

34 a 15, 30, 45, 60, 75, 90, ... 255, 270, 285, 300; multiples of 15

b 24, 48, 72, 96, 120, ... 528, 552, 576, 600; multiples of 24

SECTION 10.2

Problem-Set Notes

for page 500

- **37** combines the ideas of divisibility tests (covered in the next section) and the Fundamental Counting Principle (from the previous section).
- **38** and **39** introduce a common type of algebraic word problem.
- **41** builds on problem 20 from the previous section.

Checkpoint

In problems **1–4**, classify each number as *prime* or *composite*.
1. 9 **2.** 17 **3.** 53 **4.** 86

In problems **5** and **6**, list the first six natural-number multiples of the given number.
5. 8 **6.** 11

In problems **7** and **8**, find the least common multiple of each pair of numbers.
7. 12 and 15 **8.** 18 and 24

9. Find all natural numbers so that $\frac{32}{n}$ is a natural number.

10. Find $\frac{1}{12} + \frac{8}{15}$.

Answers:
1. Composite **2.** Prime **3.** Prime **4.** Composite **5.** 8, 16, 24, 32, 40, 48 **6.** 11, 22, 33, 44, 55, 66 **7.** 60 **8.** 72 **9.** 1, 2, 4, 8, 16, 32 **10.** $\frac{37}{60}$

Additional Answers

for page 500

Investigation Four other picture frames can be made using all of these tiles in addition to the 6 × 9 frame shown: 3 × 12, 4 × 11, 5 × 10, and 7 × 8. Many others can be made if you don't use all of the tiles. Arrangements such as 2 × 13 are possible, too, but they leave no interior room, so they probably shouldn't count.

P, R **36** The least common multiple of two different numbers a and b is a. What is true about b? b is a factor of a

◀ LOOKING BACK **Spiral Learning** LOOKING AHEAD ▶

P, R **37 a** How many three-digit multiples of 5 can be created using each of the digits 1, 2, 3, 4, 5, and 6 no more than once? 20
 b How many three-digit multiples of 5 can be created using each of the digits 0, 2, 3, 4, 5, and 6 no more than once? 36

P, R, I **38** Gil can mow $\frac{2}{15}$ of a large lawn in one hour. Juan can mow $\frac{1}{12}$ of the same lawn in an hour.
 a What fraction of the lawn can be mowed in an hour if Gil and Juan work together? $\frac{13}{60}$
 b What fraction of the lawn can Gil mow in three hours? $\frac{2}{5}$
 c What fraction of the lawn can Juan mow in three hours? $\frac{1}{4}$
 d After Gil and Juan have worked together for three hours, what fraction of the lawn is mowed? $\frac{39}{60}$
 e What percent of the lawn still needs to be mowed after Gil and Juan have worked together for three hours? 35%

R **39** If Bill is able to paint a fence in six hours, how long must he work to complete 40% of the job? $2\frac{2}{5}$ hr

P, R **40** Let $\#(x)$ be equal to the sum of all the factors of x. Find
 a $\#(28)$ 56
 b $\#(17)$ 18

P, R **41** How many 2-by-2 matrices can be formed where the elements are different prime numbers less than 20? 1680

R **42** How many four-digit numbers can be formed using the digits 0, 1, 2, 3, 4, 5, 6, 7, 8, and 9 once? 4536

R **43** Chuck can shuck about 80 ears of corn in a half hour. How long would it take Chuck to shuck 1000 ears of corn? 375 min, or 6 hr 15 min

Investigation

Tile Arrangements Raúl took his tiles and made a picture frame as shown. How many different picture frames can he make with his tiles?

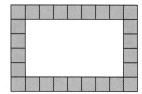

Chapter 10 Topics of Number Theory

SECTION 10.3

Investigation Connection
Section 10.3 supports Unit E, *Investigation* 2. (See pp. 426A–426D in this book.)

10.3 Divisibility Tests

TESTS FOR 2, 3, 4, 5, 6, 8, 9, 10, AND 12

We can use a calculator to test whether a number is divisible by another number. Sometimes, however, it is preferable to use a divisibility test.

A natural number n is divisible by
- 2 if the last digit of the number is even
- 3 if the sum of the digits of the number is divisible by 3
- 4 if the number formed by the last two digits of the number is divisible by 4
- 5 if the last digit of the number is either 0 or 5
- 6 if the number is divisible by both 2 and 3
- 8 if the number formed by the last three digits of the number is divisible by 8
- 9 if the sum of the digits of the number is divisible by 9
- 10 if the last digit of the number is 0
- 12 if the number is divisible by both 3 and 4

Example 1
Test the number 6,392,124 for divisibility by
a 2   b 3   c 4   d 5   e 6   f 8   g 9   h 10   i 12

Solution
a The number is divisible by 2 because the last digit, 4, is even.
b The number is divisible by 3 because the sum of the digits, 27, is divisible by 3.
c The number is divisible by 4 because the number formed by the last two digits, 24, is divisible by 4.
d The number is not divisible by 5 because the last digit is not 0 or 5.
e The number is divisible by 6 because the number is divisible by both 2 and 3.
f The number is not divisible by 8 because the number formed by the last three digits, 124, is not divisible by 8.
g The number is divisible by 9 because the sum of the digits, 27, is divisible by 9.
h The number is not divisible by 10 because the last digit is not 0.
i The number is divisible by 12 because the number is divisible by both 3 and 4.

Objective
- To apply divisibility tests for 2, 3, 4, 5, 6, 8, 9, 11, and 12

Class Planning

Schedule 1 day
Resources
In *Chapter 10 Student Activity Masters* Book
Class Openers 25, 26
Extra Practice 27, 28
Cooperative Learning 29, 30
Enrichment 31
Learning Log 32
Interdisciplinary Investigation 33, 34
Spanish-Language Summary 35

In *Tests and Quizzes* Book
Quiz on 10.1–10.3 (Honors) 263

Class Opener

Simplify each fraction.
1. $\frac{57}{129}$ 2. $\frac{176}{231}$ 3. $\frac{333}{360}$ 4. $\frac{165}{220}$

Answers: 1. $\frac{19}{43}$ 2. $\frac{16}{21}$ 3. $\frac{37}{40}$ 4. $\frac{3}{4}$

Lesson Notes

- Divisibility tests form the basis of a wide variety of challenging and interesting questions about numbers, such as those in problems 25, 29, 30, 32, 33, and 34 in the problem set. These kinds of problems also appear frequently in math contests.

- While this material may be somewhat less important than it was before calculators, some students will find it interesting, and it still plays a role in thinking about numbers. You might ask students to try and figure out why the rules work. They will probably be unable to justify them all, but they should have fun trying, and they will certainly learn something along the way.

SECTION 10.3

Lesson Notes continued

- The sample problem ties together many of the ideas in this section. Some discussion in class would probably be useful.
- Students should pay close attention to the calculator symbols next to the problems in this section. More of the problems than those marked could be done using a calculator, but doing so would prove to be more tedious than simply applying the rules for divisibility.

Additional Examples

Example 1
Test the number 5,481,962 for divisibility by
a. 2   b. 3   c. 4   d. 5   e. 6
f. 8   g. 9   h. 10   i. 12

Answers:
a. Yes b. No c. No d. No e. No
f. No   g. No h. No i. No

Example 2
Is 58,419,686,444,001 divisible by 11?

Answer: No

Cooperative Learning

- Have students review the rules for divisibility by 2, 4, and 8, then work in groups to determine a divisibility rule for 16 and 32. Have students try to come up with a general rule for divisibility by any power of 2 and explain why their rule works.

Possible answer:
A number is divisible by 2^n if the last n digits form a number divisible by 2^n.

- Problem 20 and Investigation are appropriate for cooperative learning.

You may have noticed that divisibility tests for 7 and 11 have not been included. A divisibility test for 11 follows.

DIVISIBILITY TEST FOR 11

Example 2
Is 31,829,172,691,807 divisible by 11?

Solution
The number is too large to be displayed on most calculators, so use of the calculator is limited. Here is a test for the divisibility of the number 31,829,172,691,807 by 11.

List the digits of the number in order.	3 1 8 2 9 1 7 2 6 9 1 8 0 7
Add every other digit, beginning with the first digit on the left (the odd-positioned digits).	$3 + 8 + 9 + 7 + 6 + 1 + 0 = 34$
Add the remaining digits.	$1 + 2 + 1 + 2 + 9 + 8 + 7 = 30$
Find the absolute value of the difference of the two sums.	$\|34 - 30\| = 4$
Check whether the absolute value of the difference is divisible by 11.	Since 4 is not divisible by 11 31,829,172,691,807 is not divisible by 11.

What if the final absolute value for some number n was 0? Would the number n be divisible by 11? Since 0 is divisible by 11, the number n would be divisible by 11.

> A natural number is divisible by 11 if the absolute value of the sum of the odd-positioned digits minus the sum of the remaining digits is divisible by 11.

Sample Problem

Problem Find a value for n so that the number 619,4n2 will be divisible by
a 2   b 3   c 4   d 5   e 6

SECTION 10.3

Solution
a 619,4n2 will be divisible by 2 if n = 0, 1, 2, 3, 4, 5, 6, 7, 8, or 9. Since the last digit of the number is 2, the number will always be an even number.
b 619,4n2 will be divisible by 3 if $6 + 1 + 9 + 4 + n + 2 = 22 + n$ is divisible by 3. This happens if n = 2, 5, or 8, because 24, 27, and 30 are divisible by 3.
c 619,4n2 will be divisible by 4 if the two-digit number $n2$ is divisible by 4. This will happen if n = 1, 3, 5, 7, or 9, because 12, 32, 52, 72, and 92 are divisible by 4.
d 619,4n2 will be divisible by 5 if the number ends in 0 or 5. This will never happen, so there is no value of n that will be a solution.
e 619,4n2 will be divisible by 6 if the number is a multiple of both 2 and 3. The only replacements of n that work for both 2 and 3 are 2, 5, and 8.

Think and Discuss

P **1** Is 2992 divisible by 2? Explain your answer.
P **2** Is 3516 divisible by 6? Explain your answer.
P **3** Explain how to test for divisibility on a calculator.
P **4** Why is it not always possible to use a calculator to test for divisibility?
P **5** The divisibility test for 3 says "add the digits and see whether the sum is divisible by 3." The divisibility test for 9 says "add the digits and see whether the sum is divisible by 9." Guess a divisibility test for 27 (which is 3^3). Does it work?
P **6** Which of the divisibility tests are easiest to use?   2, 5, 10
P **7** Is the statement *true* or *false?*
  a If a number is divisible by 4, then it is divisible by 8.   False
  b If a number is divisible by both 2 and 6, then it is divisible by 12.   False
  c If a number is divisible by 3 and 5, then it is divisible by 15.   True
  d If a number is divisible by 2, 3, and 5, then it is divisible by 5.   True
P **8** In part **a** of the sample problem on the previous page, why do all values of n work?

Problem-Set Notes

for page 503

- 3 and 4 should be discussed—there is more to each answer than one might think.
- 5 is an example of a pattern that fails.

Additional Practice

1. Which numbers are divisible by 3?
  a. 4691   **b.** 8145   **c.** 15,147
2. Which numbers are divisible by 4?
  a. 81,416   **b.** 27,492   **c.** 18,406
3. Which numbers are divisible by 8?
  a. 15,800   **b.** 41,512   **c.** 171,494
4. Which numbers are divisible by 9?
  a. 81,909   **b.** 512,706   **c.** 479,871
5. Which numbers are divisible by 12?
  a. 6039   **b.** 101,676   **c.** 1,265,712

Answers:
1. b, c **2.** a, b **3.** a, b **4.** a, c
5. b, c

Additional Answers

for page 503

1 Yes. A number is divisible by 2 if its last digit is even.

2 Yes. A number is divisible by 6 if it is divisible by both 2 and 3.

3 Possible answer: Use the trial-and-error approach.

4 Possible answer: The calculator's display may not have enough places.

5 Possible answer: A number is divisible by 27 if the sum of the digits is divisible by 27; no

8 Since the last digit of the number is 2, the number will always be even. Therefore, the number will always be divisible by 2.

Problems and Applications

9. Is the number 7890 divisible by
   a. 2? Yes   b. 3? Yes   c. 4? No   d. 5? Yes

10. Which numbers are divisible by 3? By 9?
    a. 108   b. 355   c. 828   d. 633   e. 552   f. 981   g. 3744

11. Is the number 4,316,280,948 divisible by
    a. 2? Yes   b. 3? Yes   c. 4? Yes   d. 5? No   e. 6? Yes
    f. 8? No   g. 9? Yes   h. 10? No   i. 11? No   j. 12? Yes

12. Which numbers are divisible by 2?
    a. 8924 Yes   b. 1735 No   c. 2190 Yes   d. 4000 Yes   e. 2481 No
    f. 5225 No   g. 2222 Yes   h. 7×10^3 Yes   i. $\sqrt{144}$ Yes

13. Which numbers are divisible by 3?
    a. 779,623,556,395 No   b. 779,623,556,396 No   c. 779,623,556,397 Yes
    d. 779,623,556,398 No   e. 779,623,556,399 No

14. Which numbers are divisible by 11?
    a. 891 Yes   b. 121 Yes   c. 585 No   d. 743 No
    e. 9284 Yes   f. 8778 Yes   g. 5353 No   h. 573,375 Yes

15. Which of the numbers are prime numbers?   1871
    a. 7713   b. 4705   c. 6318   d. 9427   e. 1871

16. If a number n is divisible by 36, by what other number is n divisible?

17. Is each of the three numbers 1257, 345, and 234 divisible by
    a. 2? No   b. 3? Yes   c. 5? No   d. 9? No   e. 11? No

18. **Communicating**   Is the area of the rectangle divisible by 12? Explain your answer.

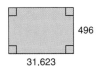

19. **Communicating**   Explain why the sum of two multiples of a number must be a multiple of the number.

20. **Communicating**   The numbers a and b are divisible by 2.
    a. Is the sum of a and b divisible by 2? Explain your answer.
    b. Is the difference of a and b divisible by 2? Explain your answer.
    c. Is the product of a and b divisible by 2? Explain your answer.
    d. Is the quotient of a and b divisible by 2? Explain your answer.

21. Which numbers are divisible by 9?
    a. 25,727,577,360,705 Yes   b. 25,727,577,360,706 No
    c. 25,727,577,360,707 No   d. 25,727,577,360,717 No
    e. 25,727,577,361,717 No

SECTION 10.3

P **22 Communicating** Explain why 32,768,325 is divisible by 15.

P, R **23** Find the values of n so that the number $32n5$ is divisible by 15.

P, R **24** The number $823n45$ is divisible by 11.
  a What is n?   8
  b Is the number divisible by 3?   Yes

P **25** Write a six-digit number in which all the digits are not the same. Then rearrange the digits of the number you wrote to form another six-digit number. Subtract one of the numbers from the other. Is your answer divisible by 9?   Yes

P, R **26** Of the three-digit numbers having three different digits from the set {1, 2, 3, 4}, which are divisible by 3?

P **27** The number $413n$ is divisible by 3 and it is an even number. What is the value of n?   4

P, R, I **28** On a coordinate system, graph the lattice points (x, y) that make the number $x3y527$ divisible by 11. (The point (x, y) is a lattice point if both x and y are integers.)

P, R **29 Communicating** Explain why 5! is divisible by 30.

P, R **30** Of the four-digit numbers having four different digits from the set {1, 2, 3, 4}, which numbers are divisible by 4?   1324, 1432, 3124, 3412, 4132, 4312

P **31** Suppose that a number divided by 4 has a remainder of 2.
  a List five such numbers.
  b Show that when these numbers are divided by 2, the quotient is odd.

P, R **32** Find all values of n that make the number $1nnnnnnnn1$ divisible by 11.

P **33 Communicating** The numbers 3625 and 5263 are a palindromic pair of numbers, since reversing the order of the digits of one number gives the other number. Explain why in a palindromic pair, if one number is divisible by 9, then so is the other number.

P, R **34 a** A number is divisible by 11. The sum of all the odd-positioned digits of the number is 22. The sum of all but one of the even-positioned digits is 18. What is the missing digit?   4
  b A number is divisible by 11. The sum of all the even-positioned digits is 22. The sum of all but one of the odd-positioned digits is 12. What is the missing digit?   No such number
  c A number is divisible by 11. The sum of the odd-positioned digits is 22. One of the even-positioned digits is missing. Can you find two possible values for the missing digit?   No

---

16 1, 2, 3, 4, 6, 9, 12, 18

18 Yes; 496 is divisible by 4, and 31,623 is divisible by 3. So the product is divisible by $3 \times 4 = 12$.

19 Let n = the number. Then $nk + nm = n(k + m)$, which is a multiple of n.

20 a Yes; the sum of two even numbers is always even.
  b Yes; the difference of two even numbers is always even.
  c Yes; the product of two even numbers is always even.
  d Not necessarily; for example, $2 \div 2 = 1$.

22 The sum of the digits is 36. 36 is divisible by 3. The last digit is 5. So the number is divisible by 5. A number divisible by 3 and 5 is divisible by 15.

23 $n = 2, 5,$ or 8

26 123, 213, 312, 234, 324, 423, 132, 231, 321, 243, 342, 432

28 The lattice points are (5, 8), (8, 5), (6, 7), (7, 6), (4, 9), (9, 4), (1, 1), (2, 0)

29 $5! = 5 \times 4 \times 3 \times 2 \times 1$. $30 = 5 \times 3 \times 2$, so 5! is divisible by 30.

31 a Possible answers: 6, 10, 14, 18, 22
  b If $n \div 4$ has remainder 2, then $n = 2, 6, 10, 14, \ldots$ The quotient of n and 2, then, is $1, 3, 5, 7, \ldots$

32 $|(4n + 1) - (4n + 1)| = |0|$. Since 0 is divisible by all multiples of 11, n can be any digit.

33 Possible answer: A number is divisible by 9 if the sum of the digits is divisible by 9. In a palindromic pair the digits are the same; therefore, if the digits of one add up to a number divisible by 9, so will the digits of the other.

SECTION 10.3

Problem-Set Notes

for page 506

- 38 is an arithmetic problem, but it uses the ideas of multiples and factors.
- 44 is a powerful representation of these ideas. It is both visual and numerical.

Communicating Mathematics

- Have students write a few sentences explaining how divisibility rules are important when working with fractions.
- Problems 1–5, 8, 18–20, 22, 29, 33, and 44 encourage communicating mathematics.

Checkpoint

1. Determine whether the number 17,820 is divisible by:
   a. 2 b. 3 c. 4 d. 5 e. 6
   f. 8 g. 9 h. 10 i. 11 j. 12
2. Find a number for n so that $3n4$ is divisible by:
   a. 4 b. 5 c. 9 d. 11 e. 12

Answers:
1. a. Yes b. Yes c. Yes d. Yes
   e. Yes f. No g. Yes h. Yes
   i. Yes j. Yes
2. a. 0, 2, 4, 6, 8 b. None
   c. 2 d. 7 e. 2, 8

Additional Answers

for page 506

35 $\{x: -4 \leq x \leq 7\}$

40 d 0.13

43 About 643 breaths per hour; about 15,430 breaths per day

44 $\frac{3k}{15k} = \frac{3}{15} = \frac{1}{5}$ for any k

Answer for Investigation is on page A8.

◀ LOOKING BACK **Spiral Learning** LOOKING AHEAD ▶

R **35** Describe the graph, using set-builder notation.

R **36** An ordinary die is tossed three times. Find the probability that the number on the first toss is prime, the number on the second toss is a multiple of 2, and the number on the third toss is a multiple of 3. $\frac{1}{12}$

P, R, I **37** Write an equation that generates the table. $y = 2x - 1$

x	−4	−3	−2	−1	0	1	2	3	4
y	−9	−7	−5	−3	−1	1	3	5	7

R, I **38** Beumont skis cross-country a distance of 1 kilometer in 180 strides. He takes 12 strides in 20 seconds. How long will it take Beumont to ski 12 kilometers? 3600 sec, or 60 min

R **39** a What is the least number of tiles needed to produce exactly one rectangle that is not a square? 2
b What is the least number of tiles needed to produce exactly one rectangle whose area is greater than 20? 23

R **40** Write each number in decimal form.
a $\frac{1}{6}$ $0.1\overline{6}$ b 315% 3.15   c $9\frac{1}{4}$ 9.25   d $\sqrt{(0.05)^2 + (0.12)^2}$

R **41** Evaluate each expression.
a $\frac{5}{12} - \frac{2}{3}$ $-\frac{1}{4}$ b $\frac{3}{5} + \frac{7}{4}$ $2\frac{7}{20}$ c $\frac{5}{6} - \frac{3}{8}$ $\frac{11}{24}$

R **42** When running a race, Teng takes a swallow of water every 8 minutes and changes her stride every 10 minutes. How many minutes after she begins will she have to do both at the same time? 40 min

R, I **43** Breathing normally, Nathan takes 5 breaths every 28 seconds. Find his breathing rate in breaths/hour. How many breaths will he take in a day?

R **44** Spreadsheets Make a spreadsheet consisting of 3 columns: column A—multiples of 3; column B—multiples of 15; column C—the quotient of the number in column A divided by the number in column B. Explain your results.

Investigation

Numeric Palindrome A numeric palindrome is a natural number whose digits read the same forward or backward. For example, 375,573 is a numeric palindrome, as is 8228. The number 121 is a three-digit palindrome that is divisible by 11. Find other three-digit palindromes that are divisible by 11. List the numbers from least to greatest value. Can you find all the four-digit palindromes that are multiples of 11?

SECTION 10.4

"Discover any new prime numbers lately?"

SECTION 10.4
Prime Factorization

Investigation Connection
Section 10.4 supports Unit E, *Investigations* 1, 2, 3. (See pp. 426A-426D in this book.)

PRIME-FACTORIZATION FORM

It's easy to find the prime factors of a number like 12.

$$12 = 2 \cdot 2 \cdot 3$$

But how do you factor a number like 4693? Or is 4693 prime? In this section we will introduce methods for determining the prime factorization of a natural number.

The **prime-factorization form** of a natural number shows the prime factors being multiplied. The prime-factorization form of 12 is shown above. We usually list prime factors in order from least to greatest.

Example 1
Write the prime-factorization form for the natural numbers 26, 27, 28, 29, 30, 31, and 32.

Solution

Natural Number	Prime-Factorization Form
26	$2 \cdot 13$
27	$3 \cdot 3 \cdot 3$ or 3^3
28	$2 \cdot 2 \cdot 7$ or $2^2 \cdot 7$
29	29 is prime.
30	$2 \cdot 3 \cdot 5$
31	31 is prime.
32	$2 \cdot 2 \cdot 2 \cdot 2 \cdot 2$ or 2^5

Objectives
- To express a number in prime-factorization form
- To determine whether a number is prime

Vocabulary:
prime-factorization form

Class Planning

Schedule 1 day

Resources
In *Chapter 10 Student Activity Masters* Book
Class Openers *37, 38*
Extra Practice *39, 40*
Cooperative Learning *41, 42*
Enrichment *43*
Learning Log *44*
Technology *45, 46*
Spanish-Language Summary *47*

Class Opener

Write each number as a product of only prime factors.
1. 21 **2.** 28 **3.** 60 **4.** 198

Answers:
1. $3 \cdot 7$ **2.** $2 \cdot 2 \cdot 7$
3. $2 \cdot 2 \cdot 3 \cdot 5$ **4.** $2 \cdot 3 \cdot 3 \cdot 11$

Lesson Notes

- The uniqueness of prime factorization is perhaps the fundamental fact of integer arithmetic. It is an underlying assumption for most problems about factoring, and it is at the center of most discussions of number theory. It is a rather surprising phenomenon. We certainly do not have uniqueness with regard to forming sums, so there is no idea corresponding to primes in addition. Primes are the building blocks of the integers with respect to

SECTION 10.4

Lesson Notes continued

multiplication, so most questions about multiplication (and therefore about division) are ultimately questions about prime factors.

- The factor tree provides a visual representation for prime factorization, but in actual practice, students should be encouraged to use the method of example 3.

- Example 4 gives a method for finding the prime-factorization form of large numbers. Students will probably not be required to do this very often, but they should know how to do it.

- The sample problem includes the key idea that the largest number we need to check as a factor is the largest prime less than the square root of the number. A table like the ones on page 508 may help students understand why the square root is the stopping point in the search for primes.

Additional Examples

Example 1
Write the prime-factorization form for the natural numbers 35, 36, 37, 38, 39, and 40.

Answers:
$35 = 5 \cdot 7$; $36 = 2 \cdot 2 \cdot 3 \cdot 3$; 37 is prime; $38 = 2 \cdot 19$; $39 = 3 \cdot 13$; $40 = 2 \cdot 2 \cdot 2 \cdot 5$

Example 2
Use a factor tree to find the prime-factorization form of 132.

Possible answer:

```
   132
  /  \
 2    66
     /  \
    2    33
        /  \
       3    11
```
$132 = 2^2 \cdot 3 \cdot 11$

PRIME-FACTORIZATION TECHNIQUES

In earlier mathematics classes, you may have used factor trees to find the prime-factorization form of a natural number.

Example 2
Use a factor tree to find the prime-factorization form of 120.

Solution
We can begin with any two factors of 120.
Let's use 10 and 12.

We now factor 10 and 12 into two factors each.

Now we factor 6.

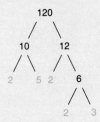

We observe that all the numbers at the ends of the factor-tree branches are prime numbers. The prime-factorization form of 120 is $2^3 \cdot 3 \cdot 5$.

Example 3
Factor 924 into its prime-factorization form without using a factor tree.

Solution
We use the divisibility rules for 3 and 4 to determine that 12 is a factor of 924.

$$924 = 12 \cdot 77$$
$$= (2 \cdot 6) \cdot (7 \cdot 11)$$
$$= 2 \cdot (2 \cdot 3) \cdot 7 \cdot 11$$
$$= 2^2 \cdot 3 \cdot 7 \cdot 11$$

GENERAL PRIME-FACTORIZATION PROCEDURE

Let's look at a process for determining the prime-factorization form of any natural number. We begin by listing the pairs of factors of 36 and 120.

The factors of 36 are	The factors of 120 are	
1 and 36	1 and 120	5 and 24
2 and 18	2 and 60	6 and 20
3 and 12	3 and 40	8 and 15
4 and 9	4 and 30	10 and 12
6 and 6		

Chapter 10 Topics of Number Theory

SECTION 10.4

We observe that each number in the last pair of factors of 36 and 120 is equal to or close in value to the square roots of 36 and 120. The numbers in the first columns of the pairs of factors are less than or equal to the square roots of the numbers 36 and 120. We will use these observations in the next example and the sample problem.

Example 4
Determine the prime-factorization form of 4693.

Solution
We will systematically check the primes, that is, 2, 3, 5, 7, 11, 13, ... , to determine whether any are factors of 4693.

We test 2, 3, 5, 7, and 11 and find that none of these prime numbers are factors of 4693.

If we check the next prime, 13, we find that it is a factor of 4693 because

$$4693 = 13 \cdot 361$$

We then check to see if 13 is a factor a second time. However, we find that 13 is not a factor of 361.

If we check the next prime, 17, we find that 17 is not a factor of 361.

However, the next prime, 19, is a factor of 361.

$$361 = 19 \cdot 19$$

Therefore, $4693 = 13 \cdot 19 \cdot 19 = 13 \cdot 19^2$.

You may be wondering how many consecutive primes we need to test. Since $\sqrt{4693} < 68.52$, we have to check whether the primes less than 68.52 are factors of 4693.

Sample Problem

Problem Show that 317 is prime.

Solution Since $\sqrt{317}$ is approximately 17.80, we need only to check whether the prime numbers less than 17.80 are factors of 317. The only such primes are 2, 3, 5, 7, 11, 13, and 17. None of these are factors of 317. Thus, 317 is a prime number.

Example 3
Factor 864 into its prime-factorization form without using a factor tree.

Possible answer:
$$\begin{aligned}
864 &= 12 \cdot 72 \\
&= (3 \cdot 4) \cdot (8 \cdot 9) \\
&= 3 \cdot (2 \cdot 2) \cdot (2 \cdot 2 \cdot 2) \cdot (3 \cdot 3) \\
&= 2^5 \cdot 3^3
\end{aligned}$$

Example 4
Determine the prime-factorization form of 6681.

Answer: $3 \cdot 17 \cdot 131$

Assignment Guide

Basic
Optional

Average or Heterogeneous
Optional

Honors
Day 1 7–11, 14–18, 20–22, 24–30, 32, 35, 36, Investigation

Strands

Communicating Mathematics	2–5, 9, 22, 28, 30, Investigation
Geometry	33, 34
Measurement	33, 34
Probability	21, 31, 35
Algebra	5, 18, 20, 22, 23, 25, 33, 34, 36
Patterns & Functions	29, Investigation
Number Sense	1–19, 21, 24, 26–30, 32
Interdisciplinary Applications	35, 36

Individualizing Instruction

The gifted learner Students could be challenged to write a computer program in BASIC to determine whether an input number is prime or composite.

SECTION 10.4

Problem-Set Notes

for pages 510 and 511

- **2** combines prime factorization with divisibility tests.
- **4** requires a bit of thought but is a useful fact. Prime factors of perfect squares must occur in pairs.
- **6** is related to a famous idea, called Goldbach's conjecture, which asserts that "every even number greater than 3 is the sum of two primes." It is a 200-year-old problem that still has not been proved true or false and is discussed further in Section 10.7.
- **7** and **18** imply that factoring or simplifying can begin right away rather than multiplying first, then dividing.
- **16** previews the idea that the greatest common factor of two numbers times their least common multiple is the product of the two numbers.
- **25** previews one of the applications of factoring to algebra—that of finding perfect square factors so that square roots can be written in simplified form.

Additional Practice

In problems **1–3**, tell which prime numbers need to be tested as possible factors in order to determine whether the given number is prime.
1. 89   **2.** 120   **3.** 247

In problems **4–8**, write the prime-factorization form of each number.
4. 150   **5.** 210   **6.** 3025
7. 936   **8.** 9504

Answers:
1. 2, 3, 5, 7   **2.** 2, 3, 5, 7   **3.** 2, 3, 5, 7, 11, 13   **4.** $2 \cdot 3 \cdot 5^2$
5. $2 \cdot 3 \cdot 5 \cdot 7$ **6.** $5^2 \cdot 11^2$
7. $2^3 \cdot 3^2 \cdot 13$ **8.** $2^5 \cdot 3^3 \cdot 11$

510

Think and Discuss

P   **1** Find the prime-factorization form of
    a 10 $2 \cdot 5$ **b** 16 2^4 **c** 23 $1 \cdot 23$ **d** 27 3^3

P   **2 a** Write the prime-factorization form of 2520. $2^3 \cdot 3^2 \cdot 5 \cdot 7$
    b Without actually doing the division, determine which of the following numbers divide evenly into 2520. Explain your answers.
    i 72   Yes; $2^3 \cdot 3^2 = 72$ **ii** 45   Yes; $3^2 \cdot 5 = 45$ **iii** 63   Yes; $3^2 \cdot 7 = 63$

P, R   **3** Faron was checking to see which numbers were factors of a given number. He used the divisibility test for 3 and found it did not work. He then announced that 3, 6, 9, 12, 15, 18, 21, 24, 27, and 30 did not divide the number. Was he correct? Explain your answer.

P, R   **4** Is the statement *true* or *false*?
    a If a prime number divides evenly into the square of a number, then it divides evenly into the number.   True
    b Explain your reasoning for part **a**.

P   **5 a** What is the greatest possible prime factor of a number n?
    b In the sample problem, why was it necessary to check only numbers less than $\sqrt{317}$ when looking for prime factors of 317?

P   **6 a** Find two positive integers that cannot be written as the sum of two primes.   Answers may vary. Possible answer: 11 and 17
    b Find two positive integers that can be written as the sum of two primes.   Answers may vary. Possible answer: 4 and 5
    c Find two positive integers that can be written as the sum of two primes in two different ways.   Answers may vary. Possible answer: 10 and 14

Problems and Applications

P   **7** Find the prime-factorization form of $9 \cdot 6 \cdot 4$. $2^3 \cdot 3^3$

P   **8** Use the tree diagrams to factor 84 into its prime-factorization form.

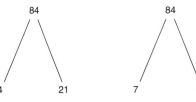
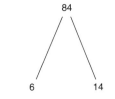

P   **9** **Communicating** Explain why $2^3 \cdot 3^2 \cdot 25^3$ is not a prime-factorization form.   25 is not prime.

P, R   **10** Completely factor each number into prime factors.
    a 4 2^2 **b** 36 $2^2 \cdot 3^2$ **c** 121 11^2 **d** 25 5^2 **e** 64 2^6 **f** 100 $2^2 \cdot 5^2$

R   **11** List all the factors of 30.   1, 2, 3, 5, 6, 10, 15, 30

510   Chapter 10   Topics of Number Theory

SECTION 10.4

P   **12** Find the prime-factorization form of
     a 286 $2 \cdot 11 \cdot 13$ **b** 301 $7 \cdot 43$ **c** 601   Prime

P   **13** Which number has a prime-factorization form of $3^2 \cdot 2^3 \cdot 5$?   360

P   **14** Use two tree diagrams to find the prime-factorization form of 380.

P   **15** Find the prime-factorization form of 8463. $3 \cdot 7 \cdot 13 \cdot 31$

P   **16** The prime-factorization form of 315 is $3^2 \cdot 5 \cdot 7$, and the prime-factorization form of 825 is $3 \cdot 5^2 \cdot 11$.
     a What is the greatest number that is a factor of both 315 and 825?   15
     b Write the answer to part **a** in prime-factorization form. $3 \cdot 5$
     c What is the smallest number that is a multiple of both 315 and 825?   17,325
     d Write the answer to part **c** in prime-factorization form. $3^2 \cdot 5^2 \cdot 7 \cdot 11$

P, R   **17 a** Copy the factor trees and fill in the missing numbers.

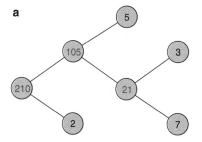
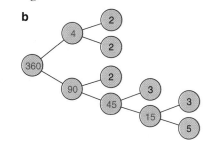

     c How could you have found the number at the beginning of the tree without filling in the entire tree?   Multiply $5 \cdot 3 \cdot 7 \cdot 2$; and $2 \cdot 2 \cdot 2 \cdot 3 \cdot 3 \cdot 5$

P, R   **18 a** Simplify the expression $\frac{3 \cdot 5 \cdot 7 \cdot 11}{5 \cdot 7 \cdot 11 \cdot 13}$. $\frac{3}{13}$
     b Express the numerator and the denominator of $\frac{2940}{945}$ in prime-factorization form.

P, R   **19 a** Find the prime-factorization forms of 225 and 150. $225 = 3^2 5^2$ and $150 = 2 \cdot 3 \cdot 5^2$
     b Find the product of the common prime factors of 225 and 150.   75
     c Find the greatest number that divides evenly into 225 and 150.   75

R   **20** Solve $2^3 \cdot 3^4 \cdot 5^2 \cdot x = 2^5 \cdot 3^4 \cdot 5^3$ for x.   20

P, R   **21** What is the probability that a number selected from the set $\{1, 2, 3, \ldots, 19, 20\}$ is prime? $\frac{2}{5}$

R, I   **22 Communicating** If $2N = 2^6 \cdot 3^5 \cdot 5^4 \cdot 7^3 \cdot 11^7$, explain why $2 \cdot 3 \cdot 5 \cdot 7 \cdot 11$ is a factor of N.

R, I   **23** If $2^a \cdot 3^b = 864$, find the values of a and b. $a = 5; b = 3$

P, R   **24** Which prime numbers need to be tested as possible factors in order to determine if 937 is prime?   2, 3, 5, 7, 11, 13, 17, 19, 23, 29

R   **25** Evaluate each expression.
     a $\sqrt{3^2}$ 3   **b** $\sqrt{3^2 \cdot 2^2}$ 6   **c** $\sqrt{5^2 \cdot 7^2}$ 35   **d** $\sqrt{9 \cdot 36}$ 18

P   **26** Find the smallest number that can be divided by three different prime numbers.   30

Section 10.4   Prime Factorization

Cooperative Learning

- Have students work in groups to find the value to substitute in each box to make the following equations true.
   a. $3! + 4! = \square (3!)$
   b. $5! + 6! = \square (5!)$
   c. $8! + 9! = \square (8!)$
   d. $9! + 10! = \square (9!)$
   e. $12! + 13! = \square (12!)$

What conclusion can be drawn?

Answers:
a. 5 **b.** 7 **c.** 10 **d.** 11 **e.** 14
Conclusion: $x! + (x+1)!$
$= x!(1) + x!(x+1) = x!(x+2)$

- Problem 29 and Investigation are appropriate for cooperative learning.

Additional Answers

for pages 510 and 511

3 Yes, if a number is not divisible by 3, it is not divisible by any multiple of 3.

4 b Students may explain by citing an example. For instance, the primes 2 and 3 both divide 6^2, or 36, because $36 = 6^2 = (2 \cdot 3)^2 = 2^2 \cdot 3^2$

5 a n

b Because when you test numbers greater than $\sqrt{317}$, you find the same pairs of factors you've already found

8 $2^2 \times 3 \times 7$

14 See students' factor trees. The prime-factorization form of 380 is $2^2 \times 5 \times 19$.

18 b $\frac{28}{9}$ or $3\frac{1}{9}$

22 2, 3, 5, 7, and 11 are all factors of N.

SECTION 10.4

Problem-Set Notes

for page 512

- **30** generalizes to "The product of *n* consecutive natural numbers is divisible by *n*." If students were successful with this problem, you might ask them if the product of three consecutive natural numbers is divisible by three.
- **32** previews greatest common factor.
- **36** previews Chapter 11.

Communicating Mathematics

- Have students write a paragraph explaining the difference between listing the factors of a number and finding the prime factorization of a number.
- Problems 2–5, 9, 22, 28, 30, and Investigation encourage communicating mathematics.

Checkpoint

In problems **1–5**, write the prime-factorization form of each number.
1. 144   2. 80   3. 625
4. 71   5. 124

6. Which numbers need to be tested as possible factors in order to tell if 901 is prime?

Answers:
1. $2^4 \cdot 3^2$ 2. $2^4 \cdot 5$ 3. 5^4
4. Prime   5. $2^2 \cdot 31$ 6. 2, 3, 5, 7, 11, 13, 17, 19, 23, 29

Additional Answers

for page 512

28 10,101 is divisible by 3, so it is not prime.

Answers for problem 34 and Investigation are on page A9.

512

P, R **27** Write the prime-factorization form of 7!. $2^4 \cdot 3^2 \cdot 5 \cdot 7$

P, R **28 Communicating**   Explain why 10,101 is not prime.

P **29** Take any six-digit number in which the first three digits are identical and in the same order as the last three digits—for example, 243,243. What prime numbers divide each of these numbers?   7, 11, and 13

P, R **30 Communicating**   Explain why the product of two consecutive natural numbers is divisible by 2.   One of the numbers must be even, and even numbers are divisible by 2.

◀ LOOKING BACK   **Spiral Learning**   LOOKING AHEAD ▶

R **31** A store sells eight different model airplanes and five choices of colored paints. How may different model planes can you make?   40

R **32 a** Find the set of all factors of 24.   {1, 2, 3, 4, 6, 8, 12, 24}
  b Find the set of all factors of 18.   {1, 2, 3, 6, 9, 18}
  c Find the intersection of the two sets in parts **a** and **b**.   **c** {1, 2, 3, 6}
  d Find the greatest number in the intersection of the sets. What is this number called?   6; greatest common factor

R **33** Find the distance between
  a A(−4, 5) and B(−2, 12) $\sqrt{53}$, or ≈ 7.28   **b** A(5, 11) and B(−4, −8) $\sqrt{442}$, or ≈ 21.02

P, R, I **34 a** Plot (10, 0), (0, 8), and (0,0).
  b Connect the points in part **a** and find the area of the triangle.
  c Find the midpoints of the sides of the triangle.
  d Connect the midpoints to form four triangles.
  e Find the area of the four triangles in part **d**.

R **35** A license plate from Bronkovia consists of two letters followed by four numbers. How many different license plates can Bronkovia have?   6,760,000

R **36 Business**   Hal O'Famer charges $5000 to appear at a sports show. He also receives $6 per autograph.
  a Write an expression to represent the amount of money Hal gets if he signs *x* autographs. $5000 + 6x$
  b If Hal made $9740 at one show, write an equation that represents the number of autographs he signed. $9740 = 5000 + 6x$
  c Solve the equation in part **b** to find the number of autographs he signed.   790

Investigation

Number Game   Find any three consecutive three-digit numbers, none of which is divisible by 3. Explain your results.

512   Chapter 10   Topics of Number Theory

SECTION 10.5

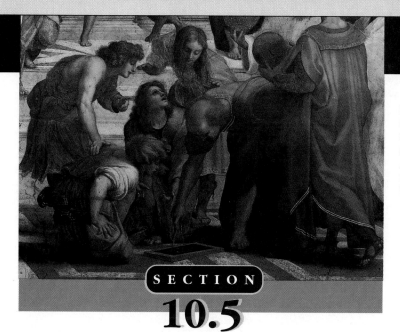

Investigation Connection
Section 10.5 supports Unit E, *Investigations* 2, 3. (See pp. 426A-426D in this book.)

SECTION 10.5
Factors and Common Factors

FACTORS OF A NATURAL NUMBER

We used tiles earlier to introduce prime numbers and composite numbers. We found the factors of a natural number n by listing the dimensions of the rectangles formed by n tiles. However, using rectangular tiles is not practical for large numbers. We need other methods of determining the factors of a number.

Example 1
Determine all the factors of 120.

Solution
Method 1
We list the factors in pairs—1 and 120, 2 and 60, 3 and 40, 4 and 30, 5 and 24, 6 and 20, 8 and 15, and 10 and 12.

Arranging the factors from least to greatest gives us

   1, 2, 3, 4, 5, 6, 8, 10, 12, 15, 20, 24, 30, 40, 60, 120

Method 2
A second method involves a tree diagram.

Show the prime-factorization form of 120 $2^3 \cdot 3 \cdot 5$

List the factors of each term 2^3 (or 8): 1, 2, 4, 8
of the prime factorization   3: 1, 3
                                      5: 1, 5

The tree diagram is shown on the next page.

Objectives
- To determine all factors of a natural number
- To determine whether a number is an abundant number, a deficient number, or a perfect number
- To determine the greatest common factor of a set of numbers

Vocabulary: abundant number, deficient number, greatest common factor, perfect number

Class Planning

Schedule 2 days
Resources
In *Chapter 10 Student Activity Masters* Book
Class Openers 49, 50
Extra Practice 51, 52
Cooperative Learning 53, 54
Enrichment 55
Learning Log 56
Technology 57, 58
Spanish-Language Summary 59

In *Tests and Quizzes* Book
Quiz on 10.4–10.5 (Honors) 264

Class Opener

1. Let A = {factors of 36} and B = {factors of 48}. Find A ∩ B.
2. Let C = {factors of 15} and D = {factors of 40}. Find C ∩ D.
3. Let E = {factors of 14} and F = {factors of 25}. Find E ∩ F.

Answers:
1. {1, 2, 3, 4, 6, 12} **2.** {1, 5} **3.** {1}

Lesson Notes

- Work through both methods of example 1 with students. While method 1 is probably shorter and simpler, method 2 might appeal

513

SECTION 10.5

Lesson Notes continued

to the visual learner and help make sense of the idea to many students in the class.

- The study of abundant, deficient, and perfect numbers has intrigued individuals for hundreds of years. It is not an essential topic for later work and may be treated accordingly. Keep an eye out, however, for the student or two whose interest is piqued by the topic.

- Greatest common factor is an important topic for subsequent work. As we mentioned in Section 10.1, try to use the words *greatest common factor* instead of GCF.

- Both methods for solving the sample problem are useful. Students will use the first method most of the time, but for theoretical and algebraic purposes, the second is important.

Additional Examples

Example 1
Determine all the factors of 90.

Answer:
1 and 90, 2 and 45, 3 and 30, 5 and 18, 6 and 15, and 9 and 10

Example 2
Find all the factors of each number. Then find the sum of the factors.
a. 2   b. 5   c. 8   d. 80

Answers:
a. 1, 2; 3   b. 1, 5; 6   c. 1, 2, 4, 8; 15
d. 1, 2, 4, 5, 8, 10, 16, 20, 40, 80; 186

Example 3
a. List the factors of 30.
b. List the factors of 50.
c. List the common factors of 30 and 50.

Now label the branches of a tree diagram. The first row of branches shows the factors of 8. The second row of branches shows the factors of 3; and the third row, the factors of 5. We multiply the numbers downward along the branches and write the products in the circles. The numbers in the circles at the end of the branches are the 16 factors of 120.

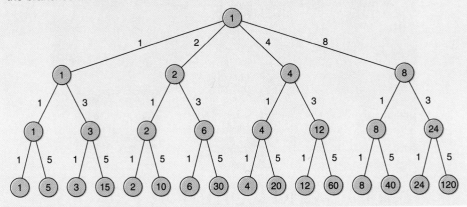

ABUNDANT, DEFICIENT, AND PERFECT NUMBERS

The early Greeks were interested in natural numbers and their related properties. The Greeks classified the natural numbers according to the sum of their factors. We will build from the Greeks' ideas.

Example 2
Find all the factors of each number. Then find the sum of the factors.
a 1   b 4   c 6   d 24   e 120

Solution

	Number	Factors	Sum of Factors
a	1	1	1
b	4	1, 2, 4	7
c	6	1, 2, 3, 6	12
d	24	1, 2, 3, 4, 6, 8, 12, 24	60
e	120	1, 2, 3, 4, 5, 6, 8, 10, 12, 15, 20, 24, 30, 40, 60, 120	360

To see whether a number n is a **perfect number**, we look at the ratio r, where

$$r = \frac{\text{Sum of the factors}}{\text{Number}}$$

If $r = 2$, the number n is a perfect number. There is only one perfect number in Example 2—the number 6 in part **c**

$$\frac{1+2+3+6}{6} = 2.$$

SECTION 10.5

If $r < 2$, the number n is **deficient.** Thus, 4 is deficient because $\frac{1+2+4}{4} = 1.75$ and $1.75 < 2$. Is 1 deficient?

If $r > 2$, the number n is called **abundant.** Thus, 24 and 120 are abundant since their values of r exceed 2. A perfect number, such as 6, is neither abundant nor deficient.

We can use a Venn diagram to show the relationship among abundant, deficient, and perfect numbers.

Natural Numbers

| Deficient Numbers | Perfect Numbers | Abundant Numbers |

GREATEST COMMON FACTORS

Look at the factors of 6 (1, 2, 3, and 6) and the factors of 9 (1, 3, and 9). Do 6 and 9 have any factors in common? Yes—both 1 and 3 are common factors of 6 and 9.

Now we will find the common factors of 45 and 60.

Example 3
a List the factors of 45.
b List the factors of 60.
c List the common factors of 45 and 60.

Solution
a The factors of 45 are 1, 3, 5, 9, 15, and 45.
b The factors of 60 are 1, 2, 3, 4, 5, 6, 10, 12, 15, 20, 30, and 60.
c The common factors of 45 and 60 are 1, 3, 5, and 15.

The **greatest common factor (GCF)** of 45 and 60 is 15.

Sample Problem

Problem Simplify the fraction $\frac{45}{60}$.

Solution **Method 1**
We will use 15, the greatest common factor of 45 and 60, to simplify the fraction.

$$\frac{45}{60} = \frac{15 \cdot 3}{15 \cdot 4} = \frac{3}{4}$$

Answers:
a. 1, 2, 3, 5, 6, 10, 15, 30 b. 1, 2, 5, 10, 25, 50 c. 1, 2, 5, 10

Assignment Guide

Basic
Optional
Average or Heterogeneous
Optional
Honors
Day 1 8–11, 13, 14, 18–24, 26, 34
Day 2 12, 15–17, 25, 28–30, 35, 36, 38, 39, 42

Strands

Communicating Mathematics	4–7, 26, 30, 31
Geometry	34, 37
Measurement	26, 37, 41, 42, 44
Data Analysis/ Statistics	17, 21, 39
Technology	27, 31, 39
Algebra	14, 16, 20, 24, 28, 30, 32, 33, 36, 40, 43, 44
Patterns & Functions	24, 31, 35
Number Sense	1–5, 7–19, 21–23, 25, 27, 29, 31, 32, 34, 35, 39, Investigation
Interdisciplinary Applications	38, 41, 42, 44

Individualizing Instruction

The visual learner Students may benefit from using the tree-diagram method to find the factors of each number in problems 8, 12, and 18. Students should be encouraged to list the factors first and then use the tree diagram to check their answers.

SECTION 10.5

Problem-Set Notes

for pages 516 and 517

- **6** refers to the fact that perfect squares will have an odd number of factors while all other natural numbers have an even number.
- **10** and **15** provide a way to help students see patterns and make generalizations.

Additional Practice

In problems **1–3**, list all the factors of each number.
1. 36 **2.** 100 **3.** 150

In problems **4–6**, find the greatest common factor of each pair of numbers.
4. 24 and 36 **5.** 18 and 24 **6.** 60 and 90

In problems **7–9**, classify each number as *abundant*, *deficient*, or *perfect*.
7. 36 **8.** 50 **9.** 30

Answers:
1. 1, 2, 3, 4, 6, 9, 12, 18, 36
2. 1, 2, 4, 5, 10, 20, 25, 50, 100
3. 1, 2, 3, 5, 6, 10, 15, 25, 30, 50, 75, 150 **4.** 12 **5.** 6 **6.** 30
7. Abundant **8.** Deficient
9. Abundant

Cooperative Learning

- Have students work through the following problems:
1. Find the number of factors and the prime-factorization form for each of the following: 4, 27, 32, 625.
2. What do you notice about the number of factors and the exponent in each case?
3. Find the number of factors and the prime-factorization form of the following numbers: 20, 36, 40.

Method 2
We will use the prime-factorization forms of 45 and 60 to simplify the fractions.

$$\frac{45}{60} = \frac{3^2 \cdot 5}{2^2 \cdot 3 \cdot 5} = \frac{3 \cdot 3 \cdot 5}{2 \cdot 2 \cdot 3 \cdot 5} = \frac{15 \cdot 3}{15 \cdot 4} = \frac{3}{4}$$

Think and Discuss

P, R **1 a** Name the one-digit natural numbers that have exactly two factors. 2, 3, 5, 7
   b What are these numbers called? Prime numbers

P, R **2** Name the one-digit natural numbers that have
   a Exactly three factors 4, 9
   b Exactly four factors 6, 8, 10
   c Exactly one factor 1

P **3 a** Are all numbers less than 6 deficient? Yes
   b Are all numbers greater than 6 abundant? No

P **4** Explain how to find the common factors of two natural numbers.

P **5** How do you find the greatest common factor of two numbers?

P **6** Explain why some numbers have an even number of distinct factors and why some numbers have an odd number of distinct factors.

P **7** Some textbooks define *perfect number* as follows:

> A natural number is perfect if the sum of all its factors, except itself, adds up to the number.

Is this definition equivalent to the definition in this section? Explain your answer.

Problems and Applications

P, R **8** List all the factors of each number.
   a 75   1, 3, 5, 15, 25, 75   **b** 130   **c** 80

P, R **9** Make a tree diagram of all the factors of 18.

P **10** Copy and complete the table.

GCF	12	15	24
16	4	1	8
31	1	1	1
42	6	3	6

Chapter 10 Topics of Number Theory

SECTION 10.5

P, R **11** Find the GCF and LCM of each pair of numbers.
    a 20 and 32   4; 160   **b** $2^2 \cdot 3^3 \cdot 5$ and $2^4 \cdot 3^2 \cdot 5^3$ 180; 54,000

R **12** List all the factors of each number.
    a 48   **b** 84   **c** 484   **d** 848

P, R **13** Find the LCM and GCF of the prime numbers 71 and 73.   5183; 1

P **14** Classify each number as *abundant, deficient,* or *neither.*
    a 16   Deficient   **b** 24   Abundant   **c** 80   Abundant

P, R **15** Copy and complete the table.

LCM	12	15	24
6	12	30	24
8	24	120	24
10	60	30	120

R **16** Simplify each fraction.
    a $\frac{15}{21}$ $\frac{5}{7}$ **b** $\frac{156}{168}$ $\frac{13}{14}$ **c** $\frac{28}{40}$ $\frac{7}{10}$

P, R **17** The prime-factorization form of 36 is
 $2 \cdot 2 \cdot 3 \cdot 3$. The prime-factorization form
 of 40 is $2 \cdot 2 \cdot 2 \cdot 5$. You may find the
 diagram helpful.
    a Find the GCF of 36 and 40.   4
    b Find the LCM of 36 and 40.   360

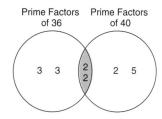

P, R **18** List all the factors of 320 that are multiples of 5.   5, 10, 20, 40, 80, 160, 320

P, R **19** Find the factors of 56 by using
 the factor tree. What percent
 of the factors are prime?
 1, 2, 4, 7, 8, 14, 28, 56; 25%

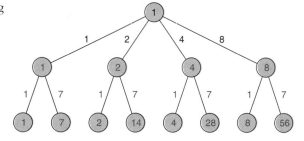

P, R **20** Let $r = \frac{\text{Sum of the factors of } n}{n}$ be the "*r*-value" of *n*. Find the "*r*-value" of *n*
 if *n* is
    a 6   2   **b** 9 $\frac{13}{9}$ **c** 28   2   **d** 120   3

P, R **21** The prime-factorization form of 32 is
 $2 \cdot 2 \cdot 2 \cdot 2 \cdot 2$. The prime-factorization form
 of 40 is $2 \cdot 2 \cdot 2 \cdot 5$. You may find the
 diagram helpful.
    a Find the LCM of 32 and 40.   160
    b Find the GCF of 32 and 40.   8

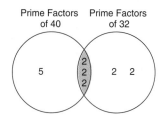

Section 10.5 Factors and Common Factors

---

4. Write a rule that relates the number of factors to the exponents in the prime-factorization form. (Hint: $2^2 \times 5$ can be written as $2^2 \times 5^1$.)

Answers:
1. 3 factors, 2^2; 4 factors, 3^3; 6 factors, 2^5; 5 factors, 5^4
2. The number of factors is one more than the power of the number. **3.** 6 factors, $2^2 \times 5$; 9 factors, $2^2 \times 3^2$; 8 factors, $2^3 \times 5$ **4.** The number of factors can be found by adding 1 to each of the exponents and multiplying the resulting numbers.

■ Problem 33 is appropriate for cooperative learning.

Additional Answers

for pages 516 and 517

4 Possible answer: List all the different factors of each number. Identify the common factors.

5 Possible answer: List all the factors of each number and pick the largest factor that they have in common.

6 Factors come in pairs, but for perfect squares, one pair is made up of the same number. Thus, perfect squares have an odd number of factors.

7 Yes; If *k* is the sum of all the factors of *n*, then according to this definition, $k - n = n$. But then $k = 2n$ and $\frac{k}{n} = 2$, which is the definition given on page 514.

8 b 1, 2, 5, 10, 13, 26, 65, 130
 c 1, 2, 4, 5, 8, 10, 16, 20, 40, 80

9 See students' tree diagrams. $18 = 2 \cdot 3 \cdot 3$.

12 a 1, 2, 3, 4, 6, 8, 12, 16, 24, 48
 b 1, 2, 3, 4, 6, 7, 12, 14, 21, 28, 42, 84
 c 1, 2, 4, 11, 22, 44, 121, 242, 484
 d 1, 2, 4, 8, 16, 53, 106, 212, 424, 848

SECTION 10.5

Problem-Set Notes

for pages 518 and 519

- **22** is an interesting problem that often pops up in math contests. The solution—find the LCM and add 2—is usually not obvious to students.

- **24** helps to generalize the concept of multiple.

- **27** illustrates the Euclidean algorithm, which has been used for 2000 years to find the greatest common factor of two numbers. Have your students try to think through the algorithm before you discuss it.

- **29** combines the Fundamental Counting Principle with prime factorization to arrive at a method for determining how many factors a number will have.

- **30** provides a visual model for example 1 (method 2) when there are only two prime bases.

- **36** encourages students to think about LCM another way. Does it make sense to ask a similar question about GCF (i.e., find all values of x such that GCF(x, 15) = 5)?

Reteaching

Another method for finding the greatest common factor of two numbers is to (1) divide the smaller number into the larger, (2) divide the remainder into the previous divisor, and (3) continue dividing in this manner until the remainder is zero. The last divisor is the greatest common factor. This method is called the Euclidean algorithm, after the mathematician Euclid.

P, R   **22** Find the smallest natural number that has a remainder of 2 when divided by 3, 5, or 7.   107

P   **23** Use the diagram to help you list the factors of 54.

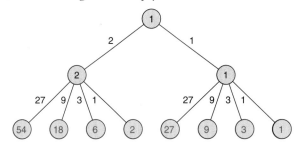

R   **24** If a is a multiple of 7, then what are the next four multiples of 7 greater than a? $a + 7, a + 14, a + 21, a + 28$

P   **25** Is the greatest common factor of two numbers always a factor of their
     a Sum?   Yes   **b** Difference?   Yes   **c** Product?   Yes   **d** Quotient?   No

R   **26 Communicating**   How can you use these rectangles to draw a picture that will convince someone that 4 is a factor of 24 + 40 = 64?

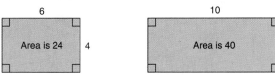

P, R   **27 Spreadsheets**   Using a spreadsheet, enter +B1 in cell A2 and +A1–B1*@INT(A1/B1) in cell B2. Then copy row 2 to rows 3–17.
     a Enter 75 in cell A1 and 27 in cell B1. As you look down column A, what is the last nonzero number you see?   3
     b Find the GCF of 75 and 27.   3
     c Try entering the other pairs of numbers in A1 and B1. What do you notice?   The last nonzero number in column A is GCF.

P, R, I   **28** Make a table of positive integer x-values and y-values that solve the equation $xy = 30$.

P, R   **29** If $600 = 2^3 \cdot 3 \cdot 5^2$, any factor of 600 must be a factor of $2^3 \cdot 3 \cdot 5^2$. How many factors does 600 have? A factor of the number 600 can have zero, one, two, or three 2's (four choices), zero or one 3's (two choices), and zero, one or two 5's (three choices). Therefore, the possible number of factors of 600 is $4 \cdot 2 \cdot 3$, or 24. Use this example to find the number of factors of each of the following numbers.
     a 70   8   **b** $2^4 \cdot 3^2 \cdot 5^7$ 120   **c** $5^4 \cdot 11^3 \cdot 13$ 40   **d** 36   9

SECTION 10.5

P, R **30 Communicating** Copy and fill in the multiplication matrix. Explain why the entries determine all possible factors of 144.

$$\begin{array}{c|ccc} & 1 & 3 & 3^2 \\ \hline 1 & 1 & 3 & 9 \\ 2 & 2 & 6 & 18 \\ 2^2 & 4 & 12 & 36 \\ 2^3 & 8 & 24 & 72 \\ 2^4 & 16 & 48 & 144 \end{array}$$

P, R **31 Spreadsheets** Create a spreadsheet in which column A contains the numbers from 1 to 100, cell C2 contains factors you want to identify, and cell B2 contains the formula $C$2/A2. Copy this formula down column B for 100 cells.
  a Describe how this spreadsheet can help you find factors.
  b Find the factors of
    i 140 **ii** 725 **iii** $2^3 \cdot 5^3 \cdot 7^3$

P, R **32** Simplify each expression.
  a $\frac{5}{12} \cdot \frac{5}{12}$
  b $-\frac{5}{6} - \frac{3}{4} - \frac{2}{3} - 2\frac{1}{4}$

R **33** There are a certain number of candies in a pile.
  a Tom eats one, divides the pile into thirds, takes one third for himself, and leaves the rest.
  b Bill takes the remaining pile, eats one candy, divides the pile into thirds, and takes one third for himself.
  c Sue follows the same procedure as Tom and Bill.
  d Juanita follows the same procedure as Sue and the others.
  e Ralph takes the remaining candies.
  Write expressions that represent the number of candies in the original pile and the number of candies each person has.

R **34** The length of segment AB is 126. Find the lengths of segments AC, CD, and DB if their measures are three consecutive multiples of 14.   AC = 28; CD = 42; DB = 56

◀ LOOKING BACK **Spiral Learning** LOOKING AHEAD ▶

R, I **35** Let $g(n)$ represent the greatest prime factor of n. Find
  a $g(44)$ 11
  b $g(68)$ 17
  c $g(49)$ 7

R, I **36** Find all values of x so that
  a LCM(x, 14) = 42   3, 6, 21, 42
  b LCM(x, 7) = 35   5, 35

R, I **37** The red car traveled 6 hours at 60 miles per hour from point A to point C. The blue car traveled 3 hours at 50 miles per hour from point A to point B. Find BC.
  390

Section 10.5 Factors and Common Factors

Stumbling Block

Remind students to include the numbers 1 and n as factors of n. Point out the differences between the factors of a number and the prime-factorization form of a number.

Multicultural Note

The main goal of the early Greek study of mathematics was to understand humanity's place in the universe according to a rational scheme. Mathematics helped people to find order in chaos, arrange ideas in logical chains, and find fundamental principles.

Additional Answers

for pages 518 and 519

26 Possible answer:

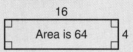

Area is 64

28

x	1	2	3	5	6	10	15	30
y	30	15	10	6	5	3	2	1

30 The entries represent all of the ways that the powers of the prime factors can be multiplied.

31 a Possible answer: Each whole number in column B and the corresponding number in column A form a pair of factors of the number in C2.

b i 1, 2, 4, 5, 7, 10, 14, 20, 28, 35, 70, 140

ii 1, 5, 25, 29, 145, 725

iii Answer on page A9.

33 Possible answer:
number in original pile = x;
Tom = $\frac{1}{3}(x-1)$; Bill = $\frac{1}{3}(\frac{2}{3}x - \frac{5}{3})$;
Sue = $\frac{1}{3}(\frac{4}{9}x - \frac{19}{9})$; Juanita = $\frac{1}{3}(\frac{8}{27}x - \frac{65}{27})$; Ralph = $\frac{2}{3}(\frac{8}{27}x - \frac{65}{27})$

SECTION 10.5

Checkpoint

In problems 1–3, list all the factors of each number.
1. 24   2. 75   3. 120

In problems 4–6, find the greatest common factor of each pair of numbers.
4. 30 and 36   5. 40 and 90
6. 80 and 120

In problems 7–9, classify each number as *abundant*, *deficient*, or *perfect*.
7. 17   8. 45   9. 90

Answers:
1. 1, 2, 3, 4, 6, 8, 12, 24   2. 1, 3, 5, 15, 25, 75   3. 1, 2, 3, 4, 5, 6, 8, 10, 12, 15, 20, 24, 30, 40, 60, 120   4. 6   5. 10   6. 40
7. Deficient   8. Deficient
9. Abundant

Communicating Mathematics

- Have students write a letter to an older brother or sister, a friend, or a relative. The letter should explain whether the number that represents that person's age is abundant, deficient, or perfect, and why.
- Problems 4–7, 26, and 30 encourage communicating mathematics.

Additional Answers

for page 520

40 c $\{x: -2 < x < 18\}$

44 2:30 P.M.

Investigation Two numbers are relatively prime if they have no common factors other than 1. If the numerator and denominator of a fraction are relatively prime, the fraction cannot be reduced further.

R **38** Barry's box contains 84 small marbles, of which 25% are blue. Bradley's box contains 240 large marbles, of which 40% are blue. Barry and Brad dump their boxes of marbles in a bucket. What percent of the marbles in the bucket will be blue?   ≈36%

I **39** In set A are the factors of a number. In set B are the factors of another number.
  a Find A ∩ B.   {2, 2, 3, 7}
  b Find A ∪ B.   {2, 2, 2, 3, 3, 7, 7, 11}
  c What are the two numbers?   504; 6468
  d What is the GCF of the numbers?   84
  e What is the LCM of the numbers?   38,808

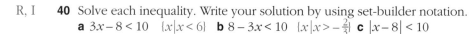

R, I **40** Solve each inequality. Write your solution by using set-builder notation.
  a $3x - 8 < 10$ $\{x \mid x < 6\}$ **b** $8 - 3x < 10$ $\{x \mid x > -\frac{2}{3}\}$ **c** $|x - 8| < 10$

R **41** Silky ran the first $1\frac{1}{4}$ miles of a race at a speed of 30 miles per hour. How long did it take Silky to run this distance?   2.5 min

R **42** Lashaun observed that a group of bricklayers could lay 8 rows of bricks all the way around a building in one day. The walls were already 48 rows high, and she estimated that the building would require 710 rows. How many working days will be needed to finish the job? $82\frac{3}{4}$ days

R **43** Solve for all possible values of x.
  a $(x - 6)(x + 7)(x + 3)(2x) = 0$ 6, −7, −3, or 0
  b $4x - 9 < 18$, x is an integer   6, 5, 4, 3, …
  c $\frac{3}{4}x = \frac{7}{8}$ $1\frac{1}{6}$

R, I **44** Bill leaves point A at 9 A.M., driving 45 miles per hour. Tom leaves point A at 10 A.M. traveling in the same direction as Bill at 55 miles per hour. At what time will Bill catch up to Tom?

Investigation

Relatively Prime Numbers For two numbers to be "relatively prime," they don't have to be prime numbers at all. For example, the following pairs of numbers are relatively prime:

  a 4, 9   **b** 5, 6   **c** 8, 5   **d** 12, 7   **e** 10, 13

But the following pairs of numbers are *not* relatively prime.

  a 4, 6   **b** 9, 3   **c** 8, 12   **d** 15, 10   **e** 21, 28

What do you think relatively prime means? Where could you apply this concept?

Chapter 10 Topics of Number Theory

SECTION 10.6

SECTION 10.6
LCM and GCF Revisited

USING VENN DIAGRAMS

In this section we will further our understanding of the least common multiple (LCM) and the greatest common factor (GCF), and we will apply both to our work with fractions. We will look at several methods for determining the LCM or the GCF of two numbers.

Example 1
Find both the LCM and the GCF of 48 and 72.

Solution
We will investigate two methods of finding the LCM and the GCF of 48 and 72.

Method 1
Write the prime-factorization form of each number.

$$48 = 2 \cdot 2 \cdot 2 \cdot 2 \cdot 3$$
$$72 = 2 \cdot 2 \cdot 2 \cdot 3 \cdot 3$$

The GCF of 48 and 72 is the product of the prime factors common to both numbers: $2 \cdot 2 \cdot 2 \cdot 3$, or 24.

The LCM of 48 and 72 is the product of the GCF and the remaining prime factors of the two numbers: $2 \cdot 3 \cdot$ GCF $= 2 \cdot 3 \cdot 24$, or 144.

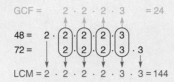

The second method is on the next page.

Objectives
- To use Venn diagrams to determine the least common multiple and the greatest common factor of a set of numbers
- To determine whether numbers are relatively prime
- To use LCM and GCF to simplify fractions and solve equations

Vocabulary: relatively prime numbers

Investigation Connection
Section 10.6 supports Unit E, *Investigation* 2. (See pp. 426A-426D in this book.)

Class Planning

Schedule 1 day

Resources
In *Chapter 10 Student Activity Masters* Book
Class Openers *61, 62*
Extra Practice *63, 64*
Cooperative Learning *65, 66*
Enrichment *67*
Learning Log *68*
Technology *69, 70*
Spanish-Language Summary *71*

Transparencies *40 a–d*

Class Opener

Let $x = 60$ and $y = 84$. Determine each of the following:
1. The least common multiple of x and y.
2. The greatest common factor of x and y.
3. The product of x and y.
4. The product of your answers to 1 and 2.

Answers:
1. 420 **2.** 12 **3.** 5040 **4.** 5040

Lesson Notes

- This section pulls together the number theory ideas developed in the first five sections. Understanding the methods in this section provides considerable

SECTION 10.6

Lesson Notes continued

insight into the natural numbers and enables students to solve a wide variety of problems quickly and efficiently.

- Methods 1 and 2 of example 1 both rely on the ideas of intersection, union, and prime factorization. Getting students to understand the ideas behind these methods may require you to work through a couple of examples.

- Note the definition of relatively prime in the solution to example 2. The definition will be used in the problems.

- Examples 3 and 4 and the sample problem are applications of the ideas in the chapter. The method given for solving example 4 will prove useful for any equation involving fractions.

Additional Examples

Example 1
Find the least common multiple and greatest common factor of 36 and 84.

Answer: GCF: 12; LCM: 252

Example 2
Find the least common multiple and greatest common factor of 35 and 48.

Answer: GCF: 1; LCM: 1680

Example 3
Simplify the expression
$\frac{1}{2} \cdot \frac{11}{25} + \frac{4}{5} \cdot \frac{10}{15}$

Answer: $\frac{113}{150}$

Example 4
Solve $\frac{2}{5}x + \frac{1}{3} = \frac{5}{6}$ for x.

Answer: $x = \frac{5}{4}$

522

Method 2
We can draw a Venn diagram. Fill in region II with the prime factors common to 48 and 72: 2, 2, 2, and 3. Fill in region I with the remaining factor of 48—2. Fill in region III with the remaining factor of 72—3.

The GCF is the product of the primes in region II, the intersection of the two circles. Thus, the GCF is 24.

The LCM is the product of the primes in regions I, II, and III—the union of the two circles. Thus, the LCM is 144.

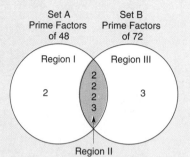

Product of factors in $A \cap B = 2^3 \cdot 3^1$, the GCF
Product of factors in $A \cup B = 2^4 \cdot 3^2$, the LCM

Example 2
Find the GCF and the LCM of 25 and 56.

Solution
The prime-factorization form of 25 is 5^2, and the prime-factorization form of 56 is $2^3 \cdot 7$. The Venn diagram shows that there are no common prime factors in region II. Since 25 and 56 have no prime factors in common, the GCF is 1. The LCM of 25 and 56 is $25 \cdot 56$, or 1400.

If the GCF of two numbers is 1, the two numbers are **relatively prime.** Thus, 25 and 56 are relatively prime. When the GCF of two numbers is 1, the LCM is the product of the two numbers.

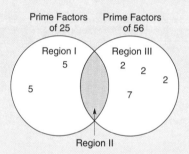

SIMPLIFYING FRACTIONS

The concepts of LCM and GCF are helpful when working with fractions.

Example 3
Simplify the expression $\frac{1}{2} \cdot \frac{7}{20} - \frac{3}{5} \cdot \frac{7}{10}$.

Solution

$$\frac{1}{2} \cdot \frac{7}{20} - \frac{3}{5} \cdot \frac{7}{10}$$

Multiply first
$$= \frac{7}{40} - \frac{21}{50}$$

LCM of 40 and 50 is 200
$$= \frac{5}{5} \cdot \frac{7}{40} - \frac{4}{4} \cdot \frac{21}{50}$$

$$= \frac{35}{200} - \frac{84}{200}$$

GCF of 49 and 200 is 1
$$= -\frac{49}{200}$$

Since the GCF of 49 and 200 is 1, the fraction cannot be further simplified.

SECTION 10.6

EQUATIONS CONTAINING FRACTIONS

If an equation contains a fraction, it may be helpful to multiply both sides of the equation by the LCM of the denominators of the fractions.

Example 4
Solve $\frac{1}{2} + \frac{2}{3}x = \frac{3}{4}$ for x.

Solution
The denominators of the fractions are 2, 3, and 4. The LCM of 2, 3, and 4 is 12. We multiply both sides of the equation by 12.

$$12\left(\frac{1}{2} + \frac{2}{3}x\right) = 12\left(\frac{3}{4}\right)$$
$$(12)\left(\frac{1}{2}\right) + (12)\left(\frac{2}{3}x\right) = 12\left(\frac{3}{4}\right)$$
$$6 + 8x = 9$$
$$-6 \qquad -6$$
$$8x = 3$$
$$x = \frac{3}{8}$$

Sample Problem

Problem Painter Phil estimates that he can paint a certain fence by himself in 6 hours. He estimates that his son Scott can paint the fence by himself in 10 hours. Can they finish the job if they work together for $3\frac{1}{2}$ hours?

Solution Phil can do $\frac{1}{6}$ of the job in one hour. Scott can do $\frac{1}{10}$ of the job in one hour. Each person is scheduled to work $3\frac{1}{2}$ hours.

Phil completes $\left(3\frac{1}{2}\right)\left(\frac{1}{6}\right) = \frac{7}{2} \cdot \frac{1}{6} = \frac{7}{12}$ of the job.

Scott completes $\left(3\frac{1}{2}\right)\left(\frac{1}{10}\right) = \frac{7}{2} \cdot \frac{1}{10} = \frac{7}{20}$ of the job.

Together they paint $\frac{7}{12} + \frac{7}{20} = \frac{35}{60} + \frac{21}{60} = \frac{56}{60}$ of the fence, which is a little short of completing the job.

Think and Discuss

P **1 a** Find the greatest common factor of 12 and 8.   4
  b Find the least common multiple of 12 and 8.   24

Assignment Guide

Basic
Optional

Average or Heterogeneous
Optional

Honors
Day 1 9, 11–13, 15–18, 20, 23, 26, 31, 35, 36, 38, 39

Strands

Communicating Mathematics	2, 4, 6
Geometry	39
Measurement	18, 20, 25, 27, 32, 39
Data Analysis/ Statistics	9, 13
Technology	17, 23, 32
Probability	31, 36
Algebra	10, 14, 16, 22, 24, 25, 28–30, 38, Investigation
Patterns & Functions	Investigation
Number Sense	1–10, 12–19, 21, 26, 28–31, 33, 37–39, Investigation
Interdisciplinary Applications	11, 12, 18, 20, 21, 26, 27, 32, 34

Individualizing Instruction

The visual learner Students may benefit from drawing a diagram to represent problem 32. Students might show a clock for each passing hour and illustrate the corresponding distance between the trucks.

SECTION 10.6

Problem-Set Notes

for pages 524 and 525

- **2–5** should help students clarify some important ideas in number theory.
- **8** describes the way most people find the LCM for small numbers.
- **12** is a familiar application of LCM.

Checkpoint

1. Use the diagram to find the least common multiple and greatest common factor of 90 and 120.

Prime Factors of 90 / Prime Factors of 120

3 | 2, 3, 5 | 2, 2

2. a. Find the least common multiple of 4, 6, and 18.
b. Use your answer to part **a** to help you solve the equation $\frac{3}{4} - \frac{5}{6}x = \frac{7}{18}$.

3. Solve the equation for x. $\frac{4}{5}x + \frac{7}{9} = \frac{31}{36}$.

In problems **4** and **5**, decide if the numbers are relatively prime.

4. 24 and 45   **5.** 16 and 27

Answers: **1.** LCM: 360; GCF: 30 **2. a.** 36 **b.** $x = \frac{13}{30}$ **3.** $x = \frac{5}{48}$ **4.** No **5.** Yes

Communicating Mathematics

- Have students write a paragraph explaining the difference between least common multiple and greatest common factor.
- Problems 2, 4, and 6 encourage communicating mathematics.

P **2 a** Explain why two consecutive integers greater than 1 will always be relatively prime.
  b Explain why three consecutive integers greater than 1 will always contain a multiple of 3.

P, R **3** Is it possible for the sum of the GCF and the LCM of two numbers to be odd? Yes

P **4** Is the statement true *always, sometimes,* or *never?* If two numbers are relatively prime, then any factor of the first number and any factor of the second number will also be relatively prime. Explain your answer.

P **5** If you multiply two numbers, is the result always a common multiple of the two numbers? Is it always the LCM of the two numbers? Yes, No

P, R **6** Explain why the product of three consecutive natural numbers is divisible by 6.

P **7** Is the statement true *always, sometimes,* or *never?*
  a The number 2 is relatively prime to an odd number. Always
  b The number 2 is relatively prime to an even number. Never
  c Two even numbers are relatively prime. Never
  d An odd number and an even number are relatively prime. Sometimes
  e Two odd numbers are relatively prime. Sometimes
  f Two integers with a last digit of 5 are relatively prime. Never
  g Two integers with a last digit of 7 are relatively prime. Sometimes

P, R **8** Martin was asked to find the LCM of 12 and 9. He started by listing the multiples of 12 (12, 24, 36, …) until he found one, 36, that was divisible by 9. Is 36 the LCM of 12 and 9? Will this method always work?

Problems and Applications

P **9** The prime-factorization form of 120 is $2 \cdot 2 \cdot 2 \cdot 3 \cdot 5$. The prime-factorization form of 75 is $3 \cdot 5 \cdot 5$. You may find the diagram helpful. Find the LCM and the GCF of 120 and 75.   600; 15

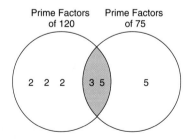

P **10 a** Find the LCM of 3, 6, and 15.   30
  b Use the answer to part **a** to help you solve the equation $\frac{4}{5} + \frac{1}{6}x = \frac{7}{15}$.   −2

P **11 Business** Machine A can make 500 CD's per hour. Machine B can make 760 CD's per hour. The company has an order for 900,000 CD's. At 8 hours per day, how many days will it take to complete the order?
About 89 days

524 Chapter 10 Topics of Number Theory

SECTION 10.6

P **12** Suppose that there are ten hot dogs to a package and eight hot dog buns to a package. What is the fewest number of packages of hot dogs you can purchase so that you can purchase a whole number of packages of buns and have exactly the same number of hot dogs and buns? 4

P **13** The prime-factorization form of 45 is $3 \cdot 3 \cdot 5$. The prime-factorization form of 68 is $2 \cdot 2 \cdot 17$. You may find the diagram helpful. Find the LCM and the GCF of 45 and 68.
3060; 1

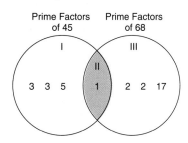

P, R **14** Solve each equation for x.
 a $\frac{1}{6}x + \frac{1}{4} = 1$ $\frac{9}{2}$
 b $\frac{1}{6}x + \frac{1}{4}x - \frac{1}{3}x = 1$ 12

P **15 a** Copy and fill in the factor trees to find all the factors of 75 and 60 if $75 = 3 \cdot 5^2$ and $60 = 2^2 \cdot 3 \cdot 5$.

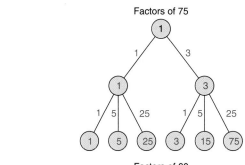

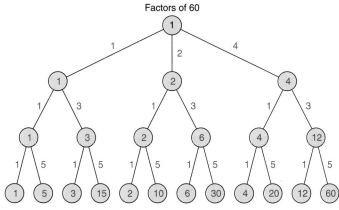

 b Find the GCF and the LCM of 75 and 60. 15; 300

P, R **16** Solve each equation for x.
 a $\frac{2}{3} + \frac{3}{5}x = \frac{5}{7}$ $\frac{5}{63}$
 b $1\frac{2}{3} + \frac{5}{9}x = 8$ $11\frac{2}{5}$

P, R **17** Find the LCM and the GCF of 7! and 8!. 40,320; 5040

Additional Answers

for pages 524 and 525

2 a Because they're only 1 unit apart, the greatest common factor that two consecutive integers can have is 1. Therefore, they're relatively prime.

b Every third number is a multiple of 3, so any set of three consecutive integers must contain a multiple of 3.

4 Always. If A and B represent the sets of prime factors of two relatively prime numbers, then $A \cap B = \{\}$. Since no subsets of A and B can intersect, any factors of the two numbers will be relatively prime.

6 For 3 consecutive factors, one is even and is therefore divisible by 2. One is a multiple of 3 and is therefore divisible by 3. A number divisible by 2 and 3 is divisible by 6.

8 Yes; yes

SECTION 10.6

Problem-Set Notes

for pages 526 and 527

- **29** is atypical in that instead of being asked to find the GCF and LCM, students are asked to find the numbers that have a given GCF and LCM.
- **35** previews another type of algebraic word problem.
- **38** may seem difficult, but all it requires is an understanding of prime factorization.

Additional Practice

1. Use the diagram to find the least common multiple and greatest common factor of 24 and 40.

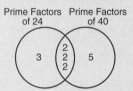

Prime Factors of 24 | Prime Factors of 40

2. a. Find the least common multiple of 12, 18, and 24.
b. Use your answer to part **a** to help you solve the equation $\frac{5}{12}x - \frac{7}{18} = 1\frac{1}{24}$.

3. Solve the equation for x.
$\frac{1}{9}x - \frac{1}{6} = \frac{35}{36}$

4. Suppose there are ten hot dogs to a package and twelve hot dog buns to a package. What is the fewest number of packages of hot dogs you can purchase so that you can have exactly the same number of hot dogs and buns?

In problems **5** and **6**, decide if the numbers are relatively prime.
5. 22 and 27   **6.** 18 and 35

Answers:
1. LCM: 120; GCF: 8
2. a. 72 **b.** $x = \frac{103}{30}$ **3.** $x = \frac{41}{4}$ **4.** 6
5. Yes **6.** Yes

P, R, I **18** Pump 1 can pump 80 gallons of water per minute, pump 2 can pump 60 gallons of water per minute, and pump 3 can pump 45 gallons of water per minute. A flooded basement has approximately 2000 gallons of water that needs to be pumped out. What percent of the water will be pumped out per minute if the three pumps are used together?   9.25%

P **19** Suppose that you are going to add $\frac{7}{36}$ and $\frac{19}{45}$.
  a Why do you need to know the LCM of 36 and 45?
  b Use the LCM to add the fractions. $\frac{111}{180} = \frac{37}{60}$
  c What is the GCF of the numerator and the denominator before you simplify the sum? After you simplify the sum?   3; 1

P, R **20** Kim can clean a fish in 2 minutes, while Ping Yo can clean 25 fish in an hour. If they have 180 fish to clean, how long will it take? $3\frac{3}{11}$ hr

P, R, I **21** In 1992, February 29 was on a Saturday. What is the next year that February 29 will be on a Saturday? What does this have to do with LCM and GCF?   2020

◀ LOOKING BACK **Spiral Learning** LOOKING AHEAD ▶

R **22** If 80 is 40% of the quantity $x + 12$, what is x?   188

R **23** **Spreadsheets**   Make a spreadsheet with three columns. Column A should show years, column B should show Tony's age if he was born in 1972, and column C should show the age of Tony's sister if she was born in 1982. In what year is Tony three times as old as his sister?   1987

R **24** Solve each equation for x.
  a $0.5x + 0.7x = 1.68$ 1.4
  b $-0.8x + 0.2x = 2.46$ -4.1

R, I **25 a** Convert $186,000 \frac{\text{miles}}{\text{second}}$ to $\frac{\text{feet}}{\text{hour}}$.   **b** Simplify $\frac{25 \text{ miles}}{10 \frac{\text{miles}}{\text{hr}}}$

R **26** **Science**   Gear 1 has 40 teeth, gear 2 has 24 teeth, and gear 3 has 60 teeth. All of the teeth are the same size. How many revolutions must each gear make before the arrows are again lined up?

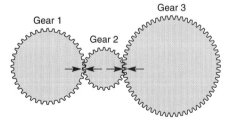

P, R **27** If 40 textbooks $2\frac{1}{2}$ inches thick can be stored on a shelf, then how many textbooks $1\frac{1}{4}$ inches thick can be stored on the same shelf?   80

P, R **28** Solve $\frac{2}{3} + \frac{3}{2}x = \frac{5}{6}$ for x. $\frac{1}{9}$

R **29** The LCM of p and g is 48. The GCF of p and g is 4. Find all possible values of p and g.   12 and 16; 4 and 48

526 Chapter 10 Topics of Number Theory

SECTION 10.6

R **30** Simplify each expression.
  a $\frac{2}{3} \cdot \frac{3}{4} + \frac{5}{6} \cdot \frac{6}{10}$ 1   **b** $3.75 + 2\frac{1}{4}$ 6   **c** $15\% + \frac{2}{5}$ $\frac{11}{20}$, or 55%

R **31** You are playing a game with a pair of dice. You win if the numbers on the two dice are relatively prime; otherwise, you lose. What percent of the dice throws should be winning ones?   About 64%

P, R, I **32** Fran drives her delivery truck at an average speed of 50 miles per hour and Fred drives his delivery truck at an average speed of 45 miles per hour.
  a At noon, the two trucks leave company headquarters and head in opposite directions. How far apart will the trucks be after $2\frac{1}{2}$ hours?   237.5 mi
  b If the trucks head back to headquarters at 2:30 P.M., at what time will each truck arrive?   5:00
  c How many miles will each truck have traveled?   250 mi; 225 mi

R **33** What are the first five perfect-square multiples of 5?   0, 25, 100, 225, 400

R **34** *Finance* A car depreciates at a rate of 17% per year for the first four years after it is purchased. If the car originally cost $17,549, what is its value after four years? $8328.46

R **35** Ten gallons of punch that is 60% ginger ale is mixed with 15 gallons of punch that is 70% ginger ale. What percent of the mixture is ginger ale?   66%

R **36** How many different three-digit numbers can be formed using three different digits?   648

P, R **37** Find the GCF and the LCM of each pair of numbers.
  a 6 and 8   2; 24   **b** 12 and 16   4; 48   **c** 18 and 24   6; 72

P **38** Let p be a prime number. List the factors of p^2. $1, p, p^2$

R **39** What is the smallest square that can be made using rectangular 3-inch-by-4-inch tiles?   12" by 12"

Investigation

Lattice Points Lattice points are points with coordinates that are both integers. Plot the lattice points that have coordinates that are relatively prime. Describe some patterns that you observe.

Multicultural Note

The Babylonians developed a place-value system based on the number 60. Although modern numerals are base-ten, the Babylonian system is reflected in our units for measuring time and angles. There are 60 seconds in 1 minute and 60 minutes in both 1 hour and 1 degree of angle measurement.

Additional Answers

for pages 526 and 527

19 a You need to find a common denominator.

21 The GCF of 7 (the number of days per week) and 1461 (the number of days between consecutive February 29s) is 1. Therefore, 7 × 1461 days (or 28 years) must pass before February 29 falls on the same day of the week. 7 × 1461 is the LCM of the two numbers.

25 a 3,535,488,000,000 ft/hr
b 2.5 hr

26 Gear 1: 3 revolutions; gear 2: 5 revolutions; gear 3: 2 revolutions

Investigation

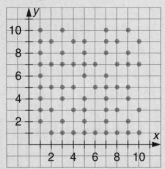

Students may note the following:
1. There are no points along the line $y = x$ or on the axes.
2. The pattern is symmetric with respect to the line $y = x$.

SECTION 10.7

Objectives
- To identify symmetric prime numbers
- To apply Goldbach's first and second conjectures
- To identify figurate numbers
- To identify Pythagorean triples

Vocabulary: figurate numbers, Pythagorean triple, symmetric prime numbers

Class Planning

Schedule 1 day

Resources
In *Chapter 10 Student Activity Masters* Book
Class Openers 73, 74
Extra Practice 75, 76
Cooperative Learning 77, 78
Enrichment 79
Learning Log 80
Using Manipulatives 81, 82
Spanish-Language Summary 83

Class Opener

Write each even number as the sum of two prime numbers and each odd number as the sum of three prime numbers.

1. 24   2. 30   3. 102
4. 17   5. 55   6. 103

Possible answers:
1. 11 + 13   2. 17 + 13
3. 5 + 97   4. 13 + 2 + 2
5. 11 + 13 + 31   6. 47 + 13 + 43

Lesson Notes

- The content of this section is not in the mainstream of required mathematics, but it makes some connections between natural numbers and geometry and raises interesting questions for further

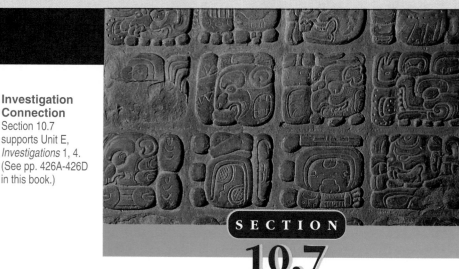

Investigation Connection
Section 10.7 supports Unit E, *Investigations* 1, 4. (See pp. 426A-426D in this book.)

SECTION 10.7
Number-Theory Explorations

SYMMETRIC PRIMES

The number 25 is midway between the prime numbers 19 and 31. The numbers 19 and 31 are **symmetric primes** (also called Euler Primes) of 25 because both numbers are six units from 25 on the number line.

Example 1
Find all other pairs of symmetric primes of 25.

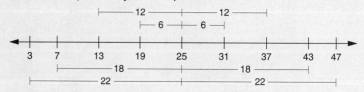

Solution
Other pairs of symmetric primes of 25 are 13 and 37, 7 and 43, and 3 and 47.

GOLDBACH'S CONJECTURES

Does every natural number have a pair of symmetric primes? The answer is no. The numbers 1, 2, and 3 do not have a pair of symmetric primes. Do all natural numbers greater than 3 have a pair of symmetric primes? The answer to this question is unknown at this time. If you could solve this problem, you would be able to prove the following conjectures by the Prussian mathematician Christian Goldbach.

528   Chapter 10 Topics of Number Theory

SECTION 10.7

▶ **Goldbach's First Conjecture** Every even number greater than or equal to four can be represented as the sum of two primes.

▶ **Goldbach's Second Conjecture** Every odd number greater than or equal to seven can be expressed as the sum of three primes.

Example 2
Test Goldbach's conjectures by expressing each number as the sum of primes.
a 42   b 23

Solution
a Since 42 is an even number greater than 4, we can express it as the sum of two primes, 5 + 37.
b Since 23 is an odd number greater than 7, we can express it as the sum of three primes, 3 + 7 + 13.

FIGURATE NUMBERS

The ancient Greeks were interested in ***figurate numbers***—special numbers associated with geometric figures. The answers to many problems that require counting are figurate numbers. Triangular numbers, square numbers, pentagonal numbers, and hexagonal numbers are examples of figurate numbers.

Example 3
a The first four triangular numbers are 1, 3, 6, and 10.

What is the next triangular number?

b The first four square numbers are 1, 4, 9, and 16.

What is the next square number?

---

exploration through special projects. Your enthusiasm will help students be more inclined to pursue the topics in this section.

▪ The figurate numbers, particularly the triangular numbers and the square numbers, occur frequently in pattern problems.

▪ Knowing some Pythagorean triples and being able to recognize them in problem situations can help make solutions easier to find.

▪ The sample problem is usually presented as a theorem in geometry.

Additional Examples

Example 1
Find the symmetric primes of 36.

Answers: 31 and 41, 29 and 43, 19 and 53, 13 and 59, 11 and 61, and 5 and 67

Example 2
Test Goldbach's conjectures by expressing each number as the sum of primes.
a. 36   b. 59

Possible answers:
a. 17 + 19   b. 11 + 19 + 29

Example 3
Three rectangular numbers are 2, 6, and 12.

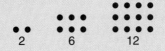

What is the next rectangular number?

Answer: 20

Example 4
Which of the following are Pythagorean triples?

a. 13, 84, 85   b. 15, 110, 111
c. 17, 144, 145 d. 18, 180, 181

Answer: **a** and **c**

SECTION 10.7

Assignment Guide

Basic
Optional

Average or Heterogeneous
Optional

Honors
Day 1   10–13, 15–23, 25, 27, 30, 31, 37

Strands

Communicating Mathematics	16, 19, 20
Geometry	27, 33, 37
Measurement	27, 31, 33, 37
Data Analysis/ Statistics	30
Technology	21
Probability	36
Algebra	12, 20, 22, 23, 25–39
Patterns & Functions	15, 18, 23, 25
Number Sense	1–15, 17, 18, 20, 22–25, 36, Investigation
Interdisciplinary Applications	16, 19, 31

Individualizing Instruction

The gifted learner Students might have their curiosity piqued by problem 18, which remains an unsolved problem in mathematics. Students might be encouraged to do some research into other mathematical problems that remain unsolved and to present their findings to the class.

---

c The first four pentagonal numbers are 1, 5, 12, and 22.

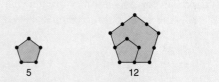

What is the next pentagonal number?

d The first four hexagonal numbers are 1, 6, 15, and 28.

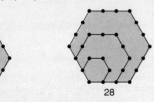

What is the next hexagonal number?

Solution

a The fifth triangular number is 10 + 5, or 15.
b The fifth square number is 16 + 9, or 25.
c The fifth pentagonal number is 22 + 13, or 35.
d The fifth hexagonal number is 28 + 17, or 45.

PYTHAGOREAN TRIPLES

Study the data displayed in the table.

a	b	c
1	0	1
3	4	5
5	12	13
7	24	25
9	40	41
11	60	61

The numbers in the chart satisfy the well-known equation $a^2 + b^2 = c^2$, which is associated with the Pythagorean Theorem. Except for the first line of data, (1, 0, 1), each line is a trio of natural numbers called **Pythagorean Triples**.

Chapter 10   Topics of Number Theory

SECTION 10.7

Example 4
What pattern can be used to describe the numbers in columns **a, b**, and **c** in the preceding table?

Solution
Here is one description of the pattern:

In column **a**, the consecutive numbers differ by two. In column **b**, the numbers are multiples of 4 and the consecutive numbers differ by 4, 8, 12, 16, and 20. Each number in column **c** is one more than the corresponding value in column **b**.

Can you find other ways to describe the number relationships in columns **a, b,** and **c**?

Sample Problem

Problem Count the number of diagonals in each drawing. How do the number of diagonals in each geometric figure relate to the triangular numbers?

Triangle   Quadrilateral   Pentagon   Hexagon

Solution We can organize the answer in a table.

Figure	Number of Diagonals
Triangle	0
Quadrilateral	2
Pentagon	5
Hexagon	9

The number of diagonals in each figure is one less than a triangular number.

$$0 = 1 - 1$$
$$2 = 3 - 1$$
$$5 = 6 - 1$$
$$9 = 10 - 1$$

Cooperative Learning

- Have students work together to discover some relationships between the figurate numbers. For example, can figurate numbers be combined somehow to form other figurate numbers?

Possible answers: Two successive triangular numbers form a perfect square; a triangular number and the next square number form a pentagonal number (i.e., the third triangular number and the fourth square number form the fourth pentagonal number—6 + 16 = 22); a triangular number and the next pentagonal number form a hexagonal number.

- Problems 16–24 and Investigation are appropriate for cooperative learning.

SECTION 10.7

Problem-Set Notes

for pages 532 and 533

- 16–24 are exploratory in nature and are therefore appropriate for students to work on together.

Additional Practice

1. What two pairs of primes are symmetric primes of 10?
2. Express 24 as the sum of two prime numbers.
3. Express 51 as the sum of three prime numbers.
4. What is the sixth pentagonal number?
5. Is 10, 12, 13 a Pythagorean triple?

Answers: **1.** 7, 13; 3, 17
2. Possible answer: 5 + 19
3. Possible answer: 17 + 17 + 17 **4.** 51 **5.** No

Multicultural Note

In the first century, Greece became the center of the civilized world. The Greeks adopted the most advanced scientific methods from earlier civilizations in Egypt and Mesopotamia. Two of the most famous Greek mathematicians, Thales and Pythagoras, traveled extensively in these regions.

Think and Discuss

P **1** The prime numbers 5 and 23 are symmetric to what number? 14

P **2** What two pairs of primes are symmetric primes of 9? 7 and 11; 5 and 13

P **3** Express 6 as the sum of two primes. 3 + 3

P **4** Refer to Example 2 on page 529.
  a Find other ways to express 42 as the sum of primes.
  b Find other ways to express 23 as the sum of primes.

P **5** Express 14 as the sum of two primes in two different ways. 3 + 11; 7 + 7

P **6** Express 11 as the sum of three primes. 3 + 3 + 5

P **7** What is a triangular number?

P **8** What are figurate numbers?

P **9** Which of the following are Pythagorean triples?
  a 6, 8, 10   **b** 5, 11, 12   **c** 8, 15, 17   a and c

Problems and Applications

P **10** Write 86 as the sum of two prime numbers in three different ways.

P **11** Write 137 as the sum of three prime numbers in three different ways.

P, R, I **12** The LCM of a and b is 12. The GCF of a and b is 4.
  a Find ab. 48
  b Find the LCM of $2a$ and $2b$. 24
  c Find the GCF of $2a$ and $2b$. 8
  d Find the LCM of $3a$ and $3b$. 36
  e Find the GCF of $3a$ and $3b$. 12

P **13** Find all pairs of symmetric primes of 30.

P, R **14** In parts **a** and **b,** determine whether the statement is true *always, sometimes,* or *never.*
  a The product of two numbers divided by their LCM is their GCF. Always
  b The product of two numbers divided by their GCF is their LCM. Always
  c Find the LCM of 24, 36, and 45. 360

P **15** The figurate numbers 1, 5, 12, 22, and 35 are called pentagonal numbers. Represent each pentagonal number greater than 1 as the sum of a square number and a triangular number.

532   Chapter 10   Topics of Number Theory

SECTION 10.7

R **16 Communicating** How might the rules for baseball change if the baseball field had five bases instead of four bases?
Possible answers: Players would have to run an extra 90 ft to score. Also, a grand slam home run would account for 5 runs.

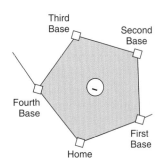

P **17** The numbers 20, 21, and 29 represent a Pythagorean Triple. Can you find other Pythagorean Triples in which the two smaller numbers differ by one? By two? Answers may vary. Possible answers: 3, 4, 5; 6, 8, 10

P, R **18** Copy the diagram and circle the primes. If the pattern were continued forever, would there be a prime in every row after the first row?
Answers will vary. This is an unsolved problem in mathematics.

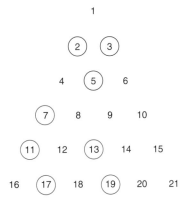

P, R **19 Sports** How would bowling change if there were 15 pins instead of 10? Possible answers: A different method of scoring would be necessary. Instead of the "7-10 split" we would have the "11-15 split."

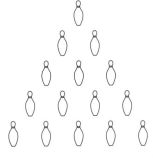

P **20** Observe that 41 and 61 are symmetric primes of 51. Also, $51 = \frac{41 + 61}{2}$ and $102 = 41 + 61$. How does this relate to Goldbach's First Conjecture?

P, R **21** Find the sum of the numbers in the fifteenth row of the triangle. 3375

```
        1
      3   5
    7   9   11
 13   15   17   19
```

Communicating Mathematics

- Have students write a paragraph explaining why Goldbach's First Conjecture includes only even numbers greater than 4 and not odd numbers. Alternately, students might explain this orally.

- Problems 16, 19, and 20 encourage communicating mathematics.

Additional Answers

for pages 532 and 533

4 a Possible answer: 11 + 31

b Possible answer: 5 + 7 + 11

7 Possible answer: A number that corresponds to the number of dots in a triangular array.

8 Possible answer: A figurate number can be represented by dots arranged in a pattern to look like a particular geometric figure.

10 Possible answers: 3 + 83; 7 + 79; 13 + 73

11 Possible answers: 11 + 13 + 113; 17 + 17 + 103; 17 + 59 + 61

13 29, 31; 23, 37; 19, 41; 17, 43; 13, 47; 7, 53

15 5 = 4 + 1; 12 = 9 + 3; 22 = 16 + 6; 35 = 25 + 10

20 For any even number n, the sum of pairs of symmetric primes of $\frac{n}{2}$ will satisfy Goldbach's First Conjecture.

SECTION 10.7

Problem-Set Notes

for pages 534 and 535

- **23** is Ulam's Conjecture, which asserts that following the given algorithm will eventually result in 1 for any natural number input. The number of steps required to arrive at 1 is called the Ulam number of the input number. The greatest Ulam number for a two-digit number is 118 (for the number 97).

- **37** is a geometric representation of the Pythagorean Theorem. Be sure your students see its significance.

Stumbling Block

Students may think that the fourth hexagonal number should be 18 rather than 28, because there are 18 dots around the perimeter. Point out that each successive hexagon is built from the previous one, so the dots in the interior need to be counted.

Reteaching

Students may have difficulty finding symmetric primes of a number n. Have students place n on a number line, then test numbers 1 unit away, 2 units away, 3 units away, and so on. Alternately, students may use a spreadsheet to display potential symmetric primes.

P **22** Two primes of the form n and $2n-1$ are "almost-double primes" (ADP). Find the first five pairs of almost-double primes.

P **23** Pick a natural number.
Step 1: Divide the number by 2 if the number you chose is even. Repeat this procedure as long as the resulting number is even.
Step 2: Multiply the number by 3, add 1, and divide by 2 if the number you chose is odd.
Continue to repeat Step 1 or Step 2 on the resulting number. What happens? *Operations result in the repeated results 1, 2, 1, 2, 1, 2, ...*

P **24** Pick a prime number greater than 3, and double it. Now, look at the two integers that are one greater than and one less than your product. Can both numbers be prime? Explain.

P **25** Let $f(n)$ = the sum of the natural-number factors of n. Find $f(n)$ for each of the following.
 a $f(24)$ 60
 b $f(31)$ 32

◀ LOOKING BACK **Spiral Learning** LOOKING AHEAD ▶

R **26** Evaluate each expression.
 a $(-4)^2$ 16
 b -4^2 -16
 c $-(-4)^2$ -16
 d $-(4^2)$ -16

R **27 a** Write an expression for the perimeter of the triangle. $12x$
 b Find the perimeter if $x = 2.4$.   28.8

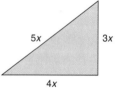

R **28** Multiply $\begin{bmatrix} -2 & 6 \\ -1 & 0 \end{bmatrix} \cdot \begin{bmatrix} 5 & -4 \\ 2 & -3 \end{bmatrix} \cdot \begin{bmatrix} 2 & -10 \\ -5 & 4 \end{bmatrix}$

P, R **29** Solve $\frac{x-6}{4} = \frac{1}{2}$.   8

R **30 a** How many points did Rick score?   9
 b What is the mean number of points scored by the six players?   9.5

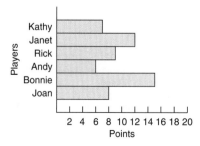

534   Chapter 10   Topics of Number Theory

SECTION 10.7

R, I **31** A.J. can mow the lawn in 30 minutes and J.R. can mow the lawn in 45 minutes. How long will it take if they work together? 18 minutes

R **32** Solve $|x - 4| = 9$ for x. $x = 13$ or $x = -5$

R **33**
   a Find the length of $\overline{AB}$. 5
   b Find the length of $\overline{BC}$. 9
   c Find the area of the rectangle. 45

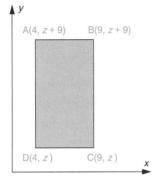

R, I **34** Two years ago, Mary was half her sister's age. If Mary's sister is 12 years old, how old is Mary? 7

R **35** Solve $x + 4 = 3x + 6$ for x. $x = -1$

R, I **36** What is the probability of randomly selecting a natural number from $(-4, -3, -2, -1, 0, 1, 2, 3, 4)$ that is 3 units or less from -2? .1

R, I **37** Find the area of square C. 289

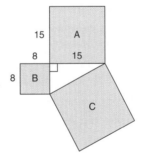

R **38** Solve each equation for x.
   a $3x = 0$ $x = 0$
   b $3(x - 4) = 0$ $x = 4$
   c $(x - 3)(x - 4) = 0$ $x = 3$ or $x = 4$

R **39** Graph the equation $2x - 4 = 1$.

Investigation

Prime Numbers Find all two-digit prime numbers that remain prime when the digits are reversed. Find a three-digit prime number that remains prime no matter how you rearrange the digits.

Checkpoint

1. The prime numbers 11 and 71 are symmetric to what number?
2. Express 12 as the sum of two prime numbers.
3. Express 29 as the sum of three prime numbers.
4. What is the sixth triangular number?
5. Is 8, 6, 10 a Pythagorean triple?

Answers:
1. 41 2. 5 + 7 3. Possible answer: 3 + 7 + 19 4. 21 5. Yes

Additional Answers

for pages 534 and 535

22 2 and 3; 3 and 5; 7 and 13; 19 and 37; 31 and 61

24 For any three consecutive integers, one must be a multiple of 3. So if n is prime and $n > 3$, then $2n$ won't be a multiple of 3; so either $2n + 1$ or $2n - 1$ will be.

39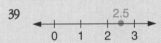

Investigation The two-digit primes that remain prime when reversed are 11, 13, 17, 31, 37, 71, 73, 79, and 97.

A three-digit number that remains prime no matter how the digits are arranged is 113.

SUMMARY

CHAPTER 10

Summary

CONCEPTS AND PROCEDURES

After studying this chapter, you should be able to
- Apply the Fundamental Counting Principle to solve problems (10.1)
- Identify prime numbers and composite numbers (10.2)
- Identify natural-number multiples of a number (10.2)
- Identify the least common multiple of natural numbers (10.2)
- Apply divisibility tests for 2, 3, 4, 5, 6, 8, 9, 11, and 12 (10.3)
- Express a number in prime-factorization form (10.4)
- Determine whether a number is prime (10.4)
- Determine all factors of a natural number (10.5)
- Determine whether a number is an abundant number, a deficient number, or a perfect number (10.5)
- Determine the greatest common factor of a set of numbers (10.5)
- Use Venn diagrams to determine the least common multiple and the greatest common factor of a set of numbers (10.6)
- Determine whether numbers are relatively prime (10.6)
- Use LCM and GCF to simplify fractions and solve equations (10.6)
- Identify symmetric prime numbers (10.7)
- Apply Goldbach's first and second conjectures (10.7)
- Identify figurate numbers (10.7)
- Identify Pythagorean triples (10.7)

VOCABULARY

abundant number (10.5)	perfect number (10.5)
composite number (10.2)	prime-factorization form (10.4)
deficient number (10.5)	prime number (10.2)
factor (10.2)	Pythagorean triple (10.7)
figurate numbers (10.7)	relatively prime numbers (10.6)
Fundamental Counting Principle (10.1)	symmetric prime numbers (10.7)
greatest common factor (10.5)	tree diagram (10.1)
least common multiple (10.2)	

CHAPTER 10

Review

1. Write the prime-factorization form of each number.
   a 27   **b** 16   **c** $16 \cdot 27$ **d** $8 \cdot 9 \cdot 5$ **e** 720

2. Coach Simon has eight pitchers and three catchers. How many combinations of one pitcher and one catcher can he select?   24

3. Camille has four blouses, five skirts, and six sweaters that can be mixed and matched. How many different outfits can Camille assemble?   120

4. Determine the perimeters of all rectangles with area 80 if the lengths of the sides of the rectangles are natural numbers.   162, 84, 48, 42, 36

5. When a ball comes down from the top of a pinball machine, it goes left 80% of the time and right 20% of the time. It then hits a bumper. The bumper on the left sends it to the left 30% of the time and to the middle 70% of the time. The bumper on the right sends it to the right 90% of the time and to the middle 10% of the time. Find the probability that, after the ball first hits a bumper, it goes to the
   a Left   .24   **b** Middle   .58   **c** Right   .18

6. The diagram represents the probabilities connected with Bart shooting two free throws.
   a Find the probability that Bart will make at least one free throw.   .92
   b Find the probability that Bart will make exactly one free throw.   .44

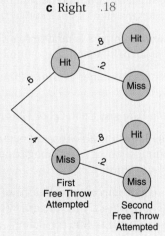

7. Boards come in 8-foot, 10-foot, and 12-foot lengths. You need seven 6-foot boards and eight 5-foot boards. What size boards should you buy to minimize waste?

8. Suppose x is a prime factor of 30 and y is a composite factor of 56.
   a What is the number of possible values for the volume of the box?   15
   b What is the greatest possible volume of the box?   2800
   c What is the least possible volume of the box?   80

Assignment Guide

Basic
Optional

Average or Heterogeneous
Optional

Honors
Day 1   1–14, 16–20, 22–27, 30

Strands

Communicating Mathematics	26
Geometry	4, 8, 12
Measurement	4, 7, 8, 12
Data Analysis/ Statistics	6, 20
Technology	9, 26
Probability	2, 3, 5, 6, 16, 24, 30
Algebra	8, 10, 11, 14, 21, 31
Patterns & Functions	26
Number Sense	1, 8–12, 14–18, 20, 22, 23, 25–29
Interdisciplinary Applications	5–7, 13, 19, 23

Additional Answers

for page 537

1a 3^3
b 2^4
c $3^3 \times 2^4$
d $2^3 \times 3^2 \times 5$
e $2^4 \times 3^2 \times 5$

7 One 8-ft board, four 10-ft boards, three 12-ft boards

REVIEW

Additional Answers

for pages 538 and 539

22 The symmetric primes of 20 are 3 and 37, 11 and 29, 17 and 23

25 a 1, 2, 4, 373, 746, 1492
b 1, 2, 3, 4, 6, 8, 12, 16, 24, 37, 48, 74, 111, 148, 222, 296, 444, 592, 888, 1776
c 1, 3, 647, 1941

26 b 180 is the LCM of 6, 15, 9, and 12

30 123, 132, 213, 231, 312, 321

---

9 Use a spreadsheet to find the factors of 2477. How far down column B do you need to copy the formula to be sure that you have found all the factors?   To cell B49

10 The five-digit number $n5n5n$ is divisible by 11. Determine the values of n. $n = 7$

11 Find all single-digit values of n so that $|n^2 - 3|$ is a multiple of 11. $n = 5$ or $n = 6$

12 The area of a rectangle is 140 and the base and height are both natural numbers. Find all possible values of the height if the height is greater than the base.   14, 20, 28, 35, 70, 140

13 The PTO has $1000 for scholarships. Each of the people receiving a scholarship must receive the same amount of money. What scholarship amounts can be given to each person if the amount of the scholarship must be at least $50? $1000, $500, $250, $200, $125, $100, $50

14 a List the natural-number factors of 18.   1, 2, 3, 6, 9, 18
   b Find the sum of the reciprocals of the factors in part **a**. $2\frac{1}{6}$

15 a Find the least common denominator of $\frac{1}{4}$, $\frac{1}{8}$, and $\frac{5}{6}$.   24
   b Subtract the sum of the last two fractions in part **a** from the first fraction. $-\frac{17}{24}$

16 If a pair of numbers is randomly selected from 5, 18, 20, 27, and 36, what is the probability that the two numbers are relatively prime? $\frac{2}{5}$

17 Classify each number as *abundant, deficient,* or *perfect*.
   a 28   Perfect   **b** 58   Deficient   **c** 36   Abundant

18 Find the LCM and the GCF of 120 and 135.   1080; 15

19 Printing Press 1 is able to print 6000 magazines per hour. What percent of a printing job of 75,000 magazines has been completed after $3\frac{1}{2}$ hours of running time?   28%

20 How many complete revolutions of Gear 2 are necessary for the arrows of the two gears to line up again?   12

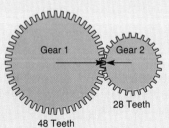

Gear 1 — 48 Teeth
Gear 2 — 28 Teeth

21 Evaluate each expression.
   a $\frac{12}{20} - \frac{9}{15}$ 0
   b $\frac{8}{12} + \frac{6}{9}$ $1\frac{1}{3}$
   c $\frac{5}{20} + \frac{6}{12}$ $\frac{3}{4}$

REVIEW

22 Locate the symmetric primes of 20 on a number line.

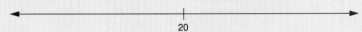

23 Bill comes to work at 6 A.M. and works for eight hours. He takes a five-minute break every three hours. Sue comes to work at 10 A.M. and works eight hours. Sue takes a break every two hours for five minutes. When will Bill and Sue have a break?   At noon

24 How many different combinations of three segments can be selected from the group shown?   20

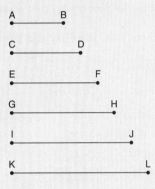

25 List all the factors of each number.
   a 1492   **b** 1776   **c** 1941

26 Make a spreadsheet consisting of these four columns:
A (multiples of 6)
B (multiples of 15)
C (multiples of 9)
D (multiples of 12)

   a What is the smallest number that appears in all four columns?   180
   b Explain why it is the first number that appears in all four columns.

27 **a** How many factors does 3^4 have?   5
   b How many factors does 2^3 have?   4
   c How many factors does $2^3 \cdot 3^4$ have?   20

28 **a** Factor 1001 into primes. $7 \cdot 11 \cdot 13$
   b Factor 1,002,001 into primes. $7^2 \cdot 11^2 \cdot 13^2$

29 List the perfect-square factors of 1800 if the prime-factorization form of 1800 is $2^3 \cdot 3^2 \cdot 5^2$.   1, 4, 9, 25, 36, 100, 225, 900

30 List the set of all three-digit lock combinations that use only the digits 1, 2, and 3 without using any of these digits more than once.

31 Evaluate each expression.
   a $\frac{1}{4} + \frac{1}{2.5}$ $\frac{13}{20}$ **b** $\frac{2}{1.5} + \frac{1.5}{2.5}$ $1\frac{14}{15}$ **c** $\frac{2}{3} + \frac{3}{2}$ $2\frac{1}{6}$

Chapter Review

CHAPTER 10

Test

1. A restaurant offers four types of sandwiches, three types of salads, and four types of soup. How many different meals could Judy order if she wants a sandwich, a salad, and a bowl of soup?   48

2. On each of the next three days, the probability of snow is listed as 20%. Calculate the probability that it snows exactly two out of three days.   .096

In problems 3–5, identify the number as *prime* or *composite*.

3. 34   Composite
4. 71   Prime
5. 292   Composite

In problems 6–8, find the LCM of each pair of numbers.

6. 3 and 13   39
7. 4 and 18   36
8. 42 and 36   252

9. Check the number 27,165,879,348 for divisibility by 2, 3, 4, 5, 6, 8, 9, 10, 11, and 12.   It is divisible by 2, 3, 4, 6, and 12.

10. What is n if the four-digit number $513n$ is divisible by 4?   2 or 6

In problems 11 and 12, determine the prime-factorization form of each number.

11. 18,000 $2^4 \cdot 3^2 \cdot 5^3$
12. 560 $2^4 \cdot 5 \cdot 7$

In problems 13–15, determine whether each number is *abundant*, *deficient*, or *perfect*.

13. 28   Perfect
14. 42   Abundant
15. 84   Abundant

In problems 16 and 17, simplify each fraction.

16. $\frac{112}{196}$ $\frac{4}{7}$
17. $\frac{1}{2} \cdot \frac{2}{5} + \frac{3}{4} \cdot \frac{1}{5}$ $\frac{7}{20}$

18. One hose can add 40 gallons of water per hour to an empty pool. Another hose can add 30 gallons of water per hour. If the pool can hold 12,000 gallons of water, what percent of the pool will be filled per hour if both hoses are used together?   .58%

19. Solve $\frac{5}{6}x + \frac{1}{4}x - \frac{2}{3}x = 3$ for x . . . $\frac{36}{5}$

20. Is 391 a prime number?   No

21. Use one of Goldbach's conjectures to express 110 as the sum of primes.

22. What is the seventh triangular number?   28

23. What is a pair of symmetric primes of 36?   31 and 41

24. What number completes the Pythagorean triple (7, ___, 25)?   24

PUZZLES AND CHALLENGES

1 Create a 4-by-4 Magic Square. That is, put the integers 1 to 16 in the squares so that the sums of the four numbers in each row, each column, and each four-square diagonal are the same.

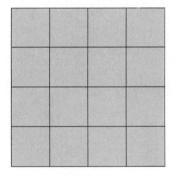

2 A palindrome is a word that reads the same forwards as backwards. Examples are tot, noon, radar, and hannah. See how many four-, five-, and six-letter palindromes you can find which are legitimate English words.

3 Palindromes can also be numbers (for example, 343 or 6776). In the multiplication problem (11)(181) = 1991, the product of a two-digit palindrome and a three-digit palindrome is a four-digit palindrome. Try to find other multiplication problems in which the product of a two-digit palindrome and a three-digit palindrome is a four-digit palindrome.

4 Some letters are missing from the word list below. Form a complete word by replacing each set of asterisks with a man's first name. For example, ad***tage plus *van* gives the complete word *advantage*.

- **a** r****serie
- **b** k****knack
- **c** b****yard
- **d** ***ichoke
- **e** ****etplace
- **f** aspara***
- **g** flot***
- **h** f*****
- **i** ***ural
- **j** ***esman
- **k** *****le
- **l** m***ge

binary equivalents of base 10 numbers. Explain how to add two binary numbers. Why do computers use binary numbers?

8. Section 10.3 does not include a divisibility test for the number 7. Explain why this is so. Include examples with your explanation.

9. Extend example 4 on page 509 as follows: Write a computer program that tests whether a number is prime. Test your program with some two-digit numbers. Once you've worked out the bugs, use your program to find the four greatest four-digit primes. Write a brief explanation of how your program works.

10. Extend problem 34 on page 527 as follows: Use a spreadsheet to make a table showing the value of the car given in the problem as it depreciates at a rate of 17%. After how many years is the car worth half of its original value? Is the value of the car ever zero? Explore what factors affect the rate at which a car depreciates. Do cars ever appreciate in value?

11. Extend problem 23 on page 534 as follows: Write a computer program that performs the calculations described for any input number. Run the program using various values less than 1000. Describe your results.

Additional Answers

for Chapter 10 Test

21 Possible answer: 3 + 107

Answers for Puzzles and Challenges are on page A9.

OVERVIEW

LONG-TERM INVESTIGATIONS

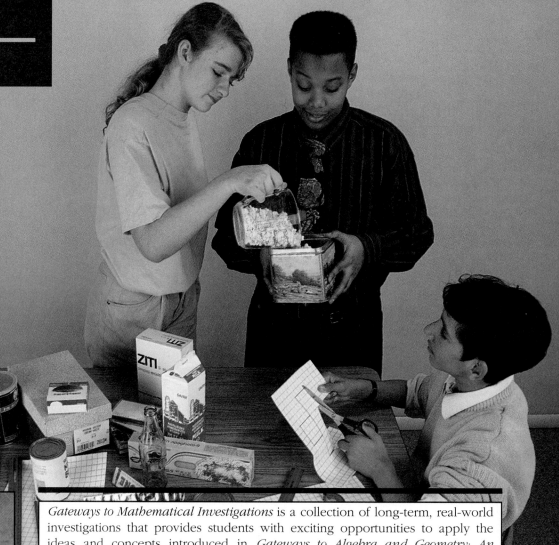

See *Investigations* book, Unit F, pages 134-159.

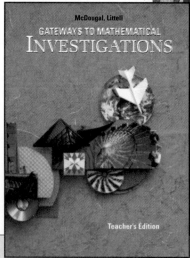

Gateways to Mathematical Investigations is a collection of long-term, real-world investigations that provides students with exciting opportunities to apply the ideas and concepts introduced in *Gateways to Algebra and Geometry: An Integrated Approach*. Unit F: An Enterprising Venture consists of five investigations that develop the idea of inventing, marketing, packaging, and financing a new product. This unit is designed so that you can pick and choose the order and number of investigations you would like to do with your students. Each investigation is an exploration that takes several days to complete. Refer to the menu at the right to learn more about the investigations that make up this unit.

As your students work through the Unit F investigations, use Chapters 11 and 12 of *Gateways to Algebra and Geometry: An Integrated Approach* for concept and skill development, and as a source of rich problems, exercises, and applications. When appropriate, refer students to additional supportive sections in *Gateways to Algebra and Geometry* as well.

UNIT F: An Enterprising Venture

MENU OF INVESTIGATIONS

OVERVIEW: Groups apply mathematical concepts and strategies from Chapters 11 and 12 within the context of inventing, marketing, packaging, and financing a new product.

Developing a Product

(3-5 days)
Students discuss ideas for a new product; they create and conduct a market research survey, and display their findings using spreadsheets, graphs and equations; they analyze their findings.
STRANDS
Functions, Algebra, Discrete Mathematics, Statistics and Probability
UNIFYING IDEAS
Multiple Representations, Patterns and Generalizations

The Container

(3-5 days)
Students conceptualize their products and consider packaging requirements for them. They examine various containers: define, describe, classify, model, and arrange them in size order. They design containers for their own products.
STRANDS
Geometry, Measurement, Logic and Language
UNIFYING IDEAS
Proportional Relationships, Multiple Representations

What's My Container?

(2-3 days)
Students play a logical reasoning game. The object of the game is to solve riddles about the volume and surface area of polyhedra and solids with circular cross-sections.
STRANDS
Geometry, Algebra, Logic and Language
UNIFYING IDEAS
Multiple Representations, Patterns and Generalizations

The Design Team Takes Over

(3-4 days)
Students discuss the concept of trademark design and create their own designs using geometric iterations.
STRANDS
Geometry, Functions, Measurement
UNIFYING IDEAS
Multiple Representations, Proportional Relationships

Approaching the Bankers

(3-4 days)
Students investigate bank accounts and loans. They use tables, graphs, and iterations to analyze and make decisions about bank accounts and loans.
STRANDS
Functions, Algebra, Logic and Language, Statistics
UNIFYING IDEAS
Multiple Representations, Patterns and Generalizations

1. Developing a Product Pages 137-140*	**3-5 Days**	• To create survey instruments; to conduct a survey; to represent and analyze results of the survey using tables, graphs, and equations
2. The Container Pages 141-144*	**3-5 Days**	• To investigate polyhedra and solids with circular cross sections; to design a container for a hypothetical new product
3. What's My Container? Pages 145-148*	**2-3 Days**	• To play a game where students determine the shape of a "mystery package" by solving riddles involving polyhedra, solids with circular cross-sections, and their volumes and surface areas
4. The Design Team Takes Over Pages 149-153*	**3-4 Days**	• To use geometric iterations to create trademark designs
5. Approaching the Bankers Pages 154-158*	**3-4 Days**	• To investigate the concept of iterations within the context of investing and borrowing money; to express an iterative process in a variety of ways including diagrams, function notation, tables, and graphs

*Pages refer to *Gateways to Mathematical Investigations* Teacher's Edition.

PLANNING

LONG-TERM INVESTIGATIONS

• Family Letter, 1 per student • Calculator, 1 per group • Grid paper, several sheets per student • Coordinate grid paper, 1-2 sheets per student • Cut-out paper squares, 30-50 per student (optional) • Blank spreadsheet, 1 per group • Computer, spreadsheet software (optional) • Resource book of population statistics for local, city, state, and country, 1 for the class • *Gateways to Mathematical Investigations* Student Edition, pp. 50-51, 1 of each per student	**Gateways to Algebra and Geometry** Section 11.2, pp. 553-558: Equations with Variables on Both Sides Section 11.4, pp. 566-572: Equations with Several Variables *Additional References* Section 9.7, pp. 469-476: Graphs of Equations Teacher's Resource File: Student Activity Masters
• Grid paper, several sheets per student • Isometric dot paper, 2-3 sheets per student • Boxes and cans, different sizes and shapes, several per group • Geometric solids and volume blocks, 1 set per group • Beans, popcorn, or other filler, enough to fill the largest container per group • Measuring container, 1 per group; string, 1 piece per group • Scissors, tape, ruler, 1 of each per student • *Gateways to Mathematical Investigations* Student Edition, pp. 52-54, 1 of each per student	**Gateways to Algebra and Geometry** Section 11.5, pp. 573-580: Polyhedra Section 11.6, pp. 581-588: Solids with Circular Cross Sections *Additional References* Teacher's Resource File: Student Activity Masters
• Calculator, 1 per student • Grid paper and isometric dot paper, several sheets per student • Geometric solids and volume blocks, 1 set per group • Two containers: a cylinder and a cone; each with the same base and height (optional) • Two containers: a prism and a pyramid, each with the same base and height (optional) • Sand or water, enough to fill cylinder and prism (optional) • *Gateways to Mathematical Investigations* Student Edition, pp. 55-56, 1 of each per student	**Gateways to Algebra and Geometry** Section 11.1, pp. 546-552: Presenting Solutions to Problems Section 11.3, pp. 559-565: More Word Problems Section 11.4, pp. 566-572: Equations with Several Variables Section 11.5, pp. 573-580: Polyhedra Section 11.6, pp. 581-588: Solids with Circular Cross Sections *Additional References* Section 2.3, pp. 64-70: Geometric Formulas Teacher's Resource File: Student Activity Masters
• Students' containers from Investigation 2, 1 per group • Blank white paper, grid paper, and isometric dot paper, several sheets per student • Iterated equilateral triangle, 1 per student • Compass protractor, and ruler, 1 of each per student • Colored pencils, 1 set per group; Scissors and tape, 1 of each per student • Assorted magazines, 3-4 per group • Computer, LOGO software (optional) • *Gateways to Mathematical Investigations* Student Edition, pp. 57-58, 1 of each per student	**Gateways to Algebra and Geometry** Section 12.1, pp. 604-611: Iterations Section 12.2, pp. 612-619: More About Iteratons Section 12.3, pp. 620-624: Geometric Iterations Section 12.5, pp. 634-639: Introduction to Fractals *Additional References*
• Students' work from Investigations 1,2,and 4 • Calculator with iterating capabilities, 1 per pair • Grid paper, coordinate grid paper, several sheets per student • Ruler, 1 per student • Local newpaper, 1 per group • Blank spreadsheet, 1 per student (optional) • Computer, spreadsheet software, 1 per group (optional) • Cardboard box, 1 (optional) • Index cards, several (optional) • *Gateways to Mathematical Investigations* Student Edition, pp. 59-60, 1 of each per student	**Gateways to Algebra and Geometry** Section 12.1, pp. 604-611: Iterations Section 12.2, pp. 612-619: More About Iterations Section 12.4, pp. 625-633: Convergent Iterations *Additional References* Teacher's Resource File: Student Activity Masters

See pages 602A and 602B: Assessment for Unit F.

CHAPTER 11

11 Solving Problems with Algebra

Contents

11.1 Presenting Solutions to Problems
11.2 Equations with Variables on Both Sides
11.3 More Word Problems
11.4 Equations with Several Variables
11.5 Polyhedra
11.6 Solids with Circular Cross Sections
11.7 Reading Information from Graphs

Pacing Chart

Days per Section

	Basic	Average	Honors
11.1	*	2	2
11.2	*	2	2
11.3	*	3	3
11.4	*	3	3
11.5	*	2	2
11.6	*	2	2
11.7	*	2	2
Review	–	1	1
Assess	–	2	2
	–	19	19

* Optional at this level

Chapter 11 Overview

■ Applying algebraic principles to real-world situations is a major challenge to students in algebra. Chapter 11 begins by posing several problems that can be interpreted algebraically. Several non-algebraic solutions are examined to broaden students' views of how to solve problems; then, algebraic solutions are introduced to these and other problems.

■ Many of the problems in the chapter involve geometry. We explore algebraic solutions to these problems in detail. The rest

CHAPTER 11

WATER ENGINEERING A water engineer gets involved in the management and distribution of supplies of fresh water from sources such as wells, rivers, and lakes. Needs for fresh water vary depending on whether the use is for home, industry, or agriculture.

Some water engineers devote their efforts to the problem of converting sea water to fresh water. This is called desalination.

Water engineers also deal with the purity of fresh water, by learning how to test for, and how to remove pollutants.

INVESTIGATION

Water Conservation Obtain a copy of a water bill. Use it to estimate the quantity of water used by a family in a year. What is the cost for a year's supply of water? List some ideas for conserving water, and estimate how much water each of your ideas would conserve if implemented.

of the chapter is devoted to applying algebraic formulas to the geometry of three-dimensional solids and to the relationships among equations, applications, and coordinate graphs.

Career Connection

Water and sewage treatment are important to every community. Water engineers and technicians are employed by federal, state, and local governments and in private industry. Population growth, droughts, and the ever-growing concern about water pollution mean increased opportunities for water engineers, technicians, and construction workers on water-management projects.

Video Connection

For more information about the career discussed here, please refer to the Futures videocassette #4, program A: *Water Engineering*, featuring Jaime Escalante. For more information or to order a copy of the program, call PBS VIDEO at 800/344-3337.

Investigation Notes

The cost of water varies widely from region to region. In the western United States, for example, where water is more scarce, it is more expensive than in the southeast, where it is more plentiful. The average family of four uses about 300 gallons of water per day at home (about 100,000 gallons per year). Water-conservation ideas include installing faucet aerators and low-water-use toilets and taking shorter showers.

CHAPTER 11

Authors' Note

These eight thought-provoking problems can be solved without the use of algebra. Use them to emphasize to students that there is usually more than one way to solve a problem. Effective strategies for problems such as these include:

- Use a spreadsheet
- Guess and check
- Use physical models
- Draw a picture or graph
- Find a pattern
- Solve a simpler problem

Answers

for pages 544 and 545

PROBLEM 1
It took Heath 40 minutes to catch Cliff.

PROBLEM 2
Smith's melon weighed the most at 64 pounds.

PROBLEM 3
The full 24-oz bottle contained 23.04 oz of sugar syrup.

PROBLEM 4
Jason is 23 and Freddy is 11.

Mathematics in Words

One reason mathematics is interesting and exciting is that there are many ways to solve most math problems. Solve each of the following problems, using logic, a drawing, or any other method that makes sense to you. See if you can find more than one way to solve each problem. Be prepared to discuss with the class how you reached your solutions.

PROBLEM 1: A Distance Problem

Cliff left his house at 8 A.M., walking at an average rate of 3 miles per hour. Two hours later, his sister, Heath, left home to catch her brother, bicycling at 12 miles per hour. How long did it take Heath to catch Cliff?

PROBLEM 2: Overlapping Data

At the annual watermelon convention, three farmers, Smith, Jones, and Squash, had the three largest melons. When the melons were weighed in pairs, each pair weighed more than 100 pounds.

- Smith's and Jones's melons together weighed 110 pounds.
- Jones's and Squash's melons together weighed 104 pounds.
- Smith's and Squash's melons together weighed 122 pounds.

Whose melon was the heaviest and how much did it weigh?

PROBLEM 3: Percent of Concentration

Mikey looked at the bottle of pancake syrup that his mother put on the table. The list of ingredients showed that the pancake syrup was 3% pure maple syrup and 1% butter. Sugar syrup was the only other ingredient. If the full bottle contained 24 ounces, how much of the full bottle was sugar syrup?

PROBLEM 4: An Age Puzzle

Five years ago, Jason was three times as old as Freddy. When Freddy was four years old, Jason was four times Freddy's age. How old are Jason and Freddy now?

Chapter 11 Solving Problems with Algebra

CHAPTER 11

PROBLEM 5: Counting Coins in a Fountain

There is a fountain in the center of a city. People like to toss coins in the fountain for luck. At the end of each day, the city cleans the fountain and donates the money to charity. On a given day, one fourth of the people tossed in a quarter, 40 people tossed in a dime, and the rest tossed in a penny. If there was a total of 200 coins tossed in the fountain that day, how much money did the city donate to charity?

PROBLEM 6: Work Rates

a A cold-water faucet fills two thirds of a water tank in one hour. How long must the faucet run in order to fill the whole tank?
b A hot-water faucet takes two hours to fill the same tank. What portion of the tank does it fill in one hour?
c If both faucets are used together, how long will it take them to fill the tank?

PROBLEM 7: A Combination Problem

Phylo won a $100 gift certificate that can be used to purchase tapes and CDs. He goes to a store where each tape costs $9 and each CD costs $12. He wants to use as much of the gift certificate as possible. How many tapes and CDs should he buy?

PROBLEM 8: A Range Problem

A small radio station is three miles from a straight highway. The station has a broadcasting range of five miles. If a car is driving down the highway, how long is the section of highway on which the car can receive the radio signal?

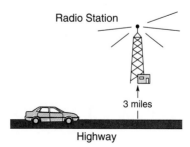

PROBLEM 5
The city donated $17.60 to charity.

PROBLEM 6
a The cold-water faucet must run for $1\frac{1}{2}$ hours to fill the tank.
b The hot-water faucet fills $\frac{1}{2}$ of the tank in 1 hour.
c Together, the faucets will fill the tank in $\frac{6}{7}$ hour, or about $51\frac{1}{2}$ minutes.

PROBLEM 7
Phylo can spend at most $99. He could purchase either 11 tapes and 0 CDs, 7 tapes and 3 CDs, or 3 tapes and 6 CDs.

PROBLEM 8
The car can receive the radio signal along an 8-mile section of the highway.

Mathematics in Words

SECTION 11.1

Objective
- To develop strategies for solving problems

Class Planning

Schedule 2 days

Resources

In *Chapter 11 Student Activity Masters* Book
Class Openers 1, 2
Extra Practice 3, 4
Cooperative Learning 5, 6
Enrichment 7
Learning Log 8
Technology 9, 10
Spanish-Language Summary 11

Class Opener

A train leaves a station and travels at an average speed of 40 mph. A second train leaves one hour later from the same station and travels along a parallel track at an average speed of 50 mph. Try to figure out how long it will take for the second train to catch up to the first.

Answer: 4 hours

Lesson Notes

- The main point of this section is that there are several ways to solve most problems. Students should not be taught a single method, but rather be exposed to many problem-solving strategies and be allowed time to practice using them. The text suggests that students should try each sample problem again, using a different method. This might make a good group activity, after which groups could share their methods with the class.

Investigation Connection
Section 11.1 supports Unit F, *Investigation* 3. (See pp. 542A-542D in this book.)

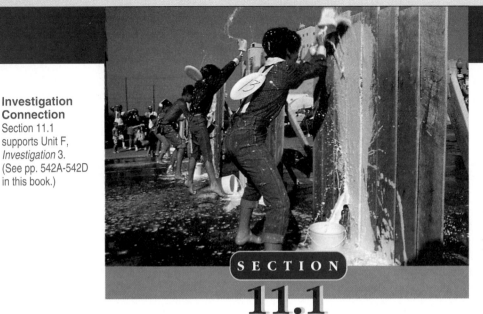

SECTION 11.1
Presenting Solutions to Problems

Each time you come across a problem, you need to decide on a strategy to solve the problem. Remember,

▶ No one strategy works for all problems.
Most problems can be solved in many ways.

As you solve problems, you will develop your own strategies. We will show you some techniques that you may not have thought of trying. Try to solve each problem before you read the method we use. Then read our solution and see how it differs from yours.

Sample Problems

Problem 1
A Distance Problem
Carlos left home at 10 A.M. on a daylong bike trip, traveling at an average speed of 9 miles per hour. One hour later, his mother realized that Carlos had forgotten his lunch. She hopped on her moped and followed him, traveling 15 miles per hour. Did she catch up with him in time for lunch?

Solution
Carlos travels 9 miles per hour. Traveling at 15 miles per hour, his mother closes the gap by 6 miles each hour. Since she left at 11 A.M., by noon she has gained 6 miles. Carlos is still 3 miles ahead.

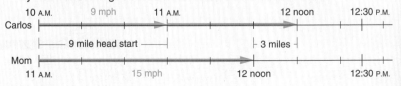

546

It will take another half hour for his mother to catch him at 12:30. You can decide if this was in time for lunch, since we don't know what time he usually eats lunch.

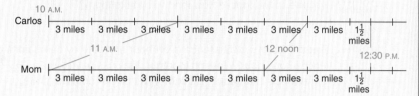

Problem 2
An Age Problem

When I was born, my father was eleven times as old as my sister, who had just turned three. How old was I when my father was
a Six times as old as my sister?
b Four times as old as my sister?
c Twice as old as my sister?

Solution

A spreadsheet provides a useful way of solving this problem. After the starting ages of 0, 3, and 33 are entered, each number in columns A, B, and C is one more than the preceding number. Each ratio in column D is found by dividing the number in column C by the number in column B.

a When my dad's age was six times my sister's, I was 3 years old.
b When my dad's age was four times my sister's, I was 7 years old.
c When my dad's age was twice my sister's, I was 27 years old.

	A	B	C	D
1	Me	Sister	Dad	Dad/Sister
2	0	3	33	11
3	1	4	34	8.5
4	2	5	35	7
5	3	6	36	6
6	4	7	37	5.28571429
7	5	8	38	4.75
8	6	9	39	4.33333333
9	7	10	40	4
10	8	11	41	3.72727273
11	9	12	42	3.5
12	10	13	43	3.30769231
13	11	14	44	3.14285714
14	12	15	45	3
15	13	16	46	2.875
16	14	17	47	2.76470588
17	15	18	48	2.66666667
18	16	19	49	2.57894737
19	17	20	50	2.5
20	18	21	51	2.42857143
21	19	22	52	2.36363636
22	20	23	53	2.30434783
23	21	24	54	2.25
24	22	25	55	2.2
25	23	26	56	2.15384615
26	24	27	57	2.11111111
27	25	28	58	2.07142857
28	26	29	59	2.03448276
29	27	30	60	2
30	28	31	61	1.96774194

Assignment Guide

Basic
Optional

Average or Heterogeneous
Day 1 8–11, 13, 15, 16, 22, 24, 26, 30
Day 2 12, 14, 17–20, 23, 25, 28

Honors
Day 1 8–11, 14–16, 20, 21, 24, 25, 28
Day 2 12, 13, 17–19, 22, 23, 26, 27, 29, 30

Strands

Communicating Mathematics	1–3, 7, Investigation
Geometry	10, 18, 30
Measurement	3–5, 10, 18, 30
Data Analysis/ Statistics	6
Technology	22
Probability	8, 12, 14, 17
Algebra	3, 8–12, 14, 15, 18–26, 28–30
Patterns & Functions	6, 9
Number Sense	9, 13, 14, 16, 17, 19, 21, 26, 27, 29
Interdisciplinary Applications	1–7, 9, 11, 15, 30, Investigation

Individualizing Instruction

Limited English proficiency
Students may have a tough time with the fine distinctions between meanings of words in problems 1, 2, and 7. You may need to provide some additional examples so that misinterpretations can be minimized.

SECTION 11.1

Cooperative Learning

Have students work together to solve the problems in the text. Heterogeneous ability groups may work well, as some students may have a hard time getting started with some problems. Emphasize the importance of persistence and of trying a variety of strategies. Be sure that students show how they solved the problems and that you commend students who come up with novel strategies.

Checkpoint

1. A point is chosen at random from the segment shown.

   $\xleftarrow{\bullet\!\!-\!\!\!+\!\!\!+\!\!\!+\!\!\!+\!\!\!+\!\!\!+\!\!\!+\!\!\!+\!\!\!+\!\!\!+\!\!\!+\!\!\!+\!\!\!\bullet\!\!\!\rightarrow}$
   $\quad-7\quad\;-2\quad\quad\;\,6$

   What is the probability that the point chosen is more than three units from -2?

2. Peter can mow a yard in 2 hours. Paul can mow the same yard in 4 hours. How long will it take if they work together?

3. Dinah works every 6 days. Don works every 4 days. Elyse works every 8 days. If all three worked April 1, when will they all work on the same day again?

Answers:
1. $\frac{7}{13}$ 2. 1 h 20 min
3. April 25

Problem 3
Working Together

Tom knows that Ben can whitewash a fence in six hours. Johney can do the job in eight hours. Tom talked them both into working together to whitewash the fence. How long will it take them?

Solution Ben can paint $\frac{1}{6}$ of the fence in an hour and Johney can paint $\frac{1}{8}$ of the fence in one hour. Working together, they can paint $\frac{1}{6} + \frac{1}{8} = \frac{4+3}{24} = \frac{7}{24}$ of the fence in an hour. Think of $\frac{7}{24}$ fence per hour as a rate.

$$(\text{Rate})(\text{time}) = \text{amount of work done}$$

$$\text{Time} = \frac{\text{amount of work done}}{\text{rate}}$$

$$= \frac{1\;\text{fence}}{\frac{7}{24}\;\frac{\text{fence}}{\text{hour}}}$$

$$= 1\;\text{fence} \cdot \frac{24}{7}\;\frac{\text{hour}}{\text{fence}}$$

$$= \frac{24}{7}\;\text{hours}$$

Together they can paint the fence in a little less than $3\frac{1}{2}$ hours.

Problem 4
Probability

A 50-foot wire is strung between the tops of two 35-foot telephone poles. If the wire breaks at a random point between the poles, what is the probability that one end of the wire will hit the ground?

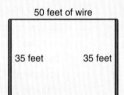

Solution Diagrams and number lines help model this problem.

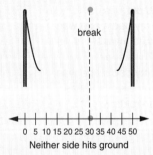

Neither side hits ground

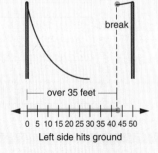

Left side hits ground

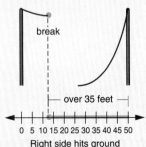

Right side hits ground

SECTION 11.1

> If the wire breaks within 15 feet of either pole, then one end of the wire will hit the ground. The "winning" length is 15 + 15 = 30, out of the total of 50 feet.
>
>
>
> The probability that the wire will hit the ground is $\frac{30}{50}$, or $\frac{3}{5}$.

As you can see, there are different types of word problems and different ways to solve them. Were the methods you used different from the methods we showed? Try each of the sample problems again, using a method different from the one we used or the one you used before.

Think and Discuss

P **1** Ms. Word gave her class the following problem:

> Moe can mow the lawn in 3 hours. Len can mow the lawn in 2 hours. How long will it take if they work together?

Adam said, "*Together* means add. So I added 2 and 3 and got 5 hours." Was Adam correct? Explain your answer.

P **2** The next day, Ms. Word gave her class another problem.

> Mark walks five blocks north, turns left, and goes three blocks. How far has he walked?

Sal said, "Two blocks." Darren said, "That doesn't make any sense." Sal answered, "The problem uses the word *left*, so I subtracted." What is wrong with Sal's thinking?

P **3** A freight train, traveling 50 miles per hour, left Chicago at 9 A.M. and headed south. An express train, traveling 65 miles per hour, left Chicago at noon and headed south on the same track. At what time will the express train catch up to the freight train? Explain how to solve this problem in at least two different ways.

P **4** Jack went uphill at 6 miles per hour to meet Jill. After he reached the top of the hill, he fell downhill at 12 miles per hour. What was his average speed for the entire trip up and down?   8 mph

P **5** Jack jogged up a hill at 10 miles per hour to meet Jill. How fast must Jack go downhill so that his average speed for the trip will be 20 miles per hour?   There is no downhill speed that will make the average 20 mph.

Problem-Set Notes

for page 549

- **1, 2,** and **7** point out the trouble with using "key" words to solve word problems instead of paying attention to the meanings of words within the contexts of the problems. Discuss these problems with your students.

- **4** and **5** show that "average speed" is not the average of the average speeds, but the total distance divided by the total time. In both problems, the distance traveled is arbitrary. For example, if the distance up the hill were 12 mi, the solution to **4** could be given as

$$\frac{12\text{ mi} + 12\text{ mi}}{2\text{ hr} + 1\text{ hr}} = \frac{24\text{ mi}}{3\text{ hr}} = 8\text{ mi/hr.}$$

If the distance were 6 mi, the solution would remain

$$\frac{6\text{ mi} + 6\text{ mi}}{1\text{ hr} + 0.5\text{ hr}} = \frac{12\text{ mi}}{1.5\text{ hr}} = 8\text{ mi/hr.}$$

Communicating Mathematics

- Have students complete the following sentence and expand their answers either in writing or orally.
"The kind of word problem I find easiest is . . . because . . . ."

- Problems 1–3, 7, and Investigation encourage communicating mathematics.

Additional Answers

for page 549

1 No; it will take Moe and Len 1 hr 12 min to mow the lawn if they are working together.

2 *Left* doesn't always mean "subtract." Sal took a rule that works in special cases and tried to use it in all cases.

3 10 P.M.; possible answers: Use a number line; set up an equation.

SECTION 11.1

Problem-Set Notes

for pages 550 and 551

- 13 can be solved efficiently by using a spreadsheet.

Additional Practice

1. Joanne bicycled uphill at a rate of 4 miles per hour. She then turned around and bicycled back downhill at a rate of 12 miles per hour. What was her average rate for the entire trip up and down the hill?

2. Martha left home at 9:00 A.M., traveling at a rate of 45 mph. Her brother Matthew left home at noon, following Martha's route. If Matthew travels at 60 mph, when will he overtake Martha?

3. When I was two years old, my sister was triple my age, and my father was six times my sister's age. How old was I when my father was seven times my sister's age?

Answers:
1. 6 mph 2. 9:00 P.M.
3. 1 year old

Multicultural Note

Hypatia, an Egyptian woman born in A.D. 370, was a professor at the university in Alexandria, Egypt, at a time when this center of learning attracted many scholars from Africa, Asia, and Europe. In the field of algebra, Hypatia wrote about the work of an earlier Egyptian mathematician named Diophantus, who had studied problems that led to equations with more than one solution.

P **6** In the last part of 1992, Nancy is one-half of Sally's age. Copy and complete the chart to determine when this will happen again.

		1992	1993	1994	1995	1996
Jan. 1 – May 3	Sally	47				
	Nancy	23				
May 4 – June 29	Sally	48				
	Nancy	23				
June 30 – Dec. 31	Sally	48				
	Nancy	24				

Make a similiar chart and decide when you will be half of your mother's age. Will this occur more than once?

P **7** Athos said, "I have five racks of three-piece suits. How many suits do I have?" Porthos said, "Fifteen." Aramis said, "How did you get that?" Porthos replied, "*Of* means times, and five times three equals fifteen." Aramis objected, "That is nonsense." Why was Porthos wrong?

Problems and Applications

P, R **8 a** What is the probability that a point chosen on the segment shown is within five units of 8? $\frac{10}{13}$

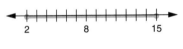

  b What is the probability that the chosen point is within five units of one of the endpoints of the segment? $\frac{10}{13}$

P **9** When I was two years old, my brother was double my age and my father was seven times my brother's age. How old was I when my father was three times my brother's age? 10

P, R **10** The area of the triangle is 45. Solve for x. $x = 13$

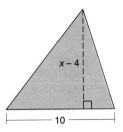

P, R **11** A quart of 4% butterfat milk is mixed with 3 quarts of 1% butterfat milk.
  a What fraction of a quart of the 4% milk is butterfat? $\frac{4}{100}$
  b What fraction of a quart of the 1% milk is butterfat? $\frac{1}{100}$
  c What fraction of the three quarts of 1% milk is butterfat? $\frac{1}{100}$
  d What percent of the mixture is butterfat? 1.75%

550 Chapter 11 Solving Problems with Algebra

SECTION 11.1

P, R **12** A point is chosen at random on the segment shown.
  a What is the probability that the point chosen is more than four units from −3? $\frac{3}{7}$
  b What is the probability that the coordinate of the chosen point makes the inequality $2x - 5 \leq 1$ true? $\frac{12}{15}$, or $\frac{4}{5}$

P **13** Charlie's Chihuahua barks 42 times per minute. Lucia's Lhasa apso barks 28 times per minute. Silvie's springer spaniel barks 10 times per minute, and Lunyee's Labrador retriever barks 4 times per minute. If all the dogs start barking at the same time
  a When will they all bark simultaneously? 30 sec later
  b When will Lucia's and Lunyee's dogs bark at the same time? Every 15 sec

P, R **14** Two numbers, x and y, are chosen at random. If $0 \leq x \leq 10$ and $0 \leq y \leq 7$, what is the probability that both numbers are at most 5? $\frac{25}{70}$, or $\frac{5}{14}$

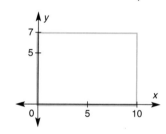

P **15** Michael can paint a wall in three hours. Angelo can paint the same wall in two hours. How long will it take if they work together? 1 hr 12 min

P **16** The planets Tragedy, Comedy, and Variety orbit the star Serious. Every so often the planets line up on one side of Serious in a formation called a syzygy. Tragedy revolves around Serious once every 12 days, Comedy once every 18 days, and Variety once every 30 days. How many days is it from one Serious syzygy to the next? 180 days

P, R **17** What is the probability that a number selected from the set {2, 3, 5, 7, 11, 13, 17, 19} is a prime number? 1

P, R **18** Find the area of the triangle. 30

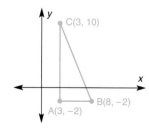

P, R **19** How many integers are there between points A and B? 14

Reteaching

Students could use manipulatives to estimate the solution to sample problem 3. Fraction wedges could be used to show how Ben and Johney completed the job.

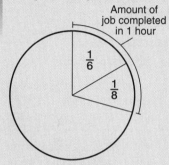

You could easily segue from the model to the solution given in the text by having students note that $\frac{1}{6} + \frac{1}{8} = \frac{7}{24}$.

Additional Answers

for pages 550 and 551

6

		'92	'93	'94	'95	'96
Jan. 1–	S	47	48	49	50	51
May 3	N	23	24	25	26	27
May 4–	S	48	49	50	51	52
Jun. 29	N	23	24	25	26	27
Jun. 30–	S	48	49	50	51	52
Dec. 31	N	24	25	26	27	28

Nancy's age will be half of Sally's from May 4 to June 29, 1994. Answers will vary for the second part of the problem; check students' charts.

7 Porthos multiplied the wrong numbers. He needs to multiply 5 racks times the number of suits per rack. The fact that a number appears in a problem does not necessarily mean it will be used.

SECTION 11.1

Problem-Set Notes

for page 552

- **22** and **25** preview the next section.
- **28** reviews solving equations with variables on one side.
- **Investigation** relates to problem 16.

Stumbling Block

Students may overgeneralize the methods shown for solving the sample problems to problems that require a different approach. Emphasize the uniqueness of each problem and praise students for pointing out subtle but significant differences between seemingly similar problems.

Additional Answers

for page 552

21 a (0, 10), (1, 7), (2, 4), (3, 1)
b (1, 7), (2, 4), (3, 1)

Investigation Syzygy is a rarely used term referring to an alignment of three heavenly bodies, such as the earth, moon, and sun. You might hear it used in reference to a solar or lunar eclipse.

◀ LOOKING BACK **Spiral Learning** LOOKING AHEAD ▶

R **20** Evaluate each expression.
 a $\frac{-3}{5} + \frac{1}{2} \quad -\frac{1}{10}$ **b** $3.6 - 9.4 \quad -5.8$ **c** $-0.25 + \frac{1}{3} \quad \frac{1}{12}$ **d** $\frac{-3}{2} + \frac{1}{4} \quad -\frac{5}{4}$

R **21** Given the equation $3x + y = 10$,
 a Find all the pairs of whole numbers (x, y) that solve the equation.
 b Find all the pairs of natural numbers (x, y) that solve the equation.

R **22** **Spreadsheets** Use a spreadsheet to find the value of x for which $5x - 40 = 3x - 15$. $12\frac{1}{2}$

23 Solve each equation for x.
 a $-6x = 30 \quad x = -5$ **b** $-30 = 6x \quad x = -5$

R, I **24 a** For what values of x and y is $xy = 0$? $x = 0$ or $y = 0$
 b Solve the equation $(2x + 18)(x - 12) = 0$ for x. $x = -9$ or $x = 12$

R **25** Suppose that $9x + 6 = 5x - 30$. Find the equation that results from performing each of the following operations.
 a Adding 30 to each side of the equation $9x + 36 = 5x$
 b Subtracting 6 from each side $9x = 5x - 36$
 c Subtracting $5x$ from each side $4x + 6 = -30$
 d Subtracting $9x$ from each side $6 = -4x - 30$

R, I **26** Find all integer values of n for which $\frac{11}{2n+3}$ is an integer. $-7, -2, -1, 4$

R **27** Find the LCM of 6, 8, and 15. 120

R **28** Solve each equation for x.
 a $2(x + 17) = 24 \quad x = -5$ **b** $1571(x - 9.3) = 1571 \quad x = 10.3$
 c $2x + 4\frac{1}{2} = 1\frac{1}{4} \quad x = -\frac{13}{8}$

R **29** Find the value(s) of n for which the five-digit number $52n36$ is divisible by 24. $n = 5$

R **30** Philip José, a farmer, has 430 feet of fencing. He wants to use it to fence in three sides of a rectangular field that borders a straight stretch of river.
 a What are the dimensions of the largest field that can be fenced?
 b What is its area? **a** 107.5 ft by 215 ft **b** 23,112.5 ft²

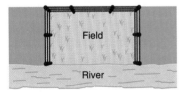

Investigation

Language Look up the word *syzygy* and explain why it makes sense to use the word in problem 16. When would you be likely to read or hear about syzygy in the news?

SECTION 11.2

Investigation Connection
Section 11.2 supports Unit F, *Investigation* 1. (See pp. 542A-542D in this book.)

SECTION 11.2
Equations with Variables on Both Sides

Example 1
Aacme Rent-an-Aardvark charges $30 plus $5 per hour to rent an aardvark. Lletme Rent-a-Llama charges $44 plus $3 per hour to rent a llama. Zzezze was planning a birthday party and found that it would cost the same amount to rent either animal for the time he needed them. For how long did Zzezze need an animal?

Solution
Let h represent the number of hours that Zzezze rents an animal. The rental charge from Aacme will be $30 + $5 · h. The rental charge from Lletme will be $44 + $3 · h. Since the total cost is the same, we set the two expressions equal to each other and solve the resulting equation, $30 + 5h = 44 + 3h$.

As we have done before when solving equations, we add the same quantity to both sides of the equation. In this case, we add $-3h$ to both sides.

Horizontal Method	Vertical Method
$30 + 5h = 44 + 3h$	$30 + 5h = 44 + 3h$
$30 + 5h - 3h = 44 + 3h - 3h$	$\quad\quad -3h \quad\quad -3h$
$30 + 2h = 44 + 0$	$30 + 2h = 44 + 0$
$2h + 30 - 30 = 44 - 30$	$\quad -30 \quad\quad -30$
$2h = 14$	$2h = 14$
$h = 7$ hours	$h = 7$ hours

If Zzezze rented an aardvark or a llama for 7 hours, the rental would cost

$\quad 30 + 5(7) = \$65 \quad\quad$ or $\quad\quad 44 + 3(7) = \$65$

Objective
- To solve equations with variables on both sides

Class Planning

Schedule 2 days
Resources
In *Chapter 11 Student Activity Masters* Book
Class Openers *13, 14*
Extra Practice *15, 16*
Cooperative Learning *17, 18*
Enrichment *19*
Learning Log *20*
Using Manipulatives *21, 22*
Spanish-Language Summary *23*

Class Opener
For each pair of expressions, find the value of x that gives the expressions the same value:
1. $x + 1$ and $2x - 3$
2. $x + 1$ and $4x - 5$
3. $2x - 3$ and $4x - 5$

Answers:
1. 4 2. 2 3. 1

Lesson Notes
- Examples 1 and 2 show two methods for solving equations. Discuss both methods with students. Neither is more difficult than the other, so we recommend that you allow students to select the method that they are more comfortable with.

- Sample problem 1 shows a correct use of cross multiplication. You may find that some students want to set the numerators equal and the denominators equal. Use a number example to show that this does not work.

SECTION 11.2

Lesson Notes continued
- The use of Algebra Tiles or a balance scale might be helpful in this section.

Additional Examples

Example 1
The Blue Cab Company charges $1.60 plus $0.15 per tenth of a mile. The Green Cab Company charges $2.00 plus $0.13 per tenth of a mile. Tex E. Ryder took a Blue cab one way and a Green cab on the return trip. Each ride cost the same amount. How far did he travel one way?

Answer: 2 miles

Example 2
Solve the equation $3(5 - x) = 4x + 1$.

Answer: $x = 2$

Communicating Mathematics
- Have students write a few sentences explaining why it is important to check the solution of an equation—especially one that contains a variable in the denominator.
- Problem 23 and Investigation encourage communicating mathematics.

Stumbling Block
In sample problem 1, for $\frac{x-6}{x} = \frac{2}{3}$, some students may solve $x - 6 = 2$ and $x = 3$. Point out that x may have only one value for this particular equation. Point out that ratios do not have to have equal numerators and denominators to be equal. Give students numerical examples of this principle.

The strategy we use for solving equations with variable terms on both sides of the equal sign is similar to strategies we have used earlier—add or subtract as necessary in order to have all variable terms on one side of the equation and all numerical terms on the other side of the equation.

Example 2
Solve the equation $2(19 - x) = 3 + 5x$.

Solution

$$2(19 - x) = 3 + 5x$$

Apply the distributive property $\quad 38 - 2x = 3 + 5x$

Now there are two methods we can use to solve the equation.

Method 1

$$38 - 2x = 3 + 5x$$

Add $2x$ to each side $\quad 38 - 2x + 2x = 3 + 5x + 2x$

$$38 = 3 + 7x$$

Subtract 3 from both sides $\quad 38 - 3 = 3 + 7x - 3$

$$35 = 7x$$

Divide both sides by 7 $\quad \frac{35}{7} = \frac{7x}{7}$

$$5 = x$$

Method 2

$$38 - 2x = 3 + 5x$$

Subtract $5x$ from both sides $\quad 38 - 2x - 5x = 3 + 5x - 5x$

$$38 - 7x = 3$$

Subtract 38 from both sides $\quad 38 - 7x - 38 = 3 - 38$

$$-7x = -35$$

Divide both sides by -7 $\quad \frac{-7x}{-7} = \frac{-35}{-7}$

$$x = 5$$

Either way, the solution is the same: $x = 5$. The variable can be eliminated from either side of the equation. Some students prefer to eliminate x terms from the right side so that the solution turns out in the form "$x =$." Others prefer to eliminate the negative terms. The choice is yours.

Sample Problems

Problem 1 Find the value of x if $\frac{x-6}{x} = \frac{2}{3}$.

Solution The equation says that the ratio $\frac{x-6}{x}$ is equal to the ratio $\frac{2}{3}$. Thus, the equation is a proportion, and we can cross multiply.

SECTION 11.2

$$\frac{x-6}{x} = \frac{2}{3}$$

$$3(x-6) = 2x$$
$$3x - 18 = 2x$$
$$\underline{+ 18 + 18}$$
$$3x = 2x + 18$$
$$\underline{-2x -2x}$$
$$x = 18$$

Check the solution: $\frac{18-6}{18} = \frac{12}{18} = \frac{2}{3}$.

Problem 2 The square and the rectangle shown have equal perimeters. Find the dimensions of both figures.

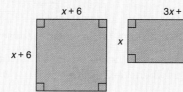

Solution

The perimeter of the square is
$4(x + 6)$
$= 4x + 24$

The perimeter of the rectangle is
$2x + 2(3x + 5)$
$= 2x + 6x + 10$
$= 8x + 10$

The two perimeters are equal, so we can write the equation $4x + 24 = 8x + 10$.

$$4x + 24 = 8x + 10$$
$$\underline{-4x -4x}$$
$$24 = 4x + 10$$
$$\underline{-10 -10}$$
$$14 = 4x$$
$$\frac{14}{4} = x$$
$$3\tfrac{1}{2} = x$$

The side of the square is
$x + 6 = 3\tfrac{1}{2} + 6$
$= 9\tfrac{1}{2}$

The length of the rectangle is
$3x + 5 = 3(3\tfrac{1}{2}) + 5$
$= 10\tfrac{1}{2} + 5$
$= 15\tfrac{1}{2}$

The rectangle is $3\tfrac{1}{2}$ by $15\tfrac{1}{2}$. Check to be sure that the perimeters are equal.

Assignment Guide

Basic
Optional

Average or Heterogeneous
Day 1   5, 7–10, 13, 17, 20 (a, c), 24, 26
Day 2   6, 11, 12, 14–16, 18, 20 (b, d), 28, 29

Honors
Day 1   5, 7, 9, 10, 12, 13, 17, 18, 20, 21, 23, 24 (a, c, e), 25, 28
Day 2   6, 8, 11, 14–16, 19, 22, 24 (b, d, f), 26, 27, 29

Strands

Communicating Mathematics	23, Investigation
Geometry	10–12, 15, 17, 27, 29
Measurement	10–17, 27, 29
Technology	11, 16
Algebra	1–25, 27, 28
Number Sense	4, 26, 28
Interdisciplinary Application	13, 14, 16, 18, 19, 22, Investigation

Individualizing Instruction

The kinesthetic learner Students may benefit from using manipulatives to model the equations in this section. For example 1, you might label a certain type of block with an h to represent the variable and use counting blocks to represent the 30 and the 44. Students can then see how taking away three h blocks from each side affects the equation.

SECTION 11.2

Problem-Set Notes

for pages 556 and 557

- 4 should be explored through numerical examples and counterexamples.
- 10–12, 15, 17, 27, and 29 review some geometric ideas.
- 18 and 19 could also be done by using a spreadsheet.

Cooperative Learning

- Refer students to problem 13. Have them graph the equations $y = 50x + 75$ and $y = 75x + 50$ on the same coordinate grid, where x = the number of hours worked and y = the total charge for services. Instruct students to use $\frac{1}{4}$-hour intervals along the x-axis and $25 intervals along the y-axis.
  1. What do the points where the lines intersect the y-axis represent?
  2. What does the point where the two lines intersect represent?
  3. When someone charges a lot for services, we say that their fees are "steep." How does this relate to the graph?

Answers:
1. The service fees 2. The time at which the total charges for Bob and Rob will be the same. 3. Bob charges more per hour than Rob, and the line representing his charges is steeper than the one representing Rob's.

- Problems 16 and 18 are appropriate for cooperative learning.

Think and Discuss

P 1 Solve $3(x + 2) = 12(x - 3)$ in two different ways.

P 2 Copy the following steps, filling in the missing coefficients and constants, to solve $28 + 4k = 7k - 14$ for k.

Step 1	$28 + 4k + \_\_\_\_k = 7k - 14 + \_\_\_\_k$	$-7, -7$
Step 2	$28 + \_\_\_\_k = -14$	-3
Step 3	$28 + \_\_\_\_k - \_\_\_\_ = -14 - \_\_\_\_$	$-3, 28, 28$
Step 4	$\_\_\_\_k = \_\_\_\_$	$-3, -42$
Step 5	$k = \_\_\_\_$	14

P 3 Solve the equation $9x + 11 = 6x - 4$ in several ways.

P 4 Indicate whether the statement is *true* or *false*.
  a If $\frac{a}{b} = \frac{2}{3}$, then $a = 2$ and $b = 3$. **b** If $\frac{a}{b} = \frac{2}{3}$, then $a = \frac{2}{3}b$.

Problems and Applications

P, R 5 Solve each equation for x.
  a $\frac{2x-4}{4} = \frac{x-2}{2}$ **b** $\frac{3}{x+4} = \frac{2}{2x+8}$

P, R 6 Solve each equation for x.
  a $\frac{x-6}{2} = \frac{x}{3}$ 18 **b** $\frac{2}{x-6} = \frac{3}{x}$ 18 **c** $\frac{x}{x-6} = \frac{3}{2}$ 18

P 7 If $2x + y = 8(x + y) + 13$ and $y = 5$, what is the value of x? -8

P 8 Copy the following steps, filling in the missing coefficients and constants, to solve $28 + 4k = 7k - 14$ for k.

Step 1	$28 + 4k + \_\_\_\_k = 7k - 14 + \_\_\_\_k$	$-4, -4$
Step 2	$\_\_\_\_ = 3k - 14$	28
Step 3	$\_\_\_\_ + \_\_\_\_ = 3k - 14 + \_\_\_\_$	$28, 14, 14$
Step 4	$\_\_\_\_ = 3k$	42
Step 5	$\_\_\_\_ = k$	14

P, R 9 Solve the proportion $\frac{2x-3}{x+5} = \frac{3}{4}$ for x. 5.4

P, R 10 The perimeter of a square with a side $2x$ is equal to the perimeter of an equilateral triangle with a side length of $x + 10$. Solve for x. 6

P, R 11 The perimeter of the square equals the circumference of the circle. Find the diameter of the circle. ≈ 4.66

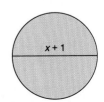

556 Chapter 11 Solving Problems with Algebra

SECTION 11.2

P, R **12** The circumference of circle A is three times the circumference of circle B. Find the radii of the circles.   Circle A: 15; circle B: 5

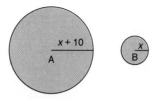

P **13** **Business**   For routing roots from a sewer pipeline, Bob the plumber charges a $50 service fee per visit plus $75 per hour. Rob the router charges a $75 service fee plus $50 per hour for the same job. Is it cheaper to hire Rob? How long must a job take in order for it to be cheaper to hire Rob?

P **14** A stream is flowing at 4 miles per hour. A small motorboat can go $5y$ miles downstream in 3 hours. It takes 5.9 hours for the boat to return upstream to its starting place. How fast does the boat travel in still water?

P, R **15** $\triangle ABC$ is similar to $\triangle DEF$. Find the length of $\overline{AB}$.   AB = 10

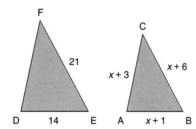

P **16** **Consumer Math**   At Scratch and Dents Rent-a-Car, it costs $34.95 a day plus $0.23 per mile to rent a car. At Rent-a-Lemon, the charge is $25.00 a day plus $0.31 per mile. If you need to rent a car for three days, how many miles must you drive for a car from both agencies to cost the same amount?   ≈ 373 mi

P, R **17** The area of the triangle shown is $8x - 20$. Find the value of x.   70

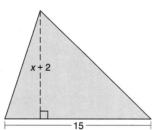

P, R **18** Candy Barr wanted to buy some big chocolate kisses that cost $0.50 each. She took them to the checkout counter and had just enough money for the kisses but forgot about the 8% tax. She put two kisses back and had just enough money. How many kisses did she buy?   25

P, R **19** **Business**   Lo was hired as a salesperson. She was given her choice of two salary plans. She could earn either $300 per week plus 10% commission or $200 per week plus 20% commission. How much merchandise does Lo have to sell in order for the $200 plan to be in her best interest? $1000

P **20** Solve each equation.
   a $3x - 9 = 5x + 11$ −10
   b $-3x - 9 = 5x + 11$ −2.5
   c $-3x - 9 = 5x - 11$ $\frac{1}{4}$
   d $-3x - 9 = -5x - 11$ −1

Additional Practice

Solve each equation for x.
1. $\frac{2x+3}{5} = \frac{3x-1}{4}$
2. $\frac{3}{2x+1} = \frac{4}{3x-1}$
3. $3(2x - 3) = 6(4x - 5)$
4. $\frac{2x+8}{3x+1} = \frac{4}{3}$
5. $2x + 3y = 6(3x - 7)$ and $y = 4$
6. $7x - 9 = 4x + 21$

7. One health club charges a $50 annual membership fee and $2 per workout session. Another health club charges a $30 annual membership fee and $3 per workout session. Raymond estimates he will work out 40 times per year. Which club should he join?

Answers:
1. $x = \frac{17}{7}$ 2. $x = 7$
3. $x = \frac{7}{6}$ 4. $x = \frac{10}{3}$
5. $x = \frac{27}{8}$ 6. $x = 10$
7. The first health club

Additional Answers

for pages 556 and 557

1 Two possible methods are:

(1) $\frac{3(x+2)}{3} = \frac{12(x-3)}{3}$
   $x + 2 = 4(x - 3)$
   $x + 2 = 4x - 12$
   $2 = 3x - 12$
   $14 = 3x$
   $x = \frac{14}{3}$

(2) $3x + 6 = 12x - 36$
   $6 = 9x - 36$
   $42 = 9x$
   $\frac{42}{9} = x$
   $x = \frac{14}{3}$

3 Possible answer: $x = -5$

4 a False **b** True

5 a x = any number
 b No solution

13 It's cheaper to hire Rob for any job over 1 hr.

14 About 12.28 mph

SECTION 11.2

Problem-Set Notes

for page 558

- **22** could also be done by using a spreadsheet.
- **24** reinforces the difference between expressions and equations.

Reteaching

Have students use the following line of reasoning for example 1: One base rate is $30, the other is $44. This is a difference of $14. The difference in rates per hour is $5/hr – $3/hr = $2/hr. Therefore, it takes $14 ÷ $2/hr = 7 hours to make up the difference in the base rates.

Checkpoint

Solve each equation for x.
1. $\frac{3x-6}{2} = \frac{2x+5}{2}$
2. $\frac{7}{2x+1} = \frac{3}{x-5}$
3. $6(2x-3) = 12(3x+1)$
4. $5x + 1 = 4x - 7$
5. $3x + y = 5(6x - 2)$ and $y = 3$
6. $\frac{4x-7}{2x+3} = \frac{1}{2}$

Answers:
1. $x = 11$ 2. $x = 38$
3. $x = -\frac{5}{4}$ 4. $x = -8$
5. $x = \frac{13}{27}$ 6. $x = \frac{17}{6}$

Additional Answers

for page 558

25 a $x =$ any real number
26 $-3.5 \times 10^{-3}, -3.5 \times 10^{-4}, 3.5 \times 10^{-4}, 3.5 \times 10^{-3}$

Answer for Investigation is on page A9.

558

P, R **21** Find the value of n if the three-digit number $nn2$ is 197 more than the three-digit number $2n5$. $n = 4$

P, R **22 Business** Willie was hired as a door-to-door salesperson. He had a choice of two salary plans. He could take a straight salary of $375 per week or he could take a salary of $200 per week plus 10% commission on his sales. How much merchandise would he have to sell in a week in order for the commission plan to be in his best interest?   More than $1750 worth

R **23** Simplify each fraction. Then explain your results.
  a $\frac{2+7}{3 \cdot 3}$ 1   **b** $\frac{8+5}{15-2}$ 1   **c** $\frac{48}{50-2}$ 1

◀ LOOKING BACK   **Spiral Learning**   LOOKING AHEAD ▶

R **24** Solve each equation and simplify each expression.
  a $3x - 5 = 9x + 4$ $x = -1.5$ **b** $(3x - 5) + (9x + 4)$ $12x - 1$
  c $(3x - 5) - (9x + 4)$ $-6x - 9$ **d** $3(x - 5) = 9(x + 4)$ $x = -8.5$
  e $3(x - 5) = 9x + 4$ $x = -\frac{19}{6}$ **f** $3(x - 5) - 9(x + 4)$ $-6x - 51$

R **25** Solve each equation.
  a $3(x - 4) + 2x = 5x - 12$ **b** $2(3x + 1) = 6x - 4$ No real solution

R **26** List the following values in order, from smallest to largest.
$$3.5 \times 10^{-4}, 3.5 \times 10^{-3}, -3.5 \times 10^{-3}, -3.5 \times 10^{-4}$$

P, I **27** The area of the trapezoid is 80. Find the value of h.   5

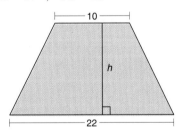

R **28** Solve each equation for x.
  a $x + \frac{2}{3} = \frac{4}{5}$ $x = \frac{2}{15}$ **b** $x + \frac{4}{5} = \frac{2}{3}$ $x = -\frac{2}{15}$
  c $\frac{4}{5}x = \frac{2}{3}$ $x = \frac{5}{6}$ **d** $\frac{2}{3}x = \frac{4}{5}$ $x = \frac{6}{5}$

R, I **29** The smallest angle of a triangle measures 36 degrees. Find the average of the other two angles.   72°

Investigation

Balance Look up the meaning of *balance beam*, *balance scale*, and *fulcrum balance*. Explain the similarities and the differences. Also, explain how each device is used.

558   Chapter 11   Solving Problems with Algebra

SECTION 11.3

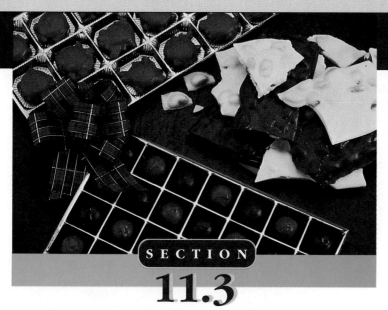

Investigation Connection
Section 11.3 supports Unit F, *Investigation* 3. (See pp. 542A–542D in this book.)

SECTION 11.3
More Word Problems

Earlier in this chapter, we worked with several types of word problems. Now we will look at some additional types of word problems.

Example 1
Ye Olde Chocolate Shoppe makes chocolate bars that measure 5 centimeters by 12 centimeters by 0.5 centimeter. Wilbur Wanna, a chocolate maker, pours melted chocolate into molds at a rate of 10 milliliters per second. How long does it take Wilbur to fill 100 chocolate-bar molds?

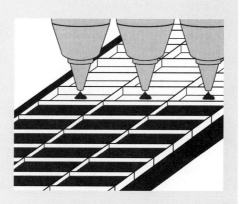

Solution
A milliliter is approximately equal to one cubic centimeter. The volume of each mold is 5(12)(0.5), or about 30 cubic centimeters, or 30 ml. Since Wilbur pours chocolate at a rate of 10 ml per second, it takes him 3 seconds to fill each mold. Wilbur fills 100 molds in 300 seconds, or 5 minutes.

Objective
- To develop strategies for solving problems

Class Planning

Schedule 3 days
Resources
In *Chapter 11 Student Activity Masters* Book
Class Openers 25, 26
Extra Practice 27, 28
Cooperative Learning 29, 30
Enrichment 31
Learning Log 32
Technology 33, 34
Spanish-Language Summary 35

In *Tests and Quizzes* Book
Quiz on 11.1–11.3 (Average) 273
Quiz on 11.1–11.3 (Honors) 274

Class Opener
A regular octagon has the same perimeter as a regular hexagon, though its sides are 3 units shorter. Find the length of a side of each figure.

Answer: Octagon: 9 units, hexagon: 12 units

Lesson Notes
- Each of the four problems given in this section have only one method of solution shown. Challenge your students to come up with alternative methods. Those with an [ANS] key on their calculators may be able to do the spreadsheet solution to the sample problem with a calculator.

Section 11.3 More Word Problems

SECTION 11.3

Additional Examples

Example 1
Chuck Layt makes bars of fudge from molds that are 3 cm by 10 cm by 0.6 cm. Chuck can fill molds at a rate of 180 ml per minute. How long does it take to fill 70 molds?

Answer: 7 minutes

Example 2
Mr. Z. Rocks said that the $950 price of a copier included the 7% sales tax. How much was the copier without tax?

Answer: $887.85

Example 3
The areas of the two rectangles are equal. Find the perimeter of each.

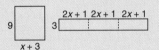

Answers: 28 and 36

Problem-Set Notes

for page 561

- 1 and 2 are good discussion problems. The answers might surprise students.
- 3 reminds students that lengths of sides of figures must be positive.

Additional Answers

for page 561

1 Yes, if the number is negative.

Example 2
Minny Olta bought a camera. Mr. Cannon, the salesperson, agreed to sell her the camera for $50, including the 6% sales tax. He had trouble figuring out what amount to enter into the computerized cash register so that it would add the tax to the price of the camera and produce a total of exactly $50. What should the price of the camera have been before tax?

Solution
The total cost, C, of the camera should equal the selling price, P, plus 6% of P in sales tax.

$$C = P + (0.06)P$$
$$50 = P(1 + 0.06)$$
$$50 = 1.06P$$
$$\frac{50}{1.06} = \frac{1.06P}{1.06}$$

Priced to the nearest cent $\qquad 47.17 \approx P$

Check the solution $\qquad \text{Tax} = 0.06(47.17)$
$$\approx 2.83$$

Total cost $\qquad \$47.17 + \$2.83 = \$50.00$

Example 3
The areas of the two rectangles are equal. Find the perimeter of each rectangle.

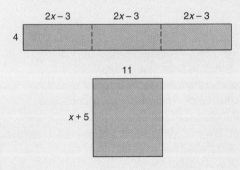

Solution
The area of the top rectangle is $3[4(2x - 3)]$.
The area of the bottom rectangle is $11(x + 5)$.

The areas are equal

$$3[4(2x - 3)] = 11(x + 5)$$
$$12(2x - 3) = 11(x + 5)$$
$$24x - 36 = 11x + 55$$
$$\underline{-11x \qquad\qquad -11x}$$
$$13x - 36 = 55$$
$$\underline{+36 \qquad +36}$$
$$13x = 91$$
$$\frac{13x}{13} = \frac{91}{13}$$
$$x = 7$$

560 Chapter 11 Solving Problems with Algebra

SECTION 11.3

The perimeters of the rectangles are

$$P_{top} = 2[4 + 3(2x - 3)]$$
$$= 2[4 + 3(2 \cdot 7 - 3)]$$
$$= 2[4 + 3(11)]$$
$$= 2(37)$$
$$= 74$$

$$P_{bottom} = 2[11 + (x + 5)]$$
$$= 2[11 + 7 + 5]$$
$$= 2(23)$$
$$= 46$$

Sample Problem

Problem Scrooge wants to use $2500 for two investments. One of the investments earns 4% interest per year. The other investment earns 9% interest per year. Scrooge decides to put half his money in each account. How much interest does he earn after one year? How much money does he have after 3 years, with interest compounded annually?

Solution A spreadsheet is helpful. If we don't have a spreadsheet, we can use a calculator and organize the results in a table similar to a spreadsheet's.

	A	B	C	D	E	F
		Investment at 4%		Investment at 9%		TOTAL
1	Time	Interest	Total $	Interest	Total $	ACCUMULATED
2	0 yr	0.00	1250.00	0.00	1250.00	2500.00
3	1 yr	50.00	1300.00	112.50	1362.50	2662.50
4	2 yr	52.00	1352.00	122.63	1485.13	2837.13
5	3 yr	54.08	1406.08	133.66	1618.79	3024.87

After one year, Scrooge has earned interest of $162.50. After 3 years, he has $3024.87.

Think and Discuss

P, R **1** Is it possible for 15% of a number to be greater than the number itself?

P, R **2** When a store discounts an item, does it matter whether the discount is computed before the tax is added or after the tax is added? Explain your answer. No; there is no difference in the final cost.

P, R **3** What restrictions must be placed on x so that the dimensions of the rectangle will make sense? $x > 15$

(Rectangle with dimensions $2x - 30$ and $x + 5$)

Section 11.3 More Word Problems

Assignment Guide

Basic
Optional

Average or Heterogeneous
Day 1 5, 6, 9, 15, 16, 28, 32
Day 2 8, 11, 13, 21, 22, 25, 27, 30
Day 3 14, 17, 18, 23, 24, 26, 29, 33

Honors
Day 1 4–7, 10, 12, 13, 15, 16, 18–20
Day 2 8, 9, 11, 14, 17, 21, 22 (a, b)
Day 3 22 (c, d), 23–30

Strands

Communicating Mathematics	2
Geometry	3, 4, 6–8, 10, 13, 16, 19, 21, 23, 25, 30, 31
Measurement	3, 4, 6–10, 12, 13, 15, 16, 18, 19, 21, 25–27, 30, 31, Investigation
Data Analysis/ Statistics	32
Technology	8, 9, 24, 27, 30
Probability	10
Algebra	1, 3, 5–9, 11, 13–17, 19–23, 25, 26, 28, 30–33, Investigation
Patterns & Functions	23, 27, 33
Number Sense	1, 3, 9, 32
Interdisciplinary Applications	2, 4–6, 9, 12, 14, 15, 17, 18, 24, 27–29, Investigation

Individualizing Instruction

Limited English proficiency
Students will benefit from having the problems read to them and having unfamiliar words clarified before attempting to solve. You might prefer to have students work in small groups.

SECTION 11.3

Problem-Set Notes

for pages 562 and 563

- **12** assumes that the price of a can of soda remains the same.
- **13, 14,** and **18** are a little more challenging.

Stumbling Block

In problems like the sample problem, students may confuse interest compounded annually with simple interest. Remind students of the differences between the two methods for calculating interest. If simple interest were used in the sample problem, the interest reported in columns B and D would stay the same each year.

Reteaching

Example 1 is a good review of the principles of dimensional analysis introduced in Chapter 4. Set up the problem as follows:

100 molds · $\frac{30 \text{ ml}}{\text{mold}}$ · $\frac{1 \text{ sec}}{10 \text{ ml}}$

= 300 sec

Students may find it easier to deal with all of the units by using this method.

Problems and Applications

P **4** A one-cubic-foot container will hold approximately 7.5 gallons of water. How much water will a 16-inch-by-12-inch-by-8-inch aquarium hold? ≈$6\frac{2}{3}$ gal

P **5** **Business** At Buster Block's Video, Le, the owner, wants to charge $3.25 for video rental. If the price must include an 8% tax, what is the pre-tax price for each rental? (Round your answer to the nearest cent.) $3.01

P **6** The edge of a cube is 4 inches. Kuba contracts to make 1000 such cubes out of plastic. If each cube weighs 0.5 ounce per cubic inch, how many pounds of plastic must Kuba buy to fill the order? 2000 lb

P, R **7** If the sides of a square are doubled in length, the perimeter increases by 40 centimeters.
  a How long are the sides of the original square? 10 cm
  b What is the perimeter of the original square? 40 cm

P, R **8** The shaded area is what percent of the total area of the rectangle? ≈21.46%

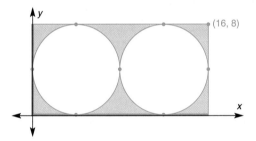

P **9** A machine produces 400 CD's per hour. A second machine produces 520 CD's per hour. On a given day, the first machine was started at 8 A.M. and the second machine was started at 11 A.M. Both machines were shut down at the same time, and they produced the same number of CD's.
  a How long did each machine run? 13 hr; 10 hr
  b What was the total number of CD's produced? 10,400

P, R **10** A circular disc with a diameter of 6 is tossed onto a flat surface so that at least part of the disc is inside the circle. What is the probability that the disc lies entirely inside the circle? $\frac{49}{169}$, or about 0.29

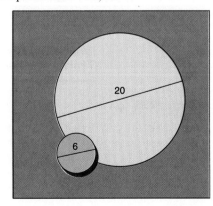

P **11** A pile contains 900 coins. Each day Penny Nichols takes 25% of the coins from the pile. Approximately how many coins remain in the pile after 10 days?

Chapter 11 Solving Problems with Algebra

SECTION 11.3

P **12 Business** The manufacturer of Kukka Koola, a soft drink, decides to change from 12-ounce cans to $11\frac{1}{2}$-ounce cans. By what percent will the manufacturer's income increase for every 120 ounces of beverage sold?

P, R **13** Harold walked around a circle. Then he walked 5 feet farther from the center of the circle and walked around another circle. All together, he walked 2000 feet. What is the radius of the inner circle?   ≈156.26 ft

P, R **14 Finance** Ole and Lars, twins from Minnesota, each inherited $10,000 from their great-uncle Calvin. Lars invested part of his money at 3% interest and part at 8%. Ole divided his money the same way Lars did, but invested one part at 4% and the other part at 6%. They ended the year with the same amount of interest. How did they divide their money?

P **15** Ace Plumbing charges $50 for a house call and $15 per hour for repairs. Deuce Plumbing charges $30 per visit and $17.50 per hour. Trey needed a plumber. Amazingly, both Ace and Deuce would have charged the same amount for the number of hours required to fix the problem. How many hours did it take to fix Trey's plumbing?   8 hr

P, R **16** The perimeter of the triangle is twice the perimeter of the square. Find the value of x.   10

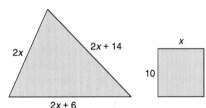

P **17 Business** Mavis sells real estate. She recently sold a house for $225,000. That is 15% more than the house was worth last year. How much was the house worth last year? $195,652.17

P **18** At the Tex-Mex horseshoe-throwing contest, the winner is determined by the sum of the lengths of three tosses. A bonus of 10 feet is given for each ounce over 16 ounces that a horseshoe weighs. Red throws a $2\frac{1}{2}$-pound horseshoe and Slats throws a 20-ounce horseshoe. Both tied for first place. What is the difference, in feet, in the sum of their tosses?   200 ft

P, R **19** Triangle ABC is similar to triangle DEF. Find the length of segment EF.   20

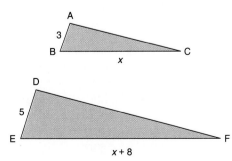

Cooperative Learning

- Have students use a circle drawn on a piece of paper, plus a coin, to model problem 10. You might choose to modify the problem a bit to have students also record whether the coin landed completely out of the circle. Have students calculate the probability that the coin will lie entirely in the circle if at least part of it is in the circle and compare this with their results.

- Problems 10 and 27 are appropriate for cooperative learning.

Communicating Mathematics

- Have students complete the following sentence and expand their answers either in writing or orally: "The kind of word problem I find the most difficult is . . . because . . . ."

- Problem 2 encourages communicating mathematics.

Additional Answers

for pages 562 and 563

11 Possible answer: 50 coins

12 ≈4.35%

14 Lars invested $6,666.67 at 3% and $3,333.33 at 8%; Ole invested $6,666.67 at 4% and $3,333.33 at 6%. The other possibility is that Lars invested $5714.29 at 3% and $4285.21 at 8% and that Ole invested $5714.29 at 6% and $4285.21 at 4%.

Section 11.3 More Word Problems

SECTION 11.3

Problem-Set Notes

for page 564

- **24** can be extended by asking students what the total percentage discount is.

Additional Practice

1. Linda has $5000 to invest. She puts $2000 in an account that pays 5% interest per year and the rest in an account that pays 6% interest per year. How much money does she have after three years if the interest is compounded annually?

2. A vendor sells a balloon for $2. If this price includes an 8% sales tax, what is the cost of the balloon?

3. The areas of the rectangles are equal. Find the perimeter of each.

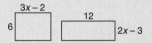

Answers:
1. $5888.30 2. $1.85 3. 32; 34

Multicultural Note

Scientists use a method called *indirect measurement* to measure such things as the height of a mountain or the weight of the moon. One of the earliest and best-known indirect measurements was a calculation of the circumference of the earth made by the Greek mathematician Eratosthenes. To make his estimate, Eratosthenes used rays of sunlight and some basic geometric principles.

P **20** Find the value of W for which $5W = W + 40$. $W = 10$

P, R **21** The area of the triangle is $28x$. What is the height of the triangle? 12

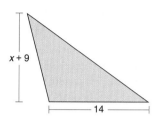

P, R **22** Solve each equation.
  a $3x - 15 = 23 + 7x$ $x = -9.5$
  b $16 - 4y = 4y - 8$ $y = 3$
  c $\frac{3}{4}a + 6 = \frac{1}{4}a - 2$ $a = -16$
  d $6k + 8 = 8(k + 10)$ $k = -36$

P, R, I **23** Write a function for the area A of a trapezoid with bases b_1 and b_2 and height h.
$A(b_1, b_2, h) = \frac{1}{2}h(b_1 + b_2)$

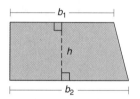

P **24** **Consumer Math** A dress originally priced at $67.95 is marked down three successive times. The first markdown is 15%, the second is 8%, and the third is 15%. What is the final cost of the dress? $45.17

P, R **25** The area of the front of the box in the drawing is twice the area of the bottom of the box. What is the volume of the box? 1296

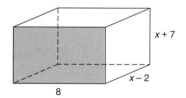

P **26** Sue gets paid $7.50 an hour. One week she notices that she worked the same number of hours as her age. If this had happened four years ago, in order to make the same amount of money, she would have had to earn $8.25 per hour. How old is Sue? 44

◄ LOOKING BACK   **Spiral Learning**   LOOKING AHEAD ►

R, I **27** **Science** A ball is dropped from 6 feet above the ground. It is allowed to bounce continuously. On each bounce, the ball rebounds to a height 85% as high as the previous bounce.
  a What is the maximum height of the ball after the fifth bounce? ≈2.66 ft
  b What is the maximum height of the ball after the ninth bounce? ≈1.39 ft
  c What is the total distance that the ball has traveled by the end of the ninth bounce? (Hint: Use a spreadsheet.) ≈58.25 ft

SECTION 11.3

R   **28 Consumer Math**   Billi paid $15 down on a CD player and then paid $12 a month until she had paid a total of $231 for the player. How many monthly payments did she make?   18

R   **29 Business**   Kirk started a business with a total investment of $6000. Then, friend Spock bought a 25% share in Kirk's business enterprise. How much did Spock invest? $1500

R, I   **30** The ratio of AB to AC is 3:5. What is the length of the diameter of the circle?   ≈11.66 cm

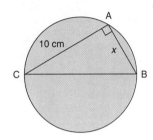

R   **31** Find x so that the area of triangle ABC is 72.   15

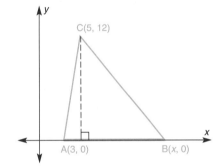

R   **32** Region I contains 20 elements.
Region II contains x elements.
Region III contains $2x$ elements.
How many elements are in Region III if $A \cup B$ contains 74 elements?   36

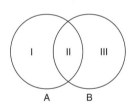

R, I   **33** If $f(x) = 3x + 8$ and $g(x) = x^2 - 4$, what is
    a $f(2)$?   14   **b** $g(-2)$?   0
    c $f(3)$?   17   **d** $g(3)$?   5

Investigation

The Teeter-totter Principle   Weights A and B are put on the ends of the scale to make it balance. What can you determine about the relationship between A and B?

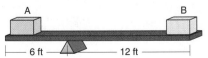

Section 11.3   More Word Problems

Checkpoint

1. The price of a rubber ball, including a 6% sales tax, is $1.68. What is the price of the rubber ball without tax?

2. The perimeters of the rectangles are equal. Find the area of each rectangle.

$3x - 5$ $5x - 10$
$2x - 3$ $x - 6$

3. If the sides of a square are tripled in length, the perimeter increases by 56 in. How long are the sides of the original square?

Answers:
1. $1.58   **2.** 247; 60   **3.** 7 in.

Additional Answers

for page 565

Investigation   Package A must weigh twice as much as package B.

SECTION 11.4

Objective
- To solve equations with several variables

Class Planning

Schedule 3 days
Resources
In *Chapter 11 Student Activity Masters* Book
Class Openers 37, 38
Extra Practice 39, 40
Cooperative Learning 41, 42
Enrichment 43
Learning Log 44
Interdisciplinary
 Investigation 45, 46
Spanish-Language Summary 47

Class Opener

Use 6 cm and 10 cm as the measurements needed to find each of the following.

1. Perimeter of a rectangle
2. Area of a parallelogram
3. Area of a triangle
4. Perimeter of an isosceles triangle
5. Perimeter of a scalene triangle

Answers:
1. 32 cm 2. 60 cm² 3. 30 cm²
4. 22 cm or 26 cm 5. Cannot be done; a third measure is needed.

Lesson Notes

- Our intent in the three examples and three sample problems is to expose students to many different types of problems, have them delve into the information given, and see what they can do to solve the problems. We do not prescribe that a single method be taught for each type of problem presented. For example,

Investigation Connection
Section 11.4 supports Unit F, *Investigations* 1,3. (See pp. 542A–542D in this book.)

SECTION 11.4
Equations with Several Variables

We have worked with formulas and functions that involve several inputs. For example, the area, A, of a triangle with base b and height h is found with the function

$$A(b, h) = \tfrac{1}{2}bh$$

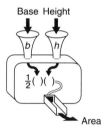

The perimeter of a rectangle is determined by its length, ℓ, and its width, w. The function that gives the perimeter of a rectangle is

$$P(\ell, w) = 2\ell + 2w$$

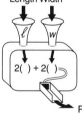

Example 1
The perimeter of a rectangle is 58 centimeters. The length of the rectangle is twice its width. Find the dimensions of the rectangle.

Solution

Formula for perimeter	$P = 2\ell + 2w$
Substitute 58 for P	$58 = 2\ell + 2w$
Since $\ell = 2w$, we can substitute $2w$ for ℓ	$58 = 2(2w) + 2w$
	$58 = 4w + 2w$
	$58 = 6w$
	$\dfrac{58}{6} = \dfrac{6w}{6}$
	$\dfrac{29}{3} = w$

566 Chapter 11 Solving Problems with Algebra

Since $w = 9\frac{2}{3}$ and $\ell = 2w$, $\ell = 2\left(\frac{29}{3}\right) = \frac{58}{3}$, or $19\frac{1}{3}$. The length of the rectangle is $19\frac{1}{3}$ centimeters, and the width is $9\frac{2}{3}$ centimeters.

Example 2
A stamp collector collects $2 stamps and $3 stamps. He has 12 more $2 stamps than $3 stamps. The total value of the stamps is $944. How many $3 stamps does the collector have?

Solution
Let a represent the number of $2 stamps and b represent the number of $3 stamps.
The value of a $2 stamps is $2a$. The value of b $3 stamps is $3b$.
The total value of the stamps is $2a + 3b = 944$.
Since he has 12 more $2 stamps than $3 stamps, $a = b + 12$.

We can substitute $(b + 12)$ for a

$$2(b + 12) + 3b = 944$$
$$2b + 24 + 3b = 944$$
$$5b + 24 = 944$$
$$\underline{-24 \quad -24}$$
$$5b = 920$$
$$b = 184$$

The collector has 184 three-dollar stamps (and 196 two-dollar stamps). Can you think of a way to do this problem without using algebraic equations?

Example 3
In a balance problem, the product of the mass on one end of the beam and its distance from the balance point (fulcrum) must equal the product of the mass at the other end of the beam and its distance from the fulcrum.

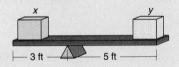

Mass x is 10 kilograms heavier than mass y. Mass x is 3 feet from the fulcrum, and mass y is 5 feet from the fulcrum. Find the mass of the heavier object.

Solution
The balance equation is $3x = 5y$.
Since mass x is 10 more than y, $x = y + 10$.

We can substitute $(y + 10)$ for x

$$3(y + 10) = 5y$$
$$3y + 30 = 5y$$
$$\underline{-3y \qquad -3y}$$
$$30 = 2y$$
$$15 = y$$

The other object, x, has mass

$$y + 10 = (15) + 10$$
$$= 25$$

The heavier object has a mass of 25 kilograms.

---

example 1 could also be done as a demonstration on a spreadsheet, with one column representing widths, a second column, set to twice the values of the first column, representing the lengths, and a third showing the various perimeters. Students could guess the widths and "zoom in" on the approximate correct value.

Additional Examples

Example 1
The two congruent sides of an isosceles triangle are each twice as long as the third side. The perimeter of the triangle is 40 in. Find the length of each side of the triangle.

Answers: 8 in., 16 in., and 16 in.

Example 2
Tickets for a show cost children $1 each and adults $4 each. Twenty-five more adults than children attended the show, and ticket sales totaled $170. How many children attended the show?

Answer: 14

Example 3
Mass y is 8 kg heavier than mass x. Find the mass of the heavier object.

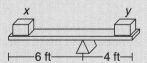

Answer: 24 kg

SECTION 11.4

Assignment Guide

Basic
Optional

Average or Heterogeneous
Day 1   7–9, 11, 13, 19, 25, 29
Day 2   10, 14, 15, 20, 24, 26–28, 33
Day 3   12, 16–18, 21–23, 30–32

Honors
Day 1   7–9, 11–13, 16, 21
Day 2   10, 14, 15, 17–19, 27
Day 3   20, 22–26, 28–33, Investigation

Strands

Communicating Mathematics	1, 15
Geometry	1, 2, 9, 13, 17, 19, 23, 28, 30, 31
Measurement	1, 2, 9, 11, 13, 17, 19, 23, 31
Technology	13
Probability	14
Algebra	1–13, 15, 17, 19, 20, 22, 23, 25–30, 32
Patterns & Functions	3, 5, 8, 20, 25, 30
Number Sense	4, 6, 7, 10, 15
Interdisciplinary Applications	11, 14, 16, 18, 20–22, 26, 27, 33, Investigation

Individualizing Instruction

The visual learner In a problem such as sample problem 2, students might be aided by graphing the line that represents the combinations for each possible value of the denominator. Since the line shows the solution set of the equation, students can find pairs of natural-number solutions by reading the graph, instead of by using trial and error.

Sample Problems

Problem 1 The area of the rectangle is numerically equal to its perimeter.
a Write an equation to describe this situation.
b Find the length of the rectangle if the width is 12 inches.
c Solve the equation you wrote in part **a** for ℓ in terms of w.

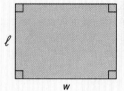

Solution The formula for the area of a rectangle is $A = \ell w$.
The formula for the perimeter of a rectangle is $P = 2\ell + 2w$.

a Since it is given that the area is numerically equal to the perimeter, $2\ell + 2w = \ell w$.

b If the width is 12 inches,
$$2\ell + 2(12) = \ell(12)$$
$$2\ell + 24 = 12\ell$$
$$-2\ell \qquad\qquad -2\ell$$
$$24 = 10\ell$$
$$2.4 = \ell$$

c To solve for ℓ in terms of w, we must have all of the ℓ terms on the same side of the equation.

$$2\ell + 2w = \ell w$$
$$2\ell - 2\ell + 2w = \ell w - 2\ell$$

Distributive property $\qquad 2w = \ell(w - 2)$
Divide both sides by
$(w - 2)$ to isolate the ℓ $\qquad \dfrac{2w}{w-2} = \ell$

The variable ℓ is now expressed in terms of w.

Problem 2 Find all pairs of natural numbers (x, y) for which $\dfrac{14}{x + 2y}$ is a natural number.

Solution In order for the fraction to be a natural number, $x + 2y$ must be a factor of 14. It must be either 1, 2, 7, or 14. We organize our trials:

$x + 2y = 1$	No pairs of natural numbers work.
$x + 2y = 2$	If $y = 1$, then $x = 0$. No pairs of natural numbers work.
$x + 2y = 7$	(1, 3), (3, 2), and (5, 1) each work.
$x + 2y = 14$	(2, 6), (4, 5), (6, 4), (8, 3), (10, 2), (12, 1)

There are nine different pairs of natural numbers that satisfy the condition.

SECTION 11.4

Problem 3 Each side of square SUAE is 7. Find the restrictions on the values of x and y for which rectangle RECT exists and for which Q is on $\overline{US}$ and T is outside of the square.

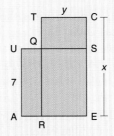

Solution In order for the rectangle to exist as shown, both x and y must be positive and y must be smaller than 7 and x must be larger than 7. In summary, $x > 7$ and $0 < y < 7$.

Think and Discuss

P, R **1** A rectangle has length 12 and width 8. Which of the following is a correct method of computing the perimeter of the rectangle? Explain your choice(s).
  a $12 + 8 + 12 + 8 = 40$
  b $24 + 16 = 40$
  c $2(20) = 40$
  d $(12 - 8) \cdot 10 = 40$

P **2** The perimeter of a rectangle is 20 millimeters. If the length of the rectangle is increased by 2 millimeters and the width is decreased by 2 millimeters, the perimeter of the rectangle is still 20 millimeters. Find the length and width of the rectangle.   Any dimensions that add up to 10 mm

P **3 a** If $Q(x, y) = 3x + 5y$, what is $Q(4, 3)$?   27
  b If $Q(x, y) = 3x + 5y$, what is $Q(3, 4)$?   29
  c Could $Q(a, b)$ ever be equal to $Q(b, a)$?   Only if $a = b$

P **4** For what values of x will $\frac{43}{x+9}$ be an integer? $-52, -10, -8, 34$

P **5** If $G(x, y) = 3x + 4y$ and $x = y + 5$, find $G(x, y)$ for
  a $y = 3$ 36
  b $y = 0$ 15
  c $x = 2$ -6
  d $y = 2x$ -55

P **6** For what pairs of natural numbers is $\frac{24}{3x + 3y}$ a natural number? What do you notice about your answer?

Problems and Applications

P **7** For what natural-number values of n is $\frac{n+7}{n}$ a natural number?   1 or 7

P **8** If $f(x, y) = x + 2y$, what is
  a $f(3, 5)$?   13
  b $f(5, 3)$?   11
  c $f(a, b)$? $a + 2b$
  d $f(b, a)$? $b + 2a$

Section 11.4 Equations with Several Variables   569

Problem-Set Notes

for page 569

- **1** has three correct answers: choice **a** for those who choose to "walk" around the perimeter, choice **b** for those who use the formula $P = 2\ell + 2w$, and choice **c** for those who walk halfway around and then double the measure.

- **3, 5,** and **8** can be related to the function machines on page 566.

Cooperative Learning

- Students can construct a simple balance beam using the following materials: 2 paper cups, a ruler, a dowel rod or pen, a paper clamp, and two paper clips.

- Students can experiment by hanging objects of different masses from the "hooks" and adjusting the position of the fulcrum to see if the relationship in example 3 holds.

- Problems 1 and 25 are appropriate for cooperative learning.

Additional Answers

for page 569

1 a Add all four sides
b Double the lengths of the sides, then add
c Double the distance halfway around
d Not a correct method

6 (1, 7), (2, 6), (3, 5), (4, 4), (5, 3), (6, 2), (7, 1), (2, 2), (3, 1), (1, 3), (1, 1). The sum of x and y is 2, 4, or 8.

SECTION 11.4

Problem-Set Notes

for pages 570 and 571

- **23** could be done as a discovery problem using the Geometer's Sketchpad.
- **25** previews the next chapter.

Additional Practice

1. If $Q(x, y) = 7x + 6y$, what is $Q(-1, 3)$?
2. For what pairs of natural numbers (x, y) is $\frac{18}{2x+3y}$ a natural number?
3. Tickets for a water slide are $5 for adults and $3 for children. Lakeisha bought $65 worth of tickets. How many tickets of each type could she have bought?
4. One box weighs 15 pounds more than the other. What must the weight of each box be for the scale to balance?

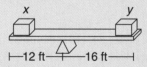

Answers:
1. 11 2. (3, 1), (3, 4), (6, 2)
3. 1 adult, 20 children; 4 adults, 15 children; 7 adults, 10 children; 10 adults, 5 children; or 13 adults 4. 45 pounds, 60 pounds

Reteaching

In example 3, if mass x is 10 kilograms heavier than mass y, then mass y is 10 kilograms lighter than mass x. Thus $y = x - 10$. Have students use this equation for the substitution into $3x = 5y$ to obtain $3x = 5(x - 10)$. Students can then solve directly for the heavier mass.

P **9** The perimeter of the rectangle is 80 centimeters.
    a Find $x + y$. $x + y = 26$
    b Graph the lattice points (x, y) in Quadrant I. (A *lattice point* is a point whose coordinates are both integers.)

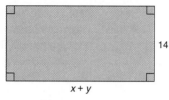

P **10** Find all ordered pairs of natural numbers (x, y) for which $\frac{21}{x+2y}$ is a natural number.

P **11** Science One of the boxes weighs 15 kilograms more than the other.
    a Which box is heavier, A or B? A
    b What must be the weight of each box so that the scale balances? A: 60 kg; B: 45kg

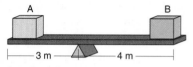

P **12** Mr. Coyne has some dimes and quarters. He has twice as many dimes as quarters. The total value of his coins is $8.10. How many dimes does he have? 36

P **13** The length of a rectangle is three times its width. The perimeter of the rectangle is 348. Find its length. 130.5

P, R **14** Sarah Alice has 14 identical diamond rings to display in the window of Diamonds Are Us. The window display has 12 sections. Sarah wants to place at least one ring in each section. In how many different ways can Sarah display the 14 rings? 78 ways

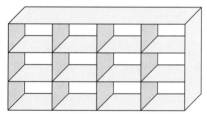

P **15** If $\frac{18}{x+2y}$ is a natural number, and x and y are natural numbers, explain why $x + 2y$ must be equal to either 3, 6, 9, or 18.

P, R **16** At the Children's Little Theatre, admission is $8 for adults and $5 for children. Adrian spent $47 on tickets. How many tickets of each type did she buy? 3 children; 4 adults

P, R **17** The perimeter of an isosceles triangle is 48 inches. The length of the base of the triangle is 4 inches greater than the length of either leg. What is the length of the base of the triangle? $18\frac{2}{3}$ in

P, R **18** Pip buys 24 cans of soda at 25¢ each and 5 liters of soda at 75¢ each. How much did Pip pay for the soda? $9.75

P, R **19** Suppose the perimeter of the rectangle and the perimeter of the triangle are equal.
    a Write an equation in terms of x to represent this situation.
    b Solve the equation for x. $x = 3.5$
    c Find the length of the rectangle. 6.5
    d Find the length of each side of the triangle. 7

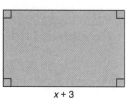
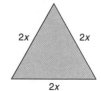

Chapter 11 Solving Problems with Algebra

SECTION 11.4

P, R, I **20 Sports** A National Football League kicker scores three points for each field goal made and one point for each extra point made.
  a Write a function for the number of points, P, a kicker gets for f field goals and e extra points. $P(f, e) = 3f + e$
  b During the 1990 season, Nick Lowery made 34 field goals and 37 extra points. How many points did he score? 139
  c During the 1990 season, Pete Stoyanovich scored 100 points, of which 37 were extra points. How many field goals did he kick? 21

P **21 Sports** Jump Shot, a basketball player, can make either two-point baskets or three-point baskets. In his last game, he scored 20 points with no free throws. How many two- and three-point baskets could Jump Shot have made?

P **22** A six-pack of cans of Sparkly soda costs $3.75. An eight-bottle carton of Sparkly soda costs $4.25. Juanita bought some six-packs and some cartons of the soda. All together, she had a combination of 48 cans and bottles at a cost of $27.75. How many six-packs of cans and how many eight-bottle cartons did she buy? Four 6-packs and three 8-packs

P, R, I **23** The length of segment RE is 2 and the length of segment TR is 3. Find the restrictions on the radius r of circle C so that the area of the shaded region is at least one third the area of rectangle RECT. $\sqrt{\frac{8}{\pi}} \leq r$

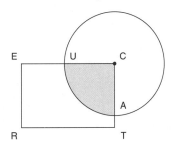

◀ LOOKING BACK **Spiral Learning** LOOKING AHEAD ▶

R **24** Find the least common multiple and the greatest common factor of 18 and 24. LCM = 72; GCF = 6

R, I **25** Suppose that $f(x) = 3x + 2$. Peter let $x = 2$ and determined $f(2)$. He took his answer and put it back into the function. He repeated the process several times, using each output as the next input.
  a Find the first ten outputs of the function if $x = 2$ is the initial input.
  b Find the first ten outputs of the function if $x = 1$ is the initial input.

R **26 Consumer Math** Connie bought identical ice-cream cones for herself, Van, and Illa. The total bill came to $4.41, including 5% sales tax. How much was each ice-cream cone? $1.40

R **27** Mike can mow the lawn in two hours and Sherry can mow the lawn in three hours. How long will it take to mow the lawn if they work together? $1\frac{1}{5}$ hr, or 1 hr 12 min

Communicating Mathematics

- Have students answer the following questions in writing: Is it possible to solve an equation with more than one variable? Why or why not?
- Problems 1 and 15 encourage communicating mathematics.

Additional Answers

for pages 570 and 571

9 b The graph will display these lattice points: (1, 25), (2, 24), (3, 23), (4, 22), (5, 21), (6, 20), (7, 19), (8, 18), (9, 17), (10, 16), (11, 15), (12, 14), (13, 13), (14, 12), (15, 11), (16, 10), (17, 9), (18, 8), (19, 7), (20, 6), (21, 5), (22, 4), (23, 3), (24, 2), (25, 1).

10 For a divisor of 21: (1, 10), (3, 9), (5, 8), (7, 7), (9, 6) (11, 5), (13, 4), (15, 3), (17, 2), (19, 1); for a divisor of 7: (1, 3), (3, 2), (5, 1); for a divisor of 3: (1, 1)

15 For $\frac{18}{x+2y}$ to be a natural number, $x + 2y$ must be a factor of 18.

19 a $2(4) + 2(x + 3) = 2x + 2x + 2x$

21

2 pt	3 pt
1	6
4	4
7	2
10	0

25 a When $x = 2$, outputs 1–10 are 8, 26, 80, 242, 728, 2186, 6560, 19682, 59048, and 177146.

b When $x = 1$, outputs 1–10 are 5, 17, 53, 161, 485, 1457, 4373, 13121, 39365, and 118097.

Section 11.4 Equations with Several Variables

SECTION 11.4

Problem-Set Notes

for page 572

- 31 previews the next section and gives students a model that shows faces, edges, and vertices.

Stumbling Block

In sample problem 2, some students may miss some of the ordered pairs that make the expression a natural number. Be sure that students set the denominator equal to each of the factors of the numerator. For each equation, have students use either graph paper or dot paper to graph the solution.

Checkpoint

1. If $Q(x, y) = 4x - 3y$, what is $Q(2, -5)$?
2. The perimeter of a rectangle is 48 centimeters. The length is 6 centimeters longer than the width. Find the dimensions of the rectangle.
3. Find all ordered pairs of natural numbers (x, y) for which $\frac{15}{2x+y}$ is a natural number.
4. One of the boxes weighs 12 kg more than the other. What must the weight of each box be for the scale to balance?

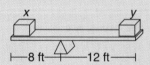

Answers:
1. 23 2. 9 cm × 15 cm
3. (1, 1), (1, 3), (2, 1), (1, 13), (2, 11), (3, 9), (4, 7), (5, 5), (6, 3), (7, 1) 4. 24 kg, 36 kg

Answers for problems 28 and Investigation are on page A9.

572

R, I **28** Draw the image resulting from
  a A 90° clockwise rotation about (0, 0)
  b A 90° clockwise rotation about (1, 2)

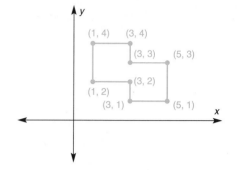

R **29** Solve each equation and simplify each expression.
  a $-\frac{2}{3}x + 4x^2 - 6x^2 + \frac{1}{4}x$ $-2x^2 - \frac{5}{12}x$ **b** $\frac{3}{5}x + \frac{4}{9} = -\frac{1}{3}$ $x = -\frac{35}{27}$
  c $9\left(\frac{1}{3}x - \frac{1}{6}\right) = 4\left(\frac{1}{2}x + \frac{1}{4}\right)$ $x = 2.5$ **d** $-9\left(\frac{1}{3}x - \frac{1}{6}\right) - 4\left(\frac{1}{2}x - \frac{1}{4}\right)$ $-5x + \frac{5}{2}$

R **30 a** Write a function for the volume, V, of the box in terms of ℓ, w, and h. $V(\ell, w, h) = \ell w h$
  b Write a function for the surface area of the box. $A(\ell, w, h) = 2\ell w + 2\ell h + 2wh$

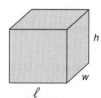

R, I **31** Suppose the figure is cut out and folded to make a box.
  a Find the area of the box.   376 in.²
  b Find the volume of the box.   480 in.³

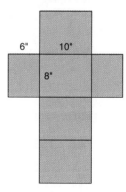

R **32** Multiply $-5\begin{bmatrix} -4 & 9 & -8 \\ 0 & 7 & -3 \end{bmatrix}$ $\begin{bmatrix} 20 & -45 & 40 \\ 0 & -35 & 15 \end{bmatrix}$

R **33** At Parker High School, the students raise money for the production of the school play by selling cookies. If the students receive 15% of total sales, how much money will they raise by selling $34,500 worth of cookies? $5175

Investigation

Chemistry A very well known equation in chemistry is $PV = nRT$. Find out what the variables P, V, n, R, and T represent.

572 Chapter 11 Solving Problems with Algebra

SECTION 11.5

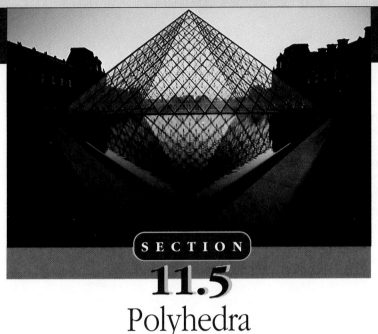

Investigation Connection
Section 11.5 supports Unit F, *Investigations* 2, 3. (See pp. 542A-542D in this book.)

SECTION 11.5
Polyhedra

PRISMS

In this section we will study **polyhedra**—solids with flat faces. The word *polyhedra* means "many faces." The singular form of the word is *polyhedron*.

A rectangular box is a polyhedron. It has 6 **faces**, 12 **edges**, and 8 **vertices**. Each face is a polygon. Each edge is formed by the intersection of two faces. Each vertex is the point of intersection of several edges. We can compute the volume of a rectangular box by using the formula

Volume = (length)(width)(height)

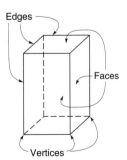

The rectangular box is an example of a type of polyhedron called a **prism**—a figure with two congruent parallel polygonal bases. Any cross section of a prism cut parallel to the bases is also congruent to the bases. Bases do not need to be on the "bottom" and "top" of a prism.

Bases are shaded green.

Objectives

- To find the volume and the total surface area of a prism
- To find the volume and the total surface area of a pyramid

Vocabulary: edge, face, lateral area, lateral face, polyhedra, prism, pyramid, slant height, vertex

Class Planning

Schedule 2 days

Resources
In *Chapter 11 Student Activity Masters* Book
Class Openers *49, 50*
Extra Practice *51, 52*
Cooperative Learning *53, 54*
Enrichment *55*
Learning Log *56*
Using Manipulatives *57, 58*
Spanish-Language Summary *59*

In *Tests and Quizzes* Book
Quiz on 11.4–11.5 (Average) *275*
Quiz on 11.4–11.5 (Honors) *276*

Transparencies *46, 47*

Class Opener

1. Evaluate $V(\ell, w, h) = \ell w h$ at $V(\frac{1}{2}, 4, 3)$.
2. Evaluate $V(A, h) = Ah$ at $V(40, 8)$.
3. Evaluate $V(A, h) = \frac{1}{3}Ah$ at $V(9, \frac{1}{3})$.
4. Evaluate $S(\ell, w, h) = 2(\ell w + \ell h + wh)$ at $S(3, 5, 10)$.

Answers: **1.** 6 **2.** 320 **3.** 1 **4.** 190

Lesson Notes

- Handling and building models of polyhedra can help students visualize three-dimensional figures. The more hands-on you make this section, the better your students will understand the

Section 11.5 Polyhedra 573

SECTION 11.5

Lesson Notes continued

concepts. For example, pouring sand into a rectangular pyramid, then emptying it three times into a rectangular prism with the same base area and height will help students remember the formula for volume.

Additional Examples

Example 1
Find the volume and the total surface area of the triangular prism.

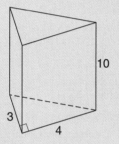

Answer: Volume: 60 units^3, Surface Area: 132 units^2

Example 2
Find the slant height of the pyramid in example 2 on page 575.

Answer: 5 feet

The formula for the volume of any prism is

$$\text{Volume}_{\text{prism}} = (\text{area of base})(\text{height})$$

The total surface area of a prism can be found by adding together the areas of the faces.

The faces of the prism that are not bases are called **lateral faces.** Often we need to know the surface area of the "sides" of a prism. The **lateral area** of a prism is the sum of the areas of the lateral faces—the surface area minus the area of the bases.

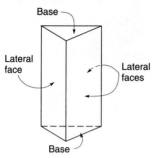

Example 1
a Find the volume of the triangular prism.
b Find the total surface area of the triangular prism.

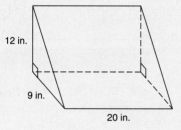

Solution

a The base of this triangular prism is a right triangle that has an area of $\frac{1}{2}(9)(12) = 54$ square inches. Since we know the area of the base, we can find the volume by using the formula

$$\begin{aligned}\text{Volume} &= (\text{area of base})(\text{height}) \\ &= (54)(20) \\ &= 1080\end{aligned}$$

The volume of the triangular prism is 1080 cubic inches.

b The total surface area is equal to the sum of the areas of the two bases and the lateral area.

$$\begin{aligned}\text{Area of bases} &= \tfrac{1}{2}(12)(9) + \tfrac{1}{2}(12)(9) \\ &= 54 + 54 \\ &= 108\end{aligned}$$

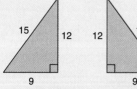

The area of the two bases of the triangular prism is 108 square inches.
The lateral area is equal to the sum of the areas of the three rectangular faces of the prism.

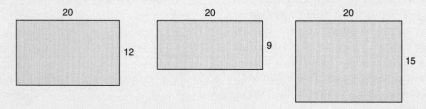

574 Chapter 11 Solving Problems with Algebra

$$\text{Lateral area} = (9)(20) + (12)(20) + (15)(20)$$
$$= 180 + 240 + 300$$
$$= 720$$

The lateral area of the triangular prism is 720 square inches.

The total surface area of the triangular prism is the sum of the areas of the two bases and the lateral area.

$$\text{Total surface area} = 720 + 108$$
$$= 828$$

The total surface area of the triangular prism is 828 square inches.

PYRAMIDS

A **pyramid** is a polyhedron with several lateral faces, but only one base. The square-based pyramid shown has 4 lateral faces, 8 edges, and 5 vertices. The height of the pyramid is measured perpendicular to the base. A ***slant height*** is the height of one of the lateral faces.

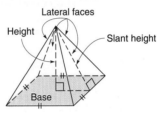
Square-Based Pyramid

The surface area of a pyramid is the sum of the areas of the lateral faces and the area of the base.

Clearly, the volume of a pyramid is less than the volume of a prism with the same base and height. In fact, the volume of a pyramid is one third the volume of a prism with the same base and height.

$$\text{Volume}_{\text{pyramid}} = \tfrac{1}{3}(\text{area of base})(\text{height})$$

Example 2

A solid stone pyramid is 4 feet tall and has a square base with a 6-foot side. What is the weight of the pyramid if the stone weighs 300 pounds per cubic foot?

Solution

We will use the formula for the volume of a pyramid.

$$V = \tfrac{1}{3}(\text{area of base})(\text{height})$$
$$= \tfrac{1}{3}(6^2)(4)$$
$$= 48$$

The volume of the pyramid is 48 cubic feet. Since the weight of the material is 300 lb/ft³ the weight of the solid pyramid is 48 ft³ · $\left(\dfrac{300 \text{ lb}}{1 \text{ ft}^3}\right)$, or 14,400 pounds.

Assignment Guide

Basic
Optional

Average or Heterogeneous
Day 1 8, 9, 12, 13, 16, 21, 23–25
Day 2 10, 15, 17, 19, 20, 22, 26, 28

Honors
Day 1 8, 9, 11, 14, 15, 20, 23, 24, 28
Day 2 10, 12, 13, 16–19, 21, 22, 25, 26

Strands

Communicating Mathematics	4–7, 11, 18, 19, 24
Geometry	1–13, 15–17, 19–21, Investigation
Measurement	2, 8–10, 12–21, 28
Data Analysis/ Statistics	7, 22
Technology	16, 21, 25
Probability	23
Algebra	12, 16, 17, 24, 26–28, Investigation
Patterns & Functions	22, 24, Investigation
Number Sense	14, 15, 27
Interdisciplinary Applications	16, 18, 21, 28

Individualizing Instruction

The gifted learner Students could be challenged to research the dimensions of the Washington Monument, then find its volume and total surface area.

SECTION 11.5

Problem-Set Notes

for pages 576 and 577

- **3** and **7** check to see whether students are correctly interpreting the dotted lines in diagrams of three-dimensional figures.

Stumbling Block

Students may confuse *height* and *slant height*. Use physical models or the diagrams in problem 15 to point out the differences between the two measures. Have students note that in figure I for problem 15, the slant heights are all equal; in figure II, two pairs of slant heights are equal and one pair is equal to the height; and in figure III, the slant heights are all different.

Multicultural Note

Hypatia, the fourth-century Egyptian mathematician, is also remembered for her research on conic sections (figures that can be formed by slicing a cone). She was quite ahead of her time, in that the importance of studying conic sections was not recognized until the 1600s, when astronomers used them to model the orbits of planets.

Sample Problems

Problem 1 The pentagonal base of the pyramid has an area of 12 square millimeters, and the pyramid has a height of 8 millimeters. Find its volume.

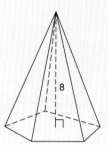

Solution The formula for the volume of a pyramid is

Volume = $\frac{1}{3}$(area of base)(height)

= $\frac{1}{3}$(12)(8)

= 32

The volume of the prism is 32 cubic millimeters.

Problem 2 A tower is built with a rectangular base. Find the volume of the tower.

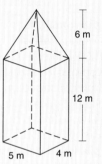

The easiest way to solve this problem is to "divide and conquer." First find the volume of the rectangular prism. Then find the volume of the pyramid. Finally, add the two volumes.

Volume$_{prism}$ = (4)(5)(12)
= 240

The volume of the prism is 240 cubic meters.

Volume$_{pyramid}$ = $\frac{1}{3}$(20)(6)
= 40

The volume of the pyramid is 40 cubic meters.

Total volume = 240 + 40
= 280

The total volume of the tower is 280 cubic meters.

Think and Discuss

P **1** Is a rectangular box a prism? Yes

P **2** In part **b** of Example 1 on page 574, what does the number 15 represent in the calculation of the lateral area of the triangular prism?

576 Chapter 11 Solving Problems with Algebra

SECTION 11.5

P 3 What do the dashed lines in the figure represent? Edges of the polyhedron that would not be visible from the given point of view.

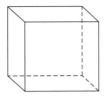

P 4 Explain why the figures in column I are prisms and why the figures in column II are not prisms.   Prisms are made up of two congruent bases connected by lateral faces; the figures in column II do not have these characteristics.

Column I   Column II

P 5 Explain the difference between the height of a pyramid and the slant height of a pyramid.

P 6 How would you calculate the surface area of a pyramid?

P 7 Explain the difference between figure I and figure II.   Figure I is an open box (bottom and four sides but no top), whereas figure II is a rectangular prism.

I   II

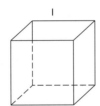 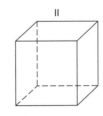

Cooperative Learning

- Give groups of students various prisms, pyramids, and figures such as the one in sample problem 2. Have them find the volume and surface area of each one.
- Investigation is appropriate for cooperative learning.

Additional Answers

2 15 is the length (or width) of one of the faces of the triangular prism.

5 The *height* is the shortest distance from the top (peak) to the plane containing the base. A *slant height* is the height of one of the lateral faces.

6 The surface area of a pyramid is the sum of the areas of the lateral faces and the area of the base.

Problems and Applications

P 8 Find the volume of each figure.

a 70

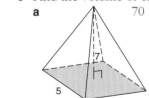

b 210

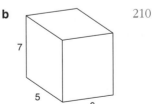

c 210

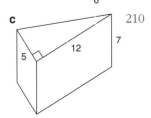

d 70

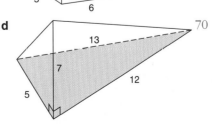

Section 11.5 Polyhedra

SECTION 11.5

Problem-Set Notes

for pages 578 and 579

- **15** is counterintuitive in that they all have equal volume.
- **16** reviews proportions. You might extend this problem by having students first find the volume of the model and then compare the ratio of the volumes to the ratio of the linear dimensions.
- **19** focuses on the different units for volume and surface area.

Additional Practice

1. Find the volume of a rectangular prism whose length is 7, width is 8, and height is 9.
2. Find the length of a rectangular prism whose volume is 264 cm³, width is 4 cm, and height is 6 cm.
3. Find the height of a square-based pyramid whose volume is 84 in.³ and whose base is 6 in. on each side.
4. Find the total surface area of the triangular prism in problem 10 on page 578.
5. Find the volume of a square-based pyramid whose base is 16 on each side and whose height is 6.
6. Find the total surface area of a rectangular prism whose length is 2, width is 1.5, and height is 1.

Answers:
1. 504 units³   2. 11 cm
3. 7 in.   4. 920 units²
5. 512 units³   6. 13 units²

P   **9** For the triangular prism, find the
   a Area of one base   24
   b Area of each lateral face   90, 120, 150
   c Lateral area   360
   d Total surface area   408
   e Volume   360

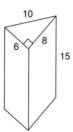

P, R   **10** Find the lengths of edges $\overline{BC}$ and $\overline{DF}$.   15, 17

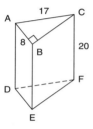

P, R   **11** **Communicating** What does the diagram appear to represent?   Possible answers: A quadrilateral with both diagonals drawn; a top view of a quadrilateral-based pyramid.

P, R, I   **12** A rectangular prism has edges $(x-8)$, x, and $(x+2)$.
   a Write a formula for the volume of the prism in terms of x. $V = x(x+2)(x-8)$
   b What is the volume of the prism for $x = 10$? $x = 8$? $x = 5$?

P, R   **13** In the square-based pyramid, each edge of the base is 16, the height is 15, and the slant height is 17.
   a Find the area of the base.   256
   b Find the lateral area.   544
   c Find the total surface area.   800
   d Find the volume.   1280

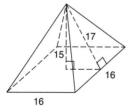

P, R   **14** Bill knows the maximum number of marbles that can fit inside a pint jar. How would he determine the maximum number of marbles that can fit inside a gallon jar?   Multiply by 8 for an estimate.

P   **15** Which of the square-based pyramids shown has the greatest volume? The least volume?   All have the same volume, since they have the same base and height.

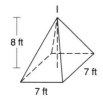 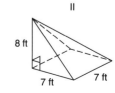 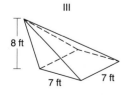

578   Chapter 11   Solving Problems with Algebra

SECTION 11.5

16 The Great Pyramid of Cheops measured 775.75 feet on each side of its square base, and it had a height of 481.4 feet.
  a Find the volume of the pyramid. ≈96,566,924 ft³
  b Tutt wants to build a model of the pyramid. He wants the model to have a height of 4 feet. How long should he make each edge of the base? About 6.4 ft

17 Find the volume of each triangular prism.
  a 288
  b $P = 30$, $A = 50$ 450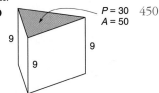

18 **Science** Which weighs more, a cubic foot of ice or a cubic foot of water? Explain your answer. Water, since its density is greater than that of ice

19 Find the volume and the surface area of the box. Explain why the volume and the surface area are not really equal. 216 in.²; 216 in.³; square inches and cubic inches measure different attributes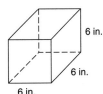

20 Find the volume of the solid. 1400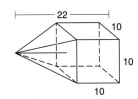

21 **Science** A hexagonal fish tank is filled with water. How much does the water weigh? ≈40.5 lb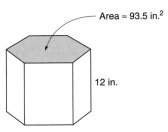

22 How many paths are there from A to B? The arrows indicate one-way paths.
  a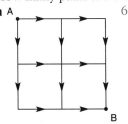
  b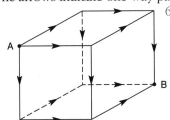

Communicating Mathematics

- Have students write a paragraph explaining the similarities and differences between prisms and pyramids.
- Problems 4–7, 11, 18, 19, and 24 encourage communicating mathematics.

Reteaching

Another way to find the lateral area of a prism is to multiply the perimeter of the base by the height. This method can be verified by the Distributive Property of Multiplication over Addition.

Additional Answers

for pages 578 and 579

12 b 240; 0; does not exist

Section 11.5 Polyhedra 579

SECTION 11.5

Checkpoint

1. Find the volume of the figure in problem 10 on page 578. (Hint: BC = 15.)
2. Find the volume of a square-based pyramid whose base is 5 cm on each side and whose height is 9 cm.
3. Find the total surface area of the figure in problem 8 b on page 577.
4. Find the total surface area of the figure in problem 8 c on page 577.

Answers:
1. 1200 units³   2. 75 cm³
3. 214 units²   4. 270 units²

Additional Answers
for page 580

24 e Since the input equals the output—that is, $f(\frac{1}{2}) = \frac{1}{2}$—we say that $\frac{1}{2}$ is fixed by the function f.

Investigation Possible answer:

Polyhedron	F	E	V
Cube	6	12	8
Octahedron	8	12	6
House	7	15	10
Right triangular prism	5	9	6

$F - E + V = 2$

P, R 23 Two polyhedral dice are shaped like an octahedron and a dodecahedron. The faces of each die are numbered with consecutive natural numbers, starting with 1. If the two dice are tossed, what is the probability that the sum of the two top faces is

a 2? $\frac{1}{96}$ b 7? $\frac{1}{16}$ c 18? $\frac{1}{32}$

Octahedron

Faces numbered 1–8

Dodecahedron

Faces numbered 1–12

◀ LOOKING BACK **Spiral Learning** LOOKING AHEAD ▶

R, I 24 In parts **a–d,** evaluate $f(x) = 3x - 1$ for each input.

a $f(4)$ 11   b $f(-2)$ -7 c $f(\frac{1}{3})$ 0   d $f(\frac{1}{2})$ $\frac{1}{2}$

e Explain why $\frac{1}{2}$ could be called a "fixed point" of function f.

R 25 At Abie's CD's, you can choose 5 songs from a list of 20 songs, and the store will make you a custom-made CD. How many different custom-made CD's can be made?   1,860,480

R, I 26 Solve each equation for x.

a $\frac{x-4}{7} = \frac{x}{5}$ $x = -10$ b $\frac{4}{x} = \frac{x}{9}$ $x = 6$ or $x = -6$

R 27 Find all ordered pairs of natural numbers (x, y) for which the expression $\frac{18}{x + 5y}$ is a natural number.   (1, 1), (4, 1), (13, 1), (8, 2), (3, 3)

R 28 **Science** Find the value of y so that the scale balances.   12 ft

Investigation

Polyhedra Make a table like the one shown, listing the number of faces (F), edges (E), and vertices (V) for a number of polyhedra, including the three pictured.

	F	E	V
Cube	6	12	8
Octahedron			
"House"			
.			
.			
.			

Examine your results and find a relationship between F, E, and V.

580   Chapter 11   Solving Problems with Algebra

SECTION 11.6

Investigation Connection
Section 11.6 supports Unit F, *Investigations* 2, 3. (See pp. 542A-542D in this book.)

SECTION 11.6
Solids with Circular Cross Sections

Objectives
- To find the volume and the total surface area of a cylinder
- To find the volume and the total surface area of a cone
- To find the volume and the total surface area of a sphere

Vocabulary: cone, cylinder, sphere

Class Planning

Schedule 2 days

Resources
In *Chapter 11 Student Activity Masters* Book
Class Openers 61, 62
Extra Practice 63, 64
Cooperative Learning 65, 66
Enrichment 67
Learning Log 68
Using Manipulatives 69, 70
Spanish-Language Summary 71

Transparencies 14, 48, 49

Class Opener

Let $A(r, h) = 2\pi r^2 + 2\pi rh$.
1. Evaluate $A(3, 4)$.
2. Evaluate $A(10, 5)$.
3. Evaluate $A(2, 1)$.

Answers:
1. ≈131.95   2. ≈942.48   3. ≈37.70

Lesson Notes

- Example 3 can lead to a lively discussion if students think about all of the air that is in most ice cream. This would be a great problem to test in the lab. We discussed this problem at length and left it as you see it but would be interested in hearing from you if you actually perform the experiment. You may want to

CYLINDERS

A **cylinder** is a geometric figure with two parallel congruent circular bases. You can think of a cylinder as being a prism with circular bases.

The formula for the volume of a cylinder is the same as the formula for the volume of a prism.

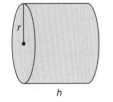
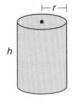

Volume = (area of base)(height)

Since the base of a cylinder is a circle, we can use the formula for the area of a circle to find the area of the base of a cylinder. Therefore, we can rewrite the formula for the volume of a cylinder as

$$\text{Volume}_{\text{cylinder}} = \pi r^2 h$$

The total surface area of a cylinder is found by adding the areas of the circular bases of the cylinder and the lateral area of the cylinder. If we think of a cylinder as a can, the lateral area is the area of the label.

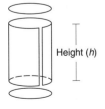

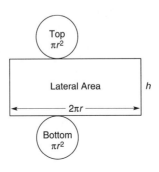

SECTION 11.6

Lesson Notes continued

coordinate the activity with your science teachers, especially if it is a hot day in May and you can get tubs of ice cream for your classes.

Additional Examples

Example 1
The height of a can is 5.6 in., and the length of a diameter is 4.2 in.
a. How much liquid will the can hold?
b. Find the area of the label that covers the side of the can.

Answers:
a. ≈77.58 in.³, or ≈$\frac{1}{3}$ gal
b. ≈73.89 in.²

Example 2
A cone has a diameter of 1.8 in., a height of 4.0 in., and a slant height of 4.1 in.
a. Find the volume of the cone.
b. Find the lateral area of the cone.

Answers:
a. ≈3.39 in.³ b. ≈11.59 in.²

Example 3
A spherical scoop of frozen yogurt 1.8 inches in diameter is placed in the cone in the previous example. If the yogurt melts into the cone, will it overflow the cone?

Answer: No; the volume of the scoop is only ≈3.05 in.³.

When we lay the label flat, we see that it is a rectangle with a length equal to the circumference of the can and a width equal to the height of the can.

$$\text{Lateral area}_{\text{cylinder}} = (\text{circumference})(\text{height})$$
$$= 2\pi r h$$

$$\text{Total surface area}_{\text{cylinder}} = \text{lateral area} + \text{area}_{\text{base 1}} + \text{area}_{\text{base 2}}$$
$$= 2\pi r h + 2(\pi r^2)$$

Example 1
The height of a can is 6.4 inches, and the length of a diameter is 4.8 inches.
a How much liquid will this can hold?
b Find the area of the label of the can.

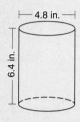

Solution
a First we will find the volume of the can. Since the radius of a circle is half the diameter, the radius of the base is 2.4 inches.

$$\text{Volume}_{\text{cylinder}} = \pi r^2 h$$
$$= \pi (2.4^2)(6.4)$$
$$\approx 116$$

The volume of the can is about 116 in.³. Since a gallon is about 231 in.³, this can holds about half a gallon.

b The label covers the lateral area of the can.

$$\text{Lateral area}_{\text{cylinder}} = 2\pi r h$$
$$= 2\pi (2.4)(6.4)$$
$$\approx 96.5$$

The lateral area of the can is about 96.5 in.².

CONES

Just as a cylinder is like a prism with circular bases, a **cone** is a pyramid with a circular base. The formula for the volume of a cone is like the formula for the volume of a pyramid. However, the area of the base of a cone is the area of a circle.

$$\text{Volume}_{\text{cone}} = \tfrac{1}{3}(\text{area of base})(\text{height})$$
$$= \tfrac{1}{3}\pi r^2 h$$

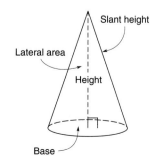

582 Chapter 11 Solving Problems with Algebra

SECTION 11.6

To find the surface area of a cone, we first need to determine the lateral area of the cone.

$$\text{Lateral area}_{\text{cone}} = \pi rL, \text{ where } L \text{ is the slant height}$$

The total surface area of the cone is the sum of the area of the circular base and the lateral area.

Example 2
An ice-cream cone has a diameter of 2.4 inches and a height of 4.5 inches. The slant height is about 4.65 inches.
a Find the volume of the cone.
b Find the lateral area of the cone.

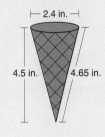

Solution
a The radius of the cone's base is $\frac{2.4}{2}$, or 1.2, inches.

$$\begin{aligned}\text{Volume}_{\text{cone}} &= \tfrac{1}{3}\pi r^2 h \\ &= \tfrac{1}{3}\pi (1.2^2)(4.5) \\ &= 2.16\pi \\ &\approx 6.8\end{aligned}$$

The volume of the cone is about 6.79 in.^{3}.

b
$$\begin{aligned}\text{Lateral area}_{\text{cone}} &= \pi rL \\ &= \pi (1.2)(4.65) \\ &\approx 17.5\end{aligned}$$

The lateral area of the cone is about 17.5 in.^{2}.

SPHERES

A common geometric shape is a **sphere**. Baseballs, soap bubbles, and planets are all approximately spherical in shape. A sphere has a radius and a diameter, but it does not have any edges. The formula for the volume of a sphere is

$$\text{Volume}_{\text{sphere}} = \tfrac{4}{3}\pi r^3$$

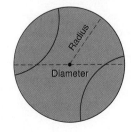

The formula for the total surface area of a sphere is

$$\text{Total surface area}_{\text{sphere}} = 4\pi r^2$$

Section 11.6 Solids with Circular Cross Sections

Assignment Guide

Basic
Optional

Average or Heterogeneous
Day 1 6–8, 11, 12, 15, 18
Day 2 9, 10, 13, 14, 16, 19, 20

Honors
Day 1 6, 9–11, 14, 15, 19
Day 2 7, 8, 12, 13, 16, 17, 20, 21

Strands

Communicating Mathematics	1, 3, 15
Geometry	1–20, 22, Investigation
Measurement	1–14, 16–20, 22, 28, Investigation
Technology	2, 4–6, 8–10, 12–17, 19, 20, 26
Algebra	1, 4, 7, 18, 21–27, Investigation
Patterns & Functions	26, 27
Number Sense	1, 21, 26, 28
Interdisciplinary Application	1–3, 6–10, 12, 13, 15, 16, 19, 20, 23, Investigation

Individualizing Instruction

The gifted learner Students could create a spreadsheet application to see how the volume of a sphere is affected by increases in radius. Have students look for patterns, such as what happens when the radius is doubled or tripled.

SECTION 11.6

Reteaching

Students may not readily accept that the volume of a cone is one third the volume of a cylinder. You might want to demonstrate this by filling a cone with sand or water and pouring it into a cylinder that has the same dimensions. Students will see that the cone needs to be filled three times in order to fill the cylinder.

Cooperative Learning

■ Have small groups of students consider the following questions.

1. If the radius of the base and the height of a cylinder must have a sum of 15 (using only whole numbers), what combination will yield the greatest volume?

2. What if the radius and the height total 30?

3. Predict what combination will maximize the volume if the sum is allowed to be 45. Verify your prediction.

Answers:
1. $r = 10$, $h = 5$
2. $r = 20$, $h = 10$
3. $r = 30$, $h = 15$

■ Problems 10 and 16 are appropriate for cooperative learning.

Example 3

A spherical scoop of ice cream has a diameter of 2.4 in. If it is placed in the cone described in Example 2 and melts into the cone, will the melted ice cream overflow the cone?

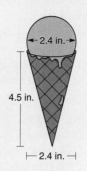

Solution

In Example 2, we found that the volume of the cone is 6.79 in.³. The radius of the scoop of ice cream is the same as the radius of the cone—1.2 in.

$$\text{Volume}_{\text{sphere}} = \tfrac{4}{3}\pi r^3$$
$$= \tfrac{4}{3}\pi(1.2^3)$$
$$= \tfrac{4}{3}\pi(1.728)$$
$$= 7.24$$

The volume of the scoop of ice cream is about 7.24 in.³. Since the volume of the ice cream is greater than the volume of the cone, the ice cream will overflow the cone when it melts.

Sample Problem

Problem

A cone-shaped paper cup has a height of 8 cm and a radius of 4 cm. If you fill the cup with water to half of its depth, what portion of the volume of the cup is filled with water?

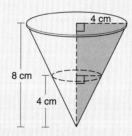

Solution

In the drawing, the smaller triangle is similar to the larger triangle. The ratio of the corresponding sides is 1:2. Therefore, the cone of water has half the height and half the radius of the cup. We will find the ratio of the volume occupied by the water to the volume of the whole cone.

$$\frac{\text{Volume of water}}{\text{Volume of cone}} = \frac{\tfrac{1}{3} \cdot \pi \cdot 2^2 \cdot 4}{\tfrac{1}{3} \cdot \pi \cdot 4^2 \cdot 8} = 0.125$$

584   Chapter 11   Solving Problems with Algebra

SECTION 11.6

Think and Discuss

P **1** A tube is just large enough to contain three tennis balls. Which is greater, the height of the tube or the circumference of the tube? Explain your answer.
The circumference, since it is about π(diameter of ball), whereas the height is about 3(diameter of ball), and π > 3.

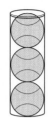

P, R **2** If the full glass of water is poured into the rectangular container shown, will the water overflow the container? Yes (barely)

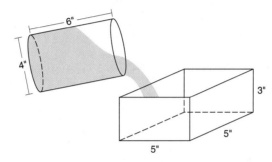

P **3** An egg is not a sphere. What are some possible ways to find the volume of an egg?

P **4** A sphere has a radius of 6 feet.
  a Find the volume and the total surface area of the sphere. ≈904.8 ft^3; ≈452.4 ft^2
  b If the volume and the surface area of a sphere are numerically equal, what is the sphere's radius? 3 units

P **5** A cone is 10 cm in diameter and 12 cm in height.
  a Find the lateral area. ≈204.2 cm^2
  b Find the area of the base. ≈78.5 cm^2
  c Find the total surface area. ≈282.7 cm^2
  d Find the volume. ≈314.2 cm^3

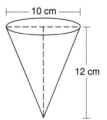

Problems and Applications

P **6** A farmer carries liquid fertilizer in a cylindrical tank that is 12 feet long and 6 feet in diameter.
  a Find the volume of the tank in cubic feet.
  b Find the number of gallons that the tank will hold. (There are about 7.48 gal per ft^3.)

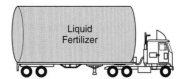

Section 11.6 Solids with Circular Cross Sections   585

Problem-Set Notes

for page 585

- **2** would be great to do as an experiment.

Communicating Mathematics

- Have students write a paragraph explaining the difference between lateral area and total surface area. Do all figures have both lateral area and total surface area? Explain.
- Problems 1 and 15 encourage communicating mathematics.

Stumbling Block

Students may accidentally use diameter in a formula that calls for radius. Remind students to check which measurement is given and which is needed when presented with a problem.

Additional Answers

for page 585

2 The volume of the pan is 75 cubic inches; the volume of the glass is about 75.4. The pan will overflow, but only by a small amount. (Surface tension may keep the water from overflowing if you are careful.)

3 Possible answer: Immerse the egg in water and find the volume of the water displaced.

6 a 108π, or ≈339.29, cubic feet
b ≈2538 gallons

SECTION 11.6

Problem-Set Notes

for pages 586 and 587

- **8** can be demonstrated physically with paper, tape, and sand.
- **15** makes connections to science. You might ask students how crushing a cube affects the rate of cooling.
- **16** calls on students to apply several concepts.

Checkpoint

1. Find the volume and total surface area of a cylinder with height 15 and radius 8.
2. Find the volume and total surface area of a spherical softball that has a radius of 2 inches.
3. Find the volume of a cone with height 12 and radius 8.

Answers:
1. ≈3015.9 units³; ≈1156.1 units²
2. ≈33.5 in.³; ≈50.3 in.²
3. ≈804.2 units³

P, R **7** A cone is partially filled with water. The radius of the cone is 12". The height of the cone is 18". The radius of the surface of the water is 4".
  a How deep is the water in the cone? 6 in.
  b What fraction of the cone is filled with water? $\frac{1}{27}$

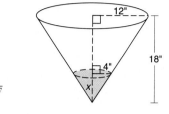

P **8** A 4" by 8" sheet of paper can be rolled into a cylinder in two ways. Circular caps can close off the ends.
  a Which cylinder has the greater lateral area? They are the same.
  b Which cylinder has the greater total surface area? A
  c Which cylinder has the greater volume? A

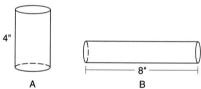

P **9** A scoop of ice cream has a diameter of 4.2 cm. It is placed in a cone with a diameter of 4 cm and a height of 6 cm. If the ice cream melts into the cone, will it overflow the cone? Yes

P **10** One thousand gallons of water is pumped into a holding tank that is 20 ft in diameter and 30 ft tall. What is the approximate depth of the water in the tank? ≈0.43 ft

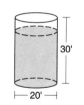

P, R **11** Consider the two cylinders shown.
  a Find the ratio of the radii of the two cylinders. $\frac{1}{3}$
  b Find the ratio of the heights of the two cylinders. $\frac{1}{3}$
  c Find the ratio of the volume of the smaller cylinder to the volume of the larger cylinder. $\frac{1}{27}$

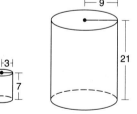

P **12** **Science** The earth is almost a sphere, with a diameter of 7920 miles. About 29% of the surface of the earth is land.
  a Find the volume of the earth. ≈2.6 × 10¹¹ mi³
  b Find the total surface area of the earth. ≈1.97 × 10⁸ mi²
  c Find the approximate land area on the surface of the earth. ≈5.71 × 10⁷ mi²

P, R **13** In Wet City each person uses approximately 100 gallons of water per day. The city reservoir is a cylinder 100 ft across and 30 ft deep. For how many people will the reservoir provide water in a day? ≈18,000 people

586   Chapter 11   Solving Problems with Algebra

SECTION 11.6

P **14** Which has the larger volume, a cube with sides of 3 units or a sphere with a radius of 2 units?   The sphere has the larger volume.

P **15 Science** A cubical piece of ice and a spherical piece of ice have the same volume. The pieces are submerged in two identical glasses of Cool-Aid. The cooling rate of a piece of ice depends on the surface area of ice exposed to the liquid. Which piece do you think will cool the drink faster? Explain your response.

P **16** An ice machine makes 20-mm cubes that have a hole through the center. Each hole has a 6 mm diameter.
  a Find the total surface area of the ice cube shown, including the inside walls of the hole.   ≈2720 mm²
  b Find the volume of the ice in the cube shown.   ≈7434.5 mm³

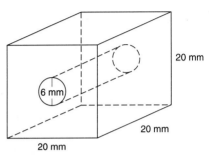

P **17** Can 50 marbles, each with a radius of 1 cm, be melted down to form a sphere with a radius of less than 4 cm?   Yes

P, R **18** The cylinder, the hemisphere, and the cone have the same radius r and height h. Find the ratio

$$\text{Volume}_{\text{cylinder}}:\text{Volume}_{\text{hemisphere}}:\text{Volume}_{\text{cone}} \quad 3:2:1$$

P, R **19 Science** The fuel for this rocket takes up 40% of the volume of the rocket. The diameter of the rocket is 12 ft. (Hint: In this problem, divide and conquer.)
  a Find the volume of the rocket that can be used for the payload.   ≈5067 ft³
  b Find the lateral area of the nose cone.   ≈188.5 ft²

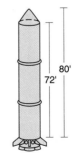

P **20** A piston cylinder has a diameter of 6 cm. If 60 cm³ of oil is pumped into the cylinder, how far does the oil push the piston rod?   ≈2.122 cm

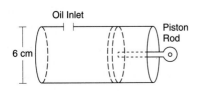

Additional Practice

1. Find the volume and total surface area of an aluminum can with height 14 cm and diameter 6 cm.

2. Find the lateral area of a cone with slant height 15 and radius 9.

3. Find the volume and total surface area of a globe with a diameter of 10 inches.

Answers:
1. ≈395.8 cm³; ≈320.4 cm²
2. ≈424.1 units²
3. ≈523.6 in.³; ≈314 in.²

Additional Answers

for page 587

15 Possible answer: The cube has more, so it will cool the drink faster. For example, a cube with a volume of 1000 units³ has a surface area of 600 units², but a sphere with a volume of 1000 units³ has a surface area of about 483 units².

17 Yes; the volume of the large sphere $(64 \cdot \frac{4}{3}\pi)$ is greater than the combined volumes of the small spheres $(50 \cdot \frac{4}{3}\pi)$.

SECTION 11.6

◀ LOOKING BACK **Spiral Learning** LOOKING AHEAD ▶

P **21** Find all whole number values of x for which $\frac{33}{2x-1}$ is a whole number.

P, R **22** A pyramid with a square base has a height of 6 and a volume of 32. What is the length of the sides of the square? 4

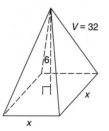

R **23** Deb U. Tant wanted to rent a room for a dance. Holly's Day Inn charges $200 plus $80 per hour. Sherry's Tonn charges $300 plus $60 per hour. Amazingly, at either place the total cost would have been the same. For how many hours did Deb plan to rent the room? 5 hours

R, I **24** Solve $x = 3y + 5$ for (x, y) if $6x - 2 = 4y$. $(-1, -2)$

R **25** Solve each equation for x.
a $2(x + 3) = 13 - 5x$ $x = 1$
b $x - 2x + 3x - 4x + 5x - 6x + 7x = 8x + 20$ $x = -5$

R **26** a Evaluate $\sqrt{.04}$ 0.2
b Evaluate $\sqrt{\sqrt{.04}}$ ≈ 0.4472
c Evaluate $\sqrt{\sqrt{\sqrt{.04}}}$ ≈ 0.6687
d Continue the same process, taking the square root of the answer to each previous problem. What value do your answers seem to approach? 1

R, I **27** If $f(1) = 7$ and $f(n + 1) = f(n) + 4$, what is the value of each of the following?
a $f(1)$ 7
b $f(2)$ 11
c $f(3)$ 15
d $f(4)$ 19
e $f(5)$ 23

R **28** Arrange the following from least to greatest by weight.
a A two-liter bottle of cola
b A two-liter bottle of diet cola
c A two-liter bottle of water   They weigh about the same amount.

Investigation

Ratio Find the ratio of the diameter of a 12-ounce soft drink can to the height of the can. Do the same for other cans. Are the ratios the same? What is the range of ratios that you find?

SECTION 11.7

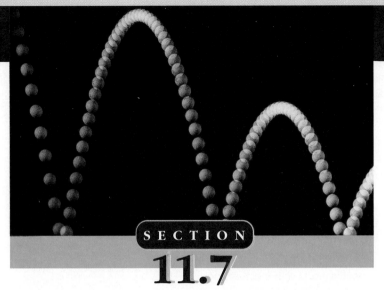

SECTION 11.7
Reading Information from Graphs

As you have seen in earlier chapters, graphs can help us see information quickly. Let's look at some more examples.

Example 1
Solve the equation $2x - 8 = -4x + 7$ in the following two ways.
a By graphing the two equations $y = 2x - 8$ and $y = -4x + 7$
b Algebraically

Solution

a We can sketch the graphs of the two equations $y = 2x - 8$ and $y = -4x + 7$. Where do the two lines seem to intersect? They seem to intersect at $(2.5, -3)$. Thus, $x = 2.5$. We must test the value in the equation in order to verify that it is correct.

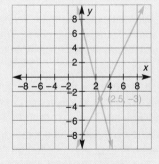

Left Side	**Right Side**
$2x - 8$	$-4x + 7$
$= 2(2.5) - 8$	$= -4(2.5) + 7$
$= 5 - 8$	$= -10 + 7$
$= -3$	$= -3$

Since the left side is equal to the right side, $x = 2.5$. Can you describe what the -3 represents in the ordered pair $(2.5, -3)$?

b
$$2x - 8 = -4x + 7$$
$$\underline{+ 4x \qquad\quad + 4x}$$
$$6x - 8 = \qquad 7$$
$$\underline{\quad + 8 \qquad + 8}$$
$$6x \quad = \quad 15$$
$$x = \frac{15}{6} = 2.5$$

Objective
- To read information from graphs correctly

Class Planning

Schedule 2 days
Resources
In *Chapter 11 Student Activity Masters* Book
Class Openers 73, 74
Extra Practice 75, 76
Cooperative Learning 77, 78
Enrichment 79
Learning Log 80
Interdisciplinary
  Investigation 81, 82
Spanish-Language Summary 83

Class Opener

A piggy bank contains $0.95 in nickels and dimes. You shake it, but you cannot determine the number of coins. Could it contain
1. 9 coins?   2. 11 coins?
3. 13 coins? 4. 18 coins?

Answers:
1. No   2. Yes: 8 d, 3 n
3. Yes: 6 d, 7 n   4. Yes: 17 n, 1 d

Lesson Notes

- Graphing both sides of an equation on a coordinate plane and looking at the points of intersection, (x, y), gives us the x values that satisfy the equation. The y values are the values of each side when an x value is substituted into each expression. A spreadsheet with columns representing the varying x values and the left- and right-side expressions might also help students see when the left side equals the right side. If the spreadsheet has

Section 11.7 Reading Information from Graphs

SECTION 11.7

Lesson Notes continued

graphing capabilities, you can show the point of intersection of the graphs.

- Both graphing and algebraic solutions are given in example 1. Be sure that students see the connection between the graph and the solution of the original equation.

- In example 2, since we are dealing with discrete numbers of coins, a scattergram of coins and values of coins is a better representation, though this may not always be practical.

- Sample problem 2 is an example of a problem that necessitates the use of technology. Either a graphing calculator or a graphing program on a computer may be used.

Additional Examples

Example 1
Solve the equation $x - 5 = 3x + 1$ by graphing and by algebra.

Answer:

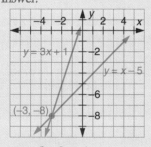

$$x - 5 = 3x + 1$$
$$-2x - 5 = 1$$
$$-2x = 6$$
$$x = -3$$

Example 2
Joe has 9 coins. Each coin is either a penny or a nickel. He has a total of 29 cents. How many pennies and nickels does Joe have?

Answer: 5 nickels and 4 pennies

590

Example 2
Cashe Box has a group of nickels and dimes with a total value of 55 cents. Cashe has 8 coins. How many nickels and dimes does Cashe have?

Solution
Although we could use trial and error to get an answer, let's look at another way to model this problem:

Let D = the number of dimes.
Let N = the number of nickels.
We can now write the equation $D + N = 8$.

One way to solve this is with 8 dimes and no nickels. We graph this as (8, 0). He could also have 8 nickels and no dimes. We graph this as (0, 8). We then connect the two points. This is the graph of $D + N = 8$.

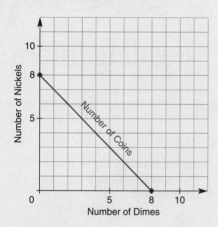

Now let's represent the value of the coins. Each dime is worth 10 cents and each nickel is worth 5 cents. The total value of the dimes is $10D$ and the total value of the nickels is $5N$. The total value of the coins is $10D + 5N = 55$.

If $D = 0$, there must be 11 nickels ($5N = 55$). We can graph this as (0, 11). If $N = 0$, there must be 5.5 dimes ($10D = 55$). Even though Cashe can't have half a dime, we can still use the point (5.5, 0) as a point on the graph.

The intersection of the two lines is (3, 5). This means that 3 dimes and 5 nickels are the 8 coins with a total value of 55 cents.

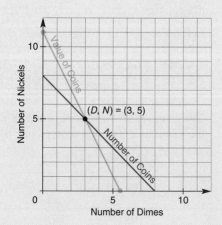

590   Chapter 11   Solving Problems with Algebra

SECTION 11.7

What if there were 9 coins? We can adjust the graph. Just slide the graph of the number of coins to the points (0, 9) and (9, 0) and look for the solution.

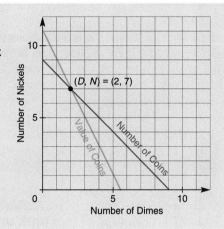

Calculators with graphing capabilities and computers with a graphing program enable us to examine graphs that are too complicated to draw.

Sample Problems

Problem 1 Graph the equation $h = 6 + 64t - 16t^2$. This equation gives the height in feet, h, of a rock that is tossed in the air at a time in seconds, t, after the rock is tossed.
  a What is the maximum height that the rock will reach?
  b The rock will reach a height of 54 feet twice, once going up and once coming down. Identify those two times (in seconds).
  c After how many seconds will the rock hit the ground?
  d From what height was the rock tossed?

Solution With the aid of a graphing calculator or a computer graphing program, we can find the information.
  a The maximum height is 70 feet.
  b The rock reaches 54 feet at 1 second and at 3 seconds.
  c The rock will hit the ground in approximately 4.1 seconds.
  d The rock was tossed from a height of 6 feet.

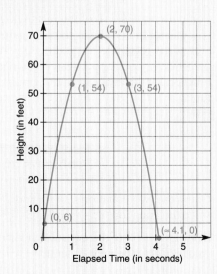

Assignment Guide

Basic
Optional

Average or Heterogeneous
Day 1 6, 8, 12, 15, 18–20, 29
Day 2 9, 10, 13, 16, 17, 21, 23, 26

Honors
Day 1 6, 10, 12, 13, 19–23
Day 2 7–9, 17, 18, 24–26, 28, 29

Strands

Communicating Mathematics	9, 12, 18
Geometry	4, 5, 15, 19–22, 25
Measurement	4, 5, 7, 10, 14, 15, 17, 19–22, 25
Data Analysis/ Statistics	12, 16, 18
Technology	1, 5, 22, 25, Investigation
Algebra	1–11, 13–17, 19, 20, 22, 23, 26–30
Patterns & Functions	5, 9, 13, 17, 29
Number Sense	12, 21, 24, 27, 30
Interdisciplinary Applications	10, 12, 14, 16–18, 29, Investigation

Individualizing Instruction

The gifted learner Students might be challenged to verify the results of sample problem 1 by creating a spreadsheet application that relates time and height. Students might use intervals of 0.01 between 4.00 and 4.20 seconds to come up with a more exact time for when the rock hits the ground.

SECTION 11.7

Problem-Set Notes

for pages 592 and 593

- 1–5 could all be done by using technology. Using computers or graphing calculators will help most students—especially the visual learners—understand the concepts.
- 10 and 11 require that students use a graphing calculator or computer.
- 12 c should be discussed, as there may be many valid answers. Predictions are not certainties.

Additional Practice

1. Solve the equation $3x + 1 = -x + 5$.
2. Solve the equation $3x - 4 = -2x + 1$.
3. Craig invented a card game in which each card is worth either 5 points or 9 points. Write an equation showing the number of cards F worth five points and the number of cards N worth nine points that will yield 75 points. Find all ordered pairs (F, N) that will give the 75 points.

Answers:
1. $x = 1$ 2. $x = 1$
3. $5F + 9N = 75$; $(6, 5), (15, 0)$

Checkpoint

1. Solve the equation $3x = -x + 4$.
2. Two angles are supplementary. The measure of one angle is three times the measure of the other. Find the measure of the two angles.
3. Movie tickets cost \$6 for adults and \$4 for children. Write a function that represents the total cost of A adults'

Problem 2 Solve the equation $2^x = 2x$.

Solution Graph the two equations $y_1 = 2^x$ and $y_2 = 2x$. There are two points of intersection of the graphs. We are looking for the x-coordinates. They are at $x = 1$ and at $x = 2$.

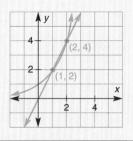

Think and Discuss

P 1 Use a graphing calculator or a computer to create the graphs of the equations $y = x^2$ and $y = 2^x$. Then use the trace or zoom feature to find the three points of intersection. $(2, 4), (4, 16),$ and $(\approx -0.77, \approx 0.59)$

P 2 On graph paper, draw the graph of the equation $y = \frac{1 + x}{4 + x}$ for the whole-number values $x = 0, 1, 2, \ldots, 10$. For what values of x is $y \geq 0.6$? $x \geq 4$

P 3 Box-seat tickets cost \$7 each, and grandstand seats cost \$4 each.
   a Write a formula that represents the cost of B box seats and G grandstand seats. $C = 7B + 4G$
   b Find all combinations (B, G) that could be bought with exactly \$100.

P, R 4 Two angles are complementary. The measure of one angle is 4 times the measure of the other.
   a Write two equations to describe the two facts. $x + y = 90, y = 4x$
   b Graph the two equations on the same coordinate system and find the measures of the two angles.   18 and 72

P 5 Compute the area, A, of the rectangle for values of x greater than 0. Then sketch a graph of the area of the rectangle by plotting the ordered pairs (x, A). Check your result by writing a function for the area, $A(x)$, and graphing the function with a graphing calculator or computer.
   $A(x) = (10 - x)(10 + x)$

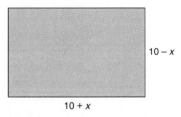

Problems and Applications

P 6 Graph the two equations $y = 6x$ and $y = -2x + 16$. Use the graphs to solve the equation $6x = -2x + 16$. $x = 2$

P 7 Draw the graph of the equation $A = \pi r^2$. (Hint: You may need to find values of A to the nearest tenth.)

592   Chapter 11   Solving Problems with Algebra

SECTION 11.7

P **8** For the equation $y = 16 - 4t - t^2$,
  a Find the values of y when $t = 0, 1, 2, 3, 4, 5,$ and 6. Graph the seven points (t, y).   16; 11; 4; −5; −16; −29; −44
  b What value of t gives the greatest value of y? $t = 0$

P, I **9** **Communicating** Explain how a graph can be used to determine the value in cents, y, of a number of dimes, x.

P, I **10** **Science** I shot an arrow into the air, with an initial vertical velocity of 48 feet per second. It left my hand at a height of about 5 feet. Its height-versus-time equation is $h = -16t^2 + 48t + 5$.
  a What is the maximum height the arrow will reach?   ≈41 ft
  b After how many seconds will it hit the ground?   ≈3.1 sec

P **11** Graph the two equations $y = 2^x - 3$ and $y = 3x + 1$.
  a Use the graph to solve the equation $2^x - 3 = 3x + 1$ for x.
  b Verify the result by substituting the value of x into the equation.

P, R **12** **Business** A company's profits for the years 1975 through 1990 are shown on the graph.

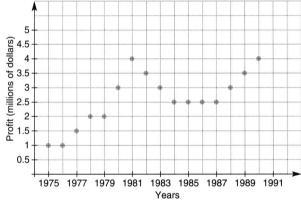

  a Between what years were the profits declining?   Between 1981 and 1984
  b In what years were the profits at an all-time high?   1981, 1990
  c Use the graph to project what the profits will be in the next two years. Explain your answer.

P, R **13** In the game of LegBall, it is possible to score 8 points with a touch-up and 5 points with a parkgoal.
  a Write a function for the value, V, of t touch-ups and p parkgoals.
  b In the Colossal Bowl, a team scored 34 points. Write an equation to represent this, and graph the result. $8t + 5p = 34$
  c What point on the graph represents the kinds of scores made?   (3, 2)

P **14** Let x be the number of gallons in a batch of punch. Ginger ale makes up 60% of the punch.
  a Write a formula to describe the number of gallons of ginger ale in x gallons of punch. $y = 0.6x$
  b Draw the graph of the equation.

Section 11.7 Reading Information from Graphs

---

tickets and C childrens' tickets. Find all combinations (A, C) that could be bought with exactly $50.

Answers:
1. $x = 1$ 2. 45°; 135°
3. $f(A, C) = 6A + 4C$; (1, 11), (3, 8), (5, 5), (7, 2)

Additional Answers
for pages 592 and 593

2
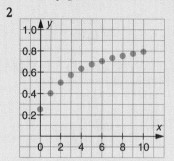

3 b
Box seats	0	4	8	12
Grandstand	25	18	11	4

4 b
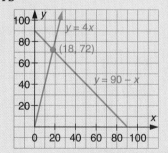

Answers for problems 5–7, 8 a, 9, 11, 12 c, 13 a and b, and 14 b are on pages A9 and A10.

CHAPTER 11.7

Problem-Set Notes

for pages 594 and 595

- **15** can be also done by using a spreadsheet.
- **18** requires some creativity. You might suggest that the y-axis represents distance traveled.
- **24** previews the next chapter.

Multicultural Note

During the Middle Ages, the mathematicians of Europe adopted the word *algebra* to mean the study of methods of finding the numbers that satisfy conditions which are expressible as equations. Now, we more accurately call this the "theory of equations."

Cooperative Learning

Have students work together to solve the more challenging homework problems. Again, heterogeneous ability groups may work well. If any students own a graphing calculator or know how to run a graphing program, have them demonstrate how the software can be used to solve problems 10 and 11.

Communicating Mathematics

- Have students write a paragraph discussing how making a graph can be useful in solving problems. For what types of problems is it a good strategy?
- Problems 9 and 12 encourage communicating mathematics.

R, I **15** The length of the rectangle shown is 20 more than the width. The perimeter of the rectangle is 80.
   a Write two equations to describe the given conditions. $\ell = w + 20,\ 80 = 2\ell + 2w$
   b Graph the equations you wrote in part **a** and find ℓ and w. $\ell = 30,\ w = 10$

R, I **16** The Very Little Theater sold 17 tickets for a performance and took in $3.00. Adult tickets cost $0.30 and children's tickets cost $0.15. The manager, Joey, wrote the equations $A + C = 17$ and $30A + 15C = 300$, where A represents the number of adults and C represents the number of children. Joey used a graph to answer the following questions.
   a How many of each kind of tickets were sold to produce total sales of $3.00?   14 children, 3 adults
   b The next day Joey also made $3.00 but sold only 12 tickets. How many of each kind of ticket were sold?   4 children, 8 adults

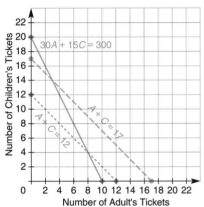

P, R **17** **Science** Use the values $F = 32, 41, 50, 59, 68, 77, 86, 95,$ and 104 to draw the graph of the equation $C = \frac{5}{9}(F - 32)$. What does this equation represent?

P **18** **Communicating** Write a story that describes the graph shown.   Answers will vary.

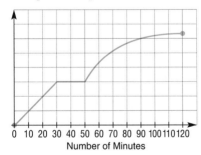

◀ LOOKING BACK **Spiral Learning** LOOKING AHEAD ▶

R **19** A cone has a radius of 5 and a slant height of 13.
   a Find the height of the cone.   12
   b Find the lateral area of the cone. 65π, or ≈ 204.2
   c Find the volume of the cone. 100π, or ≈ 314.2

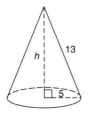

594 Chapter 11 Solving Problems with Algebra

CHAPTER 11.7

R **20** There are approximately 7.48 gallons in a cubic foot. How many gallons can a cylindrical barrel hold if it is 4 feet tall and 3 feet in diameter?

R **21** Which has the greatest volume, cylinder A, cylinder B, or cone C?   B

 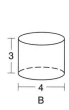 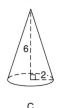

R   **22** A cylindrical drainpipe is 60 feet long and has a diameter of 4 inches.
    a Find the lateral area of the drainpipe. 20π, or ≈ 62.8, ft^2
    b What volume of water can the pipe hold at one time? $\frac{5}{3}\pi$, ≈ 5.24 ft^3

R **23** Solve each equation for x.
    a $-2x - 7 = -4x + 13$ 10
    b $-\frac{2}{3}x + 5 = -\frac{1}{3}x + 10$ -15

R **24** Find the value of $\sqrt{\sqrt{\sqrt{\sqrt{\sqrt{\sqrt{\sqrt{\sqrt{3}}}}}}}}$ to the nearest thousandth. ≈ 1.004

R **25** Find the surface area of a hemisphere that has a radius of 12 cm. (Hint: Be sure to include the flat face.) 432π, or ≈ 1357, cm^2

R **26** Coina Clecter has 50 dimes for every 7 nickels she has. What is the value of her coins if she has 228 coins? $21.40

R, I **27** Find all pairs of whole numbers (x, y) for which the value of $\frac{17}{3x + 4y}$ is a whole number.   (3, 2)

R **28** Solve each equation for x.
    a 20% of x = 12% of $(x + 20)$ $x = 30$
    b $12 - 3x = 15\left(1 + \frac{1}{3}x\right)$ $x = -\frac{3}{8}$

R, I **29** The total number of squares in an n-by-n checkerboard is given by the formula $f(n) = \frac{n(n+1)(2n+1)}{6}$. How many squares are in

    a A 4-by-4 checkerboard?   30
    b An 8-by-8 checkerboard?   204
    c A 12-by-12 checkerboard?   650

R **30** Evaluate each expression.
    a $(-3)(-4) + 5(-6)$ -18
    b $\left(-\frac{1}{2}\right)\left(-\frac{2}{3}\right)\left(-\frac{3}{4}\right)\left(-\frac{4}{5}\right)\left(-\frac{5}{6}\right)$ $-\frac{1}{6}$
    c $\frac{-3}{5} + \frac{4}{7}$ $-\frac{1}{35}$
    d $\frac{1}{2} + \frac{1}{3} + \frac{1}{4}$ $\frac{13}{12}$

Investigation

Graphs Interview people in various lines of work to find out how they use graphing calculators or computer graphics in their jobs.

Reteaching

Use logical reasoning and guess-and-check to reteach example 2. Start by having students notice that since 55 is not a multiple of 10, Cashe must have an odd number of dimes. Since Cashe has 8 coins, he could have 1, 3, 5, or 7 nickels. If Cashe had 1 nickel, he'd have 7 dimes, which totals more than 55¢. Similarly, 3 nickels and 5 dimes is more than 55¢. Five nickels and 3 dimes is the answer.

Additional Answers

for pages 594 and 595

15 b

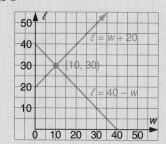

17

F	32	41	50	59	68	77	86	95	104
C	0	5	10	15	20	25	30	35	40

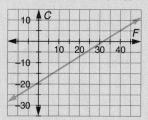

Converts temperature in degrees Fahrenheit to degrees Celsius.

20 211.5 gal

Investigation Answers will vary. Many people in fields related to science, engineering, product design, or data analysis will use either or both.

SUMMARY

CHAPTER 11

Summary

CONCEPTS AND PROCEDURES

After studying this chapter, you should be able to
- Develop strategies for solving problems (11.1, 11.3)
- Solve equations with variables on both sides (11.2)
- Solve equations with several variables (11.4)
- Find the volume and the total surface area of a prism (11.5)
- Find the volume and the total surface area of a pyramid (11.5)
- Find the volume and the total surface area of a cylinder (11.6)
- Find the volume and the total surface area of a cone (11.6)
- Find the volume and the total surface area of a sphere (11.6)
- Read information from graphs (11.7)

VOCABULARY

cone (11.6)
cylinder (11.6)
edge (11.5)
face (11.5)
lateral area (11.5)
lateral face (11.5)

polyhedra (11.5)
prism (11.5)
pyramid (11.5)
slant height (11.5)
sphere (11.6)
vertex (11.5)

REVIEW

CHAPTER 11

Review

1. If one of your friends didn't believe that the formula for the volume of a cone should include the factor $\frac{1}{3}$, how could you convince your friend that $V = \frac{1}{3}\pi r^2 h$ is correct?

2. The price of garbanzo beans increased 20%, to a cost of $45 per ton. What was the cost before the price increase? $37.50 per ton

3. A rectangle has a length of x meters and a width of 12 meters. A square with the same perimeter has a side of 17 meters. Find the length of the rectangle.   22 m

4. Mark Kett invests $1000 in each of two stocks. The value of the first stock increases 3% per month, and the value of the second increases 2% per month. What is the total value of the stocks for each month up to one year? (Hint: Use a spreadsheet.)

5. Find the value of x if the perimeters of the rectangle and the triangle are equal. $x = 7$

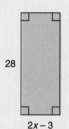
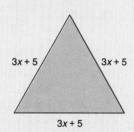

6. Freddy leaves his home on the corner of Maple Street and heads east on Forest Street. Jason leaves camp and heads west on Forest Street at exactly the same time, traveling 6 mph. They meet at Oak Street. How fast was Freddy traveling?   4 mph

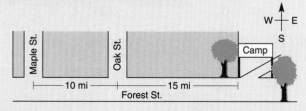

7. Simplify each expression and solve each equation.
   a. $50 - 3x = 5(x + 18)$ $x = -5$
   b. $50 - 3x - 5(x + 18)$ $-40 - 8x$
   c. $\frac{1}{2}(2x + 8) + \frac{1}{4}(8x - 20)$ $3x - 1$
   d. $\frac{1}{2}(2x + 8) = \frac{1}{4}(8x - 20)$ $x = 9$

8. Video Heaven rents movies for $3 per night. Video Heaven also offers a video club plan. The plan costs $100 per year and allows movie rentals at $1 per night plus two free rentals per month. How many movies must you rent per year to make the video club worthwhile?   38

Chapter Review   597

Assignment Guide

Basic
Optional

Average or Heterogeneous
Day 1   2, 4, 6–9, 13, 15, 16, 18, 19, 21, 24

Honors
Day 1   1–8, 10–15, 17, 19–21, 23, 24

Strands

Communicating Mathematics	1
Geometry	1, 3, 5, 11, 14, 16–20, 22
Measurement	3, 5, 6, 11, 14–17, 19, 20, 22, 24
Data Analysis/ Statistics	6, 9
Technology	4, 10, 24
Algebra	1–3, 5–9, 11–15, 18, 21–24
Number Sense	12
Interdisciplinary Applications	2, 4, 6, 8, 9, 13, 16, 23

Additional Answers

for page 597

1. Possible answer: A cone is $\frac{1}{3}$ of a cylinder, which has a volume of $\pi r^2 h$.

Answer for problem 4 is on the following page.

597

REVIEW

Additional Answers
for pages 597, 598, and 599

4

	Stock 1	Stock 2	Total
Jan	$1030.00	$1020.00	$2050.00
Feb	1060.90	1040.40	2101.30
Mar	1092.73	1061.21	2153.94
Apr	1125.51	1082.43	2207.94
May	1159.27	1104.08	2263.35
Jun	1194.05	1126.16	2320.21
Jul	1229.87	1148.69	2378.56
Aug	1266.77	1171.66	2438.43
Sep	1304.77	1195.09	2499.86
Oct	1343.92	1218.99	2562.91
Nov	1384.23	1243.37	2627.60
Dec	1425.76	1268.24	2694.00

12 (12, 2), (9, 4), (6, 6), (3, 8)

19 a Volume: 904.8 units^3, Surface Area: 452.4 units^2
b Volume: 339.3 units^3, Surface Area: 282.7 units^2
c Volume: 1005.3 units^3, Surface Area: 628.3 units^2

23 a Possible answers: Make a table, guess and check, solve an equation.

---

9 Katrina earns $12.68 per hour for up to 8 hours a day. She earns time and a half for additional hours.
 a How much does she earn per hour of overtime? $19.02
 b If she works for 8 hours at regular pay and then works 3 hours overtime, what is her total earning for that day? $158.50
 c If she works the 11-hour day in part **b**, what is her average hourly wage for that day? $14.41

10 When I was a lad (9 years old), I worked for my dad (he was 5 times as old as I). My sister was born that year. Use a spreadsheet to determine
 a How old I was when my dad was 10 times as old as my sister 14
 b How old I was when my dad was twice as old as I 36

11 Triangle ABC is similar to triangle DEF. Use a proportion to find the value of x. $x = 13\frac{1}{3}$

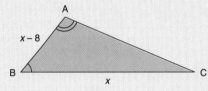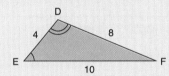

12 Find all pairs of natural numbers (x, y) for which $2x + 3y = 30$.

13 Cardy collects sets of baseball cards. He has 12 sets of 1992 Bottoms and 5 sets of 1992 Flairs. Their combined worth is $400. He buys 6 more sets of Flairs and sells 2 sets of Bottoms. The value of the cards he owns now is $470. Find the value of a set of Flairs cards. $20

14 The perimeter of a rectangle is 80 cm. Find the rectangle's length if its width is 300% of the length. 10 cm

15 Find the value of x so that the lever will balance. 30

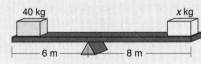

16 A house has the shape of a rectangular prism with a triangular prism on the top. Find the interior volume of the house.
13,440 ft^3

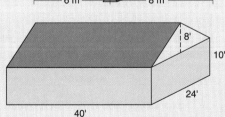

17 A pyramid has a square base with a side of 6 and a slant height of 5.
 a Find the lateral area of the pyramid. 60
 b Find the total surface area. 96
 c Find the volume. 48

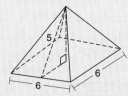

REVIEW

18 Use the trapezoidal pyramid shown.
  a How many vertices, V, does the pyramid have? 5
  b How many edges, E, does the pyramid have? 8
  c How many faces, F, does the pyramid have? 5
  d Is the equation $F - E + V = 2$ *true* or *false*? True

19 Find the volume and the total surface area of each of the following solids. If necessary, round your answer to the nearest tenth.
  a Sphere **b** Cylinder **c** Cone

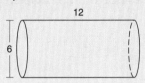

20 Find the ratio of the volume of the cone to the volume of the cylinder. 1:3

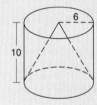

21 Graph the equations $y = 2x + 7$ and $y = -3x - 3$ and find the point where the graphs intersect. $(-2, 3)$

22 An open-top box has a square base.
  a Find the lateral area of the box. 560 in.^{2}
  b Find the total surface area. 756 in.^{2}
  c Find the volume. 1960 in.^{3}
  d Convert the volume from cubic inches to cubic feet. About 1.13 ft^3

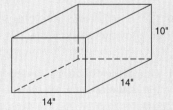

23 Mr. Mint has one more dime than he has quarters. The total value of his dimes and quarters is $4.30.
  a How many methods can you think of to find the number of dimes and quarters Mr. Mint has? Answers will vary.
  b How many dimes and quarters does he have? 13 dimes, 12 quarters

24 Use a graphing calculator or a computer to graph the equations $y = 3^x$ and $y = x^3$. Find the points of intersection of the graphs in order to solve $3^x = x^3$. If necessary, find your answer to the nearest tenth. $x = 2.5$ or $x = 3$

Chapter Review

CHAPTER 11

Test

In problems 1–3, solve each equation.

1 $3x - 8 = 7x - 20$ 3
2 $\dfrac{x-4}{3x} = \dfrac{4}{9}$ -12
3 $2(3x - 1) = 2x$ $\dfrac{1}{2}$

4 Today, Rita's father is three times her age. Six years ago, he was four times her age. How old are Rita and her father today?   Rita: 18; father: 54

In problems 5–7, find the volume of each figure.

5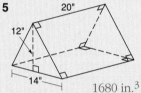
1680 in.³

6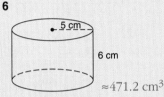
≈471.2 cm³

7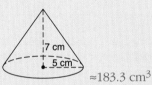
≈183.3 cm³

8 Ginny bought two tires for $120, including sales tax. If sales tax is 5%, what was the price of the tires before tax? Round your answer to the nearest cent. $114.29

9 The areas of the two rectangles are equal. Find the perimeter of each rectangle.   59, 58

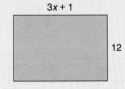
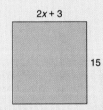

10 David decided to invest his $10,000 inheritance. He put half of the money in a savings account that pays 5% interest and the other half in a CD that pays 8% interest. How much interest will David earn in one year? $650

11 Graph the equations $y = 4x + 3$ and $y = 3x + 1$. Use the graphs to solve the equation $4x + 3 = 3x + 1$. $x = -2$

In problems 12–14, find the total surface area of each figure.

12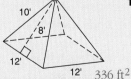
336 ft²

13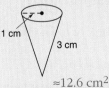
≈12.6 cm²

14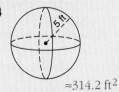
≈314.2 ft²

15 Juanita's piggy bank contains only quarters and dimes. The value of the coins is $5.75. If she has 32 coins, how many are quarters and how many are dimes?   17 quarters; 15 dimes

PUZZLES AND CHALLENGES

1 Eric, Dave, and Carl are triplets. Each participates in two of the following six sports: baseball, football, soccer, basketball, hockey, and swimming. No two are in the same sport. From the facts below, deduce which two sports each participates in.
a Eric and the football player went to watch the basketball game.
b Dave doesn't wear any special kind of protective equipment in either of his sports, just special attire.
c The football player and the hockey player watched their brother set a swimming record.
d Soccer and football are played at the same time.

2 Can you read the two familiar proverbs printed at the right?

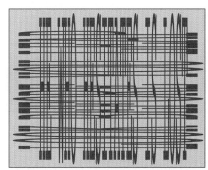

3 Match each item on the left with the one on the right that has a similar or related meaning. The relationships are disguised by the use of double meanings or altered spacing between words. For example, the word *avenue* (a venue) goes with "one location" since *venue* means "location." The word *holster* goes with "arm rest" since it is a place that an arm (gun) gets placed (rested). These puzzles, called "Mind Flexers" by their inventor, psychology professor Morgan Worthy, are meant to improve mental ingenuity. He says, "Do not take them too seriously, and you will quickly improve at seeing the relationships."

1	Avenue	A	Laugh at pigs' home
2	Pantry	B	Large flower
3	Mango	C	Bad ratio
4	Hasty	D	Promote music
5	Bumper	E	Attempt to cook
6	Meant	F	Adequate golfer
7	Fortune	G	I'm an insect
8	Parking	H	One location
9	Maximum	I	Do more work
10	Restless	J	He left
11	Holster	K	Arm rest

8. Extend problem 11 on page 562 as follows: Create a spreadsheet application to solve the problem. Does Penny ever remove all of the coins from the pile? Why or why not? Suppose that Penny had $900 in pennies instead of 900 coins. How could you change your spreadsheet to account for this change? Do the results change? Explain what happens.

9. Example 2 on page 567 challenged you to come up with a nonalgebraic method for solving the problem. Show another method for solving the problem. Discuss the advantages and disadvantages of your method compared to the one shown.

10. Extend problem 31 on page 572 as follows: Design figures that could be folded to form a triangular prism, a rectangular prism, and a square-based pyramid. Construct each of your figures out of posterboard or sturdy paper. Find the volume and lateral area of each figure.

11. Extend problem 26 on page 588 as follows: Design a spreadsheet application that performs the indicated computations with any input number. Use your spreadsheet to explore what happens as you vary the input number from 0 to 1000. Explain your results and create a diagram that shows what happens.

Additional Answers

for Puzzles and Challenges

1 Eric, hockey and soccer; Dave, basketball and swimming; Carl, football and baseball.

2 He who hesitates is lost. Look before you leap.

3 1–H, 2–E, 3–J, 4–A, 5–C, 6–G, 7–D, 8–F, 9–B, 10–I, 11–K

ASSESSMENT

LONG-TERM INVESTIGATIONS

UNIT F: An Enterprising Venture

Suggestions for various ways to assess students' performance are integrated throughout the investigations in this unit. These forms of assessment include observing the students while they work, interviewing students, looking at students' self-evaluations and group evaluations, observing peer review/response situations, evaluating students' products, and evaluating students' portfolios.

Refer to the chart below to preview the variety of assessment opportunities that are integrated throughout this unit.

Assessment Opportunities	Investigations						
	Before Unit	1	2	3	4	5	End of Unit
Form A: Observation Notes **Bank A/B:** Observation Guide		✘		✘		✘	
Form B: Observation Evaluation Record **Bank A/B:** Observation Guide				✘			
Form C: Interview Record **Bank C:** Interview Guide		✘					
Form D: Product Evaluation Record **Bank D:** Product Evaluation Guide			✘		✘	✘	
Form E: Rubric Chart **Bank E:** Sample Rubric			✘				
Form F: Portfolio Item Description	✘						✘
Form G: Portfolio Notes **Bank G:** Portfolio Evaluation Criteria	✘						✘
Form H: Self-evaluation			✘				✘
Form I: Group Evaluation				✘		✘	
Form J: Peer Review/Response					✘		

Rubrics

Rubrics are used to evaluate students' products. You can develop a scoring rubric for a specific investigation by adapting the blackline masters shown below, which are provided in the *Gateways to Mathematical Investigations* Teacher's Edition. You will probably want to develop a rubric with students at least once per unit. The chart on the opposite page suggests that for Unit F: An Enterprising Venture, you complete a rubric before Investigation 2. Remind students that, at times, they may be asked to revise their work to improve it and to make it complete.

Bank D, page 211

Bank D: Product Evaluation Guide

The statements below can be used as a guide to assess students' products. Adapt these statements as needed to evaluate specific assignments. Record your adaptations on a clean copy of **Form D: Product Evaluation Record** (p. 210). You may also wish to create and record your own criteria on **Form D.** After you have adapted **Form D** for use with a specific assignment, make copies for each student or group as needed.

Mathematical Reasoning Skills

- The student includes examples and/or counter examples to support the mathematical argument.
- The student synthesizes and summarizes own or group's thinking.

Mathematical Understanding

- The student identifies all the important information of the situation or problem.
- The student shows understanding of the situation's mathematical ideas. (List important mathematical concepts and skills for the assignment on Form D.)

Form E, page 212

Form E: Blank Rubric Chart

NAME (S) ______________ ______________ GROUP __________
______________ ______________ DATE __________

STANDARDS FOR STUDENT'S WORK

LEVEL	Standard of performance to be achieved for each specified level
4 Exceptional/ Fully Acceptable	

Bank E, page 213

Bank E: Sample Rubric

STANDARDS FOR STUDENT'S WORK

LEVEL	Standard of performance to be achieved for each specified level
4 Exceptional/ Fully Acceptable	Your presentation fully accomplishes the purpose of the task. Your work is complete, well organized, and clearly explained. You illustrate ideas using clear diagrams and models and communicate them effectively to the audience. You identify all the important information of the situation and show a thorough understanding of the mathematical ideas and procedures. You use strong arguments to support findings, including examples (and/or counterexamples). You demonstrate that you fully grasp the central mathematical ideas developed in this task. You demonstrate an orignial extension or applicatrion of the mathematical concepts.
3 Satisfactory/ Acceptable	Your presentation substantially accomplishes the purpose of the task. Your work is complete, reasonably organized, and explained. You use an appropriate diagram or model and communicate ideas effectively to the audience. You identify most of the important information of the situation and show understanding of the mathmatical ideas and procedures. You present arguments to support findings, including examples (and/or counterexamples). You understand and use the central mathematical ideas developed in this task. You demonstrate and apply the mathematical concepts.
2 Incomplete/ Revise	Your presentation partially accomplishes the purpose of the task. However, your work may be incomplete, or misdirected, or not clearly organized and/or explained. Your diagram or model may be inappropriate or unclear. You may communicate ideas inadequately to the audience. You may identify some of the important information of the situation and you may show only partial and/or limited understanding of the mathematical ideas and procedures. Your arguments to support findings may be weak or incomplete. You may misunderstand or misuse the central mathematical ideas developed in this task. You may have difficulty demonstrating and applying the mathematical tasks. You need to revise this work. (Your teacher will provide suggestions for a satisfactory or competent rating and a due date for your revised work.)
1 Inadequate/ Redo	Your presentation shows little or no understanding of the assignment. Your work is barely, if at all, comprehensible. Your explanations are not understandable. You show little or no understanding of the mathematical ideas and procedures. You made serious computational mistakes and misused mathematical terms. You should meet with your classmates or your teacher to be sure you fully understand this assignment. Then you need to revise this work. (Your teacher will provide suggestions for a satisfactory or competent rating and a due date for your revised work.)

Gateways © McDougal, Littell & Company   Bank E: Sample Rubric   213

Form F, page 214

Form F: Portfolio Item Description

NAME ______________ DATE ______________

Complete this form each time you add an item to your portfolio.

Type of Portfolio Entry: ______________

Why did you choose to place this item in your portfolio?

____

Form G, page 215

Form G: Portfolio Notes

NAME ______________ CLASS ______________

Date	Observations/Comments	Recommendations

Bank G, page 216

Bank G: Portfolio Evaluation Criteria

At the start of the school year, have a class discussion on the goals and uses of students' portfolios. Discuss with the class some ideas regarding what to include in a portfolio. Then, use the portfolio evaluation criteria which follows to develop a rubric with the class that communicates the criteria you will use to assess their portfolios. Have students record this criteria on copies of **Form E: Rubric Chart.** Have students store this completed form in their portfolios so that they can refer to it throughout the year.

Each time students add items to their portfolios, have them complete copies of **Form F: Portfolio Item Description.**

Use copies of **Form G: Portfolio Notes** to record your observations, comments, and recommendations each time you evaluate students' portfolios.

Ideas for Types of Portfolio Entries

- a portfolio table of contents
- a letter to the reader of the portfolio explaining each item
- teacher-completed checklists/evaluations
- a mathematical autobiography
- notes from interviews, teachers, peers, etc.
- papers that show the student's corrections or errors or misconceptions
- photos or sketches of the student's work with manipulatives or mathematical models
- excerpts from the student's daily journal or reflective writing samples
- tables, charts, and diagrams created by the student
- artwork supporting a problem such as drawings
- designs, coordinate pictures, scale drawings, maps, etc.
- student-created problems with or without solutions
- group projects or individualized reports
- student-developed surveys and results
- self-evaluations

Suggestions for Portfolio Evaluation Criteria

Mathematical Reasoning Skills
The Student:
- analyzes problem situations.
- is flexible and comfortable in using several different problem solving approaches.
- synthesizes and summarizes the individual's or group's thinking.
- provides supporting statements to verify findings.
- shows original and creative thought.

Mathematical Tools and Techniques
The Student:
- uses a variety of diagrams, models, technology, manipulatives, and other resources appropriately.
- uses displays (e.g. graphs, tables, diagrams, sketches, and models) to represent mathematical ideas.

Mathematical Understanding
The Student:
- shows understanding of mathematical ideas and procedures. (List specific skills/concepts on **Form G.**).

Mathematical Communication
The Student:
- organizes work in a logical and clear way.
- expresses mathematical ideas clearly and appropriately.
- uses mathematical language and notation appropriately.

Mathematical Dispositions
The Student:
- shows development as as group participant.
- shows development of a positive attitude.
- shows growth in learning.
- includes revised and complete work.
- shows consistent effort.
- includes a variety of work in his or her portfolio.

216   Assessment   *Gateways* © McDougal, Littell & Company

Portfolios

At the beginning of each unit, you may wish to review with the students the criteria they should use when placing work in their portfolios. Have students complete Form F: Portfolio Item Description each time they wish to place an item in their portfolios.

Periodically, you may wish to evaluate students' portfolios using the blackline masters shown above. For Unit F: An Enterprising Venture, the chart on the opposite page suggests that you evaluate students' portfolios at the end of the unit.

CHAPTER 12

Contents

12.1 Iterations
12.2 More About Iterations
12.3 Geometric Iterations
12.4 Convergent Iterations
12.5 Introduction to Fractals

Pacing Chart

Days per Section

	Basic	Average	Honors
12.1	*	*	2
12.2	*	*	2
12.3	*	*	2
12.4	*	*	2
12.5	*	*	2
Review	–	–	2
Assess	–	–	2
	–	–	14

* Optional at these levels

Chapter 12 Overview

- This chapter connects many of the concepts learned in the previous eleven chapters. It is an enrichment chapter that many of your students will find exciting and not overly difficult. We strongly recommend that if there is time, you introduce your students to the concept of iteration. Both spreadsheets and some calculators have the ability to use preceding outputs as inputs and should be used throughout the chapter. Geometric iterations are used to introduce fractals, a fairly new branch of mathematics.

- There is some software available to enrich the lessons in this chapter. For the Macintosh, there is *Discover Form* and *Fractal Attraction*. For IBM-compatible computers, there is *Chaos: The Software*. These software packages make it possible to display iterations and fractals.

12 Looking Ahead: Iterations and Fractals

CHAPTER 12

SOUND ENGINEERING A sound engineer is chiefly concerned with three aspects of sound: the frequency, the amplitude (or loudness), and the quality.

The frequency is the number of vibrations per second. Electronic equipment can produce sounds with a given frequency, and can vary the intensity. Sound engineers are continually working on ways to improve the quality of music, whether it be a live orchestra, a rock concert greatly amplified, or music recorded on a tape or a CD.

If you like both music and electronics, you could be headed for a career related to sound engineering.

INVESTIGATION

Music What is harmony? Explain in mathematical terms why some musical notes sound good when played together, and some combinations sound less pleasant.

Career Connection

There are many opportunities for sound engineers and technicians, and not only in the music industry. Convention centers and hotels, for example, often require sound equipment for various nonmusical programs. A basic understanding of acoustics and the electronic processes used to create sound can help students increase their appreciation of music. It is important for students to realize that mathematics often plays a vital role in artistic creation.

Video Connection

For more information about the career discussed here, please refer to the Futures videocassette #5, program B: *Sound Engineering,* featuring Jaime Escalante. For more information or to order a copy of the program, call PBS VIDEO at 800/344–3337.

Investigation Notes

Harmony is the combination of musical pitches, or notes, to produce chords. Consonant, or "pleasant," harmonies include combinations of notes in which the ratios of the frequencies of the sound waves are simple, such as an octave (e.g., C to c, 1:2), a third (e.g., C to E, 4:5), or a fifth (e.g., C to G, 2:3). Dissonant, or "unpleasant," harmonies do not have such simple frequency ratios. They result in beats, or interference, between pulsations of sound waves, which make the combinations of notes less pleasant.

SECTION 12.1

Objectives
- To recognize iterative processes
- To interpret and construct iteration diagrams

Vocabulary: iteration, iteration diagram

Class Planning

Schedule 2 days

Resources
In *Chapter 12 Student Activity Masters* Book
Class Openers 1, 2
Extra Practice 3, 4
Cooperative Learning 5, 6
Enrichment 7
Learning Log 8
Technology 9, 10
Spanish-Language Summary 11

Transparencies 7, 50

Class Opener

Find the next four numbers in each pattern, and state what the pattern is.
1. 2, 5, 11, 23, ...
2. 3, 8, 23, 68, ...

Answers:
1. 47, 95, 191, 383; add 1 to twice the preceding number
2. 203, 608, 1823, 5468; subtract 1 from three times the preceding number

Lesson Notes

- The iteration diagram is an extension of the function machine that was presented in Section 11.4. Point out to students that the initial input number is the zero term, not the first term.
- To use a calculator that has iteration capability to complete example 1, "seed" the first value

604

Investigation Connection
Section 12.1 supports Unit F, *Investigations* 4, 5. (See pp. 542A–542D in this book.)

SECTION 12.1
Iterations

REPEATED CALCULATIONS

The following example will help you to understand the meaning of *iteration*.

Example 1
If Mary puts $500 in a bank account that gives 6.5% interest per year figured once a year, how much will she have in the bank at the end of ten years?

Solution
Let's first just figure out how much she has after one year. This will be the original $500 plus the interest earned on the $500 calculated by multiplying the rate as a decimal 0.065 times $500.

Times	Amount in Bank at That Time
Start of Year 1	$500
End of Year 1	$500 + 0.065 ($500) = $532.50

Now we can figure the amount at the end of the second year by repeating the same calculations as used for the first year, only starting with $532.50 in the bank. That is, our result from the previous year is used as the starting value for the next year. We now use this idea to complete our table for ten years.

Year	Amount	End of Year	Amount in Bank at End of Year
1	$500	1	500.00 + 0.065 (500.00) = $532.50
2	532.50	2	532.50 + 0.065 (532.50) = 567.11
3	567.11	3	567.11 + 0.065 (567.11) = 603.97
4	603.97	4	603.97 + 0.065 (603.97) = 643.23
5	643.23	5	643.23 + 0.065 (643.23) = 685.04
6	685.04	6	685.04 + 0.065 (685.04) = 729.57
7	729.57	7	729.57 + 0.065 (729.57) = 776.99
8	776.99	8	776.99 + 0.065 (776.99) = 827.50
9	827.50	9	827.50 + 0.065 (827.50) = 881.29
10	881.29	10	881.29 + 0.065 (881.29) = 938.57

604 Chapter 12 Looking Ahead: Iterations and Fractals

SECTION 12.1

We see that Mary will have $938.57 in the bank at the end of ten years. **Note:** If you try these calculations, you might get slightly different answers because of rounding differences.

The repetitive process in Example 1 is called iteration. It is used when the next calculation to be done is exactly the same as the previous calculation, except that the next calculation uses the results of the previous calculation as its starting value.

ITERATION DIAGRAMS

One good way to model iterations is by means of an ***iteration diagram*** like the following.

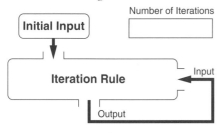
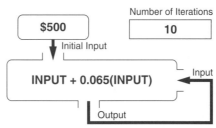

Notice the characteristics of an iteration diagram.

- There is always an initial input value. In Example 1, it was the $500 initially placed in the bank.
- There is an iteration rule to enable us to take any input and compute the output. In Example 1, it was INPUT + 0.065(INPUT).
- There is an arrow indicating that the previous output should be used as the next input. In Example 1, the money in the account at the end of each year is used as the input for the next year.
- There is a number of iterations, indicating the number of times to repeat the process. In Example 1, there were ten iterations, corresponding to the ten years.

Example 2
Consider the iteration diagram shown.
  a Calculate the outputs for the indicated number of iterations.
  b How is this problem different from the problem in Example 1?

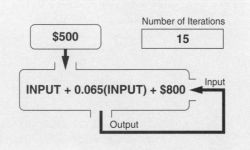

by entering it and pressing EXE. Then enter ANS + 0.065 × ANS and press EXE ten times. Some calculators without the ANS key still have this capability. The copy function gives spreadsheets the ability to iterate.

- You may recognize that both sample problems involve geometric sequences, but we don't recommend that you introduce the term—the intent here is to have students iterate and discover relationships.

- Bringing in a super ball and having students watch it bounce may be a motivational introduction to the lesson.

- The graphs in the sample problems were initially generated by a spreadsheet. If your spreadsheet has graphing capability, you may want to allow students to use it to generate similar graphs.

Additional Examples

Example 1
John deposits $800 into an account that pays 4.75% interest per year, calculated once a year. How much will he have at the end of six years?

Answer: $1056.85

Example 2
Suppose John deposits $200 more at the beginning of each new year. How much will be in his account at the beginning of the fifth year?

Answer: $1822.00

SECTION 12.1

Assignment Guide

Basic
Optional

Average or Heterogeneous
Optional

Honors
Day 1 4, 6–10
Day 2 5, 11–16

Strands

Communicating Mathematics	3, 11, 13, Investigation
Geometry	2, 9, 12, 15
Measurement	2, 9, 11, 12, 14, 15
Technology	3, 6, 8, 11, 16
Algebra	1–16
Patterns & Functions	1–16, Investigation
Number Sense	2
Interdisciplinary Applications	3, 5–8, 10, 11, 13, 14, 16, Investigation

Individualizing Instruction

The gifted learner Students might be familiar enough with spreadsheets by this time to recognize that creating one for many of the problems could be helpful. Encourage students to follow through with this idea.

Solution

a Initial input value: $500
Iteration rule: INPUT + 0.065(INPUT) + $800
Continue iterations 15 times.

Iteration	Amount ($)
0	500
1	1332.5
2	2219.113
3	3163.355
4	4168.973
5	5239.956
6	6380.553
7	7595.289
8	8888.983
9	10266.77
10	11734.11
11	13296.82
12	14961.12
13	16733.59
14	18621.27
15	20631.66

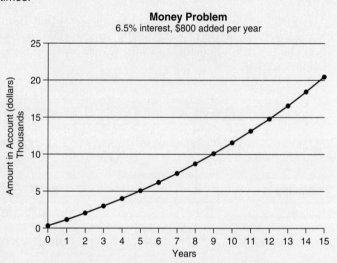

b This is the same problem as Example 1, only Mary adds an additional $800 to her account at the end of each year, and the iterations are done for 15 years instead of 10 years.

Sample Problems

Problem 1 A ball is dropped from a height of 12 feet, and each time it rebounds to a height that is 85% of the previous height. On which rebound will the ball first fail to reach a height of 1 foot?

Solution Initial input value: 12 feet
Iteration rule: 0.85(INPUT)
Continue until the output is less than 1 foot.

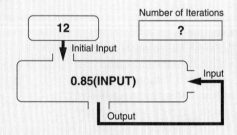

Spreadsheets are especially useful in solving problems using iteration. Here is a spreadsheet solution to the problem.

SECTION 12.1

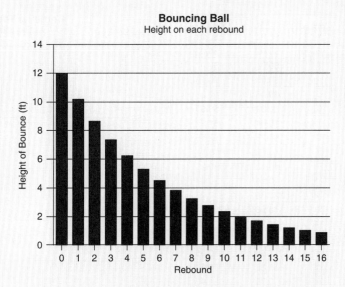

Bouncing Ball — Height on each rebound

Bounce	Height (ft)
0	12
1	10.2
2	8.67
3	7.3695
4	6.264075
5	5.324464
6	4.525794
7	3.846925
8	3.269886
9	2.779403
10	2.362493
11	2.008119
12	1.706901
13	1.450866
14	1.233236
15	1.048251
16	0.891013

On the sixteenth rebound the ball first bounces less than 1 foot high.

Note: Many graphing calculators have an [ANS] key, and this can be used to do iteration. For this problem, you would do the following:

Enter	Display
12 [EXE]	12
0.85 [×] [ANS] [EXE]	10.2
[EXE]	8.67
[EXE]	7.3695
.	.
.	.
.	.

On some calculators, the enter key or the [=] works like the [EXE] key.

Problem 2

Jake Whitehorse has just won a radio music quiz. The station gives Jake two possible options for his prize.

Option 1: A flat sum of $10,000
Option 2: $0.01 today, $0.02 tomorrow, $0.04 the third day, and so on for 24 days

Solution Which option should Jake take?

Initial value: $0.01
Iteration rule: 2(INPUT)
Continue iterations 23 times.

We use a spreadsheet to help us quickly and painlessly get our answer. We make a third column to show the total of the money received by each day.

Cooperative Learning

- In example 1, suppose the bank has two other investment plans. Option 2 pays 3.25% interest per half year ($\frac{1}{2} \times 6.5\%$), and Option 3 pays 1.625% interest per quarter ($\frac{1}{4} \times 6.5\%$). Have students work in groups to investigate which option Mary should choose if she wants to have the greatest possible amount of money after three years.

Answer:
Option 1: $603.97; option 2: $605.77; option 3: $606.70; should choose Option 3

- Problem 16 is appropriate for cooperative learning.

Communicating Mathematics

- Have students write a paragraph that explains what iteration is and how it applies to mathematics.

- Problems 3, 11, 13, and Investigation encourage communicating mathematics.

Multicultural Note

Many mathematicians of the twentieth century have been concerned with the philosophical foundations of mathematics. Bertrand Russell (1872–1970), a British philosopher and mathematician, attempted to show that all pure mathematics follows from strictly logical premises.

SECTION 12.1

Problem-Set Notes

for pages 608 and 609

- **1** reviews graphing, tables, and writing equations.
- **2** and **3** could be done in small groups.

Checkpoint

1.

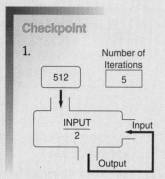

Copy and complete the following table for the iteration diagram above.

Term	0	1	2	3	4	5
Output						

2. Carlos placed $1000 in a savings account. He earns 5.5% interest per year. How much money will he have at the end of five years?

3.

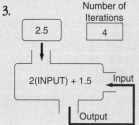

Copy and complete the following table for the iteration diagram above.

Term	0	1	2	3	4
Output					

Answers:

1.

Term	0	1	2	3	4	5
Output	512	256	128	64	32	16

2. $1306.96

3.

Term	0	1	2	3	4
Output	2.5	6.5	14.5	30.5	62.5

608

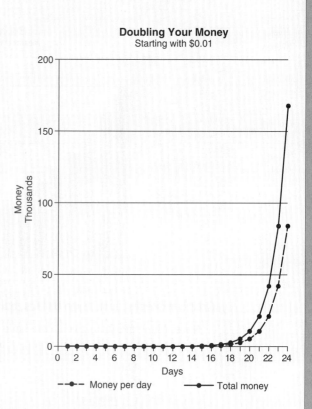

Day	Money/day	Total
1	$0.01	$0.01
2	$0.02	$0.03
3	$0.04	$0.07
4	$0.08	$0.15
5	$0.16	$0.31
6	$0.32	$0.63
7	$0.64	$1.27
8	$1.28	$2.55
9	$2.56	$5.11
10	$5.12	$10.23
11	$10.24	$20.47
12	$20.48	$40.95
13	$40.96	$81.91
14	$81.92	$163.83
15	$163.84	$327.67
16	$327.68	$655.35
17	$655.36	$1,310.71
18	$1,310.72	$2,621.43
19	$2,621.44	$5,242.87
20	$5,242.88	$10,485.75
21	$10,485.76	$20,971.51
22	$20,971.52	$41,943.03
23	$41,943.04	$83,886.07
24	$83,886.08	$167,772.15

Jake should definitely take option 2. (There is an interesting pattern involving the numbers in the second column and those in the third column. Describe this pattern.)

Think and Discuss

P **1**

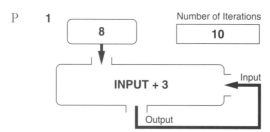

a Copy and complete the following table for the iteration diagram above.

Term	0	1	2	3	4	5	6	7	8	9	10
Output	8	11	14	17	20	23	26	29	32	35	38

b Make a graph on a coordinate system for the table in part **a**.

c Write an equation that goes with the table and graph. $y = 3x + 8$

608   Chapter 12   Looking Ahead: Iterations and Fractals

SECTION 12.1

P **2** A lily pad on a pond doubles its area every day. If it will completely cover the pond at the end of 100 days, after how many days will it cover 50% of the pond? 99 days

P **3** There is a legend about a king who agreed to grant any wish to the subject who slew a giant. The subject who did this asked for a grain of wheat for the first square of a checkerboard, two for the second square, four for the third square, and so on, with the number of grains of wheat being doubled each time until all 64 squares of the checkerboard had been accounted for.
  a Draw an iteration diagram for this situation.
  b How many grains would go on the last square? 2^{63}, or $\approx 9.22 \times 10^{18}$
  c Explain why the king could not fulfill his promise.

Problems and Applications

P **4**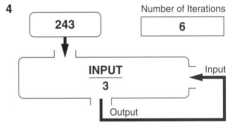

  a Copy and complete the following table for the iteration diagram above.

Term	0	1	2	3	4	5	6
Output	243	81	27	9	3	1	$0.\overline{3}$

  b At what term will the output be less than 1?   The sixth

P **5** **Business** In 1960, the cost of a certain computer was $750,000. Every five years since then, the cost of an equivalent computer has been cut in half.
  a Draw an iteration diagram for this situation.
  b Make a table showing the cost of this computer every five years from 1960 to 2000.
  c When would the cost of this computer be less than $50?   In 2030

P, R **6** **Spreadsheets** In Example 1, Mary put $500 in a bank account that gives 6.5% interest per year for ten years.
  a How much money would she have if her original deposit had been $800 instead of $500? $1501.71
  b How much would she have if the bank had paid 7.5% interest on her $500 for ten years? $1030.52

Reteaching

You might reteach example 1 by using the iteration rule 1.065(INPUT). Point out that using this rule requires only one calculation rather than two. You can verify this rule for students by applying the distributive property:
  INPUT + 0.065(INPUT)
  = (1)(INPUT) + (0.065)(INPUT)
  = (1 + 0.065)(INPUT)
  = (1.065)(INPUT)

Additional Answers

for pages 608 and 609

1 b
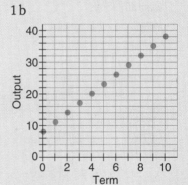

3 a Initial input: 1; iteration rule: 2(INPUT); no. of iterations: 63

c Possible answer: Fulfilling the promise would require more wheat than there is in the world.

5 a Initial input: $750,000; iteration rule: INPUT ÷ 2

b

Year	Amount
1960	$750,000.00
1965	375,000.00
1970	187,500.00
1975	93,750.00
1980	46,875.00
1985	23,437.50
1990	11,718.75
1995	5,859.38
2000	2,929.69

SECTION 12.1

Problem-Set Notes

for pages 610 and 611

- **9 b** and **15** preview Section 12.4.

Additional Practice

1.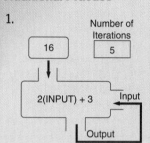

   Copy and complete the following table for the iteration diagram above.

Term	0	1	2	3	4	5
Output						

2. Deena bought a machine for work. The machine's value each year is 90% of its value the preceding year. If the machine originally cost $15,000, when will its value be less than $1000?

3. Gene put $1500 in a savings account that pays 6% interest per year. How much money will he have at the end of eight years?

Answers:

1.
Term	0	1	2	3	4	5
Output	16	35	73	149	301	605

2. After 26 years
3. $2390.77

Stumbling Block

In example 2, students may try to add the $800 to INPUT first, and then multiply by 0.065. Remind students that the $800 has not yet been in the account for a year, so it may not yet earn interest.

P **7** A theater has 20 rows of seats and has 23 seats in the first row. In each successive row there are 4 more seats than in the row in front of it.
  a Draw an iteration diagram to describe this situation.
  b How many seats are in the twentieth row? 99 seats

P **8** **Banking** Ann puts $1000 into a bank account. She gets 5.1% interest per year. Each year, she adds another $600 to the account. How much will she have after ten years? $9226.53

P, R **9**

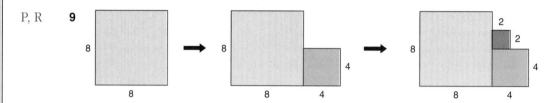

  a What is the area of the smallest square when there are 12 squares in the figure? $\approx 1.526 \times 10^{-5}$
  b Continuing the pattern, what would be the total area of the entire figure after 12 iterations? $85\frac{1}{3}$

P **10** **Science** S. Cargo started out with 32 snails in his aquarium. He discovered that every month the number of snails increased by one fourth of the number he had the previous month. For example, at the end of the first month he had $32 + \frac{1}{4}(32)$, or 40, snails. Assume that no snails die.
  a Draw an iteration diagram for the number of snails S. Cargo has in any given month.
  b When would S. Cargo have 1,000,000 snails in his aquarium? In 47 months

P **11** A toy car is 12 feet from a wall. At each successive second it travels one half of the way to the wall.
  a Copy the table, extending and completing it for the first 10 seconds.

Seconds	0	1	2	3	...
Distance from wall					

  b When is the car less than 3 inches from the wall? In 6 sec
  c Does the car ever reach the wall? Explain.
  d Describe the speed of the car as time increases.

P, R **12**

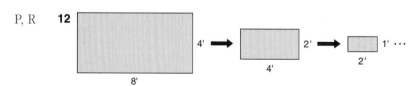

  a Make an iteration diagram to figure the area of each successive rectangle in the pattern shown.
  b What is the area of the tenth rectangle? $\approx 1.22 \times 10^{-4}$ ft^2

610   Chapter 12   Looking Ahead: Iterations and Fractals

SECTION 12.1

P **13** **Consumer Math** Larry takes out an auto loan for $9000 at 9.5% interest per year. He makes payments of $350 per month.
  a What is the monthly interest rate?   ≈0.79%
  b The following expression represents the amount Larry still owes after making his first $350 payment.
  $$9000 + \frac{0.095}{12}(9000) - 350$$
  Explain why this is the correct formula.
  c Make an iteration diagram for the amount owed at the end of any month.
  d How many months must Larry make payments?   29 months

P **14** **Science** A swimming pool has had chlorine added to it to raise the level of chlorine in the water to 3 parts per million. Each day 15% of the chlorine that was present the previous day is lost. A swimming pool must have at least 1 part per million of chlorine for safety. How many days will go by before more chlorine must be added?   6 days

P, R **15**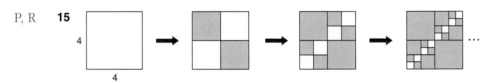
Find the shaded area of the sixth figure in the pattern shown above.   15.5

P **16** **Consumer Math** Tom takes out a $9000 loan for 48 months at 9.5% yearly interest. He makes monthly payments of $200. The iteration diagram can be used to find the amount of the loan remaining each month.
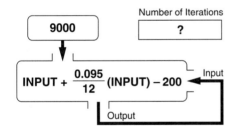
  a Why is the 0.095 divided by 12?
  b How much is owed at the end of one year? Two years? Three years? Four years?
  c Experiment with changing the monthly payment until you find the correct monthly payment to pay off the loan in exactly four years. $226.11

Investigation

The Calendar Suppose that your birthday falls on a Saturday this year. On what day will your birthday fall next year? Explain why.

Additional Answers

for pages 610 and 611

7 a Initial input: 23; iteration rule: INPUT + 4; no. of iterations: 19

10 a Initial input: 32; iteration rule: INPUT + $\frac{1}{4}$(INPUT)

11 a

Seconds	Distance from Wall
0	12
1	6
2	3
3	1.5
4	0.75
5	0.375
6	0.188
7	0.094
8	0.047
9	0.023
10	0.012

c In theory no, but in reality the distances get too small to be meaningful.
d Speed is decreasing.

12 a Initial input: 32; iteration rule: 0.25(INPUT)

13 b The first month's interest (one twelfth of the yearly interest) is charged on $9000, then $350 is subtracted from this total.
c Initial input: 9000; iteration rule: INPUT + $\frac{0.095}{12}$(INPUT) − 350

16 a 9.5% is the yearly interest rate. Tom is making monthly payments; therefore, yearly interest must be divided by 12.
b $7385.92; $5611.65; $3661.28; $1517.35

Investigation Often the answer is Sunday, because 365 days equals 52 weeks and 1 day. However, when a leap-year day (Feb. 29) falls between the birthdays, the answer is Monday. That is because there are 366, not 365, days between the birthdays. (If your birthday happens to be February 29, then you have a birthday only every four years, and this problem doesn't apply.)

SECTION 12.2

Objectives
- To recognize situations in which an iterative calculation involves the number of preceding iterations
- To express an iterative process in function notation

Class Planning

Schedule 2 days

Resources

In *Chapter 12 Student Activity Masters* Book
Class Openers 13, 14
Extra Practice 15, 16
Cooperative Learning 17, 18
Enrichment 19
Learning Log 20
Using Manipulatives 21, 22
Spanish-Language Summary 23

Transparencies 7, 50

Class Opener

Let $f(0) = 0$. Find $f(1)$ through $f(5)$, using the following equations.
1. $f(1) = f(0) + 3(1) - 2$
2. $f(2) = f(1) + 3(2) - 2$
3. $f(3) = f(2) + 3(3) - 2$
4. $f(4) = f(3) + 3(4) - 2$
5. $f(5) = f(4) + 3(5) - 2$

Answers:
1. 1   2. 5   3. 12   4. 22   5. 35

Lesson Notes

- Point out to students the connections between the examples and problems in this section and the ones in the first chapter. Here, students are introduced to notation that describes what they did earlier. The function notation should be a natural extension of their previous experience.

Investigation Connection Section 12.2 supports Unit F, *Investigations* 4, 5. (See pp. 542A–542D in this book.)

SECTION 12.2
More About Iterations

RULES INVOLVING THE ITERATION COUNTER

In many iteration problems the iteration rule depends on both the input value and the number of iterations that have been performed. The following diagrams show such a situation.

Figure	Number of Dots in Bottom Row	Total Number of Dots
•	1	1
△	2	1 + 2 = 3
△	3	3 + 3 = 6
△	4	6 + 4 = 10
△	5	10 + 5 = 15
△	6	15 + 6 = 21

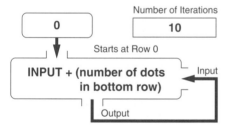

Notice that the total number of dots in each triangle depends on two things:
1. The number of dots in the previous triangle.
2. The number of dots in the new row.

Many iteration rules depend on two conditions such as those above.

612   Chapter 12   Looking Ahead: Iterations and Fractals

SECTION 12.2

FUNCTION NOTATION FOR ITERATION

It is convenient to use function notation to describe iteration problems. Consider the following example.

Example 1
If $f(0) = 10$ and $f(n) = f(n-1) + n$, what is $f(5)$?

Solution
We proceed one step at a time. We make a table of values for n, enter 10 as the first function entry, and then use the iteration formula $f(n) = f(n-1) + n$ to calculate each successive function entry.

n	$f(n)$
0	$f(0) = 10$
1	$f(1) = f(1-1) + 1 = f(0) + 1 = 10 + 1 = 11$
2	$f(2) = f(2-1) + 2 = f(1) + 2 = 11 + 2 = 13$
3	$f(3) = f(3-1) + 3 = f(2) + 3 = 13 + 3 = 16$
4	$f(4) = f(4-1) + 4 = f(3) + 4 = 16 + 4 = 20$

Notice that the function operates just like the iteration diagram: $f(0) = 10$ is the initial input, and the counter, n, starts at 0. The iteration rule, $f(n) = f(n-1) + n$, indicates that the next output is the sum of the previous output, $f(n-1)$, and the current value of the counter, n.

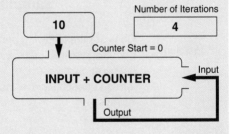

Example 2
Write an iterative function to describe the following situation:

Mary deposits $500 in a bank account that gives 6.5% interest per year. At the start of each year thereafter, she deposits an additional $300 in the account.

Solution
$M(0) = \$500$ — Money at start of first year.
$M(n) = M(n-1) + M(n-1)(0.065) + \$300$ — Money at end of each year equals money at start of that year, plus interest earned on that money for the year, plus $300 deposit.

- A note about iteration diagrams: f(counter start value) is always equal to the initial input. In example 1, $f(0) = 10$. The counter increases by 1, then, both when the initial value first "drops" into the iteration diagram and at each new input.

- The addition of a counter value to an iteration rule means that students will no longer be able to iterate simply by using the [ANS] and [EXE] keys on a calculator. Have students keep track of iterations by creating tables similar to the one in example 1 (with n in one column and $f(n)$ in the other).

- You might bring in toothpicks so that students can model sample problems 1 and 2 before they use spreadsheets and function notation to find the answers.

Additional Examples

Example 1
If $f(0) = 30$ and $f(n) = f(n-1) - n$, what is $f(6)$?

Answer: 9

Example 2
Write an iterative function to describe the following situation: Fred deposits $1000 in a bank account that pays 6% interest per year. At the start of each year, he withdraws $100 from the account.

Answer: $M(0) = 1000$; $M(n) = M(n-1) + M(n-1)(0.06) - 100$

Example 3
Write a function for the following information.

Initial input: 10
Iteration rule:
  INPUT + 3(COUNTER)
Counter start: 0

Answer: $f(0) = 10$, $f(n) = f(n-1) + 3n$

Section 12.2 More About Iterations

SECTION 12.2

Assignment Guide

Basic
Optional

Average or Heterogeneous
Optional

Honors
Day 1 5, 6, 8, 9, 11, 12, 14, 15
Day 2 7, 10, 13, 16–20, Investigation

Strands

Communicating Mathematics	14, 16
Geometry	3, 8, 9, 11, 13, 18
Measurement	11, 18
Data Analysis/ Statistics	3, 4, 7, 9–11, 13–16
Technology	19
Algebra	1–20, Investigation
Patterns & Functions	1–13, 15, 17, 18–20, Investigation
Number Sense	14, 17, Investigation
Interdisciplinary Applications	7, 19, 20, Investigation

Individualizing Instruction

The gifted learner Students could be challenged to write iterative functions that generate the hexagonal numbers, the heptagonal numbers, and the octagonal numbers. You could offer a hint that they might look for a pattern, starting with the triangular numbers, square numbers, and pentagonal numbers.

Example 3
Convert the iteration diagram into function notation.

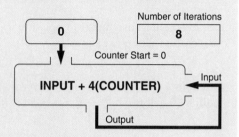

Solution
The initial input is 0, so we write $f(0) = 0$. Each time through the iteration diagram, the counter, which counts the number of iterations done so far, increases by 1. The iteration rule states that we find each output by adding 4 times the current value of the iteration counter to the preceding output. In function notation we write $f(n) = f(n-1) + 4n$. The complete function is $f(0) = 0$ and $f(n) = f(n-1) + 4n$.

Sample Problems

Problem 1 Consider the following pattern of angles. The next picture is formed from the previous picture by adding one more ray having the same endpoint as the previous rays. Count the number of different angles formed in each picture. How many angles would be formed with 11 rays?

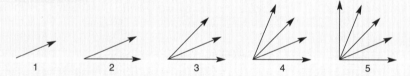

Solution A good way to solve many of these problems is to jump right into the middle and think about how many new angles are formed when we add an additional ray.

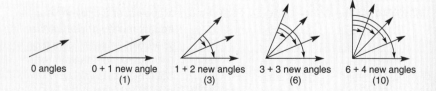

SECTION 12.2

The additional ray makes a new angle with each of the rays that were in the preceding diagram. Therefore,

$A(1) = 0$ — With one ray there are no angles.

$A(n) = A(n-1) + (n-1)$ — With n rays there are all the angles previously there, plus the $(n-1)$ new angles formed by adding the nth ray.

Number of Rays	Number of Angles
1	0
2	1
3	3
4	6
5	10
6	15
7	21
8	28
9	36
10	45
11	55

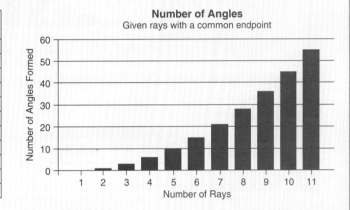

Number of Angles
Given rays with a common endpoint

Problem 2

It takes 4 toothpicks to make the 1 × 1 square shown. It takes 12 toothpicks for the 2 × 2 figure and 24 toothpicks for the 3 × 3 figure. How many toothpicks will it take for an 8 × 8 figure?

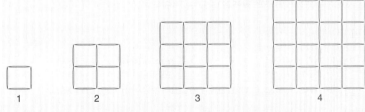

1   2   3   4

Solution

Again we jump in the middle and try to see how many toothpicks must be added to a given figure to produce the next figure.

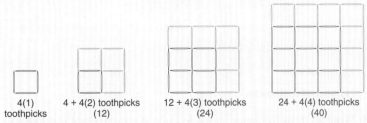

4(1) toothpicks   4 + 4(2) toothpicks (12)   12 + 4(3) toothpicks (24)   24 + 4(4) toothpicks (40)

We need to add $4n$ toothpicks, where $n \times n$ is the size of the figure. Therefore, for the next figure we need all the toothpicks for the previous figure plus $4n$ more. In function notation we write $S(1) = 4$ and $S(n) = S(n-1) + 4n$. For an 8 × 8 figure, 144 toothpicks are needed.

Additional Practice

1. If $f(0) = 3$ and $f(n) = 3 \cdot f(n-1) + 7$, what are $f(1), f(2), f(3), f(4)$, and $f(5)$?

2. Write the iteration shown in the diagram in function notation.

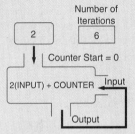

3. Brett decided to start with 5 push-ups. Each day he does 5 more than the day before. When will Brett be doing more than 100 push-ups in one day?

Answers:
1. 16, 55, 172, 523, 1576
2. $f(0) = 2$, $f(n) = 2 \cdot f(n-1) + n$
3. On the 21st day

Multicultural Note

In the late 1930s, a group of mathematicians, most of whom were French, began to publish an influential series of books under the pen name Nicolas Bourbaki. The group's purpose was to present mathematics in a contemporary and original fashion and to illustrate the structure of modern mathematics.

SECTION 12.2

Problem-Set Notes

for pages 616 and 617

- **2** can be done on any scientific calculator by repeatedly pressing the square root key.
- **3** might be better visualized if you had some toothpicks available for students to use.
- **8** and **13** are two representations of essentially the same problem.

Reteaching

To reteach the iteration process, you might use a BASIC computer program to generate the results in a problem such as example 1.

```
10 LET N = 0
20 LET F(N) = 10
30 PRINT F(N)
40 IF N = 4 THEN END
50 LET N = N + 1
60 LET F(N) = F(N – 1) + N
70 GOTO 30
```

The outputs for the program are 10, 11, 13, 16, 20.

Cooperative Learning

- Have students act out the iteration process in example 1. You'll need eleven people. One person will be the number cruncher, and the others will be the *n* values from 0 to 4 and the *f(n)* values from *f*(0) to *f*(4). One student should be labeled "*n* = 0" and another "*f*(0) = 10," and the others should simply be labeled "*n* = " and "*f*( ) = ." The number cruncher should fill in values as the simulation proceeds. The product of the simulation should be a table of values similar to the one on page 613. You may also want to have

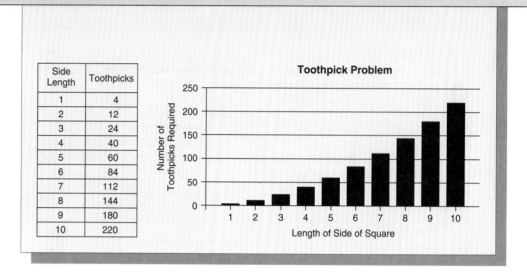

Side Length	Toothpicks
1	4
2	12
3	24
4	40
5	60
6	84
7	112
8	144
9	180
10	220

Think and Discuss

P   **1** Consider the following two functions.

$f(1) = 100$ $g(1) = 100$
$f(n + 1) = 2 \cdot f(n)$ $g(n + 1) = g(n) - 100$

Notice that as the value of *n* increases, the value of $f(n)$ increases rapidly and the value of $g(n)$ decreases rapidly. If $h(1) = 100$ and $h(n + 1) = 2 \cdot h(n) - 100$, does the value of $h(n)$ increase or decrease as the value of *n* increases? The value of $h(n)$ is constant.

P, R   **2** Suppose that $f(n) = \sqrt{f(n-1)}$.
    a Find $f(n)$ for $n = 1, 2, 3, …, 12$ if $f(1) = 16$.
    b Find $f(n)$ for $n = 1, 2, 3, …, 12$ if $f(1) = 0.16$.

P   **3** Suppose we connect triangular numbers with toothpicks.

•
0   3   9   18

    a Copy and complete the following table.

Length of side	0	1	2	3	4	5	6
Number of toothpicks	0	3	9	18	30	45	63

    b Write an iterative function to describe the number of toothpicks needed. $f(0) = 0, f(n) = f(n-1) + 3n$
    c How many toothpicks would be needed for a triangle with 15 toothpicks on each side?   360

616   Chapter 12   Looking Ahead: Iterations and Fractals

SECTION 12.2

P **4 a** Write an iterative function for this diagram.
b Make a table of outputs for the indicated number of iterations.

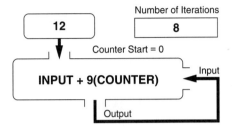

Problems and Applications

P **5** If $f(0) = 8$ and $f(n) = 2 \cdot f(n-1) - 5$, what are $f(1), f(2), f(3),$ and $f(4)$? 11; 17; 29; 53

P **6** If $f(1) = 5$ and $f(n+1) = \frac{1}{2}\left(f(n) + \frac{1}{f(n)}\right)$, what are $f(2), f(3),$ and $f(4)$? 2.6; ≈1.49; ≈1.08

P **7 Consumer Math** The iteration diagram is for a loan repayment where payments are made monthly.
 a What is the original amount of the loan? $8750
 b What is the yearly interest rate? 8.6%
 c What is the monthly payment? $250
 d Write the iteration in function notation.
 e How many months will it take to repay the loan? 41 months

P, R **8** How many diagonals does an 18-sided polygon have? (Hint: Look at the pattern for several polygons. Write an iteration diagram for this pattern.) 135 diagonals

9

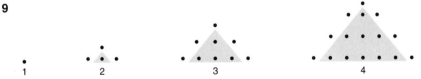

Write an iterative function to represent the total number of dots in the nth triangle. $f(1) = 1, f(n) = f(n-1) + (2n-1)$

P **10**

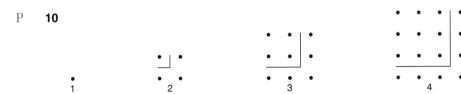

If $f(1) = 1$, which of the following functions describes the iteration pictured in the figures above? **d**
a $f(n+1) = n^2$ **b** $f(n+1) = 2n + f(n)$ **c** $f(n+1) = 2n + 1 + f(n)$
d $f(n+1) = 2n - 1 + f(n)$ **e** $f(n+1) = n + 1 + f(n)$

Section 12.2 More About Iterations 617

SECTION 12.2

Problem-Set Notes

- **14** is Ulam's Conjecture in an iteration diagram. You might have students compare this problem with problem 23 in Section 10.7.
- **18** involves some ideas introduced in Section 3.7. You may want to refer students back to pages 136 and 137.

Checkpoint

1. If $f(0) = 8$ and $f(n) = f(n-1) + 3n$, what is $f(4)$?
2. Write the iteration shown in the diagram in function notation.

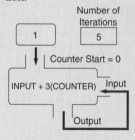

3. Describe the iteration pictured below, using function notation.

   1 2 3 4

Answers:
1. 38 2. $f(0) = 1$, $f(n) = f(n-1) + 3n$
3. $f(1) = 2$, $f(n) = f(n-1) + 2n$

Stumbling Block

Students may have a tough time writing iterative functions. Remind them that every iterative function must have a starting value (e.g., $f(0) = 10$) and must have $f(n)$ on one side of the equation and $f(n-1)$ on the other.

618

P, R **11** It takes 12 toothpicks glued together to make a $1 \times 1 \times 1$ box.

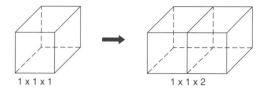

  a How many toothpicks are needed to make a $1 \times 1 \times 2$ box? A $1 \times 1 \times 3$ box? A $1 \times 1 \times 4$ box?   20; 28; 36
  b Write an iterative function for a $1 \times 1 \times n$ box. $f(1) = 12, f(n) = f(n-1) + 8$
  c Use your answer to part **b** to find the number of toothpicks needed for a $1 \times 1 \times 12$ box.   100

P **12** Suppose that $f(1) = 1, f(2) = 4, f(3) = 9, f(4) = 16, f(5) = 25$, and $f(6) = 36$.
  a Write a normal function for the pattern. $f(n) = n^2$
  b Write an iterative function for the pattern. $f(1) = 1, f(n) = f(n-1) + (2n-1)$

P **13**

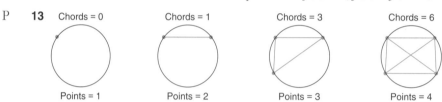

Chords = 0, Points = 1; Chords = 1, Points = 2; Chords = 3, Points = 3; Chords = 6, Points = 4

Write an iterative function for the total number of chords connecting n points on a circle. (Hint: When you add one more point, how many additional chords are added?) $f(1) = 0, f(n) = f(n-1) + (n-1)$

P **14** Try using the numbers 4, 16, 11, 18 and 49 as initial inputs for the iteration diagram shown. What results do you get? It has been conjectured that for any input, this diagram will eventually produce an output of 1. Do you agree?

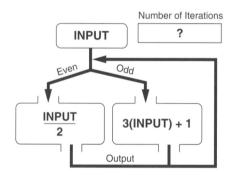

P **15** Copy the following table and use the iteration diagram to fill in the missing entries. What do you notice about the outputs?   They are all square numbers.

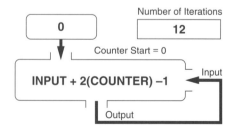

Counter	1	2	3	4	5	6	7	8	9	10	11	12
Output	1	4	9	16	25	36	49	64	81	100	121	144

618   **Chapter 12**   Looking Ahead: Iterations and Fractals

SECTION 12.2

P **16 Communicating**
a Make a table for this iteration diagram.
b Evaluate $1 \cdot 2 \cdot 3 \cdot 4 \cdot \ldots \cdot 10$.
c Compare the result of the tenth iteration in part **a** with your answer in part **b**. Why does this occur?

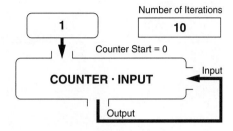

P **17** If $f(1) = 100$ and $f(n + 1) = 2 \cdot f(n)$, the value of $f(n)$ increases each time the value of n increases. If $g(1) = 100$ and $g(n + 1) = g(n) - 150$, the value of $g(n)$ decreases each time the value of n increases. If $h(1) = 100$ and $h(n + 1) = 2 \cdot h(n) - 150$, what happens to the value of $h(n)$ as the value of n increases? It decreases.

P, R **18**

 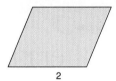

　　　　1　　　　　　　　2　　　　　　　　3

a What is the sum of the measures of the angles in figure 1? 180
b What is the sum of the measures of the angles in figure 2? 360
c What is the sum of the measures of the angles in figure 3? 540
d When a new side is added to the previous figure, by how much does the sum of the angle measures increase? By 180
e Write an iterative function to describe the sum of all the angles of an n-sided figure. $f(3) = 180$, $f(n) = f(n-1) + 180$
f Copy and complete the table.

Number of sides	1	2	3	4	5	6	7	8	9	10
Sum of angles	0	0	180	360	540	720	900	1080	1260	1440

P **19 Consumer Math** At age 16, Bill deposits $10,000 in a bank account that earns 6.5% interest per year, and he adds another $5000 each year thereafter. At what age will Bill have a quarter of a million dollars in the bank? At age 38

P, R **20** Each day a writer writes 5% more than she did on the preceding day. If she writes 2 pages the first day, when will she be writing more than 35 pages per day? On the sixtieth day

Investigation

Mental Math Approximately how many years will it take to double the money you invest if interest is compounded annually at a rate of
a 4%?　　　b 6%?　　　c 8%?　　　d 12%?

Find out what the "rule of 72" is and how it will allow you to solve the above problems mentally.

Section 12.2　More About Iterations

SECTION 12.3

Objective
- To perform iterative procedures on geometric figures

Class Planning

Schedule 2 days

Resources

In *Chapter 12 Student Activity Masters* Book
Class Openers 25, 26
Extra Practice 27, 28
Cooperative Learning 29, 30
Enrichment 31
Learning Log 32
Using Manipulatives 33, 34
Spanish-Language Summary 35

In *Tests and Quizzes* Book
Quiz on 12.1–12.3 (Honors) 297

Transparencies 7, 50

Class Opener

Draw a square with sides 8 in. in length. Locate the midpoint of each side, and connect these points to form a second square. Connect the midpoints of the sides of this second square to form a third square. Repeat the process three more times.

Lesson Notes

- Geometric iterations are a natural extension of the number-pattern iterations done in the previous two sections. If you or your students know Logo, these examples can be done graphically. Students will benefit from seeing patterns emerge visually from geometric iterations.

- The solution to the sample problem can be worked out on a spreadsheet or by using function notation.

Investigation Connection
Section 12.3 supports Unit F, *Investigation* 4. (See pp. 542A-542D in this book.)

SECTION 12.3
Geometric Iterations

In addition to iterating with numbers, it is possible to iterate with shapes. Consider the equilateral triangle shown, with points marked at the midpoint of each side.

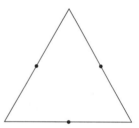

For our first iteration, we will connect the midpoints and mark the midpoints of the segments we drew. Notice that the triangle we just drew is similar to the original triangle.

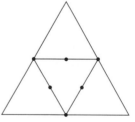

Theoretically, you can repeat this process forever, but in actual practice, after a few times the triangles become too small to draw.

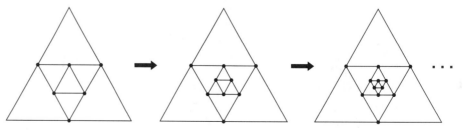

Chapter 12 Looking Ahead: Iterations and Fractals

SECTION 12.3

Example 1
Start with a square, ABCD. Locate a point one third of the way from A to B. Locate points in the same way on $\overline{BC}$, $\overline{CD}$, and $\overline{DA}$. Connect the points to form another square, and keep iterating in the same manner (clockwise).

Solution
We follow the instructions to form square ABCD and the square inside it.

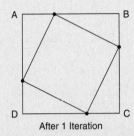

After 1 Iteration

Then we repeat the procedure with the inner square, selecting and connecting points one third of the way between consecutive vertices, in a clockwise pattern. We could repeat the process for as long as space allows.

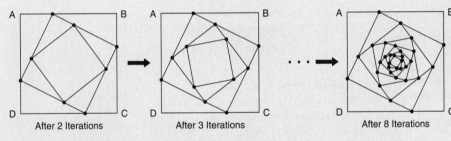

What would happen if this iteration were tried on a triangle or a pentagon? What if the distance were something other than one third of the way to the next point?

In Example 1, we started with a polygon. We can also iterate with segments.

Example 2
Begin with a horizontal segment, $\overline{AB}$, as shown. At the right-hand endpoint (B), draw two segments that are the same length as $\overline{AB}$ and that form congruent (120°) angles all the way around, then draw segments parallel and congruent to $\overline{AB}$ at the right-hand endpoints of these two new segments. Repeat the process for a few iterations.

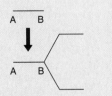

Additional Examples

Example 1
Start with a rhombus. Locate the midpoints of its sides and connect them to form a rectangle. Iterate several more times.

Answer:

Example 2
Start with two congruent segments with a common endpoint, one slanting 30° above horizontal, the other 30° below horizontal. From each nonshared endpoint, draw two more segments identical to the first pair. Repeat this process for several iterations.

Answer:

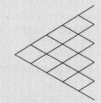

Cooperative Learning

- Have students work in groups to write an iterative function for the area of the *n*th square in example 1. Assume that AB = 30.

  Answer: $f(1) = 900$, $f(n) = \frac{5}{9}[f(n-1)]$

- Problems 6 and 11 are appropriate for cooperative learning.

SECTION 12.3

Assignment Guide

Basic
Optional

Average or Heterogeneous
Optional

Honors
Day 1 5, 8, 10
Day 2 6, 7, 9, 11

Strands

Communicating Mathematics	6
Geometry	1–10, Investigation
Measurement	1–10, Investigation
Algebra	2, 5, 6, 9–11
Patterns & Functions	1–11, Investigation
Interdisciplinary Applications	Investigation

Individualizing Instruction

The gifted learner Students might be challenged to write a computer program that uses the computer's graphic capability to perform the iteration in the class opener or example 1.

Communicating Mathematics

- Have students write a paragraph comparing numeric iteration with geometric iteration.
- Problem 6 encourages communicating mathematics.

Solution
This time, the figure eventually gets too big to fit on the paper, not too small to draw.

Second Iteration → Third Iteration ...

How many new segments will there be in the twelfth iteration? How many segments will there be all together after 12 iterations?

Sample Problem

Problem On a coordinate system, start at the origin, and draw a segment 1 unit long on the x-axis. Then turn left 90° and draw a segment 2 units long. For each subsequent iteration, turn left and draw a segment 1 unit longer than the one you drew last time. Continue for a while.

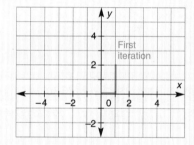

Solution The diagram shows the result of the first 5 iterations. Can you predict how long the segment will be that is added on at the eleventh iteration? Can you predict in which quadrant(s) it will lie? Can you predict the total length of the figure after 11 iterations? Will the fifteenth segment be horizontal? Will it be drawn from left to right?

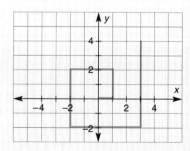

622 **Chapter 12** Looking Ahead: Iterations and Fractals

SECTION 12.3

Think and Discuss

P   **1** We are going to alter Example 2 by modifying the iteration rule. We will begin with a horizontal segment as before, and draw new segments as before, only this time their length will be half the length of the original segment. Do you think the segments will ever come together, as they did before? Will there be the same number of them as before? Will it stretch off the page if we do it enough times? No; no; no

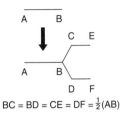

$BC = BD = CE = DF = \frac{1}{2}(AB)$

P, R   **2** Suppose that the perimeter of the triangle shown at the beginning of this section is 3 units.
     a What is the perimeter of the triangle made on the first iteration? The second iteration? The third iteration? 1.5; 0.75; 0.375
     b Write an iterative function to describe the perimeter on the nth iteration. $f(0) = 3, f(n) = f(n-1) \div 2$

P, R   **3** Draw a circle. Draw a diameter. Draw two circles, each having a diameter half that of the original circle, so that the new circles are tangent to each other and to the original circle. Iterate this process several times.

P, R   **4** Try the iteration procedure in Example 1 on an equilateral triangle. Start with an equilateral triangle ABC, select a point one third of the way from A to B, and repeat on $\overline{BC}$ and $\overline{CA}$. Connect these three points. Iterate several times.

Problems and Applications

P, R   **5** Draw a rectangle. Locate the midpoints of all four sides. Connect them in order.
     a What shape do you get? Rhombus
     b Now iterate several more times. What shapes do you get?

P   **6 Communicating** Consider this iteration performed on a coordinate system.
     a List the next six vertices of the right angles in the iteration.
     b What is the length of the first line segment? The second line segment? The third line segment? The fifty-fourth line segment?
     c Does the ninety-seventh line segment slant / or \ ?
     d Is the ending vertex of the ninety-seventh line segment in the first quadrant? Explain.

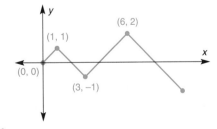

Section 12.3 Geometric Iterations   623

Problem-Set Notes

for page 623

- **1** previews convergence. Have students do this problem with a ruler before you discuss it.
- **3** has several possibilities—see how many your students can come up with. You'll probably need to let students know that tangent circles are circles that intersect at exactly one point.
- **5 a** could be turned into a communicating mathematics exercise by asking students to justify their answers.

Additional Practice

1. The first triangle in the diagram below is an equilateral triangle with a perimeter of 6.

What are the perimeters of the second and third triangles formed?

2. List the next four vertices in the iteration shown below.

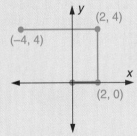

3. In problem 2, will the twentieth segment be a vertical or a horizontal segment? In which quadrant will the ending vertex of the twentieth segment lie?

Answers:
1. 3; 1.5   **2.** (−4, −4), (6, −4), (6, 8), (−8, 8)
3. Vertical; Quadrant III

Answers for problems 3, 4, 5 b, and 6 a, b, and d are on page A10.

SECTION 12.3

Problem-Set Notes

for page 624

- **8** can be shared with the art teacher at your school.

Checkpoint

1. Draw a regular pentagon. Connect the midpoints of each side to form a new pentagon. Continue this iteration three more times.

2. On a coordinate grid, start at (1, 1) and iterate as shown. All of the angles are right angles. Find the coordinates of the next six points.

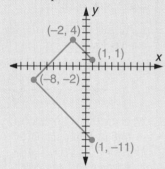

3. What is the length of the fourth segment of the figure produced in problem 2?

Answers:

1.

2. (13, 1), (−2, 16), (−20, −2), (1, −23), (25, 1), (−2, 28)

3. ≈16.97 units

Answers for problems 7, 8, 10 a, b, and d, 11, and Investigation are on page A10.

P   **7** Try the iteration scheme that generated the triangles at the beginning of this section, only this time begin with an isosceles right triangle instead of an equilateral triangle.

P   **8** On graph paper make a large square 24 units on a side. Locate eight points on the sides that are one third of the distance from one vertex to the next vertex. Connect every other one of these to form another square. Connect the remaining four to form still another square. These two new squares intersect at eight new points. Connect every other one to form another new square and the remaining four to form yet another square. Continue this iteration process. Color the picture to form an interesting mosaic.

P, R   **9** Starting at the origin of a coordinate system, draw a segment 1 unit long on the x-axis to (1, 0). Turn left 90° and draw a segment 2 units long. Turn left 90° and draw another segment 2 units long. Each iteration consists of 2 segments and 2 turns. The subsequent iterations increase the length of the segments by 1 unit. Draw ten iterations (20 segments). Which lattice points will be missed?   (0,1)

P, R   **10** On a coordinate system, start at (0, 0) and iterate as shown. All the angles are right angles.

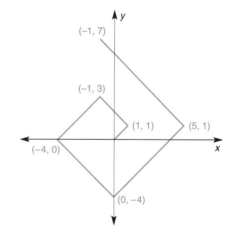

   a Find the coordinates of the next six points that are the vertices of the right angles.
   b What is the length of the first segment? The second segment? The third segment? The tenth segment?
   c In which direction does the seventeenth segment slant?
   d What is the total length of the first ten segments? Write an iteration diagram for this part of the problem.
   e What is the average length of the first ten segments? $5.5\sqrt{2}$, or ≈7.78

P   **11** In this section's sample problem, which lattice points in the first quadrant does the figure pass through? Which lattice points in the first quadrant are missed? Can you think of a way to iterate that would pass through all of the first-quadrant lattice points? All of the lattice points?

Investigation

Snowflake Curve   Find out about the Koch snowflake curve. Show how it is iterated from a single line segment.

624   Chapter 12   Looking Ahead: Iterations and Fractals

SECTION 12.4

12.4 Convergent Iterations

Investigation Connection
Section 12.4 supports Unit F, *Investigation* 5. (See pp. 542A-542D in this book.)

CONVERGENCE

Iterations that we have done thus far have been of two distinct types—those that approached closer and closer to a particular value and those that didn't. In this section, we will look more closely at those that approach a particular value. These are called **convergent** iterations.

Example 1

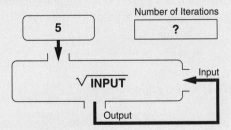

Consider the iteration diagram above. Each successive output is calculated by taking the square root of the previous output. In function notation, we can write

$$f(1) = 5$$
$$f(n) = \sqrt{f(n-1)}$$

What are the values of $f(1)$ through $f(21)$?

Objectives

- To recognize convergent iterations
- To identify the limiting value of a convergent iteration

Vocabulary: convergent, limiting value

Class Planning

Schedule 2 days

Resources

In *Chapter 12 Student Activity Masters* Book
Class Openers 37, 38
Extra Practice 39, 40
Cooperative Learning 41, 42
Enrichment 43
Learning Log 44
Technology 45, 46
Interdisciplinary
  Investigation 47, 48
Spanish-Language Summary 49

Transparencies 7, 50

Class Opener

Find the next four numbers in each pattern. What value do the terms get closer and closer to?

1. $9, 3, 1, \frac{1}{3}, \ldots$
2. $\frac{3}{2}, \frac{4}{3}, \frac{5}{4}, \frac{6}{5}, \ldots$

Answers:
1. $\frac{1}{9}, \frac{1}{27}, \frac{1}{81}, \frac{1}{243}; 0$
2. $\frac{7}{6}, \frac{8}{7}, \frac{9}{8}, \frac{10}{9}; 1$

Lesson Notes

- Spreadsheets and calculators are tools that can help a student to see and understand convergence better. Almost all calculators will show the convergence of the square root, just by pressing the key repeatedly. Spreadsheets or calculators with graphing capability allow students to visualize the convergence.

SECTION 12.4

Additional Examples

Example 1
To what value does the iteration $f(1) = 9$ and $f(n) = \frac{2}{3} \cdot f(n-1)$ converge?

Answer: 0

Example 2
A tennis ball is dropped from a height of 5 feet, and each time it bounces, it rebounds to a height 50% of its preceding height. Find the total distance the ball travels.

Answer: 15 feet

Reteaching

Reteach convergence and limiting values by giving students nonmathematical examples of the concepts. For example, during the chase scene of a movie, the good guys converge on the bad guys. In bowling, the score 300 is a limiting value. You might also discuss some nonexamples of convergence, such as the balance of a bank account and the population of the world.

Solution
This could be done with a calculator, a graphing calculator, or a spreadsheet. Below are results from a spreadsheet.

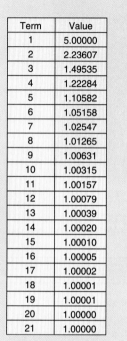
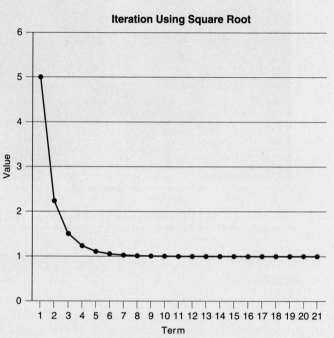

Term	Value
1	5.00000
2	2.23607
3	1.49535
4	1.22284
5	1.10582
6	1.05158
7	1.02547
8	1.01265
9	1.00631
10	1.00315
11	1.00157
12	1.00079
13	1.00039
14	1.00020
15	1.00010
16	1.00005
17	1.00002
18	1.00001
19	1.00001
20	1.00000
21	1.00000

Notice that as we continue iterating, the values get closer and closer to the number 1. In fact, the number we use for our initial input does not matter as long as we choose a number that has a square root. (Try this iteration with several other starting values.) We say that the iteration converges to the number 1.

We now give two intuitive ideas for convergence.

Convergence of Iterations

1. An iteration converges if after a while each output looks almost identical to the preceding output.
2. We also say an iteration converges if the input and the output become almost identical—in other words, if INPUT ≈ f(INPUT).

In some cases the convergence may happen very quickly, and in others it takes many iterations for this to occur. In some cases the iteration does not converge at all.

Chapter 12 Looking Ahead: Iterations and Fractals

SECTION 12.4

LIMITING VALUES

When an iteration converges, the number that the iteration converges to is called the ***limiting value*** of the iteration. The limiting value of the iteration in Example 1 was 1.

Example 2
A ball is dropped from a height of 6 feet, and each time it bounces, it rebounds to a height 75% of its preceding height.
a If we iterate the height of the bounce, does this height converge? If so, what is its limiting value?
b If we iterate the total (vertical) distance the ball travels, does this distance converge? If so, what is the limiting value?

Solution
a If we draw an iteration diagram for this problem, we see that each output is produced by multiplying the input by 0.75. We can use a spreadsheet to make a table for the first 25 bounces.

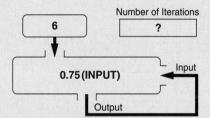

Bounce	Height (ft)
0	6.00
1	4.50
2	3.38
3	2.53
4	1.90
5	1.42
6	1.07
7	0.80
8	0.60
9	0.45
10	0.34
11	0.25
12	0.19
13	0.14
14	0.11
15	0.08
16	0.06
17	0.05
18	0.03
19	0.03
20	0.02
21	0.01
22	0.01
23	0.01
24	0.01
25	0.00

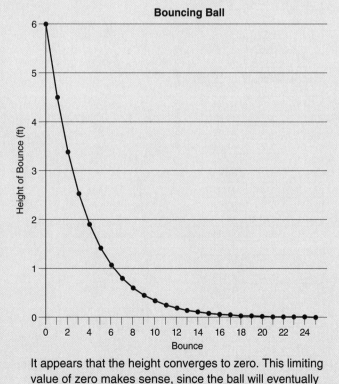

It appears that the height converges to zero. This limiting value of zero makes sense, since the ball will eventually stop bouncing.

Assignment Guide

Basic
Optional

Average or Heterogeneous
Optional

Honors
Day 1 6–9, 11, 16, 21
Day 2 10, 12–15, 19, 20

Strands

Communicating Mathematics	8, 12, Investigation
Measurement	7, 12, 13, 16, 18, 20
Data Analysis/ Statistics	1–3, 11, 19
Algebra	1–7, 9–13, 16, 17, 19–21
Patterns & Functions	1–21
Interdisciplinary Applications	7–9, 12, 13, 16, 17, Investigation

Individualizing Instruction

The gifted learner Students might be interested in finding out what an asymptote is and how it relates to convergence and limiting values.

Multicultural Note

Fractal geometry was named by the French mathematician Benoit Mandelbrot. This branch of geometry was created as Mandelbrot began to study how to measure the lengths of common borders and coastlines. This became of interest when a British scientist, Lewis Richardson, discovered that two countries sharing a border had significantly different measurements for the length of that common border.

SECTION 12.4

Stumbling Block

Students may stop the iteration process too soon when trying to find the value to which an iteration converges. Use the tables in example 2 to show how the values converge and when students can stop the iteration process.

Cooperative Learning

- Have students discuss the following problem in groups: A $100 bill is pinned to the floor. The pin is exactly in the center of the bill. You are standing 100 ft from the bill. Every 10 minutes you are allowed to move to the midpoint of the segment created by the pin and the tips of your shoes. You may not claim the $100 bill until you are standing on it. How much time will elapse before you can claim the $100 bill?

Answer:
Either 90 or 100 minutes, depending on the position of the bill.

- Problems 9 and 15 are appropriate for cooperative learning.

b The ball travels 6 feet when it is dropped. On any given bounce, it has traveled all the distance so far plus twice the height of that particular bounce.

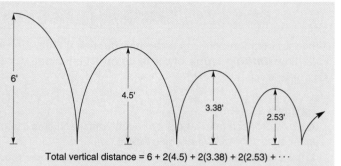

Total vertical distance = 6 + 2(4.5) + 2(3.38) + 2(2.53) + ···

We can add a third column to our table in part **a** to describe the results. The formula +C2+2*B3 was used to obtain the value 15.00 in cell C3.

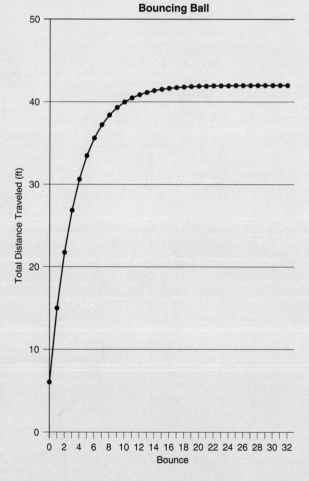

Bounce	Height (ft)	Distance (ft)
0	6.00	6.00
1	4.50	15.00
2	3.38	21.75
3	2.53	26.81
4	1.90	30.61
5	1.42	33.46
6	1.07	35.59
7	0.80	37.19
8	0.60	38.40
9	0.45	39.30
10	0.34	39.97
11	0.25	40.48
12	0.19	40.86
13	0.14	41.14
14	0.11	41.36
15	0.08	41.52
16	0.06	41.64
17	0.05	41.73
18	0.03	41.80
19	0.03	41.85
20	0.02	41.89
21	0.01	41.91
22	0.01	41.94
23	0.01	41.95
24	0.01	41.96
25	0.00	41.97
26	0.00	41.98
27	0.00	41.98
28	0.00	41.99
29	0.00	41.99
30	0.00	41.99
31	0.00	42.00
32	0.00	42.00

We see that the iterations seem to converge to 42 feet. The ball has a limiting distance of travel of 42 feet.

Chapter 12 Looking Ahead: Iterations and Fractals

SECTION 12.4

Sample Problem

Problem Suppose you start with a ball of string at point A and unroll it along the segments. If the pattern continues in the same manner, how many meters of string must be in the ball for you to unroll it over the entire figure?

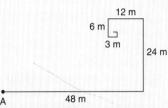

Solution At any step, we need the amount of string we used before plus one half of the amount of string we used on the last straight piece we traced. Let's start by making a table.

Iteration	String for That Section	Total String Used
1	48 meters	48 meters
2	24 meters	72 meters = 48 + 24
3	12 meters	84 meters = 72 + 12
4	6 meters	90 meters = 84 + 6
.	.	.
.	.	.
.	.	.

Segment	Length (m)	Total Length (m)
1	48.000	48.000
2	24.000	72.000
3	12.000	84.000
4	6.000	90.000
5	3.000	93.000
6	1.500	94.500
7	0.750	95.250
8	0.375	95.625
9	0.188	95.813
10	0.094	95.906
11	0.047	95.953
12	0.023	95.977
13	0.012	95.988
14	0.006	95.994
15	0.003	95.997
16	0.001	95.999
17	0.001	95.999
18	0.000	96.000
19	0.000	96.000

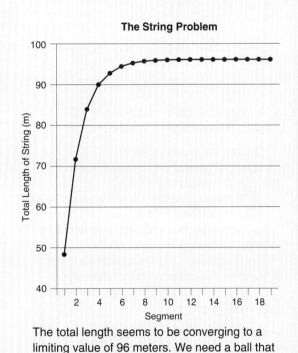

The total length seems to be converging to a limiting value of 96 meters. We need a ball that contains 96 meters of string.

Communicating Mathematics

- Have students answer the following questions in writing or orally: Do the concepts of convergence and limiting value apply to geometric iterations? If so, how?

- Problems 8, 12, and Investigation encourage communicating mathematics.

Additional Practice

1. To what number does the function $f(0) = 4$ and $f(n) = \frac{1}{2} \cdot f(n-1)$ converge?

2. Frannie invested $100 at 6% annual interest. She deposited another $100 at the beginning of each of the next four years. How much did she have at the end of the fifth year?

3. A ball is dropped from a height of 16 feet, and each time it bounces, it rebounds to a height 80% of its preceding height. What is the total vertical distance traveled by the ball before it stops bouncing?

Answers:
1. 0 **2.** $597.53 **3.** 144 feet

Section 12.4 Convergent Iterations

SECTION 12.4

Problem-Set Notes

for pages 630 and 631

- **1** might require you to suggest to students that they substitute a perfect square for 8 in the iteration rule.
- **4** shows that not all iterative functions converge.
- **7, 13,** and **16** make connections to science.
- **10** uses iteration to guess and check. This is a relatively new problem-solving approach, made possible by more sophisticated calculators and spreadsheets.

Additional Answers

for pages 630 and 631

1 a $2.8\overline{3}$, 2.828431, 2.828427; following values remain constant at 2.828427.

b Outputs converge to same result as in part **a**.

c Outputs converge to 8.

d Answers will vary.

e It computes $\sqrt{a}$, where $\frac{1}{2}(\text{INPUT} + \frac{a}{\text{INPUT}})$ is the iteration rule.

2 The values converge to the negative square root of the number. That is, they converge to $-\sqrt{a}$, where the iteration rule is $\frac{1}{2}(\text{INPUT} + \frac{a}{\text{INPUT}})$.

3 b Yes

5 a

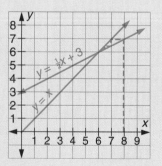

Think and Discuss

P **1 a** Use the iteration diagram to determine the outputs for ten iterations.
  b Repeat part **a** for a different positive initial input.
  c Change the 8 in the iteration rule to 64 and repeat parts **a** and **b**.
  d Try using various other positive numerators in place of the 8.
  e What does this iteration compute?

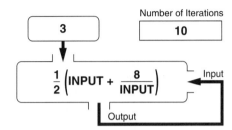

P **2** Repeat problem 1, but this time use negative numbers as the initial inputs. What do you notice about the convergence now?

P **3 a** Perform iterations as shown in the diagram until you see what the limiting value is. To what value do the iterations converge? 100
  b Repeat part **a** with other initial inputs. Do the iterations converge to the same value as in part **a**?

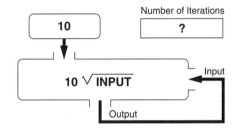

P **4** To what number does the following iteration function converge? It does not converge.

$$f(1) = 4$$
$$f(n) = f(n-1) + 3$$

P **5 a** Duplicate this diagram on a sheet of graph paper. Then follow these steps:

  1. Put your pencil at (8, 0).
  2. Move vertically (drawing a segment) to the graph of $y = \frac{1}{2}x + 3$.
  3. Move horizontally (drawing a segment) to the graph of $y = x$.
  4. Repeat steps 2 and 3 for four iterations.

  Note: The figure of segments that you have just drawn is called a *web diagram*.

  b Make a table of the x and y values of the ordered pairs on the two lines that you "visited" in part **a**.

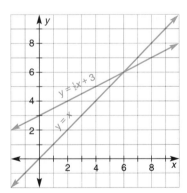

630 **Chapter 12** Looking Ahead: Iterations and Fractals

SECTION 12.4

Problems and Applications

P **6** Repeat parts **a** and **b** of problem 5, using this diagram.

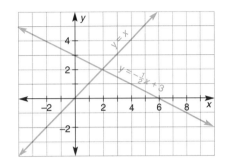

P, R **7** **Science** A swimming pool has had chlorine added to bring the chlorine concentration to 3 parts per million (3 ppm). Fifteen percent of the chlorine present on any day is lost by the next day.
  a Iterate to see how much chlorine is left each day.
  b The ideal concentration of chlorine in a pool is 2 ppm. If there are 3 ppm now, how much chlorine should be added each day to have the amount of chlorine converge to 2 ppm? 0.3 ppm

P **8** **Communicating** These are the directions on a bottle of shampoo:
  1. Put a small dab of shampoo in your hand.
  2. Work the shampoo into your hair.
  3. Rinse.
  4. Repeat.

  Explain what these directions mean iteratively and what would happen if they were taken literally.

P, R **9** **Consumer Math** Bill borrows $7000 to be paid back monthly over 24 months. The yearly interest rate is 9.5%.
  a What is the monthly interest rate? ≈0.79%
  b We do not know what his monthly payment is, so we do the following. We make a table of monthly payments and how much of the loan is left at the end of 24 months. Copy and complete the table.

Monthly payment	$100	$150	$200	$250	$300
Amount left at end of 24 months	$5826.69	$4510.82	$3194.96	$1879.09	$563.23

  c Make a graph, using the data from part **b**. Use this graph to estimate the value of the monthly payment needed to repay the loan in 24 months. About $320
  d Use iteration to check your result in part **c**. Actual payment = $321.40

P, R **10** A man spent the first third of his life unmarried. After he had been married for 50 years, his wife died. Six years later, he died. How old was the man when he died? (Hint: Guess any age for an initial input and then iterate $\frac{1}{3}$(INPUT) + 50 + 6.) 84 years old

Section 12.4 Convergent Iterations

---

b

x	8	8	7	7	6.5	6.5	6.25
y	0	7	7	6.5	6.5	6.25	6.25

x	6.25	6.125	6.125
y	6.125	6.125	6.0625

6a

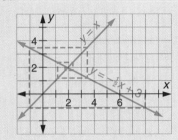

b

x	8	8	−1	−1	3.5	3.5	1.25	1.25
y	0	−1	−1	3.5	3.5	1.25	1.25	2.375

x	2.375	2.375	1.8125	1.8125
y	2.375	1.8125	1.8125	2.09375

7a

Day	Chlorine (ppm)
0	3
1	2.55
2	≈2.17
3	≈1.84
4	≈1.57
5	≈1.33
6	≈1.13
7	≈0.96

8 Step 4 says repeat, but does not tell you how many times to repeat, so you would be continuously washing and rinsing your hair.

SECTION 12.4

Problem-Set Notes

for pages 632 and 633

- 14 and 15 could be a class project—try many different rectangles and look for a pattern.
- 18 is a variation of 14.

Checkpoint

1. To what number does the iteration $f(0) = 200$ and $f(n) = \frac{1}{4} \cdot f(n-1)$ converge?

2. Jason invested $1500 at 8% annual interest. He deposited another $500 at the beginning of each of the next three years. How much did he have at the end of the fourth year?

3. A ball is dropped from a height of 10 feet, and each time it bounces, it rebounds to a height 60% of its preceding height. What is the total distance traveled by the ball before it stops bouncing?

Answers:
1. 0   2. $3793.79   3. 40 feet

P **11** Does this iteration converge? If so, what is the limiting value?   Yes; 1.618033989

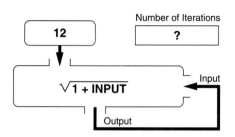

P **12** The imaginary substance antgrav rebounds to a height that is 115% of its height on the previous bounce. A ball of antgrav is dropped from a height of 5 feet.
   a When will it touch a ceiling that is 9 feet above the floor on which it bounces?   On the fifth bounce
   b Does this iteration problem converge? Why or why not?

P, R **13** **Science**   A certain prescription says to take two 500-milligram pills every 4 hours. If every 4 hours 80% of the medicine that is in the body is used up, what is the amount of medicine that remains in the body over a period of several days? (That is, what does the medicine level converge to?)   1250 mg

P **14** In each diagram, a ball begins at point A and is shot diagonally. At which point will it exit—A, B, C, or D?

a 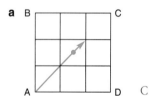   C

b 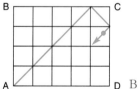   B

c

d 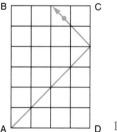   D

   C

P **15** Draw some other diagrams like those in problem 14 and look for patterns. Try to predict conditions that will lead the ball to exit at point A, point B, point C, or point D.

P, R **16** **Science**   A doctor prescribes an initial dosage of 2000 milligrams of a certain medicine. If 75% of the medicine is gone from the body one day later, what daily dosage should be given after the initial dosage in order to maintain an effective level of 1260 milligrams of medicine in the patient?   945 mg

SECTION 12.4

P, R **17** Sally wants to borrow $6000 and pay it back in payments of $275 per month. What is the greatest possible interest rate she could have and be able to repay the loan in 24 months?   9.3% annually

P **18** Each side of the triangle is divided into equal thirds (trisected). A ball at point K is shot toward B and bounces toward F and so on. Trace the entire path of the ball. What happens?

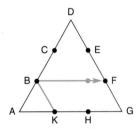

P **19** In the iteration diagram shown, to what value do the iterations converge?   ≈2.293166287

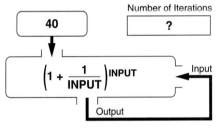

P **20** If the pattern of segments continues, how much string is needed to trace the entire diagram?   1000 ft

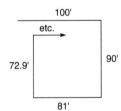

P **21** Consider the following two functions.

$f(1) = 47{,}000$
$f(n) = 0.8 \cdot f(n-1) + 0.15 \cdot g(n-1)$

$g(1) = 64{,}000$
$g(n) = 0.85 \cdot g(n-1) + 0.2 \cdot f(n-1)$

a Copy and complete the following table for values of n from 1 to 15.

n	1	2	3	4	5	6	
f(n)							...
g(n)							

b After many iterations, what do $f(n)$ and $g(n)$ converge to?   47,571 and 63,429

Investigation

Consumer Math What is a mortgage? If a bank loans you money to buy a house, who is the mortgagor and who is the mortgagee? Obtain a copy of a mortgage payment schedule that shows principal and interest payments, and explain it to the class.

Section 12.4 Convergent Iterations 633

Additional Answers

for pages 632 and 633

12 b No, because it rebounds to a height greater than its height on the previous bounce

15 Let (x, y) represent the dimensions of ABCD (x the horizontal component, y the vertical). Let $f(x, y)$ represent the exit point of a ball shot diagonally from point A. The following properties hold:

x	y	f(x, y)
odd	even	B
even	odd	D
odd	odd	C
even	even	not A

18 K → B → F → H → C → E → K; the process will repeat, following the same pattern.

21 a

n	f(n)	g(n)
1	47,000	64,000
2	47,200	63,800
3	47,330	63,670
4	47,414	63,586
5	47,469	63,531
6	47,505	63,495
7	47,528	63,472
8	47,543	63,457
9	47,553	63,447
10	47,559	63,441
11	47,563	63,437
12	47,566	63,434
13	47,568	63,432
14	47,569	63,431
15	47,570	63,430

Investigation A mortgage is a pledge of property as security for a loan. Thus, since the buyer is the one who pledges the property as security to the bank that loans him or her the money, the buyer is the mortgagor, and the bank is the mortgagee.

SECTION 12.5

Objective

- To identify the characteristics of a fractal

Vocabulary: fractal, self-similarity

Class Planning

Schedule 2 days

Resources

In *Chapter 12 Student Activity Masters* Book
Class Openers 51, 52
Extra Practice 53, 54
Cooperative Learning 55, 56
Enrichment 57
Learning Log 58
Interdisciplinary
  Investigation 59, 60
Spanish-Language Summary 61

Transparency 7

Class Opener

Draw a square with each side 8 in. long. Connect the midpoints of opposite sides to form four squares. Shade the upper right and lower left squares. In each of the two unshaded squares, connect the midpoints of opposite sides to form four squares. Again, shade the upper right and lower left squares. Repeat this process two more times.

Lesson Notes

- Even students with minimal programming experience can produce Sierpinski's triangle by using a programmable calculator or a programming language such as Logo, BASIC, or Pascal. One such program is given in the answer to Investigation.

- This section provides only a brief introduction to fractals.

634

SECTION 12.5

Introduction to Fractals

We now return to geometry for a look at a new and fascinating branch of mathematics. For our first example, we begin with an equilateral triangle and connect the three midpoints of its sides. This time, we also shade the center triangle.

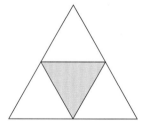

After the first iteration, there are three unshaded triangles. We now repeat the process, locating the midpoints of the sides of each unshaded triangle, connecting them, and shading each center triangle.

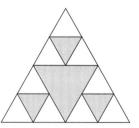

After the second iteration, there are nine unshaded triangles, so our third iteration will be done on each of these nine triangles.

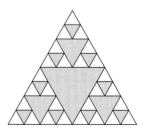

634   Chapter 12   Looking Ahead: Iterations and Fractals

SECTION 12.5

There are a few things to notice about the process.

- It is an iterative process.
- We could continue until the next set of triangles would be too small to draw. In our minds, it could continue forever. Theoretically, we could iterate infinitely many times.
- Part of the figure that is added each time an iteration is completed is similar to the previous picture. This property is called *self-similarity*. (Notice that the red portion of the third iteration is similar to the entire figure of the second iteration.)

In general, if we continue this process on a figure that has the three properties listed above, the figure converges to a figure that is called a *fractal*. If we continue iterating, eventually the figure shown here appears. This figure is called the Sierpinski triangle, after the Polish mathematician, Waclaw Sierpinski, who first introduced it.

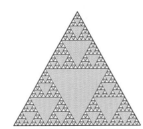

It is also possible to generate fractal curves by iterating line segments. For our example, we begin with a segment that has been divided into three congruent parts.

For our initial step, we remove the center segment and replace it with two segments that are equal in length to the segment removed. We place them as shown in the figure, to form two sides of an equilateral triangle. The segment we removed would have been the third side.

Now we iterate by repeating the same procedure on each of the four segments in the new diagram.

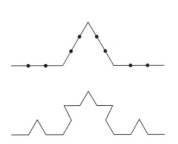

Section 12.5 Introduction to Fractals

SECTION 12.5

Problem-Set Notes

for page 637

- **1** may result in several different diagrams, depending on how the pattern is interpreted. Have students share their results.

Communicating Mathematics

- Have students write a paragraph explaining the concept of self-similarity.
- Problems 8 and 9 encourage communicating mathematics.

Additional Practice

1. The first iteration of an equilateral triangle is shown. Continue for two more iterations.

2. Start with a 6-inch-by-6-inch square. Divide it into nine squares, shading them as shown. Iterate on the unshaded squares. What does the total shaded area converge to?

Answers:
1.

2. 36 square inches

As before, we continue to iterate. Notice that the three properties of a fractal are present.

If we had started with an equilateral triangle and had iterated each side of the triangle in the same manner as the single segment, the curve generated would be the Koch snowflake, named after the Swedish mathematician, Helge von Koch.

The study of fractals began in about 1900 but did not become a major topic until the early 1970's, when computers made it possible to draw the complicated pictures in a reasonable period of time. Fractal geometry is closely connected with chaos theory, a rapidly developing branch of science.

The best way to learn about fractal geometry is to create your own fractal curve. Try this one. Draw a long line segment down the center of a sheet of paper. Find its midpoint and draw two segments from the midpoint, each half as long as the segment and making 120° angles with the bottom half of the segment. You should now have four congruent segments. Repeat the iterative process on each of these four, and then again on the new segments produced. Continue for as long as you can.

Sample Problem

Problem What is the pattern for the number of shaded triangles in the Sierpinski triangle?

Solution Going back and looking at successive pictures, we notice that the first few numbers are 1, 4, and 13. We could just play with the patterns as we did in Chapter 1. Perhaps a more effective approach is to think about what is happening each time we iterate.

We start with 1 triangle shaded and 3 unshaded. Each of the 3 unshaded triangles yields 1 shaded and 3 unshaded triangles. The shaded total is 1 + 3, or 4. The unshaded total is 3 · 3, or 9. At the next iteration, each of the 9 unshaded triangles again yields 1 shaded and 3 unshaded triangles. The shaded total is 4 + 9, or 13. The unshaded total is 3 · 9, or 27.

	Shaded	Total Shaded	Total Unshaded
Begin	1	1	3
Iterate	+3	4	9
Iterate	+9	13	27
Iterate	+27	40	81
Iterate	+81	121	243

SECTION 12.5

It looks difficult to generate a single formula, but we can generate an iterative function for the number of shaded triangles after n iterations.

$$f(1) = 1$$
$$f(n) = f(n-1) + 3^n$$

Let's put it on a spreadsheet and see what happens. Notice that after only a few iterations, there is quite a large number of triangles. Perhaps you guessed a formula for the number of triangles after n iterations.

	A	B	C
1	Step	Shaded	Unshaded
2	1	1	3
3	2	4	9
4	3	13	27
5	4	40	81
6	5	121	243
7	6	364	729
8	7	1093	2187
9	8	3280	6561
10	9	9841	19683
11	10	29524	59049
12	11	88573	177147
13	12	265720	531441
14	13	797161	1594323
15	14	2391484	4782969
16	15	7174453	14348907
17	16	21523360	43046721
18	17	64570081	129140163
19	18	193710244	387420489

Checkpoint

1. Begin with a square. Divide it into four congruent squares and shade as shown. Iterate on the empty squares two more times.

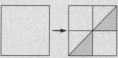

2. Draw a rectangle. Divide it into four congruent rectangles and shade them as shown. Iterate on the unshaded rectangles one more time.

Answers:

1.

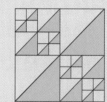

2.

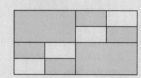

Think and Discuss

P **1** Begin with the legs of an isosceles right triangle, $\overline{AB}$ and $\overline{BC}$. Locate the midpoints, M and N. Replace ∠MBN with ∠MDN as shown. Iterate on all the remaining right angles in the same manner. Repeat for at least two more iterations. Draw the resulting diagrams. If AB = 4, find the total length of the segments in each iteration. *In each case, the length is 8.*

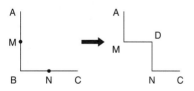

P **2** Draw a square. Divide it into nine congruent squares and color four of them as shown in the diagram. Iterate on the remaining noncolored squares. Do two more iterations.

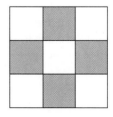

Additional Answers

for page 637

2

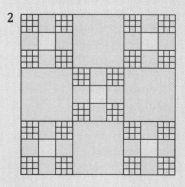

SECTION 12.5

Problem-Set Notes

for pages 638 and 639

- **4** and **5** can be time-consuming. You might want to assign half the class to each problem and have them share their results.
- **6** reviews circumference of a circle.
- **10** geometrically shows a convergent iteration.

Reteaching

Reteach fractals by cutting a 12-inch square of poster board. Have students cut squares with the following side lengths out of colored paper: 6", 3", $1\frac{1}{2}$", and $\frac{3}{4}$". Have students place the 6" squares on the poster board as follows:

Have students position the correct-sized squares to complete the next three iterations.

Cooperative Learning

- Have students work in groups to create a diagram representing the first 16 rows of Pascal's triangle. Instead of actual sums, however, have students use only O's (for odd numbers) and E's (for even numbers). Then have students shade the O's. What patterns appear? How is this related to the concept of self-similarity?

Answers:
The figure looks very similar to the one at the top of page 635. The figure made by

Problems and Applications

P **3** **a** Draw a regular pentagon. (Hint: Start with a circle and divide it into five equal parts.)
   b Draw all five diagonals. Notice that you get another pentagon formed in the center of the original pentagon.
   c Iterate by drawing the diagonals of this inner pentagon. Repeat one more time. Is this self-similar?   No

P **4** An L-shaped figure can be divided into four congruent L shapes as shown. Shade one of the four pieces as shown and iterate the unshaded L-shaped figures. Repeat several times. Is the figure self-similar?   Yes

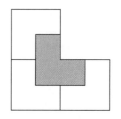

P **5** Do the Sierpinski triangle iteration scheme on an isosceles right triangle. Do three iterations.

P **6** Begin with a semicircle. Replace it with two semicircles as shown. If the length of the first semicircle is 16π, find the length of the curve formed on each iteration for the first five iterations.   In each case, the length is 16π.

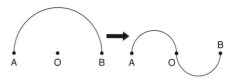

P **7** Do the Koch snowflake iteration using squares instead of triangles. The first iteration is shown.

P, R **8** **a** Find the missing sums.
      b How are the numbers to the left of the equal signs related to the Sierpinski triangle?

$1 = 1$
$4 = 1 + 3$
$13 = 1 + 3 + 3^2$
$\underline{40} = 1 + 3 + 3^2 + 3^3$
$\underline{121} = 1 + 3 + 3^2 + 3^3 + 3^4$
$\underline{364} = 1 + 3 + 3^2 + 3^3 + 3^4 + 3^5$

P, R **9** **a** Find the numbers in the fifth level of the tree diagram.
      b Find the sum of the numbers at each level.
      c How are the sums in part **b** related to the Sierpinski triangle?

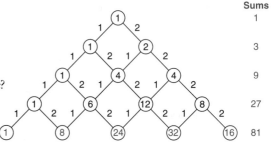

638 Chapter 12 *Looking Ahead: Iterations and Fractals*

SECTION 12.5

P **10** On graph paper begin at (0, 0) and make a path in the following manner: Move 16 units up, turn right, move 27 units right, turn right, move 8 units down, turn right, move 9 units left, turn right, move 4 units up, turn right, and move 3 units right. Continue in this pattern. What unique point does this pattern spiral toward? $\left(20\frac{1}{4}, 10\frac{2}{3}\right)$

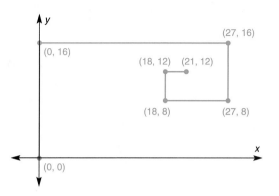

P, R **11** Start with an 8-inch-by-8-inch square. Divide it into four squares as shown, coloring the upper left square red and the lower right square blue. Iterate in the same manner on the remaining two uncolored squares.

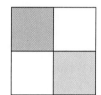

a Copy and complete the following table.

Iteration	0	1	2	3	4
Blue area shaded on this iteration	0 in.^{2}	16 in.^{2}	8 in.^{2}	4 in.^{2}	2 in.^{2}
Total blue area shaded so far	0 in.^{2}	16 in.^{2}	24 in.^{2}	28 in.^{2}	30 in.^{2}
Red area shaded on this iteration	0 in.^{2}	16 in.^{2}	8 in.^{2}	4 in.^{2}	2 in.^{2}
Total red area shaded so far	0 in.^{2}	16 in.^{2}	24 in.^{2}	28 in.^{2}	30 in.^{2}

b What does the total blue area converge to? The total red area? What happens to the picture as the blue and red areas converge?

P **12** If the largest equilateral triangle in the diagram is 1 inch on each side and if the "spiral" shown is continued indefinitely, what will be the total length of the spiral? 1 in.

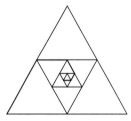

Investigation

Technology Talk to someone who knows how to program computers. See if you can come up with a program that can be used to generate the Sierpinski triangle, and demonstrate the program to the class.

Section 12.5 Introduction to Fractals 639

---

rows 1–4 is repeated eight times in rows 5–16. The figure made by rows 1–8 is repeated twice in rows 9–16.

■ Problem 11 is appropriate for cooperative learning.

Additional Answers
for pages 638 and 639

3 a–c

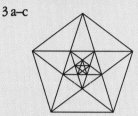

c No

5

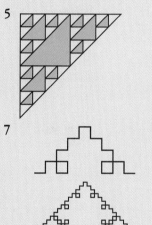

7

8 b The numbers in the left column are the numbers of shaded triangles after successive iterations of the Sierpinski triangle.

9 c The sums can represent either the numbers of unshaded triangles at successive iterations or the increases in the number of shaded triangles at successive iterations.

11 b 32 in.^{2}; 32 in.^{2}; it looks like a square with a red half and a blue half and no white space.

Answer for Investigation is on pages A10 and A11.

SUMMARY

CHAPTER 12

Summary

CONCEPTS AND PROCEDURES

After studying this chapter, you should be able to
- Recognize iterative processes (12.1)
- Interpret and construct iteration diagrams (12.1)
- Recognize situations in which an iterative calculation involves the number of preceding iterations (12.2)
- Express an iterative process in function notation (12.2)
- Perform iterative procedures on geometric figures (12.3)
- Recognize convergent iterations (12.4)
- Identify the limiting value of a convergent iteration (12.4)
- Identify the characteristics of a fractal (12.5)

VOCABULARY

convergent (12.4)
fractal (12.5)
iteration (12.1)
iteration diagram (12.1)
limiting value (12.4)
self-similarity (12.5)

Review

CHAPTER 12

1 a Use the iteration diagram to make a table of values for the indicated number of iterations.
  b Change the initial input to +25 and repeat part **a**.
  c What changes when you switch from a negative initial input to a positive initial input? Why?

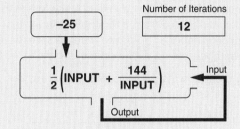

2 At the beginning of the quarter, Mr. Tuff Teacher lets his students choose how much homework they will be assigned. "You have two options," he says. "I will give you 45 minutes of homework daily, or I will give you 1 second of homework tonight, 2 seconds of homework tomorrow, 4 seconds the next day, and so forth." If Mr. Teacher's class meets 35 times per quarter, how many hours of homework would his students have under each option? Option 1: $25\frac{1}{2}$ hr; option 2: 4,772,185 hr

3 A new automobile decreases in value each year by about 13% of its value the previous year. Assume that a new Mitsundai car costs $17,500.
  a Make an iteration diagram to describe this situation.
  b Make a table to show the value of a Mitsundai each year for ten years.
  c When will a Mitsundai be worth only 50% of its original value?   In about 5 years

4 The following function can be used to calculate the amount of a loan that remains to be repaid.

$$f(0) = 5000$$
$$f(n) = f(n-1) + \frac{0.0875}{12} \cdot f(n-1) - 257$$

  a What is the amount of the loan? $5000
  b What is the yearly interest rate?   8.75%
  c What is the monthly payment? $257
  d How many payments are required to entirely pay off the loan?   21, plus $14.80

5 For this iteration diagram, make a table showing the relationship between the counter number and the output.

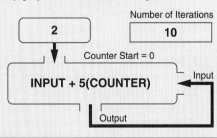

REVIEW

6 Each time a ball bounces, it rebounds to 75% of its height on the previous rebound. The ball was initially dropped from the top of a building 35 feet high.
  a Make a table of bounce number versus rebound height for this problem. When does the ball bounce less than 1 foot high?
  b Make a table to describe the total distance that the ball has traveled for a given number of bounces. (Hint: A spreadsheet might be useful here. Remember that on each bounce, the ball goes both up and down.)

7 Study the following pattern of toothpick figures.

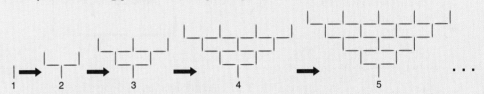

  a Make a table showing the relationship between the number of the figure and the number of toothpicks required to make the figure.
  b How many toothpicks would be required for the fifteenth figure? 330 toothpicks
  c Write an iterative function to describe the number of toothpicks necessary to make each figure. $f(1) = 1, f(n) = f(n-1) + (3n-2)$

8

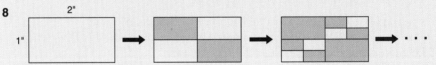

  a Iterate the figure for two more iterations.
  b Does this process satisfy the three conditions necessary for a fractal iteration? Explain your answer.
  c If the iteration is continued indefinitely, what will be the sum of the areas of the shaded sections? 2 in.^{2}

9 Consider the iterations shown. Each horizontal segment is exactly one half as long as the preceding horizontal segment. Each vertical segment is exactly one third as long as the preceding vertical segment.
  a List the coordinates of the next three vertices of the figure. $(20, 14); (20, 13\frac{1}{3}); (22, 13\frac{1}{3})$
  b How long will the tenth horizontal segment be? The tenth vertical segment? $\frac{1}{16}; \frac{2}{2187}$
  c What does the sum of the lengths of the horizontal segments converge to? What does the sum of the lengths of the vertical segments converge to? If the iterations continued forever, what would be the total length of all the segments in the diagram? 64; 27; 91
  d To what point does the diagram converge? $(21\frac{1}{3}, 13\frac{1}{2})$

REVIEW

7 a

Figure	1	2	3	4	5	6	7	8
Toothpicks	1	5	12	22	35	51	70	92

8 a

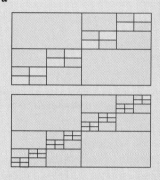

10 a Copy the following table, using the iteration diagram to complete it.

Iteration	Output
0	25
1	30
2	32
3	32.8
4	33.12
5	≈33.25
6	≈33.30
7	≈33.32

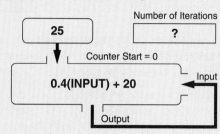

b To what number do the iterations converge if continued forever? $33.\overline{3}$

b Yes, it satisfies the conditions for a fractal iteration—it is an iteration process that could be continued an infinite number of times, and it has the property of self-similarity.

11 In the game Tower of Hanoi, the disks on peg A must be moved from that peg to peg C, using peg B to help. The only rules are (1) you may move only one disk at a time and (2) no disk may be placed on top of a smaller disk.

The number of moves needed to complete a game of Tower of Hanoi depends on the number of disks initially placed on peg A. For one disk, it takes only one move, so $f(1) = 1$. For n disks, we must move $n - 1$ disks from peg A to peg B, then the lowest disk from peg A to peg C, and then the $n - 1$ disks from peg B to peg C—that is, $f(n) = 2 \cdot f(n-1) + 1$.

a Copy the following table, using the iterative function described above to complete it.

Number of disks	1	2	3	4	5	6	7
Total number of moves	1	3	7	15	31	63	127

b Write the iterative function as an iteration diagram.

11 b Initial input: 1; iteration rule: 2(INPUT) + 1

12 b $f(0) = 60{,}000$; $g(0) = 42{,}000$; $f(n) = 0.8[f(n-1)] + 0.16[g(n-1)]$; $g(n) = 0.84[g(n-1)] + 0.2[f(n-1)]$

12 There are 60,000 people in Algebraville and 42,000 people in Geometry Town. Each year, 20% of the population of Algebraville moves to Geometry Town, and 16% of the population of Geometry Town moves to Algebraville. Assume that the rest of the people do not move at all.
a Copy and complete the table.
b Write iterative functions for both Algebraville and Geometry Town.
c At what population does each town converge? Algebraville: 45,333; Geometry Town: 56,667

Year	Population of Algebraville	Population of Geometry Town
0	60,000	42,000
1	54,720	47,280
2	51,341	50,659
3	49,178	52,822
4	47,794	54,206
5	46,908	55,092

Chapter Review 643

CHAPTER 12

Test

1. Paul puts $1000 into a bank account that pays 5.5% interest per year.
   a. Make an iteration diagram for this situation.
   b. How much money will be in the account after eight years? $1534.69

2. A ball is dropped from a height of 16 feet. Each time it bounces, it rebounds to a height that is 75% of its previous height.
   a. After what bounce will the ball first fail to reach a height of 2 feet? The eighth
   b. What is the total distance the ball will travel before coming to rest? 112 ft

3. Peggy was offered a job for which she would be paid $0.01 the first day, $0.03 the second day, $0.09 the third day, and so forth. On which day would she first be paid more than $20.00? The eighth

4.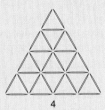

   1   2   3   4

   It takes 3 toothpicks to make the first figure, 9 to make the second, 18 to make the third, and 30 to make the fourth. How many toothpicks are needed to make the eighth figure? 108

5. Stephen puts $700 in a bank account that pays 5.5% interest per year. At the start of each year thereafter, he adds $500 to the account.
   a. Write an iterative function to describe this situation.
   b. How much money will be in the account at the start of the seventh year? a $f(0) = \$700, f(n) = f(n-1) + f(n-1)(0.055) + \$500$ b $4409.22

6. If $f(0) = 3$ and $f(n) = f(n-1) + n^2 + 2$, what is $f(8)$? 223

7. Draw a square, ABCD. Then connect midpoints of adjacent sides to form another square. Iterate on the innermost square two more times. If ABCD has a side length of 20 cm, what is the area of the innermost square after the three iterations? 50 cm^2

8. If $f(1) = 5$ and $f(n) = \frac{f(n-1)}{n}$, do the outputs of the function converge? If they do, what is the limiting value? Yes; 0

9. Start at the origin of a coordinate system. For the first iteration, draw a segment 1 unit long along the negative y-axis, then turn right. For each subsequent iteration, draw a segment 2 units longer than the preceding iteration's, then turn right. Continue iterating for a while. How long will the segment drawn on the fifteenth iteration be? At what point will that segment end? 29 units; (−15, 16)

PUZZLES AND CHALLENGES

A kind of word puzzle you may not be familiar with is called the **cryptic crossword clue.** In this kind of clue, a word is both defined and given a wordplay clue. The number in parentheses is the number of letters in the word.

Examples

Clues:
- **a** To desire half a loaf is no good (4)
- **b** Weird tone is live on video receiver (10)
- **c** Sounds like buckets turn ashen (5)

Answers:
- **a** LONG ("To desire" is a definition of "long." The wordplay part is "half a loaf" or *lo* followed by *ng* for "no good.")
- **b** TELEVISION ("Weird" implies an anagram of "tone is live" and a video receiver is a definition of television.)
- **c** PALES (Buckets are pails. "Sounds like" implies a homonym. "Turn ashen" means pales.)

See if you can figure out these cryptic clues:

1. Mixed-up life wears down fingernails (4)
2. Auto weighing 2000 pounds in a package (6)
3. Silly lab gear in a mathematical discipline (7)
4. Chief school person is basic, I hear (9)
5. Live outdoors first half of run for office (4)
6. Proper speech and weird beam of light give meaning (10)
7. Sounds like totals includes minute measures occasionally (9)
8. Book of maps holds up the world (5)

Try to make up some cryptic clues of your own.

---

payment come out the same as in the ad? Explore how the interest rate and term of the loan (the number of months you pay) affect the total amount paid for the car. Write a brief report explaining your findings.

8. Find out more about fractals. Research the following questions: Who made up the term *fractal*? Who are the pioneers in the field of fractal study? What applications do fractals have? What interesting properties do fractals have? Find pictures of fractals that you think are particularly interesting. If possible, find a computer program that generates fractals. Demonstrate the program for your class.

9. As suggested in Section 12.5, the best way to learn about fractal geometry is to create your own fractal curve. Design several fractal curves. Continue the iterative process as long as you can. Display your fractals for your classmates to see.

Additional Answers

for Chapter 12 Test

1a Initial input: $1000; iteration rule: INPUT + 0.055(INPUT)

for Puzzles and Challenges

1. File
2. Carton
3. Algebra
4. Principal
5. Camp
6. Dictionary
7. Sometimes
8. Atlas

HANDBOOK OF ARITHMETIC SKILLS

1: PLACE VALUE

The value that a digit represents depends on where the digit appears. For example, the 7 in 375 represents 70, and the 3 represents 300. In expanded notation, we can write

$$375 = 3(100) + 7(10) + 5.$$

We can also write decimals in expanded notation.

$$38.64 = 3(10) + 8 + 6(0.1) + 4(0.01)$$

Refer to the following place-value chart as you study examples 1–3. The chart shows the place value for each digit in the number 1,369,028.457.

Place-Value Chart

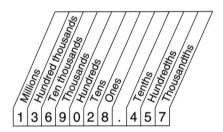

Example 1   Find the place value of the 4 in the number 14,982.

Solution   The 4 is in the thousands place. Its value is 4,000.

Example 2   Find the place value of the 3 in the number 25.73.

Solution   The 3 is in the hundredths place. Its value is 0.03.

Example 3   Use expanded notation to write 12,674.8.

Solution $12{,}674.8 = 1(10{,}000) + 2(1{,}000) + 6(100) + 7(10) + 4 + 8(0.1)$

2: ROUNDING

The *Morning Herald* newspaper said that the city's education budget was $76 million. The exact amount of the school budget was $76,329,104. The paper had rounded the figure to the nearest million.

• HANDBOOK OF ARITHMETIC SKILLS •

To round to a certain place value, look at the digit to the right of that place. If the digit is 5 or greater, round the number up by increasing the digit in the rounding place by 1 and changing all numbers to the right of it to 0. If the digit is less than 5, round down; that is, leave the digit in the rounding place unchanged and include zeros to the right of the rounding place as needed.

Example 1 Round 47,508 to the nearest thousand.

Solution The digit in the thousands place is 7.
The digit to the right of the thousands place is 5.
5 fits the rule for rounding up. It is 5 or greater.
Answer: 47,508 rounded to the nearest thousand is 48,000.

Example 2 Round 8.123 to the nearest tenth.

Solution The digit in the tenths place is 1.
The digit to the right of the tenths place is 2.
Since 2 is less than 5, round down.
Answer: 8.123 rounded to the nearest tenth is 8.1.

3: SIMPLIFYING FRACTIONS AND MIXED NUMBERS

Equivalent Fractions

A fraction shows a part of something. If the numerator and denominator of a fraction have a common factor, then the fraction can be reduced to an equivalent fraction that has a smaller denominator. When the numerator and denominator have no common factors other than 1, the fraction is in lowest terms, or simplest form.

Example 1 The total floor area of a school is 16,000 square feet. The library has a floor area of 2,000 square feet. In lowest terms, what fractional part of the school's floor space does the library take up?

Solution $\dfrac{\text{Library area}}{\text{Total area}} = \dfrac{2{,}000}{16{,}000} = \dfrac{2000(1)}{2000(8)} = \dfrac{1}{8}$

Handbook of Arithmetic Skills

HANDBOOK OF ARITHMETIC SKILLS

Example 2   Reduce $\frac{75}{45}$ to lowest terms.

Solution   15 is a common factor of 75 and 45.

Dividing by the common factor: $\frac{75}{45} = \frac{75 \div 15}{45 \div 15} = \frac{5}{3}$

The fraction $\frac{5}{3}$ is in lowest terms.

Converting mixed numbers and improper fractions

A mixed number is a number that includes both a whole number and a fraction, such as $2\frac{3}{5}$. An improper fraction is a fraction, such as $\frac{13}{5}$, in which the numerator is larger than the denominator. Mixed numbers can be converted to improper fractions, and improper fractions can be converted to mixed numbers.

Example 3   Alicia is preparing quarter-pound packages of cheese in the food market. She uses $3\frac{1}{4}$ pounds of cheese to prepare 13 quarter-pound packages. Show that these two amounts are equal.

Solution   Converting from improper fraction to mixed number:

13 quarters is $\frac{13}{4}$.

Convert by dividing. $\frac{13}{4} = 3\frac{1}{4}$

Converting from mixed number to improper fraction:

There are 4 quarters in each pound.

So $3\frac{1}{4} = \frac{(3 \times 4 + 1)}{4} = \frac{13}{4}$.

4: ADDING AND SUBTRACTING FRACTIONS AND MIXED NUMBERS WITH COMMON DENOMINATORS

When you add fractions with common denominators, the denominator remains the same for the sum. Add the numerators, then simplify the resulting fraction if possible.

To add mixed numbers, add the whole number parts and the fraction parts separately and then simplify the resulting mixed number.

• HANDBOOK OF ARITHMETIC SKILLS •

Example 1 The first part of a triathlon is a swim of $1\frac{3}{10}$ miles. The second part is a run of $3\frac{5}{10}$ miles. The third part is a bike ride of $20\frac{7}{10}$ miles. What is the total distance of the race?

Solution We add to find the answer.

$$\begin{array}{r} 1\frac{3}{10} \\ 3\frac{5}{10} \\ + 20\frac{7}{10} \\ \hline 24\frac{15}{10} \end{array} = 24 + 1\frac{5}{10} = 25\frac{5}{10} = 25\frac{1}{2} \text{ miles}$$

To subtract fractions with common denominators, you keep the same denominator and subtract the numerators. Then, if possible, simplify the resulting fraction. With a mixed number, if you cannot subtract the fraction part, then you must rename the number. To rename, you take 1 from the whole number, convert 1 to a fraction and add it to the fraction part of the mixed number. In the following example, $\frac{5}{8}$ is less than $\frac{7}{8}$, so we rename $13\frac{5}{8}$ as $12\frac{13}{8}$ so we can subtract.

Example 2 Beth was framing a picture. She had a piece of glass $13\frac{5}{8}$ inches wide. She cut off a piece that was $2\frac{7}{8}$ inches wide. How wide was the remaining piece?

Solution You must subtract fraction from fraction and whole number from whole number:

$$\begin{array}{rl} 13\frac{5}{8} \rightarrow 12\frac{13}{8} & \text{Rename. } 1 = \frac{8}{8} \text{ and } \frac{8}{8} + \frac{5}{8} = \frac{13}{8} \\ -2\frac{7}{8} \rightarrow -2\frac{7}{8} & \text{Subtract whole numbers and fractions.} \\ \hline 10\frac{6}{8} = 10\frac{3}{4} & \text{inches, the width of the remaining piece.} \end{array}$$

5: ADDING AND SUBTRACTING FRACTIONS AND MIXED NUMBERS WITH DIFFERENT DENOMINATORS

To add or subtract fractions with different denominators, one or more of the fractions must be converted so that the fractions have a common denominator.

The least common denominator (LCD) of a set of fractions is the smallest number that is a multiple of all of the denominators. The LCD of two fractions can be found by writing the multiples of the larger denominator and finding the smallest multiple that is divisible by the smaller denominator.

HANDBOOK OF ARITHMETIC SKILLS

Example 1 Find the least common denominator and add the two fractions. Simplify the answer.

$\frac{1}{2} + \frac{3}{5} = ?$ The LCD for 2 and 5 is 10.

Solution
$\begin{aligned} \frac{1}{2} &\to \frac{5}{10} \\ +\frac{3}{5} &\to +\frac{6}{10} \\ & \frac{11}{10} = 1\frac{1}{10} \end{aligned}$

$\frac{1}{2} \cdot \frac{5}{5} = \frac{5}{10}$
$\frac{3}{5} \cdot \frac{2}{2} = \frac{6}{10}$

Example 2 On Monday Carlos swims $\frac{5}{8}$ of a mile and on Tuesday he swims $\frac{3}{4}$ of a mile. What is the total distance he swims during the two days?

Solution

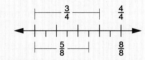

The LCD is 8. So convert $\frac{3}{4}$ to eighths: $\frac{3}{4} = \frac{3 \cdot 2}{4 \cdot 2} = \frac{6}{8}$.

$\frac{3}{4} + \frac{5}{8} = \frac{6}{8} + \frac{5}{8} = \frac{11}{8}$ or $1\frac{3}{8}$ miles

Example 3 Find the least common denominator and add the two fractions. Simplify the answer.

$2\frac{5}{12} + 5\frac{7}{8} = ?$ The LCD for 12 and 8 is 24.

Solution
$\begin{aligned} 2\frac{5}{12} &\to 2\frac{10}{24} \\ +5\frac{7}{8} &\to +5\frac{21}{24} \\ & 7\frac{31}{24} = 8\frac{7}{24} \end{aligned}$

$\frac{5}{12} \cdot \frac{2}{2} = \frac{10}{24}$
$\frac{7}{8} \cdot \frac{3}{3} = \frac{21}{24}$

You can check addition of fractions by using a calculator to evaluate both sides of the equation. For Example 3, $(2 + \frac{5}{12}) + (5 + \frac{7}{8}) = 8.2916\overline{6}$, and $8 + \frac{7}{24} = 8.291\overline{6}$.

When subtracting either fractions or mixed numbers with different denominators, you should find the least common denominator, convert the fractions to equivalent fractions with the LCD, then do the subtraction.

Example 4 Subtract $2\frac{5}{6}$ from $7\frac{2}{5}$.

Solution The least common denominator of 6 and 5 is 30. Each fraction must be converted to an equivalent fraction with denominator 30.

$$7\frac{2}{5} \rightarrow 7\frac{12}{30} \rightarrow 7\frac{12}{30} = 6 + \frac{30}{30} + \frac{12}{30} = 6\frac{42}{30}$$
$$-2\frac{5}{6} \rightarrow -2\frac{25}{30} \rightarrow \qquad\qquad\qquad -2\frac{25}{30}$$
$$\qquad\qquad\qquad\qquad\qquad\qquad\qquad\qquad 4\frac{17}{30}$$

You can check this answer by doing the subtraction on a scientific calculator, then evaluating the answer as a decimal to see that the result is the same.

$$\left(7 + \frac{2}{5}\right) - \left(2 + \frac{5}{6}\right) = 4.5\overline{6}$$
$$4 + \frac{17}{30} = 4.5\overline{6}$$

6: MULTIPLYING FRACTIONS AND MIXED NUMBERS

To multiply fractions, multiply numerator by numerator and denominator by denominator.

Example 1 Of the land in Fairfax County, $\frac{5}{8}$ is used for farming. Of the farmland, $\frac{3}{4}$ is used for dairy farms. What fractional part of the county is used for dairy farming?

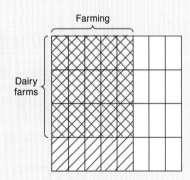

Solution $\frac{5}{8} \times \frac{3}{4} = \frac{15}{32}$

If there are common factors in the numerators and denominators, you can either multiply the fractions, then simplify, or divide out common factors, then multiply.

• HANDBOOK OF ARITHMETIC SKILLS •

Example 2   Multiply $\frac{9}{2} \times \frac{4}{15}$.

Solution   **Method 1** $\frac{\overset{3}{\cancel{9}}}{\underset{1}{\cancel{2}}} \times \frac{\overset{2}{\cancel{4}}}{\underset{5}{\cancel{15}}} = \frac{6}{5} = 1\frac{1}{5}$

            Method 2 $\frac{9}{2} \times \frac{4}{15} = \frac{36}{30} = \frac{6}{5} = 1\frac{1}{5}$

To multiply mixed numbers, convert to improper fractions, then multiply.

Example 3   Multiply $5\frac{3}{4} \times 2\frac{1}{3}$.

Solution $5\frac{3}{4} \times 2\frac{1}{3} = \frac{23}{4} \times \frac{7}{3} = \frac{161}{12} = 13\frac{5}{12}$

The results of multiplication can be checked using a calculator.

$$\left(5 + \frac{3}{4}\right) \times \left(2 + \frac{1}{3}\right) = 13.41\overline{6}$$

$$13\frac{5}{12} = 13.41\overline{6}$$

7: DIVIDING FRACTIONS AND MIXED NUMBERS

Dividing by a number is the same as multiplying by the reciprocal of the number. The reciprocal of a number is the number that, when multiplied by the original number, will equal 1. The reciprocal of $\frac{4}{5}$ is $\frac{5}{4}$, so dividing by $\frac{4}{5}$ is the same as multiplying by $\frac{5}{4}$.

You can see this if you consider the following equations and expressions.

$$80 \div 2 = 40 \qquad 80 \times \frac{1}{2} = 40$$

$$50 \div \frac{1}{3} \text{ is equivalent to } 50 \times 3.$$

To divide one fraction by another, multiply by the reciprocal of the divisor.

• HANDBOOK OF ARITHMETIC SKILLS •

Example 1 The eighth-grade class is dividing 15 pounds of candy into $\frac{3}{4}$-pound packages. How many packages will the class have?

Solution Divide 15 by $\frac{3}{4}$.

$15 \div \frac{3}{4} = 15 \times \frac{4}{3}$ Multiply by the reciprocal.

$\phantom{15 \div \frac{3}{4}} = 20$

The class will have 20 packages.

8: RATIO AND PROPORTION

A ratio is a comparison of two numbers that uses division. A ratio may be written in different ways. The ratio of 2 to 3 may be written 2:3 or $\frac{2}{3}$.

Example 1 Write the ratio 42 to 56 as a fraction in lowest terms.

Solution $\frac{42}{56} = \frac{3}{4}$

A proportion shows the equality of two ratios. For example, $\frac{3}{4} = \frac{6}{8}$. Notice that if $\frac{a}{b} = \frac{c}{d}$, then $ad = bc$. This is called cross-multiplying. When you cross-multiply with the proportion $\frac{3}{4} = \frac{6}{8}$, you get $6 \times 4 = 8 \times 3 = 24$.

When three out of four terms in a proportion are known, you can find the fourth term by solving the equation.

Example 2 Solve the proportion $\frac{5}{12} = \frac{x}{18}$.

Solution

$5 \cdot 18 = 12x$ Cross multiply.
$90 = 12x$ Solve.
$x = 7.5$

Handbook of Arithmetic Skills

HANDBOOK OF ARITHMETIC SKILLS

Example 3   A recipe calls for $3\frac{1}{2}$ cups of flour to make 20 muffins. How much flour is needed to make 50 muffins?

Solution   We use a proportion that relates cups of flour to muffins.

$$\frac{3.5}{20} = \frac{x}{50}$$

$$3.5 \cdot 50 = 20x$$

$$20x = 175$$

$$x = 8.75 \quad \text{To make 50 muffins you will need } 8\frac{3}{4} \text{ cups of flour.}$$

9: CONVERTING FRACTIONS, DECIMALS, AND PERCENTS

Any rational number can be written as a fraction, a decimal, or a percent.

For example:
$$\frac{1}{2} = 0.5 = 50\%$$

$$\frac{5}{8} = 0.625 = 62.5\%$$

$$2 = \frac{2}{1} = 2.0 = 200\%$$

Converting a fraction to a decimal and to a percent

To convert a fraction to a decimal, divide the numerator by the denominator and round to the desired place value.

Example 1   Convert to decimals, then to percents.
(a) $\frac{3}{4}$ (b) $\frac{2}{3}$

Solution   (a) $\frac{3}{4}$ means $3 \div 4$.

Thus, $\frac{3}{4} = 0.75 = 75\%$.

(b) $\frac{2}{3} = 0.66\ldots = 66.66\ldots\%$

The answer can be written as $66.\overline{6}\%$, where the bar means that the 6 repeats. Or you can round the answer to 66.67%, 66.7%, or 67% depending on the amount of precision you want.

• HANDBOOK OF ARITHMETIC SKILLS •

A mixed number can be converted to a decimal by separating the whole number from the fraction, converting the fraction to a decimal, then combining the whole number and the decimal. Since *percent* means "parts out of 100," a decimal is converted to a percent by multiplying by 100. An easy way to do this is to move the decimal point two places to the right. This procedure creates a number that is 100 times larger than the decimal number.

Example 2 Convert $1\frac{3}{7}$ to a decimal rounded to the nearest thousandth, and convert the decimal to a percent.

Solution $1\frac{3}{7} = 1 + \frac{3}{7} = 1 + 0.4286 = 1.429$

$1.429 = 142.9\%$ Multiply the decimal by 100.

Converting a percent to a decimal and a fraction

Percent means "parts out of 100," so 8% means 8 out of 100. This percent can be written as a decimal, 0.08. You can convert any percent to a decimal by removing the percent sign and dividing by 100.

Example 3 Convert to decimals, then to fractions.
(a) 25%   (b) 7.5%   (c) 300%

Solution (a) 25% → $0.25 = \frac{25}{100} = \frac{1}{4}$

(b) 7.5% → $0.075 = \frac{75}{1000} = \frac{3}{40}$

(c) 300% → $\frac{300}{100} = 3$

Equivalent fractions, decimals, and percents

The table shows some equivalent fractions, decimals, and percents. Working with percents will be much easier for you if you are familiar with most of these basic conversions.

Fraction	Decimal	Percent
$\frac{1}{100}$	0.01	1%
$\frac{1}{10}$	0.1	10%
$\frac{1}{8}$	0.125	12.5%
$\frac{1}{5}$	0.2	20%
$\frac{1}{4}$	0.25	25%
$\frac{1}{3}$	$0.\overline{3}$	$33\frac{1}{3}\%$
$\frac{1}{2}$	0.5	50%

Handbook of Arithmetic Skills

10: USING PERCENT

Finding a percent of a number

To find a percent of a number, change the percent to a decimal and multiply the decimal by the number. A method that works for all percent problems is to use the linear equation $pN = A$. In this equation, p is the percent rate, N is the base number, and A is the percent amount. If p is a percent rate less than 100%, then A will be less than N.

Example 1 Find the amount of sales tax on an item selling for $30 if the sales tax rate is 7.5%.

Solution The percent rate, p, is 7.5%. The base number, N, is 30. Find A.
$$p \times N = A$$
$$0.075 \times 30 = 2.25 \quad \text{The sales tax is \$2.25.}$$

Example 2 Find 110% of 400.

Solution The percent rate, p, is 110%. The base number, N, is 400. Find A.
$$p \times N = A$$
$$1.1 \times 400 = 440$$

Finding what percent one number is of another

You have seen how the percent equation can be used to find the percent amount. The same equation can be used to find the percent rate when the base number and the percent amount are known. The solution will be in decimal form. You then convert this decimal to a percent.

Example 3 Ian had 56 items correct on a test with 80 items. What percent of the items did he have correct?

Solution The percent amount is 56. The base number is 80. Find p.
$$p \times N = A$$
$$p \times 80 = 56$$
$$p = \frac{56}{80} = 0.7 \to 70\%$$
56 is 70% of 80. Ian had 70% of the items correct.

Example 4 What percent of 78 is 120?

Solution
$$p \times N = A$$
$$p \times 78 = 120$$
$$p = \frac{120}{78} = 1.54 \rightarrow 154\%$$

Finding the number when a percent amount is known

The same equation used to find the percent and percent amount can be used to find the base number. You substitute values for p and A, and then you solve the equation.

Example 5 94.5 is 125% of what number?

Solution The percent rate is 125%. The percent amount is 94.5. Find N.
$$p \times N = A$$
$$1.25 \times N = 94.5$$
$$N = 94.5 \div 1.25 = 75.6$$

11: SQUARE ROOTS

Because 6 × 6 = 36, we say that 6 is the *square root* of 36. Square roots are always positive numbers.

To find a square root on some scientific calculators, you press the square-root key after pressing the number. On other scientific calculators, you press the square-root key and then the number. When you are taking the square root of a sum or difference, it is important to use calculator grouping symbols so that you will find the correct square root.

Example 1 Find $\sqrt{78}$.

Solution Using a calculator, you will find that $\sqrt{78} \approx 8.83$ or 8.8.

Example 2 Find the value of $\sqrt{26^2 + 30^2}$.

Solution Use the appropriate grouping symbols.
$$\sqrt{26^2 + 30^2} = \sqrt{(26^2 + 30^2)} \approx 39.7$$

• HANDBOOK OF ARITHMETIC SKILLS •

12: EXPONENTS

In the expression 3^4, we call 3 the base and 4 the exponent.

Example 1 Write each expression using exponents.
(a) $2 \cdot 2 \cdot 2 \cdot 2 \cdot 2$ (b) $a \cdot a \cdot a$

Solution (a) $2 \cdot 2 \cdot 2 \cdot 2 \cdot 2 = 2^5$ (b) $a \cdot a \cdot a = a^3$

Example 2 Write each expression without using exponents.
(a) t^5 (b) $3s^4$ (c) $(3s)^4$

Solution (a) $t^5 = t \cdot t \cdot t \cdot t \cdot t$ (b) $3s^4 = 3 \cdot s \cdot s \cdot s \cdot s$
(c) $(3s)^4 = (3s)(3s)(3s)(3s) = 81 \cdot s \cdot s \cdot s \cdot s$

Notice that the exponent in Example 2(b) applies only to the variable, s.

A quick way to remember powers of 10 is that the exponent shows the number of zeros that will follow the 1. Therefore, $10^2 = 100$, $10^4 = 10,000$, and so on.

Scientific notation is a shorthand method of using exponents to write large numbers. In scientific notation a decimal number between 1 and 10 is multiplied by a power of 10. For example, $6.75 \times 10^{11} = 6.75 \times 100,000,000,000 = 675,000,000,000$.

Example 3 Write the number 9.125×10^8 in standard notation.

Solution 9.125×10^8 means $9.125 \times 100,000,000$.
Moving the decimal point eight places gives 912,500,000.

Example 4 Write 49,000,000,000 in scientific notation.

Solution There are 10 places after the 4, therefore
$49,000,000,000 = 4.9 \times 10^{10}$

Note that 4.9 is a number between 1 and 10.

HANDBOOK OF ARITHMETIC SKILLS

Your scientific calculator uses scientific notation when a number is very large. Usually the calculator shows a number between 1 and 10 with a decimal part followed by E and an integer. The E stands for *exponent* and the integer represents a power of 10. For example, $9.3 \times 10^7 = 9.3 \text{ E}7 = 93,000,000$.

The calculator shows a negative exponent for very small numbers. For example, 3.24 E–4 means 3.24×10^{-4}.

Example 5
a Write 3.24E–4 in standard form.
b Write 0.000082 in scientific notation.

Solution
a $3.24 \text{ E}{-4} = 3.24 \times 10^{-4} = 0.000324$
b $0.000082 = 8.2 \times 10^{-5}$

13: ORDER OF OPERATIONS

The rules for evaluating arithmetic expressions include performing mathematical operations in the following order:

1. Work within grouping symbols, starting from the innermost symbols.
2. Evaluate powers and roots.
3. Do multiplication and division from left to right.
4. Do addition and subtraction from left to right.

Example 1 Evaluate.
(a) $2(3 + 4)$
(b) $2(3 + 5) + 3(5 - 1)$

Solution
(a) $2(3 + 4)$
$= 2 \times 7 = 14$

(b) $2(3 + 5) + 3(5 - 1)$
$= 2 \times 8 + 3 \times 4$
$= 16 + 12 = 28$

Example 2 Evaluate $5[6 + 4(4 - 7) - 8 \div (9 - 5)]$.

Solution
$5[6 + 4(4 - 7) - 8 \div (9 - 5)]$
$= 5[6 + 4(-3) - 8 \div 4]$
$= 5[6 - 12 - 2]$
$= 5(-8) = -40$

Example 3 Evaluate $1 + n^2$ if $n = 7$.

Solution $1 + n^2 = 1 + 7^2 = 1 + 49 = 50$

TABLES AND CHARTS

Table of Measurements

English Units

Length
1 mile = 5280 feet
1 yard = 3 feet
1 foot = 12 inches

Mass and Weight
1 ton = 2000 pounds
1 pound = 16 ounces

Capacity (Liquid Measures)
1 gallon = 231 cubic inches
1 gallon = 4 quarts
1 quart = 2 pints
1 pint = 16 fluidounces

Area
1 square mile = 640 acres
1 acre = 4840 square yards
1 square yard = 9 square feet
1 square foot = 144 square inches

Volume
1 cubic yard = 27 cubic feet
1 cubic foot = 1728 cubic inches

Metric Units

Length
1 kilometer = 1000 meters
1 hectometer = 100 meters
1 dekameter = 10 meters
1 decimeter = 0.1 meter
1 centimeter = 0.01 meter
1 millimeter = 0.001 meter

Mass and Weight
1 kilogram = 1000 grams
1 hectogram = 100 grams
1 dekagram = 10 grams
1 decigram = 0.1 gram
1 centigram = 0.01 gram
1 milligram = 0.001 gram

Capacity (Liquid Measures)
1 kiloliter = 1000 liters
1 hectoliter = 100 liters
1 dekaliter = 10 liters
1 deciliter = 0.1 liter
1 centiliter = 0.01 liter
1 milliliter = 0.001 liter

Area
1 square kilometer = 1,000,000 square meters
1 square meter = 10,000 square centimeters
1 square centimeter = 100 square millimeters

Volume
1 cubic meter = 1,000,000 cubic centimeters
1 cubic centimeter = 1,000 cubic millimeters

English-Metric Conversions

Length
1 mile ≈ 1.61 kilometers
1 yard ≈ 0.914 meter
1 foot = 30.48 centimeters
1 inch = 2.54 centimeters

Mass and Weight
1 pound ≈ 0.453 kilogram
1 ounce ≈ 28.35 grams

Capacity
1 gallon ≈ 3.785 liters
1 fluidounce ≈ 29.57 milliliters

Area
1 mi^2 ≈ 2.58 km^2
1 yd^2 ≈ 0.836 m^2
1 ft^2 ≈ 0.093 m^2
1 in.^{2} ≈ 6.45 cm^2

Volume
1 yd^3 ≈ 0.765 m^3
1 ft^3 ≈ 0.028 m^3
1 in.^{3} ≈ 16.39 cm^3

Densities of Selected Substances

Substance	Density g/cm³	lb/ft³
Air	0.0012	0.075
Aluminum	2.7	168.5
Copper	8.89	555
Gasoline	0.66	41.2
Glass	2.6	162
Gold	19.3	1204
Helium	0.00018	0.01123
Ice	0.922	57.5
Lead	11.3	705
Milk	1.03	64.27
Silver	10.5	655
Steel	7.8	486.7
Water	1.0	62.4
Wood (balsa)	0.13	8.11
Wood (oak)	0.72	44.9
Wood (pine)	0.56	34.9

The First 100 Prime Numbers

2	3	5	7	11	13	17	19	23	29
31	37	41	43	47	53	59	61	67	71
73	79	83	89	97	101	103	107	109	113
127	131	137	139	149	151	157	163	167	173
179	181	191	193	197	199	211	223	227	229
233	239	241	251	257	263	269	271	277	281
283	293	307	311	313	317	331	337	347	349
353	359	367	373	379	383	389	397	401	409
419	421	431	433	439	443	449	457	461	463
467	479	487	491	499	503	509	521	523	541

Table of Geometric Formulas

Perimeter Formulas

The perimeter of any polygon is the sum of the lengths of the sides of the polygon.

Rectangle

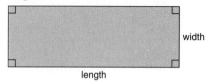

Perimeter = 2 · (length) + 2 · (width)

Circle

The perimeter of a circle is called the circle's **circumference**.

Circumference = π · (diameter)   Circumference = π · 2 · (radius)

Area Formulas

Rectangle   **Square**

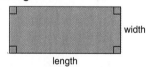

Area = (length) · (width)   Area = (side) · (side)

Triangle   **Parallelogram**

    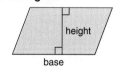

Area = $\frac{1}{2}$ (base) · (height)   Area = (base) · (height)

Trapezoid   **Circle**

    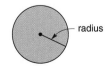

Area = $\frac{1}{2}$ · height · (base$_1$ + base$_2$)   Area = π · (radius)^2

Surface Area

Prism

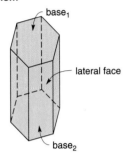

Total surface area =
(area of base$_1$) + (area of base$_2$) +
(sum of the areas of the lateral faces)

Pyramid

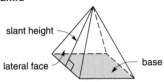

Total surface area = (area of base) + (sum of the areas of the lateral faces)

Cylinder

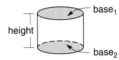

Total surface area =
(area of base$_1$) + (area of base$_2$) +
(circumference of either base · height)

Cone

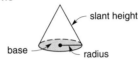

Total surface area =
(area of base) + (π · radius · slant height)

Sphere

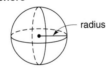

Total surface area = $4 \cdot \pi \cdot$ (radius^2)

Volume

Prism

Volume = (area of base) · (height)

Pyramid

Volume = $\frac{1}{3} \cdot$ (area of base) · (height)

Cylinder

Volume = (area of base) · (height)

Cone

Volume = $\frac{1}{3} \cdot$ (area of base) · (height)

Sphere

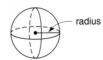

Volume = $\frac{4}{3} \cdot \pi \cdot$ (radius^3)

Symbols Used in Mathematics

Symbol	Meaning
Geometry	
$\overleftrightarrow{AB}$	line AB
$\overrightarrow{AB}$	ray AB
$\overline{AB}$	segment AB
AB	length of $\overline{AB}$
$\angle$	angle
$m\angle A$	measure of angle A
∟	right angle
△	triangle
⊙	circle
~	is similar to
≅	is congruent to
⊥	is perpendicular to
∥	is parallel to
⇉	parallel lines
Sets	
{ }	set
∩	intersection
∪	union
⊂	is a subset of
Computer Operations	
*	multiplication
/	division
^	power

Symbol	Meaning
Algebra	
+	addition
−	subtraction
×, ·	multiplication
÷, —	division
$\sqrt{}$	root
$|x|$	absolute value of x
$n!$	n factorial
=	is equal to
≈	is approximately equal to
≠	is not equal to
<	is less than
≤	is less than or equal to
>	is greater than
≥	is greater than or equal to
Other Symbols	
°	degrees
'	feet
"	inches
π	pi
%	percent

Tables and Charts

Table of Squares and Square Roots

n	n²	√n	n	n²	√n	n	n²	√n
1	1	1.000	51	2,601	7.141	101	10,201	10.050
2	4	1.414	52	2,704	7.211	102	10,404	10.100
3	9	1.732	53	2,809	7.280	103	10,609	10.149
4	16	2.000	54	2,916	7.348	104	10,816	10.198
5	25	2.236	55	3,025	7.416	105	11,025	10.247
6	36	2.449	56	3,136	7.483	106	11,236	10.296
7	49	2.646	57	3,249	7.550	107	11,449	10.344
8	64	2.828	58	3,364	7.616	108	11,664	10.392
9	81	3.000	59	3,481	7.681	109	11,881	10.440
10	100	3.162	60	3,600	7.746	110	12,100	10.488
11	121	3.317	61	3,721	7.810	111	12,321	10.536
12	144	3.464	62	3,844	7.874	112	12,544	10.583
13	169	3.606	63	3,969	7.937	113	12,769	10.630
14	196	3.742	64	4,096	8.000	114	12,996	10.677
15	225	3.873	65	4,225	8.062	115	13,225	10.724
16	256	4.000	66	4,356	8.124	116	13,456	10.770
17	289	4.123	67	4,489	8.185	117	13,689	10.817
18	324	4.243	68	4,624	8.246	118	13,924	10.863
19	361	4.359	69	4,761	8.307	119	14,161	10.909
20	400	4.472	70	4,900	8.367	120	14,400	10.954
21	441	4.583	71	5,041	8.426	121	14,641	11.000
22	484	4.690	72	5,184	8.485	122	14,884	11.045
23	529	4.796	73	5,329	8.544	123	15,129	11.091
24	576	4.899	74	5,476	8.602	124	15,376	11.136
25	625	5.000	75	5,625	8.660	125	15,625	11.180
26	676	5.099	76	5,776	8.718	126	15,876	11.225
27	729	5.196	77	5,929	8.775	127	16,129	11.269
28	784	5.292	78	6,084	8.832	128	16,384	11.314
29	841	5.385	79	6,241	8.888	129	16,641	11.358
30	900	5.477	80	6,400	8.944	130	16,900	11.402
31	961	5.568	81	6,561	9.000	131	17,161	11.446
32	1,024	5.657	82	6,724	9.055	132	17,424	11.489
33	1,089	5.745	83	6,889	9.110	133	17,689	11.533
34	1,156	5.831	84	7,056	9.165	134	17,956	11.576
35	1,225	5.916	85	7,225	9.220	135	18,225	11.619
36	1,296	6.000	86	7,396	9.274	136	18,496	11.662
37	1,369	6.083	87	7,569	9.327	137	18,769	11.705
38	1,444	6.164	88	7,744	9.381	138	19,044	11.747
39	1,521	6.245	89	7,921	9.434	139	19,321	11.790
40	1,600	6.325	90	8,100	9.487	140	19,600	11.832
41	1,681	6.403	91	8,281	9.539	141	19,881	11.874
42	1,764	6.481	92	8,464	9.592	142	20,164	11.916
43	1,849	6.557	93	8,649	9.644	143	20,449	11.958
44	1,936	6.633	94	8,836	9.695	144	20,736	12.000
45	2,025	6.708	95	9,025	9.747	145	21,025	12.042
46	2,116	6.782	96	9,216	9.798	146	21,316	12.083
47	2,209	6.856	97	9,409	9.849	147	21,609	12.124
48	2,304	6.928	98	9,604	9.899	148	21,904	12.166
49	2,401	7.000	99	9,801	9.950	149	22,201	12.207
50	2,500	7.071	100	10,000	10.000	150	22,500	12.247

GLOSSARY

A

absolute value (6.7) The distance of a number on the number line from zero. For example, |3| = 3, and |–3| = 3.

abundant number (10.5) A number whose r ratio is greater than two.

acute angle (3.3) An angle with a measure between 0 and 90.

adjacent angles (7.1) Angles that have the same vertex, have a side in common, and do not overlap.

adjacent sides (7.2) Sides of a polygon that have a common endpoint.

angle (3.5) A representation of a rotation using two rays meeting at a vertex.

area (2.3, 4.1) The size of a surface, usually expressed in square units.

artistic graph (5.3) A graph that displays information pictorially.

Associative Properties of Addition and Multiplication (9.5) Grouping does not affect the sum or product of three numbers. For numbers a, b, and c: $(a + b) + c = a + (b + c)$, and $(ab)c = a(bc)$.

attribute (4.1) A characteristic that distinguishes an object from other objects.

average (5.4) A value that tells something about the center of a set of data. See *mean*; *median*; *mode*.

axes (6.8) In the rectangular coordinate system, the two number lines that intersect to form right angles.

B

bar graph (5.2) A graph that displays data as vertical or horizontal parallel bars.

base (2.2) The number that is raised to a power. In the expression 5^2, 5 is the base.

base of a triangle (2.3) The side to which the line segment for the height of the triangle is drawn.

base unit (4.2) An expression of rate as a fraction with a denominator of 1.

bases of a trapezoid (2.3) The two parallel sides of a trapezoid.

bisect (7.5) To divide into two congruent parts.

box-and-whisker plot (5.5) A graph used to identify the middle 50% of a set of data.

C

capacity (4.1) The amount (volume) that containers can hold.

cell (1.5) A spreadsheet "box," identified by its column letter and row number.

center of rotation (7.4) A point about which a figure is rotated.

center of symmetry (7.3) A point about which a figure is rotated until it fits on its original position exactly.

centi- (4.1) A prefix that indicates 0.01 unit.

centimeter (4.1) A unit of measurement equal to 0.01 meter.

circle graph (5.1) A graph in which pie-shaped parts of a circle represent data. Also called *pie chart*.

circumference (2.3, 9.4) The distance around a circle, equal to π times the diameter.

coefficient (8.1) The number multiplying the variable. In the term $2x$, 2 is the coefficient.

common factor (10.5) A number that is a factor of two or more numbers. The number 5 is a common factor of 15 and 20.

Commutative Properties of Addition and Multiplication (9.5) The order in which numbers are added or multiplied does not change the sum or product. For any numbers a and b: $a + b = b + a$ and $ab = ba$.

complement (3.5, 9.2) (1) One of two angles that have measures with a sum of 90. Each angle is the complement of the other. (2) The complement of set A is the set of all elements that are not in A.

complementary angles (3.5) Two angles with measures that add up to 90.

composite number (10.2) A natural number greater than 1 that has more than two factors.

cone (11.6) A three-dimensional figure with a circular base and a vertex.

congruent (3.1) Having exactly the same size and shape.

convergent (12.4) Approaching a value.

conversion factors (4.4) Ratios used to change units of measure.

convex polygon (3.7) A polygon in which any two interior points can be connected by a segment that is entirely within the polygon.

coordinate (6.1) The number associated with a point on a number line. Two coordinates identify a point in the coordinate plane.

counting numbers (9.3, 10.1) The set of natural numbers, {1, 2, 3, ... }.

cross multiplying (4.5) A method used to solve proportions by multiplying the numerator of each ratio by the denominator of the other.

cylinder (11.6) A three-dimensional figure with two parallel congruent circular bases.

D

data analysis (5.1) A process for collecting, organizing, displaying, and interpreting data.

deficient number (10.5) A number with an r ratio less than 2.

degree (3.5) A unit of measure for an angle.

density (4.3) The ratio of the mass of a substance to its volume.

diagonal (3.1) A segment that joins two nonconsecutive vertices of a polygon.

diameter (2.3) A line segment that passes through the center of a circle and has endpoints on the circumference of the circle.

dimensional analysis (4.4) A technique that is used to convert a measurement from one type of unit to another.

dimensions of a matrix (5.6) The number of rows and columns; a 3 × 4 matrix has 3 rows and 4 columns.

disjoint (5.1) Able to be separated into categories that do not overlap.

disjoint sets (9.3) Two sets that have an intersection that is empty.

distance formula (9.6) The distance between $X(a,b)$ and $Y(c,d)$ is given by $XY = \sqrt{(c-a)^2 + (d-b)^2}$.

Distributive Property (8.1) For numbers a, b, and x, $ax + bx = (a + b)x$ and $ax - bx = (a - b)x$.

E

edge (11.5) In a polyhedron, the line of intersection of two faces.

encasement (7.6) A procedure used to find the area of a triangle.

equal matrices (5.6) Matrices with corresponding entries that are equal.

equation (1.6) A sentence stating that the values before and after an equal sign are equal.

equiangular (3.7) Figures in which all angles are congruent.

equilateral triangle (3.7) A triangle with all sides congruent.

equivalent expressions (8.1) Two expressions that produce the same value.

equivalent fractions (10.2) Two or more fractions that represent the same number.

estimate (3.3) To find the approximate value.

evaluate (2.1) To find the value of an expression.

exponent (2.2) A number used to indicate the repeated multiplication of a number. In the expression 10^4, 4 is the exponent.

extremes (5.5) The least and greatest value for a given set of data.

F

faces (11.5) The flat surfaces (polygons) of polyhedral solids.

factor (10.2) A number that divides into another number and leaves a remainder of zero.

Fibonacci sequence (8.1) A pattern in which the first and second numbers are both 1 and each following number is the sum of the two numbers that precede it (1, 1, 2, 3, 5, 8, 13, 21, ...).

figurate numbers (10.7) Special numbers associated with geometric figures.

final ray (3.5) A ray that represents the final position in a rotation.

fractal (12.5) A figure resulting from iterations, in which each new part is similar to the previous figure.

function (8.7) A relationship in which one value depends on the other values. (See *input*; *output*.)

G

gram (4.1) The basic metric unit of mass.

greatest common factor (GCF) (10.5) The greatest number that is a factor of two or more numbers.

grouping symbols (2.1) Symbols that show which operation is to be done first.

H

height of a triangle (2.3) A perpendicular segment from a triangle's vertex to the opposite side.
heptagon (3.1) A polygon with seven sides.
hexagon (3.1) A polygon with six sides.
histogram (5.3) A graph that shows the number of times a value or range of values occurs in a set of data.
hypotenuse (2.3) In a right triangle, the side opposite the right angle.

I

image (7.4) A figure obtained through the transformation of an original figure (preimage).
inequality (6.2) A statement that two quantities are not equal.
initial ray (3.5) A ray that represents the initial direction of a rotation.
input (8.7) For a function, the value substituted for the variable. For example, let $f(x) = x + 1$. If $x = 5$ is the input, then 6 is the output.
integers (6.1, 9.3) The set I = { ... , –3, –2, –1, 0, 1, 2, 3, ... }.
intersection (9.2) The set that contains all common elements of two given sets.
inverse operations (3.4) Operations that reverse the effect of each other. For example, addition and subtraction are inverse operations.
irrational numbers (9.3) Real numbers that cannot be written as the ratio of two integers and cannot be written as terminating or repeating decimals.
isosceles trapezoid (7.5) A trapezoid in which the nonparallel sides are congruent.
isosceles triangle (7.5) A triangle with two congruent sides.
iteration (12.1) A process used to do repetitive calculations. Each calculation uses the results of the previous calculation as its starting value.
iteration diagram (12.1) A diagram used to model iterations.

K

kilo- (4.1) A prefix that indicates 1000 units.
kilogram (4.1) A unit of mass equal to 1000 grams.
kilometer (4.1) A unit of distance equal to 1000 meters.
kite (7.5) A quadrilateral that has two pairs of congruent adjacent sides.

L

lateral area (11.5) The area of the surfaces, other than bases, of a prism, pyramid, cylinder, or cone.
lateral faces (11.5) The faces of a prism or pyramid that are not bases.
least common denominator (10.2) The least common multiple of the denominators of the fractions being compared.
least common multiple (LCM) (10.2) The smallest number other than zero that is a multiple of two or more numbers.
legs (2.3) In a right triangle, the sides that form the right angle.
like terms (8.1) Terms that have the same variable raised to the same power.
limiting value (12.4) A number to which an iteration converges.
line A straight path of points that extend infinitely in opposite directions.
linear equation (9.7) An equation for which the graph is a line.
linear measurement (3.1) A measure of length or distance.
line graph (5.2) A graph that displays data as points that are connected by segments.
line of symmetry (7.3) A line that divides a figure into two halves that are mirror images of each other.
line plot (5.3) A histogram that has data graphed on a number line.
line segment Two points on a line, plus all the points in between. The two points are the endpoints of the segment.
liter (4.1) The basic metric unit of capacity.
lower quartile (5.5) The middle term of the lower half of a set of data.

M

magic square (6.6) A matrix in which the numbers in each row, column, and diagonal have the same sum.
matrix (5.6) A two-dimensional table of rows and columns.
mean (5.4) A type of average. It is the sum of data items divided by the number of data items.

measures of central tendency (5.4) The term given to various averages, such as mean, median, and mode, because they each measure the center of a set of data.

median (5.4) The middle term for a set of data that is arranged in order from least to greatest.

meter (4.1) The basic unit of length in the metric system. A meter is equal to 100 centimeters.

midpoint (6.8) A point that is halfway between two given points.

midpoint formula (9.6) If (a, b) and (c, d) are any two points, then M, the point midway between these two points, has coordinates given by $M = \left(\frac{a+c}{2}, \frac{b+d}{2}\right)$.

midquartile range (5.5) The difference between the upper and lower quartile. The middle 50% of a set of data.

milli- (4.1) A prefix that indicates 0.001 unit.

milligram (4.1) A unit of measurement equal to 0.001 gram.

milliliter (4.1) A unit of measurement equal to 0.001 liter.

millimeter (4.1) A unit of measurement equal to 0.001 meter.

mode (5.4) The value that occurs most often in a given set of data.

multiple (10.2) The product of a number and any whole number. The multiples of 3 are 0, 3, 6, 9, …

N

natural numbers (10.1) The set of counting numbers, {1, 2, 3, … }.

natural-number multiples (10.2) A set that contains the result of multiplying a number by each of the natural numbers. For example, the natural-number multiples of 5 are 5, 10, 15, 20, …

negative numbers (6.1) Numbers that are to the left of zero on a number line.

nonagon (3.1) A polygon with nine sides.

number line (6.1) A line used to represent the set of real numbers.

numerical term (8.1) In an equation or expression, the terms without the variable.

O

obtuse angle (3.3) An angle with a measure between 90 and 180.

octagon (3.1) A polygon with eight sides.

opposites (6.1) Numbers that are the same distance from zero on a number line.

opposite sides (7.2) The sides of a quadrilateral that do not have a common endpoint.

opposite vertices (7.2) The vertices of a quadrilateral that are not consecutive.

ordered pair (6.8) A pair of numbers, with specified order, that can be used to designate a point in the coordinate plane.

order of operations (2.1, 2.2) The order in which mathematical operations are performed.

origin (6.1) On a number line, zero; in the coordinate plane, the point (0, 0).

outliers (5.5) Data that are above the upper quartile or below the lower quartile by more than 1.5 times the midquartile range.

output (8.7) For a function, the result after evaluating for given input. For example, let $f(x) = x + 1$. If $x = 5$ is the input, then 6 is the output.

P

parallel (7.1) Two lines that are everywhere the same distance apart and never intersect.

parallelogram (7.2) A quadrilateral in which both pairs of opposite sides are parallel.

pattern (1.1) Any set (numbers, letters, figures, etc.) that seems to be in a certain order, from which you could predict the next element in the set.

pentagon (3.1) A polygon with five sides.

percent (2.4) Means "in each 100."

perfect number (10.5) A number whose factors, except itself, add up to the number.

perimeter of a polygon (2.1) The sum of the lengths of the sides.

perpendicular (7.1) Intersecting at right angles.

pi (π) (2.3, 9.4) The ratio of the circumference to the diameter of a circle. Pi is approximately 3.1416.

pictograph (5.3) A graph that uses symbols to represent data.

pie chart (5.1) See *circle graph*.
polygon (3.1) A closed figure made up of line segments that intersect only at their endpoints.
polyhedron (11.5) A three-dimensional figure in which each face is a polygon.
positive numbers (6.1) Numbers that are to the right of zero on a number line.
power (2.2) A number indicating repeated multiplication. The number 1000 is a power of 10 because $10^3 = 1000$.
preimage (7.4) The original figure in a transformation.
prime-factorization form (10.4) A number expressed as a unique product of prime factors usually multiplied in order from least to greatest.
prime number (10.2) A natural number greater than 1 and with only two factors, 1 and itself.
prism (11.5) A polyhedron with two congruent parallel bases.
probability (1.4) The ratio of the number of favorable outcomes to the number of possible outcomes.
proportion (4.5) An equation stating that two ratios are equal.
protractor (3.5) An instrument used to measure angles.
pyramid (11.5) A polyhedron with one base that is a polygon, several lateral faces that are triangles, and a vertex.
Pythagorean Theorem (2.3) States that in any right triangle,
$$(\text{leg}_1)^2 + (\text{leg}_2)^2 = (\text{hypotenuse})^2.$$
Pythagorean triple (10.7) Three natural numbers, such as 3, 4, and 5, that satisfy the equation $a^2 + b^2 = c^2$

Q

quadrants (6.8) The four regions in a rectangular coordinate system.
quadrilateral (7.2) A polygon with four sides.
quotient (6.4) The answer to a division problem.

R

radicand (2.2) The quantity inside a root symbol.
radius of a circle (2.3) A line segment with one endpoint at the center of a circle and the other endpoint on the circumference.
range (5.5) The difference between the greatest value and the least value for a given set of data.
rate (4.2) A ratio that compares the measurements of two different attributes, for example, miles per hour.
ratio (4.2) A way of comparing two values. The ratio of 3 to 5 can be written as 3:5 or as $\frac{3}{5}$.
rational number (9.3) A number that can be written in the form a/b where a and b are integers and $b \neq 0$.
ray (3.5) A straight path of points that start from one endpoint and extend indefinitely.
real numbers (9.3) The union of the sets of rational and irrational numbers; the numbers on a number line.
reciprocal (8.2) For any nonzero number $\frac{a}{b}$, the reciprocal is the number $\frac{b}{a}$. Zero has no reciprocal.
rectangle (7.2) A quadrilateral with four right angles and opposite sides that are congruent.
rectangular coordinate system (6.8) Two perpendicular number lines intersecting at their zero points, producing a plane in which each point can be identified by an ordered pair.
reflection (7.4) A kind of transformation that makes the new figure a mirror image of the original figure.
reflection symmetry (7.3) A type of symmetry in which a line can be drawn that divides the figure into two halves that are mirror images of each other.
regular polygon (3.1) A polygon in which all sides are congruent and all angles have the same measure.
relatively prime (10.6) Two numbers that have a GCF of 1. The numbers 20 and 27 are relatively prime.
rhombus (7.2) A quadrilateral with four congruent sides and opposite sides that are parallel.
right angle (3.3) An angle with a measure of 90.
Right angle property (9.4) If $\overline{AC}$ is a diameter of a circle and B is any point on the circumference of the circle other than A or C, then $\angle ABC$ is always a right angle.
right triangle (2.3) A triangle with one right angle.

root (2.2) The base that corresponds to a given power. For example, the third root of 8 is 2, because $2^3 = 8$.

root index (2.2) The number used to indicate what root to find.

rotation (7.4) A transformation that rotates a shape a certain number of degrees about a point.

rotation symmetry (7.3) Symmetry in which a figure can be rotated 180° or less in order to lie atop its original position exactly.

rounding (3.3) Using an approximate value, according to certain rules. If 297 is rounded to the nearest hundred, it becomes 300.

S

scalar (6.4) The number used to multiply each element in a matrix.

scalar multiplication (6.4) The multiplication of each element of a matrix by the same number.

scale (4.3) The ratio of distances on a model or map to real-world distances.

scale factor (4.6) The ratio that corresponds to the amount of enlargement or reduction of objects.

scattergram (5.3) Data plotted as points in a plane with vertical and horizontal axes. The data cannot be connected by a line.

scientific notation (3.2) A number expressed as the product of a number between 1 and 10 and a power of 10.

self-similarity (12.5) A property in which part of the figure that is added each time an iteration is completed is similar to the previous part.

set (9.1) A collection of objects.

side (3.1, 3.5) (1) A segment of a polygon. (2) A ray of an angle.

similar (4.6) Having the same shape but possibly a different size.

slant height (11.5) The height of one of the lateral faces of a pyramid or the distance from the vertex of a cone to the circumference of the cone's base.

slide (7.4) See *translation*.

solution (3.4) The value(s) of the variable(s) for which an equation or inequality is a true statment.

sphere (11.6) A round three-dimensional figure in which each point on the surface is the same distance from the center.

spreadsheet (1.5) A powerful data-analysis tool that creates tables and manipulates data in them.

square (7.2) A quadrilateral with four congruent sides and four right angles.

square root (2.2) One of two equal factors of a number.

stem-and-leaf plot (5.5) A numerical diagram used to organize data.

subset (9.1) Set A is a subset of set B if every element in A is also in B. This can be written as $A \subset B$.

supplement (3.5) One of two angles that have a sum of 180. Each angle is the supplement of the other.

supplementary angles (3.5) Two angles with measures that add up to 180.

surface area (11.5) The total area of the surface(s) of a solid figure.

symmetric prime numbers (10.7) A pair of prime numbers that are the same distance away from a given number on the number line.

symmetry (7.3) See *reflection symmetry; rotation symmetry*.

T

table (1.5) A chart that displays data.

tessellation (1.1) A design that repeats without leaving gaps and without overlapping.

transformation (7.4) A way a figure is moved so that the position of the figure changes, but its size and shape do not. See *reflection; rotation; translation*.

translation (7.4) A transformation that slides a figure from one position to another. Also called a *slide*.

translation numbers (7.4) Numbers used to find the image in a translation.

trapezoid (7.2) A quadrilateral with only one pair of parallel sides.

tree diagram (10.1) A diagram used to show all possible combinations in order to solve a problem.

U

union (9.2) Given sets A and B, the set of all elements that are in set A or in set B (or both).

unit cost (4.2) The cost of one unit of a product.

upper quartile (5.5) The middle term of the upper half of a set of data.

V

variable (3.4) A symbol that represents numbers.

Venn diagram (9.2) A diagram that shows relationships among sets.

vertex (3.1, 3.5, 11.5) (1) A point where two sides of a polygon intersect. (2) The center of rotation. (3) In a polyhedron, the intersection of two edges.

vertex matrix (7.6) A matrix that organizes coordinates of the vertices of a polygon.

vertical angles (7.1) Two nonadjacent angles formed when two lines intersect.

vinculum (2.1) A horizontal bar used as a grouping symbol.

volume (4.1) The amount of space an object takes up, usually expressed in cubic units.

W

whole numbers (9.3) The set of natural numbers and the number zero, {0, 1, 2, 3, … }.

X

x-axis (6.8) The horizontal axis in a rectangular coordinate system.

x-coordinate (6.8) The first number of an ordered pair. It indicates how far the point is to the right or the left of the y-axis.

x-intercept (9.7) A point at which a graph crosses the x-axis.

Y

y-axis (6.8) The vertical axis in a rectangular coordinate system.

y-coordinate (6.8) The second number of an ordered pair. It indicates how far the point is above or below the x-axis.

y-intercept (9.7) A point at which a graph crosses the y-axis.

Z

Zero Product Property (9.5) For numbers a and b, if $a \times b = 0$, then either $a = 0$ or $b = 0$.

ANSWERS
TO ODD-NUMBERED PROBLEMS

Chapter 1: Patterns

1.1 A Tiling Problem pages 5–9
1 36 **3a**
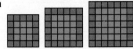
b 4, 8, 12, 16, 20 **c** 0, 1, 4, 9, 16 **5** 100
7a $\frac{3}{4}$ **b** $\frac{7}{8}$ **c** $\frac{15}{16}$ **d** $\frac{31}{32}$ **e** $\frac{63}{64}$ **9** 51
11a
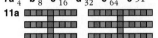

b Red: 15, 15, 21, 21; Blue: 14, 20, 20, 26
13a Layers: 4, 8, 16; Length of each layer: 3", 1.5", 0.75" **b** Direction of folds different in each case.
15a 12, 14, 36, 2, 144, 48, 72 **b** 0, 2, 0, 0, 0, 12, 0
17 Angles 1 and 4 **19a** $222.75 **b** $74.25

1.2 Number and Letter Patterns pages 11–15
1 21, 28, 36, 45 **3** 6, 10 **5** –1, –2 **7** JC, RR, GB
9
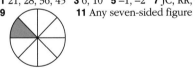
11 Any seven-sided figure
13 65, 129, 257, 513 **15**

17 27 **19** 2.5
21
  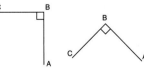
23a Count the smallest squares and multiply by 4.
b 64 **25a** 15, 60, 135, 240, 375, 540, 735
b 1,058,400
27

1	2	3	4	5	6
3	5	8	12	17	23
5	10	18	30	47	70
7	17	35	65	112	182
9	26	61	126	238	420
11	37	98	224	462	882
13	50	148	372	834	1716
15	65	213	585	1419	3135

29a $\frac{3}{7}$ **b** $\frac{5}{7}$

31 Not possible **33a** 56 **b** 93

1.3 Practical Patterns pages 18–22
1 13 **3a** False **b** True **c** True **5** 6: $8.00, 7: $6.86, 8: $6.00; $6.86 **7a** $500 **b** $625 **c** $750
9   **11** 10 units **13**
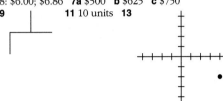

15 32, 64, 128, 256 **17** 72, 63, 56, 51 **19** A vehicle that sometimes accelerates, slows down, travels at a constant speed, and is at rest. **21a** 1, 3, 6, 10
b 15; 21 **23** XVI **25** (6, 36), (7, 49), (8, 64)
27a

#	3	4	6	7	8
3	3	12	6	21	24
4	12	4	12	28	8
6	6	12	6	42	24
7	21	28	42	7	56
8	24	8	24	56	8

b Least common multiple **29** 343 **31** 8 **33** $\frac{160}{100}$
35 6 **37** $11.52 **39** $1\frac{1}{3}$ gal

1.4 Probability pages 24–28
1 $\frac{1}{2}$ **3** Always **5** A fraction, a quotient, etc.
7 Notes in a song; answers to a multiple-choice test
9 28; 12; 245; 80 **11** $\frac{4}{13}$ **13a** $\frac{1}{4}, \frac{1}{4}$ **b** $\frac{1}{6}, \frac{2}{3}$
15   **17a** $\frac{9}{10}$ **b** $\frac{7}{10}$ **c** $\frac{1}{2}$
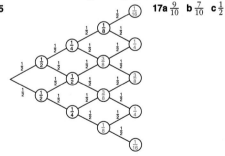

19a $\frac{1}{27}$ **b** $\frac{2}{9}$ **c** $\frac{4}{9}$ **d** $\frac{8}{27}$ **e** 0 **21** 9 **23** 3
25 0.3, 33%, $\frac{1}{3}$ **27** 6; 35

1.5 Tables and Spreadsheets pages 31–35

1a B2, B3, B4, C4, D4 **b** 20, 23, 71 **c** A2+3, A3+3, A4+3 **3a** 169 **b** 7 **5**

A	B	C	D
1	12	3	2
2	15	4	4
3	18	6	8
4	21	9	16
5	24	13	32
6	27	18	64

7 $1.90 **9** 35 **11**

Miles	Amount (dollars)
10	2.80
20	5.60
30	8.40
40	11.20
50	14.00
60	16.80
70	19.60
80	22.40
90	25.20
100	28.00
110	30.80
120	33.60
130	36.40
140	39.20
150	42.00
160	44.80
170	47.60
180	50.40
190	53.20
200	56.00

13a 9 **b** 27 **c** 81 **d** 243 **e** 729
15a 1980 **b** 1919 **c** Any reasonable answer accepted. **17** 124 ft
19 Possible answers:

	A	B	C	D
1	1	2	2	2
2	2	3	3	3
3	A2 + 1	A2 + B2	C1 + C2	D1 + D2
4	A3 + 1	A3 + B3	C2 + C3	D3 + D3
5	A4 + 1	A4 + B4	C3 + C4	D4 + D4
6	A5 + 1	A5 + B5	C4 + C5	D5 + D5

21a $\frac{1}{9}, \frac{1}{36}, \frac{1}{12}, \frac{1}{72}$ **b** $\frac{55}{72}$ **23a** 49 sq units **b** 343 cu units
25 25
27

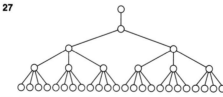

29a 11, 12, 33, 42, 45 **b** Two digits are the same or differ by less than 3.

1.6 Communicating Mathematics pages 38–41

1 The center is the point of intersection of the diagonals. **3a** 13 **b** 2×A2+5; 2×A3+5; 2×A4+5; 31; 67; 139 **5** 56.23 **7** $\frac{3}{2}$ is greater than 1 **9a** 32°F or 0°C **b** 212°F or 100°C **11a** 64 sq units **b** 512 cu units **13** 2.828427125; nearly 8 **15** Add the second number to twice the first. **17** Multiply the area of a face by the length of an edge.

19 A1→5, A2→A1+2, etc. **21** Enter the numbers 1-7 in cells A1-A7. Enter 1 in cell B1, A2×B1 in cell B2, then copy the formula in cell B2 to cells B3-B7.
23a 180 **b** $\frac{1}{100}$ **c** $\frac{5}{9}$ **d** A1→0, A2→A1+1, etc; B1→32, B2→B1+9÷5, etc. **25** 5 × 8, or 8 × 5
27a 1320 **b** 906 **c** 23
29a

Square	1	2	3	4	5	6
Side length	1 cm	1 cm	2 cm	3 cm	5 cm	8 cm
Perimeter	4 cm	4 cm	8 cm	12 cm	20 cm	32 cm
Area	1 cm²	1 cm²	4 cm²	9 cm²	25 cm²	64 cm²

b 55 cm, 220 cm, 3025 cm² **31** $\frac{1}{4}$

Chapter 1 Review pages 43–45

1a 46.21 **b** 157.675 **c** 38.79 **d** ≈11.46 **3a** $17.50 **b** $8.75 **c** $5.85 **d** $4.40 **5** A sequence of items in which each item is generated by a rule from the preceding item **7** Any reasonable answer is accepted. **9** $\frac{3}{4}$ **11** $\frac{22}{25}$ **13a** True **b** True **c** False **d** True **e** False **f** False **15**

	A	B	C	D
1	1	1	2	3
2	2	3	5	5
3	3	6	11	16
4	5	11	22	38
5	8	19	41	79

17a $\frac{1}{4}$ **b** $\frac{1}{13}$ **c** $\frac{1}{52}$ **d** $\frac{4}{13}$ **19a** $19.25 **b** $28.00
c $18.25 **21** d, E, F
23

Length	Width	Perimeter	Area
6	6	24	36
9	4	26	36
3	12	30	36
18	2	40	36
36	1	74	36

25 G, K, L

Chapter 2: Formulas and Percent

2.1 Operations on Numbers pages 53–56

1 Possible answer: So that everyone will assign the same value to the expression **3** The operations were not performed from left to right. **5** The numbers 3 and 5 were added. However, 3 and (12-7) should be multiplied first. **7a** True **b** True **9a** 29 **b** 69 **c** 37 **d** 117 **11a** $\frac{1}{2}$ **b** $5\frac{5}{14}$ **c** $2\frac{4}{7}$ **13** 6
15 0 **17** 1 **19** Possible answer: two five-dollar bills and six one-dollar bills **21** 6 **23** 0 **25** 16
27 18.84 **29** 36 **31** $\frac{16}{15}$; ≈ 1.0667 **33a** Tom: 75.3, Serena: 92, Jamie: 74, Ivan 85.3, Mary 75.3
33b E2: (B2 + C2 + D2)/3 etc **c** 80.4
d (E2 + E3 + E4 + E5 + E6) ÷ 5 **35** Possible answer: $209.85 **37a** 16,807 **b** 117,649 **c** 279,936 **d** 6561
39 False **41** True **43** Always **45a** 49 **b** 125 **c** 81
47 Greek letter, symbol stands for ratio of a circle's circumference to its diameter.

2.2 Powers and Roots pages 60–63

1a If s is the length of the side of a square, s^2 is its area. **b** If e is the length of the edge of a cube, e^3 is its volume. **3a** True **b** True **c** False **d** False

5a Sometimes **b** Sometimes **7** 1135.87 ft/sec
9a 4 **b** ≈ 3.3556 **c** 5 **11** ≈ 2.236 **13** ≈ 29.275
15a 36 **b** 180 **17a** True **b** False **c** False **d** True
19 320 in.^{2} **21** True **23a** 729; 2187; 6561 **b** 3^A7;
3^A8, 3^A9 **23c** 3 × B6; 3 × B7; 3 × B8 **25** 0
27 Willie **29** 25.92 sq units **b** 12.96 sq units **31** False
33 True **35** ≈2.320 **37a** 3 + 5 · (6 + 7) = 68
b (3 + 5) · 6 + 7 = 55 **c** (3 + 5) · (6 + 7) = $\underline{104}$
39a True **b** False **41** 13 lbs **43** 0.6 **45** $4.\overline{3}$ **47** Yes

2.3 Geometric Formulas pages 67–70

1a 16 ft **b** 16 ft^2 **c** The units differ **3a** 28 **b** $\underline{21}$
c 112 **d** 60 **5a** Sometimes **b** Always **7** 6 **9** $\sqrt{12}$,
or ≈ 3.46 **11** 75 **13** 50 **15** $a = 10$, $b = 9$, $c ≈ 11.18$,
$d ≈ 8.66$, $e ≈ 5.29$ **17** $x = \sqrt{2}$, or ≈1.41; $y = \sqrt{3}$, or
≈1.73 **19a** 5(4 + 6); 5(4) + 5(6) **b** 50 **21a** $5^2\pi \sim 3^2\pi$
b ≈50.265 cm^2 **23a** $\sqrt{10}$ cm, or ≈3.16 cm
b $2\sqrt{20} + 2\sqrt{10}$ cm, or ≈15.27 cm **25** 7 **27** $\frac{7}{10}$
29 131 **31** 13 **33** $448 **35** $\frac{250}{100}$ **37a** $2775
b $21,275

2.4 Percent pages 74–77

1a 45 **b** 200 **c** 46.98 **d** 9.4356 **e** 25% **f** 2 **g** 750
3a False **b** True **c** True **5a** $33.60 **b** 52%
7 37.5% **9** $33\frac{1}{3}$% **11** 0.6, 66%, $\frac{2}{3}$ **13** $\frac{153}{1000}$; 0.153
15 $\frac{3}{5}$; 0.6 **17** d **19** $788.40 **21** 52 g **23** Bob: $336
a week; Carol: $336 a week; Ted $350 a week
25a $151.80 or more **b** Answers will vary.
27a $\sqrt{125}$, or ≈11.18; $\sqrt{5}$, or ≈2.24 **b** $\frac{5}{1}$
29 D5: +(A5+B5+C5)/3, D6: +(A6+B6+C6)/3; $≈15.\overline{6}$,
$≈42.\overline{6}$ **31** 0.1, 0.01 **33a** $\frac{4}{3}$ **b** $\frac{4}{3}$ **c** 5 **d** 7

2.5 More About Spreadsheets pages 81–85

1

A	B	A + B	Percent
15	20	0.75	75%
3	15	0.2	20%
130	200	0.65	65%
4	25	0.16	16%
18	12	1.5	150%
6	30	0.2	20%

3 Values become smaller, toward 1.73205 or $\sqrt{3}$.
5 8 **7** True **9** 1490
11

	A	B	C
1	Prices	20% Discount	25% Discount
2	$ 3.00	$ 2.40	$ 2.25
3	$ 5.00	$ 4.00	$ 3.75
4	$ 7.00	$ 5.60	$ 5.25
5	$ 9.00	$ 7.20	$ 6.75
6	$11.00	$ 8.80	$ 8.25
7	$13.00	$10.40	$ 9.75
8	$15.00	$12.00	$11.25
9	$17.00	$13.60	$12.75
10	$19.00	$15.20	$14.25
11	$21.00	$16.80	$15.75

13 Always **15a** $\frac{5}{8}$ **b** $\frac{3}{5}$ **c** Cardinals **17** 9 hours tells
time, not distance **19** 36 in., or 3 ft **21** 56 mm,
or 5.6 cm **23** $\sqrt{\frac{13}{2}}$, $\sqrt{4 + 9}$, $\sqrt{4} + \sqrt{9}$ **25** Always
27 100% **29a** +A3·10 **b** +A3/2.54 **c** +C3/12

Chapter 2 Review pages 87–89

1a 118 **b** 200 **c** 528 **d** 118 **e** 83 **f** $5.\overline{5}$ **3** $A = 5.22$;
$P = 11.4$ **5** ≈23% **7** 12 **9** 33.75 **11** 75% **13** 100%
15a 8 × 185 + 30 **b** $1510 **17a** 1728 **b** 756 **c** 1728
19 38.5°C **21**

	A	B	C	D
1	Width	Length	Perimeter	Area
2	1	2	6	2
3	2	4	12	8
4	3	6	18	18
5	4	8	24	32
6	5	10	30	50
7	6	12	36	72

23 60 **25** (4 + 4) · (4 + 4) · 4 = 256
27 (8 − 3) · (4 + 5) = 45 **29a** ≈497.18 **b** ≈79.04
c 6.29 **31** For any (x, y), find x% of y. **33** Values
become smaller, toward 1.41421 or $\sqrt{2}$

Chapter 3: Measurement and Estimation

3.1 Linear Measurements pages 96–100

1 $2\frac{3}{4}$ in. **3** 24 posts **5** Always **7** Always **9** Always
11a About $2\frac{1}{8}$ in. **b** About 54 mm, or 5.4 cm
13a True **b** False **c** True **d** False **15** 2 ft
17 15 mm **19** 23,401 **21** Measure drawings
23 Yes **25a** $\frac{1}{6}$ **b** $\frac{1}{6}$ **c** $\frac{1}{6}$ **d** 0 **27a** 86 **b** 216
29 10% of 1 m **31** 1 m **33** $\frac{9}{16}$ **35** $1\frac{5}{8}$ in. **b** 180 in.,
or 15 ft **37a** 31.4159265 **b** 314.159265
c 3141.159265 **d** 0.314159265 **e** 0.0314159265
f 0.00314159265 **39a** 4 **b** 1 **c** 0.25 **d** 0.0625; The
results get smaller as the exponents decrease in
value. **41** 2^6, greater by 28 **43a** ≈224% **b** ≈112%
45a ≈240 mi **b** Possible answer: ≈780 mi
47a In each cell, the cell named in the formula
changes to the cell above. **47b** Place 1 in A1. Enter
+A1*2 in A2 and copy down.

3.2 Scientific Notation pages 103–106

1 For a shorter way to record numbers, etc.
3a −2 **b** 10^{-2} **c** 1.2 × 10^6 **5** True **7** 4.3 × 10^4
9a 10,100,000 **b** 1,000,000,000,000 **c** 100 **d** $\frac{1}{100}$
e −9,900,000 **f** 9,900,000 **11** 3.8512 × 10^2
13 5.647 × 10^3 **15** $\overline{BC}$ **17** 5.2 × 10^5 **19** 0.234
21 87.643 **23** ≈1.539 × 10^{-6} **25a** 25% **b** 10% **c** 1%
27 2.310679612 × 10^{32} **29** 3.18253 × 10^{-4}
31a 1.7 × 10^{27} **b** 6 × 10^{53} **c** 4 × 10^{27} **33** 3
35 Measure drawings. **37** $\sqrt{0.1296}$ **39** Between 9
and 10 **41** About 6.9 **43** About 2 **45** No. He will
need to increase his body fat from 6.5%. **47** 1000,
125, 27, 8, 8, 1, 1, 1, 1 **49** 10 in. **51a** 36 **b** 32 **c** 4
d 2 **53a** $\frac{3}{8}$ **b** $\frac{2}{3}$ **55a** 4 + 12 + (7 × 9) **b** 4 × (12 − 7) + 9
c (4 + 12) ÷ (7 + 9) **d** 4 ÷ 12 × 7 × 9

3.3 Estimation and Rounding pages 109–113
1 (Answers will vary.) **3** (Answers will vary.)
5 2 **7** 18 **9** Possible answer: To make measurement look as if it was not an estimate **11** 4.2600
13a 7 packages **b** 10 boxes **c** 3 bags
d 11 newspapers **15** Possible answer: The sale is designed to move goods. If the buyer does not meet the terms of the sale, he does not benefit.
17 475.39 **19** 132.62 **21** 263.11 **23 c** **25** 4
27 52,500 to 53,499 **29** 49 movies **31a** ≈0.8 sec
b 72 beats per min **33a** 7 hr **b** 13.6 gal **c** $17.95
35a Possible answer: A: line 34; B: line 32 **b** The numbers are so small that they are rounded to zero.
37 The greatest depth between the sea bottom and the sea surface is 3363 ft. **39** East **41** $3\frac{3}{7}$ **43a** 44, 22, 11 **b** If the length and width are both divided in half, so is the perimeter. **c** +B3/2 **45a** 42,200
b ≈7.1154 **c** 31,800 **d** 1.924×10^8 **47** 52.3
49 5,230,000 **51** False **53** True **55** True

3.4 Understanding Equations pages 117–120
1 7 **3** (0, 10), (1, 9), (5, 5) **5** 45 **7** 4.75 **9** 1.8
11 23 **13** 5184 **15** 113.6 **17** 9.4 **19** ≈6.3
21 Possible answers: $x = -1, y = -6; x = 12, y = \frac{1}{2}$
23 32.47 in. **25** −2 and 5 **27** 10 **29a** 7.5 cm
b Check drawing. **c** $\frac{2}{3}$ **31** 2.96 **33a** 2 **b** $\frac{1}{4}$ **c** 75
35a 2 **b** 6 **c** No such value. **37** 0 **39a** 4.2 cm, ≈6.1 cm, ≈6.1 cm **b** Check drawing. **c** ≈20.6 cm
41 25 **43** Greater than 6.6 cm

3.5 Angle Measurement pages 125–128
1 West **3** Possible answer: S is the vertex of each angle. **5** Initial ray EF, final ray ED, or initial ray ED, final ray EF **7** 50° to the left **9** m∠DEF = 110, obtuse **11** m∠XYZ = 30, acute **13** 90, right
15 130, obtuse **17** Check drawing. **19** Check drawing. **21** ∠ADB, ∠BDC, ∠ADC **23a** 116
b 26 **25** Check drawing. **27** Check drawing.
29 Check drawing. **31** 63 **33a** B **b** About $45,000,000 **35** 30 **37a** 85.8 **b** 15.3 **c** 97.8
d 550.8 **39** 60 **41** 3.92×10^{-2} **43** 3.156×10^3
45 13 **47** 899 **49** 8.397

3.6 Constructions with Ruler and Protractor
pages 132–135

1 Both have four sides. Only one pair of opposite sides of a trapezoid is parallel. **3a** Angles are supplementary. **b** Angles are supplementary.
5a A Geometric theorem says if two lines are cut by a transversal and the corresponding angles are congruent, the lines are parallel. **5b** Use congruent alternate interior angles, congruent alternate exterior angles, or perpendicular lines. **7** Check drawing. **9** Check drawing. **11** Check drawing.
13a True **b** True **15a** 1:1 **b** Square **17** $\frac{2}{11}$
19a 5 **b** 7 **21a** $\frac{3}{8}$ **b** 37.5% **c** $\frac{3}{8}$ **23a** 2; 2 **b** 1.6; 2.4

c

	A	B	C
1	Percent of antifreeze	Gallons of water	Gallons of antifreeze
2	40	2.4	1.6
3	50	2	2
4	60	1.6	2.4
5	70	1.2	2.8
6	80	0.8	3.2
7	90	0.4	3.6
8	100	0	4

25 About 12.7
27 About 3.5 **29** $\frac{1}{2}$ **31a** Always **b** Never **c** Never
d Always **e** Always **33a** $\frac{3}{4}$ **b** $\frac{5}{4}$ **c** $\frac{9}{10}$

3.7 Geometric Properties pages 138–141
1a 45 **b** 85 **c** 33 **d** 37 **3** 360 **5** 1080 **7** Always
9 Always **11** Never **13a** $\overline{AC}$ **b** $\overline{BC}$ **15a** True
b False **c** False **d** True **e** False **17** ∠1:77; ∠2:68, ∠3:33 **19** m∠CBA = 100; m∠CBD = 80; m∠ABE = 80; m∠DBE = 100; m∠FEB = 100; m∠BEH = 80; m∠FEG = 80; m∠HEG = 100
21 Parallelogram **23a** $\frac{3}{41}$ **b** 7 **25** Areas are equal.
27a $\frac{2}{3}$ **b** $\frac{2}{3}$ **29a** $\frac{3}{2}$ **b** $\frac{3}{5}$ **c** 150% **d** 60% **31** 15 **33** $\frac{1}{3}$

Chapter 3 Review pages 143–145
1a 40.5 **b** 3.25 **c** 76.5 **d** 1053 **3** 4.1758×10^{-3}
5 15 **7** 644,000 **9** Opposite sides are congruent. Opposite angles are congruent. Consecutive angles are supplementary. Opposite sides are parallel.
11 300 **13** ≈12 **15** ≈50 **17** 9.3×10^8 cm³ **19** 22
21 10^5 **23** 0.111 **25** $30 **27** The census, next year's enrollment, population of earth, etc. **29** ≈3.9 **31** 15
33 The measure of the other pair of opposite angles is 120. **35a** $\frac{1}{2}$ **b** $\frac{1}{2}$ **37a** 45 **b** 22.5 **c** 135 **39** 138
41a ≈153.9384 **b** 33.59 **c** 6.998217242

Chapter 4: Ratio and Proportion

4.1 A Closer Look at Measurement
pages 153–157

1 cm **3** mm **5** qt **7** g **9a** 10^2 **b** 10^{-1} **c** 10^{-3} **d** 10^3
e 10^{-2} **f** 10^1 **11a** Tape measure **b** Meterstick, ruler, or tape measure **c** Tape measure **d** Odometer
e Ruler **f** Meterstick or ruler **13** A league is about 3 mi so title refers to a 60,000-mi undersea journey in a submarine **15** $\frac{4}{1}$ **17** $\frac{2}{3}$ **19a** 15 ft 3 in. **b** $15\frac{3}{12}$ or $15\frac{1}{4}$
21a $\frac{1}{3}$ **b** $\frac{1}{9}$ **c** $\frac{1}{27}$ **23** $\frac{3}{2}$ **25** $\frac{2}{3}$ **27** 46 ft, or $15\frac{1}{3}$ yd
29 300 in.³ **31** 3 yd, or 9 ft **33a** $\frac{2}{3}$ **b** Distance covered in 2 hr 40 min at 42 mi/hr **35a** $\frac{9}{10}$ **b** $\frac{28}{27}$
37a The diagonal measurement **b** The amount of pull the line has consistently borne without breaking.
39 Enter +A3+1 in cell A4, and copy B3 to B4. Then copy A 4−B 4 to A102−B102. **41a** 427.2 **b** 1495.2
c 1.8 **d** 513,280.8 **e** −427.2 **f** −1495.2
43a $\frac{2}{7}$ **b** $\frac{a}{d}$ **45** 1979 **47a** 3 **b** 3 **c** x **d** x

4.2 Ratios of Measurements pages 161–164

1a 40 mi/hr **b** $15/book **c** 24 mi/gal **3a** False
b False **c** True **d** True **5** Shreds; Ingredients,
quality, brand loyalty **7** $\frac{5}{9}, \frac{4}{7}, \frac{3}{5}, \frac{2}{3}$ **9a** $0.45 **b** $11.25
11 About 720 bricks **13** 1512.5 mi **15** About
22,200 bushels **17** About 6.15 hr, or 6 hr 9 min
19 4:5 **21** 2 min. **23** Mayflower **25** $\frac{309}{173}$ **27** 30%
29a 1 **b** 13 **c** 0 **d** −4 **31** About 32%
33 About $11 **35** $\frac{2}{3}$

4.3 More Applications of Ratios pages 167–171

1 About 200 mi **3a** 1:240 **b** ≈2.9m × ≈4.8 m;
≈3.6 m × ≈4.6m **5** About 264.5 persons/sq mi
7 ≈0.14 lb/in.^{3} **9** 10 ft 8 in. **11** 1:10,000,000
13 About 2.5 kg **15a** ≈360 mi **b** ≈165 mi **17** The
map with scale 1:3,500,000. **19a** Oil floats on water.
b 2400g or 2.4kg; 2190g or 2.19kg **21** 40 in.
23 2150.4 kg **25** Aluminum, oak, and glass balls
27a About 11 persons/sq mi; about 675 persons/sq mi
b About 74,200,000 **c** About 7 persons/sq mi; about
755 persons/sq mi **29** Sandy **31** 290:375 or 58:75
33 8 P.M. **35** $x = \frac{5}{3}$ **37** $x = 3$

4.4 Working with Units pages 174–177

1 a, c, and d **3a** 3600 sec **b** 360 in. **5a** 180 sec;
$\frac{1}{15}$ mi/sec **b** 30 min; $1\frac{2}{15}$ mi/min **7a** $\frac{7}{3}$ **b** $\frac{7}{3}$ **c** $\frac{7}{3}$
d $\frac{7}{3}$ **9** About 310.6 lengths **11a** 20 qt **b** 48 pt
c 40 oz **13** 3 in./sec **15** ≈551.8 ft^3 **17** 16 pitchers
19a 112 mi **b** About 17 days 7 hr **21** About 306 km
23 5.4 hr, or 5 hr 24 min **25** $2\frac{2}{3}$ in. **27a** 3:21:36 P.M.
b About 4:2:7 **29** 1 **31** $x = 3$ **33** 10 ft 5 in.
35 $x = 51, y = 64, z = 19$ **37a** 655,360 bytes
b $1024 = 2^{10}$, near 1,000.

4.5 Proportions pages 181–184

1 $\frac{3}{4}, \frac{4}{3}$, or 12 **3** 10 in. **5** 27 in., or $2\frac{1}{4}$ ft **7** $x = 16$
9 $x = 10$ **11** $x = 60$ **13a** 25 **b** 8:25 **15** About 9 gal
17 $x = 7\frac{1}{5}$ **19** $z = 12$ **21a** 10 **b** 9 **c** 10% **23** 189 in.^{2}
25 About 17.14 **27** Fall short; 500 units **29a** $y = 6$
b $y = 6$ **c** $y = 8$ **d** $y = 6.71$ **31** 5:1 **33** $1\frac{1}{2}$ hr
35 The angle formed by the two segments is a right
angle and the sum of the measures of the other two
angles is 90. **37a** $1.\overline{6}$
b 13 **c** $17.\overline{3}$

4.6 Similarity pages 187–191

1 Possible answer: Check that corresponding
angles are congruent or that the ratios of the
lengths of corresponding sides are equal. **3** 12, 15
5 2 ft 1 in. **7a** Yes **b** No **c** No **9a** 1:3 **b** 1:3
c 1:3 **d** 9 **11a** 4:3 **b** 16:9 **c** 4:3 **13a** 6 in.
b About 16.67 in. **15a** 12 in. **b** 16 in. **c** 12 ft, 16 ft
17a About 3.43 **b** About 11.67 **19** 70 sq units **21** 15

23 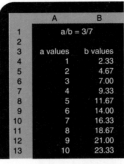 **25** Slugger

27 $w ≈ 3.17$ **29** $w ≈ 7.74$ **31** Electric: about 271.9
milion barrels; residential and commercial: about
512.2 million barrels; industrial: about 1561.9 million
barrels; transportation: about 3977.5 million barrels.

Chapter 4 Review pages 193–195

1a Yards **b** Teaspoons **c** Quarts **d** Fluidounces
3b Daisies are 25¢ each in each case. **5** $x = 9.6$
7 $x = 9.6$ **9** 210 in., or 17 ft 6 in. **11** $6.\overline{6}$ **13** $\frac{1}{8760}$,
about 1.14×10^{-4} **15a** 38 m^2 **b** About 408.6 ft^2
17 Height, texture, weight, color, volume, girth;
height, weight, volume, girth **19** About 200 mi
21 About 35.76 mi/hr **23a** 60 ft^3 **b** 3450 lb
25 $810 **27** 34.5 **29** The two rectangles are not
similar. **31a** About 2.35 acres **b** $0.011 /yd^2
33 60; 135

Chapter 5: Data Analysis

5.1 Circle Graphs pages 203–206

1 An entire quantity divided into groups. **3** 360
5 So that the percents add up to 100%. **7** Answers
may vary. Discuss $\frac{42}{177} ≈ 24\%$ versus $\frac{42}{154} ≈ 27\%$
9 No **11** The "rock" pie segment is much larger
than the corresponding segment in problem 9. The
"country" pie segment is much smaller than the
corresponding segment in problem 9. **13** A: ≈155;
B: ≈65; C: ≈80; D: ≈60 **15** About 3,983,868,000
barrels **17** The answer cannot be determined.
Even though the percent went down, the total
amount consumed in 1970 is not known.
19b 139,500,000 **c** Less than 1 year: 5.6%; 1–5 years:
34.1%; 5–10 yrs: 30.6%; 10–15 yrs: 19.9%; 15 yrs or
older: 9.8% **d** Less than 1 year: 20°; 1–5 yrs: 123°;
5–10 yrs: 110°; 10–15 yrs: 72°; 15 yrs or older: 35°
e  **21** Possible answer:
Each number is 9
more than the
preceding number.

23 5 yds **25** ≈162° **27** About 74% **29** 4

5.2 Bar Graphs and Line Graphs pages 209–212

1 The Richter-Scale readings would appear along the vertical axis, and the countries would appear along the horizontal axis. **3** It appears that in the year 2000, the women's high jump record will reach 7 ft. **5a** Data for which we know both the total and the breakdown of the total **b** Data that are used to compare different conditions
7a 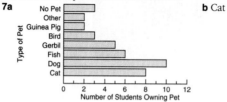 **b** Cat
9a Price per share in U.S. dollars **b** Increased $\frac{7}{8}$ dollar or about $.88 **c** $37\frac{1}{8}$ to $33\frac{1}{2}$ **d** Tuesday
e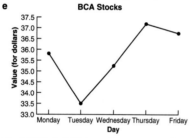
11a 0.85% **b** 1.06% **c** 1% **d** Cannot determine. The answer would depend on the population in that year. **e** 3.86; 3.72; 2.08; 2.04; 2.02
13 Possible answer: Plane takes off, circles the airport, lands, takes off again
15 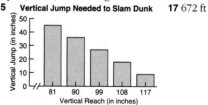 **17** 672 ft
19 7.5 gal **21a** 0.059 kg

5.3 Other Data Displays pages 216–220

1 In a histogram, a mark is made for each individual piece of data. **3** Possible answer: Use a scattergram if data is given in pairs of numbers and you want to determine a relationship. Use a bar graph if you know the number of items in each category and you want to determine a trend.
5 It is easier to compare information on a bar graph.
7 About 3.2 years **9** 15:41, or $\frac{15}{41}$
11 Possible answer: Long goods are more fun to eat.

13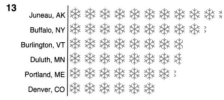
15 About 1.36×10^{10} gal; About 1.72×10^{10} gal
17 1970 to 1980 **19**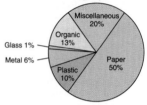
21 Answer depends on graph drawn in problem **20**.
23 0 **25** $\frac{1}{12}$ **27a** $1:\frac{1}{2}$; 2:3; 3:$\frac{15}{2}$; 4:14; 5:17; 6:20; 7:21
b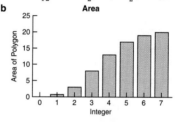
29a The perimeter of rectangle 2 is half the perimeter of rectangle 1. **b** The area of rectangle 2 is one fourth the area of rectangle 1. **31a** Less than 1 year: ≈2.5 million; Between 1 and 5 years: ≈23.8 million; Between 5 and 10 years: ≈17.2 million; Between 10 and 15 years: ≈6 million; more than 15 years: ≈1 million. **b** Less than 7; between 7 and 35; between 35 and 70; between 70 and 105; more than 105. **33a** $\frac{61}{100}$ **b** $\frac{21}{50}$

5.4 Averages pages 224–227

1 35; 48; $48\frac{1}{3}$ **3** Since each of the three averages may yield a different number, it is important to identify the average so that the data are not incorrectly interpreted. **5a** 136.2; 130 **b** ≈148.8; 133 **c** Change in mean is about 12.6; Change in median is 3. **7a** 72.46 **b** 65 **c** 95; No **d** Possible answer: No, the data items vary in value.
9a 74 **b** 0 **11a** 8 **b** 15 **c** 11 **13** Since you don't know the number of women or the number of pieces in the "5 or more" category, the mean and median can't be found. The mode is zero.
15a ≈$9.23 **b** 0 **c** 0 **17a** 71
b,c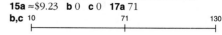

19a ≈30.7; 21.5
b
```
 6   21.5  30.7                                        120
```
21a 233 **b** ≈21 **c** Health-care costs; Skill levels and aging work force **23** −56; (−7)(8) = −56
25a The larger the waist size, the larger the neck size **b** About 25 cm **c** A scale beginning at zero would not be a feasible set of numbers. **27a** No **b** Possible answer: Other teams in Oakland's division may be weak. **c** Cincinnati won the World Series in 1975, 1976, and 1990. Los Angeles won the World Series in 1981 and 1988.

5.5 Organizing Data pages 230–233

1a
```
 6 | 3 9
 7 | 9
 8 | 2 8
 9 | 0 1 3 4 5 7
10 | 6
 :
35 | 3    where 6|3 means 6.3
```
b 35.3 for Mt. Washington; 35.3 is more than 1.5 times the midquartile range above the upper quartile.

3 25% **5** Percentiles divide the data into 100 parts, and quartiles divide data into four parts.
7a 83 **b** 69 **c** 68 **d** 60 **e** 84
f
```
15        60 69    84    98
```
9a
```
2 | 3 3
3 | 3 3
4 | 2 4 8
5 | 1 5
6 |
7 | 4    where 2|3 means 2.3
```
b 4.3; 3.3; 5.1; 2.3 and 7.4; 5.1
c
```
2.3   3.3 4.3  5.1        7.4
```
d No; none of the data are 1.5 times the midquartile range from the upper or the lower quartile.
11a 6, 9, 11, 16, 17, 21, 24, 38, 56, 80, 96, 106, 113, 126, 149, 157, 174, 191, 287, 336, 496, 651, 1356, 2639 **b** 109.5; 22.5; 239; 6 and 2639 **c** Most of the data are closer in value to median than mean.
d
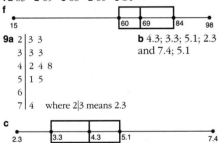

e 651, 1356, 2639 **f** Answers will vary. Fish are a main source of food in Alaska; Alaska has the longest coastline. **13** ≈4.7 cm **15a** False **b** True **c** False **d** True **e** False

5.6 Matrices pages 236–241

1 A matrix is a two dimensional table of rows and columns. **3** 2 × 3 **5** $\begin{bmatrix} 12 & 18 \\ 15 & 20 \end{bmatrix}$ **7** $\begin{bmatrix} 2 & 3 & 4 & 5 \\ 3 & 4 & 5 & 6 \\ 4 & 5 & 6 & 7 \\ 5 & 6 & 7 & 8 \end{bmatrix}$

9a $e_{1,2}, e_{2,1}, e_{2,3}$ **b** $e_{1,2}, e_{1,3}, e_{2,3}$
11
Perimeter	Area
6	2
12	8
18	18
24	32
30	50
36	72

13 $w = 6$; $x = 9$; $y = 6$; $z = -1$; $t = 2$

15 Possible answer: $\begin{bmatrix} 1 & 2 & 3 \\ 2 & 4 & 6 \\ 3 & 6 & 9 \end{bmatrix}$ **17** Job A: Arti; Job B: Tria; Job C: Rita

19
	A	B	C	D
A	0	3	1	1
B	1	0	1	1
C	3	3	0	1
D	2	3	2	0

21
Years Experience	Bagger	Stockboy	Cashier
1	4.80	5.07	7.84
2	5.23	5.60	8.54
3	5.71	6.24	9.28
4	6.08	6.56	10.14
5	6.67	7.20	11.15

23a
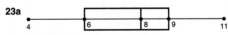

b 8 **c** $\frac{2}{5}$ or 0.4 **d** about 120 **25** 1 **27** 3 **29** 7
31a Kix: $182.00; Trox: $93.50; Soggies: $259.25 **b** Store A: $173.75; Store B: $93.25, Store C: $101.75; Store D: $166 **c** Store A **33** ≈0.11 **35** LCM
```
40
30
 6
42
10
24
```

37 8 **39** 8

Chapter 5 Review pages 243–246

1a ≈4.2 **b** ≈7.5 **c** 2:1 **d** Possible answer: Students do better in school if their attendance is better. **3** $\begin{bmatrix} 5 & 7 \\ 9 & 11 \end{bmatrix}$ **5** Answers will vary.

7a Bill: $188.75; Sue: $219; Harry: $246.25
b 6 hr as a bagger, 7 hr as a clerk, or 4 hr as a cashier
9
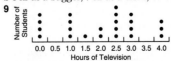

11a East: Boston; West: Oakland **b** East: Toronto; West: Oakland **c** Texas **d** Oakland
13 Answers will vary. **15a** North Dakota and Texas **b** $\frac{1}{20}$ **c** Colorado and New Mexico

Chapter 6: Signed Numbers

6.1 The Number Line pages 256–259

1 0 **3** Possible answer: The integers consist of the natural numbers, their opposites, and zero. **5** 45 sec **7a** 3.6 **b** –102 **c** 0 **9a** A: –2 or 8; B: 11 or 19 **b** 13 units (–2 to 11); 21 units (–2 to 19); 3 units (8 to 11); 11 units (8 to 19) **11** 16° **13a** –11 **b** –7 **15a** –14 **b** –π **c** $\sqrt{2}$ **d** 0 **e** 1 **f** $\frac{3}{4}$ **17a** 2 **b** –9 **c** –8 **d** –4 **e** –3.5 **f** –4$\frac{2}{3}$ **19a** 2 **b** 1 **c** 0 **d** –1 **e** –2 **f** –3 **g** –4 **21**

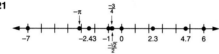

23a

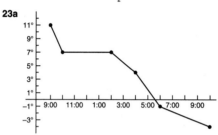

b –4°C **25a** –2 **b** –2 **27a** 26 **b** 10 **c** –2 **d** 2 **29a** 6 **b** 3 **c** 0 **d** –3 **e** –6 **f** –9 **31** 0 **33** $206.25 **35a** 16 **b** 16 is the midpoint of the segment connecting x to y. **37** In parentheses

6.2 Inequalities pages 263–266

1a 3 **b** 5 **c** 1 **d** 4 **e** 2 **3** A shaded endpoint indicates that the endpoint is included in the set. **5** An arrowhead represents the continuation of the set in that direction. **7** Possible answer: An *and* inequality represents a graph that is shaded between two numbers; an *or* inequality represents a graph that has two or more separate sections. **9a** –4 **b** Less than **11** Yes **13a** 6 and 7 **b** –7 and –6 **15a**, **b**, **c**

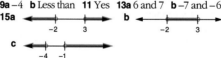

17a This inequality implies that 3 < –3. **b** Number < –3 or number > 3 **19** 0 < L + W < 25 **21a** Possible answer: –2 < x ≤ 3 **b** Possible answer: y ≥ –1 **c** Possible answer: z < 1 **23** The graph of x < –2 or x > –1 is shaded to the left of –2 and to the right of –1. The graph of x < –2 and x > –1 has no shading because no numbers satisfy both conditions. **25a** Yes, but it is customary to put the negatives to the left of zero and the positives to the right. **b** No, because the distance from 0 to 1 establishes the scale, and this scale must be maintained along the number line.

27 **29a** 13 **b** 7 **c** 6

31 12, 6, 0, –6, –12, –18

33a, **b**

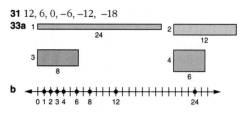

6.3 Adding Signed Numbers pages 269–274

1 No, the dimensions of the matrices must be the same. **3** $\frac{19}{24}$ **5** Possible answer: The sum of two positive numbers is positive; the sum of two negative numbers is negative. **7a** 8 **b** 3 **c** –7 **d** –2 **9** –3 + 8 + (–2) = 3 **11** Yes **13a** Never **b** Always **c** Sometimes **15** $\begin{bmatrix} -11 & -10 & 5 \\ -39 & 21 & 13 \end{bmatrix}$ **17** 1.54

19a $\frac{1}{4}$ **b** $\frac{1}{2}$ **c** $\frac{1}{4}$ **21a** 15 **b** 15 **c** 6.3 **d** 6.3 Possible answer: Adding the opposite of a number gives the same answer as subtracting that number. **23** $\frac{1}{2}$ **25a** $\frac{57}{21}$, or ≈2.7 **b** $\frac{17}{12}$, or ≈1.4 **c** $\frac{-119}{24}$, or ≈–4.96

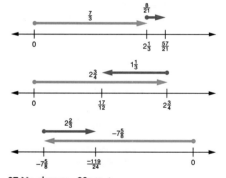

27 No change **29a** Balance
$65.53
33.00
8.82
–7.60
–13.09
–32.34

b

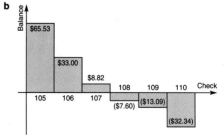

31 11.5 and −5.3 **33** $\begin{bmatrix} -4 & 0 \\ -10 & 0 \\ -18 & 0 \end{bmatrix}$

35 No; the negative numbers should *increase* from left to right. **37a** 20 **b** ≈9.22 **c** ≈4.47
39a −28 **b** −24 **c** 20 **d** −28 **e** −24 **f** 20
41a

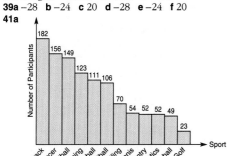

b Some athletes may have participated in more than one sport. A circle graph may be used only when each item belongs to exactly one category.

6.4 Multiplying and Dividing Signed Numbers
pages 278–281

1a −18 **b** −18 **c** −18 **d** −18 **3a** Possible answer: A scalar is a number by which a matrix is multiplied. **b** Possible answer: Because a scalar changes the "scale" of a matrix by increasing or decreasing each element by the same factor **5a** 4 **b** −4 **c** Because the square of −2 is the opposite of the opposite of the square of 2 **d** −8 **e** −8 **f** Because the cube of −2 and the opposite of the cube of 2 are both negative **7** Possible answer: Division by a number is equivalent to multiplication by the number's reciprocal.
9a −20 **b** −15 **c** −10 **d** −5 **e** 0 **f** 5 **g** 10 **h** 15 **i** 20

Possible answer: The products increase in increments of 5. **11a** −8 **b** 8 **c** 8 **d** −8 **13a** −29 **b** 9
15a Possible answer: $1.07 \begin{bmatrix} 4.95 \\ 5.65 \\ 4.85 \\ 5.75 \end{bmatrix}$ **b** Raul: $5.30; Pablo: $6.05; Sue: $5.19; Melissa: $6.15

17 −4 **19a** $\begin{bmatrix} 6 & -8 \\ 18 & -36 \\ -12 & 4 \end{bmatrix}$ **b** $\begin{bmatrix} 4.2 & 21 \\ -11.9 & 31.5 \\ -38.85 & 0 \end{bmatrix}$

21a $0.86 \begin{bmatrix} 175,000 \\ 98,000 \\ 112,000 \\ 145,000 \\ 205,000 \end{bmatrix}$ **b** $\begin{bmatrix} 150,500 \\ 84,280 \\ 96,320 \\ 124,700 \\ 176,300 \end{bmatrix}$

23a ←——|——→ **b** ←——|——→
       3   3
c ←——|——→ **d** ←——|——→
       3   3
e ←——|——→ **f** ←——|——→
       3   3
g ←——|——→ **h** ←——|——→
       3   3

25 $\frac{2}{3}$ **27a** 80 posts **b** 56 posts **c** 66 posts

6.5 Matrix Multiplication pages 285–289

1a $\begin{bmatrix} -7 & 11 \\ -8 & 14 \end{bmatrix}$ **b** $\begin{bmatrix} 0 & -10 \\ -1 & 7 \end{bmatrix}$ **c** Possible answer: The order in which matrices are multiplied affects the answer.
3 Possible answer: 2 × 3 and 3 × 4 **5** 4 × 3 **7a** [22]
b $\begin{bmatrix} -6 & -8 \\ 21 & 28 \end{bmatrix}$ **9a** $\begin{bmatrix} 65 \\ 62 \end{bmatrix}$ **b** $\begin{bmatrix} 25 \\ -46 \end{bmatrix}$ **11** $\begin{bmatrix} 2 & 2 \\ 1 & 4 \end{bmatrix} \cdot \begin{bmatrix} 5 & -3 \\ -2 & -3 \end{bmatrix}$

13 Sometimes **15a** $\begin{bmatrix} 24 & 26 \\ 38 & 40 \end{bmatrix}$ **b** $\begin{bmatrix} 24 & -26 \\ -38 & 40 \end{bmatrix}$

17 Clyde: $70.80; Chloe: $71.20 **19a** Booth 1: $45,475; Booth 2: $47,766.10; Booth 3: $40,608.80 **b** $133,849.90 **c** Possible answer: Because they are responsible for more wear on the roadway
21a −156 **b** 156 **23** −1.468 × 10²⁹ **25** 36π, or ≈113.1
27a −19 **b** −75 **c** 30 **29** 10 **31a** −24 **b** −24 **c** −42

6.6 Subtracting Signed Numbers
pages 292–295

1 Possible answer: By multiplying the number by −1
3a The opposite of negative 100 **b** 100 **5a** 3 **b** 1
c 4 **d** 2 **7** $\begin{bmatrix} -6 & 1 & 7 \\ 6 & -2 & -2 \end{bmatrix}$ **9a** −3 **b** −13 **c** 13 **d** 3

11

x	−8	−6	−4	−2	0	2	4	6	8
y	12	10	8	6	4	2	0	−2	−4

13a 4 **b** −6 **c** 37 **d** −4.5 **e** −12.56 **f** 23.89 **15a** 11
b −2 **c** −12 **17a** 12 **b** 20 **c** 201 **d** 101
19
```
 36   -84  126  -126   84   -36
120  -210  252  -210  120
330  -462  462  -330
792  -924  792
1716 -1716
3432
```
21a −28 **b** −16 **c** −19 **d** 50

23 40 units **25a** 35 **b** 35 **c** 12 **d** 12
27a Ted [12] Bill [12] Laura [10] Kim [19] **b** Ted [3.0] Bill [2.0] Laura [2.0] Kim [≈3.2] **29** 1, −18; 2, −9; 3, −6; 6, −3; 9, −2; 18, −1

31a True **b** True **c** True **d** True

6.7 Absolute Value   pages 298–301

1a 5 **b** 0 **c** 6.23 **d** $3\frac{2}{3}$ **3a** 10, –10 **b** 1.3, –1.3
c 0 **5a** 4 **b** 20 **7** Both expressions stand for the distance between the points whose coordinates are x and y. This distance is unique, so the expressions must be equal. **9a** 4 **b** –8
11 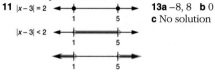   **13a** –8, 8 **b** 0 **c** No solution

15 $|-7| - |18|, |18 + (-7)| = |-18 + 7| = |7 - 18| = |-7 - (-18)|, |-7 - 18|$ **17a** –5 or 5 **b** –5 or 5
c 4 or 14 **19a** 12 **b** 12 **c** 18 **d** 18 **21** –4, –2, 0
23a Possible answers: $|3 + 2| = |3| + |2|; |-3 + (-2)| = |-3| + |-2|; |3 + 0| = |3| + |0|$ **b** Possible answers: $|3 + (-2)| < |3| + |-2|; |-3 + 2| < |-3| + |2|; |-1 + 1| < |-1| + |1|$ **c** The statement is never true.
25 –7 and 15 **27** 0.1 **29a** 9 **b** 21; 21 **c** $|x - 9| = 2$
31a $|x - 3| = 12$ **b** $|x + 5| = 8$ **33** Mean: ≈–5.58; median: –2; mode: –12 **35** –65 40 –40 **37** $\begin{bmatrix} -32 \\ 38 \end{bmatrix}$
–105 80
–185

39a –4 **b** –10 **c** –12 **41**

*	–3	–7	+4	–2
+2	–6	–14	8	–4
–5	15	35	–20	10
–6	18	42	–24	12

6.8 Two-Dimensional Graphs   pages 305–309

1 (2, 4); (0, 1); (–2, 2); (–3, 0); (–3, –4); (3, –1)
3 Possible answer: Quadrant I ordered pairs have the form (+, +); Quadrant II ordered pairs, (–, +); Quadrant III ordered pairs, (–, –); and Quadrant IV ordered pairs, (+, –). **5** (2.5, 1)
7

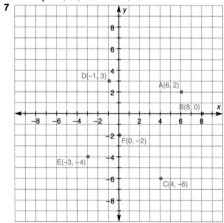

A: Quadrant I; B: No quadrant; C: Quadrant IV;
D: Quadrant II; E: Quadrant III; F: No quadrant
9 (–5, 1); (–3, 2); (1, 1); (0, –3); (–4, –2)

11 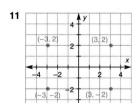   Possible answer: The points are the vertices of a rectangle.

13 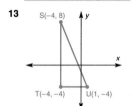   ST = 12; TU = 5; SU = 13

15a R(–6, 5); E(2, 5); C(2, –1); T(–6,–1) **b** 48 **c** 10
17a Possible answers: With the bottom side and the left side lying on the x- and y-axes; with the bottom side lying on the x-axis and the origin at its midpoint; with the origin at the figure's center
b Possible answer: With the bottom side and the left side lying on the x- and y-axes
c 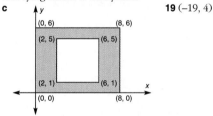   **19** (–19, 4)

21 (5, –2); (8, –2); (8, –6) **23a** 88 **b** ≈13.6%
25 88 ft/sec **27** ◄——●————●——►
                              –8   8

29 (–1, 3); (–5, –1); (–2, –2) **31** –5394 **33** 36.7
35 2 or –2 **37** 4 or –2 **39** Possible answer: $|x| < 9$

Chapter 6 Review   pages 311–313

1a –153 **b** –786 **c** 58 **d** 6 **e** 4956 **f** –8 **3a** 8 units
b $3\frac{1}{2}$ units **c** 7 units **d** 5.34 units **5a** $x \geq -4$
b $-5 \leq x < 1$ **7** 18 ft **9a** –10 **b** –6 **c** 16 **d** 4 **e** –4
f –12 **g** 4096 **h** –2 **11** $w = -80; x = 100; y = -9; z = 9$ **13a** –27 **b** –64 **c** –125 **d** 81 **e** 256 **f** 625
15 $u = -13; v = 8; x = -3; y = 6; z = -20$
17a –6(5 + 2) = –42 **b** –9(5) + (–9)(2) = –63
c –3[4 + (–9)] = 15 **19a** 45 **b** A'(–3, 1), B'(–3, 11), C'(6, 6) **21** –42 **23a** False; –2 · –3 = 6 **b** False; –2 + –3 = –5 **25a** –64 **b** –64 **c** 64 **d** 64 **27a** 14
b –8 **c** 16 **29a** 13, –3 **b** –13, 3 **c** $-3 \leq w \leq 13$

Chapter 7: Exploring Geometry

7.1 Line and Angle Relationships
pages 320–324

1 Possible answers: Top edge and bottom edge of desk are parallel; top and side of chalkboard are perpendicular. **3a** 33 **b** 147 **c** 147 **5** Not adjacent; the angles do not have a common vertex. **7** Adjacent **9a** 51 **b** 129 **c** 51
11 Possible answer:

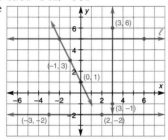

13 No **15** $\overline{QR} \parallel \overline{SE}$; $\overline{QS} \parallel \overline{RE}$; $\overline{QR} \perp \overline{RE}$; $\overline{RE} \perp \overline{ES}$; $\overline{ES} \perp \overline{SQ}$; $\overline{SQ} \perp \overline{QR}$ **17a** 90 **b** ≈101 **c** ≈27 **d** ≈63 **19a** m∠1 = 60; m∠2 = 150; m∠3 = 120; m∠4 = 60; m∠5 = 30; m∠6 = 150; m∠7 = 120; m∠8 = 60 **b** ∠1, ∠4, ∠8 are congruent. ∠2, ∠6 are congruent. ∠3, ∠7 are congruent. **c** ∠1 and ∠5, ∠4 and ∠5, ∠8 and ∠5 **d** ∠1 and ∠3, ∠1 and ∠7, ∠3 and ∠4, ∠7 and ∠4, ∠8 and ∠3, ∠7 and ∠8, ∠5 and ∠2, ∠5 and ∠6 **21a** 6 units **b** 6 **c** 20 units **d** 20 **23** 5; 5; 5; 5 **25**

Figure	Number of Rays	Number of Angles
A	3	3
B	4	6
C	5	10
D	6	15
E	12	66

27 5 and 6 **29** −3 and −2

7.2 Quadrilaterals pages 328–331

1 Always **3** Never **5** Sometimes **7** A square is a rectangle because adjacent sides are perpendicular. A square is a rhombus because all four sides are congruent. A square is a parallelogram because opposite sides are parallel. A square is not a trapezoid because it has two pairs of parallel sides. **9a** Square **b** Possible answer: They could be moved the same distance to the left. **c** Possible answer: They could be moved the same distance upward.
11

Quadrilateral	Trapezoid	Parallelogram	Rhombus	Rectangle
a	✓			
b	✓		✓	✓
c	✓		✓	✓
d	✓	✓		

13a Adjacent **b** Adjacent **c** Opposite **d** Adjacent
15 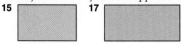 **17**

19a Square **b** Possible answer: They could be moved the same distance to the right or left. **c** Possible answer: They could be moved the same distance up or down. **d** Possible answer: D could be moved to the left, and U could be moved the same distance to the right. **21** (−3, 4) **23** m∠F = 50, m∠R = 130, m∠E = 50, m∠D = 130; possible answer: opposite angles are congruent and consecutive angles are supplementary. **25** Possible answer: Wyoming and Colorado appear to be shaped like rectangles. **27** 3 parallelograms **29a** −8 **b** 200% **c** 25% **31** |x−2| < 4 **33b** = −28 **35** 90; 56

7.3 Symmetry pages 335–338

1 Both **3** Rotation **5a** Sometimes **b** Always **c** Always **7** **9**

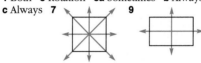

11a True; an equilateral triangle is one example. **b** False for all finite figures
13 **15** 180° rotation symmetry about the midpoint of $\overline{PL}$

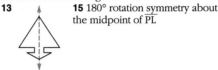

17 **19**

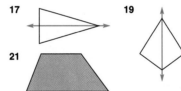

21

23 Impossible **25** Impossible **27** Impossible **29a** Possible answer: An open-ended wrench can be slipped onto it from 6 directions instead of just 4. **b** Possible answer: They are too "round," so that a wrench might slip on them, or wear away the corners. **c** Possible answer: Opposite sides are not parallel, and an open-ended wrench could not be used on them. **d** Pentagonal bolt heads are found on fire hydrants so unauthorized persons may not turn on the water. **31a** (−1, 1); (−4, 0); (−1, −1); (0, −4); (1, −1) **b** Yes; (0, 0); 90°
33 Possible answer: Many kinds of flowers have radial (rotation) symmetry; some other kinds, as well as many kinds of leaves, have bilateral (reflection) symmetry. The arrangement of leaves on plant stems is often bilaterally symmetric. Some lower animals (starfish, coral polyps) have radial symmetry, whereas most higher animals (including human beings) are bilaterally symmetric.
35 56 **37** −6 **39** The triangles are congruent right triangles. **41** The triangles are congruent isosceles triangles. **43** 2.45 m **45** ≈3.05 m

Answers to Odd-Numbered Problems

7.4 Rigid Transformations pages 343–346

1 At 6, 8, and 9, respectively

3a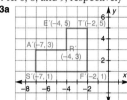

b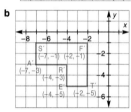

c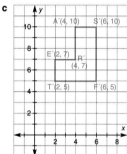

d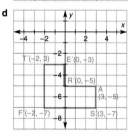

5 Possible answer: A figure is symmetric if it can undergo a transformation (reflection or rotation of 180° or less) without its position changing.

7a 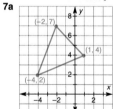   **b**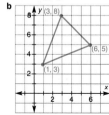

9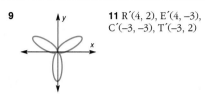

11 R′(4, 2), E′(4, −3), C′(−3, −3), T′(−3, 2)

13 130   **15** Parallelogram

17a 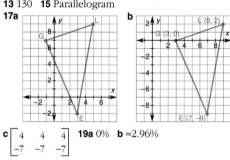   **b**

c $\begin{bmatrix} 4 & 4 & 4 \\ -7 & -7 & -7 \end{bmatrix}$ **19a** 0%   **b** ≈2.96%

21 True   **23** True   **25** True   **27a** $\frac{2}{5}$ **b** $\frac{3}{4}$

7.5 Discovering Properties pages 349–353

1 Possible answer: All four sides are congruent, so ABCD is a rhombus.   **3a** Reflect △NST over $\overline{TN}$, forming △T′S′N′, △TSS′ will be equilateral.
b SN = $\frac{TS}{2}$ **5** Possible answers: One diagonal ($\overline{KT}$) bisects the other ($\overline{EI}$). The diagonals are perpendicular. One diagonal ($\overline{KT}$) bisects a pair of opposite angles. A pair of opposite angles (∠E and ∠I) are congruent.
7 74   **9** (5, 7); (11, 2)   **11** 9
13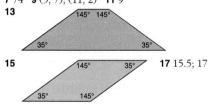

15   **17** 15.5; 17

19 68; 68   **21** 50   **23** 34; 24; 21.2; 20   **25** (10, 0); (0, 8)
27 Square and rectangle

29

	A	B	C
1	k	2k	AB
2	1	2	2.236
3	2	4	4.472
4	3	6	6.708
5	4	8	8.944
6	5	10	11.180
7	6	12	13.416
8	7	14	15.652
9	8	16	17.889
10	9	18	20.125
11	10	20	22.361
12	11	22	24.597
13	12	24	26.833
14	13	26	29.069
15	14	28	31.305
16	15	30	33.541
17	16	32	35.777
18	17	34	38.013
19	18	36	40.249
20	19	38	42.485
21	20	40	44.721
22	21	42	46.957
23	22	44	49.193
24	23	46	51.430
25	24	48	53.666

31 32° F **33** −25.6° F **35a** 20 mm **b** 100 mm² **c** No. The answer to **a** is a linear measure in mm, while the answer to **b** is a measure of area in mm².

7.6 Transformations and Matrices
pages 357–361

1 c **3a** $\begin{bmatrix} 0 & 10 & 14 \\ 6 & 16 & 4 \end{bmatrix}$ **b** 80 **c** 1:4

5 Possible answer: $\begin{array}{c} \text{P} \quad \text{Q} \quad \text{I} \\ \begin{bmatrix} -3 & 4 & -4 \\ 3 & -1 & -2 \end{bmatrix} \end{array}$ **7** A′(0, −7), B′(4, −5), C′(6, 5)

9a The x-coordinates change to their opposites; the y-coordinates remain the same. **b** $\begin{bmatrix} 6 & -5 & 8 \\ 4 & 3 & 2 \end{bmatrix}$

11a $\begin{array}{c} \text{Y}' \quad \text{E}' \quad \text{S}' \\ \begin{bmatrix} 6 & -4 & -10 \\ 10 & 8 & 10 \end{bmatrix} \end{array}$

b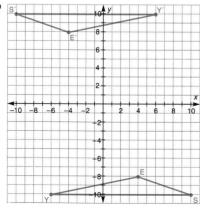

A rotation of 180° about the origin

13a $\begin{bmatrix} 4 & 2 & 7 \\ 9 & 6 & 5 \end{bmatrix}$ **b** $p = -3$; $q = -11$; S′(1, −2); A′(−1, −5)

15 H′(−1, 6), A′(2, 4), T′(−5, 3) **17** H′(−1, 2), A′(1, 5), T′(2, −2) **19a** $\begin{array}{c} \text{R} \quad \text{E} \quad \text{V} \\ \begin{bmatrix} 0 & 0 & 3 \\ 4 & 0 & 0 \end{bmatrix} \end{array}$

b $\begin{array}{c} \text{R}' \quad \text{E}' \quad \text{V}' \\ \begin{bmatrix} 0 & 0 & -3 \\ 4 & 0 & 0 \end{bmatrix} \end{array}$

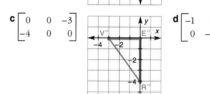

c $\begin{bmatrix} 0 & 0 & -3 \\ -4 & 0 & 0 \end{bmatrix}$ **d** $\begin{bmatrix} -1 & 0 \\ 0 & -1 \end{bmatrix}$

e $\begin{bmatrix} 0 & 0 & -3 \\ -4 & 0 & 0 \end{bmatrix}$; possible answer: Rotating a figure 180° is the same as reflecting it over the y-axis, then reflecting the image over the x-axis.

21 $\begin{bmatrix} 2 & -1 & 1 \\ 3 & -4 & 6 \end{bmatrix}$ The product is identical to the multiplicand.

23a 18 **b** 18 **25** −3.585 **27** 398.068 **29** P′(−2, −3); P″(2, −1); P‴(6, 1); P⁗(10, 3); P⁗′(14, 5)

Chapter 7 Review pages 363–365

1 Each is 55. **3** D′(−2, 5), E′(3, 1), F′(1, 0), G′(0, −5)
5 Both

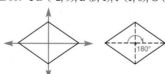

7 Both

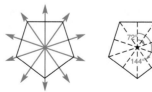

9a Rectangle **b** Move 4 units left. **c** Move 4 units up. **d** Move 4 units down.

11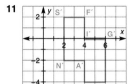

13a Move A to (8, 0). **b** Move A to (8, 0), or move H to (–3, 0), or move E to (3, 5).

15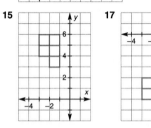

17

19a Similarities: 4 congruent sides, diagonals are perpendicular bisectors of each other. Differences: The angles of a rhombus need not be right angles; the diagonals of a rhombus need not be congruent. **b** Similarities: Opposite sides are congruent and parallel; opposite angles are congruent; consecutive angles are supplementary; diagonals bisect each other. Differences: The angles of a parallelogram need not be right angles; the diagonals of a parallelogram need not be congruent. **21** 100 **23a** $K'(-1, -5)$, $I'(1, -7)$, $T'(3, -5)$, $E'(1, 0)$ **b** $K'(1, 5)$, $I'(-1, 7)$, $T'(-3, 5)$, $E'(-1, 0)$

Chapter 8: The Language of Algebra

8.1 Simplifying Expressions pages 374–377

1 No **3** $8x + x = (8 + 1)x = 9x$ **5** Susie; Like terms have the same variable and the same power.
7a $135 = 135$ **b** Right **c** $540 = 540$ **d** Left **e** When it is easier to add first; when it is easier to multiply first
9 $3x + 4x = 7x$ is true for every value of x, but $x + 5 = 8$ is true only when $x = 3$. **11a** $14x$ **b** 0 **c** $11x + 7y$ **d** $-6x - 3$ **e** $-4x^2 - 5x$ **f** $46x - 9y$
13 $4xy + 9xy - 5xy = 8xy$ **15a** False **b** True **c** True **d** False **17a** $-\frac{1}{4}$ **b** $-\frac{1}{4}x$ **c** Cannot be simplified.
19a $4x + 4$ **b** $3y + 1$ **c** $x = 3, y = 5$ **d** AB = 28, CD = 28 **21** $0x = 0$, so 0 is the simpler form.
23a $13x + 3$ **b** 4.3

25a $11x - 2$
b

x	Side TR	Side RI	Side TI	Sum of Sides
1	1	5	3	9
2	5	7	8	20
3	9	9	13	31
4	13	11	18	42
5	17	13	23	53
6	21	15	28	64
7	25	17	33	75
8	29	19	38	86
9	33	21	43	97
10	37	23	48	108
11	41	25	53	119
12	45	27	58	130
13	49	29	63	141
14	53	31	68	152
15	57	33	73	163
16	61	35	78	174
17	65	37	83	185
18	69	39	88	196
19	73	41	93	207

27 $x = 6 - y$; $y = 6 - x$ **29a** True **b** False **31a** –17 **b** 2 **33a** 99; –11 **b** –495; 55 **c** –297; 33 **35a** > **b** = **c** <

8.2 Using Variables pages 381–384

1a $8 + (-8)$; 0 **b** $\frac{2}{3} \cdot \frac{3}{2}$; 1 **c** $x + (-x)$; 0 **d** $y\left(\frac{1}{y}\right)$; 1
3 The sentence is true when you substitute any value for x. **5a** $31.20 **b** $3b + (0.12)d$ **7** $(2x + 1) + (8x - 3) + (15x + 2)$, or $25x$ **9a** $n - 6$ **b** Half of a number **c** $2(n + 3)$ **d** A number subtracted from 5
11 $C = P + 0.06P$ or $C = 1.06P$ **13a** $4(6) + 0.09($850)$, or $100.50 **b** $4(H) + 0.09(D)$ **15** Possible answers: $M = 14 + Y$, $Y = 3C$, $63 = C + (3C) + (14 + 3C)$
17a $a^2 = b^2 + c^2$ **b** $(t + 7)^2 = (k - 2)^2 + (m + 3)^2$
19 $4x + 17 = 2x + 21$ **21a** $33,000 - 400(1)$ **b** $33,000 - 400(2)$ **c** $33,000 - 400(3)$ **d** $33,000 - 400(4)$ **e** $33,000 - 400(m)$ **23a** Yes; $2x + 4y = 6x + 3$
b No; not unless the trapezoid is isosceles **25a** $-2x - 5$ **b** $11x + 11y$ **27** 55 **29a** < **b** < **c** < **d** < **e** > **f** >
31a Substitute 4 for x. Multiply 4 by 3, then add 7. **b** Subtract 5 from 35 to get 30. Divide 30 by 2 to get 15.

8.3 Solving Equations pages 388–391

1 Possible answer: Raising the shade, then opening the window, then closing the window and lowering the shade **3a** ii **b** i **c** iv **d** iii; Each is an inverse operation. **5a** $7 = n + 3$ **b** $n + 7$ or $7 + n$ **c** $3n + 7$ **d** $3n + 7 = 12$ **e** $3 + 7 = n$ **7a** $x = 5.5$ **b** $x = 5.25$ **c** $x = -8$ **9a** $3x + 7$ **b** $x = -1$ **c** $x = 3$ **d** $-17 - 7x$
11a $x = 4.5$ **b** $x = -193$ **c** $4a - 3b$ **d** $y = 146.51$ **e** $t = 27$ **f** $k = 19.3$ **g** $k = -13$ **h** $2m - 392$ **i** $5x - 10x^2$ **j** $d = 321$ **k** $t = 27$ **l** $15 - 3x$ **13** 135
15a $57 = 36 + 7x$ **b** $x = 3$ **17a** $23x + 19 = 180$ **b** $x = 7$ **c** $m\angle A = 89$, $m\angle B = 78$, $m\angle C = 13$
19a $a + 4 = 15$ **b** $a = 11$ **c** 5; 5 **d** 6 **21a** $\frac{x^2}{36} = .4$ **b** $x \approx 3.8$ **23a** 3 **b** 3 **c** 3 **d** –3 **25a** 4 **b** 33

27a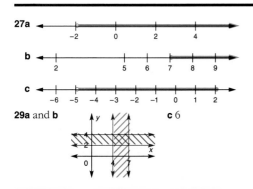

b

c

29a and b   c 6

8.4 Some Special Equations pages 395–398

1a $x = 17$ or $x = -17$ **b** $x = 17$ or $x = -25$ **c** $x = -4$
d No solution **e** Part a is absolute value of x, Part b is absolute value of a sum. **f** 4 has only one additive inverse, -4. **3** Kwamie; the first step is to simplify each side of the equation.
5a Sometimes **b** Sometimes **c** Always **d** Never
7a 1; 256 **b** 2; 4 and -4 **c** 2; 1 or 0 **d** All values
9a $x = 81$ **b** No solution **c** $x = 3$ or $x = -3$
11a $x = 5.5$ ft **b** Possible answer: Divide 162 by 12 and subtract 8 **13** $x = 2$ **15a** C = $5(x^2) + 8(4)x^2$
b $259 **17** $a = 7$ or $a = -7$ **19** $x - 10$, x, $x + 5$, $x + 75$, $x + 100$ **b** 195 **21** $x = 3$ **23a** $-3 \le x \le 7$
b $-5 < x \le 5$ **25** The values of $|x|$ and $\sqrt{x^2}$ are the same. **27a** True **b** True **c** True **d** True **e** False
f False **29a** 40 **b** C' = (11, 2), D' = (3, 2) **c** 16
31a $x = 2$ **b** $x = 5$ **33** A' = $(-1, 5)$; B' = $(2, -2)$; C' = $(10, 2)$

8.5 Solving Inequalities pages 402–405

1 Parts d and h are the only operations that reverse order. **3** Both decrease the value by 5. **5a** $x > -0.75$
b $13 + 4(7 - 2x) < 47$ means the length 13 plus the length $4(7 - 2x)$ is shorter than 47 units. **7a** False
b True **c** False **d** True **9** $3 \ge 5$ **11a** 10 **b** -13 **c** -15
13 $x > -13$ **b** $x \ge 14$ **c** $x < -50$ **d** $x < 3$
15a $3(x - 2) + 8 < 39$ or $3x + 2 < 39$ **b** $x < \frac{37}{3}$ or $x < 12.\overline{3}$
c   **17** $x \ge 8.75$

19a $k > 4$ **b** $k < -6$ **c** $k > 8$ **21** $x \ge 6$
23   **25** She sold more than 18.

27a True **b** True **c** False **d** False **e** False **f** False
g True **h** True **29** She always added the same amount to each side. **31** $\begin{bmatrix} -5 & 1 \\ -30 & 46 \end{bmatrix}$

33a B = 6 or B = -10 **b** A = -4 or A = 4, B = -4 or B = 4 **35a** $\overline{AB} \parallel \overline{CD}$, $\overline{AD} \parallel \overline{BC}$ **b** $\overline{AB} \perp \overline{BC}$ and $\overline{AD}$, $\overline{DC} \perp \overline{DA}$ and $\overline{CB}$, $\overline{AC} \perp \overline{DB}$

8.6 Substitution pages 409–412

1a $x = 6$ **b** $NUM = 6$ **c** $a + b = 6$ **d** $5x - 9 = 6$ **e** $x = 3$
3a Possible answer: A store substitutes one brand product for another; substitutes a generic prescription drug for a brand-name prescription drug.
b A nonmathematical substitute can take the place of something and not be equivalent. In mathematics, the substituted value must be equivalent to the value it replaces. **5** Possible answer: The calculator stored at least one more decimal place than it showed. **7** $x = 3$
9 $a = 20$ **11** $x = -0.5$ **13** $5(2x - 7)$a $10x - 35$
15 b, d, f, g, h **17** 60, 90, 90, 120 **19** Edna substituted 1 for $\frac{6}{6}$, then multiplied. **21a** $x = -4$ **b** $11x + 66$
c $x \ge -4$ **d** 0 **e** All values of x **23** 36 **25** P = 1
27 $24,000, $24,000, and $30,000 **29a** $x \ge -12$
b $x > -3$ **c** $x \le -397.21$ **d** $x \le 5.5$ **31a** $2x + 7$
b $y = 1.5$ **c** $3z - 7$ **d** $a < 1.5$ **e** $110q$ **f** $q \approx 0.91$
33 85 **35a** $18x$ **b** $13x + 12$ **c** $18x = 13x + 12$; Tom caught up with Bill. **d** $x = 2.4$; Tom will catch up with Bill in 2 hr 24 min **37a** 0; 1; 3; 6 **b** 10

8.7 Functions pages 416–419

1 A function uses a formula to assign an output value to an input value. **3a** The parentheses indicate multiplication. **b** The parentheses enclose the input of a function. **c** The parentheses indicate the order of operation—addition before multiplication. **5** 16
7a 1 **b** 1 **c** 7 **d** 11 **9a** 19 **b** 30 **c** 19 **d** 15 **e** 3.75 **f** 5
g 0 **h** 4 **11** $360 **13a** $P(s) = 4s$ **b** $P(l, w) = 2l + 2w$
c $P(b, s) = 2b + 2s$ **d** $P(s_1, s_2) = 2s_1 + 2s_2$ **e** $P(b_1, b_2, s) = b_1 + b_2 + 2s$ **15** $V(l, w, h) = lwh$ **17a** $A(s) = s^2$
b $A(b, h) = \frac{1}{2}bh$ **c** $A(b, h) = bh$ **d** $A(d_1, d_2) = \frac{1}{2}d_1 d_2$
e $A(b_1, b_2, h) = \frac{1}{2}h(b_1 + b_2)$ **19a** 48; 144 **b** $\frac{1}{3}$, 1
c 12; 36 **d** 0 **21** $f(x, y) = 2x + y$ **23a** $y = 20$ **b** $y > 5$
25a 72 **b** 8 **c** ≈ 51.8 **d** ≈ 139.2 **27** $b = 28$ **29** $x = 12$, $x = 9$ **31** The points on $\overline{BE}$ **33** 10.5, 21, 28, 31.5
35a $\frac{15}{88}$ **b** $-\frac{1}{10}$ **c** $-\frac{2}{5}$ **d** $-\frac{17}{12}$ or $-1\frac{5}{12}$ **37** 15

Chapter 8 Review pages 421–423

1a 137 **b** 137 **c** 110 **d** $\frac{49}{3}$ **3** $\begin{bmatrix} 5x & 5y & 5z \\ 5x & 5y & 5z \\ 5x & 5y & 4z \end{bmatrix}$

5 b, c, and **d** **7a** $5x + 8$ **b** $3x^2 + x + 3$ **9a** 9 more than 3 times a number **b** 3 times the sum of a number and 9 **c** 5 divided by a number is $\frac{2}{3}$
11 $P = 6x + 18y + 20$ **13a** $80 **b** $30 + 20b$ **15** 86
17 60, 80, 100, 120 **19** P(AB, AC, BC) = $13x + 11$
21a $x = 5$ or $x = -5$ **b** $x = 5$ or $x = -11$ **c** No solution
d $x = -5$ or $x = -11$ **23** $x = 6$ or $x = -6$ **25a** < **b** >
c = **d** < **27a** $2x + 33 > 99$ **b** $x > 33$ **29** Treat $(2x + 4)$ as a unit and combine like terms, or multiply to remove parentheses, then combine like terms.
31a 5 **b** -17 **c** 16 **d** 16 **e** 9 **f** $\frac{1}{9}$ **g** 0 **h** -33
33 a is decreased by 6

Answers to Odd-Numbered Problems

Chapter 9: Real Numbers

9.1 Sets pages 431–434

1 The set of all points on the number line between -8 and 2 inclusive; $\{x: -8 \leq x \leq 2\}$; because there are infinitely many elements. **3a** True **b** False **c** True **d** True **5** No, since the set contains all real numbers greater than 4 (including, for example, 4.1)
7a The set of even integers between -4 and 16, inclusive
b
c $\{x: -4 \leq x \leq 16$ and x is an even integer$\}$ **d** Possible answer: Graphing **9** Possible answer: $\{(0, 0), (4, 0)\}$; the set of points consisting of the origin and $(4, 0)$ **11** $\{5, 10, a\}$ **13** False **15** True **17** $\{\angle CAN, \angle CAP, \angle CAT, \angle PAT, \angle PAN, \angle TAN\}$ **19** 100 **21** 1 **23** 0 **25** $\{-1, 0, 1, 2, 3\}$ **27** 2 **29** 9 **31** Yes **33** $\{x: -5 < x < 15$ and x is an integer$\}$ **35** $\{9, 10, 11, 12\}$ **37a** $\{1, 2, 4, 8, 16, 32, 64\}$ **b** $\{2, 4, 8, 16, 32, 64\}$ **39a** 6 **b** $\{3, 6, 9, 12, 15\}$, $\{3, 6, 9, 12, 15, 18\}$, $\{3, 6, 9, 12, 18\}$, $\{3, 6, 9, 12\}$ **41** 4 **43** 13 from math club, 4 from science club **45** 40 **47a** $Z \subset X$ **b** $\{$Multiples of 12$\}$

9.2 Intersection and Union pages 438–441

1 It might refer to the members of the band and the members of the orchestra, or it might refer only to those who are members of both the band and the orchestra. **3** Intersection **5** The intersection is a region common to both roads. **7a** $\{1, 2, 3\}$ **b** $\{4, 5, 6, 7\}$ **c** $\{1, 2, 3, 4, 5\}$ **d** $\{3\}$ **e** $\{6, 7\}$ **9** $B \cap C$ **11a** $\{3, 6, 9, 12, 15, 18, 21, 24, 27, 30, 33, 36, 39\}$ **b** $\{4, 8, 12, 16, 20, 24, 28, 32, 36, 40\}$ **c** $\{12, 24, 36\}$ **d** Possible answer: $\{x: x = 12n$, where $n = 1, 2, 3\}$ **13a** B **b** A **15a** 27 **b** 141 **17a** Interior of rectangle BCFG **b** Interior of rectangle ADEH **19a** $\{x: -4 \leq x \leq 4\}$ **b** $\{x: -6 \leq x \leq 6\}$
21a, **b**, **c**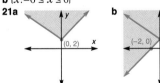
23a $\{$Rectangles measuring 2 cm by 3 cm$\}$ **b** $\{$Rectangles measuring 1 cm by 6 cm, 2 cm by 3 cm, and 1 cm by 4 cm$\}$
25a $\{4\}$ **b** $\{1, 3, 4, 5\}$ **c** $\{1, 3, 4, 5, 7, 10\}$ **d** $\{4\}$ **e** $\{\}$ **f** $\{1, 2, 3, 4, 5, 6, 7, 10\}$ **g** $\{1, 3, 4, 5\}$ **h** $\{4\}$ **27a** $\frac{3}{10}$ **b** $\frac{1}{3}$ **c** $\frac{3}{20}$ **d** $\frac{46}{5}$ **29a** $\{x: x < 7\}$ **b** $\{x: x \geq -1\}$ **31a** $\{-6, 6\}$ **b** $\{1296\}$ **c** $\{\}$ **d** $\{x: x \geq 0\}$ **33a** Possible answer: $\frac{1}{2}$, 1, $\sqrt{30}$ **b** $\{x: 0 < x < 3\}$

9.3 Important Sets of Numbers pages 444–447

1a True **b** False **c** True **d** False **3 a, b, d,** and **e** **5 a, b, c,** and **d** **7** No **9** Possible answer: Because $1 = \frac{1}{3} + \frac{2}{3} = 0.\overline{3} + 0.\overline{6} = 0.\overline{9}$ **11a** $\frac{7}{20}$ **b** $\frac{1}{3}$ **c** $\frac{5}{9}$ **d** $\frac{2}{3}$ **13 a, b,** and **c** **15** $\frac{100}{99}$, $\frac{11}{10}$, 1.111…, $\sqrt{5} \div 2$ **17** $(1, 9), (2, 8), (3, 7), (4, 6), (5, 5), (6, 4), (7, 3), (8, 2), (9, 1)$
19a True **b** False **c** False **d** True
21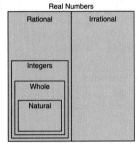
23 a, b, and **d**
25a $\sqrt{13}$ **b** 4 **27** $\begin{bmatrix} -18 & 32 \\ 20 & -34 \\ -33 & 61 \end{bmatrix}$ **29a** The set of positive multiples of 12 **b** $\{12\}$
31 $\{3, 6, 12, 15, 30, 60\}$

9.4 Pi and the Circle pages 451–454

1 Yes; for example, $\frac{\pi}{\pi} = 1$ **3** Rational
5a Rectangle, because it has four right angles and opposite sides congruent **b** Possible answer: Inscribe a rectangle in the circle and draw the rectangle's diagonals. Their intersection is the center of the circle. **7a** False **b** False **c** True **9** $\frac{1}{16}$
11a 49π, or ≈ 153.9 **b** 16π, or ≈ 50.3 **c** 9π, or ≈ 28.3 **d** 25π, or ≈ 78.5 **13** 20 ft **15** $56 + 8\pi$, or ≈ 81 **17** $(12, 6), (12, -6)$ **19a** 5 **b** AB **21** 3π, or ≈ 9.4 ft **23** $24 + 6\pi$, or ≈ 42.8 **25a** The rear-wheel gear **b** $\frac{24}{7}$, or ≈ 3.4, revolutions **27** 4
29, **31**, **33** $\frac{409}{33}$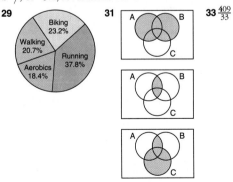

9.5 Properties of Real Numbers pages 458–461

1 It doesn't matter—the expressions are equivalent (by the Associative Property of Multiplication).
3a $=$ **b** $<$ **c** $<$ **d** $=$ **5a** Yes **b** Yes **c** Yes **d** Yes

Answers to Odd-Numbered Problems

7 Yes **9a** Commutative Property of Multiplication **b** Associative Property of Addition **c** Zero Product Property **d** Distributive Property of Multiplication over Addition **e** Commutative Property of Addition **11a** = **b** = **c** > **d** < **13** {−2, −1, 0}, {−1, 0, 1} or {0, 1, 2} **15a** Yes **b** No **c** Yes **d** No **17** According to the Associative Property of Multiplication, $\left(\frac{1}{2} \cdot 6\right) \cdot 11 = \frac{1}{2} \cdot (6 \cdot 11)$ **19a** No **b** Yes **21a** $\begin{bmatrix} 0 & 0 \\ 0 & 0 \end{bmatrix}$ **b** No

23a $A(x) = (6 - x)(x - 2)$ **b** $x = 6$ and $x = 2$ **c** {$x : 2 < x < 6$} **25a** 5 **b** 70 **c** 9 **d** $-\frac{1}{2}$ **27a** 1; 8

b 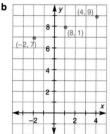 **c** (1, 8) is the midpoint of the segment joining (4, 9) and (−2, 7)

29 24, 60, 96 **31a** 4 **b** ≈8.06 **c** ≈25.32 **33a** 2 **b** 2 **c** 6 **d** 6 **e** 24 **f** 24 **g** 120 **h** 120 **35** $x = 11$, $y = -2$

9.6 Two Coordinate Formulas pages 465–468

1 No, because $(x_2 - x_1)^2$ and $(y_2 - y_1)^2$ have the same value no matter in which order the coordinates are substituted. **3** The coordinates of the midpoint are the average of the x-coordinates and the average of the y-coordinates. **5a** (8, −3) **b** Possible answer: Since the right triangles are congruent, AM = MB. **7a** ≈9.90 **b** ≈9.90 **c** ≈9.90 **d** No matter what the values of a and b, the difference between a and $a + 7$ and between b and $b + 7$ is always 7. **9a** (12, −7) **b** (4, 7.5) **c** (1.9, −0.1) **d** (−0.4, 2) **11a** 5 **b** 5 **c** 5 **d** 5 **13a** 3000 ft **b** 4243 ft **c** 4123 ft **d** 5000 ft **15** AB = √32; BC = √80; AC = √80; isosceles **17a** (0,4) **b** (0,0.5)

19 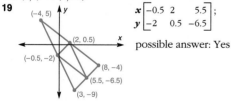 $x \begin{bmatrix} -0.5 & 2 & 5.5 \\ \end{bmatrix}$; $y \begin{bmatrix} -2 & 0.5 & -6.5 \end{bmatrix}$ possible answer: Yes

21a Parallelogram **b** (0, 2) **c** (0, 2) **d** √73, or ≈8.5 **e** √125, or ≈11.2 **23a** (5, 2.5) **b** (5, 2.5) **c** The midpoints are the same. **25** $19.22 **27a** 13 **b** −7 **c** 6 **d** 2 **29** 4

31a Commutative Property of Addition **b** Associative Property of Multiplication **c** Commutative Property of Addition **d** Distributive Property of Multiplication over Addition **e** Commutative Property of Multiplication **f** Commutative Property of Multiplication and Association Property of Addition

9.7 Graphs of Equations pages 473–476

1 a and **c** **3** Possible answer: 3 points to be safe **5** Each value of x is multiplied by 5. **7a** $625 **b** $4

c
Mugs sold	0	1	2	3	4	5	6
Profit (dollars)	−625	−621	−617	−613	−609	−605	−601

d $p = 4m - 625$ **e** 257

9
x	−5	−4	−3	−2	−1	0	1	2	3	4
y	−13	−9	−5	−1	3	7	11	15	19	23

11a 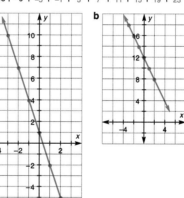 **b**

13a 0.5 **b** −0.5 **c** $y = 0.5x - 0.5$ **15a** $y = 3x + 6$ **b** $y = -3x - 2$

17a, b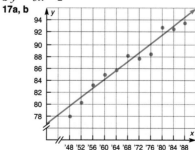

c Possible answer: About 96 or 97 in.

19a
x	−3	−2	−1	0	1	2	3
y	−15	−10	−5	0	5	10	15

b
x	−3	−2	−1	0	1	2	3
y	9	9	9	9	9	9	9

21a $y = x-1$ **b** $y = 2x + 3$
23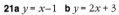

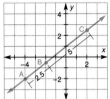

25a ≈5.31 **b** It is half of AC. **27** 32 **29** –15
31 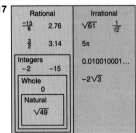 **33a** A′(2, –4); A″(7, –12) **b** (2, –4)

Chapter 9 Review   pages 478–481

1a {$x:x < 5$} **b** {$x:x < 0.2$} **c** {$x:x \leq 1$} **3a** {1} **b** $\sqrt{26}$
5 Possible answers: **a** $12 − (7 − 2) \neq (12 − 7) − 2$
b $36 \div (6 \div 2) \neq (36 \div 6) \div 2$ **c** $7 − 4 \neq 4 − 7$
d $81 \div 3 \neq 3 \div 81$ **7a**

x	–3	–2	–1	0	1	2
y	4	6	8	10	12	14

b

x	–3	–2	–1	0	1	2
y	5	2	–1	–4	–7	–10

9 {1, 3, 5, 7} **11** $\frac{36}{25\pi}$, or ≈.46 **13a** 13 **b** 6.5 **c** ≈132.7
d ≈40.8 **15a** { }, {□}, {@}, {□, @} **b** { }, {B}, {E}, {A}, {R}, {B, E}, {B, A}, {B, R}, {E, A}, {E, R}, {A, R}, {B, E, A}, {B, E, R}, {B, A, R}, {E, A, R}, {B, E, A, R}
17 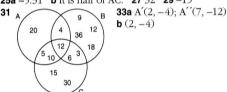 **19** {0, 4, $\frac{3}{2}$, $\frac{-7}{3}$}

21

x	–4	–3	–2	–1	0
y	$\frac{8}{3}$	3	$\frac{10}{3}$	$\frac{11}{3}$	4

23a {Positive integer divisors of 6} **b** 6
25 ≈.611 **27** 13π **29a** 11 **b** Infinitely many
31 About 2:1 **33a**

x	–3	–2	–1	0	1	2	3
y	4	6	8	10	12	14	16

$y = 2x + 10$

b

x	–3	–2	–1	0	1	2	3
y	5	3.5	2	0.5	–1	–2.5	–4

$y = -1.5x + 0.5$

Chapter 10: Topics of Number Theory

10.1 Techniques of Counting   pages 489–493

1 8 **3** If one event can happen in m ways and another event can happen in n ways, then the number of ways both events can happen is $m \cdot n$.
5a 37.5% **b** 37.5% **7a** 12 **b** 9 **c** Possible answer: Neither the numbers in column C nor those in column D are relatively prime.
9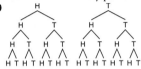

$P(\text{exactly 3 heads}) = \frac{4}{16} = \frac{1}{4} = .25$

11 (tree diagram A, B, C with P Q R S branches) **13** 12 **15** 24 **17** 900

19 15 **21** 16.8% **23** 32 **25** Possible answer: How many choices does Jane have for lunch if she can choose milk or juice and either a sandwich, a pizza, or a salad? **27** 15 **29a** $\frac{1}{6}$ **b** $\frac{1}{2}$ **c** $\frac{1}{12}$ **d** $\frac{1}{12}$ **e** $\frac{1}{6}$ **f** $\frac{8}{9}$
31 $576.00
33a Column 1, Minutes: 1, 2, 3, … 60
   Column 2, Machine A: 8, 16, 24, 32, … 480
   Column 3, Machine B: 12, 24, 36, 48, … 720
b 24, 48, 72, 96, 120, 144, 168, 192, 216, 240, 264, 288, 312, 336, 360, 384, 408, 432, 456, 480
c They are the common multiples of 8 and 12.
35 18, 24, 42

10.2 Primes, Composites, and Multiples   pages 497–500

1 2, 3, 5, 7, 11 **3** 10 **5** All prime natural numbers; also 1 **7** 4, 6, 8, 9, 10, 14 **9a** $10 \times 1, 5 \times 2$ **b** 11×1
c $9 \times 1, 3 \times 3$ **11a** 16, 32, 48 **b** 24, 48, 72 **c** 48
13 30, 60, 90, 120, 150 **15a** Prime **b** Composite
c Composite **d** Prime **17a** 35 **b** 6 **c** 143 **19a** 35
b 12 **c** 24 **21a** 24 **b** 60 **23** 1, 2, 3, 4, 6, 12
25 20×1; 10×2 **27a** 7, 14, 21, 28,…
b 0, 7, 14, 21, 28,… **c** …, –14, –7, 0, 7, 14, 21,…
29a 24 **b** 24 **31a** 36 **b** $\frac{4}{36}$, $\frac{3}{36}$ **c** $\frac{1}{36}$
33a 144 **b** $\frac{136}{144} − \left(\frac{40}{144} + \frac{33}{144}\right)$ **c** $\frac{7}{16}$ **35** About 2:25 P.M.
37a 20 **b** 36 **39** $2\frac{2}{5}$ hr **41** 1680 **43** 375 min, or 6 hr 15 min

10.3 Divisibility Tests   pages 503–506

1 Yes **3** Possible answer: Divide the larger number by the smaller number; If the answer is a whole number, the numbers are divisible. **5** Possible answer: A number is divisible by 27 if the sum of the digits is divisible by 27; No **7a** False **b** False **c** True **d** True **9a** Yes **b** Yes **c** No **d** Yes
11a Yes **b** Yes **c** Yes **d** No **e** Yes **f** No **g** Yes **h** No **i** No **j** Yes **13** Only **c** **15** Only **e** **17a** No **b** Yes **c** No **d** No **e** No

19 Possible answer: According to the distributive property, $ax + bx = (a + b)x$, so the sum of two multiples of any number x will always have a factor of x. **21** Only **a** **23** $n = 2$, $n = 5$, or $n = 8$ **25** Yes **27** $n = 4$ **29** Possible answer: $51 = 1 \cdot 2 \cdot 3 \cdot 4 \cdot 5$, so it is a multiple of 2, 3, and 5, and therefore a multiple of $(2 \cdot 3 \cdot 5)$.
31a Possible answer: 6, 10, 14, 18, 22 **b** $4n + 2 = 2(2n + 1)$. $2n + 1$ is odd since $2n$ is even, and 1 is added to an even number.
33 Possible answer: A number is divisible by 9 if the sum of the digits is divisible by 9. In a palindromic pair the digits are the same; therefore, if the digits of one add up to a number divisible by 9, so will the digits of the other. **35** $\{x : -4 \leq x \leq 7\}$ **37** $y = 2x - 1$ **39a** 2 **b** 23 **41a** $-\frac{1}{4}$ **b** $2\frac{7}{20}$ **c** $\frac{11}{24}$ **43** About 643 breaths/hour; about 15,430 breaths/day

10.4 Prime Factorization pages 510–512

1a 2.5 **b** 2^4 **c** 1.23 **d** 3^3 **3** Possible answer: Yes; if a number is not divisible by 3, it is not divisible by any multiple of 3. **5a** n **b** Because when you test numbers greater than $\sqrt{317}$, you find the same pairs of factors you've already found **7** $2^3 \cdot 3^3$ **9** 25 is not prime. **11** 1, 2, 3, 5, 6, 10, 15, 30 **13** 360 **15** $3 \cdot 7 \cdot 13 \cdot 31$ **17a** From right to left: 21, 105, 210 **b** From right to left: 15, 45, 90, and 4,360 **c** Multiply $5 \cdot 3 \cdot 7 \cdot 2$ and $2 \cdot 2 \cdot 2 \cdot 3 \cdot 3 \cdot 5$.
19a $225 = 3^2 \cdot 5^2$; $150 = 2 \cdot 3 \cdot 5^2$ **b** 75 **c** 75 **21** $2\frac{2}{5}$ **23** $a = 5$; $b = 3$ **25a** 3 **b** 6 **c** 35 **d** 18 **27** $2^4 \cdot 3^2 \cdot 5 \cdot 7$ **29** 7, 11, 13 **31** 40 **33a** $\sqrt{53}$, or ≈ 7.28 **b** $\sqrt{442}$, or ≈ 21.02 **35** 6,760,000

10.5 Factors and Common Factors pages 516–520

1a 2, 3, 5, 7 **b** Prime numbers **3a** Yes **b** No **5** Possible answer: List all the common factors, and pick the largest of them. **7** Yes
9

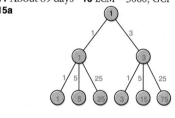

11a 4; 160 **b** 180; 54,000 **13** 5183; 1
15

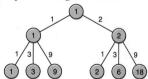

17a 4 **b** 360

19 1, 2, 4, 7, 8, 14, 28, 56; 25% **21a** 160 **b** 8 **23** 1, 2, 3, 6, 9, 18, 27, 54 **25a** Yes **b** Yes **c** Yes **d** No **27a** 3 **b** 3 **c** The last nonzero number in column A is always the GCF of the two numbers. **29a** 8 **b** 120 **c** 40 **d** 9

31a Possible answer: Each whole number in column B and the corresponding number in column A form a pair of factors of the number in C2. **b i** 1, 2, 4, 5, 7, 10, 14, 20, 28, 35, 70, 140 **ii** 1, 5, 25, 29, 145, 725 **iii** 1, 2, 4, 5, 7, 8, 10, 14, 20, 25, 28, 35, 40, 49, 50, 56, 70, 98, 100, 125, 140, 175, 196, 200, 245, 250, 280, 343, 350, 392, 490, 500, 686, 700, 875, 980, 1000, 1225, 1372, 1400, 1715, 1750, 1960, 2450, 2744, 3430, 3500, 4900, 6125, 6860, 7000, 8575, 9800, 12,250, 13,720, 17,150, 24,500, 34,300, 42,875, 49,000, 68,600, 85,750, 171,500, 343,000
33 Possible answer: original pile = x; Tom = $\frac{1}{3}(x - 1)$; Bill = $\frac{1}{3}\left(\frac{2}{3}x - \frac{5}{3}\right)$; Sue = $\frac{1}{3}\left(\frac{4}{9}x - \frac{19}{9}\right)$; Juanita = $\frac{1}{3}\left(\frac{8}{27}x - \frac{65}{27}\right)$; Ralph = $\frac{2}{3}\left(\frac{8}{27}x - \frac{65}{27}\right)$
35a 11 **b** 17 **c** 7 **37** 390 **39a** {2, 2, 3, 7} **b** {2, 2, 2, 3, 3, 7, 7, 11} **c** 504 and 6468 **d** 84 **e** 38,808 **41** 2.5 min **43a** 6, −7, −3, or 0 **b** 6, 5, 4, 3, … **c** $1\frac{1}{6}$

10.6 LCM and GCF Revisited pages 523–527

1a 4 **b** 24 **3** Yes **5** Yes; no **7a** Always **b** Never **c** Never **d** Sometimes **e** Sometimes **f** Never **g** Sometimes **9** LCM = 600; GCF = 15 **11** About 89 days **13** LCM = 3060; GCF = 1
15a

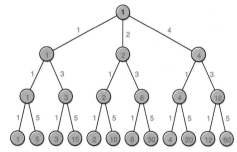

b GCF = 15; LCM=300 **17** LCM = 40,320; GCF = 5040
19a You need to find a common denominator. **b** $\frac{111}{180} = \frac{37}{60}$ **c** 3; 1 **21** 2020 **23** 1987 **25a** 3,535,488,000,000 ft/hr **b** 2.5 hr **27** 80 **29** 16 and 12; 4 and 48 **31** About 64% **33** 0, 5, 100, 225, 400 **35** 66% **37a** 2; 24 **b** 4; 48 **c** 6; 72 **39** 12" × 12"

10.7 Number-Theory Explorations
pages 532–535

1 14 **3** 3 + 3 **5** 3 + 11; 7 + 7 **7** Possible answer: A number that corresponds to the number of dots in a triangular array. **9 a** and **c** **11** Possible answer: 11 + 13 + 113, 17 + 17 + 103, 17 + 59 + 61 **13** 29, 31; 23, 37; 19, 41; 17, 43; 13, 47; 7, 53 **15** 5 = 4 + 1; 12 = 9 + 3; 22 = 16 + 6; 35 = 25 + 10 **17** Possible answers: 3, 4, 5; 6, 8, 10 **19** Possible answer: A different method of scoring would be necessary. **21** 3375 **23** The operations eventually result in the repeated results 1, 2, 1, 2, 1, … **25a** 60 **b** 32 **27a** $12x$ **b** 28.8 **29** 8 **31** 18 min **33a** 5 **b** 9 **c** 45 **35** –1 **37** 289 **39** [number line with point at 2.5 between 0 and 3]

Chapter 10 Review pages 537–539

1a 3^3 **b** 2^4 **c** $2^4 \cdot 3^3$ **d** $2^3 \cdot 3^2 \cdot 5$ **e** $2^4 \cdot 3^2 \cdot 5$ **3** 120 **5a** .24 **b** .58 **c** .18 **7** One 8-ft board, four 10-ft boards, three 6-ft boards **9** To cell B49 **11** $n = 5$ or $n = 6$ **13** $1000; $500; $250; $200; $125; $100, $50 **15a** 24 **b** $-\frac{17}{24}$ **17a** Perfect **b** Deficient **c** Abundant **19** 28% **21a** 0 **b** $1\frac{1}{3}$ **c** $\frac{3}{4}$ **23** At noon **25a** 1, 2, 4, 373, 746, 1492 **b** 1, 2, 3, 4, 6, 8, 12, 16, 24, 37, 48, 74, 111, 148, 222, 296, 444, 592, 888, 1776 **c** 1, 3, 647, 1941 **27a** 5 **b** 4 **c** 20 **29** 1, 4, 9, 25, 36, 100, 225, 900 **31a** $\frac{13}{20}$ **b** $1\frac{14}{15}$ **c** $2\frac{1}{6}$

Chapter 11: Solving Problems with Algebra

11.1 Presenting Solutions to Problems
pages 549–552

1 No **3** 10 P.M.; Possible answers: Use a number line or set up an equation. **5** There is no downhill speed that will make the average 20 mph. **7** Portnos multiplied the wrong numbers. He needs to multiply the 5 racks times the number of suits per rack. Just because a number appears in a problem does not mean it will be used. **9** 10 **11a** $\frac{4}{100}$ **b** $\frac{1}{100}$ **c** $\frac{1}{100}$ **d** 1.75% **13a** 30 sec later **b** Every 15 sec **15** 1 hr 12 min **17** 1 **19** 14 **21a** (0, 10), (1, 7), (2, 4), (3, 1) **b** (1, 7), (2, 4), (3, 1) **23a** $x = -5$ **b** $x = -5$ **25a** $9x + 36 = 5x$ **b** $9x = 5x - 36$ **c** $4x + 6 = -30$ **d** $6 = -4x - 30$ **27** 120 **29** $n = 5$

11.2 Equations with Variables on Both Sides
pages 556–558

1 $x = \frac{14}{3}$ **3** $x = -5$ **5a** x = any number **b** No solution **7** $x = -8$ **9** $x = 5.4$ **11** ≈4.66 **13** It is cheaper to hire Rob for any job over 1 hour. **15** AB = 10 **17** 70 **19** $1000 **21** $n = 4$ **23a** 1 **b** 1 **c** 1 **25** x = any number **b** No real solution **27** 5 **29** 72°

11.3 More Word Problems pages 561–565

1 Yes, if the number is negative **3** $x > 15$ **5** $3.01 **7a** 10 cm **b** 40 cm **9a** 13 hr; 10 hr **b** 10,400 **11** Possible answer: 50 coins **13** ≈156.26 ft **15** 8 hr **17** $195,652.17 **19** 20 **21** 12 **23** $A(b_1, b_2, h) = \frac{1}{2}h(b_1 + b_2)$ **25** 1296 **27a** ≈2.66 ft **b** ≈1.39 ft **c** ≈58.25 ft **29** $1500 **31** 15 **33a** 14 **b** 0 **c** 17 **d** 5

11.4 Equations with Several Variables
pages 569–572

1 a, b, c **3a** 27 **b** 29 **c** Only if $a = b$ **5a** 36 **b** 15 **c** –6 **d** –55 **7** 1 or 7 **9a** $x + y = 26$

b

x	1	2	3	4	5	6	7	8	9	10	11
y	25	24	23	22	21	20	19	18	17	16	15
x	12	13	14	15	16	17	18	19	20	21	22
y	14	13	12	11	10	9	8	7	6	5	4
x	23	24	25								
y	3	2	1								

11a A **b** A: 60 Kg; B: 45 Kg **13** 130.5 **15** For $\frac{18}{x + 2y}$ to be a natural number, $x + 2y$ must be a factor of 18 in order that the quotient is a natural number. **17** $18\frac{2}{3}$ in. **19a** Possible answer: $2(x + 3) + 2(4) = 3(2x)$ **b** 3.5 **c** 6.5 **d** 7 **21** (1, 6), (4, 4), (7, 2), (10, 0) **23** $\sqrt{\frac{8}{\pi}} \le r$ **25a** When $x = 2$, outputs 1-10 are 8, 26, 80, 242, 728, 2186, 6560, 19682, 59048 and 177146. **b** When $x = 1$, outputs 1-10 are 5, 17, 53, 161, 485, 1457, 4373, 13121, 39365, 118097. **27** $1\frac{1}{5}$ hrs, or 1 hr 12 min **29a** $-2x^2 - \frac{5}{12}x$ **b** $x = -\frac{35}{27}$ **c** $x = 2.5$ **d** $-5x + \frac{5}{2}$ **31a** 376 in.² **b** 480 in.³ **33** $5175

11.5 Polyhedra pages 576–580

1 Yes **3** Edges of the polyhedron that would not be visible from the given point of view. **5** The height of the pyramid is a segment from the vertex and perpendicular to the plane containing the base. The slant height is on the surface of the pyramid, extending from the vertex perpendicular to the base edge. **7** Figure I is an open box. Figure II is a rectangular prism. **9a** 24 **b** 90, 120, 150 **c** 360 **d** 408 **e** 360 **11** Possible answer: A quadrilateral with both diagonals drawn; a top view of a quadrilateral-based pyramid. **13a** 256 **b** 544 **c** 800 **d** 1280 **15** All have the same volume. **17a** 288 **b** 450 **19** Surface area: 216 in.²; Volume: 216 in.³; The surface area and volume are not equal because the units in.² and in.³ measure different attributes of the polyhedron. **21** ≈40.5 lb **23a** $\frac{1}{96}$ **b** $\frac{1}{16}$ **c** $\frac{1}{32}$ **25** 1,860,480 **27** (1, 1), (4, 1), (13, 1), (8, 2), (3, 3)

11.6 Solids with Circular Cross Sections
pages 585–588

1 The circumference, since it is π × (diameter of ball) whereas the height is about 3 × (diameter of ball) and $\pi > 3$. **3** Possible answer: Immerse the egg in water and find the volume of the water displaced. **5a** ≈ 204.2 cm² **b** ≈ 78.5 cm² **c** ≈ 282.7 cm² **d** ≈ 314.2 cm³ **7a** 6 in. **b** $\frac{1}{27}$ **9** Yes **11a** $\frac{1}{3}$ **b** $\frac{1}{3}$ **c** $\frac{1}{27}$ **13** $\approx 18,000$ people **15** Possible answer: The cube has more surface area so it will cool the drink faster. **17** Yes **19a** ≈ 5067 ft³ **b** 188.5 ft² **21** 1, 2, 6, 17 **23** 5 hr **25a** $x = 1$ **b** $x = -5$ **27a** 7 **b** 11 **c** 15 **d** 19 **e** 23

11.7 Reading Information from Graphs pages 592–595

1 ($\approx -0.77, \approx 0.59$), (2, 4), (4, 16) **3a** $C = 7B + 4G$

b
Box seats	0	4	8	12
Grand stand	25	18	11	4

5
x	1	2	3	4	5	6	7	8	9
A	99	96	91	84	75	64	51	36	19

$A(x) = (10 - x)(10 + x)$

7
r	1	2	3	4	5
A	≈3.1	≈12.6	≈28.3	≈50.3	≈78.5

9 Possible answer: Let $y = 10x$. Draw the graph. **11a** $x = 4$ or $x \approx -1.2$ **b** $2^4 - 3 = 13$; $3 \cdot 4 + 1 = 13$; $2^{-1.2} - 3 \approx 2.6$; $3(-1.2) + 1 = -2.6$ **13a** $V = 8t + 5p$ **b** $8t + 5p = 34$;
t	0	1	2	3	4
p	6.8	5.2	3.6	2	0.4
c (3, 2)

15a $\ell = w + 20$; $80 = 2\ell + 2w$ **b** $w = 10$; $\ell = 30$

17
F	32	41	50	59	68	77	86	95	104
C	0	5	10	15	20	25	30	35	40

Converts temperature in degrees Fahrenheit to degrees Centigrade. **19a** 12 **b** 65π or ≈ 204.2 **c** 100π or ≈ 314.2 **21** B **23a** 10 **b** -15 **25** 432π cm² or ≈ 1357 cm² **27** (3,2) **29a** 30 **b** 204 **c** 650

Chapter 11 Review pages 597–599

1 Possible answer: A cone is $\frac{1}{3}$ of a cylinder whose volume is $\pi r^2 h$. **3** 22m **5** 7 **7a** -5 **b** $-40 - 8x$ **c** $3x - 1$ **d** $x = 9$ **9a** $19.02 **b** $158.50 **c** $14.41 **11** $x = 13\frac{1}{3}$ **13** $20 **15** 30 **17a** 60 **b** 96 **c** 48 **19a** $V = 904.8$ Total Surface Area = $144\pi \approx 452.4$ **b** $V = 108\pi \approx 339.3$ TSA = $90\pi \approx 282.7$ **c** $V = 320\pi \approx 1005.3$ TSA = 200π, or ≈ 628.3 **21** $(-2, 3)$ **23a** Possible answers: Make a table; guess and check; solve an equation. **b** 13 dimes, 12 quarters

Chapter 12: Looking Ahead: Iterations and Fractals

12.1 Iterations pages 608–611

1a
Term	0	1	2	3	4	5	6	7	8	9	10
Output	8	11	14	17	20	23	26	29	32	35	38

c $y = 3x + 8$

3a Initial input: 1; Iteration rule: 2(INPUT); Number of Iterations: 63 **b** 2^{63} or $\approx 9.22 \times 10^{18}$ **c** Possible answer: checkerboard square could not hold this many grains of wheat

5a Initial input: $750,000; Iteration rule: $\frac{\text{INPUT}}{2}$; Number of Iterations: 8

b
Year	Amount
1960	$750,000
1965	$375,000
1970	$187,500
1975	$ 93,750
1980	$ 46,875
1985	$ 23,476
1990	$ 11,719
1995	$ 5,859
2000	$ 2,920

c Year 2030

7a Initial input: 23; Iteration rule: INPUT + 4; Number of Iterations: 19 **b** 99 seats **9a** $1.526 \times 10^{-5} = \frac{1}{256} \cdot \frac{1}{256}$ **b** $85.\overline{3}$

11a
Seconds	0	1	2	3	4	5	6	7	8	9	10
Distance from wall	12	6	3	1.5	.75	.375	.188	.094	.047	.023	.012

b In 6 seconds **c** In theory, no, but in reality, the distances get too small to be meaningful **d** speed is decreasing **13a** $\approx 0.79\%$ **b** The first month's interest is charged on $9000, then $350 is subtracted from this total **c** Initial input: 9000; Iteration rule: INPUT + $\frac{0.095}{12}$(INPUT) $- 350$ **d** 29 months **15** 15.5

12.2 More About Iterations pages 616–619

1 The value of $h(n)$ is constant.

3a
Length of side	0	1	2	3	4	5	6
Number of toothpicks	0	3	9	18	30	45	63

b $f(0) = 0$; $f(n) = f(n-1) + 3n$ **c** 360 **5** $f(1) = 11$; $f(2) = 17$; $f(3) = 29$; $f(4) = 53$ **7a** $8750 **b** 8.6% **c** $250 **d** $f(0) = 8750$; $f(n) = f(n-1) + f(n-1) \cdot \left(\frac{0.086}{12}\right) - 250$ **e** 41 months **9** $f(1) = 1$; $f(n) = f(n-1) + (2n-1)$ **11a** 20; 28; 36 **b** $f(1) = 12$; $f(n) = f(n-1) + 8$ **c** 100 **13** $f(1) = 0$; $f(n) = f(n-1) + (n-1)$

15
Counter	1	2	3	4	5	6	7	8	9	10	11	12
Output	1	4	9	16	25	36	49	64	81	100	121	144

The answer is the square of the counter. **17** The value of $h(n)$ decreases. **19** At age 38

12.3 Geometric Iterations pages 623–624

1 No; no (for a finite number of iterations); no **3** One possible answer is given. Others exist.

5a Rhombus **b** Alternates between a rhombus and a rectangle

7   9

(0, 1)

11 lattice points hit: (1, 1), (1, 2), (3, 1), (3, 2), (3, 3), (3, 4) lattice points missed: (1, 3), (2, 1) (2, 2), (2, 3) From the origin, go right one unit, turn left 90°, go up one unit, and turn left 90°. Each iteration consists of two line segments and two left turns. Increase the lengths of the segments by one unit each iteration.

12.4 Convergent Iterations pages 630–633

1a $2.8\overline{3}$, 2.828431, 2.828427; Following values remain constant at 2.828427 **b** Outputs converge to same result as part **a**. **c** Outputs converge to 8. **d** Answers will vary. **e** It computes $\sqrt{a}$, where $\frac{1}{2}\left(\text{INPUT} + \frac{a}{\text{INPUT}}\right)$ is the iteration rule. **3a** 100 **b** yes

5b
x	8	7	7	6.5	6.5	6.25	6.25	6.125	6.125	6.0625
y	7	7	6.5	6.5	6.25	6.25	6.125	6.125	6.0625	6.0625

7a 2.55 ppm, 2.17 ppm, 1.84 ppm, 1.57 ppm, 1.33 ppm **b** 0.3 ppm **9a** ≈0.79% **b** $5826.69; $4510.82; $3194.96; $1879.09; $563.23 **c** About $320 **d** Actual payment is $321.40 **11** Yes; 1.618033989 **13** 1250 mg **15** Let (x, y) represent the dimensions of ABCD (x the horizontal component, y the vertical). Let $f(x, y)$ represent the exit point of a ball shot diagonally from point A. The following properties hold:

x	y	f(x, y)
odd	even	B
even	odd	D
odd	odd	C
even	even	not A

17 9.3% annually **19** 2.293166287
21a
n	f(n)	g(n)
1	47000	64000
2	47200	63800
3	47330	63670
4	47414	63586
5	47469	63531
6	47505	63495
7	47528	63472
8	47543	63457
9	47553	63447
10	47559	63441
11	47563	63437
12	47566	63434
13	47568	63432
14	47569	63431
15	47570	63430

b ≈47,571 and ≈63,429

12.5 Introduction to Fractals pages 637–639

1 In each case, the length is 8.
3a,b   **c** No **5**

7

9a 1, 8, 24, 32, 16 **b** 1, 3, 9, 27, 81 **c** They are the numbers of shaded triangles for the iterations of the Sierpinski triangle.

11a
Iteration	0	1	2	3	4
Blue area shaded on this iteration	0 in.²	16 in.²	8 in.²	4 in.²	2 in.²
Total blue area shaded so far	0 in.²	16 in.²	24 in.²	28 in.²	30 in.²
Red area shaded on this iteration	0 in.²	16 in.²	8 in.²	4 in.²	2 in.²
Total red area shaded so far	0 in.²	16 in.²	24 in.²	28 in.²	30 in.²

b 32 in.²; 32 in.²; It looks like a square with a red half and a blue half and no white space.

Chapter 12 Review pages 641–643

1a,b
iteration	output	output
0	−25	25
1	−15.38	15.38
2	−12.3714	12.37140
3	−12.0055	12.00557
4	−12.0000	12.00000
5	−12	12
6	−12	12
7	−12	12
8	−12	12
9	−12	12
10	−12	12
11	−12	12
12	−12	12

c It converges on the opposite of −12. The results are the same; the signs of the numbers are just different.

3a Initial input: 17,500; Iteration rule: INPUT − INPUT(0.13)

b
year	value
0	$17,500
1	$15,225
2	$13,246
3	$11,524
4	$10,026
5	$ 8,722
6	$ 7,588
7	$ 6,602
8	$ 5,744
9	$ 4,997
10	$ 4,347

c In about 5 years

5
Counter	0	1	2	3	4	5	6	7	8	9	10
Output	2	7	17	32	52	77	107	142	182	227	277

7a
Figure number	1	2	3	4	5	6	7	8
Number of toothpicks	1	5	12	22	35	51	70	92

b 330 toothpicks **c** $f(1) = 1; f(n) = f(n−1) + (3n−2)$
9a $(20, 14)$ $\left(20, 13\frac{1}{3}\right)$ $\left(22, 13\frac{1}{3}\right)$ **b** $\frac{1}{16}, \frac{2}{2187}$ **c** 64; 27; 91
d $\left(21\frac{1}{3}, 13\frac{1}{2}\right)$

11a
Number of disks	1	2	3	4	5	6	7
Total number of moves	1	3	7	15	31	63	127

b Initial input: 0; Iteration rule: 2(INPUT) + 1

INDEX

A

Absolute value, 296–297
Abundant number, 515
Accuracy, degree of, 107
Acute angle, 110
Addition
    Associative Property of, 455
    Commutative Property of, 456
    Distributive Property of Multiplication over, 373
    of fractions, 6, 648-651
    as inverse operation, 115
    of like terms, 371–372
    of matrices, 269
    of signed numbers, 267–268
    to solve inequalities, 399
Adjacent
    angles, 319
    sides, 325
Agriculture, 218
Algebra
    combining like terms, 371–372
    evaluating expressions, 50–52
    functions, 413–414, 613–615
    graphing equations, 469–471, 589–592
    inequalities, 260–262, 399–401
    investigation, 63
    matrices, 234–236, 276, 282–284
    properties of real numbers, 455–457
    signed numbers, 252, 267–268, 275-278, 290–291
    simplifying expressions, 371–372
    solving equations, 385–387, 392–394, 553–555, 566–568
    variable, 114
Angle(s), 121–124
    acute, 110
    adjacent, 319
    central, 387
    complementary, 124
    congruent, 320
    corresponding, 186–187
    measuring, 122–123
    naming, 122
    obtuse, 110
    right, 110
    straight, 124
    supplementary, 124
    vertex of, 121
    vertical, 319
Applications
    agriculture, 218
    architecture, 169, 184, 324, 433
    aviation, 165, 389
    banking, 272, 274, 610
    budgeting, 270
    business, 210, 233, 240, 280, 287, 381, 390, 396, 404, 411, 417, 473, 491, 512, 524, 557, 558, 562, 563, 565, 593
    car maintenance, 22
    communicating, 6, 9, 12, 20, 26, 27, 35, 54, 56, 62, 74, 83, 98, 106, 110, 119, 127, 133, 139, 140, 155, 157, 171, 176, 184, 204, 211, 219, 225, 227, 232, 233, 264, 265, 271, 272, 273, 274, 279, 286, 306, 307, 321, 329, 338, 346, 353, 359, 361, 376, 377, 382, 384, 389, 390, 391, 398, 404, 410, 432, 433, 434, 441, 447, 452, 459, 492, 504, 505, 510, 511, 512, 518, 519, 533, 578, 593, 594, 619, 623, 631
    consumer math, 15, 20, 22, 32, 54, 55, 56, 63, 67, 70, 74, 83, 99, 110, 111, 154, 155, 156, 161, 162, 163, 170, 189, 191, 237, 259, 281, 287, 391, 321, 382, 459, 468, 557, 564, 565, 571, 611, 617, 619, 631, 633
    design, 346
    environmental science, 76, 104, 191, 287
    estimating, 110, 111, 127, 128, 134, 164, 167, 168, 169, 174, 182
    family tree, 16–17
    finance, 75, 416, 527, 563
    geography, 100, 330, 468
    health, 73, 105, 390
    history, 33, 128, 162,
    industrial arts, 337
    language arts, 241, 330, 438
    matrices, 269, 278, 282, 284, 287
    mental math, 56, 377, 619
    nutrition, 75
    postal rates, 29, 33
    ratios, 165–166
    sales tax, 43
    science, 18, 39, 40, 53, 60, 102, 103, 104, 111, 168, 169, 177, 210, 212, 217, 218, 232, 238, 256, 353, 384, 398, 404, 453, 475, 492, 499, 526, 564, 570, 572, 579, 580, 586, 587, 593, 594, 610, 611, 631, 632
    signed numbers, 253, 255
    social science, 211
    sports, 15, 69, 205, 233, 257, 288, 336, 416, 474, 491, 533
    spreadsheets, 7, 8, 13, 14, 19, 28, 32, 34, 35, 38, 40, 55, 61, 75, 76, 82, 83, 85, 100, 106, 111, 112, 118, 134, 135, 139, 156, 157, 163, 190, 191, 205, 224, 225, 265, 272, 280, 324, 330, 353, 397, 411, 418, 446, 493, 499, 506, 518, 519, 526, 552, 609
    stock, 277–278, 280, 284
    surveying, 466
    technology, 22, 26, 177, 259, 447, 461, 493, 639
    time zones, 255
Approximating π, 454
Architecture, 169, 184, 324, 433
Area, 150
    of circle, 64
    lateral, 574
    of parallelogram, 64
    of rectangle, 64
    of square, 61
    surface, 574
    of trapezoid, 66, 564
    of triangle, 64, 117, 455, 566
Artistic graph, 215
Associative Property of Addition, 455
Associative Property of Multiplication, 455
Attribute, 150
Auto trip, 171
Averages, 221–223
Aviation, 165, 389
Axes, 207–208, 302

B

Balance, 558
Banking, 272, 274, 610
Bar graph, 207
Base
    of cone, 582

in exponential expression, 57
of parallelogram, 64
of prism, 574
of pyramid, 575
of rectangle, 64
of trapezoid, 66
of triangle, 64
Baseball, 227
Base unit, 159
Binary digit, 492
Binary number, 492
Bisect, 348
Box-and-whisker plot, 229–230
Budgeting, 270
Business, 210, 233, 240, 280, 287, 381, 390, 396, 404, 411, 417, 473, 491, 512, 524, 557, 558, 562, 563, 565, 593

C

Calculator
  adding signed numbers, 254, 268
  decimal forms, 77
  exponents, 58
  expressions, 50–51
  factorial, 461, 491
  graphing, 591
  iterations, 607, 626
  percents, 71
  scientific notation, 102
  square roots, 58
Calendar, 611
Capacity, 150
Career Connection
  agriculture, 485
  aircraft designer, 149
  architect, 93
  astronauts, 3
  automotive design, 369
  cartography, 251
  fashion, 49
  optics, 427
  sound engineering, 603
  sports performance, 317
  statistician, 199
  water engineering, 543
Car maintenance, 22
Cell, 30
Celsius, 40, 53, 353
Center of rotation, 121, 340
Centi-, 151, 153
Centimeter, 94, 151
Central angle, 387

Central tendency, measures of, 222
Chemistry, 572
Chord, 448
Circle(s)
  area of, 64
  center of, 448
  chord of, 448
  circumference of, 64
  concentric, 451
  diameter of, 448
  radius of, 448
  right angle property of, 449
Circle graph, 200–202
Circumference, 64
Coefficient, 371
Combining like terms, 371–372
Common factor, 515
Communicating mathematics, 6, 9, 12, 20, 26, 27, 35, 36–37, 54, 56, 62, 74, 83, 98, 106, 110, 119, 127, 133, 139, 140, 155, 157, 171, 176, 184, 204, 206, 211, 219, 225, 227, 232, 233, 264, 265, 271, 272, 273, 274, 279, 286, 306, 307, 321, 329, 338, 346, 353, 359, 361, 376, 377, 382, 384, 389, 390, 391, 398, 404, 410, 432, 433, 434, 441, 447, 452, 459, 492, 504, 505, 510, 511, 512, 518, 519, 533, 578, 593, 594, 619, 623, 631
Communications, 206
Commutative Property of Addition, 456
Commutative Property of Multiplication, 456
Complement
  of angle, 124
  of set, 437
Complementary angles, 124
Composite number, 494
Compound interest, 73, 561, 619
Computer. *See* Spreadsheets, Technology
Concentric circles, 451
Cone, 582–583
Congruent
  angles, 320
  segments, 96
Consecutive vertices, 327
Constructions, 129–132

Consumer math, 15, 20, 22, 32, 54, 55, 56, 63, 67, 70, 74, 83, 99, 110, 111, 154, 155, 156, 161, 162, 163, 170, 189, 191, 237, 259, 281, 287, 382, 391, 459, 468, 557, 564, 565, 571, 611, 617, 619, 631, 633
Convergent iterations, 625
Convex polygon, 139
Coordinate, 254
Corresponding angles, 186–187
Corresponding sides, 186–187
Counting numbers, 442, 486
Cross multiplying, 178–179
Cylinder, 581

D

Data analysis
  box-and-whisker plot 229–230
  graphs, 200–202, 207–208, 213–215
  investigation, 220, 476
  measures of central tendency, 222
  stem-and-leaf plot, 228–229
Decimals
  fraction and percent equivalents, 72, 654–655
  repeating, 77, 443
  terminating, 77, 443
  vs. fractions, 412
Deficient number, 515
Degree (measure), 123
Degree of accuracy, 107
Denominator, 51
Density, 166, 169
Design, 346
Diagonal, 95, 327
Diameter, 448
Digraph, 236
Dimensional analysis, 172
Dimensions, 234
Discount, 73, 391
Disjoint
  groups of data, 200
  sets, 443
Distance, 297
Distance formula, 463
Distributive property, 373
Divisibility, 501–503
Division
  of fractions, 652–653
  as inverse operation, 115

in order of operations, 51
of signed numbers, 277

E

Early civilization, 128
Edge, 573
Element (of set), 428
Empty set, 429
Encasement, 357
Endpoints, inequalities, 260
English system, 94, 100
Enlargement, 185–186
Environmental science, 76, 104, 191, 287
Equal matrices, 235
Equations, 114–117
   absolute value, 392
   containing fractions, 523
   estimating solutions, 115
   graphing, 469–472, 589–592
   proportions, 178–180
   representing problems, 379–380
   with several operations, 386–387
   with several variables, 566–568
   simplifying to solve, 393–394
   solution of, 114
   solving, 385–387
   with square roots, 393
   substitution, to solve, 406–407
   with variables on both sides, 553–555
Equiangular, 137
Equilateral triangle, 137
Equilibrium, 266
Equipment design, 317
Equivalent
   expressions, 370–371
   fractions, decimals, percents, 72, 654–655
Escape velocity, 3
Escher, M.C., 346
Estimating, 107
   applications, 110, 111, 127, 128, 134, 164, 167, 168, 169, 174, 182
   measure of angles, 122
   solutions, 115
Euler primes, 528
Evaluating expressions, 50–52
Exponent, 57, 658–659

Expressions
   adding and subtracting, 370–372
   to describe phrases, 378–380
   equivalent, 371
   simplifying, 371–372
Extremes (data), 228

F

Face, 573
Factor, 494
   common, 515
   greatest common, 515, 521–522
Factorial, 461, 491
Factorization, prime, 507–509, 521–522
Fahrenheit, 40, 53, 353
Family tree, 16–17
Farming, 485
Fibonacci sequence, 377
Figurate numbers, 529
Final ray, 121
Finance, 75, 416, 527, 563
Floor plan, 184, 309
Formulas
   area, 64, 117, 566, 568
   circumference, 64
   for converting Celsius to Fahrenheit, 353
   for converting Fahrenheit to Celsius, 53
   density, 166
   distance, 463
   interest, 415
   midpoint, 464
   perimeter, 52, 413, 566
   Pythagorean Theorem, 65
   surface area, 381
   volume, 381, 573
Fractal, 635
Fraction bar, 51
Fractions. *See also* Ratio
   addition of, 6, 649–650
   converting improper, 647–648
   converting mixed numbers, 647–648
   as decimal, 72, 654–655
   decimal and percent equivalents, 72, 655
   division of, 652–653
   equations containing, 523
   equivalent, 647–648
   investigation, 447
   multiplication of, 43, 651–652

   as percent, 71–72, 654–655
   as probability, 23
   as proportion, 178
   as rate, 159
   as ratio, 158
   simplifying, 515, 522, 647–649
   subtraction of, 649–651
Fuel efficiency, 369
Function(s)
   defined, 413
   evaluating, 414–415
   input, 413
   notation for iteration, 613
   output, 413
Fundamental Counting Principle, 486

G

Gasoline mileage, 301
Geography, 100, 330, 468
Geometric properties, 136–137
Geometry
   acute angle, 110
   adjacent angles, 319
   area formulas, 64
   bisect, 348
   central angle, 387
   chord, 448
   circumference, 64
   complementary angles, 124
   concentric circles, 451
   cone, 582–583
   congruent angles, 320
   congruent segments, 96
   constructions, 129–132
   convex polygon, 139
   corresponding angles, 186–187
   corresponding sides, 186–187
   cylinder, 581
   diameter, 448
   equilateral triangle, 137
   geometric properties, 136–137
   hypotenuse, 65
   investigations, 41, 135, 141
   isosceles trapezoid, 349
   isosceles triangle, 347
   kite, 349
   lateral area, 574, 581–582
   leg, 65
   midpoint, 304
   obtuse angle, 110
   parallel lines, 318
   parallelogram, 326
   perimeter, 52, 95, 413, 566

perpendicular lines, 318
polyhedra, 573, 580
prism, 573
protractor, 122
pyramid, 575
Pythagorean Theorem, 65
quadrilaterals, 95, 325–328
radius, 448
ray, 121
rectangle, 36, 52, 64, 413, 566
regular polygon, 96
rhombus, 326
right angle, 110
right triangle, 65
similar polygons, 186–187
sphere, 583–584
square, 61, 325
straight angle, 124
supplementary angles, 124
surface area, 574, 581–582
tessellations, 9, 331
trapezoid, 327
vertical angles, 319
Goldbach's Conjectures, 528–529
Golden rectangle, 93
Gram, 151
Graphing
    equations, 469–472, 589–592
    geometric figures, 303–304
    inequalities, 260–262
    table of values, 469–471
    two equations, 589–592
Graphs (in data analysis)
    artistic, 215
    bar, 207
    circle, 200–202
    histogram, 214
    line, 208
    pictograph, 214
    scattergram, 213
Graphs (investigation), 595
Greatest common factor (GCF), 515, 521–522
Grouping symbols, 51
Guess and check, 115

H

Health, 73, 105, 390
Height of a triangle, 64
Heptagon, 95
Hexagon, 95, 137
Histogram, 214
History, 33, 106, 128, 162, 199
Hypotenuse, 65

I

Image, 340
Index, root, 58
Industrial arts, 337
Inequalities
    graphing, 260–262
    solving, 399–401
    symbols, 261
Infinite sets, 434
Initial ray, 121
Input, 413
Integers, 254. *See also* Signed numbers
Interest
    compound, 73, 561
    formula, 415
Intersection of sets, 435
Inverse operations, 115, 385–386
Investigations
    algebra, 63
    approximating π, 454
    architecture, 324
    auto trip, 171
    balance, 558
    banking, 274
    baseball, 227
    calendar, 611
    chemistry, 572
    communications, 206
    consumer math, 391, 633
    data analysis, 220, 476
    decimals, 77
    decimals *vs.* fractions, 412
    early civilization, 128
    English system, 100
    equilibrium, 266
    equipment design, 317
    escape velocity, 3
    farming, 485
    Fibonacci sequence, 377
    floor plans, 184, 309
    fractions, 447
    fuel efficiency, 369
    gasoline mileage, 301
    geography, 468
    geometry, 41, 135, 141,
    Golden rectangle, 93
    graphs, 595
    history, 106, 199
    infinite sets, 434
    in the news, 113
    language, 461, 552
    language arts, 241
    lattice points, 527
    leap year, 28
    lift-to-drag ratios, 149
    limits, 441
    logos, 338
    lotteries, 493
    M.C Escher, 346
    map coloring, 251
    mass and weight, 157
    mental math, 56, 619
    merchandising, 120
    music, 603
    number game, 512
    numeric palindrome, 506
    origami, 361
    optical illusion, 427
    patterns on a telephone, 15
    physics, 281
    playing-card symmetries, 353
    polyhedra, 580
    prime numbers, 535
    radar guns, 164
    ratio, 588
    relatively prime numbers, 520
    rent costs, 419
    science, 384, 398
    selling prices, 49
    similar polygons, 191
    snowflake curve, 624
    sports, 85, 212
    spreadsheets, 35
    statistics, 233
    technology, 22, 639
    teeter-totter principle, 565
    temperature scales, 259
    tessellations, 9, 331
    tile arrangements, 500
    time differences, 295
    triangle inequality, 405
    using a matrix, 289
    volume, 70
    water conservation, 543
    water consumption, 177
Irrational number, 443
Isosceles trapezoid, 349
Isosceles triangle, 347
Iteration
    convergent, 625
    diagram, 605
    fractal, 635
    function notation for, 613
    geometric, 620–622
    limiting value of, 627
    rule, 613

K

Kilo-, 151, 153
Kilogram, 151
Kilometer, 151
Kite, 349
Koch, 636

L

Language, 461
Language arts, 241, 330, 438
Lateral area, 574, 581–582
Lateral faces, 574
Lattice points, 527
Leap year, 28
Least common denominator, 496, 649–650
Least common multiple (LCM), 495–496, 521–522
Leg (of triangle), 65
Legend, 165
Lift-to-drag ratios, 149
Light-year, 102
Like terms, 371
Limiting value, 627
Limits, 441
Linear measurement, 94
Line
  graph, 208
  plot, 214
  segment, 36
  of symmetry, 333
Lines, parallel and perpendicular, 318
Liter, 151
Logos, 338
Lotteries, 493
Lower quartile, 229

M

Map coloring, 251
Mass and weight, 157
Matrices
  addition of, 269
  definition of, 234
  equal, 235
  investigation, 289
  multiplication of, 282–285
  multiplication of, by scalar, 276
  subtraction of, 291
  vertex, 355
Matrix multiplication, 282–285, 355–356
Mean, 221
Measurement
  angle, 121–124
  English system of, 94
  linear, 94
  metric system of, 94, 151
  operations with, 152
  of perimeter, 52, 95
  ratios of, 158–160
  of volume, 150
Measures of central tendency, 222
Median, 221
Mental math, 56, 377, 619
Merchandising, 120
Meter, 151
Metric system, 94
Metric units, 151
Midpoint, 304
Midpoint formula, 464
Midquartile range, 230
Milli-, 151, 153
Millimeter, 151
Mode, 222
Multiples, 495–496
Multiplication
  Associative Property of, 455
  Commutative Property of, 456
  Distributive property, 373
  of fractions, 43, 651–652
  as inverse operation, 115
  of matrices, 282–285, 355–356
  in order of operations, 50–51
  powers of signed numbers, 276
  scalar, 276
  of signed numbers, 275
Music, 603

N

Natural-number multiples, 495
Natural numbers, 442, 486
Negative signed numbers, 252
News, in the, 113
Nonagon, 95
Notation, scientific, 101–102, 658–659
Number(s)
  abundant, 515
  binary, 492
  composite, 494
  counting, 442, 486
  deficient, 515
  figurate, 529
  integers, 254, 428
  irrational, 443
  natural, 442, 486
  negative signed, 252
  perfect, 514
  positive signed, 252
  powers of signed, 276
  prime, 494, 535
  rational, 442
  real, 443
  relatively prime, 520, 522
  translation, 341
  whole, 442
Number game, 512
Number line, 252–254
Numerator, 51
Numerical terms, 371
Numeric palindrome, 506
Nutrition, 75

O

Obtuse angle, 110
Octagon, 95
Operation symbols, 52
Opposites, 253
Opposite sides, 325
Opposite vertices, 327
Optical illusion, 427
Ordered pair, 302
Order of operations, 51, 59, 659
Origami, 361
Origin, 254, 334
Outliers, 230
Output, 413

P

Palindrome, numeric, 506
Parallel lines, 318
Parallelogram, 326
Patterns, 10–11, 16–17
Patterns on a telephone, 15
Pentagon, 95
Percent, 71
  calculating mentally, 72, 74
  of change, 278
  changing fraction to, 71
  compound interest, 73
  discount, 73
  fractional and decimal equivalents, 72, 656–657
  rewriting as decimal, 71
  rewriting as fraction, 71
  solving problems with, 72–73
  using, 656–657
Perfect number, 514
Perimeter,
  of polygon, 95
  of rectangle, 52, 413, 566
Perpendicular lines, 318
Physics, 281
Pi, 65, 448, 454
Pictograph, 214
Pie chart, 200

Place value, 646
Playing-card symmetries, 353
Polygon(s)
    classification of, 95
    convex, 139
    diagonal of, 95
    perimeter of, 95
    quadrilaterals, 325–328
    regular, 96
    sides of, 95
    similar, 186–187, 191
    sum of angles of, 136–137
    vertex of, 95
Polyhedra, 573, 580
Positive signed numbers, 252
Postal rates, 29, 33
Power, 57
Powers of signed numbers, 276
Preimage, 340
Prime factorization, 507–509, 521–522
Prime number, 494, 535
Prism, 573
Probability, 23–24
Property
    Associative, of Addition, 455
    Associative, of Multiplication, 455
    of circle, 449–450
    Commutative, of Addition, 456
    Commutative, of Multiplication, 456
    Distributive, 373
    geometric, 137–138, 347–348
    of real numbers, 455–457
    right angle, 449
    zero product, 456–457
Proportion, 178–180, 653–654
Protractor, 122
Pyramid, 575
Pythagorean Theorem, 65
Pythagorean triples, 530–531

Q

Quadrant, 302
Quadrilaterals, 325–328
    isosceles trapezoid, 349
    kite, 349
    parallelogram, 326
    rectangle, 36, 52, 64, 325, 413, 566
    rhombus, 326
    square, 325
    trapezoid, 327
Quartile, 229
Quotient, 277

R

Radar guns, 164
Radical sign, 58
Radicand, 58
Radius, 448
Range, 40, 228
Rate, 158–159
Ratio, 158, 653–654
    applications, 165–166
    investigation, 588
    as probability, 23–24
    as proportion, 178
    scale factor, 185–186
Rational number, 442
Ray, 121
Real numbers, 443
Reciprocal, 381
Recommended Daily Allowance (RDA), 73
Rectangle, 36, 325
    area of, 64
    perimeter of, 52, 413, 566
Rectangular coordinate system, 302
Reduction, 185–186
Reflection, 339, 355–356
Reflection symmetry, 333
Regular polygon, 96
Relatively prime numbers, 520, 522
Rent costs, 419
Rhombus, 326
Right angle, 110
Right angle property, 449
Right triangle, 65
Root, 58
Root index, 58
Rotation, 340, 355–356
Rotation symmetry, 334
Rounding, 37, 107–108, 646–647

S

Sales tax, 43
Scalar, 276
Scalar multiplication, 276
Scale, 165
Scale factor, 186
Scattergram, 213
Science, 18, 39, 40, 53, 60, 102, 103, 104, 111, 168, 169, 177, 210, 212, 217, 218, 232, 238, 256, 353, 384, 398, 404, 453, 475, 492, 499, 526, 564, 570, 572, 579, 580, 586, 587, 593, 594, 610, 611, 631, 632
Scientific notation, 101–102, 658–659
Self-similarity, 635
Selling prices, 49
Set(s)
    complement of, 437
    definition of, 428
    disjoint, 443
    element of, 428
    empty, 429
    finite, 434
    infinite, 434
    intersection of, 435
    set-builder notation, 428
    solution, 430
    subset of, 429
    union of, 435
    Venn diagram, 436
Set-builder notation, 428
Side(s)
    adjacent, 325
    of angle, 122
    corresponding, 186–187
    opposite, 325
    of polygon, 95
Signed numbers
    addition of, 267–268
    calculator and, 254, 268
    division of, 277
    multiplication of, 275
    negative, 252
    positive, 252
    powers of, 276
    subtraction of, 290–291
Similarity
    defined, 186
    enlargement, 185–186
    geometric figures, 186–187
    reduction, 185–186
    scale factor, 186
Similar polygons, 191
Simplifying
    expressions, 371–372
    fractions, 515, 522, 647–649
Slant height, 575
Slide (transformation), 341, 355
Snowflake curve, 624
Social science, 211
Solution, 114
Solution set, 430
Solving equations, 114
    with absolute value, 392
    estimating solutions, 115

using inverse operations, 385–386
with several variables, 566–568
with square roots, 393
using substitution, 406–407
with variables on both sides, 553–555
Speed, 159
Sphere, 583–584
Sports, 15, 69, 85, 205, 212, 237, 257, 288, 336, 416, 474, 491, 533, 571
Spreadsheets, 7, 8, 13, 14, 19, 28, 30–31, 32, 34, 35, 38, 40, 55, 61, 75, 76, 78–80, 82, 83, 85, 100, 106, 111, 112, 118, 134, 135, 139, 156, 157, 163, 190, 191, 205, 224, 225, 265, 272, 280, 324, 330, 353, 397, 411, 418, 446, 493, 499, 506, 518, 519, 526, 552, 609
Square, 61, 325
Square root, 58, 657
Statistics, 233. *See also* Data analysis
Stem-and-leaf plot, 228–229
Stock, 277–278, 280, 284
Straight angle, 124
Subset of sets, 429
Substitution, to solve equations, 406–407
Subtraction
 of fractions, 648–649
 as inverse operation, 115, 117
 of like terms, 371–372
 of matrices, 291
 of signed numbers, 290–291
 to solve inequalities, 399
Supplement (of angle), 124
Supplementary angles, 124
Surface area, 381, 574, 581–582
Surveying, 466
Symbols
 absolute value, 296
 angle, 122
 circle, 448
 grouping, 51
 inequalities, 37, 261, 401
 intersection of sets, 435
 "is approximately equal to", 58
 "is congruent to", 133, 320
 line, 320
 line segment, 187
 operation, 52

parallel, 318
percent, 71
perpendicular, 318
ray, 319
similarity, 185–187
square root, 58
subset, 429
union of sets, 435
variable, 114
Symmetric primes, 528
Symmetry, 332
 line of, 333
 reflection, 333
 rotation, 334
System of equations. *See* Graphing, two equations

T

Table (chart), 29
Table of values, 469, 472
Technology, 22, 26, 177, 259, 447, 461, 493, 639
 See also Calculator, Iterations, Spreadsheets
Teeter-totter principle, 565
Temperature
 converting from Celsius to Fahrenheit, 353
 converting from Fahrenheit to Celsius, 53
 of human body, 53
 scales, 259
Terminating decimals, 77
Terms
 combining like, 371–372
 like, 371
 numerical, 371
Tessellations, 9, 331
Tile arrangements, 500
Tiling, 4–5
Time differences, 295
Time zones, 255
Transformations
 reflection, 339, 355–356
 rotation, 340, 355–356
 translation, 341,355
Translation, 341, 355
Translation numbers, 341
Trapezoid, 66, 327
Tree diagram, 487
Triangle(s)
 area of, 64, 117, 566
 equilateral, 137
 isosceles, 347
 right, 65
 similar, 186–187

sum of angles in, 136
Triangle inequality, 405

U

Union of sets, 435
Unit(s)
 base, 159
 costs, 159
 cubic, 150
 metric, 151
 square, 150
 working with, 172–173
Upper quartile, 229
Using a matrix, 289

V

Value
 of expression, 50
 of functions, 414
 place, 646
Variable, 114
Venn diagram, 436, 522
Vertex (Vertices)
 of angle, 121
 consecutive, 327
 of image, 354–355
 opposite, 327
 of polygon, 95
 of polyhedron, 573
Vertex matrix, 355
Vertical angles, 319
Vinculum, 51
Volume, 150
 of cone, 582
 of cylinder, 190, 581
 investigation, 70
 of prism, 574
 of pyramid, 575
 of rectangular solid, 150, 381, 573
 of sphere, 583

W

Water conservation, 543
Water consumption, 177
Web diagram, 630
Whole numbers, 442
Word chain, 367
Word problems, 544–548, 559–561
 age, 526, 547, 550
 averages, 222–225
 balance, 567
 banking, 269
 bar graph, 207–211

circle graph, 201–202, 204
coin, 590
cost, 282–283, 396, 553, 560
distance, 412, 546
dollar gain or loss, 277, 284
earnings, 415, 472
interest, 73, 561, 604, 619
line graph, 210
matrices, 312–313
percent, 72–73, 201–202
percent of change, 277–278
probability, 23–24, 490, 550
proportion, 182
rate, 160, 163
ratio, 169
temperature, 268, 316, 342
work, 523, 548

X

x-axis, 302
x-coordinate, 302

Y

y-axis, 302
y-coordinate, 302

Z

Zero Product Property, 456–457

Credits

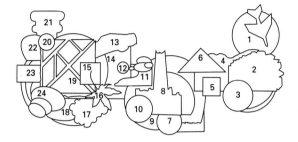

Cover

Collage by Carol Tornatore

1 The Earth as seen from space. © Telegraph Colour Library/FPG International; **2** Ferris Wheel. © SUPERSTOCK; **3** French Crown paperweight (blue cane with pink and green and blue and red ribbon twists) glass, diam.: 8.2 cm. The Art Institute of Chicago. Bequest of Arthur Rubloff, 1988.541.265; **4** Death Valley Sand Dunes. © Richard Pharaoh /International Stock Photography; **5** Quilt. Debora J. Bass; **6** Glass Pyramid at the Louvre, Paris, by I.M. Pei. © Telegraph Colour Library/FPG International; **7** Compact Disc. © Pete Saloutos/The Stock Market; **8** Golden Gate Bridge, San Francisco. © 1989 Luis Castañeda/The Image Bank. **9** Fiber Optics Bundle. © TSW; **10** French, Paperweight (green, red, and white swirls), Clichy, glass, diam.: 6.6 cm. The Art Institute of Chicago. Gift of Arthur Rubloff, 1977.825. Photograph by Christopher Gallagher; **11** Rippling water. © Comstock, Inc.; **12** Contemporary Marble. © David Young Wolff/PhotoEdit; **13** Silver Porsche 959. © 1989 Ron Kimball Photography; **14** *Fall*. 1988. Hand printed paper. Margaret Ahrens Sahlstrand; **15** *Rainbow Stars* pieced quilt. Jean Ray Laury. Photograph by Stan Bitters; **16** Orange Leaf. © Telegraph Colour Library/FPG International; **17** Mola, from Cuna Indians of Panama © Carmine Fantasia; **18** Epcot Center. © Carmine Fantasia; **19** Honey Bees. © Hans Pfletschinger/Peter Arnold, Inc.; **20** Antique Marble. Courtesy, Paul C. Baumann, Author, *Collecting Antique Marbles;* **21** Ionic Column, New Orleans. © W. Cody/Westlight; **22** Chambered Nautilus Shell Cross Section. © Louise K. Broman/Photo Researchers; **23** Circuit Board. © John Michael/International Stock Photography; **24** Solar Energy Collectors, Australia. © 1987 Otto Rogge/The Stock Market.

Fine Art and Photographs

ix Rippling Water. © Comstock, Inc.; **vi** (above) *Rainbow Stars* pieced quilt. Jean Ray Laury. Photograph by Stan Bitters; **vi** (below) Zodiac. Venice, Italy. © SUPERSTOCK; **vii** Glass Pyramid at the Louvre, Paris, designed by I.M. Pei. © Telegraph Colour Photography/FPG International; **viii** Space shuttle astronaut during space walk. NASA; **x** (above) The Earth as seen from space. NASA; **x** (below) Circuit Board. © John Michael/International Stock Photography; **xi** Taj Mahal, Agra, India. © Telegraph Colour Library/FPG International; **xii** (above) Death Valley Sand Dunes. © Richard Pharaoh/International Stock Photography; **xii** (below) Circuits printed on Computer Chip (200% enlargement). © Phil Degginger/TSW; **xiii** Fiber Optics. © TSW; **xiv** Compact Disc. © Robert George Young /Masterfile; **xv** Solar Energy Collectors, Australia. © Otto Rogge/The Stock Market; **xvi** St. Louis Arch. © Tom Till/ International Stock Photography; **2–3** *Rainbow Stars* pieced quilt. Jean Ray Laury. Photograph by Stan Bitters; **3** (inset, right) Space shuttle astronaut during space walk. NASA; **3** (inset, left) Space shuttle astronaut in pilot's station. NASA; **4** Tile mosaics in a mosque, Casablanca, Morocco. © Lisl Dennis/The Image Bank; **9** Rice Packets. © Guido Alberto Rossi/The Image Bank; **10** *Numbers in Color*, (detail) 1958–59, Jasper Johns. Albright-Knox Art Gallery, Buffalo, New York. Gift of Seymour H. Knox; **16** Bees on Honeycomb. © TSW; **21** Ancient Ruins at the Forum, Rome. © SUPERSTOCK; **23** from *Treatise on the Game of Chess* (Persian Manuscript 211). Royal Asiatic Society; **28** Aztec Calendar cartoon. © 1992 Nick Downes; **29** Stamp Mosaic. © Adrienne McGrath; **36** Crazy Quilt. © Adrienne McGrath; **41** Spider Web. © 1990 Tom Bean/The Stock Market. **48–49** Computer-aided-design workstation. Courtesy, Gerber Garment Technology; **49** (inset, left) Computer-aided-design program. © Art Montes de Oca; **49** (inset, right) Pattern cutting in clothing factory. © Bill Bachmann/Uniphoto; **50** Foreign currency. © Viesti Associates, Inc./The International Exchange Photo Library; **57** Wind generators, Altamont Pass, California. © Kevin Schafer/Peter Arnold, Inc.; **64** Pythagorean Theorem. © Adrienne McGrath; **66** Aerial photo of the Orange Bowl, Miami. © Walter Iooss, Jr./The Image Bank; **71** Varieties of apples. © Michel Viard/Peter Arnold, Inc.; **77** Sale sign. © Peter Southwick/Stock Boston Inc.; **78** Compact discs. © Ted Horowitz/The Stock Market; **84** Jim Abbott. © M. Grecco/Stock Boston Inc.; **92–93** Houston, Texas cityscape. © Superstock International Inc.; **93** (inset) *View of an Ideal City*, late 15th century, artist unknown. Central Italian School. The Walters Art Gallery, Baltimore; **94** Roger Bannister. © AP/Wide World Photos; **101** The Pleiades. © Hale Observatories/ Photo Researchers; **107** Fiesta, San Antonio, Texas. © Bob Daemmrich/Stock Boston Inc.; **114** Man with yoke on path beside rice paddies. © Margaret Gowan/TSW; **121** Hang gliding, Aspen, Colorado. © D. Brownell/The Image Bank; **129** *Cube and Four Panels*, 1975, Ron Davis. Albright-Knox Art Gallery, Buffalo, New York. By exchange, National Endowment for the Arts Purchase Grant and Matching Funds, 1976; **135** Photographed from Oliver Byrne's *The Elements of Euclid*, published by William Pickering, 1847; **136** Kite. © Bill Ross/Westlight; **141** Cotton harvest, Arizona. © Jim Sanderson/The Stock Market; **148–149** United States Air Force Thunderbirds F-16 Fighting Falcons flying in formation. © 1991 Jordan Coonrad; **149** (inset, left) Boeing 737 jetliners, Boeing assembly plant, Renton, Washington. © 1982 David Barney/The Stock Market; **149** (inset, right) Electronic fuel control for aircraft. © 1982 Gabe Palmer/The Stock Market; **150** Attribute tiles; **158** *Out of This World* video game board. Courtesy, Brian Fargo, Interplay Productions; **165** Map of Nigeria and surrounding countries; **172** Beakers, flasks, and graduated cylinders. © 1988 Blair Seitz/Photo Researchers; **177** Sears Tower, Chicago. © Frank Cezus/FPG International; **178** Mural. Courtesy, Mexican Fine Arts Center Museum, Chicago; **179** Mac art of Mural. Joe Terrasi; **185** Russian nesting dolls. © Carmine Fantasia; **198–199** Democratic National Convention, Atlanta. © 1988 Christopher Morris/Black Star; **199** (inset, left) Harry Truman holding erroneous Dewey victory newspaper. The Bettmann Archive; **199** (inset, right) Exit polling. © Bob Daemmrich/Stock Boston Inc.; **200** Pizza segments. © Lois Ellen Frank/Westlight; **207** Earthquake photos. © Bill Ross/Westlight; **213** Aerial shot of cars in parking lot. © Ronn Maratea/International Stock Photography; **215** Map: *Rain Forest Priorities*, Dave Herring. Courtesy, *World Monitor: The Christian Science Monitor Monthly* © 1991; **218** Trash illustration by Pierre Mion. © 1991 National Geographic Society; **221** Group of blue men, one white. © Bruce Rowell/Masterfile; **228** Boxes of fish, Farmer's Market, Seattle. © Morton Beebe/The Image Bank; **233** © 1989 by Sidney Harris from *Einstein Simplified*, Rutgers University Press; **234** Attribute blocks; **250–251** Earth wind-speed patterns from space. © Telegraph Colour Library/ FPG International; **251** (inset, left) Multi-colored world map with time zones. © TSW; **251** (inset, right) Man plotting data on a sonar-produced ocean map. © Franz Edson/TSW; **252** Roller coaster. © M. Timothy O'Keefe/Tom Stack & Associates; **260** *Blue Package with Ostrich Eggs*, 1971, Claudio Bravo. Courtesy, Staepfli Gallery, New York; **267** *Collection of Southern Signs*, 1924, Paul Klee. Washington University Gallery of Art, St. Louis; **275** Electronic stock ticker board. © C.P. George/H. Armstrong Roberts; **282** Double cheeseburger. © Karen McCunnell/Leo de Wys, Inc.; **289** Electronic brain for fuel injection. © Brownie Harris/The Stock Market; **290** Soo Canal locks, Sault Ste. Marie, Michigan. © 1992 Adrienne S. McGrath; **296** Weather instrument. Measures rainfall, temperature, wind direction, and wind speed. © Barry L. Runk/Grant Heilman Photography, Inc.; **302** Blue subway car reflection. © Joe Marvullo; **316–317** Multiple exposure bicycle photograph. © 1989 Globus Brothers/The Stock Market; **317** (inset) Wheelchair racing. © David Young Wolff/PhotoEdit; **318** Building based on crystal shapes, Poitiers, France. © Tony Craddock/TSW; **325** Lumber. © 1989 Albert Normandin/The Image Bank; **332** Abstract computer graphic. © Larry Keenan Assoc./The Image Bank; **339** Skyline with reflections. © Mark Stephenson/Westlight; **347** *Composition 8*, 1923, Vassily Kandinsky. Solomon R. Guggenheim Museum, New York. Gift, Solomon R. Guggenheim, 1937. © The Solomon R. Guggenheim Foundation, New York; **354** Origami. Kimi Igarashi; **368–369** Electric car dealership sign. © Comstock, Inc.; **369** Car engineer working on computer-aided designs. © Andrew Sacks/TSW; **369** Anecoic test chamber, Japan. © Ken Straiton/The Stock Market; **370** Tomb paintings, Valley of the Kings, Deir EL Bahari, Egypt. © Superstock International Inc.; **378** Basketball game with scoreboard. © D. Logan/H. Armstrong Roberts; **385** *Far Side* cartoon by Gary Larson. Reprinted by permission of Chronicle Features, San Francisco; **391** Matrix; **392** *Screwarch Bridge (State II)*, 1980, Claes Oldenburg. The Museum of Modern Art, New York. Etching and aquatint 203 11/16 x 50 3/4 in. Gift of Klaus G. Perls and Heinz Berggruen in memory of Frank Perls, art dealer (by exchange). Photograph © 1992 The Museum of Modern Art, New York; **399** Pyramids, Egypt. © Camerique; **406** Triazzle. Courtesy Dan Gilbert and DaMert Company © 1991; **413** Oyster pearl beds, Japan. © Comstock, Inc.; **426–427** Prism. © Ted Mahieu/The Stock Market; **427** (inset, top left) Ophthalmoscope. © 1991 Larry Gatz/The Image Bank; **427** (inset, bottom right) Bar code. © David Young Wolff/PhotoEdit; **428** Sand buckets. © David Young Wolff/PhotoEdit; **435** *Several Circles* (detail), 1926, Vassily Kandinsky. Solomon R. Guggenheim Museum, New York. Gift, The Solomon R. Guggenheim, 1941. © The Solomon R. Guggenheim Foundation, New York. Photograph by David Heald; **442** Tape measure. © Murray Alcosser/The Image Bank; **448** *The Red Sun Gnaws at the Spider*, 1948, Joan Miró. Private collection, Japan; **455** Commuters at rush hour, New York. © J. Ramey/The Image Bank; **462** The Isometric Map of Midtown Manhattan. © 1989 The Manhattan Map Co. All Rights Reserved; **469** The Romerberg facade, Frankfurt, Germany © Terry Williams/ The Image Bank; **484–485** Aerial view of wheat harvest. © Andy Sacks/TSW; **485** (inset) Produce market, Bali, Indonesia. © Steve Satushek/The Image Bank; **486** Pond study at outdoor education camp, Texas. © Bob Daemmrich/ Stock Boston Inc.; **494** Tiles; **501** Pastry slices. © Nino Mascardi/The Image Bank; **507** © 1977 by Sidney Harris from *What's So Funny About Science?* William Kaufmann, Inc.; **513** *School of Athens* (detail), Raphael. Vatican Rooms, Scala/Art Resource, New York; **521** Watch gears. © Bernard Van Berg/The Image Bank; **528** Mayan Glyphs, Chiapas, Mexico, A.D. 600–900. © Robert Frerck/Odyssey, Chicago; **542–543** Glen Canyon Dam, Arizona. © L. Jacobs/H. Armstrong Roberts; **543** (inset, lower left) Aerial view, sewage treatment plant. © Steve Proehl/The Image Bank; **543** (inset, upper right) Engineer testing water at sewage treatment plant. © 1982 Chris Jones/The Stock Market; **546** Fence painting contest. Courtesy Hannibal, Missouri, Visitor's Bureau; **553** Llama. © Kahl/Bruce Coleman, Inc.; **559** Cherry chocolate turtles and almond bark. © Skip Dean/The Image Bank; **566** Postage stamps from various countries. ©1989 Ed Bohon/The Stock Market; **573** Glass pyramid at the Louvre, Paris by I.M. Pei. © Walter Iooss, Jr./The Image Bank; **581** Cultivated flowers against geometric skyline. © Pete Turner/The Image Bank; **589** Golf balls. © The Harold E. Edgerton 1992 Trust. Courtesy of Palm Press, Inc.; **602–603** Recording studio sound-mixing console. © Ted Horowitz/The Stock Market; **603** (inset, left) Pythagoras' discovery of arithmetic ratios in stringed instruments, 1492. Woodcut. The Granger Collection; **603** (inset, right) Recording studio. © E. Masterson/H. Armstrong Roberts; **604** Ahu Akivi statues, Easter Island. © Harald Sund/The Image Bank; **612** *Government House Fort and Dalhousie Barracks*, Calcutta. The Granger Collection; **620** *Cubic Space Division* (detail), 1952 by M.C. Escher. Cornelius Van S. Roosevelt Collection. © 1992 National Gallery of Art, Washington, D.C.; **625** *Michigan Avenue with View of the Art Institute*, 1984, Richard Estes, American, b. 1937, oil on canvas 91.4 x 121.9 cm, Gift of the Capital Campaign Fund, 1984.177. © 1991 The Art Institute of Chicago. All Rights Reserved; **634** Fractal. © Homer W. Smith/Peter Arnold, Inc.

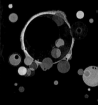

435

ADDITIONAL ANSWERS

Chapter 1

Page 7

6

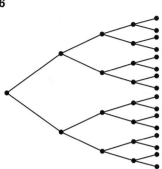

7 a $\frac{3}{4}$ **b** $\frac{7}{8}$ **c** $\frac{15}{16}$ **d** $\frac{31}{32}$ **e** $\frac{63}{64}$

Possible pattern: The denominator of each sum is the greater of the denominators of the addends, and the numerator of each sum is 1 less than the sum's denominator.

8 Possible answer: Add to the preceding figure one tile at the right end of each horizontal row and one tile in the middle of the vertical column.

11 a

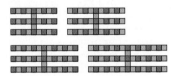

12 Possible answer: In the *n*th figure, there are *n* columns of tiles, with *n* tiles in the first column and with the number of tiles in each succeeding column increasing by 1; 92 tiles

Pages 20 and 21

20 a Possible answers: $0.10–$0.29; $0.30–$0.49; $0.50–$0.69; $0.70–$0.89; $0.90–$1.09

b Possible answer: Multiply the number of whole dollars in the purchase price by 0.05. Then use the answers to part **a** to determine how much more tax is owed on the remaining cents in the purchase price.

22 Possible answer: The number of green tiles in the *n*th figure is $(n-1)^2$. The number of yellow tiles in the *n*th figure is $4 \times n$. The total number of tiles in the *n*th figure is $(n+1)^2$.

27 b Possible answer: The expression *m # n* seems to represent the value of the least common multiple of the numbers *m* and *n*.

Page 22

30 Possible answer: Arnold was right, since we usually do not say that a person is 19 years old until the person has had a 19th birthday.

Investigation We read from top left to bottom right, so it makes sense for the numbers on a phone to be in the order that we read. The calculator, on the other hand, displays the numbers in order from bottom to top. This practice probably dates back to the old adding machines which, for mechanical reasons, had the numbers arranged in ascending order.

Page 28

Investigation Most years have 365 days, with 28 in February. Approximately every fourth year, February 29 is added, and the year has 366 days. Such years are called leap years. Every year that is divisible by 4 is a leap year, except for century years (1700, 1800, etc.), which are not leap years (except for century years divisible by 400, which are!). There are 24 leap years in the 100 years from 1800 to 1899, so the probability is .24.

Page 35

27

Investigation Answers will vary. For example, in a bicycle shop, an employee could use a spreadsheet to identify the following: item, description of item, manufacturer, quantity, price, discount, and cost of item.

Page 39

12 Possible answers: Because when you divide both the numerator and the denominator of $\frac{3}{6}$ by 3, you get $\frac{1}{2}$; because if you cut one pizza in half and an identical pizza into sixths, three of the sixths are the same amount of pizza as one of the halves

14 a Possible answer: The postal service's transportation and delivery costs are much the same for letters of any weight, so these costs must be included in the charge for a letter of the least weight.

b

Weight (ounces)	Cost (dollars)
1	0.35
2	0.60
3	0.85
4	1.10
5	1.35
6	1.60
7	1.85
8	2.10

c Combine them into a 7-ounce letter, which costs $1.85. Sending them separately would cost $1.95.

16 Possible answer: Multiply the length of a side by itself.

17 Possible answer: Multiply the area of a face by the length of an edge.

18 Possible answer: Enter 1 in cell A1 and A1 + 1 in cell A2, then copy the formula in A2 to A3–A7.

19 Possible answer: Enter 5 in cell A1 and A1 + 2 in cell A2, then copy the formula in A2 to A3–A7.

20 Possible answer: Enter 0.02 in cell A1 and 2 × A1 in cell A2, then copy the formula in A2 to A3–A7.

21 Possible answer: Follow the procedure given in the answer to problem 18 to get the whole numbers 1–7 in cells A1–A7, then enter 1 in cell B1, enter A2 × B1 in cell B2, and copy the formula in B2 to B3–B7.

Page 47

1 $9.00; The jeweler can open each link of the three-link piece (for $3.00) and then connect the remaining three lengths with those links (for $6.00).

2 35; Label each vertex of the pentagon and each point of intersection of the diagonals of the pentagon. Make an organized list of the different triangles in the figure.

3 Pick a marble from the bag labeled Blue/Green.

4

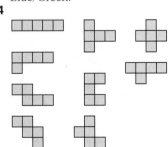

A1

ADDITIONAL ANSWERS

5 a 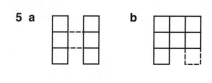 b

Chapter 2

Page 91

1 A Fool on the hill
2 A Stepladder
3 A What goes up must come down.
1 B Wide body
2 B Life of ease
3 B Search high and low
1 C Look both ways before crossing.
2 C Sidestep
3 C Life after death
1 D The plot thickens.
2 D Mixed-up kid
3 D The rise and fall of the Roman Empire
1 E Top of the morning
2 E Three blind mice
3 E Circles under the eyes
1 F Stairway to heaven
2 F Less is more.
3 F Bad investment

Chapter 3

Page 100

Investigation Possible answer: Inches and feet were used by the ancient Greeks and Romans. The word *inch* comes from the Latin word *uncia*, which means "twelfth part." The Greek word *pous* means "foot," which measured about 12.1 modern inches. Later, the French phrase *pied du roi* (foot of the king) measured about 12.8 inches. The word *yard* is derived from the middle English *yerde*, a measuring rod.

Pages 104 and 105

26 2.8455×10^{42}
31 a 1.7×10^{27} b 6×10^{53} c 4×10^{27}
34–36 See students' work.
39 Possible answer: Between 9 and 10
40 Possible answer: About 950
41 Possible answer: About 6.9
42 Possible answer: About 30
43 Possible answer: About 2
44 Possible answer: About 25
46 Expressions a, c, and d; Possible explanation: The simplified form of each expression results in the same number.
48 a Yogi: 50%; Ellie: 25%; Mickey: ≈16.7%

Page 113

Investigation Answers will vary. Some examples follow: Rounded number: 120,000 attend an outdoor concert. Exact number: a round-trip airline ticket is advertised at $775. Cannot be determined: Hotel accommodations would have cost $160 per night; a voter-registration drive signed up 60 new voters.

Page 147

1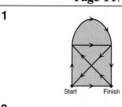

2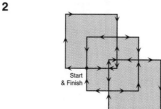

3 Cannot be done
4 5
5 13
6 Cheatem

Chapter 4

Pages 156 and 157

40 Possible answers given. Proposed situations will vary.
  a A very small unit of length (1×10^{-10} m) used to measure such things as wavelengths of light and distances on the atomic scale
  b A unit of capacity (4 pecks or 8 dry quarts) used mainly to measure amounts of grain and produce
  c A unit of heat most familiarly used as a measure of the energy-producing potential of foods
  d A unit of weight (200 mg) used as a measure of the size of diamonds and other gemstones
  e A unit of volume (128 ft^3) used to measure quantities of firewood
  f A unit of intensity most familiarly used to measure the loudness of sounds
  g A unit of length (6 ft) used mainly by mariners as a measure of water depth
  h A unit of length (220 yd) most familiarly used as a measure of lengths of horse races
  i A unit used to measure the amount of light passing through a given region of space
  j A unit of capacity (see part b above for equivalents) used mainly to measure small amounts of produce
  k A unit used to measure the intensity of x-rays and gamma radiation
  l A unit of work (about 1.34×10^{-3} horsepower) most familiarly used to measure the amount of electricity required to operate an electrical device
41 a 427.2 b 1495.2 c 1.8 d 513,280.8
   e −427.2 f −1495.2
44 Possible answer: To add the fractions, she had to express them as numbers of equal parts (that is, to rewrite them with a common denominator). Since multiplying a number by 1 does not change its value, she was able to change the denominators of the fractions by multiplying each by an appropriate form of 1.

Investigation The weight of an object depends on the force of gravity; for example, you would weigh less on the moon than you do on Earth, and in outer space (far from any stars or planets) you weigh almost nothing. Mass, on the other hand, indicates an amount of matter and is independent of any gravitational force. On the surface of Earth, mass and weight are usually interchangeable; if your mass is 60 kg, you weigh 60 kg. But on the moon, while your weight would drop to about 10 kg, your mass remains 60 kg.

Page 191

Investigation Students may observe that when triangles are similar the corresponding angles are equal, and the corresponding sides are proportional. With nontriangles, even if the angles are the same, the figures may not be similar (e.g., a square and a rectangle). All squares are similar, because corresponding angles are equal (they're all 90°) and corresponding sides are in proportion.

ADDITIONAL ANSWERS

Chapter 5

Page 203

10 Consumer Spending on Albums, Tapes and Compact Discs

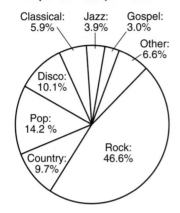

11 The Rock sector is much larger than the corresponding sector in problem 9. The Country sector is much smaller than the corresponding sector in problem 9.

Pages 210 and 211

3 It appears that in the year 2000, the women's high-jump record will reach 7 feet. This conclusion follows a pattern, but there is no guarantee that the seven-foot mark will be reached.

5 a Data for which we know both the total and the breakdown of the total are best displayed in a circle graph.
b Data that are used to compare different conditions are best displayed in a bar graph. When we construct a bar graph, we do not need to know the total number of data.

6
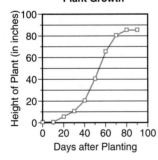

7 a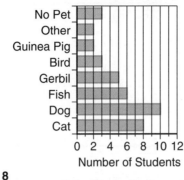

8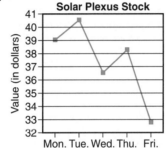

9 a Price per share in U.S. dollars
b The stock increased $\frac{7}{8}$ dollar or about $0.88.
e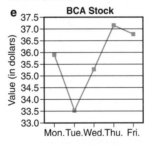

10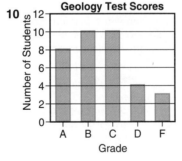

11 d From the information given, you cannot determine whether more people got married in 1970 or in 1975. The answer would depend on the size of the population in each of those years.

12 a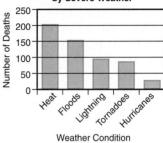

b Because the total number of deaths caused by weather conditions is unknown.

13 Possible answer: Plane takes off, circles an airport, lands, takes off again.

14 b The number of farms decreased.
c Possible answer: Modern equipment allows fewer farmers to produce more crops.

15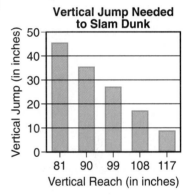

Page 220

29 a The perimeter of rectangle 2 is half the perimeter of rectangle 1.
b The area of rectangle 2 is one fourth the area of rectangle 1.

31 a Less than 1: ≈2.5 million; between 1 and 5: ≈23.8 million; between 5 and 10: ≈17.2 million; between 10 and 15: ≈6.3 million; more than 15: ≈1.0 million
b Less than 7; between 7 and 35; between 35 and 70; between 70 and 105; more than 105

Investigation It is a common practice when presenting information graphically to vary the vertical axis to show only part of the range. This can mislead a careless reader. For example, if school enrollment is 1800 in 1991, 1792 in 1992, and 1780 in 1993, the decline in enrollment is not very dra-

ADDITIONAL ANSWERS

matic percentage-wise. But it can be made to appear more dramatic in a graph by limiting the range to, say, 1700 to 1800, instead of 0 to 1800.

Page 230

where 6 | 3 means 6.3

Pages 232 and 233

9 a
```
2 | 3 3
3 | 3 3
4 | 2 4 8
5 | 1 5
6 |
7 | 4
```
where 2 | 3 means 2.3
b 4.3; 3.3; 5.1; 2.3 and 7.4; 5.1
c

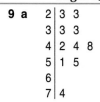

d No. The midquartile range is 2.3, and there are no data that are 1.5 times the midquartile range from the upper or lower quartile.

10 a

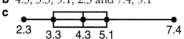

d

11 a 6, 9, 11, 16, 17, 21, 24, 38, 56, 80, 96, 106, 113, 126, 149, 157, 174, 191, 287, 336, 496, 651, 1356, 2639
b 109.5; 22.5; 239; 6 and 2639
c Most of the data are closer in value to the median than the mean.
d

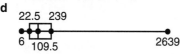

e 651, 1356, and 2639; these numbers are more than 1.5 times the midquartile range from the upper quartile.
f Possible answer: Alaska has the longest coastline.
14 > means "is greater than" and < means "is less than."
16 e Sizes 8–16, with the most dresses in size 10
Investigation Answers will vary.

Pages 240 and 241

23 a

31 b Store A: $173.75; Store B: $93.25; Store C: $101.75; Store D: $166.00
Investigation The most general meaning of *matrix* is "that from which something originates, develops, or takes form." In model making, it can refer to a die or mold; in geology, the material in which a fossil is embedded; in printing, a metal mold for casting the face of type; and in anatomy, the intercellular substance in which living cells are embedded. In mathematics, the rectangular array might suggest a basic structure from which mathematical ideas can develop via calculations. Note that spreadsheets are in matrix form.

Page 249

1 Possible answers:
$6 = ((4 + 4) \div 4) + 4$; $7 = 4 + 4 - (4 \div 4)$; $8 = (4 + 4) \div (4 \div 4)$; $9 = 4 + 4 + (4 \div 4)$
2 $99 + (99 \div 99)$ or $(9)(9) + 9 + 9 + (9 \div 9)$
3
```
8 1 6
3 5 7
4 9 2
```
4 a 36 **b** 27 **c** 25 **d** 54 **e** 29, 38, 47, 56

Chapter 6

Pages 264 and 265

2 An equation is a mathematical sentence that contains the symbol =. The equation indicates that the quantities on both sides of = have the same value. An inequality is a mathematical sentence that uses one of the following symbols: <, ≤, >, ≥, and ≠. An inequality symbol indicates that the quantity on the left side and the quantity on the right side of the symbol do not have to be equal.
3 A shaded endpoint indicates that the endpoint is included in the set.
4 An open endpoint indicates that the endpoint is not included in the set.
5 An arrowhead represents the continuation of the set in that direction.
6 The symbol < indicates that the set does not include the number that it precedes, whereas ≤ indicates that the set includes the number that it precedes.
7 Possible answer: An *and* inequality represents a graph that is shaded between two numbers; an *or* inequality represents a graph that has two or more separate sections.
8 Possible answers:
a It makes it easier and quicker to use a continuous measuring tape to measure all dimensions of the box.
b One person, unassisted, can lift the package.
c The box must fit the typical length of a delivery-truck bed.
10

11 Yes. The three USP conditions are met.
12 a

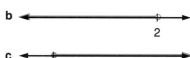

b

c

d

15 a 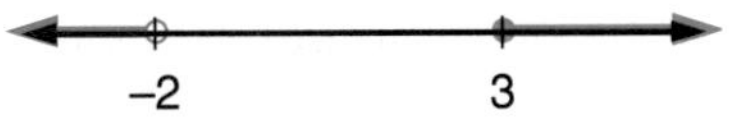

b 

c 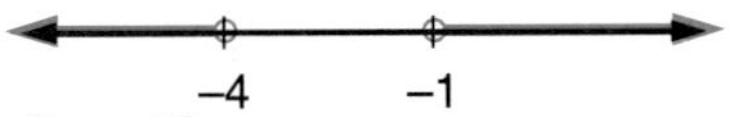

16 a $-2 < x \le 3$
b $x \ge -1$
c $x < 1$
17 a This inequality implies that 3 < –3.
b Number < –3 or number > 3
20 No. Although the 130-inch combined girth and length is acceptable, with a girth of 10 inches, the total length of the package is 120 inches, which exceeds the 108-inch maximum length.
22 a 

b 

c 

23 The graph of $x < -2$ or $x > -1$ is shaded to the left of –2 and to the right of –1. The graph of $x < -2$ and $x > -1$ has no shading because no numbers satisfy both conditions.
24 a

	A	B
1	x	SQRT (9–x^2)
2	–6	ERROR
3	–5	ERROR
4	–4	ERROR
5	–3	0
6	–2	2.236
7	–1	2.828
8	0	3
9	1	2.828
10	2	2.236
11	3	0
12	4	ERROR
13	5	ERROR
14	6	ERROR

c 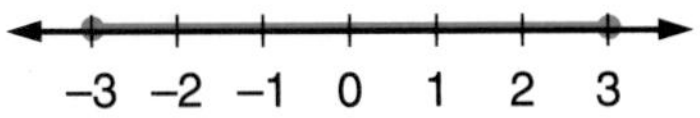

25 a Yes, but it is customary to put the negatives to the left of zero and the positives to the right.
b No, because the distance from 0 to 1 establishes the scale, and this scale must be maintained along the number line.
27 

Pages 280 and 281

22 a

	A	B	C	D	E
1	Original	5%	10%	12%	15%
2	Score	Increase	Increase	Increase	Increase
3	97	101.85	106.7	108.64	111.55
4	84	88.2	92.4	94.08	96.6
5	69	72.45	75.9	77.28	79.35
6	42	44.1	46.2	47.04	48.3
7	79	82.95	86.9	88.48	90.85
8	84	88.2	92.4	94.08	96.6
9	86	90.3	94.6	96.32	98.9
10	77	80.85	84.7	86.24	88.55
11	62	65.1	68.2	69.44	71.3
12	71	74.55	78.1	79.52	81.65
13	55	57.75	60.5	61.6	63.25

24 a $\begin{bmatrix} -4 & -4 & 12 \\ 10 & 2 & 2 \end{bmatrix}$

b See students' drawings.
c Original area: 16 square units; final area: 64 square units, or four times as much as original area.
28 Alf: \$13.25; Betty: \$8.20; Grandma: \$7.25; Del: \$9.90
Investigation Half-life is the length of time it takes for half of the atoms in a radioactive substance to disintegrate. The half-life is constant for a given substance but varies greatly from one substance to another. A graph showing time on the horizontal axis and the remaining amount of radioactive substance on the vertical axis will show a curve that declines with time and approaches zero. For example, a typically shaped curve might pass through (1, 1), $(2, \frac{1}{2})$, $(3, \frac{1}{4})$, $(4, \frac{1}{8})$, etc.

Pages 300 and 301

23 a Possible answers:
$|3 + 2| = |3| + |2|$;
$|-3 + (-2)| = |-3| + |-2|$;
$|3 + 0| = |3| + |0|$
b Possible answers:
$|3 + (-2)| < |3| + |-2|$;
$|-3 + 2| < |-3| + |2|$;
$|-1 + 1| < |-1| + |1|$
c The statement is never true.

24 a 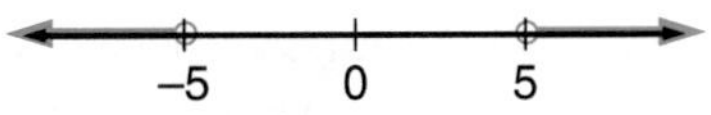

b 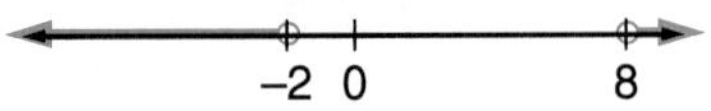

c 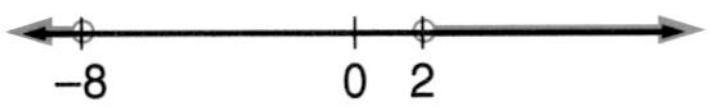

35
–65 40 –40
–105 80
–185

Investigation The region will often be circular in shape, the radius being the distance you can travel on one tankful. For a car that holds 15 gallons and gets 20 miles per gallon, this is about 300 miles.

Page 314

6 c $\begin{bmatrix} 12 & -8 & 24 \\ 0 & -16 & 28 \end{bmatrix}$

9 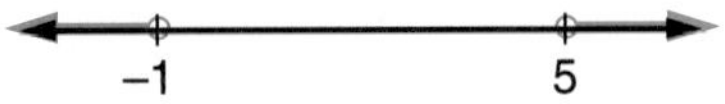

10 a–e 

18 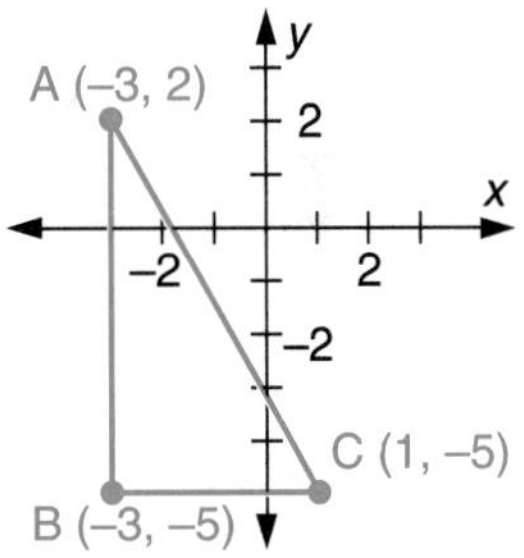

Page 315

1 77
2 White
3 Possible answer:

1		11	8	17
10	7	16	3	12
15	2	9	18	21
6	19	22	13	4
	14	5	20	23

Chapter 7

Page 324

26

a	AC	CB	AB	Perimeter	Area
1	4.123	4.123	8	16.246	4
2	4.472	4.472	8	16.944	8
3	5.000	5.000	8	18.000	12
4	5.657	5.657	8	19.314	16
5	6.403	6.403	8	20.806	20
6	7.211	7.211	8	22.422	24
7	8.062	8.062	8	24.125	28
8	8.944	8.944	8	25.889	32
9	9.849	9.849	8	27.698	36
10	10.770	10.770	8	29.541	40
11	11.705	11.705	8	31.409	44
12	12.649	12.649	8	33.298	48
13	13.601	13.601	8	35.203	52
14	14.560	14.560	8	37.120	56
15	15.524	15.524	8	39.048	60
16	16.492	16.492	8	40.985	64
17	17.464	17.464	8	42.928	68
18	18.439	18.439	8	44.878	72
19	19.416	19.416	8	46.833	76
20	20.396	20.396	8	48.792	80

Investigation Since vertical lines will actually meet at the center of the earth, they are not parallel; the distance between them increases slightly the higher you go. Thus, two vertical walls will be a little farther apart at their tops than at their bases. While the differences may be too little to measure in most buildings, vertical supports that can be miles apart—say, in a suspension bridge—are measurably farther apart at their tops than at their bases.

Pages 330 and 331

22 Quadrilateral is derived from the Latin prefix *quadri-* ("four") and the Latin word *latus* ("side") and refers to a polygon with four sides. Rectangle is derived from the Latin words *rectus* ("right") and *angulus* ("angle") and refers to a quadrilateral all of whose angles are right angles. Parallelogram is derived from the Greek words *parallélé* ("parallel") and *grammé* ("line") and refers to a quadrilateral with two pairs of parallel sides.

26

a	AB	BC	AD	DC	Perimeter	Area
1	4.472	8	1	9.849	23.321	18
2	4.472	8	2	8.944	23.416	20
3	4.472	8	3	8.062	23.534	22
4	4.472	8	4	7.211	23.683	24
5	4.472	8	5	6.403	23.875	26
6	4.472	8	6	5.657	24.129	28
7	4.472	8	7	5.000	24.472	30
8	4.472	8	8	4.472	24.944	32
9	4.472	8	9	4.123	25.595	34
10	4.472	8	10	4.000	26.472	36
11	4.472	8	11	4.123	27.595	38
12	4.472	8	12	4.472	28.944	40
13	4.472	8	13	5.000	30.472	42
14	4.472	8	14	5.657	32.129	44
15	4.472	8	15	6.403	33.875	46
16	4.472	8	16	7.211	35.683	48
17	4.472	8	17	8.062	37.534	50
18	4.472	8	18	8.944	39.416	52
19	4.472	8	19	9.849	41.321	54
20	4.472	8	20	10.770	43.242	56

Investigation Tiles that are shaped like squares, rectangles, rhombuses, or parallelograms can be used to tile a floor without leaving gaps or overlapping (i.e., those figures, among many others, will always tessellate). A trapezoid will tessellate nicely if, for example, alternate pieces are turned upside down. A regular hexagon will tessellate, while a regular pentagon will not.

Pages 336 and 337

6 a If the numerals marking yardage lines and lettering on the field are not taken into account, a football field has reflection symmetry about the 50-yard line and about a line through the midpoints of the end lines. It also has 180° rotation symmetry about the midpoint of the 50-yard line.
b The infield of a baseball field has reflection symmetry about a line through home plate and second base. In some cases, the entire field of play is symmetric about the same line.
c A tennis court has reflection symmetry about the net and about a line through the midpoints of the baselines. It also has 180° rotation symmetry about the midpoint of the net.
d Golf courses are asymmetric.

11 a True; for example, an equilateral triangle
b False; just as a circle can have only one center, so any figure can have only one center of rotation.

29 a Possible answer: An open-ended wrench can be slipped onto it from 6 directions instead of just 4.
b Possible answer: They are too "round," so that a wrench might slip on them, or wear away the corners.
c Possible answer: Opposite sides are not parallel, and an open-ended wrench could not be used on them.
d Pentagonal bolt heads are found on fire hydrants so that unauthorized persons may not turn on the water.

Page 353

29

k	2k	AB
1	2	2.236
2	4	4.472
3	6	6.708
4	8	8.944
5	10	11.180
6	12	13.416
7	14	15.652
8	16	17.889
9	18	20.125
10	20	22.361
11	22	24.597
12	24	26.833
13	26	29.069
14	28	31.305
15	30	33.541
16	32	35.777
17	34	38.013
18	36	40.249
19	38	42.485
20	40	44.721
21	42	46.957
22	44	49.193
23	46	51.430
24	48	53.666

35 c No; a length and an area cannot be added, since they are measurements of different attributes.

Investigation Many cards have rotation symmetry, so that they are never upside-down in your hand. However, note that some cards (such as the A, 3, 5, 6, 7, and 9 of any suit but diamonds) don't actually have true rotation symmetry, since the symbol(s) in the center must be oriented either up or down. The symmetry of picture cards may vary depending on their design.

Page 359

11 b

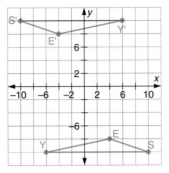

A rotation of 180° about the origin

Pages 360 and 361

15 H'(−1, 6), A'(2, 4), T'(−5, 3)
16 H'(−2, 1), A'(−5, −1), T'(2, −2)
17 H'(−1, 2), A'(1, 5), T'(2, −2)

19 a R   E   V
$$\begin{bmatrix} 0 & 0 & 3 \\ 4 & 0 & 0 \end{bmatrix}$$

b R' E' V'
$$\begin{bmatrix} 0 & 0 & -3 \\ 4 & 0 & 0 \end{bmatrix}$$

c R" E" V"
$$\begin{bmatrix} 0 & 0 & -3 \\ -4 & 0 & 0 \end{bmatrix}$$

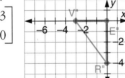

d $\begin{bmatrix} -1 & 0 \\ 0 & 1 \end{bmatrix}$

e $\begin{bmatrix} 0 & 0 & -3 \\ -4 & 0 & 0 \end{bmatrix}$

Possible answer: Rotating a figure 180° is the same as reflecting it over the y-axis, then reflecting the image over the x-axis.

21 $\begin{bmatrix} 2 & -1 & 1 \\ 3 & -4 & 6 \end{bmatrix}$

It is identical to the multiplicand.

22 One is 8, the other is 10.

28 a There are five reflection symmetries, over each of the five lines through a vertex and the midpoint of the opposite side.
b There are two rotation symmetries, of 72° and 144°. The center of symmetry is the point of intersection of the lines connecting the vertices to the midpoints of the opposite sides.
c Possible answers: By reflecting the pentagon over its lines of symmetry (or by rotating about its center of symmetry), one can make any of the figure's angles match up with any of the others, so all the figure's angles are congruent; by the same reflections or rotations, it can be seen that all the figure's diagonals are congruent.

29 P'(−2, −3); P"(2, −1); P"'(6, 1), P""(10, 3), P""'(14, 5)

Investigation Origami is the art of folding paper to form a figure, such as a boat or an animal. It has been practiced for centuries, more extensively in Japan, Korea, and China, but to some degree in most countries around the world. The artist starts with a square piece of paper, probably because a square has more lines of symmetry than do other rectangles. Various folds will result in slides, reflections, or rotations.

Page 367

1 Possible answers:
a Mast, past, pest, peat, seat, sear, spar
b Cent, dent, dint, dine, dime
c Fast, last, lost, loot, boot, blot, slot, slow
d Work, cork, cook, coot, clot, cloy, clay, play
e Shave, shake, slake, flake, flame, blame, blase, blast, blest, bleat, cleat, clean
f Meat, feat, fest, fist, fish
g Wok, won, ton, toy, try, fry
h Wish, wise, wipe, ripe, rope, hope

2 Answers will vary. Some students may conclude that the least possible number of moves is 18.

Chapter 8

Pages 390 and 391

20 b R = 0.75(220 − a)
c

	A	B
1		Target
2	Age	Heart Rate
3	10	157.5
4	11	156.75
5	12	156
6	13	155.25
7	14	154.5
8	15	153.75
9	16	153
10	17	152.25
11	18	151.5
12	19	150.75
13	20	150
14	21	149.25
15	22	148.5
16	23	147.75
17	24	147
18	25	146.25
19	26	145.5

29 a, b

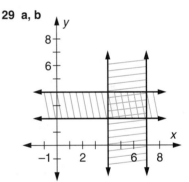

Investigation Many people are surprised to learn that both methods give the same answer. Let x be the list price. If you calculate the 20% discount first, and then the tax, you get (0.80x)(1.05) = 0.84x. If you calculate the 5% tax first, then the discount, you get (1.05x)(0.80) = 0.84x.

ADDITIONAL ANSWERS

Page 425

1. Possible answer: Facetious
2. Possible answer: Bookkeeper
3. 1 6 15 20 15 6 1
   1 7 21 35 35 21 7 1
4. Possible answer: Fill the 4-quart bottle and pour it into the 7-quart bottle. Fill the 4-quart bottle again and use it to fill the 7-quart bottle, which leaves one quart of water in the 4-quart bottle. Empty the 7-quart bottle and pour the one quart of water in the 4-quart bottle into the 7-quart bottle. Completely fill the 4-quart bottle again and empty it into the 7-quart bottle. The 7-quart bottle now contains 5 quarts of water.
5. a All-star team
   b Circle the wagons
   c The rain in Spain
   d Safe on first

Chapter 9

Page 434

Investigation Examples of finite sets: $\{A, B, C, \ldots, Z\}$, {all people}; examples of infinite sets: $\{0, 2, 4, 6, \ldots\}$, {integers}, {points on a circle}; $\overline{AB}$ has a midpoint, M_1; then $\overline{AM_1}$ has a midpoint, M_2; and so on, thus identifying an infinite set of points, $\{M_1, M_2, M_3, \ldots\}$, all on $\overline{AB}$.

Pages 440 and 441

18. Possible answer: A = {1, 2} and B = {1, 2}
20. Possible answers: {squares}, {rectangles}, {trapezoids}, {rhombuses}
21. a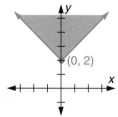
    b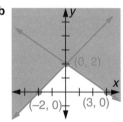

24. a, b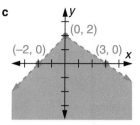
    c, d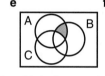
    e, f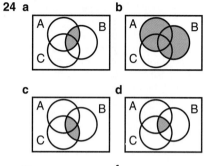

25. a {4} b {1, 3, 4, 5} c {1, 3, 4, 5, 7, 10}
    d {4} e { } f {1, 2, 3, 4, 5, 6, 7, 10}
    g {1, 3, 4, 5} h {4}
26. {2, 3, 5, 7}, {2, 3, 5}, {2, 3, 7}, {3, 5, 7}, {2, 5, 7}, {2, 3}, {2, 5}, {2, 7}, {3, 5}, {3, 7}, {5, 7}, {2}, {3}, {5}, {7}, { }
30. a $A = 9\pi \approx 28.27$, $C = 6\pi \approx 18.85$
33. a Possible answer: $\frac{1}{2}$, 1, $\sqrt{30}$
    b $\{x: 0 < x < 3\}$

Investigation Finding the square root always brings you closer to 1, whether you start out with a positive number less than or greater than 1. (Many students are surprised to learn that, for numbers less than 1, the square root is actually larger.) Thus, no matter where you start, repeatedly pressing the square root key will generate a sequence that approaches 1 as a limit.

Pages 446 and 447

22. a $\{(x, y): -2 \leq x \leq 5; -3 \leq y \leq 2$ where x and y are integers$\}$
24. {(3, 4), (4, 3), (5, 12), (12, 5), (6, 8), (8, 6), (9, 12), (12, 9)}

Investigation If the fraction has been reduced and the denominator contains only 9s, then the number of 9s in the denominator gives the number of digits that repeat in decimal form. Thus, $\frac{13}{9} = 1.\overline{4}$, $\frac{13}{99} = 0.\overline{13}$, $\frac{13}{999} = 0.\overline{013}$, etc.

Page 461

Investigation Possible answer: Commuting to and from school or work; associates as friends or working partners. Just as commuting involves traveling the same distance in opposite directions, so the commutative property states that the same answer will result when the operations are reversed. The common meaning of the word *associates* as "individuals who are part of a group" relates clearly to the associative property, which refers to grouping. In Latin, *commutare* means "to change" (as in, change for another); *associare* means "to join to."

Page 483

1.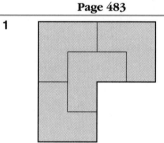
2. 5
3.
   ```
     37091
   + 37091
     74182
   ```
4. 3

Chapter 10

Page 506

Investigation The three-digit palindromes divisible by 11 are 121, 242, 363, 484, 616, 737, 858, and 979. All four-digit palindromes are divisible by 11.

ADDITIONAL ANSWERS

Page 512

34 a, d

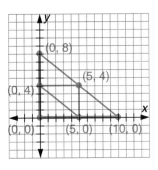

b 40
c (0, 4), (5, 0), (5, 4)
e 10, 10, 10, 10

Investigation It is impossible to find three consecutive three-digit numbers, none of which is divisible by 3. A simple way to show this is to list some natural numbers in order and circle the multiples of 3. Clearly, every third number is a multiple of three. So in every set of three consecutive numbers, exactly one of them will be a multiple of 3.

Page 519

31 b iii 1, 2, 4, 5, 7, 8, 10, 14, 20, 25, 28, 35, 40, 49, 50, 56, 70, 98, 100, 125, 140, 175, 196, 200, 245, 250, 280, 343, 350, 392, 490, 500, 686, 700, 875, 980, 1000, 1225, 1372, 1400, 1715, 1750, 1960, 2450, 2744, 3430, 3500, 4900, 6125, 6860, 7000, 8575, 9800, 12250, 13720, 17150, 24500, 34300, 42875, 49000, 68600, 85750, 171500, 343000

Page 541

1

16	2	3	13
5	11	10	8
9	7	6	12
4	14	15	1

2 Possible answers: Anna, Otto, Madam, toot, deed
3 Possible answers: (22)(131) = 2882; (11)(171) = 1881

4
a otis   **b** nick   **c** rick
d art   **e** mark   **f** gus
g sam   **h** allen   **i** nat
j sal   **k** chuck   **l** ira

Chapter 11

Page 558

Investigation
Balance beam: a narrow wooden beam supported in a horizontal position approximately four feet above the floor. It is used for performing gymnastic feats.
Balance scale: a beam that is supported freely in the center and has two pans of equal weight suspended from its ends; it is used for measuring mass.
Fulcrum balance: a fulcrum is a support about which a lever pivots; its balance is at the point about which the lever moves as force is applied.

Page 572

28 a

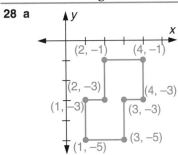

b

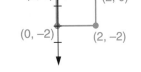

Investigation The formula represents the Ideal Gas Law, where
P = pressure
V = volume
n = number of moles
R = ideal gas constant
T = temperature

Pages 592 and 593

5

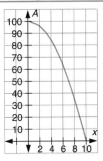

6

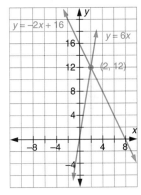

7

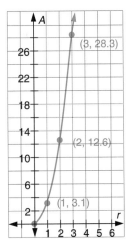

8 a

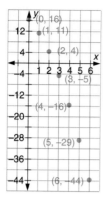

9 Possible answer: Let $y = 10x$. Draw the graph.

11 a

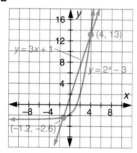

b $2^4 - 3 = 13$, $3 \times 4 + 1 = 13$;
$2^{-1.2} - 3 \approx -2.6$; $3(-1.2) + 1 = -2.6$

12 c Between $2.3 million and $5 million; explanations should refer to trends shown on the graph

13 a $V = 8t + 5p$
b

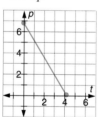

14 b

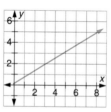

Chapter 12

Page 623

3 One possible answer is given; others exist.

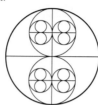

4

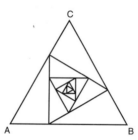

5 b Alternates between a rhombus and a rectangle

6 a (10, –2); (15, 3); (21, –3); (28, 4); (36, –4); (45, 5)
b $\sqrt{2}$; $2\sqrt{2}$; $3\sqrt{2}$; $54\sqrt{2}$
d Yes; the ending vertices of all the odd-numbered segments are in Quadrant I.

Page 624

7

8

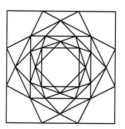

10 a (–8, 0); (0, –8); (9, 1); (–1, 11); (–12, 0); (0, –12)
b $\sqrt{2}$; $2\sqrt{2}$; $3\sqrt{2}$; $10\sqrt{2}$
d $55\sqrt{2}$; initial input: 0; iteration rule: INPUT + $\sqrt{2}$ (COUNTER); counter start: 0; iterations: 10

11 Some lattice points hit: (1, 1); (1, 2); (3, 1); (3, 2); (3, 3); (3, 4)
Some lattice points missed: (1, 3); (2, 1); (2, 2); (2, 3)
From the origin, go right one unit, turn left 90°, go up one unit, and turn left 90°. Each iteration consists of two line segments and two left turns. Increase the lengths of the segments by one unit each iteration.

Investigation The Koch snowflake curve is a fractal formed by iterating the sides of an equilateral triangle. For each iteration, the line segments that make up the figure are divided into three equal parts. Then the center part is removed and replaced with two segments of equal length to form two sides of an equilateral triangle. See pages 635 and 636 for a diagram.

Page 639

Investigation The Sierpinski triangle is a graphical introduction to a recursive process in chaos theory. It is an example of how order forms out of randomness. To begin, pick three points—the vertices of a triangle—and then pick a fourth point as a starting point. Randomly choose one of the three vertices of the triangle and calculate the midpoint between the starting point and this vertex. Plot this midpoint, then repeat the process using the midpoint you calculated as a new starting point. Repeat this until a figure appears.

The program on the following page, for the TI-81 graphics calculator, will create the Sierpinski triangle on the graphics screen.

Note: Before executing the program, set the range as follows: XMin=0; XMax=10; Scl=0; YMin=0; YMax=10; Scl=0. Also, clear or deselect all equations from the Y= screen. To "break" the program, press ON, then 2.

ADDITIONAL ANSWERS

Actual program steps	Comments
Prgm2 : SRPINSKI	Enter the program as program 2
: 0→A	
: 0→B	First vertex
: 10→C	
: 0→D	Second vertex
: 0→E	
: 10→F	Third vertex
: 5→G	
: 5→H	Starting point
: ClrDraw	Find command in Draw menu
: DispGraph	Find command in Prgm menu
: Lbl C	Find command in Prgm menu
: PT−On (G, H)	Find command in Draw menu
: Int(3∗Rand)→I	Find commands in Math menu
: If I = 1	Find commands in Prgm and Test menus
: Goto A	
: If I = 2	
: Goto B	
: (E+G)/2→G	
: (F+H)/2→H	
: Goto C	
: Lbl B	
: (C+G)/2→G	
: (D+H)/2→H	
: Goto C	
: Lbl A	
: (A+G)/2→G	
: (B+H)/2→H	
: Goto C	